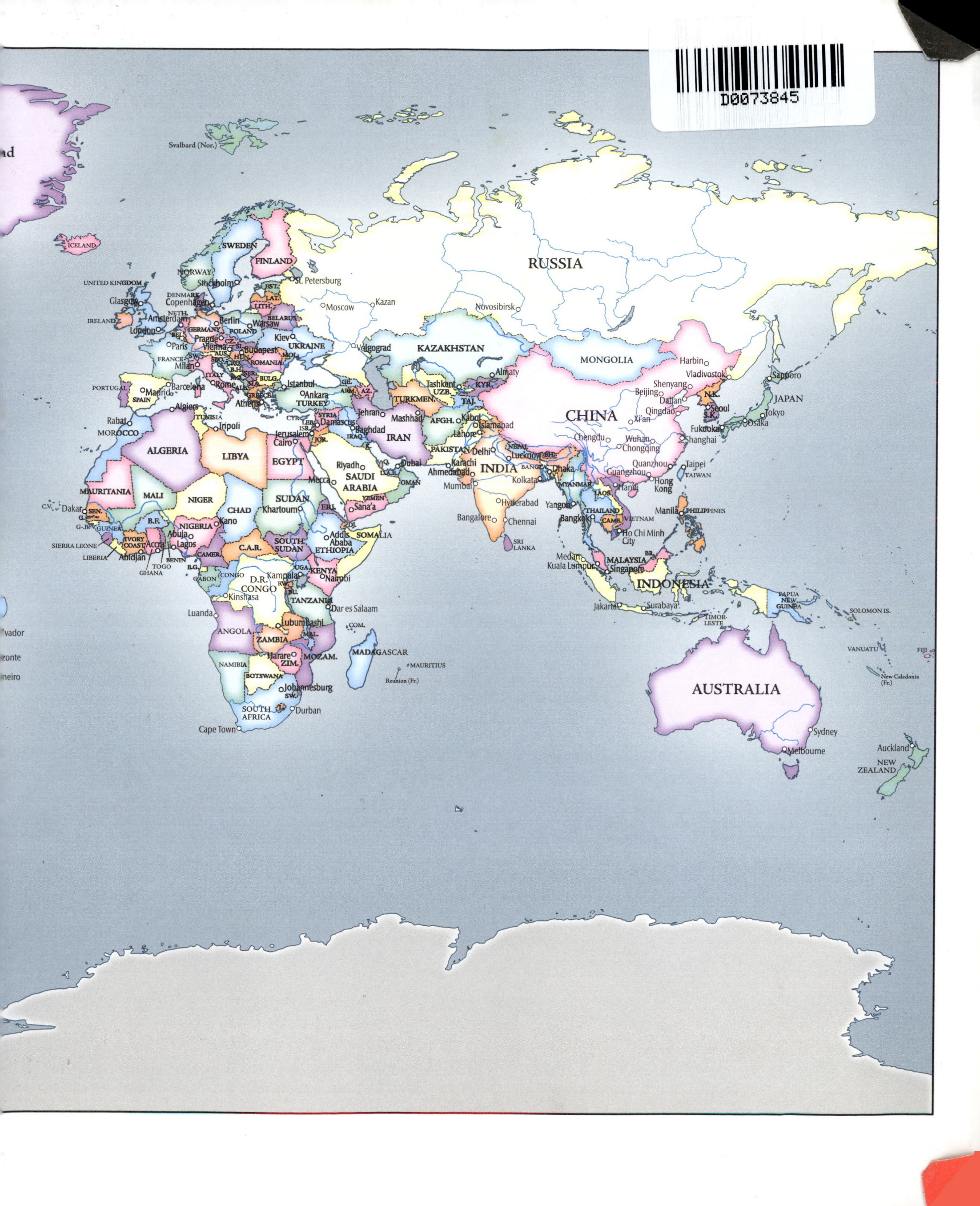

D0073845
Svalbard (Nor.)
ICELAND
SWEDEN
FINLAND
NORWAY
UNITED KINGDOM
Stockholm
St. Petersburg
RUSSIA
Glasgow
DENMARK
Copenhagen
EST.
LAT.
LITH.
Moscow
Kazan
Novosibirsk
IRELAND
NETH.
Amsterdam
Berlin
BELARUS
Warsaw
London
GERMANY
POLAND
Paris
Prague
Kiev
UKRAINE
Volgograd
KAZAKHSTAN
MONGOLIA
Harbin
FRANCE
Vienna
Budapest
Milan
ROMANIA
Vladivostok
Sapporo
PORTUGAL
Madrid
Barcelona
ITALY
Rome
BULG.
Istanbul
Almaty
Shenyang
Beijing
N.K.
JAPAN
SPAIN
Athens
Ankara
TURKEY
GREECE
Tashkent
UZB.
KYR
Dalian
Seoul
S.K.
Tokyo
Algiers
TURKMEN.
TAJ.
Qingdao
Osaka
Rabat
TUNISIA
Tripoli
CYP.
SYRIA
Damascus
Tehran
Mashhad
AFGH.
Kabul
Islamabad
CHINA
Xi'an
MOROCCO
LEB.
ISR.
Jerusalem
Baghdad
IRAQ
IRAN
Lahore
Fukuoka
ALGERIA
LIBYA
Cairo
JOR.
Chengdu
Wuhan
Chongqing
Shanghai
EGYPT
PAKISTAN
Delhi
NEPAL
Lucknow
Riyadh
Dubai
U.A.E.
Karachi
INDIA
Dhaka
Quanzhou
Taipei
TAIWAN
Ahmedabad
Kolkata
Guangzhou
SAUDI ARABIA
Mecca
OMAN
Mumbai
MYANMAR
Hanoi
Hong Kong
MAURITANIA
MALI
NIGER
YEMEN
LAOS
CHAD
SUDAN
Sana'a
Hyderabad
Yangon
C.V.
Dakar
SEN.
Khartoum
ERI.
THAILAND
Manila
PHILIPPINES
B.F.
Bangkok
CAMB.
VIETNAM
Bangalore
Chennai
G.-B.
GUINEA
NIGERIA
Kano
DJ.
Addis Ababa
SOMALIA
Ho Chi Minh City
IVORY COAST
Accra
Abuja
SIERRA LEONE
Lagos
C.A.R.
SOUTH SUDAN
ETHIOPIA
SRI LANKA
LIBERIA
Abidjan
TOGO
BENIN
CAMER.
BR.
GHANA
E.G.
UGA.
KENYA
Medan
MALAYSIA
Kuala Lumpur
Singapore
GABON
CONGO
D.R. CONGO
Kampala
Nairobi
INDONESIA
Kinshasa
RW.
BU.
TANZANIA
PAPUA NEW GUINEA
Dar es Salaam
Jakarta
Surabaya
SOLOMON IS.
Luanda
Lubumbashi
TIMOR-LESTE
COM.
ANGOLA
ZAMBIA
MAL.
VANUATU
FIJI
Harare
ZIM.
MOZAM.
MADAGASCAR
MAURITIUS
NAMIBIA
Reunion (Fr.)
New Caledonia (Fr.)
BOTSWANA
AUSTRALIA
Johannesburg
SW.
SOUTH AFRICA
Durban
Cape Town
Sydney
Auckland
Melbourne
NEW ZEALAND

Fundamentals of World Regional Geography 4e

Joseph J. Hobbs
University of Missouri

Cartography by Andrew Dolan

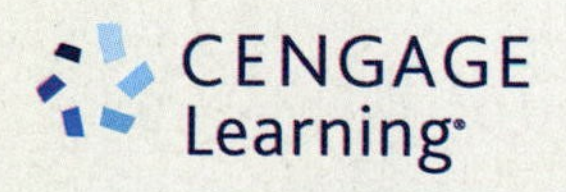

Australia • Brazil • Mexico • Singapore • United Kingdom • United States

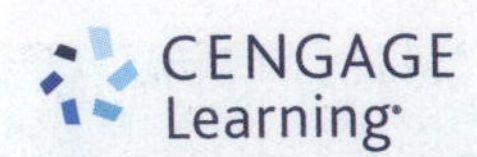

***Fundamentals of World Regional Geography,* Fourth Edition**
Joseph J. Hobbs

Product Manager: Morgan Carney
Senior Content Developer: Jake Warde
Associate Content Developer: Kellie N. Petruzzelli
Product Assistant: Victor Luu
Market Development Manager: Ana Albinson
Content Project Manager: Hal Humphrey
Senior Designer: Michael C. Cook
Manufacturing Planner: Becky Cross
Project Manager: Lori Hazzard, MPS Limited
Photo Researcher: Veerabhagu Nagarajan, Lumina Datamatics
Text Researcher: Manjula Subramanian
Copy Editor: Jill Pellarin
Cartographer: Andrew Dolan
Cover Image: Shivji Joshi/National Geographic Creative
Compositor: MPS Limited

Library of Congress Control Number: 2015938979

Student Edition:

ISBN: 978-1-305-57826-5

Cengage Learning
20 Channel Center Street
Boston, MA 02210
USA

Cengage Learning is a leading provider of customized learning solutions with employees residing in nearly 40 different countries and sales in more than 125 countries around the world. Find your local representative at **www.cengage.com**.

Cengage Learning products are represented in Canada by Nelson Education, Ltd.

To learn more about Cengage Learning Solutions, visit **www.cengage.com**.

Purchase any of our products at your local college store or at our preferred online store **www.cengagebrain.com**.

Printed in the United States of America
Print Number: 05 Print Year: 2018

Brief Contents

Contents

Photos courtesy of Joe Hobbs

Maps

Preface

To appreciate how our complex world works today, it is vital to have a solid grounding in the environmental, cultural, historic, economic, and geopolitical contexts of the world's regions and nations. *Fundamentals of World Regional Geography* establishes that foundation and offers you an opportunity to explore the events, issues, and landscapes of the world in more detail.

Chapters 1 through 3 provide the basic concepts, tools, and vocabulary of world regional geography. In the first chapter, geography's uniquely spatial approach to the world is introduced along with some of the discipline's milestone concepts and its considerable career possibilities—especially those growing from the "geospatial revolution." The second chapter covers the essential characteristics of the world's physical processes and how human activity has altered some of them. Climate change and the treaties to control it have a prominent role in that chapter. Chapter 3 traces the modification of landscapes by human actions, describes trends and projections of population growth, and considers agendas to slow destructive trends in resource use.

Then come eight chapters exploring the world's regions through a consistent, thematic approach focusing in turn on five elements: Area and Population, Physical Geography and Human Adaptations, Cultural and Historical Geographies, Economic Geography, and Geopolitical Issues. The final section of each chapter, entitled "Regional Issues and Landscapes," contains a selection of short studies of critical problems in global affairs and exemplary or important problems in human or physical geography.

The book is built for use in either a one- or two-semester course. If time is limited, the five thematic elements of each chapter may be a priority, with limited use of the case studies in the "Regional Issues and Landscapes" section. After the three introductory chapters, it does not matter what order the regional chapters are read in; no regional chapter presumes that any other regional chapter has been read. However, a unique cross-reference system allows a theme or issue introduced in one region to be tied immediately to other regions.

MindTap for Fundamentals of World Regional Geography, 4e implements the cross-reference system through easy-to-use links so readers can instantly navigate to the related theme or issue. Read more about the powerful learning tools made available in MindTap in the Course Support section of the Preface.

New to the Fourth Edition

Both longtime and first-time users of *Fundamentals of World Regional Geography* should be pleased with this edition. Much of the critical content of the previous editions is retained, but this is the most extensive revision to date, and a number of new elements are introduced here.

- Almost all of the maps are new or newly designed. Cartographer Andrew Dolan and I have worked to tie the maps tightly to the content.
- The climate and biome classification and mapping schemes have been revised to be consistent with the Köppen system and the World Wildlife Fund ecoregions data.
- Pie charts have replaced population cartograms.
- The language and religion maps in each regional chapter are revised using better and more consistent data.
- The book is more thematic and conceptual than in previous editions. Fritz Gritzner's big geographic question, "What is where, why there, and why care?," leads us to critical thinking about the concepts and themes that span the world's regions.
- The new *Geographic Spotlight* feature depicts geographers' methods of capturing, analyzing, and depicting geographic information.
- The definitive "18 Standards of Geography" authored by the National Council for Geographic Education (NCGE), and presented in Chapter 1, served as a constant reference in writing the book. At the end of the course, the book's reader will be able to claim confidently, "My geographic understanding has been informed by all the standards. I can match each standard with content in the book."
- Extensive use of the NCGE Standards is one of many elements that recommend this book in preparing for the AP Human Geography test.
- Regional chapters have a new feature entitled "Life in . . . ," where a resident of that region discusses land and life in a particular country. The only exception is the chapter on the United States because most of the book's readers know life in that country.
- In the introductory Chapter 3 and in each regional chapter, economic geography has been given more attention. Globalization has shifted wealth from more to less developed countries, lifting hundreds of millions of people out of poverty while gutting middle class jobs in developed countries and widening inequalities in many societies. These issues, along with the crosscurrents of post-Great Recession economic growth and the impacts of China's economic slowdown, are tied together throughout the book.
- The geographies of fresh and marine waters have more attention than in previous editions.
- Urban geography is much more prominent than in earlier editions.
- Geopolitical issues are more important than ever. Here the reader is brought up to date on geopolitical problems while also given a geographic foundation for understanding post-publication current events.
- Geopolitical instability and other climate change impacts on human systems are covered throughout.
- *Think Critically* questions are raised with many figure captions, challenging the student to use the text information in a thoughtful manner.

- Graphs have a new clean look and are more accessible.
- A thorough Study Guide useful for both instructors and students concludes each chapter.

Chapter-Specific Changes

- Chapter 1, "Objectives and Tools of World Regional Geography," introduces NCGE's "Six Essential Elements" and "18 Standards of Geography." Two new objectives of the book are presented: *To become geographically literate,* and *To use geographic critical thinking to understand the world*. Tobler's First Law of Geography is introduced. There is a figure showing "how to lie with maps." The AAG's "Strategic Research Directions" for students and scientists of geography are highlighted as thematic elements covered in the book.
- Chapter 2, Physical Processes and World Regions, contains global maps using the new world climate and biome data. There is more discussion of ocean resources, with a new map of global shipping lanes. Data, modeling, and treaties related to climate change are updated, with a global map of changes observed to date. IPCC conclusions structure the discussion.
- Chapter 3, Human Processes and World Regions, includes a new map and discussion of global human migrations, based on DNA analysis. The Geography of Economic Development section introduces the dramatic shifts in wealth across the globe, and the factors behind them. The new discussion of *globalization and development* with five outcomes of globalization is especially useful. The Human Development Index, Fragile States Index, and Human Values Index are mapped and discussed at the global scale, as is "land grabbing," which re-emerges in many of the regional chapters. There is new discussion of freshwater, including the concepts of "virtual water" and "water footprints," along with a new map of the world's aquifers. A unique graphic of global human migrations serves as a reference in many regional chapters. There are new maps and discussions of urban geography and the globalization of industrial supply chains.
- Chapter 4 on Europe covers the region's growing immigration crisis and the related rise of populist and nationalist parties, and its "crisis with Islam." Recent developments in the devolution of Scotland and Catalonia are covered. There is a vivid new map of the Columbian exchange and a thematic map of Lampedusa as a migrants' stepping stone to Europe. A new graph and discussion depict the decline of Europe's primary and secondary economic sectors and the rise of its tertiary services sector. "Life in France" is the guest essay. Problems in the Eurozone and European Union (EU), including the Greek financial crisis, are described, along with Germany's linchpin status. European reactions to Russia's invasion of Crimea are discussed. Environmental Perception is illustrated with English landscape tastes. Attributes of the "global city" with a worldwide map of global cities are introduced here. A new map of Iceland's and Norway's EEZs shows where their whaling takes place. The geopolitical evolution of the East European shatterbelt is depicted in five maps.
- Chapter 5, Russia and the Near Abroad, sees Russia falling into economic crisis (with "Dutch Disease") while Vladimir Putin consolidates his hold on power in the wake of Russia's invasion and intervention in Ukraine. Future actions by Russia, especially in "frozen conflicts" like that in Transnistria, and where Russian irredentism is strong, are considered. Russia's power to use pipelined energy as a political tool is analyzed and mapped. The country's demographic crisis is updated and mapped with a comparison of neighboring countries' demographics. The Chelyabinsk and Tunguska meteor events are in a new discussion of near-earth objects. A Cold War map depicts communism globally in 1980. "Life in Moldova" is the guest essay. Eurasia's status as the "cockpit of history" girded by the "Iron Silk Road" is proposed. The Armenian Genocide is discussed and mapped.
- Chapter 6, the Middle East and North Africa, is filled with new features. The region's enduring economic importance is seen in a map of global oil reserves, alongside global oil consumption. The tense standoff over Nile Basin waters is discussed and mapped, and the *Geography of Water* feature takes a closer look at the importance of virtual water and land grabbing for many of the countries. The region's geopolitical importance as a shatterbelt with numerous chokepoints is underscored in view of emerging threats, particularly terrorism. The evolution of Islamic State in Iraq and Syria (ISIS) from Syria's civil war and America's engagements in Iraq are traced and mapped, and US options for confronting ISIS are considered. There is discussion and a map of the refugee camps of externally and internally displaced Syrian refugees. There is an in-depth comparison of al-Qa'ida and ISIS, and a backgrounder on what Salafist movements are seeking. There is a map of the distribution of Islamist terrorist movements across Eurasia and Africa. An extraordinary graph shows the tangled web of allies and adversaries within and beyond the region. American geopolitical interests in the region are detailed. US options in dealing with Iran's nuclear ambitions are considered. A sequence of maps shows the evolution of political borders in Mesopotamia and raises the prospect of Iraq's fragmentation. The previous edition's "Arab Spring" is bookended here by the "Arab Fall." The Saudi intervention in Houthi-dominated Yemen is discussed. Three maps of Freedom House ratings through time depict the region's lack of democratic institutions. Dubai's troubled artificial islands are a "problem landscape." A graphic shows how the Gulf's traditional architectural "wind catchers" work. "Life in Egypt" is the guest essay.
- Chapter 7 on South and East Asia is the book's longest because it covers the most populous world region. It discusses the revision of China's one-child policy. A new graph shows the ecological succession associated with slash-and-burn

cultivation. In this chapter and others, we consider Edward Glaeser's ideas about cities, including their slums, as engines of innovations. The economic geography text and maps have expanded to consider the International Monetary Fund's characterization of China as the world's largest economy; the interdependence of the Chinese and American economies; the "onshoring" of US jobs previously offshored to China, as wages rise there; China's gaps between rich and poor, rural and urban; the vulnerabilities of China's real estate sector and overall economy; and the domestic fallout and international repercussions of China's economic slowdown. Geopolitical Hot Spots in the Western Pacific describes and maps each of the conflicting maritime claims in this area. China's strategic "string of pearls" is discussed and mapped. China's growing frustration with North Korea is discussed. Maps and prose show the evolution of colonial India into India, Pakistan, East Pakistan, and disputed Kashmir, and the emergence of the Naxalite "Red Corridor" in India. A map shows the distortions between genders, due largely to abortions, in India's states. We consider the future of Afghanistan, the "Graveyard of Empires," in the wake of the US drawdown. A new map shows the contributions of palm oil plantations and other developments to deforestation in Indonesia and Malaysia. Myanmar earns a dedicated ethnic map and discussion of growing democracy and new relationships with the US. Indochina's political evolution is described and mapped. The benefits and drawbacks of golden rice are discussed. The guest essay is about "Life in Vietnam."

- Chapter 8 takes us across Oceania and Antarctica. Helping to illustrate New Zealand's position in tectonic movements, there is a new map of the mostly submerged continental fragment of Zealandia. Christchurch's 2011 earthquake is discussed in the accompanying text and in the guest essay on "Life in New Zealand," by a Maori woman geographer. The Hawaii-Emperor Seamount Chain map reveals how the Hawaiian Islands and others were born. We reconsider the simple parable of Easter Island's decline. There is more insight into how control over natural resources spawned conflict in the Solomon Islands. Renewed US military interests in the Pacific as a counterweight to China are discussed. The titanic Castle Bravo atomic test is seen in the context of foreign military uses of the region. Australia's "Pacific Solution" for its unwanted immigrants is introduced. Aboriginal and Torres Strait Islander efforts to reclaim territories and resources are brought up to date with a recent Supreme Court ruling in Australia. There is a *Geographic Spotlight* on countermapping. Recent research on climate change related to Antarctica accompanies a new map of the continent.
- Chapter 9 on Sub-Saharan Africa portrays the continent in its most hopeful state in decades. The fight against HIV/AIDS continues with new hope as ARV drugs reach more people. The sudden epidemic of Ebola and the success in fighting it are discussed. Maps and prose describe the devastating trafficking of ivory. There are discussions of better governance, less conflict, more foreign investment, and diversification away from commodities toward manufacturing and services, all helping to grow economic prospects. Africa's "durable strengths and resources" are listed. "Conflict metals" have become problematic. We see land grabbing and the continuing "sustained looting" of Africa's resources. We consider China's infrastructure-for-minerals swaps with African governments, and the different priorities of Chinese and American aid. The growth of Islamist terrorism is viewed as one reason for renewed Western geopolitical interest in the region. The Sustainable Development Goals and the Millennium Development Goals preceding them are described in the African context. Current and planned urban projects across the region are introduced. The historic geography of race relations in South Africa and a map of the "homelands" are introduced to provide more understanding of the current situation. "Life in Tanzania" is the guest essay.
- Chapter 10 on Latin America includes a completely revised section on economic geography that reveals the region's growing relationship with China (including impacts of China's economic slowdown), disturbing economic inequalities, the drive to diversify into manufacturing and services, and the wide adoption of neo-liberalism and related free trade agreements (amidst pushback by Brazil and the "Bolivarian socialist countries"). A new map depicts minerals and mining. The status of NAFTA and other FTAs is updated. There is a town plan of the colonial city and discussion of life in the region's informal settlements. The reestablishment of diplomatic relations between Cuba and the US is marked. There are new insights and maps in the "Geography of Drug Trafficking" about smuggling routes, the drugs value chain, and the role of cartels. The Nicaragua Canal is discussed and mapped. Students are invited to contribute to humanitarian mapping projects in Latin America and elsewhere, including through MapGive and Tomnod. The guest essay is "Life in Amazonia for Uncontacted Tribes."
- Chapter 11 is the most thorough revision to date of the United States and Canada. Current debates over legal and illegal immigration in both countries are discussed. There are new maps of physiographic regions and natural hazards. The section on "Cultural and Historical Geographies" includes new discussions of early peoples' migrations into North America, ecological and cultural impacts of Europeans, European settlers and settlements, and peoples of the US and Canada today. There is a new map of the ethnic composition of selected US metropolitan areas. There is a new discussion of the geographic underpinnings of the region's prosperity. Using a new map of the region's energy resources and routes, readers are encouraged to weigh in on the Keystone XL Pipeline controversy. The revised economic geography section depicts the American "energy revolution," the convergence of new oil drilling and falling oil prices, new insights into alternative energies and agricultural technologies, growth in information technologies and how they have displaced traditional workers, projected

sectors of employment, and growing social and economic inequalities. America's infrastructure woes are described. Geopolitical issues include expressions of war-weariness while the US finds itself on a perpetual war footing. Saul Cohen argues that the US is no longer a superpower. Readers are asked to consider six great geopolitical challenges for the US. New patterns of settlement in small, midsized, and large cities are discussed, along with urban "smart growth." There is a much-expanded discussion of issues related to Colorado River waters and drought in the West.

Acknowledgments

I am grateful to everyone who encouraged me throughout, especially my wife, Cindy; daughters, Katie and Lily; and my Mom. Brothers Greg and Will offered valuable help with Chapter 11 content, as did Andy Dolan. In addition to drafting outstanding maps, Andy considered every word in the book for accuracy, timeliness, and relevance, and researched all the numbers. Andy also contributed substantially to Section 11.3, and to the climate and biome discussions of each chapter. I have never had a better and more supportive editor than Jake Warde. Lori Hazzard worked tirelessly and kindly through all the book's stages.

The Cengage Learning team was headed up by Product Manager Morgan Carney and included Content Developer Jake Warde, Associate Content Developer Kellie Petruzzelli. Lori Hazzard of MPS Limited, Editorial Assistant Victor Luu, Marketing Manager Ana Albinson, Content Developer Digitization Project Manager Jennifer Chinn, Intellectual Property Analyst Christine Myaskovsky, Art Director Michael Cook, Photo Research Manager Veerabaghu Nagarajan of Lumina Datamatics and Text Permissions Manager Manjula Devi Subramanian, also of Lumina Datamatics.

Course Support Resources

The text is accompanied by a number of ancillary publications to assist instructors and enhance student learning, including full *MindTap* course support!

MindTap for *Fundamentals of World Regional Geography*, 4e

MindTap is a personalized, fully online digital learning platform of authoritative content, assignments, and services that engages your students with interactivity while also offering you choice in the configuration of coursework and enhancement of the curriculum via web apps known as MindApps. MindApps range from ReadSpeaker (which reads the text out loud to students) to Kaltura (allowing you to insert inline video and audio into your curriculum), to ConnectYard (allowing you to create digital "yards" through social media—all without "friending" your students). *MindTap for Fundamentals of World Regional Geography*, 4e provides the following unique features to enhance your course:

- An interactive eBook with highlighting, note taking, and an interactive glossary
- Unparalleled content cross-referencing so students can make important connections across the regions of the world
- Interactive mapping exercises based on the high-quality maps in the text
- Global Geoscience Watch, an ideal one-stop site for current events and research projects for all things geography
- Pre-tests and post-tests for each chapter that are auto-graded in MindTap and include helpful hints for students

Instructor Resources

Instructor Companion Site

Everything you need for your course in one place! This collection of book-specific lecture and class tools is available online via www.cengage.com/login. Instructors can access and download preassembled Microsoft® PowerPoint® lecture slides, the instructor's manual, the image library, animations, videos, blank maps, test banks, and more.

Cengage Learning Testing Powered by Cognero

This flexible online system allows the instructor to author, edit, and manage test-bank content from multiple Cengage Learning solutions; create multiple test versions in an instant; and deliver tests from an LMS, a classroom, or wherever the instructor wants.

About the Author

Joe Hobbs received his B.A. at the University of California Santa Cruz and his M.A. and Ph.D. at the University of Texas at Austin. He is a geography professor at the University of Missouri, where he also serves as director of the Vietnam Institute. He is a mainly a geographer of the Middle East, with many years of field research on Bedouin peoples and natural environments in Egypt's deserts. Joe's interest in the region grew from his boyhood in Saudi Arabia. His profession in geography grew out of life abroad with his Mom and Dad, all of his travels, and especially his being a "wayfellow" of Saleh Ali, a Bedouin of the Ma'aza tribe. His research in Egypt has been supported by grants from Fulbright, the American Council of Learned Societies, the American Research Center in Egypt, and the National Geographic Society Committee for Research and Exploration. He served as the team leader of the Bedouin Support Program, a component of the St. Katherine National Park project in Egypt's Sinai Peninsula, and led an effort to establish a national plan for environmental management in the United Arab Emirates. His most recent field research has been with a team, funded by the Norwegian Research Council, studying the interactions between nomadic pastoralists and acacia trees in Egypt and Sudan. Upcoming work deals with impacts of climate change on cultures and livelihoods in the Lower Mekong Basin and with best practices in environmental management for Ajman, United Arab Emirates.

Joe is the author of other books including *Bedouin Life in the Egyptian Wilderness* and *Mount Sinai* (both University of Texas Press), co-author of *The Birds of Egypt* (Oxford University Press), and co-editor of *Dangerous Harvest: Drug Plants and the Transformation of Indigenous Landscapes* (Oxford University Press).

Joe has taught graduate and undergraduate courses in world regional geography, geopolitics, environmental geography, the geography of the Middle East, the geography of caves, the geography of global current events, and the geographies of drugs and terrorism, as well as a field course on the ancient Maya geography of Belize. He has received the University of Missouri's highest teaching award, the Kemper Fellowship, and awards for leadership in international education at MU. He has led adventure tours to remote areas in Latin America, Africa, the Indian Ocean, Asia, Europe, and the High Arctic. Joe lives in Missouri with his wife, Cindy; daughters, Katie and Lily; and turtles, lizards, cats, and dogs.

In loving memory of Tommy, Jack,
Elizabeth, and Avantika

Above: Maps are the primary way geographers visualize spatial information of all kinds. Above is a selection of maps from this book, each focusing on a different relationship between people, places, and the environment.
Left: Paddling the Perfume River, central Vietnam. Joe Hobbs

1 Objectives and Tools of World Regional Geography

We are living in the era of the geographer.

—HAL MOONEY, STANFORD ECOLOGIST[1]

Welcome to world regional geography. What an important and useful field of study! In recent times the world has seemed endangered on so many fronts: great powers struggle for control in Ukraine; violent Islamists threaten the social and political fabric of the Middle East; China exerts its power over the marine territories of less powerful Asian countries; Ebola ravages West Africa and threatens other regions, for example. What on Earth is going on? But buried by the worrisome headlines are remarkable stories of breakthroughs in technology, communications, and agriculture as well as advancements in the eradication of disease and hunger. What are those all about? Where are we, Earth's peoples, headed?

Chapter Outline

Chapter Objectives

This chapter will enable you to

- Learn about the scope of geography as an academic discipline.
- Get acquainted with the essential themes, elements, and standards of geography.
- Learn some key concepts in geography.
- Appreciate the book's overall objectives.
- Learn the basic language of maps.
- Explore the "geospatial revolution," geographic information systems (GIS), and remote sensing.
- See how geographic knowledge is put to work in the job market.

1.1 What Is Where, Why There, and Why Care?

In studying world regional geography, we seek to understand what is going on, and why, and especially where. How are we doing? Many findings suggest "not very well." A study carried out recently by professors in three Ivy League colleges revealed that only one in six adult Americans could accurately locate Ukraine on a world map. Asked to locate Ukraine on a world map with only country borders drawn in, the 2066 respondents were literally all over the map, placing Ukraine on every continent except Antarctica, which was not depicted. Ukraine turned up in a number of US states, especially in Alaska. A number of respondents put it in Greenland.[2]

What difference does it make? Who cares if you know where Ukraine is, much less Greenland? Long ago *geography* earned a reputation for mind-numbing memorization of state capitals, and for driving students away (**•Figure 1.1**). Netflix's description of a 2014 film called *Geography Club* reads in part: "Looking for a haven from the social hell of [high] school, the teens in this dramedy form a social club they know no one else will join."[3] That's not very funny to geographers like me, but I understand it. The truth is, by itself, a piece of knowledge like where Ukraine is probably means little. But geography is all about context and connections. Understanding *where* things are makes it much easier to appreciate and answer the *who, what, when, why,* and *how* questions in life, at every scale—from your daily activities to world affairs. Geography always starts with the *where* question, but it is far more interesting and important than its old reputation for memorizing places suggests. Helping you to understand contexts and relationships, geography can help you make better-informed judgments and decisions. My geographer colleague Fritz Gritzner coined this definition of geography, which also serves as a methodology and as a challenge for us to think critically: "***What is where, why there, and why care?***"[4]

To illustrate the importance of geographic insight, let's drill down a little deeper into that Ukraine study, which was conducted after Russia annexed the Crimean Peninsula in 2014 and appeared ready to take eastern Ukraine by force. The professors who conducted the study found that the farther away from the actual location of Ukraine the survey participants guessed Ukraine was, the more likely they were to support US military intervention in Ukraine. Should that kind of disconnect concern us as we think about expending American "blood and treasure" in the world's hotspots?

Most of us using this book are Americans, and our collective experience in recent decades has prompted us to say this to our politicians: We are tired of getting it wrong, and we can't afford to get it wrong. Our decision makers are responding. Here is what the former US Secretary of Defense Robert Gates told cadets at the US Military Academy at West Point in 2011:

> Any future defense secretary who advises the president to again send a big American land army into Asia or into the Middle East or Africa should "have his head examined," as General [Douglas] MacArthur so delicately put it . . . Just think about the range of security challenges we face right now beyond Iraq and Afghanistan: terrorism and terrorists in search of weapons of mass destruction, Iran, North Korea, military modernization programs in Russia and China, failed and failing states, revolution in the Middle East, cyber-piracy proliferation, natural and man-made disasters, and more. And I must tell you, when it comes to predicting the nature and location of our next military engagements, since Vietnam, our record has been perfect. We have never once gotten it right.[5]

In his second term in office, President Obama depicted his foreign policy motto as "don't do stupid stuff."[6]

If only American presidents were advised by geographers . . . Geography is all about "getting it right" and "doing smart stuff" when it comes to understanding how the world works. Geographic knowledge of the *where, who,*

• **Figure 1.1** Geography used to be associated with memorizing mind-numbing facts. Not anymore!

what, when, why, and *how* can help guide informed decision making at all scales, from whether and how the United States should commit troops to a ground war to how you can get from point A to point B in your own community. Geographic insight has the power to transform our lives and contribute to the welfare of our communities and our countries.

By the end of this chapter, you will know what geography is, recognize the benefits you can gain from learning world regional geography, understand the organization and objectives of this book, and learn some of the key concepts and tools of geography.

Before You Go On . . .

I have been teaching world regional geography (WRG) for more than 25 years, and I know the challenges you face as a student in taking on such a large and important subject as the world. Knowing the Earth gets a lot easier when you recognize the patterns that repeat themselves in different places, and also when you recognize the key points of what you are reading. So, before you continue reading, you need to know about some important features of the book that help you with WRG recognition. The first is its *cross-referencing system*. The book is written with global interconnections in mind. "Globalization" is understandable as a concept, but how exactly does it work? The page and figure numbers in the book's margins (and the hyperlinks in the ebook) tie the diverse strands of global issues together. For example, when you read in Chapter 3 how countries running low on productive agricultural land become "land grabbers" in other countries, page numbers in the margin lead you to the places where land grabbing is occurring (go to **page 65** to see what I mean). As you read about China's economic growth and its appetite for raw materials, you are likewise directed to places around the world where these forces come into play (see **page 65**). I put a lot of effort into making these connections for you, and I hope you will use this feature often and learn much from it.

I also want to draw your attention to features that will help you know what the most important points in the book are. My WRG students often ask me that famous question: "Do I need to know that for the test?" I cannot tell you what your professor or TA will put on your test or quiz, but I can help you recognize the ideas, issues, concepts, themes, and information that are *fundamental* to world regional geography (*fundamental* means "of central importance") and that are worthy of testing. I encourage you to use the Study Guide at the end of each chapter. It highlights the chapter's most important points and issues. If you want to double its usefulness, I recommend that you read the Study Guide even before you read the chapter, and use it more thoroughly after your reading and when you are preparing for the test. Another device that I am fond of as a writer and that should be useful to you is the *topic sentence* or *phrase* introducing or summarizing the main point or content of a given passage. Usually my topic sentence is at the beginning or end of a paragraph, but not always. Want a quick read of the chapter to get up to speed? Follow the topic sentences like highway signs.

What Is Geography?

Geography, a term first used by the Greek scholar Eratosthenes in the 3rd century BCE,[7] literally means "description of the Earth" but is probably best characterized as "the study of the Earth as the home of humankind." Focusing on interactions between people and the environments in which we live, the modern academic discipline of geography has its roots in the Greek and Roman civilizations and the Scientific Revolution in Europe.

Geography has unique properties as a scientific discipline. These traits are articulated especially well in the set of **National Geography Standards**, composed by the National Council for Geographic Education (NCGE) and promoted by the National Geographic Society.[8] The standards are based on the NCGE's **six essential elements of geography.** Each of the six elements has a subset of geographic knowledge standards, eighteen in all, that "represent the most current conception of what it means to be geographically literate." These eighteen standards represent the substantive content of the field of geography, and they also underpin this book's contents. You should be able to take any issue discussed in the text and match it with one or more of the eighteen standards. The standards are presented in •**Table 1.1**.

In this book, I have worked mainly behind the scenes to ensure that your geographic literacy is informed by these NCGE standards. The book's three introductory chapters employ all eighteen standards to set the world stage for you, and the chapter outline of each regional chapter reflects (but does not mirror) the six essential elements.

Another conceptual summary of geography's distinctive properties is known as the **Five Themes of Geography.** The National Council for Geographic Education and the **Association of American Geographers (AAG)** developed these themes. Because of their clarity and easy use, many geographers prefer them for teaching, and I encourage you to try them out for yourself (see Try It, **page 6**). Your prof may wish to use this set instead of or alongside the six essential elements and their eighteen standards. The Five Themes of Geography are listed here:[9]

1. Location
2. Place
3. Human–Environment Interaction
4. Movement
5. Region

Table 1.1 The Six Essential Elements and 18 Standards of Geography

1 *The World in Spatial Terms.* Geography studies the relationships among people, places, and environments by mapping information about them into a spatial context (*spatial* means "of or relating to space").

- Standard 1: How to use maps and other geographic representations, tools, and technologies to acquire, process, and report information.
- Standard 2: How to use mental maps to organize information about people, places, and environments.
- Standard 3: How to analyze the spatial organization of people, places, and environments on Earth's surface.

2 *Places and Regions.* The identities and lives of individuals and peoples are rooted in particular places and in human constructs called "regions."

- Standard 4: The physical and human characteristics of places.
- Standard 5: That people create regions to interpret Earth's complexity.
- Standard 6: How culture and experience influence people's perception of places and regions.

3 *Physical Systems.* Physical processes shape the Earth's surface and interact with plant and animal life to create, sustain, and modify ecosystems.

- Standard 7: The physical processes that shape the patterns of Earth's surface.
- Standard 8: The characteristics and spatial distribution of ecosystems on Earth's surface.

4 *Human Systems.* People are central to geography; human activities, settlements, and structures help shape the Earth's surface, and humans compete for control of the Earth's surface.

- Standard 9: The characteristics, distribution, and migration of human populations on Earth's surface.
- Standard 10: The characteristics, distributions, and complexity of Earth's cultural mosaics.
- Standard 11: The patterns and networks of economic interdependence on Earth's surface, process, patterns, and functions of human settlement.
- Standard 12: The process, patterns, and functions of human settlement.
- Standard 13: How forces of cooperation and conflict among people influence the division and control of Earth's surface.

5 *Environment and Society.* The physical environment is influenced by the ways in which human societies value and use the Earth's physical features and processes.

- Standard 14: How human actions modify the physical environment.
- Standard 15: How physical systems affect human systems.
- Standard 16: The changes that occur in the meaning, use, distribution, and importance of resources.

6 *Uses of Geography.* Knowledge of geography enables people to develop an understanding of the relationships among people, places, and environments over time—that is, of the Earth as it was, is, and might be.

- Standard 17: How to apply geography to interpret the past.
- Standard 18: To apply geography to interpret the present and plan for the future.

Source: National Council for Geographic Education, 2012. Geography for Life: National Geography Standards, 2012. http://education.nationalgeographic.com/standards/national-geography-standards/.

Try it The Geography of Anyplace

Try using the Five Themes of Geography to characterize any place. Here is an example to work from, using Ground Zero in Manhattan.

Geographic Characteristics of Ground Zero

Location: Lower Manhattan, New York City (with an exact location of latitude: 40 degrees, 42 minutes, 43 seconds N; and longitude: 74 degrees, 00 minutes, 49 seconds W (later in the chapter, we will look at latitude and longitude).

Place: Formerly, office buildings and firms at the heart of one of the world's great financial centers (a reason it was targeted for destruction); now, a place of historical significance and collective grief for people of the United States.

Human–Environment Interaction: Lower Manhattan occupies low-lying ground that once was marshy swampland. Construction of the twin towers of the World Trade Center, as well as the buildings erected after the 9/11 attacks, required special foundations to keep the Hudson River's water from pouring in.

Movement: Before 9/11, the daily comings and goings of office workers in the World Trade Center; on 9/11, the diversion of airplanes to target the buildings; after 9/11, the flow of tourists and construction crews to the site.

Region: Situated in region of the United States known as the Northeast, in a humid subtropical climate region (in the next chapter, we look at such physical regions).

You can use the five themes to appreciate any place geographically, from the Great Pyramids of Egypt to where you are now. Try it.

The National Geographic Society's educational division recommends, "While the five themes are still used, essential geography content knowledge for students is best described in the National Geography Standards, which were updated in 2012."[10]

The five themes, the six elements, and the eighteen standards cover a lot of ground. In its scope of interests, geography is the most all encompassing of the social sciences (a point of pride for us geographers). Broadly, the discipline has two major branches, **physical geography** and **human geography**, each of which has roots and relationships with other disciplines in the social and physical sciences (•**Figure 1.2**). Although we are classified as social scientists, we geographers often bridge the social and natural sciences and even the humanities in our research, publication, and teaching (another point of pride for us).

As you can see in the center of Figure 1.2, where all the components of the discipline converge, *geography is almost always concerned with the theme of* ***human–environment interaction***. This concern has put geographers at the cutting edge of science and policy in the 21st century because so many of the Earth's most pressing problems—climate change, population growth, and hunger, for example—involve the coupling of human and environmental systems.

Geographers' interests in human–environment interaction, and especially in the ways in which people are changing the face of the Earth, go way back. The great German geographer Alexander von Humboldt (1769–1859) began geography's modern era in a series of classic studies on this theme. From field observations in Venezuela, he concluded, "Felling the trees which cover the sides of the mountains provokes in every climate two disasters for future generations: a want of fuel and a scarcity of water."[11] A century and a half later, we are von Humboldt's future generations. Look at some of the most pressing global environmental issues that concern us today: they include deforestation and shortages of fresh water.

In Humboldt's wake, other geographers in Europe and the United States wrote about environmental changes due to deforestation and the expansion of agriculture and industry. The American geographer Carl Sauer (1889–1975) wrote, "We have accustomed ourselves to think of ever expanding productive capacity, of ever fresh spaces of the world to be filled with people, of ever new discoveries of kinds and sources of raw materials, of continuous technical progress operating indefinitely to solve problems of supply. Yet our modern expansion has been affected in large measure at the cost of an actual and permanent impoverishment of the world."[12] These words have a modern ring to them, but Sauer, a geographer at the University of California–Berkeley, wrote them in 1938. Sauer focused geographers' attention on how the forces of nature and culture shape the **landscape**—the collection of physical and human geographic features on the Earth's surface—and in particular the roles that human ideas, activities, and cultures play in modifying the landscape. Sauer is credited with founding the **landscape perspective** in

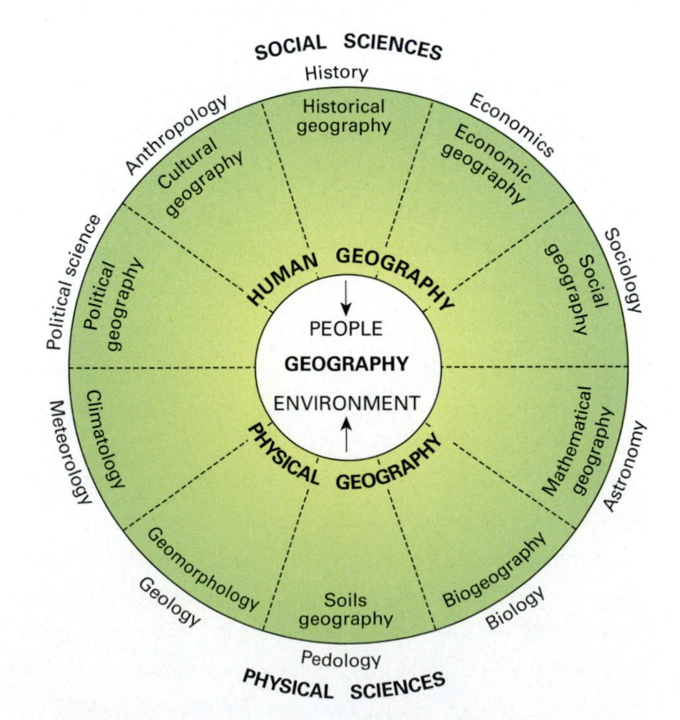

• **Figure 1.2** Selected subfields of geography. These are the main subject areas in human geography and physical geography and their links with the most closely related disciplines in the social and natural sciences.

American geography, based on the method of studying the transformation through time of a **natural landscape** to a 55 **cultural landscape.** Essentially, Sauer challenged us to think of what the world would look like without people and then understand what people have done to reshape the world through time.

Culture—the system of values, beliefs, and attitudes that shapes and influences perception and behavior—underlies many of our decisions about how to use and modify the landscape.[13] That is why geographers are so concerned with cultural features such as ethnicity, language, and religion, and why you will learn much about them in this book.

The World Regional Approach to Geography

The **world regional approach to geography** ranges across the human and physical subfields of geography, synthesizing, simplifying, and characterizing the human experiences of Earth as home. It is impossible to deal with something as large and diverse as our planet without an organizing framework. World regional geography simplifies the task by dividing the world into **regions** (•**Figure 1.3** and •**Table 1.2**). These subdivisions of space are human constructs, not "facts on the ground." People create and draw boundaries around regions that share relatively similar characteristics. A region is simply a convenience and a generalization, helping us become acquainted with the world and preparing us for more detailed insights. This WRG book recognizes eight world regions; others have more or less.

Three types of regions are recognized by geographers. Each is helpful in its own way in conveying information about different parts of the world:

- A **formal region** (also called a **uniform** or **homogeneous region**) is one in which all the population shares a defining trait or set of traits. A good example is a political unit such as a county or a state, where the regional boundaries are defined on a map. Figure 4.2 on **page 91** is a formal region map showing the countries of Europe.
- A **functional region** (also called a **nodal region**) is a spatial unit characterized by a central focus on some kind of activity (often an economic activity). At the center of a functional region, the activity is most intense, whereas toward the edges of the region the defining activity becomes less important. A good example is the distribution area for a metropolitan newspaper, with the highest numbers of subscribers in the city and diminishing numbers at growing distances from the city.
- A **vernacular region** (or **perceptual region**) is a region that popularly exists in people's minds but has no definitive boundaries This region may play an important role in

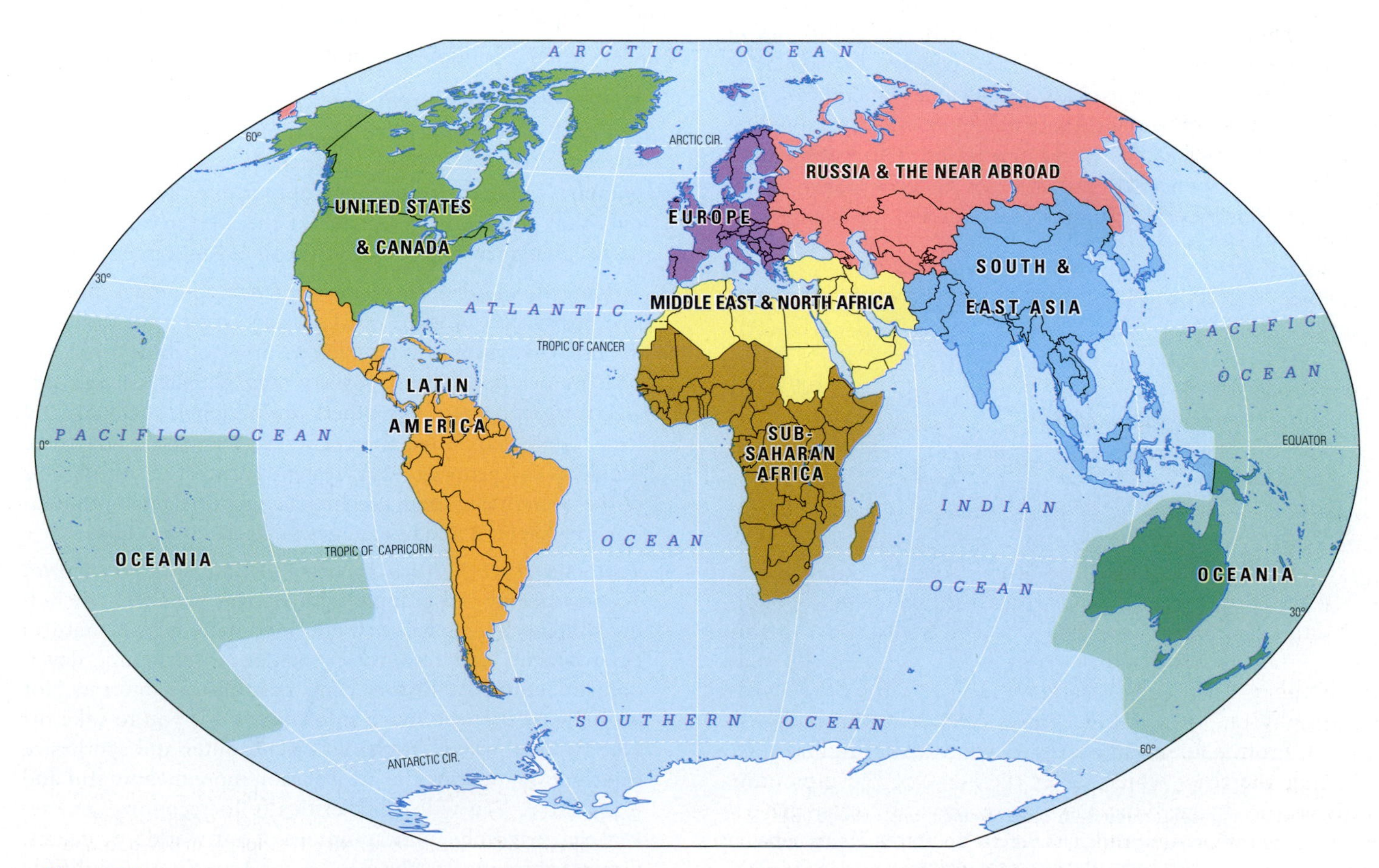

• **Figure 1.3** World regions as identified and used in this book.

Table 1.2 The Major World Regions: Basic Data

Political Unit	Area (sq mi, thousands)	Area (sq km, thousands)	Population (millions)	Rate of Natural Increase (%)	Urban Population (%)	Population Under Age 15 (%)	Agricultural Workers (%)	Per Capita GDP (PPP) ($US)	GDP ($US, billions)	Oil Production (million bbl/day)	Literacy, Female (%)	Literacy, Male (%)	HDI
World	**52,485.9**	**135,935.9**	**7,227.7**	**1.2**	**52**	**27**	**18**	**11,900**	**107,405**	**84**	**79**	**87**	**0.669**
Europe	1,959.3	5,072.0	532.2	0.1	73	16	3	32,400	19,145	3	98	98	0.878
Russia and the Near Abroad	8,533.2	22,100.8	284.5	0.4	65	19	9	12,900	5,068	13	99	98	0.745
Middle East and North Africa	5,416.1	14,027.8	531.2	1.9	63	30	10	11,600	9,150	29.5	77	87	0.694
South and East Asia	8,265.3	21,407.1	3,950.7	1.1	45	25	14	7,900	41,074	7.4	79	88	0.646
Oceania	3,306.8	8,564.5	38.4	1.1	70	24	10	31,200	1,311	0.5	86	89	0.808
Sub-Saharan Africa	8,655.2	22,417.2	919.5	2.6	37	43	61	2,600	3,275	5.7	52	69	0.462
Latin America	7,946.2	20,580.7	618	1.2	78	27	16	14,000	9,336	9.3	88	89	0.734
North America	8,403.8	21,765.8	353.2	0.4	81	19	2	51,800	19,046	15.3	99	99	0.913

Sources: World Population Data Sheet, Population Reference Bureau, 2014; Human Development Report, United Nations, 2014; World Factbook, CIA, 2014.

cultural identity but does not necessarily have official or clear-cut borders. Good examples are the South, the Bible Belt, and the Rust Belt in the United States (•**Figure 1.4**). These regional terms have economic and cultural connotations, but ten people might have ten different definitions of the qualities and boundaries of these regions. Vernacular or perceptual regions, created by individuals and cultures, represent the regional identities that help us organize, simplify, and make sense of the world around us. This book's eight regions are vernacular regions: not all geographers agree which countries make up the Middle East, for example. In

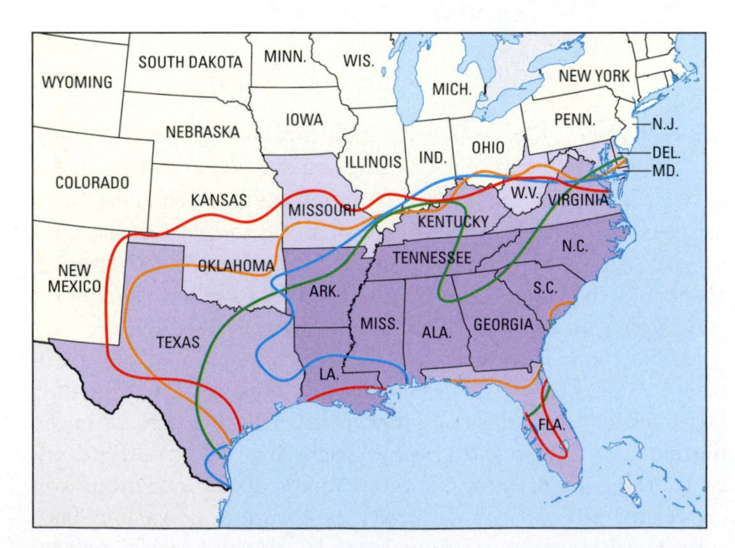

• **Figure 1.4** Definitions of a vernacular region, the American South. Purple shading represents three state-based delineations; colored lines delimit various religious, linguistic, and cultural "Souths." These are just a few of the many different interpretations of the region.

introducing each region, I will tell you what characteristics I chose to define it.

The Objectives of This Book

I have written this book to help you achieve five objectives:

1. *To become geographically literate.* This book will empower you with a comprehensive geographic vocabulary and an advanced command of the "language" of world regional geography. Using the framework of world regions, this book puts the "meat on the bones" of the 18 geographic standards, giving you all you need to achieve geographic literacy.
2. *To understand Earth's problems and their potential solutions.* Like geography broadly, WRG is concerned with problems in human–environment interaction. Some of these problems, such as overpopulation, poverty, and climate change, are global in scope, whereas others are national, regional, and local, or are manifested at these scales. We will see how these problems can be made less threatening and even solved. One of the overarching ones, climate change, first gets our attention in Chapter 2 and re-emerges in all the other chapters.
3. *To use geographic critical thinking to understand the world.* To understand and grapple with Earth's problems, including climate change, we must consider many factors: natural environments and resources, population, economic development, ethnicity, history, and geopolitical interests, for example. Is that too much information for you to take in? No. You will use the tools of WRG to filter and synthesize information, making the information more meaningful and memorable. You will think critically to recognize and reveal the geographic underpinnings of our world's problems. Critical thinking is "the process of actively and skillfully

conceptualizing, applying, analyzing, synthesizing, and evaluating information to reach an answer or conclusion."[14] Using geography's holistic and integrative approach in a regional framework, you will synthesize information, techniques, and perspectives from both the natural sciences and the social sciences. You will tread into the grounds of political science, history, economics, anthropology, sociology, geology, atmospheric science, and other areas. Pulling these issues and perspectives together, thinking critically and finding the links among them is doing geography. Doing this synthesis within a regional framework is doing world regional geography. Thinking critically in this framework, you *will* be able to understand Earth's problems and potential solutions.

Growing your habit of geographic critical thinking will be rewarding for you. Your overall university experience will be richer as you connect the dots between your diverse courses. As you carry on through life, your insight and wisdom may reward you both professionally and personally. More complete knowledge of the world—good geography—is also good business. In the competitive environment of the global economy, better understanding of cultures and environments throughout the world helps boost the "bottom line." You may be surprised how much your geographic knowledge, enhanced by your ability to produce insight and advice from it, will help you in your career, whatever it turns out to be.

4. *To understand the geography of current events.* This book is carefully written to set the stage of world events for you. With the book and your professor's guidance, you should become able to read and view news with a much better understanding of the issues underlying world events. Incidents like earthquakes and tsunamis in the western Pacific, disease epidemics originating in southern China, and Russia's invasions of neighboring countries are not random, unpredictable events. They are rooted in consistent, recognizable problems that have geographic dimensions. You will find it satisfying to be "pre-informed" about a problem that suddenly appears in the news. You will become somewhat of an expert on **geopolitics**, the struggle for space and power played out in a geographical setting.[15]
5. *To develop the ability to interpret places and "read" landscapes.* In doing geography, you will be concerned both with **space** (the exact placement of locations on the face of the Earth) and with **place** (the imprecise but important physical and cultural contexts of a location). Place is much more subjective than space because, like a vernacular region, it often is defined by the meanings of a particular location. For example, your perceptions of New York City may be very different from those of your friend and may be shaped by personal experience in the "Big Apple" or by photographs or movies you have seen. In this book, there is much discussion of the "sense of place" that individuals and groups have about locations and regions. Perception of place can have a very strong influence on how we make decisions and interact with others. Perception of place can even have a strong impact on world events. For example, in Chapter 6 on the Middle East, you will see how Jewish and Muslim perceptions of sacred places located within a few meters of each other in Jerusalem play crucial roles in conflict and peacemaking in the Middle East and beyond. With •**Figure 1.5**, let's consider an example of how you can use your critical thinking skills to define and identify place.[16] As you work forward through your book and course, you will get better at identifying the many elements of place identity, including climate, vegetation, and landforms of the physical environment and the language, religion, history, and livelihoods of the people living in that environment. Your skill in interpreting places will even help make you a better traveler.

Joe Hobbs

• **Figure 1.5** Study this photograph, and name the country—or at least the region—where it was taken. What clues in the physical and human geographies of this place help you locate it? For more clues and the place identification, see note 16 on page 23.

1.2 The Language of Maps

We turn now to the most important tool that geographers use to explore and explain relationships on our planet: the map. As geographers study people, places, and environments, we usually (but not always) collect and depict information that can be mapped. In other words, we are interested in the **spatial** context of the things. As noted in the first essential element of geography, **spatial** means "of or relating to space."

A **map** is a representation of various phenomena over all or a part of the Earth's surface, usually rendered on a flat surface such as paper or a computer monitor. The science of making maps is called **cartography**. There are two basic types of maps: reference maps and thematic maps. **Reference maps** are concerned mainly with depicting the locations of various features, both natural and human-made, on the Earth's surface (road atlases are a good example, as are the opening maps for each regional chapter, such as Europe in **Figure 4.1**). **Thematic maps** show the spatial distribution of one or more attributes

Insights Mental Maps

When someone asks you, "Could you draw me a map of how to get there?" you might quickly draw some lines, write down some street names, talk about some familiar landmarks, and apologize for how crude your map is. Your map would probably end up looking very different from that of another person asked the same question. Our understanding of location is not completely objective. Each of us has a personal sense of space and place and associations with them.

A **mental map**, like a vernacular region, is a collection of personal geographic information that each of us uses to spatially organize the images and facts we have about places, both local and distant. We constantly draw upon that geographic information to make our way through daily life, and are always revising and updating that information as we succeed or fail on our way. Sometimes we use that information to create actual maps. These maps are not accurate, precise, or scientific, but they portray useful information and tell us much about the individuals and cultures that create them.

across a given area. Thematic maps can be divided into two categories: quantitative and qualitative. Quantitative thematic maps show the spatial distribution of numerical information (such as population density or income levels, as in Figure 3.7 on **page 56**), whereas qualitative thematic maps display non-numeric data (such as the distribution of climates or languages, as in Figure 2.4 on **page 30**).

As maps are an essential tool in the study of world regional geography, it is important that you know how to read them. The main elements of the "language of maps" are *scale, coordinate systems, projections*, and *symbolization.*

Scale

A map is a reducer; it shrinks an area to the manageable size of a chart, piece of paper, or computer monitor. The amount of reduction appears on the map's **scale**, which shows the actual distance on Earth as represented by a given linear unit on the map. A common way to depict scale is with a fraction or ratio, such as 1:10,000 or 1:10,000,000. In the fraction, one linear unit on the map (for example, 1 inch or 1 centimeter) represents 10,000 or 10,000,000 such real-world units on the ground. A **large-scale map** is one with a relatively large representative fraction (for example, 1:10,000 or even 1:100) that portrays a relatively small area in more detail. A **small-scale map** has a relatively small representative fraction (such as 1:1,000,000 or 1:10,000,000) that portrays a relatively large area in more generalized terms. Compare the two maps in **• Figure 1.6**. Figure 1.6a is a small-scale map showing San Francisco and surrounding parts of the Bay Area. Figure 1.6b is a large-scale map that "zooms in" on part of San Francisco. *Remember this inverse relationship: a small-scale map shows a large area, and a large-scale map shows a small area.*

(a)

(b)

• Figure 1.6 **(a)** Small-scale and **(b)** large-scale maps of San Francisco and environs.

How good are you at reading maps? What about imagining or drawing them? (See the Try It feature on this page.) Did you ever wish you were better at it? This section should help you—and it will be useful as you navigate this book.

Mental Maps

Try it

You have mental maps in your mind. Try this: without referring to this book or any other source, draw your map of the world. It does not need to be detailed. But try to get outlines of the continents on your map, with their rough shapes and relative sizes. Then compare yours with a world map in the book or elsewhere. How did you do? Yes, you can laugh at yourself. This is not something that most of us are proficient at. Did you lean toward a certain projection you might be familiar with, like the Mercator with its large polar land areas (see **page 16**)? It is very likely that if you try this again when you are finished with the course, your mental world map will be much improved.

Coordinate Systems

Maps cannot convey the subjective meanings associated with place, but they are very effective in conveying information about space and location. In this book, you will be concerned with two kinds of **location:** relative location and absolute location.

Relative location *defines a place in relationship to other places.* You can derive this kind of information from many maps. Relative location is one of the most basic reference tools of everyday life; you might say you live south of the city, just west of the shopping mall, or next door to a good friend. As you proceed through the book, relative location will become part of your basic geographic knowledge and your critical thinking about geography. You might look at Figure 5.33, on **page 193**, to see, for example, how tantalizingly close the legally-Ukrainian port of Sevastopol is to Russia's southwestern border. Despite its vast size, Russia has few ports in warm waters accessible for seafaring throughout the year. Understanding the implications of relative location will prove quite useful for you in following world affairs; in this case, you can easily appreciate one of the reasons why Russia asserted control over the Ukraine's Crimean Peninsula, which juts into the Black Sea (see **page 180** and Geographic Spotlight, **page 12**).

Absolute location refers to a point on the Earth's surface. Also known as **mathematical location**, absolute location is essential in reference maps, but not always in thematic maps. **Coordinate systems** are used to determine absolute location. These coordinate systems use a network of grids consisting of horizontal and vertical lines covering the entire globe. The intersections of these lines create addresses in a global coordinate system, giving each location a specific, unique, and mathematical placement (as appears, for example, as a "waypoint" in the common global positioning system (GPS) device).

The most common coordinate system uses **parallels** of **latitude** and **meridians** of **longitude**. The term *latitude* denotes position with respect to the Equator and the poles (see **•Figure 1.7**). Latitude and longitude are measured in **degrees** (°), **minutes** (′), and **seconds** (″). Each degree of latitude, which is made up of 60 minutes, is about 69 miles (111 km) apart; these distances vary a little because Earth is a slightly flattened ("oblate") sphere or ellipsoid. Each minute of latitude, which is made up of 60 seconds, is therefore roughly a mile apart. The **Equator**, which circles the globe east and west midway between the poles, has a latitude of 0°. All other latitudinal lines are parallel to the Equator and to each other, which is why they are called *parallels*.

Every point on a parallel has the same latitude (for example, places on the Equator in both South America and Africa are located at 0° latitude). Places north of the Equator are in **north latitude**. Places south of the Equator are in **south latitude**. The highest latitude a place can have is 90°N (the **North Pole**) or 90°S (the **South Pole**). Places located between the **Arctic Circle** at 65.56°N and the North Pole, and between the **Antarctic Circle** at 65.56°S and the South Pole, form the most commonly recognized boundaries of the **high latitudes**. Places located between the **Tropic of Cancer** and the **Tropic of Capricorn**, at 23.44°N and 23.44°S, respectively, are said to be in **low latitudes**. Places occupying an intermediate position with respect to the poles and the Equator are said to be in the **middle latitudes**. Incidentally, there are no universally accepted definitions for the boundaries of the high, middle, and low latitudes. The northern half of the Earth between the Equator and the North Pole is called the **Northern Hemisphere**, and the southern half between the Equator and the South Pole is the **Southern Hemisphere**.

Meridians of longitude are straight lines connecting the poles (see **Figure 1.7b**). Every meridian runs due north–south. All the meridians converge at the poles and are farthest apart

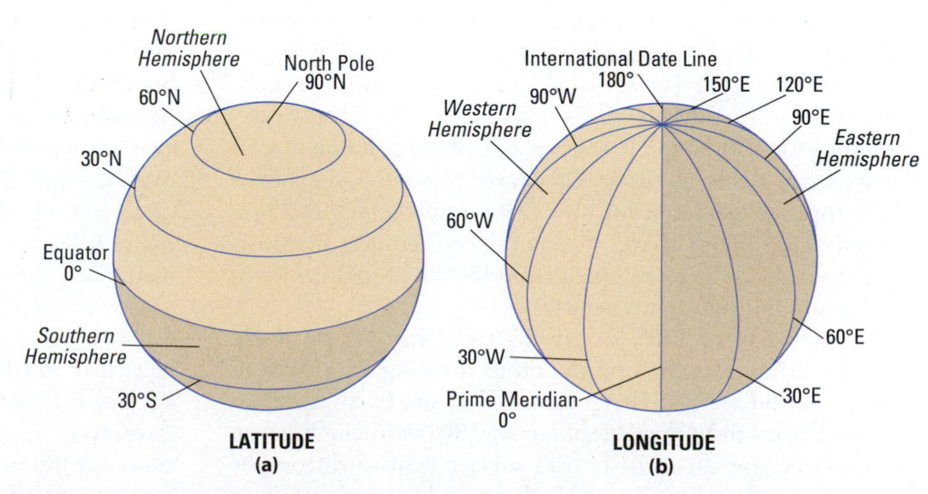

• Figure 1.7 **(a)** Earth's lines of latitude (parallels) in increments of 30 degrees, from the Equator (0 degrees) to the North Pole (90 degrees north latitude). **(b)** Earth's lines of longitude (meridians) in increments of 30 degrees.

Geographic Spotlight Tobler's First Law of Geography

Relative location is at the heart of a geographic axiom known as ***Tobler's First Law of Geography***: *"Everything is related to everything else, but near things are more related than distant things."* This observation is especially useful in the quantitative realm of spatial data analysis, such as in maps and GIS. But it is useful in the qualitative or subjective realm as well. As we explore Russia, for example, we will peel back the layers of Russia's geopolitical concerns and see that Russia's periphery, or "Near Abroad," including Ukraine, is most relevant to its foreign policy.

177

As the Swiss-American geographer Waldo Tobler himself admitted, his observation may be more a "principle" than a "law." In any case, as you practice geography, you will find that it is more often true than not.[17]

at the Equator. Lines of longitude are not the same distance from one another across the globe, so their values vary. At the Equator, the distance between lines of longitude is about 69 miles (111 km), whereas at the Arctic Circle it is only about 28 miles (45 km). Just as there must be a zero reference line for lines of latitude—the Equator—there must be a zero reference line for lines of longitude. Known as the **Prime Meridian**, it has a longitude of 0 and serves as the reference line from which longitude east and west is measured. Places east of the Prime Meridian are in **east longitude**; places west of it are in **west longitude.** The Prime Meridian is also known as the **Greenwich Meridian** because it passes through the Royal Astronomical Observatory in Greenwich (pronounced "Gren-ich"), England.

Why on Earth Greenwich? By 1884, with the Industrial Revolution in high gear and global economic activities becoming increasingly interconnected, there needed to be universally accepted reference points of zero for both longitude and time. Conferring in Washington, DC, in 1884, 23 major world powers voted overwhelmingly for Greenwich, mainly because most nautical charts at the time already used Greenwich as Prime Meridian. The Royal Astronomical Observatory in Greenwich had state-of-art equipment, including the Transit Circle telescope whose crosshairs would exactly define the longitude 0° for the world. The 1884 conference also established the world's 24-hour time system, with all time zones referring back to Greenwich Mean Time (GMT; also referred to as Coordinated Universal Time, or UTC) on the Prime Meridian. You can straddle the Prime Meridian line in the Observatory so that one foot is in the Eastern Hemisphere and one foot in the Western. Being in the world center of time and space is an appropriate metaphor for how Britain saw herself in 1884. Others, especially rival and frequent enemy France, saw Britain differently. France abstained from the vote on Greenwich. (see Geographic Spotlight, page 13)

The meridian of 180°, exactly halfway around the world from the Prime Meridian, is the other dividing line between places east and west of Greenwich. All of the Earth's surface eastward from the Prime Meridian to 180° is in the **Eastern Hemisphere**, and all of the Earth's surface westward from the Prime Meridian to 180° is in the **Western Hemisphere.** This meridian of 180° has another purpose in addition to complementing the Prime Meridian. It serves as the **International Date Line**, where the beginning of one day and the end of another day meet. The Earth has 24 time zones, and there must be a line where the Earth's clock begins and ends. The line has a few zigzags in it for political and practical reasons (especially to fit a country or part of a country in a single time zone; see Figure 8.2, page 370). The date west of the line is one day ahead of the date east of the line. The person traveling west across the International Date Line gains a day (crossing from Monday to Tuesday, for example); and traveling east, loses a day (crossing from Monday to Sunday).

You now have the ability to create and interpret the absolute location of any spot on Earth. Try it—see the feature on page 14.

Projections

The truest and most reliable cartographic representation of Earth is a globe. But you can't be carrying a globe to class, and to depict Earth we unfortunately almost always have to flatten it out. Think of peeling an orange and pressing the peel, representing the surface of the Earth, onto a piece of paper (as in Figure 1.8f on page 15); the spread-out peel represents the sphere but cannot recreate its shape. How do cartographers render our 3D planet in two dimensions? Because the Earth is spherical, any map created on a flat surface will inevitably have some distortion. **Map projections** are mathematical applications to minimize this distortion. There are thousands of projections, but there is no single "correct" or "perfect" projection for any particular map. All flat maps have varying amounts of distortion among the four basic properties of a globe: area, shape, distance, and direction. On large-scale maps (such as a map of a city), the distortions are small enough that they may be disregarded for most purposes, but on maps of smaller scales (countries, continents, the entire world) distortion becomes a greater problem as the scale becomes smaller.

Most projections work by transferring the features of the spherical Earth onto a "developable surface" (a geometric surface that can be flattened into a plane without tearing or stretching) such as a plane, cylinder, or cone. These projections are referred to as **azimuthal**, **cylindrical**, and **conic**, respectively (there are also notable subsets of these categories, such as pseudocylindrical and polyconic). See **•Figure 1.8** for examples of these projections. A few projections are

Geographic Spotlight Core Location and Peripheral Location

Among the geographical concepts used throughout this book are those of **core location** and **peripheral location** (the region of Europe is subdivided along these lines, for example). Some locales have greater importance in local, regional, or world affairs because they have a central, or core, location relative to others. Other, peripheral, locales are less important because they are situated farther from "where the action is." A comparison of two countries, the United Kingdom (UK) and New Zealand, provides a good example (**• Figure 1.A**). Both are island countries. Their climates are remarkably similar, although they are in opposite **hemispheres** (half spheres) and are about as far apart as two places on Earth can be. Westerly winds blow off the surrounding seas, bringing abundant rain and moderate temperatures throughout the year to both.

But there are important differences. The United Kingdom is located in the Northern Hemisphere, which has the bulk of the world's land (it is sometimes described as Earth's **land hemisphere**) and most of its principal centers of population and industry; New Zealand is on the other side of the Equator, surrounded by the vast expanses of water in the Southern Hemisphere (sometimes known as Earth's **water hemisphere**) and off the beaten track of the globe's economic activity.

As Figure 1.A illustrates well, the United Kingdom is located near the center of the world's landmasses. Only a narrow channel separates it from the densely populated industrial areas of western continental Europe. Many major oceanic commercial routes converge on this western seaboard area of Europe (see Figure 2.11). For centuries, the United Kingdom has played a major role in the economic and political development of northwestern Europe. New Zealand, meanwhile, has been a far outpost of that development and history. The United Kingdom has a central, or core, location in the modern framework of human activity on Earth, whereas New Zealand has a peripheral location.

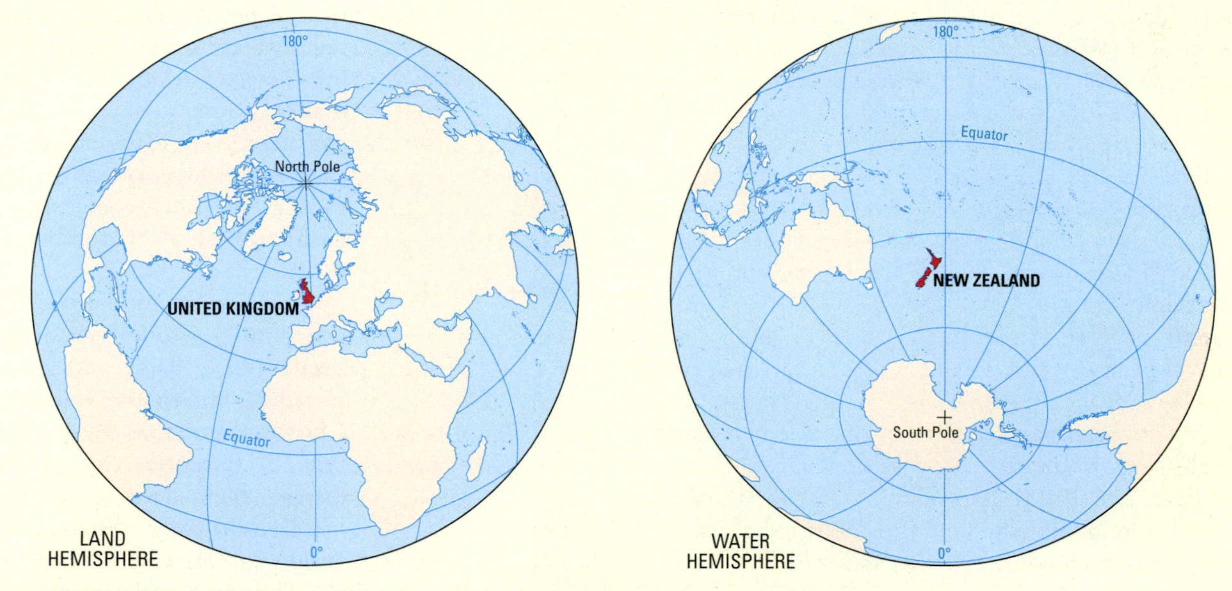

• Figure 1.A In the left map, note how the major landmasses are grouped around the margins of the Atlantic and Arctic Oceans. The British Isles and the northwestern coast of Europe lie in the center of the "land hemisphere," which constitutes 80 percent of the world's total land area and has about 90 percent of the world's population. In the map on the right, New Zealand lies near the center of the opposite hemisphere, or "water hemisphere," which has only 20 percent of the land and about 10 percent of the population.

"interrupted," instead of contiguous; the most well known of these is Goode's Homolosine, which has a "peeled orange" look (see **Figure 1.8f**).

Projections are also classified by which metric property they preserve (or distort the least). **Conformal projections,** which include the Mercator projection discussed in the feature on **page 16**, preserve shapes well. **Equal-area projections,** as the name suggests, preserve area (but no equal-area projection can preserve shapes, and no conformal projection can preserve area). **Equidistant projections** preserve distance from a specific point to all other points. Map projections that do not preserve any one metric, or try to distort all properties about equally for aesthetic purposes (making the map "look right"), are called **compromise projections.** The Winkel Tripel projection used for the world maps in this book (**Figure 1.3** is an example) is a compromise projection.

Another important property of a map is its **orientation,** the relationship between the direction on the map and the corresponding compass directions in reality. Almost all maps place north at the top, but there are sometimes reasons to orient a

Try it Latitude and Longitude

Here is a useful exercise to ensure that you understand how latitude, longitude, and absolute location work.

On the map in • **Figure 1.B**, the latitude of Madrid, Spain, is approximately 41 degrees north latitude, 4 degrees west longitude (41°N, 4°W).

What are the approximate latitude and longitude coordinates of Oslo, Norway? In which hemispheres (north/south; east/west) is Oslo located?[18]

The answer is in note 18 on **page 23**. Understanding absolute location is a simple and indispensible part of the language of maps.

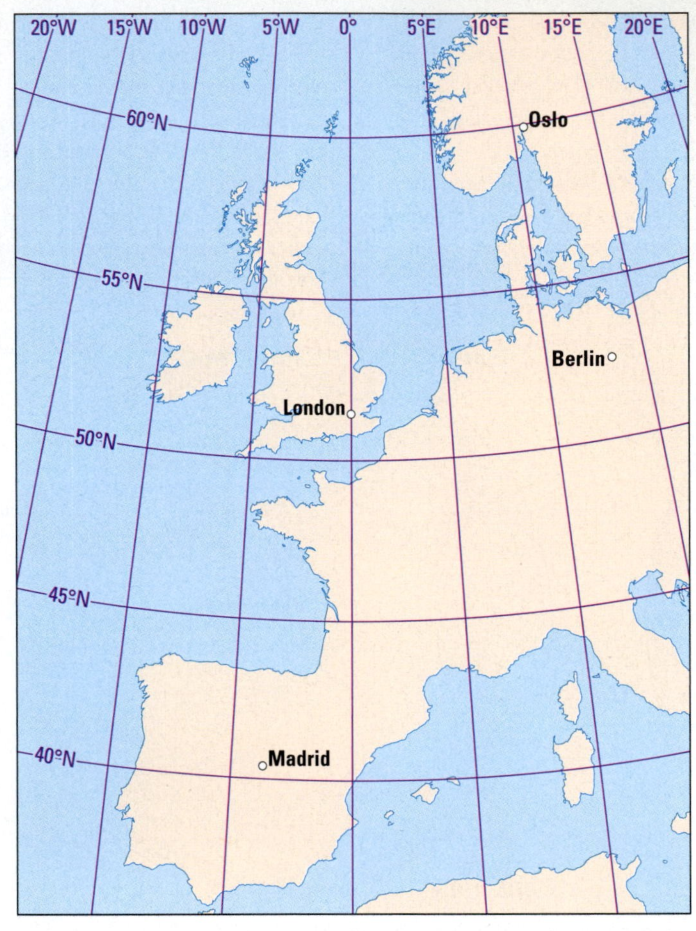

• **Figure 1.B** What are the approximate latitude and longitude coordinates of Oslo, Norway? (The answer is in note 18 on page 23.)

map differently, and it is possible to present a different perception of geographical space by changing a map's orientation. There are some radical orientations that literally turn the world upside down. One of the most interesting is Australian—the "What's Up? South" world map with south at the top, that you may find easily in an Internet search.

Symbolization

Maps allow us to get information, to see patterns of distribution, and to compare these patterns with one another. But no map can be a complete record of a given area. In a process called "cartographic abstraction," the map's cartographer chooses important details to convey the map's information. As a map user, you must keep in mind that no map can be complete and that many details must be simplified or omitted to keep a map legible. All maps must "lie" to some degree to inform their readers (a truth that is told in an interesting book entitled *How to Lie with Maps*). In some instances, you should look at maps very critically, in case details are altered or left out in order to deceive you. Could a "hidden agenda" be lurking in a map you're looking at? See •**Figure 1.9** on **page 15**, and the issue of Israel not appearing on some maps in note 19.

Symbolization refers to the need for geographic features shown on maps to be represented by symbols, such as lines, fills, shapes, colors, and type. For example, the political map of Europe in Figure 4.2 indicates the relative importance of cities by varying the size and boldness of the type, makes different countries easier to tell apart by using different fill colors for each, and shows coastlines and rivers as solid blue lines, contrasting with the dashed gray lines that indicate country boundaries.

Thematic maps often (but not always) use just one type of symbol to display their data, and they can be classified by which symbol they use. **Choropleth maps**, the most common type of thematic map in this book, display their data by filling in political units with differing colors. A good example of a choropleth map is the Human Development Index map (Figure 3.8 on **page 57**). **Isarithmic maps** do not use political units, but instead use lines or bands of color to join points of equal value across the mapped area. A topographic map showing contour lines is an example of an isarithmic map, as is the map of world precipitation in Figure 2.3. **Graduated symbol maps** use simple symbols, such as circles, scaled proportionally to the quantity of the attribute being mapped. See **Figure 4.50**, showing the distribution of the Roma people across Europe, for an example. **Dot density maps** (such as Figure 3.21, the map of world population, on **page 75**) use dots to represent a stated amount of some phenomenon within a political unit (for example, if one dot equaled 1000 people, twelve dots in an area would indicate 12,000 people). **Flow maps** use arrows of various widths to indicate the movement of people or goods from one area to another (see Figure 6.41, the map of the movements of Palestinian refugees, on **page 251**).

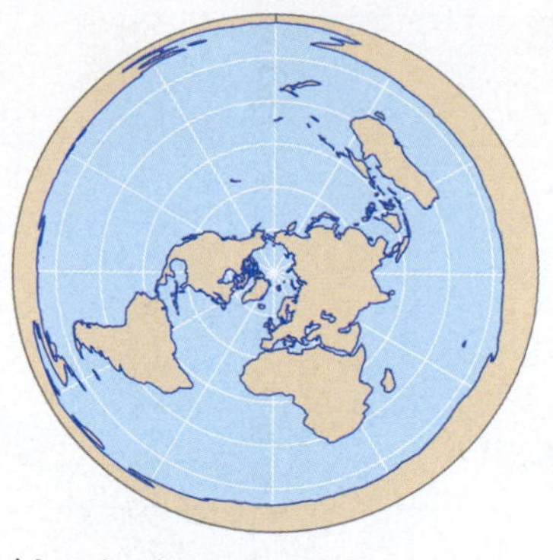

(a) An azimuthal projection (North Pole Azimuthal Equidistant)

(b) A cylindrical projection (Miller)

(c) A pseudocylindrical projection (Mollweide)

(d) A conic projection (Albers Equal-Area Conic)

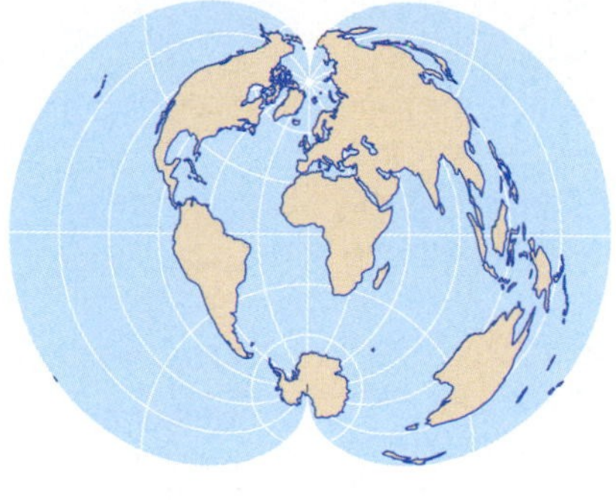

(e) A polyconic projection (American Polyconic)

(f) An interrupted projection (Goode Homolosine)

• **Figure 1.8** Examples of different map projections.

Enbridge Map of Douglas Channel

Kitimat

(a)

Actual Map of Douglas Channel

(b)

• **Figure 1.9** Maps are not always objective renditions of the landscape. Some have an agenda, and it is possible to "lie" with maps. **(a)** Depicts an open, easy access to the sea through the Douglas Channel, while **(b)** depicts a passage with islands that require ships to navigate with caution.

Source: http://sumofus.org/campaigns/enbridge/?sub=fb.

THINKING CRITICALLY: The Canadian energy company Enbridge wants to ship oil out of Kitimat, British Columbia. A straight channel without islands would minimize the prospect of a catastrophic oil spill. Such geographic conditions, as suggested in the Enbridge map of this area **(a)**, would help win public and regulatory approval for this energy project. **(b)** Includes more islands in the passage, reflecting site and situation more accurately. What's going on here? To turn Fritz Gritzner's phrase, "What is not where, why not there, and why care?"

1.3 Geographic Technologies and Careers

To close this chapter, we turn to some of the most innovative tools and breakthroughs in geography and consider how you (or someone you know) might become part of the action in this growing field. This section will also open your eyes to some of the newest tools for gaining both an easier and deeper understanding of the world's regional geographies. The opportunities that may open to you by studying geography are abundant and exciting.

The Geospatial Revolution

In recent decades, technological advances in the field of geography and related sciences have accelerated so quickly and with such impact that they can truly be called "revolutionary." These advances are not confined to the laboratory or library. Like

Geographic Spotlight The Mercator Projection

The most widely recognized map projection is the one developed by Gerardus Mercator in 1569. Mercator (1512–1594) developed his cylindrical, conformal **Mercator projection** (**Figure 1.C**) mainly for navigational use: it was designed to show lines of constant compass bearing (known as "rhumb lines") as straight lines, which was very important to assist sailing vessels in charting their courses. This was a significant improvement over previous projections, and Mercator projections are still used for water navigation today. However, the projection's usefulness at sea makes it largely unsuitable for other purposes, including reference world maps. In order for the rhumb lines to be shown straight, the projection must continually increase the spacing between the parallels away from the Equator. This results in enormous distortions of size approaching the polar areas (in fact, the poles themselves cannot 108 be shown on a Mercator map as they lie at infinity). On a Mercator map, Greenland and Africa appear similarly sized, whereas in reality Africa is about 14 times larger than Greenland!

Despite the objections of cartographers for many decades, the Mercator world map projection is still common in classrooms, TV newscasts, and online mapping services such as Google's. Its straight lines and convenient rectangular shape help make it an attractive "go-to" default map of the world. But think critically about this map projection. On many map images, the Mercator projection depicts the United States or Europe, despite their northern locations, as being essentially at the "center of the world." This is done by removing most of Antarctica but keeping much of the Arctic in the frame. Indeed, the Mercator projection's exaggeration of mid-northern latitudes made it very popular in the West during the age of European colonialism. The Mercator projection was long accepted as being "authoritative," and intentionally or otherwise conveyed a geographic sense of Western dominance of the world. Although some people are slow to abandon it for less distorted projections, the Mercator projection is fading in part because of such perceived biases. Most atlases and textbooks no longer use Mercator. But it is still prevalent enough in other media that it is important to recognize this projection's drawbacks and to think critically about how a map projection can influence our perceptions of the world.

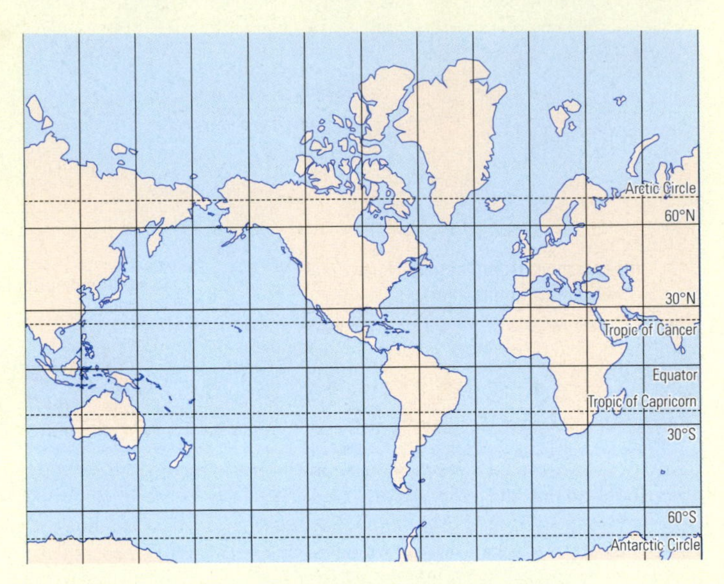

• **Figure 1.C** The Mercator Projection.

THINKING CRITICALLY: Examine the Mercator projection of the world in Figure 1.C, and compare it with other projections in Figure 1.8. What issues do you see with Mercator and the similar cylindrical projection in Figure 1.8b? What about that issue of the size of Greenland? Do you see why we favor the Winkel Tripel projection used in Figure 1.3?

information on the Web, they have been democratized: you can use the power of geographic tools on your own computer, tablet, or smartphone, and even accept the challenge of *Wikimapia: "Let's Describe the Whole World!"* (http://wikimapia.org). Digital cartography has given you access to newer, better, and more accurate maps than the world has ever seen. Lost perhaps are the romance and mystery of *terrae incognitae*, the unknown lands inscribed across sometimes vast regions of old maps of South America and Africa. As recently as the 1980s, I worked in the deserts of Egypt with very old, small-scale paper maps where large wilderness areas were blank. My nomadic Bedouin desert companions and I spent much time walking, talking, and filming to fill in these blanks. We could not have foreseen that 30 years later technology would allow us to probe the remotest corners of their vast territory. Recently, one of these old Bedouin friends, now settled down, gave me a tour of his desert, using Google Earth on his laptop! (•**Figure 1.10**)

A relatively new term relates to most of these technological advances: **geospatial**. This adjective *geospatial* means "pertaining to the geographic location and characteristics of natural or constructed features and boundaries on, above, or below the Earth's surface, especially referring to data that is geographic and spatial in nature."[20] Even with simple paper maps, geographers have always worked with geospatial information. But the word *geospatial* has connotations of powerful computer hardware, software, and information-gathering tools including satellites. These technologies are evolving rapidly. The geospatial revolution is still underway, and it will touch your life in many ways. Chances are that you have used or seen a GPS device in a car or on a mobile device in hand. You have probably looked up directions on Google Maps or MapQuest. Underlying these devices and applications are enormous and sophisticated technologies. Two of the most important geospatial technologies are geographic information systems (GIS) and remote sensing.

GIS—the familiar acronym for **geographic information systems**—is a computerized data management system that allows people to create, capture, retrieve, manipulate,

• **Figure 1.10** Abdel Zaahir Sulimaan 'Awda of the Khushmaan Ma'aza Tribe in his home in Hurghada, on Egypt's Red Sea Coast. I was astonished when he pulled out his laptop, powered it up, and used Google Earth to show me around the rugged desert wilderness of the Ma'aza. Thirty years earlier, when his family was nomadic, I walked with him in that desert, and we used primitive paper maps to find our way to the summit of mainland Egypt's highest mountain, Shaayib al Banaat (2187 m). And now we correspond on Facebook!

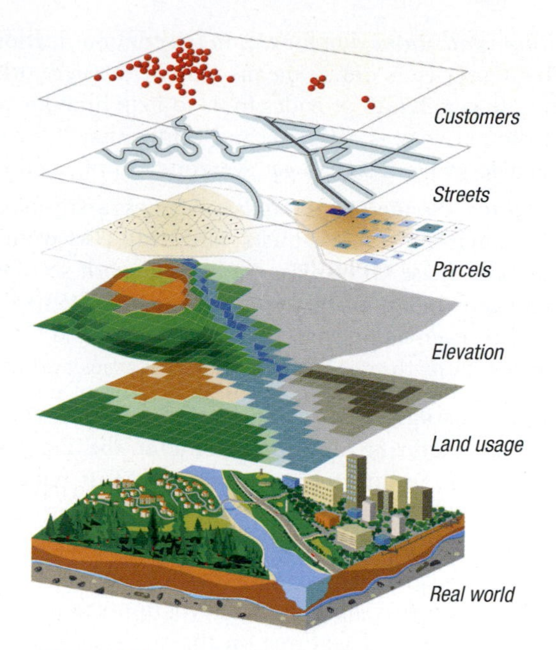

• **Figure 1.11** Geographic information systems (GIS) create and use layers of spatial data. GIS data, images, and models have an enormous range of applications.

analyze, view, and display spatial information. GIS is our discipline's leading area of growth and employment because of its value in addressing so many real-world problems. Like a conventional paper map, a GIS view can offer you a road map. But do a mouse click on the road, and you unleash the power of GIS: Up comes a box displaying the road's name; its length and width; when it was last repaved; its speed limit; whether it is owned by the city, county, or state; the amount of traffic it receives in a certain time frame; and so forth. These different types of nonspatial information about geographic features in the GIS, linked to those geographical features by unique identifiers stored in tables, are called "attributes." People spend an enormous amount of time and energy to compile these data from a variety of sources.

GIS data are created and displayed in "layers"—databases storing the locations and attributes of features belonging to a single theme (for example, with layers showing a company's potential customers, the streets on which those customers live, their land ownership, the land's elevation, and land uses in their community, as in •**Figure 1.11**). A GIS can have dozens of layers displaying all kinds of geospatial data for a given area. GIS users can select and highlight certain attributes in various layers that relate to their interests or analysis, while turning off those attributes that are not needed. The ability to query and selectively view certain types of geospatial data, as well as to add other data (such as from a handheld GPS device, from a digitized paper map, or from an image taken by an orbiting satellite), allows users to see spatial relationships with extraordinary ease and clarity. The knowledge they gain from understanding geographic relationships is extremely valuable for decision making and explains why GIS is known as a "critical thinking technology" (see Geographic Spotlight GIS in Action on **page 19**).

GIS makes it possible to see geographic patterns, problems, and connections easily and efficiently—a capability that Esri, the world's largest GIS software maker, calls "The Geographic Advantage."[21] One of the best ways to see what GIS can do is to visit Esri's website "gis.com," where you will see many ways that GIS is being used all around you—and all around the world. Here are a few examples of the geographic advantage as defined by Esri.

- *Natural resource management.* In forestry, caring for existing and future trees ensures a steady supply of wood for the world's building needs. GIS provides tools to help determine where to cut today and where to seed tomorrow, while minimizing negative impacts.
- *Business.* Every day, businesses deliver goods and services to clients all around a city. Each truck driver needs a route of how to most efficiently visit each client. GIS provides tools to create efficient routes that save time and money and reduce pollution.
- *Defense.* In the military, leaders need to understand terrain to make decisions about how and where to deploy their troops, equipment, and expertise. They need to know which areas to avoid and which are safe. GIS provides tools to help get personnel and materials to the place where they can best do their job.
- *Emergency preparedness and response.* During floods and hurricanes, emergency response teams save lives and property. GIS provides tools to help locate shelters, distribute food and medicine, and evacuate those in need.

- *Communications and media*. In telecommunications, when phone service is out, it means part of the network may be disconnected. GIS provides tools to help find out what part of the network is affected, and brings that information to the field so workers can get everyone talking again.
- *Planning*. Planners of all kinds—business analysts, city planners, environmental planners, and strategists from all organizations—use GIS to lay out a framework so that growth can occur in a managed way. As you will see in Chapter 7, GIS can even help us plan for a warmer world in which rising sea levels threaten to flood coastal cities and farms.

346

Remote sensing, also known as *Earth observation*, is the science of acquiring information about the Earth's surface without being in direct contact with it. Most remote sensing data are obtained by sensors on Earth-orbiting satellites or by aerial photography (cameras mounted on airplanes taking pictures of the ground). Remote sensing is not limited to cameras that capture visible light; much important information about processes and features on the surface or in the atmosphere is gleaned from satellite sensors that can "see" other parts of the electromagnetic spectrum, such as infrared and microwave wavelengths. Radar (radio detecting and ranging, which measures the reflection of satellite-emitted radio waves bouncing off of ground features) and lidar (light detecting and ranging, which uses light in the same way) are also examples of remote sensing.

Remote sensing is an exceptionally good tool for helping geographers understand how people and natural processes modify the Earth. This book uses remote sensing to introduce you to many places, patterns, and problems. "A picture is worth a thousand words," the adage says, and remote sensing images like those you see in Google Earth go far in describing natural and cultural landscapes (see Geographic Spotlight on **page 20**, and Try It on **page 21**).

YOUR MOM SAID YOU SHOULD MAJOR IN SOMETHING THAT WILL GET YOU A GOOD JOB. YOU REALLY DO WANT A GOOD JOB AFTER YOU GRADUATE. BUT DON'T YOU WANT TO DO SOMETHING YOU LOVE? WHAT IF YOU COULD DO BOTH? WHAT IF YOU COULD ENJOY YOUR WORK, GET PAID FOR IT, AND HAVE A REAL IMPACT ON THE WORLD? AFTER ALL, WE ALL WANT TO MAKE A DIFFERENCE.

YOU REALLY DO KNOW WHERE YOU WANT TO GO.

GEOGRAPHY

CAN TAKE YOU THERE.

For more information about Careers in Geography, go to www.aag.org.

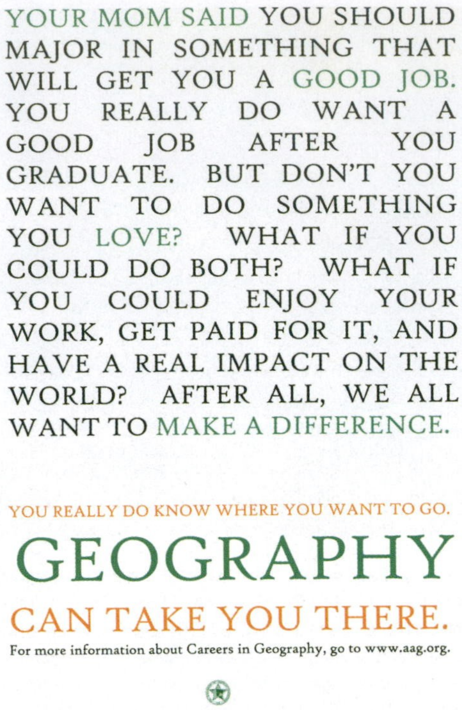

• **Figure 1.12** Geography is awesome.
Source: Dr. Patricia Solís, Association of American Geographers, copyright registered 2004. Reprint permission granted for educational and dissemination purposes only; please do not reprint, translate, or otherwise alter without express written permission by the author.

Careers in Geography

This striking poster in • **Figure 1.12** was produced by the Association of American Geographers, using data about how young people in the United States choose their career paths.[25] The employment trends in favor of geography since 2000 have been remarkable. The US Department of Labor has identified geospatial technology as one of the most important emerging and evolving fields in the technology industry. This agency reports that the geospatial market is growing at an annual rate of almost 35 percent, with the commercial subsection of the market expanding at the rate of 100 percent each year.[26]

Much of this job growth is in professions in which people feel like they can have a positive impact. The Association of American Geographers had these observations of global trends that are contributing "to a renaissance of geography and its potential for making a difference in society and the world":

> These include globalization at an increasing pace and scale, phenomena that compel greater understanding of the world, places, people, and natural systems that affect us as a planet and as global citizens and consumers. It includes a recent proliferation of geographic technologies, once fairly obscure and now pervasive in our daily lives, such as GPS in cell phones and cars, online mapping at your fingertips, cable news reports using spatial visualizations, and many more applications in modern business and government services that underlie operations, planning, and progress in all sectors everywhere we live and work. It also includes an academic trend toward greater interdisciplinarity, especially a renewed focus on big questions that matter but that require a breadth of knowledge and multiple fields to tackle. Geography's long-standing intellectual traditions in crossing those usual disciplinary boundaries are now better understood, increasingly seen as relevant and more widely respected in scholarly circles. These trends have produced unprecedented growth in the field.[27]

More and more students are graduating with geography degrees—bachelors, masters, and doctoral—and more of them are finding jobs. At my *alma mater*, the University of Texas at Austin, recent B.A. grads in geography have become a Nature Conservancy preserve manager, a transportation planner, a natural science teacher, and a geospatial engineer. Masters students became a US Geological Survey water specialist, the chief executive officer of an insurance company in the West Indies, and a vice president for market research in real estate. Many Ph.D. graduates became geography professors (ask your prof which of the 60 specialty groups of the AAG she or he belongs to), whereas others took executive positions in GIS departments and companies or became decision makers in foundations and research institutions all around the world.[28] Every

Geographic Spotlight GIS in Action

The geographic advantages of the geospatial revolution can be deployed to peer into the future and to reveal the past, and to help us understand the spatial affiliations of both modern, complex, and globalized peoples and of traditional peoples practicing a subsistence economy. One of the most remarkable applications of GIS, and one that clearly employs the geographic advantage, is the documentation of indigenous geographical knowledge. The Inupiat people of Alaska's North Slope are dealing with unprecedented changes in the Arctic environment that, as we will see in the next chapter, are related to global processes 40 of climate change. Anxious to preserve their highly detailed and previously undocumented environmental knowledge for future generations, they have asked geographers to create a GIS database of indigenous knowledge. The response is the **Arctic Cultural Geography Project**, funded by the National Science Foundation. In videotaped interviews, more than 50 Inupiat elders, hunters, and berry pickers discussed their activities, landforms, environmental changes, and other issues, and identified related locations of satellite images and topographic maps (**• Figure 1.D**). Geographers categorized and created GIS attributes on appropriate layers. Endowed with this detailed and deep body of knowledge, geographers and other scientists are using these data to understand climate change, landform processes, and other problems. The Inupiat are using this body of knowledge for educational and resource management tools.

Dr. Wendy R. Eisner

• Figure 1.D Interviewing an Inupiaq elder in the Arctic Mapping Project.

In many places around the world where indigenous and local peoples retain traditional knowledge and practices, there are opportunities like this to render indigenous mental maps as geospatial data and put them to practical, beneficial use. It is vital at the same time to ensure this information is not used for nefarious purposes. In the Inupiat case, you may view the GIS at http://www.arcticmapping.org, but only after explaining your interest in a short email.

year we see growing numbers of undergraduate students in our Geography Department's courses positioning themselves for careers in geospatial intelligence (GEOINT) and human intelligence (HUMINT) in the federal government's intelligence agencies.

In this chapter, we have seen clearly that Geography is *a lot* more than rote memorization of state capitals. What do you think are the most important questions facing the world today? Chances are that some of yours will overlap with the AAG's eleven "strategic research directions" for students and scientists in Geography.[29]

1. How Are We Changing the Physical Environment of Earth's Surface?
2. How Can We Best Preserve Biological Diversity and Protect Endangered Ecosystems?
3. How Are Climate and Other Environmental Changes Affecting the Vulnerabilities of Coupled Human–Environment Systems?
4. How and Where Will 10 Billion People Live on Earth?
5. How Will We Sustainably Feed Everyone in the Coming Decade and Beyond?
6. How Does Where People Live Affect Their Health?
7. How Is the Movement of People, Goods, and Ideas Transforming the World?
8. How Is Economic Globalization Affecting Inequality?
9. How Are Geopolitical Shifts Influencing Peace and Stability?
10. How Might We Better Observe, Analyze, and Visualize a Changing World?
11. What Are the Societal Implications of Citizen Mapping and Mapping Citizens?

These are some of geography's big questions, and we will begin to answer *all* of them in the next two chapters and beyond. By examining and answering these questions in the context of world regional geography, you will gain deep and lasting insights into what cosmologist Carl Sagan described as that "pale blue dot" in which "we float like a mote of dust in the morning sky":[30]

> Look again at that dot. That's here. That's home. That's us. On it everyone you love, everyone you know, everyone you ever heard of, every human being who ever was, lived out their lives. The aggregate of our joy and suffering, thousands of confident religions, ideologies, and economic doctrines, every hunter and forager, every hero and coward, every creator and destroyer of civilization, every king and peasant, every young couple in love, every mother and father, hopeful child, inventor and explorer, every teacher of morals, every corrupt politician, every "superstar," every "supreme leader," every saint and sinner in the history of our species lived there—on a mote of dust suspended in a sunbeam.

Geographic Spotlight Google Earth

There are two tools that can greatly enhance your understanding and enjoyment of world regional geography. One is a globe. As we saw earlier, flat maps distort representations of the Earth, but a globe puts a representation of the true world in your hands. If you keep a globe handy and use it as you work your way through the book, it will help you learn about the Earth and remember what you learn.

The other tool is a virtual globe, map, and geographic information software program like Google Earth or NASA's World Wind (• **Figures 1.E** and **1.F**). If you have not already done so (and if you have permission to do so), on the PC or Mac, navigate to *earth.google.com* (apps are also available for most smartphones and tablets). Download the free Google Earth application, open it, bookmark it, and use it often—*PLEASE!* With Google Earth you are in control of an easy-to-use but incredibly powerful geospatial set of tools. It is so powerful, in fact, that it enables you to become a rather effective spy: you can see not only your home or apartment, but also top secret North Korean and Iranian nuclear facilities. A Russian intelligence official said of Google Earth, "Terrorists don't need to reconnoiter their target. Now an American company is working for them."[22]

By the way, when I photograph a bridge, power plant, or dam (as I do often, thinking I might use the photo in one of my classes or for this book), I am sometimes detained and questioned by security guards (this has happened in Ethiopia, Egypt, Lebanon, and even at the Navajo power plant in Arizona (see **page 545**); and so far, I have thankfully been released, without having to give up any photos). That is ironic, considering what Google Earth can do.

Google Earth's base map is composed of many thousands of remotely sensed images, from both aerial photography and orbiting satellites, put together as a giant mosaic. And although Google Earth is not a full-fledged GIS (it can display geographic information, but cannot analyze or modify that data), its "Layers" sidebar works in a similar fashion: you can turn on and off certain features like country boundaries and road networks. With the "weather" layer, you can get real-time views of storm systems and temperatures. How hot or cold is it in Arabia or Siberia right now? It's easy to find out. You might also be interested in the "global awareness" layer that lets you explore some of the world's environmental problems. And then there is the "Street View" layer for many locations, infamous for news stories like this: "A woman, checking out a female friend's house on Google Maps, was surprised to see her husband's Range Rover out front. A divorce is underway."[23]

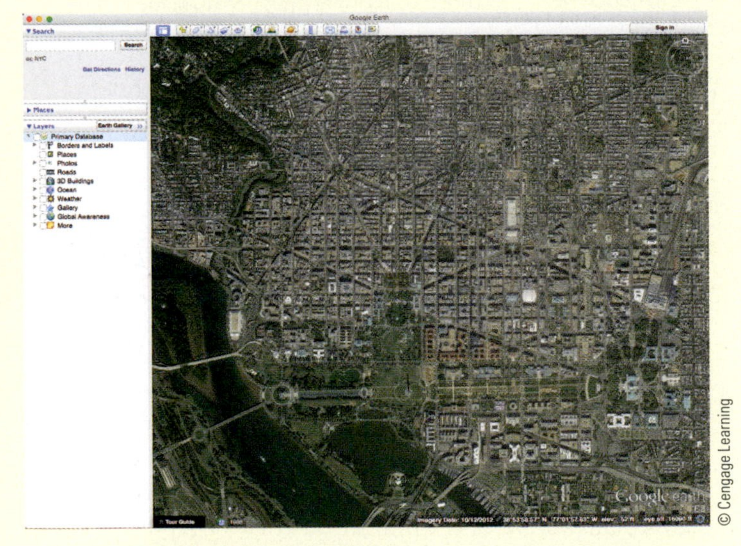

© Cengage Learning

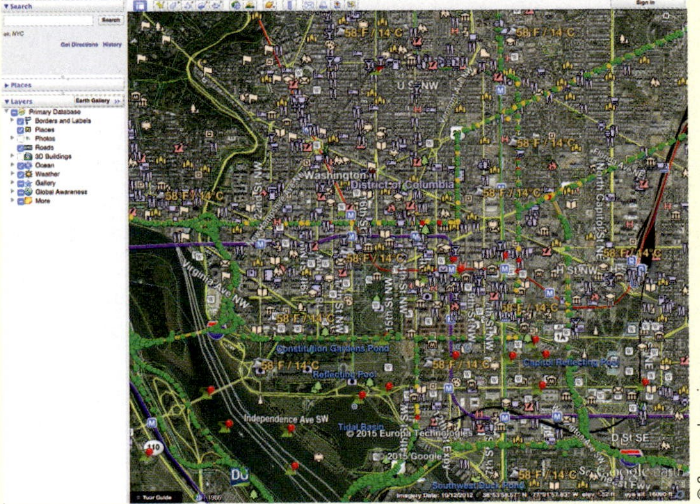

© Cengage Learning

• **Figures 1.E and 1.F** Opening Google Earth without (E) and with (F) layers checked to view Washington, DC.

> The Earth is a very small stage in a vast cosmic arena. Think of the endless cruelties visited by the inhabitants of one corner of this pixel on the scarcely distinguishable inhabitants of some other corner, how frequent their misunderstandings, how eager they are to kill one another, how fervent their hatreds. Think of the rivers of blood spilled by all those generals and emperors so that, in glory and triumph, they could become the momentary masters of a fraction of a dot.
>
> Our posturings, our imagined self-importance, the delusion that we have some privileged position in the Universe, are challenged by this point of pale light. Our planet is a lonely speck in the great enveloping cosmic dark. In our obscurity, in all this vastness, there is no hint that help will come from elsewhere to save us from ourselves.
>
> The Earth is the only world known so far to harbor life. There is nowhere else, at least in the near future, to which our species could migrate. Visit, yes. Settle, not yet. Like it or not, for the moment the Earth is where we make our stand.
>
> There is perhaps no better demonstration of the folly of human conceits than this distant image of our tiny world. To me, it underscores our responsibility to deal more kindly with one another, and to preserve and cherish the pale blue dot, the only home we've ever known.
>
> —Carl Sagan

Reading the Landscape with Remote Sensin

Remote sensing images have a way of bringing out the detective in us. Have a look at the before and after remote sensing images of Sendai, a town in northeastern Japan. • **Figure 1.G** is an image taken on April 4, 2010, by a satellite about 450 miles above the Earth. • **Figure 1.H** shows how nature reclaimed Sendai. On March 11, 2011, a magnitude 9.0 earthquake thrust a section of the Earth's crust upward from the seafloor about 80 miles from Sendai. The displaced water roared ashore as devastating tsunami waves as high as 40 feet. Compare Figures 1.G and 1.H, examining an ... the details of the cultural landscape and its transformation. Use the "before" remotely sensed image of Sendai (**Figure 1.G**) to make some of your own observations about the Japanese cultural landscape.

Use the "after" image of Figure 1.H to observe changes that this natural disaster wrought on Sendai. I have written some of my own observations in note 24.

• **Figures 1.G and 1.H** The shoreline of Sendai, Japan, before and after the gigantic earthquake and tsunami on March 11, 2011.

Study Guide

Summary

- Recent studies suggest that US citizens generally have poor knowledge of world geography. More and better geographic knowledge would serve us well in many contexts.
- There are six essential elements of the national geography standards: the world in spatial terms, places and regions, physical systems, human systems, environment and society, and the uses of geography. Each of these has a subset of standards, totaling 18.
- The five themes of geography are: Location, Place, Human–Environment Interaction, Movement, and Region.
- Geography means "description of the Earth" and is also defined as "the study of the Earth as the home of humankind."
- Five main objectives of the text are for readers (1) to become geographically literate, (2) to understand Earth's problems and their potential solutions, (3) to use geographic critical thinking to understand the world, (4) to understand the geography of current events, and (5) to develop the ability to interpret places and "read" landscapes.
- Maps are the geographers' most basic tools. The basic language of maps includes the concepts and terms of scale, coordinate systems, projection, and symbolization. Maps can depict spatial data in a variety of ways.
- Individuals and cultures generate their own unique "mental maps." Regions are in effect mental maps that help us make sense of a complex world.
- Modern geographic thought derives from a long legacy of interest in how people interact with the environment. The dominant approach has been to understand how people have changed the landscape or face of the Earth.
- The discipline of geography may be divided into regional and systematic specialties, with the systematic fields having the most followers. Their concerns overlap many disciplines in the natural and social sciences. Geographers are employed in many private and public capacities. The strongest growth area with the most jobs is in geographic information systems (GIS).

Review Questions

1. What is geography? Is it usually classified as a natural or a social science? What are some of its characteristic approaches?
2. What are the six essential elements of geography as defined by the National Council for Geographic Education? What does each element indicate about geography's concern with space, place, or the environment? How do the eighteen standards help inform geographic literacy?
3. What does *spatial* mean, and how does geography's interest in space differentiate it from other disciplines?
4. What geographic features make the United Kingdom and New Zealand different?
5. What are the major terms and concepts associated with scale, coordinate systems, projections, and symbolization?
6. Why is a map made with the Mercator projection more suitable for navigation than a map made with a compromise projection, such as the Winkel Tripel?
7. What is the difference between a dot density map and a choropleth map?
8. What is a mental map?
9. What is GIS, and what typically makes it different from old-fashioned manual cartography? What are some applications of GIS and remote sensing?
10. What do geographers study, and what do they do for a living?

Key Terms + Concepts

Notes

1. Personal communication between Hal Mooney and Tom Wilbanks, cited in *Understanding the Changing Planet: Strategic Directions for the Geographical Sciences* (2010). National Academies Press.
2. 2014 Americans: Ukraine is in . . . Greenland? http://www.bbc.com/news/blogs-echochambers-26943479, BBC Online, April 8, 2014 (on misplacing Greenland).
3. Netflix.com description of film *Geography Club.*
4. Charles F. Gritzner, "Defining Geography: What Is Where, Why There and Why Care?" AP Central, accessed September 10, 2014, http://apcentral.collegeboard.com/apc/members/courses/teachers_corner/155012.html.
5. Secretary of Defense Speech, US Department of Defense, http://www.defense.gov/speeches/speech.aspx?speechid=1539.
6. David Rothkopf, "Obama's 'Don't Do Stupid Shit' Foreign Policy." June 4, 2014. http://foreignpolicy.com/2014/06/04/obamas-dont-do-stupid-shit-foreign-policy.
7. The abbreviation BCE stands for "before the Common Era," which is a reference to the dating system invented by European Christians that sets the birth of Jesus Christ as year 1. In Christian cultures, dates before that year are expressed as BC, meaning "before Christ," and later years are identified as AD, which stands for *anno Domini* (Latin, "in the year of our Lord"). Religion-neutral dating systems such as the one used in this

book employ BCE ("before the Common Era") and CE ("Common Era"), respectively, but the years are numbered the same in the two systems.

8. National Council for Geographic Education. The Eighteen National Geography Standards, http://www.ncge.org/publications/tutorial/standards/.
9. To learn more about the Five Themes, visit the National Geographic Society's Education FAQ at http://education.nationalgeographic.com/education/faq/?ar_a=1.
10. http://education.nationalgeographic.com/education/faq/?ar_a=1.
11. Quoted in Geoffrey J. Martin and Preston E. James, *All Possible Worlds: A History of Geographical Ideas* (New York: Wiley, 1993), p. 150.
12. Quoted in Andrew Goudie, *The Human Impact on the Natural Environment* (Oxford: Blackwell, 1986), p. 6.
13. Kathleen A. Dahl, "Culture," https://docs.google.com/viewer?a=v&pid=sites&srcid=ZW91LmVkdXxrYXRobGVlbi1kYWhsfGd4OjMyOTM4NDFlNjdhZTJhZWU.
14. Dictionary.com, "critical thinking," in Dictionary.com's 21st Century Lexicon, accessed June 22, 2013, from http://dictionary.reference.com/browse/criticalthinking.
15. This definition is by Robert Strausz-Hupé, *Geopolitics: The Struggle for Space and Power. Foreign Affairs* (New York: Arno Press, October 1942).
16. This is a rehearsal for a community cultural event in the city of Ha Giang in northern Vietnam. By looking at faces, you may conclude this place is in Asia. Other clues would call on your knowledge of Vietnam's recent past, when North Vietnam's communist forces, led by Ho Chi Minh, prevailed in a war with the US-backed South Vietnamese government (see page 344). "Uncle Ho's" name appears on the right in Vietnamese language, which has a "romanized" script. Ho Chi Minh also strikes a heroic prose in the statuary. The predominant ethnic group here is Hmong, and you can see Hmong flutes being waved by a number of the participants. The luxuriant vegetation around this park points to a humid region, as do the peoples' umbrellas.
17. Waldo Tobler, "On the First Law of Geography: A Reply," *Annals of the Association of American Geographers*, 94(2), 2004, pp. 304–310.
18. Using this map, you should determine Oslo's location as about 60 degrees north, 11 degrees east (60°N, 11°E). Oslo is therefore in the Northern Hemisphere, the Eastern Hemisphere, and the land hemisphere.
19. Activist David Eaves proposes that Enbridge is "lying with maps;" see his blog at http://eaves.ca/2012/08/15/lying-with-maps-how-enbridge-is-misleading-the-public-in-its-ads/. In 2015, the publisher HarperCollins was widely criticized for omitting the political name "Israel" in a world atlas designed to be used in the Arab Gulf States of the Middle East. It is still common in the Arab World to see maps omitting the name "Israel," instead leaving the land area blank or using the label "Palestine." This reflects regional animosities dating back to the 1967 Arab-Israeli War. HarperCollins quickly pulled the Atlas from circulation when this controversy emerged.
20. Defined in www.dictionary.com.
21. Esri, at www.gisday.com/cd2009/fliers/what_is_gis.pdf.
22. Lieutenant General Leonid Sazhin, Federal Security Service, quoted in Katie Hafner and Saritha Rai, "Google Offers a Bird's-Eye View, and Some Governments Tremble," *New York Times*, December 20, 2005, p. 1.
23. http://gawker.com/5191459/cheating-husband-said-caught-via-google-street-view.
24. Each person may see different features in these images. Here are some of the things I see; what about you? Figure 1.E is a remotely sensed snapshot of a Japanese cultural landscape. Just inland from the concrete wave breakers and scalloped shoreline, people have built homes between a greenbelt of trees and an irrigation canal. The canal is a straight and uniform feature—a good indication that it is the handiwork of people rather than of nature. Bridges and streets also stand out as built features. Continuing westward (left) across the image, you see many more homes, including some nestled against a patch of trees (the Japanese prize their forests and go to great lengths to protect them). Finally, at the far left of the image, the long brown rows of empty land are fields for wet rice, which would become green when planted and irrigated from the nearby canal. This satellite image provides an excellent snapshot of how the people of Japan have modified nature to suit their needs. Figure 1.F shows in breathtaking fashion how nature reclaimed Sendai: almost every feature on the landscape, both natural and cultural, has been destroyed or transformed by the tsunami waves. You can describe this destruction feature by feature, using the terms *before* and *after*.
25. http://www.aag.org/cs/jobs_and_careers/geography_can_take_you_there.
26. http://www.doleta.gov/brg/indprof/geospatial_profile.cfm.
27. Committee on Strategic Directions for the Geographical Sciences in the Next Decade, National Research Council, *Understanding the Changing Planet: Strategic Directions for the Geographical Sciences* (Washington, DC: The National Academies Press, 2010).
28. Information courtesy of Bill Doolittle, University of Texas at Austin.
29. National Academies Press. *Understanding the Changing Planet: Strategic Directions for the Geographical Sciences* (Washington, DC: National Academies Press, 2010).
30. Carl Sagan, *Pale Blue Dot: A Vision of the Human Future in Space*, http://www.goodreads.com/author/show/10538.Carl_Sagan.

Global Geoscience Watch

Global Geoscience Watch is your portal into the full GREENR database (Global Reference on the Environment, Energy, and Natural Resources). For starters, take a tour of the interactive Google Map found at the top of the GREENR database home page. Click on the "World Map" menu item found on the Global Geoscience Watch homepage. Chapters 4–11 of the text cover eight major regions of the world. Refer to the Table of Contents and select one of these regions to explore on the interactive map. Explore that region on the map by identifying and reviewing maps of countries in the region. Click some of the "pins" on the county maps to see the extensive information available within GREENR database. Write a paragraph about what you were able to learn from this exercise. Include the names of the countries you explored in your tour of the database.

Online Resources

For access to MindTap and additional study materials visit www.cengagebrain.com. Read your textbook, take notes, complete activities, take practice quizzes and more.

Above: The Goosenecks of the San Juan River in southeast Utah, where an entrenched meander cuts 1000 feet deep through limestone, siltstone, sandstone, and shale beds lain down by ancient seas. Joe Hobbs
Left: Here is an aerial view of the Goosenecks. It's always good to sit by the plane window! Joe Hobbs

2 Physical Processes and World Regions

I see Earth! It is so beautiful.

—Yuri Gagarin, Soviet cosmonaut, first person in space[1]

Many issues in world regional geography relate to the interaction of people and the natural environment, and how these relationships have changed the face of the Earth. Physical processes are the main focus in this chapter, and human processes the focus in Chapter 3. But these processes are so intertwined that human forces receive much attention in this chapter, and natural forces much attention in the following chapter.

This discussion of physical geography introduces you to the four "spheres" that make up the Earth's habitable environment: the **lithosphere**, its outer "rind" of rock; the **hydrosphere**, made up of all the world's water features; the **atmosphere**, the layer of gases surrounding the Earth; and the **biosphere** (also known as the "ecosphere"), which is the global ecological system including all the relationships played out among the lithosphere, hydrosphere, and atmosphere. Earth's lithosphere is a work in progress, and you will see how its changing surfaces provide both opportunities and threats to people. We will consider the climate and

Chapter Outline

Chapter Objectives

This chapter will enable you to

- Understand the tectonic forces behind some of the world's major landforms and natural hazards.
- Recognize consistent global patterns in the distribution of vegetation types and climates.
- Identify the natural areas most threatened by human activity and explain how natural habitat loss may endanger human welfare.
- Appreciate the important roles of the world's oceans.
- Describe the potential impacts of global climate change and international efforts to prevent them.

vegetation types that play such large roles in human activities, and appreciate the rich diversity of wild plant and animal species. We will look briefly at the planet's often-overlooked oceans and the resources they hold. Finally, we will examine how the climate is changing, what these changes may mean for life on Earth, and what people can do to adapt to or mitigate some of the most dangerous climatic changes.

2.1 Geologic Processes

The Earth's surface is in motion. Those of you living on the West Coast of the United States have probably experienced the unforgettable sensation of the ground moving beneath you in a mild "temblor" or even a large earthquake. These events reflect processes that are constantly changing the face of the Earth. Here we explore just one set of geological processes—arguably the most important of them all because of its role in shaping global landforms—plate tectonics.

Plate Tectonics

About a century ago, a German geologist named Alfred Wegener came up with what seemed to many like an outlandish theory known as **continental drift.** Pointing to such things as the jigsaw puzzle–like geometry of Africa's west coast and South America's east coast, he proposed that the continents were once joined in a supercontinent (which he named Pangaea) but that they "drifted apart" over time. However, Wegener was not able to explain the forces behind these movements, and he was derided by a scientific community that seemed unwilling to accept anything other than a terra firma that was truly firm. Detractors denounced his conclusions as "utter damned rot" and "mere geopoetry."[2] But later discoveries in deep-sea science established Wegener's basic proposition as factual, and today a good deal is known about how continental drift occurs.

The Earth's lithosphere, which varies from 50 to 125 miles in thickness, is made up of about a dozen giant and several smaller sections of rock called **plates.** These plates move in various directions in processes known collectively as **plate tectonics** (•**Figure 2.1**). On the ocean floors in places such as the Mid-Atlantic Ridge and the East Pacific Rise, new lithosphere is "born" as molten material rises from the Earth's mantle and cools into solid rock (•**Figure 2.2**). Plate tectonics are often explained by the analogy of a "conveyor belt" (the convection cell in Figure 2.2) in constant motion. On either side of the long, roughly continuous ridges, the two young plates move away from one another, carrying islands with them; this process of **rifting** or pulling apart is called **seafloor spreading.**

Seafloor spreading has few impacts on people where it originates, but at the other end of the conveyor belt, where Earth's plates collide, there are often huge consequences. **Tectonic forces** that cause the lithosphere's plates to diverge, converge, and subduct create some of the planet's greatest natural hazards. The most dangerous tectonic forces are related to **seismic activity** (seismic refers to Earth vibrations, mainly earthquakes) that causes earthquakes and **tsunamis**, and the **volcanism** (movement of molten earth material, especially in volcanoes) often associated with plate movements. The globe's most active and deadly realm of tectonic activity is the so-called **Ring of Fire** on the rim of the Pacific Ocean (see **Figure 2.1** and **page 374**).

Plates collide and converge in different ways and with different consequences; among the most notable consequences are the building of Earth's great mountain ranges. In some parts of the world—off the east coast of Japan, as we saw in Chapter 1, and off the coast of the US Pacific Northwest, for example—one plate "dives" below another in a tectonic process known as **subduction** (see Figure 2.2).

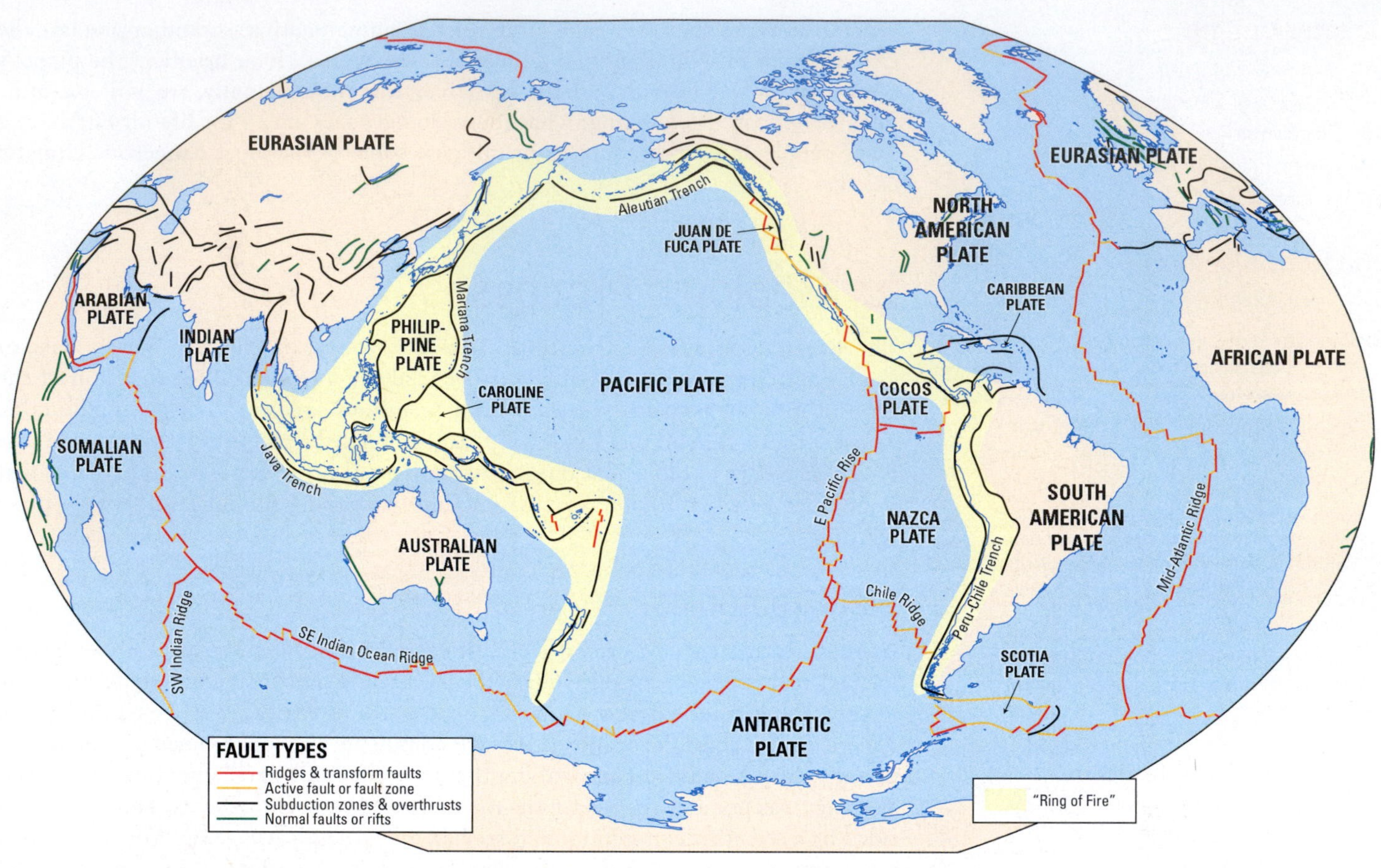

• **Figure 2.1** Major tectonic plates and their general direction of movement. Earthquakes, volcanoes, and other geologic events are concentrated where plates separate, collide, or slide past one another. Where they separate, rifting produces very low land elevations (well below sea level at the Dead Sea of Israel and Jordan, for example) or the emergence of new crust on the ocean floor (in the middle of the Atlantic Ocean, for example). Note the "Ring of Fire" around the edges of the Pacific Plate.

Source: Adapted from NASA, "Global Tectonic Activity Map of the Earth," DTAM-1, 2002.

The descending lithosphere is melted again as it dives into the Earth's superheated mantle along a deep linear feature known as a **trench**. One of these is the Mariana Trench in the western Pacific, where the deepest place on Earth lies at a crushing 35,814 feet (10,916 m) below sea level (see **Figure 2.1**). Subduction is another stage along the "conveyor belt" process that will eventually see this material recycled as newborn lithospheric crust.

Subduction can release enormous amounts of energy. Periodically, the great stress of one plate pushing beneath another is released in the form of an earthquake. The world's largest recorded earthquakes—registering 9.5 (Chile, 1960), 9.2 (United States, 1964), 9.1 (Indonesia, 2004), and 9.0 (Japan, 2011), respectively, on the **moment magnitude scale (MMS)** (a replacement of the Richter scale) which measures the strength of the earthquake at its source—have struck along these subduction zones. This sudden displacement of a section of oceanic lithosphere is also what triggers a tsunami, one of nature's most powerful and destructive processes. See **pages 21, 338,** and **375** for more insight into tsunamis, and **page 28** for geography's professional concerns with natural hazards.

In other places where they meet, lithospheric plates grind and slide along one another, as in California and Turkey. The processes of rock crowding together or pulling apart along these fracture lines is known as **faulting**. Movement along various kinds of **faults**—places where sections of the Earth's crust are in contact—causes earthquakes, the emergence of new landforms, and other geologic consequences.

Volcanism generally takes place along and near **subduction zones** (see Figure 2.2) and also in the world's several dozen **geologic hot spots** where molten material has broken through the crust as a "plume" (as in Yellowstone National Park and in Hawaii; see **page 373**). Despite posing a host of natural hazards to people living on their slopes or downwind—including pyroclastic flows (fast-moving currents of rock fragments, hot gases, and ash), lava flows, ashfalls, and other dangers—volcanoes have beneficial qualities. Volcanic rock usually breaks down to form fertile soils (as in Ethiopia and the Nile Valley), and the range of climate and vegetation types on their slopes creates opportunities for growing crops and raising livestock.

240, 404

We turn now to global patterns of climate and vegetation.

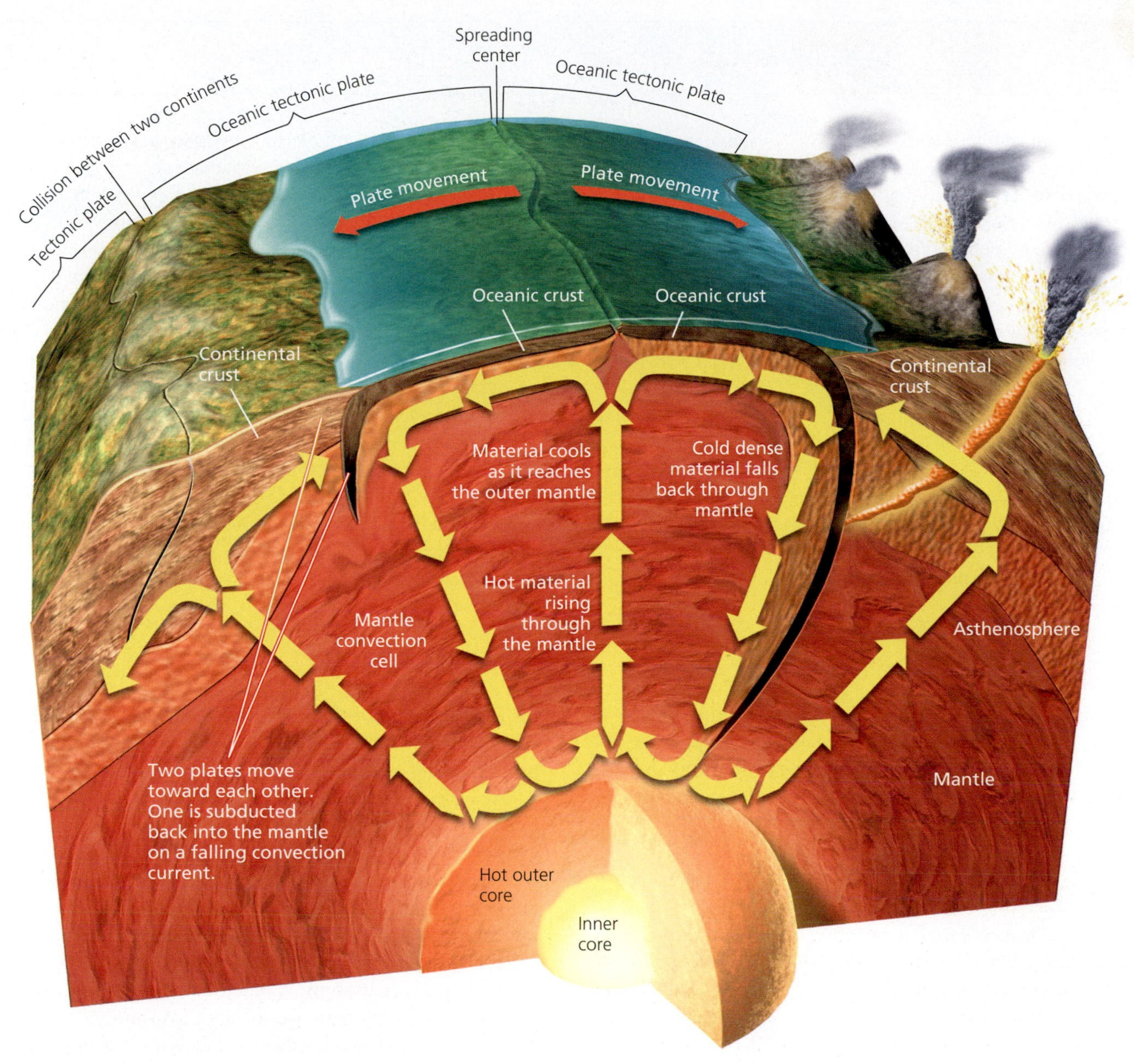

• **Figure 2.2** The Earth is composed of core, mantle, and crust. Dynamic forces within the core and mantle have major effects on what happens in the crust and on the surface, including impressive but dangerous events like earthquakes and volcanoes. These forces create many of Earth's landforms. The yellow arrows illustrate how the lithosphere is "recycled."

2.2 Patterns of Climate and Vegetation

"Everybody talks about the weather, but nobody does anything about it," Mark Twain reportedly said. We see in this chapter that we have actually done a lot about the weather and even more about the climate. First we will look at natural patterns in vegetation and climate that present different opportunities for people around the world. You will see these patterns repeated in the regions you are about to explore.

As you experience a warm, dry, cloudless summer day or a cold, wet, overcast winter day, you are encountering **weather**—the atmospheric conditions occurring at a given time and place. **Climate** is the average weather of a place over a long time period. Dr. J. Marshall Shepherd of the National Weather Service offered this analogy: "Weather is your mood; climate is your personality."[6] Along with surface conditions such as elevation and soil type, climatic patterns have a strong correlation with patterns of natural vegetation and, in turn, with human opportunities and activities on the landscape.

Precipitation and temperature are the key variables in weather and climate. Warm air holds more moisture than cool air, and **precipitation**—rain, snow, sleet, and hail—results

Geographic Spotlight Natural Hazards

In the not too distant future, you will hear about it: a quake in California injuring people and causing costly damage to infrastructure, a catastrophic "twister" grinding a Midwestern neighborhood to its foundations, or a typhoon (hurricane or cyclone) pounding a Pacific island community with roaring winds and rains. These impressive and unforgiving phenomena are some of Earth's **natural hazards**, defined by the three geographers who pioneered their study as "those elements of the physical environment, harmful to man and caused by forces extraneous to him."[3] People are part of the definition of natural hazards: a volcanic eruption that does not impact people would be a natural phenomenon, not a natural hazard. Natural hazards represent one of the most important and useful subfields of geography. Geographers equipped with the intellectual and material tools of natural hazards research are in high demand in planning and in emergency response offices at all levels of civil administration. One of the 65 specialty groups of the AAG is Hazards, Risks, and Disasters. It falls right in the crosshairs of geography's scope of human–environment interaction. An excellent resource on natural hazards research is the University of Colorado–Boulder's Natural Hazards Center, where Boulder's Geography Department plays a prominent role.[4]

6

A question that geographers of natural hazards commonly address is why people build where they do. Usually there is an economic rationale: the most affordable dwelling may be the flimsy subsidized "projects" housing situated in a floodplain that has flooded and will flood again. On the flip side, there is often a perverse economic rationale: because the location is beautiful or convenient, wealthy people will pay a premium to live on a hazard-prone site (for example, on landfill that will "liquefy" near a fault line, or in a forest that will burn), where they can afford to lose and rebuild. "The attraction of risky places can be strong; they can be as beautiful as they are deadly," wrote a journalist in the wake of the 2014 landslide that obliterated the little town of Oso, Washington.[5] In this book, we present several case studies of natural hazards and examine several situations in which poverty and income inequality play a role in natural hazards.

471, 502

from processes that cool the air to release moisture. Water is, of course, essential for life on Earth. Some geographers would tell you that the map of the world's precipitation—here in • **Figure 2.3**—is the most important map of all for understanding life on Earth.

Climate

Geographers have created a number of systems for categorizing climates, but the most widely used is the **Köppen climate classification system (KCC)**. Devised by the German scientist Wladimir Köppen early in the 20th century, it has been modified several times since then. Köppen's climate types are based on measurements of monthly and yearly temperatures and precipitation. There are about 30 different climate types and variants in the KCC, designated by letters such as *BSk* and *Dfa*. These labels are not very user-friendly, so for easier use and for the sake of correlation with biome types, this book distills the KCC into 11 climate types (see • **Figure 2.4**).

Three climates are designated "tropical," with their defining characteristic being consistent warm to hot temperatures; every month has an average temperature of 18°C (64°F) or higher. Looking at Figure 2.4, you can see that all three tropical climates are found in the Amazon basin of South America, in the following sequence poleward from the Equator. The **tropical rainforest climate** is hot, humid, and rainy year-round, with little seasonal variation in temperature or precipitation. The tropical monsoon climate has a short dry season, when the otherwise heavy rains taper off for one or more months. The **tropical savanna climate** (also called the tropical wet-dry climate) has marked wet and dry seasons, typically lasting roughly half the year. The savanna's extended dry season results in lower annual precipitation amounts than the other tropical climates.

Two climates are classified by their aridity, which is defined as precipitation amounts that are less than the **potential evaporation (PE)**. PE is a measure of how much water would be evaporated in a given area if the amount of water available were unlimited. Hot areas have higher PE than cold areas; low precipitation in low latitudes (close to the Equator) results in sparse flora, while the same amount of precipitation at higher latitudes can support much more vegetation. The **desert climate** is very dry year-round, with low annual precipitation. The Sahara in Africa has the world's largest contiguous desert climate region. The **semiarid climate** receives more precipitation than do deserts and is commonly a transition zone between deserts and more humid climates. A good example is the Sahel, the orange stripe just south of the Sahara in Figure 2.4. Although arid climates are usually thought of as hot, desert and semiarid regions are also found in colder, higher latitudes, such as in the Gobi Desert of northern China and southern Mongolia.

Often adjacent to arid climates along coastlines is the **Mediterranean climate** (or dry summer subtropical climate), which is more humid than arid climates but has a pronounced summer dry season with little or no precipitation. Winters are mild, and summers are generally hot. Most of California enjoys this pleasant climate. Poleward of the Mediterranean climates lies the **oceanic climate** (also known as the **maritime or marine west coast climate**). Proximity to the ocean moderates the climate, preventing it from getting too hot in summer or too cold in winter. This factor, along with the higher latitudes it is found in, makes the oceanic climate relatively

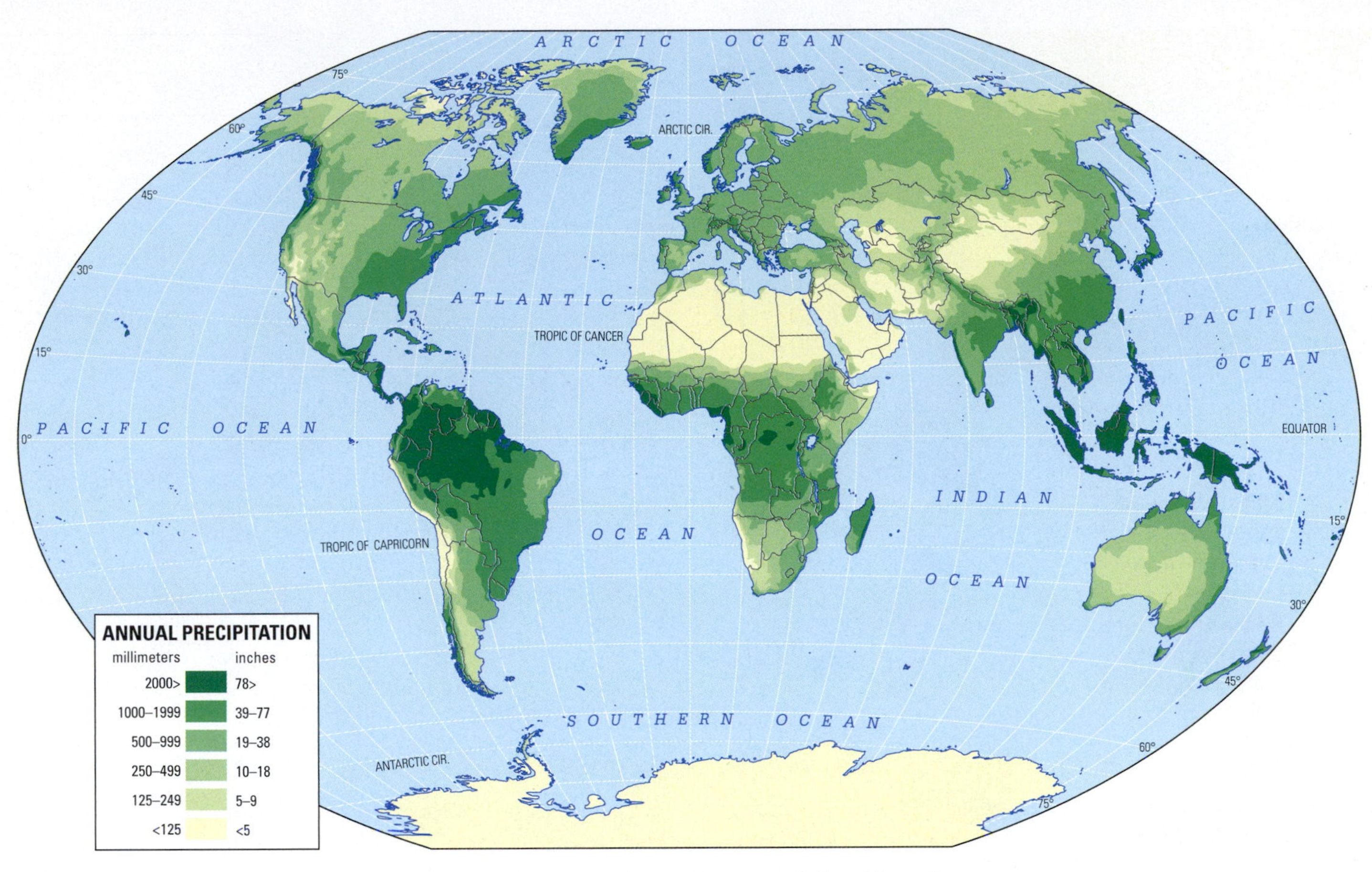

• **Figure 2.3** World precipitation map. Is this, as some authorities say, the most important of all world maps?

mild year-round, with precipitation amounts roughly equal in each season. Much of Europe has this climate. Some mountainous areas in tropical and subtropical regions also have an oceanic climate, thanks to the cooling effects of higher elevations.

The **humid subtropical climate** receives abundant precipitation, spread roughly evenly throughout the year, although some areas with this climate experience a monsoonal, dry-winter precipitation pattern. The humid subtropical climate features hot and muggy summers, and mild winters with occasional cold snaps. This is the climate of the southeastern United States and the southern Midwest. The **humid continental climate** also receives year-round precipitation, usually in lesser amounts than in the subtropical and oceanic climates. The humid continental climate has large temperature swings between its cold winters and the often subtropical-like hot summers. This is the climate of much of the northeastern United States and the northern Midwest.

Found exclusively in the Northern Hemisphere, the **subarctic climate** (also called the subpolar, boreal, or taiga climate) has short, mild summers and long, cold, severe winters. Most of Russia's Siberia region has a subarctic climate. Annual precipitation amounts tend to be low in subarctic areas; however, this climate is more humid than true deserts as cold air cannot hold as much moisture as warm air, so less evaporation occurs (its PE is low). Low PE values are also the reason why the **polar climate**, despite very low precipitation totals, can support numerous lakes, rivers, and wetlands in tundra areas and permanent ice cover elsewhere. In polar climates, no month has an average temperature above 10°C (50°F); in ice-covered areas, the average temperature is always below freezing. As we will see in this chapter, rising temperatures across the polar realms are thawing ice and the ground it lies on, with serious implications for climate change (see **page 41**).

On our climate maps, mountainous areas of the tropics and midlatitudes are designated as **undifferentiated highland**. These are not climate and biome zones per se, but are regions where no single climate or vegetation dominates. Instead, these rugged locales, such as in the Rocky Mountains and the Himalayas, have many small microclimates, which vary greatly depending on elevation and exposure to wind, precipitation, and sun.

Biomes

As with climate, scientists have developed a number of ways to categorize Earth's ecosystems. This book uses the 14 **biomes**—regions distinguished by their associations of climate, flora, and fauna—developed by the World Wide Fund for Nature (World Wildlife Fund [WWF]), well known by its panda logo

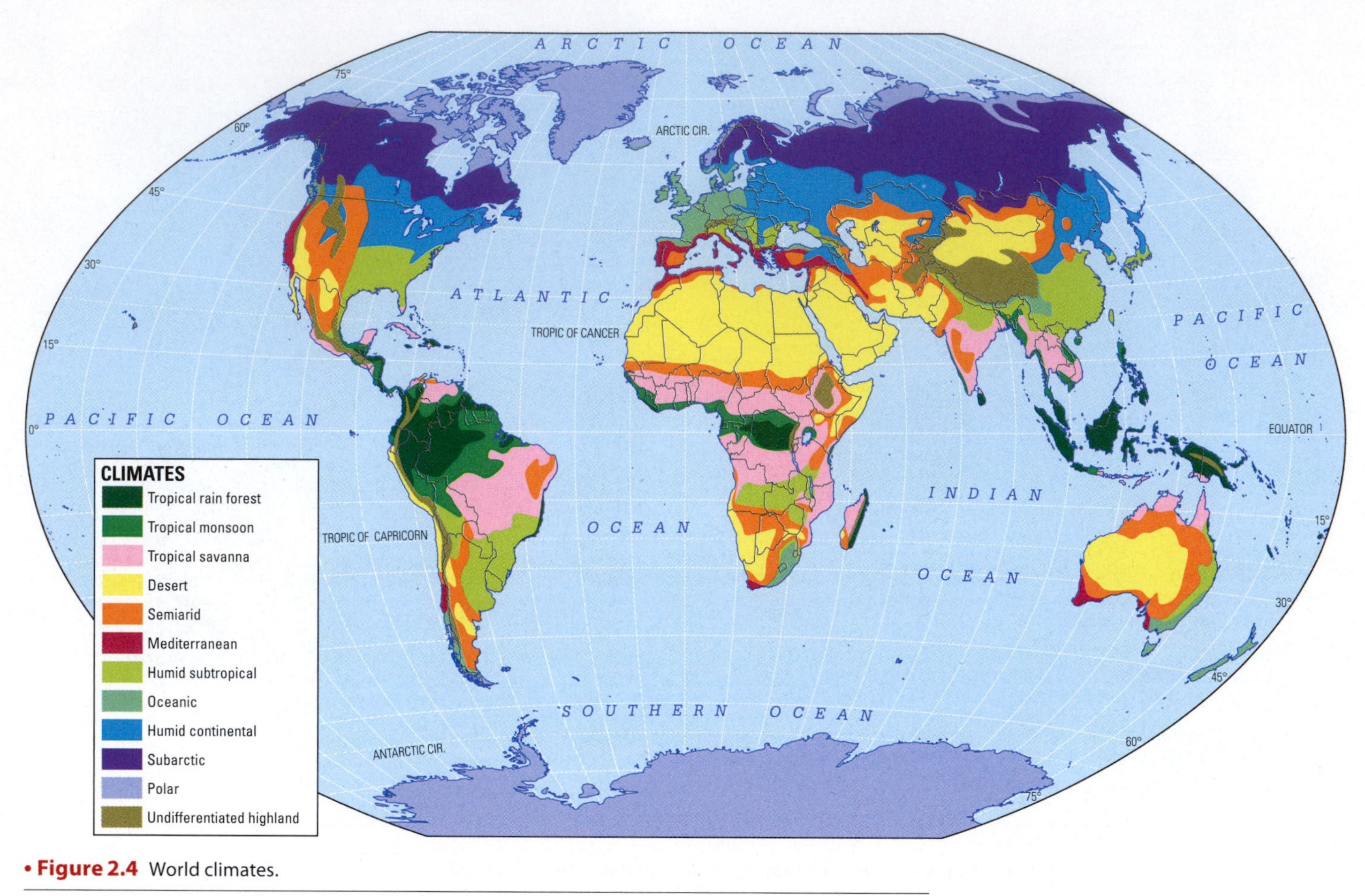

• **Figure 2.4** World climates.

THINK CRITICALLY: Are Figures 2.4 and 2.5 large-scale or small-scale maps?

(•**Figure 2.5**). Temperature and precipitation are major influences on the types of vegetation found around the world, and biomes group these varieties of vegetation together based on common aspects of plant morphology, such as the dominant basic plant structure (trees or grasses, for example), water requirements, and leaf types.

As you use Figure 2.5, it is important to keep in mind that it depicts vegetation potential—the flora that would dominate in the absence of human activity. For example, Figure 2.5 depicts a vast, unbroken area of temperate broadleaf and mixed forest in the eastern United States, but of course much of this region has been transformed by agriculture and urbanization. The "human footprint" in Figure 3.6 on **page 55** is a small-scale map showing what people have done to impact biomes and change the face of the Earth. By comparing it with Figure 2.5, you can, for example, see the cities and transportation corridors that have broken up the forests of the eastern United States. In some sources, maps of the "human footprint" are labeled as "natural capital degradation." Maps of land use and/or land cover also depict human activities and impacts on the landscape.

Climate plays the main role in determining the distribution of biomes, but differing soils and landforms host different types of vegetation where the climatic pattern is essentially the same. Biome and climate types are sufficiently related that many climate types take their names from vegetation types—for example, the desert climate and the Mediterranean climate. You may easily see the geographic links between climate and vegetation by comparing the maps in Figures 2.4 and 2.5. The spatial distributions of climate and vegetation types do not overlap perfectly, but there is a high degree of correlation in many cases (see Try It on **page 32**).

The highest biodiversity in the world is found in the **tropical and subtropical moist broadleaf forest biome**. The forests in this biome are mainly a thick cover of broadleaf evergreen trees; this includes the tropical rainforests (the "jungle" of popular imagination; •**Figure 2.6a**). It is estimated that as many as half of all species on Earth inhabit tropical rainforest areas, the largest of which is in the Amazon Basin. In humid areas of higher latitudes, the **temperate broadleaf and mixed forests** biome dominates, where broad swathes of deciduous forests are peppered with evergreen needle-leaf trees. These forests cover much of Europe and the eastern United States (•**Figure 2.6b**).

The **tropical and subtropical dry broadleaf forest** occurs in hot areas with pronounced wet and dry seasons, as in central India and parts of mainland Southeast Asia. To conserve water during the dry season, deciduous trees drop their leaves, just as

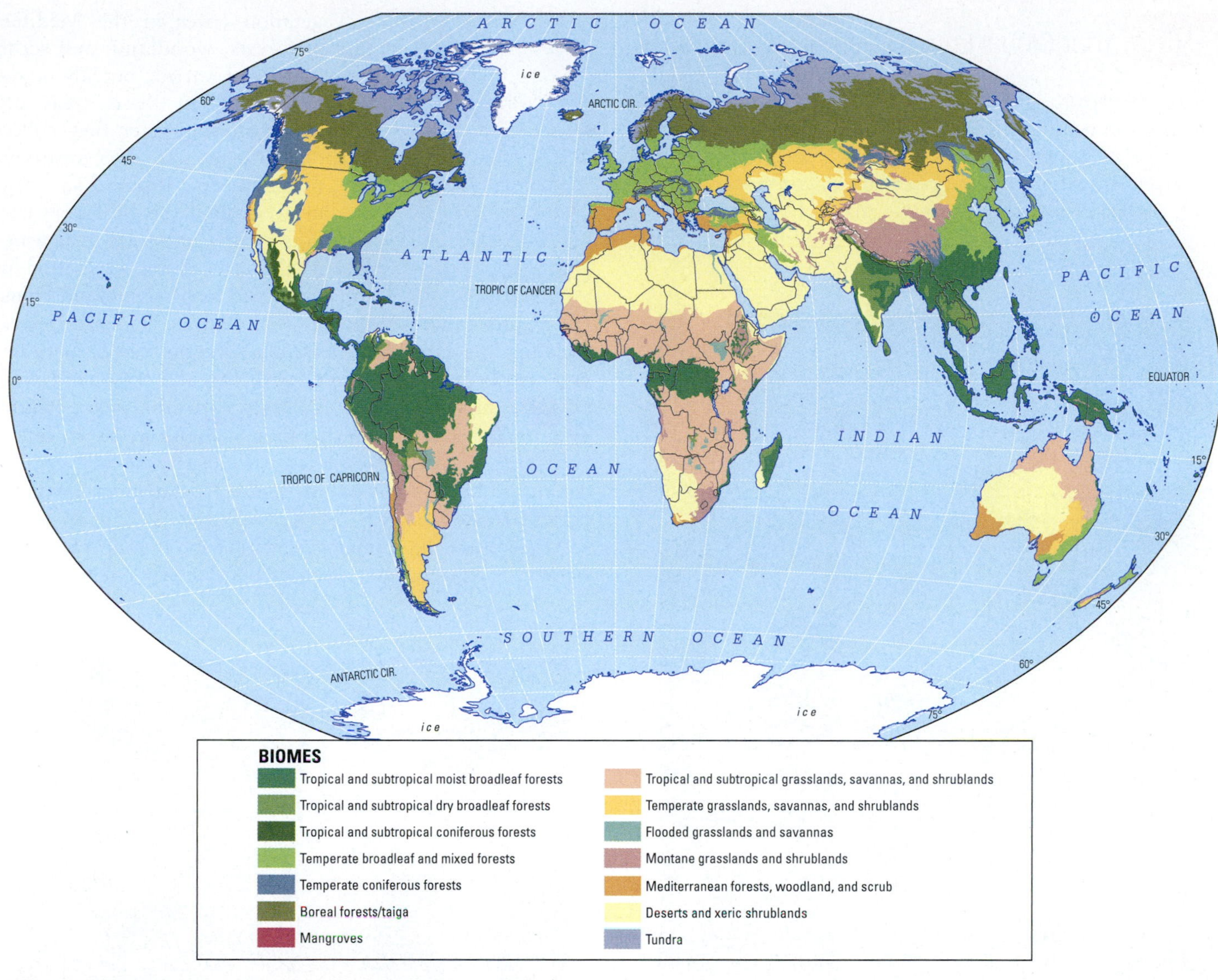

• Figure 2.5 World biomes (natural vegetation) map.
Source: Based on World Wildlife Fund ecoregions data, 1999.

• Figure 2.6a Tropical moist broadleaf forest (tropical rain forest) at the ancient Maya site of Tikal, northern Guatemala.

temperate deciduous trees do before the onset of the cold winter (**•Figure 2.6c**). Wooded tropical areas with lower amounts of precipitation and more pronounced seasonal temperature changes make up the **tropical and subtropical coniferous forest** biome, as found in mountainous regions of Mexico and Guatemala (**•Figure 2.6d**). **Coniferous trees** are better adapted to poor soils and cool, dry conditions than broadleaf trees are.

Needle-leaf trees also predominate in the temperate coniferous forest biome, which is found in both mild, humid coastal climates like the southeastern United States and in cooler, drier inland and mountainous regions (**•Figure 2.6e**). Farther north, in snowy, cold-winter areas lie the vast, nearly unbroken tracts of the coniferous **boreal forest/taiga** biome (**•Figure 2.6f**). Russia's Siberian region has mainly taiga vegetation. Evergreen needle-leaf trees predominate in both of these biomes, but tracts of broadleaf forests grow among them.

Try it Your Global Environmental Counterparts

Examine the maps in Figures 2.4 and 2.5. In each map, find approximately where you live. What kind of climate region is this? What is the characteristic biome or natural vegetation there? Now find one or more other areas in the world that have the same climate and biome types—what places are most like where you live? Remember, these maps are greatly generalized, so you may need to be forgiving about the details.

Vegetation cover in the **Mediterranean forests, woodland, and scrub** biome is quite varied, but all of the plants growing in these areas are adapted to storing water they collect in cool, wet winters in order to survive hot, dry summers (**•Figure 2.6g**). Vegetation all around the Mediterranean Sea is the archetype. Minimizing water loss is also crucial for vegetation in the **desert and xeric shrubland** biome (*xeric*, pronounced "zeeric," means "extremely dry"). Deserts including the Sahara and Australia's interior can be both hot and cold because of latitude, elevation, and seasonal as well as day or night variations. All deserts are arid, with potential evaporation exceeding precipitation. Plant life in deserts can be surprisingly rich when water is available; a single rainfall event can transform a barren surface into a carpet of green and other colors (**•Figure 2.6h**).

Grasses are the primary vegetation cover in a number of biomes. In warm climates wetter than deserts but too dry to

Joe Hobbs

• Figure 2.6b Temperate broadleaf and mixed forest, northwest Florida.

Joe Hobbs

• Figure 2.6d Tropical coniferous forest in highland Guatemala, here the setting for modern Maya rituals.

Joe Hobbs

• Figure 2.6c Tropical dry broadleaf forest, Angkor Wat, Cambodia.

Joe Hobbs

• Figure 2.6e Temperate coniferous forest by the Pine River, a fisherman's paradise in southern Colorado's Weminuche Wilderness.

• **Figure 2.6f** Boreal forest in Bowron Lake Provincial Park, British Columbia, Canada.

• **Figure 2.6g** Mediterranean scrub and woodland of California's coastal range. The Calaveras Fault lies at the boundary of San Jose's eastern suburbs with the hills beyond.

• **Figure 2.6h** Desert shrubland annuals in full glory following winter rains, Anza Borrego Desert State Park, southern California.

support extensive forests, exists the **tropical and subtropical grasslands, savannas, and shrubland** biome (including the African savanna so familiar to viewers of nature documentaries; •**Figure 2.6i**); its cooler counterpart, the **temperate grasslands, savannas, and shrubland** biome, is found in places like the steppes of Mongolia and the prairies of the Great Plains (•**Figure 2.6j**). The **montane grasslands and shrubland** biome

• **Figure 2.6i** Tropical grassland (savanna), southern Kenya.

• **Figure 2.6j** Temperate grassland (prairie), Flint Hills, eastern Kansas.

exists in mountainous and highland areas, such as on the vast plateau of Tibet, with plant cover varying by elevation (•**Figure 2.6k**). The **flooded grasslands and savannas** biome is present in tropical and subtropical areas on very low, flat lands with soggy soils that are seasonally flooded from rivers or are perpetually covered by marshes and other wetlands. Examples are the large Sudd region of South Sudan and Florida's Everglades (•**Figure 2.6l**).

Restricted to tropical coasts, **mangroves** are saltwater swamps dominated by salt-tolerant mangrove species (•**Figure 2.6m**). Largely protected from large waves and rough surf, mangroves themselves also provide the vital ecosystem service of protecting vulnerable coastlines from erosion. Mangroves make up the world's smallest biome; the total worldwide area is about the size of Arkansas. Commercial aquaculture and coastal developments have taken a big toll on mangroves in recent decades.

Tundra is Earth's coldest biome. Trees cannot grow in most tundra locations because of the low temperatures and widespread permafrost (frozen soil) just below the surface. Far northern Canada, Scandinavia, and Russia have tundra vegetation, which includes grasses, shrubs, and moss (•**Figure 2.6n**). In Franz Josef Land, a Russian archipelago just 600 miles from the North Pole, I was delighted to find colorful flowers growing from a thin layer of melted permafrost.

Polar areas such as Antarctica and most of Greenland are outside of any biome, as they are covered by glaciers or a more stationary **ice cap** year-round, where no vegetation can grow at all (•**Figure 2.6o**). Perpetual ice in the form of glaciers may be found at very high elevations in the lower latitudes (even in equatorial regions) as well as in the polar realms. Geographers have a name for those parts of Earth where the surface is frozen: it is the **cryosphere** (from the Greek word root meaning "cold"): the area covered by ice sheets and glaciers, permafrost, and sea areas periodically covered by ice. Numerous biomes, for example, from ice cap down through boreal and deciduous forests to desert, occur in the steep mountainous areas depicted as "undifferentiated highland" on many climate and biome maps (•**Figure 2.6p**).

Earth's biomes are not static. As we will see later in this chapter, climate change appears to be shifting biotic zones toward the poles and upward in elevation. The process of biotic

© Daniel Prudek/Shutterstock.com

• **Figure 2.6k** Montane grassland on the Tibetan Plateau.

© Vilainecrevette/Shutterstock.com

• **Figure 2.6m** Mangrove habitats above and below water, Caribbean Sea, Panama.

Mike Goldwater/Alamy

• **Figure 2.6l** Flooded grassland and savanna in the vast Sudd region of the White Nile, South Sudan.

Joe Hobbs

• **Figure 2.6n** Tundra, northern Norway.

Joe Hobbs

• Figure 2.6o Ice cap, western Svalbard (Spitsbergen), Norway.

Joe Hobbs

• Figure 2.6p Undifferentiated highland vegetation, San Juan Mountains, Colorado, United States. Note the timberline above which trees do not grow. Here, biomes change with elevation, and so the vegetation climate types have to be characterized as "undifferentiated" on small-scale maps.

change may be seen in striking fashion in the world's coniferous forests, for example. Recently, from Russia to the Rockies, pine and spruce beetles, which lay eggs under the bark, disrupting the flow of nutrients and water in the tree, have laid waste to these forests (**•Figure 2.7**). The culprit appears to be climate change: winters have not been cold enough to kill off the beetles, which are also now able to live at much higher elevations. Mortality of the trees is as much as 90 percent, creating vast reservoirs of fuel for forest fires. This habitat change affects wildlife, and where people live amongst these dead trees, they must contend with the potential natural hazard of fire.

2.3 Biodiversity

All of Earth's natural regions have remarkable life forms. But scientists recognize the exceptional importance of some biomes because of their **biological diversity** (or **biodiversity**)—the

Jean Hobbs

• Figure 2.7 Spruce beetles have killed up to half of this stand of coniferous forest near Ouray, Colorado. Nearby stands have been completely killed off.

number of plant and animal species present and the variety of genetic materials these organisms contain.

As mentioned above, the most diverse biome is the tropical rain forest. From a single tree in Peru's Amazon region, entomologist Terry Erwin collected about 10,000 insect species (he uses a gigantic version of a household flea or roach fogger, along with large funnels to collect the falling insects). From another tree several yards away, he counted another 10,000 species, many of which differed from those in the first tree. Alwyn Gentry of the Missouri Botanical Garden recorded 300 tree species in a 1-hectare (2.47-acre) plot of the Peruvian rain forest. Less than 50 years ago, scientists calculated that there were 4 million to 5 million species of plants and animals on Earth. Now, however, their estimates are much higher, in the range of 40 million to 80 million. This startling revision is based on research, still in its infancy, on species inhabiting the rain forest. Research like Erwin's and Gentry's also raises important questions about what is at stake in destruction of the world's rain forests and other critical habitats.

The Importance of Biodiversity

Such diversity is important in its own right, but it also has vital implications for nature's ongoing evolution and for people's lives on Earth. Humankind now relies on a handful of crops as staple foods. In our agricultural systems, the trend in recent decades has been to use biotechnology to develop genetically modified high-yield varieties of grains and to plant them as vast **monocultures** (single-crop plantings). This trend, which is the cornerstone of the so-called **Green Revolution**, is controversial. On the one hand, it puts more food on the global table. But on the other, it may render agriculture more vulnerable to pests and diseases and thus pose long-term risks of famine. In evolutionary terms, the Green Revolution has reduced the natural diversity of crop varieties

116, 312, 314

that allows nature and farmer to turn to alternatives when adversity strikes. At the same time, while we remove tropical rain forests and other natural ecosystems to provide ourselves with timber, agriculture, and living space, we may be eliminating the foods, medicines, and raw materials of tomorrow even before we have collected them and assigned them scientific names. "We are causing the death of birth," lamented biologist Norman Myers.[7]

Places on Earth that are both biologically rich and deeply threatened by human activities are known as **biodiversity hot spots**, which scientists believe deserve immediate attention for study and conservation (these are not to be confused with the geologic hot spots defined on **page 26**). The world's 35 biodiversity hot spots as identified by Conservation International are depicted in **•Figure 2.8**. Referring to the map of the Earth's biomes in Figure 2.5, you see that most of these hot spots are in tropical and subtropical areas. Many are islands that tend to have high biodiversity because species on them have evolved in isolation to fulfill special roles in these ecosystems, and because human pressures on island ecosystems are particularly intense. Efforts are underway in most of these hot spots to establish national parks and other protected areas. Conservationists are also turning their attention increasingly to the state of the world's oceans, which play critical roles in the Earth's physical processes and sustain many people.

375, 443

2.4 The World's Oceans

Life on Earth would be impossible without the roles and resources of the hydrosphere, which includes both oceans and freshwater sources such as lakes and rivers. In this book, beginning in the next chapter, you will see many examples of how the most precious resource of fresh water is increasingly imperiled and contested. Here the focus is on the world's oceans.

67, 239, 421

Why Should We Care about Oceans?

A remarkable feature of life on Earth is that over half of the world's people live within 60 miles (100 km) of a coastline. Ten percent of the world's people live within 6 miles (10 km) of the shore. You can see these spatial characteristics in the map of the world's population on **page 75**. After decades living as one of those lucky 10 percent, I now live in Missouri, deprived of the sea's wonders. Whether or not we enjoy a personal relationship with the sea, there are many reasons to appreciate it. It may be that people are neurologically hardwired to benefit from the ocean and other waterways. In his book *Blue Mind*, Dr. Wallace Nichols argues that "being near, in, on, or under water can make you happier, healthier, more connected, and better at what you do."[8] Here are other essential qualities of the sea and its resources:

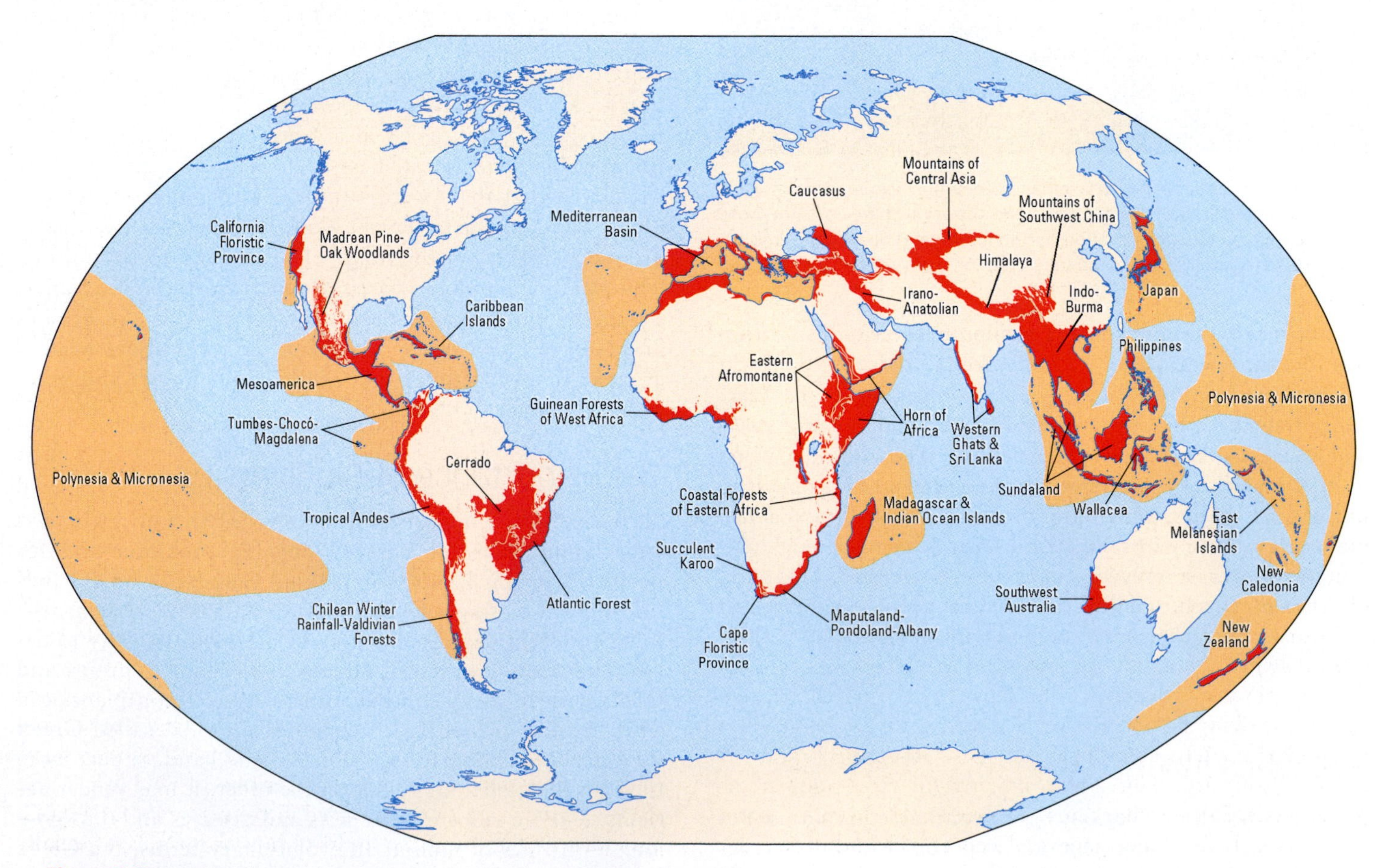

• Figure 2.8 World biodiversity hot spots as recognized by Conservation International.

Joe Hobbs

• Figure 2.9 Seattle's Pike Place Market. Humanity's demand for fish and seafood is growing, but what about supplies? Stocks are falling, and in many species, plummeting.

Joe Hobbs

• Figure 2.10 Shark fins at a Hong Kong Market. There are several streets in this section of the city filled with shops selling shark fins and edible birds' nests. Both are specialty items in high demand among ethnic Chinese people throughout the world. Hong Kong is the main distribution center for the world's shark fin trade, with at least 10 million shark fins moving through Hong Kong each year. The shark fins pictured here sell for over $300 per pound.

It's a watery world. About 70 percent of the world's surface is comprised of water. It is important to consider how water shapes life on "the Blue Planet." Oceans have the largest role in the **hydrologic cycle**, which is the process, powered by the sun's energy, that moves water between the oceans, the sky, and the land.[9] Without the seas, Earth's usable freshwater resources would be extremely limited.

The oceans feed us. About a billion people, or 15 percent of the world's population, rely mainly on fish for their protein (**•Figure 2.9**). Growing numbers of people, increased buying power among the middle classes in developing countries, and technological improvements in the fishing industry are putting these resources under unprecedented pressure.[10] Since 1980, growth in worldwide demand for seafood has exceeded 40 percent. Some scientists have issued a stern warning on the future of the world's fisheries: Unless major steps are taken quickly, there will be a "global collapse" of all species currently fished by 2050.[11]

Defining *collapse* as a population less than 10 percent of its previous levels, these authorities noted that 30 percent of the fish species currently exploited had already collapsed. The declines are most extreme among stocks of large predatory fish like sharks, tuna, and swordfish. Up to 73 million sharks are killed each year (mainly by Spanish and Indonesian fishermen) to provide shark fin soup (which can cost up to $100 per bowl) to the world's ethnic Chinese, who believe it bestows health and virility (**•Figure 2.10**). An uncertain but large percentage of these sharks are "finned," a cruel practice in which the fins are cut off the living fish, which are then thrown back into the sea to die. Fortunately, PR campaigns against the consumption of shark fin soup are credited with slowing down the trade in fins in East Asia. Sales of shark fins in Hong Kong, the main center for the shark fin trade, have declined significantly in recent years. What about other fish in high demand? Populations of the bluefin tuna—a magnificent, huge fish built for speed and endurance—are under pressure because their flesh is prized for Japanese sushi and sashimi. In 2013, a Japanese sushi restaurateur paid a record $1.76 million for a single tuna weighing 488 lbs. (222 kg). Not surprisingly, Japan blocked efforts to put the bluefin on a United Nations list of protected species for many years. 134

Fortunately these grim realities and scenarios can be changed by action on several fronts, including reducing the number of unwanted fish caught in nets and cracking down on overfishing. Governments need to intervene. Legislation like the United Nations (UN) Convention on the Law of the Sea (see **page 317**) can encourage marine conservation and include mechanisms to enforce the law. Like the climate change treaties discussed later, it is easy for a country to sign an international agreement. But then this legislation needs to be ratified, and most importantly, it needs to be enforced.

Just as land can be farmed, so can waters: **aquaculture** (the cultivation of aquatic organisms for food, including **fish farming**) has the potential to increasingly substitute for wild-caught fish stocks. Fish farms already supply half of the fish consumed around the world. There has been explosive growth in aquaculture in China (which leads the world, producing 70 percent of the world's farmed fish). Some aquaculture is what may be described as "sustainable seafood," including the unlikely harvest of shrimp raised in well water deep in the Arizona desert.[12] But much commercial fish farming is controversial because of inputs of antibiotics, pesticides, fertilizers, and hormones, and because of the prospect that genetically modified salmon and other fish (dubbed "Frankenfish" by detractors) would escape their confines and alter the genetic heritage of wild fish. These issues mirror many of those raised in conventional land-based agriculture. There are also limits to how many fish, especially salmon, can be farmed. 116, 314

Oceans provide energy and other raw materials for human use. As terrestrial sources of conventional energy dwindle, our fossil fuel–addicted economies create demand for new supplies. Deep seas are the final frontier for energy exploration. Under the US territorial waters of the Gulf of Mexico, there are as many as 45 billion barrels of petroleum (enough to supply the US oil demand for about five years). But getting to that oil is problematic: much of it lies beneath more than 10,000 feet (3000 m) of water and then 5 miles (8 km) of rock, salt, and sand. Deep-water drilling presents special engineering challenges and environmental risks. Human error in operating the *Deepwater Horizon* well, which was pumping oil from an unbelievable depth of about 35,000 feet (10,000 m) in the Gulf of Mexico, led to a historic oil spill and cleanup effort in 2010. 522 Similar opportunities and perils lie in the depths of the Arctic and Atlantic Oceans (see **pages 484** and **538**). Meanwhile, there is enormous potential to capture unconventional energy supplies from the sea, especially by using the power of surface winds and waves and of rising and falling tides to generate electricity. Europeans are leading the way in using these forms 113 of energy. Prospects are increasing for the deep-sea mining of other minerals, including gold, silver, and the copper, cobalt, nickel, and "rare earth" minerals (many of which are used in high-tech devices) held within manganese nodules strewn across much of the world's seafloor.

318, 439

Oceans play important roles in trade and commerce (• **Figure 2.11**). "They that go down to the sea in ships and occupy their business in great waters" are still among us (Psalm 107:23). The seafaring days are far from over. A remarkable 90 percent of global trade is seaborne. About 98 percent of the Chinese goods seen everywhere in America travel across the Pacific on cargo ships.[13] There is an economic rationale for this. Airfreight can cost 20 times as much as sea freight. In her fascinating book *Ninety Percent of Everything*, Rose George diagnoses most people as being "seablind," and opens our eyes with some amazing accounting: shipping is so cheap that it makes more financial sense for Scottish cod to be sent 10,000 miles to China to be filleted, then sent back to Scottish shops and restaurants, than to pay Scottish filleters to do their work—now there's one of globalization's many twists![14] As we will see in many cases, protecting the world's sea-lanes from pirates and military adversaries has long been one of the world's most important geopolitical concerns. These concerns are greater now than at any time in history, particularly where oil is involved (see, for example, **pages 236** and **321**).

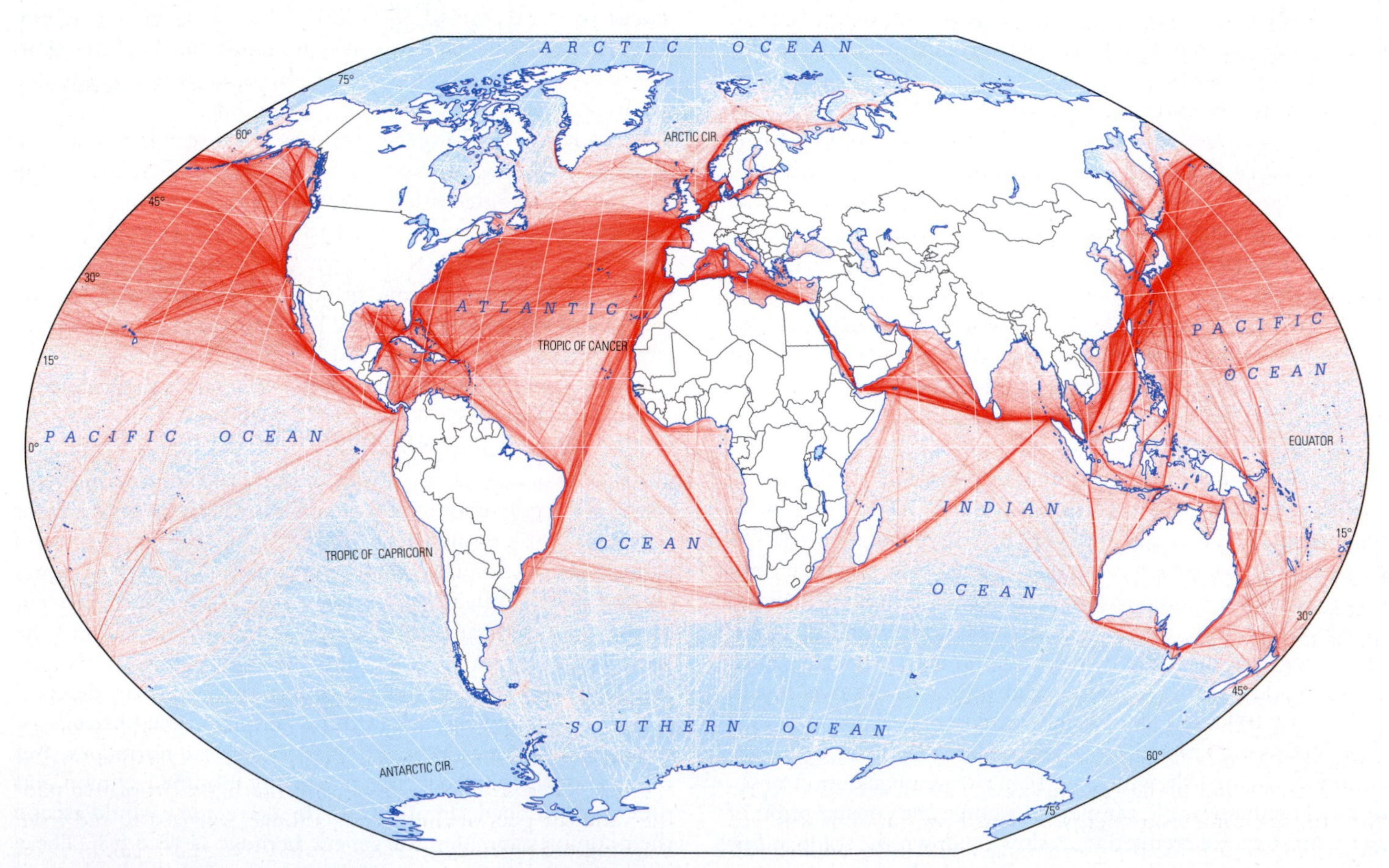

• **Figure 2.11** World commercial shipping routes, as mapped by the National Center for Ecological Analysis and Synthesis. Over 3300 commercial and research vessels were tracked for a year to develop this map showing where human activity on the oceans is highest.

2.5 Global Climate Change

In the atmosphere—the layer of oxygen, nitrogen, and other gases extending from the Earth's surface to about 60 miles (100 km) above it—changes are occurring that are having profound effects on natural and human systems. Many of these changes are attributable to human activities, and the final section of this chapter is a prelude to the following chapter on human processes affecting the planet. "Climate change is the greatest challenge of our time," said Thomas Stocker, co-chair of the United Nations' Intergovernmental Panel on Climate Change. "It threatens our planet, our only home."[15] Now we will examine closely the threats posed by climate change and the promising agendas for combating them. Because so much of the world is being affected by climate change, this detailed overview will help you as you proceed through the regional chapters. You will have the foundation to think critically about mitigation and adaptation choices facing individual countries and will be able to see what future environments may look like around the world (last but not least, National Geography Standard 18 is "To apply geography to interpret the present and plan for the future").

Until about 2000, there was considerable uncertainty in the science of climate change. Most scientists had insisted that human activities were responsible for a documented warming of the Earth's surface (by 1.4°F, or about 0.8°C, since the late 19th century), but a significant minority argued either that warming was not occurring or, if it was, that a natural climatic cycle was responsible. Now there is far less scientific uncertainty about the human role in climate change; by 2014, only an estimated 3 percent of scientists were "climate change skeptics."[16]

The remaining 97 percent of the roughly 2500 atmospheric scientists from more than 130 countries who make up the United Nations–sponsored **Intergovernmental Panel on Climate Change (IPCC)** have concluded that global warming is unequivocal, and that human activities are causing the temperature changes. According to the IPCC's *Climate Change 2014: Impacts, Adaptation, and Vulnerability* report, there was a 95 to 100 percent chance that human production of **greenhouse gases (GHGs)** such as **carbon dioxide (CO_2)** has been responsible for most of the warming of recent decades (• **Figure 2.12**).[17]

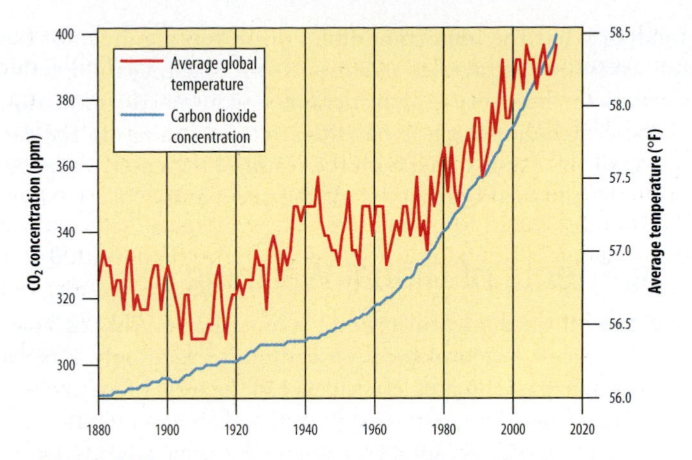

• **Figure 2.12** Industrialization, the burning of tropical forests and other factors have produced a steady increase in carbon dioxide emissions. Most scientists believe that these increased emissions explain the corresponding steady increase in the global mean temperature.

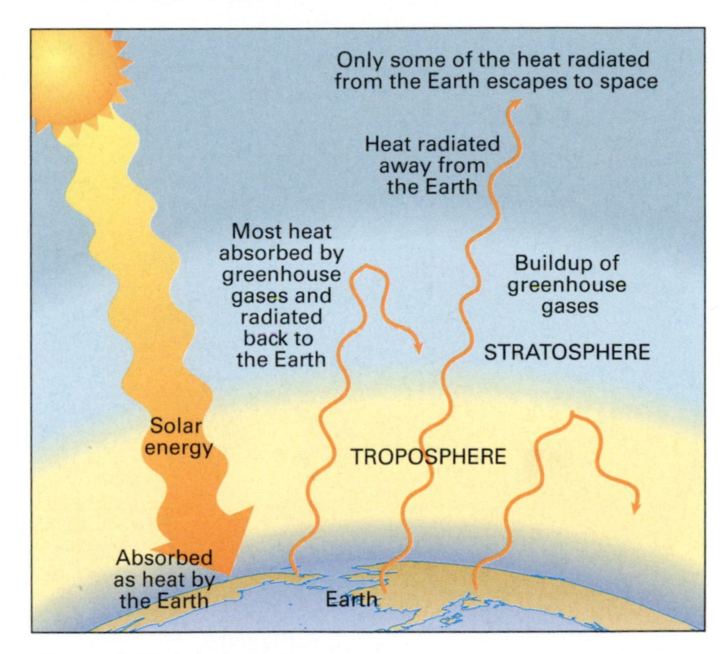

• **Figure 2.13** The greenhouse effect. Some of the solar energy radiated as heat (infrared radiation) from the Earth's surface escapes into space, while greenhouse gases trap the rest. Naturally occurring greenhouse gases make the Earth habitable, but carbon dioxide and greenhouse gases emitted by human activities accentuate the greenhouse effect, making the planet unnaturally warmer.

The Greenhouse Effect

In 1827, a French mathematician named Jean-Baptiste Joseph Fourier (who had distinguished himself as a member of Napoleon Bonaparte's scientific survey of Egypt), established the concept of the **greenhouse effect.** This term is a metaphor of Earth's atmosphere acting like the transparent glass cover of a greenhouse (• **Figure 2.13**; you can also think of the atmosphere as being like a car's windshield). Visible sunlight passes through the glass to strike the planet's surface. Oceans and land, like the floor of the greenhouse or the car's upholstery, reflect the incoming solar energy back as heat (invisible infrared radiation). Acting like the greenhouse glass or car windshield, the Earth's atmosphere traps some of that heat.

The greenhouse effect is not a bad thing. In fact, if not for naturally occurring greenhouse gases such as carbon dioxide and water vapor, the Earth would be too cold for most forms of life. What is worrisome is what happens when human activities add newer and greater quantities of greenhouse gases into the atmosphere, trapping unnatural amounts of heat; these are the so-called "anthropogenic" (made by human) causes of climate change. Carbon dioxide released into the atmosphere from the burning of coal, oil, and natural gas is the greatest source of concern, but **methane** (from rice paddies and the guts of ruminating animals like cattle, and from thawing permafrost), **nitrous oxide** (from the breakdown of nitrogen fertilizers), and **chlorofluorocarbons (CFCs)** (used as coolants and refrigerants) are also greenhouse gases resulting from human activities. CFCs also destroy stratospheric ozone, a gas that has the important effect of preventing much of the sun's harmful ultraviolet radiation from reaching the Earth's surface. In the car-as-Earth metaphor, continued production of these greenhouse gases has the effect of rolling up the car windows on a sunny day, with the result of increased temperatures and physical overheating of the occupants.

The Effects of Global Warming

The trend toward a warmer world is remarkable. Although formal records of meteorological observations began only about a century ago, past climates left evidence in the form of marine fossils, corals, glacial ice, fossilized pollen, and annual growth rings in trees. These indirect sources, along with formal records dating back to 1861, indicate that the 20th century was the warmest of the past six centuries. The ten warmest years on record have all happened since 1998. In descending order, they are 2014, 2010, 2005, 1998, 2013, 2003, 2002, 2006, 2009, and 2007.[18]

And what is the forecast? The IPCC uses a number of different "emission scenarios" that consider such variables as rates of population and economic and industrial growth, as well as efforts that might be made to curb greenhouse gas emissions. IPCC scientists predict that if emissions continue unabated, global mean temperatures between the years 2080 and 2100 will be 6.7°F to 8.6°F (2.6°C to 4.8°C) higher than today. Not long ago we spoke of climate change mainly in futuristic terms, but the IPCC is telling us that climatic changes and their consequences are happening all around us now: "In recent decades, changes in climate have caused impacts on natural and human systems on all continents and across the oceans."[19] Following are the most prominent changes that are anticipated or underway in this warming world (see **•Figure 2.14**).

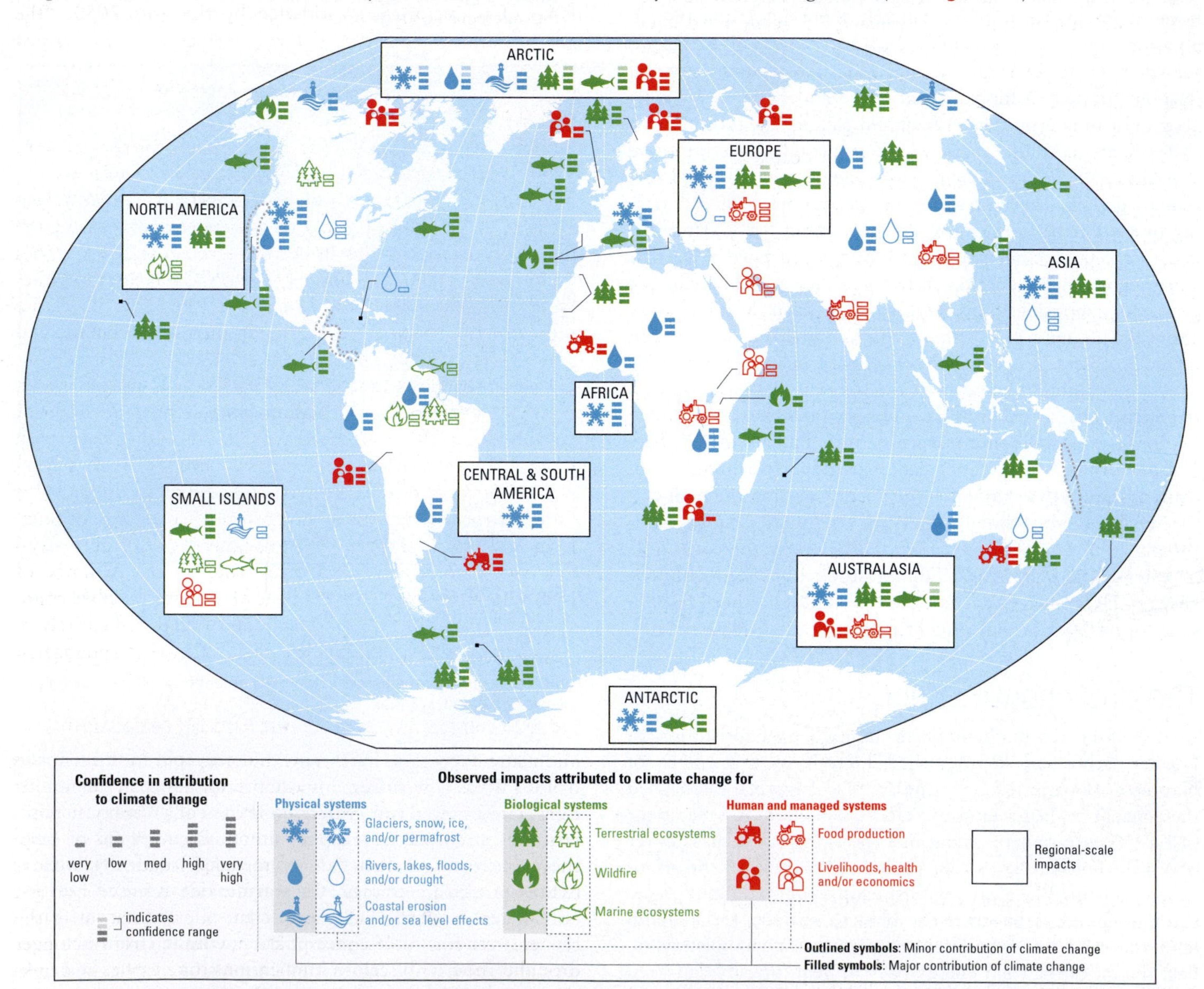

• Figure 2.14 Global patterns of impacts in recent decades attributed to climate change. Impacts are shown at a range of geographic scales. Symbols indicate categories of attributed impacts, the relative contribution of climate change (major or minor) to the observed impact, and confidence in attribution.

Source: The UN Intergovernmental Panel on Climate Change, and the map may be accessed at *http://ipcc-wg2.gov/AR5/images/uploads/WG2AR5_SPM_FINAL.pdf*.

A Warmer Climate Overall, but Not Warmer Everywhere

As geographers, we want to know *where* the anticipated changes will occur and what they will be. Computer models suggest that there will not be a uniform temperature increase across the globe. Increases will instead vary spatially and seasonally. Some places will actually become cooler. Warming will be greatest *in the polar and tropical regions.*

Geographically, the impacts of climate change are expected to be greatest at the higher latitudes, especially in the polar realms. Permafrost, the frozen ground typical of the polar regions, has been melting at an alarming rate. Among other things, this has been forcing Arctic peoples to relocate from 504 traditional settlements as their buildings sink into muck. Impacts on marine environments are also profound in these high latitudes. The IPCC estimates that the average coverage of Arctic sea ice has shrunk about 3 percent per decade since 1978, with summertime ice decreasing 7 percent per decade. Aboard a Russian icebreaker, I saw open water at the North Pole in August 1996. The vessel's Russian "Ice Master" said that in 30 years of working on icebreakers in the Arctic, he had never seen anything like this: vast expanses of the Arctic Ocean completely ice-free or merely dotted with sea ice and icebergs. Ever since that summer, there have been reports of unprecedented change in Arctic waters. The trend is self-perpetuating: the darker ocean waters that replace the white ice cover absorb ever more solar radiation. (Think of cars here again: How hot would the surface of a dark blue car feel to your touch on a summer day compared with the surface of a white car?) That warming melts even more ice, creating more dark surface water and more warming (**• Figure 2.15**). This is known as **polar amplification** and is an example of a positive feedback loop, which represents change in a natural system. Scientists fear that beyond certain thresholds some climatic feedbacks could become irreversible. The theoretical point at which an impact becomes irreversible is known as a **tipping point.**

• Figure 2.15 From the safety of an icebreaker, I photographed this polar bear feeding on a seal in the Arctic Ocean. The sea ice seen here deflects the sun's rays, but the dark blue water absorbs them, contributing to warmer water and more ice melt. The bear's scientific name, *Ursus maritimus*, meaning "marine bear," aptly suggests that this great predator spends much of its time at sea. But it must come up on the sea ice to feed, sleep, hibernate, and bear young. As sea ice retreats, there are fewer places for the bear to eat and rest, and there are increasing accounts of bears swimming to their death in open water.

Joe Hobbs

In some corners of the world, melting ice is seen as good news as it means more money. A further retreat of polar ice would bode well for maritime shipping through the long-icebound Northwest Passage across Arctic Canada, and through the Northern Sea Route across the top of Russia, which until recently has been navigable only in summer and only with the aid of icebreakers. 198, 538

Recent science at the University of Hawaii–Manoa has concluded that warming will also be more rapid in the Earth's tropical regions, where there is a normal baseline of year-round warmth. In their report in the journal *Nature*, the Hawaii scientists also painted this uncomfortable scenario of temperatures worldwide by the year 2050: "the coldest year in the future will be warmer than the hottest year in the past."[20]

Rising Sea Levels

As global temperatures rise, so do sea levels, for two reasons: melting glacial ice on land is pouring more water into the sea, and as seawater warms, it occupies more volume ("thermal expansion," just as mercury in a thermometer does, causing the mercury to climb the glass tube). Sea levels rose 6 to 9 inches (15 to 23 cm) in the 20th century. The IPCC forecasts a further rise of 12 to 35 inches (31 to 88 cm) by 2100. Some of this rise will be fueled by increased melting in the vast Greenland and Antarctic ice sheets.

Decision makers in coastal cities throughout the world, in island countries, and in nations with important urban or agricultural lowland areas adjacent to the sea are worried about the implications of rising sea levels. The low-lying coastal countries of the Netherlands and Bangladesh have for many years been outspoken advocates of reduced greenhouse gases. Their natural allies are the roughly 40 vulnerable island 387 countries around the world who make up the **Alliance of Small Island States.** The president of alliance member country Maldives, an Indian Ocean country comprised entirely of islands barely above sea level, pleaded, "We are an endangered nation!" (**• Figure 2.16**).[21] 329

More Precipitation Overall, but Also More Drought

Higher temperatures will cause more evaporation from the world's oceans, resulting in more precipitation, but it will be distributed irregularly. What the IPCC calls "heavy precipitation events" will be more common. Even unusually large snowfall trends are expected, and the magnitude of hurricanes (whose strength increases with warmer ocean surface waters) will increase. The IPCC forecasts more precipitation in the 346 higher latitudes, but less precipitation, with intense and longer droughts, and with serious implications for supplies of drinking water, in the lower latitudes. "In many regions, changing precipitation or melting snow and ice are altering hydrological systems, affecting water resources in terms of quantity and quality," reports the IPCC.[22]

Peter Essick/Aurora/Getty Images

• **Figure 2.16** Malé (pronounced "mah-lay"), capital city of the Maldives. Some sources describe this as the world's most densely populated city. For more on the Maldives, see **page 329**.

THINK CRITICALLY: How is Malé vulnerable to some of the climate change problems described by the IPCC (**pages 40–43**)? How would you describe its geographical situation?

Shifting Biomes, with Impacts on Plant and Animal Species

With warming, the distribution of climatic conditions typical of biomes is shifting poleward around the world and upward in mountainous regions. Many animal species will be able to migrate to keep pace with changing temperatures, but plants, being stationary, will not. Conservationists who have struggled to maintain islands of habitat as protected areas are particularly concerned about rapidly changing climatic conditions. The IPCC reports that these impacts have already been observed: "*Many terrestrial, freshwater, and marine species have shifted their geographic ranges, seasonal activities, migration patterns, abundances, and species interactions in response to ongoing climate change.*"

Shifting Biomes, with Impacts on Agriculture

Agriculture will also be affected as growing zones move poleward. And although the right temperature and rainfall combinations for productive agriculture might emerge in new areas, the right soil conditions may not be there (the hard rock of the Canadian Shield will never make good farmland, for example). The IPCC reports on agricultural impacts so far: "Based on many studies covering a wide range of regions and crops, negative impacts of climate change on crop yields have been more common than positive."

Why would there be positive agricultural impacts from climate change? One answer might be because carbon dioxide is a natural fertilizer, and its increase in the atmosphere should increase crop yields. Factoring this in their models, scientists are disappointed to find that potential gains in crop yields are offset by the losses incurred by heat waves and drought.[23] The IPCC predicts that climate change will reduce crop production by as much as 2 percent each decade through 2100. The numbers are concerning because population and consumption growth are projected to increase the demand for food by as much as 14 percent each decade through 2100. Couldn't more land be farmed? Yes, but in many parts of the world this would involve clearing and burning forests, contributing more carbon dioxide emissions and reducing the carbon storage function of flora. Couldn't existing farmland be made more productive? Yes. Agriculture can be adapted to new conditions, including a warming climate. Crops can be planted earlier, and new varieties more resilient to climatic stress can be developed. This would of course require more use of genetically modified foods, the potential downsides of which were cited earlier.

36

More Climate-Related Extreme Events, with Significant Impacts on People

These impacts are material and psychological. According to the IPCC, "impacts from recent climate-related extremes, such as heat waves, droughts, floods, cyclones, and wildfires . . . include alteration of ecosystems, disruption of food production and water supply, damage to infrastructure and settlements, morbidity and mortality, and consequences for mental health and human well-being."

Geopolitical Instability and Other Impacts on Human Systems

The IPCC reports that "climate-related hazards exacerbate other stressors, often with negative outcomes for livelihoods, especially for people living in poverty." Climate-related hazards affect poor people's lives directly through impacts on livelihoods, reductions in crop yields, or destruction of homes and indirectly through, for example, increased food prices and food insecurity.

Ironically, many of the countries most vulnerable to climate change are among the least responsible for greenhouse gas emissions. Bangladesh, a fertile, low-lying country the size of Iowa, with a population of 158 million, is a good example; its climate change problems are described on **page 330**. There is consensus that some of the poorest and most populous nations will be hit hardest by climate change and that political and economic instability will result. The IPCC forecasts that rising sea levels will flood tens of millions of poor people out of their homes each year. By 2080, 1 billion to 3 billion people in developing countries (as much as 10–30 percent of the world's population) will face shortages of freshwater related to growing aridity. Epidemics of malaria and other killer diseases will be more widespread, the IPCC warns.

In the next chapter we consider many of humanity's biggest challenges, including the gulf between the haves and have-nots, population growth, poverty, migration, and refugees. With climate change added to this mix, risks rise to alarming levels. The 2014 IPCC report asserts that climate change could so destabilize the world's food system that rising hunger or even mass starvation could result: by 2080, possibly 200 million to 600 million people will face starvation because of failing crops.[24] Such dramatic impacts on food security could lead to social unrest. We may have had a preview of this in 2007 and 2008, when a series of poor grain harvests worldwide (in part due to extreme weather events) fell far short of global demand. Grain prices more than doubled, and several countries halted grain exports. There was panic buying, and food riots occurred in more than 30 countries.

Environmental crises like these would presumably lead to movements of **climate change refugees**—people on the move because of untenable conditions at home—and to conflicts over resources, upsetting the fragile balance of power between countries. In a published report, US military thinkers concluded that climate change

> presents significant national security challenges to the United States . . . Projected climate change will seriously exacerbate already marginal living standards in many Asian, African and Middle Eastern nations, causing widespread political instability and the likelihood of failed states. The chaos that results can be an incubator of civil strife, genocide and the growth of terrorism. The United States may be drawn more frequently into these situations.[25]

What Can We Do about Global Climate Change?

This is scary. Can Earth's future really be that bleak? What if those 97 percent of scientists are wrong? The IPCC scientists acknowledge that there are many uncertainties in their modeling and forecasting. However, they argue that where possibly dangerous, irreversible, or catastrophic events are involved, policy should be based on the **precautionary principle**: adverse events should be preempted or prevented by action. Uncertainty is not a reason for inaction. We need to act (• **Figure 2.17**).

• **Figure 2.17** Protestors calling for action on negotiating the climate change treaties in Copenhagen in 2009. Copenhagen went down as a huge disappointment for advocates of a binding treaty.

Motivated by the scientific consensus, policy makers are considering the steps required to blunt or cope with climate impact scenarios. Scientists are urging them to take actions to limit future warming of the Earth's atmosphere to no more than 2°C (3.6°F) above the level of preindustrial times (well below the IPCC forecast of nearly 4.8°C (8.6°F) that unchecked emissions could bring by 2100). That temperature increase of 2°C (3.6°F) is the level above which scientists think the most dangerous effects of climate change would begin to occur.

There are essentially two approaches to confronting climate change: mitigation and adaptation. **Mitigation** measures aim to avoid the adverse impacts of climate change in the long term; these include steps like switching from coal to cleaner fossil fuels to produce electricity, replacing fossil fuels with other energy resources, reducing energy consumption, and removing greenhouse gases from the atmosphere by boosting photosynthesis (for example, by planting more trees, which absorb carbon dioxide in the photosynthetic process of producing plant tissue and oxygen). **Adaptation** measures are designed to cope with and reduce the unavoidable impacts of climate change in the short and medium terms; these include building sea walls and dikes to prevent flooding related to rising sea levels, relocating people from flood-prone areas to higher ground, and developing crop varieties that are more suited to expected changes in precipitation and temperature. Generally, mitigation is seen as a problem that the world's richer countries plus China should deal with because they are the main producers of greenhouse gases. Adaptation is typically seen as a problem for the poorer countries, which are expected to suffer the most from climate change. Many of the world's largest population clusters are in low-lying river deltas in poor countries: Vietnam's Mekong, Bangladesh's Brahmaputra/Ganges, and Egypt's Nile, for example. These are the broad associations of adaptation and mitigation. In reality, most poor and wealthy countries alike will have to take measures to both adapt to and mitigate climate change.

Following are some of the main approaches to mitigating climate change and thereby avoiding the kinds of almost-unthinkable adaptations that would be needed in vulnerable countries like Vietnam (see feature on **page 346** in Chapter 7). This discussion excludes the untested prospect of **geoengineering**, using technology to intervene in the climate system in an effort to combat global warming. Launching Earth-orbiting mirrors to deflect sunlight back into space is one technique under consideration. Another is "to create a 'space sunshade' by shooting reflective particles into the stratosphere that block out a small but significant fraction of incoming solar radiation."[26]

Negotiate and Implement International Treaties to Reduce Greenhouse Gas Emissions

Most scientists and policy makers agree that the best way to confront global climate change is to implement international treaties to reduce emissions (see Geography of Energy, page 45, and Figure 2.14). Countries have proven they can unite in effective action against greenhouse gases. Thanks to the **Montreal Protocol** and its amendments, signed by 37 countries in the late 1980s, the production of CFCs worldwide was reduced in phases to zero by 2010. As a result there have been reductions in the size of stratospheric ozone "holes" that have been observed seasonally over the southern and northern polar regions since about 1985. CFC molecules have very long life spans, but scientists predict that the ozone layer should recover to pre-1980 levels by 2050. This recovery will also depend on the successful phaseout of other ozone-destroying refrigerants known as hydrochlorofluorocarbons (HCFCs). The other landmark treaty to combat climate change is the Kyoto Protocol described on page 45.

Cut emissions through market-based incentives. Compared with richer nations, the less developed countries (LDCs), other than China and India, release relatively little CO_2 into the atmosphere. Because the LDCs produce relatively little carbon dioxide, policy makers are coming up with innovative mechanisms to keep them from doing so, and to encourage them to develop clean industrial technologies. It is very expensive for the already highly energy-efficient, more developed country (MDC) to further cut its greenhouse gas emissions, but relatively inexpensive for the rich country to invest in emission cuts in a poorer, energy-inefficient country. The Kyoto Protocol requires the MDC Annex I countries to reduce greenhouse gas emissions, but not necessarily at home. Mechanisms in the agreement offer these countries flexibility in how they meet their reductions targets, providing market-based mechanisms working on the principle that the world benefits from reduced emissions wherever they are reduced. In these mechanisms, carbon dioxide is a commodity that is traded in the "carbon market," with one credit equaling 1 ton of CO_2.

With the **clean development mechanism (CDM)**, a wealthier Annex I country can earn emissions units by investing in emission-reduction projects in a poorer country, and use those certified emission reduction (CER) credits to meet part of its overall emission reduction target. A rich country like France, for example, can build a solar panel farm for rural electrification in the African country of Mali. Mali thereby benefits by clean, sustainable energy development, whereas France exercises a cost-effective way to meet its treaty obligations. In such a scheme, both countries apparently benefit, as does the global atmosphere. Critics however say that the CDM ultimately lets the big polluters in the industrialized countries "off the hook," by letting them avoid the hard work of cutting emissions further at home.

Joint Implementation (JI) allows an Annex I country to earn emission reduction units (ERUs) by investing in emission reduction in another Annex I country. For example, Japan has a project to capture methane from Ukraine's coal mines before it is emitted into the atmosphere. Japan thereby earns emission reduction units, each equaling 1 ton of CO_2, that Japan can use to meet its emission cut targets or sell on the open carbon market. Joint implementation offers Japan a flexible and cost-efficient means of fulfilling its treaty commitment, while the host party Ukraine benefits from foreign investment and technology transfer. The main difference between the CDM and JI lies in their application, as JI projects can only be hosted by countries like Japan with emission reduction or limitation commitments to treaty obligations.[27]

Finally, there is the mechanism of the marketplace where emissions credits are bought and sold. The **international emissions trading** (also called **cap and trade**) mechanism lets countries that have an excess of emission units (emissions permitted to them, but not used) to sell this excess capacity to countries that are over their targets. Russia is one country making a lot of money with this mechanism: its industrial sector is flagging, so it is not polluting as much as the Kyoto Protocol allows it to. It is selling its unused emission credits to 175 countries like Germany whose industries are flourishing and producing more emissions than allowed by treaty. No net gain in GHG emissions is involved. The country that is greener, either by accident or by design, profits, while the browner country saves money it would have to spend to reduce its own emissions.

Increase Carbon Sequestration

Carbon sequestration refers mainly to the natural capture and long-term storage of carbon in forests, farmlands, or in the oceans, so that the buildup of carbon dioxide in the atmosphere will reduce or slow (there are also artificial means of carbon sequestration, for example, pumping carbon dioxide deep into underground repositories).[29] In the process of photosynthesis, plants absorb carbon dioxide. Collectively, the world's forests, farmlands, and seas (which contain carbon-absorbing phytoplankton) thus serve as giant **carbon sinks**. In international climate-change negotiations, forest- and farm-rich countries want to receive credit for having these natural buffers against climate change.

Climate change can also be mitigated by reducing deforestation, especially in poorer countries. People in these countries rely heavily on trees and other vegetation for fuel. When 68 burned, trees release the CO_2 that they had stored in the process of photosynthesis. At the same time, in being burned they cease to exist as organisms that, in the process of photosynthesis, remove CO_2 from the atmosphere. Deforestation for agricultural expansion, conversion to pastureland, infrastructure development, logging, and other purposes accounts for about 20 percent of global greenhouse gas emissions (more than the global transportation sector and second only to the energy sector.)[30] Since 2008, the United Nations has been developing a program called **Reducing Emissions from Deforestation and Forest Degradation (REDD)**. This is an effort to create a financial value for the carbon stored in forests, offering incentives for developing countries to reduce carbon emissions from forested lands and invest in low-carbon development. REDD has other potential benefits, like protecting biodiversity, alleviating poverty, and enhancing the status of indigenous peoples (however some groups protest that REDD will close forests off to them, and represents new agents of land grabbing and marginalization), and protecting biodiversity. REDD is emerging as part of the carbon-trading scheme described above, 64, 66

Geography of Energy The Kyoto Protocol and Beyond

Climate change is a global problem, ultimately linked to energy consumption, that demands the attention of all the countries. Fortunately, the struggle to confront climate change has been more than just "hot air." In 1992, at the United Nations Conference on Environment and Development (the "Earth Summit") in Rio de Janeiro, 155 countries signed a treaty called the **United Nations Framework Convention on Climate Change (UNFCC)**. This was a nonbinding treaty; although all these countries acknowledged that climate change was a problem, the UNFCC did not require any specific commitments. The main reason the treaty was made "toothless" was that the United States, then responsible for one-quarter of global CO_2 emissions, would not sign a treaty that required it to reduce emissions. A global climate treaty without the biggest emitter on board would be meaningless, so the treaty was watered down so that the United States would sign. It was hoped that later negotiations would secure specific commitments from the United States and other countries.

Those negotiations came in 1997. In Kyoto, Japan, at a United Nations conference on climate change, 84 of the 160 countries present signed the **Kyoto Protocol**, a landmark international treaty for climate change. This treaty committed the countries that signed and ratified it to binding (mandatory) emission reduction targets. Recognizing that developed countries are principally responsible for the high levels of GHG emissions in the atmosphere as a result of more than 150 years of industrial activity, the Protocol places a heavier burden on developed nations under the principle of "common but differentiated responsibilities." The agreement required 38 more developed countries known as ("Annex I countries") to reduce carbon dioxide emissions to 5 percent or more *below their 1990 levels* by the year 2012. The United States (with Vice President Al Gore signing) pledged to cut its emissions to 7 percent below 1990 levels by that date. The European Union made a promise of 8 percent below 1990 levels, and Japan promised a 6 percent reduction below 1990 levels.

Meeting these pledges would require substantial legislative, economic, and behavioral changes in these more affluent countries. Transportation and other technologies that use fossil fuels would have to become more energy efficient, making these technologies at least temporarily more expensive. Gasoline prices would rise, so consumers would feel the pinch. Advocates of the protocol argued that the initial sacrifices would soon be rewarded by a more efficient and competitive economy powered by cleaner and cheaper sources of energy derived from the sun. Opponents, however, felt that higher fuel prices would be too costly to bear. In 2005, after the European Union countries, Japan, and finally Russia ratified the Kyoto Protocol, it went into effect for the countries that ratified it.

The United States signed but did not ratify the Kyoto Protocol during the Clinton administration. Even then, partly because of the strength of the fossil fuel industry lobbies, the political will for ratification did not exist. When President George W. Bush assumed office in 2001, he rejected the Kyoto Protocol outright. The Bush administration had two objections: the potentially high economic cost of implementing the treaty and the fact that China, along with all the world's less developed countries, was not required by the Kyoto Protocol to take any steps to reduce greenhouse gas emissions. China argued then, as it still does now, that while it is a major greenhouse gas producer, it is a developing country whose growth should not be hampered by pressure from countries that themselves became rich by burning fossil fuels.

Efforts to reduce the production of greenhouse gases focus on the world's wealthiest nations, but there is a good reason for the growing pressure on China to join the fight. By far the greatest producers of carbon dioxide emissions, China and the United States together account for roughly 40 percent of the entire global emissions output. As recently as 1994, China was predicted to catch up with the United States' CO_2 emissions levels by 2019. The fact that China achieved this distinction in 2007 instead is testimony to the dizzying pace of that country's development—a theme that echoes throughout this book. China's thinking about 310 climate change has changed a lot in recent years, especially because its increasingly affluent people do not want to suffer the choking effects of China's growth. In 2014, the two "great powers of climate change" came to an agreement (but not a treaty): the United States would cut net greenhouse gas emissions to 26–28 percent below 2005 levels by 2025. China would reach its peak CO_2 emissions around 2030 and increase the non–fossil fuel share of all energy sources to around 20 percent by 2030.

As the Kyoto Protocol approached expiration in 2012, an agreement was reached to extend it to 2020. The terms of a successor document to be developed by 2015 would go into effect when the Kyoto Protocol expires in 2020. Attention turned to exactly what new emission-cut goals should be set. The IPCC advocated a "carbon budget" for humanity—a limit on the amount of carbon dioxide that can be produced by industrial activities and the clearing of forests. No more than 1 trillion metric tons of carbon could be burned and the resulting gases released into the atmosphere, the panel found, for planetary warming to be kept below that 3.6°F (2°C) above the level of preindustrial times. Over a half-trillion tons have already been burned since the beginning of the Industrial Revolution, and unless there is a reduction in the rate at which energy consumption is growing, the trillionth ton will be burned sometime around 2040.[28] Keep your eyes on what world powers do or don't do to keep the planet below that critical threshold of a 3.6°F (2°C) increase.

allowing richer countries to purchase carbon credits from poorer countries. But in the long run, the richer countries may also be expected to donate funds to REDD in the interest of the planet's welfare.

As we move through the world's regions, we will see how individual countries and supranational organizations are dealing with this "greatest challenge" of climate change. The next chapter explores how the world's developed and less developed countries confront problems like climate change, population growth, and resource use differently. Like this chapter, it will ask you to consider specific ideas about how to deal with some of the most pressing issues of our time.

Study Guide

Summary

- The Earth's three layers of habitable space are the hydrosphere, atmosphere, and lithosphere. The lithosphere is made up of separate plates that are in motion, a process known as plate tectonics. These movements result in mountain building, volcanic activity, earthquakes, and other consequences that represent natural hazards when they affect people.
- Weather refers to atmospheric conditions prevailing at one time and place. Climate is a typical pattern recognizable in the weather of a region over a long period of time. Climatic patterns have a strong correlation with patterns of vegetation and in turn with human opportunities and activities in the environment.
- Geographers group local climates into major climate types, each of which occurs in more than one part of the world and is associated with other natural features, particularly vegetation. Geographers recognize 10 to 20 major types of ecosystems or biomes, which are categorized by the type of natural vegetation. Vegetation and climate types are so sufficiently related that many climate types are named for the vegetation types.
- Some biomes are particularly important because of their biological diversity—the number of plant and animal species and the variety of genetic materials these organisms contain. Regions where human activities are rapidly depleting a rich variety of plant and animal life are known as biodiversity hot spots, places scientists believe deserve immediate attention for study and conservation.
- Oceans cover about 70 percent of the Earth's surface. They play the key role in the hydrologic cycle, sustain large numbers of people through the protein in fish and seafood, and contain valuable mineral resources. The bulk of the world's cargo trade is by sea.
- Most of the sun's visible short-wave energy that reaches the Earth is absorbed, but some of it returns to the atmosphere in the form of infrared long-wave radiation, which generates heat and helps warm the atmosphere. This is the Earth's natural greenhouse effect.
- The scientific community represented by the United Nations–sponsored Intergovernmental Panel on Climate Change (IPCC) is convinced with 95 percent or greater certainty that human activities, particularly the production of carbon dioxide and other greenhouse gases, are responsible for global warming. Computer-based climate change models use various emissions scenarios for atmospheric carbon dioxide and indicate that the mean global temperature might warm up by an additional 6.7°F to 8.6°F (2.6°C to 4.8°C) by 2100, unless action is taken. The scientific consensus is that with global warming, the distribution of climatic conditions typical of biomes will shift poleward and upward in elevation, sea levels will rise, and mean global precipitation will increase (but with drought intensified in some areas). According to the IPCC, these trends are already observable.
- Polar amplification, with melting of ice that opens up more blue water and melts more ice, represents a positive feedback loop that could become irreversible. The point at which an impact become irreversible is a tipping point.
- There are two approaches to confronting climate change: Mitigation measures aim to avoid the adverse impacts of climate change in the long term, and are usually associated with more developed countries (MDCs). Adaptation measures are designed to cope with and reduce the unavoidable impacts of climate change in the short and medium terms, and are usually associated with less developed countries (LDCs).
- In the policy arena, the European Union countries tend to exert the strongest leadership in fighting climate change. The United States has often been slow or unwilling to take strong, specific actions. China has tried to be exempted from mandatory steps but, along with the United States, has recently pledged to do more.
- The Kyoto Protocol was an international agreement requiring the industrialized countries that ratified it to make substantial cuts in their carbon dioxide emissions to reduce global warming. Mechanisms to make cuts include the Clean Development Mechanism, Joint Implementation, and Cap and Trade. The United States dropped its support for the treaty. It went into effect for the ratifying countries after Russia ratified it in 2004. Set to expire in 2012, the treaty was extended to 2020. Efforts continued to reduce carbon dioxide emissions, especially in the MDCs and China, and to maximize ways to reduce deforestation, especially in the LDCs. The goal is to keep the mean global temperature below a 3.6°F (2°C) increase.

Review Questions

1. What are the three "spheres" of habitable life on Earth?
2. What are plate tectonics, and what are some of the main consequences of tectonic activity?
3. What is the difference between weather and climate? What are the main forces that produce precipitation and aridity?
4. What are the major climate types and their associated biomes? Where do they tend to occur on Earth?
5. Where are the biodiversity hot spots? In what kinds of locations and biomes do many of them occur?

6. What important roles do the world's oceans play, and what are their major resources? In what ways are these resources threatened?
7. What are the most prominent changes that are anticipated or underway in the warming world? What steps are being taken or considered to avert these changes through mitigation and through adaptation?
8. Why are China and the United States particularly important in issues related to climate change, and how do they characterize their abilities to modify greenhouse gas emissions?
9. According to the IPCC, what specific cap on greenhouse gas emissions would prevent the worst consequences of global warming?
10. What are the market-based incentives that could help treaty members meet their emission reduction targets?

Key Terms + Concepts

adaptation (p. 43)
Alliance of Small Island States (p. 41)
aquaculture (p. 37)
atmosphere (p. 24)
biodiversity (p. 35)
biodiversity hot spots (p. 36)
biological diversity (p. 35)
biomes (p. 29)
 boreal forest/tiaga (p. 31)
 desert and xeric shrubland (p. 32)
 flooded grasslands and savannas (p. 34)
 Mediterranean forests, woodland, and scrub (p. 32)
 montane grasslands and shrubland (p. 33)
 temperate grasslands, savanna, and shrublands (p. 33)
 temperate broadleaf and mixed forest (p. 30)
 tropical coniferous forest (p. 31)
 tropical and subtropical dry broadleaf forest (p. 30)
 tropical and subtropical grasslands, savannas, and shrubland (p. 33)
 tropical and subtropical moist broadleaf forest (p. 30)
biosphere (p. 24)
cap-and-trade system (p. 44)
carbon sequestration (p. 44)
carbon sink (p. 44)
clean development mechanism (CDM) (p. 44)
climate (p. 27)
 desert (p. 28)
 humid continental (p. 29)
 humid subtropical (p. 29)
 ice cap (p. 34)
 mangroves (p. 34)
 maritime or marine west coast (p. 28)
 Mediterranean (p. 28)
 oceanic (p. 28)
 polar climate (p. 29)
 semiarid (p. 28)
 subarctic (p. 29)
 tropical rain forest (p. 28)
 tropical savanna (p. 28)
 tundra (p. 34)
 undifferentiated highland (p. 29)
climate change refugees (p. 43)
coniferous trees (p. 31)
continental drift (p. 25)
cryosphere (p. 34)
fault (p. 26)
faulting (p. 26)
fish farming (p. 37)
geo-engineering (p. 43)
geologic hot spot (p. 26)
Green Revolution (p. 35)
greenhouse effect (p. 39)
greenhouse gases (GHGs) (p. 39)
 carbon dioxide (CO_2) (p. 39)
 chlorofluorocarbons (CFCs) (p. 39)
 methane (p. 39)
 nitrous oxide (p. 39)
hydrologic cycle (p. 37)
hydrosphere (p. 24)
Intergovernmental Panel on Climate Change (IPCC) (p. 39)
international emissions trading (p. 44)
joint implementation (JI) (p. 44)
Köppen climate classification system (KCC) (p. 28)
Kyoto Protocol (p. 45)
lithosphere (p. 24)
mitigation (p. 43)
moment magnitude scale (MMS) (p. 26)
monoculture (p. 35)
Montreal Protocol (p. 44)
natural hazard (p. 28)
plates (p. 25)
plate tectonics (p. 25)
polar amplification (p. 41)
potential evaporation (PE) (p. 28)
precautionary principle (p. 43)
precipitation (p. 27)
Reducing Emissions from Deforestation and Forest Degradation (REDD) (p. 44)
rifting (p. 25)
Ring of Fire (p. 25)
seafloor spreading (p. 25)
seismic activity (p. 25)
subduction (p. 25)
subduction zone (p. 26)
tectonic forces (p. 25)
tipping point (p. 41)
trench (p. 26)
tsunami (p. 25)
United Nations Framework Convention on Climate Change (UNFCC) (p. 45)
volcanism (p. 25)
weather (p. 27)

Notes

1. http://www.goodreads.com/author/show/690500. Yuri Gagarin.
2. Wolf Roder, "Three Near Misses: Are These Science or Pseudoscience?" *Cincinnati Skeptics Newsletter*, June 1996. Accessed November 14, 2006, from http://www.cincinnatiskeptics.org/newsletter/vol5/n5/misses.html.
3. This definition of natural hazards is from http://www.oas.org/DSD/publications/Unit/oea66e/ch01.htm.
4. http://www.colorado.edu/hazards/about/.
5. http://www.nytimes.com/2014/03/29/us/governments-find-it-hard-to-restrict-building-in-risky-areas.html?_r=0.
6. On Face the Nation, February 16, 2014; transcript at http://www.cbsnews.com/news/face-the-nation-transcripts-february-16-2014-mccrory—demint/.
7. Norman Myers, *Primary Source: Tropical Forests and Our Future* (New York: Norton, 1984), p. x.
8. http://www.wallacejnichols.org/122/bluemind.html.
9. This definition of the hydrologic cycle is from http://geography.about.com/od/physicalgeography/a/hydrologiccycle.htm.
10. Review the United Nations' *Annual Report of the State of the World's Fisheries and Aquaculture* at http://www.fao.org/sof/sofia/index_en.htm.
11. Boris Worm et al., "Impacts of Biodiversity Loss on Ocean Ecosystem Services," *Science*, November 3, 2006, pp. 787–790.
12. See the Monterey Bay Aquarium's guide to sustainable seafood, "Seafood Watch," at www.mbayaq.org/cr/seafoodwatch.asp.

13. Amy Roach Partridge, "Global Trucking Woes," *Global Logistics*, November 2006, accessed September 27, 2007, from http://www.inboundlogistics.com/cms/article/tms-soothes-transportation-woes/.

14. Dwight Garner, "Life on Ships That Make the World Go Round," October 17, 2013, pp. C1, 7.

15. Stocker quoted in http://www.globalresearch.ca/climate-change-the-greatest-challenge-of-our-time/5360852.

16. Dana Nuccitelli, "Survey Finds 97% of Climate Science Papers Agree Warming Is Man-Made," *The Guardian*, May 16, 2013; accessed April 20, 2015, from http://www.theguardian.com/environment/climate-consensus-97-per-cent/2013/may/16/climate-change-scienceofclimatechange.

17. Intergovernment Panel on Climate Change, "Climate Change 2014: Impacts, Adaptation, and Vulnerability," accessed January 7, 2015, from http://www.ipcc.ch/report/ar5/wg2/.

18. For the warmest years recorded, see http://www.nasa.gov/content/goddard/nasa-finds-2013-sustained-long-term-climate-warming-trend/index.html#.VBxia0vYMdJ. There is an animation of the warming planet here.

19. http://report.mitigation2014.org/spm/ipcc_wg3_ar5_summary-for-policy makers_approved.pdf.

20. Justin Gillis, "By 2047, Coldest Years May Be Warmer Than Hottest in the Past, Scientists Say," *New York Times*, p. A9.

21. http://www.earth-policy.org/plan_b_updates/2008/update76.

22. In the following paragraphs, quotes from the IPCC report come from http://report.mitigation2014.org/spm/ipcc_wg3_ar5_summary-for-policy makers_approved.pdf.

23. Justin Gilles, "A Jolt to Complacency on Food Supplies," *New York Times*, November 12, 2013. p. D3; and Justin Gilles, "Climate Change Seen Posing Risk to Food Supplies," *New York Times*, November 2, 2013. pp. A1, 8.

24. See this video: http://www.ipcc.ch/report/ar5/wg2/.

25. See "U.N. Climate Panel Endorses Ceiling on Global Emissions." http://www.nytimes.com/2013/09/28/science/global-climate-change-report.html. For an assessment of impacts of climate change on the United States, see http://www.globalchange.gov/what-we-do/assessment.

26. Martin Weitzman, "The World in 2114," *Financial Times*, February 15, 2014.

27. http://einstitute.worldbank.org/ei/course/clean-development-mechanism-and-joint-implementation-navigating-kyoto-project-based-mechanism.

28. Incidentally, more than 3 trillion tons of carbon are still left in the ground as fossil fuels. The IPCC is encouraging governments to insist that once the trillion-ton budget is exhausted, companies that want to keep burning fossil fuels will have to capture carbon dioxide and store it underground. Under fire, the Obama administration pressed ahead with rules requiring that technology be upgraded, which would be expensive for the coal-burning power plants that supply most of America's electricity. During President Obama's two terms, by executive action and without congressional approval he advanced a "climate action plan" for the United States. It would cut US carbon emissions to 17 percent below 2005 levels by 2020, on par with one target advocated by the European Union. But unbound by treaty, such a goal would be easily overturned in a new political environment. And how is the leading agent of climate change agents faring? Like the United States, China has consistently averted any binding commitments to reduce carbon emissions.
45
29. For carbon sequestration, see www.carbonventures.com/policy/article.php?list=The%20Kyoto%20Protocol&id=4467&link=Carbon%20Sequestration/.

30. Source on deforestation and carbon: www.unredd.org/AboutREDD/tabid/582/Default.aspx/.

Global Geosciences Watch

Go to the Environment and Ecology Portal. Select the Kyoto Protocol subcategory and read the Overview document beginning at the top of this page. Write two paragraphs. In the first paragraph describe briefly why the proponents of the agreement think it is a landmark environmental and political achievement. In the second paragraph describe the points that opponents use to criticize the agreement.

Online Resources

For access to MindTap and additional study materials visit www.cengagebrain.com. Read your textbook, take notes, complete activities, take practice quizzes and more.

Above: Bedouin on the move in Egypt. The Bedouins' environmental knowledge is great and their environmental "footprint" small. Joe Hobbs
Right: A young European visitor to an Indian camp near Iquitos, Peru. Joe Hobbs

3

Human Processes and World Regions

The key problem facing humanity is how to bring a better quality of life—for 8 billion or more people—without wrecking the environment entirely in the attempt.

—EDWARD O. WILSON[1]

This chapter continues your introduction to geography's basic vocabulary, focusing on how people have interacted with the environment to change the face of the Earth, with an emphasis on the human roles. You will see how modern trends in human–environment interactions are products of revolutionary changes in the past: the arrival of agriculture and of industrialization. The chapter uses some of the concepts of economic geography to explain where rich and poor countries are located on the Earth's surface and account for some of these patterns of prosperity and poverty. As you explore our eight world regions, you will understand where and why populations are increasing, and what the implications of that growth are. Finally, with your geographic perspectives you will consider ideas about how to solve some of the most important global problems of our time.

Chapter Outline

Chapter Objectives

This chapter will enable you to

- Gain a historical perspective on the capacity of human societies to transform environments and landscapes.
- Understand why some countries are rich and others poor, and recognize the geographic distribution of wealth and poverty.
- Recognize trends associated with globalization, including the decline of poverty worldwide.
- Explain the simultaneous trends of falling population growth in the richer countries and rapid population growth elsewhere.
- Explore the principles of sustainable development.

3.1 Two Revolutions That Have Changed the Earth

The geographer's approach to understanding a current landscape—in almost all cases, a cultural landscape that has been shaped or touched by human activity—is often historical, involving study of its transformation from the pre-human or early human natural landscape. The current spatial patterns of our relationship with the Earth[7] may be seen as products of two "revolutions": the Agricultural Revolution that began in the Middle East about 10,000 years ago and the Industrial Revolution that began in 18th-century Europe. Each of these revolutions transformed humanity's relationship with the natural environment. Each increased substantially our capacity to consume resources, modify landscapes, grow in number, and spread in distribution. Why should we consider these historic processes? The answer borrows from the theme our colleagues in the Geological Society of America explored in a recent annual meeting: "The past is the key to the future."

Hunting and Gathering

Until about 10,000 years ago, our ancestors lived by **hunting and gathering** (also known as **foraging**). We were quite good at it—it served us well for more than 100,000 years until we began experimenting with the revolutionary technologies of agriculture.

Foraging was quite different from the farming and industrial ways of living that succeeded it. Joined in small bands of extended family members, hunters and gatherers were nomads with no villages, homes, or other fixed dwellings. They moved to take advantage of changing opportunities on the landscape. These foragers scouted large areas to locate foods such as seeds, tubers, foliage, fish, and game animals (**• Figure 3.1**). Moving their small group from place to place, they had a relatively limited impact on the natural environment, especially compared with the impacts left by agricultural and industrial societies.

Hunters and gatherers may have been the **original affluent society.**[2] Many scholars have seen in these pre-agricultural people an apparent harmony with the natural world in both economies and spiritual beliefs. After short periods of work to collect

Joe Hobbs

• Figure 3.1 Until the relatively recent past—just a few thousand years ago—people were exclusively hunters and gatherers. This rock art in Egypt's Sinai Peninsula was created over a long time span, as it depicts Neolithic period hunting of ibex, later uses of camels and horses, and writing from the Nabatean period (1st century CE).

the foods they needed, they enjoyed long stretches of leisure time. Studies of those few hunter-gatherer cultures that lingered into modern times, such as the San (Bushmen) of southern Africa and several Amerindian groups of South America, 464 suggest that although their life expectancy was low, they suffered little from the debilitating social and psychological problems found in industrialized societies.

Hunters and gatherers did modify their landscapes. These foragers were not always at peace with one another or with the natural world. With upright posture, stereoscopic vision, opposable thumbs, an exceptionally large brain relative to the rest of the body, and a year-round mating "season," *Homo sapiens* (Latin for "wise man") became, after our emergence in southern Africa about 200,000 years ago, an **ecologically dominant species**—defined as one that competes more successfully than other organisms for nutrition and other essentials of life, or that exerts a greater influence than other species on the environment. Before they started farming, our wise ancestors were people on the move. They came out of Africa and crossed the Isthmus of Suez or southern Red Sea about 125,000 years ago, reaching South Asia by 50,000 years ago, Europe by 43,000 years ago, Australia by 40,000, East Asia by 30,000 years ago, North America by 14,000 years ago (possibly 30,000 years ago), and South America 11,000 years ago (new research may revise these dates to much earlier). There are several remarkable projects that have traced human migrations by DNA pathways, including the Genetic Atlas of Human Admixture and the National Geographic's Genographic Project[3] (•**Figure 3.2**).

All across their growing range, our pre-agricultural ancestors shaped the face of the land on a vast scale relative to their small numbers. They used fire to flush out or create new pastures for the game animals they hunted. Many of the world's prairies, savannas, and steppes, where grasses now prevail, developed as hunters and gatherers repeatedly set fires. These people also overhunted and in some cases eliminated animal species. The controversial **Pleistocene overkill hypothesis** states that rather than being at harmony with nature, hunters and gatherers of the late Pleistocene Epoch (a geological period from about 2 million to 10,000 years ago) hunted many species to extinction, including the elephant-like mastodon of North America.

Farming: Welcome to the Anthropocene

Despite these excesses, the environmental changes that hunters and gatherers could cause were limited. Humans' power to modify landscapes took a giant step with **domestication**, the controlled breeding and cultivation of plants and animals. Domestication brought about the **Agricultural Revolution**, also known as the **Neolithic Revolution** or the **Food-Producing Revolution**.

This was the dawn of a new age, the likes of which the world had never experienced. In conventional geologic time scales, the Agricultural Revolution that began about 13,000 years ago coincided with a great retreat of glacial ice sheets and the onset of the Holocene Epoch in which

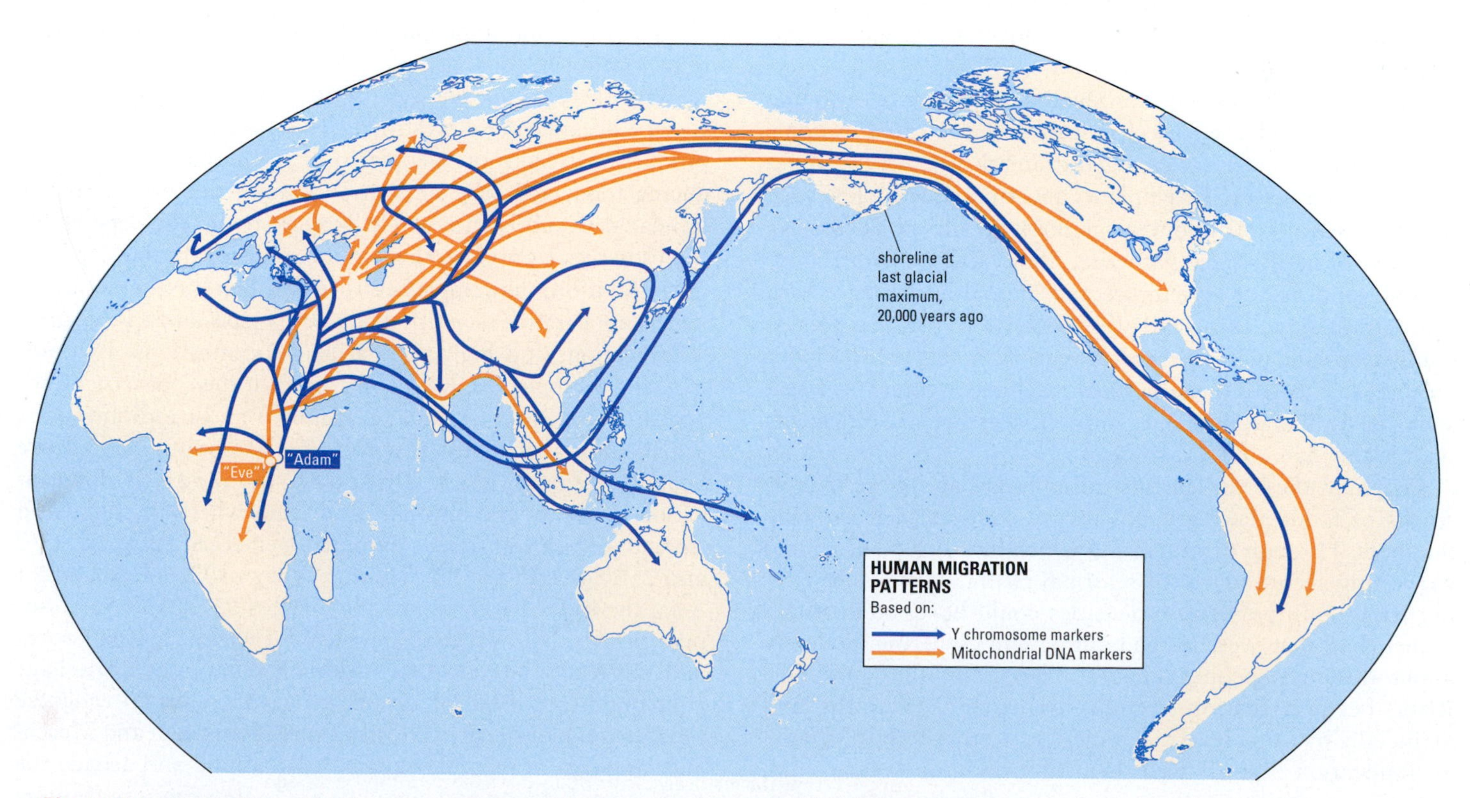

• **Figure 3.2** DNA pathways trace human migrations out of Africa and across the Earth. Scientists have identified the human lineages of the world descended from 10 sons of a genetic Y-chromosome figurative "Adam" and 18 daughters of a mitochondrial figurative "Eve".

we now live (Holocene comes from Greek words meaning "entirely new"). There is a fascinating scientific effort underway to rewrite our understanding of the age in which we live. This process involves renaming the Holocene the **Anthropocene Epoch**, or simply "the Anthropocene," meaning "the age of humans." Peoples' impacts on the atmosphere, lithosphere, and hydrosphere have been so monumental in the last 13,000 years—and especially within the last 250 years—that they rank alongside the great natural forces that defined past ages.

It all began with agriculture. Why people started to produce rather than continue to hunt and gather plant and animal foods—first in the Middle East and later in Asia, Europe, Africa, and the Americas—is uncertain. Although there are many ideas about this process, surprisingly little hard evidence exists to explain it confidently. Two theories are most often put forward. One is that the climate changed. Increasing drought and reduced plant cover may have forced people and wild plants and animals into smaller areas, where people began to tame wild herbivores (including the progenitors of sheep and goats) and sow wild seeds (including the progenitors of wheat and barley) to produce a more dependable food supply.

A more widely accepted theory is that their own growing populations in areas originally rich in wild foods forced people to find new food sources, so they began sowing cereal grains and breeding animals. This process began about 13,000 years ago in the Zagros Mountains of what is now Iran and in neighboring areas of what are now northern Iraq and southern Turkey. The culture of domestication spread outward from there but also developed independently in several world regions.

299, 412, 460

Now that humans were producing as well as consuming foods, their landscape uses, cultures, social organizations, and other characteristics changed dramatically. Among other things, in choosing to breed plants and animals, people settled down. Gradually abandoning the nomadic life and **extensive land use** of hunting and gathering, they came to favor the **intensive land use** of agriculture and animal husbandry. With this pattern of intensive land use, people could sow and harvest crops in specific places year after year. With less need to move around, they began living in fixed dwellings, at least on a seasonal basis. These developed into villages, small settlements with fewer than 5000 inhabitants.

People in these settlements raised larger and more reliable stocks of food, making it possible to support their growing numbers. Through **dry farming**, which involved planting and harvesting according to the seasonal rainfall cycle (and without irrigation), population densities could be 10 to 20 times higher than they were in the hunting and gathering mode. By about 4000 BCE, people along the Tigris, Euphrates, and Nile Rivers began **irrigation** of crops—bringing water to the land artificially by using levers, channels, and other technologies—an innovation that allowed them to grow crops year round, independent of seasonal rainfall or river flooding (•**Figure 3.3**). Irrigation allowed even more people to make a living off the land; irrigated farming yields five to six times more food per unit area than dry farming. In ecological terms, the expanding food surpluses of the Agricultural Revolution raised the Earth's **carrying capacity**, the size of a species' population (in this case, humans) that an ecosystem can support. When domestication of plants and animals began about 13,000 years ago—very recently on Earth's timeline—there were probably fewer than 5 *million* people on Earth. Today, there are more than 7 *billion*.

Joe Hobbs

• **Figure 3.3** The Tigris River in southeastern Turkey. The brown areas are rainless in the long, hot summers and are capable of producing only one crop a year through dry (unirrigated) farming. The green areas along the river are irrigated and can produce two or three crops a year. Irrigation was thus a revolutionary technology that greatly increased the number of people the land could support.

With these conditions culture became more complex, and society became more stratified. The steep increase in food production freed more people from the actual work of producing food, and they undertook a wide range of activities unrelated to subsistence needs. Irrigation and the dependable food supplies it provided thus set the stage for the development of **civilization**, the complex culture of urban life characterized by the appearance of writing, economic specialization, social stratification, and high population concentrations. By 3500 BCE, for example, 50,000 people lived in the southern Mesopotamian city of Uruk, in what is now Iraq. In addition to the Tigris-Euphrates Valleys that comprise Mesopotamia, **culture hearths**—regions where civilization followed the domestication of plants and animals, and from where the **diffusion** (spread) of cultural traits originated—emerged between 8000 and 2500 BCE in the Wei-Huang Valley of China, Southeast Asia, the Indus River Valley, the Ganges River Valley, Mesopotamia, the Nile Valley of Egypt, West Africa, Mesoamerica, and the Andes (•**Figure 3.4**). Although religion and language have important roles in culture, geographers often recognize separate religious hearths (for example, Jerusalem and Mecca) and language hearths (for example, in China and astride the Black Sea). Religion and language are so important in framing the characteristics of world regions that we give them special treatment in each chapter.

220

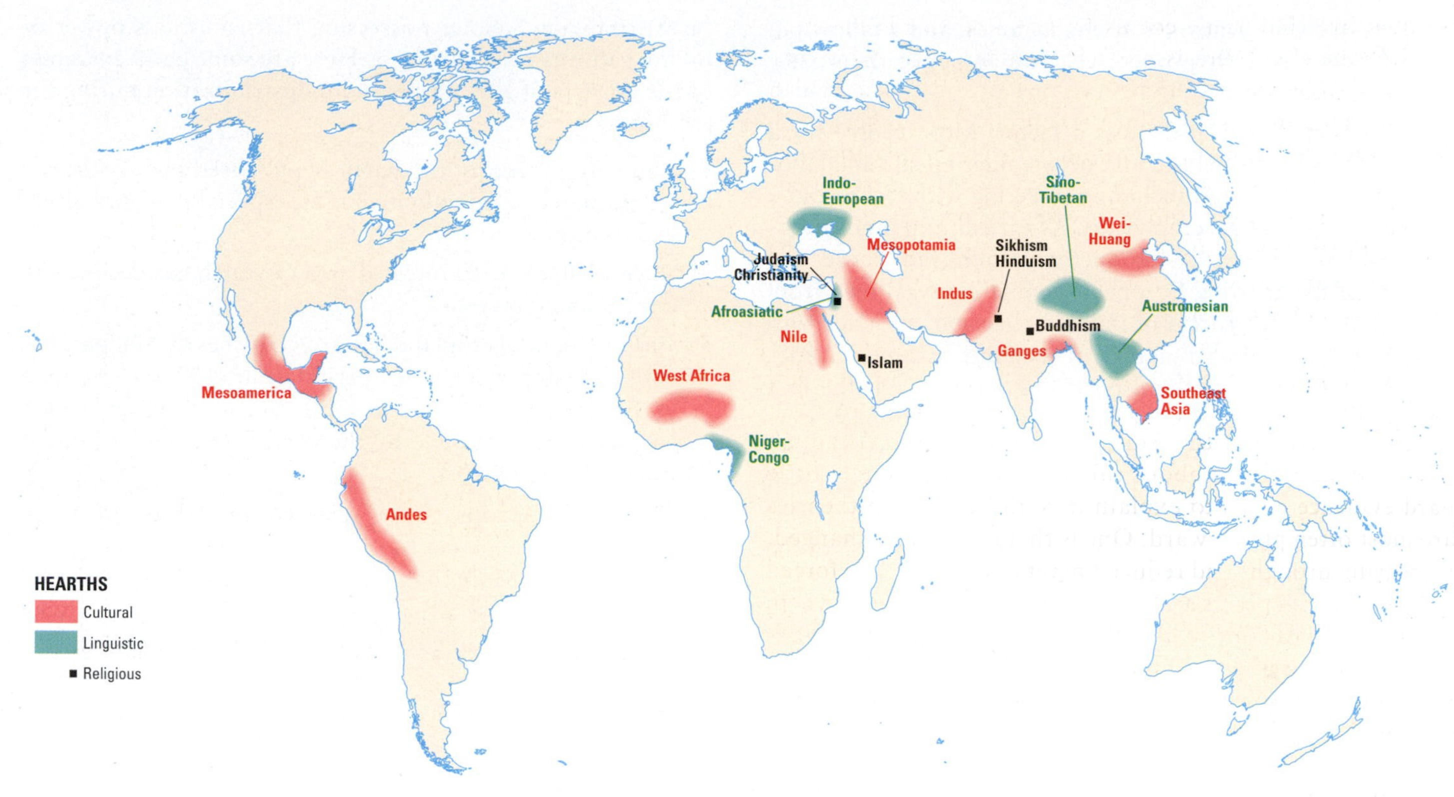

• **Figure 3.4** Cultural, linguistic, and religious hearths worldwide. All areas are approximate.

Human impacts on the natural environment increased. The agriculture-based urban way of life that spread from these culture hearths had larger and more lasting impacts on the natural environment than either hunting and gathering or early agriculture. Acting as agents of humankind, domesticated plants and animals proliferated at the expense of the wild species that people came to regard as pests and competitors. For example, *Bos primigenius*, a wild bovine that was the ancestor of most of the world's domesticated cattle, was hunted to extinction in Europe by 1627. Farmers who resented the animals' raids on their crops were probably among the most ardent hunters. Agriculture's permanent and site-specific nature magnified the human imprint on the land, and the pace and distribution of that impact increased with growing numbers of people.

The Industrial Revolution

The human capacity to transform natural landscapes took another giant leap with the **Industrial Revolution**, which began in Europe around 1750 CE. New patterns of human–land relations emerged from innovations in technology. Several factors came together to spark these breakthroughs (•**Figure 3.5**):

- Western Europe had the economic capital necessary for experimentation, innovation, and risk. Much of this money came from the lucrative trade in gold, undertaken initially in the Spanish and Portuguese empires after 1400.
- Significant improvements in agricultural productivity took place in Europe after about 500 CE. New tools such as the heavy plow, as well as more intensive and sustainable use of farmland, led to increased crop yields. The **three-field system of crop rotation**, with land rotated

Leemage/Historical/Corbis

• **Figure 3.5** The Industrial Revolution got underway with few environmental safeguards. Here a 19th-century artist depicts effluent from copper mills in Cornwall, England pouring directly into waterways.

between cereals, nitrogen-fixing legumes, and a fallow period, was especially productive. Human populations grew correspondingly.

- Population growth itself was a factor. More people freed from work in the fields could devote their talents and labor to trying new things, including tinkering with technologies that would improve crop yields. As agricultural innovations spread and industrial productivity improved, more and more of the growing European population was freed from farming. In the process of **industrialization**, societies with a mainly agricultural (agrarian) economy became societies manufacturing goods and providing services. For the first time in history, a region (Europe) emerged in which city-dwellers producing and consuming those goods and services outnumbered rural folk. Having now spread to nearly all parts of the world, the process of industrialization continues to promote urbanization today (see The Triumph of the City feature on **page 78**).

Most geographers see population growth today as a drain on resources, but the Industrial Revolution illustrates that given the right conditions, more people do create more resources. 80 Industrializing Europeans used innovations such as the steam engine to tap the vast energy of fossil fuels—initially coal and later oil and natural gas—and boost the manufacturing of goods and the productivity of agriculture (note that our synthetic nitrogen fertilizers are made from natural gas). Fossil fuels, which are the stored-up carbon products created by photosynthesis in ancient ecosystems, allowed the Earth's carrying capacity for humankind to be raised again, this time into the billions.

Industrialization, Colonization, and Environmental Change

As they began to deplete their local supplies of resources needed for industrial production, Europeans started to look for these materials abroad. As early as their **Age of Discovery**, also known as the **Age of Exploration**, which began in the 15th century, Europeans probed ecosystems across the globe to feed a growing appetite for innovation, economic growth, and political power. The process of European **colonization**—the extension of European countries' political and economic control over foreign areas—was thus linked directly to the Industrial Revolution. Mines and plantations from such faraway places as central Africa and India supplied the copper and cotton that fueled economic growth in colonizing countries such as Belgium and England. 308, 417

No longer dependent on the foods and raw materials they could procure within their own political and ecosystem boundaries, Europeans of the Industrial Revolution had an impact on the natural environment that was far more extensive and permanent than that of any other people in history. The world today is heir to their legacy: we tend to think of countries and economies being "industrialized" or "industrializing," either possessing that enormous power or on the pathway to acquiring it. Here are some good examples of the impacts of agriculture and industrialization during the Anthropocene:[4]

- About 40 percent of the Earth's land-based photosynthetic output is dedicated to human uses, especially in agriculture and forestry.
- Since 1750, the total forested area on Earth has declined by more than 20 percent.
- Since 1750, total cropland has grown by nearly 500 percent, with more expansion in the period from 1950 to today than in the century from 1750 to 1850. Much of this expansion has been made possible by the use of fertilizers produced from fossil carbon.
- Since 1750, people have released quantities of fossil carbon that the planet took hundreds of millions of years to store away. More than a half-trillion tons of greenhouse gases (GHGs) have already been burned since then, giving people a commanding role in the planet's carbon cycle (see **page 39**).
- There have been five mass species extinctions in Earth's 3.8-billion-year history. In her book *The Sixth Extinction*, science writer Elizabeth Kolbert argues that the consequences of industrialization have put us on course to eliminate 20–50 percent of all species on Earth by 2100 (remember that most species are small, like those collected in Terry Irwin's funnels; **page 35**). This sixth extinction episode is the only one overwhelmingly attributable to human activity.

The global impacts of human activities are well summarized on the map of the "human footprint" in **•Figure 3.6**. It shows uneven distributions of human impacts across the globe. These are the results of economic, population, and other processes we will now explore.

3.2 The Geography of Economic Development

One of the most striking characteristics of human life on Earth is the large disparity between wealthy and poor people, both within and between countries. At a high level of generalization, the world's countries can be divided into *haves* and *have-nots* (**•Figure 3.7** and **•Table 3.1**). Writers refer to these distinctions variously as *developed* and *underdeveloped*, *developed* and *developing*, *more developed* and *less developed*, *industrialized* and *industrializing* or *industrialized* and *nonindustrialized*, and *North* and *South*, or *Global South*, based on the concentration of wealthier countries in the middle latitudes of the Northern Hemisphere and the abundance of poorer nations in the Southern Hemisphere.

I use the terms **more developed countries (MDCs)** and **less developed countries (LDCs)**. This framework is an introductory

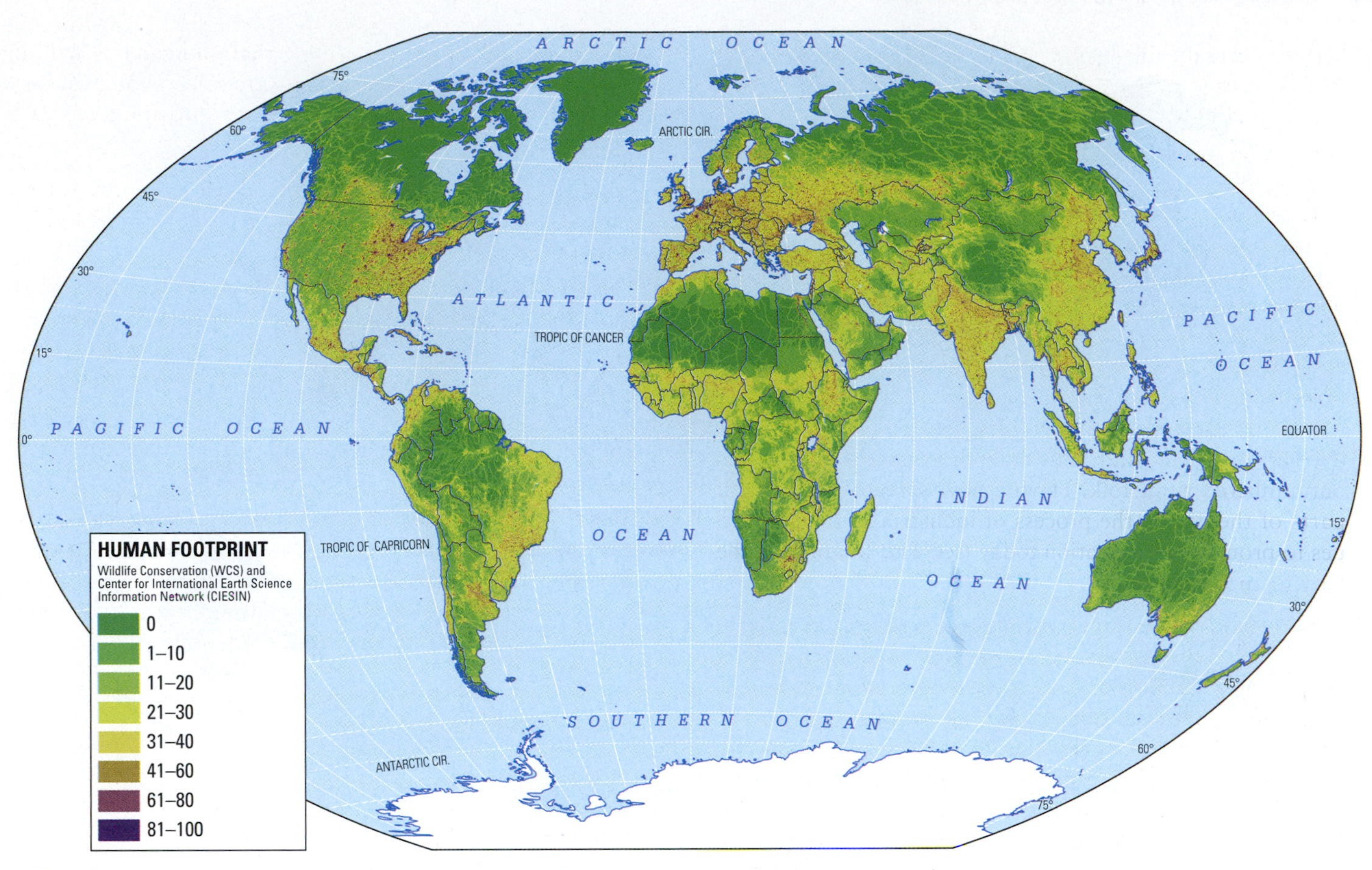

• **Figure 3.6** The human footprint. The biome map on **page 31** depicts what the world's vegetation would look like without human activity. Here we see how strongly people have changed the natural environment. This isarithmic map of the human footprint is a quantitative analysis of human impacts on the Earth's biomes. A score of 1 indicates the least human influence in a given biome. A higher number means greater impact. However, because each biome has its own independent scale, a score of 1 in a tropical rainforest might reflect a different level of human activity than in a broadleaf forest.
Sources: Wildlife Conservation Society (WCS) and Center for International Earth Science Information Network (CIESIN).

THINK CRITICALLY: What are some of the main factors related to higher levels of human impact? Use **Figure 3.21** (population density), **Figure 3.C** (world aquifers), and **Figure 3.E** (urban populations), and discussion of overpopulation on **pages 81–83.**

TABLE 3.1 Generalized Characteristics of More Developed Countries (MDCs) and Less Developed Countries (LDCs)

Characteristic	MDC	LDC
Per capita GDP and income	High	Low
Percentage of population in middle class	High	Low
Percentage of population involved in agriculture	Low	High
Energy use	High	Low
Percentage of population living in cities	High	Low
Birth rate	Low	High
Population growth rate	Low	High
Percentage of population under age 15	Low	High
Percentage of population that is literate	High	Low
Amount of leisure time available	High	Low
Life expectancy	High	Low

tool and cannot account for the tremendous variations and continuous changes in economic and social welfare that characterize the world today. Some countries, including those known as the "Asian Tigers," are best described as **newly industrializing countries (NICs)** because they do not fit the MDC or LDC idealized types. The relevant regional chapters describe these cases.

Measuring Development

There is no single, universally acceptable standard for measuring wealth and poverty on the global scale. But you are likely to encounter the following indexes and issues as you read about development.

Annual Per Capita Gross Domestic Product

Gross domestic product (GDP) is the total output of goods and services that a country produces for home use in a year. Divided by the country's population, the resulting figure of

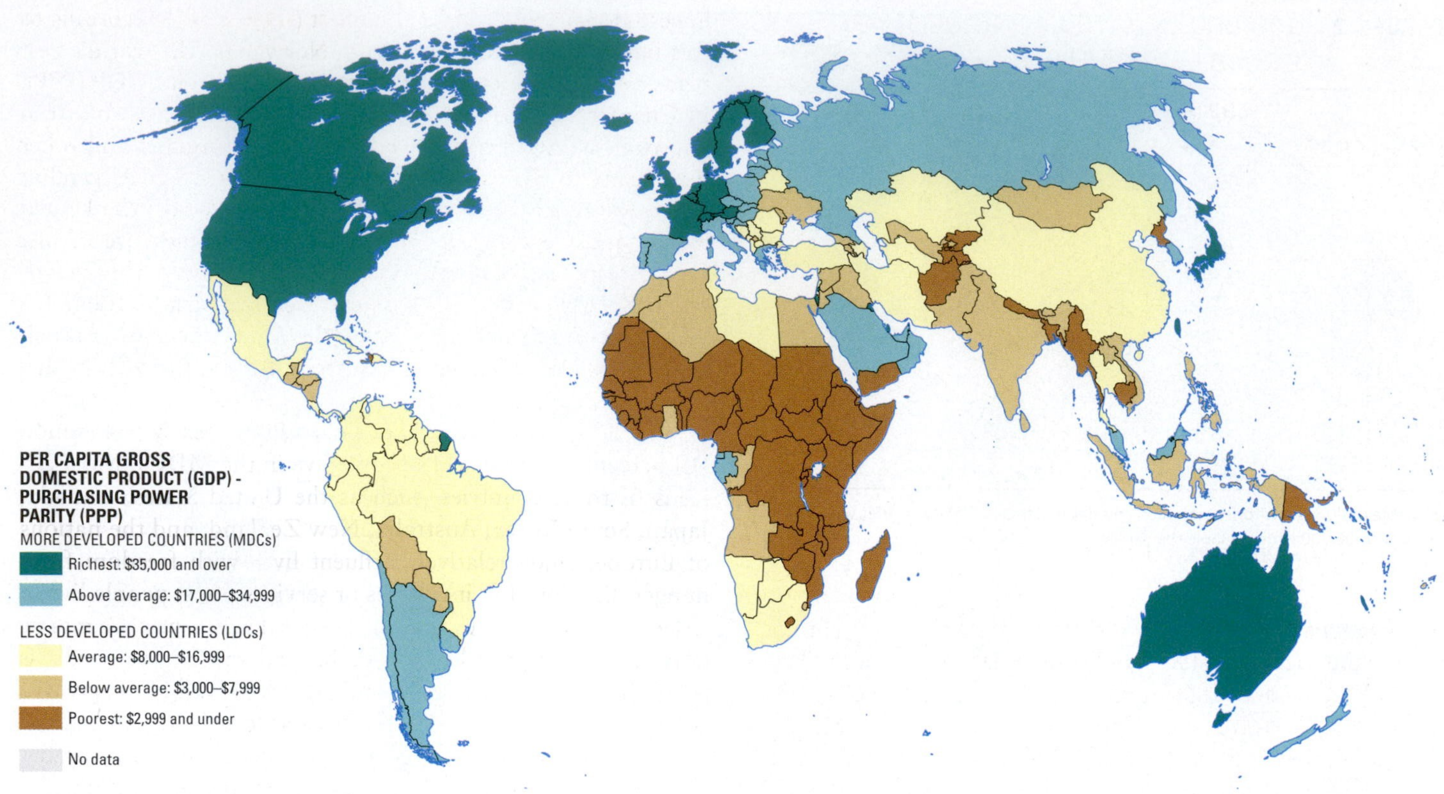

• Figure 3.7 Wealth and poverty by country. Note the concentration of wealth in the middle latitudes of the Northern Hemisphere.
Source: Data from World Factbook, CIA, 2015.

per capita GDP is one of the most commonly used measures of economic well-being and is the one used in this book. Closely related measures are per capita **gross national product (GNP)**, which includes foreign output by domestically owned producers, and **gross national income (GNI)**, which includes gross domestic production plus income from abroad from sources such as rents, profits, and labor.

Purchasing power parity (PPP) conversion factors consider differences in the relative prices of goods and services, providing a better overall way of comparing the real value of output between different countries' economies. **Gross domestic product purchasing power parity (GDP [PPP])**, used to compare countries in this book, is measured in current "international dollars," which indicate the amount of goods and services one could buy in the United States with a given amount of money. Definitions vary, but in this book, an MDC is a country with an annual per capita GDP (PPP) of $17,000 or more; all others are considered LDCs.

Statistics like GDP (PPP) are useful for measuring the relative strengths of national economies, but they are not necessarily representative figures for the "average citizen" of a particular country. For example, the small, oil-rich African nation of Equatorial Guinea has a relatively high GDP (PPP) of $25,700; this is the amount each person in that country would earn each year if the GDP were divided equally among all its citizens. However, over 60 percent of Equatorial Guineans live on less than $2 a day, indicating the country's oil wealth is being concentrated in the hands of a very few people while most of its citizens live in poverty. This is an aspect of the typical "resource curse" afflicting all too many African nations; this problem is described on **page 424** in Chapter 9 on Africa.

The gulf between the world's richest and poorest countries is striking (**•Table 3.2**). The average per capita GDP (PPP) in the MDCs is about five times greater than in the LDCs. With a per capita GDP (PPP) of $102,100, Qatar is the world's richest country, and the Democratic Republic of the Congo is the world's poorest with a per capita GDP (PPP) of $400. Thirty percent of the world's people live on less than $2 per day, the World Bank's average poverty line for the LDCs in 2015. These raw numbers suggest that economic productivity and income alone characterize **development**, which, according to a common definition, is the process of improvement in the material conditions of people through diffusion of knowledge and technology.

The Human Development Index

Definitions like this, and statistics such as per capita GDP (PPP), reveal little about important measures of well-being such as gender equality, literacy, and life expectancy. Recognizing the shortcomings of strictly economic definitions, the United Nations Development Programme created the **Human**

TABLE 3.2 Top and Bottom Countries in Per Capita GDP (PPP) (U.S. Dollars, 2014)*

Top Countries	GDP (PPP) Per Capita	Bottom Countries	GDP (PPP) Per Capita
Qatar	102100	Guinea	1100
Luxembourg	77900	Madagascar	1000
Singapore	62400	Malawi	900
Norway	55400	Niger	800
Switzerland	54800	Liberia	700
Brunei	54800	Central African Rep.	700
United States	52800	Burundi	600
Netherlands	43300	Somalia	600
Canada	43100	Zimbabwe	600
Australia	43000	Congo, Dem. Rep.	400

*This table excludes territories, colonies, and dependencies. Only countries with populations over 100,000 with available data are listed.

Source: CIA World Factbook, 2015.

Development Index (HDI), a scale that takes quality-of-life features into account. This book uses HDI in the Basic Data tables of the chapter that introduce each region (for example, Table 4.1 on page 92). The HDI scale is computed using the criteria of life expectancy, years of schooling, and GDP (PPP), and then indexed on a scale of 0 to 1, with 0 representing the lowest global value and 1 the highest (•Figure 3.8). According to this index, with a rating of 0.944, Norway is "the world's best place to live," although it ranks sixth in per capita GDP (PPP). In Chapter 4, we will see how the "social compact" between governments and peoples has allowed Norwegians and other Europeans to live so well. Following Norway, in descending order, are Australia, Switzerland, the Netherlands, the United States, and Germany. In HDI terms, with a measure of just 0.337, Niger is the world's lowest-rated country; the Democratic Republic of the Congo, the Central African Republic, Chad, and Sierra Leone fare only slightly better. Note that all five of these low-rated countries are in Africa, for reasons discussed in Chapter 9.

On the basis of per capita GDP (PPP), nearly 1.4 billion (19 percent) of the world's people live in the MDCs. Most citizens of these countries, such as the United States, Canada, Japan, South Korea, Australia, New Zealand, and the nations of Europe, enjoy relatively affluent lives with freedom from hunger. Employed in industries or services, most people live in cities rather than in rural areas. Disposable income, the money that people can spend on goods beyond subsistence needs, is generally high. There is a large middle class. Population growth is low as a result of low birth rates and low death rates (these demographic terms are explained in Section 3.3 on population). Life expectancy is long, and the literacy rate is high.

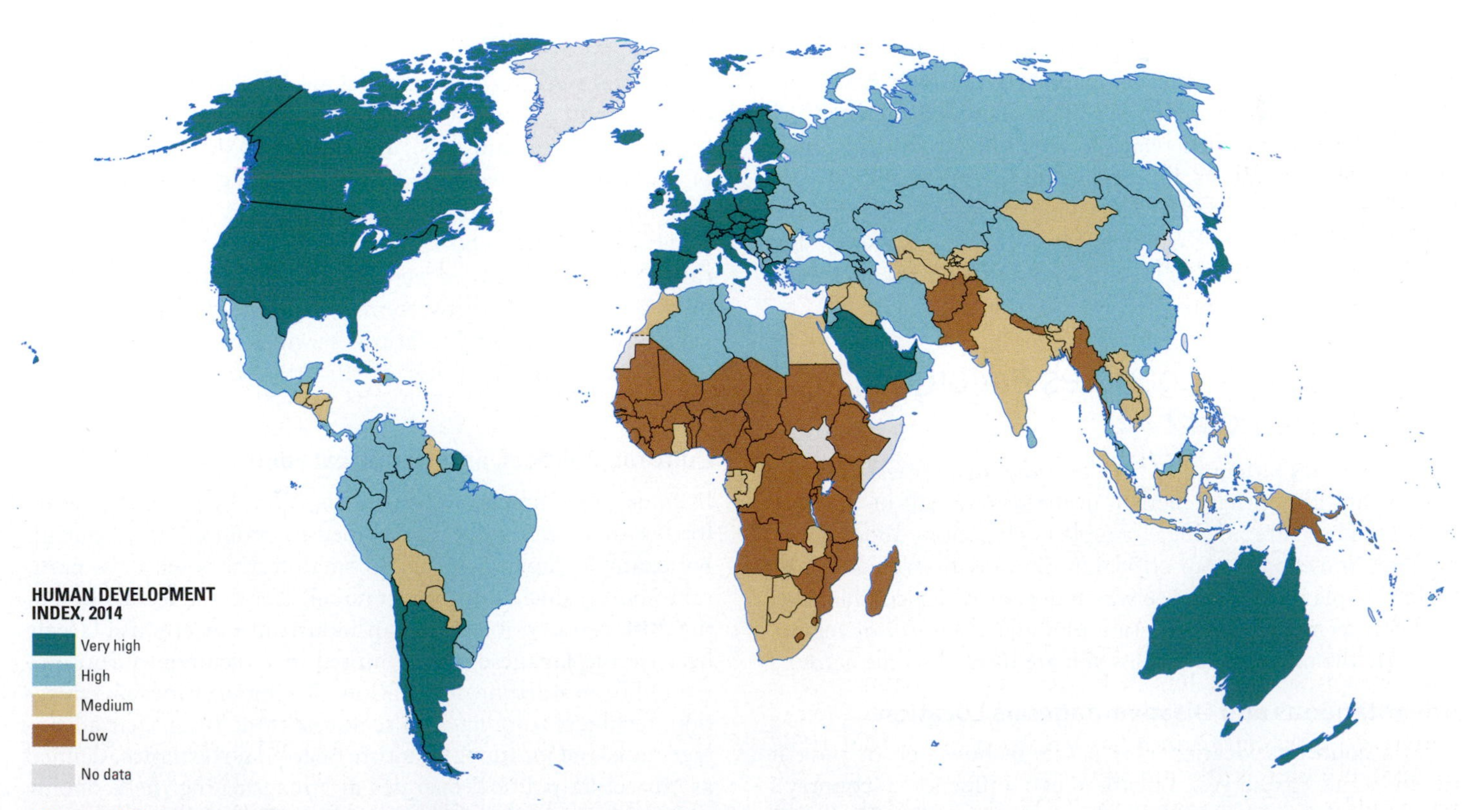

• **Figure 3.8** The Human Development Index classifies its scores into four categories: 1.0 to 0.8 is "very high" development; 0.79 to 0.7 is "high"; 0.69 to 0.55 is "medium"; and scores of 0.54 and under are "low."

Life for a very large number of the planet's other 81 percent, or about 5.8 billion people, is quite different. In the LDCs, which prevail in Africa, Latin America, and Asia, poverty and food insecurity are commonplace. Of particular concern is the bottom billion, the world's most impoverished people, living mainly in Africa. Subsistence agriculture is important; only recently has industry become a significant component of economies in all of these regions except Africa. Most people live in rural areas, but that statistic will change soon; 48 percent of the LDCs' population is urban. There is an enormous economic gulf between the poor and a small but wealthy elite who own most of the private landholdings in the countryside. The middle class was very small until recently, but is now growing rapidly. With relatively high birth rates and falling death rates, population growth is high relative to that in MDCs. Compared with MDCs, life expectancy is shorter, and the literacy rate is lower.

LDCs are not destined to be in poverty forever. Effective development policies and practices have changed the outlook for many countries. Doomsayers predicted for decades that the colossal country of India would one day collapse under the weight of overpopulation and poverty. As we will see in Chapter 7, India is instead emerging as a global economic power to be reckoned with, effectively using its human capital to generate wealth from software and outsourced services (see Insights feature 313 on **page 313**). A few decades ago, China was a desperately poor country, but in 2014 it edged past the US to become the world's largest economy measured by PPP (China also has the world's largest population, so on a per capita basis it still lags far behind the US, which by most measures is the world's largest economy). Still, because of their large populations, China and India face many challenges—including big gaps between their haves and have-nots—and neither ranks as a more developed country.

With four-fifths of the world's people living in the poorer, less developed countries, it is important to understand the root causes of underdevelopment and to appreciate how wealth and poverty affect the global environment in very different but equally profound ways. With this knowledge, it becomes possible to envision solutions.

Why Did Some Countries Become Rich and Others Poor?

Many theories attempt to explain the disparities between MDCs and LDCs that emerged in the second half of the 20th century. There are no single, simple explanations about development. You can use your critical thinking skills to consider the various explanations and see which one, or which combination of them, seem to fit the situation of a given country or region. Here are the main explanations you are likely to come across.

Advantageous and Disadvantageous Location

Does a country's location play a role in how rich or poor it is? In some cases, yes. Location can influence a country's economic fortunes. For example, as discussed in Chapter 1 (**page 13**), because it is situated close to a great mainland with which to trade, the island of Great Britain enjoys a core location favorable for economic development. Japan has a similar location relative to the Asian landmass. In contrast, landlocked nations such as Bolivia in South America and numerous nations in Africa have locations unfavorable for trade and economic development, and they have not overcome this disadvantage. But it is important to recognize that geographic location is never the sole decisive factor in development. Like Japan and Britain, Madagascar and Sri Lanka are island nations situated close to large mainlands, but for a variety of (mainly political) reasons discussed in the relevant chapters, neither has experienced prosperity. Switzerland, which is very prosperous, is landlocked.

Resource Wealth or Poverty

Having or lacking a diversity or abundance or **natural resources**—naturally occurring materials that people can use—plays a significant role in development. Superabundance of an especially valuable resource (for example, oil in the Persian Gulf countries) or a diversity of natural resources has helped some countries become more developed than others. The former Soviet Union and the United States achieved superpower status in the 20th century in large part by using the enormous natural resources of both countries. Resource wealth does not guarantee development however: some "failed states," better known as **fragile states**—low-income countries characterized by poor governance and/or weak state capacity, leaving citizens vulnerable to a range of shocks—have squandered their abundant natural resources and again illustrate the "resource curse," as described on **page 424** (**•Figure 3.9**).[5] The resource-rich Democratic Republic of the Congo is one of these.

Natural resources, which include inexhaustible resources (like solar and wind energy), renewable resources (such as water, plants, and animals), and nonrenewable resources (like oil and gas), are one component of **natural capital**, defined as "the natural resources and natural services that keep us and other species alive and support human economies."[6] **Natural services** (also called **ecosystem services**), the processes provided by healthy ecosystems, represent the other component of natural capital. Natural services include the natural processes of air and water purification and the renewal of topsoil.

Cultural, Political, and Historical Factors

In some cases, human industriousness has helped compensate for resource limitations and helped to promote development. For example, Japan has a rather small territory with few natural resources (including almost no oil). Yet in the second half of 355 the 20th century, it became an industrial powerhouse largely because the Japanese people united in a common purpose to rebuild from wartime devastation, placing priorities on education, technical training, and seaborne trade from their advantageous island location. A culture of **good governance**, defined as "directing political energies at strengthening the economy rather than trying to cement power and keep down the opposition," is important too, promoting prosperity even in the most unlikely geographical settings.[7] Singapore, a small island at the

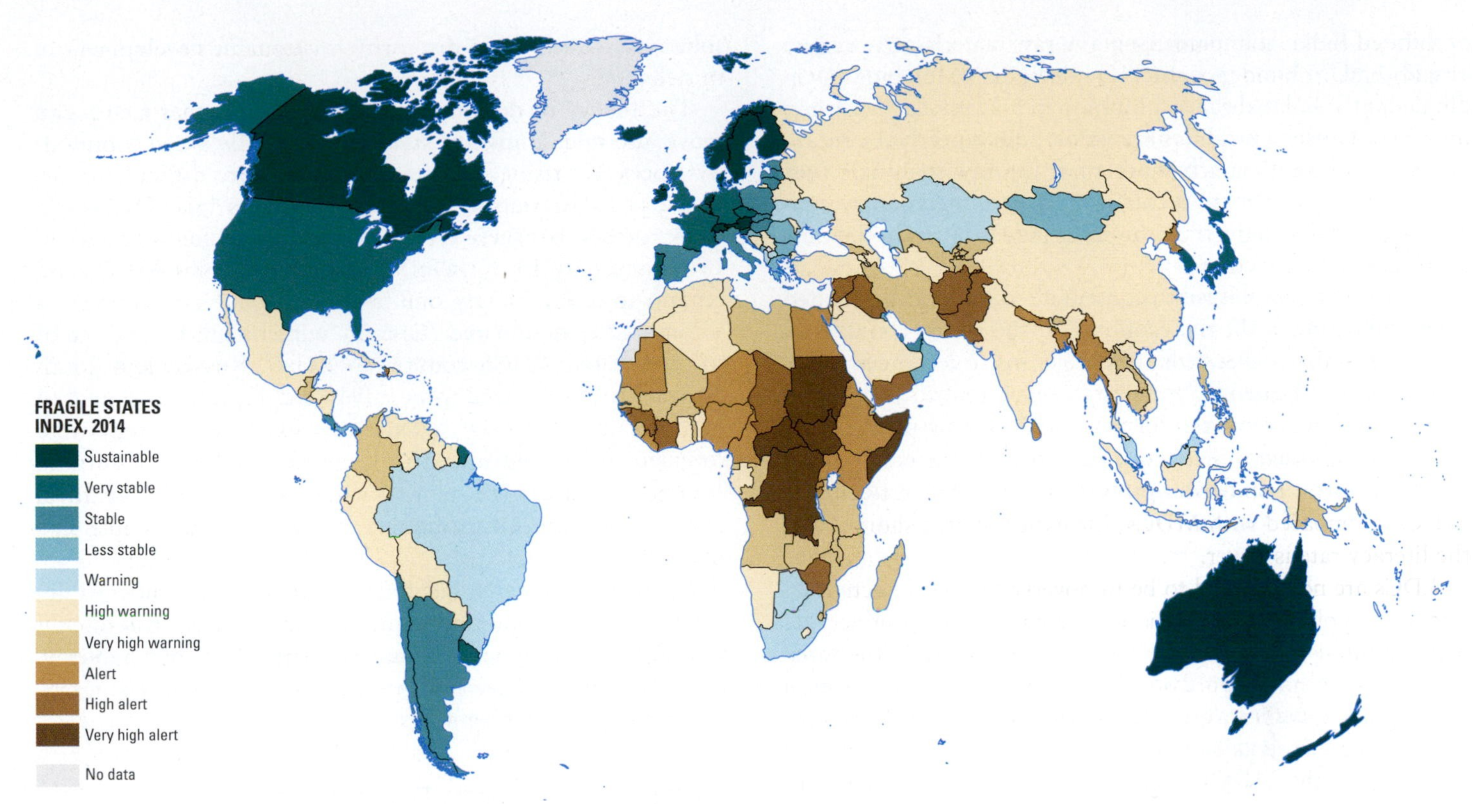

• **Figure 3.9** Fragile states are low-income countries characterized by poor governance and/or weak state capacity, leaving citizens vulnerable to economic shocks, natural hazards, and climate change. Stable states are the opposite of fragile states. The most stable states are depicted in this classification as "sustainable states."

THINK CRITICALLY: What patterns do you recognize by comparing this map with **Figures 3.7, 3.8,** and **3.16**?

tip of the Malay Peninsula (see Figure 7.2 on **page 285**) has few natural resources. It does, however, have a Chinese-influenced culture of productive trade, and practices of good governance. Conversely, cultural or political problems like corruption and ethnic factionalism can hinder development in a resource-rich nation, again as in the mineral-wealthy Democratic Republic of Congo.

Embraced most strongly in the LDCs, **dependency theory** as an explanation for the rich–poor divide argues that the worldwide economic pattern established by the Industrial Revolution and by the forces of colonialism that accompanied it persisted in the postcolonial world. In his book *Ecological Imperialism*, historian and geographer Alfred Crosby explains how dependency led to the rich–poor divide, depicting the two very different ways in which European powers used foreign lands during the Industrial Revolution.[8] In the pattern of **settler colonization**, Europeans sought to create new Europes, or **neo-Europes**, in lands much like their own: temperate middle-latitude zones with moderate rainfall and rich soils where they could raise wheat and cattle. And so, between 1630 and 1930, more than 50 million Europeans emigrated from their homelands to create European-style settlements in what are now Canada, the United States, Argentina, Uruguay, Brazil, South Africa, Australia, and New Zealand. Many of these lands would later become some of the world's wealthiest countries.

In contrast to their preference to settle familiar mid-latitude environments, Europeans viewed the world's tropical lands mainly as sources of raw materials and markets for their manufactured goods. The environment was too different from home to make settlement attractive. In establishing a pattern of **mercantile colonialism**, Crosby explains, Europeans were less inhabitants than conquering occupiers of the colonies, overseeing indigenous peoples and resettled slaves in the production of primary commodities: sugar in the Caribbean; rubber in Latin America, West Africa, and Southeast Asia; and gold and copper in southern Africa, for example. Colonialism required huge migrations of people to extract the Earth's resources, including 30 million slaves and contracted workers from Africa, India, and China to work in mines and plantations around the globe.

In the mercantile system, the colony provided raw materials to the ruling country in return for finished goods; this meant, for example, that people in India would purchase clothing made in Britain from the raw cotton they themselves had harvested. The relationship was most advantageous to the colonizer. Britain, for example, would not allow its colony India to purchase finished goods from any country but Britain. It prohibited India from producing any raw materials the empire already had in abundance, such as salt (India's Mohandas Gandhi defiantly violated this prohibition in his famous "March to the Sea"). Finished products are **value-added products**, meaning they are worth much more than the raw materials they are made from, so manufacturing in the ruling country concentrated wealth there while limiting industrial and economic development in the colony.

The colony was obliged to contribute to, but was prohibited from competing with, the economy of the ruling country—a relationship that dependency theorists insist continues today in some cases. Dependency theory asserts that to participate in the world economy, the former colonies, now independent countries, continue to depend on exports of raw materials to and purchases of finished goods from their former colonizers and other MDCs, and this disadvantageous position kept them poor. Dependency theorists call this relationship **neocolonialism**. With independence, the former colonies needed revenue. To earn that money, they continued to produce the goods for which markets already existed—generally the same unprocessed primary products they supplied in colonial times. Dependency theorists argue that when the former colonies try to break their dependency by becoming exporters of manufactured goods, the MDCs impose trade barriers and quotas to block that development (see Insights, **page 526**).

Some developing countries continue to rely heavily on income from the export of a handful of **primary commodities**, which are raw materials such as fruits and ores whose extraction or harvest needs little processing before use: examples are oil, rubber, bananas, and sugar. This reliance makes them vulnerable to the whims of nature and the world economy. The economy of a country heavily dependent on rubber exports, for example, may suffer if an insect pest or disease wipes out the crop or if a foreign laboratory develops a synthetic substitute. When demand for rubber rises, that country may actually harm itself trying to increase its market share by producing more rubber, because in doing so, it drives down the price. (Consider oil: members of the Organization of Petroleum Exporting Countries [OPEC], cartel of oil-producing countries, drove oil prices to all-time lows in the mid-1980s when they overproduced oil in a bid to earn more revenue.) If the rubber-producing country withholds production to shore up rubber's price, it provides consuming countries with an incentive to look for substitutes and alternative sources. (Again, look at oil: after OPEC embargoed shipments of oil to the United States in the 1970s, the United States began developing domestic oil supplies and becoming more energy efficient, thus temporarily reducing oil prices and OPEC's revenues.) The developing country is in a dependent and disadvantaged position—even if it produces something as vital as oil.

In this discussion we have provided the context of the so-called **poverty traps**, which are factors that impede development. The four poverty traps are conflict, natural resources, landlocked situations, and bad governance.[9] Oxford University economist Paul Collier uses the concept of poverty traps to diagnose the problems of what he calls the "bottom billion"—the poorest people on Earth—and begin to prescribe solutions, as we will in discussing sustainable development in this chapter.

The theory of **development traps** argues that LDCs can move into middle-income status rather easily when commodity prices are rising, but find it much more difficult to advance to MDC status if commodity prices fall. The World Bank reveals that there were 101 middle-income countries in 1960, and only 13 of them became high-income MDCs and remain so today.[10] Only one of these, Equatorial Guinea, is a commodity-dominated economy with its oil. Even three of the celebrated BRICS countries—Brazil, Russia, and South Africa—have appeared to be advancing rapidly on course to join the advanced MDCs. On closer examination they were revealed to have advanced in large part because of surging commodity prices (of oil in Russia, for example, up until 2014), with an unsustainable foundation for continued development.

Ominously, research has found an association between dependence on primary commodity exports and the risk of civil war. We will examine this connection in Africa in Chapter 9. 423 Here we have the opportunity to acknowledge some encouraging trends in development.

Globalization and Development

Recent trends in globalization and development suggest that the gap between MDCs and LDCs is closing. **Globalization** is defined as "the spread of free trade, free markets, investments, and ideas across borders, and the political and cultural adjustments that accompany this diffusion." *The World Is Flat* journalist Thomas Friedman entitled his book on the subject. He observed, ironically, that in his search for India, Christopher Columbus proved the world was round, but in visiting India, Friedman himself found the world to be flat: the "playing field" of the international economy had been leveled: "It is now possible for more people than ever to collaborate and compete in real time with more people on more different kinds of work from more different corners of the planet and on a more equal footing than at any previous time in the history of the world—using computers, email, fiber-optic networks, teleconferencing, and dynamic new software."[11]

Globalization has strong supporters and detractors. Supporters argue that the newly emerging global economy will bring increased prosperity to the entire world. Innovations in one country will be transferred instantly to another country; productivity will increase; and standards of living will improve. With the elimination of barriers to free trade, free enterprise will prosper, pumping additional capital into national economies and raising incomes for all (some of these barriers are discussed on **page 526** in Chapter 11). Many of the inputs into globalization come from **multinational companies** (also called **transnational companies**, meaning companies with operations outside their home countries).

Opponents of globalization argue that the process will actually increase the gap between rich and poor countries, and even within countries; a selected few developing nations (or people within a country) will prosper from increased foreign investment and resulting industrialization, but the hoped-for global wealth will bypass other countries (or people within a country) altogether. And the increasing interdependence of the world economies will make all the players more vulnerable to economic and political instability. The multinational companies will recognize huge profits at the expense of poor wage laborers. In addition, environments will be harmed if environmental regulations are reduced to a lowest common denominator (for example, the high standards of air quality demanded by the US Clean Air Act would be deemed noncompliant with World Trade Organization (WTO) rules because they make it harder for countries with "dirtier" technologies to compete in the marketplace). Such concerns often lead to massive protests against "corporate-led globalization," especially at meetings of the WTO, the International Monetary Fund (IMF), and the World Bank. Typical protesters' demands are that working conditions be improved in the foreign sweatshops, where textiles and other goods are produced at low cost for US corporations, and that corporations like Starbucks should sell only Fair Trade coffee beans bought at a price giving peasant coffee growers a living wage rather than at the "exploitive" price typically paid. Corporations like Starbucks have often been quick to respond to the demands of the environmentally and socially conscious consumers (**•Figure 3.10**). 472

Are the protestors' concerns warranted? Let's look at five recent trends and effects of the globalization process.

First Is the Breathtaking Reduction of Poverty and Growth of the Middle Classes

The world's poverty rate has been cut by half since 1990. According to the World Bank, 42 percent of the world's people lived in poverty in 1990, but just 21 percent did by 2010. The trend is remarkable, and we need to consider the reasons for it. At the same time we must remember that it is a huge number of people, 1.5 billion—mainly in Sub-Saharan Africa, that remains poor. Incidentally, mapping poverty is a critical tool in reducing poverty.[12]

Thanks to the redistributive powers of globalization, the ranks of the middle class have grown in LDCs in recent years. The so-called BRICS countries—Brazil, Russia, India, China, and South Africa—have been the "poster children" of this process. The world's middle classes are projected to overtake the poor by 2022 and to at least double their 2014 population to as many as 5 billion by 2030, with Asians representing two-thirds of the total (**•Figure 3.11**). Improved and more widely available 174, 310, 470

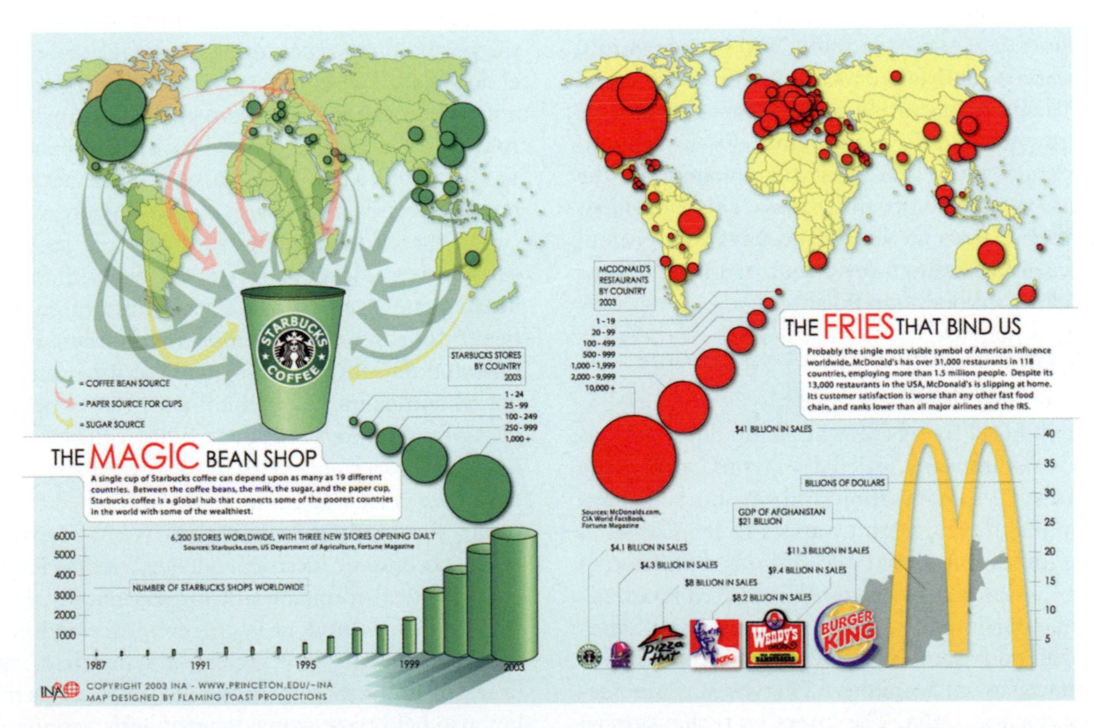

• Figure 3.10 Globalization of American food. This image shows growth in the number of worldwide franchises, amount of profit, and their franchise locations as well as ingredient sources.

Source: Copyright 2003 INA—www.princeton.edu/. (Map designed by Flaming Toast Productions.) Reproduced with permission.

THINK CRITICALLY: The AAG poses four questions about these maps:

- The images of McDonald's and Starbucks reflect the globalization of American products. What does that mean? Can you think of other examples?
- How does each of these products foster international trade connections?
- Which company, McDonald's or Starbucks, has more stores worldwide?
- How does Starbucks support the economies of other world nations?

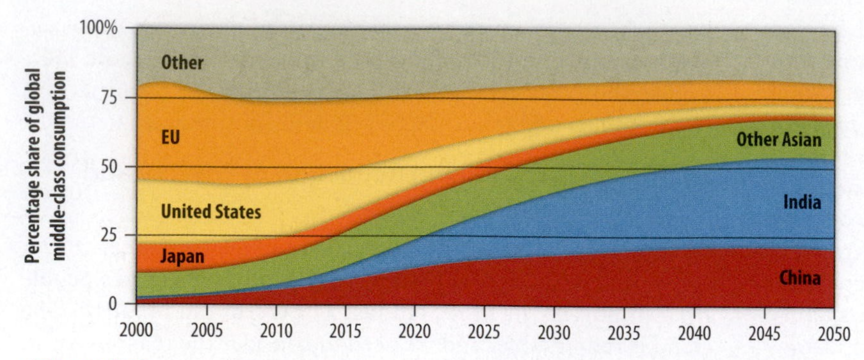

• **Figure 3.11** The global middle-class wave. Global middle-class consumption will shift heavily toward China, India, and other Asian countries (including Japan) as the high-income countries see their share decrease.

Source: Based on CS Monitor. http://www.csmonitor.com/World/2011/0517/Surging-BRIC-middle-classes-are-eclipsing-global-poverty.

education, much of it facilitated by graduate study abroad (especially in the US), has generated a surge of human capital fit for placement in the "knowledge economies" that generate a growing portion of the world's economic capital. Your Asian classmates testify to this trend.

Globalization is credited for this transformation, with more economic as well as human capital entering the system: taxes have been lowered while wages and foreign investments have grown in the BRICS and other "emerging market" economies. India and Brazil should overtake Germany and France to join China, the United States, and Japan as the world's five largest economies by 2050. A note of caution is important here: events can force the new members of the middle class downward into poverty. It is also noteworthy that while the super-rich ("the 1 percent") in MDCs have also benefitted from this movement of capital, the middle classes of the MDCs have lost ground economically, as have the bottom 5 percent of people in the LDCs, concentrated in Sub-Saharan Africa.

Second Is the Growth in Inequality Between Some of the Socioeconomic Groups

Globalization moves jobs to where wages are low, and so growth in the LDC middle classes is tied directly to losses in the MDC middle classes. "Mechanization and outsourcing have made Americans jobless," writes John Gapper in the *Financial Times*. "The same forces that have hollowed out manufacturing and clerical jobs in the US and Europe have lifted hundreds of millions of people out of poverty in China and India. They have made Western economies far more unequal, while tilting the global balance toward equality. The winners are factory workers in China and India. The losers [are] the western middle class."[13] These factories are in cities, and the processes of globalization and urbanization are tied together closely (see **page 78**). For more on outsourcing, see **page 313**.

Income inequalities certainly are growing in the MDCs, and especially in the United States. Seventy-five percent of wealth is owned by the richest 10 percent of people in the US. Among MDC peers, Sweden and Switzerland also are in the 70 percent, but most are in the 50th percentile. Among LDCs, wealth concentrations are especially high, also in the 70th percentile, in India, Indonesia, South Africa, and Chile.[14]

Third Is the Potential for Conflict

These income and wealth inequalities are bad news: they erode social cohesion and increase the scope for internal conflict. According to the World Economic Forum's *Global Risks Report*, income disparities represent the worldwide risk most likely to be manifested through 2025. Social protests against inequality and corruption to date have included the Occupy Wall Street movement in the US and demonstrations in Brazil and India (the *B* and *I* of the BRICS), Spain, Turkey, and Egypt. In many cases, these protests have been joined or tinged by elements often promoted as "populist," but representing what Robert Kaplan describes as "toxic nationalism:" anti-immigrant, nationalist, and fascist forces. "We truly are in a battle between two epic forces," Kaplan writes: "those of integration based on civil society and human rights, and those of exclusion based on race, blood and radicalized faith."[15] Radical geographer David Harvey paints a grim picture of violent responses to inequalities in rich and poor countries alike: riots and protests in Turkey, Egypt, Brazil, and Sweden "look more and more like the prior tremors for a coming earthquake that will make the postcolonial revolutionary struggles of the 1960s look like child's play. The longer it goes on, the less I think that there is a possibility that it will be a peaceful transition."[16] We should recall that climate change may also be increasing the potential for conflict (see **page 43**), so the intersection of natural and human stresses may be especially dangerous.

On the surface, globalization would seem to foster peaceful interdependence. Defying this logic, though, the economic interdependence of globalization has not guaranteed peace. In forcefully contesting the maritime borders of neighbors in the East and South China Seas, China, for example, appears willing to accept the economic losses that would be incurred by disrupting its interdependence with Taiwan, South Korea, Vietnam, and other countries in its critical supply chains. "In this 316 world the economic competition unleashed by globalisation becomes the handmaid of an assertive nationalism that in turns trumps economic interdependence," writes Philip Stephens, the chief political commentator for the *Financial Times*.[17]

The ironies and crosscurrents are striking: globalization is credited with "leveling the playing field" and lifting hundreds of millions of people out of poverty around the world, but also held responsible for troubling economic discrepancies that spark conflict. And although the income gap in the MDCs is growing, global inequality is shrinking.[18] Is globalization "good" or "bad?" You have to evaluate that, keeping in mind that there are winners and losers.

Fourth Are Geopolitical Changes

To further set the stage for our exploration of world regions, it is important that we recognize major changes underway in the status of nation-states and their citizens. Before the Soviet Union fractured in 1991, we lived in a "bipolar" world in which two great countries, the Union of Soviet Socialist Republics (USSR or Soviet Union) and the United States, and their allies, vied for dominance. There was a simple competition, the Cold War, between these two poles. When the Soviet Union collapsed a unipolar world evolved, in which the US stood as the sole superpower. Francis Fukuyuma famously called it "the end of history," arguing that the end of the Cold War left Western liberal democracy as the "final" form of government.

119, 150, 172, 533

History did not end of course, and power shifted in new ways geographically. Globalization weakened the unipolar dominance of the United States in economic, political, military, and cultural matters, and through a momentous tilt of power, we have entered a new multipolar world, according to many observers. Fareed Zakaria, a thoughtful geopolitical analyst you may want to follow (on CNN, iTunes, or Twitter), calls the current age "the post-American world," an era of greater multipolarity, "a genuinely global system where every part of the global system has countries that are rich and vibrant and participating."[19]

Fifth Are Changes in Technology, Knowledge, and People Power

If not peace among nations, then what can we look forward to? One very hopeful trend accompanying globalization is that the flow of information is becoming more democratized. Rapid developments and falling prices are making smartphones accessible in even the poorest parts of the world. These bring more knowledge and greater opportunities to shape events—and in the most extreme cases, literally provide instructions about when and where to turn up and overturn oppressive governments, as we will see in Egypt. Philip Stephens observes a crucial change in relations between peoples and their governments around the world: power is shifting from governments to peoples, especially because of the growth of the middle classes and their challenges to the governing elites, and because ordinary peoples for the first time in history have empowering access and influence through social media.[20] Countries that restrict access to the Internet, notably China, will face growing pressures from citizens, including those educated abroad, who are eager for more diverse and true information. According to the social scientists who author the *World Values Survey*, growth of a "knowledge society" is correlated with increasing standards of living (see Insights, **page 64**).

258

306

Even where there are no restrictions on access to information, the rapid growth and spread of computer and wireless technologies are not yet benefiting everyone; there is still a **digital divide** to overcome, both between and within countries. Internationally, there is a divide in between the minority of countries leading in **information technology (IT)** and the majority of nations less able to create, purchase, or use new technologies. The fear in the technology-lagging countries is that the growth of e-commerce will concentrate the wealth generated by that commerce in the technology leadership countries. This would enable the leaders to make even further advances in technology and realize even bigger economic gains from a so-called **knowledge economy** based on innovation and services, while the technology laggards fail to catch up and simply become poorer (**•Figure 3.12**). Efforts underway to empower the world's poor through access to information technology include the One Laptop per Child program, which seeks to provide the world's children with rugged, low-cost, low-power, Internet-connected laptops (see http://onelaptop.org).

426

Another plus of globalization is that it has given people everywhere access to the cultures of others. Not long ago the prevailing view was that globalization would result in the "McDonaldization" of the world, eliminating the diversity of local cultures in favor of a monolithic and mainly Western worldview. This is not happening. Considering just the geography of food as an example, hamburgers have not conquered the world. Local cuisines flourish and become accessible through the Web and television to people in the West curious to explore foods of the world. We can now listen to world music, the name given to all music not native to the West. The migrations of peoples across the world introduce us to a variety of cultures that usually retain their distinctiveness even after many generations in a new land. It is then up to us to welcome and celebrate this diversity.

Or not. The same revelation of cultural wealth that invites us to appreciate others also makes it possible to revile others in thought (especially through writing on the Web and social media) and in deed. Recently preparing lectures for my "Geography of the Middle East" course, I read rafts of anti-Islamic writings on the Web; not anti–Islamist terrorism but anti-Islam, the religion. I also read through a multitude of ISIS

Joe Hobbs

• Figure 3.12 Staying connected in Ho Chi Minh City, Vietnam. Information technology—especially in the form of cell phones, computers, and the Internet—is spreading rapidly around the world. Some people believe that there is now an unfolding Information Revolution comparable to the Agricultural and Industrial Revolutions in its capacity to change humanity's relationship with the Earth.

Insights Mapping World Values

Are you interested in the intersections of geography and culture? You may be interested in the research and findings of the *World Values Survey*, a project undertaken since 1981 to "help social scientists and policy-makers better understand worldviews and changes that are taking place in the beliefs, values and motivations of people throughout the world."[21] The survey's basic finding is that with an increase in standards of living, and the transition from developing country status to post-industrial "knowledge society" status, a country tends to move from poor to rich accompanied by a transition from "survival values" to "self-expression values," and a transition from "traditional values" to "secular-rational values." According to the researchers, value differences around the world show a pronounced *culture zone pattern* that they have mapped (**•Figure 3.A**). "The strongest emphasis on traditional values and survival values is in Islamic societies of the Middle East," the researchers find. "The strongest emphasis on secular-rational values and self-expression values is found in the Protestant societies of Northern Europe. These culture zone differences reflect different *historical pathways* of how entire groups of societies entered modernity."[22]

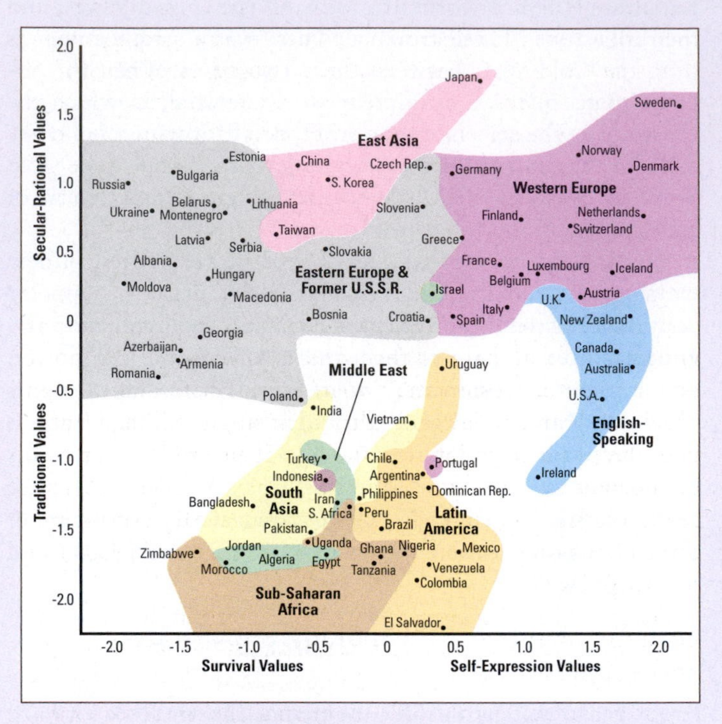

• Figure 3.A World Values Survey.

THINK CRITICALLY: What do you think of the World Values Survey? What examples seem to hold up or undermine its basic finding? Does it suggest that some values and countries are "better," or is it a true correlation of observed phenomena?

(Islamic State of Iraq and Syria), al-Qa'ida, and other Islamist terrorist invective against Jews, Christians, and even other Muslims, with instructions on how and where to kill these innocent peoples. Information, like globalization broadly, is a double-edged sword.

Environmental Impacts of Underdevelopment

As geographers, we are very interested in the environmental impacts of relations between MDCs and LDCs, especially on the poorer countries. LDCs generally lack the financial resources needed to build roads, dams, energy grids, and other infrastructure critical to development. Then they turn to the World Bank, IMF, and other institutions of the MDCs, as well as increasingly to China, to borrow funds for these projects. Many borrowers are unable to pay even the interest on these loans, which is sometimes huge; debtor nations have been known to spend as much as 40 percent of annual revenues on interest payments to the lenders, or more than they spend on education and health combined.

This debt burden can be very harmful to natural environments. When lender institutions threaten to cut off assistance, borrowing countries often try to raise cash quickly to avoid this prospect, generally by one or both of two methods. One method is to dedicate more land to the production of **cash crops** (also known as **commercial crops**). These are the primary commodities such as coffee, tea, sugar, coconuts, and bananas exported to the MDCs, where they are luxuries or staples (illegal drug crops are also cash crops, of course, but fall outside most statistics and many of the general associations made here). Governments or foreign corporations often buy out, force out, or otherwise displace subsistence food farmers in the search for new lands on which to grow these commercial crops. In this process, known as **marginalization**, poor subsistence farmers are pushed onto fragile, inferior, or marginal lands that cannot support crops for long and end up depleted by cultivation. In Brazil's Amazon Basin, for example, peasant migrants arrive from Atlantic coastal regions, where government and wealthy private landowners cultivate the best soils for sugarcane, soybeans, and other cash crops. The newcomers to Amazonia slash and burn the rain forest to grow rice and other crops that exhaust the soil's limited fertility in a few years. Then they move on to cultivate new lands, and in their wake come cattle ranchers whose land use further degrades the soil. Other marginalized peasant farmers throughout the world are "pushed" into the cities to join the urban poor. These are very common, important problems in the LDCs, and you will see many examples of them in the following chapters. One of the most outstanding examples, involving every continent except Antarctica, involves **land grabbing** (see Insights on page 66). 243, 417, 457

Remember the IPCC's warnings about climate change? The already marginalized of the world are and will continue to be the most vulnerable to the impacts of climate change. "People who are socially, economically, culturally, politically, institutionally, or otherwise marginalized are especially vulnerable to climate change and also to some adaptation and mitigation responses," writes the IPCC. This heightened vulnerability is rarely due to a single cause. "Rather, it is the product of intersecting social processes that result in inequalities in socioeconomic status and income as well as in exposure. Such social processes include, for example, discrimination on the basis of gender, class, ethnicity, age, and (dis)ability."[23] 41

After stepped-up production of commodities, the second way for LDC debtor countries to raise cash quickly is to simply sell off their natural assets—as many have done to the land grabbers, for example. Who is buying them? The wealthier countries, and increasingly and most strikingly, China—an issue that this book covers often. LDCs often face difficult choices between using natural resources for immediate gain or for longer term, more sustainable growth. In most cases, they feel compelled to take short-term profits, by cash cropping and other strategies, a choice that can have devastating environmental consequences. Most LDCs have resource-based economies that rely not on industry but on stocks of productive soils, forests, and fisheries. The countries' long-term economic health could benefit through protection and careful use of these assets. But to pay off international debts and meet other needs, the LDCs often draw down their ecological capital faster than nature can replace it. In ecosystem terms, they exceed **sustainable yield** (also known as the **natural replacement rate**), the highest rate at which a **renewable resource** can be used without decreasing its potential for renewal. Worldwide, twice as many trees are cut down as are planted or replaced with new growth, and that ratio is far higher in the forest-rich LDCs Brazil, Congo, Indonesia, Peru, and India (after these five, Russia, Canada, the United States, China, and Australia round out the top 10 in global forest reserves, making up two-thirds of the total). 427, 468

Based on deforestation and other indexes of natural resource wealth, numerous LDCs are in a state of **ecological bankruptcy**, the exhaustion of environmental capital. These imbalances are tracked by a number of nongovernmental organizations (NGOs), including the Global Footprint Network.[29] This environmental poverty plays itself out in several damaging ways. First, there are negative effects on the health and well-being of the people living in such countries. People in the LDCs feel the impacts of environmental degradation more directly than people in the MDCs (•**Figure 3.13**).

Joe Hobbs

• **Figure 3.13** The Thu Bon River in Vietnam. In the LDCs, it is very common for people to rely on polluted water sources for drinking, bathing, cooking, and washing their clothes and eating utensils. Health effects can be severe; one of the most tragic is infant diarrhea, typically a waterborne illness that kills an estimated 2 million youngsters worldwide every year.

People in the LDCs are more dependent on nature's abilities to replenish and cleanse itself, and so they suffer more when those abilities are diminished (see Insights, page 68). Let us consider the most important resource of water. Shortages of water represent a shortcut to ecological bankruptcy and geopolitical crisis. Water is a "zero sum" commodity: one individual's gain is another's loss in the global supply. Many of the world's poor drink directly from untreated water supplies; an estimated 1 billion people lack access to clean water. The World Commission on Water estimates that by 2025, 4 billion people will live under conditions of severe water stress.[30] Some of the most serious imbalances are in the Middle East and other arid regions. Yemen has been forecast to be the world's first country that will literally run out of water. With respect to the world's forests, in the LDCs people tend to cook their meals with fuelwood rather than with fossil fuels (see Insights, page 68). 239

The fuelwood crisis is a problem that governments and societies can confront successfully. A combination of new technologies, government policies, and sound economic growth can halt and even reverse deforestation. The biogas digester, a simple household device that can be used in warm climates, ferments animal manure for the production of methane cooking gas and thus spares fuelwood. Governments can protect forest while also allowing people to earn their livelihoods in them from **nontimber forest products (NTFPs)**, such as handicrafts made from bamboo (see www.ntfp.org). The process of economic development can itself lead to less pressure on forest resources. Despite the historic loss of forests worldwide, noted earlier (see page 54), the world's wealthier countries have more forest now than they did in 1990, and forest cover has actually increased in about half of the world's 50 most forested

Insights Land Grabbing

Although we celebrate the growing middle classes and the emergence of so many people from poverty worldwide, we need to keep an eye on the price paid for this growing affluence. The impacts of population growth and, just as importantly, the dietary preferences among the middle classes toward more meat have increased pressure on land and freshwater resources around the world. In the wake of the frenzy surrounding grain shortages and skyrocketing grain prices in 2007, which resulted from reduced supply with poor harvests and increased demand, including for biofuels, a recent agent of marginalization accelerated: **land grabbing**, which refers to rich countries and individuals buying up land and water resources overseas to protect their future food security and to turn profits. Among 62 documented "grabbers" are governments, corporations, state interests, private investors, and speculators mainly from Britain, the United States, China, the United Arab Emirates, South Korea, South Africa, Israel, India, and Egypt. The 41 countries with farmlands "grabbed" are mainly in Africa and Asia, and notably in Congo, Sudan, Indonesia, Tanzania, Mozambique, and Ethiopia (•**Figure 3.B**). The main local impact is the displacement of subsistence farmers and herders on a vast scale—again, marginalization.

It is estimated that in a single decade (2002–2012), 2 million square kilometers, an area the size of Mexico, were "grabbed," two-thirds of this in Africa. Grabbers often want very large land tracts, for example, 50,000 hectares (123,000 acres), which in many countries are available only as commonly owned, ancestral tribal lands.[24] International monitors report that many of these deals were closed "with limited consultation of the local population, without adequate

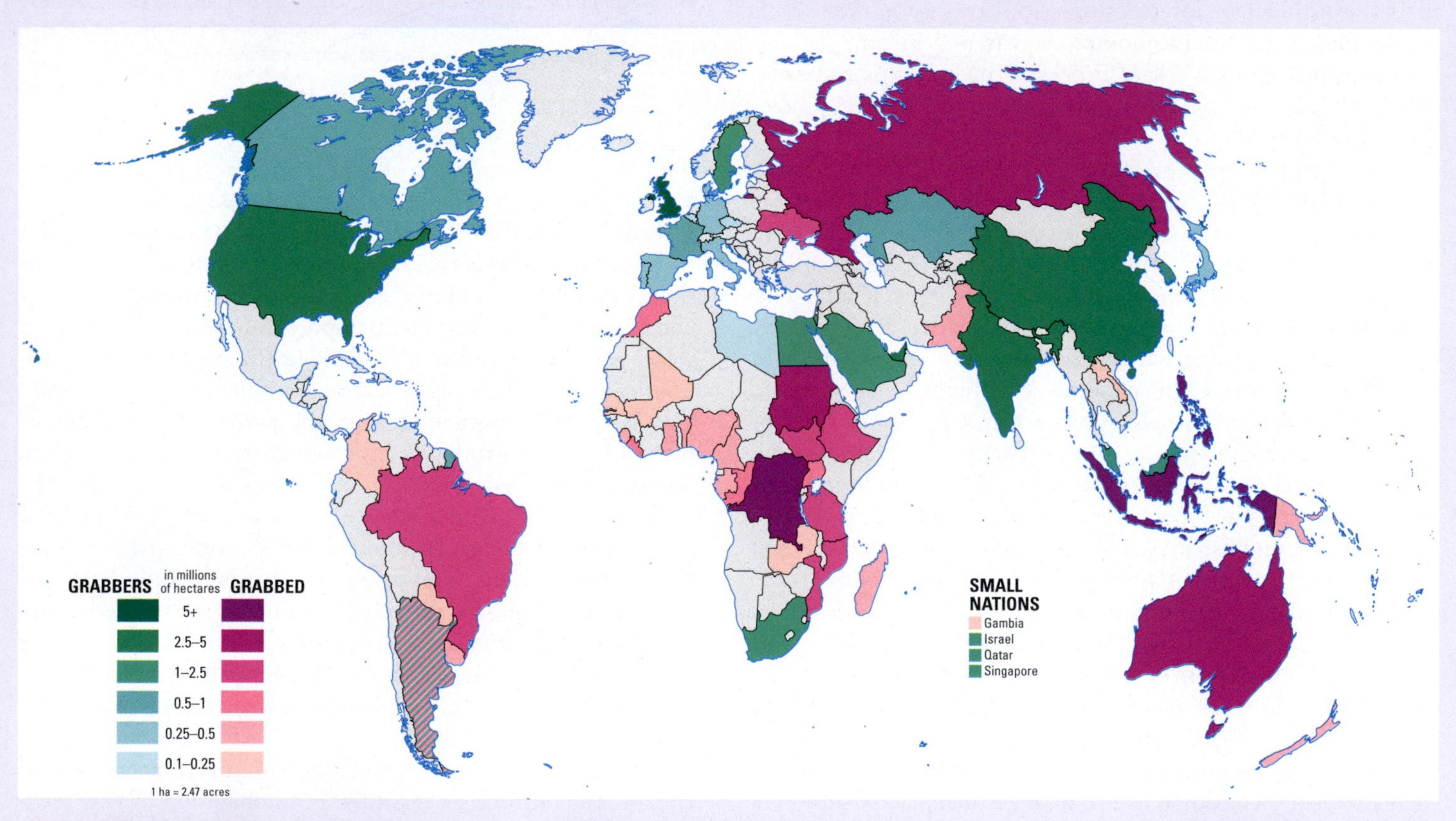

• **Figure 3.B** Global map of land grabbing.

THINK CRITICALLY: Using your knowledge of the issues underlying MDCs and LDCs, what patterns do you recognize? What other factors might be at work that make some countries grabbers and others grabbed?

countries since that time. A key challenge for the wealthier forested countries is to slow their demand for timber from the LDCs, and thus help break one of the cycles of deforestation, debt, and poverty.

Political and social crises can emerge from the ecological bankruptcy process. Revolutions, wars, and refugee migrations in developing nations often have underlying environmental causes. Such problems, as related in the earlier discussion of climate change refugees in Chapter 2, are central to the national interests of the United States and other countries far beyond the affected nations. Deforestation in Mexico and Haiti, for example, is among the root causes

43

compensation of the previous land users, and without seeking opportunities to create new jobs or enhance environment sustainability."[25]

Despite economic growth in the LDCs generally, there is still much reason for concern about hunger and shortages of freshwater, especially in the least developed countries. Scientists describe land grabbing as a "a new form of colonialism," and report that "the per capita volume of grabbed water often exceeds the water requirements for a balanced diet and would be sufficient to improve food security and abate malnourishment in the grabbed countries."[26] Although on the surface this is a problem about land, how the land is used has important implications for the global redistribution of water resources (•**Figure 3.C**).

Land grabbing is a rush not only for land but also for the freshwater resources available therein. The production of all food commodities (except fish) requires, directly or indirectly, both land and water. Because about 86 percent of the human appropriation of freshwater resources is used to sustain agricultural production, land grabbing is mainly a grabbing of freshwater resources, including both rainwater and irrigation water.[27]

Land grabbing is thus related to the issues of **virtual water**, the total volume of water needed to produce and process a commodity or service, which is of particular importance in the Middle East, Australia, and other arid regions of the world; and the **water footprint**, an indicator of freshwater use that looks at both direct and indirect water use of a consumer or producer (see **page 243**).[28]

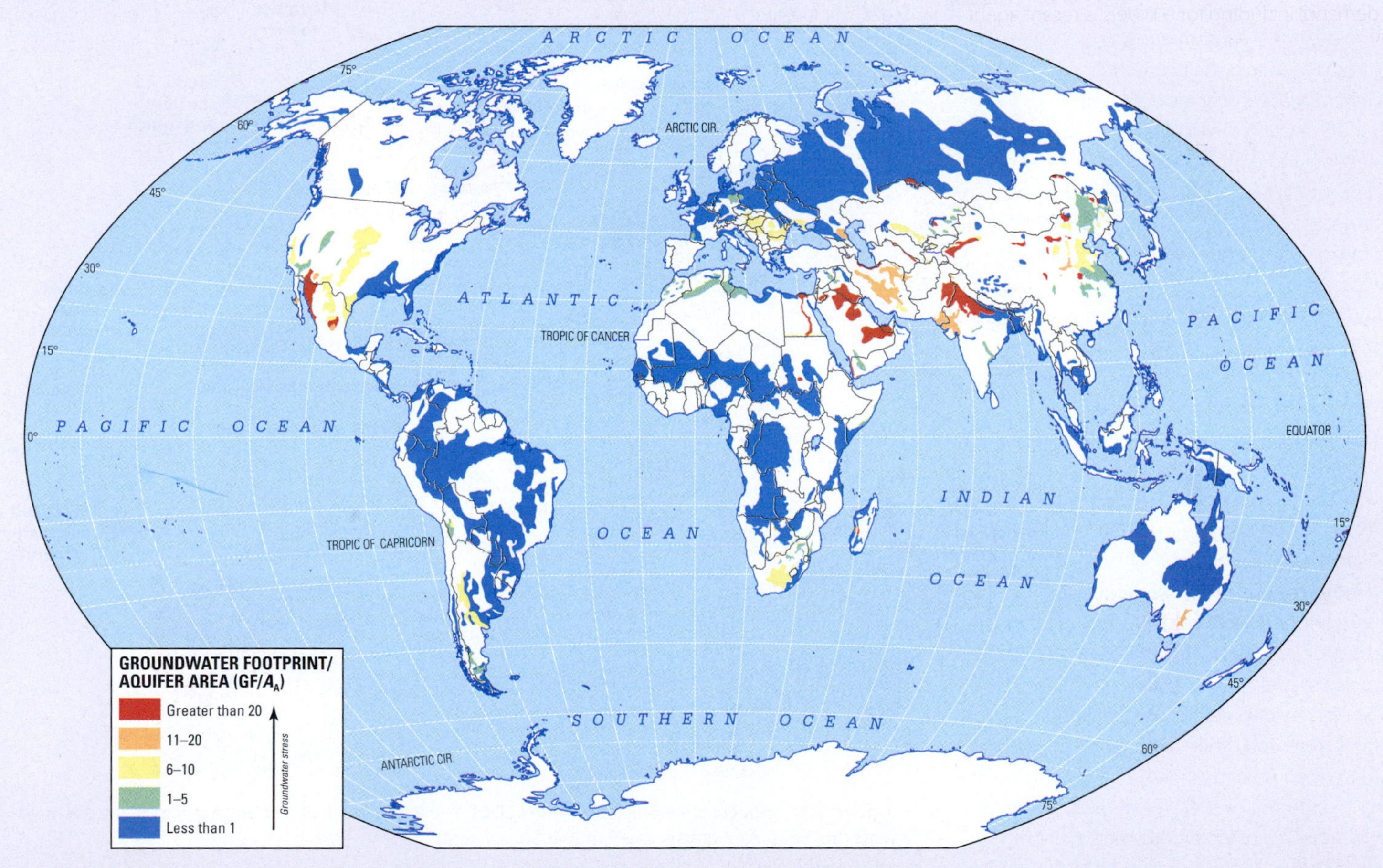

• **Figure 3.C** The status of the world's aquifers. A higher groundwater footprint per aquifer area represents greater stress on this vital water resource.

Source: Tom Gleeson, Yoshida Wada, Marc F. P. Bierkens and Ludovicus P. H. van Beek. 2012. *Nature* 488:197–200 (August 9), http://www.nature.com/nature/journal/v488/n7410/full/nature11295.html.

why rural people from those countries try to move to the United States, legally or otherwise. When their home environments are so degraded that they simply cannot support shifting farming as they once did, people will move. This brings us to the geographical issues of population and migration.

3.3 The Geography of Population

The welfare of humanity and of the planet's other species and natural habitats is tied closely to two related issues: the number of people there are and the rates at which we consume resources. In some quarters, there are fears that the human

Insights Deforestation and the Fuelwood Crisis

Deforestation—the removal of tree cover—has many detrimental effects in the LDCs. You can see how these relationships play out in **•Figure 3.D**. As people remove trees to use as fuel, for construction, or to make room for crops, their existing crop fields lose protection against the erosive force of wind. Less water is available for crops because, in the absence of tree roots to funnel water downward into the soil, it runs off quickly. Increased salinity (salt content) generally accompanies increased runoff, causing the quality of irrigation and drinking water downstream to decline. Eroded topsoil resulting from reduced plant cover can choke irrigation channels, reduce water delivery to crops, raise floodplain levels, and increase the chances that floods will destroy fields and settlements. As reservoirs fill with silt, hydroelectric generation, and therefore industrial production, is diminished. Upstream, where the problem began, fewer trees are available to use as fuel.

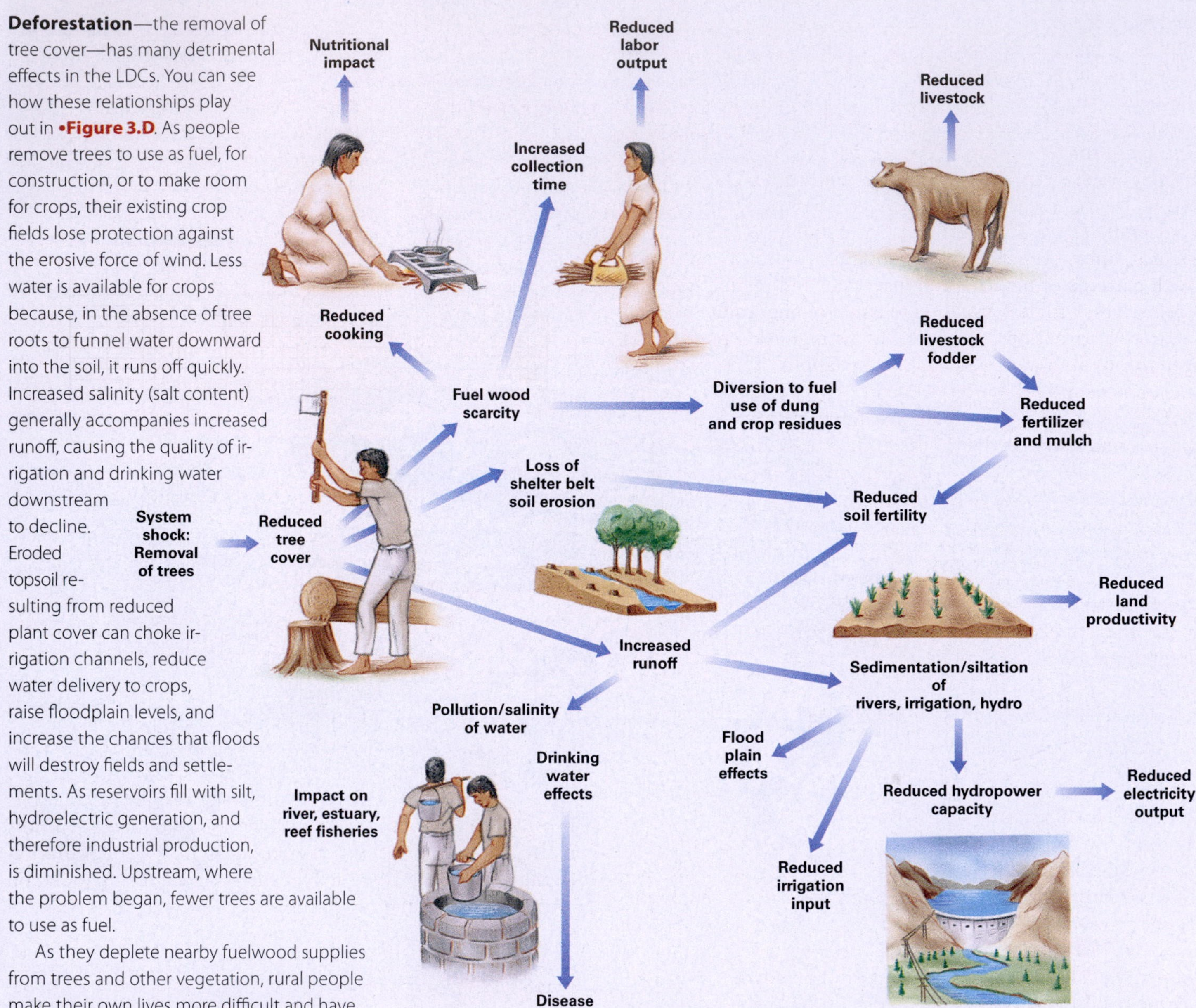

• Figure 3.D Impacts of deforestation in the LDCs. The initial shock of tree removal has many negative effects on the human–environment interaction system.

As they deplete nearby fuelwood supplies from trees and other vegetation, rural people make their own lives more difficult and have to change their activity patterns. Women and children, in almost every society the main fuelwood collectors, must walk farther, and then farther again, to gather fuel. Using animal dung and crop residues for fuel can solve one problem but create another: it deprives the soil of the fertilizers these poor people often use when farming. Reduced food output is the result. A family may eventually give up one cooked meal a day or put up with colder temperatures in their homes because fuel is lacking—steps that have a negative impact on the family's health. These impacts of fuelwood depletion are known collectively as the **fuelwood crisis** (sometimes the term refers only to the impact of women and children walking farther for fuel).

population explosion the world has experienced since 1800 will lead to a precipitous crisis: a massive disease outbreak, famine, or some other kind of catastrophe triggered by too many people living too close together without sufficient food and other resources to sustain them. On the other hand, there is optimism that the human population growth rate is slowing, and our numbers are expected to stabilize—ideally at a population that can be sustained with minimal risk of famine or other suffering.

Meanwhile, large numbers of people move across and within political borders, usually by choice, but in some cases by force. Migrants bring new cultures, ideas, and opportunities with them, but their reception is not always warm: tension and violence have characterized relations between majority and migrant minority groups in recent years in Europe and Australia, for example, and in the United States there is a heated debate about "immigration reform" directed at illegal migrants, especially from Latin America. Migration is one of the major themes discussed throughout this book.

95, 221, 353, 385, 434, 449, 494

Geography's interests in population are shared by many other fields, including biology, sociology, anthropology, and political science. The study of population is known as **demography**. The field of demography is most concerned with patterns of birth, death, marriage, and related issues in themselves, with less attention to issues of migration and population distributions. What most distinguishes **population geography** is its focus on spatial variations. This section of the book examines how many people have lived on Earth and how many we may be in the future, mindful of who these people are and why they have been so few or so many, and pays especially close attention to the fundamentally geographic issue of *where* they are. This discussion will be a very important reference for you as you work your way forward through the world regions. I hope you will find it interesting, and I recommend that you explore the best Internet resource on world population (where you can also download a free world population data table): the Population Reference Bureau site at prb.org.

How Many People Have Ever Lived on Earth?

Around 125,000 years ago, and perhaps earlier, our *Homo sapiens* ancestors came out of Africa and began to populate Eurasia. By around 10,000 years ago, after plants and animals began to be domesticated, there were probably about 5.3 million humans in the world—roughly the current number of residents of the Detroit metropolitan area or the country of Finland. By 1 CE, humans numbered between 250 and 300 million, slightly less than the population of the United States today. The first billion was reached around 1800. Then a staggering population explosion occurred in the wake of the Industrial Revolution. The second billion came in 1930, the fourth in 1975, the sixth in 1999, and the seventh in 2011 (**•Figure 3.14**). *Homo sapiens* is now by far the most populous midsize mammal on Earth and has succeeded where no other animal has in extending its range to the world's farthest corners. Using a 24-hour time period to represent the 200,000 years of our species' history, we reached our first billion within just the last 3 minutes. The people alive today represent about 6 percent of all the people who have ever lived on Earth!

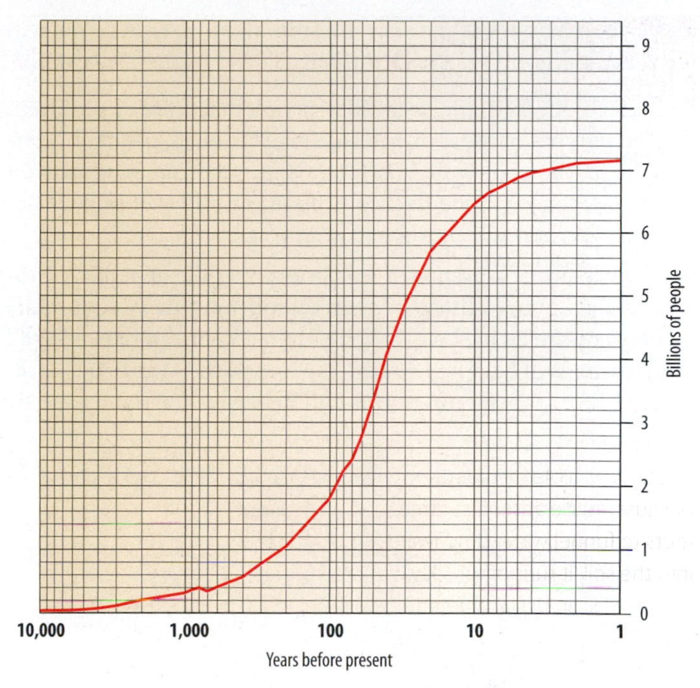

• Figure 3.14 Human population over the last 10,000 years. This graph uses a logarithmic scale to make the exponential population growth of the last 100 years easier to see.

Figure 3.14 illustrates very dramatically that right after about 1800, the human population surged. What happened after tens of thousands of years that made our numbers suddenly skyrocket? In addition to the factor, addressed earlier, of greater food surpluses that made it possible to feed more people, we answer that question by considering some of the basic vocabulary of population geography.

How Can We Measure Population Changes?

Excluding the issue of migration for now, two main variables determine population change in a given village, city, or country—or the entire planet: birth rate and death rate. The **birth rate** is the annual number of live births per 1000 people in a population. The **death rate** is the annual number of deaths in that same sample population of 1000. The **population change rate**—the figure that is often called the "population growth rate" but that may represent either growth or loss—is the birth rate minus the death rate in that population.

We can put these measures to work to appreciate the Earth's population now. Suppose there were a perfect sample of 1000 people representing the world's population in 2014. Their birth rate was 20 per 1000 and their death rate was 8 per 1000. This means that by the end of that year, among the 1000 people, 20 babies had been born and 8 people had died, resulting in a net growth of 12. That figure, 12 per 1000, or 1.2 percent, represents the 2014 population change rate for the world. The numbers for 2015 were identical.

We may now look at some of different combinations of birth rates, death rates, and population change rates, and the various forces behind them.

What Determines Family Size?

Many factors affect birth rates, which tend to be much higher in the less developed countries of the world and among the poorer residents of more affluent countries. Some of the motivations and circumstances may seem unfamiliar at first, but you will see there are good reasons for them:

- Better-educated and wealthier people have fewer children. The parents of most of you reading this book probably considered the economic cost of raising you and sending you to college and made decisions about family size in part because of their education and yours.
- Conversely, less educated and poorer people generally want and have more children. One reason for this is economic: poorer parents in the LDCs are often convinced that "extra" children will help bring more income to the family by working in fields or factories and will help care for them in their old age.
- People in cities tend to have fewer children than those in rural areas.
- In the LDCs, those who marry earlier generally have more children (their reproductive life span is longer).
- Couples with access to and understanding of contraception generally have fewer children.
- Value systems and cultural norms play very important roles. Even where contraception is available and understood, a couple may decide not to interfere with what they perceive as God's will or may seek the social status associated with a larger family.

What Determines Death Rates?

Death rates are correlated mainly with health factors, particularly the level of nutrition and level of medical care available. Improvements in food production and distribution help reduce death rates. Better sanitation, better hygiene, and cleaner drinking water eliminate fatal diseases such as infant diarrhea, a common cause of infant mortality in the less developed countries. The availability of antibiotics, immunizations, insecticides, and other improvements in medical and public health technologies also correlate with death rates.

Human death rates overall have been on a steady trend of decline for decades. This translates into an explosion in population, but despite more people, the quality of their lives is generally improving. However, death rates sometimes rise, of course, especially with the outbreak of epidemics such 408 as HIV/AIDS and Ebola. Large proportions of the world's population have, in fact, been killed in disease epidemics and natural disasters. The "Black Death" (caused by bubonic plague) in Europe and Asia killed about 15 percent of the world's population between 1334 and 1349; at least 40 percent of Europe's population died. An influenza (flu) pandemic (widespread epidemic) in 1918–1919 killed at least 20 million people worldwide. We cannot assume that such devastation will never happen again. Earlier this century, there were fears that the "bird flu" might kill as many as 150 million people if it mutated to a human strain, and more recently the worry was that Ebola could become highly contagious as well as highly infectious.

Closely related to the measure of death rates is that of **life expectancy**, the number of years a person may expect to live in a given environment (typically defined as a country [see **•Figure 3.15**] and differentiated between women, who usually live longer, and men). As death rates fall, life expectancy increases, and the reverse is also true. Life expectancy worldwide has risen 50 percent since 1900, mainly because of improvements in health technologies and education in the LDCs. Life expectancy at birth for the world as a whole rose from 64.8 years in the period 1990–1995 to 70.0 years in the period 2010–2015. LDCs made substantial progress, gaining 8.9 years of life expectancy over the same period. With exceptions mainly in Sub-Saharan Africa, life expectancies continue to rise. In the United States in 2015, life expectancy for women was 81 years, and for men, 76 years. In Botswana, a southern African country hit hard by the HIV/AIDS epidemic, a woman could expect to live 47 years and a man 48 years. But as we will see in Chapter 9, that is an improvement for Botswana as it and other Sub-Saharan African countries begin to make headway against the scourge of AIDS. 408

What Determines the Population Change Rate?

Throughout history, natural disasters, diseases, and wars have taken huge bites out of our numbers. Overall, however, with birth rates higher than death rates, the trend line has been one of growth—and since 1800, of spectacular growth. In 1968, the rate of population growth hit an all-time high of 2.0 percent. To appreciate how rapid that growth rate was, you can calculate the **doubling time**. Applying the often-used "Rule of 70," in which 70 is divided by the growth rate, the doubling time is the number of years required for the human population to double (assuming that the rate of growth would be unchanged over the entire period). In 1968, the human population was growing at a rate that would, if unchanged, have doubled in 35 years. That rate did slow down, however, illustrating the limited usefulness that doubling time has in projecting future growth. At the 2015 population change rate of 1.2 percent, our numbers would double in 58 years.

Doubling time may be only approximate, but it is a good tool for comparisons among countries and also illustrates that human populations have the potential for exponential growth; that is, not an incremental or arithmetic increase from 1 to 2 to 3 to 4, and so on, but geometric growth from 2 to 4 to 8 to 16 for example. (Exponential growth in the context of the famous Malthusian scenario is discussed on **page 80**.) The annual rate of population change worldwide is depicted by country in **•Figure 3.16**.

Why Has the Human Population "Exploded?"

If both birth rates and death rates are high (as they were when our ancestors were hunters and gatherers and as recently as the dawn of the Industrial Revolution), population growth

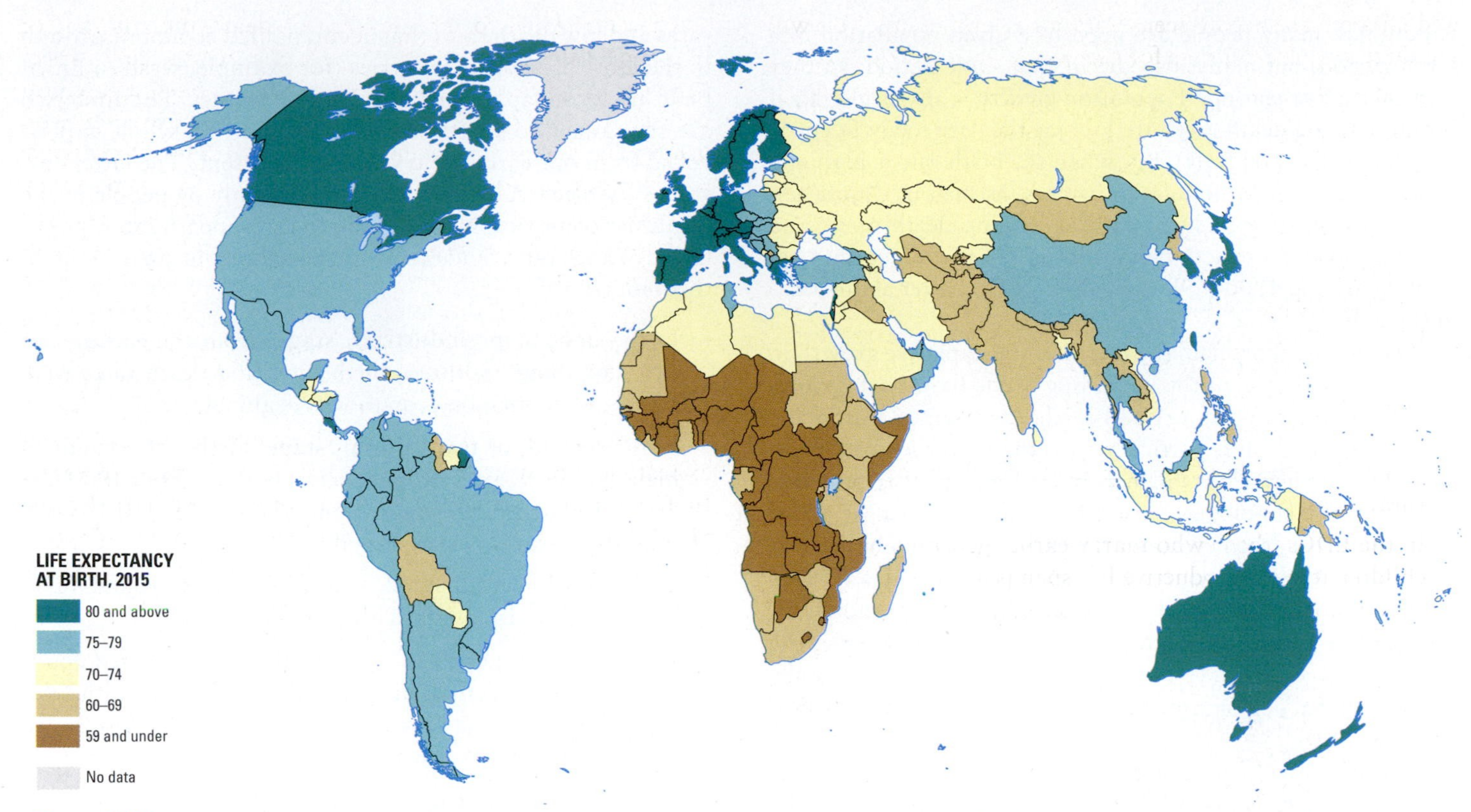

• **Figure 3.15** Life expectancy is closely tied to economic well-being; people live longer where they can afford the medicines and other amenities and technologies that prolong life.

THINK CRITICALLY: What patterns do you recognize by comparing this figure with **Figures 3.7, 3.8,** and **3.9?**

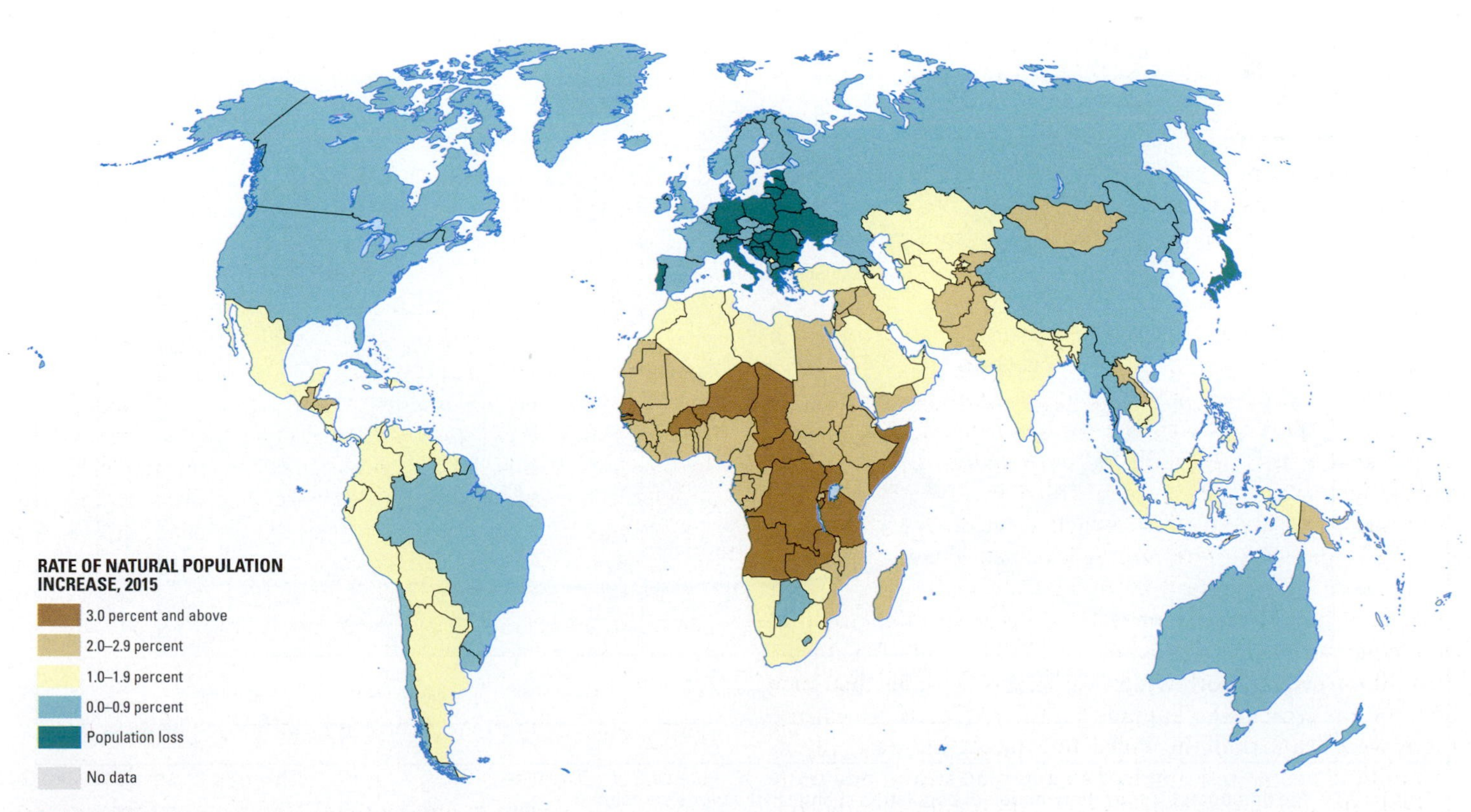

• **Figure 3.16** Population change rates are highest in the countries of Africa and other regions of the developing world and lowest in the more affluent countries.

is minimal: many people are born to a given population in a given period, but many also die in that same period, so they "cancel each other out." Population growth is also negligible if both birth and death rates are low (as they are today in countries like Japan and Italy). But when the birth rate is high and death rates are low, population surges, as seen in •**Figure 3.17**. This scenario of high birth rates, plunging death rates, and surging growth is exactly what played out for our species beginning around 1800 in Western Europe and after about 1950 in the LDCs.

It is vital to appreciate the fact that the explosive growth in world population since the beginning of the Industrial Revolution is the result not of a rise in birth rates, but of a dramatic decline in death rates, particularly in the less developed countries. The death rate has fallen as improvements in agricultural and medical technologies have diffused from the richer to the poor countries. Until recently, however, there were no strong incentives for people in the less developed countries to have fewer children. With birth rates remaining high and death rates falling quickly, the population has grown sharply; the LDCs are generally in or emerging from stage 2 of an important model that demographers call the **demographic transition** (see **Figure 3.17**).

In its entirety, from stages 1 to 4 as seen in Figure 3.17, the demographic transition model depicts, and is defined as, the change from high birth rates and high death rates to low birth rates and low death rates that accompanied economic growth in the more developed countries (for example, western European nations, Japan, and the United States). The first two are the same two stages that humankind as a whole experienced from our earliest days until the present. The latter two stages have generally been experienced only by people in the wealthier countries. Note how birth rates, death rates, population change rates, and economic development correspond in this model:

- In the **first**, or **preindustrial, stage** (from the earliest humans to about 1800 CE), birth rates and death rates were high, and population growth was negligible.
- In the **second**, or **transitional, stage**, birth rates remained high, but death rates dropped sharply after about 1800 due to medical and other innovations of the Industrial Revolution. Population growth surged.
- In the **third**, or **industrial, stage**, beginning around 1875, birth rates began to fall as affluence spread.
- Finally, after about 1975, some of the industrialized countries entered the **fourth** or **post-industrial stage**, with both low birth rates and low death rates and therefore (once again, as in stage 1), low population growth.

In this model, the United States, with population growth of about 0.4 percent per year, is in the early years of the

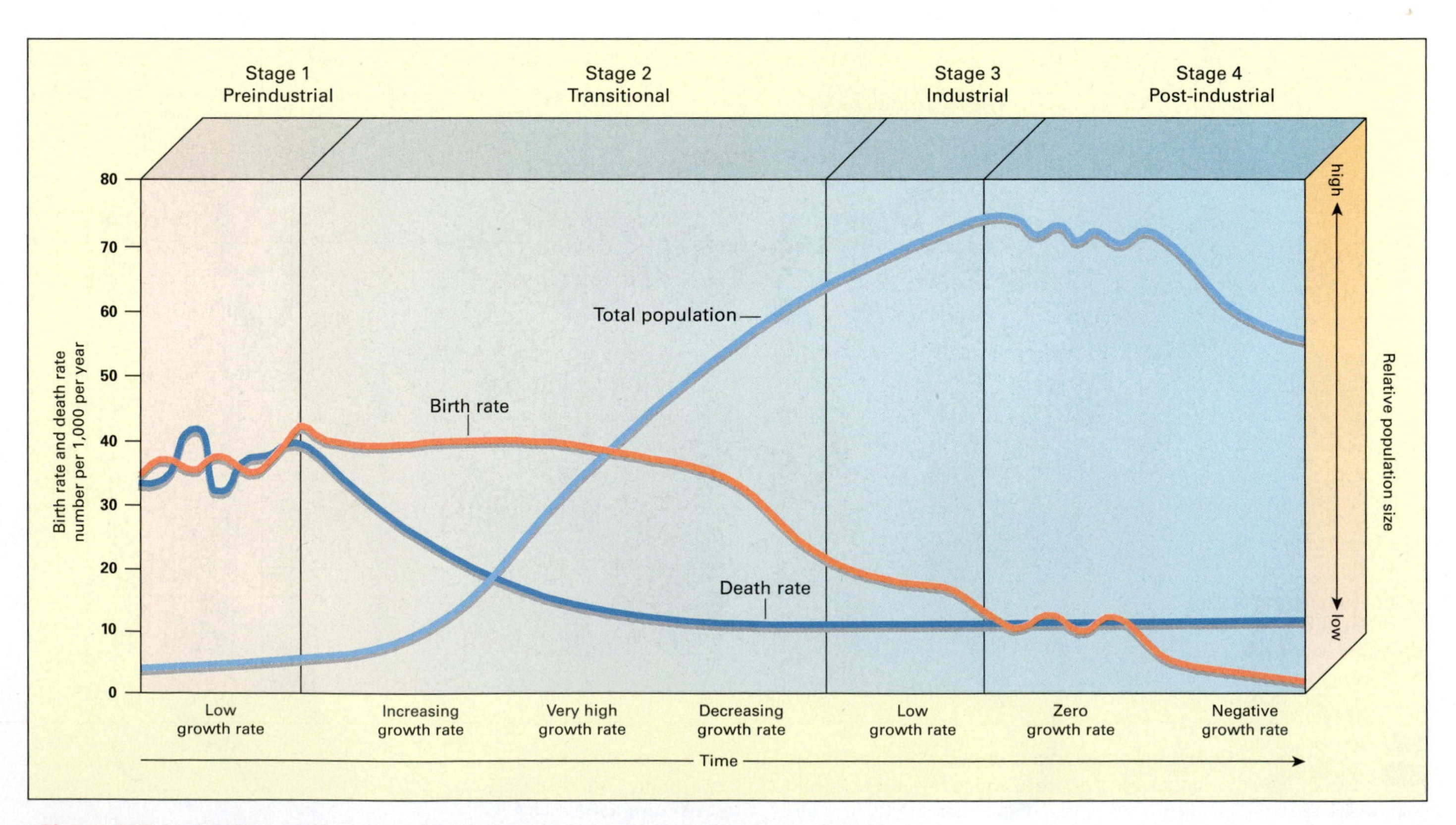

• **Figure 3.17** The demographic transition models of population change in the world's wealthier countries. Note how the population surged in the wake of the Industrial Revolution as death rates fell while birth rates remained high, but then leveled out and began to decline as economic development advanced.

post-industrial stage. Other industrialized, affluent countries have made their way well into that stage. Some wealthier countries, including Austria and the Czech Republic, have in recent years officially registered **zero population growth** (ZPG) (in which birth rates and death rates are equal), and many other European countries are close to ZPG. Because of the growing desire of women in some countries—including Japan and Germany—to pursue their own careers and postpone marriage, birth rates in those countries have fallen below death rates, meaning that these countries are actually losing population or even experiencing a **population implosion**. The populations of more than 40 countries are projected to decrease between now and 2050. The largest absolute declines should occur in China, Germany, Japan, Poland, Romania, Russia, 91, 289, 357 Serbia, Thailand, and Ukraine. In most of these cases, declines will be due to the fertility rate of post-industrial countries falling below the **population replacement level**, the number of new births required to keep the population steady (generally calculated as 2.1 children per woman in the MDCs). Other countries, notably Russia, that qualify as MDCs on the basis of per capita GDP (PPP) have experienced population losses 156 owing to rising death rates as well as falling birth rates. Trends of this sort can be most easily appreciated by looking at a very useful and informative device, the age structure diagram.

The Age Structure Diagram

Both current and projected rates of population growth are distributed quite unevenly between the poorer and richer nations. This phenomenon is apparent in the age structure diagrams typical of these countries. An **age structure diagram** (often called a **population pyramid**) classifies a population by gender and by 5 year age increments (•**Figure 3.18**). One important index these profiles show is the percentage of a population under age 15. A country with a broad-based age structure diagram like Niger in Figure 3.18 is typically poor and faces the prospect of increasing poverty because so many new jobs, food, and other resources will have to be created to meet the demands of those children as they mature and have their own children. About 9 of every 10 babies born in the world today are in the poorer countries. The bottom-heavy age structure diagram also suggests a continued surge in population as those children grow to enter their reproductive years. A large, youthful population—known as the **youth bulge**—in which competition for jobs, education, and land is intense is a social environment ripe for discord. A study by Population Action International found that 80 percent of the civil conflicts of recent decades took place in countries where at least 60 percent of the population was younger than 30.[31] Disaffection among huge populations

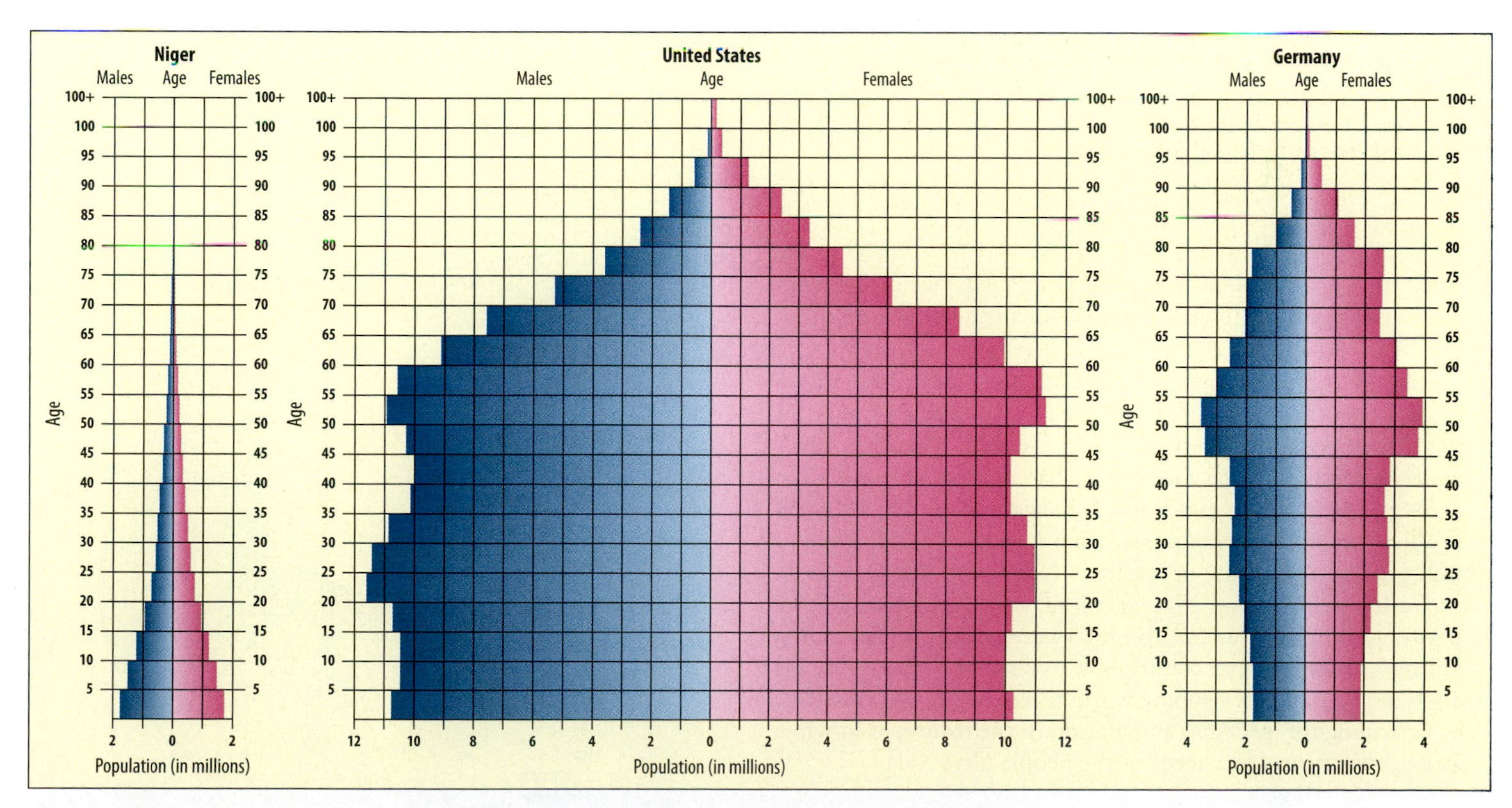

• **Figure 3.18** Population by age and gender in representative countries. Relative to older age cohorts, Niger's base of population under age 15 is broad and represents half of the country's population. This pyramid-like age structure diagram is typical of the poorer LDCs. In contrast, wealthier US and Germany have a narrow base under age 15 relative to their older age cohorts. Niger has a relatively high population growth rate, while the US (excluding immigration) has a slow growth rate. Germany and some other post-industrial countries have "negative growth," with declining populations.

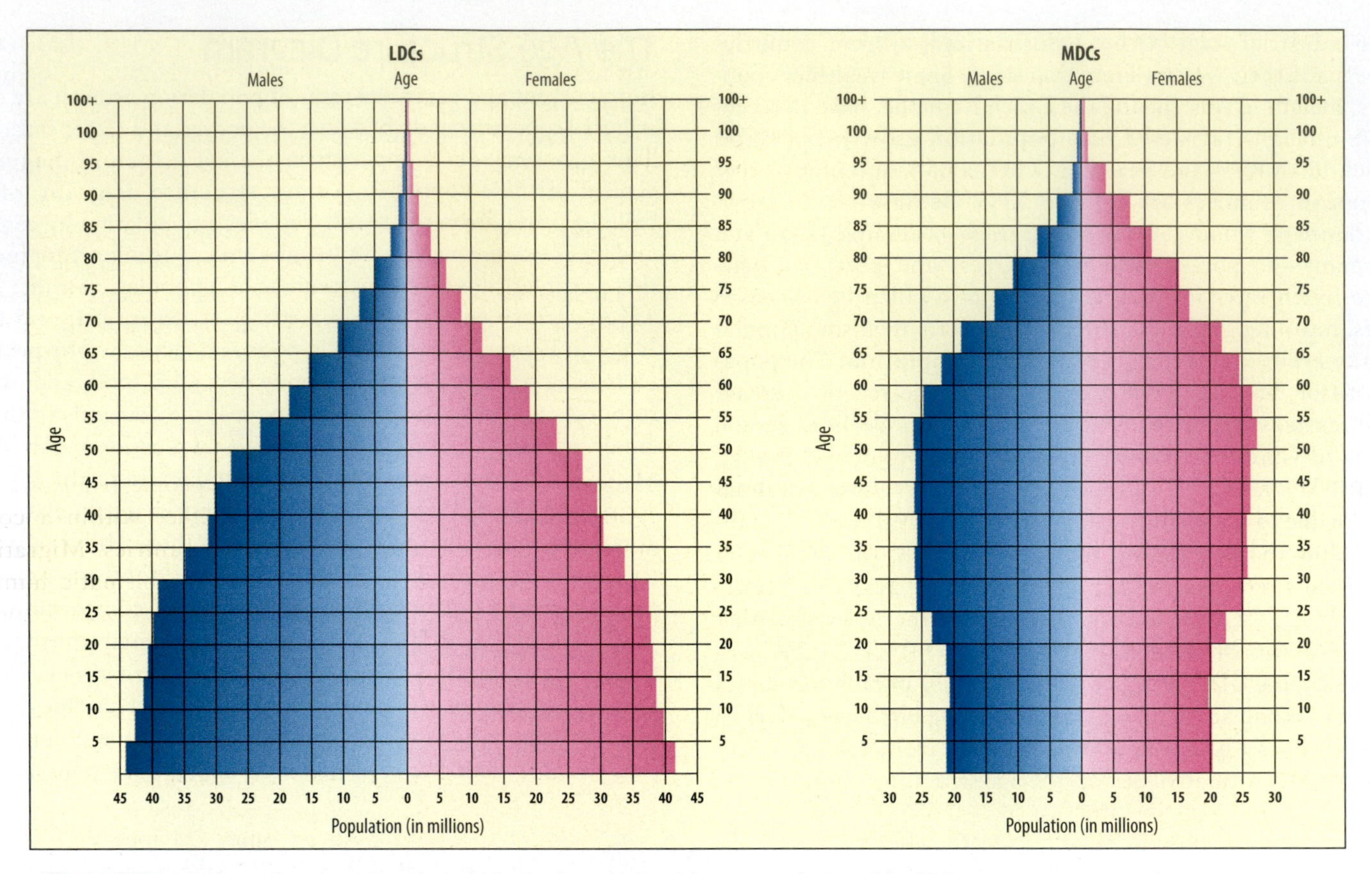

• **Figure 3.19** Population by age and gender worldwide. Note the much larger population of the LDCs and the enormous "youth bulge" present in many of those countries. By contrast, MDCs have far fewer people in total, and a much smaller proportion of young people.

THINK CRITICALLY: You could look at the pyramid-shaped structure of the LDCs and say, "these countries are in trouble." Knowing what you do about underdevelopment, what troubles may be underway or may lie ahead for these countries?

of impoverished, unentitled young people was one of the main forces that brought down several authoritarian governments in 256 the "Arab Spring" of 2011.

The bottom-heavy, pyramid-shaped age structure diagram of Niger contrasts markedly with the more square structures of the United States and Germany in Figure 3.18. These wealthier countries have a much more even distribution of population through age groups, with a modest share under age 15. Such profiles suggest that, not considering migration, their population growth is low (US) or negative (Germany).

Collectively, about 29 percent of the population of the poorer countries is under age 15, whereas the corresponding figure for the wealthier countries is 16 percent (•**Figure 3.19**). As a whole, then, the developing world faces the critical challenge of providing for a surging population in the future, even while struggling to meet the needs of the people alive today.

Where Do We Live?

Where do all these people live? The world population pie chart in •**Figure 3.20** shows clearly that China and India are the most populous countries, with about 1.4 and 1.3 billion people,

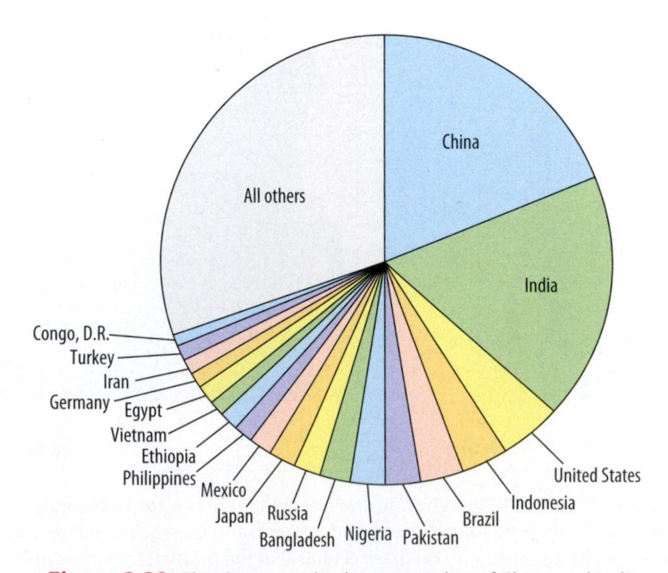

• **Figure 3.20** The demographic heavyweights of China and India stand out in this pie chart of the world's population. Half of the world's population lives in just six countries.

respectively. Think about it this way: about one-third of all people living on Earth today are either Chinese or Indian!

Why are so many people in just two countries? There are several reasons. Both countries are large. China is 20 percent larger than the contiguous 48 US states, and India is about one-third the size of China. Both have been populated since very ancient times, and successful intensive agriculture has been practiced in both for more than 4000 years. Both have large areas of productive soils and high rainfall, promoting successful farming. Both are developing countries in which birth rates remained high while death rates fell. China's population explosion prompted the government to adopt an aggressive population control policy several decades ago, and India followed suit with less forceful measures. India's higher growth rate—1.5 percent, compared with China's 0.5 percent, means India will likely overtake China in population by 2030.

Although there are many variables to consider in explaining why large numbers of people are clustered in particular countries, including cultural factors and family planning policies, the natural setting is by far the most important factor. The population densities shown in the dot density map (•**Figure 3.21**) correlate generally with agricultural and other environmental conditions. The clustering of dots representing the highest population densities are in the more humid and fertile regions of both China and India. Conversely, in the very western part of China, including the high Tibetan Plateau, very dry conditions limit agriculture to a few favored areas, and population dots are sparse. The world's highest mountains, the Himalayas, rise just north of the band of high population density in northern India. The moisture-laden winds that bring so much productive rainfall to India cannot cross that mountain barrier, which has been a divide between densely and sparsely populated regions for thousands of years. Looking at the lightly populated areas of the world, you can recognize similar environmental factors at work: in the Sahara of northern Africa, the Arctic of northern Canada, and the Amazon Basin of South America, conditions are too dry, too cold, or too wet and infertile to support large numbers of people.

The Geography of Migration

Migration refers to the movement of people from one location to another in any setting, whether within a community, within a country, or between countries. Migration is one of the most dynamic and most problematic human processes on Earth (•**Figure 3.22**). Recent studies by geographers have determined that the poorest countries are *not* the largest source of international migrants; instead, countries in transition—those whose middle classes are swelling such as China, India, and Mexico—are the largest. As countries transition to MDC status, they become destinations of migrants themselves.[32]

A migrant is always both an **emigrant** (one who moves from a place) and an **immigrant** (one who moves to a place).

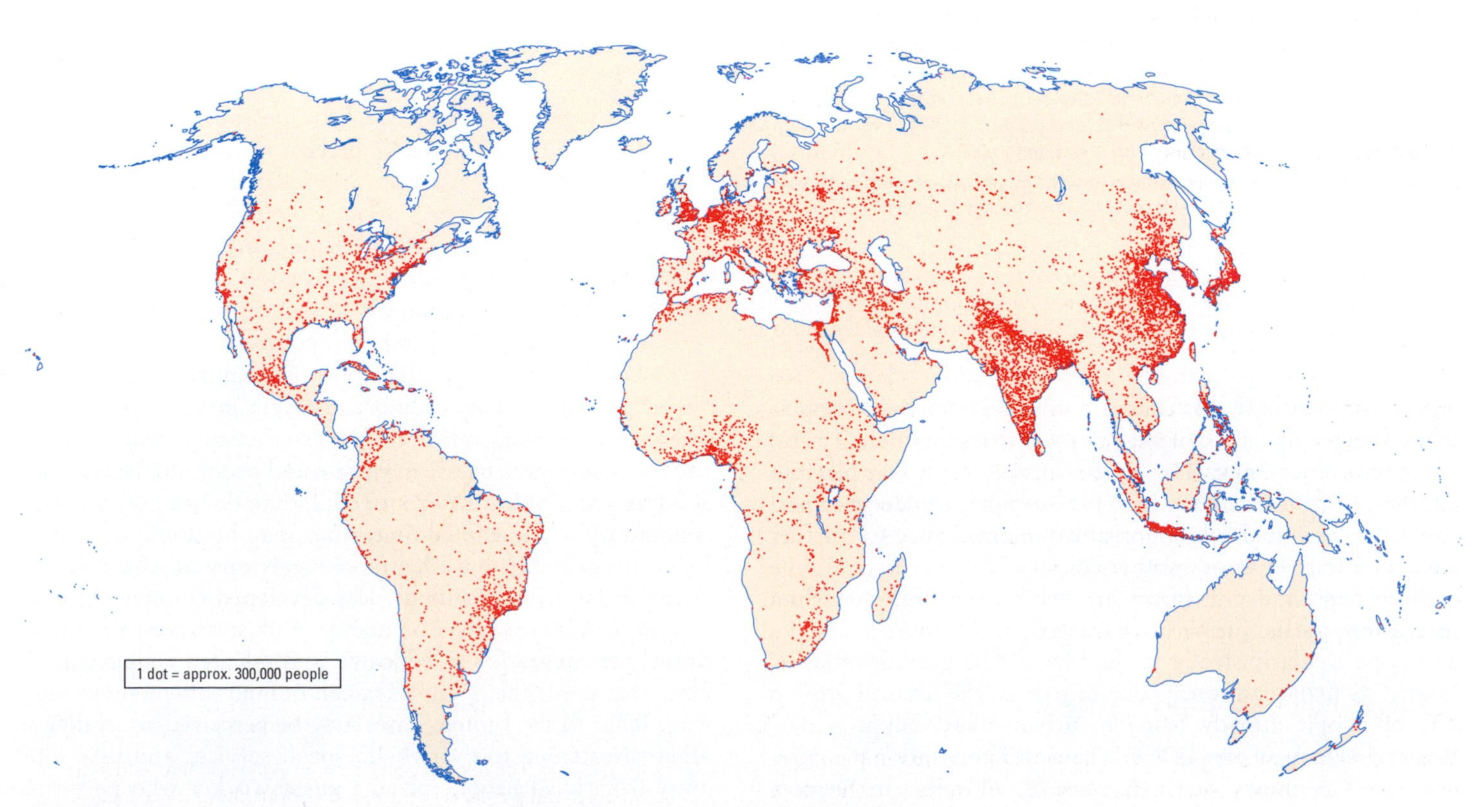

• **Figure 3.21** This dot density map shows the approximate distribution of people around the world.

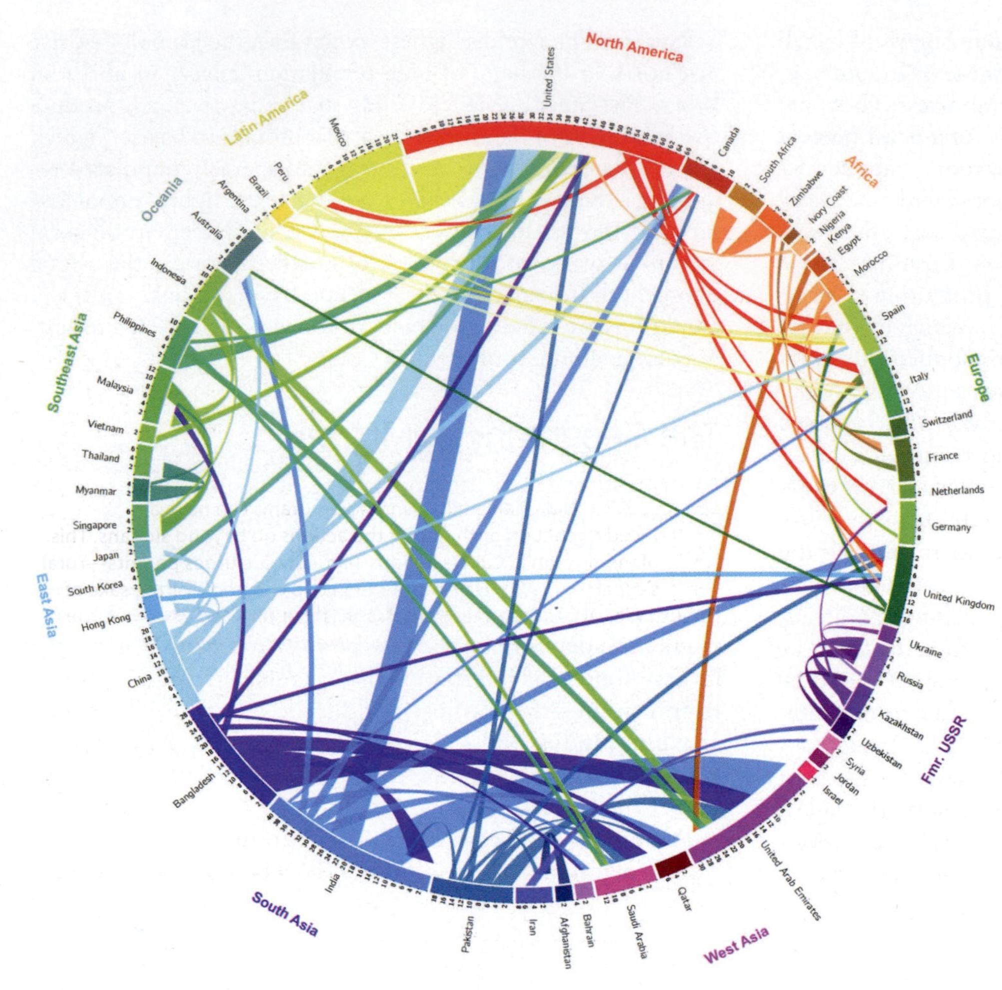

• **Figure 3.22** The global picture of people on the move, drafted by a team of geographers. The major trends are of migrants in search of work in more affluent countries and of refugees driven by warfare or environmental adversity. The bilateral flows between 196 countries are comparable across countries and capture the number of people who changed their country of residence between mid-2005 and mid-2010. The circular plot shows the estimates of directional flows between the 50 countries that sent and/or received at least 0.5 percent of the world's migrants in that period. Tick marks indicate gross migration (in + out) in 100,000s.

Source: "The Global Flow of People" by Nikola Sander, Guy J. Abel and Ramon Bauer, published online at www.global-migration.info and in Science on March 28, 2014 (Abel & Sander, "Quantifying global international migration flows." Vol. 343: 1520–1522).

THINK CRITICALLY: What single country has the largest (on the graph, widest) out-migration, and what is the destination country? What two countries are most Indian migrants going to? What is the world's largest destination country for migrants? See note 32.

Migration is usually associated with either **push factors**, as when hunger or lack of land "pushes" peasants out of rural areas into cities or warfare pushes people from one place to another, or **pull factors**, when, for example, an educated person is "pulled" by a job opportunity or the chance for further education in the city or another country.

Both push and pull forces are behind the **rural-to-urban migration** pattern that is characteristic of most countries and is particularly strong in the LDCs. The growth of cities, known as **urbanization**, is due in part to the natural growth rate of people already living in urban areas, but it is particularly strong in the LDCs because of this internal migration (see Geography of Cities, **page 78**). Whereas in Western cities urbanization typically goes hand in hand with economic growth, the tide of migrants "pushed" from rural areas into the city amounts to a situation characterized as "urbanization without industrialization" or "urbanization without modernization." Parts of LDC cities are actually being "ruralized" as country folk bring their values, customs, dress, and even small livestock into the city. Living in the hearts of modern Cairo and Hanoi, I was often awakened by crowing roosters, and saw rooftop and ground floor "barnyards" with rabbits, goats, and pigeons. 434

Within and among countries today, there are considerable movements of **refugees**, the victims of such severe push factors as persecution, political repression, and war. In their destination countries, they may be classified as "illegal immigrants" or as people granted **asylum**, meaning they are given permission to immigrate on the grounds that they would be harmed or persecuted in their country of origin. Immigration and asylum laws and quotas vary widely around the world, depending on host countries' political, economic, and social systems. Among the most disadvantaged of the world's peoples are the **internally displaced persons (IDPs)** who are dislodged and impoverished by strife in their home country but have little prospect of emigrating. Over 60 percent of internally displaced persons are located in just 5 countries: Syria, Colombia, Nigeria, Sudan, and the Democratic Republic of the Congo. In this book you will read many 265 accounts of others moved by forces beyond their control.

Migration is almost never clearly "good" or "bad," but is almost always a mixed blessing and often elicits strong anti-immigrant sentiments. Migration is 94 bad for the country whose most talented people are leaving but good for the country they go to; the Indian doctor benefits your community in the United States but may be sorely missed in India, for example, but even more sorely missed would be the doctor from a less populous, less developed country, such as Ghana or Afghanistan. The doctor is illustrative of the **brain drain**, the emigration of educated and talented people from a 128, 129, 157, 175 place that needs them. The Mexican immigrant who does low-wage labor in the United States may be perceived as an **illegal alien** threatening to overwhelm social services and take jobs away from local people, or as a **guest worker** who performs services that few others want to do but that are critical to the 93, 497 country's economic well-being. Some host-country peoples are more welcoming of new cultures that enrich their ethnic mosaic (Americans are generally perceived as among the most accommodating cultures in the world), whereas others fear losing their ethnic majorities and privileges (Japanese people generally want to keep their homogeneity, and official policies 356 discourage immigration).

How Many People Will Live on Earth?

Population geographers are fairly confident in their calculations of how many people have lived on Earth at various times in the past. Projecting future numbers is another matter. There are many uncertainties. Will birth rates fall faster than expected in the developing world? Will death rates surge because of HIV/AIDS or some other epidemic? These are among the wild cards in the population deck.

In asking how many people will live on Earth in the future, we are essentially asking, "Will the poorer countries of the world go through the demographic transition?" Will the current, relatively high birth rates in the less developed countries continue their present slow decline? In 1970, Kenya had a birth rate of 51 and Bangladesh had a birth rate of 45; by 2015, Kenya's birth rate had dropped to 34 and Bangladesh's to 20. Considering what might have been, given the country's huge population, China's decline may be the most impressive of all, from a rate of 38 in 1965 to 12 in 2015. The Chinese government's generally harsh, recently amended **one-child policy**, which uses a combination of incentives and punishments, has been the main reason for the decline. But the demographic transition model also suggests to us that many increasingly affluent 289 Chinese couples are making the choice to limit their children. In Kenya and Bangladesh, growing literacy and the slow but steady economic progress of women have brought down the birth rate.

Family planning policies and levels of education and economic well-being have played various roles in the poorer countries but have collectively combined to bring birth rates down steadily since 1968 (**•Figure 3.23**). If the processes of increasing economic and social development continue in the LDCs as a whole, they should make steady progress through the demographic transition, and the Earth's population should cease to grow and should stabilize.

But will the poorer countries complete the transition successfully? An unsavory but possible scenario would see death rates rise dramatically (due to disease or famine, for example) in at least some of the countries, in effect pushing them back to stage 1 of the demographic transition, where both birth and death rates are high. Others could remain in stage 2, with high birth rates and falling death rates, long enough to bring unexpectedly high numbers of people into the world. With such different scenarios in mind, the United Nations prefers to use a widely respected model with three projections: high, medium, and low growth (**•Figure 3.24**).

In view of declining birth rates worldwide between 2007 and 2010, the United Nations revised its medium projection for population growth downward. The agency had predicted 9.3 billion by 2050 (half a billion fewer than it had estimated in earlier projections) and stabilization—the maximum number of people that will ever live on Earth at one time—at 10 billion in 2100. The UN's 2014 projection revised that medium figure *upward* to 9.6 billion by 2050, with 10.9 billion by 2100. Instead of leveling off and stabilizing, the population will continue to grow beyond 2100. During the period 2015–2050, nine countries are expected to account for more

• Figure 3.23 National family planning programs can have a pronounced impact on birth rates if the actions go beyond slogans. This family planning sign in Cairo's famous Tahrir Square urges parents—rural parents especially—to have no more than two children "for the sake of a better life." The government's message is that the ideal family of four can afford amenities a large family could not, including the CD player/radio. Egypt's birth rate fell from 40 in 1970 to 31 in 2015. It is interesting to note that with growing religiosity in Egypt, the government was pressured recently to drop the term "family planning," used here, because "planning" suggests that people can override God's ultimate say on family size.

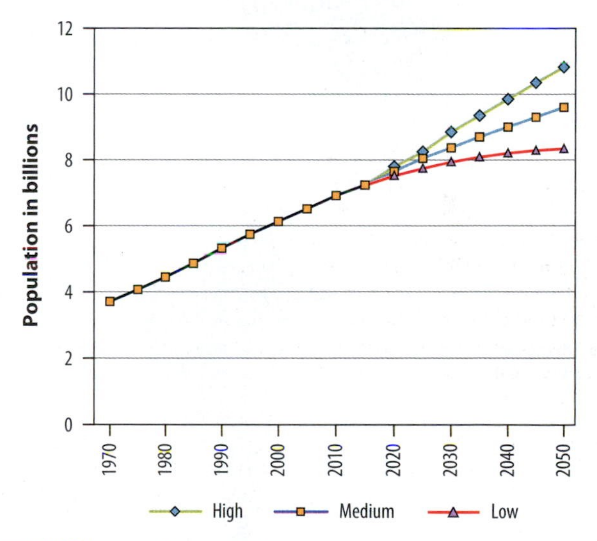

• Figure 3.24 United Nations' 2014 projection for population growth. Most users of these data prefer to cite the medium projection of 9.6 billion by 2050.

Geography of Cities The Triumph of the City

In 2008, the number of city-dwellers exceeded the number of rural peoples worldwide for the first time, and this trend is not likely to reverse. Fifty-three percent of the world's population was urban in 2015; the UN forecasts that that figure will grow to 66 percent by 2050, with 90 percent of that growth taking place in Asia and Africa (•**Figure 3.E**). The percentage and speed of urbanization differ between world regions. North America and Latin America are near 80 percent urban; Europe, Oceania, Russia, and the Near Abroad, and the Middle East and North Africa are about 70 percent urban; although urbanization lags in South and East Asia at 46 percent, and Sub-Saharan Africa at 40 percent. Africa and Asia will have the most rapid rates of urbanization through 2050, in part because of their relatively low levels now. Urban growth in South and East Asia is spectacular. Over half of the world's 34 **megacities**—those with populations greater than 10 million—are in this region. The number of megacities worldwide is projected to reach 41 by 2030, with Asia leading the way.[33]

Triumph of the City is an appropriate title for Harvard economist Edward Glaeser's book on urbanization. "It turns out that the world isn't flat, it's paved," Glaeser writes.[34] Whatever negative images come to mind—urban blight, urban decay, urban crime—there are compelling arguments that cities have a lot going for them. They are engines of economic growth, in fact, producing roughly 80 percent of global GDP. There is an almost perfect correlation between prosperity and urbanization. In MDCs, urbanization and industrialization first grew hand in hand. That

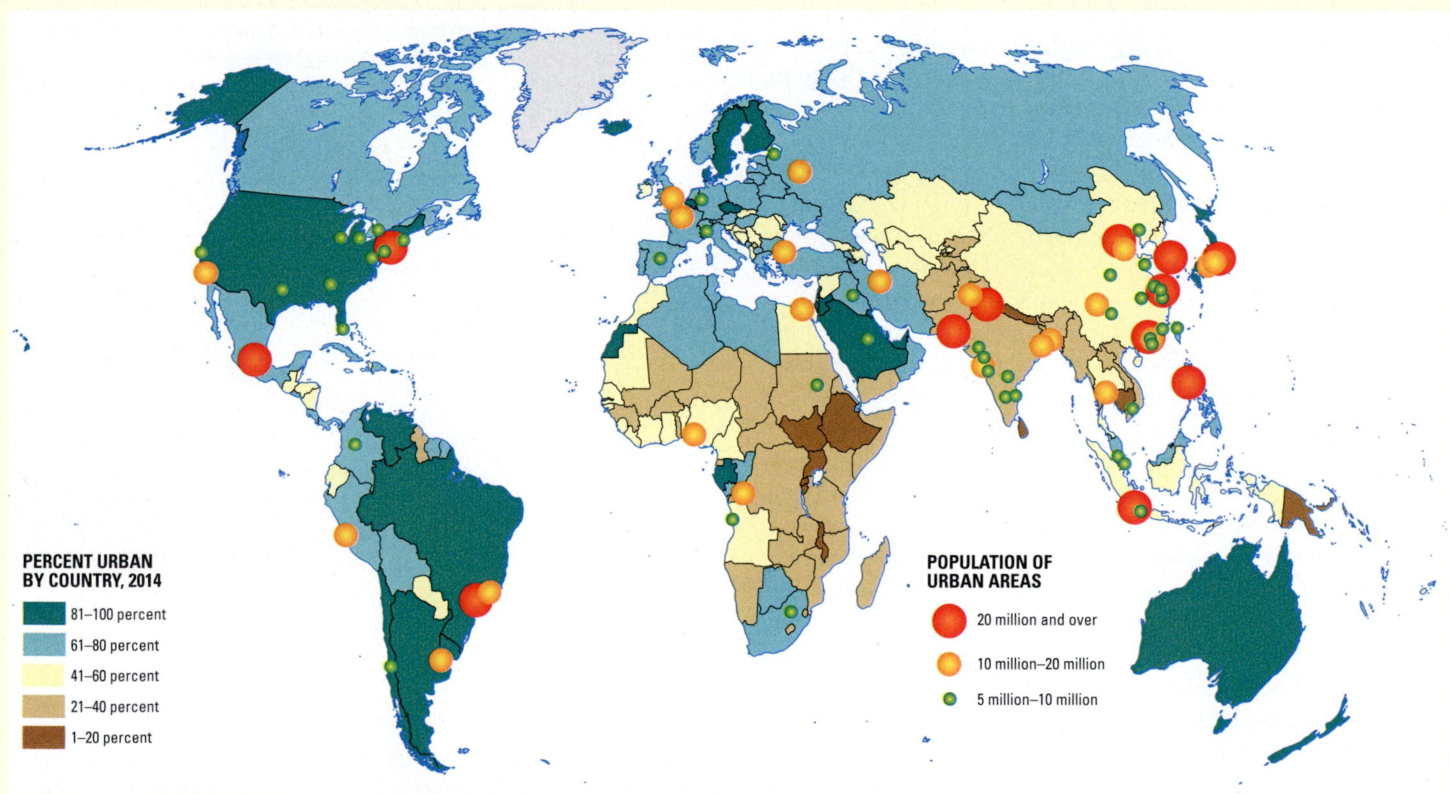

• **Figure 3.E** This map shows the percentage of the population in each country living in urban areas, and also shows the locations of all metropolitan areas with populations of 5 million or more.

than half of the world's projected increase: the Democratic Republic of the Congo, Ethiopia, India, Indonesia, Nigeria, Pakistan, Tanzania, the United States, and Uganda.[36] This upward revision dampened the rash of popular and academic articles proclaiming that "the population explosion is over" and even some essays arguing there would soon be too few people on Earth, particularly in the richer countries. Some experts have cautioned that it is too soon to declare the population bomb defused. "World population growth turned a little slower," a Population Institute report argues. "The difference, however, is comparable to a tidal wave surging toward one of our coastal cities. Whether the tidal wave is 80 or 100 feet high, the impact will be similar."[37] Population growth rates have been declining since 1968—from 2.1 percent per year to 1.2 percent per year in 2015—but the population base is so large that 82 million are added to the world's population correlation has faded with the rise of globalization, which killed off manufacturing in many a city. The modern Western city's economy instead prizes and in many cases thrives on innovation. Think of the iPhone and Apple's trademark phrase "Designed by Apple in California." Although the IT professionals designed Apple's lifestyle-changing devices there in Silicon Valley, the actual manufacturing of an iPhone is outsourced mainly to Chinese cities, where components from a supply chain of 30 other countries go into it (•**Figure 3.F**, a fine illustration of how urbanization and globalization also go hand-in-hand).[35] In addition to information services, major sectors of the urban economy in MDCs include financial services, health services, retail, tourism, and media. Urban employment in LDCs is not correlated exclusively with manufacturing by any means. India has its own "Silicon Valley," a city called Bangalore or Bengaluru, where information services such as call centers are important sectors.

Glaeser subtitled his book about cities *How Our Greatest Invention Makes Us Richer, Smarter, Greener, Healthier, and Happier.* The notion of "green cities" may seem intuitively absurd, but it is easy to appreciate that the urbanite has a smaller "carbon footprint" than the suburbanite who burns a lot of fuel getting to and from work downtown. Your book *Fundamentals of World Regional Geography* will acquaint you with many concepts related to urbanization, among them "smart" cities, megacities, urban decay, gentrification, and informal settlements (slums).

60

313

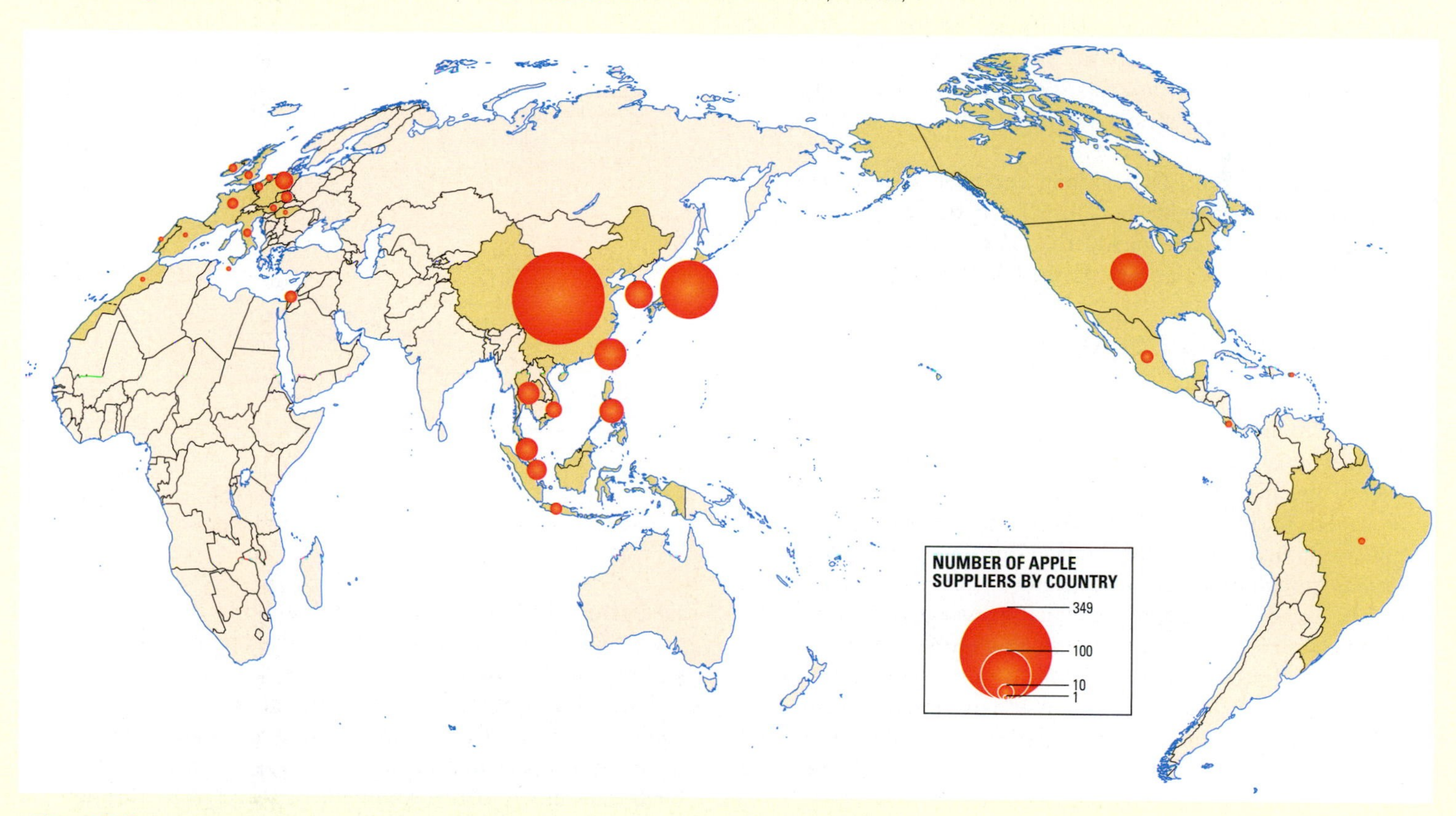

• **Figure 3.F** Over 700 suppliers around the world contribute to the design and development, sourcing, manufacturing, warehousing, and distribution of Apple's iPhone. The components come from cities and are assembled in cities; the processes of globalization and urbanization are linked.
Source: Apple Supplier List 2014.

each year. Of those, 54 percent are in Asia and 33 percent in Africa. Those ratios will shift, however, as growing prosperity reduces population growth in Asia and leaves the lion's share of population growth in the world's least developed region, Sub-Saharan Africa. The UN projects that by 2050 80 percent of the global increase will take place in Africa, with just 12 percent in Asia. Reminding us that the figures are not predictions, but projections, the United Nations has pledged to do its best to help stabilize the global population at a lower number than now projected, by lowering the birth rate. Like most organizations concerned with bringing down the birth rate, the UN does not focus on birth control. Instead, it calls for and funds efforts that both bring down birth rates and promote development in the world's LDCs. These goals can be reached most effectively by increasing the education and employment of women.

329, 266

As birth rates fall and life expectancies rise, the world's population is aging. Older people (over age 60) were just 9 percent of the world's population in 1994. They represented 12 percent in 2014 and are expected to reach 21 percent by 2050. Many MDCs already have very low **old-age support ratios**, the number of working-age adults for each older person. Germany, Italy, and Japan, for example, have only two and a half to three working-age adults for each older person. European countries cluster at the lower end of old-age support ratios; Latin American countries are mainly in the middle range; and with their larger families, countries of South and East Asia and Sub-Saharan Africa have relatively high old-age support ratios.[38]

Shouldn't we celebrate that people are living longer and population growth rates are falling? Yes, but population aging also poses a number of challenges. Stresses on families are especially strong in Western MDCs, where there generally is not a culture of a small nuclear family caring for the elderly at home. Instead, the elderly go to "old folks homes." These institutions and health care systems, along with pension and retirement systems, represent a growing cost burden that younger people must bear.

Generally in the LDCs, strong extended family networks help look after the elderly. When I tell friends in Malaysia, Egypt, or Madagascar that elderly people in America live apart from their families, they are stunned, and ask, "Why?"

The Malthusian Scenario

Even while the birth rate falls, the world's population increases. Already large and growing numbers pose fundamental questions: Can the Earth sustain so many people? Will we exceed the planet's carrying capacity, and what will happen if we do? Early in the Industrial Revolution, an English clergyman named Thomas Malthus (1766–1834) postulated that human populations, growing geometrically or exponentially, would exceed food supplies, which grow only arithmetically or linearly. He predicted a catastrophic human die-off as a result of this irreconcilable equation (•**Figure 3.25**). He could not have foreseen that the exploitation of new lands and resources, including tapping the energy of fossil fuels, would permit food production to keep pace with or even outpace population growth for at least the next two centuries.

This **Malthusian scenario** of the lost race between food supplies and mouths to feed is still debated today. On one side of the argument are optimists, the so-called **technocentrists** or **cornucopians**, who argue that through technological ingenuity people have always been able to conquer food shortages and other problems, and therefore they always will be able to in the future (•**Figure 3.26**). Julian Simon, a University of Maryland economist, argued that far from being a drain on resources, additional people create additional resources. Technocentrists therefore insist that people can raise the Earth's carrying capacity indefinitely and that the die-off that Malthus predicted will always be averted. Our more numerous descendants will instead enjoy more prosperity than we do.

In contrast, the **neo-Malthusians** (heirs of the reasoning of Thomas Malthus) argue that although we have been successful so far, we cannot increase the Earth's carrying capacity indefinitely. There is an upper limit beyond which population growth cannot occur, with calculations ranging from 8–40 billion people. Neo-Malthusians insist that the poorer countries cannot remain indefinitely in the second stage of the demographic revolution. Either they must intentionally bring birth rates down further and make it successfully through the demographic transition, or they must unwillingly suffer nature's solution, a catastrophic increase in death rates. Thus by either the **birth rate solution** or the **death rate solution** to the problem, the neo-Malthusians argue, the LDCs must confront their population crises (see **pages 72–74**).

There is more "grist for the mill" than ever for neo-Malthusians: Craig Hanson, Director of Food, Forests and Water Programs at the World Resources Institute, points to studies that say feeding more than 9 billion people in 2050 will require 70 percent more calories than the world's population consumes today.[39] As we have seen in Chapter 2, although food demand is rising at a rate of 14 percent per decade, climate change is predicted to reduce crop outputs by as much as 2 percent per decade. The Malthusian race is on . . . or at least 42 the debate is still timely and lively.

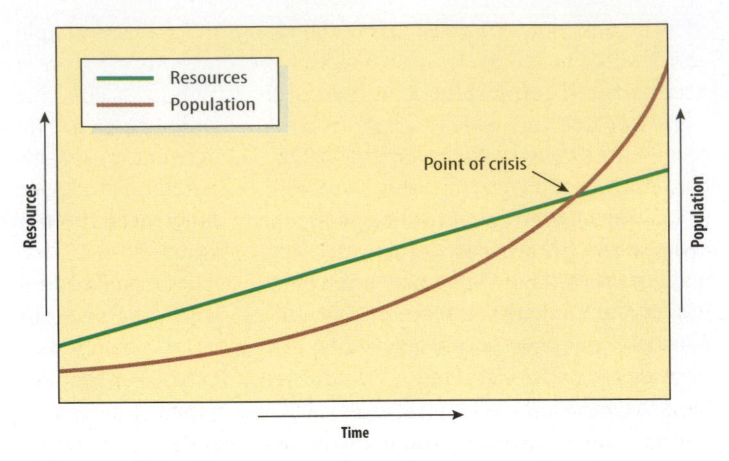

• **Figure 3.25** Malthus envisioned a race between people and resources, in which people lost.

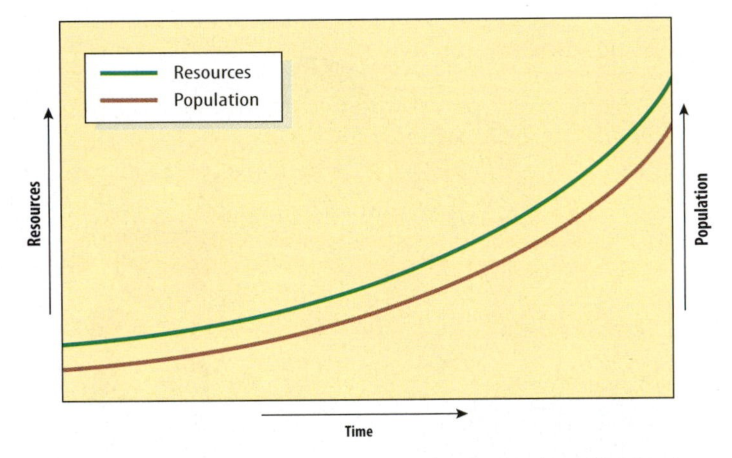

• **Figure 3.26** The technocentrists reason that production of food and other resources will always stay ahead of population growth.

Whereas technocentrists view the equation between people and resources passively, insisting that no corrective action is needed, neo-Malthusians tend to be activists who describe terrible scenarios of a death rate solution in order to motivate people to adopt the birth rate solution. For decades the biologist Paul Ehrlich insisted that we are increasingly vulnerable to a Malthusian catastrophe, particularly as viruses diffuse around the globe with unprecedented speed. "The only big question that remains," Ehrlich wrote, "is whether civilization will end with the bang of an all-out nuclear war, or the whimper of famine, pestilence and ecological collapse."[40] Such dire warnings have earned many neo-Malthusians the reputation of being "gloom-and-doom pessimists." The neo-Malthusians insist that there are simply too many people already for the world to support, especially in the LDCs, where there are not enough resources to support them. But what does "too many people" mean? How many are too many, and based on what criteria?

What Is "Overpopulation"?

It is probably most useful to think about two distinct types of overpopulation, one characteristic of the poor countries and the other of the rich (•Figure 3.27). **People overpopulation** is an apparent problem in the poorer countries. The environmental problems characteristic of LDCs are intensified by the relatively high rates of human population growth in those countries. More people cut more trees, hunt more wildlife, and otherwise use more resources. Many persons, each using a small quantity of natural resources daily to sustain life, have a great collective impact on the environment and may add up to too many people for the local environment to support. Common consequences are malnutrition and even the famine emergencies in which richer countries are called upon to provide relief.

Consumption overpopulation is characteristic of the MDCs. In the wealthier countries, there are fewer persons, but each uses a large quantity of natural resources from ecosystems around the world. Their collective impacts also degrade the environment, and even their smaller numbers may be "too many" at such unsustainably high levels of consumption, particularly with their habit of feeding so high on the food chain; much more energy is required to produce meat than to produce grains (see Insights on page 82). With less than 5 percent of the world's population, the United States may be regarded as the world's leading over-consumer. According to a useful measure known as the **ecological footprint**—the amount of biologically productive land needed to sustain a person's consumption and

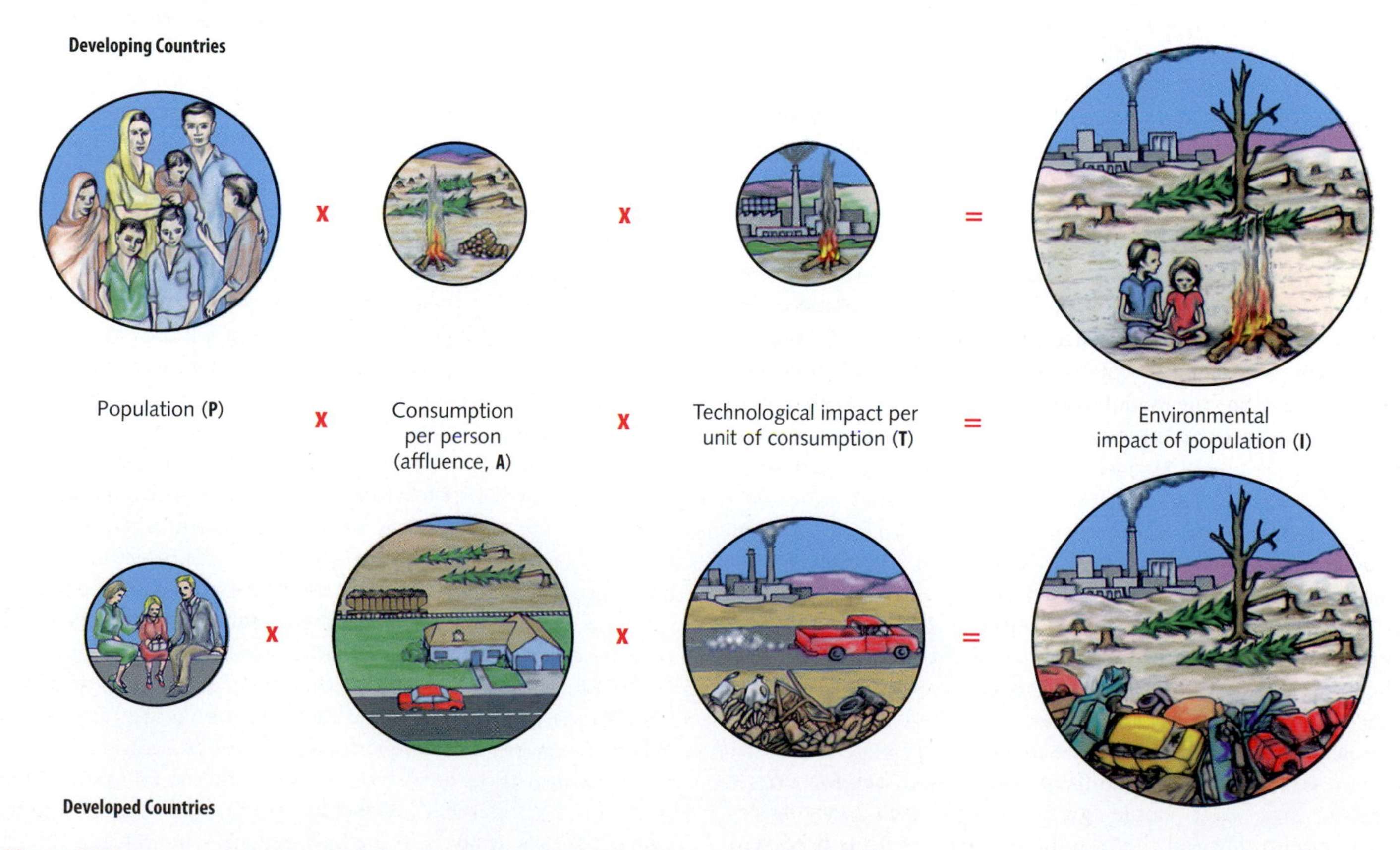

• **Figure 3.27** Two types of overpopulation, calculated according to this formula: number of people × number of units of resources used per person × environmental degradation and pollution per unit of resource used = environmental impact. Circle size shows the relative importance of each factor. People overpopulation is caused mostly by growing numbers of people and is typical of LDCs. Consumption overpopulation is caused mostly by growing affluence and is typical of MDCs.

Insights Population and Food Energy

This book discusses energy a lot, and for good reason: energy supply and demand issues impact our lives at many scales. But energy does not refer just to fossil fuels. Issues of food energy, which is made available to ecosystems through the process of photosynthesis, and of how that energy flows through ecosystems, are central to the question "How many people can the world support?"

67 **Food chains** (for example, in which a deer eats grass and a mountain lion eats a deer) are short, seldom consisting of more than four "trophic (feeding) levels." The collective weight (known as **biomass**) and absolute number of organisms decline substantially at each successive trophic level (there are fewer leopards than antelopes). The reason is a fundamental rule of nature known as the **second law of thermodynamics**, which states that the amount of high-quality usable energy diminishes dramatically as the energy passes through an ecosystem. In living and dying, organisms use and lose the high-quality, concentrated energy that green plants produce and that is passed up the food chain. As an organism is consumed in any given link in the food chain, about 90 percent of that organism's energy is lost, in the form of heat and feces, to the environment. The amount of animal biomass that can be supported at each successive level thus declines geometrically, and little energy remains to support the top carnivores. So, for example, there are many more deer than there are mountain lions in a natural system.

The second law of thermodynamics has important consequences and implications for human uses of resources. The higher we feed on the food chain (especially the more meat we eat), the more energy we use. The question of how many people the Earth's resources can support thus depends very much on what we eat. The affluent consumer of meat demands a huge expenditure of food energy because that energy flows from grain (producer) to livestock animal (primary consumer, with a 90 percent energy loss) and then from livestock animal to human consumer (secondary consumer, with another 90 percent energy loss). It has been estimated that each year, in eating meat products (beef, pork, poultry) the average US citizen consumes the food energy equivalent of two-thirds of a ton of grain. Earlier in this chapter we looked at the similar concept of "virtual water" consumed by livestock and farming. Strictly speaking, by skipping meats and eating just the grains (or something else on that trophic level) used to feed these animals, many people could live on the energy required to sustain just one meat-eater. The point is not to feel guilty about eating meat but to realize that there is no answer to the question "How many people can the world support?" The question needs to be rephrased: "How many people can the world support based on people consuming—?" (with the blank filled in with a specific number of calories per day or with a general lifestyle; an adult body needs roughly 2200 calories per day).

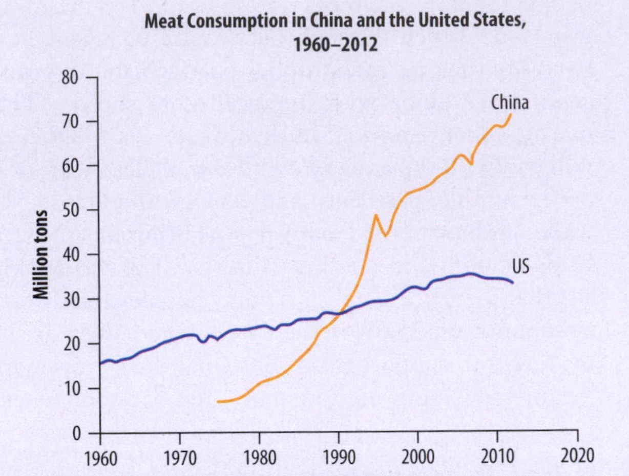

• **Figure 3.G** China overtook the United States as the world's leading meat consumer in 1992, and by 2012 the Chinese ate more than twice the meat that Americans did (but on a per capita basis, they lag behind Americans). China's consumption of meat continues to rise, influencing the nature of agriculture around the globe.

THINK CRITICALLY: At the "big picture" level, imagine all the world's peoples consuming food and other resources at the level Americans do. What are some of the implications of that?

Development means a more affluent lifestyle, including eating more meat (feeding higher on the food chain). Some analysts fear the burden that the changing dietary preferences accompanying development will pose to the planet's food energy supply. In 1994, Lester Brown of the WorldWatch Institute wrote a provocative essay titled "Who Will Feed China?"[41] Mindful of the second law of thermodynamics, he argued that if Chinese people continue on a course of growing affluence, moving away from reliance on rice as a dietary staple to a diet rich in meats and beer, the entire planet will shudder. How could over a billion people be sustained so high on the food chain?

Brown's concerns were prescient. China's overall meat consumption overtook that of the US about when he wrote his essay, and by 2012 it was twice the America's and is continuing to rise (•**Figure 3.G**). Of course, with the absolute numbers of people and ranks of the middle class surging worldwide, it's not just China that is concerning. A global population growth of less than 30 percent is projected to double the demand for animal products. Meat is expensive everywhere, and its production and consumption favor those with means. As almost always, the poorest of the poor will lose ground: "Paradoxically, as increasing numbers of people can afford to eat well, food for the poor will become scarcer, because demand for animal products will surge, and they require more resources like grain to produce," writes Mark Bittman. "But there is not the land, water or fertilizer—let alone the health care funding—for the world to consume Western levels of meat."[42] The "land grabbing" described earlier in this chapter (**page 66**) nevertheless represents efforts by China and other countries to secure the land, water, and other resources needed to sustain growing appetites.

absorb his or her wastes—for every acre needed to support an average Ethiopian, 69 acres are needed to support an average American. This disparity suggests that if Ethiopia is "underdeveloped," the United States is "overdeveloped".

If the vast majority of the world's population were to consume resources at the rate that US citizens do, the environmental results would certainly be ruinous. Even if consumption levels in the LDCs do not rise substantially, the sheer increase in numbers of people in those countries suggests that degradation of the environment will accelerate in the coming decades.

3.4 An Action Plan for Global Problems

The cornucopian view is comforting: we have nothing to worry about. By keeping up the good, ingenious work as we have in the past, our futures will be secure. If, however, one diverges modestly from this premise or even accepts the most dire neo-Malthusian view and promotes the "death-rate solution" to overpopulation (as did ecologist Garrett Hardin, whose famous *Living on a Lifeboat* essay advocated to "let nature take its course" and allow people imperiled by famine or other catastrophe to perish; "We can't cure a shortage by increasing the supply," Hardin wrote), it is appropriate to ask what can be done to prevent or solve some of the world's critical problems involving natural resources and human population numbers.[43]

The Birth Rate Solution and Sustainable Development

In recent decades, new concepts and tools for managing the Earth and its resources in an effective, long-term way have emerged. Known collectively as **sustainable development** (also **ecodevelopment**), these ideas and techniques consider what both MDCs and LDCs can do to avert the possible Malthusian dilemma and improve life on the planet. By promoting the "birth rate solution" and other concrete actions, sustainable development offers an activist agenda without the peril of doom depicted by the neo-Malthusians.

The World Conservation Union defines sustainable development as "improving the quality of human life while living within the carrying capacity of supporting ecosystems."[44] Sustainable development seeks to modify the current pattern of unsustainable development, whereby economic growth is based in large part on excessive depletion of natural capital. Sustainable development challenges the perverse incentives of conventional development. For example, a country that depletes its resource base for short-term profits gained through deforestation increases its GDP and appears to be more "developed" than a country that protects its forests for a long-term harvest of sustainable yield. Deforestation appears to be beneficial to a country because it raises GDP through the production of pulp, paper, furniture, and charcoal. However, GDP growth does not measure the negative impacts of deforestation, such as erosion, flooding, siltation, and malnutrition. These consequences are **external costs**, or **externalities**, harmful "hidden costs" that are not included in the market prices of goods and services. Carbon dioxide emissions are another good example: they are pumped into the global "commons" with costly but unaccounted for consequences, including those of global climate change. Efforts are sometimes made to monetize external costs; the Obama administration, for example, calculated the social costs of carbon at $36 per ton, and used that figure to regulate coal.[45] Advocates of sustainable development argue that these externalities should be added, or "internalized," before a good or a service is marketed, because the true higher costs of producing these goods and services would be recognized. The lower true costs of less destructive practices would then be evident, providing stronger incentives for individuals, companies, and nations to invest in sustainable practices and technologies.

Sustainable development is a complex assortment of theories and activities, but its proponents call for eight essential changes in the way people perceive and use their environments:

1. People need to change their worldviews and value systems, recognizing that many resources are finite and reducing their expectations to a level more in keeping with the Earth's environmental capabilities. Proponents of sustainable development argue that this change in perspective is needed especially in the MDCs, where instead of trying to "keep up with the Joneses," people should try to enjoy life through more social rather than material pursuits.
2. People should recognize that development and environmental protection are compatible. Rather than viewing environmental conservation as a drain on economies, we should see it as a guarantor of future economic well-being. This is especially important in the LDCs, with their resource-based economies.
3. People all over the world should consider the needs of future generations more than we do now. Much of the wealth we generate is in effect borrowed or stolen from our descendants. Our economic system values current environmental benefits and costs far more than future benefits and costs, and so we try to improve our standard of living today without regard to tomorrow.
4. Communities and countries should strive for self-reliance, particularly through the use of appropriate technologies. For example, remote villages are relying increasingly on off-the-grid solar power for electricity without having to tie into larger networks of coal-burning plants.
5. LDCs need to limit population growth as a means of avoiding the destructive impacts of people overpopulation. Advances in the status of women, improvements in education and social services, and effective family planning technologies can help limit population growth.
6. Governments need to practice land reform, particularly in the LDCs. Poverty is often not the result of too many people on too little total land area, but of a small, wealthy minority, including corporate farming interests holding a disproportionately high share of quality land. To avoid the environmental and economic consequences of marginalization

(as discussed on page 64), a more equitable distribution of land is needed. Increased yields from commercial agricultural landholdings may not be the key to warding off global hunger. Peasant agriculture is arguably more efficient than the industrial model, according to the Canadian research group ETC: the industrial food chain uses 70 percent of agricultural resources to provide 30 percent of the world's food, and the "peasant food web" produces the remaining 70 percent using only 30 percent of the resources.[46]

7. Economic growth in the MDCs should be slowed or adjusted to reduce the effects of consumption overpopulation. If economic growth, understood as the result of consumption of natural resources, continues at its present rate in excess of sustainable yield, the Earth's natural capital will continue to be depleted, with potentially serious consequences.
8. Wealth should be redistributed between the MDCs and LDCs, particularly the least developed countries. The growth of the middle class has not reached all countries. Because poverty is a fundamental cause of environmental degradation, the spread of a reasonable level of prosperity and security to the poorest of the LDCs is essential. Proponents argue that this does not mean that rich countries should give cash outright to poor countries. There are a number of ways of helping the poor. One way to help redistribute global wealth is to reduce or eliminate **trade barriers**—government-imposed restrictions on the free international exchange of goods or services—that MDCs raise against LDCs. The lending institutions of MDCs can forgive some existing debts owed by LDCs or use such innovations as **debt-for-nature swaps**, in which a certain portion of debt is forgiven in return for the borrower's pledge to invest that amount in national parks or other conservation programs. Recently, there have been a number of efforts to reduce the debt burdens of LDCs, particularly those identified as most crippled by debt. Since 2000, the World Bank and the IMF have identified about three dozen countries—all but a few of them in Sub-Saharan Africa—as "heavily indebted poor countries" (HIPCs) and forgiven major portions of their debts. We explore the UN's Sustainable Development Goals in the African context on page 433.

Some geographers and other scientists believe that sustainable development will bring about the **"Third Revolution,"** a shift in human ways of interacting with the Earth so dramatic that it will be compared with the origins of agriculture and industry. Dealing as they do with some of the most critical issues of our time, the formidable changes called for in sustainable development are well worth our attention.

With these first three chapters we have laid the groundwork for exploring the world as geographers. Let's put out knowledge to work in the context of world regions. Europe is first up, but it is not necessary to start there. Let's go!

Study Guide

Summary

- A useful way to begin to appreciate the current spatial patterns of our relationship with the Earth is to view these patterns as products of two "revolutions" in the relatively recent past: the Agricultural, or Neolithic (New Stone Age), Revolution and the Industrial Revolution. Each transformed humanity's relationship with the natural environment. After the Industrial Revolution, the human impact on Earth grew dramatically.
- The world is markedly divided between the haves and the have-nots, characterized at the largest scale by contrasts between more developed and less developed countries (MDCs and LDCs). Explanations for these disparities include dependency theory, cultural factors, geographic location, and natural resource base.
- LDC economies often rely on the export of a few raw materials or commercial crops and tend to draw down their natural "capital" quickly, with profound impacts on the environments.
- LDC economies in debt often step up commercial agriculture and displace or "marginalize" peasant farmers to land that cannot sustain and is degraded by agriculture (for example, rain forest, semi-arid rangeland, steep slopes). Some countries are involved in "land grabbing," or securing land for commercial agriculture in other countries; this also marginalizes local subsistence farmers.
- There is disagreement on whether globalization is beneficial or detrimental to poorer countries and especially to the poorer people who live in those countries. However, globalization is largely credited for dramatically reducing global poverty in recent decades. Globalization's associations include growing middle classes in the LDCs; the transfer of jobs from MDCs to LDCs, especially in manufacturing; and growing income inequality within MDCs, with associated social unrest.
- Geopolitically, recent decades have seen profound changes from a "bipolar" power rivalry between the US and USSR to the "unipolar" reign of the US, to the "post-American" and "multipolar" present. Current trends include stronger assertions of nationalism, the democratization of information through digital technology, the growth of "people power," and the ability of distinct cultures to flourish rather than succumb to leveling forces of "Westernization."
- Population growth is measured by birth rates, death rates, and migration. Growth rates tend to be higher in the LDCs due to relatively high but falling birth rates and much lower and

falling death rates. The demographic transition, a model of what happened to populations in the MDCs, shows the wealthier countries passing from high birth rates and death rates to low birth rates and death rates.

- The explosive growth in world population since the beginning of the Industrial Revolution is the result not of a rise in birth rates, but of a dramatic decline in death rates, particularly in the LDCs. High rates of human population growth intensify environmental problems characteristic of the LDCs. More people use more resources, a phenomenon that suggests there is a problem of people overpopulation in the poorer countries. Consumption overpopulation is more characteristic of the MDCs, where fewer people use large quantities of natural resources from around the world.
- Human demands for more food and other photosynthetic products are growing, and it is questionable whether supplies can keep pace. The Malthusian scenario, which has not been realized so far, insists that a catastrophic die-off of people will occur when their numbers exceed food supplies.
- New concepts and tools for managing the Earth and its resources in an effective, long-term way have emerged in the past two decades. Known collectively as sustainable development or ecodevelopment, these ideas and techniques consider what humanity can do to avert the Malthusian dilemma and improve life on the planet.

Review Questions

1. What were the Agricultural and Industrial Revolutions? In what ways did they initiate important changes in human–Earth relationships?
2. What are the typical differences between MDCs and LDCs?
3. According to dependency theory, what are the causes of disparities between MDCs and LDCs? What other factors may explain global wealth and poverty?
4. What are the five principal recent trends and effects of globalization, both positive and negative? In what ways are the economies of MDCs and LDCs intertwined?
5. Geopolitically, how has power been rearranged in recent years? What are some of the characteristics of the theorized "post-American multipolar world"?
6. How have relations between citizens and their governments changed with increased access to digital information? What forces have given rise to social unrest, or may do so in the future? Have local cultures succumbed to or survived the proliferation of information and possible "McDonaldization"?
7. What is the fuelwood crisis? What are some of the effects of deforestation on human lives and economies in the LDCs?
8. What are the two types of overpopulation, and how do they differ?
9. What has been the main cause of the world's explosive population growth since the beginning of the Industrial Revolution?
10. What variables distinguish the four stages of the demographic transition, and what explains them?
11. Which world regions have the most and the fewest people, and what factors might account for these differences?
12. How do technocentrists and neo-Malthusians view the balance between people and resources?
13. What are the goals and methods of sustainable development?

Key Terms + Concepts

Age of Discovery (p. 54)
Age of Exploration (p. 54)
age structure diagram (p. 73)
Agricultural Revolution (p. 51)
Anthropocene Epoch (p. 52)
asylum (p. 76)
biomass (p. 82)
birth rate (p. 69)
birth rate solution (p. 80)
brain drain (p. 76)
carrying capacity (p. 52)
cash crops (p. 64)
civilization (p. 52)
colonization (p. 54)
commercial crops (p. 64)
consumption overpopulation (p. 81)
cornucopians (p. 80)
culture hearth (p. 52)
death rate (p. 69)
death rate solution (p. 80)
debt-for-nature swap (p. 84)
deforestation (p. 68)
demographic transition (p. 72)
first (preindustrial) stage (p. 72)
second (transitional) stage (p. 72)
third (industrial) stage (p. 72)
fourth (post-industrial) stage (p. 72)
demography (p. 69)
dependency theory (p. 59)
development (p. 60)
development traps (p. 60)
diffusion (p. 52)
digital divide (p. 63)
domestication (p. 51)
doubling time (p. 70)
dry farming (p. 52)
ecodevelopment (p. 83)
ecological bankruptcy (p. 65)
ecological footprint (p. 81)
ecologically dominant species (p. 51)
ecosystem services (p. 58)
emigrant (p. 75)
extensive land use (p. 52)
external costs (p. 83)
externalities (p. 83)
food chain (p. 82)
Food-Producing Revolution (p. 51)
foraging (p. 50)
fragile states (p. 58)
fuelwood crisis (p. 68)
globalization (p. 60)
good governance (p. 58)
gross domestic product (GDP) (p. 55)
gross domestic product purchasing power parity (GDP [PPP]) (p. 56)
gross national income (GNI) (p. 56)
gross national product (GNP) (p. 56)
guest worker (p. 76)
Human Development Index (HDI) (p. 57)
hunting and gathering (p. 50)
illegal alien (p. 76)
immigrant (p. 75)
Industrial Revolution (p. 53)
industrialization (p. 54)
information technology (IT) (p. 63)
intensive land use (p. 52)

internally displaced persons (IDPs) (p. 76)
irrigation (p. 52)
knowledge economy (p. 63)
land grabbing (p. 65)
less developed countries (LDCs) (p. 54)
life expectancy (p. 70)
Malthusian scenario (p. 80)
marginalization (p. 64)
megacities (p. 78)
mercantile colonialism (p. 59)
migration (p. 75)
more developed countries (MDCs) (p. 54)
multinational companies (p. 60)
natural capital (p. 58)
natural replacement rate (p. 65)
natural resource (p. 58)
natural services (p. 58)
neocolonialism (p. 60)
neo-Europes (p. 59)
Neolithic Revolution (p. 51)
neo-Malthusians (p. 80)
newly industrializing countries (NICs) (p. 55)
nontimber forest products (NTFPs) (p. 65)
old-age support ratio (p. 80)
one-child policy (p. 77)
original affluent society (p. 50)
people overpopulation (p. 81)
per capita GDP (p. 56)
Pleistocene overkill hypothesis (p. 51)
population change rate (p. 69)
population explosion (p. 68)
population geography (p. 69)
population implosion (p. 73)
population pyramid (p. 73)
population replacement level (p. 73)
poverty trap (p. 60)
primary commodities (p. 60)
pull factors (p. 76)
purchasing power parity (PPP) (p. 56)
push factors (p. 76)
refugees (p. 76)
renewable resource (p. 65)
rural-to-urban migration (p. 76)
second law of thermodynamics (p. 82)
settler colonization (p. 59)
sustainable development (p. 83)
sustainable yield (p. 65)
technocentrists (p. 80)
"Third Revolution" (p. 84)
three-field system of crop rotation (p. 53)
trade barriers (p. 84)
transnational companies (p. 60)
urbanization (p. 76)
value-added products (p. 60)
virtual water (p. 67)
water footprint (p. 67)
youth bulge (p. 73)
zero population growth (ZPG) (p. 73)

Notes

1. E. O. Wilson quoted in http://www.worldpopulationbalance.org/quotes.
2. Marshall B. Sahlins, *Stone Age Economics* (Chicago: Aldine-Atherton, 1972).
3. Genetic Atlas of Human Admixture: http://www.policymic.com/articles/82595/interactive-map-presents-a-new-way-to-look-at-human-history-using-cutting-edge-science; National Geographic's Genographic project: https://genographic.nationalgeographic.com/human-journey/.
4. Many of these points come from this special edition: "The Geology of the Planet: Welcome to the Anthropocene," *The Economist*, May 26, 2011.
5. See the discussion of fragile states at the Fund for Peace website, http://ffp.statesindex.org/faq.
6. G. Tyler Miller and Scott Spoolman, *Living in the Environment*, 18th ed. (Belmont: Cengage, 2015), p. 6.
7. Tyler Cowen, "Why Emerging Markets Should Look Within," *New York Times,* February 9, 2014, page 6.
8. Alfred W. Crosby, *Ecological Imperialism: The Biological Expansion of Europe, 900–1900* (New York: Cambridge University Press, 1986).
9. For Paul Colliers' poverty traps, see http://makewealthhistory.org/2008/12/08/why-some-countries-remain-poor-paul-colliers-four-poverty-traps/.
10. These 13 are Equatorial Guinea, Greece, Hong Kong, Singapore, Ireland, Israel, Japan, Mauritius, Portugal, Spain, Puerto Rico, South Korea, and Taiwan.
11. Thomas Friedman, *The World Is Flat: A Brief History of the Twenty-First Century* (New York: Picador, 2005).
12. See, for example, http://web.worldbank.org/WBSITE/EXTERNAL/TOPICS/EXTPOVERTY/EXTPA/0,,contentMDK:20239110~menuPK:462078~pagePK:148956~piPK:216618~theSitePK:430367~isCURL:Y,00.html.
13. John Gapper, "Capitalism: In Search of Balance," *Financial Times*, December 23, 2013.
14. *Credit Suisse Global Wealth Databook*, 2013, p. 146.
15. Robert Kaplan, "The Return of Toxic Nationalism," *The Wall Street Journal*, December 23, 2012.
16. Scott Carlson, "Mapping a New Economy: The Geographer David Harvey Says Fixing Inequality Will Take More Than Tinkering," *Chronicle of Higher Education Review*, May 12 2015, http://chronicle.com/article/Mapping-a-New-Economy/146433/.
17. Philip Stephens, "How the Best of Times Is Making Way for the Worst," *Financial Times*, March 27, 2014, http://www.ft.com/intl/cms/s/0/19bb96ec-b418-11e3-a102-00144feabdc0.html#axzz3ZRogLxDs.
18. Gapper, "Capitalism: In Search of Balance."
19. Nora Dunne, "How the American Dream Went Global: Interview with Fareed Zakaria," *Christian Science Monitor*, May 23, 2011.
20. On power shifting to people, see Stephens, "How the Best of Times Is Making Way for the Worst."
21. For the World Values Survey, see http://www.iffs.se/wp-content/uploads/2012/12/WVS-brochure-web.pdf.
22. From http://www.worldvaluessurvey.org/WVSContents.jsp.
23. See https://ipcc-wg2.gov/AR5/images/uploads/IPCC_WG2AR5_SPM_Approved.pdf.
24. Fred Pearce, *The Land Grabbers: The New Fight over Who Owns the Earth* (Boston: Beacon Press, 2013).
25. Maria Cristina Rulli, Antonio Saviori, and Paolo D'Odorico, "Global Land and Water Grabbing," *Proceedings of the National Academies of Science of the USA*, 110(3): 892–897.
26. Ibid.
27. Ibid.
28. For the water footprint, see http://www.waterfootprint.org/?page5files/Glossary.
29. For the Global Footprint Network, see http://www.footprintnetwork.org/en/index.php/GFN/page/earth_overshoot_day/.
30. Sources on water shortages include http://www.waterencyclopedia.com/St-Ts/Survival-Needs.html-ixzz3DmJphc4k and http://www.waterencyclopedia.com/St-Ts/Survival-Needs.html-ixzz3DmJ9YxZE. Reference for 2025 statistic: http://www.env-edu.gr/Documents/World%20Water%20in%202025.pdf.
31. Celia W. Dugger, "Very Young Populations Contribute to Strife, Study Concludes," *New York Times,* April 4, 2007, p. A6.
32. The findings reported in the journal *Science* may be read at http://qz.com/192440/where-everyone-in-the-world-is-migrating-in-one-gorgeous-chart/. The graph shows that the largest migration is from Mexico to the US. Most Indian migrants go to the United Arab Emirates and the US. The US is the largest of all migrant destinations.

33. These comparative figures and projections on urbanization are based on United Nations, Concise Report on the World Population Situation in 2014, http://www.un.org/en/development/desa/population/publications/pdf/trends/Concise%20Report%20on%20the%20World%20Population%20Situation%202014/en.pdf.

34. Edward Glaeser, *Triumph of the City: How Our Greatest Invention Makes Us Richer, Smarter, Greener, Healthier, and Happier* (New York: Penguin, 2012).

35. An excellent detailed look at the iPhone supply chain is at http://comparecamp.com/how-where-iphone-is-made-comparison-of-apples-manufacturing-process/.

36. These 2014 UN projections are from United Nations, Concise Report on the World Population Situation in 2014, http://www.un.org/en/development/desa/population/publications/pdf/trends/Concise%20Report%20on%20the%20World%20Population%20Situation%202014/en.pdf.

37. Cited in Steven A. Holmes, "Global Crisis in Population Still Serious, Group Warns," *New York Times,* December 31, 1997, p. A7.

38. For population aging, see United Nations, Concise Report on the World Population Situation in 2014, http://www.un.org/en/development/desa/population/publications/pdf/trends/Concise%20Report%20on%20the%20World%20Population%20Situation%202014/en.pdf.

39. Pilita Clark, "UN Study Says Climate Change Already Hitting Food Supply, " *Financial Times*, March 31, 2014, p. 4.

40. Paul R. Ehrlich, "Populations of People and Other Living Things," in *Earth '88: Changing Geographic Perspectives*, ed. Harm De Blij (Washington, DC: National Geographic Society), p. 309. Paul Ehrlich and Julian Simon made a famous Malthusian wager back in 1980. Always a champion of human abilities to improve resource security, Simon said that Ehrlich could choose any five commodity metals and that these would be cheaper, not more expensive, 10 years later. Guess who won? Julian Simon won. But neo-Malthusians hasten to add that in most 10-year periods of the last century, Simon would have lost.

41. Lester R. Brown, "Who Will Feed China?" *WorldWatch Magazine*, September–October 1994, p. 10.

42. Mark Bittman, "How to Feed the World," *New York Times*, October 30, 2013, p. wk9.

43. Garrett Hardin, "The Tragedy of the Commons," *Science*, 162, (1968), pp. 1243–1248; Garrett Hardin, "Lifeboat Ethics: The Case against Helping the Poor," *Psychology Today*, September 1974, p. 6; and Garrett Hardin, "Living on a Lifeboat," *BioScience*, 24, 1974, p. 568. I learned recently that Hardin had ties to racist organizations, and so I removed the *Lifeboat* discussion from the main body of this book. However, if used critically and carefully the *Lifeboat* model raises valid and important issues worth discussing. Here is the setup for that discussion.

    Confronted with the Malthusian scenario, one option is the death rate solution, to "let nature take its course" and allow people imperiled by famine or other catastrophe to perish. The neo-Malthusian ecologist Garrett Hardin introduced "lifeboat ethics," the question of whether or not the wealthy should rescue the "drowning" poor—in a distressing and challenging essay about the Malthusian scenario. "People turn to me," wrote Hardin, "and say, 'My children are starving. It's up to you to keep them alive.' And I say, 'The hell it is. I didn't have those children.'" He described the world not as a single "spaceship *Earth*" or "global village" with a single carrying capacity, as many environmentalists do, but as a number of distinct "lifeboats," each occupied by the citizens of single countries and each having its own carrying capacity. Each rich nation is a lifeboat comfortably seating a few people. The world's poor are in lifeboats so overcrowded that many fall overboard. They swim to the rich lifeboats and beg to be brought aboard. What should the passengers of the rich lifeboat do? The choices pose an ethical dilemma for you to consider.

    Hardin set out the following scenario. There are 50 rich passengers in a boat with a capacity of 60. Around them are 100 poor swimmers who want to come aboard. The rich boaters have three choices. First, they could take in all the swimmers, capsizing the boat with "complete justice, complete catastrophe." Second, as they enjoy an unused excess capacity of 10, they could admit just 10 from the water. But which 10? And what about the margin of comfort that excess capacity allows them? Finally, the rich could prevent any of the doomed from coming aboard, ensuring their own safety, comfort, and survival—and ensuring the others would die.

    As a critical thinking exercise, you may think about these options, with some concrete examples in mind. As an occupant of the rich lifeboat *United States*, for example, what would you do for the drowning refugees from lifeboat *Haiti* or lifeboat *Somalia*? Help bail out everyone in need? Assist just a few? Ignore them altogether? Why did you make this choice?

    This can be a difficult ethical question. I do this exercise in my world regional geography and environmental geography classes, with everyone making their choices anonymously on a piece of paper. I tally up the results of "save" and "drown," and we discuss as much as the students want to.

    Hardin's choice was the third: drowning. To preserve their own standard of living and ensure the planet's safety, the wealthy countries must cease to extend food and other aid to the poor and must close their doors to immigrants from poor countries. "Every life saved this year in a poor country diminishes the quality of life for subsequent generations," Hardin concluded. "For the foreseeable future, survival demands that we govern our actions by the ethics of a lifeboat." Incidentally, Hardin jumped off his lifeboat, committing suicide in 2003.

44. World Wildlife Fund, *Sustainable Use of Natural Resources: Concepts, Issues and Criteria* (Gland, Switzerland: World Wildlife Fund, 1995), p. 5.

45. Robin Harding, "A High Price for Ignoring the Risks of Catastrophe," *Financial Times*, February 19, 2014, p. 9.

46. Mark Bittman, "How to Feed the World," *New York Times*, October 30, 2013, p. wk9.

Global Geoscience Watch

Go to the Forests and Deforestation Portal and next to the Statistics heading click "View All." On this page, click on "Share of Tropical Deforestation, 2000–2005." Choose one of the countries shown in the graphic and research the deforestation in this country further. (Tip: use the World Map feature.) Write a brief report on your findings, and include any information you find about possible solutions to the problem in the country you researched.

Online Resources

For access to MindTap and additional study materials visit www.cengagebrain.com. Read your textbook, take notes, complete activities, take practice quizzes and more.

Above: An elder of Bergen, Norway. Europe's population is aging, leaving doors open to youthful immigrants and the controversy over whether immigration is the solution to Europe's demographic problems. Joe Hobbs
Left: One of the world regions most transformed by human activities, prosperous Europe is aglow at night. C. Mayhew & R. Simmon (NASA/GSFC), NOAA/NGDC, DMSP Digital Archive

4 Europe

Europe is so well gardened that it resembles a work of art, a scientific theory, a neat metaphysical system. Man has re-created Europe in his own image.

—ALDOUS HUXLEY (1894–1963)[1]

Many culture hearths have influenced peoples and landscapes around the world, but none more impressively than Europe. For more than 500 years, European nation building, scientific and technological achievements, and colonization reshaped cultures and natural systems around the world. Europe is not as strong as it once was, but the region still has enormous economic and political influence on the rest of the world. These influences come especially from the 28-country-strong **European Union (EU)**, which participates in the global system as a single unit. This chapter introduces the human and physical geographies of this influential world region, and explores some of its characteristic and unique problems.

Chapter Outline

Chapter Objectives

This chapter will enable you to

- See how Europe has become one of the most heavily modified landscapes on Earth.
- Recognize Europe as a post-industrial region with a wealthy, declining population
- Trace Europe's emergence from wartime divisions to supranational unity within the European Union (EU) and know why the EU is important.
- Consider the "social compact" that most European governments have with their heavily taxed citizens.
- Appreciate the trend toward greater unity under the EU and the North Atlantic Treaty Organization (NATO) and the concurrent trend of devolution of power from central governments to provinces.
- Become acquainted with the origins and issues of some of Europe's persistent ethnic and political struggles.
- Understand the "malaise" affecting European institutions and peoples, and the nationalistic anti-immigration, anti-euro, and anti-EU countercurrents to European integration.

4.1 Area and Population

Europe is often classified as one of the world's seven continents. But looking at the map of Europe in •**Figure 4.1**, you will see that Europe is not a distinct landmass like Australia or Africa. It is more accurately a subcontinent of the world's greatest landmass, **Eurasia** (a *portmanteau* combining the names Europe and Asia). The popular and scholarly designation of Europe as a region distinct from the rest of Eurasia is a time-honored and useful organizing device however, and is used in this book.

Europe in this book is a political region made up of the countries of Eurasia lying west of Turkey, Russia, and three of the former republics of the Soviet Union: Belarus, Ukraine, and Moldova (the traditional *physical* dividing line between Europe and Asia is drawn from the Ural Mountains down to the Caucasus, technically placing a small part of Turkey, a large part of Russia, and the three former republics within Europe). Defined by these boundaries, Europe is a great peninsula, fringed by smaller peninsulas and islands and bordered on its seaward sides by the Arctic and Atlantic Oceans, the Mediterranean Sea, and the Black Sea (•**Figures 4.1** and **4.2**). Think of it as a "Peninsula of Peninsulas."

Europe's Subregions

In Chapter 1, you learned the concept of core and periphery. This text recognizes Europe as having four subregions: a core and three peripheral areas (•**Table 4.1**). 13

The **European core** spreads across Northwestern and North Central Europe. This subregion includes some of the world's most geopolitically and economically significant countries (see Table 4.1, **page 92**). They include the United Kingdom, Ireland, France, the **"Benelux"** countries (a *portmanteau* for Belgium, the Netherlands, and Luxembourg), Switzerland, Austria, and Germany. These countries have the largest populations and the most important economic and political roles in Europe. Within the European core are also three tiny political entities, known as **microstates**: Andorra, Monaco, and Liechtenstein. Following are Europe's peripheral areas:

- Northern Europe is made up of Denmark, Iceland, Norway, Sweden, and Finland.
- Southern Europe includes Portugal, Spain, Italy, Greece, Malta, and Cyprus.
- Eastern Europe includes Estonia, Latvia, Lithuania, Poland, the Czech Republic, Slovakia, Hungary, Romania, Bulgaria, Albania, Serbia, Kosovo, Montenegro, Bosnia and Herzegovina, Croatia, Macedonia, and Slovenia. In many sources these are known as the countries of Central and Eastern Europe, with the acronym CEE.

Small but Powerful Europe

Europe's influence on the world has long been disproportionately large, given its size. Europe is only about half as large as the United States' "lower 48" (the contiguous United States, a comparison easily seen in Figure 4.5). The average European country is only 50,000 square miles (130,000 sq km) in area, or about the size of Arkansas.

Within Europe, some countries far outsize the others in population and influence (illustrated most clearly by the pie chart in •**Figure 4.3b**). Four countries—Germany, the United Kingdom, France, and Italy—are very populous. Their populations are about 81 million in Germany, 64 million in both France and the United Kingdom, and 61 million in Italy. These four countries together make up over half of Europe's population (and for comparison are equal to about 85 percent of the population of the United States). Europe's population of 542 million is about 1.7 times that of the United States.

• **Figure 4.1** The physical geography of Europe.

Europe has one of the world's great clusters of human population (**Figure 4.3a**); see also **Figure 3.21** and **Table 4.1**. You can clearly see this in the images of Earth from space at night, in the chapter-opening photo on **page 88**. One of every 13 people in the world is a European, living in a space half the size of the United States. Within this great cluster of people, there are some remarkable variations; some European countries like Norway are lightly populated, and others like the Netherlands have large, dense concentrations of people.

Urban Populations and Industries in Europe

Most of Europe's largest and most densely populated modern cities are located where they are for a reason: their geographical situation favored them (see Geographic Spotlight on **page 94** and the major cities in **Table 4.2**). Cities like Manchester in England, Essen in Germany, Prague in the Czech Republic, and Lille in France are situated near historical sources of coal and hydroelectric power. These energy resources were critical in the

• **Figure 4.2** The political geography of Europe.

location and rapid development of many European cities during and in the wake of the Industrial Revolution (see Figure 4.31 on **page 124**, and the Industrial Revolution on **page 53**). As you can see in Figures 4.3a and 4.31, Europe has a discontinuous, roughly triangular region of high population density and industrialization, from England in the west to Poland in the east, to northern Italy in the south.

This urban-industrial region covers only a small portion of Europe but produces more economic goods and services than the rest of Europe combined. There are only three other areas on Earth that resemble Europe's urban-industrial belts; these are in eastern North America, Japan, and China.

Why Is Europe's Population Declining?

There are several outstanding characteristics of Europe's population: it is urban, aging, and—except where migration is factored in—shrinking.

Europeans are one of the most urbanized peoples on Earth, with nearly one-third of them living in cities of more than

Table 4.1 Europe: Basic Data

Political Unit	Area (sq mi, thousands)	Area (sq km, thousands)	Population (millions)	Rate of Natural Increase (%)	Urban Population (%)	Population Under Age 15 (%)	Agricultural Workers (%)	Per Capita GDP (PPP) ($US)	GDP ($US, billions)	Trade Balance ($US, billions)	Oil Production (million bbl/day)	Literacy, Female (%)	Literacy, Male (%)	HDI
Europe	**1959.3**	**5072.0**	**532.2**	**0.0**	**73**	**16**	**3**	**32,400**	**19,145**	**—**	**3**	**98**	**98**	**0.878**
European Core														
Andorra	0.2	0.5	0.1	0.5	90	15	14	37,200	3	−1	—	99	99	0.830
Austria	32.4	83.8	8.5	0	67	14	1	42,600	387	−2	*	99	99	0.881
Belgium	11.8	30.5	11.2	0.2	99	17	1	37,800	467	−17	—	99	99	0.881
France	212.9	551.2	64.1	0.3	78	18	2	35,700	2,587	−55	*	99	99	0.884
Germany	137.8	356.7	80.9	−0.2	73	13	1	39,500	3,621	228	*	99	99	0.911
Ireland	27.1	70.1	4.6	0.9	60	22	2	41,300	224	54	—	99	99	0.899
Liechtenstein	*	*	*	0.4	15	16	8	89,400	3	2	—	99	99	0.889
Luxembourg	0.1	2.6	0.6	0.4	83	17	0	77,900	50	−9	—	99	99	0.881
Monaco	*	*	*	−0.1	100	13	0	85,500	6	0	—	99	99	—
Netherlands	15.8	40.9	16.9	0.2	67	17	3	43,300	798	64	*	99	99	0.915
Switzerland	15.9	41.1	8.2	0.2	74	15	1	54,800	444	55	—	99	99	0.917
United Kingdom	94.5	244.6	64.5	0.3	80	18	1	37,300	2,435	−183	0.8	99	99	0.892
Northern Europe														
Denmark	16.6	43	5.6	0.1	87	17	1	37,800	248	7	0.2	99	99	0.900
Faroe Islands (Den.)	*	*	*	—	42	20	10	30,500	1	0	—	99	99	—
Finland	130.6	338.1	5.5	0.1	85	16	3	35,900	221	5	—	99	99	0.879
Iceland	39.8	103	0.3	0.7	95	21	6	41,300	14	0	—	99	99	0.895
Norway	125.1	323.8	5.1	0.4	80	18	2	55,400	339	59	1.8	99	99	0.944
Sweden	173.7	449.7	9.7	0.2	84	17	2	40,900	434	20	—	99	99	0.898
Southern Europe														
Cyprus	3.6	9.3	1.2	0.6	67	17	3	24,500	25	−4	—	98	99	0.845
Greece	51	132	11	−0.1	73	15	3	23,600	284	−25	—	99	99	0.853
Italy	116.3	301.1	61.3	−0.1	68	14	2	29,600	2,066	52	—	99	99	0.872
Malta	0.1	0.2	0.4	0.2	100	15	2	29,200	13	−1	—	91	93	0.829
Portugal	35.5	91.9	10.4	−0.2	61	15	3	22,900	276	−10	—	94	97	0.822
San Marino	*	*	*	0.3	94	15	0	—	2	1	—	95	97	—
Spain	195.4	505.9	46.5	0.1	77	15	3	30,100	1,534	−21	—	99	99	0.869
Vatican City	*	*	*	—	100	—	0	—	*	—	—	—	—	—
Eastern Europe														
Albania	11.1	28.7	3	0.5	54	19	18	10,700	30	−3	*	95	98	0.716
Bosnia and Herzegovina	19.7	51	3.8	−0.1	46	16	8	8,300	38	−6	—	99	99	0.731
Bulgaria	42.8	110.8	7.2	−0.5	73	14	5	14,400	123	−63	—	98	98	0.777
Croatia	21.8	56.4	4.2	−0.3	56	15	4	17,800	87	−8	*	99	99	0.812
Czech Rep.	30.4	78.7	10.5	0	74	15	3	26,300	299	12	—	99	99	0.861
Estonia	17.4	45	1.3	−0.1	68	16	3	22,400	35	−1	—	99	99	0.840
Hungary	35.9	92.9	9.9	−0.4	69	14	3	19,800	239	3	*	99	99	0.818
Kosovo	4.2	10.8	1.8	1.1	38	14	13	7,600	16	−3	—	87	96	—
Latvia	24.9	64.4	2	−0.4	68	14	5	19,100	48	−3	—	99	99	0.810
Lithuania	25.2	65.2	2.9	−0.4	67	15	4	22,600	79	−4	—	99	99	0.834
Macedonia	9.9	25.6	2.1	0.2	65	17	10	10,800	27	−3	—	96	99	0.732
Montenegro	5.4	14	0.6	0.3	64	19	1	11,900	9	−2	—	97	99	0.789
Poland	124.8	323.1	38.5	−0.1	61	15	4	21,100	941	2	*	99	99	0.834
Romania	92	238.1	20	−0.3	54	16	12	14,400	386	−9	*	97	98	0.785
Serbia	29.9	77.4	7.1	−0.5	59	14	8	11,100	90	−6	*	97	99	0.745
Slovakia	18.9	48.9	5.4	0.1	54	15	3	24,700	150	4	—	99	99	0.830
Slovenia	7.8	20.2	2.1	0.1	50	15	2	27,400	60	1	—	99	99	0.874

* Less than 0.1. — Data not available or not applicable.

Sources: World Population Data Sheet, Population Reference Bureau, 2014; Human Development Report, United Nations, 2014; World Factbook, CIA, 2014.

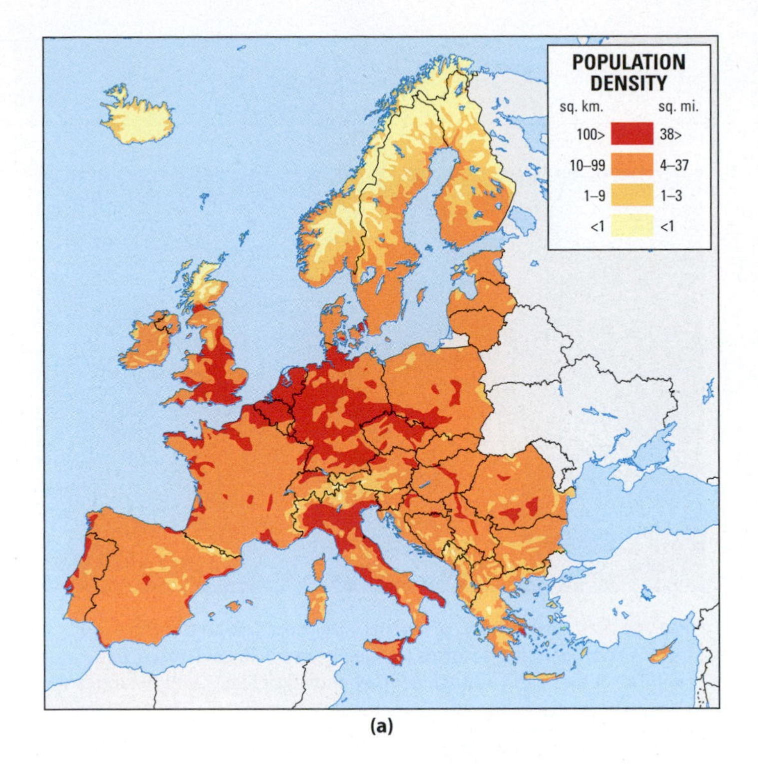

(a)

Table 4.2 Largest Cities in Europe

City	Population
Paris, France	10.8
London, United Kingdom	10.2
Düsseldorf-Cologne-Ruhr, Germany*	8.8
Madrid, Spain	6.2
Milan, Italy	5.2
Barcelona, Spain	4.7
Berlin, Germany	4
Rome, Italy	3.9
Naples, Italy	3.7
Athens, Greece	3.5
Lisbon, Portugal	2.6
Rotterdam, Netherlands	2.6
Manchester, United Kingdom	2.6
Birmingham, United Kingdom	2.5

* The *Ruhr* is the name for the urban area made up of the cities of Dortmund, Essen, Duisburg, and surrounding areas.
Population in millions.
Source: Demographia World Urban Areas, 2015.

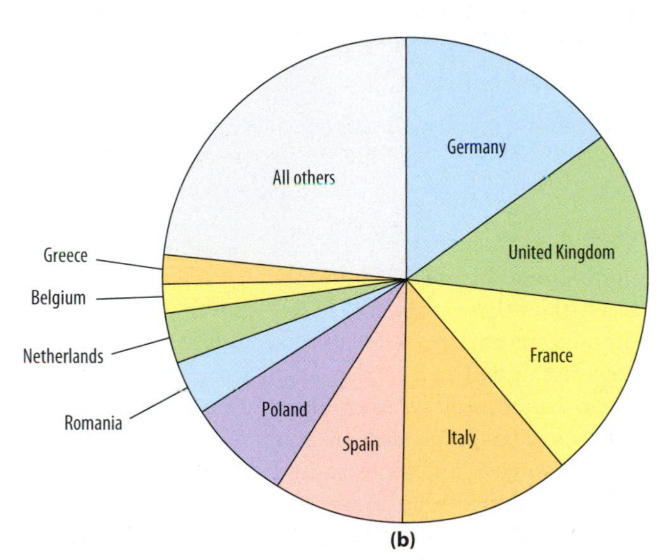

(b)

• **Figure 4.3** **(a)** Population distribution and **(b)** population pie chart of Europe. The continent's demographic heavyweights are Germany, the United Kingdom, France, and Italy.

1 million people. This large urban population is also declining, as the region has entered the final stage of the demographic transition. It is useful to think of Europe as the model of the demographic transition enacted in real life. This region has followed a trajectory from preindustrial high birth rates and high death rates to post-industrial low birth rates and low death rates (see Figure 3.17, **page 72**). Over a period of many decades, Europe recovered from the demographic setbacks of the two world wars. Poland lost an astonishing 19 percent of its people in World War II, and France more than 3 percent in World War I, for example. Europe's population peaked in 1997 and then began a slow, steady decline to its current situation of **"birth dearth"**. Today, birth rates are low in this region of relative affluence and high urbanization, where increasing numbers of employed and educated women are saying, "There is no time for motherhood, and children are too expensive anyway."[2]

Europe's fertility rate is in fact below population replacement level, with no European country maintaining its population through births. 73 Have a look at Table 4.1: you can see that most countries are hovering right around or below zero population growth. Birth rates in Eastern Europe are particularly low. This is due mainly to the economic challenges that came with the fall of communism in that subregion. Many of the communist governments provided free housing, education, and child care to their citizens. But as they transitioned to more democratic and free market systems, these governments cut the state subsidies that provided a "safety net" for their citizens. Couples decided rationally to delay childbirth or not have children at all. Perceiving better working opportunities elsewhere, people of the east also began moving westward in search of education and work. Western Europe is a great magnet for immigration from many more distant places as well. More than 20 million people of non-European origin now live in Europe. The major groups are Turks, Kurds, Arabs, Indians, Pakistanis, Sri Lankans, and a variety of peoples from Sub-Saharan Africa, the Caribbean, and Latin America.

During periods of solid economic growth (especially 1997–2007), immigrants often provided much-needed labor and were generally welcomed in Europe. "Guest workers" were encouraged to come to industrial and agricultural centers that were not able to meet their labor needs with domestic workers. West 75

Geographic Spotlight Site and Situation

Site and situation are related but distinct dimensions of place and space, and are crucial to understanding the geography of Europe and the world's other regions. **Site** refers to the physical properties of a piece of land on which something is or will be located: Rome is known as "the city of seven hills." For a contrast within Italy, consider the site of Venice on flat ground, penetrated, and often inundated by the Adriatic Sea. Known as *La Serenissima*, "the Most Serene [city]," Venice is in fact imperiled by its site; see **•Figure 4.A**. **Situation** is the larger geographical context of the site; the saying "all roads lead to Rome" indicates that "the Eternal City" was situated as a transportation hub. Our Venice example focuses on the city's situation as an Adriatic seaport. In studying Europe, you will see many examples of cities that grew up at the crossroads of transportation routes or where coal, hydropower, and other resources could be obtained nearby; these are examples of how situation plays an important role in urban and economic geography. Sometimes situation and site correspond in the most advantageous ways for human settlement. In other cases, however, the perfect situation is an imperfect site: Venice is a good example (arguably the best is New Orleans, situated near the mouth of the Mississippi, a vital transportation artery, but on a site precariously at and below sea level, inviting flooding by storm surges or by sea level rise associated with the warming Earth; see **page 41**). Venetians are fighting back with the "Moses Project," a system of 78 mobile floodgates to block all three inlets to the city's lagoon and thwart the kind of devastating flooding that Venice suffered in 1966.

Will Hobbs

• Figure 4.A There are no problems with the situation of Venice, but its site is another matter. Storms and sea level rise imperil this urban jewel. This is the island and 9th century church of San Giorgio Maggiore, in Venice's lagoon.

THINK CRITICALLY: What are your city's site and situation? Which features are advantageous and which are not? Were these characteristics different when the site was first selected?

Indians from British Caribbean territories came to Britain to help with labor shortages in the 1950s, for example. Germany invited Turks as guest workers during its economic boom years from 1961 to 1997, and 2 million Turks responded to that opportunity. Spain welcomed Latin Americans, Moroccans, Romanians, and others when it was enjoying its property and construction boom in the 1990s (with the benefit of hindsight, we can now see that was a classic real estate bubble, similar to the one that inflated in the United States during those same years).

European relationships with immigrant and minority communities have long been strained in times of economic hardship, however, and recent years have been no exception. Shockwaves of the great economic recession that began in 2007, growing domestic unemployment, tax burdens, and terrorist threats and attacks have focused native attitudes against immigrants. Austerity and unemployment have energized political parties known kindly as "populist" and less so as "fascist." European historian Andrea Mammone reminds us that ever since World War I, far-right European nationalists have singled out for persecution those who are different—Jews, Gypsies, homosexuals, and the disabled in particular. Germany's fascist Nazi party was the worst, but even today France's National Front, the Italian Social Movement, Denmark's People's Party, the Netherland's Party for Freedom, and Greece's Golden Dawn, to name a few, are fiercely anti-immigrant. In the 2014 European Parliament elections, about a fifth of these seats in Brussels went to members of such anti-immigrant parties. As we will see, many of them are not only against outsiders but also against the institution of the European Union. Today, Mammone argues, a philosophy of 110 **Euro-Fascism** has resprouted from old roots. This time targets include Jews again but focus most intently on immigrants—especially Muslims—from Africa, the Middle East, and Asia[3] (see Regional Perspective on **page 96**, and the introduction to Islam on **page 229**). Strategic analyst Robert Kaplan describes these feverish sentiments as the "return of toxic nationalism" to Europe, and recognizes similar forces in South and East Asia and in the Middle East.[4]

Europe's shrinking population is aging faster than that of any other world region. The median age in Italy is projected to rise from 44 in 2015 to 53 by 2050. The populations of the Czech Republic and Spain are expected to contract by about 20 percent by 2050, and the Netherlands may shrink by 10 percent. As we will see in many of the more developed countries (MDCs), a birth dearth can have significant economic impacts, creating labor shortages and a greater burden on the young to support the elderly. How is it possible to get more 157, 357 European babies? The Italian government has tried to reverse its population shrinkage by offering parents a **"baby bounty"** of about $1200 for each child after their first—but only if the mother is Italian or carries a European Union passport. France has offered similar cash incentives for third children, and Germany for second children. In Denmark, there is a hilarious and naughty TV advertising campaign called "Do It for Denmark," encouraging Danes to take romantic getaways and make babies. But the quickest surefire way of counteracting Europe's population shortage is to welcome immigrants.

Bring on the Immigrants?

Europe's shrinking, aging population could use a youthful boost from immigration, which is already the main source of population growth in most European countries. If birth rates remain at their current low level, the European Union will have a shortfall of 20 million workers by 2030. The EU countries would need an annual inflow of about 3 million migrants a year to prevent the future support ratio from dropping sharply. But that prospect raises a number of concerns in the individual countries and for the European Union itself.

Governments in the past were reluctant to impose harsh measures that would restrict migration. Borders have been porous, particularly because of the Schengen Agreement on freedom of movement (described on **page 121**). But Europe's open door attitudes and policies are changing, especially because of growing unemployment at home and the fallout from surging numbers of illegal immigrants. European views of immigrants shifted markedly when the global economic crisis struck Europe in 2007. Most Europeans since then have seen immigrants as a financial burden on society and as competing with them for jobs and benefits; immigrants threaten to unravel the social safety net of the European welfare state, and they live outside of mainstream European society instead of becoming integrated within it (see **page 113**). Germany's Prime Minister Angela Merkel went as far as to declare multiculturalism a "total failure" in her country. She said that the so-called "multikulti" concept, in which people would "live side-by-side happily," was not working, and that immigrants needed to do more to integrate—including learning to speak German.[10]

In some countries, notably Denmark and the Netherlands, generous welfare benefits were a magnet for immigration. But Denmark has pushed back, passing tough anti-immigration laws and reducing welfare benefits for new arrivals. In the Netherlands, increasing numbers of Dutch nationals are expressing weariness with their tax burden to support literacy, technical training, and health care costs for immigrants, who now make up 10 percent of the country's population.

An unknown number of migrants into Europe slip across borders undetected and without legal documentation, or overstay tourist or student visas. Like the US, individual EU countries had been preoccupied with what to do about their "illegals," as Europeans call their undocumented immigrants. In 2015, the United Kingdom had nearly a million, and Italy, Germany, and France each had over 400,000. But in 2015, these concerns were sidelined, as Europe found itself dealing with the worst refugee crisis since the end of World War II. War and deteriorating economic conditions in Afghanistan, the Middle East, North Africa, and Sub-Saharan Africa created a tidal wave of immigration into Europe. In just the 8-month period of January to August 2015, more than 350,000 migrants were counted at EU borders, about two-thirds of them in Italy and Greece.

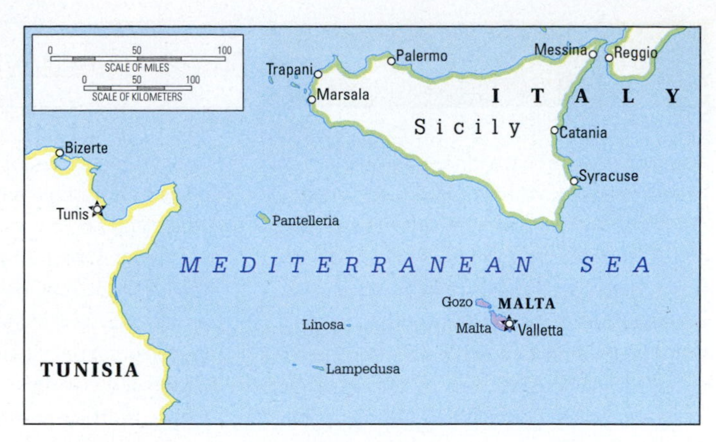

• Figure 4.4 By virtue of its location and political status, Lampedusa is a primary destination for migrants wanting to reach Europe from Africa by sea.

Geographic proximity is the decisive factor behind the migrant arrivals in Italy and Greece. Italy's Mediterranean island of Lampedusa is actually closer to North Africa's shore than it is to mainland Italy (**• Figure 4.4**). Immigrants from many countries on the African continent and Middle East funnel toward Libya, boarding vessels there to make the dangerous passage to Lampedusa. Waves of Syrian refugees in particular struggle to reach European shores as their civil war drags on. Heartbreaking shipwrecks and the dumping of human cargo by traffickers have killed thousands of Middle Easterners and Africans trying to make the passage. European leaders reevaluated their policy of not patrolling the Mediterranean to rescue migrants; they had feared that would only encourage more immigration. Germany, which receives the most asylum applications in the EU, expected to absorb 800,000 refugees in 2015, and appealed to other EU countries to deal with the humanitarian crisis by adopting their own quotas to accommodate migrants.

When they no longer wanted immigrants, European governments, particularly those of Italy and Spain, were able to use generous foreign aid to effectively pay Libya, Tunisia, and Morocco to keep would-be immigrants to Europe bottled up within those countries or to repatriate them to their home countries. As North African regimes toppled in the Arab Spring of 2011, those guarantees disappeared, and tens of thousands of immigrants—mainly Somalis, Eritreans, Senegalese, and Nigerians—began making their way to mainland Europe through Lampedusa. Italian authorities held them for two days in Lampedusa, then classified them in mainland Italy as either refugees seeking political asylum or "economic immigrants" seeking work. And then, much to the chagrin of France, Germany, and other destination countries, Italy issued them national residence permits allowing them to move on elsewhere in Europe. Italy hoped the migrants would exploit the passport-free Schengen area to travel on to countries where many have family members. (Under the EU 121 law known as the "Dublin Regulation," the country where

Regional Perspective Europe's Crisis with Islam

Jews and Muslims have long suffered in Europe. In 1492, the same year of Columbus's New World discovery, Catholic Spain began the systematic expulsion of Muslims and Jews from Spain. This was a domestic counterpart to the **Crusades**, violent efforts to restore Christendom to the Holy Land during Europe's Middle Ages. Given the opportunity to do so, many Europeans today would like to see both of these peoples pushed out or worse. Anti-Semitism is on the upswing again in Europe, where it has long had profound consequences. Here the focus is on anti-Muslim sentiment in Europe and the deadly force that feeds it most: terrorism carried out by a small sliver of Muslims who have hijacked Islam to commit murder.

108

Europeans are worried about the migration of Muslims into their countries and the growing cultural presence of their religion, Islam. Deeply seated cultural divides periodically incite the proverbial "clash of civilizations." In 2005, a Danish newspaper published satirical cartoons of the Prophet Muhammad (among other things, wearing a bomb in his turban). For all Muslims except some Shi'ite groups, it is heresy to depict the Prophet or God in any form. The Danish caricatures thus touched a raw nerve, and a wave of demonstrations against Denmark, some of them violent, erupted across the Muslim world. On one side of the abyss of misunderstanding stood a secular European media, an "equal-opportunity offender" poking fun at a variety of religions and public figures, and on the other side, believers bound by their faith to defend the Prophet's honor.

Many "native" Europeans view Muslims as inflexible and intolerant people who cannot be accommodated within their value systems and societies. Germany's Interior Minister was quoted saying that Islam is not a part of the German way of life.[5] Switzerland has outlawed the construction of Islamic minarets. French and Belgian governments have prohibited Muslim women and girls from wearing scarves (*hijab*) in schools and government offices, and wearing face-covering veils (*niqab* and *burqa*) in public places. Some governments cite both security concerns (it being important to see a person's face in numerous circumstances, such as in banking and retail selling or wherever presenting identification is required) and long-standing bans on any religious symbols in secular society. France's secular political system insists on strict separation of church and state. It would be unthinkable for a French politician to take the oath of office using a Bible, for example. Christian students in France are prohibited from wearing crosses, and Jewish students from wearing skullcaps. The French government has banned any "conspicuous signs of religious affiliations" in public schools. But Muslims feel that the restrictions are in reality a form of discrimination aimed directly at them.

Britain and France have taken different positions on accommodating Muslim and other immigrants. Britain favors the multicultural model of encouraging immigrants to hang onto the cultural practices and values of their homelands. France, in contrast, wants immigrants to fall in line with French customs and values. Neither of these two approaches proves to be especially successful.[6]

Writers and editorialists on the right have depicted the transformation of the cultural landscape from Europe to **Eurabia**. Islam is "patiently conquering Europe's cities, street by street," writes British columnist Christopher Caldwell.[7] He forecasts a Europe swamped by a tidal wave of Muslims, whose birth rates far exceed those of native Europeans and whose radical interpretations of Islam threaten the continent's security. Such sentiments testify to a growing wave of **Islamophobia**—a fear of Islam and the Muslims who practice this faith—across Europe. Is such fear justified?

Polls and statistics reveal a rather benign picture of European Muslims, and the emergence of a Muslim majority in any western European country as highly unlikely. Like most native Europeans, Muslims in Europe are far more preoccupied with their economic well-being than with militant politics, and their birth rates are falling. "If you drew up a composite profile of the average French European Muslim today," wrote journalist Simon Kuper, "it would look something like this: She has two or three children, who attend non-religious schools. She is relatively poor but generally content, although angry about discrimination. She feels more religious than a decade ago, but doesn't wear a headscarf, although she has friends who do. She opposes terrorism, although she probably knows terrorist sympathizers. She votes socialist, and worries about economic issues more than about anything in the Middle East."[8]

migrants first arrive is responsible for deciding their status as immigrants, refugees, or asylum seekers. A protocol signed by the European countries gives everyone the right to seek asylum in Europe and gives refugees the right not to be sent to a place where they may be tortured or persecuted.)

When migrants reached their destination countries—Spain is one case—a few were flown back to their home countries, but most found themselves destitute and living in squalid camps in the countryside. Spain had prided itself on being immigrant friendly, especially during its economic boom/bubble years of 1997–2007. These immigrants contributed to Spain's economic success. But in the depths of the financial crisis after 2007, Spain struggled to halt the surge of immigrants. Generally, the southern and eastern EU economies feel too overwhelmed to accommodate migrants; northern European countries take in far more asylum seekers. Latvia, Estonia, Luxembourg, Switzerland, Slovenia, and Sweden all have immigrants making up over 10 percent of the population. As elsewhere in Europe, anti-immigrant politics have come to the fore, even in Sweden, known previously for its generous accommodation of migrants.

4.2 Physical Geography and Human Adaptations

Europe has diverse and impressive physical features. Among its most distinctive physical characteristics are its irregular shape, its jagged coastal outline, its high latitude, and its temperate climate. The main peninsula of Europe is fringed by a number

But terrorism undoes the rational sense of well-being and evokes fear. In 2004, a Dutch citizen of Moroccan descent murdered Theo van Gogh, a Dutch filmmaker (and distant relative of the artist Vincent van Gogh) who had criticized the status of women in Muslim societies. The London subway bombings of 2005 and the London and Glasgow car bombing attempts of 2007 were attributed to Muslim dissidents of Pakistani and Middle Eastern origin. Saying it was in retribution for British killings of Muslims, two Nigerian converts to Islam killed a British Army soldier in London in 2013. The flow of young European recruits to join ISIS in Syria and Iraq and the prospect of terrorists trained in the Middle East and then returning to Europe with the tide of Syrian refugees aroused deep-seated fear. Then those fears were realized.

It happened twice in Paris in 2015. In November, two gunmen, brothers of Algerian descent, armed with automatic weapons and a rocket-propelled grenade, burst into the 245 office of a satirical magazine called *Charlie Hebdo* during a staff meeting. They opened fire, killing 12. Why? *Charlie Hebdo* had depicted the Prophet Muhammad in a variety of irreverent situations. In apparent retribution, several months earlier the magazine *Inspire*, published by al-Qa'ida in the Arabian Peninsula (AQAP), listed one of the *Charlie Hebdo* journalists on its roster of "most wanted" and called for his death. AQAP later claimed responsibility for the attack on *Charlie Hebdo*. Within days, a massive dragnet led to the men being trapped and killed by French security forces. Simultaneously, a gunman inspired by al-Qa'ida's rival organization, ISIS, killed several Jewish shoppers at a kosher supermarket in Paris. He too was killed by French police.

The death toll of 17 civilians stunned France and indeed people around the world, who saw it above all as an attack on freedom of speech. More than a million people turned out at a rally in Paris, holding pens high in a statement that "the pen is mightier than the sword" (**• Figure 4.B**).

• Figure 4.B *Charlie Hebdo* magazine's answer to the slaying of its staff was "The pencil is mightier than the sword." Soon after the January 2015 terrorist attack about 4 million people, including 40 world leaders, marched in unity rallies across France "for freedom of expression." At this rally in Lyon, as elsewhere, demonstrators held up pens and pencils, and with the slogan "I am Charlie" showed solidarity with those who lost their lives.

In November 2015, ISIS carried out a far bloodier assault on civilian targets in Paris, killing more than 120 and promising more attacks on Western cities.

It is difficult to predict where public sentiment and military thinking in the West, especially in France, will go from here. The French showed remarkable restraint in the wake of the attack on *Charlie Hebdo*; anti-Islamic expression was absent from that huge rally. But after the November attack French President Hollande declared that "France is at war" with ISIS, and his people began to reassess their situation.

France has Europe's largest Muslim population, an estimated 6 million, and the backdrop of anti-immigrant sentiment and nationalist fervor make France and Europe ripe for backlash. *Daily Beast* Foreign Editor Christopher Dickey spoke of the dangers of "intolerance of the intolerant."[9] Meanwhile the same freedom of speech that *Charlie Hebdo* exercised has given militant Islamists a platform to incite violence against Jews and Christians. Asking themselves whether it is safe to live in Europe, Jews are once again on the move, many of them exercising their right of return to Israel (see **page 251**).

of smaller peninsulas, notably the Scandinavian, Jutland, Iberian, Italian, and Balkan peninsulas (see **Figures 4.1** and **4.2**). Offshore the European "peninsula of peninsulas," there are numerous islands, including Great Britain, Ireland, Iceland, Sicily, Sardinia, Corsica, and Crete. The island of Cyprus lies far in the eastern Mediterranean. Around the indented shores of Europe, arms of the sea penetrate the land in the form of estuaries (the tidal mouths of rivers), and harbors offer protection for shipping. This complex mingling of land and water provides many opportunities for maritime activity, and much of Europe's history has focused on seaborne trade, sea fisheries, and sea power.

On the map in **• Figure 4.5**, you can see that much of Europe lies north of the latitudes of the 48 conterminous United States. The British Isles is at the same latitude as Hudson Bay in Canada. Athens, Greece, is only slightly farther south of the latitude of Saint Louis, Missouri. You may expect Europe to be cold, considering those facts. But generally, it is rather warm.

Why Is Europe So Warm?

Europe's mild climates, especially in winter, are particularly surprising given its high latitudes. London, for example, has about the same average temperature in January as Richmond, Virginia, which is 950 miles (1500 km) farther south. The interaction between surface winds and the moderating influence of ocean waters, even in winter, accounts for Europe's warmth. The **Gulf Stream** and its appendage the **North Atlantic Drift** (**• Figure 4.6**), which originate in the western

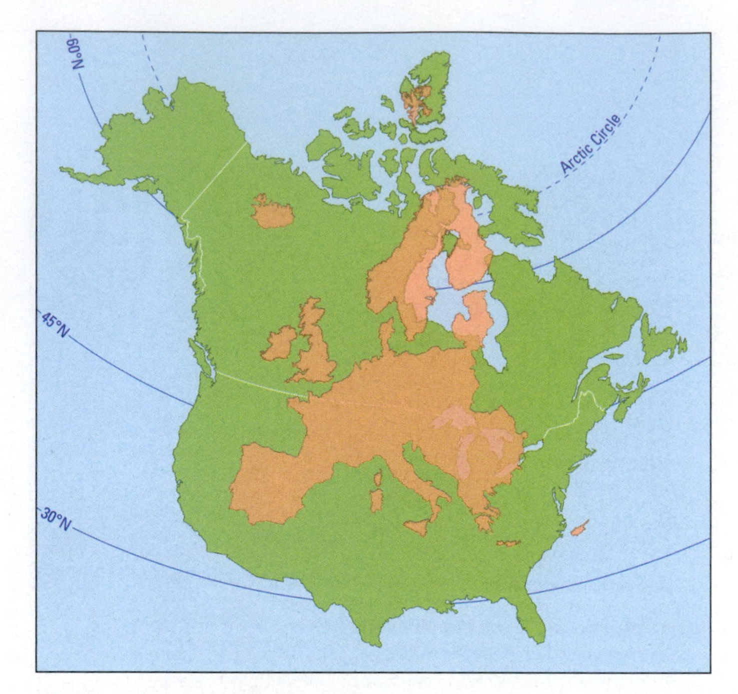

• **Figure 4.5** Europe in terms of latitude and area compared with the United States and Canada.

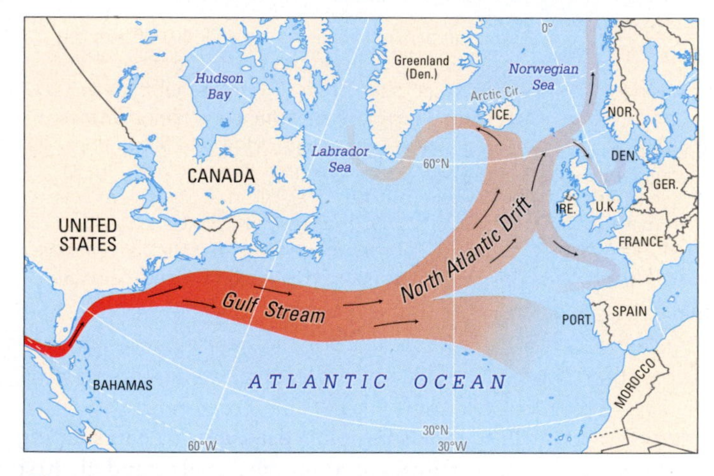

• **Figure 4.6** The westerly winds carry moderating influences from the Gulf Stream and North Atlantic Drift waters ashore in the winter to warm the land, and in the summer to cool the land.

parts of the Atlantic Ocean, flow to the north and east and lap Europe's shores. The capacity of the sea to retain warmth, more than the tropical ocean warmth itself, plays a critical role. With **westerly winds** (*westerlies*) blowing from sea to land along the Atlantic coast of Europe, the moving air in winter absorbs relative warmth from the ocean and transports it to the land. This makes winter temperatures abnormally mild for this latitude, especially along the coast. In the summer, the climatic roles of water and land are reversed: instead of being warmer than the land, the ocean is now cooler, so the air brought to the land by the westerlies in the summer has a cooling effect. These influences are carried as far north as the Scandinavian Peninsula and into the Barents Sea, making Norwegian ports generally ice-free in winter and giving the Russians an essentially warm-water port at Murmansk, even though it lies above the Arctic Circle at a latitude equivalent to Canada's far north. 198

Northern Europeans—the Norwegians, Swedes, Finns, and Icelanders—enjoy their long summer days. If you are visiting this region from the middle or lower latitudes, your body clock is out of sorts for a number of days until you can accustom yourself to local rhythms. At midnight, the sky is still bright, and you have to find or create darkness to sleep. The locals enjoy this "midnight sun" and celebrate the summer solstice with great fervor at "midsummer" festivals. They revel because they know winter, with its long, dark nights, is coming.

The same winds that bring warmth in winter and coolness in summer also bring abundant moisture. Most of this falls as rain, although the higher mountains and more northerly areas have considerable snow. Abundant, well-distributed, and relatively dependable moisture has always been one of Europe's major assets (see the world precipitation map, Figure 2.3 on **page 23**). The average precipitation in most places in the European lowlands is 20 to 40 inches (50 to 100 cm) per year. This is enough for a wide range of crops, partly because mild temperatures and high atmospheric humidity reduce the rate at which plants lose their moisture.

Human Settlement on Europe's Varied Landscapes

With plains, plateaus, hills, mountains, and water bodies, Europe's topographic features are diverse. Enriched by the human associations of an eventful history, Europe has many of our Earth's most beautiful and storied landscapes and affords residents and visitors the chance to enjoy distinctive and often highly scenic landscapes.

One of the most prominent surface features of Europe is an undulating, or rolling, plain that extends without a break from near the French-Spanish border, across western and northern France, central and northern Belgium, the Netherlands, Denmark, northern Germany, and Poland, and far into Russia. Known as the **North European Plain** (see Figure 4.1), it has outliers in Great Britain, the southern part of the Scandinavian Peninsula, and southern Finland.

Geographic characteristics of the North European Plain have had important influences on human settlement. The plain includes the largest part of Europe's farmland, much of it on soils formed from deposits of **loess** (a windblown soil of generally high fertility). It is underlain in some places by deposits of coal, iron ore, potash (used in the production of soap and glass), and other minerals that were important in the region's industrial development. This broad lowland provides an important natural transportation route skirting the highlands to the south. Thanks to these geographic characteristics, many of the largest European cities, including London, Paris, and Berlin, developed on the plain. From northeastern France eastward, there is a band of especially dense population extending 124 along the southern edge of the plain; you can see this "plainly" in Figure 4.3a.

Europe's highlands are very important in the character of European landscapes and peoples. South of the North European Plain, Europe is mainly mountainous or hilly, with scattered plains, valleys, and plateaus. The hills are geologically old, and as in America's Appalachian Mountains, erosive forces have worn down their elevations over long periods of time. But many mountains in Southern Europe are geologically young and often high and ruggedly spectacular, with jagged peaks and snowcapped summits. The highest is Mount Blanc (15,771 feet [4807 m]) on the French-Italian border, but the most iconic is Switzerland's Matterhorn, at 14,690 feet (4478 m) (**•Figure 4.7**). Another prominent mountain chain, the Pyrenees, forms a wall up to 10,695 feet (3404 m) high along the border of Spain and France. With three dozen elephants brought from Africa, the army of the famed Carthaginian general Hannibal crossed both the Pyrenees and the Alps to confront Rome in the 3rd century BCE, beginning the second Punic War (**Figure 4.8**). The Transylvanian Alps and the Carpathian Mountains, mountain chains evoking legends of vampires and mysterious woodlands, meet further to the east in Romania.

Glaciation—in this case a geologic and climatologic process in which great ice sheets formed in the polar regions and advanced toward the Equator—played a major role in shaping Europe's landscapes. Continental ice sheets descended, first sweeping the Scandinavian Peninsula and Scotland (**•Figure 4.9**), during the **Ice Age** of the Pleistocene Epoch (2.5 million to 10,000 years ago). Evidence suggests that there were four major periods of widespread glacial coverage. Today, landscapes of **glacial scouring**—the erosive action of ice masses in motion—characterize most of Norway and Finland, much of Sweden, parts of the British Isles, and Iceland. By creating precipitous drops for water, these landscape changes created many favorable sites for hydroelectric installations (**•Figure 4.10**). **Glacial deposition**—the process in which the glacier's movements offload rock and soil—had a major effect on the present landscape. Glacial deposits were scattered across most of the North European Plain and are productively farmed today. The weight of the glaciers was so great, and their retreat so recent in geologic terms, that parts of the Scandinavian Peninsula are still rising—rather like a sponge from which a rock has been removed. This process, called

UniversalImagesGroup/Getty Images

• Figure 4.8 In 218 BC, Hannibal's Carthaginian army of 100,000 men set out from North Africa with the aim of sacking Rome. During winter this force crossed Europe's greatest mountain barriers with about three dozen elephants. They had multiple intended uses, including to carry men and matériel, and to trample and terrorize the enemy. However, few of the animals survived their odyssey to actually engage Roman forces. One of the enduring mysteries is whether these were African or Asian elephants. African elephants are most likely given the proximity of their range to Carthage; however, their temperaments are not well suited to handling, as Asian elephants are. Only one elephant lived through the entire campaign, Hannibal's favored mount, Surus.

Joe Hobbs

• Figure 4.7 The Matterhorn rises above the Swiss resort of Zermatt, a model of sustainable and low-impact tourism. What a day this was! I flew out of balmy Cairo early in the morning, arriving in Geneva and transferring to a train that operated like a Swiss watch. By late afternoon, I was hiking in falling snow in one of the prettiest places I'd ever seen. I worked for several days at the offices of the International Union for the Conservation of Nature and Natural Resources (IUCN) in Gland and then took the train to Milan. On that journey, the train traveled mostly inside the Alps rather than over them. With their tunnels and bridges, European rails and roads are engineering marvels.

• **Figure 4.9** The maximum extent of glaciation in Europe about 18,000 years ago.
Source: After Hoffman, 1990.

• **Figure 4.10** The Sima plant on Norway's Eidfjord is one of Europe's largest hydropower plants. A reservoir 940 m straight up from this spot holds meltwater from the Hardangerjokulen glacier. Water is released from the reservoir into artificial conduits excavated in the rock. Falling downward with tremendous force, the water spins turbines and generates electricity for distribution through the network of power lines seen here.

isostatic rebound, has the curious effect of raising some of Scandinavia's harbor infrastructure and forcing their relocation. These are some of the few places on Earth where rising sea levels are not a concern!

The North European Plain is bordered by glaciated lowlands and hill lands in Finland and eastern Sweden, and by rugged, ice-scoured mountains and spectacular fjords in western Sweden and most of Norway (•**Figure 4.11**). The British Isles also have extensive areas of glacially scoured hill country and low mountains, along with lowlands where glacial deposition occurred.

• **Figure 4.11** Although shrinking, Norway's glaciers continue to scour its landscapes, leaving lakes and fjords in their wake. This glacier caps the Hallingskarvet Plateau.

Wherever they call home, Europeans tend to have a good relationship with nature, with particularly strong bonds to the seas and the mountains. Europe is relatively small; it has an outstanding transportation infrastructure of roads, rails, and ferries; it has city-dwellers who long for the weekend or seasonal getaway; and it has beautiful beaches and mountains. Even where Europeans don't live close to natural wonders, they can reach them quickly. Europe's natural gems inevitably succumb to their own successes: some beach and mountain resorts are notoriously crowded and, many would say, spoiled; but others have managed to maintain their special charm through careful planning and management. Switzerland's alpine resort of Zermatt is one example (**Figure 4.7**).

Diversity of Climate and Vegetation

Despite its relatively small size, Europe has remarkable climatic and biotic diversity (•**Figure 4.12a, b**); see also **Figures 2.4** and **2.5**. An oceanic climate extends from the coast of Norway to northern Spain and inland to the center of Europe. Comparable to the climate of the US Pacific Northwest, the main characteristics are mild winters, cool summers, and ample rainfall, with many drizzly, cloudy, and foggy days. Throughout the year, changes of weather follow each other in rapid succession as different air masses temporarily dominate or collide with each other along weather fronts. In lowlands, winter snowfall is light, and the ground is seldom covered for more than a few days at a time (in the first decade of this century, however, there were several freakish, massive snow events that blanketed much of the British Isles and the European continent). Summer days are longer, brighter, and more pleasant than the short, cloudy days of winter, but even in summer there are many chilly and overcast days. The frost-free season of 175 to 250 days

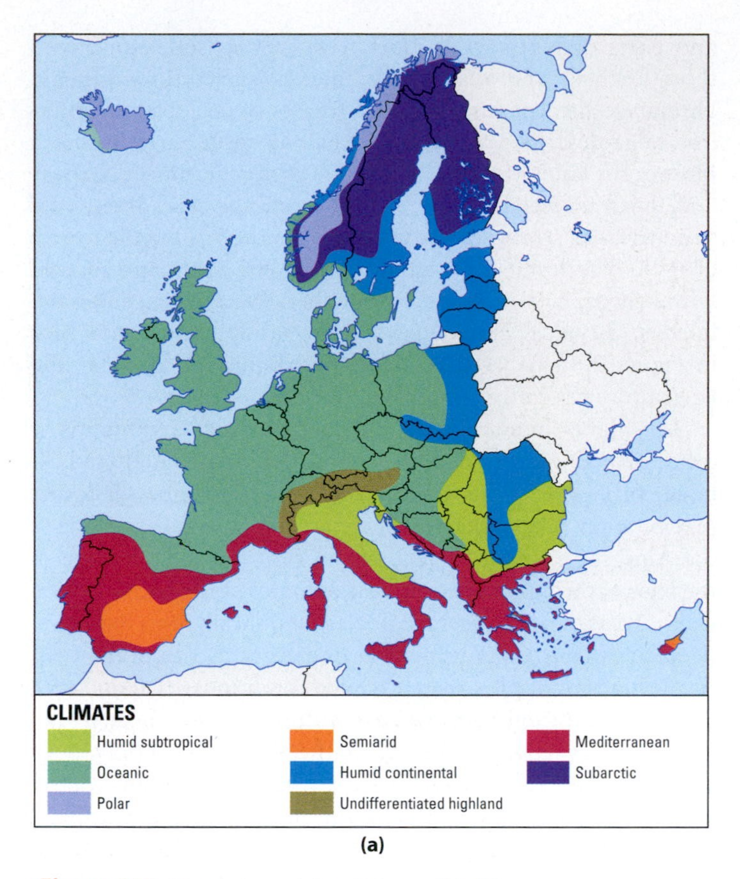

(a)

BIOMES
Temperate broadleaf and mixed forests
Temperate coniferous forests
Boreal forests/taiga
Temperate grasslands, savannas, and shrublands
Mediterranean forests, woodland, and scrub
Tundra

(b)

• **Figure 4.12** Climate types **(a)** and biomes **(b)** of Europe.

is long enough to support typical middle-latitude crops such as wheat, barley, and potatoes, but most areas have summers that are too cool for heat-loving crops like corn (maize). Agricultural and other land uses are depicted in •**Figure 4.13**.

Inland from the coast, in Western and Central Europe, the marine climate gradually changes. Relative to the coast, winters become colder and summers are hotter; cloudiness and annual precipitation decrease. Influences of maritime air masses from the Atlantic are replaced by continental air masses from inner Asia. Conditions are different further inland, with two distinctive climate types: the humid continental climate in the north (ranging from central Sweden through the Baltics and Poland, to the mountainous areas of Romania and Bulgaria), similar to the climate of the Midwestern United States, and the humid subtropical climate in the warmer south (covering much of Italy to Serbia and the coast of the Black Sea), which is similar to the Mid-Atlantic US, though with lesser precipitation amounts. The natural vegetation is mostly temperate mixed forest, with varying soil quality. Among the best soils are those formed from alluvium (soils deposited by rivers and streams) and loess along the roughly 1800-mile (3000-km) valley of the Danube River in Central and Eastern Europe (see **Figure 4.1** and **page 103**).

Southernmost Europe has the distinctive Mediterranean climate, with a pattern of dry summers and wet winters (this is the archetype of Mediterranean climates worldwide, like those found in California; see **Figure 2.4** for their distribution).

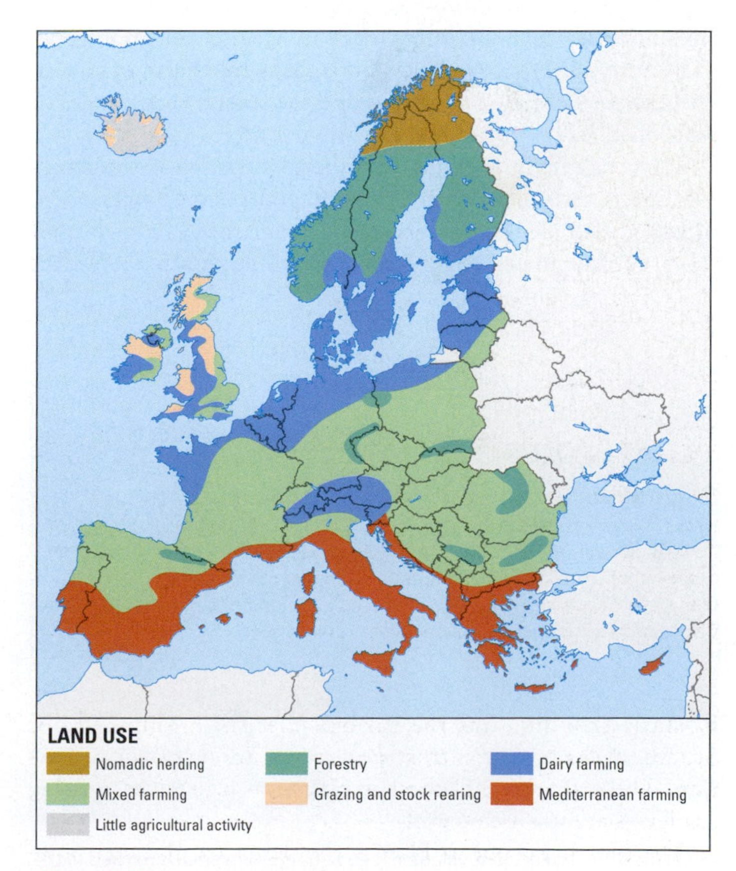

• **Figure 4.13** Land use in Europe.

This pattern results from seasonal shifts in atmospheric belts. In winter, the belt of westerly winds shifts southward, bringing precipitation. In summer, a belt of high atmospheric pressure over the Sahara shifts northward, bringing extremely dry and warm or hot conditions. Winters are mild and frosts are rare.

Drought-resistant trees originally covered Mediterranean lands, but little of that forest remains, having been depleted by centuries of human use—especially the clearing of land for farming, and the pressure of goats and other grazing livestock. George Perkins Marsh, one of the early great American geographers, wrote about these landscape changes in the Mediterranean in his classic 19th-century book *Man and Nature*. He was particularly scornful of Spanish stewardship:

> The laws of almost every European state more or less adequately secure the permanence of the forest; and I believe Spain is the only European land which has not made some public provision for the protection and restoration of the woods—the only country whose people systematically war upon the garden of God.[11]

Through human activities, forests have been extensively replaced by the wild scrub that the French call *maquis* (called *chaparral* in the United States and Spain) or by cultivated fields, orchards, and vineyards. Much of the land consists of rugged, rocky, and eroded slopes where thousands of years of deforestation, overgrazing, and excessive cultivation have taken their toll. The region's subtropical temperatures can support a wide variety of crops, but irrigation is a must in the dry summers. This climate is also a draw for summer travelers, and tourism has immense economic importance in Mediterranean Europe. "The South of France" evokes both the rural charm of Aix en Provence in Cezanne's paintings and the seaside indulgences of the Côte d'Azur (**•Figure 4.14**).

Some northerly sections of Europe experience the harsh conditions associated with subarctic and polar climates. The subarctic climate, characterized by long, severe winters and short, cool summers, covers most of Finland, much of Sweden, and parts of Norway. A short frost-free season, along with thin, leached, and acidic soils, makes agriculture difficult. Human settlement is sparse, and forests of needleleaf conifers (the taiga described on **page 31**), such as spruce and fir, cover most of the land. In the polar climate of northernmost Norway and much of Iceland, cold winters combine with brief, cool summers and strong winds to create conditions hostile to tree growth. The result is an open, windswept landscape covered with lichens, mosses, grass, low bushes, dwarf trees, and wildflowers (a typical Norwegian tundra landscape may be seen in Figure 2.6n on **page 34**). Human inhabitants are few, and agriculture generally impossible, on the tundra.

The higher mountains of Europe, like high mountains in other parts of the world, have an undifferentiated highland climate, varying with elevation and exposure to sun, wind, and precipitation. The variety can be startling. The Italian slope of the Alps, for example, ascends from subtropical conditions at the base of the mountains to polar climates at the highest elevations. Areas of permanent ice, such as mountain tops, have temperatures that average below freezing every month of the year. At the higher latitudes, these extreme conditions are found even at sea level. Europe's glaciers are following the almost worldwide trend of retreat in the past half century due mainly, most climatologists believe, to increasing levels of greenhouse gases in the Earth's atmosphere and the accompanying warming. Probably because of their interior continental locations, the Alpine glaciers are melting much faster than the world average (**•Figure 4.15 a, b**).

330, 457

Rivers and Waterways

Europe's rivers are famous around the world for their historical and scenic associations: the Blue Danube, the left and right banks of the Seine in Paris, and the castles on the Rhine are examples. These waterways are important for transport, water supply, electricity generation, and recreation. Rivers were a critical part of Europe's transport system at least as far back as Roman times, and they still make it possible to move cargo at low cost. The most important rivers for transportation are those of the highly industrialized areas in the core area of Europe (see **Figure 4.29**). An extensive system of canals connects and supplements the rivers. Some are quite old; the Romans built canals throughout Northern Europe and Britain, mainly for military transport.

The Dutch developed the pound (pond) lock for canals in the late 14th century, leading to the expansion of regional connections of waterways through canals and to the use of canals to link farmlands with the coasts. Starting in the late 1700s, the British built canals to move raw materials toward factory locations. By the mid-19th century, the expanding influence of railroads and the steam engine competed with canals. Although canals continue to play a significant transport role, since the 1950s railroads and highways have been the main means to transport goods throughout Europe.

Important seaports developed along the lower courses of many rivers, and some became major cities. London on the Thames, Antwerp on the Scheldt, Hamburg on the Elbe, and

iStockphoto.com/ACMPhoto

• Figure 4.14 The ancient village of Les-Baux-de-Provence is nestled in the classic Mediterranean landscape of Provence in southern France.

Joe Hobbs

• **Figure 4.15a** The author took this photo of Briksdal Glacier in Norway in 1984.

Ellen van Bodegom/Getty Images

• **Figure 4.15b** This photo of Briksdal Glacier was taken in 2008. The pace at which glaciers are melting in most areas of the world is anything but glacial.

Rotterdam in the **delta** (the usually triangular-shaped alluvial area at the mouth of a river) of the Rhine are outstanding examples.

The Rhine and the Danube are particularly important European rivers, touching or crossing the territory of many countries (see **Figure 4.2**). The Rhine countries include Switzerland, Liechtenstein, Austria, France, Germany, and the Netherlands. Highly scenic for much of its course, the Rhine is Europe's most important inland waterway. Along and near it, an axis of intense urban industrial development and high population density developed with the Industrial Revolution. At its North Sea end, the Rhine connects to world commerce at one of the world's most active seaports, Rotterdam, in the the Netherlands. The Danube, on its journey from the Black Forest of southwestern Germany to the Romanian shore of the Black Sea, touches or crosses more countries than any other river in the world. The river is an important artery for the flow of goods and is also held in deep esteem by the peoples that share its waters. The Danube delta region of eastern Romania is one of the world's most important wetlands for waterfowl, and draws growing numbers of ecotourists.

4.3 Cultural and Historical Geographies

Europe's cultural diversity is impressive: a train ride of a few hours in any direction on the Continent takes passengers across a number of ethnic regions and cultural landscapes. Today, Europe's peoples coexist in a state of peace, following a terrible legacy of war.

Ethnic and Linguistic Geographies of Europe

Ethnicity

Modern humans have lived in Europe for at least 40,000 years. Agriculture expanded out of the heart of the Middle East, reaching Southeastern Europe by approximately 6000 BCE, and most of the rest of the continent by 3000 BCE through migrations and along early trade routes. The domestication of plants and animals led to the first villages which, with the aid of such innovations as copper, later mixed with tin to create bronze, eventually grew into cities, and then civilizations.

Mass migrations of peoples known as Proto-Indo-Europeans radiated outward from the steppes of what is now Ukraine starting around 4000 BCE. Over the next two millennia, they settled across much of Europe and parts of western, central, and southern Asia, and their descendants diversified into a number of ethnicities speaking the wide variety of languages in the Indo-European family.

105

Greece was the site of the first European civilizations, including the Minoan culture on the island of Crete and the Indo-European culture of the Mycenaean's on the Greek mainland, whose people mingled with subsequent invaders and settlers over time to form the ancient Greek civilization. Although often separated into city-states (such as Athens and Sparta) that frequently warred with each other, the ancient Greeks developed previously unsurpassed heights of philosophical inquiry, literary and artistic expression, science and mathematics, and representative government. Greek colonists spread across the shores of the Mediterranean Sea, bringing their civilization and language with them. This was most notable in the Greek colonies on the Italian peninsula, which greatly influenced the rapidly expanding empire of Rome.

Another wave of Indo-Europeans settled in Italy and diversified to become the Italic tribes. The Latins would emerge as the most powerful Italic tribe after their capital city of Rome, founded in the 8th century BCE, conquered all of Italy by the 3rd century BCE. By the 2nd century CE, the Roman Empire stretched from Great Britain to modern-day Kuwait, encompassing both sides of the Mediterranean Sea (the Roman name for which was *Mare Nostrum*, "our sea"). Approximately one-fifth of humanity lived within the empire's bounds at its greatest extent. Rome's influence was most strongly felt in the empire's lesser-developed northern and western fringes, whereas the more established lands in the eastern Mediterranean retained more of their cultural identities. Greece remained Greek, even after centuries of Roman rule, while north of Greece, Roman settlers mixing with indigenous tribes became the ancestors of present-day Romanians.

In the 3rd century, Rome was increasingly beset by political corruption, invasions, civil wars, and economic troubles. In the 4th century, the Roman Empire split into eastern and western sections ruled by separate emperors. The poorer, less stable western half of the empire, ruled from Rome, collapsed entirely the following century; the wealthier eastern half (usually referred to as the Byzantine Empire), ruled from Constantinople, would survive until 1453. The legacy of Roman times—arts, religion, architecture and engineering, and government, building upon the knowledge of the ancient Greeks—were felt throughout Europe for centuries after the empire dissolved.

In the 1st millennium BCE, around the same time as the Italic tribes were forming, the **Celts**, another Indo-European group, were settling in Central Europe. They spread westward into modern-day France (where they were known to the Romans as Gauls), Spain, and the British Isles. Celtic peoples blended with Romans during the time of the empire but were increasingly pushed to the western margins of the continent by invasions and mass migrations of Germanic tribes from the north and east as the empire declined.

Descendants of an ancient Proto-Indo-European migration, the Germanic (or Teutonic) tribes originated in modern-day Germany and southern Scandinavia around 1500 BCE. The Romans could never conquer the Germanic "barbarians," and as the empire weakened, Germanic tribes rapidly expanded westward and southward. The Germanic people known as the Franks expanded into the Roman-Celtic region of Gaul (modern France and parts of Belgium, western Germany, and northern Italy), where they mixed with the local population; this Frankish kingdom ("Francia") was the forerunner of modern France. Similarly, Roman-Celtic Britannia was invaded by the Angles and Saxons of northern Germany; the resultant Anglo-Saxon kingdoms eventually coalesced into the country of England. Other migrating Germanic groups, such as the Visigoths, Ostrogoths, and Lombards, affected the development of Spain and Italy.

As the Germanic tribes were expanding west and south from their hearth, the Slavs were radiating outward from their ancestral lands of Ukraine and surrounding areas. The Slavic peoples moved into depopulated areas of Eastern and Central Europe and invaded the Balkan Peninsula, conquering or assimilating the indigenous population during the 6th century. Most early Slavic nations were small and soon came under the domination of non-Slavic kingdoms and empires; only a few were able to establish a long-lasting identity. North of the Carpathian Mountains, the Polan tribe absorbed several surrounding Slavic tribes to form Poland in the 10th century, which grew to be a large, powerful kingdom in the Middle Ages but was erased from the map in the 18th century, not to reappear again until after World War I. Bulgaria too was a strong nation founded in the 7th century, but it was absorbed into the Ottoman Empire in 1396 and did not regain its independence until five centuries passed.

The origin of the Albanian people is unclear. They are of Indo-European descent but are not closely related to any other ethno-linguistic group. Also of uncertain origin are the Balts, of which the Latvians and Lithuanians are the only surviving groups. They are distant relatives of Slavic peoples, but when the Baltic and Slavic ethnicities diverged is uncertain. The approximately 4 million **Roma (Gypsies)**, who live across Europe and have no country or region of their own, are an Indo-European group that probably migrated from India into Southeastern Europe around or before 1100 CE.[12]

142

The Finno-Ugric peoples make up the largest European ethno-linguistic group that is not descended from the original Indo-Europeans. They probably emerged around the Ural Mountains of Russia (the languages they speak are classified as Uralic) in or before the 4th millennium BCE. The Finns, Estonians, and Sami (Laplanders) began migrating toward their present locations soon after. Hungarian tribes took a different path; they lived in various parts of western Russia before settling in the Pannonian or Carpathian Basin (centered in what is now Hungary) around 900 CE, displacing or absorbing the various Germanic, Slavic, and Turkic tribes that were already living there.

The Basques of northern Spain and southwestern France are unique in Europe. Their origins are uncertain, but Basques are certainly a very old ethnic group; they are unrelated to any other existing ethnicity. They may be the last surviving descendants of a Stone Age race that lived over 30,000 years ago, when small bands of Neanderthals still lived across Europe, and glacial ice covered Scandinavia and the British Isles.

Language

Europe's diverse linguistic landscape is the legacy of all of those peoples and events through the centuries. The vast majority of European languages belong to three main branches of the **Indo-European** language family: Romance, Germanic, and Slavic. Several Indo-European languages outside those branches, as well as Uralic languages, constitute nearly all the rest of the tongues spoken across Europe. Basque, a **language isolate** unrelated to any other existing tongue, is the only surviving language from before the Indo-European expansion across the region (•**Figure 4.16**).

Romance languages descended from the Vulgar Latin ("vulgar" meaning "common," as opposed to the "classical" written, or formal, Latin) spoken by citizens of the Roman Empire. Latin as an everyday tongue spread through all of Italy and beyond to areas primarily north and west as the empire

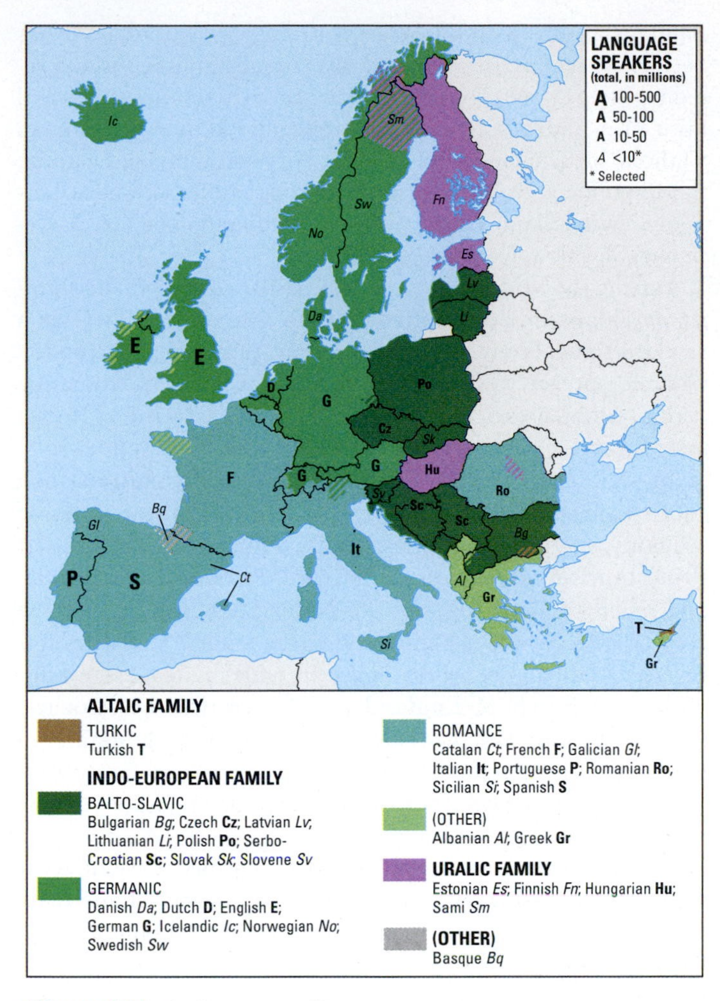

• **Figure 4.16** The languages of Europe.

expanded. After the decline and fall of the empire, the Latin dialects spoken in these areas—isolated from each other and mixing with words and structure from the local (primarily Celtic) inhabitants of the conquered lands—eventually grew into separate languages.

Spanish is by far the most widely spoken Romance language in the world, thanks to the large colonial empire Spain created in the 16th and 17th centuries. Approximately 400 million people speak Spanish as their native tongue, more than any other language except Mandarin Chinese. Over 200 million people speak Portuguese, primarily in Brazil. French has 75 million speakers worldwide, the majority of them living in France, Switzerland, and southern Belgium. In the 18th and 19th centuries, French was the primary language of diplomacy and the elite throughout Europe; it remains one of the official languages of many international organizations. Italian and Romanian (the sole Romance language found in Eastern Europe) have about 60 million and 25 million speakers, respectively, located primarily in their home countries.

Germanic languages were spread throughout Northern and Central Europe during the centuries of mass ethnic migrations after the collapse of the Roman Empire. In the first millennium BCE there were three Germanic languages, called North, West, and East Germanic. North Germanic evolved into Old Norse, the language spoken by the Vikings, and afterward differentiated into the various modern Scandinavian languages such as Danish, Norwegian, and Swedish. The branches of the East Germanic language went extinct in the Middle Ages. The West Germanic line produced German, the language with the most native speakers in Europe (about 90 million). German is the main language of Germany and Austria, and is spoken by a significant minority in several surrounding countries. German is the second-most common language in use on the World Wide Web. Closely related to German is Dutch, the language of the Netherlands and northern Belgium, spoken by over 20 million Europeans.

English is the most widely spoken Germanic language, spoken by nearly 400 million people worldwide as their first language and possibly over two billion as a second language. Originally English developed among the Anglo-Saxons that invaded England in the 5th century. In the 9th century, waves of Viking armies conquered much of England, and in 1066 the country was invaded by William the Conqueror of Normandy. These occupations brought in elements of Old Norse and French to the English language. English spread around the world from the 16th to the 19th centuries as the British colonial empire grew, from Ireland to North America, Australia and beyond; the rising power of the United States after World War II helped cement English's standing as the world's lingua franca. English is the predominant language of the Internet and scientific and technical journals, and is the sole language of international aviation and shipping.

The **Slavic** languages are divided into three branches: South, West, and East. The South Slavic branch languages are spoken in the Balkans. Twenty million people across the countries that constituted the former Yugoslavia (a word that means "Land of the South Slavs") speak Serbo-Croatian, with each of those countries having developed its own dialect of the language. Bulgarian and Slovene are other South Slavic languages. In Europe, the largest language on the West Slavic branch is Polish, with 38 million speakers, nearly all of them in Poland itself. The other West Slavic languages are Czech and Slovak, which are similar enough that a speaker of Czech can understand a Slovak speaker. East Slavic languages are spoken in former Soviet countries and are discussed in Chapter 5. 164 Related to the Slavic languages are the **Baltic** languages of Latvian and Lithuanian. Baltic and Slavic languages are classified together under the term "Balto-Slavic."

Greek occupies its own branch of the Indo-European languages. Greek has a long literary tradition stretching back millennia; its alphabet was adopted and modified by the Romans and later spread throughout most of Europe. Greek, a lingua franca of the early eastern Mediterranean region, was the language of highly educated people in Roman times and became the Byzantine (Eastern Roman) Empire's official tongue. Also occupying its own Indo-European branch is **Albanian**, the origin and history of which is largely unknown. Outside of Albania itself, Albanian is the primary language of Kosovo and is a significant minority language in Macedonia. **Celtic** languages, which ranged across Europe 2000 years ago, are now restricted to extreme western portions of Ireland, Great Britain, and France. The largest of these, Ireland's Gaelic, is spoken natively by only about 5 percent of the Irish population.

The **Uralic** languages of Europe, including Hungarian, Finnish, and Estonian, have a separate origin from the Indo-European languages. Hungarian (called *Magyar* in Hungary) is the largest, with 12 million speakers. The **Sami** language of the Sami people (Lapps) in northern Scandinavia is also of Uralic origin.

Europeans' Religious Roots

At first glance, Europe's religious geography seems relatively straightforward: one of this region's principal culture traits is the dominance of **Christianity** (see the map of Europe's faiths in •**Figure 4.17**; and the discussion of Christianity in Chapter 6, **pages 226–227**). But Christianity has a broad spectrum in Europe, where Christian encounters with other believers became part of the continent's troubled history. Initially outlawed by the Romans, Christianity was embraced as the faith of the realm by the Emperor Constantine in the 4th century, and it became a dominant institution in the late stages of the empire. It survived the empire's fall and continued to spread, first within Europe and then to other parts of the world. In total number of adherents, the **Roman Catholic Church**, led by the pope in Vatican City in Rome, is Europe's largest religious group (with 48 percent of the population), as it has been since the Christian church was first established in the region (•**Figure 4.18**). Today, Catholicism remains the principal faith of a highly secularized Europe. Predominantly Roman Catholic areas include Italy, Spain, Portugal, France, Belgium, the Republic of Ireland, large parts of the Netherlands, Germany, Switzerland, Austria, Poland, Hungary, the Czech Republic, Slovakia, Croatia, and Slovenia.

During the Middle Ages, another great center of Christianity developed in Constantinople (now Istanbul, Turkey) as a rival to Rome. From that seat of power, the Eastern Orthodox Church spread westward to Greece, Bulgaria, Romania, Serbia, Montenegro, and Macedonia, and eastward to Russia and Ukraine.

165

In the 16th century, the **Protestant Reformation** took root in various parts of Europe, and a subsequent series of religious wars, persecutions, and counter-persecutions left **Protestantism** dominant in Great Britain, Scandinavia, and much of Germany and the Netherlands. Except for the Netherlands, where a higher birth rate among Catholics has reversed the balance, these areas are still mainly Protestant, with the **Church of England** and **Calvinism** the strongest sects in Britain and **Lutheran Protestantism** dominant in Germany and Scandinavia. Among the three Baltic countries, Lutheranism is the dominant faith in Estonia and Latvia, while most Lithuanians are Catholic.

The fact that Christians form a vast majority is a somewhat misleading characterization of religion in Europe. Mainstream Christianity is in decline as Europeans have become increasingly secularized in recent decades. Historic legacy plays a strong role in the trend: over a period of centuries, Europeans paid a costly toll in lives in fighting for causes associated with religion, and they have increasingly associated peace with secularism. As a viable option to warfare, European groups with strong ethnic, religious, and linguistic identities have also increasingly called for political power to "devolve" to them (see Insights, **page 107**).

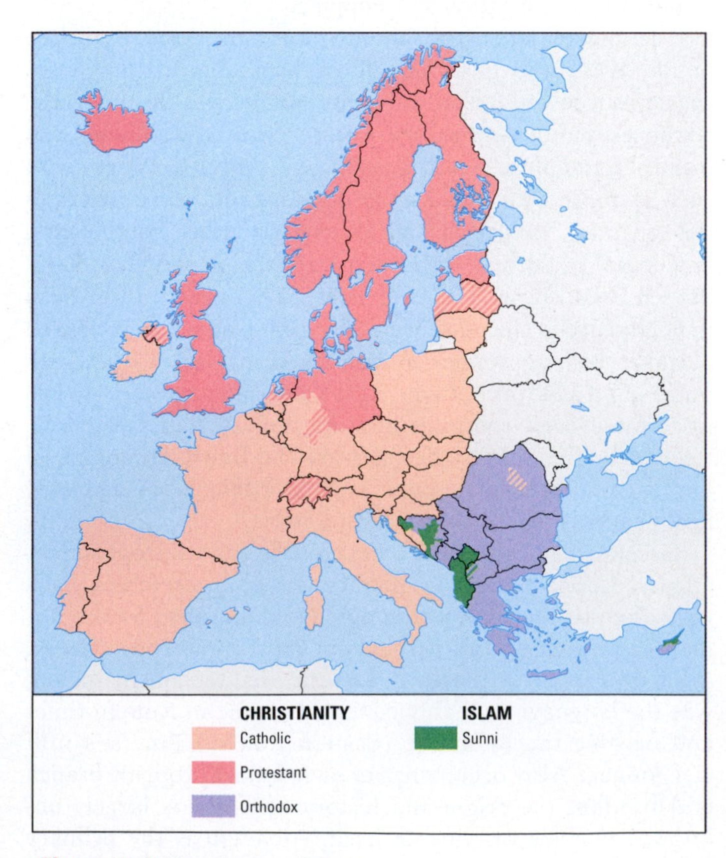

• **Figure 4.17** The religions of Europe.

• **Figure 4.18** The Vatican in Rome is the seat of the Roman Catholic Church. Christ's apostle St. Peter is said to be buried there. "Thou art Peter, and upon this rock I will build my church," Christ reportedly told Peter (Matthew 16:18). The Pope in Rome is Peter's successor.

Insights Devolution

A notable political process in modern Europe, often related to issues of linguistic, religious, and ethnic identity, is **devolution**, the dispersal of political powers to ethnic minorities and other subnational groups. Like that of many other world regions, Europe's historical geography is rife with dominant groups occupying the traditional lands of smaller minority groups. Minorities are sometimes absorbed into the larger culture and political system, but more often than not they retain some degree of distinctiveness and a desire to control their own affairs.

This book relates numerous examples from around the world of strife and war taking place because the dominant power refuses to yield any measure of authority to minority ethnolinguistic groups. But in Europe, the process of devolution has generally been occurring energetically but peacefully. This may be because the European countries, after centuries of strife, are finally self-assured of their sovereignty and well-being; they are so focused on other priorities that demands for autonomy among their minorities are no longer perceived as threats. However, the European Union, preoccupied with meeting the diverse needs of 28 member states, sees devolution as a vexing problem. Devolution is an important issue in Scotland, Northern Ireland, Spain, and elsewhere (• **Figure 4.C** is a map of the major locales where devolution has occurred or is being appealed for in modern Europe).

Arguably, the most significant push for political devolution is in Spain's northeastern province of Catalonia, where a majority strongly favors independence. Catalans share a sense of cultural distinctiveness, relative economic well-being, and historic oppression by the national government in Madrid. Spanish dictator Francisco Franco (in power 1939–1975) revoked Catalonia's considerable autonomy and suppressed Catalan identity, even outlawing use of the distinctive Catalan language in schools.

Catalonia is Spain's strongest economic engine. Representing just 16 percent of the country's population, it generates about 20 percent of Spain's GDP through manufacturing (especially of biopharmaceuticals, medical technologies, processed foods, metalworks, and ICT [information and communications technology]). Catalonia bucked the recent trend of Spain's crippling unemployment and economic crisis, making Catalan independence more inviting than ever. Catalonians insist that they contribute far more to the central government than they get back in the form of services and other benefits.

The Catalan slogan is *España nos roba* ("Spain is robbing us"), akin to the *Roma ladrona* ("Rome the robber") slogan of Italy's secessionist-leaning **Lega Nord**, or **Northern League**.

The uniquely Catalonian practice of building human towers, known as *castells* (• **Figure 4.D**), grew as a metaphor for nation building in the months leading up to a November 2014 "straw poll" on independence. Catalan authorities organized the vote in defiance of the central government in Madrid. The result was 80 percent in favor of independence, a marked contrast with Scotland's "no independence" outcome. The push for Catalonian independence bears watching because a full-on push for independence could stir others now on the sidelines.

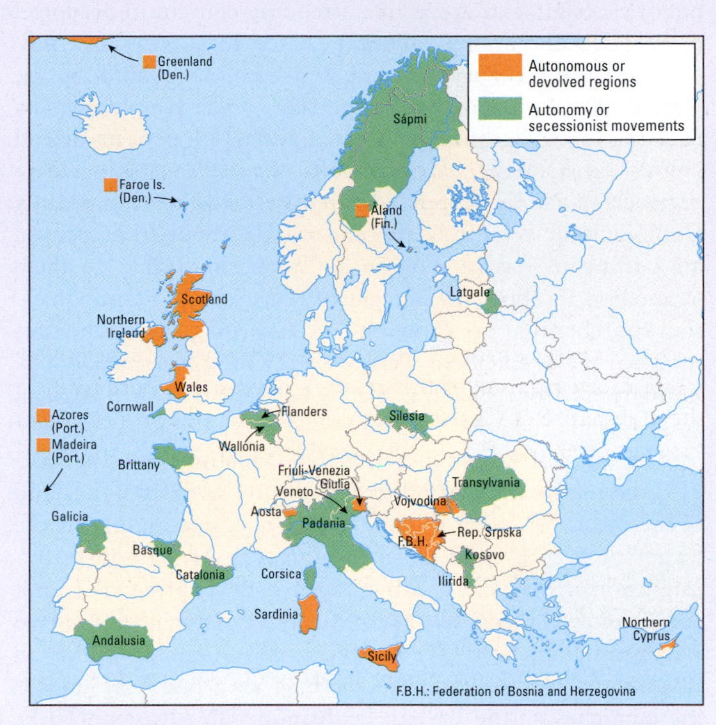

• **Figure 4.C** Devolutionary areas of Europe. Power is being devolved from central governments to regional authorities in many parts of Europe. This map depicts places like Scotland and Wales, where the transfer has taken place, and others like Corsica, where devolution is anticipated, declared, or sought.

136

Artur Debat/Getty Images

124

• **Figure 4.D** The practice of building a human tower, or *castell*, has been around for a long time in Catalonia, but recently it has become a symbol of nation building for the independence-minded Catalans.

Numbers strongly confirm the overall trend of secularization in Europe. Recent polls indicate that 51 percent of Europeans believe in God, but in most countries even larger majorities shun traditions like attending church. In predominantly Catholic France, about 1 person in 20 attends a religious service every week (compared with one in three in the United States). In Poland, 95 percent of the people identify themselves as Catholic, but only about 40 percent attend church on Sundays. Various polls suggest that somewhere between 15 and 33 percent of Italians attend church in mainly Catholic Italy. Polls indicate that overall, about 20 percent of all Europeans regard religion as "very important" to them (compared to about 60 percent of Americans). Those Christian churches that are growing—even surging—in attendance tend to be independent **Pentecostal churches** attended by immigrants from Nigeria, Sierra Leone, and other African countries.

518

The faith of **Islam** (practiced by **Muslims**) has the fastest-growing number of adherents of any world religion and is the fastest-growing religion in Europe, due mainly to the legacy of French colonialism in northern Africa and to immigration across the continent (for a description of Islam, see Chapter 6, **pages 229–232**). Muslims are a majority in two European countries (Albania and Kosovo) and a plurality in Bosnia and Herzegovina (all of these are "Balkans," meaning countries on the Balkan Peninsula). The Ottoman Turks established Islam as the faith of their extensive realm during their period of rule from the 14th to the early 20th centuries (see Figure 6.39, **page 250**). Islam was also the religion of the Muslims who brought their faith and civilization to the Iberian Peninsula in 711 CE after their rapid spread across North Africa in the wake of the Prophet Muhammad's death. In Spain, known to the **Moors** as "Andalus," a Moorish civilization flourished until the end of the 15th century, when Spain's Catholic monarchy ousted them along with Spain's Jews. Among the many Moorish legacies remaining today are the architectural wonders of the Alhambra palace in Grenada.

230

Muslims have come back to Spain and have immigrated widely across Europe. Most of Islam's recent growth on the continent has been tied to immigrant communities in Western Europe. Estimates vary, but France has nearly 6 million Muslims (representing 7.5 percent of the population), followed by Germany with 4 million, the United Kingdom with almost 3 million, and Italy and Spain each with over 1 million.

As related in the earlier discussion of immigration, many "native" Europeans treat Muslims with fear, suspicion, or hostility. But nothing can compare with the experience of the **Jews** as a minority in Europe. One of humankind's greatest tragedies is the story of their plight on the continent (see also Chapter 6, **page 226**). The Romans forcibly expelled these followers of **Judaism** to Europe from Palestine in the 1st century CE, and over time, Jews became more numerous in Europe than anywhere else. In 1880, before many emigrated to the United States, 90 percent of the world's Jews lived in Europe. In 1939, as World War II erupted, there were 9.7 million Jews, 60 percent of the world total, in Europe. By the end of the war, systematic murder by the Nazis in the **Holocaust** reduced their numbers to 3.9 million. About half of those subsequently emigrated to Israel and elsewhere, and there are about 1 million Jews in Europe today.

96

226

In the 1930s anti-Semitism became state policy in Germany under the Nazis, led by Adolf Hitler. Many German Jews fled to the United States, and others emigrated to Palestine in support of the Zionist movement which, since the late 1890s, had aimed at establishing a Jewish homeland in Palestine, with Zion (a sacred hill in Jerusalem serving as a synonym for Jerusalem) as its capital. Most Jews were not as fortunate as the Zionist emigrants. Within the boundaries of the Nazi empire that conquered most of continental Europe during World War II, Hitler's regime implemented its "Final Solution" to the "Jewish problem" by transporting Jews to a network of camps where they were processed to die in gas chambers, do hard labor, or become victims of cruel medical experiments. The Nazi Germans and their allies killed an estimated 6 million Jews, along with millions of other minorities they deemed "inferior," including Roma (Gypsies) and homosexuals. It was this Holocaust that prompted the victorious allies of World War II, from their powerful position in the newly formed United Nations, to create a permanent homeland for the Jewish people in Palestine. A seemingly endless cycle of violence, discussed beginning on **page 249**, ensued in the heart of the Middle East.

138

European Colonialism and Its Consequences

Far more than any other region, Europe has shaped the human geography of the modern world. Before the late 15th century, Europe played only a minor role in world trade patterns: goods moved from southern France and northern Italy northwest to Britain, for example, and more active trade connected Italy with the margins of the Mediterranean Sea and beyond to southwestern Asia. But the most important global trade routes were much farther east and south. The longest link was the 5000-mile (8000-km) **Silk Road** along which goods moved overland (and on the Mediterranean Sea) from Venice to Chang-an (Xi'an) in China.

190, 327

The balance of world affairs started to shift to Europe with the beginning of the Age of Discovery in the 15th century. European sailors, missionaries, traders, soldiers, and colonists tracked out across the world. Although these agents had different motives, they shared a conviction that European ways were superior to all others, and by various means they sought to bring as much of the world as possible under their sway. Economic benefits for the European homelands were among their main motivations. By the end of the 19th century, Europeans had shaped a self-serving world in which they were economically dominant and exercised great influence on faraway indigenous cultures.

54

The process of exploration and discovery by which Europeans filled in their world maps began with Portuguese expeditions down the west coast of Africa. In 1488, a Portuguese expedition headed by Bartholomeu Dias rounded the Cape of Good Hope at the southern tip of Africa and opened the way for European voyages eastward into the Indian Ocean. In 1492, North America was brought into contact with Europe when a Spanish expedition commanded by a Genoese (Italian) man named Christopher Columbus (Cristóbal Colón) crossed the Atlantic to the Caribbean Sea. 463 Less than half a century later, these early feats of exploration culminated in the first circumnavigation of the globe by a Spanish expedition commanded initially by a Portuguese man, Ferdinand Magellan. Worldwide exploration continued apace for centuries, with the Dutch, French, and English eventually wresting leadership from the Spanish and Portuguese.

Where Did That Come From? — Try it

Consider the momentous impacts of the Columbian Exchange.

Imagine the United States and the rest of the Western Hemisphere without coffee; there was no "morning Joe" for anyone in the Americas before 1492. Imagine Europe, Asia, and Africa without cigarettes; just about 500 years ago, there was no tobacco except in the Americas.

Species by species of plant and animal, the Columbian Exchange profoundly impacted the world's cultural and economic geographies. Every single one of the plants and animals we depend on for our sustenance and indulgence had roots in geographic space; its ancestors thrived in a particular environment in ancient times and then were domesticated and diffused through human activities. 51

You can easily appreciate the origins and diffusion of the Columbian Exchange by considering one or more of your favorite foods (plant or animal), pets, or domestic animals. Where did it originally come from? ("Old World" and "New World" cover a lot of territory). If you have a sweet tooth, what about chocolate or sugar? Where did the crops that yield them originate from? Answering these questions will take you into the geographic realms of biogeography, diffusion, natural and cultural landscapes, and historical geography. Try it.

The explorers were the vanguards of a global European invasion that would in turn bring the missionaries, soldiers, traders, settlers, and administrators. They gradually built the frameworks of European colonization of the Americas, Africa, Asia, and the Pacific. The colonial patterns are described on **pages 54–60** in Chapter 3 and are not repeated in detail here. Briefly, their main manifestations were the transfer of wealth from the colonies to Europe, the movement of great numbers of indigenous laborers to serve European interests in other lands, the settlement of Europeans in agriculturally productive colonies (where their descendants are great majorities or significant minorities today), and the attempted subjugation of indigenous cultures to European ones. These are among the most prominent issues in world geography; every chapter of this book deals with the impacts of European colonization on cultures and landscapes beyond Europe.

Of great importance in reshaping the world's **biogeography** (the distribution of living things on the Earth) was the so-called **Columbian Exchange**, the transfer of plants and animals from one region to another, following Europe's conquest of the Americas. Major examples include the introduction of hogs and cattle and the reintroduction of horses to the Americas from Europe; of tobacco and corn (maize) from the Americas to Europe and other parts of the world; of rubber from South America to Asia; of the potato from the Americas to Europe, where it became a major food; and of coffee from Africa to Latin America, where it became the main cash crop in a number of countries. More examples are mapped in **•Figure 4.19**, which shows some of the most significant foodstuffs diffused from the New World to the Old World after 1492 and the major Old World foods introduced to the New World after that time (see the Try It feature above). Europeans also, for the most part unwittingly, introduced virulent diseases, including smallpox, to the New World, with devastating results. Over time, the New World's "gift" of tobacco to the Old World became a major killer. 465, 508

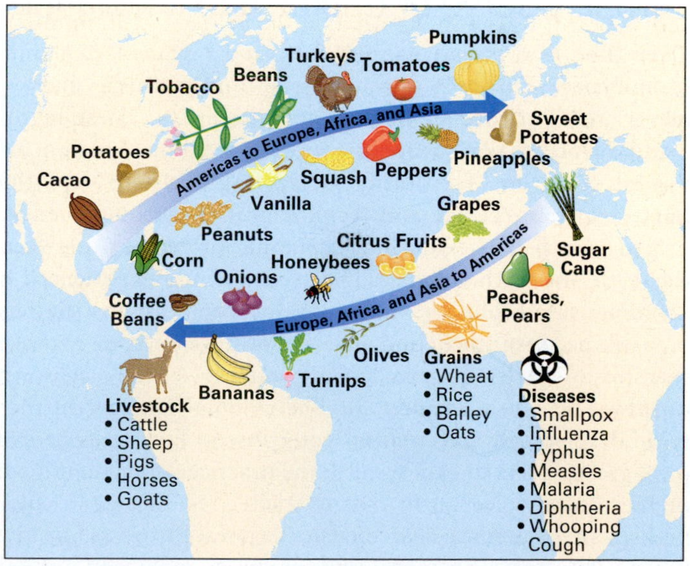

• Figure 4.19 The Columbian Exchange. Europe served as the gateway for New World goods that diffused throughout the Old World.

THINK CRITICALLY: The "exchange" sounds benign, reciprocal, and symbiotic. Would it be more appropriate to think of the "Columbian Revolution," on the scale of the original "Agricultural Revolution?" Pick an animal or plant, such as cattle going to the New World or tobacco to the Old World, and consider some of its impacts on land and life in its new setting.

4.4 Economic Geography

Europe has been troubled lately, and it is important to consider the economic underpinnings of some of the region's problems we have explored so far in this chapter. The biggest danger posed by Europe's unemployment crisis—described above in the context of immigration (**page 94**)—is its rapid contamination of the European Union's foundations. High unemployment is generating support for the populist and "Euro-Fascist" parties that denounce immigration and the currency of the euro (see **page 94**). "The crisis is threatening key pillars of EU integration such as the free movement of people, the free movement of goods and a common currency," wrote Robert Kaplan. "The greatest threat to the European Union's future is not Greece leaving the eurozone, or Portugal defaulting on its debt, or even Spain's banks collapsing; it is persistently high unemployment."[13]

Unemployment has crippled Europe since the beginning of the economic crisis in 2008. It rose from 7 percent in 2008 for all the EU countries to 10 percent in 2014. The most serious concerns are in the European periphery, where in Spain and Greece unemployment rose in those years from 8 to 27 percent. Joblessness among young people is even higher, prompting migration within passport-free Schengen countries: Young people from Greece and Portugal for example are moving north to Germany and Britain in search of work. Many are taking this discontentment and their anti-EU, anti-euro, and anti-immigration sentiments with them. Let's see how this region got into its "Euromess."

Europe's Surge of Capital and Innovation

Over time Europe has gained, lost, and regained economic prominence. Europe's own material and cultural endowments were enriched by its colonial enterprises. These ingredients helped make Europe the world's wealthiest region for several centuries. The European capitalist system, and the huge strides in science and technology that accompanied it, helped launch the region into economic supremacy. Between the 13th and 18th centuries, parts of Europe experienced a "commercial revolution" in which the capitalist institutions of banking, insurance, and investment developed to manage transfers of wealth from trade and conquest (see the discussion on **pages 59–60** in Chapter 3). The colonial era saw further rapid development of economic enterprise in Europe as people took greater risks to gain wealth; the practices of mercantilism or mercantile colonialism that characterized European colonialism are sometimes described as the quest for wealth by any means necessary. Energetic entrepreneurs mobilized capital into large companies and used (or exploited) both European and overseas labor.

Early in the Age of Discovery, Europeans reached a level of technology generally superior to that of non-Europeans with whom they came in contact. Achievements in shipbuilding, navigation, and the manufacture and handling of weapons gave them decided advantages. Powers such as China, once technologically superior, fell quickly behind in terms of the power and geographical reach they projected (as we will see throughout the book, China has changed that balance in recent decades). Europe gained almost free rein in control and expansion of technological innovation for several centuries. The foundations of modern science were also constructed almost entirely in Europe, and the great names in science between about 1500 and 1900 were overwhelmingly European.

These material, cultural, and intellectual forces combined to launch the Industrial Revolution in Europe, which was the first world region to evolve from an agricultural into an industrial society. In the first half of the 18th century, primarily in Britain, industrial innovations made water power and then coal-fueled steam power increasingly available to turn machines and drive gears in new factories. The Scotsman James Watt's invention of the steam engine in 1769 made coal a major resource and greatly increased the amount of power available. New processes and equipment, including the development of coke (coal with most of its volatile constituents burned off), made iron smelting cheaper and more abundant. The invention of new industrial machinery made of iron and driven by steam engines multiplied the output of manufactured products, initially textiles. Then, in the 19th century, British inventors developed processes that allowed steel, a metal superior to iron in strength and versatility, to be made inexpensively and on a large scale for the first time. These developments soon diffused to new hearths of massive, mechanized industrial production and brought unprecedented changes to landscapes and ecosystems around the world (for more on the revolutionary nature of the Industrial Revolution, see Chapter 3, **page 53**). Among the most immediate and most significant processes accompanying industrialization were rapid population growth and a dramatic shift of people from rural areas into cities. England's population surged from 9 million in 1801 to about 16 million by 1841, a 77 percent increase. Cities grew at an even faster rate; in this same time frame, London's population increased 89 percent, Birmingham's by 149 percent, and Manchester's by 169 percent. This was the first "population explosion." There had never been such growth or pressure on urban services to accommodate so many people.

53

70, 78

By 1900, the people and machines of western European cities created about 90 percent of the world's manufacturing output. Europe led the world in the production of textiles, steel, ships, chemicals, and other industries. But in the 20th century (especially after 1960), Europe's preeminence in world trade and industry diminished to about 25 percent of the world's manufacturing output, for several reasons. A principal one is that like most MDCs, those of Europe have transitioned into post-industrial societies in which the wealth generated by the manufacturing sector (the "secondary sector") has been overtaken by the service sector (the "tertiary sector" that includes entertainment, government, health care, financial services, education, retail services, and information technology).

Europe Displaced

Several factors eroded Europe's economic supremacy in the 20th century. One was warfare (•**Figure 4.20**). Europe suffered enormous casualties and damage in World Wars I and II, which were initiated and fought mainly on that continent. Western Europe eventually recovered from the wars, in large part with assistance from the United States, but did not regain its former supremacy. "War is mainly a catalogue of blunders," wrote the then British Prime Minister Winston Churchill. He knew from experience: as First Lord of the Admiralty during World War I, in what is known as the Battle of Gallipoli, he doomed more than 50,000 young men trying to achieve his strategic ambition of securing the Turkish straits between the Black and Adriatic Seas. 237, 391

Another factor disempowering Europe was the loss of its colonies. During the 20th century, rising **nationalism**—in this context, referring to the quest by colonies and ethnic groups to possess their own homelands—brought the European colonial empires to an end. Taking advantage of a weakened Europe and mounting disapproval of colonialism in the world at large, one European colony after another gained independence quickly in the decades following the end of World War II. Ironically, opposition to continued European control was often spearheaded by colonial leaders who had been educated in Europe and had acquired nationalistic ideas there. Later, we will meet some of these leaders, including Vladimir Lenin of Russia and Ho Chi Minh of Vietnam.

Europe's predominance was also seriously eroded by the rising economic and political stature of the United States and the Soviet Union. Both far larger than any European nation, these countries outpaced Europe in military power, economic resources, and world influence, especially in the postwar years 1950–1970. 171, 520

A major shift in global manufacturing patterns also weakened Europe. Europe once enjoyed a near monopoly in exports of manufactured goods, but manufacturing accelerated in many countries outside Europe in the 20th century. The United States reached industrial maturity by 1950 and competed vigorously with Europe. More recently, the Asian countries of China, Japan, South Korea, and Taiwan became Europe's industrial competitors in markets all over the world. 309

Finally, Europe's leadership was weakened by a new dependence on outside sources of energy. The region's traditional reliance on coal was superseded by technological developments and environmental concerns, and the energy balance shifted to petroleum and natural gas. Despite the development of North Sea oil and gas resources (belonging mainly to the United Kingdom and Norway), Europe became and remains highly dependent on imported Middle Eastern and Russian 132

Edmund O. Rhodes

• **Figure 4.20** In the 20th century, two devastating world wars originated in and centered on Europe, causing enormous suffering and eclipsing the region's global economic dominance. These are the ruins of the Reichstag in Berlin in 1945, photographed by chemical engineer Edmund Rhodes, my grandfather. The US Army sent him to Germany twice to assess the conditions and potential reconstruction of German industry and technology. He worked closely with the Soviet authorities who reached Berlin before the Americans did. Soviet treatment of German civilians, especially women, was atrocious as the war drew to a close.

oil and gas. The geopolitics of Europe's dependence on Russian fossil fuels has constrained Europe's ability to confront Russian ambitions in Ukraine, Georgia, and elsewhere.

An Imbalance of Wealth

A striking pattern in Europe's economic geography is that Western Europe is wealthier than Eastern Europe; Europe's four largest economies are (in descending order) Germany, France, the United Kingdom, and Italy. Additionally, the European core, with Germany as its anchor, is much wealthier than its periphery. These trends date to at least the 1870s, when per capita incomes in the West were twice those in the East. The gap grew even wider throughout the 20th century. After World War II, eastern European countries were in effect colonized by the Soviet Union, serving as vassal states that gave up human and material resources to service the motherland. With the dissolution of the Soviet Union in 1991 and the admission of eastern European countries to the European Union, have the eastern countries been able to hope that they may eventually catch up with their western counterparts? Poland, Central Europe's largest economy, has been particularly successful; although not prosperous, it did not suffer the near collapse that gripped some of the southern periphery members such as Spain and Greece in recent years, and has attracted much new investment. Membership in the European Union is structured to help redistribute wealth from west to east through mechanisms such as agricultural subsidies that could help to breathe new life into the faltering farms and factories of the east. As we will see, however, the newer member states have not been automatically granted all the privileges enjoyed by the older, wealthier members.

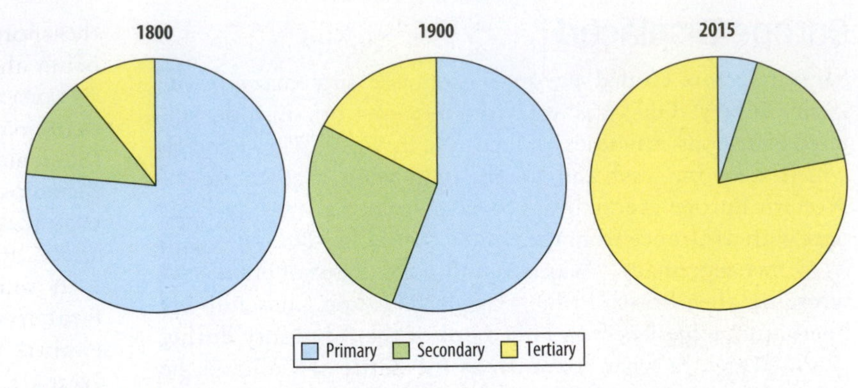

• **Figure 4.21** Change in time of European workers employed in the three major sectors of the economy. The primary sector is mostly agriculture and resource extraction (like mining). The secondary sector is primarily manufacturing. The tertiary sector is made up of "white-collar jobs," jobs involving providing services of some kind to a customer. As Europe has grown wealthier and more technologically advanced, the proportion of its people working in the primary sector has plummeted, whereas a large majority now work in the tertiary sector.

Living off Land and Sea: Europe's Primary Sector

Despite the dominance of a post-industrial, urban economy, some European countries continue to rely significantly on the productivity of land and sea, the so-called "primary sector" of any economy. Agriculture was the original foundation of Europe's economy, and it is still important (•**Figure 4.21**), though only 5 percent of Europeans work in agricultural jobs (down from 76 percent in 1800 and 56 percent in 1900). Food provided by the region's agriculture allowed Europe to become a populous area at an early time. After about 1500, a period of steady agricultural improvement began. Introduction of important new crops such as the potato played a part, but so did such practical improvements as new systems of crop rotation and scientific advances that produced better knowledge of the chemistry of fertilizers. The continuing expansion of industrial cities provided growing markets for European farmers, who received protection through **tariffs** (tax penalties imposed on imports) or direct **subsidies** (payments made to domestic producers) to encourage production and support rural incomes (these are also discussed on **page 526** in Chapter 11, in the context of similar protective measures by the United States). As a result of European Union policies, there are often surpluses of many products, including grains, butter, cheese, olive oil, and table wines (see the discussion of the Common Agricultural Policy or CAP on **page 117**). But Europe is not agriculturally self-sufficient and imports 60 percent more agricultural products than it exports (see Insights, **page 116**).

Europe, like the United States, is finding itself increasingly at odds with the rest of the world as global free-trade agreements insist that its doors be open to more (and generally cheaper) food imports. Europe wants to protect its farmers, particularly those producing what it calls the "sensitive" products (including beef, chicken, and sugar). It does this with high tariffs on imported foods and with subsidies that allow farmers the **dumping** of their surplus production onto the world market at very low prices. The European Union says it is doing enough to open its doors to imports because it imposes no tariffs on imports from the world's 50 poorest countries, but free-trade advocates insist that more trade barriers must fall.

Throughout history, fishing has been an important part of the European food economy, and the coasts still have many busy fishing ports (•**Figure 4.22**). At times, control of fishing grounds has been a major commercial and political objective of nations, even resulting in warfare. Fisheries are most productive in shallow seas that are rich in plankton, the principal food of herring, cod, and other commercially valuable fish. The Dogger Bank (to the east of Great Britain) in the North Sea and the waters off Norway and Iceland are major historical fishing areas. Norway, Iceland, and Denmark are Europe's leading nations in total catch. Fishing has long been especially crucial to Norway and Iceland, which have little agriculturally productive land. However, overfishing of cod, the world's most popular food fish, sometimes known as the "beef of the sea," has grown so severe, particularly in the

Jeffrey L. Rotman/Terra/Corbis

• **Figure 4.22** Cod fishing in the North Atlantic.

North Sea, that since 2006 the European Union and Norway have legislated cuts of cod fishing by as much as 25 percent and have periodically closed off sections of the North Sea during crucial spawning seasons. Environmentalists have cried foul; organizations, including the World Wildlife Fund and Greenpeace, have insisted on a complete ban on cod fishing. Already, plummeting fish catches have caused economic hardship in many traditional fishing communities on North Sea shores. 37, 537

Post-Industrialization

Europe's economy today, like its population trend, is best characterized as **post-industrial**, defined as more reliant on tertiary service activities than secondary manufacturing activities. In the European context, this means that since about 1960, the region has experienced **deindustrialization**, shifting by choice and by economic necessity away from energy-hungry, labor-costly, and polluting industries toward an economy based on production of high-tech goods and on services. The high-technology products include electronic devices, data processing hardware, **robotics**, and telecommunications equipment. Other industries like drug and pesticide manufacturing make use of these technological products to create their own, so there is an important interconnectedness centered on high technology. Europe is also renowned for the production and export of high-quality, expensive luxury brands, such as Ferrari automobiles and Rolex watches. In the tertiary service sector, growing proportions of Europeans are employed in transportation, health care, banking, retailing, wholesaling, advertising, legal services, consulting, information processing, research and development, education, and tourism.

Deindustrialization has had important spatial consequences for Europe. As we have seen, cities and industries grew up where important natural resources like coal and iron ore were located. Now most of these geographical correlations and barriers have become obsolete. With the capabilities of the Internet and the advantages of fast and convenient transport of goods, entrepreneurs can start up new enterprises almost anywhere. European cities are generally cleaner than they once were and, as in North America, neglected and decaying cores of formerly industrial cities are being rehabilitated and gentrified. 542 Many European cities are exceptionally attractive places to live and visit, providing amenities that encourage socializing, good health, and enjoyment of the museums and historical legacies that endow each city and give it its unique character. English literary buffs, for example, flock to locations like Whitby on Yorkshire's eastern coast (•**Figure 4.23**), which features prominently in Bram Stoker's *Dracula*.

Neither high-tech nor service industries employ as many people as the old manufacturing sector did, and like the United States, the European countries are grappling with problems resulting from industrial unemployment. There are concerns over where new jobs will come from and how to make the transition from manufacturing to service roles without too much social dislocation. Many European nations for a long time fit the model of the **welfare state**, using resources collected through high taxation rates to provide generous social services to citizens (see Life in France: The Fraying Safety Net, **page 114**). That "social model" of the European state is, however, under tremendous pressure from the forces of globalization, the costs of absorbing immigrants, the lingering effects of the global financial crisis of 2008–2009, and the debt crisis that erupted in Europe soon after that. Europe recently has been looking less like the relatively dynamic United States and more like the stagnant Japan of the 1990s. 356 Demand within Europe for European products is declining, especially as the population ages (as it is doing in Japan). Competition for export markets is becoming more intense. Ideally, the pooling of resources in a unified effort should help solve such problems.

Katie Hobbs

• **Figure 4.23** Whitby, England. On the hilltop stand St. Mary's church (left) and the ruins of Whitby Abbey (right), which was shelled by German warships in 1914. Bram Stoker writes that Dracula came ashore here as a black dog, sheltering in the Abbey and feeding on townsfolk until a priest drove a wooden stake into the vampire's heart.

life in FRANCE

The Fraying Safety Net

Let's say you are a French woman or man categorized economically as a middle-class worker. Your lifestyle is not extravagant, but simply by being a French citizen you enjoy a lot of amenities that the middle-class American workers (for example) would not have. You can call these the benefits of the welfare state or the strands of a "safety net."

See if you like these living and working conditions: workers are required to take a five-week vacation every year. Most higher education costs are paid for by the government, and this benefit is available to all citizens. The same goes for child care and, perhaps most important of all, for health care. In case of losing her or his job (an all-too-common scenario in Europe today), the worker has two years of government-paid unemployment insurance to bridge the gap between jobs. A generous government pension begins with retirement at age 62. Parents get a monthly payment for each child after their first, helping to cover the expenses of child rearing but also incentivizing couples to have children on a continent where low birth rates have economic consequences (from the government point of view, including not enough people to pay the taxes to keep the safety net intact).

Very high taxes, especially on employers, pay for such cradle-to-grave benefits. Every employee forks over 22 percent of his or her paycheck in payroll taxes for the benefits. The employer pays even more, as much as 50 percent. But France's government is having a hard time raising taxes high enough to pay for all the benefits. The welfare state is also disadvantaged in the worldwide marketplace of innovation and competition. Companies looking for places to start up or expand will consider the high taxes and deep layers of government regulations, and look elsewhere.

My friend Claire De Circourt, a small business owner in her native Paris, wrote of the burdens threatening to unravel France's safety net. It is a common refrain across Europe:

> Someone has to pay for the cost of all that seems so wonderful in France! It is too often forgotten. The people paying are the entrepreneurs, the artisans and all the self-employed who pay a heavy tribute for not being on the receiving end, since they chose to be their own person. Having no safety net to fall onto, [entrepreneurs] work all the time and cannot afford to be sick. They have too much to lose . . . As a result, the entrepreneurs, those who dare, those who do not wait to be fed, to be taken care of and protected for beyond what they should for their own good, are weakened and unjustly discarded when they are unable to resist the pressure anymore—it eventually happens with illness or aging.
>
> The whole country is weakened from it. France starts to feel the foundations shaking, for too many people live on the strength of too few. The State does not see that; it would mean recognizing that their generosity has little to do with them! They buy votes by giving away money they have not earned, but the real richness—the creativity, the work power, the courage of people—is disappearing. Either those who manage to hold on until then are getting older and going on retirement, or, if they are young, many leave the country in order to utilize their energy and their ideas, and their wish to work and get things by themselves. France is in a bad shape. It is the next Lehman Brothers.
>
> Maybe you'll understand why the French never smile: some are prisoners of their privileges, while the others are slaves.

How difficult would it be to break this social compact that most European governments have with their citizens? Very. One Frenchman puts it this way: "You cannot take away guns from Americans, and in the same way you cannot take away social benefits from French."[14]

The European Union, Built on a Market

The European Union, which has its headquarters in the Belgian capital of Brussels, is Europe's most important **supranational organization** (an organization in which member countries are united beyond the authority of any single national government and are planned and controlled by a group of nations). A good way to think of the European Union is as "virtual" United States of Europe, with a federation of nations similar to the states of the US. The 28 countries of the European Union are mapped in • **Figure 4.24**.

The European Union's roots go back to 1957, when the Treaty of Rome created the **European Economic Community (EEC)**, also known as the **Common Market**, made up of France, West Germany, Italy, Belgium, Luxembourg, and the Netherlands. By 1996, three years after it acquired the name "European Union," the organization had nine more members: the United Kingdom, Denmark, Ireland, Greece, Portugal, Spain, Austria, Finland, and Sweden.

Much of the European Union's raison d'être is economic. The EEC was designed, as the European Union is today, to secure the benefits of large-scale production by pooling the resources—natural, human, and financial—and markets of its members. Tariffs on goods moving from one member state to another were eliminated, and restrictions on the movement of labor and capital between member states were eased (in theory, a citizen of any EU country may work in any other EU country, but the process is not simple; a work permit must be obtained from the host country). Monopolies that restricted competition were discouraged. A common set of external tariffs was established to regulate imports from other countries,

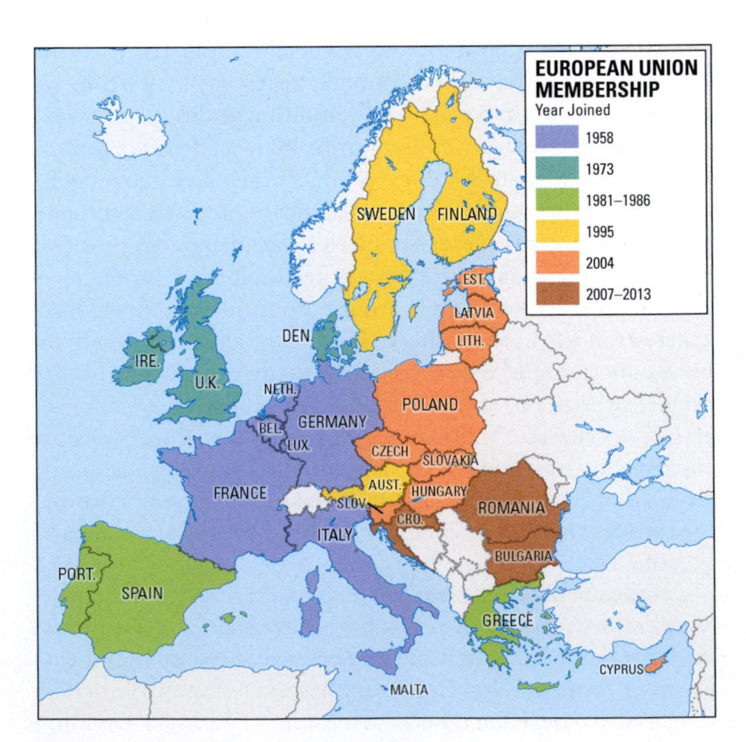

• **Figure 4.24** Members of the European Union.

and a common system of price supports for agriculture replaced the individual systems of member states (see Insights on **page 116** for the EU policy on importing genetically modified foods). The founders expected that free trade within such a populous and highly developed bloc of countries would stimulate investment in mass-production enterprises, which could sell freely to all member countries. This trade would encourage productive geographic specialization, with each part of the union expanding lines of production for which it was best suited. Each country might thereby achieve greater production, larger exports, lower costs to consumers, higher wages, and a higher standard of living than it could achieve on its own.

Bring on the Euro

The European Union, created under provisions of the **Maastricht Treaty on European Union** that went into force in 1993, began pursuing even greater new steps toward European unification. A single EU currency, the **euro**, was launched in 1999 as the centerpiece of the **European Economic and Monetary Union** (EMU; •**Figure 4.25a**). This common currency was based on economic theories suggesting that sharing a currency across borders had many advantages, such as lower transaction costs, more certainty for investors, enhanced competition, and more consistent pricing. A common currency could theoretically restrain public spending, reduce debt, and tame inflation. In practical terms, consumers are able to spend money in euro-using countries just as Americans do at home; they can compare the price of a car, for example, in Germany and Italy, and decide to buy the cheaper car, just as an American might buy a car in Utah instead of Colorado. Besides sharing a currency, all euro-using countries agree to allow the European Central Bank to make critical decisions on issues such as interest rates (similar to the role of the Federal Reserve in the US), which must be identical for all euro-using nations.

A certain amount of euphoria—shall we call it "europhoria?"—accompanied the creation of the euro: this would be yet another symbol of a strong and united Europe. But from the beginning the common currency had detractors. The Swedes, Danes, and British rejected the euro. Sweden feared a loss of financial and monetary independence. Danes worried that the euro would threaten their generous welfare benefits or even their national unity. When Britain had to decide whether to use the common currency, its economy was stronger than those of Germany and France, giving the "Brits" confidence in shunning the euro. The new eastern European members of the European Union decided to adopt the euro, but had to make difficult economic reforms to meet the requirements of adopting this common currency. The number of euro-using nations grew to 19 of the 28 EU states; these countries comprise the so-called **Eurozone** (•**Figure 4.25b**; five other countries and microstates use the euro but are not in the Eurozone).

Like the European Union itself, the Eurozone has had growing pains: a major crisis developed in the Eurozone in 2010 in part because of the economic burdens imposed by the expansion of the common currency and of the European Union itself. Problems have plagued the Eurozone and the EU ever since then.

pitchr/Shutterstock.com

(a)

(b)

• **Figure 4.25** The "euro" **(a)** and the countries that use it **(b)**. The euro is the common currency of 19 of the European Union's 28 countries. Greece leads the way of countries that want to drop it, sparking fears of a greater unraveling of the European Union. Most conspicuous by its absence, the United Kingdom was among three countries refusing to join the Eurozone to begin with. The euro was worth US $1.21 on January 1, 2015, the date that Lithuania joined the Eurozone as the last Baltic country to do so.

Europe's "Big Bang"

Embracing the Less Wealthy

Ten eastern European nations joined the European Union all at once in what was known as the **"big bang"** of 2004. The newcomers were Poland, the Czech Republic, Hungary,

Insights Genetically Modified Foods and EU/US "Food Fights"

Joe Hobbs

• **Figure 4.E** Biotechnological processes can manipulate the genetic makeup of mung beans (seen here in Vietnam with University of Missouri plant scientists Jerry Nelson, left, and Henry Nguyen) and other food crops to make them more productive, increasing yields—and also, say many Europeans, endangering local plant varieties and evolutionary processes. 118, 120

The United States is Europe's leading trading partner, but the exchange is not always open or friendly, especially when it comes to food. Countries often pay lip service to the concept of free trade while actually limiting trade or imposing tariffs on some products to protect their own industries and interests. Nations on both sides of the Atlantic use these trade barriers, sometimes even precipitating **trade wars**, in which the countries try to damage one another's trade. In the late 1990s, for example, the United States wanted to sell bananas that US corporations grew in Latin America to the countries of the European Union. The Europeans, who wanted to protect their investments in banana farming in former European colonies in Africa, the Caribbean, and the Pacific, refused to buy, and a so-called **"banana war"** ensued.

An ongoing trade dispute involves **genetically modified (GM) foods**, also called **genetically modified organism (GMO) foods**, produced in the United States. These products, such as rice, soybeans, corn, and tomatoes, have been genetically manipulated to be more productive and resilient through biotechnology (defined as the use of living systems and organisms to develop or make useful products) (•**Figure 4.E**). Genetically modified crops can promote nutrition and pest resistance, and tolerance of herbicides, cold, drought, and salt. European farmers and consumers, with their long-established food cultures, generally disdain GM foods, citing unintended harm to other organisms, reduced effectiveness of pesticides, gene transfer to non-target species, unknown effects on human health, and allergenicity (creating allergic reactions in people). For a long time, the European Union practiced a "zero tolerance" policy toward GM foods. Some biotechnology firms, St. Louis–based Monsanto in particular, have been vilified as evil and greedy multinational corporations decimating ecosystems and suing small farmers who violate their intellectual seed-stock property. Monsanto, however, replies that it is practicing its motto of "Doing well by doing good."

The United States argued that the EU ban on GM foods violated the free-trade principles of the World Trade Organization (WTO). Under pressure in international trade talks, the European Union gradually made concessions, allowing some GM foods and livestock feeds to be imported from the United States and elsewhere. Under the rubric of the proposed, highly controversial trade agreement known as the Transatlantic Trade and Investment Partnership (TTIP), the United States and Canada are pressuring EU member states to open their markets to imports of GM foods.

The European Union insists that GM foods must be labeled (in the US, they have usually not been, but some state referendums have resulted in labeling). And European consumers, who derisively refer to GM food as **"Frankenfoods"** (as in Frankenstein), have not shown much appetite for these imports. A handful of European countries, notably France, Germany, and Italy, have defied EU guidelines and imposed outright bans on GM foods and crops.

Globally, the Europeans may find themselves increasingly isolated on the issue of GM foods and less able to compete in the international agricultural marketplace, especially if China and other large food producers and consumers embrace GM products. The heated GMO debate is likely to continue. We will see it in other regional contexts in sometimes surprising ways. Some African nations, for example, have actually refused to accept international food aid in the form of genetically modified corn. 314, 344, 484

Slovakia, Estonia, Latvia, Lithuania, Slovenia (the first former Yugoslav state to join the European Union), Malta, and officially all of Cyprus, although de facto EU membership applies only to the Greek portion of the divided island nation. 127 Turkey's application to join was rejected. The big bang created a mega-Europe of 450 million people, stretching from the Atlantic to Russia, with an economy valued at almost $10 trillion, nearly as strong as the US economy. The European Union grew in population by about 20 percent (75 million people, almost half of them in Poland) and in geographic size by about one-third (with a 40 percent growth in farmland). A more modest expansion took place in 2007, when

Romania and Bulgaria joined as the European Union's poorest member nations.

The overall near doubling to 28 countries was accompanied by some clear advantages for the European Union and its new members, but also by some problems. Some thorny issues had to be resolved before the dozen new countries could join, and even thornier problems may lie ahead.

Consistent with Europe's historical patterns of west and east, and core versus periphery, the most outstanding difference between the old and new members is in their economies. The old EU member nations have some of the highest per capita gross domestic product purchasing power parity (GDP [PPP]) levels in the world. The new states, by contrast, are much less affluent, and some are classified statistically as less developed countries (LDCs) with a GDP (PPP) of less than $17,000 (see **Figure 3.7** and **Table 4.1**). Bulgaria and Romania both have a per capita GDP (PPP) of just $14,400. The old EU countries have 95 percent of the continent's wealth, and the new ones only 5 percent. When the "big bang" countries joined in 2004, the European Union's average wealth per person fell by 13 percent.

Naturally, questions arose about the European Union's ability to afford its new members. There is particular concern with the costs of the European Union's **Common Agricultural Policy (CAP)**. The European Union provides infamously generous agriculture subsidies ("direct farm payments") to its member states. Many farmers in the European Union would be unable to farm profitably without EU subsidies. The CAP protects domestic farmers by taxing agricultural imports and subsidizing exports. When the market prices of agricultural products fall below an agreed-upon target price, the EU government steps in to buy that produce. This protects farmers but also provides a perverse incentive for them to overproduce, thereby driving market prices ever lower. The results have included "mountains" of butter, "lakes" of milk, and other metaphors for over-the-top outputs. Overall, the CAP makes EU foods among the most expensive in the world.

At long last, there are serious reforms underway in the CAP. These have reduced subsidies from 50 to 40 percent of the EU's budget, and that ratio will drop further. With its fiscal house in disorder, the European Union simply cannot continue to be so lavish with agriculture, which produces only 1.6 percent of the EU's GDP. As in America, the numbers of farmers are in steady decline and make up only 5 percent of the population (reflecting the broader demographic trend, they are also getting older, so reforms aspire to attract younger farmers). Also, as in the United States, subsidies tend to favor large agribusinesses rather than small farmers, so the CAP helps make the rich richer.

The CAP meanwhile fails to help the world's poor farmers, particularly in Africa, who cannot compete with the low prices of sugar, dairy, cereals, and other produce exported worldwide by the large European (and American) agribusinesses. Their dumping, or the sale on world markets of goods at prices well below their costs of production, is neither "free trade" nor "fair trade," and is an excellent example of one of the dark sides of globalization. The European Union's lavish quotas for domestic sugar production is especially hard on the world's poorer agricultural exporters.[15]

421, 526

Feeding the PIGS

One way the European Union will cut back on these expensive subsidies is by not providing them in full to the newer members. The older member states argue that the new countries do not have the high production costs and quality controls of the old countries. Mainly, though, the older European Union members fear that providing full subsidies to the new countries would bankrupt the organization. Germany is especially fearful. It is the wealthiest member of the European Union and carries the largest cost burden. But Germany had its own economic pains from absorbing the costs of reunion and integration with the poor former East Germany and is concerned about this far more expensive integration on a much grander scale. Despite hopes among many in the newly joining states that membership would bring them prosperity, there has been at least short-term pain. Food prices have risen sharply while salaries have increased slowly. Many western Europeans fear that the lower salaries and taxes of the new eastern member countries will lure industries and jobs away from the West, contributing to the decline of their home economies.

130

A similar set of problems revealing the strains within a united Europe has plagued the Eurozone. In this case, the European Union's strongest economies, those of Germany (which is seen as the "economic locomotive" of the continent) and France, have come under pressure to help out the weaker countries—especially Portugal, Ireland, Greece, and Spain, known collectively by the unfortunate acronym **PIGS**. After joining the common currency, these countries embarked on spending and borrowing sprees, building up huge deficits and debts (apparently hoping that in case of any troubles, the larger economies of the Eurozone would bail them out). In the good years of 2000 to 2008, Germany was the manufacturing giant at Europe's core while the PIGS on the European periphery used easy credit to consume what Germans produced. In the debt crisis that followed, beginning in 2010, a number of economic bubbles in these countries burst—most loudly in Spain's real estate sector.

With economic growth replaced by economic stagnation, the euphoric vision of the euro faded away, and new words crept into common usage: Euromess, Europhobia, Eurocrisis, and especially the **European debt crisis** or "sovereign debt crisis." The European debt crisis began early in 2010, when it appeared that Greece might default on its loans. The Eurozone and the European Union more broadly faced the same stark economic choice that had stared down the United States in 2008: austerity versus demand (or austerity versus growth). Should economies tighten belts and slash spending, or try to stimulate the economy through spending (by borrowing even more, or "deficit spending")? The United States chose the stimulus model, which also involved slashing interest rates to encourage investment and spending. The Eurozone, led by Germany, chose austerity, but it was a bitter pill: longer working hours, shorter vacations, and fewer of the amenities 114 of Europe's welfare state.[16] Austerity programs included salary cuts, leading to even more unemployment, poverty, and social unrest.

Europeans have been there before, especially when suffering from crippling rates of inflation; you may have heard of Germans pushing around wheelbarrows filled with cash to buy everyday items in the post–World War I era. The European Central Bank (ECB) was established in Germany precisely because the European Union knew that Germany would ensure that inflation would never grow out of control. Recently, however, the European Union has become vexed by **deflation**, a sustained decline in prices that can lead to higher unemployment and that is quite difficult to reverse. High levels of indebtedness have prompted Europeans to put their money into debt repayment rather than new purchases that would stimulate the economy. Reduced demand leads to lower prices, and inflation turns to deflation.[17]

Although the United States appears to have got it right with deficit spending, emerging slowly but surely from its crisis, Europe's austerity formula was not as successful. By 2014, the Eurozone was turning away from austerity and belatedly considering the American deficit spending model. The ECB cut interest rates to near zero and in 2015 tried to stimulate the economy through "quantitative easing" by buying financial assets of commercial banks. There is a psychological factor in quantitative easing: concerned that inflation will rise, people will spend now rather than risk paying higher prices in the future, and thus they stimulate the economy.[18] This strategy helped both Britain and the United States to recover from their financial crises, but there are fears that the Eurozone has acted too late. With all of Europe's social fault lines so close to the surface, there is a real danger of unrest and even conflict on the streets of Europe.

In theory, the European Union and the Eurozone should provide strength through unity during times of crisis. The reality is far different. The burdens are shared unequally, and the Germans see quantitative easing as a form of wealth transfer to countries that have been irresponsible: feeding the PIGS again. The Eurozone is not nimble in times of crises: there is no way to address a problem quickly because all 19 members have to be consulted for input. The ECB lacks much authority to make binding decisions for all its members. Individual countries in the Eurozone have different goals and expectations. The wealthier ones, like Germany, do not want their taxes used to bail out countries like Greece that are heavy borrowers. The poorer indebted countries like Greece do not want wealthier outsiders like the Germans to dictate how their government taxes their citizens and spends their money.

Overall, the European Union tends to function smoothly in times of prosperity, but during times of economic weakness the fault lines of its member states and the complaints of social movements within them are activated.[19] Somehow, Europe's economies need to start growing again; the Transatlantic Trade and Investment Partnership (TTIP) between the European Union and the United States, discussed below, could help make this happen. If economic stagnation or deflation continues, forces opposed to the EU could prevail. 116, 120 One or more member states might leave the European Union, threatening the grand experiment. The conventional view was that problem-riddled Greece might bow out. But a much more consequential candidate for withdrawal emerged when Britain's Conservative (Tory) Party, led by Prime Minister David Cameron, prevailed in the UK general election in 2015: Cameron had promised that, if elected, by the end of 2017 he would hold a referendum on the UK pulling out of the European Union. Along with the prospective devolution of Scotland, Wales, and Northern Ireland, this could leave England as the "rump state" of a once great power.

4.5 Geopolitical Issues

In the past 100 years, Europe's geopolitical situation changed more profoundly and more violently than that of any other world region. Europe's two world wars wrought unprecedented devastation. World War I (1914–1918), known until 1939 as the Great War, saw Europe fracture along fault lines of military alliances. Germany and the Ottoman Empire (based in what is now Turkey) were solidly defeated by an alliance led by Britain, France, Russia (early on), and the United States (later). The war accomplished little and destroyed much. The peace treaty at war's end placed a costly burden of reparations on the defeated Germany.

The economic and social humiliation of Germany helped plant the seeds for the rise of Adolf Hitler's fascist Nazi (National Socialist) Party and the outbreak of World War II (1939–1945). Hitler's Germany surged in strength and allied with Italy, the Soviet Union (early), and Japan (later) in a military quest to establish European and Far Eastern empires. Early in the war, Germany advanced swiftly on the European mainland and Scandinavia, and set its sights on an occupation of Britain. In 1941, when Japan bombed the US naval base at Pearl Harbor in the Hawaiian Islands, the United States was drawn into the war firmly on Britain's side. Soon joined by the Soviet Union, these forces, assisted by 534 many smaller ones, began to turn the tide against Germany. In 1945, with German cities in ruin and two Japanese cities nearly obliterated by the first atomic bombs, the war ended. In the European theater, about 50 million people had died. "What is Europe now?" Prime Minister Churchill asked. "It is a rubble heap, a charnel house, a breeding ground of pestilence and hate."[20]

Postwar Europe

The wounds of that gigantic conflagration have healed through a number of decisive steps toward unity. The European Union is the latest and largest of the various postwar European supranational organizations. There are many reasons, largely the economic ones discussed above, for this push toward greater unity. One of the most important but understated reasons is simply that Europeans want their continent to be war-proof, with its member countries so intertwined in economics, foreign policy, and other ways that war among them would be all but impossible.

Europe is still recovering and reorganizing from another momentous geopolitical shift: the end of the **Cold War**, which came with the collapse of the Soviet Union in 1991. When World War II ended, Soviet forces and political institutions moved in on most of the countries of Eastern Europe, including part of defeated Germany. As Churchill memorably described it, an **"Iron Curtain"** had descended from Poland southward to the Balkan Peninsula. West of it, the capitalist democracies led by Britain, France, West Germany, Italy, and the United States prevailed; to the east, across a 4000-mile (6400-km) line of concertina wire, walls, minefields, and guard towers, the Soviet Union controlled authoritarian, state-run regimes in the countries that came to be known as its European "satellites," including East Germany, Poland, 138 Czechoslovakia, Hungary, Romania, and Bulgaria. The United States nurtured Western Europe's interests and from 1947 to 1952 pumped billions of dollars into the reconstruction of the devastated European infrastructure in a program known as the **Marshall Plan**.

The United States was also the linchpin of the military alliance known as the **North Atlantic Treaty Organization (NATO)**, founded in 1949 between the US, Canada, most of the European countries west of the Iron Curtain, and Turkey. NATO faced off against the **Warsaw Pact**, an alliance of the Soviet Union and its eastern European satellites. On more than one occasion, notably in the Cuban Missile Crisis of 1962 and in the Arab-Israeli War of 1973, these military alliances came all too close to apocalyptic 239, 474 nuclear war.

In 1989, the Berlin Wall, which separated the German city that epitomized the Cold War, was dismantled in a popular and peaceful uprising (**•Figure 4.26**). The Soviet grip on East Germany and its other satellites had been slipping for years, and now it crumbled. The Soviet Union itself disbanded in 1991 (see Chapter 5, **page 173**). An exhilarated public throughout Eastern Europe embraced the prospects of democracy, capitalism, and material improvement. The Warsaw Pact was dissolved. The nuclear arsenals of the respective alliances were reduced substantially. Plans were made to turn the path of the Iron Curtain into the **European Greenbelt**, a mosaic of national parks and other protected areas stretching across 18 countries, from the Barents Sea to the Black Sea. There is an analogous place in the world, where someday an official protected natural area may emerge from a former war front: the demilitarized zone (DMZ) separating North and 358 South Korea.

Since the end of the Cold War, NATO has periodically had a significant peacemaking role. In the Balkans during the 1990s, NATO forces reversed the advance of the Serbs against their neighbors in the former Yugoslavia (see **page 141**). More recently, NATO members deployed their military assets against the regime of Muammar Gaddafi in Libya. In that conflict, the United States, in an effort labeled "leading from behind," orchestrated a leading role for NATO, especially to prevent the United States from sustaining a reputation of warmongering in the Middle East. NATO's membership has grown to 28 with 257 the addition of 9 former Soviet allies (Croatia, Estonia, Latvia, Lithuania, Slovakia, Slovenia, Romania, and Bulgaria; see **•Figure 4.27**). Russia resented this Western expansion into its former sphere of influence, viewing it as encirclement by potential adversaries, but for awhile its opposition was tempered

• Figure 4.26 The fall of the Berlin Wall, November 10, 1989. East German border guards watch peacefully at the Branderburg gate as history unfolds below them.

• Figure 4.27 European members of the North Atlantic Treaty Organization (NATO).

by its own growing alignment with Western defense, security and economic concerns. Along with most of the former Soviet countries, Russia joined NATO's "Partnership for Peace" program, and for a time there was even talk of Russia joining NATO!

All of Russia's growing engagements with Europe and Europe's allies came to a screeching halt with Russia's annexation of Ukraine's Crimea Peninsula in 2014 (see **pages 180–184**). NATO, which had maintained a quiet position on the sidelines, was jolted into action and new strategic thinking. Russian President Putin vowed to protect the interests of ethnic Russians wherever they lived. Countries with significant Russian minorities felt threatened by what they saw as Russian 180 provocation. Three of these were NATO members: Latvia, Estonia, and Lithuania. These three called for and received an immediate (although small) infusion of NATO troops, sending a message that Russia should not contemplate invading these former members of the Soviet Union. NATO's relevance in the post–Cold War world was made clear; Russia did not dare invade a NATO member. After Russia's invasion, Ukraine's Western-oriented government lamented that it had not pushed for NATO membership.

Russia's invasion of Crimea and agitation in eastern Ukraine sent shock waves through Europe. Besides the political destabilization, Russia upset the economic order. Western European economies counted on a steady and secure supply of 111 Russian natural gas through a network of pipelines. To punish Russia for invading Ukrainian territory, the United States called for tough economic sanctions, especially in the sector that would most hurt Russia: energy. The problem, of course, was that any sanction against Russian energy would also punish European consumers. Geography presented no easy alternatives: it would be difficult, costly, and time-consuming for gas-rich Iran, Qatar, or the US, for example, to step into the breach.

American officials found it generally difficult to persuade their European counterparts to be proactive against Russia. Detractors in the United States shared this refrain: the Europeans were enjoying their "vacation from history," counting on the United States to carry the economic, military, and political burdens of countering Vladimir Putin in Eastern Europe and 533 ISIS (Islamic State of Iraq and Syria) in the Middle East.[21] For their part, European powers felt increasing unease that the United States was not providing its traditionally strong and reliable leadership, and instead was seeking to offload many burdens on its shoulders.

An unexpected shock shook up European geopolitics. On July 17, 2014, pro-Russian rebels in Eastern Ukraine shot down a civilian Malaysian airliner (Flight MH 17) flying from Amsterdam to Kuala Lumpur, killing all on board. Most of the passengers were Dutch. This finally led the Europeans to take a tougher stand against Russia, including by tightening some of those economic sanctions that would also hurt Europe. More details are related on **page 183** in Chapter 5.

Another piece on the geopolitical chessboard between the European Union and the United States, is the Transatlantic Trade and Investment Partnership (TTIP). Proponents say that, if agreed upon, the pact would be highly beneficial to EU economies, particularly to Spain, Sweden, and the United Kingdom, boosting such key exports as cars, steel, and chemicals. The pact would also remove tariffs and other trade barriers that have long hindered trade between the European Union and the United States.

But the TTIP wavered and approached collapse, yet another victim of many of the problems plaguing Europe and EU–US relations. It was negotiated in 2011 and 2012 when the euro was endangered and the European Union distracted by internal problems. In 2013, a young man named Edward Snowden quit his job at the American National Security Agency (NSA) and soon took asylum in Russia. Working with journalists, he made public reams of classified information about NSA spying, including embarrassing revelations about American eavesdropping on Germany's Angela Merkel and other European leaders. Another wave of wide and deep anti-American sentiment—including reactions against the TTIP and agreements with the United States—swept Europe.

Using the analogy of the EU as a kind of "United States of Europe," it makes sense that along with its other common institutions the European Union should have a contiguous and effectively borderless subregion in which its citizens can travel freely. It does: this is "Schengenland," an internal geopolitical bloc that has helped those EU countries that are part of it function more efficiently, but also raises serious questions about immigration and broader geopolitical challenges including terrorism (see Regional Perspective, **page 121**).

Differences between Europeans and Americans

Many of us instinctively equate the United States with Europe when we speak about global patterns; after all, we share many common cultural roots, carry a great deal of the world's wealth, and have many of the same measures of quality of life. But there are differences, and these seem to have grown in recent years. "Americans are from Mars and Europeans are from Venus," wrote policy analyst Robert Kagan.[22]

Here are a few differences—and please keep in mind that these are not hard facts, but generalizations reflecting policies and public opinions. The concept of social justice, in which steps are taken to reduce unemployment and poverty among the least privileged, is a key value in Europe, whereas it lags in the United States. European governments and people have historically had a stronger sense of the social compact between the state and its citizens. Governments have been expected to provide "free" public services like education and health care, and citizens have understood they must foot the bill for these "free" services by paying very high taxes. (Incidentally, undergraduate public education in most European nations is free. However, per capita spending on public education is higher in the United States, and an American graduate degree is widely

Regional Perspective Europe's Borders: Welcome (?) to Schengenland

In theory, the European Union would like to move toward a situation in which there were no passport, visa, or other control issues at any internal land, sea, and airport frontiers of its member countries. This would be an extraordinary geopolitical bloc, considering Europe's fractious history. A framework that began outside the European Union and now operates within it attempts this kind of integration. Known as the **Schengen Agreement**, it allows free circulation of people between the nations that signed the agreement; these are the EU countries (except for Britain, Ireland, Croatia, Romania, and Bulgaria), 95 plus non-EU Iceland, Norway, and Switzerland (**• Figure 4.F**). To maintain internal security within this greater **Schengenland**, the member states are supposed to exercise common visa, asylum, and other policies at their external borders. Some countries, notably the United Kingdom, are skeptical of the effectiveness of control of external borders and have stayed out of Schengen. Some people within Schengen countries have challenged their membership; for example, Denmark's rightist Danish People's Party, part of the country's governing coalition, wanted to reimpose border controls to keep a tighter lid on immigration and terrorism. Threatened by expulsion from Schengenland, Denmark did not reinstate border controls but did step up security spot-checks.

Truly open borders across Europe are probably still far in the future. Anti-immigrant fears that rose with the exodus of refugees from North Africa during 2011's Arab Spring, and with the rise of ISIS, have caused some countries, notably France and Denmark, to backpedal on their commitment to the principle of open borders. The main concern is that Europeans who have joined ISIS in the Middle East could return with their fighting expertise, travelling freely across Schengenland to wage terror against Europe. Western European fears of a tidal wave of cheap eastern labor, terrorists, or human traffickers have produced special rules for the eastern countries of the European Union, long regarded as abodes for these shady activities. Overall, like the euro, Schengenland is one of the major European goals that has not been fully achieved.

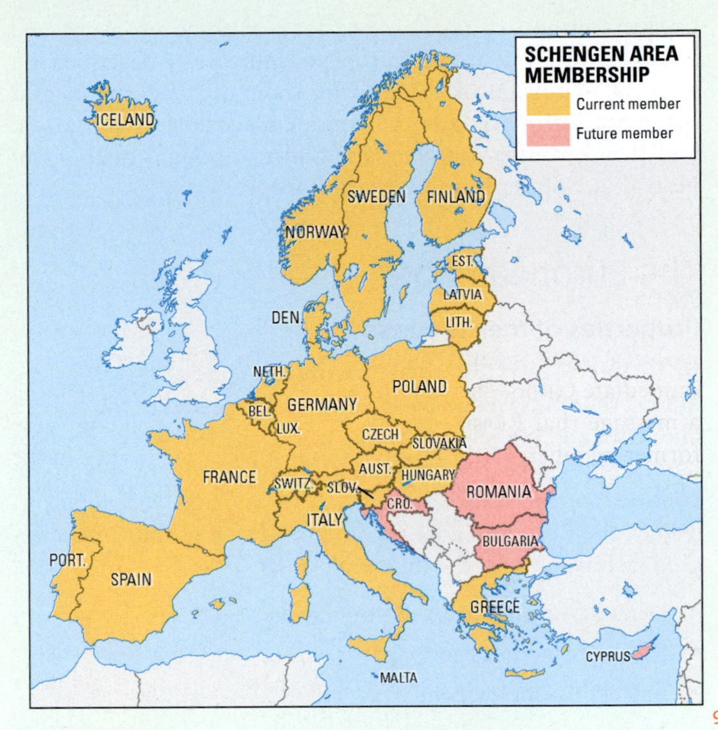

97

• Figure 4.F Schengenland.

THINK CRITICALLY: What are the benefits and drawbacks of open borders in Europe? What roles do immigration and terrorism play?

regarded as superior to that obtainable in Europe.) European peoples and their politicians accept high taxes on gasoline, to pay for a variety of services, as an incentive to drive small cars and conserve energy, whereas for an American politician, advocating higher fuel taxes is political suicide; as a result, gas prices in Europe are roughly double those in the United States. Europeans complain that US cultural industries such as Hollywood films degrade other cultures and marginalize languages other than English. For awhile there was strong resistance, especially in France, to the proliferation of McDonald's and other American fast-food restaurants in Europe; but that has largely faded, and McDonald's has "Europeanized" its restau- 61 rants in both menus and architecture. Europeans are less inclined than Americans to allow questions of spirituality into political debate; in fact, they are repelled by personal expressions of religiosity in public, which they consider indiscreet. Europeans abhor the death penalty, which is outlawed in all EU countries.

Europeans tend to apply the precautionary principle—the notion that it is appropriate to attempt to reduce or eliminate any risky practice—to things they think may be harmful. Thus there may be risks from GM foods, so many people would like to ban them; climate change is probably occurring, so most people want to take action to reduce greenhouse gases. The United States has adopted more of a "show me" attitude—for example, with respect to global warming; compared with Europeans, a larger percentage of Americans think there is not enough scientific proof, so it is better not to impose reductions in carbon dioxide emissions. There are differences on the geopolitical front too. For Americans, September 11, 2001, was the watershed time when it reevaluated its place in the world and began to feel less secure. For Europeans, the fall of the Berlin Wall in 1989, leading to the collapse of the Soviet bloc, was the watershed that made them start to feel more secure. Americans and Europeans are different. *Vive la différence!*

4.6 Regional Issues and Landscapes

In this section, we take a close look at some problems and issues of interest in the European subregions described on **page 89**. We begin by recognizing some of the elements that made the European core the continent's strongest subregion, and then examine characteristic and exceptional qualities of Northern, Southern, and Eastern Europe.

The European Core

Properties of the European Core

In Chapter 1, we saw that core and peripheral locations are among the simplest and most common properties of the Earth's countries and regions (see **page 13**). Europe has a recognizable core, a subregion that has long played a dominant role in the continent's political, economic, and cultural development. Today, the European core is recognized as a prominent subregion of Europe with the following traits:

- Densest, most urbanized population
- Most prosperous economy
- Lowest unemployment
- Most productive agriculture
- Most conservative politics
- Greatest concentration of highways and railroads
- Highest levels of crowding, congestion, and pollution[23]

Defined in this way, the core consists mainly of three of Europe's major countries and of smaller nations in the British Isles and the west central portions of the European mainland (**•Figures 4.28** and **4.29**). The United Kingdom and the Republic of Ireland share the British Isles. The countries of West Central Europe are Germany and France and their smaller neighbors, the Netherlands, Belgium, Luxembourg, Switzerland, and Austria. Also sharing these traits of the core are roughly the northern half of Italy, Denmark, southern Sweden, and parts of Poland, but other characteristics of these countries place them more appropriately in the European periphery, discussed later. Because they are nestled among the core countries, the microstates of Liechtenstein, Monaco, and Andorra are included here. The European core includes some of the world's most geopolitically and economically significant countries (see Table 4.1, **page 92**).

As we saw in Chapter 1, geography is very concerned with appreciating how people have reshaped the face of the Earth. This European subregion became prosperous in part through the exploitation of its own natural resources. Today, it has one of the world's most transformed landscapes, shaped by thousands of years of productive agriculture and hundreds of years of industrial development tempered by a preference for sightliness (**•Figure 4.30**). European countries do not have the natural advantage enjoyed by one of their affluent peers, the United States: vast size. As we will see in North America in

• Figure 4.28 The countries comprising Europe's core.

Chapter 11, there is actually more forest in the United States now than there was when European settlement began (see **page 505**). Europeans have not enjoyed the luxury of allowing vast parts of their hinterlands to keep or reclaim something similar to their primeval conditions (see Geographic Spotlight, **page 125**).

The European core is one of just four regions in the world classified as a **major cluster of continuous settlement**, meaning that all habitations lie no more than 3 miles (5 km) from other habitations in at least six different directions. In addition, roads or railroads lie no farther than 10–20 miles (16–32 km) away in at least three directions.[26]

Europe's cultural landscape is unusually intricate. Different layers of history appear at every turn, and there are sharp differences in human geographies and national perspectives of its peoples. Environments and resources also vary widely. Local resources, especially coal (fossilized carbon) and iron ore, and transportation opportunities were the basis for industrial and urban development in past centuries—that correlation is seen in **•Figure 4.31**—but the modern European core has a pronounced post-industrial geography. There are distinct areas 113 of declining and growing industrial activity. Resources and industries that put many major cities on the map are played out or are in serious decline, in many cases replaced by information technologies, financial services, and other new economic trends.

This process of deindustrialization hit Britain especially hard, but other regions in the European core, notably the Ruhr in Germany, also suffered. High unemployment is a major consequence of this trend, but a positive result is that the cities are cleaner and more livable. Many cities are replacing old industrial areas with greenbelts and attractive architecture.

• **Figure 4.29** Political geography of the European core.

Improved environments should lure in high-tech and other clean industries, financial and other services, and tourists that would drive a new economy (*should* and *would* because these trends were underway before being set back for years by the economic crisis that erupted in 2007). In addition, Europe's high-tech sector suffers from a shortage of skilled engineers and its fragmented markets. Only nine of the world's top 100 high-tech firms were headquartered in Europe in 2014. Not enough future leaders in the Information and Communication Technology (ICT) sector are in the European pipeline now; just 17 percent of EU students are in the STEM fields (science, technology, engineering, and math), compared with 25 percent in the United States, 29 percent in South Korea, and 31 percent in China.[27] Europe does, however, have its share of IT zones, and these have a smaller footprint than the old industrial areas. In a geographical play on words evoking California's Silicon

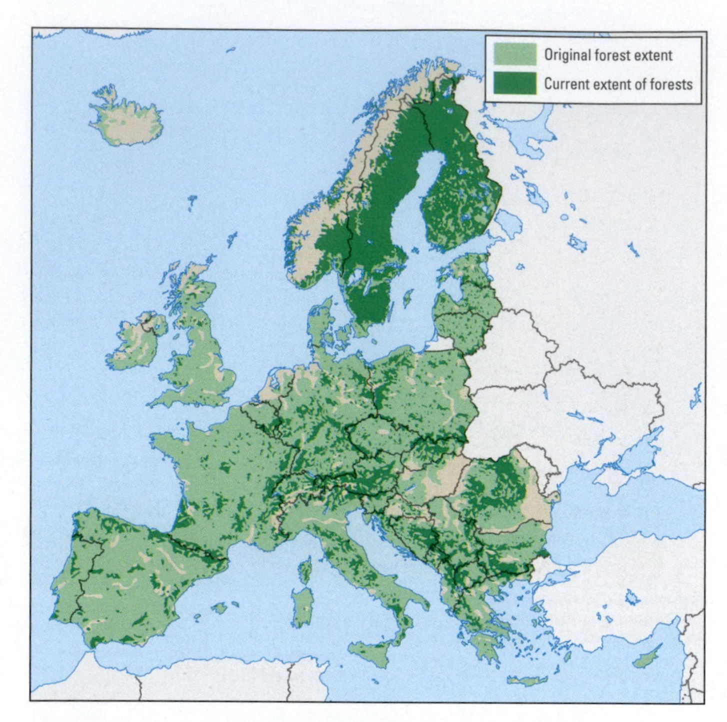

• **Figure 4.30** Human activities over long periods of time have transformed the natural landscapes of Europe. This map depicts "pre-settlement" temperate mixed and coniferous forests, and the distribution of forests currently.

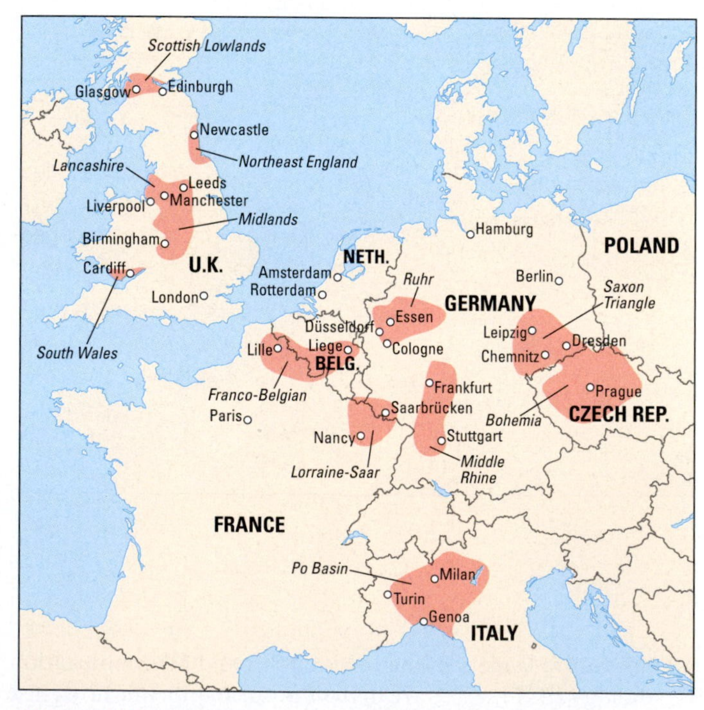

• **Figure 4.31** Historical industrial concentrations, cities, and seaports of the European core. Older industries such as coal mining, heavy metallurgy, heavy chemicals, and textiles clustered in these congested districts. Local coal deposits provided fuel for the Industrial Revolution in most of these areas, which have shifted increasingly to newer forms of industry as coal has lost its economic significance and older industries have declined.

Valley, parts of the British Isles have been dubbed "Silicon Bog," "Silicon Glen," and "Silicon Fen," and the continent boasts Germany's "Silicon Saxony," France's "Scientific City," and northern Italy's "Techno City."[28]

Great Britain: Aging Seat of World Power

The islands of Great Britain and Ireland, off the northwestern coast of Europe, are known as the British Isles. They are home to the Republic of Ireland, with its capital at Dublin, and the much larger United Kingdom of Great Britain and Northern Ireland (often referred to simply as "Britain" or the UK), with its capital at London. Altogether, the United Kingdom encompasses about four-fifths of the area and over 90 percent of the population of the British Isles. The United Kingdom is made up of the entire island of Great Britain with its political units of England, Scotland, and Wales; the northeastern corner of Ireland, known as Northern Ireland; and most of the smaller outlying islands.

Lying only 21 miles (34 km) from France at the closest point, England is the largest of the United Kingdom's four main subdivisions. England was originally an independent political unit and twice conquered mainly Catholic Ireland, including Northern Ireland. The six counties that make up Northern Ireland separated from Ireland in 1922 and maintained their links with the United Kingdom, while the remaining counties achieved full independence as the Republic of Ireland.

Scotland was first joined to England when a Scottish king inherited the English throne in 1603, and the two became one kingdom under the Act of Union in 1707. Scots, however, are extremely proud of their distinctive culture, history, and identity, and bristle when they are identified as "English," as happens all too often. In 1997, nearly 300 years after Scotland joined the union, the UK government yielded to increasing Scottish demands and allowed for the devolution of many administrative powers (including agriculture, education, health, housing, taxation, and lawmaking, but excluding defense and social security) to a new Scottish parliament. In a groundbreaking 2007 election, Scottish nationalists won political dominance in the parliament and promised a referendum on whether Scotland should secede from the United Kingdom and declare independence. In 2014, this question appeared on the ballot for all Scots age 16 and older: "Should Scotland be an independent country?" The results disappointed nationalists: 45 percent of the electorate, shy just six points of the simple majority needed, voted for independence (•**Figure 4.32**). The referendum was a nail-biter, with profound implications had Scotland opted for independence. Economic concerns were decisive in the defeat. Many Scots worried that even with their wealth of North Sea oil, they might not be able to maintain their quality of life. Another decisive consideration was that before the vote, UK politicians promised Scots much stronger control of their own affairs even if the referendum on independence failed. The issue of outright Scottish independence has not faded, however; in fact, a strong showing by the Scottish National Party (SNP) in the UK's 2015 General Election virtually guaranteed that there would be another push for

107

Geographic Spotlight Environmental Perception

Environmental perception is the concept that people, societies, and cultures perceive and use the environment in different ways, depending on their age, gender, socioeconomic standing, language, religion, nationality, and other variables. This is an important concept in geography because it helps explain why people modify landscapes as they do.

Environmental perception is essential for understanding the cultural landscape. In this book we describe the landscapes of Europe and provide photos and maps of them. But these features are superficial, and to understand the landscape—to explain the orderly arrangements and parklike qualities of some European landscapes, for example—we need to drill more deeply and shed light on peoples' values and preferences. With environmental perception, René Dubos, the French humanist and environmentalist who coined the maxim "Think globally, act locally," uses environmental perception to "read" European landscapes:

> I realize that there is little original wilderness left in Europe. Even the highest European mountains have become humanized. Except for their glaciers, the Alps and Pyrenees have acquired a gigantic parklike quality. All over Europe, man has created a semiartificial landscape by converting primeval nature into an orderly arrangement of farmlands, pastures, and wooded areas, all of which are linked by roads and trails which may extend beyond the timberline and onto the glaciers. Much of Europe is so carefully groomed that it resembles a work of art. Man has recreated it in his own image. The wonderful harmony that now exists in many parts of Europe among the various components of nature cannot therefore be regarded as a spontaneous expression of wilderness; it is instead the outcome of a continuous and intimate collaboration between man and the site on which he lives—the "wooing of the earth." At its best, the European landscape is a creation of peasants, painters, and poets.[24]

• **Figure 4.G** Pitmidden Garden, established in 1675. Although in Scotland, it reflects the characteristic "tidiness" of English landscape tastes as described by David Lowenthal and Hugh Prince.

Joe Hobbs

Dubos wrote about the continent as a whole. Geographers David Lowenthal and Hugh Prince tell us why the English have groomed their characteristic landscapes as they have (• **Figure 4.G**):

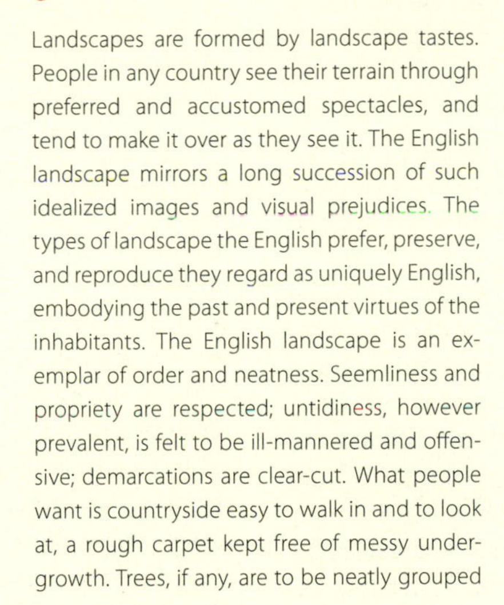

> Landscapes are formed by landscape tastes. People in any country see their terrain through preferred and accustomed spectacles, and tend to make it over as they see it. The English landscape mirrors a long succession of such idealized images and visual prejudices. The types of landscape the English prefer, preserve, and reproduce they regard as uniquely English, embodying the past and present virtues of the inhabitants. The English landscape is an exemplar of order and neatness. Seemliness and propriety are respected; untidiness, however prevalent, is felt to be ill-mannered and offensive; demarcations are clear-cut. What people want is countryside easy to walk in and to look at, a rough carpet kept free of messy undergrowth. Trees, if any, are to be neatly grouped or trimmed so as not to detract from the grassy expanse. Most downland and much other pastureland in England share with private gardens and landscape parks this amalgam of clarity, order, and accessibility. The scraggly growths of briar and bramble, the motley groundcover, the untended seedlings that clutter backyards, roadsides, and derelict agricultural lands in America would be tolerated by few English landowners, private or public.[25]

Environmental perception, on the humanistic side of geography, is as fundamental to the discipline as GIS is. Environmental perception plays a major role in answering those fundamental geographic questions of *"What is where, why there, and why care?"*

an independent Scotland. Devolution of more power to both Wales and Northern Ireland is yet another likely outcome of the election.

England conquered Wales in the Middle Ages, but like the Scots, the Welsh managed to preserve their cultural distinctiveness and hold it high—especially in staying true to their Welsh language (a member of the Celtic subfamily). Any thought that the Welsh are English withers when one considers this Welsh place name (celebrated as the longest place name in the world): Llanfairpwllgwyngyllgogerychwyrndrobwllllantysiliogogogoch, meaning "The Church of Saint Mary in the Hollow of White Hazel Trees near the Rapid Whirlpool by Saint Tysilio's of the Red Cave." A 1997 referendum voted into existence a Welsh assembly dedicated to promoting Welsh interests. In 2011, the Assembly assumed primary lawmaking powers independent of the United Kingdom. Growing numbers of Welsh are calling for a stronger devolution of power that would transfer more authority from the national government to a full-fledged parliament, as Scotland has now.

Between the defeat of Napoleonic France in 1815 and the outbreak of World War I in 1914, the United Kingdom was the world's most powerful country. Its overseas empire covered a quarter of the Earth at its maximum extent; "the sun never sets on the British Empire" was a proud boast. The influence of English law, education, and culture spread still farther. The British take great pride in much of that legacy; the

Dylan Martinez/Reuters

• Figure 4.32 Waving Scotland's flag (the "Saltires" or "St. Andrew's Cross"), supporters of Scottish independence made an eleventh-hour "yes" pitch for support on Glasgow's streets. The "nos" won the next day, September 18, 2014, keeping Scotland within the UK, at least temporarily.

©Liubov Terletska/Shutterstock.com

• Figure 4.33 Fronting the Thames River in London, the Palace of Westminster complex encompasses Britain's Houses of Parliament and the iconic clock tower of Big Ben. Behind Big Ben is Westminster Abbey. Since construction of this Anglo-French Gothic style church began in 1245, it has been the burial place of British monarchs, writers, artists, scientists, and other eminent citizens. Royal weddings and coronations also take place there.

Magna Carta, an English charter written in 1215, granted certain human rights that centuries later would be enshrined in the constitutions of other democracies, including that of the United States. However, Britain was one of the main forces behind the cruel industry of the enslavement of Africans. Until the late 19th century, the United Kingdom was the world's leading manufacturing and trading nation. Britain's Royal Navy dominated the seas, and its merchant marine moved half or more of the world's ocean trade.

London became the center of a free-trade and financial system that invested its profits from industry and commerce around the world (**•Figure 4.33**). London continues to be one of the world's greatest financial centers (see Geography of Cities, **page 127**). Unlike many of the country's other large cities, London did not owe its existence to the proximity of nearby coal or other natural mineral resources. The Thames River has provided London's port with access to worldwide ocean trade. The city has long been a magnet for entrepreneurs, artists, immigrants, tourists, and vagabonds.

The United Kingdom's economic and political decline began late in the 19th century. World War I damaged its free-trade and financial system, and the downward trend accelerated after World War II. All of the large colonies of the former British Empire gained independence by the early 1980s, and Britain's remaining overseas possessions are scattered and small.

The UK is, however, still a very important country on the world stage, in large part because of its lasting impacts from colonial times. It plays a major role in the European Union and is associated with many of its former colonies in the worldwide **Commonwealth of Nations.** The Commonwealth is a voluntary association of 54 countries that nominally recognizes the British monarch as its head. Its faithful alliance with the United States, even in times of great adversity such as during the wars in Iraq and Afghanistan, is another of Britain's characteristic traits in the international arena. Britain's involvement in those wars was, however, costly in lives and in money, and it deepened the fallout on the country from the global economic financial crisis. There is a growing sense among Britons that their country is experiencing "declinism," with Britain taking a more passive place in world affairs.[30] That perception mirrors Americans' uneasiness that the United States too is losing ground.

Much world culture has British roots. Global dominance of the English language has been discussed on **page 105**. In the 1960s, the Beatles, four fab lads from Liverpool (a city recently crowned a "European Capital of Culture"), revolutionized music in their home country, moved on to conquer a generation in the United States, and left an indelible mark on music everywhere. British royalty, pop culture, and great historical and literary legacies draw fans and visitors from around the world. For the foreigner, images and artifacts of Britain may be distilled into something homogeneous, but a Briton of Anglo origin is quick to speak about his or her unique piece of the country and what makes it distinct. And then there is multicultural Britain, in large part a legacy of its colonial past, with a bit of Jamaica, India, Kenya, and scores of other lands infused in every corner of the country.

Ireland: Struggles and Resilience on the Emerald Isle

Ireland (known as *Eire,* pronounced like "air" or "air-a," in the indigenous Celtic language) is a land of hills and lakes, marshes and peat bogs, cool dampness, and the verdant grassland that earned its epithet the "Emerald Isle" (**•Figure 4.34**). The island consists of a central plain surrounded on the north, south, and west by hills and low rounded mountains.

Ireland ranked low among western European powers for many decades, but recent economic growth changed this

Geography of Cities Global Cities

Urban geography classifies London as a **"global city,"** also known as a "world city" or "alpha city," as a city that is a center of economic and political power in the global economy. A. T. Kearney's authoritative study of global cities observes that "globally integrated cities are intimately linked to economic and human development. By creating an environment that spawns, attracts, and retains top talent, businesses, ideas, and capital, a global city can generate benefits that extend far beyond municipal boundaries."[29] • **Figure 4.H** depicts the world's leading global cities in 2014, with London a close second to New York.

Global cities share these attributes:

- A variety of international financial services, notably in finance, insurance, real estate, banking, accountancy, and marketing
- Headquarters of several multinational corporations
- The existence of financial headquarters, a stock exchange, and major financial institutions
- Domination of the trade and economy of a large surrounding area
- Major manufacturing centers with port and container facilities
- Considerable decision-making power on a daily basis and at a global level
- Centers of new ideas and innovation in business, economics, culture, and politics
- Centers of media and communications for global networks
- Dominance of the national region with great international significance
- High percentage of residents employed in the services sector and information sector
- High-quality educational institutions, including renowned universities, international student attendance, and research facilities
- Multifunctional infrastructure offering some of the best legal, medical, and entertainment facilities in the country

Appearing just as dots on a map, these cities are in fact at the dynamic crossroads of the forces of globalization. Constantly in motion with the tides of economic and geopolitical events, these cities can gain or lose their rankings over time. It can well be imagined, for example, that 17th-ranked Moscow could fall off the top 20 list of global cities as its economy, reflecting Russia's, sinks.

174

• **Figure 4.H** Global cities of the world. The factors used to determine a city's score are its level of business activity, human capital, information exchange, cultural amenities, and political engagement. Source: A.T. Kearney, "Global Cities, Present and Future," 2014.

Katie Hobbs

• **Figure 4.34** View of Ireland's characteristic green landscape from the Rock of Cashel (St. Patrick's Rock) in County Tipperary.

pattern. High-tech industries such as electronic products and software boosted Ireland's economy. The size of the economy doubled in the 1990s, earning Ireland the nickname **"Celtic Tiger"** and placing it on par with the United Kingdom in per capita GDP (PPP) (see Table 4.1, **page 92**). Unemployment dropped sharply. Ireland's government sparked the country's surging development with a program attracting foreign-owned industries with inexpensive labor, tax concessions, and help in financing plant construction. Hundreds of industrial factories, owned mainly by US, British, and German companies, began to produce computers, electronics, foods, textiles, office machinery, organic chemicals, and clothing, and Ireland's economy became more than 50 percent reliant on exports of these products to overseas markets. Most of these plants sprang up around the two main cities, Dublin and Cork, where they make up an economic region dubbed "Silicon Bog," and near Shannon International Airport in western 124 Ireland.

Unfortunately, however, Ireland is the "I" in the PIGS economies described on **page 117**. Much of Ireland's strong economic growth—with an annual rate of about 7 percent between 1999 and 2007—was the result of a borrowing and spending spree prompted by very low interest rates. Housing and banking bubbles developed, and inevitably—as elsewhere in the Eurozone, but with greater consequences—they popped. In 2008, the country adopted economic austerity measures including increased taxes and cuts in salaries and 117 budgets. Ireland fell into a deep recession as industrial and service companies shuttered completely or cut way back on staff. As happened so many times in Ireland's history, the Irish began fleeing their home island in search of better lives abroad. But this time, the Irish emigrants took skills in IT, law, and financial services with them—they were among Ireland's best and brightest, not menial laborers looking for construction jobs abroad (this is one manifestation of the "brain drain" problem described in Chapter 3; see **page 76**). Despite this loss of human capital, the deep sacrifices made by the Irish helped the country come out of recession rather quickly. Low wages and prices attracted more foreign investment, and exports surged. The Celtic Tiger clawed its way back up to having the EU's highest annual growth rate (7 percent again) in 2014.

Ireland's Troubles

In the past, Ireland had an agrarian economy that made the country ripe for suffering. England conquered Ireland in the 17th century and designated it part of the United Kingdom. Ireland was in effect an English colony. Most of the land was expropriated (seized from its owners) and divided among English landlords into large estates, where Irish peasants worked as tenants. Most of the landlords were absentees, living not in Ireland, but in England, and income from their enterprises generally found its way to Britain. During the 17th and 18th centuries, British authorities forbade Irish Catholics, who made up about 80 percent of the population, from owning land, voting, getting an education, and living within 5 miles of a town. Under British rule, Ireland became a land of poverty, sullen hostility, and periodic violence. Worse yet, a crop disease known as blight triggered the notorious **potato famine** (*Gorta Mor*, the "Great Hunger") of 1845–1851. Nearly 10 percent of Ireland's pre-famine population of 8 million died of starvation or disease during these years. While the Irish died, under British military guard Ireland exported diverse and large amounts of crop and livestock products to England and Scotland, with profits going the same directions. In his 2013 book entitled *The Famine Plot*, prominent Irish historian Tim Pat Coogan makes the case that the Irish potato famine was genocide carried out under British colonial rule.

During the famine, large numbers of Irish emigrated to England, Wales, Scotland, and, in "coffin ships," to Australia and North America. Between emigration and starvation, Ireland's population declined by 20–25 percent during the Potato Famine. The large Irish American community in the United States traces much of its ancestry to this migration. Today, Americans of Irish descent are just a few of the eager consumers abroad of Irish culture, especially dance and music.

Northern Ireland, now a majority Protestant part of the United Kingdom, has had a particularly difficult struggle that also has colonial roots. One outcome of the second English conquest of Ireland, in the 17th century, was the settlement of English and Scottish Presbyterians and other Protestants in the north, where they became numerically dominant. These newcomers were supposed to form a loyal population in an otherwise hostile, conquered, largely Roman Catholic country. In 1921, when the Irish Free State (later to become the Republic of Ireland) was established in the southern, mainly Catholic part of the island, for economic and religious reasons the predominantly Protestant north chose to remain with the United Kingdom. Northern Ireland, known in Britain as Ulster, was given much autonomy, including its own parliament in an estate called Stormont in Belfast (Stormont is a metonym for the Government of Northern Ireland). Northern Ireland's Catholic minority felt increasingly marginalized by the pro-British Protestant majority and in the late 1960s began agitating—often violently—for change. The British responded by imposing **direct rule**—in which the UK central government in London made all major policy decisions for Northern Ireland—and Catholic discontent soon focused on the perceived British occupation. The **Irish Republican Army (IRA)** began a campaign of bombings, shootings, and arson, sometimes into the heart of England, ostensibly designed to drive the British army from Northern Ireland. **The Troubles**, as the locals refer to this period of strife, involved an often violent struggle between **Catholic Republicans** and **Protestant Unionists** in Northern Ireland.

In 1998, by which time more than 3500 Irish on both sides had died in the conflict, the US-brokered **Good Friday Agreement** was signed. Approved by the IRA (and its political counterpart, **Sinn Fein**, pronounced "shin fayn") and the Unionists, the agreement called for the devolution of British power to a Northern Irish legislature. Direct rule by Britain was to be replaced by the **Northern Ireland Assembly**, in which Protestants and Catholics would share power. Violent challenges to the peace agreement persisted from both sides, however, leading the British to halt devolution and reimpose direct rule in 2002. In 2005, the IRA conceded to permanently cease military operations. Despite the refusal of three small splinter groups, including one calling itself the "Real Irish Republican Army," to lay down arms, Unionists and the IRA did settle into a power sharing arrangement. The two sides continue to negotiate a more enduring agreement. Theoretically there could even be a referendum for Northern Ireland's independence from the UK.

If it lasts, peace could help bring prosperity to Northern Ireland. From 1997 to 2007 Ulster's economic growth was high, second only to London's in the UK. Northern Ireland's economy was based previously on a strong manufacturing sector with shipbuilding and linen textile industries concentrated around Belfast, on the east coast. Set back by "the Troubles," its own burst property bubble, and the high British tax rate that discourages investment, Northern Ireland's economy has recently rebounded. About 70 percent of its GDP comes from services, and its manufacturing is now centered on high-tech and capital-intensive industries, including aerospace.

Paris as a Primate City

A country's **primate city** is a city that, according to a convenient definition, has a population at least twice as large as its next largest city (sometimes it is defined as more populous than the second and third largest cities combined). Paris, for example, has a metropolitan area population of 11 million, far exceeding the doubled population of France's next largest cities, Lyon (with 1.5 million people) and Marseille (with 1.4 million). In practical terms, a primate city dwarfs all others in its demographic, economic, political, and cultural importance. The primate city produces and consumes a disproportionately high share of the country's goods and services, and many national resources—such as bureaucratic offices and educational opportunities—are available only in the primate city. The primate city therefore acts as a great magnet for rural-to-urban migration and tends to absorb more investment than other communities.

Urban primacy is often bad for a country's development because funds that could be spent to improve services and quality of life in smaller cities and rural areas are diverted to the primate city instead. With greater opportunities available in the primate city than in the smaller cities of the hinterland, the most educated and talented people abandon those centers for the primate city. This internal brain drain contributes further to underdevelopment outside the primate city. These situations are typical of the LDCs, where development in smaller cities and in rural areas lags. Cairo, Karachi, and Mexico City are exemplary primate cities of Africa, Asia, and the Americas.

Paris is the greatest urban and industrial center of France, a primate city overshadowing all other cities in both population and economic activity. This alpha city is by far the largest city on the mainland of Europe. Paris is located at a strategic point on the Seine River relative to natural lines of transportation; therefore, its geographic *situation*, as described earlier in the chapter, was important in the city's development. During the Middle Ages, Paris became the capital of a succession of kings who gradually extended their control over all of France. As the French monarchy became more absolute and centralized, its seat of power grew in size and came to dominate the cultural and political life of France. As Paris city grew in population and wealth it also became an increasingly large and rich market for goods. Paris' growth was in large part the product of the growth and centralization of the French government and the transportation system it created. As national road and rail systems were built, their trunk lines were laid out to connect Paris with outlying regions. The result was a radial pattern with Paris at the hub, and this is essentially the pattern that exists today. It would be appropriate to say of France that "all roads lead to Paris" (see **• Figure 4.35**).

Like London, which is also a primate city, Paris has no major natural resources for the industry in its immediate vicinity, yet it is the greatest industrial center of the country. Today, Paris generates a third of the country's wealth and represents the world's sixth largest urban economy. It is also one of the world's costliest cities in which to buy goods and services; all tourists experience sticker shock.

Paris' local market, plus transportation advantages and proximity to government offices, provided the foundations for its huge industrial complex. This development involved two major classes of industries. Paris became the main producer of the high-quality luxury items (fashions, perfumes, cosmetics, and jewelry, for example) for which that city and France have long been famous. Paris also became the country's leading center of engineering industries, secondary metal manufacturing, and diversified light industries, concentrated in a ring of industrial suburbs that sprang up in the 19th and 20th

• Figure 4.35 Paris is the primate city at the hub of France's transportation network.

centuries. Typically of post-industrial Europe, Paris' economy shifted away from manufacturing and today is dominated by the service sector, with business, professional, technical, and other services employing more than 80 percent of the city's workforce.

Paris inverts the usual pattern of the industrial city by maintaining a core of low-profile historic monuments, while the skyscrapers of the modern economy tower away from the city center (this layout is clearly visible in **•Figure 4.36**). Parisian urban planners have carefully cultivated and protected the "crown jewels" of this beautiful city's architecture, and they have succeeded admirably. Paris maintains its international reputation as the romantic and monument-studded **"City of Light."** It is the world's number 1 urban tourist destination, and France attracts more tourists than any other country in the world.

• Figure 4.36 The Champs de Mars leads to the Eiffel Tower in Paris. In the distance is the skyline of the city's modern commercial district.

Divided and Reunified Germany

One of the most important geopolitical events of the late 20th century—and one that has lingering effects in the new millennium—was the reunification of the two Germanys. If you go on to study Korea in the book, you will see that one of the big questions for this century is whether the divided Koreas might follow the German example. In both cases, economic issues play a prominent role; in each case, there is a stronger, wealthier country to bear much of the cost of merging with a weaker and poorer counterpart.

The Federal Republic of Germany (**•Figure 4.37**) reappeared on the map of Europe as a unified country in 1990. Between 1949 and 1990, it had been divided into two nations: democratic West Germany (known as the German Federal Republic), formed from the zones occupied by Britain, France, and the United States after Germany's defeat in World War II in 1945, and Communist East Germany (the so-called German Democratic Republic), formed from territory occupied after the war by the Soviet Union. The former capital, Berlin, was similarly divided. West Berlin was a part of West Germany for those decades, even though the city was entirely surrounded by East Germany. The withdrawal of Soviet support for East Germany

• Figure 4.37 Principal features of Germany. Note which parts of the country belong to what formerly were East and West Germany.

in 1989 and the destruction of the Berlin Wall led to the rapid collapse of its Communist government in 1990 and to the jubilant reunification of Germany's 16 states. 119

Reunified Germany is Europe's dominant country in many respects. Its population of 81 million is much greater than that of any other nation in the region. Politically it is seen, along with France, as the cornerstone of the European Union. Economic considerations are even more important; Germany has the world's fourth largest economy, after China, the United States, and Japan. Germany is among the world's top three countries, along with China and the United States, in exports of goods. Germany typically bears an outsized burden in paying for the economic problems of the Eurozone and the European Union. 117

The former West Germany was Europe's leading industrial and trading country during the latter half of the 20th century. The former East Germany was an advanced country by Soviet bloc standards but was an economic disaster by West German standards. Many billions of dollars and much time have been required to rehabilitate Germany's east, and the task is still incomplete.

Germany's urban geography is unusual. Industrial and urban centers are widely scattered. Many cities are large, but none is as nationally dominant as London or Paris. This dispersed pattern came about because Germany was divided for a long time into petty states, many of which eventually became internal states (*länder*) within German federal structures. These often had their own capitals, which would in time become Germany's large cities.

During World War II the western Allies, led by the United States, heavily bombed many of these industrial cities, which also had large civilian populations. Perhaps the most tragic bombing was of Dresden in Germany's Saxony region, in 1945. The rationale for targeting this city, when Germany's demise was already assured, was questionable. The firestorm produced by the incendiary bombing of this city, known as the "Florence on the Elbe" because of its cultural treasures, was immortalized in Kurt Vonnegut's novel *Slaughterhouse Five*.

Allied bombing did not eliminate Germany's industrial capacity. West Germany was reconstructed rapidly in the postwar years, thanks to several factors. The massive US-funded aid package known as the Marshall Plan supplied billions of dollars in capital goods between 1948 and 1952 (see **page 119**). This aid was strategically aimed at rebuilding the country to stave off the Communist threat from the east. Skilled workers were eager to work, even for low wages, and markets were hungry for the industrial products that Germans could make. Germany's economic growth continued through the 1960s. Foreign workers, especially Turks and Kurds from Turkey, filled labor vacancies created by upwardly mobile Germany workers. 94

East Germany, Now Eastern Germany

East Germany was Communist ruled and Soviet dominated, but it was still German. Even Soviet exploitation and bureaucratic inefficiency were unable to destroy the traditional German ethics of hard work, efficiency, attention to detail, and high standards. East Germany was the most productive of the Soviet satellite countries that lay between Russia and Western Europe, but its economy lagged far behind West Germany's. When East Germany collapsed in 1989 and the country was unified, a major migration stream began flowing from east to west. More than 4 million people fled westward from what Germans call the "new states" of the former East Germany. The emigrating easterners left behind what one German professor described as "a series of towns and enclaves for senior citizens."[31] The urban landscapes of the east changed to accommodate an aging population. Hundreds of the old Soviet-style apartment blocks were demolished, and new parks and other green spaces developed in their wake. The ongoing urban architectural experiment in eastern Germany, known as *Rueckbau*, is a cross between demolition and reconstruction.

The pool of young and educated people that remained in the east began to attract employers looking for lower-wage workers, and the region's economy rebounded. In a striking counter-migration, more than 3 million people have moved from the west to the east to take advantage of new economic opportunities. There is a growing semiconductor industry in what local promoters like to call "Silicon Saxony" (one of those plays on words, modeled on Silicon Valley, as described on **page 124**), especially in Dresden and Leipzig. The automakers Volkswagen and BMW have opened plants in Dresden and Leipzig, respectively. Employing skilled and inexpensive workers, scores of US chip-making companies have set up operations in the region.

Billions of dollars of investment are helping to raise economic standards in the east and to clean up the severe environmental damage inflicted during Communist rule there. The former West Germans are bearing the lion's share of the tax and social burdens of bringing the former East Germans up to the standards long enjoyed in the west. More than 25 years after reunification, western Germans (known as "Wessies" by the east Germans) continue to resent carrying the load of this program of "fiscal equalization." They believe that easterners (whom they call "Ossies") are ungrateful for their help and have developed a culture of dependency.[32]

Economically, Germany has been among those European welfare states subsidizing its citizens generously. 113 The costs for that support contributed to Germany's economic downturn early in the 2000s, when economic growth was the slowest of all EU countries and unemployment surged to 10 percent. In the area of former East Germany, conditions were worse, with the rate of unemployment more than twice that in the west. Germany adopted a course of economic austerity as it set about its painful transition from a welfare state to one in which citizens increasingly pay for health care and other vital services. Other difficult measures included cutting jobs, 131 extending work hours without extra pay, and outsourcing work to other countries. These measures helped Germany's export-driven economy to rebound after 2005 and grow even more sharply after 2009 in the wake of the global recession. Germany's climb out of its economic hole has been remarkable, and it raised hopes for an economic rebound throughout the European Union.

But is Germany really committed to the European community and to its partnership in the Atlantic alliances, including NATO? These became nagging questions after 2010. Germany is still considered the economic locomotive of Europe, and other EU member countries look to Germany for leadership and aid (many of the poorer EU countries came to believe that Germany was their "paymaster" in times of need). Germany is, however, increasingly cultivating economic partnerships with countries to its east. Fewer German manufactured goods like precision machinery, cars, chemicals, and electronics are going to France, whereas more are being exported to China, India, and Russia. France is foremost among the EU countries feeling nervous that Germany will "decouple," going its own way and serving its own interests rather than those of the EU. Germany also broke ranks politically with its NATO allies, including France, the United Kingdom, and the United States, in refusing to support military action against Muammar Gaddafi's Libya in 2011.

267

Europe Lights the Way on Alternative Energy

An EU poll known as the "Eurobarometer" reveals that 50 percent of EU citizens regard climate change as one of the world's most serious problems (by comparison, 28 percent of Americans hold this view). Reflecting public sentiment, the European Union has taken the official stance that global climate must be confronted through dramatic changes in the ways energy is produced and consumed (**• Figure 4.38**). Both to help mitigate global climate change and reduce its

Joe Hobbs

(a)

Joe Hobbs

(b)

©Kiev.Victor/Shutterstock.com

(c)

Joe Hobbs

(d)

• Figure 4.38 Europe has a wide variety of energy sources, with Poland almost completely reliant on coal to produce electricity and France nearly so on nuclear power. Traditional and modern sources of energy in Europe include **(a)** peat from a Scottish bog, **(b)** coal in Poland, and windmills, both old **(c)** (in the Netherlands) and new **(d)** (in Denmark).

THINK CRITICALLY: With your knowledge from this book and other sources, do you think that Europeans are "tilting at windmills" and paying too much for alternative energies? Do these technologies make economic sense for the United States? What role does climate change have in your views?

dependence on fossil fuel imports, the EU has a target of deriving 20 percent of its energy consumption from renewable sources (including nuclear) by the year 2020 (by the Kyoto Protocol GHG benchmarks described on page 45, that will require cuts in greenhouse gas emissions to 20 percent *below* their 1990 levels). This is the overall 2020 target for the 28 countries; individual country targets for renewables' shares are variable, from 50 percent in Sweden to 10 percent in Malta.

Profound changes are underway to turn the clock back on the relentless forward movement of Europe's fossil fuel–driven economies and lifestyles. These include increasing fuel efficiency in automobiles, encouraging even more use of public transportation, expanding the cap-and-trade mechanisms of carbon emissions trading, and bringing a wide array of alternative energies—hydroelectric power, solar power, tidal power, wave power, geothermal power, nuclear power, and even trash power—online.

One of nuclear power's advantages is that it produces no greenhouse gas emissions, but EU states differ in using this most dangerous energy source. France is the world champion of nuclear energy, getting about 80 percent of its electricity from nuclear power. Until 2011, Germany was a powerhouse of nuclear energy, deriving 25 percent of its electricity from this source, but now it has vowed to phase out its nuclear power program completely by 2022. What happened? The devastating tsunami of 2011 in Fukushima, Japan, and the series of nuclear power plant accidents in its wake occurred. Awed by what happened in Japan, Germany's leaders and citizens decided that nuclear power was simply too risky to maintain. France, on the other hand, is too deeply committed to nuclear power to abandon this resource.

Germany's decision to forsake nuclear power puts it in a difficult position to attain its target of getting 18 percent of its total energy from renewables by 2020. Germany is rising to this challenge with the world's most creative and successful renewable energy sector in the world, and already by 2009 earned recognition as "the world's first major renewable energy economy."[33] Germany's ambition is to be the world's first industrial power to use 100 percent renewable energy by 2050, using a mix of wind, solar, geothermal, hydropower, and biomass.

One of the most popular renewable energies in Europe is wind power. Germany has established itself as the world's leading manufacturer of wind turbines. More than 20,000 of these mills dot Germany itself, providing about 8 percent of the country's electricity needs. That share is projected to rise to 20 percent by 2020, after 40 offshore wind farms as far as 70 miles from the coast come online. Denmark uses wind power to meet almost 30 percent of its electricity needs—the highest rate in the world from this source. Denmark aims to be 50 percent reliant on wind energy for electricity by 2020 and is racing against Germany to be fossil fuel–free by 2050. To generate 10 percent of its electricity by 2020, Britain is building more than 6,000 wind turbines off England's coast, mainly in the Thames estuary, the east coast area, and along the northwest coast.

Wind energy produces no greenhouse gas emissions, and you might think Greens would embrace it. But environmentalist Britons, recalling William Blake's image of the "dark satanic mills" of early English industry, lament that the "silver satanic blades" will sully the lands of the storied Lake District of the northwest.[34] The **NIMBY** ("not in my back yard") attitude often wins the day. This viewpoint, combined with the fact that Europeans are very short on land (compared with Americans especially), as well as the characteristic geography of Western Europe's numerous islands, peninsulas, and seas, and enormous amount of coastline, makes a perfect geographic prescription for offshore wind power.

Some of the technical challenges of providing wind power are impressive. Have a look at the North Sea in Figure 4.1 on page 90. Its waters can be turbulent, with winds often howling at 60 miles per hour and waves cresting higher than 30 feet (9 m). Many of Britain's 6,000 gigantic wind turbines will be installed in these treacherous North Sea waters. Fortunately, at their innovation hub in Aberdeen, Scotland, British engineers have already developed the know-how of installing oil rigs in the North Sea and in other deep and daunting marine environments around the world. The plan is to put that fossil fuel experience to work on alternative energies. Innovation is the name of the game in European alternative energies, with wave power perhaps the most innovative. Ingenious wave energy converters are already at work in Portuguese and Scottish waters.

Can Europe afford this renewable energy future, especially given its rather bleak economic situation? Although the cost of renewables has been declining, they are still not competitive with fossil fuels, and they are heavily subsidized through taxes on Europeans. Yet the Eurobarometer polls reveal that even Greece and other countries hit hardest by the downturn, people want to push ahead with renewables. Generally, European peoples, governments, and industries see renewables as an investment that will lower their costs in the long term, whereas in the United States there are lingering perceptions that moving away from fossil fuels will be too costly—and, of course, climate change itself has influential doubters.

The European Periphery

Properties of the European Periphery

Just as there is a recognizable European core, where the region's demographic, economic, and political weight is centered, there is also a **European periphery**. This "rimland" is 89 made up of countries whose interests are tied closely to those of the core and are strongly influenced by the core countries. But they have less clout in the political and economic systems than the core countries do and tend to be dependent on the core. As we have seen in the context of the European Union, for example, many of the countries of the periphery hope that the core countries Germany and France will bail them out of their economic troubles.

Here we use the tool of geographic regions to subdivide these countries into three categories: those of Northern,

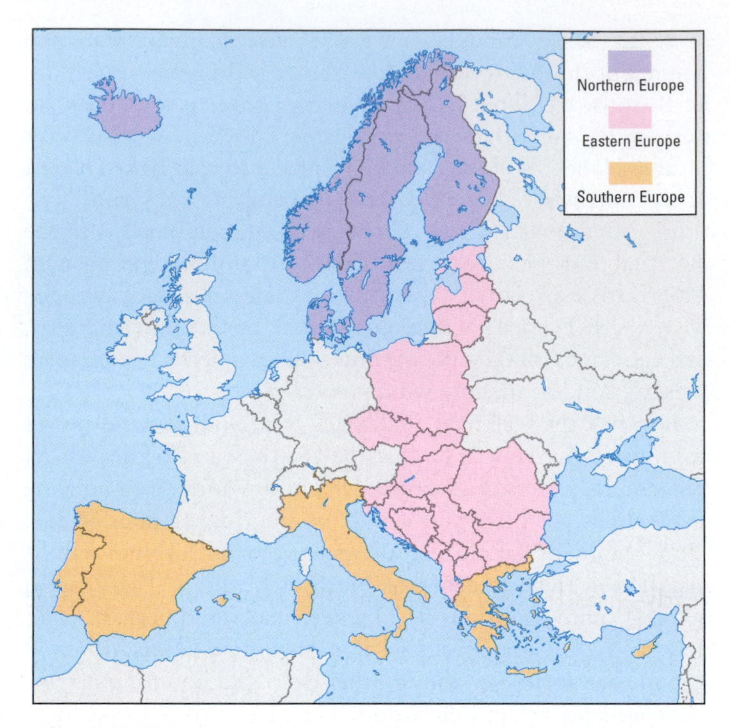

• **Figure 4.39** The subregions and countries of the European periphery.

Southern, and Eastern Europe (•**Figure 4.39**). Within these subgroups, the countries have distinct national identities, histories, and experiences. In this section, we explore some problems and issues of these countries outside the European core. The major traits of each nation are listed in Table 4.1 on **page 92**.

Northern Europe

Save the Whales . . . for Dinner One reason that both Norway and Iceland (see •**Figure 4.40**) have refused to join the European Union is that both have economically vital fishing industries, and they fear that common EU policies on fishing might diminish their fishing profits. These two countries, along with Japan, also are distinguished by their unique perspective on another marine resource: the world's whales. People in all three countries maintain that populations of some whale species, especially the minke (pronounced "ming-key"), have rebounded to levels that should allow regular, limited harvesting for human consumption. One of their arguments is that growing populations of these and other whales they hunt (fin, Bryde's, sei, humpback, and sperm) would dent the fishing industry by feeding on huge amounts of commercially important fish stocks. Whale meat has also long been a prized food for people in these countries. "Save the whales for the dinner table" is the barbed Norwegian response to the popular Western bumper sticker "Save the whales."

In a landmark agreement honored by most of the world's marine countries, the International Whaling Commission (IWC) banned commercial whaling worldwide in 1986. The Norwegians never accepted the ban, calling it "cultural imperialism," and defiantly continue to take about 500 to 600 minkes each year, well short of their self-imposed quota of

• **Figure 4.40** Whaling takes place within the exclusive economic zones (EEZs) of Norway and Iceland.

about 1300 whales per year (•**Figure 4.41a**). Invited recently into a family home in Bergen, I was honored with a hefty serving of lightly singed, almost uncooked, very dark minke meat and a side of potatoes, salad, and red wine (•**Figure 4.41b**). It did not taste like chicken. Norwegians eat whale meat without remorse, arguing that the rest of the world is foolish in its insistence that the whale is a "sacred cow." Even Norway's famously staunch environmentalists see no problem with the whale harvest. But with video footage of water stained red by the blood of writhing and apparently suffering whales, environmentalists outside the country have stepped up pressure on Norway to stop whale hunting.

Although Norway simply defies the IWC whaling ban and since 1993 has hunted whales commercially for consumption in Norway and for export abroad (especially to Japan), the other whaling countries at least pretend to be considerate of the ban. The 1986 ban forbids commercial whaling but allows whales to be killed for scientific purposes and for subsistence (the latter clause allows indigenous peoples in

AP Images/Svein Andersen/HO

• **Figure 4.41a** Norwegian whalers haul in a minke whale.

Joe Hobbs

• **Figure 4.41b** A Norwegian host preparing minke meat; it is only lightly singed before serving.

Alaska, for example, to hunt whales). Iceland and Japan have exploited the loophole of the scientific purpose and by this means put whale meat on the table. They argue that there is a legitimate need for the science: Iceland estimates that its territorial waters are inhabited by more than 40,000 whales that eat perhaps 1 to 2 million tons of fish each year. Advocates insist that only by studying the whales' stomach contents can reliable estimates of these mammals' impacts on Iceland's vital fishing industry be evaluated. The studies accomplish two other goals: they reduce the pressure on the fish resource, and they allow whale meat to find its way to markets for ordinary consumers.

Killing whales does carry some external costs. Iceland resumed commercial whaling in 2006, with a 2013–2018 annual quota of 229 minkes and 154 fin whales each year. Most Icelanders support the hunt, but others argue that whale watching—which previously had drawn 90,000 tourists

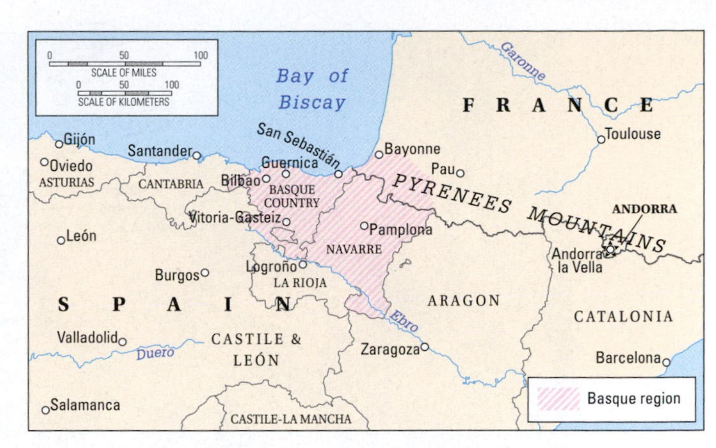

• **Figure 4.42** The Basque country of Spain and France.

yearly—was the more economically sensible way to manage the resource. As the European economy hit hardest by the global economic downturn, Iceland could ill afford to lose tourists. And with "Save the whales" an international mantra, Iceland, Norway, and Japan are subjected to intense criticism from around the world. On the global scale the anti-whaling movement has been enormously successful, reducing the number of these cetaceans taken worldwide from about 40,000 annually in the 1960s to about 2000 per year since the ban took effect in 1986.

Southern Europe

The Basque Country The 2 million **Basques** of Spain and 250,000 Basques of France (see Figure 4.C and •**Figure 4.42**) have a unique ethnicity and culture unrelated to those of their host country majorities. In both countries, the Basques 104 have worked hard at retaining their unique customs. At home and in diasporas around the world, Basques have been known as sheepherders and ranchers, fishermen, and merchants. Like other minorities, they have often been the targets of discrimination and violence. This is especially true in Spain where, during the Spanish civil war in 1937, a German bombing (called for by Spanish fascists) led to tragic loss of life among Basque civilians in the city of Guernica. Artist Pablo Picasso painted a mural of the suffering of Guernica's civilians that brought international attention to the Spanish civil war and ever since has served as a symbol of the tragedy of war (•**Figure 4.43**).

By the 1960s, Basque desire for independence from Spain led to the emergence of the militant group **ETA** (an acronym for **Basque Homeland and Liberty** in Euskara, the Basque language). Spanish authorities gradually allowed considerable autonomy for the Basques, and they now control their own affairs in taxation, education, and policing. For more than four decades, these concessions were not enough for ETA, which carried out attacks on Spanish interests and triggered a vigorous crackdown by Spanish security forces. But in 2011, ETA declared a permanent ceasefire and an end to violence. The European Union and the United States regard ETA as a terrorist organization and are wary of this

Ullstein Bild/The Granger Collection

• Figure 4.43 During the Spanish Civil War, Spanish Republicans commissioned Pablo Picasso to create an image of the suffering inflicted by Spanish Nationalists on Spanish civilians. Picasso's Guernica went on display at the 1937 World's Fair in Paris.

ceasefire because of ETA's history of breaking earlier ones. It is not clear whether in the long term ETA will settle for what they have now, a devolutionary status comparable to Scotland's autonomy within the United Kingdom, or will push for outright independence as Kosovo did (as described 124 later in this chapter). Independence for Basques in Spain or France seems unlikely, and aside from ETA most Basques are not pushing for it or expecting it. French Basques would like to have the kind of autonomy their counterparts in Spain have achieved.

North versus South in Italy Italians are united by their pride in the heritage bestowed by civilization from the Romans to the Renaissance and beyond, and Italians enjoy some of the most picturesque and storied landscapes on Earth (see **•Figure 4.44**). But within Italy, there is a long-standing vernacular distinction between the north and the south. Northerners tend to see themselves as more sophisticated and cosmopolitan than the more earthy and provincial southerners. Southern Italians, whose region is known as the **Mezzogiorno** (a fine example of a vernacular or perceptual region, as described on **page 7** in Chapter 1), often acknowledge their agrarian roots as the source of their superior kinship values and enjoyment of life. Statistics underscore the economic discrepancies between the two Italys—for example, northern Italy has labor shortages while the south has unemployment; northern labor and industries (such as luxury automobile, jewelry, textile, leather, ceramic, and furniture industries) are more productive, and income levels in the north are much higher.

• Figure 4.44 North and South in Italy, as perceived by Italians. This is a vernacular map, so there are no "true" boundaries separating north and south. The green striped regions are sometimes, though not always, considered part of Padania (northern Italy).

The feeling among many northerners that they would be better off if they did not have to subsidize the south has even led to a secessionist movement, championed by the Northern League (Lega Nord), that calls for the creation of an autonomous or independent state called **Padania** in the north (see Figure 4.B). The Northern League wants to maintain the isolated affluence of northern Italy in part by keeping out immigrants, particularly Muslims and Africans. An entity of the Northern League, called **Liga Veneta**, wants to create an autonomous region of Veneto, centered on the affluent city of Venice.

94

In the toe of Italy's boot, many illegal immigrants, mainly black Africans, live in squalid conditions. They began coming to Italy to pick oranges and tangerines when farming was still profitable because of EU agricultural subsidies. But by 2010, when EU subsidies for farming were curbed, there was no work for them in a weakening agricultural sector. The xenophobic Northern League denounced the "excessive tolerance" of illegal immigration in Italy. In Italy's toe, some white Italians engaged in the "hunting of blacks." Both whites and blacks ignited ugly race riots that took many lives and destroyed property. As elsewhere in Europe, the climate of economic hardship provided a perfect breeding ground for this kind of racism and violence.

North versus South in Cyprus Keep your eyes on Cyprus. It is another one of those situations where there are divisions between north and south, and questions about reunification or continued divisiveness. In the case of Cyprus, there is much at stake beyond the confines of this Mediterranean island. What happens there may play a big role in the question of whether Turkey will join the European Union, changing the character and complexity of the European Union, much as the "big bang" did. 115

The large Mediterranean island of Cyprus (see **• Figure 4.45**, and Figure 4.1 for its location relative to the continents), located near southeastern Turkey, came under British control in 1878 after centuries of Ottoman Turkish rule. In 1960, it gained independence as the Republic of Cyprus. The critical problem on the island is the division between the Greek Cypriots, who are Greek Orthodox Christians and make up about three-fourths of the population of 1.2 million, and the Turkish Cypriots, who are Muslims and make up about one-fourth of the population. Agitation by the Greek majority for union (*enosis*) with Greece led in the 1950s to widespread terrorism and guerrilla warfare by Greek Cypriots against the occupying British. Violence also erupted between Greek advocates of *enosis* and the Turkish Cypriots, who feared a transfer from British to Greek sovereignty.

• Figure 4.45 Cyprus is vexed by its divisions.

A major national crisis erupted in 1974 when a short-lived coup by Greek Cypriots, led mainly by Greek military officers, temporarily overthrew Cyprus's President Makarios, who had embraced a conciliatory policy toward the Turkish minority. Turkey then launched a military invasion that overran the northern part of the island. Cyprus was soon partitioned between the Turkish north and the Greek south. A buffer zone (the **Attila Line**, or Green Line) sealed off the two sectors from each other, and even the main city of Nicosia was divided. A separate government was established in the north, and in 1983 the Turkish Republic of Northern Cyprus was proclaimed. Only Turkey recognizes this state. Meanwhile, the internationally recognized Republic of Cyprus has functioned in the Greek Cypriot sector, which makes up about three-fifths of the island. Both republics have their capitals in Nicosia.

The north had dominated the economy prior to the partitioning of Cyprus, but after that the north had economic problems while Greek Cyprus prospered from tourism, financial services, and shipping. Most nations have refused to trade directly with Turkish Cyprus ever since the invasion, and Turkey has not been able to afford boosting the north's prospects. The depressed north remains tied to Turkey, and the Greek sector makes effective use of economic aid from Greece, Britain, the United States, and the United Nations. A member of the Eurozone, Greek Cyprus suffered the fallout from the Eurozone's financial and banking crisis in 2012–2013 and was given a bailout by the European Commission, the European Central Bank, and the IMF. 118

Events leading to the EU's expansion in the big bang of 2004 provided a strong incentive for the two sides to resolve their differences and join the union. The United Nations devised a plan for the two halves to vote separately in a referendum on reunification. Had both sides voted in favor of reunification, a united Cyprus would have taken up membership in the European Union. However, the United Nations and European Union agreed that should either the north or the south, or both, reject the referendum, only the Greek south would actually join the European Union. In the vote, the Turkish north voted overwhelmingly in favor of reunification, and the Greek south overwhelmingly rejected it. The somewhat perplexing result is that in EU terms, "Cyprus" nominally refers to both parts of the island, but only the Greek south is a de facto member of the European Union. This allows the United Nations, the European Union, and other international bodies to continue their policy of not officially recognizing the north as a legitimate, separate political entity. Turkey, long seen as the obstacle to reunification of Cyprus, won rare acclaim in Europe because of the Turkish Cypriots' strong vote in favor of reunification. 273

However, relations between Turkey and the European Union soured in 2011, when Turkey threatened to send warships against the Republic of Cyprus. The Republic had come to an agreement with Israel over seabed resources in the eastern Mediterranean and began exploratory drilling for oil and gas. Turkey and the Turkish Cypriots argued that any maritime agreements or drilling should take place only after reunification of the island. Turkey's support of Northern Cyprus has diminished its prospects of joining the European Union; as an EU member the Republic of Cyprus has blocked further Turkey–EU negotiations until the issue of the island's division is resolved.

Eastern Europe

Wrenching Reforms in the Shatterbelt of Eastern Europe

The eastern European countries are in the long process of reinventing themselves after more than four decades of direct or indirect control by the Soviet Union. Prior to German reunification in 1990, the former East Germany was also part of Eastern Europe, and it is brought into the present discussion where appropriate. Three of the countries—Estonia, Latvia, and Lithuania—were incorporated into the Soviet Union during World War II, and many others became **"Soviet satellites,"** countries where local Communist governments were effectively 119 controlled from Moscow. Exceptions to that subjugation were Albania and the former Yugoslavia (consisting of Slovenia, Croatia, Bosnia and Herzegovina, Serbia, Montenegro, and Macedonia), where national Communist resistance forces took power on their own as German and Italian powers collapsed with the end of World War II.

Majority Slavic ethnicity, former Communist status, and subjugation to Soviet interests were among the few unifying themes of the region prior to the end of the Cold War. Now the true complexity of the region is more apparent.

The extension of Soviet power into Eastern Europe after World War II was a replay of history. In the Middle Ages, several peoples in the region—Poles, Czechs, Hungarians, Bulgarians, and Serbs—enjoyed political independence for long periods and at times controlled extensive territories outside their homelands. Their situation deteriorated as stronger powers—Germans, Austrians, Ottoman Turks, and Russians—pushed into East Central Europe and carved out empires. These empires frequently collided, and the local peoples were caught in wars that devastated great areas, often resulted in a change of authority, and sometimes brought about large transfers of populations from one area to another. In geopolitical terms, Eastern Europe is a classic **shatterbelt**—a large, strategically located region composed of conflicting states caught between the conflicting interests of great powers (see the political geographic evolution of this region in •**Figure 4.46**). Other shatterbelts are discussed here in this book, including the Middle East. 239

Peoples of this region have been on the move in large numbers since the beginning of World War II. Jews, Germans, Poles, Hungarians, Italians, and others were uprooted by Nazi and Soviet authorities, often without notice, losing all their possessions, and were dumped as refugees in so-called "homelands" that many had never seen. During the war, Nazi Germany systematically killed approximately 6 million Jews and perhaps as many as 1.5 million Gypsies (Roma) and other "undesirables" in the Holocaust. The prewar populations of 108 Germans in Poland and Czechoslovakia were expelled at the end of World War II and forced into East and West Germany. Overseen by the Soviet Union, many died in this process of forced migration. Ethnic minorities now make up only 2 percent of the population in Poland, which transferred most of its German population to Germany and whose territories with populations of ethnic Lithuanians, Russians, and Ukrainians were absorbed by neighboring countries. A number of countries still have large minorities, including the Roma (see **page 142**). These population transfers, often involving **ethnic cleansing** (the forced emigration or murder of one ethnic group by another within a certain territory), helped create an ethnic map of Eastern Europe that was overwhelmingly and purposefully Slavic.

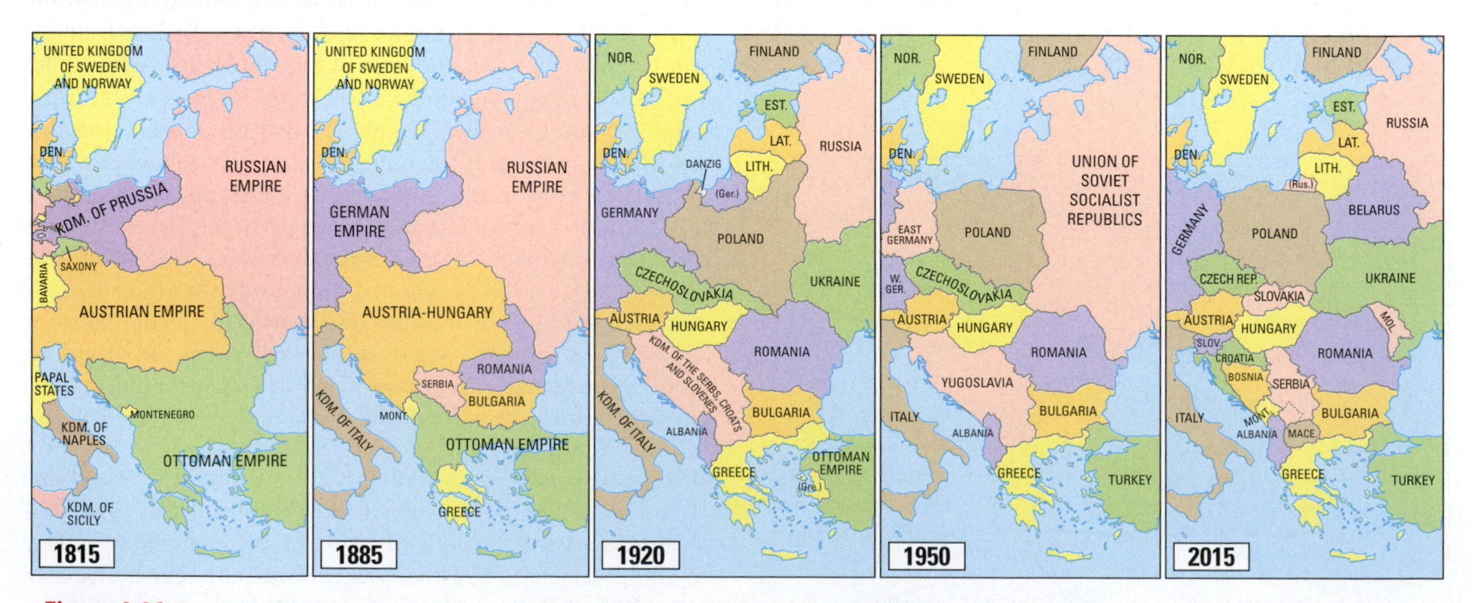

• **Figure 4.46** Positioned between stronger powers to the east and west, Eastern Europe is a classic shatterbelt, with a tumultuous past reflected in its shifting borders.

Communism in Ideals and Practice

In becoming satellites of the Soviet Union, the nations of Eastern Europe were transformed by **communism**. You need to consider the principal traits of communism in order to understand their legacy in today's geography of Eastern Europe and of the former Soviet Union, which is subject of the next chapter. Communism had these principal traits:

- One-party dictatorial governments
- National economies planned and directed by organs of the state
- Abolition of private ownership (with some exceptions) in the fields of manufacturing, mining, transportation, commerce, and services
- Abolition of independent trade unions
- Varying degrees of **socialization** (state ownership) of agriculture and industry

Soviet military force crushed attempts to break away from Soviet control by East Germany in 1953, Hungary in 1956, and Czechoslovakia in 1968. In agriculture, the new communist governments liquidated the remaining large private holdings and in their place introduced programs of collectivized agriculture on the Soviet model. Some farmland was placed in large state-owned farms on which the workers were paid wages, but most was organized into collective farms owned and worked jointly by peasant families who shared the proceeds after operating expenses of the collective had been met. This **collectivization**—in the farming context, the bringing together of individual landholdings into a government-organized and government-controlled agricultural unit—met with strong 171 resistance.

Conversion back to private farming after four decades of collectivization presented painful obstacles, and progress has been slow. This process is still ongoing, which is why it is important to understand the legacies of communism and socialism. Some eastern European members of the European Union, notably Romania, have seen what many observers describe as "land grabs" by multinational agribusinesses, banks, and private investment funds, and in general small farmers have lost ground (see **page 66** for the problem of "land grabbing"). Corporate farming has boosted overall yields considerably in the eastern EU countries, which are benefitting from the scaling up of EU CAP 117 subsidies. Some crops such as soybean have declined where EU bans on genetically modified foods have prevented the planting of the previously prolific "Roundup Ready" variety. Despite recent attempts to diversify and intensify agriculture, in Eastern Europe agriculture remains focused on a relatively narrow range of products, especially corn, wheat, pigs, and sheep.

In industry, as it did in agriculture, communism in Eastern 170 Europe inaugurated a new era based on a Soviet model. Most industries were taken out of private hands, and national economic plans were developed. Communist planners did not aim at a balanced development of all types of industry. Instead, they emphasized those industries deemed essential to the development of a greater Soviet empire: mining, iron and steel, machinery, chemicals, construction materials, and electric power, for example. These were favored while consumer-type industries were neglected.

The central planning agencies of the Communist governments maintained rigid controls over individual industries. Plant managers were ordered to produce certain goods in quantities determined by state governmental planners. The plant's success was judged by its ability to meet production targets rather than its ability to sell its products competitively and at a profit. Plant managers were expected to conform to the central plan and not to ask questions. As you can imagine, this was not an economic environment that fostered creativity and entrepreneurship. There were no visionaries like Apple's Steve Jobs—or if there were, they would have been branded as nonconformist and been alienated or worse.

This system resulted in shoddy goods that were in short supply or in mountainous oversupply because production was not driven by consumer demand and satisfaction. The planned expansion of mining and industry in Eastern Europe increased the output of minerals, manufactured goods, and energy. But it also produced inefficient, overstaffed industries that were ill-prepared for competition in the world markets of the post-Communist period. This was a problem shared by all the economies of all the successor states of the Soviet Union, as we will also see in the next chapter.

Under communism, the economies of Eastern Europe were closely tied to that of the Soviet Union. During the 1970s and 1980s, however, the region's trade relationship with the Soviet Union weakened. Trade and financial relations with Western countries, companies, and banks expanded rapidly. With relatively little to export, the eastern European countries borrowed massively from Western governments and banks to pay for imports from the West. Poland and Romania pursued this course especially strongly. New industrial plants and equipment acquired in this way were supposed to be paid for by goods that would be produced for export to the West by industries that had learned efficient production and marketing techniques from the West.

But then the Western economies began to slump, lowering the demand for imports. This led to a situation by the 1980s in which large debts to Western governments and banks needed to be repaid if countries were to maintain any credit at all. However, with the help of mismanagement by Communist bureaucrats, exports with which to pay were not being produced or, if produced, could not be sold. The results were severe impediments to economic expansion and falling standards of living as the governments squeezed out the needed money from their people. In Poland, social action through the formation of an independent trade union called **Solidarity** paralyzed the country and threatened to upset Communist dominance. These conditions, along with the Soviet Union's own growing economic and political distress, set the stage for the withdrawal of Soviet control and de-communization in 1989 and 1990.

As late as the early summer of 1989, East Germany and the eastern European countries seemed to be firmly in the control of totalitarian Communist governments. Then, suddenly, the Communist order began to crumble. Public demands for freedom, democracy, and a better life gathered momentum in one country after another. Communist dictators who had ruled for many years were forced out, and reformist governments took charge. The only violent removal of a dictator in the 1989 revolutions occurred when Romanians hunted down and executed their brutal leader Nicolae Ceaușescu. In 1989 and 1990, the Soviet Union withdrew its backing for the region's Communist order. That support was too costly for the Soviet Union, whose leaders were forced to recognize the inevitable victory of "people power" in Eastern Europe. By mid-1991, democratic multiparty elections had been held in all the countries. This liberalizing process continued throughout the 1990s and into the 2000s.

Governments in Eastern Europe have been struggling ever since 1989 to build democracies with capitalist economies out of the economic wreckage of unproductive, unprofitable, uncompetitive, state-owned enterprises. Many countries pursued reforms aggressively to meet qualifications for joining the European Union. A number of countries formed a free trade agreement known as the **Central European Free Trade Agreement (CEFTA)** to position themselves for EU membership. Many of these countries have left CEFTA and joined the European Union, and most CEFTA members today are the Balkan countries hoping to join (see **• Figure 4.47**).

• Figure 4.47 Present and former members of CEFTA. For eastern European countries, joining CEFTA is a precursor to EU membership.

Economic restructuring has required many difficult steps, including the **privatization** (shift to nongovernmental ownership) of state-owned enterprises, an increase in the efficiency of state-run enterprises by allowing noncompetitive enterprises to fail and removing support of poor performers, and encouragement of the development of new private enterprises, often with foreign capital and management playing a role. In addition, economic restructuring ended price controls so that prices reflect competition in the market; the development of new institutions required by a market-oriented economy (such as banks, insurance companies, stock exchanges, and accounting firms); the fostering of joint enterprises between state-owned firms and foreign firms; and the elimination of bureaucratic restrictions on the private sector. Internationally convertible currencies have also been created to increase trade and enhance competition, and access to international communications media of all sorts has been expanded to help local entrepreneurs learn from foreign examples.

One of the most remarkable economic trends has been western European and Chinese **outsourcing** to take advantage of relatively cheap labor in Eastern Europe. Also known as "offshore outsourcing" or "offshoring," outsourcing is the flight of technology and other jobs from countries with high manufacturing and service costs to countries with low manufacturing and service costs. In the late 1990s, Austrian and German electronics firms established large factories in East Central Europe, especially in Hungary. Even China opened television and other electronics factories in Eastern Europe, because China's labor costs were rising.

311

The eastern European countries enjoyed strong economic growth for about a decade (1997–2007). But when the global economic crisis hit Europe in 2007, it hit the former communist countries hardest. They are coming back now. The "Steve Jobses" or perhaps more accurately the "Eric Schmidts" (of Google fame) *have* emerged in this new environment; in fact, in 2014 Google launched a high-tech talent search dubbed the "New Europe 100" throughout the region. Eastern Europe's post-Soviet educational system has prized the STEM fields, producing a skilled labor force that has been creating its own labors. Estonia is particularly well-known for its start-ups, including Skype and the money-transfer service TransferWise. Once a Soviet backwater, Estonia has become a world-leading digital economy with one of the world's highest rates of broadband use. Many eastern European businesses that are expanding move their headquarters and sales teams to the United States and the United Kingdom but keep their developers at home. Programmers' salaries are cheaper in Eastern Europe, where it is easier to keep the best talent from being "poached" by the competition, as might happen in Silicon Valley. Since Russia's invasions of Crimea and eastern Ukraine in 2014, the eastern European countries, especially Estonia, have also been attracting tech start-ups from Russia and Ukraine looking for more open and stable societies than their own.[35]

Why They Call It "Balkanization"

You may have the terms "balkanize" or "balkanization" in geographic or political contexts. This word comes from the real-life geopolitical events in a part of Southeastern Europe known as the Balkans or the Balkan Peninsula. Here we will see what these events were and what their impact has been on the modern geography of the Balkan states. Some of the world's youngest countries were born here.

Although economic progress characterized most of Eastern Europe in the 1990s, Yugoslavia (the outlines of which appear in the 1950 map of the Balkan Peninsula in **Figure 4.46**) was plagued by ethnic warfare after the Yugoslav federal state dissolved in 1991. Previously, under the Communist regime of Marshal Josip Broz Tito during World War II, Yugoslavia ("Land of the South Slavs") had been organized into six "socialist people's republics" (Serbia, Croatia, Slovenia, Bosnia and Herzegovina, Montenegro, and Macedonia) and two "autonomous provinces" within Serbia (Kosovo and Vojvodina).

As long as Tito and his successors could retain a firm grip, these groups could coexist within the artificial boundaries of a single nation (rather like Iraq under Saddam Hussein; see **page 265**). However, this apparent unity belied many underlying tensions and conflicts based on the linguistic, ethnic, and religious distinctions you read about earlier in the chapter: Orthodox Serb versus Muslim Kosovar, Catholic Croat versus Orthodox Serb, and so on.

As the Iron Curtain dissolved across Europe, Yugoslavia began to fracture along its ancient ethnic fault lines (•**Figure 4.48**). In 1988, Serbia took direct control of Kosovo and Vojvodina as part of its push (under an elected Communist president, Slobodan Milosevic) for greater influence in federal Yugoslavia. At first the quasi-independent states within Yugoslavia demanded a looser federation with more autonomy for each republic. But in 1991 and 1992, Croatia, Slovenia, Macedonia, and Bosnia insisted on independence. When Croatia declared its independence, fighting broke out between ethnic Croats and Serbs in the new country. Croatian Serbs took over one-third of Croatia and embarked on a policy of ethnic cleansing. Although Bosnia's Muslims (known as *Bosniaks*) and Croats voted for independence from Yugoslavia, ethnic Serbs in Bosnia and the Yugoslav government of Slobodan Milosevic opposed this.

• **Figure 4.48** Ethnic composition of Yugoslavia's successor states.

War erupted in 1992 between the Bosnian government and local Serbs. Supported by Serbia and Montenegro, the Bosnian Serbs fought to partition Bosnia along ethnic lines and join Serb-held areas to a "greater Serbia." The ethnic Serbian quest for greater Serbia is an excellent example of **irredentism**, a policy advocating annexation of territories administered by another state by reason of common ethnic, linguistic, cultural, or historical ties, or of prior historical possession, genuine or alleged. Bosnian Serbs laid siege to the mainly Bosniak city of Sarajevo. The United Nations withdrew recognition of Yugoslavia because of the government's failure to halt Serbian atrocities against non-Serbs in Croatia and Bosnia. In 1994, the Bosniaks and Croats agreed to create the joint federation of Bosnia and Herzegovina.

In 1995, the presidents of Bosnia and Herzegovina, Croatia, and Serbia signed the **Dayton Accord** (it was brokered by US diplomats at an airbase near Dayton, Ohio). This peace agreement retained Bosnia and Herzegovina's international boundaries and created a joint multiethnic, democratic government responsible for conducting foreign and economic policies. The agreement also recognized a second tier of government composed of two similarly sized entities charged with overseeing internal functions: the Bosniak-Croat Federation of Bosnia and Herzegovina, and the Bosnian Serb-led Serb Republic (Republika Srpska). To help implement and monitor the agreement, NATO fielded a 60,000-strong implementation force known as IFOR. This was NATO's first major post–Cold War world engagement.

While the Bosnian situation quieted, in 1998 Serbia turned its sights on Kosovo, a majority ethnic Albanian region with a minority of ethnic Serbs in the north (this was in part a religious conflict; as a reminder, most Albanians are Muslims, whereas most Serbs are Orthodox Christians). Here again, irredentist Serbs aspired to ethnically cleanse Kosovo of non-Serbs. Serbs asserted that Kosovo was the cradle of their culture and the site of the decisive battle of Kosovo Polje in 1389, which Serbs lost to the Ottoman Turks. NATO responded to the Serb offensive with 11 weeks of bombing. When the bombing stopped, the United Nations took control of Kosovo and oversaw a tense truce between its Serb and Muslim inhabitants. The fate of more than 200,000 ethnic Serbs who fled during the conflict from Kosovo to Serbia and Montenegro has not been resolved. Nominally still a part of Serbia, but with a population of only 5 percent ethnic Serb and 95 percent ethnic Albanian, Kosovo came under United Nations jurisdiction.

Kosovo declared its independence in 2008, naming itself the Republic of Kosovo. The ethnic Albanian majority poured out into the streets of the capital Pristina to joyously proclaim their country as "newborn" (•**Figure 4.49**). Soon after, Kosovo was recognized as independent by more than 80 countries (now grown to over 100), including the US, most EU member countries, and many other nations. United Nations membership is the ultimate stamp of a country's sovereignty, but Russia and China (which have veto power in the United Nations Security Council) uphold Serbia's claim that Kosovo is a Serbian province, so Kosovo may never receive that recognition. In addition to Russia and China, dozens of other countries do not recognize Kosovo; many of them (such as Morocco, Israel, Georgia, and Spain) have issues with independence movements or calls for regional autonomy in their own borders. In this book, we recognize Kosovo as an independent nation.

The ouster of Serbian leader Milosevic in 2000, and his capture and subsequent transfer to The Hague to stand trial for war crimes (he died before the trial was completed), opened the way for a period of stable and peaceful borders in the 2000s. The international community insisted on no further redrawing of Balkan boundaries. The region had undergone a profound and violent process of **balkanization**, which is political-geographic shorthand for fragmentation into ethnically based, contentious units that took its name from the characteristic disharmony of this region. Balkanization was not new to this area. This geopolitical term was coined early in the 19th century in reference to the Balkan's fragmentation under Ottoman control and was widely used after World War I when the Austro-Hungarian and Russian empires shattered.

The Balkan conflicts unfolded on television and other media through the 1990s. They reminded Europeans and Americans of the earlier world wars on the continent and seemed a horrible anomaly in a world region now accustomed to peace and prosperity. Western military and political institutions eventually seemed to extinguish the fires of conflict, and the Balkan countries were set on the back burners of European political concern. By all measures of quality of life and economic prosperity, Albania and the countries of the former Yugoslavia are the poor stepchildren of the continent, and most cannot expect to join the European Union for many years. The two former Yugoslav countries that are EU members, Slovenia and Croatia, are by far the most prosperous of that group.

The Roma

One of Europe's largest ethnic minorities is the Gypsies—properly known as the Roma—who number about 4 million in Europe (see •**Figure 4.50**). Romania has the highest number, 800,000 by some estimates; but estimates of the Romany population vary widely because the Roma tend to live on the road and at the margins of society.

The Roma began their great odyssey in what is now India, and their Romany language is very close to languages still spoken on the Indian subcontinent. Having in early times traveled 104 thousands of miles from their homeland, and still often moving in caravans, they have come to be the archetypal people on the move (in informal usage, we often refer to inveterate travelers as "gypsies.") Throughout the Roma realm, host governments and majority populations have for centuries regarded the Gypsies with disdain. They are typically depicted as a rootless, lawless, and violent people. They are "the other," apart, not deserving of the educational and economic opportunities offered to majority populations. Europeans have long projected their stereotypes onto the Roma, whether belittling or

AP Images/Visar Kryeziu

• **Figure 4.49** On February 17, 2008, the Albanian majority of Kosovo celebrated their self-proclaimed newborn country in the city of Pristina. Kosovo is not recognized as independent by UN Security Council members Russia and China.

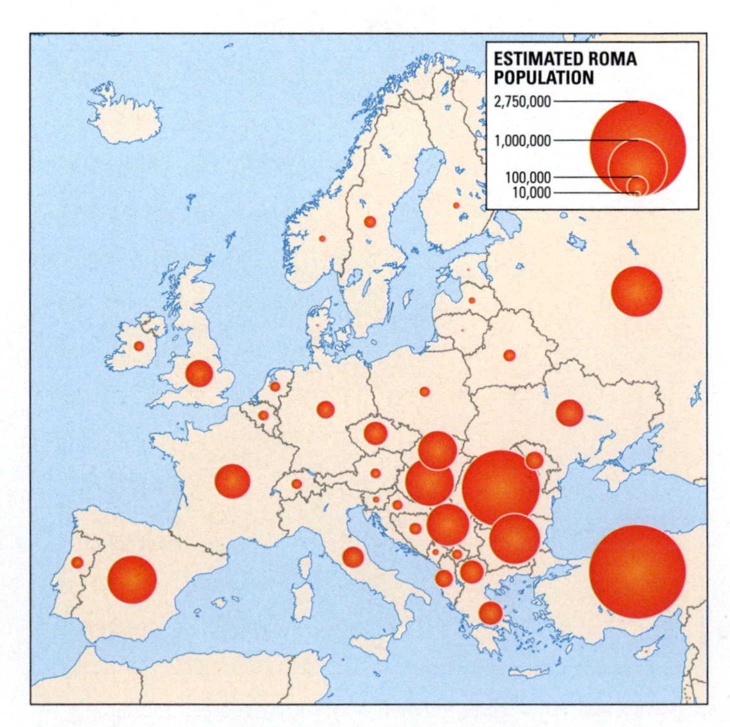

• **Figure 4.50** Distribution of the Roma across Europe.

beguiling. European men have objectified the Gypsy woman as a smoldering seductress capable of casting an irresistible spell, as does *Carmen* in the Frenchman George Bizet's opera, popular since 1875 (**•Figure 4.51a**).

Wherever they are, Roma families are larger than those of the majority populations. Sterilizations of Roma women without their consent have been reported in the Czech Republic, Slovakia, Hungary, Romania, and Bulgaria. The Roma are poorer than the majority populations; in Hungary, for example, the Roma unemployment rate is five times that of non-Roma, and in some Roma communities the unemployment rate is as high as 90 percent. Roma people were the first to lose their jobs when the eastern European countries gained economic freedom from the Communists.

The Roma typically live in shantytowns without water, sewage, or other services, and make their living on scant child benefit payments, minor mechanical work, begging, foraging for food, and petty crime (**•Figure 4.51b**). One Czech city built a wall between Gypsy and Czech neighborhoods, insisting it was needed to protect other townspeople from Roma criminals, noise, and visual squalor. Most Roma children do not attend schools because their parents do not believe that education will improve their job prospects, given the discrimination against them. Many that want to attend school are funneled into institutions for the mentally disabled. There is also violence; Gypsies are often the targets of attacks by racist "skinheads." These assaults, especially in the Czech Republic and Slovakia, have led to waves of Roma emigration to Canada, France, Britain, and Finland.

Some Europeans have begun to speak of "the Roma Problem," which has a disturbing ring to it for people old enough to remember the Holocaust (which also preyed on Roma). Chillingly, neo-Nazis have been quoted calling for "Gypsies to the gas chambers!"[36] This "Roma Problem" has grown with rates of unemployment across Europe. France's ultraright National Front founder Jean-Marie Le Pen called the Roma "smelly" and "rash-inducing," and used anti-Roma rhetoric to boost election results in favor of National Front candidates. Italy's Northern League, Greece's Golden Dawn, and other right-wing parties across Europe have stepped up their rhetoric against the Roma. Long stereotyped as child kidnappers, Roma with light-haired children have been detained and subjected to DNA testing. Many fear for the safety of their own children.[37]

Romany activists have responded to discrimination by demanding more rights to welfare, pensions, and other benefits offered to regular citizens. Already feeling overwhelmed by the

©F. JIMENEZ MECA/Shutterstock.com

(a)

Joe Hobbs

(b)

• Figure 4.51 Stereotype and reality of a Romany woman. **(a)** The Gypsy woman is often portrayed as a fortune teller or a seductress. **(b)** I photographed this Romany woman and her children begging on a street corner in Szczecin, Poland.

economic burden of their welfare services and by the migrants described on **pages 95** and **96**, the wealthier countries of the European core are worried that the new EU membership of eastern European countries will unleash a tide of Roma immigration. The 15 countries that joined the European Union most recently have about 2–3 million Roma. Some Roma have taken advantage of the free passage allowed within Schengenland to travel, especially to France, in search of work. When petty crime or unacceptable practices such as the overharvesting of mushrooms pick up in France, the usual suspects are Roma who have made their way there thanks to Schengen allowances. France responds by deporting many Roma back to their countries of origin, especially Romania. French authorities insist that their policy is that immigrants may settle in France only if they can support themselves. After three months in France, they must prove they are working or studying, and have sufficient funds and health care. If they cannot provide proof—and many Roma cannot—France can deport them, and it does. The deported Roma find themselves back in Romania with scant hope of work there—which is what drove them to France in the first place. Economic recession closed the mines and other enterprises in Romania where the Roma once found work.

When Romania and other eastern European countries joined the European Union, the European Union earmarked funds to be spent in those countries on projects that would help Roma communities. These funds got caught up in bureaucratic red tape, and most of these projects have yet to get off the ground.

The stigma associated with being Roma has made successful Roma generally reluctant to reveal their ethnic identity for fear of losing their jobs. When someone takes a public stance in favor of the Roma, that person is often silenced or dismissed. For example, at a concert in Bucharest, Romania, Madonna sang and spoke to urge greater tolerance and compassion for the Roma. She was booed.

This concludes the survey of Europe. Given the book's size limitations, we do not have the chance to explore Europe's diverse countries further, but you now have a good foundation for appreciating land and life in Europe. You are in a good position to read or view news coming out of Europe with much more understanding. You are ready to see how the pieces fit together between Europe and other regions. Those pipelines feeding natural gas to Western Europe . . . let's follow them to their origin in Russia and see what's going on there.

Study Guide

Summary

- Europe is physically part of the great continent of Eurasia but is generally labeled a continent and is treated as a separate region.
- Europe's population is about twice that of the United States, and Europe is more densely settled.
- Having passed through the demographic transition, Europe is demographically post-industrial, with a slowly declining and aging population.
- Immigration could help stimulate Europe's economic growth by welcoming a larger working-age population, but migration is controversial. There is much anti-Muslim sentiment, and anti-Semitism is also on the rise again. Issues surrounding Muslim immigration are especially intense because of terrorist attacks attributed to Islamist extremist groups. Ultranationalist political parties in a number of countries have gained popularity and power through anti-Muslim immigrant positions.
- Among Europe's most distinctive physical geographic traits are its northerly location, temperate climate, and varied topography.
- The North European Plain is a major belt of settlement and agricultural productivity. Oceanic, continental, and Mediterranean climates are characteristic of the region. Complemented by canals and other engineering works, the Rhine and the Danube have long the two most important rivers for transport, and provided geographic situations for development of many cities.
- European languages derive primarily from Indo-European roots and include Romance, Germanic, and Slavic languages. English is a Germanic language, and like most European languages has been enriched by many other tongues. The dominant religion in Europe is Christianity, with the major sects of Protestantism, Roman Catholicism, and Eastern Orthodoxy. There are significant populations of Muslims (most of them recent immigrants) and of Jews (whose population on the continent dates as far back as Roman times, and represents a small remnant of the pre-Holocaust community).
- From the beginning of the 16th century until late in the 19th century, Europe was at the center of global forces of colonization and foreign settlement, long-distance trade, and agricultural and industrial innovation. During this time, Europeans diffused crops and animals between the Old and New Worlds; this is the "Columbian Exchange."
- The Industrial Revolution originated in Europe, with energy from coal used to manufacture textiles, iron products, and a growing array of goods. Industrialization and colonization

went hand in hand to launch Europe to global economic and political supremacy.

- Recent decades have seen a global shift in power away from Europe. War dislocation, rising nationalism, the ascendancy of the United States, a shift in world manufacturing patterns, and new energy sources have combined to diminish Europe's global centrality. Europe is nevertheless a very strong force in world economic, political, and social affairs, and its peoples are among the most prosperous in the world.
- Europe's economy is post-industrial, making the transition from energy-hungry, labor-costly, and polluting industries to leaner high-tech industries and services. This shift has caused unemployment, which worsened during the recent global economic crisis. European countries imposed economic austerity measures that were generally unsuccessful, so following the American model the countries turned to more borrowing for economic stimulus. Debt and deficit spending continues to trouble European economies.
- European nations often try to protect their domestic industries and have been involved in trade wars with the United States, which likewise wants to protect its industries. Europe is under pressure from international trade agreements to reduce its trade barriers and subsidies that protect its domestic agriculture and other industries.
- Eastern Europe has long been much poorer than Western Europe.
- In recent decades, Europe has been reorganizing itself to ensure that nothing like the two world wars will happen again and to strengthen its economies. Its principal military alliance, NATO, is growing in membership and redefining its focus toward peacekeeping. NATO has found new missions in confronting apparent Russian expansionism in Ukraine and anticipating more Russian interventions. The Warsaw Pact, the military alliance counterpart of NATO in Soviet times, dissolved soon after the Soviet Union broke up.
- The most important development in Europe's postwar reorganization is the growth of the 28-member European Union, a supranational organization pooling the economic and human resources of its member countries. There have been obstacles to its achievement of common EU policies in economic policies, defense, and foreign affairs. Some countries (notably Norway) chose to stay out of the EU, and some (notably the United Kingdom) chose to not become part of the "Eurozone"—the 19 countries using the common currency of the euro.
- Overall there are several marked differences in the ways Europeans and Americans view the world and interact with their governments. Unlike the US system, where benefits or the lack of them are largely left to market forces and personal circumstances, Europeans tend to have a "social compact" with their governments in which they pay high taxes in return for extensive government services from cradle to grave. Austerity measures however eroded some of the public benefits.
- The nations making up the European core are the United Kingdom, Ireland, France, Germany, the Low Countries (the Netherlands, Belgium, and Luxembourg), Austria, and Switzerland, along with the microstates of Andorra, Monaco, and Liechtenstein. The most important countries in this region are Germany, the United Kingdom, and France. They are the main engines of economic momentum in the region.
- The United Kingdom is made up of England, Scotland, Wales, and Northern Ireland. It forms the greater part of two major islands, Great Britain and Ireland, which are known as the British Isles.
- Most of the 18th- and 19th-century industrial and urban developments of Britain were situated near sources of coal and iron ore, the two leading resources of the Industrial Revolution. London arose on the Thames estuary without nearby coal or iron ore as a local resource. It developed into a major global financial center classified as an "alpha" or "global" city.
- Paris is a primate city outranking all other French cities in its demographic, economic, political, and cultural importance. Situated on the Seine River, Paris serves as a hub for transportation and other important French networks.
- Although the Republic of Ireland has recently experienced strong economic growth based on high-tech industries, Northern Ireland's growth has been restrained by a long-standing but perhaps nearly resolved conflict between historically indigenous Roman Catholics and more recent residents of Protestant, English background.
- The EU countries have established an ambitious schedule to replace fossil fuel energy with wind and other renewable energy alternatives. Several countries seek global leadership in alternative energy technologies, especially to reduce greenhouse gas emissions; Germany for example wants to become by 2050 the world's first industrial power to use 100 percent renewable.
- Germany, restored when West Germany reunified with East Germany in 1990, has the largest economy and population in Europe. German urban centers have a much broader distribution than the industrial cities of France and Britain. As in Britain, these cities were generally situated near coal and other industrial resources and along rivers for transport. Also as in Britain, they have been deindustrializing in recent decades. Germany's economy has had to bear the costs of reunification of being a welfare state. The former East Germany remains poorer than the west, but new industrial investment is drawn there by low labor costs.
- The core region of Europe is ringed by a periphery of three subregions: Northern, Eastern, and Southern Europe. These have historically been less integrated into the core, and the eastern and southern countries have been less prosperous.
- There is an imbalance between northern and southern Italy, with the north far more prosperous. Some northerners want to develop an autonomous or independent region of Padania.
- As the result of a 2004 referendum in which both sides of divided Cyprus voted on whether they should be reunited, the Greek south (which rejected reunification) was allowed to join the European Union, whereas the Turkish north (which favored reunification) was not allowed to join.

- Eastern European nations have been shaped powerfully by struggles between stronger countries and make up a geopolitical shatterbelt. Most critical was their role as satellite states of the Communist Soviet Union between World War II and 1990.
- Collectivization under Communist rule during the Soviet era from the late 1940s until the 1980s was the agricultural model of Eastern Europe. Privatization of farmland has taken place since the fall of communism.
- Mining, iron and steel, machinery production, construction materials, and electrical power were the highlights of the Eastern Europe's industrial programs during the years of Soviet domination, with the majority of raw materials coming from Russia. The shift to a market economy left many firms uncompetitive, and many were abandoned. Both western European countries and China have taken advantage of inexpensive labor in Eastern Europe to outsource some production to the region, and many economies in this subregion have rebounded.
- The violence-ridden disintegration of the former Yugoslavia in the 1990s was the most profound political development in post-Soviet Eastern Europe. Yugoslavia was carved into six countries. Historic rivalries gave way to outright war between some ethnic groups. "Balkanization" means fragmentation into ethnically based, contentious political units and took its name from recurrent geopolitical fracturing that took place in Balkan Peninsula. In the late 1990s, NATO and the United Nations deployed military and diplomatic assets to prevent further balkanization of the countries. Kosovo declared its independence in 2008, but it is not recognized by Russia and some other countries.

Review Questions

1. Why is Europe usually treated as a separate region?
2. What terms and trends best describe Europe's population today?
3. Why is immigration so critical to Europe's demographic future, and what problems are associated with immigration?
4. What are some of Europe's major physical and environmental characteristics? Like the region's cultures, they are described as "diverse"—what are some of the attributes of this diversity? What is unusual about Europe's coastline?
5. What roles have rivers played in Europe's development, and which rivers are the most important?
6. What are the dominant languages, religions, and other ethnic traits of Europe?
7. What factors led to Europe's global dominance in economic and political affairs? What impacts did that dominance have on other peoples and environments?
8. What factors came together to spark the Industrial Revolution, and what role did natural resources play in Europe's urban geography? How does Europe's present urban spatial pattern reflect its industrial past, and what developments have made these associations less relevant today?
9. What are Europe's main economic traits? How have these changed in recent decades?
10. Why is the Common Agricultural Policy (CAP) problematic within Europe and between Europe and its trading partners?
11. What major factors weakened Europe in the 20th century? How are Europe's political and economic institutions of today trying to create a different Europe?
12. How did the Cold War affect the geopolitics and economic geography of Europe?
13. What are the main goals and principles of the European Union? What successes and difficulties have the organization had? What makes Germany such an important power within the European Union? What are some of the weaker powers within the EU, and how does the European Union struggle to accommodate the diverse needs of its 28 members?
14. What is Schengenland, and why is it so controversial now?
15. What are some of the differences between the Europeans and Americans?
16. In what ways has Europe established global leadership in alternative energy development?
17. What are the countries of the European core, and on what basis have they been designated as belonging to the core?
18. What are the political affiliations of Ireland, Northern Ireland, England, Scotland, and Wales? What is devolution, and what experiences have the UK and other countries had with this political process?
19. What are "the Troubles" of Northern Ireland?
20. What is a primate city? What examples exist in Europe and elsewhere?
21. What was the impact of German reunification on the country's economy?
22. What north–south distinctions characterize Italy and Cyprus, and what problems are associated with them?
23. How are Norway, Iceland, and Japan out of step with the rest of the world where whaling is concerned?
24. What are the major countries and ethnolinguistic groups of Eastern Europe?
25. Why is Eastern Europe considered a shatterbelt in geopolitical terms?
26. How did communism shape agriculture and industry in Eastern Europe? What economic processes have occurred there since 1990?
27. What are the main forces behind the breakup of Yugoslavia and the current borders and ethnic components of its successor countries?
28. Who are the Roma, and what are their unique attributes? Why and how have other ethnic groups in Europe discriminated against them?

Key Terms + Concepts

Attila Line (p. 137)
"baby bounty" (p. 94)
balkanization (p. 142)
"banana war" (p. 116)
Basques (p. 135)
Benelux (p. 89)
"big bang" (p. 115)
biogeography (p. 109)
"birth dearth" (p. 93)
Calvinism (p. 106)
Catholic Republicans (p. 129)
Celtic Tiger (p. 128)
Central European Free Trade Agreement (CEFTA) (p. 140)
Christianity (p. 106)
Church of England (p. 106)
"City of Light" (p. 130)
Cold War (p. 119)
collectivization (p. 139)
Columbian Exchange (p. 109)
Common Agricultural Policy (CAP) (p. 117)
Common Market (p. 114)
Commonwealth of Nations (p. 126)
communism (p. 139)
Crusades (p. 96)
Dayton Accord (p. 141)
deflation (p. 118)
deindustrialization (p. 113)
delta (p. 103)
devolution (p. 107)
direct rule (p. 129)
dumping (p. 112)
enosis (p. 137)
environmental perception (p. 125)
ETA (Basque Homeland and Liberty) (p. 135)
ethnic cleansing (p. 138)
Eurabia (p. 96)
Eurasia (p. 89)
euro (p. 115)
Euro-Fascism (p. 94)
European core (p. 89)
European debt crisis (p. 117)
European Economic and Monetary Union (p. 115)
European Economic Community (EEC) (p. 114)
European Greenbelt (p. 119)
European periphery (p. 133)
European Union (EU) (p. 88)
Eurozone (p. 115)
"Frankenfoods" (p. 116)
genetically modified (GM) foods (p. 116)
genetically modified organism (GMO) foods (p. 116)
glacial deposition (p. 99)
glacial scouring (p. 99)
glaciation (p. 99)
global city (alpha city, world city, p. 127)
Good Friday Agreement (p. 129)
Gulf Stream (p. 97)
Gypsies (p. 104)
Holocaust (p. 108)
Ice Age (p. 99)
Indo-European languages (p. 104)
Baltic languages (p. 105)
Celtic (p. 106)
Germanic languages (p. 105)
Greek (p. 105)
Romance languages (p. 104)
Slavic languages (p. 105)
Irish Republican Army (IRA) (p. 129)
Iron Curtain (p. 119)
irredentism (p. 141)
Islam (p. 108)
Islamophobia (p. 96)
isostatic rebound (p. 100)
Jews (p. 108)
Judaism (p. 108)
language isolate (p. 104)
Lega Nord (Northern League) (p. 107)
Liga Veneta (p. 136)
loess (p. 98)
Lutheran Protestantism (p. 106)
Maastricht Treaty on European Union (p. 115)
major cluster of continuous settlement (p. 122)
Marshall Plan (p. 119)
Mezzogiorno (p. 136)
microstates (p. 89)
Moors (p. 108)
Muslims (p. 108)
nationalism (p. 111)
NIMBY (p. 133)
North Atlantic Drift (p. 97)
North Atlantic Treaty Organization (NATO) (p. 119)
Northern Ireland Assembly (p. 129)
North European Plain (p. 98)
outsourcing (p. 140)
Padania (p. 136)
Pentecostal churches (p. 108)
PIGS (p. 117)
post-industrial (p. 113)
potato famine (p. 128)
primate city (p. 129)
privatization (p. 140)
Protestant Reformation (p. 106)
Protestant Unionists (p. 129)
Protestantism (p. 106)
robotics (p. 113)
Roma (Gypsies) (p. 104)
Roman Catholic Church (p. 106)
Saami (Lapps) (p. 106)
Schengen Agreement (p. 121)
Schengenland (p. 121)
shatterbelt (p. 138)
Silk Road (p. 108)
Sinn Fein (p. 129)
site (p. 94)
situation (p. 94)
socialization (p. 139)
Solidarity (p. 139)
Soviet satellites (p. 138)
subsidies (p. 112)
supranational organization (p. 114)
tariffs (p. 112)
trade wars (p. 116)
the Troubles (p. 129)
undocumented workers (p. 95)
Uralic languages (p. 106)
Warsaw Pact (p. 119)
welfare state (p. 113)
westerly winds (p. 98)
"world city" (p. 127)

Notes

1. Aldous Huxley, *Do What You Will* (London: Chatto & Windus, 1970).
2. Elisabeth Rosenthal, "European Union's Plunging Birthrates Spread Eastward." *The New York Times,* September 4, 2006, p. A3.
3. Andrea Mammone, "Europeans United, in Hating Europe," *The New York Times*, January 3, 2014, p. A17.
4. Robert Kaplan, "The Return of Toxic Nationalism," *The Wall Street Journal*, December 23, 2012. http://www.wsj.com/articles/SB10001424127887323297104578174932950587010.
5. Quoted in Judy Dempsey, "New Interior Minister Revives a Debate: Can Muslims be True Germans?" *The New York Times*, March 7, 2011, p. A5.
6. Robin Simcox, "The Battle from Algiers: Lessons from the Charlie Hebdo Attack." *Foreign Affairs*, January 12, 2015, http://www.foreignaffairs.com/articles/142763/robin-simcox/the-battle-from-algiers.
7. Quoted in Dwight Garner, "A Turning Tide in Europe as Islam Gains Ground," *The New York Times*, July 30, 2009, pp. C-1, C-4.
8. Simon Kuper, "The Myth of Eurabia," *Financial Times Life and Arts*, October 4, 2009, p. 2.
9. Interview with Brooke Baldwin on CNN, January 7, 2015.
10. "Merkel Says German Multicultural Society Has Failed," BBC News, October 17, 2010, http://www.bbc.com/news/world-europe-11559451.
11. George Perkins Marsh, *Man and Nature, or, Physical Geography as Modified by Human Action* (New York: Scribner, 1864), p. 279.
12. David Levinson, *Ethnic Groups Worldwide* (Westport, Conn.: Oryx Press, 1998), p. 1.
13. Robert Kaplan, "Europe Struggles to Fight Unemployment," *Stratfor*, October 8, 2014.
14. Alissa J. Rubin and Maia de la Baume, "Under Strain, France Examines its Safety Net: A Model of Social Democracy at Risk," *The New York Times*, November 8, 2013, pp. A4, A8.
15. http://www.bbc.com/news/world-europe-11216061. An Oxfam briefing paper called "Stop the Dumping! How EU Agricultural

Subsidies Are Damaging Livelihoods in the Developing World" is an excellent read on this topic (http://www.iatp.org/files/Stop_the_Dumping_How_EU_Agricultural_Subsidies.htm).

16. For an article on reforms by cutting subsidies, see "Europe's Economic Comeback Is Already Fading," *International Strategic Analysis*, August 18, 2014.
17. Paul Mason, "Deflation Would Be the Stuff of Nightmares for European Economies," *The Guardian*, September 7, 2014, http://www.theguardian.com/commentisfree/2014/sep/07/deflation-secular-stagnation-europe-economic-nightmare.
18. David Jolly and Jack Ewing, "E.C.B. Stimulus Calls for 60 Billion Euros in Monthly Bond-Buying," *The New York Times*, January 22, 2015, p. B1.
19. "Europe's Fragmentation Endangers a Trans-Atlantic Pact," *Stratfor*, October 23, 2014.
20. Quoted in A. A. Byatt, "What Is a European?" *New York Times Magazine*, October 13, 2002, p. 58.
21. http://www.realclearpolitics.com/video/2014/08/12/scarborough_on_europe_ignoring_world_crises_why_is_it_always_on_our_shoulders.html.
22. Robert Kagan, *Of Paradise and Power: America and Europe in the New World Order* (New York: Knopf, 2003), p. 1.
23. Terry G. Jordan and Bella Bychova Jordan, *The European Culture Area: A Systematic Geography* (Lanham, Md.: Rowman & Littlefield, 2002), p. 402.
24. René Dubos, *A God Within* (New York: Scribner), pp. 135–152.
25. David Lowenthal and Hugh C. Prince, "English Landscape Tastes," *Geographical Review*, 55(2): 186–222.
26. Jordan and Jordan, 2002, p. 163.
27. Sarah Gordon, "Europe's High-Tech Sector Losing Ground," *Financial Times*, p. 3
28. Jordan and Jordan 2002, p. 148.
29. A. T. Kearney, "2014 Global Cities Index and Emerging Cities Index: Global Cities, Present and Future," February 20, 2014, http://www.atkearney.com/documents/10192/4461492/Global+Cities+Present+and+Future-GCI+2014.pdf/3628fd7d-70be-41bf-99d6-4c8eaf984cd5.
30. Philip Stephens, "Shrunken Ambitions," *Financial Times*, April 28, 2010, p. 9.
31. Quoted in Kevin J. O'Brien, "Last Out, Please Turn Off the Lights: Poor Economy Is Driving East Germans from Home," *The New York Times*, May 28, 2004, pp. W1, W7.
32. These perceptions of Germans are discussed in Quentin Peel, "An Unequal Union," *Financial Times*, October 1, 2010, p. 11
33. Jane Burgermeister, "Germany: The World's First Major Renewable Energy Economy," Renewable Energy World.com, April 3, 2009, http://www.renewableenergyworld.com/rea/news/article/2009/04/germany-the-worlds-first-major-renewable-energy-economy.
34. Ian Cowell, "Menacing the Land, but Promising to Rescue the Earth," *The New York Times*, July 4, 2005, p. A4 (Churchill quote).
35. Sally Davies, "Start-Ups Emerge from Eastern Europe's Post-Soviet Era," *Financial Times*, June 3, 2014, http://www.ft.com/intl/cms/s/0/6c287f48-eb0a-11e3-9c8b-00144feabdc0.html#axzz3MLXRI5yS.
36. Dan Bilevsky, "Roma, Feared as Kidnappers, See Their Own Children at Risk," *The New York Times*, pp. A1, A8
37. Ibid.

Global Geoscience Watch

Use the Google Earth "World Map" tool at the top of the GREENR database home page to research the use of alternative energy in Europe. Click on a country, then click on the "pin" to access the database for this country. Search on "alternative energy." Write a brief report highlighting an interesting use of alternative energy you read about. Address such topics as the technology used, the chances for long-term success, and use in other countries. If for any reason you can't find information in the country you chose, select another country.

Online Resources

For access to MindTap and additional study materials visit www.cengagebrain.com. Read your textbook, take notes, complete activities, take practice quizzes and more.

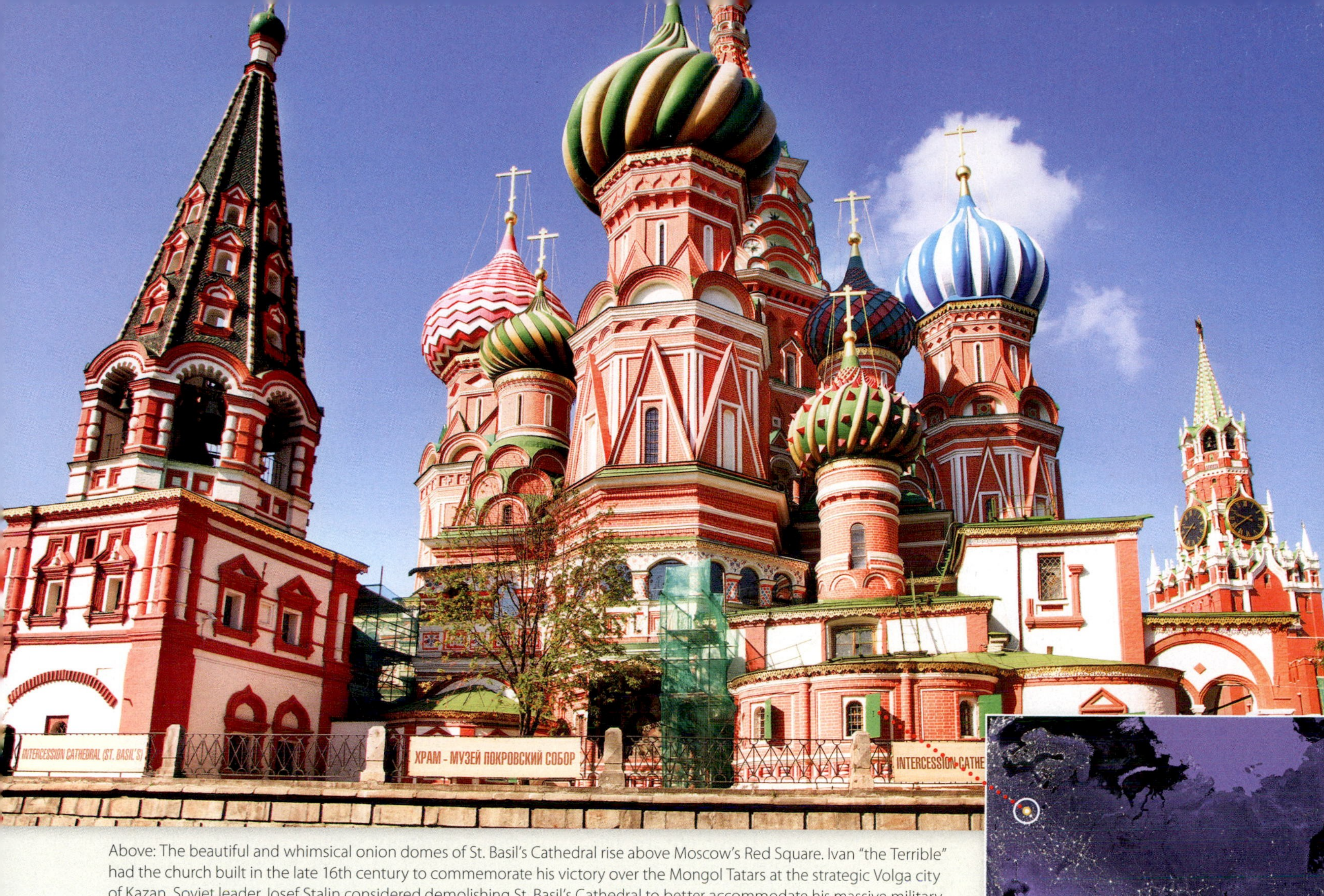

Above: The beautiful and whimsical onion domes of St. Basil's Cathedral rise above Moscow's Red Square. Ivan "the Terrible" had the church built in the late 16th century to commemorate his victory over the Mongol Tatars at the strategic Volga city of Kazan. Soviet leader Josef Stalin considered demolishing St. Basil's Cathedral to better accommodate his massive military parades on Red Square. ©Nikolay Vinokurov/Shutterstock.com

Right: The great landmass of Russia and neighboring lands at night. Note the string of lights along the Trans-Siberian Railway. C. Mayhew & R. Simmon (NASA/GSFC), NOAA/ NGDC, DMSP Digital Archive

5

Russia and the Near Abroad

Russia is a mystery wrapped inside a riddle inside of an enigma.

—WINSTON CHURCHILL

Take a moment to appreciate the size of the region you are about to study. Russia and the countries along its perimeter make up an area of breathtaking scale. Use the images within this book's front and back covers, Google Earth, a globe, or the chapter's reference maps (•**Figures 5.1** and **5.2**). You are about to explore a region that is spatially massive and of enormous consequence in world affairs, a fact that was overlooked when complacency about post–Cold War Russia shuttered many US academic programs dedicated to Russian studies. On the heels of Russia's invasion of Ukraine's Crimean Peninsula in February 2014, Angela Stent, an expert on Russian foreign policy, in a *Washington Post* essay entitled "Why American Doesn't Understand Putin," wrote, "The need for a dedicated and deep understanding of Russia—especially the motives and machinations emanating from the Kremlin—is as critical as ever. Otherwise the United States is doomed to repeat cycles of 'resets,' great expectations of better relations with Russia followed by serial disappointments."[1] This chapter will examine those motives and machinations, and kindle your interest in the issues.

Chapter Outline

Chapter Objectives

This chapter will enable you to

- Appreciate the environmental obstacles to development in vast areas of the world's largest country and nearby nations.
- Learn the significant milestones in the historical and geographic development of Russia and the Soviet Union, some of them accompanied by unimaginable loss of life.
- Recognize the differences between command and free-market economies and the post-Soviet difficulties in shifting from one to the other.
- Understand the reasons for the reversal of Russia's progress through the demographic transition.
- Come to know the geopolitical and ethnic forces threatening the unity of Russia and pitting various groups and countries within and outside the region against one another.
- Understand Russia's motivations in seizing Ukrainian territory and the country's prospective interventions in other neighboring states.
- Appreciate the importance of fossil fuels in Russia's economy, and how overdependence on natural resources poses risks to the country's development.

This vast geographic realm is made up of Russia; the kindred Slavic countries of Ukraine, Belarus, and Moldova; the decidedly non-Russian but heavily Russian-influenced "stans" of central Asia—Kazakhstan, Uzbekistan, Turkmenistan, Kyrgyzstan, and Tajikistan; and the countries of Georgia, Armenia, and Azerbaijan in the fractious Caucasus Isthmus. Nine of these 12 countries are also the member states of the **Commonwealth of Independent States (CIS),** an economic and political association anchored by Russia (Ukraine, Turkmenistan, and Georgia are not CIS members).

Geographers have not had an easy task in defining and naming the regions that emerged from the breakup of the Soviet Union. As discussed in Chapter 1, regions are organizing tools, not facts on the ground. There is no acknowledged best way to classify this region. Some geographers have chosen to cleave the central Asian "stans" from Russia, establishing them as an entirely separate region, or to include them as part of a greater Middle Eastern region. Some still consider the Baltic countries (Estonia, Latvia, and Lithuania) as part of the greater Russian realm, especially because large ethnic Russian populations live in them. The variety of geographers' names for the region suggests how unsettled the terminology is: "Post-Soviet Region," "Russia and Its Neighbors," "Russia and the Newly Independent States," "Russia and Neighboring Countries," "Former Soviet Union" (with the acronym FSU), and this book's "Russia and the Near Abroad" are among them. The must-read book on the geography of this region is Mikhail Blinnikov's *A Geography of Russia and Its Neighbors.*

From 1917 to 1991, the huge region known in this text as Russia and the Near Abroad, plus the three Baltic countries, made up a single Communist-controlled country known as the Union of Soviet Socialist Republics (USSR), or Soviet Union for short. (A *soviet*, which translates as "council" in English, is a political or economic system in which the working class in a business or political unit manages the affairs of that organization collectively.) The Russian-dominated government in Moscow controlled the affairs of many non-Russian peoples in the country, and after World War II also effectively controlled its Communist satellite countries in Eastern Europe (those satellite countries were discussed in Chapter 4). For four decades, the Cold War between the Soviet bloc of nations and the US-led Western bloc dominated world politics. These contending spheres vied in a geopolitical struggle for the allegiance of newly independent countries and other nations around the world after World War II.

The stakes in this contest were high, and at times the chilly relations of the Cold War threatened even to bring on a nuclear conflict with unimaginable consequences. Each side had massive nuclear stockpiles ready to be unleashed at a moment's notice. The 11th hour of such a nuclear showdown was reached in the Cuban Missile Crisis of 1962, when Russians and Americans prepared for the end (see **page 474**). In recently released audiotapes, American First Lady Jackie Kennedy can be heard pleading with her husband, US President John Kennedy, to be together with their children at the White House when the Soviet nuclear warheads detonated there. Each side regularly denounced the other with hostile rhetoric. American President Ronald Reagan attached moral significance to the struggle, dubbing the Soviet Union the **"Evil Empire"** in 1983.

Remarkably, just four years after President Reagan used that term, he and Soviet leader Mikhail Gorbachev stood together shaking hands and smiling in an agreement to reduce their nuclear stockpiles. And almost unbelievably, in 1991, while Gorbachev was still the Soviet leader, the giant country he ruled split into 15 independent countries (see **•Table 5.1**). Russia (officially known as the Russian Federation) remains by far the largest of these in area, population, and political and economic influence (see Insights, **page 154**).

Russia emerged from the fragmentation of the USSR as a great power, but not a superpower. Russia suddenly found itself struggling to maintain influence in the other countries that emerged from the ashes of the Soviet Union. From the Russian perspective, the other 14 countries constituted the **Near Abroad,** a zone in which Russia had to preserve and exert its special interests and influence.

Table 5.1 Russia and the Near Abroad: Basic Data

Political Unit	Area (sq mi, thousands)	Area (sq km, thousands)	Population (millions)	Rate of Natural Increase (%)	Urban Population (%)	Population Under Age 15 (%)	Agriculture Workers (%)	Per Capita GDP-PPP ($US)	GDP ($US, billions)	Trade Balance ($US, billions)	Oil Production (million bbl/day)	Literacy Female (%)	Literacy Male (%)	HDI
Russia and the Near Abroad	**8533.2**	**22100.8**	**284.5**	**0.4**	**65**	**19**	**9**	**12,900**	**5,068**	**—**	**13**	**99**	**98**	**0.745**
Core														
Belarus	80.2	207.7	9.5	−0.1	76	15	7	16,100	171	−3	*	99	99	0.786
Moldova	13.0	33.6	4.1	0	42	16	15	3,800	17	−3	—	98	99	0.663
Russia	6592.8	17075.3	143.7	0	74	16	4	18,100	3,568	196	10.1	99	99	0.778
Ukraine	233.1	603.7	42.9	−0.4	69	15	12	7,400	373	−8	*	99	99	0.734
Caucasus Region														
Armenia	11.5	29.8	3	0.5	63	19	22	6,300	24	−2	—	99	99	0.730
Azerbaijan	33.4	86.5	9.5	1.3	53	22	6	10,800	168	20	0.8	99	99	0.747
Georgia	26.9	69.6	4.8	0.2	54	17	9	6,100	34	−4	—	99	99	0.744
Central Asia														
Kazakhstan	1049.2	2717.4	17.3	1.5	55	25	5	14,100	420	40	1.6	99	99	0.757
Kyrgyzstan	76.6	198.4	5.8	2.1	34	31	19	2,500	19	−4	—	98	99	0.628
Tajikistan	55.3	143.2	8.3	2.7	26	36	27	2,300	22	−4	—	99	99	0.607
Turkmenistan	188.5	488.2	5.3	1.4	47	28	13	9,700	82	5	0.3	98	99	0.698
Uzbekistan	172.7	447.3	30.7	1.8	51	28	18	3,800	170	1	0.1	99	99	0.661

* Less than 0.1. — Data not available or not applicable.
Sources: World Population Data Sheet, Population Reference Bureau, 2014; Human Development Report, United Nations, 2014; World Factbook, CIA, 2014.

Table 5.2 Largest Cities in Russia and the Near Abroad

City	Population
Moscow, Russia	16.1
St. Petersburg, Russia	5.1
Tashkent, Uzbekistan	2.2
Kiev, Ukraine	2.2
Minsk, Belarus	1.9
Almaty, Kazakhstan	1.5
Novosibirsk, Russia	1.4
Kharkiv, Ukraine	1.4

Population in millions.
Source: Demographia World Urban Areas, 2015.

The choice to use "Russia and the Near Abroad" in this book is not meant to impose a Russian-centered view of the region, as the name might imply, but to reflect the enormous power that Russia wields (or would like to wield) there. The former Soviet countries have enduring and often uncomfortable strategic and economic associations with Russia. For example, the Soviet-era oil refinery and pipeline system still links Russia with many countries of the Near Abroad. As we will see, Russia has periodically shut off the pipelines supplying oil and natural gas, or has increased the prices of these commodities, to obtain political concessions from some of these nations. The needs of many of the Near Abroad countries to buy Russian energy or export their own energy resources across Russian territory give the countries an incentive to maintain peaceful political relations with Russia. The countries of the Near Abroad generally do not want to trigger Russian military intervention, fearing this could open the door to Russia re-exerting its control. Russia seized parts of Georgia in 2008 and Ukraine in 2014, so these are reasonable fears. In this chapter, you will get acquainted with many of these issues in their geographical context, which is simply enormous.

181

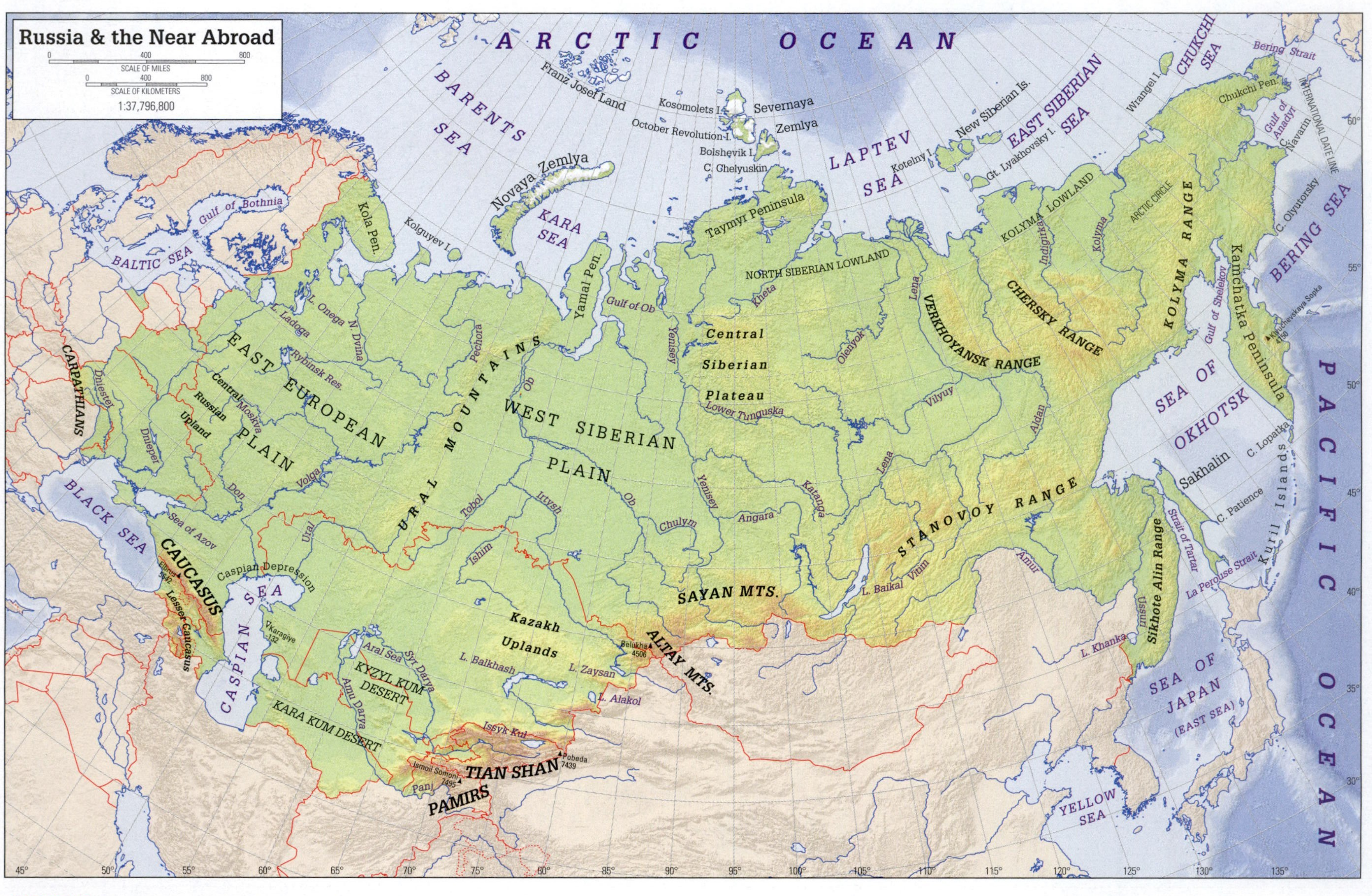

• **Figure 5.1** Physical geography of Russia and the Near Abroad.

• **Figure 5.2** Political geography of Russia and the Near Abroad.

Insights Regional Names of Russia and the Near Abroad

Here are some useful geographic terms to know when studying this region. The region of Russia and the Near Abroad consists of what used to be the Union of Soviet Socialist Republics, minus the Baltic countries of Estonia, Latvia, and Lithuania. The Soviet Union came into existence in 1922, following the overthrow of the last Romanov tsar in the Russian Revolution of 1917 and the subsequent civil war. Prerevolutionary Russia is known as Old Russia, Tsarist Russia, Imperial Russia, or the Russian Empire.

The name *Russia* now refers to the independent country of Russia, which was the largest in size and population of the 15 Soviet Socialist Republics (Union Republics, or SSRs) that made up the Soviet Union. The full name of the Russian Federation during the Communist period was the Russian Soviet Federated Socialist Republic, or RSFSR; the name appears on many older maps.

The loosely aligned Commonwealth of Independent States (CIS) was formed by 10 of the 15 former Union Republics late in 1991 (the country of Georgia would withdraw from the CIS in 2008). Estonia, Latvia, and Lithuania are not members of this organization. They distanced themselves strongly from Russia, and in 2004 joined the European Union. They are characterized as European in Chapter 4.

The area west of the Ural Mountains and north of the Caucasus Mountains has been known historically as European Russia. The Caucasus and the area east of the Urals have been called Asiatic Russia, Soviet Asia, or the eastern regions. Transcaucasia is the region of the Caucasus Mountains and the area south of them. Siberia is the general name for the huge area between the Urals and the Pacific. central Asia is the arid area occupied by the five countries with large Muslim populations immediately east and north of the Caspian Sea. They were known collectively in pre-Soviet times as "Turkestan" and are known informally today as "the Stans" because all the country names end in 120 -*stan* (Persian for "land of").

5.1 Area and Population

With an area of 8.5 million square miles (22.1 million sq km), the region of Russia and the Near Abroad is the largest in this book. Russia is about twice as big as the United States, including Alaska. You can get a sense of this comparison by looking at Figure 5.4 on **page 158**. A good indication of its staggering size is the fact that this region (and even Russia by itself) spans 10 time zones; in comparison, the United States, from Maine to Hawaii, spans 6. For most Russians, having a landmass that sprawls unbroken across 10 time zones has been a matter of national pride, rather like the way Texans or Alaskans feel about the size of their states. Many Russians were dismayed when, in 2009, their president proposed (in the interest of the country's economy) that Russia should cut back to just four time zones. This would make it easier for businesses in Moscow, for example, to communicate with offices and customers in the Russian Far East on the same business day.

The eventful geopolitical history of Russia and the Near Abroad has given this region land frontiers with 15 countries in Eurasia (Figure 5.2). Between the Black Sea and the Pacific, the region borders Turkey, Iran, Afghanistan, China, Mongolia, and North Korea. Pakistan and India also lie close by. In the Pacific Ocean, narrow water passages separate the Russian-held islands of Sakhalin and the Kurils from Japan. In the west, the region has frontiers with Romania, Hungary, Slovakia, Poland, Lithuania, Latvia, Estonia, Finland, and Norway.

Although the region of Russia and the Near Abroad is huge, it does not have many people: only about 240 million, or about three-quarters of the population of the United States. Much of this vast region is sparsely populated (**•Figure 5.3a, b**). Outside of a distinctive core region, where many of the largest cities are located (see Table 5.2), great stretches of economically unproductive periphery separate populated areas from one another. 166 The mountainous frontiers in Asia have few inhabitants. As we will see, shrinking populations pose many challenges for Russia and, to a lesser extent, some of its neighbors. Russian leadership worries that there are not enough Russians in its far eastern region along the frontier with overcrowded China. You can see the population distributions clearly in Figure 5.3a. In contrast with the lightly populated east, you can see that to the west, on the frontier between the Black and Baltic Seas, international boundaries pass through populous lowlands. Vulnerable to invaders, these have long been disputed territories between Russia and other countries.

Russia has the region's largest population, with about 143 million people. The next largest country is Ukraine, with about 43 million. Uzbekistan is the most populous central Asian nation, with about 30 million people. The region's other countries trail far behind, with populations generally under 10 million apiece. Population growth rates are highest, around 1.8 percent annually, among the predominantly Muslim populations of the central Asian countries. At the other end of the spectrum, Russia, Ukraine, and Belarus are losing population at a rate of up to 0.4 percent per year. After the collapse of the Soviet Union, Russia experienced an astonishing population loss for years, declining by 6 million people between 1991 and 2011, before the downward trend finally started reversing. Population declines such as Russia's have seldom been seen on Earth except in times of warfare, famine, and epidemic.

But there has been no war or shortage of food. The depopulation has been instead due to an imbalance between fertility and mortality, which has its roots in the declining quality of life that occurred for two decades after the breakup of the Soviet Union (see Insights, **page 156**). In an atmosphere of economic and political uncertainty, couples chose to have fewer children, and there were more broken marriages. More disturbingly, death rates soared.

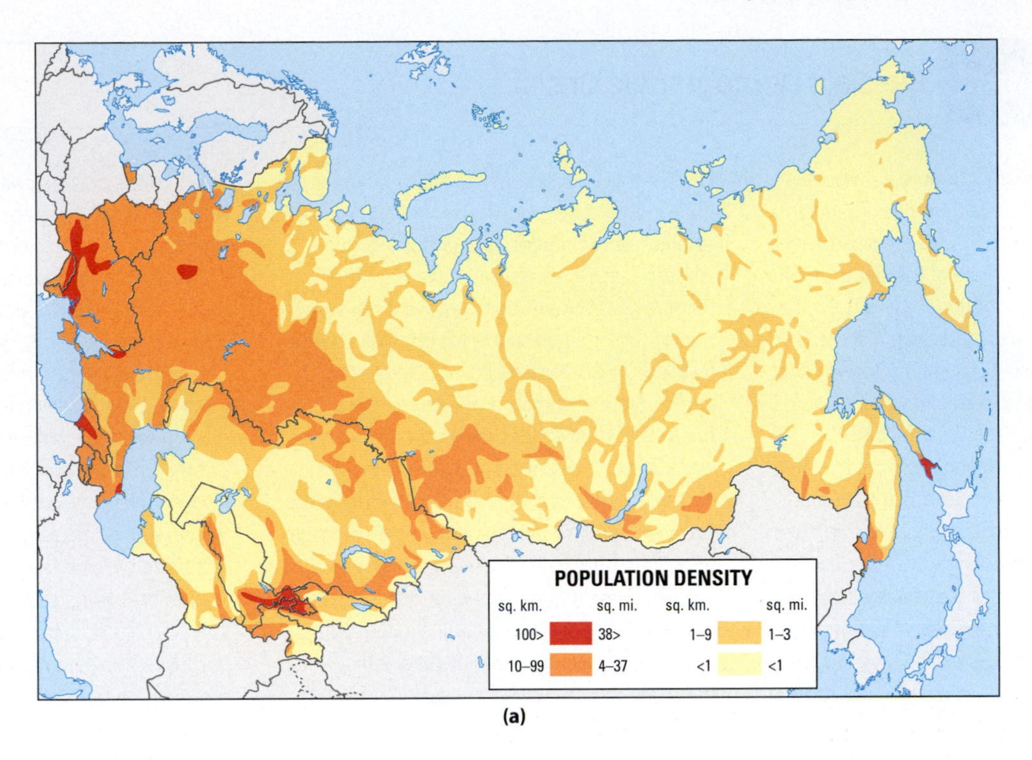

(a)

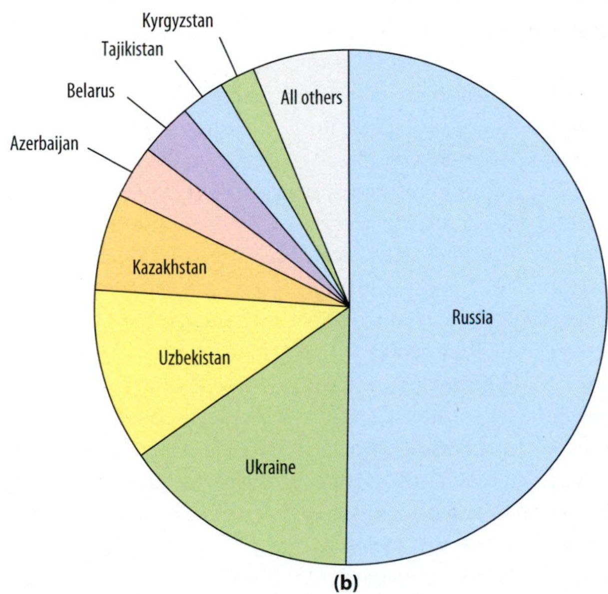

(b)

• **Figure 5.3** **(a)** Population distribution and **(b)** population pie chart of Russia and the Near Abroad.

5.2 Physical Geography and Human Adaptations

Stretching nearly halfway around the globe in northern Eurasia, the region of Russia and the Near Abroad is geographically gifted by its sheer size. But it is also geographically challenged. Most of the region has some combination of cold temperatures, infertile soils, marshy terrain, aridity, and ruggedness. These natural conditions are more comparable to those of Canada than to those of the United States. Four-fifths of the total area is farther north than any point in the conterminous United States (•**Figure 5.4**). Interaction with a demanding environment was a major theme in Russian and Soviet expansion and development, and is a hallmark of everyday life.

The Roles of the Climates and Vegetation

The region of Russia and the Near Abroad has a harsh climatic setting (•**Figure 5.5a**). Severe winter cold, short growing seasons, drought, and hot, crop-shriveling winds are major disadvantages. But there are also advantages in the form of good soils, the world's largest forests, natural pastures for livestock, diverse wild fauna, and great reserves of fossil fuels. Many of

Insights Russia's Demographic Crisis

Demographically, Russia ought to be a post-industrial country like the countries of Western Europe or like the United States. It should have passed into the fourth state of the demographic transition, where both birth rates and death rates have fallen parallel to one another (like the United 72 States in Figure 5.B). But it hasn't happened that way. Look at the trend lines for Russia in Figure 5.B, where a startling intersection took place.

Russia's economic health plummeted after the breakup of the Soviet Union, and so did human health. Along with the economic downturn came more unemployment, a decline in health and other services, increasing crime rates, and a growing sense of despair at the individual level. These gave rise to some very unhealthful habits among Russians, which have stubbornly persisted for more than two decades.

One of the biggest problems has been an increase in alcoholism. Russians do not imbibe much in beer and wine, the customary adult beverages of Western Europe. Instead they go for the harder stuff, especially vodka (**•Figure 5.A**). The figure for annual consumption of hard liquor is an eye-blearing 3.5 gallons per person (that statistic would include every man, women, and child in the country). Alcoholism is blamed for one-quarter of Russian men dying before reaching age 55, from accidents, violence, and direct alcohol poisoning. Alcohol poisoning kills an estimated 40,000 Russians (mainly men) each year, compared with fewer than 300 in the United States. One statistic after another points to alcohol-related problems. Perhaps the worst one is the Russian government's own admission that one in eight Russians die every year from causes directly related to or aggravated by alcohol.

Joe Hobbs

• Figure 5.A Alcoholism is a major killer in Russia.

Russia's government has repeatedly tried to crack down on alcohol abuse through public awareness campaigns, higher taxes, and restricted sales. It is a delicate balance, however. Russia's government does not want consumption of unregulated and untaxed vodka to rise further; the black market producing cheap vodka met an estimated one-third of the demand late in 2014. Nor does Russia's government want Russians deprived of the customary means of "taking the edge off," particularly during the economic kind of hardship the country experienced after Russia invaded Crimea and oil prices collapsed late in 2014. The solution, at least temporarily? The government announced a price cut of 16 percent for vodka in February 2015.

There are other problems too. Obesity has been rising and now affects about 25 percent of the population, with fast food taking most of the blame (more than 600 McDonald's restaurants are now in Russia). Smoking has long been a national pastime; an estimated 60 percent of Russian men smoke (compared to just 15 percent of Russian women). Organized and petty crime has resulted in an alarmingly high incidence of physical violence. Infections of HIV/AIDS, often accompanied by tuberculosis, increased 20 percent between 2005 and 2011, through the population of intravenous drug users and their sexual partners. Russia's geography is part of the problem: it is uncomfortably close to the world's largest producer of heroin, Afghanistan. Highly potent and inexpensive heroin snakes its way swiftly out of Afghanistan and through the central Asian countries into Russia.

Heroin addicts are left to their own devices to wean themselves off the drug through the rather primitive "cold turkey" method. This is not very effective, and the relapse rate is high. The West's common treatment of addiction with methadone is not practiced in Russia, where the health care system suffers from a vast array of problems. The universal system of health care under the Soviets was not efficient, but a health care system close to collapse replaced it, and hospitals were ill-equipped to help stem a rising tide of deaths from a variety of causes.

After 1990, Russia's death rate rose about 16 percent, and it has persisted as one of the world's highest levels (13 per 1000 in 2014), a rate comparable with many disease-ridden countries in Africa. Even war-torn Iraq, Syria, and Afghanistan have lower death rates than does Russia. This clearly is not the benign demographic transition in which a country's population declines because of falling birth rates that reflect growing affluence (see the graph and discussion on **pages 72–74**). Instead, Russia did something completely unprecedented in the world's demographic experience: it moved *backward* through the transition. Not only did death rates rise, but as quality of life deteriorated, birth rates also plummeted (by 43 percent between 1990 and 2000). The conditions that would promote higher fertility were in short supply. Half of Russian marriages ended in divorce in 2014, one of the highest divorce rates in the world. A physician cited "male/female estrangement and a loss of family cohesion" among the reasons for Russia's plummeting birth rate.[2] There were two abortions for every live birth in 2014.

The disturbing trend lines of increasing death rates and falling birth rates in Russia intersected in the mid-1990s, forming the grim graphic that demographers dubbed the "**Russian cross**" (**•Figure 5.B**). It was an ominous intersection; at the time, some population projections showed Russia's population 191

Birth and Death Rates
per 1,000 population

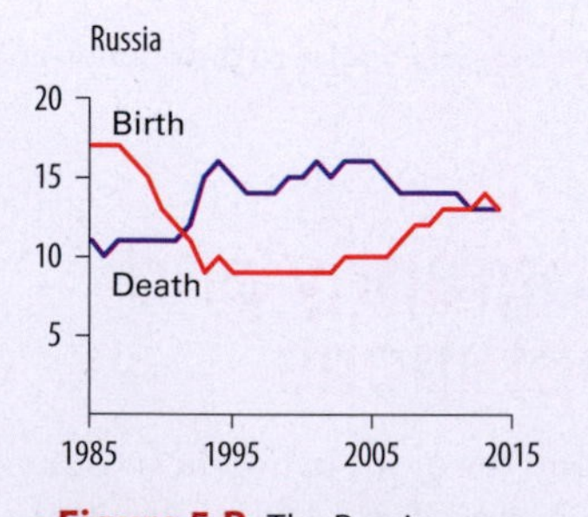

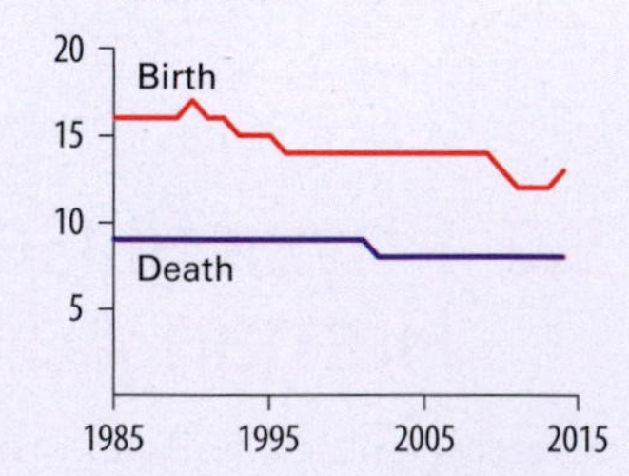

• Figure 5.B The Russian cross.
Source: Data from World Factbook, CIA, 2015.

THINK CRITICALLY: How are the demographic events depicted and discussed in this section revealed in Figure 5.C, Russia's age structure diagram?

declining from 140 million to about 100 million by 2050. The odd shape of Russia's age structure diagram in •**Figure 5.C** illuminates the problem: Russia's population is aging, and there are comparatively fewer young people. This is a common phenomenon across the countries of Russia and the Near Abroad; •**Figure 5.D** shows current 2050 population projections of some of the countries of the region, compared with Middle Eastern and African countries to the south. Countries of Eastern Europe and the former USSR are shrinking—Moldova's 2050 population may only be half of its 2015 total—while the youthful countries of Asia and Africa are projected to grow rapidly.

Alarmed, in 2006 the government in Moscow established a strategy to reverse these trends. It is trying to incentivize higher fertility through cash payouts to mothers, extended maternity leaves, and childcare benefits, similar to the incentives offered in Italy and Japan where birth rates are also very low. Russia's president introduced "Family Contact Day," a new national holiday on which Russians are encouraged to stay home and make babies. These efforts appear to be fruitful: in 2013 births exceeded deaths for the first time since 1991, and the population hemorrhage stopped. Officials hope that Russia's population can stabilize at 140 million in 2020 and then begin increasing.

This is a critical time in Russia's demographic crisis. The recent rise in birth rates and fall in death rates coincided with a boom in Russia's economy due mainly to high prices for oil, the country's principal export. With the economy's sudden downturn, will the trend lines turn in the wrong directions again? 94, 357

A major factor that could help Russia grow its population is immigration. Moscow has been trying to lure Russian expatriates home, but the economic crisis is instead growing the brain drain of Russia's educated and trained population. Immigrants are coming into Russia, but they tend to the jobless motivated by "push" factors in neighboring Mongolia and China. 75 As in Europe, there is strong anti-immigrant sentiment in Russia, and the overall ingredients for sustained demographic growth are absent.

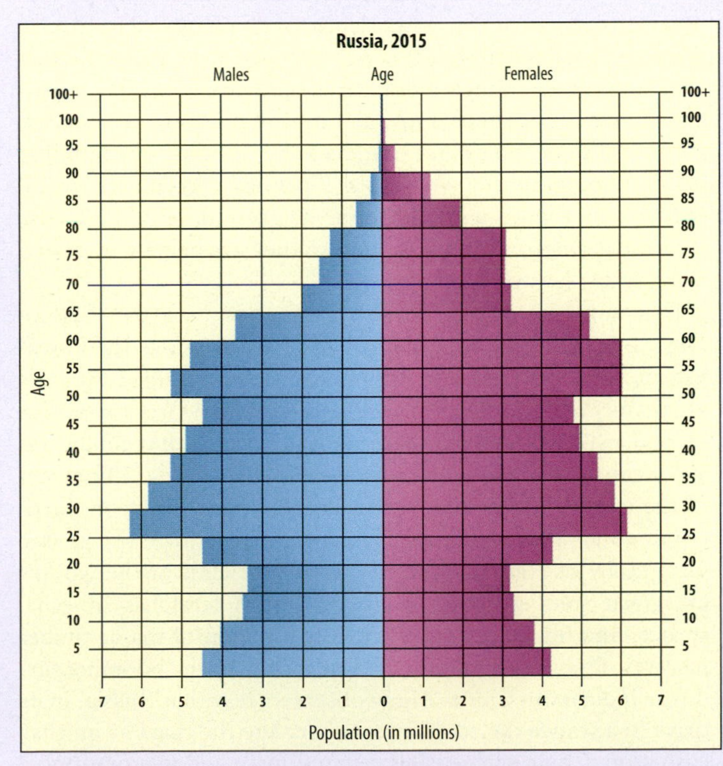

• **Figure 5.C** Russia's age structure diagram. Russia's population topped out at 148 million when the Soviet Union broke up in 1991 and has been falling ever since. Note the relatively low numbers of the country's young people. World War II's impact can be clearly seen in the imbalance between men and women at the top of the diagram; the men were killed.
Source: Based on US Census Bureau, International Database.

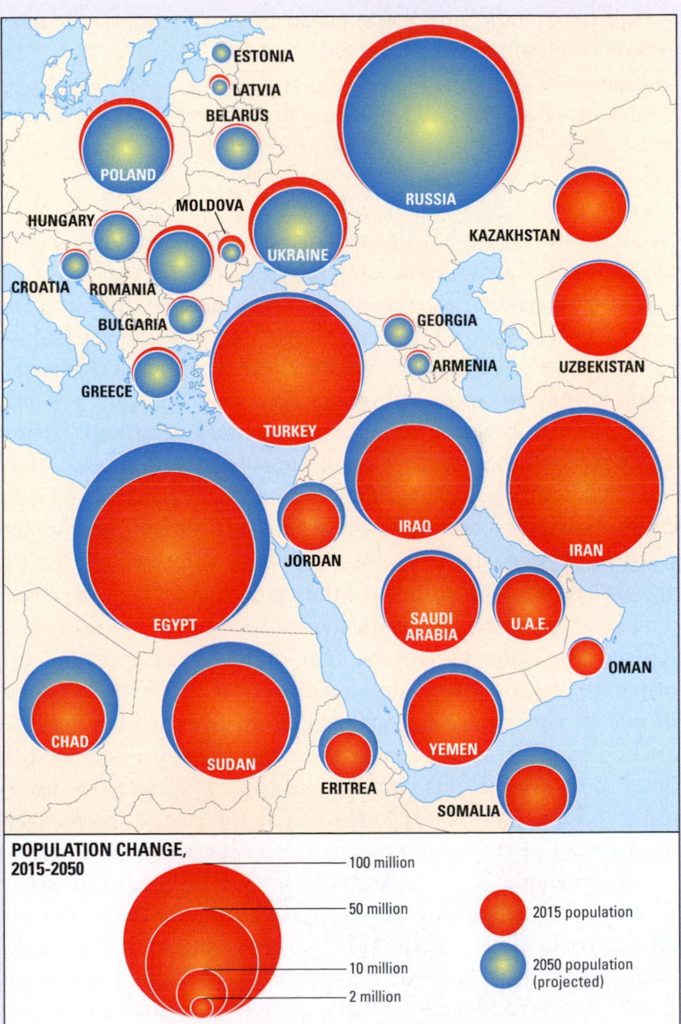

• **Figure 5.D** A comparison of the 2015 populations of selected countries, shown in red, with their projected 2050 populations, shown in blue.
Source: Based on US Census Bureau, International Database.

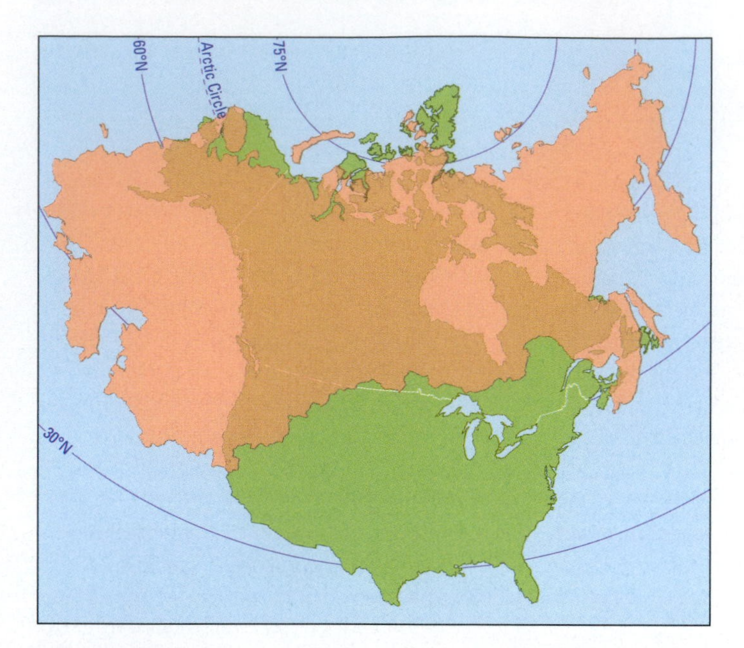

• Figure 5.4 Russia and the Near Abroad compared in latitude and area with the continental United States and Canada.

THINK CRITICALLY: What impacts do Russia and the Near Abroad's high latitude have on land and life in the region?

these resources are hard to reach because of the logistical difficulties posed by climatic extremes and vast distances.

Most parts of the region have continental climatic influences, with long, cold winters; short, warm summers; and low to moderate precipitation. Severe winters, for which Siberia—roughly the eastern two-thirds of Russia—is particularly infamous, prevail because the region is generally at a high latitude and has few of the moderating influences of oceans. As we have seen in Chapter 4, Europe benefits from those moderating influences. Westerly winds from the Atlantic reach only the western portion of Russia and the Near Abroad, and their moderating effects become weaker toward the east, where it is much colder.

97

The most extreme continental climate of the world is here. Russia can claim the lowest temperature ever recorded in a populated region (uninhabited Antarctica can be much colder): *minus 90°F* (minus 68°C), in the Siberian settlement of Verkhoyansk. That location is well inland from the shores of northern Siberia, so even in this context you can appreciate the moderating influence of ocean waters. Even the Arctic Ocean has a moderating effect on temperature, so Siberia's northern coast is not quite as cold as the region's interior, further to the south.

Russia and the Near Abroad have a large amount of arable land; Russia alone has the third largest amount of the world's land that can be farmed, behind only the United States and India. But several factors make agriculture difficult in large parts of this world region. The average frost-free season of 150 days or less in most areas (except the extreme south and west) is too short for many crops to mature. Aridity and drought create problems too. The annual average precipitation is less than 20 inches (50 cm) nearly everywhere except in the extreme west, along the eastern coast of the Black Sea and the Pacific coast north of Vladivostok, and in some of the higher mountains. Most places are relatively warm during the brief summer, and the southern steppes and deserts are hot. Thankfully, although the summer is short, the high latitudes bring longer days in the summer, and these encourage plant growth. Summer is also a time for more outdoor activities. Russians are passionate campers and sun worshipers, and they take every opportunity to "get away from it all" to enjoy the fleeting pleasures of summer. Urban residents never lose the traditional Russian passion for nature, and many retreat as much as possible to their rural cabins, called *dachas*. These have an important place in Russia's agriculture; see **page 194**.

There are five main climatic belts in the region of Russia and the Near Abroad: polar, along the Arctic shore; subarctic, which covers most of Siberia; humid continental, where most of the region's population lives; and both semiarid and desert, mostly confined to central Asia. You can see how each of these belts with its associated vegetation and soils succeeds the other from north to south in **•Figure 5.5b**. Most human activities take place in the subarctic, humid continental, and steppe climatic zones. **•Figure 5.6** depicts land uses all across the vast region of Russia and the Near Abroad.

Unlikely Sources of Global Warming in Russia's Frozen Wilderness

Much of the tundra and subarctic climatic zones are permanently frozen to an average depth of 3 meters. The frozen ground, called **permafrost,** makes construction difficult. Heat generated by buildings melts the upper layers of permafrost and causes foundations and walls to sink and tilt. Most buildings are elevated on pilings to reduce this risk, like the building in the photograph of **•Figure 5.7**. Pipelines carrying crude oil (which is hot when it comes from the ground) would likewise melt the permafrost and sink, and so they are heavily insulated or built on elevated supports.

34, 160

The subarctic climate zone corresponds with the Russian *taiga*, or northern coniferous forest (also known as the boreal forest). This is not an agriculturally productive zone; the soils are mainly acidic, infertile **spodosols** (from the Greek word for "wood ash" and known in Russian as ***podzols***). But conditions are ideal for coniferous tree growth. Representing 20 percent of the world's forested land, this is the largest forest on Earth (**•Figure 5.8**). This wilderness has witnessed strange things (see Geography of Natural Hazards, **page 161**). Half of the world's evergreen trees are here, with enormous stretches of pine, spruce, and fir. Economically, this is a region of major timber reserves. Russia exports raw logs primarily to the booming furniture industry in China. Many observers fear that Russia, in its drive to advance the economy, will deplete the taiga as an easy cash crop. China's increasing demand for wood is prompting a massive increase in illegal logging of the taiga in Russia's Far Eastern region; two to four times the amount of timber legally allowed to be cut is exported to China each year. There would be global consequences for selling off this commodity. Russia's blanket of coniferous forest acts as a major carbon dioxide "sink" that helps absorb human-made greenhouse gases, so its removal would contribute to climate change.

44

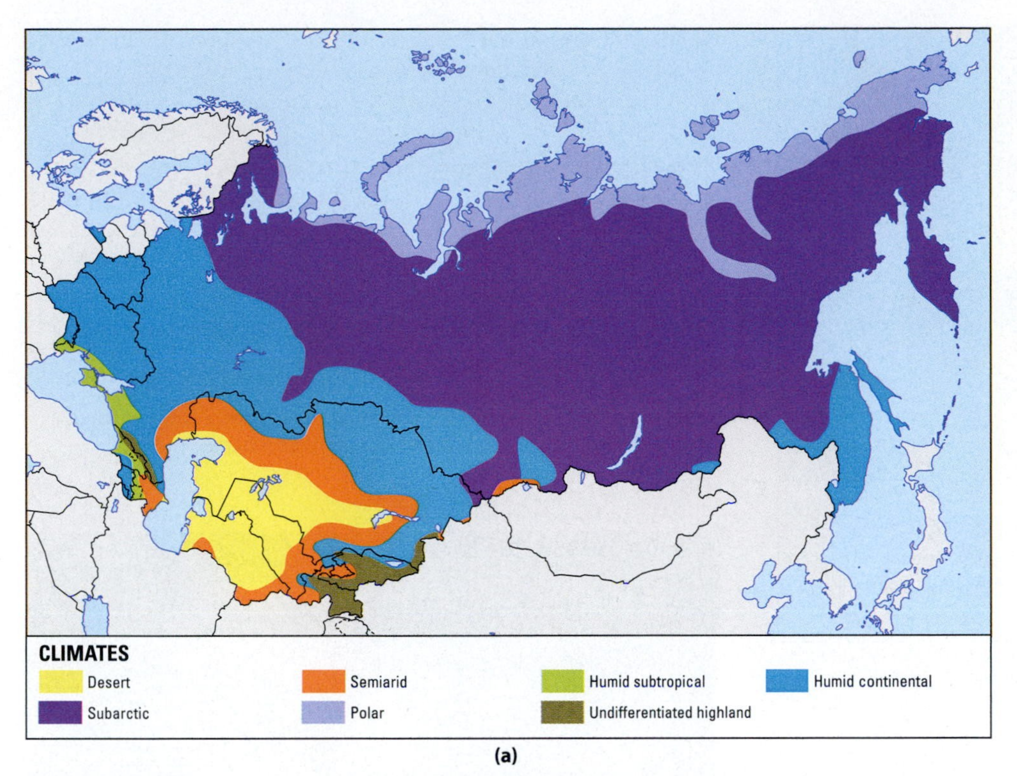

(a)

(b)

• **Figure 5.5** **(a)** Climates and **(b)** biomes of Russia and the Near Abroad.

As discussed in Chapter 2 (**page 41**), global warming is likely to be most intense in the high latitudes and polar regions. We looked at a "tipping point" scenario in which warming temperatures trigger other mechanisms that warm the atmosphere irreversibly. We saw how warm temperatures melt arctic sea ice, leading to more warming. Another positive feedback loop of this kind involves the permafrost underlying the subarctic and tundra zones, most extensively in Russia, but also in northern Europe, northern North America, and Greenland. The carbon locked up in this organic matter of permafrost is more than double the carbon already in Earth's atmosphere. If warmer temperatures thaw this frozen ground, bacteria and fungi will begin to break down the carbon contained in the organic matter and release it into the atmosphere as carbon dioxide and methane, both of which are greenhouse gases. This would lead to even more atmospheric warming.

Steppes and Good Soils

A humid continental climate (similar to the climate of the Midwestern US) lies south of the subarctic climate, stretching eastward from the region's western border to the vicinity of Novosibirsk (see Figure 5.5a). Both the climate and the soils are more favorable for agriculture in the humid continental climate than in the subarctic zone.

The northern part of the humid continental climate is forested. The more southerly regions, stretching from Ukraine and southern Russia to Kazakhstan, are dominated by steppes—vast, largely treeless plains dominated by short grass with deep roots, similar to the Great Plains of North America. Pastoralists, including the Scythians, who in the 4th century BCE produced astonishing artwork of gold in what is now Ukraine, grazed their herds across these steppes for millennia.

The steppe corresponds with the **black-earth belt,** which today is the most important area of crop and livestock production in the region of Russia and the Near Abroad. In the 1950s, the Soviet Union began extending the areas of intensive agricultural production beyond Ukraine and into the so-called **virgin and idle lands** (or "new lands")—the fertile, black-earth soils of northern Kazakhstan and southern Siberia (•**Figure 5.9**). Today, spring wheat and spring barley are the main crops in this rich agricultural zone, supplemented by sugar beets and sunflowers. Cattle are the principal livestock (•**Figure 5.10**). The main soils of this belt are known as ***chernozem,*** which is Russian for "black earth." Among the best soils to be found anywhere in the world, these are the thick and productive soil type known as **mollisols.** (A similar belt of mollisols spans the

• **Figure 5.6** Land use in Russia and the Near Abroad.

Joe Hobbs

• **Figure 5.7** In the High Arctic, buildings must be erected on pilings so that they do not melt the permafrost below. This is the Russian coal-mining settlement of Barentsburg in Norway's Svalbard (Spitsbergen) Archipelago.

©Dmitry Strizhakov/Shutterstock.com

• **Figure 5.8** A glimpse of Siberia's vast taiga near Ulan-Ude in Buryatia, southeast of Lake Baikal.

• **Figure 5.9** The belt of fertile *chernozem* (black earth) soils stretches from Ukraine through Kazakhstan into southern Siberia.

Joe Hobbs

• **Figure 5.10** A grazier with his cattle on the steppe near the Don River in southern Russia.

eastern portion of the Great Plains of North America.) Their great fertility is due to an abundance of **humus** (decomposing plant and animal matter) in the topsoil.

In central Asia, growing crops is difficult in the desert climate areas east of the Caspian Sea. Cotton is the most important crop where irrigation water is available from the Syr Darya, Amu Darya, and other rivers flowing from high mountains to the south and east. The Syr Darya and Amu Darya flow into the Aral Sea—or what is left of it; on **page 199**, there is discussion of Soviet-era management of agriculture in this area and its legacy in the disrupted ecosystems of central Asia.

The Role of Rivers

In the early history of Russia, rivers formed natural passageways for trade, conquest, and colonization. They were especially crucial in settling Siberia, which is drained by some of the greatest rivers on Earth: the Ob, Yenisey, Lena, and Kolyma, all flowing northward to the Arctic Ocean, and the Amur, flowing eastward to the Pacific. By following these rivers and their lateral tributaries, the Russians advanced from the Urals to the Pacific in less than a century.

Geography of Natural Hazards Near-Earth Objects

Thanks to nearly every Russian vehicle having front-facing dash cams, something relatively rare was well documented on February 15, 2013: shortly after sunrise, a meteor approximately 60 feet (20 m) wide streaked across the sky of the central Russian city of Chelyabinsk (**•Figure 5.E**). The object, traveling at over 40,000 miles an hour, entered the Earth's atmosphere at a shallow angle and exploded approximately 18 miles (29 km) above the ground. The blast and subsequent shock wave, which released over 20 times the energy of the nuclear bomb detonated over Hiroshima, caused building damage in an area several dozen miles across. More than 1000 people were injured after the blast, mostly from the glass fragments of shattered windows. A few dozen people even got a kind of "sunburn" from the intensely bright flash of the explosion.

The Chelyabinsk meteor was the largest object to have entered Earth's atmosphere since June 30, 1908, when a similar space rock exploded over the Tunguska region of Siberia, 1800 miles (2900 km) to the east of Chelyabinsk. These impacts bring up such questions as "What would happen if a similar object hit a populated area today?" An eyewitness to Tunguska reported:

> Suddenly in the north sky . . . the sky was split in two, and high above the forest the whole northern part of the sky appeared covered with fire . . . At that moment there was a bang in the sky and a mighty crash . . . The crash was followed by a noise like stones falling from the sky, or of guns firing. The earth trembled.[3]

The dimensions of the Tunguska object, possibly 150 feet wide, are uncertain; but like the Chelyabinsk meteor, it exploded in the atmosphere—approximately 5 miles (8 km) up—and did not directly strike the ground. The explosion generated as much energy as 1000 Hiroshima-type nuclear bombs, leaving a breathtaking signature on the landscape: an area 800 square miles (2000 sq km) of 80 million flattened and burned coniferous trees pointing directly away from an epicenter. It was still an apocalyptic scene 30 years later when a scientific expedition reached the site and took the photograph shown in **•Figure 5.F**.

These and other smaller events like them are a wake-up call for Planet Earth. There are over 10,000 near-Earth objects (NEOs), which are small asteroids and comets orbiting the sun at a similar distance to Earth. As the two Siberian events show, even "small" objects, with diameters in the dozens of meters, can have significant effects. Asteroids with diameters of 10 feet (about 3 m) strike the Earth's atmosphere nearly every year. A Tunguska-sized rock may strike Earth once every 800 years on average. Most of the Earth's surface is covered in water, and even most land areas are sparsely populated, so the likelihood of an NEO directly affecting a major city is very low. Still, if the Tunguska object had exploded over a populated area, the damage would be as if nuclear weapons had been detonated there—damage and casualties would be enormous.

Sixty-six million years ago, an asteroid roughly 6 miles (10 km) wide struck the Earth near the northern tip of Mexico's Yucatán Peninsula. Releasing energy over *5 billion* times more powerful than a Hiroshima-sized bomb, the explosion created dust and debris that lingered in the atmosphere for so many years that global net primary productivity plummeted. This brought the end of the dinosaurs and began the rise of mammals, the lineage from which we emerged. The unthinkable effects of a modern-day equivalent have scientists and decision makers around the world working on projects to prevent them; these include Europe's "NEOShield" and the NASA-supported "Spaceguard" projects. Ideally, with enough warning NASA or its European or Chinese partners could launch a space mission to knock the object off-track and, in effect, save the world. For more information on these compelling issues, visit NASA's Near Earth Object Program on the Web.[4]

AP Images

• Figure 5.E The Chelyabinsk meteor viewed from a dash cam. The 10 ton "near-Earth object" tore through Earth's atmosphere with tremendous energy. When it exploded 18 miles above the ground it generated a shock wave that shook buildings and shattered glass, leaving more than 1000 people injured.

Sovfoto/Universal Images Group/Getty Images

• Figure 5.F The Tunguska explosion of 1908 flattened great areas of surrounding taiga, with the fallen trees pointing away from the blast's epicenter. A scientist took this photo in 1938, 30 years after this celestial encounter with Earth.

THINK CRITICALLY: What "what-ifs" come to mind when you consider a meteor of this size or greater impacting a metropolitan area? What repercussions might there be in the context of globalization?

The Moscow region lies on a low upland from which a number of large rivers radiate like the spokes of a wheel (see Figure 5.1). The longest ones lead southward: the Volga to the landlocked Caspian Sea, the Dnieper to the Black Sea, and the Don to the Sea of Azov, which connects with the Black Sea through the strategic Kerch strait (see Geography of the 181 World's Great Rivers, **page 163**). Shorter rivers lead north and northwest to the Arctic Ocean and Baltic Sea. These river systems are accessible to each other by portages.

A major link in the inland waterway system is the Volga-Don Canal, tying the two rivers together where they approach each other near the city of Volgograd (see the canal in **•Figure 5.11** and its location in **•Figure 5.12**). This canal and others allowed the White Sea and Baltic Sea in the north to be linked to the Black Sea and Caspian Sea in the south in a single water transport system. On the map in Figure 5.12, you can see how these engineering works effectively allow ships to sail from the heartland of Russia to the Atlantic and Arctic oceans and the Mediterranean Sea.

• Figure 5.12 Moscow's situation near many important rivers, and the canals linking them, allows goods to be transported from Russia's heartland to the Mediterranean Sea and the Arctic Ocean.

The Role of Topography

Most of the big rivers in Russia and the Near Abroad wind slowly for hundreds or thousands of miles across large plains. Such plains and low hills compose nearly all the terrain from the Lena River to the western border of the region (see Figure 5.1). The only mountains interrupting this lowland are the Urals, a low and narrow range (with an average elevation less than 2000 ft [600 m]) separating Europe from Asia and European Russia from Siberia (remember that there is no distinct physical cleft between Europe and Asia on the great landmass of Eurasia, but the Urals range is widely acknowledged as a vernacular dividing line). The Urals trend almost due north–south. A wide lowland gap between the southern end of the Urals and the Caspian Sea allows uninterrupted east–west movement by land. Cut by river valleys offering easy passageways, the Urals are not a serious barrier to transportation.

Joe Hobbs

• Figure 5.11 A lock in the Volga-Don Canal, a vital link for trade between the Volga watershed and the Black Sea.

Between the Urals and the Yenisey River, the West Siberian Plain is one of the flattest areas on Earth. Immense wetlands, through which the Ob River and its tributaries slowly wind their way, cover much of this vast flatland. This waterlogged country, underlain by permafrost that blocks downward seepage of water, is a major barrier to land transport and discourages settlement. Tremendous floods occur in the spring when the breakup of ice in the upper basin of the Ob releases great quantities of water while the river channels farther north are still frozen and act as natural dams. Russian aircraft sometimes bomb ice jams on the rivers to prevent worse flooding.

The hilly Central Siberian Plateau, rising 1000 to 1500 feet (300 to 450 m) above sea level, sprawls across the area between the Yenisey and Lena Rivers. Mountains dominate the landscape east of the Lena River and Lake Baikal, a natural gem described on **page 196**. Extreme northeastern Siberia is an especially bleak and difficult country for people to live in.

Low mountains rim the region from Lake Baikal to the Pacific, and high mountains flank the southern frontiers of

Geography of the World's Great Rivers The Volga

Russians think of the Volga as their most important river. They call it *Matushka*—"Mother." The longest river in Europe, it rises about 200 miles (320 km) southwest of Saint Petersburg and flows 2300 miles (3700 km) to the Caspian Sea. Before railroads supplemented river traffic, the Volga was Russia's most important commercial artery. Wheat, coal, and pig iron from Ukraine; fish from the Caspian Sea; salt from the lower Volga; and oil from Baku, on the western shore of the Caspian, traveled upriver toward Moscow and the Urals. Timber and finished products moved downriver to the lower Volga and Ukraine (•**Figure 5.G**). Boatmen towed barges upstream on a 70-day journey from Astrakhan, near the Volga mouth, to Kazan, on the middle Volga. The steamboat arrived in the late 1800s, bringing an end to the way of life commemorated in the famous Russian "Song of the Volga Boatmen." But the river still carries more than half of Russia's freight, including about a quarter of its timber.

©S.Borisov/Shutterstock.com

• **Figure 5.G** Cargo moving along the Volga, Russia's "Mother" river. Russian literary depictions of life along the river are sometimes reminiscent of Mark Twain's Mississippi. Down along the Volga wharves at Kazan, Maxim Gorki found "a whirling world where men's instincts were coarse and their greed was naked and unashamed."

The Volga was difficult to navigate until the latter half of the 20th century, when the **Great Volga Scheme** transformed the river. The goal was to control the flow of the river completely with a stairway of huge reservoirs, each of which reaches upstream to the dam forming the next reservoir. These bodies of water allow shipping during the six months when the river is ice-free, and they supply hydroelectric power and water for irrigation. These reservoirs are so vast that they are often labeled as "seas" on Russian maps.

The "taming" of rivers for power, navigation, and irrigation was an important component of Soviet economic development. Large dams were raised, and reservoirs formed behind them, on many major Russian rivers. The conquest of nature through engineering feats like these was publicly celebrated and contributed to the Soviet sense of pride.

Russia and the Near Abroad from the Black Sea to Lake Baikal. In the Caucasus Mountains of extreme southern Russia, Mt. Elbrus soars to 18,510 feet (5642 m). Great mountains rise east of the central Asian deserts. Their local names suggest their majesty: the Pamir Mountains are the "Feet of the Sun," and the Tian Shan (•**Figure 5.13**) are the "Celestial Mountains" or the "Mountains of the Spirits." The Soviet Union named the highest mountain in its borders "Communism Peak," at 24,590 feet (7495 m). When it gained independence in 1991, Tajikistan renamed this peak in the Pamirs "Ismail Somoni," after a 10th-century leader of the Samanid Dynasty.

Outside of the settled southwestern core land written about in **pages 192–194**, "rugged," "wild," and "vast" describe much of the physical geography of Russia and the Near Abroad.

©Evgeny Dubinchuk/Shutterstock.com

• **Figure 5.13** The Tian Shan, the" Celestial Mountains," soar over a meadow in Kyrgyzstan.

5.3 Cultural and Historical Geographies

A Babel of Languages

The peoples of Russia and the Near Abroad speak more than 100 languages (•**Figure 5.14**), yet languages with over a million speakers are relatively few. The dominant language of the region is unquestionably Russian, spoken by about 150 million people worldwide, including 95 percent of Russian citizens. Russian is also the primary language of Belarus, with over

• **Figure 5.14** Languages of Russia and the Near Abroad. The Soviet Union had challenges trying to hold together such a large collection of ethnic groups. The Russian Federation is facing similar difficulties.

70 percent of that country's residents opting for Russian instead of Belarusian. Thanks to the decades-long process of 165 Russification, Russian is a minority language in every other country in the region. Russian is a Slavic language, on the East Slavic branch. Closely related to Russian and Belarusian is Ukrainian, the primary language of Ukraine, spoken by over 40 million people. Russian is a significant minority language in Ukraine, a cultural issue that has been a source of 181 conflict for years. Eastern Slavic languages use the Cyrillic alphabet.

Slavic languages are by far the most widely spoken tongues in the region, but there are some other Indo-European languages in use as well. Romance languages are represented by the Moldovan dialect of Romanian. On the Iranian branch, Tajikistan uses a variant of **Persian**. The ancient **Armenian** language occupies its own Indo-European branch, although some scholars have suggested it may be a distant relative of Greek.

After Indo-European languages, the **Altaic** languages are the next most widely spoken tongues throughout Russia and the Near Abroad. The relationships between the various language groups in the Altaic family (such as Turkic and Mongolic) are uncertain. The "Altaic" label is often used to group these language groups together as a convenience, but linguists are increasingly skeptical of any historical link between them.

Few **Mongolic** languages are spoken in Russia (Buryat, east of Lake Baikal, is one). **Turkic** languages, named for their relationship to Turkish, occupy much of central Asia and portions along the edges of Russia itself. In the Russia and the Near Abroad region, **Azerbaijani** and **Uzbek** are the largest Turkic languages, with about 25 million speakers each in total (including significant concentrations of speakers outside the region, like Uzbek in Afghanistan and Azerbaijani in northwestern Iran). **Turkmen, Kazakh**, and **Kyrgyz** are the smaller national Turkic languages of central Asia, collectively spoken by about 20 million people.

Several ethnic minorities in Russia speak Turkic languages. **Tatar** is the largest, with about 5 million speakers mostly in the Tatarstan and Bashkortostan republics. **Tyvan** is the language of the Tuva people along the Mongolian border. The **Yakut** language is spoken by only half a million people across the vast Sakha autonomous republic in Siberia.

There are numerous Uralic languages (related to Finnish and Hungarian) spoken in Russia, primarily in the far northwest; put together, the speakers of all these tongues total only about 5 million people. Across northern and eastern reaches of Siberia are other indigenous language families (such as the Yukaghir and Chukotko-Kamchatkan) who have so few speakers that they are in imminent danger of language death. 380

The Caucasus area—southeastern Russia, Armenia, Azerbaijan, and Georgia—is very linguistically complex. Over 50 languages are found here, in an area roughly the size of California. Many researchers argue that the rugged terrain of the Caucasus serves a natural laboratory for the evolution and isolation of distinct languages, a remarkable parallel to the topographic variations that promote the development of new plant and animal species and subspecies.

After Russian, Azerbaijani, and Armenian, the most widely spoken language in this region is **Georgian**, the primary member of the Kartvelian language family, which is found almost exclusively in Georgia. The Northwest Caucasian and Northeast Caucasian language families are comprised of numerous tongues spoken by just a few thousand people each. Several larger languages in these families are spoken in the western Georgian region of Abkhazia and southern Russian republics of Ingushetia and Dagestan, but the largest is the **Chechen** language, used by over 1 million people in Chechnya.

This is just a brief overview of the region's tongues; there are many languages spoken by smaller populations. This rich multiethnicity is part of the great cultural wealth of

Russia and the Near Abroad. This diversity has too often been a source of discord between peoples; see especially **pages 176–178**.

The cultural histories of the region's peoples have long been influenced and confronted by the Russians, whose origins reach more than a thousand years into the past. With a policy known as **Russification**, in Tsarist and Soviet times Russians tried to assert their power by having speakers of non-Russian languages give up their tongues in favor of the Russian language. Language is a powerful political and cultural tool. Settling ethnic Russians in non-Russian territories was another means of Russifying these potentially restive areas, in some cases setting the stage for Russian irredentism as we have seen 183 recently in eastern Ukraine. The story of Russia's political and cultural dominance began with Slavic peoples who colonized Russia from the west, interacted with many other peoples, stood off or outlasted invaders, and acquired the giant territory that would become Russia.

Vikings, Byzantines, and Tatars

Long before there were Russians, much of what is now western Russia was occupied by nomads and more settled tribes that had formed small, weak states on the steppes, later overrun by migrating Turkic and Slavic tribes. Slavic peoples originated in what is now Ukraine and far western Russia in the early centuries of the Christian era. During the Middle Ages, Slavic tribes living in the forested regions of western Russia came under the influence of Viking adventurers from Scandinavia known as **Varangians.** The newcomers carved out trade routes, planted settlements, and organized principalities along rivers and portages connecting the Baltic and Black Seas. In the 9th through 11th centuries, the principality of Kievan Rus' (centered on modern-day Kiev, Ukraine), ruled by a mixed Scandinavian and Slavic nobility, achieved mastery over the others and became a powerful state, reaching from the Black Sea to the Baltic Sea. The culture it developed was the foundation on which the Russian, Ukrainian, and Belarusian cultures later arose.

Contacts with Constantinople (modern Istanbul, Turkey; see its strategic location astride Europe and Asia in Figure 5.2 on **page 152**) greatly affected Kievan Rus'. Located on the straits connecting the Black Sea with the Mediterranean, Constantinople was the capital of the Byzantine (Eastern Roman) Empire. Constantinople became an important magnet for Russian trade, and the Russians borrowed heavily from its culture. In 988, the ruler of Kievan Rus', Grand Duke Vladimir I, formally accepted the Byzantine Christian faith and had his subjects baptized. Following its cleavage from the Roman Catholic Church in 1054, Orthodox Christianity became a permanent feature of Russian life and culture (**•Figure 5.15**). Moscow, established in the 12th century, eventually came to be known as the **"Third Rome"** (after Rome itself and Constantinople) for its importance in Christian affairs. Moldova, Georgia, Ukraine, and Belarus are also Orthodox Christian nations. Another important country in the history of Christianity is Armenia, which was founded in the 6th century BCE.

Joe Hobbs

• Figure 5.15 The Armenian Church of the Holy Cross on the island of Akhtamar, now in eastern Turkey. Armenia was the world's first Christian country, and its ancient borders included Mount Ararat and other land later lost to Turkey. In 2010, in a symbolic act of growing religious tolerance, Turkish authorities allowed Armenians to pray inside the Church of the Holy Cross for the first time since the collapse of the Ottoman Empire in 1915.

In 301 CE, Armenia became the first country in the world to adopt Christianity as its national religion; its Armenian Apostolic Church remains the country's official faith to this day (**•Figure 5.16**). The map in Figure 5.15 shows the region's religions.

The Bolshevik Revolution in 1917 began a 75-year period of official repression and neglect of the Orthodox Church and other religions. Since the breakup of the Soviet Union, however, there has been a renaissance in religious observance—not just among Christians but also among Muslims, Buddhists, and others—across the vast region of Russia and the Near Abroad (**•Figure 5.17**). In the decade following the breakup of the Soviet Union, half of Russia's Jews fled the country; however, there has been a small reversal in the emigration of Jews to Israel (a migration known as an *aliyeh*) and the West. Ten million Jews lived in the region before the Nazi Holocaust killed 3 million of them, and perhaps another million assimilated to escape official Soviet anti-Semitism. Today, Judaism is one of Russia's four official religions; the others are Orthodox Christianity, Islam, and Buddhism.

Yet another cultural influence reached the Russians from the heart of Asia. The steppe grasslands of southern Russia had long been the habitat of nomadic horsemen of Asian origin. During the later days of the Roman Empire and in the

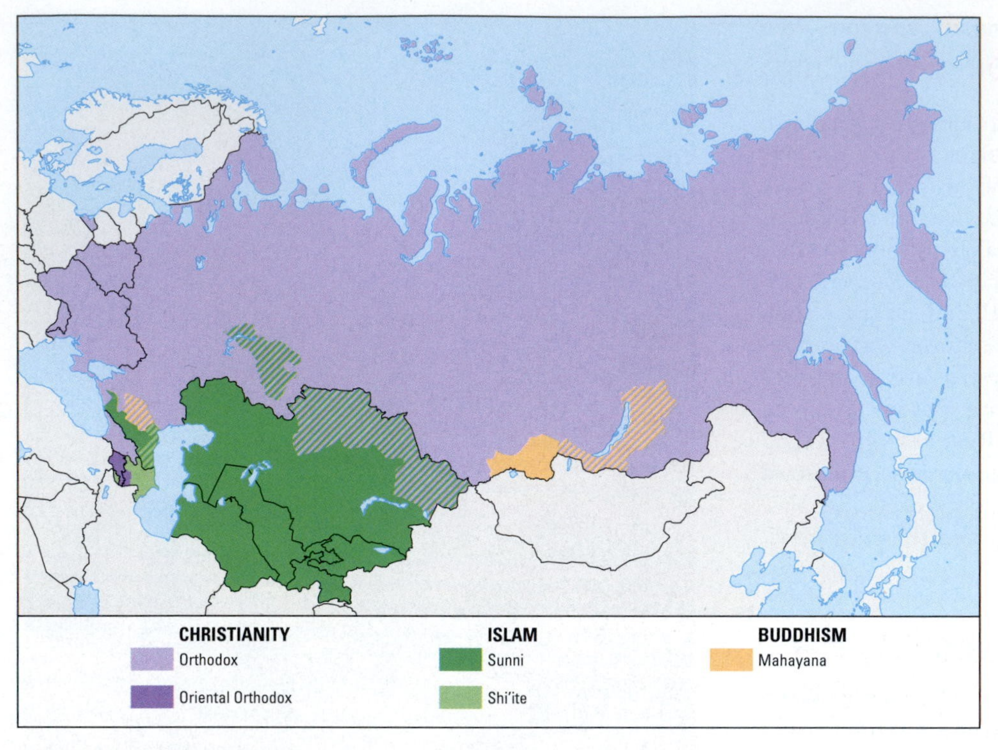

• **Figure 5.16** Religions of Russia and the Near Abroad.

Joe Hobbs

• **Figure 5.17** Some of the faithful in a Russian Orthodox Church in Vyborg, Russia. Since the breakup of the Soviet Union, there has been a resurgence of organized worship throughout the region.

Middle Ages, these grassy plains, stretching far into Asia, provided the Huns, Bulgars, and other nomads a passageway into Europe. In the 13th century, the **Tatars** (also known as the Tartars) of central Asia took this route. Many steppe and desert peoples of Turkic origins were in their ranks, with the **Mongols** in the lead. In 1237, Batu Khan ("Batu the Splendid"), the grandson of Genghis Khan, launched a devastating invasion that brought all the Russian principalities (except Novgorod, near present-day Saint Petersburg) under Tatar rule.

The Tatars collected taxes and tributes from the Russian principalities but otherwise generally allowed these units to be autonomous. Even the princes of Novgorod, who were not technically under Tatar control, paid tribute to avoid trouble. The Tatars established the khanates (governmental units) of Kazan (on the upper Volga), Astrakhan (on the lower Volga), and Crimea (on the Black Sea). Russians called the Kazan Tatars the **"Golden Horde,"** after the brightly colored tents in which they lived. When Tatar power declined in the 15th century, the rulers of the Muscovy (Moscow) principality were able to begin a process of territorial expansion that resulted in the formation of present-day Russia. Oil-rich Tatarstan, with its capital at Kazan on the upper Volga River, was annexed in the 16th century by Ivan the Terrible, and would later become one of Russia's most important "autonomous" political units. Its distinct ethnic Tatar population has proved remarkably resilient to centuries of Russian supremacy. We will have a closer look at the Tatars and their culture when we look at Russia's struggle to maintain control over diverse peoples within Russia's borders, on **page 177**.

The Empire of the Russians

From the 15th century until the 20th, the tsars of Moscow created an immense Russian peripheral empire by building onto this core. This imperialism by land, in the era when the maritime powers of Western Europe were expanding by sea, brought a host of non-Russians under tsarist control (see Insights on **page 167** to understand how Russia built a "land empire"). Russian motivations for expansion were diverse. Quelling raids by troublesome neighbors, particularly nomadic Muslim peoples of the southern steppes and deserts, was one objective. Many Russian cities, such as Volgograd (originally named Tsaritsyn and then Stalingrad) on the Volga River, were founded as fortified outposts on the steppe frontier. In the wilderness of Siberia, the search for valuable furs and minerals, especially gold, stimulated early expansion, and the missionary impulse of Orthodox priests also played a role. Land hunger and a desire to escape serfdom and taxation led to the flight of many peasants into the fertile black-earth belt of southwestern Siberia, and many landlords also moved there with their serfs.

The initial outward thrust from Muscovy under Ivan the Great (reigned 1462–1505) was mainly northward. Ivan annexed the rival principality of Novgorod, near present-day Saint Petersburg. This secured a Russian domain that extended northward to the Arctic Ocean and eastward to the

Insights Russia and Other Land Empires

Russia, and later the Soviet Union, developed as a **land empire.** Rather than establishing its colonies overseas, as imperial powers such as Spain and Britain did, Russia established colonies in its own vast continental hinterland (•**Figure 5.H**). Like most overseas European colonies, non-Russian regions of the periphery of Russia and later the Soviet Union were drawn into a relationship of economic dependence on the imperial Russian core. The central Asian republics of the Soviet Union, for example, followed Moscow's demands to grow cotton, which was shipped to Moscow to be manufactured into value-added clothing that was subsequently sold throughout the empire. Typically of colonization, this pattern contributed to the growth of the Soviet economy but often inhibited local development. Other entities characterized historically as land empires include the United States, China, and Brazil, as well as the Roman, Ottoman, Mongol, Aztec, and Inca empires.

108
200

• **Figure 5.H** The development of the land empires of Russia and the Soviet Union. Moscow has always been at its core.

Ural Mountains. Later tsars pushed the frontiers of Russia westward toward Poland and southward toward the Black Sea. Peter the Great (reigned 1682–1725) defeated the Swedes under Charles XII, to gain a foothold on the Baltic Sea, where he established Saint Petersburg as Russia's capital and its **"Window on the West."** At the expense of Turkey, Catherine the Great (reigned 1762–1796) annexed the Crimean peninsula and a frontage on the Black Sea. Ever since then it has been a foreign policy principle of Russia (and during its existence, the Soviet Union) to warn other powers against threatening its "warm-water ports" on the Black Sea—which with Russia's 2014 invasion now include the Crimean port of Sevastopol—and its access through the Turkish straits to the Mediterranean Sea and Atlantic Ocean. It is ironic that despite its 23,000-mile long coastline, Russia faces major geographic obstacles to seaborne trade and power.

182, 237

Russia's conquest of its eastern periphery was impressive. Tsar Ivan IV ("Ivan the Terrible") (reigned 1533–1584) added large new territories in conquering the Tatar khanates of Kazan and Astrakhan, thus giving Russia control over the entire Volga River (see chapter-opening photo). By the end of Ivan IV's reign, traders and pioneers were already penetrating wild and lonely Siberia. Among the new settlers were the **Cossacks,** peasant-soldiers of the steppes who originally were runaway serfs and others fleeing from the tsars' rule. They eventually gained special privileges as military communities serving the tsars (Cossacks continue to serve in the Russian military to this day). Cossack armies defeated the Khanate of Sibir (from which Siberia gets its name) in the late 16th century. Subsequent Russian advances conquered the indigenous areas farther east, subjecting indigenous peoples like the Yakut and Chukchi to slavery, forced migrations and widespread massacres that some contemporary scholars suggest were genocidal in intent.

A Cossack expedition reached the Pacific in 1639, but Russian expansion toward the east did not stop at the Bering Strait. It continued down the west coast of North America as far as northern California, where the Russian trading post of Fort Ross was active between 1812 and 1841. Russia found it difficult to maintain such distant colonies, and in 1867 Russia sold Alaska to the United States (for 2¢ per acre) and withdrew from North America. The relentless quest of Russian fur traders for sable, sea otter, and other valuable pelts brought great cultural changes to aboriginal peoples in Siberia and Pacific North America. These effects became increasingly pronounced, especially in the last decades of the Russian Empire, when millions of Russians moved into Siberia. Russia's expansionism bore similarities to the American sense of fulfilling a "manifest destiny" in acquiring territory, and to Chinese "Sinification" efforts (ongoing now) to settle ethnic Chinese in the non-Chinese areas along China's western and southern borders.

511
348, 511

There was one stumbling block in Russia's eastern expansion. In the Amur River region near the Pacific, the Russians were unable to consolidate their hold for nearly two centuries because of opposition by strong Manchu emperors claiming this territory for China. Russians were thus barred from the Siberian area best suited to growing food for people in the fur trade, and best endowed with good harbors for maritime expansion. This situation changed in 1858–1860 after the victory of European sea powers over China in the Opium 307 Wars. From two treaties, Russia was able to add the Amur region to its earlier gains in the Ob, Yenisey, and Lena basins. Finally, in a series of military actions during the 19th and early 20th centuries, Russian tsars annexed most of the Caucasus region. They also conquered "Turkestan," their name for the Muslim lands east of the Caspian Sea.

The Soviet state that succeeded tsarist Russia had to govern and interact with peoples representing scores of languages and cultures. Soviet authorities allowed different ethnolinguistic groups to keep their own languages and other elements of their traditional cultures, and even created alphabets for those that had no written language. Politically, the Soviet Union officially recognized many of the groups as non-Russian nationalities. The Soviet Union established 16 Autonomous Soviet Socialist Republics (ASSRs) as homelands for large ethnic minorities, in theory endowing them with limited autonomous (self-governing) powers. Smaller nationalities were organized into politically subordinate autonomous units, including 5 autonomous 176 regions (*oblasts*) and 10 autonomous areas (*okrugs*). The totalitarian Soviet labeling of these units as "autonomous" came back to haunt the nominally democratic Russian nation 176 in the 1990s.

Despite the official egalitarianism of the USSR, ethnic Russians were prominent in positions of responsibility in non-Russian republics, holding the majority of top posts in the Communist Party, the government, and the military. Most of the ethnic groups that the Russians tried to Russify were very resilient and carried on their own distinct cultural practices. As we will see, Russian ways had little impact on non-Slavic cultures such as the Georgians and Armenians. Even Slavic kinspeople of the Russians, notably the Ukrainians, insisted on retaining their cultural distinctiveness and their original mother tongues. This resolve helps explain why ethnic Ukrainians pushed back so hard against Russia's irredentism beginning in 2014. In general, the policy of Russification failed to achieve its objective of "converting" other peoples because it underestimated the strength of ethnicity and culture.

Russia and the Soviet Union: Tempered by Revolution and War

Russia and the Soviet Union repeatedly triumphed over powerful invaders, notably the Swedish forces led by King Charles XII in 1709, the French and their allied European forces under Napoleon I in 1812, and the German and other European forces sent by Hitler into the Soviet Union in World War II. In each case, the Russians lost early battles and much territory but eventually inflicted a crushing and decisive defeat on the invaders. The success of Old Russia and the Soviet Union in withstanding invasions by such formidable armies was due in part to the environmental rigors (particularly, the brutal Russian winter) that invaders faced, the overwhelming distances of a huge country with poor roads, and the defenders' love of their homeland. It was also due to talented Russian military leadership, and that leadership's willingness to lose great numbers of soldiers in combat. Finally, the successful defense used **scorched-earth** tactics to protect the motherland; rather than leaving Russian railways, crops, and other resources to fall into the invaders' hands, the defenders destroyed them. Napoleon fulfilled his ambition of taking Moscow, but this was a pyrrhic win because retreating Russian forces left little for his army to live on. Nazi forces pushed eastward during World War II through a scorched-earth wasteland, but in that case Moscow did not fall.

The **Russian Revolution** of 1917, which set the stage for the formation of the Soviet Union, was really two revolutions that occurred against the backdrop of World War I. In 1914, when a Serb in Sarajevo assassinated the heir to the throne of Austria-Hungary, a complicated series of alliances (some affirmed by marriages) required Tsar Nicholas II to commit Russian troops to fight along with Serbia, France, and Britain against Austria-Hungary and Germany in World War I. 118 The first revolution in Russia began early in 1917 as a general protest against the terrible sacrifices of Russian forces on the Eastern Front during this war. That revolt overthrew Nicholas II, a detached and indecisive ruler who was the last of the Romanov tsars. The Russian social hierarchy began to dissolve quickly.

The second was the **Bolshevik Revolution** that came later that year. Led by Vladimir Ilyich Lenin (1870–1924), who with German help reached Russia from his political exile in Switzerland, the Bolshevik faction of the Communist Party seized control of the government and executed the tsar and his family. The new regime made a separate peace with Germany and its allies and survived a difficult period of civil war and foreign intervention between 1917 and 1921. Lenin presided over the establishment of Russia's successor state, the Soviet Union, in 1922. To say that Lenin was the "George Washington of the Soviet Union" would only begin his biography. He was the head of the pantheon of Soviet leaders. Lenin's ideology and name were evoked at every opportunity to inspire the Soviet peoples (•**Figure 5.18**). His embalmed body remains on display to this day in a mausoleum on Moscow's Red Square. Russian President Boris Yeltsin advocated finally burying Lenin's body, but in what may be seen as another symbol of his Soviet sentiments, his successor Vladimir Putin opposed that move. Lenin's portraits and statues were omnipresent throughout the USSR. Some survive outside Russia, but when the Soviet Union fell, so did many of his likenesses—often quite savagely, with hammers—in quarters where he was resented. It was a sure sign of Ukraine's turn to the West in December 2013 when protestors destroyed the famous monument to Lenin in Kiev and replaced that with a gold-covered toilet; both actions symbolized their disgust with the toppled Yushchenko government seen as a greedy crony proxy of Russia.

Joe Hobbs

• **Figure 5.18** "Under the Banner of Lenin We Will March to Complete Victory!" This 1944 propaganda poster was designed to inspire the final push by the Red Army into the heart of the Nazi homeland.

World War II found the Russians again allied with France and Britain in a far more ferocious war against Germany. The Soviet Union's success in withstanding the German onslaught that began as **Operation Barbarossa** in June 1941 surprised many outside observers, who had predicted that the Soviets would prove too weak and disunited to resist for more than a few weeks or months. The crucially important cities of Leningrad (now Saint Petersburg) and Moscow held out through brutal sieges. Late in 1942, Soviet forces halted the push of the German *wehrmacht* eastward at the Volga River in the huge Battle of Stalingrad (see Geography of Sacred Space, **page 170**). This engagement was the turning point in the war, when Soviet and Allied forces began to reverse Nazi advances.

The failure of powerful Germany to conquer the Soviet Union was a clear indication that the USSR had been underestimated and that the country had become a major player in world affairs. The war's impact lingers to this day in the post-Soviet countries, quite noticeably in their age-structure diagrams. Note how males in the top cohorts of Russia's population pyramid in Figure 5.C are a fraction of the females. Although male life expectancy today is about 10 years shorter than female, much of this shortfall on the diagram represents death in battle. The German invasion took an estimated 23 million or more Soviet lives and caused the relocation of millions of people. It did enormous damage to settlements, factories, and livestock. As a strategic precaution during the war, Lenin's successor, Josef Stalin, directed major Soviet industries to relocate eastward away from the front, especially behind natural barriers like the Volga River, a move that still has a legacy in the region's economic geography.

So much of what happened in World War II bears on current events in this region. A majority of the Soviet war dead were ethnic Russians, but the rest were people from a variety of ethnicities that in some cases followed the "enemy of my enemy is my friend" principle, with Russian Soviets being their enemy; they worked with the Nazis. That history rose to the surface with Russia's moves on Ukraine, beginning with the occupation of Crimea in 2014. In 1941, many Ukrainians had allied with the invading Germans as potential liberators (the Baltic countries, invaded by the Soviets in 1940, also allied themselves with the Nazis against Moscow). This was in large part because less than a decade earlier (1933–1934) the Soviet government of Josef Stalin instigated a famine known as the ***holodomor*** ("hunger extermination") that killed millions of ethnic Ukrainians, mainly to blunt Ukrainian independence hopes. These events haunt relations between ethnic Ukrainians and ethnic Russians in Ukraine to this day. To garner Russian public favor for his venture in Crimea, Vladimir Putin was quick to brand pro-Western Ukrainians as "neo-Nazis," and in fact some of the forces fighting pro-Russia separatists in Ukraine described themselves as such (see Figure 5.26).[5] Historical and rhetorical patterns like these will help you understand recent Russian activities and interests near its borders.

5.4 Economic Geography

The Soviet Union's collapse dramatically changed the global political landscape and had far-reaching economic consequences. It was in fact the failing economic system that brought the giant country down. This section discusses the economic events leading up to and in the wake of the Soviet Union's demise and then focuses on the modern economy of the largest successor state, Russia. This background will help you appreciate Russia's sometimes-confounding place on the world stage today.

The Communist Economic System

The Soviet Union's Communist economic system was an attempt to put into practice the economic and social ideas of the 19th-century German philosopher Karl Marx. According to Marx, the central theme of modern history is a struggle between the capitalist class ("the **bourgeoisie**") and the industrial working class ("the **proletariat**"). He forecast that exploitation of workers by greedy capitalists would lead the

139

Geography of Sacred Space Stalingrad

You know already that Geography is a *very* wide-ranging discipline, so you will not be surprised to learn that a necrogeography—the geography of the dead—and especially the geography of cemeteries, is a lively field of study. Its practitioners glean much geographic information even by studying headstones. The cemeteries and other mortuary monuments in the Russia and Near Abroad region reveal a great deal about the geographies of both the living and the dead.

A **sacred space**, or **sacred place**, is any location that people revere. Sacred places in the realms of ordinary experience include places of worship such as synagogues, churches, and mosques. Cemeteries too are perceived and managed as sacred sites. Places where people have lost their lives defending their homeland, or where they died at the hands of an enemy, are among the most revered and enduring sacred spaces—the Gettysburg Battlefield and Manhattan's Ground Zero are examples in the United States.

For most Americans today, World War II is "ancient history," its dead and its survivors generally respected, but not revered. The situation in the post-Soviet region, especially Russia, is *very* different. The number of Soviet citizens killed in the war may never be known, and estimates vary widely, but approximately 8 million military personnel and perhaps 15 million civilians perished between 1939 and 1945.

Perhaps in no other country is the memory of a war etched so vividly in the national consciousness as it is in Russia. It is deeply moving to observe the peoples of this region visiting the cemeteries and monuments of Stalingrad (modern Volgograd), Leningrad (modern St. Petersburg), and numerous other cities. Today, the battlegrounds where those soldiers fell are sacred places. They are kept hallowed by the continuous ritual visitation of veterans, war widows, and three generations of descendants of war survivors. Even today, these pilgrims visit a large number of war monuments and cemeteries in an effort to heal deep emotional wounds and to keep alive the memory of the Soviets' costly wartime resistance. These rites of visitation and commemoration pass to each new generation. Immediately after the wedding ceremony, for example, it is common for a newlywed couple to place flowers at the local tomb of the unknown soldier. Contemporary Russian pride and nationalism have strong roots in wartime sacrifice. Russian leaders do not want war memories to fade because they represent a deep well from which they can draw to resist "the West" and specific groups like the Germans.

During the Second World War, which Russians call "the Great Patriotic War," superiors urged Russian soldiers to fight to the death for the motherland: Russia was sacred ground to be defended at any cost. Stalingrad, site of the most ferocious battle on World War II's eastern front, was the costliest to defend. Germany's Nazi leader, Adolf Hitler, had a geographic rationale for sending his forces against the city in August 1942. Located on the steppe where the Volga and Don Rivers are closest together, it was strategically situated. Take a close look at the geographical *situation* of Volgograd/Stalingrad in Figure 5.12 on **page 162**. It was a grain and livestock center with railway and water connections (through the Volga-Don Canal described on **page 162** and in the photo of Figure 5.11) to the Don Valley and the Caucasus region. It was critical to Hitler's war strategy that he seize the vital oilfields of the Caucasus—the same ones that are still prized in the global oil trade and that are described on **page 189**. This vast energy supply would fuel the German war effort. But first, Stalingrad had to be wrested from Soviet control. Although vital strategically, Stalingrad was to Hitler as much a symbolic as a military prize: Because it bore the name of the Soviet leader, its fall would be of great propaganda value to the German war effort. Equally, its salvation from the invader was a goal of nearly religious significance for the Stalin regime.

94

workers to revolt, overthrow the capitalists, and turn over ownership and management of the means of production to new workers' states. In the classless societies of these states, there would be social harmony and justice, with little need for formal government.

Marx's utopian vision did not materialize anywhere, but it did inspire anti-government revolts and Communist political systems in many countries (**•Figure 5.19**). The ideas of Marx and of Lenin, as Lenin's successor, Josef Stalin, interpreted and implemented them, provided the philosophical basis for the Soviet Union's centrally planned **command economy.** Beginning in 1928, a series of five-year economic plans demanded the fulfillment of quotas for the nation: types and quantities of minerals, manufactured goods, and agricultural commodities to be produced; factories, transportation links, and dams to be constructed or improved; and residential areas to be built for industrial workers. The goals were to abolish the old aristocratic and capitalist institutions of tsarist Russia and to develop a strong socialist state equal in stature to the major industrial nations of the West.

In the command economy, an agency in Moscow called **Gosplan (Committee for State Planning)** formulated the national plans, which were then transmitted downward through the bureaucracy until they reached individual factories, farms, and other enterprises. This was an unwieldy and inefficient process in several ways. First, the planners in Moscow were essentially required to act as CEOs of a giant corporation, effectively "USSR, Inc.," that would manage the economy of an area larger than North America—a gargantuan, impossible task. Further, the Soviet planning bureaucracy had no free market to guide it, so the system produced goods that people would not buy or failed to produce goods that people would have liked to buy. And Gosplan stated production targets in quantitative rather than qualitative terms, churning out abundant but substandard products. There was often an obsession with fulfilling huge quotas or implementing grandiose

"Not one step backward," Stalin ordered his troops. Soviet resistance at Stalingrad is an extraordinary chapter in the history of warfare. The invaders were unprepared for the resolve of the Red Army and for the ferocity of the Russian winter. The 5-month battle ended in a devastating defeat for the Germans: Two armies consisting of 24 generals, 2000 officers, and 90,000 soldiers were taken prisoner. One-quarter of the German army's war matériel was lost. Two years earlier, the Germans could not have imagined such a defeat.

But victory for the Soviet Union exacted an unimaginable cost. Stalingrad lay in ruins; Soviet authorities dubbed it a "city without an address" and, in honor of its defenders, a "hero town." Fifty years after the battle, Russian military authorities finally released figures on the number of dead. In this sacred ground lay the bodies of 3.5 *million* soldiers and civilians, of whom 2.7 million were Soviet citizens, mainly Russians.

The visitor to Volgograd cannot help but feel the pain of war that has lingered through the decades. All over the city are monuments to remind the living of the dead. The devastated shell of a mill stands as the only physical artifact of the past, but the monuments built after the war rekindle the emotional losses most strongly. Volgograd's central memorial is the complex on Mamayev Hill, a mecca for Russians. The last war veterans wear their medals as they pass through. Grandmothers lead small children to place flowers at the feet of statues. Over the mass grave of an estimated 300,000 Soviet and German soldiers, a huge hand raises an eternal torch. The honor guard changes in goose-step once each hour. This sacred place is dominated by the world's largest statue, a female sword-wielding figure called *The Russian Motherland* (**•Figure 5.I**).

Joe Hobbs

• Figure 5.I The Mamayev Hill monument to the memory of the defenders of Stalingrad. Within the hill lie the bodies of hundreds of thousands of the city's defenders and attackers. This sword-wielding figure, known as *The Russian Motherland,* is the world's largest statue, standing 279 feet (85 m) high and weighing 8000 tons. To appreciate its scale, look at the people below.

schemes, a Soviet preoccupation sometimes known as **gigantomania.** Finally, fearing reprisals from people higher up the ladder, no one wanted to suggest ways to increase quality and efficiency.

Although cumbersome, this system succeeded in propelling the Soviet Union to superpower status, improved the overall standard of living, prompted rapid urbanization and industrialization, and altered the landscape profoundly. The principal goal of Soviet national planning after 1928 was a large increase in industrial output, with emphasis on heavy machinery and other capital goods, minerals, electric power, better transportation, and military hardware. Masses of peasants were converted into factory workers. New industrial centers were founded, and old ones were enlarged. There were huge investments in defense, and the country grew strong enough to survive Germany's onslaught in World War II. After the war, the Soviets maintained large armed forces and accumulated a massive arsenal of conventional and nuclear weapons in an **arms race** against the world's only other superpower, the United States. By the final decade of the Cold War in the 1980s, 15 to 20 percent of the country's GDP was dedicated to the military (in contrast to less than 10 percent in the United States), representing an enormous diversion of investment away from the country's overall economic development.

Farmers and farming had troubled careers under the Soviet system of **collectivized agriculture.** Between 1929 and 1933, 139 about two-thirds of all peasant households in the Soviet Union were collectivized. In this process, peasant landholdings were confiscated and reorganized into two types of large farm units: the **collective farm** (***kolkhoz***) and the factory-type **state farm** (***sovkhoz***). The consolidation of individual farmsteads and villages into fewer but larger communities on the collective farms was supposed to allow the government to administer, monitor, and indoctrinate the rural population and provide services, including education, health care, and electricity, more cheaply and efficiently.

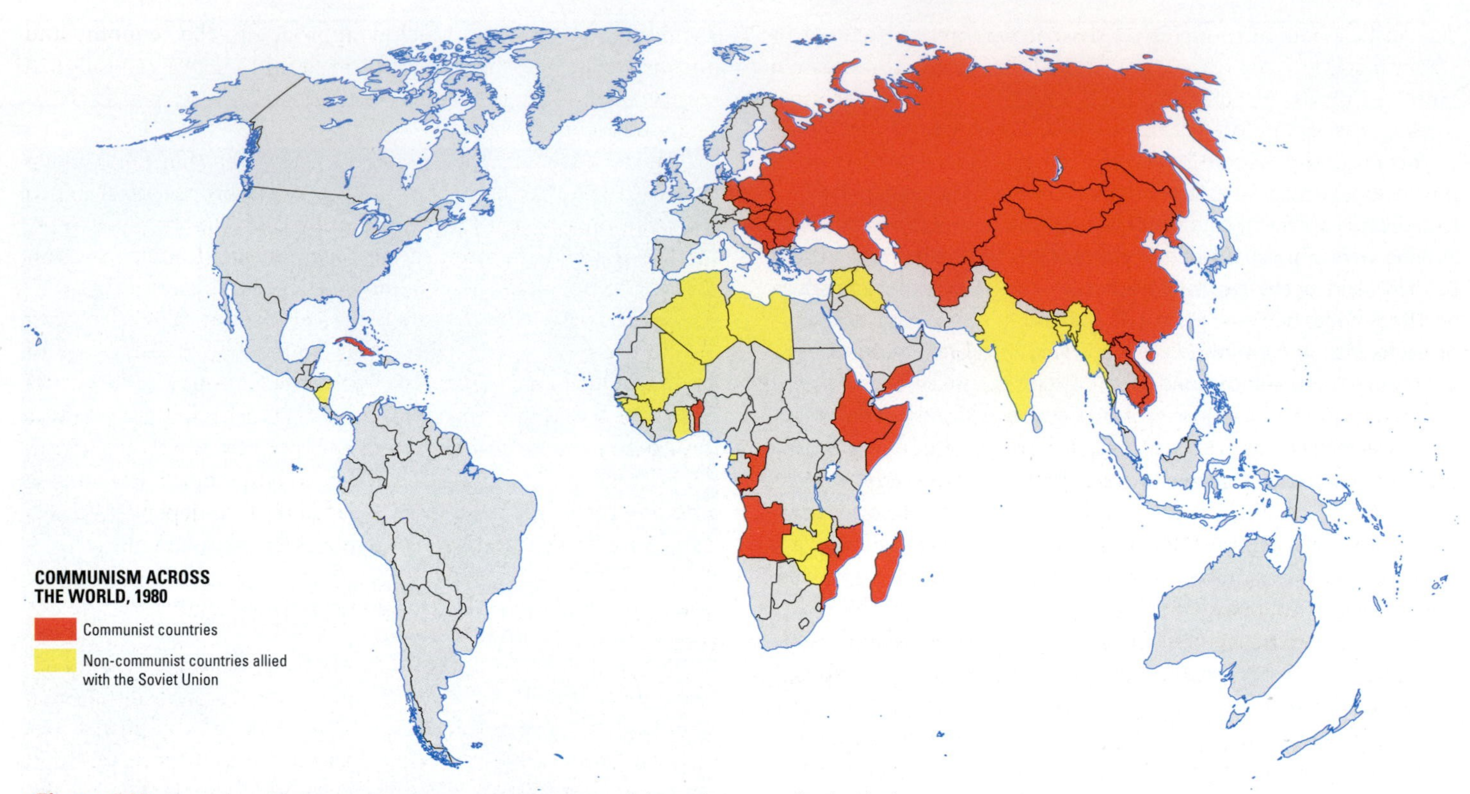

• **Figure 5.19** Countries with Communist governments in 1980 and those that were allied with the Soviet Union. Between World War II and 1991, many countries declared themselves neutral in the Cold War between Washington and Moscow, and some that did align with either the US or USSR changed allegiances over time.

But the rural people fiercely resisted collectivization. In their own version of scorched earth, peasants and nomads slaughtered millions of farm animals and burned crops to avoid turning them over to the "socialized sector." Government reprisals followed, including wholesale imprisonments and executions, together with confiscation of food at gunpoint (often including the peasants' own food reserves and seed). The more prosperous private farmers, known as *kulaks,* were killed, exiled, sent to labor camps, or left to starve. Famine induced by Soviet politics took untold millions of lives—perhaps 10 or 12 million, mainly in 1932–1933. Soviet leaders—most notoriously, Josef Stalin—disregarded these costs, and by 1940, virtually all of the Soviet Union's farmed land was collectivized. It must be noted that Ukraine figured especially strongly during these events; it was known as the "breadbasket" of the Soviet Union, mainly for its wheat. Stalin's apparatus sifted Ukrainian wheat out of Ukraine, distributing it in the vast USSR hinterland, "dumping" it at low prices abroad and using that income to fund industrialization. That historical legacy too feeds into modern anti-Russian sentiment among ethnic Ukrainians.

Soviet enterprises in both agriculture and industry harnessed the energies and resources of the entire country to achieve specific objectives. The government called on the people to sacrifice and to make the country strong, especially by working on so-called **"hero projects"** such as the construction of tractor plants, dams, railways, and land reclamation. The government in turn operated as a collectivized welfare state, providing guaranteed employment, low-cost housing, free education and medical care, and old-age pensions. As we will soon see, when the Soviet Union fell apart, so did these benefits that people counted on for sustenance. Social services were often 131 minimal but in some sectors were quite successful; the literacy rate, for example, rose from 40 percent in 1926 to 99 percent in 1959.

Economic Roots of the Second Russian Revolution

The Soviet Union's ambition to achieve and maintain military-industrial superpower status was generally achieved at the expense of the country's consumers. Many of the needs and wants of Soviet citizens were overlooked in favor of heavy metallurgy and the manufacture of machinery, power-generating and transportation equipment, and industrial chemicals. Consumers lined up in stores to purchase scarce items of clothing and everyday conveniences. The exasperation of shoppers confronted by long lines and empty shelves was an important factor generating dissatisfaction with the economic system and demands that it be reformed.

The flow of goods and services ebbed to a trickle in the 1980s. People began taking to the streets in protest. In large demonstrations and strikes, the public expressed their anger at a political and economic system that was sliding rapidly downhill. Their outpouring of dissent was unprecedented in Soviet history. They openly challenged the Communist system on the grounds that it failed to provide a good living for most people, stifled democracy, and blocked the ambitions of the country's many ethnic groups for a greater voice in running their own affairs. Meanwhile, the economy turned from bad to worse (•**Figure 5.20**).

The revamping of the economic system became an urgent priority during the regime of Mikhail Gorbachev, which began in 1985. Gorbachev had a leadership style that was also unprecedented in Soviet history: he was a reformer. He introduced new policies of ***glasnost*** ("openness") and ***perestroika*** ("restructuring") to allow a more democratic political system, more freedom of expression, and a more productive economy with a market orientation. Little did he know that opening the door slightly to allow free market, capitalist-style enterprise would lead quickly to a flood of public demands for more reforms and would unravel the Soviet empire. Under Gorbachev, the various republics and ethnic groups organized by the Soviet system into nominally "autonomous" units took advantage of new freedoms to resurrect old quarrels and demand greater control over their own affairs. Clashes between ethnic groups erupted in several republics. In all of the republics, declarations of sovereignty and in some cases outright independence challenged the authority of the central government. In 1991, the three Baltic republics of Estonia, Latvia, and Lithuania unilaterally declared their independence. The Soviet Union's disintegration had begun.

At the center of the suddenly crumbling empire, having failed to reverse the downward slide of the economy, Gorbachev faced growing public demands to scrap the command economy and move as rapidly as possible to an entirely market-oriented economy. Gorbachev's go-slow approach to the economy and his insistence that the Soviet Union should survive as a political entity, even after the Baltic republics fell away, was increasingly unpopular.

Matters came to a head with an attempted (but failed) coup against the Soviet leaders in August 1991. Gorbachev resigned his position as head of the Communist Party and began to work with Boris Yeltsin, a reform-minded political leader, to reconstruct the political and economic order. But his efforts to preserve the Soviet Union and a modified form of the Communist economic system could not stand against the growing tide of change. During the autumn of 1991, the Communist Party was disbanded, and on December 25, 1991, Gorbachev resigned the presidency. The national parliament voted the Soviet Union out of existence the following day. A powerful empire had quickly and quietly faded away, to be replaced by 15 independent countries, in what was dubbed the **Second Russian Revolution**.

Russia's Road to Misdevelopment

The unprecedented experiment to transform a colossal, authoritarian Communist state into a free-market democracy produced strange and often unfortunate results in post-Soviet Russia. President Boris Yeltsin introduced a program of rapid economic reform, known as the **economic shock therapy,** designed to replace the Communist system with a free-market economy. Formerly, the Soviet government had controlled all sectors of the economy; now, all price controls were removed, and privatization of businesses was encouraged.

Unfortunately, shock therapy proved to be too much too fast for the Russian people. Former members of the Communist Party resented the loss of their jobs and privileges. Poor people living on fixed incomes struggled to survive high prices for basic necessities. A new consumer-oriented society developed along class lines. Unemployment and homelessness increased, and the gap between rich and poor widened. There was a bewildering explosion of new goods and services that most people simply could not afford (•**Figure 5.21**).

Joe Hobbs

• **Figure 5.20** Having fallen through the cracks in Russia's transition from a command economy to free enterprise, these women were begging for help on a Saint Petersburg sidewalk. Their placards say, in essence, "Help us dear brothers and sisters. Please give us money for food, for God's sake. We are invalids with cancer. We have given of ourselves but are now forgotten by the state. We are poor and hungry, with no protection, just left to rot."

Oliver Lang/AFP/DDP/Getty Images

• **Figure 5.21** Although some beg on Russia's streets, others indulge in Western-style overconsumption. This Maserati sports car, out of reach for all but Russia's super-rich, was on display in Moscow's "Millionaire Fair."

One of the major components that emerged in Russia's new economic geography was the **underground economy,** also known as the *countereconomy* or *second economy.* Rampant black marketeering developed. Although much of this exchange was illegal, there was a general tendency to overlook such transactions because they were essential to the economy. Widespread **barter**—the exchange of goods and services instead of cash—resulted from the declining value of Russia's currency, the ruble. Many people resorted to selling personal possessions to buy high-priced food and other necessities. Most Russians grew economically worse off. Many privately owned industries were too unproductive to pay their employees, or else they paid workers "under the table" to avoid taxation. Unpaid workers could not pay taxes to the government, of course, and workers paid under the table had no incentive to pay taxes. Russia's gross domestic product plummeted, shrinking by almost half in the 1990s. This was the largest fall in production that any industrialized country had ever experienced in peacetime. Its trend followed those of rising death rates and falling birth rates seen in the "Russian cross" on page 156.

Russia's new private entrepreneurs came to include a large criminal "mafia" that preyed on government, business, and individuals. Russia became a **kleptocracy,** an economic and political system based on crime. Organized crime and corruption became pervasive and made "free enterprise" far from free. Lacking an independent judiciary and any mechanisms of accountability, corruption ran rampant. Bribes and kickbacks throughout all levels of Russian society have skyrocketed the costs of everyday necessities such as water, housing, and electricity. A company wanting to build a factory, for example, has to pay off officials at every stage of construction; without the payouts, officials' partners in organized crime would terminate the project. Anti-corruption laws are often imposed only on those who are critical of authorities. Like any other commodity, bribes are subject to inflation: in 2011, the average cost of a business-related bribe of $4000 was a 26-fold increase from the 2008 level. All of this corruption adds up, wasting 48 percent of Russia's GDP, according to the World Bank.

Tracking figures like these, some observers have classified post–Soviet Russia not as a less developed or more developed country, but as a **"misdeveloped country."** From a technical standpoint, Russia is an MDC—but just barely, with a per capita GDP (PPP) of $18,100 ($17,000 is the dividing line between MDCs and LDCs in this book). 56

As the great majority of Russians slid backward, there were those who prospered. The biggest beneficiaries of the new economic system were the so-called **oligarchs,** Russia's leading businesspeople (*oligarchy* is literally the "rule of the few"). In the 1990s, a controversial privatization program let these tycoons buy up formerly state-controlled sectors of the economy and profit handsomely from them. By some estimates, they acquired control over about 70 percent of Russia's economy.

Wielding enormous political as well as economic power, the oligarchs helped usher Vladimir Putin, a former KGB (state security) officer of the Soviet Union, into Russia's highest office, the presidency, in 2000. Putin's authoritarian style, reminiscent of former Soviet ways, proved popular with voters, who awarded him a succession of terms as president and prime minister. He enjoyed broad popular support among Russians, often with an approval rating of over 80 percent. Even with the secondary office title of prime minister, he was the "power behind the throne" of President Dmitry Medvedev between 2008 and 2012, at which time Putin regained the presidency. Putin has had help in retaining his enormous popularity: nearly all news media in Russia are state run, continuously promoting Putin while tackling political opponents. Television news is overtly hostile to the United States, other Western nations, and the Ukrainian government. Russia employs many "professional Internet trolls" who create thousands of fake profiles on Facebook, Twitter, and news websites to bombard the comment sections of Russia-related news stories with pro-Kremlin propaganda. Putin's consolidation and retention of power, with little public opposition, has seemed to reaffirm what that most powerful Soviet leader Josef Stalin often said: "The Russians need a tsar."

Although Putin reminded many people of the relative stability and security of the Soviet state, he was also credited with pulling Russia out of the economic black hole it fell into when the Soviet Union broke apart. But he may have taken his power and popularity too much for granted. In 2011, when he and President Medvedev announced Putin's plan to return to the presidency, many Russians took to the streets in protest. They wanted real rivals to Putin, and they wanted a verifiably fair election. Their appeals were soon muffled by the din of nationalistic pride that arose as Russia seized Ukraine's Crimea Peninsula. 180

Putinomics

Vladimir Putin had a simple plan for Russia. He would export the country's natural resources, especially oil, at full tilt, to flood the country with wealth, thereby starting to rebuild its economy and reinstilling a sense of national pride in Russians.

The plan worked, at least until it didn't. The steady rise in global energy demand and prices after 1999 (with a setback in 2008–2009, and a larger, faster one in 2015) was a big stimulus to the Russian economy. Russia experienced a new period of self-assurance and economic renewal. Gross domestic product and wages more than doubled in the decade of the 2000s. In 2001, Russia was famously classified by the financial firm Goldman Sachs a member of an elite group: the so-called **"BRIC economies,"** with the acronym standing for Brazil, Russia, India, and China. 60 According to the Goldman Sachs outlook, these four countries (which together had a quarter of the world's land mass and two-fifths of the world's population) would have the world's largest economies by 2050. Russia and Brazil would establish dominance by exporting raw materials, whereas India and China would do so through manufacturing and services. Putin's plan was for Russia to be world's fifth largest economy by 2020.

It didn't work out that way. BRICs turned into BICs, according to many observers, after Russia took over Crimea in 2014.[6]

Insights Russia's Dutch Disease

In economic geography terms, it appears that Russia has a bad case of "**Dutch Disease**," the link between growing exploitation of natural resources and declining performance in other sectors, like manufacturing and agriculture.

The respected journal *The Economist* coined the term "Dutch Disease" in 1977 to describe how, when production and export of the Netherlands' natural gas soared, the Dutch currency became too valuable. This made the Netherlands' other products too costly to export and made foreign investment uncompetitive. Industrialization fell, unemployment grew, and the larger economy sank. Volatility in the prices of natural resources—as has happened with Russian oil—can worsen the divergence between the natural resource and industrial sectors.[9] A big risk in this scenario is that exports of the resource could crash along with the value of the currency, as did Russia's ruble from 30 rubles per dollar to 70 in the wake of the 2014 Crimean invasion.

The natural remedy to Dutch Disease would be to increase exports of manufactured goods, but Russia does not have them and has become dependent upon imports. Russia is fortunate in having socked away some of its fossil fuel revenues for future generations in a "sovereign wealth fund." One strategy for alleviating or preventing Dutch Disease is to make the hard choice not to exploit all the country's natural resources at once. This can be quite difficult in the LDCs, particularly the heavily indebted ones . In 65 numerous LDCs, especially in Sub-Saharan Africa, we see examples of the "resource curse" or "resource paradox," another strain of Dutch Disease. 424

Russia is well endowed with industrial resources, especially minerals. Those resources, along with the country's manufacturing industries, are concentrated in several areas: the regions of Moscow and Saint Petersburg, the Urals, along the Volga, and around the city of Novosibirsk in southwestern Siberia. Among these assets are Russia's enormous energy reserves. Russia's proven oil reserves represent 5 percent of the world total; Russia also has 23 percent of the globe's proven natural gas reserves, the largest in any single country. Russia also has the world's second largest coal reserves, after the United States. Relative positions change routinely, but Russia typically ranks third (behind Saudi Arabia and the United States) in crude oil production, and second to Saudi Arabia in exports. Its reserves, however, are just a fraction of those of Venezuela, Saudi Arabia, or Canada. Russia's oil production is expected to start trailing off after 2020, and Russia will have to rely on a more diversified economy. Producing raw materials is no substitute for the kind of economic diversification that can protect a country's economy in a variety of market conditions.

The biggest danger to Russia's economy has for many years lain in excessive dependence on natural resources, including fossil fuels, metals, and timber. Energy represents nearly 70 percent of the value of Russia's exports, and oil and minerals sales fund half of Russia's federal budget. Should prices for these commodities slump—as the price of oil did in 2008 and 2015—revenues would also plummet and drag the entire economy into a recession, as appeared to be happening by 2015. If, on the other hand, revenues from oil and other natural resource exports would rebound to new highs, the profits could be rolled into manufacturing and high-tech industries, allowing the country a more stable, diversified economy. This was the scenario that Vladimir Putin was counting on, so much so that this economic rationale came to be known as "Putinomics."

Detractors have argued, however, that Putin's Russia focuses on the natural resources profits while neglecting industries, hampering the country's development. One critic, Laza Kekic, wrote that Russia was "on the way to becoming a third-world petrokleptocracy."[7] Heavy industries, which had been so vital during the Soviet era, went idle when the recent oil boom began. "We have lost our manufacturing base," said Yury Krupnov, an analyst at the Moscow-based Institute for Demography, Migration and Regional Development. "We have become addicted to an economic model centered on commodities exports."[8] These symptoms have given Russia an economic diagnosis of "Dutch Disease" (see Insights above).

For Russia there is one silver lining in its neglected industrial sector: benefiting from the Kyoto Protocol's emissions trading allowances. Russian industries are producing well below their benchmark 1990 carbon dioxide emissions, so in effect Russia can sell rather than buy its emission rights. In Chapter 2 you learned about "Joint Implementation (JI)," the climate change mitigation mechanism that allows an Annex I country to earn emission reduction units (ERUs) by investing in emission reduction in another Annex I country. In this JI case, Denmark is earning ERUs by investing in the conversion of coal-fired electricity plants to cleaner-burning gas in Siberia. 44

Russia will have an uphill climb in preparing for a future that is not driven by natural resource exports. Russia has not promoted the "knowledge economy" as post-industrial European, Asian, and North American countries have been doing while replacing heavy industry in major urban areas. Part of the problem is that much of the knowledge fled Russia after the Soviet Union collapsed, in one of the most unfortunate brain drains ever. 76

Why did the best minds start looking for the exits? Russians are justifiably proud of their past achievements in literature, art, music, and a variety of fields in the humanities, and in science and technology. The Soviet Union had a healthy lead against the United States at the start of the "space race" in the 1950s, for example. Enormous resources were deployed to advance research and development. In research parks such as "Academy Town" (Akademgorodok), the best Soviet scientists came to work. These scientists and their research and publication were esteemed, and many developing countries sent their best students to earn degrees in Soviet universities. But in the economic downdrafts that followed the breakup of the Soviet Union, knowledge was devalued. In Russia, the salary of a senior researcher with a PhD decreased by a factor of 10 between 1989 and 1999.[10] Chambermaids, car mechanics, and bus drivers earned up to twice what the average educator did. Not surprisingly, the "best and the brightest" fled the country or went into fields other than science and academia.

5.5 Geopolitical Issues

"Where did our great country go?" is a lament shared across Russian society.[11] Among Russians longing for the "good old days" of the Soviet Union has been Vladimir Putin. He described the collapse of the Soviet Union as "the greatest geopolitical catastrophe of the century."[12] There are a number of reasons to think that Putin's plan is to claw Russia's way back to geopolitical greatness, in part by reassembling pieces of the old empire.

There are three concentric spheres of major geopolitical concern in the former Soviet realm: the unity of Russia itself, Russia's relationships with its Near Abroad (the other countries that were once part of the Soviet Union), and the relations between this region and the rest of the world. There are compelling geopolitical issues to consider, and getting to know them is worth the effort. "I cannot forecast to you the action of Russia," Britain's Prime Minister Winston Churchill said in a radio address to his nation in 1939. "It is a riddle, wrapped in a mystery, inside an enigma; but perhaps there is a key. That key is Russian national interest."[13] Just what is Russia's national interest? Let's explore. Here we will find out whether, early in the 21st century, that key is to be found in Russia's geopolitical contexts.

Geopolitics within the Russian Federation

Internally, Russia is struggling to maintain a cohesive whole that is fashioned from diverse and sometimes volatile ethnic units. As mentioned earlier, the Soviet Union included a number of "autonomous" units based on ethnicity. When Russia 168 emerged from the Soviet Union, it kept those designations within its borders (•**Figure 5.22**). Of the 83 political divisions (called "subjects") of Russia, 26 are **autonomies** (nationality-based units): 21 autonomous republics, 4 autonomous *okrugs* (areas), and 1 autonomous *oblast* (region). The amount of real autonomy, or self-rule, varies somewhat between these three categories, but republics are the "most" autonomous, having the right to establish their own official languages and constitutions, and to pass laws that conflict with those of the federal government. Altogether the autonomies occupy more than 40 percent of Russia's total area and are home to about 20 percent of the country's population. Russia designated Crimea its 22nd republic after the 2014 invasion.

Nearly half of the people in Russia's autonomies are ethnic Russians. The other peoples are ethnically diverse. Some, such as the Karelians, Mordvins, and Komi, are Uralic (related to the Finns and Hungarians); others are Turkic (for example, Tatars, Tuvans, and Yakuts), Mongolian (for example, Buriats near Lake Baikal), and members of many other ethnic groups. The largest autonomous units form a nearly solid band stretching across northern Russia from Karelia, bordering Finland, to the Bering Strait. The Karelian, Komi, and Sakha (Yakut) republics are in this group. This band also includes several other large but thinly inhabited units of aboriginal peoples.

What gives some of these political and ethnolinguistic regions their exceptional geopolitical significance is the presence of major mineral resources within their boundaries: oil and

• **Figure 5.22** The autonomies of Russia, and the boundaries of the federal districts created in 2000.

gas in the Volga-Urals fields in Tatarstan and Bashkortostan, high-grade coal deposits at Vorkuta in the Komi Republic, and one-fifth of all the world's diamonds in Sakha. After the breakup of the Soviet Union, many of the former Autonomous Soviet Socialist Republics (ASSRs, see **page 168**), recalling Moscow's long-standing promises of self-rule for them, issued declarations of sovereignty, asserting their right to greater self-direction of their internal affairs and greater control over their own resources. Eighteen of Russia's republics signed the 1992 Federation Treaty granting them considerable autonomy. The treaty called for the devolution of power centralized in Moscow and for more cooperation between regional and federal governments. 107 Each republic of the Russian Federation was legally entitled to have its own official language, constitution, president, budget, tax laws and other legislation, and foreign and domestic economic partnerships. But many soon complained that Moscow did not honor the treaty, and they began looking for regional solutions to their economic and other problems.

Moscow gave in to some of these demands. The Sakha Republic, for example, won the right to sell 20 percent of its diamonds and 11 percent of its gold independently, compared with only a small fraction during the Soviet era. In turn, Sakha would pay for the government subsidies that Moscow previously gave to local industries. The taxes Sakha collected on sales of its natural resources would first be spent on salaries and improvements within the republic; Moscow would only get the remainder of the tax revenue after Sakha's internal projects were funded. Salaries and other indexes of living standards in the Sakha Republic rose after this agreement was implemented. The majority-Muslim, oil-rich Tatarstan, which voted for independence in a 1992 referendum, did not sign the Federation Treaty until further negotiations gave it an amount of sovereignty that, years later, was declared invalid by Moscow; today Tatarstan has no more autonomy than any other republic. The autonomous republic of Chechnya—of particular geopolitical importance because of its crossroads location for oil pipelines and its links with militant Islamist movements in the region and abroad—also desired independence and refused to sign the treaty. Chechnya's future, along with that of neighboring Dagestan, became critical to the integrity of the Russian Federation (see Problem Landscape, **page 178**).

As Moscow sees it, regional demands for greater self-rule threaten the unity of the Russian Federation and also could deprive Russia of valuable resources. Independence for either Tatarstan or Chechnya could inspire the oil-rich Turkic-speaking Bashkirs and the Chuvash to push for their own independence. This could begin a process that would virtually cut Russia in half. Although some analysts argue that this process of devolution in Russia may actually bring more stability and prosperity to the country, the government in Moscow is worried that Russia might break up just as the USSR did.

Putin's strong-arm tactics in both Chechnya and Ukraine were meant to blunt any push toward disintegration, as were his political-geographical changes in Russia's regional structure. To help keep the country's vast periphery tied to its core, in 2000 Putin's government created a system of eight federal districts into which all "subject" political units were placed. These are the Northwestern, Central, Southern, North Caucasian, Volga, Urals, Siberian, and Far Eastern Administrative Regions (**•Figure 5.22**). Early on, each district appointed a presidential envoy (known as a governor-general) to manage national defense, security, and justice matters. In addition, the heads of Russia's republics, formerly elected, came to be appointed by Moscow. Overall, Putin's central government rolled back the powers of autonomy and began tightly controlling regional affairs through appointments of Putin loyalists.

Blowback against this resurgent authoritarianism culminated in the anti-government protests across Russia in 2011 and 2012. Within a year, the government in Moscow appeared 174 to relent by allowing direct elections of regional leaders. However, Putin's government introduced a loophole requiring these leaders to have "voluntary consultations" with President Putin. Following these meetings Putin dismissed numerous governor-generals and republic presidents on the grounds of "loss of trust," replacing them with "acting heads" in advance of future elections.[15] Their common denominator was that they were proven Putin loyalists who would keep the lid on dissent. Putin established a new "social contract" with the regions and with the Russians: he will ensure that standards of living will rise, as long as people pay taxes and do not demand greater freedoms.[16] With Russia's economy in decline, the government's part of this contract became harder to uphold.

Geopolitics in the Near Abroad

And then there are geopolitical intrigues in the Near Abroad, which President Medvedev called Russia's "zone of privileged interests." With the collapse of the Soviet Union, Moscow took over the former property of the USSR within Russia's borders. This included the Kremlin, Russia's historical seat of power, a compound in Moscow with a stunning array of powerful architectural buildings that backs up to Red Square and the mausoleum of Vladimir Lenin (**•Figure 5.23**). Does

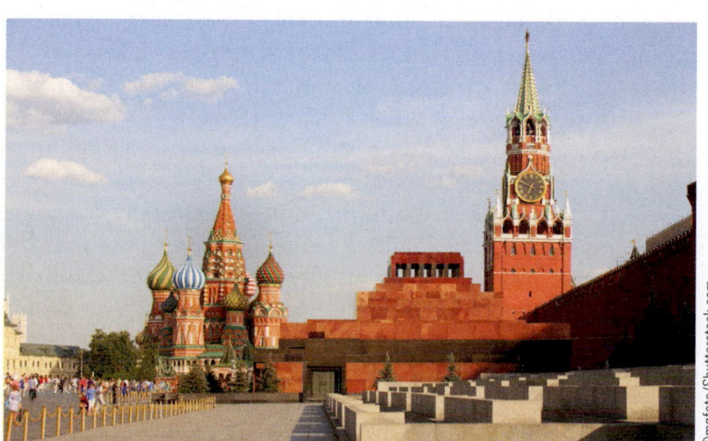

• Figure 5.23 Moscow's epicenter of power and history. The Kremlin Wall and Spasskaya (Savior) Tower at right are on the edge of the Kremlin Fortress where it adjoins Red Square. The Lenin Mausoleum is the stair-stepped structure just right of center. Left of center in the distance is St. Basil's Cathedral.

Problem Landscape Chechnya and Dagestan: Defying Russia in the Caucasus

Nominally part of the Russian Federation, the Caucasus autonomy of Chechnya (population 1 million; for its location, see Figure 5.30) insisted on independence after the Soviet Union broke up. The Chechens, who are Sunni Muslims, have long resisted Russian rule, and Moscow has fought back. Russian troops attempted but failed to seize control over Chechnya soon after its 1991 declaration of independence. Beginning a second offensive in 1994, and at the cost of an estimated 30,000 to 80,000 lives on both sides, Russian troops succeeded in exerting physical control over most of Chechnya. But in 1996, Chechen forces recaptured the capital city of Groznyy. At that time, the Chechen war was deeply unpopular among Russians, and the prospect of further Russian losses led to a peace agreement with the Chechens.

The cessation of hostilities was short-lived. Chechen rebels kidnapped and tortured hundreds of Russian citizens in southern Russia and in 1999 appeared to carry their campaign to the heart of Russia. Chechen separatists were blamed for bombings in Moscow and other civilian centers that killed more than 300 Russians. The Chechen resistance by this time had in fact become internationalized and more violent, with Islamic militant groups from abroad (including from Osama bin Laden's al-Qa'ida organization based in Afghanistan) supplying men and war matériel to their Islamic guerrilla brethren in Chechnya. No longer controlled by Chechnya's president, in August 1999 these Chechen Islamists invaded the neighboring province of Dagestan with the hope of creating an independent Islamic state there.

The invasion of Dagestan and the terrorist bombings in Moscow led to a surge in anti-Chechen sentiment among Russians and to calls for resolute action against Chechnya. Russia's new prime minister, Vladimir Putin, decided to pursue an all-out second Chechen war. In September 1999, Russian forces launched a furious attack that this time put them firmly in control of most of Chechnya and left Groznyy in ruins (•**Figure 5.J**). Russians celebrated Putin as a resolute leader who would finally protect them. But in the years since 1999, much evidence has surfaced that Russia's Federal Security Service (FSB), which succeeded the KGB, was actually responsible for the Moscow and other bombings. The contention is that the FSB sacrificed Russian civilians in a ploy to set the stage for a Chechen conflict that would consolidate Putin's power and popularity.

243

In a Russian-sponsored referendum in 2003, Chechens voted that Chechnya should remain part of Russia, and in a subsequent election appeared to favor a new president, Akhmad Kadyrov. Behind the scenes, Russia ran a rigged election to ensure the victory of their hand-picked candidate, who most Chechens revile as

Vladimir Mashatin/AFP/Getty Images

• Figure 5.J Two Russian military campaigns against Chechen insurgents left Groznyy in ruins.

President Putin or his successor intend to forcibly "inherit" much more than the monuments of the Soviet empire, including independent countries of the Near Abroad? This may be the most important geopolitical question of our time, and one with high stakes: the landscapes of the Cold War could become battlefields of a hot war in a new struggle between East and West.

What exactly does Vladimir Putin want? Some longtime Soviet Union/Russia watchers contend that Russia's diplomatic, economic, and military involvement in the countries of the Near Abroad is an effort to establish a strategic **buffer zone**—an area serving to keep conflicting powers apart—between Russia and the "Far Abroad," much as the Soviet Union had with its eastern European "satellites." The Warsaw Pact no longer exists, and so Russia has lost the important strategic depth the Soviet Union had. Russia does, however, have a military presence in more than half the former Soviet countries. Russia's incursions into parts of Ukraine, and the subsequent revitalization of NATO in Eastern Europe, have energized the Russia-anchored **Collective Security Treaty Organization (CSTO)**; see **•Figure 5.24**. Russia founded this security alliance in 1992 to replace the Warsaw Pact as a counterbalance to NATO, but its membership dwindled from nine to six, and it suffered from shortages of funding and purpose.

Other analysts conclude that above all Russia wants access to parts of the former Soviet republics because of their important natural resources and other economic assets. These include the Black Sea coast and its accommodation for Russia's naval fleet in Ukraine's Crimea (taken by Russia in 2014), the industrial resources of eastern Ukraine (many also seized by Russian proxy forces in 2014), uranium in Tajikistan, aviation plants and the vital BTC oil pipeline in Georgia, and military factories in Moldova.

Vladimir Putin spoke of Russia's urgent priority to draw the countries that were republics of the former Soviet Union closer to Russia. Such a reconfiguration could, of course, develop through **"soft power,"** a persuasive approach to international relations using diplomatic, economic, or cultural a traitor (in the first Chechen war, he fought the Russians, but he switched sides during the second war). He was assassinated the following year. His son Ramzan Kadyrov became Chechnya's president in 2007 with the blessings of Vladimir Putin. With Kadyrov, Putin implemented his policy of "Chechnization," using a handpicked and highly paid local ruler to carry out Moscow's will in Chechnya, quell the Chechen insurgency against Russia, and discourage any other would-be separatist movements.

This Russian government poured money into Kadyrov's Chechnya, investing in the reconstruction of roads and the urban infrastructure of shattered Groznyy. Chechnya's capital now boasts skyscrapers and glitzy shopping malls (**•Figure 5.K**). But underneath the glamour and the veneer of peace, Chechen resentment against Russia remains. Chechen guerrilla resistance has continued with periodic terrorist attacks against Russian targets, including civilians. Many of these are suicide bombings carried out by Chechen or Dagestani women whose fathers, sons, or brothers had been killed by Russians. These are the so-called "Black Widows" who are very effective at concealing explosive devices in their conservative Muslim clothing. They were much feared in advance of the 2014 Winter Olympics in Sochi, for which a massive Russian security infrastructure including a so-called "ring of steel" was built.

Islamist groups are still active in Dagestan (population 3 million), a volatile area with over 30 distinct and fractious ethnic groups described by one observer in these terms: "a seemingly intractable knot of clan and business rivalries, religion, separatism, crime and extortion, feuds and vendettas whose roots runs deep into its region's history."[14] As in Chechnya, but with less success, Moscow pours subsidies into Dagestan to try to keep the area loyal to Russia. Dagestan ought to be prosperous, with its Caspian Sea beaches attracting tourists, its oil and gas reserves flowing, and its fisheries and vineyards yielding harvests. But violence or the potential for violence keeps investors and tourists away and keeps resources from being utilized. You may recall that one of the Tsarnaev brothers who bombed the Boston Marathon in 2013 is suspected to have had links with violent Dagestani Islamists. The brothers' ethnic roots were in Chechnya and Dagestan.

Many Russians harbor strong prejudices against ethnic minorities and immigrants, particularly those from the Caucasus. Just as there is Islamophobia in Western Europe, so there is in Russia. About 1 million Muslims live in Moscow, where they frequently encounter discrimination. Over 10 million Russians (7 percent of the country's population) are Muslim.

96

©Alenvl/Shutterstock.com

• Figure 5.K Glitzy Groznyy of today is a showcase of Russia's will to stabilize and control Chechnya.

tools. As soon as the Soviet Union broke apart, Russia sought to serve as the anchor for all of that country's newly independent successor states. In 1991, Russia, Ukraine, and Belarus formed a loose political and economic organization, the Commonwealth of Independent States (CIS). Except for Turkmenistan and the Baltic countries of Estonia, Latvia, and Lithuania, the other former Soviet republics eventually joined the CIS. Russia took a strong leadership role in the CIS, although it was officially headquartered in Minsk, Belarus. To date, the CIS has been concerned mainly with establishing common policies on economic concerns. Russia wants its members to unite on other issues, particularly security issues like fighting Islamist terrorism and blunting NATO expansion. CIS members are also members of a free trade agreement known as **CISFTA (Commonwealth of Independent States Free Trade Agreement)**. Ukraine's partnership with Russia under this free trade banner complicated Ukraine's newer **"Deep and Comprehensive Free Trade Area" (DCFTA)** agreement with the European Union (see Figure 5.24). The pro-Russian Ukrainian government's decision to put off signing the DCFTA agreement in fact sparked the Ukrainian Revolution in 2014.

182

In its geopolitical orientation, Russia has often historically posed itself as a "Eurasian" power, neither Western nor Eastern, but distinctly unique and great. Aiming to formalize a strategic vision giving Russia the central role, in 2012 President Putin unveiled a plan to build a "Eurasian Union" of former Soviet states: Russia, Belarus, and Kazakhstan (previously associated together in a "customs union" that eliminated tariffs) would be the first members, and would later be joined by Kyrgyzstan and Armenia, both already heavily dependent upon Russian aid. President Putin said the Eurasian Union would become "a powerful supranational union capable of becoming one of the poles in the modern world."[17] Russia is especially keen to prevent any CIS countries from becoming full members in the western military alliance of NATO, although several countries, including Russia itself, are members of NATO's **Partnership for Peace (PfP)** program. "Based on a commitment to the democratic principles that underpin the

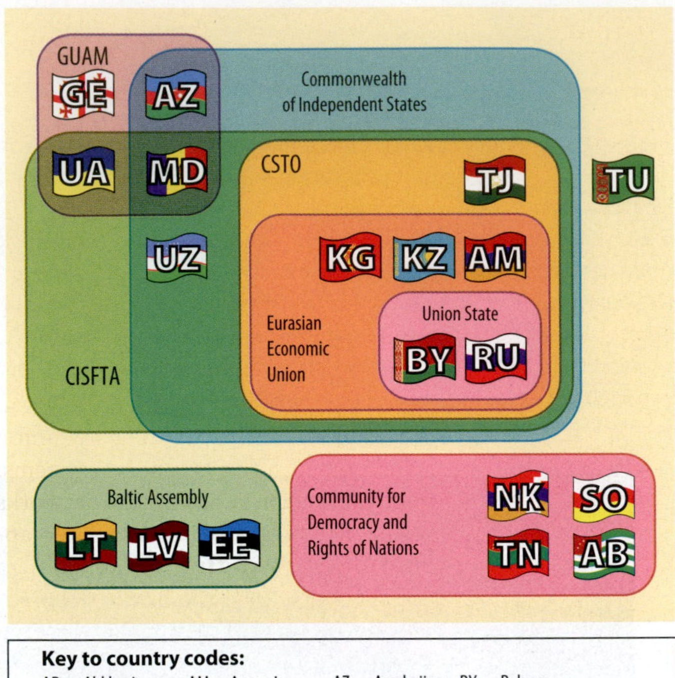

Key to country codes:

AB = Abkhazia AM = Armenia AZ = Azerbaijan BY = Belarus
EE = Estonia GE = Georgia KG = Kyrgyzstan KZ = Kazakhstan
LT = Lithuania LV = Latvia MD = Moldova NK = Nagorno-Karabakh
RU = Russia SO = South Ossetia TJ = Tajikistan TN = Transnistria
TU = Turkmenistan UA = Ukraine UZ = Uzbekistan

• **Figure 5.24** Euler diagram showing the relationships between post-Soviet multinational organizations in the region of Russia and the Near Abroad, and the three Baltic countries.

Alliance itself," PfP's mission statement reads, "the purpose of the Partnership for Peace is to increase stability, diminish threats to peace and build strengthened security relationships between individual Euro-Atlantic partners and NATO, as well as among partner countries."[18] Clearly, Russia is no longer a good match for the group.

Twelve former Warsaw Pact countries have become full members of NATO: Albania, Bulgaria, Croatia, Czech Republic, Estonia, Hungary, Latvia, Lithuania, Poland, Romania, Slovakia, and Slovenia. For years, Russia has made it clear that if Ukraine or Georgia were to become full members of NATO, they would pay a heavy price. How heavy? In 2008, Putin threatened Ukraine with nuclear annihilation if it joined NATO and deployed a US missile defense system there. "It is horrible to say and even horrible to think," he said at a joint news conference with Ukraine's pro-Western President Viktor Yushchenko, "that, in response to the deployment of such facilities in Ukrainian territory, which cannot theoretically be ruled out, Russia could target its missile systems at Ukraine. Imagine this just for a second."[19] Four years before Putin made this threat, Ukraine had given up the 2000 nuclear weapons it inherited when the Soviet Union broke apart. According to the **Budapest Memorandum** spelling out the terms of this agreement, Russia and the United States vowed to "respect the independence and sovereignty and the existing borders of Ukraine" and "refrain from the threat or use of force against the territorial integrity or political independence of Ukraine."[20] Certainly Russia would not have invaded eastern Ukraine or Crimea if Ukraine still had its nuclear deterrent, and President Putin would not have been talking about Ukraine in this way.

Ukraine is of key importance in understanding Russia's prospective bid to reestablish parts of its empire. Let's examine it closely, and then look at other geopolitical hot spots in Russia's Near Abroad.

Russia's Moves on Ukraine: Context and Pretext

Ukrainians are the second largest ethnic group in the Slavic Core and are closely related to the Russians in language and culture; "Russia writ small" is how one expert describes Ukraine.[21] The name Ukraine means "at the border" or "borderland," and this area has long served as a buffer zone between Russia and neighboring lands to the south and west (see Figure 5.33). Here armies of the Russian tsars fought for centuries against nomadic steppe peoples, Poles, Lithuanians, and Turks until the 18th century, when the Russian Empire finally absorbed Ukraine. For three centuries, Ukrainians were subordinate to Moscow, first as part of the Russian Empire and then as a Soviet republic.

Slightly smaller than Texas, Ukraine lies partly in the forest zone and partly in the steppe. On the border between these biomes, and on the banks of the Dnieper River, is the historic city of Kiev, Ukraine's political capital and a major industrial and transportation center. Ukraine is blessed with an abundance of fertile black earth soils, especially in the steppe region south of Kiev. Historically, Ukraine has been a great "bread- 159 basket"—especially of globally important wheat—along with barley, livestock, sunflower oil, beet sugar, and many other products. Ukraine also has a generous endowment of industrial raw materials, especially coal and iron ore, located mainly in the country's east, where separatism and fighting are most intense.

Ukraine's cultural geography is critical in understanding the recent political crisis. It is an ethnically divided country, in simplified terms with a West-leaning population of ethnic Ukrainians in the west and a Russia-leaning population of ethnic Russians in the east (•**Figure 5.25**). Many of these ethnic Russians are the progeny of Moscow's Russification program during Soviet times, which left about 25 million ethnic Russians in the 14 smaller Soviet states at the time of independence. Since independence in 1991, many of the governments and ethnic majorities of the post-Soviet countries have asserted their non-Russianness and tried to sweep away the remains of Moscow's Russification program. Many ethnic Russians in 165 Near Abroad countries migrated to Russia after 1991. Those who remained have often had difficulties finding housing and employment, complaining that host governments and peoples discriminate against them.

President Putin has indulged in these complaints because they offer Moscow the pretext for greater engagement and even outright control when "Russian interests" are threatened outside of Russia. In 2014, Russia's parliament, intent on justifying the Crimea invasion, gave Putin the authority to disregard

• **Figure 5.25** Ukraine is divided between a Ukrainian-speaking west and a Russian-speaking east.

THINK CRITICALLY: Compare the regions with the highest percentages of Russian speakers with the areas Russia has occupied or encouraged separatist movements on the map in Figure 5.28.

Ukraine's sovereignty in order to "protect Russia's interests and those of Russian-speakers."[22] This interest in Russian speakers casts the widest possible net for future Russian interventions: in Belarus for example, only 8 percent of the people are ethnic Russians, but over 70 percent speak Russian. Could Belarus be another target of Russia's reacquisition of former Soviet republics?

In the countries most committed to erasing Soviet legacies, governments are set on "de-Russification" and the resurrection of pre-Soviet cultures and languages. In western Ukraine, and among leaders elected from that region, the sentiment is to advocate Ukrainian language and culture while cultivating closer ties with Western Europe and the United States rather than with Russia. Ukrainian language is a required subject in Ukraine's schools, even in the country's eastern region where Russian has long been the first language for most people. In 2009, Ukraine's pro-Western President Viktor Yushchenko called for a deeper understanding of Ukrainian language as a means of distancing the country from Russia; he said, "With our native language, we preserve our culture. That greatly contributes to preserving our independence. If a nation loses its language, it loses its memory, its history, and its identity."[23]

Russia has used its fossil fuel endowment to achieve maximum political clout in the Near Abroad, especially (but not only) in Ukraine. Russian oil and gas are also vital to the former Soviet republics of Belarus, Kyrgyzstan, and the Baltic countries (as well as to Germany and other countries in Western Europe; see **page 120**). Under Vladimir Putin, the Russian government began in 2004 to increase its control of Russia's energy sector, a process of **renationalization** that reversed the trend toward privatization that had been underway. The world's largest natural gas producer is Russia's largest company, its state-owned **Gazprom**. With control of the critical energy sector, the Russian state has repeatedly wielded fossil fuel as a political weapon of state. Loyal former Soviet countries get the "carrot" of large subsidies, whereas disloyal ones get the "stick."

A good example of the carrot was that in the 1990s, Russia charged Belarus just one-third of what it charged Latvia and Estonia for natural gas. This was at a particularly good time in relations between the two countries, when they were actively pursuing a "union state" in which their economic systems, political systems, and defense networks would be integrated. Russia later applied the stick, cutting oil supplies to Belarus when its authoritarian president, Alexander Lukashenko (known as the "Europe's last dictator"), had a spat with Vladimir Putin. The Russian leader regards Belarus' independence as "contingent," suggesting that Russia would assume control of the country should Lukashenko's grip on power weaken.

Ukraine typically depends on Russia for two-thirds of its energy and so is especially vulnerable to Moscow's hand on the energy tap. Russian energy tends to flow freely when Ukrainian leaders are pro-Russian, but not otherwise. In what was widely seen as retaliation for the so-called **Orange Revolution** that brought westward-leaning President Victor Yushchenko to power in 2005, Gazprom briefly cut natural gas supplies through a pipeline to Ukraine. (Incidentally, political opponents tried to assassinate Yushchenko with polonium. He survived, but his face was grotesquely disfigured.) When a pro-Russian president from eastern Ukraine, Viktor Yanukovych, succeeded Yushchenko, the energy flowed freely again.

Russia had its eyes on parts of Ukraine's territory almost as soon as the Soviet Union fragmented. At contest was the Crimean Peninsula, a picturesque and verdant resort region on the Black Sea, with a population of 2 million that is 70 percent ethnic Russian (see Figure 5.25). Russia's Catherine the Great annexed Crimea from the Ottoman Empire in 1783, and in 1954 Soviet leader Nikita Khrushchev transferred it from the RSFSR (Russian Soviet Federated Socialist Republic) to Ukraine. This was mainly a symbolic gesture because Ukraine at the time was part of the Soviet Union. When the Soviet Union collapsed, Russian irredentists began agitating to get the Crimea back. Ukraine yielded a bit by granting the Crimea special status as an autonomous republic.

During the years when post-Soviet Russia recognized Crimea as belonging to Ukraine, it was nevertheless at odds with Ukraine over a strategic strait separating the Crimea from Russia's Taman Peninsula to the east. This Kerch Strait is the vital sea connection between the Azov Sea and the Black

Sea—and therefore between southern Russia and the wider world through the warm water access it covets (see **page 167** and Figure 5.28 on **page 183**). Russia wanted to share sovereignty of the strait with Ukraine, but Ukraine claimed most of the waters as its own and charged Russia several million dollars per year in transit fees.

Russia has a strategic interest in the Crimean port of Sevastopol, where it bases its Black Sea naval fleet. After Ukraine gained independence, it charged Russia a hefty fee to use the port. The two countries eventually struck a deal by which Ukraine allowed Russia to use the naval facility, and in exchange Russia gave Ukraine a 30 percent discount for buying Russian natural gas.

Breakout!

Although Ukraine serves as a buffer state along Russia's borderlands with Western Europe, it is also taking shape as a new eastern European shatterbelt. (As we saw in southeastern Europe's Balkan region, a shatterbelt is "a large, strategically located region composed of conflicting states caught 138 between the conflicting interests of great powers.") Ukraine is large, important, endowed, contested, and certainly shattered. The country has long been fractured by its east–west cultural regions and political orientations. In 2014, Ukraine's pro-Russian president Viktor Yanukovych, who hailed from eastern Ukraine, considered endorsing the Deep and Comprehensive Free Trade Area association agreement with the 179 European Union. With this DCFTA the EU offered significant financial assistance to help with economic, political, and social reforms in Ukraine. In return the EU hoped to secure markets for its exports, a reliable source of imported wheat and other produce, and the continued supply of (mainly Russian) natural gas through pipelines leading across the country. The agreement had enormous symbolic importance for Ukrainians, essentially representing their country's growing orientation toward the West and its democratic free market systems and Ukraine's release from the Russian orbit. During Soviet times, Lenin once said, "If we lose Ukraine, we lose our head." Fearful of losing Ukraine to the West, Russia pressured Ukraine *not* to sign the agreement and offered instead a trade pact between Russia, Ukraine, and Kazakhstan. Yanukovych was persuaded, turning his back on the EU agreement and instead signing a multi-billion dollar loan agreement with Russia. He also entered into talks over Russia's longheld desire to construct a bridge over the Kerch Strait, and Russia once again slashed the price of natural gas it supplied to Ukraine.

Yanukovych's decision not to sign the EU agreement was generally well received in Ukraine's mainly ethnic Russian east, but in the west and the capital of Kiev, people were furious. In November 2013, they took to the streets in protests that in Kiev were concentrated around the main square known as the *Maidan*. The subsequent revolution that ousted Yanukovych in February 2014 came to be known as **Euromaidan**, Ukraine's second pro-Western revolution in 10 years. Yanukovych fled to Russia, and a pro-Western interim government took office in Kiev until a pro-Western president, the billionaire "chocolate tycoon" Petro Poroshenko, was voted into office in May 2014. Poroshenko signed the EU agreement the following month.

The loss of Ukraine's pro-Russian president amidst massive demonstrations advocating closer ties to the West was, of course, deeply disturbing to Russia. Putin and his colleagues in the Kremlin greatly feared the prospect of dissent like this breaking out in Russia. Russia needed to flex its muscles, instilling nationalist fervor that would preempt dissent. But it would have to wait a bit: the Euromaidan revolution was taking place at exactly the same time Russia was enjoying the international spotlight of hosting the Winter Olympics in Sochi and showing the world what a modern, progressive and open country Russia was. Almost as soon as the Olympics closing ceremony concluded, Russia made its move on Crimea.

Following pro-Russian demonstrations in Crimea, pro-Russian forces occupied strategic positions across the peninsula. Many of these men wore military uniforms without any insignia designating what country they belonged to. These "little green men," as they came to be known, were conventional Russian military forces in disguise. In March 2014, Crimea's pro-Russian parliament hosted a referendum, seen as illegal in Ukraine's capital, asking Crimeans whether they wanted to join Russia (**•Figure 5.26**). The results were overwhelmingly in favor of doing so. The next day Crimea's parliament declared Crimea's independence from Ukraine and asked to join Russia. Crimea's separatist government signed a "treaty of accession" with Russia, formally accepting Crimea and the port of Sevastopol as part of Russia. This was a triumph for President Putin; he even took a victory lap by sailing into Sevastopol amidst great fanfare. His popularity within Russia surged higher again (**•Figure 5.27**).

Filippo Monteforte/AFP/Getty Images

• Figure 5.26 A billboard in the Crimean city of Sevastopol reads, "On March 16 we will choose either [Crimea in red with a swastika (left)] . . . or [Crimea with the colors of the Russian flag]. If Crimeans vote against joining Russia, the swastika implies, they are in league with Nazis past and present. As discussed on **page 169**, World War II's legacy is profound in the region of Russia and the Near Abroad.

THINK CRITICALLY: Swastikas are rarely seen in the West, but this one is prominently displayed in Russian-occupied Crimea. What does the swastika evoke in Russia and in the West?

• **Figure 5.27** On May 9, 2014, Vladimir Putin took a victory lap in the port of Sevastopol, Crimea, forcibly annexed by Russia a year earlier. Characteristically the Russian leader appealed to the patriotism and emotions of Russians recalling their sacrifices in World War II. May 9 is Russia's "Victory Day," marking the end of what Russians call the "Great Patriotic War." The international community joined Ukraine in denouncing Putin's visit.

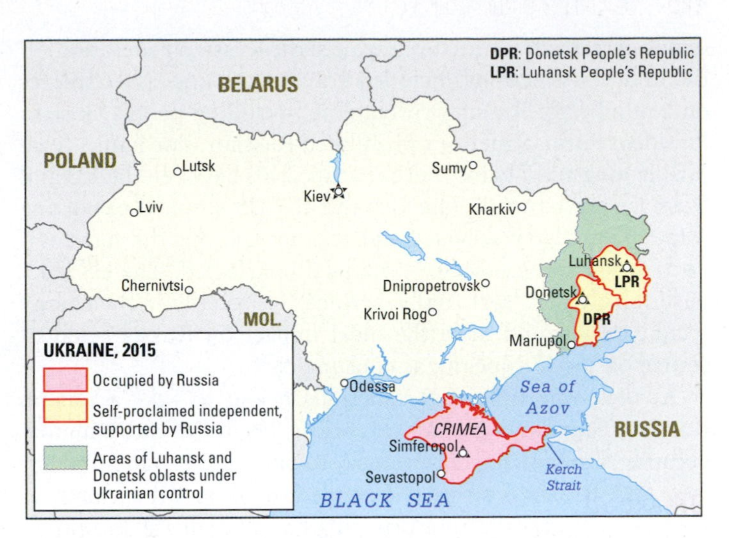

• **Figure 5.28** Russia invaded and annexed Crimea in 2014, and anti-Ukrainian separatists have proclaimed their own independent states along Ukraine's eastern border with Russia.

THINK CRITICALLY: The book you are reading needs timely and accurate cartography. After reading about Crimea and the "peoples' republics" of eastern Ukraine, how do you think these entities' political geography should be portrayed? How and why are they represented as they are on this map?

Reaction in Kiev and in most countries abroad was quite different. The United Nations, the European Union, and the United States, among many others, expressed shock and outrage that on the borders of Europe in the 21st century one country was forcibly occupying another. They called for the immediate withdrawal of Russian forces from Crimea and for the peninsula's restoration to rule from Kiev. These appeals fell on deaf ears: Russia was just beginning its tactical moves in an apparent strategy to secure eastern Ukraine, giving it overland access to the Crimean Peninsula and proxy control over eastern Ukraine's coal and other valuable natural resources. If you look at Crimea in Figure 5.28, you can recognize Russia's geographical challenge in annexing the peninsula: politically it is a virtual island cut off from Russian territory. It is almost imperative to secure a land bridge through eastern Ukraine to Crimea. A complementary move would be for Russia to secure a land bridge from the pro-Russian breakaway Transnistria region of Moldova across southwestern Ukraine to Crimea (see **page 185**).

Through the following months, the little green men fought alongside pro-Russian separatists of the self-declared **Donetsk People's Republic (DPR)** and **Luhansk People's Republic (LPR)** and against Ukrainian army forces in the Donbass region of eastern Ukraine (•**Figure 5.28**). Both President Putin and the pro-Russian separatists in eastern Ukraine began calling the area ***Novorossiya***, meaning "New Russia." This was the name of a political region in Tsarist Russia, acquired by Catherine the Great in the Russo-Turkish conflict, corresponding to what is now eastern and southern Ukraine. This resurrected geographical term is well suited to the irredentist narrative of a Ukraine that belongs with Russia.

The West views Russia's military support for Russian irredentism in Ukraine as a dangerous breach of international law and a serious threat to post–Cold War order in Europe and beyond. Hoping that squeezing Russia economically could halt its intervention, Western powers, including the EU and the US, began to impose economic sanctions on Russian individuals and institutions. From the American point of view, its Western allies, including Germany, France, and Britain, were too reluctant to use economic sanctions that would punish Russia too harshly. They had good reasons not to squeeze Russia too hard: Germany, in particular, had grown dependent upon Russian natural gas; Britain had profited from high-rolling Russian oligarchs whose penthouses and playgrounds were in London; and Russian oil money had gushed into British banks.

The catalyst that overcame Western Europe's reluctance to impose harsh economic sanctions against Russia literally fell out of the sky. On July 17, 2014, suspected pro-Russian separatists operating a Russian-supplied Buk missile system near Donetsk shot down a Malaysian Airlines jet en route from Amsterdam to Kuala Lumpur, killing all aboard. Dutch nationals were 193 of the 298 passengers on flight MH 17. Shock, grief, and anger spread from the Netherlands across Europe, and grew as pro-Russian forces on the scene made it very difficult for emergency responders and accident investigators to recover human remains and document the crash circumstances. Along with Russian President Putin, the separatists who, probably by accident, shot down the plane blamed the incident on Ukrainian forces. Meanwhile, the Kremlin continued to deny that Russia was involved in the fighting in eastern Ukraine.

The EU began to speak and act more assertively, declaring, "In response to the illegal annexation of Crimea and deliberate destabilisation of a neighbouring sovereign country, the EU has imposed restrictive measures against the Russian Federation."[24] EU and US sanctions included travel bans and asset freezes on a number of Russia's ruling elite, including people close to President Putin. Sanctions prohibited Russian state banks from raising long-term loans. Future arms deals between the EU and Russia were banned. The EU and the US stopped exporting some oil industry services and technology, but the most important natural gas sector remained unaffected. The US led a push for greater "sectorial sanctions," especially in the energy sector, that would have the most impact on Russia—and of course on the European gas consumers.[25]

At the same time that sanctions began to take a toll on Russia's economy, global oil prices began to fall (mainly because of Saudi Arabia's refusal to cut back production; see **page 175**). Russia is so very dependent upon its oil exports, as we have seen, that falling prices had an immediate and broad impact on Russia's economy. Early in 2015, President Putin stared into the abyss of his country's economy and weighed his choices. The West had offered him diplomatic "off-ramps" to de-escalate the Ukraine crisis by withdrawing Russian support. Instead, Putin drove ever-greater numbers of Russian men and matériel into eastern Ukraine. Western leaders and institutions, including NATO, considered providing heavy weaponry to Ukrainian forces. This would, of course, feed into Putin's narrative, and the broad Russian public sentiment, that NATO is encircling and endangering Russia and that Russia must blunt that threat by expanding its presence in the Near Abroad. Contesting worldviews resurrected many phantoms of the Cold War, including even the specter of nuclear war. In effect, a proxy war between Russia and NATO was taking shape. This **"New Cold War"** could heat up quickly with any miscalculation or escalation.

There were diplomatic efforts. Ukraine, Russia, and rebel leaders of the DPR and LPR in eastern Ukraine sought a regional diplomatic solution, signing the Minsk Protocol in September 2014, which called for a cease-fire and for more devolution of power from Kiev to the region. This agreement was soon in tatters as fighting intensified. Against a backdrop of growing risks, the leaders of Ukraine, Russia, Germany, and France met in Minsk once again in February 2015 to extinguish the fighting in eastern Ukraine. "Minsk II's" 12-point plan called for a cease-fire accompanied by political concessions to the Ukrainian rebels and, critically, the pullout of "foreign forces." Would Russia withdraw its little green men and their big guns?

We have considered a range of factors influencing Putin's decision to muscle in on Ukraine, none more important than Lenin's "losing our head" scenario. For Russia it's not just the head, but the belly (Ukraine's wheat) and the pot-bellied stove (Ukraine's coal, iron ore, and other industrial resources). We have considered the irredentism that is a legacy of Russification, and the long shadows of the Nazi occupation. Putin may have acted with careful, long-term strategic reasoning, or, as some have suggested, he was simply thinking and acting on the fly. It does appear that Putin was counting on the continued good luck of high oil prices to keep Russia afloat and was betting that gas-thirsty Europeans would not join an American-backed sanctions regime. Fate seems to have crossed the unlucky Malaysian Airlines with Putin's plans. The future of Putin's "go it alone against the world order" way of doing things will depend very much on Russian pocketbooks. Russian history informs us that ruinous economic decisions bring about changes in leadership.

The Frozen Hot Spots of Russia's Near Abroad

We have considered two strategic motivations for Russia's recent assertiveness in Ukraine: an interest in rebuilding parts of the former Soviet empire and an interest in building buffer zones between Russia and NATO-aligned Western Europe. Here is another explanation that helps frame Russian interventions preceding Crimea/Ukraine and may help answer the question "What are Russia's next geopolitical moves in its Near Abroad?" Russia maintains a series of "frozen conflicts" that will serve its interests in the Near Abroad and beyond without prompting all-out war or severe repercussions in the international arena (such as harsh sanctions). A **frozen conflict**, explains Serbian political scientist Nikola Vujinović, is "any prolonged ethno-political conflict that falls short of all-out war."[26]

The authoritative British journal *The Economist* offers more insights into frozen conflicts. A frozen conflict needs the support of a large power with the funds and the willingness to sustain the dispute (which Russia is good at, *The Economist* notes). The supporting power usually has a financial or strategic interest in the contested area; a good example is Russia's dependence on the heavy industries of Ukraine, including the aircraft manufacturer Antonov (builder of the world's largest plane, the behemoth six-engine An-225 Mriya). A frozen conflict can thaw and escalate into a full conflict at short notice. And finally, "If frozen conflicts are hard to end, it's probably because powerful interests don't want them to."[27]

What further moves might Russia make in its Near Abroad, beyond Crimea and eastern Ukraine? The Baltic countries of Estonia, Latvia, and Lithuania, situated between the bulk of Russian territory and the small Russian exclave of Kaliningrad on the Baltic Sea, are genuinely fearful of Russian intentions because they have suffered from Russian incursions in the past. Russia in effect colonized them, and their anti-Russian sentiment is quite strong. Russian language is banned from government offices in Lithuania, for example. People in Estonia and Latvia must pass language tests to obtain citizenship, forcing ethnic Russians in these countries to learn the local languages or suffer the consequences of not knowing them. But how much danger does Russia really pose to the three Baltic nations? It would be very risky for Russia to ignite irredentism and deploy its little green men in the Baltics, as it did in Crimea and eastern Ukraine. Why? The Baltic countries are NATO members, and a Russian incursion would automatically trigger a military response from NATO members, including the United States. The Baltics thus have virtual immunity from Russian military action.

Several other political entities are far more vulnerable to Russia's interests in maintaining frozen conflicts, igniting irredentism, and leaving doors open to Russian expansionism: Transnistria (within Moldova), Nagorno-Karabakh (within Azerbaijan), and Abkhazia and South Ossetia (both within Georgia). All are members of an organization called the **Community for Democracy and Rights of Nations**, also known as the Commonwealth of Unrecognized States. This group and other political and economic associations blocs of Russia and the Near Abroad are illustrated in Figure 5.24. Let's look at these four entities and how they fit into Moscow's strategic vision for the Near Abroad.

Moldova (known formerly as Moldavia and, in the 19th and early 20th centuries, as Bessarabia; population 4 million), made up largely of territory that the Soviet Union took from Romania in 1940, adjoins Ukraine at the southwest (•**Figure 5.29**). The Moldovan majority (about two-thirds of the country's population) is ethnically and linguistically Romanian, not Slavic. Issues of ethnic antagonism, irredentism, and political separatism involving Transnistria (population 500,000), a sliver of land the size of Rhode Island along Moldova's eastern border with Ukraine, have preoccupied Moldova since Soviet times.

In Moldova, as in the independent Baltic countries, the ethnic majorities sought to distance themselves from the Russian influences that had been imposed on them in Soviet times. Language was an important component in the reassertion of national identity. In 1989, Moldovan was made the country's official language, and the Latin alphabet supplanted the Russian Cyrillic alphabet. The Moldovan government advocated a linguistic and cultural "Moldo-Romanian" identity, and some Moldovans proposed reuniting Moldova and Romania. These measures angered the ethnic Russians of Moldova's east. When Moldova declared its independence, these separatists, wishing to remain part of the Soviet Union, proclaimed their "Pridnestrovian Moldavian Soviet Socialist Republic." Neither Moscow nor the Moldovan capital Chisinau recognized this new republic, and war soon broke out. After the fall of the Soviet Union, the goal of the separatists changed to establishing an independent country called Transnistria. A 1992 cease-fire "froze" the conflict; while Moldova continues to claim Transnistria, and no other country recognizes it as an independent entity, Transnistria is not under the control of the Moldovan government and has its own president, military, and currency.

• **Figure 5.29** Transnistria is still internationally recognized as part of Moldova.

Despite protests from the European Union and other international organizations, Russia maintains about 1000 troops in Transnistria, ostensibly for peacekeeping purposes. Russia is widely seen as being sympathetic to the Transnistrian cause, as about 30 percent of the population is ethnic Russian (as opposed to only 5 percent in the rest of Moldova). As in eastern Ukraine, the presence of Russian troops gives Moscow enormous influence in Transnistria. After Russia annexed Crimea in 2014, Transnistrian politicians petitioned Moscow to have their region join Russia as well. You can get a fine sense of how Moldovans feel squeezed by Russia, along with insights into land and life in Moldova, in Tatiana Darie's sketch of Life in Moldova on **page 186**.

Moldova has turned increasingly pro-Western in reaction to Russian assertiveness in Transnistria. Moldova is forging closer ties with various European political and economic institutions, and it is a member of the Central European Free Trade Agreement, an initial step toward possible future EU membership. Moldova faces challenges with further integration into Europe; besides the Transnistria issue, Moldova is the poorest European country (with a GDP (PPP) of just $3800) and also has the lowest Human Development Index score in Europe. Remember that this book's dividing line between the regions of Europe and Russia and the Near Abroad is a vernacular tool, and most Moldovans wants to be perceived as belonging with Europe.

With its pro-Russian leanings and ethnic Russian population, Transnistria represents "low hanging fruit" should Moscow want to annex additional post-Soviet territories. The risks of frozen conflicts heating up to all-out conflicts are greater in the three Caucasus situations. Each of them has a recent history of hostilities, with major involvement of Russia. Russia has a number of geopolitical concerns with countries and ethnic groups of the Caucasus. The Armenians embrace Russia as a strategic ally against their historic enemy, the Turks. The Georgian government is strongly anti-Russian, but several nominally Georgian provinces have populations that are strongly pro-Russian. Russia is keen to nurture these pro-Russian elements and to keep or make the national governments friendly to Russia. 199

When, after considering Crimea and mainland Ukraine, we ask whether Russia might muscle into other now-independent former Soviet republics, we need to look at what has already happened in Georgia. Ever since the Soviet Union broke apart, Georgia has wanted to loosen ties with Russia in favor of better relations with the West and its allies. Georgia wants to join the European Union (an unlikely prospect in the near future, although Georgia has signed several association agreements with the EU) and is an "aspirant" for membership in NATO (also unlikely, but NATO and the Georgian military cooperate

life in MOLDOVA

Tatiana Darie

Tatiana Darie

"I am writing this letter from a distant land. She lies somewhere between the Middle Ages and the 21st century, between nostalgia for a failed revolution and an imaginary hope called Europe . . . Was it our compulsion for traveling that first brought us together? Perhaps it was the threshold of the indeterminate that attracted us, the promise of a different becoming, even though its moment has now been lost."

This is Joanne Richardson, a Romanian-born cultural theorist narrating her journey to Moldova through 10 letters that reflect on the country's uneasy struggle between the collapse of the Soviet Union and the broken European dream. Richardson perhaps best captures the grim mood in the tiny nation, historically disputed by the east and the west.

Next year will mark 25 years since Moldova has emerged as an independent republic, following the dramatic collapse of the Soviet Union in 1991. A quarter of a century since the tiny nation, sandwiched between Romania and Ukraine, has been battling its Soviet past and fighting for a European future. The road to independence has been economically painful and politically bloody. Racked by a civil war after the Soviet breakup, the country still struggles with a conflict in the eastern breakaway region of Transnistria, the last chunk of the former Soviet empire where Moscow still keeps hundreds of troops and a large stockpile of Soviet-era weapons. The war in neighboring Ukraine sounded alarms that the "frozen conflict" could heat up and raised fears the small republic could end up bludgeoned into Russia's orbit again.

It is in the heart of Moldova's capital where you'd ask a question in Romanian and still get an answer in Russian. It is the epicenter of a geopolitical divide, split between a pro-Russian east and a pro-European west. Ironically, we share identity with both. The similarities are deeply embedded in our culture, geographically located at the crossroads of the Latin, Slavic, and other cultures. Our language has heavy influences from Russia, Romania, and Ukraine. While Romanian is widely spoken in the country, some claim Moldovan (a mix of Russian and Romanian) to be the nation's official language, while others deny its existence. But perhaps the influences are most visible in our cuisine. Culture hawks and food lovers would definitely adore Moldova's most treasured traditional dishes such as mamaliga, zeama, and plăcintă. The rest of the menu mainly originated from the cultures of many people who lived here in the past, like Ukrainians, Russians, Romanians, Jews, and others.

Only vaguely known in Europe and perhaps anonymous to the rest of the world, the landlocked country lies on a picturesque landscape with abundant vineyards, wedged between the Dniestr river to the east and the Prut to the west. A largely rural nation, once known as the garden of the Soviet Union, the republic is located on a fertile plain with a few areas of hills across the center and the northeast, with its lowest point in the Dniestr at 2 meters and its highest point on the Balanesti hill at 430 meters.

For those who love wine and are looking for unique places to visit, Moldova certainly has a lot to offer—from the world's greatest wine cellars to vast monasteries and historical complexes to landmark monuments and museums, including the house where the great Russian writer Pushkin spent his days in exile, penning some of his most famous works.

If you venture out to visit Moldova, you'll find some of the best wines in the world, that not only us, Moldovans, like to drink. Our ruby dark dry red Negru de Purcari has been long-time favored in London by her Majesty, Queen Elizabeth II. The tiny republic has consistently been among the world's top 10 wine producers, an industry that also props up its economy. A few miles away from the capital city Chisinau, you'll find one of the world's greatest wine cellars, built by Moldova's Soviet masters nearly half a century ago. Deep inside a warren of old mining tunnels with limestone hills, subterranean chambers stretch across 70 miles, housing more than a million bottles. Not far from there, you'll find another wine store, Milestii Mici. Its impressive collection of 1.5 million bottles has scored a mention in the Guinness World Records as the world's largest wine collection.

Despite being one of the poorest countries on the European continent, Moldova also has some of the most hard working and welcoming people, deeply moral and spiritual by nature. But Moldovans are bearing the brunt of a corrupt government, stagnant economy and Russia's never-ending meddling. The small nation of 3.5 million people keeps losing its population year by year as many hopefuls leave their families behind in search for a better life. Many of them abroad still hope for a chance to go back home, but they are highly aware they shouldn't hold their breath yet.

Moldovan native Tatiana Darie, a master's student in broadcast journalism at the University of Missouri-Columbia, is a freelance journalist with an emphasis on business and international news. She is pictured here above Trebujeni village, Moldova/Orheiul Vechi.

extensively). Together with Ukraine, Azerbaijan, and Moldova, Georgia has formed the regional grouping known as **GUAM** (an acronym for its four members) to promote economic integration, democratic reform, and increasing orientation toward Europe—and away from Russia (see Figure 5.25). For many years before 2012, when Russia became a member state, Georgia was able to block Russian accession to the World Trade Organization (WTO), the organization that sets the ground rules for global trade.

Russia has retaliated for Georgia's increasing orientation toward the West. In 2001 and again in 2006, when Georgia insisted that it was time for Russia to scale back its military presence in the region, Russia responded by cutting off supplies of natural gas to Georgia. Russia imposed punitive tariffs against imports of Georgian wine and other products, and threatened to send home a million Georgian guest workers whose remittances are vital to the Georgian economy. Most road and rail links between Russia and Georgia were severed, and Turkey replaced Russia as Georgia's main trading partner. Georgia also wants to reestablish its historically important position in overland trade with the central Asian countries, particularly by serving as an outlet for the export of oil from the Caspian Basin.

For many years, a critical issue in Georgia was the fate of Adjara, a province along the Black Sea that contains the city of Batumi, a vitally important oil-shipping port (•**Figure 5.30**). Adjara's Russian-backed leader Aslan Abashidze, bolstered by the presence of a Russian military base, came to power in 1991 and ruled the region almost completely independently from the national government in Tbilisi. In 2003, reformist Georgian President Mikhail Saakashvili reasserted the national government's control over all autonomous and separatist regions. Abashidze resisted Saakashvili until increasing anti-Russian protests in Batumi prompted him to abdicate and flee to Russia in 2004. Adjara reintegrated into Georgia peacefully and is now an autonomous region. The Russian military base in Adjara closed in 2007.

Although Russia wound up being unable to control Adjara, Russia supports the aspirations of the mostly Muslim South Ossetians, who would like to free themselves from Georgian control and establish an Ossetian nation (almost 100 percent of the population voted to do so in a 2006 referendum). South Ossetia declared itself independent in 1990, which led to a Georgian military operation to regain control of the area until a cease-fire was brokered by Moscow in 1992. Some South Ossetians wish to join their territory with North Ossetia, a mostly Christian region adjacent to them in southern Russia. (Incidentally, North Ossetia was the setting of the horrific deaths of nearly 200 children in Beslan in 2004, after Islamists from nearby Ingushetia seized hostages at a school.)

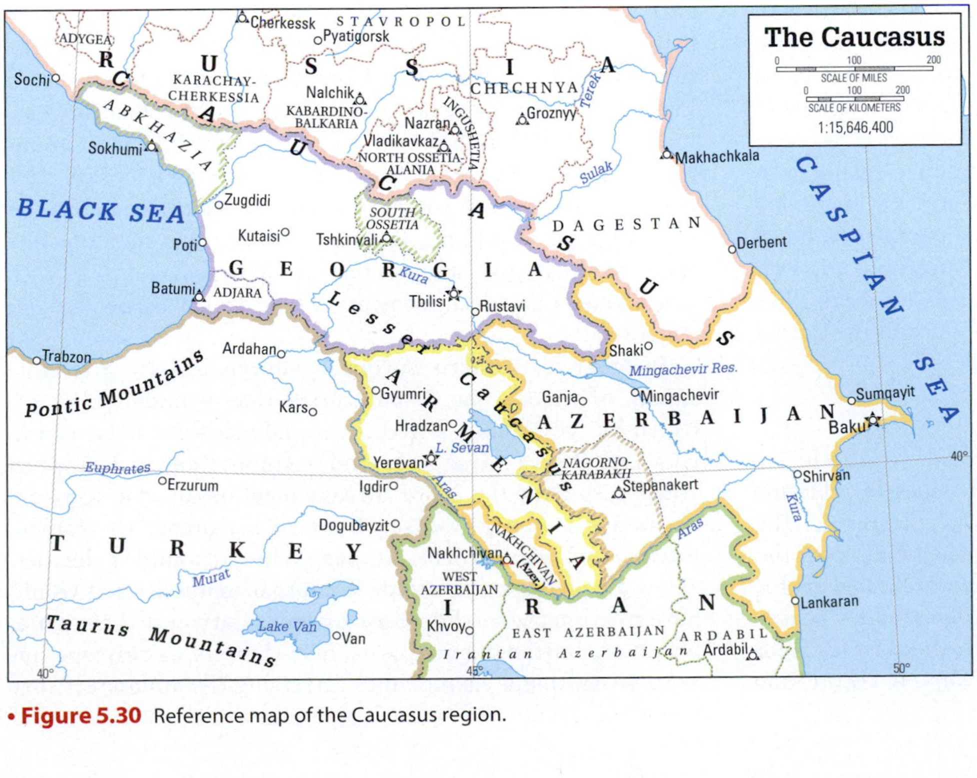

Also seeking independence from Georgia with Russian help are the Abkhazians, who make up about 5 percent of Georgia's population. In 1993, separatists captured the Georgian Black Sea port of Sokhumi in Abkhazia. Russian forces initially aided the Abkhazians, both to regain access to Black Sea resorts and to take revenge on Georgia's President Eduard Shevardnadze, a former Soviet foreign minister whom many Russians held partly responsible for the breakup of the Soviet Union. Russia was also pressuring Georgia to join the Commonwealth of Independent States, which it had initially refused to do. Shevardnadze responded by having Georgia join the CIS, agreeing to allow four Russian military bases to remain on Georgian soil and having Russian troops stationed on Georgia's border with Turkey. Russian forces then put a stop to the Abkhazian offensive but did not drive the rebels from the territory they had captured. Russia began granting Russian citizenship to the majority of Abkhazians—a step that Shevardnadze's successor, Mikhail Saakashvili, regarded as Russian annexation of Georgian territory.

For Russia, a friendly Abkhazia is a geopolitical prize, as it affords Russia an even larger area of warm waterfront on the Black Sea, from where Russia can project military and commercial power. Abkhazia is also very close to the Russian city of Sochi, the venue of the 2014 Winter Olympics. A friendly Abkhazia helped secure a wider buffer around the Sochi Olympics site and continues to assist Russia in safeguarding the vulnerable gas and oil pipelines crossing the area.

In 2008, Russia jumped on another opportunity to widen and tighten its grip in the Caucasus at Georgia's expense. Determined to assert control over South Ossetia, in August 2008 Georgian president Saakashvili sent Georgian troops against South Ossetian rebels. Directed by Russian leaders Putin and Medvedev, Russian forces invaded and quickly secured control over both South Ossetia and Abkhazia. While Georgia scrambled to secure its fractured borders, Russia recognized both regions as newly independent countries (besides Russia, only Venezuela and Nicaragua recognize the independence of these areas). This conflict was very much an old-style Cold War engagement of proxy forces: The United States and the European Union backed the Georgian side, whereas Russia propped up South Ossetia and Abkhazia.

Six years later, Russia hoped that the precedent it established in Georgia, supporting rebels and driving wedges into geographical and political integrity, would serve its interests in Ukraine. These tactics did give Russia an early advantage in Crimea and eastern Ukraine. In Cold War proxy war style, Russia has been conducting military exercises in Crimea, Abkhazia, and South

• **Figure 5.30** Reference map of the Caucasus region.

Ossetia, and the United States has been doing so in Georgia. The US military exercises in Georgia are part of a much larger "Operation Atlantic Resolve" theatre that includes Romania, Bulgaria, Hungary, the Czech Republic, Poland, Estonia, Latvia, and Lithuania. Put yourself in Russian shoes for a moment and look at a map of Russia and these countries (see Figure 5.33). You will see the Russian narrative about America's intentions to encircle, intimidate, and belittle the great country Russia. With a captive media, President Putin can fight back without much fighting, stirring the pot of Russian nationalism and enjoying sky-high approval ratings even as his economy crumbles.

One more frozen conflict needs our attention: that of Nagorno-Karabakh, a predominantly ethnic Armenian region that had been placed within the boundaries of neighboring Azerbaijan during Soviet times. In the late 1980s, ethnic Armenians in Nagorno-Karabakh began agitating for their region to be separated from Azeri territory and attached to Armenia. Azerbaijan refused those demands and stripped Nagorno-Karabakh of its autonomy. A referendum held within the region in 1991 resulted in a declaration of independence, which Azerbaijan also rejected. War broke out between Azerbaijan and Nagorno-Karabakh, with aid for the separatists coming from Armenia. Warfare in Nagorno-Karabakh took 30,000 lives and created nearly a million refugees until Russia negotiated a cease-fire in 1994. Virtually all ethnic Azeris living in Armenia (approximately 200,000) fled during the war, many winding up in refugee camps in Azerbaijan for years. Similarly, nearly all ethnic Armenians living within Azerbaijan proper also fled.

Like the situation with Transnistria in Moldova, Nagorno-Karabakh is still internationally recognized as part of Azerbaijan, but the Baku government exercises no control over the area. Since the end of the war, Nagorno-Karabakh has acted as an independent nation, albeit one that is unrecognized by any other country (including Armenia). Periodic violations of the cease-fire agreement have raised tensions in both Armenia and Azerbaijan, which have engaged in talks for years in an attempt to reach a final settlement but have yet to officially make peace with each other; indeed, both countries engaged in another round of rhetoric and increased military spending in 2014. Armenia has always enjoyed good relations with Russia and hosts several Russian military bases on its territory, but it is relatively poor and has few other allies. Azerbaijan is well liked by its other neighbors, and its increasing oil wealth has made it important to foreign powers like China and the United States (see Geography of Energy, **page 189**).

The Far Abroad

Finally, there are outstanding geopolitical issues in relations between this region and the "Far Abroad," the rest of the world. In the years immediately following the breakup of the USSR, the most crucial issues involved the development of a new model of relations between East and West and a peaceful succession to the Cold War. After 9/11, these issues were complemented by others related to the "War on Terror," as perceived by both the United States and Russia. Another phase began after 2004, with a popular and confident Vladimir Putin, buoyed especially by Russia's soaring energy revenues, putting more distance between Russia and the West and constantly reminding Russians of their greatness.

It is important to keep in mind—and Russians especially want the world to know—that Russia *is* one of the world's great powers. Russia was, between 1997 and 2014, a member of the **Group of Eight (G-8)**, the world's eight most economically powerful and politically influential countries (along with Canada, France, Germany, Italy, Japan, the United Kingdom, and the United States). Russia is a member of the UN Security Council and, along with its other permanent members (China, France, the United Kingdom, and the United States), has the enormous veto power that position affords. Therefore, Russia can (and does) have a large impact on world affairs.

Energy Issues

For years, countries of the Far Abroad have been asking serious questions about how reliable a business partner Russia is and how far it might go in projecting its power in the countries of the former Soviet Union. Now they have some answers. Russia had been advertising itself to the West as a more stable source of energy than the volatile Persian Gulf countries. But that claim rang hollow when Russia began manipulating fuel supplies and prices in its Near Abroad, with downstream effects on supplies to Western Europe. 111

Western Europe frets because the Russian gas and oil it depends on—one-quarter of its natural gas and one-quarter of its oil—flow through a cat's cradle of politically vulnerable pipelines. Remember how Russia cut off its oil to Belarus and its natural gas to Ukraine? The pipeline carrying Russian oil to Belarus 181 continues on to Germany, so Western Europe's oil supplies were also temporarily disrupted. The pipelines carrying Russian natural gas to Ukraine likewise continue on to Austria, so the suspension of gas delivery to Ukraine also meant that western European markets were not getting their supplies of natural gas.

Even before Ukraine was invaded, western European countries had begun to regard Russia as an unreliable energy supplier. These countries sought ways to reduce their reliance on Russian gas and oil and, in some cases, to find new pipeline routes for Russian oil and gas that would bypass countries with which Russia had disagreements. These routes are depicted in **•Figure 5.31**.

In order for western European nations to have a reliable supply of natural gas from Russia that would not be affected by the Kremlin's frequent squabbles with Belarus and Ukraine, Russia's Gazprom and a consortium of European companies built the Nord Stream pipeline at enormous expense across the Baltic sea floor to Germany, the largest foreign consumer of Russian gas. Why not build a cheaper, easier pipeline above ground? Because of geopolitics: it would have to cross the territories of Estonia, Latvia, or Lithuania, which are unfriendly to Russia. The share of gas Europe imports from Russia via pipelines traversing Ukrainian territory

Geography of Energy Oil in the Caspian Basin

The Caspian Sea and the land around it, making up the Caspian Basin, have some of the world's largest reserves of oil and natural gas. As you might imagine, each one of the five countries surrounding the Caspian wants to maximize its share of the fossil fuels under the seabed. Although the Caspian "Sea" is actually a lake in strictly physical geographic terms, the countries on its shores argue variously that it is a *sea* or a *lake* in political terms. The disagreement stems from the fact that legal definitions of *sea* and *lake* determine what share of the fossil fuels each country has the right to exploit. Lakebed resources are common property, so shares of natural resources would be divided equally among the five surrounding countries. There is little oil or gas in Iran's and Turkmenistan's sections of the Caspian, so those countries argue that the Caspian should be regarded as a lake. Conversely, Russia, Kazakhstan, and Azerbaijan have signed treaties recognizing the Caspian as a sea. This designation establishes separate sovereign territorial waters, with each country having exclusive access to the reserves 317 in its domain (see **•Figure 5.L**). These three countries are actively exploiting their fossil fuel reserves within their designated portions of the seabed.

Let's look at the geopolitics involved in getting the biggest share of Caspian oil to market: Azerbaijan's. Azerbaijan is landlocked, and its oil must cross the territories of other nations by pipeline to reach the Mediterranean Sea. That looks easy enough on the map, but any export route must cross territories embroiled in post-USSR political unrest.

Russia's position is clear: Azerbaijan's oil should be pumped by pipeline through Russian territory to Russia's Black Sea port of Novorossiysk. In keeping with Moscow's principle that the Near Abroad is Russia's "zone of privileged interest," this route (which can be seen in **•Figure 5.M**) would maximize Russia's influence over Azerbaijan's economic affairs. Russia would also earn revenue from transit fees it would charge to allow oil to pass through Russian territory.

Azerbaijan's leaders, however, want to build political and economic ties with the West (Europe and the United States). Western nations in turn want to be able to obtain Caspian Sea oil without relying on Russia's goodwill. Azerbaijan therefore decided to export its oil through routes that do not transit Russia at all. One route is the Baku-Supsa pipeline, which began operations in the 1990s and terminates on the Black Sea coast of Georgia. Tankers carry the oil from there to European ports. This pipeline comes very close to the separatist area of South Ossetia, 187 and although the Baku-Supsa pipeline remains in use today, political, security, and ecological concerns led to the construction of a new route in 2006: the vital Baku-Tbilisi-Ceyhan (BTC) pipeline.

The BTC pipeline, which has seven times the capacity of the Baku-Supsa pipeline, is strategically routed to block either Russian or Iranian intervention. Russia and Iran are, of course, displeased with this route. The BTC runs across Azerbaijan from Baku through Georgia (avoiding South Ossetia) and into Turkey. Note how it skirts Armenia, which is technically still at war with Azerbaijan over Nagorno-Karabakh (see **page 188**). It continues across Turkey to the Mediterranean port of Ceyhan. Why does the BTC take a sudden 90° turn in central Turkey? As you will see in Chapter 6 (on **page 269**), Turkey has a domestic security problem with its large Kurdish population, located mainly in southeastern Turkey. The BTC route detours around this volatile region.

Azerbaijan's oil exports through these pipelines have brought considerable new wealth to the country. Its capital of Baku on the Caspian Sea coast was once a bleak, windblown place with monolithic blocks of Soviet-era apartments. Now it glistens, Dubai style, with ostentatious urban wealth.

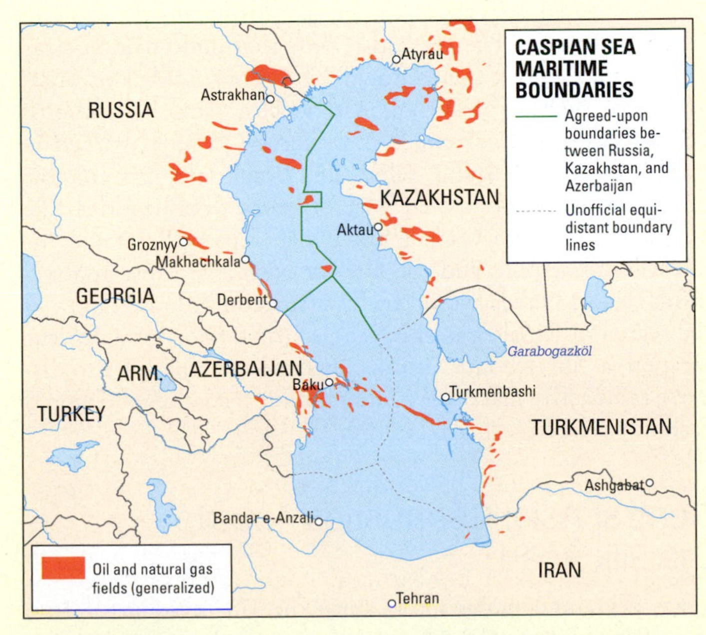

•Figure 5.L Russia, Azerbaijan, and Kazakhstan have agreed on how to divide Caspian Sea oil and gas between them. Iran and Turkmenistan want each country to have an equal share.

•Figure 5.M Oil pipeline routes from Baku.

• Figure 5.31 Is Russia a reliable provider of fossil fuels to Europe, or is it too ready to turn off the tap for political reasons?

THINK CRITICALLY: How is the geography of Russia's energy pipelines linked with Russia's geopolitical interests and with those of the countries the pipelines cross and serve? See also **Figure 5.37**.

dropped from 80 to 60 percent after the Nord Stream's completion in 2012.

For many years, Russia trumpeted a South Stream pipeline project as a companion to the successful Nord Stream; in this case, the pipeline would be built from Russia across the floor of the Black Sea to a port in Bulgaria. The route would bend noticeably southward to avoid Ukrainian territorial waters. After Russia invaded Ukraine, the EU and US put sanctions on Russia that affected the state-run Gazprom's finances. Without the necessary money to invest in the project, combined with the worsening of relations between Russia and the west in the aftermath of the invasion, the South Stream pipeline, in the early stages of construction, was canceled in 2014.

In Turkey, construction is underway on the Trans-Anatolian pipeline, which will carry natural gas from Azerbaijan. A proposal for a branch called the Trans-Adriatic pipeline would bring a reliable source of non-Russian natural gas to Europe. If completed, this route would run through Greece and Albania and terminate in southeastern Italy, where it would connect with local pipelines to bring gas to the rest of Europe.

These projects ultimately do nothing to reduce Western Europe's reliance on foreign sources of energy, especially Russian sources that the EU may sanction further. The European countries are now looking under their own feet to see whether they can find energy independence. One possibility getting much attention these days is the production of shale gas from Europe's old coal fields—the same coal fields that fueled the Industrial Revolution but were later abandoned. 124

Weapons Proliferation Issues

In addition to energy concerns, the West worries about other problems related to an adversarial or unstable Russia, such as Russia's assistance to the nuclear and would-be nuclear weapons powers North Korea and Iran; its support of the Islamist Hamas and Hezbollah organizations; and its military intervention to prop up the Assad regime in Syria.

As important as these issues are, they pale in comparison to the Cold War specter of a nuclear war of annihilation between East and West. That scenario retreated markedly after the Soviet Union collapsed. The Warsaw Pact, a military alliance formed by the Soviet Union to confront the West's North Atlantic Treaty Organization (NATO), was dissolved in 1991. 119 NATO's mission after that was not centered on a confrontation with Moscow. Both sides reduced their nuclear arsenals, with Russia taking the lead in calling for even deeper cuts. US-Russian relations warmed. In 2001, President George W. Bush described Vladimir Putin this way: "I looked the man in the eye. I found him to be very straightforward and trustworthy and we had a very good dialogue. I was able to get a sense of his soul."[28] After President Obama took office, he and his Russian counterpart agreed to "reset" US-Russian relations from a friendly new starting point. Even after Russia's intervention in Crimea and Ukraine, which poisoned the relationship, the United States has little appetite for a direct military confrontation with Russia.

Although no longer preoccupied with the prospect of a conventional nuclear war, the United States and its NATO allies are fearful that low budgets and lax security at nuclear facilities throughout Russia and the Near Abroad could help divert nuclear weapons or nuclear fuel to terrorists. There have been hundreds of thefts of radioactive substances at nuclear and industrial institutions in the former USSR. The destinations of these smuggled materials, known as **"loose nukes,"** remain largely unknown. The West also fears that underpaid nuclear scientists are selling their know-how to governments or organizations, such as al-Qa'ida, with nuclear strike intentions. Former Soviet nuclear experts are known to have worked in Iran, Iraq, Algeria, India, Libya, and Brazil, and it is possible that some have cooperated with stateless organizations like al-Qa'ida. The late Osama bin Laden repeatedly stated his intention to acquire and use nuclear weapons. Most analysts think that al-Qa'ida would try first to acquire loose nukes in the region of Russia and the Near Abroad. Short of a nuclear weapon, a "dirty bomb" using conventional explosives to disperse radioactive materials could produce casualties and would certainly terrorize. 246

Central Asia: Geopolitics Astride the Silk Road

Since 9/11, and particularly since the US invasion of Iraq in 2003, the geopolitical significance of the central Asian "stan" countries has increased enormously (see **•Figure 5.32**). This region (which like Ukraine is dispossessed of the nuclear weapons stationed there in Soviet times) borders unsettled

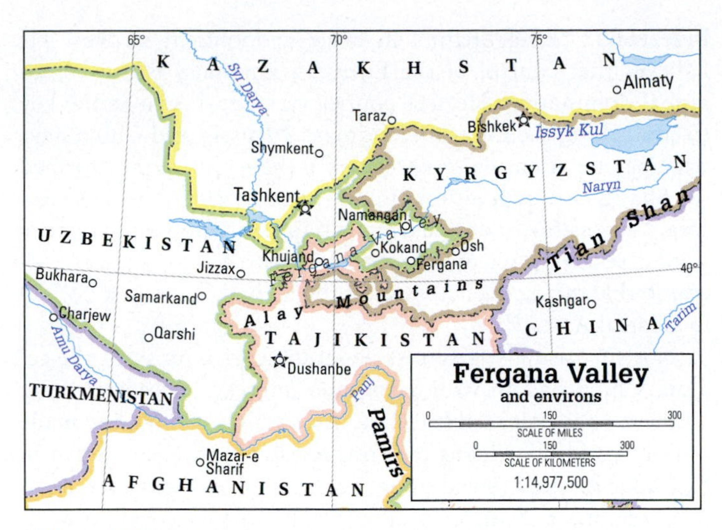

• **Figure 5.32** Reference map of the strategically important Fergana Valley region of central Asia.

Afghanistan, the potential flashpoint of Iran, and the great power of China. Central Asia is itself rich in mineral resources. Diverse kinship, spiritual, economic, and political relations exist between the central Asian Muslims and their neighbors on the other side of the international frontiers. Along with internal ethnic conflicts and dissatisfaction over living conditions, these international affiliations may make the region a major political problem area in the future. The popular uprisings of the Arab Spring did not immediately spread into non-Arab central Asia, but some of the same ingredients for the Arab Spring, such as autocratic regimes controlling large numbers of disaffected youth, are also present in central Asia.

Both Iran and Turkey are vying for increased influence in central Asia. Except in Tajikistan, support for Iran is largely lacking, in part because the majority of central Asians practice Sunni rather than Shi'ite Islam (these branches of Islam are described on **page 231**). Ethnicity is also important; the four Turkic states are inclined to orient with Turkey, with which they have already established cultural ties such as educational exchanges and shared media. Turkey has extended economic assistance and established small business ventures and large construction contracts in central Asia. Turkmenistan has good relations with both Turkey and Iran, and has built a rail link with Iran. Turkmenistan also tries to maintain good ties with Russia.

The region has become a focus for modern rivalries between the historical enemies Turkey and Russia. Russians are concerned that Turkey is winning too much influence in the young Muslim countries of central Asia. For their part, some Turks believe that, in reassembling pieces of the Soviet empire, Russia will try to retake the region the Soviets called the "Transcaucasus"—now the countries of Georgia, Armenia, and Azerbaijan—and then grab culturally Turkic central Asia. Some Turks uphold their own dream of **pan-Turkism**, uniting all the Turkic peoples of Asia from Istanbul to the Sakha Republic in Russia's Siberia region—a prospect that Moscow, of course, does not like.

The twin scourges of terrorism and narcotics—often closely linked—have put central Asia on the world map of geopolitical hot spots. Uzbekistan has erected concrete and barbed-wire barriers and sown land mines on its borders with Tajikistan and Kyrgyzstan. The Uzbek government is trying to prevent the incursion of what it sees as two huge threats: an epidemic in heroin use and a militant Islamist insurgency. Both issues have preoccupied the region's governments. The major routes for exports of heroin from Pakistan and Afghanistan are north through central Asia (primarily Tajikistan) and then into Russia and Western Europe. Along this pathway, large numbers of 156 central Asians have fallen prey to inexpensive, pure heroin. Its associated health and social problems, including HIV infection, hepatitis, and prostitution, have also taken root. Along its trade routes, heroin corrupts people at every level: it is likely that many high-ranking government officials in Tajikistan are at least indirectly associated with drug trafficking. The United Nations estimates that about 40 percent of Tajikistan's economy is drug related.

Relations between Russia and the United States improved in the aftermath of 9/11, when both powers grew alarmed about an Islamist threat to central Asia. Russia expressed great concern about a violent Islamist wave spreading from Afghanistan and Pakistan into central Asia and Russia itself. Tajikistan and Kyrgyzstan allowed Russia to deploy thousands of Russian troops in their countries, in part to stave off such a threat. Some observers, however, believe that Russia is using the Islamist threat to counter the growing US military presence and strategic footprint there since 9/11, or as a pretense to maintain its own influence in the region, or even to help build Putin's Eurasian Union by force. And human rights groups in the West 179 argue that authoritarian, oppressive regimes in central Asia exaggerate threats from Islamist militants to justify their harsh treatment of political opponents.

The United States is genuinely worried about the Islamist danger in central Asia. Its greatest fears are focused on Tajikistan, central Asia's poorest country. Islamist rebels with links to both al-Qa'ida and Afghanistan's Taliban regularly targeted US military supply lines from Uzbekistan into Afghanistan and also attacked Tajikistan's soldiers. There are fears about the reemergence of the clan-based civil war that plagued Tajikistan from 1992 to 1997.

Civil war also threatens Kyrgyzstan, where the minority ethnic Uzbeks (who make up 15 percent of the country's population) are seeking greater autonomy. They frequently engaged in bloody clashes with ethnic Kyrgyz. Uzbekistan is fearful that the dominant Kyrgyz people will ethnically cleanse the country's Uzbeks. Ethnic Kyrgyz for their part fear that Uzbekistan may take advantage of instability to occupy land in Kyrgyzstan. Russian authorities express concern that Kyrgyzstan could become "a second Afghanistan."

Tajikistan was home to a group called the **Islamic Movement of Uzbekistan (IMU)**, which has recently been pressured southward into the Afghanistan-Pakistan border region. For many years, its principal aims were to overthrow the Uzbek government of President Islam Karimov and establish an Islamic state in the heavily populated Fergana Valley, which runs through Uzbekistan, Tajikistan, and southern Kyrgyzstan (see Figure 5.32). Following a 1999 attempted assassination blamed on the rebels, Karimov cracked down hard on suspected Islamists and even began to restrict personal symbols of being Muslim, such as beards on men and headscarves on women. Tajikistan's government is also employing these repressive tactics, which many observers believe will only help grow violent Islamist and anti-government sentiments.

Anxious to repress any external support for Islamist insurgency within its borders and to curry political and economic favor with Washington, Uzbekistan initially welcomed the US military to use its land and air space in attacks against al-Qa'ida and the Taliban in 2001 and 2002. Uzbekistan shut down this access in 2005, after the United States and its European partners in NATO, citing human rights abuses, slashed economic assistance to the country. Soon, however, the US's escalating engagements in Afghanistan led Washington to overlook poor human rights records in both Uzbekistan and Kyrgyzstan. American and NATO military supplies began pouring by air and rail into Afghanistan from both countries, and Uzbekistan built a 45-mile-long rail line into the Afghan city of Mazar-e-Sharif, mainly for the transport of these military goods. The US airbase at Manas in Kyrgyzstan served as a critical airbase in the West's military campaign in Afghanistan. The Russian military also operates bases in both Kyrgyzstan and Tajikistan. In 2009, Kyrgyz president Bakiyev double-crossed Russia, promising to shut down US access to the Manas airbase in return for a large economic aid package from Russia. Once Bakiyev secured the Russian aid, he collected another large sum from the United States to renew its lease of Manas. Bakiyev was ousted in a coup the following year.

As part of the American agenda to wrap up the Afghan war, the US shut down Manas in 2014 when the Kyrgyz president refused to extend its lease. Russia, however, forgave the country's debts owed to it and in the process renewed its military presence in Kyrgyzstan. Russia has the stronger hand in its "great game" against America and the West in this country. Not long ago Kyrgyzstan was on a road to "failed state" status, but democratic reforms and economic assets, including some fossil fuels and the potential for adventure tourism, favor this country. The breathtaking Tian Shan mountains, towering over much of the country, are a major draw (see Figure 5.13).

As we consider the Far Abroad we come to that country whose influence is on the rise in every region covered in this book: China. China has sought to build security relationships with the central Asian countries, especially to blunt the aspirations of Muslim Uighur insurgents in China's far west. In a proclaimed effort to confront terrorism, Islamic extremism, and separatism, China and Russia formed a confederacy with all of the central Asian countries (except Turkmenistan) in 2001. Known as the **Shanghai Cooperation Organisation (SCO) or "Shanghai Pact,"** this confederation has the unstated aim of pushing away any aspirations NATO may have in central Asia. Iranian journalist Hamid Golpira tapped into the geopolitical thinking of former US National Security Advisor Zbigniew Brzezinski. "Zbig" paints in large geopolitical strokes. He believes that control of the Eurasian landmass is the key to global domination. He sees control of central Asia as the key to controlling the Eurasian landmass. "Russia and China have been paying attention to Brzezinski's theory since they formed the Shanghai Cooperation Organisation in 2001," writes Golpira, "ostensibly to curb extremism in the region and enhance border security, but most probably with the real objective of counterbalancing the activities of the United States and NATO in Central Asia."[29]

348

Not surprisingly, given its geographic proximity to the region, China has growing economic interests in central Asia. Chinese exports to central Asia are surging. China has made multibillion dollar loans to Turkmenistan and Kazakhstan in exchange for guaranteed future access to oil and gas supplies in these countries. Natural gas flows from Turkmenistan to western China through a 1140-mile-long pipeline crossing three countries. This pipeline is also notable as the first to run east out of the region, as China rapidly replaces Russia as the main buyer of Turkmenistan's natural gas. China is also upgrading its transportation links, both road and rail, with central Asian countries. In 2015, a freight train made the longest rail journey in the world, 8000 miles from the eastern Chinese city of Yiwu all the way to Madrid—and back, effectively opening a new Silk Road made of steel. If all roads once led to Rome, do they now to lead to China? And are they actually rails rather than roads? Consider this geopolitical perspective offered by *Quartz*'s Parag Khanna:

> The vast oceans that separate North America from the western and eastern halves of Eurasia will continue to have a major impact on the evolution of geopolitics. The pace of globalization has altered our perceptions of space and time: Communications technology inspires many to proclaim the "death of distance." Yet a contrary narrative is also emerging, one in which America's distance from Eurasia places it on the wrong side of the world from the "cockpit of history," a rapidly integrating Eurasian super-continent that is shaping its own future independently of the Western Hemisphere and the U.S. And the technology that is driving this epochal transformation is one of the most traditional: railways.[30]

434

Following are more insights into land and life in the subregions and countries that make up Russia and the Near Abroad.

5.6 Regional Issues and Landscapes

Peoples and Resources of the Coreland

You first read about the geographic concepts of core and periphery in the book's Chapter 1. Europe has distinctive traits of core and periphery, and so do Russia and the Near Abroad (see •**Figure 5.33**).

An outstanding feature of this huge region's geography is that resources are distributed unevenly, favoring development in certain subregions and countries. Agricultural and industrial resources are in fact clustered in a core region encompassing five countries partially or entirely: western Russia, northern Kazakhstan, Ukraine, Belarus, and Moldova. Nearly three-fourths of the region's people—the great majority of them Slavic—and an even larger share of the cities, industries, and cultivated land of this immense region are packed into this roughly triangular **Fertile Triangle** comprising about one-fifth of the region's total area. Also known as the **Agricultural Triangle** and the **Slavic Core,** this is the distinctive functional hub of the region (Figure 5.9, **page 160**).

Slavic peoples are the dominant ethnic groups in most of the region of Russia and the Near Abroad in numbers and in political and economic power. The major groups are the Russians (discussed extensively earlier in the chapter; the Slavs and other ethnolinguistic groups are introduced on **page 164**), Ukrainians, and Belarusians. Most of these Slavs live in the core region extending from the Black and Baltic Seas to the neighborhood of Novosibirsk in Siberia. Moscow, with a population of about 16 million, and Saint Petersburg (called Petrograd and then Leningrad during Soviet times), with 5 million people, are the largest cities, followed by 2 million people living in the Ukrainian capital, Kiev.

Notes on Ukraine

By almost every measure, Ukraine is the second-most prominent country in this world region, after Russia. Much of the discussion of Ukraine that you would expect to find here is in the earlier section about Russia's moves on Ukraine, beginning on **page 180**.

• **Figure 5.33** Political geography of the coreland of Russia and the Near Abroad.

As a reminder, Ukraine's capital of Kiev is on the banks of the Dnieper River on the border between the forest and steppe biomes. Kiev is a major industrial center, and is one of Russia and the Near Abroad's most important industrial regions. Most of the country's industrial resources—coal and iron ore especially—are in the east, and most of the factories that convert those raw materials into manufactured goods are in the west. The conflict that began in 2014, pitting east against west, is economically disastrous for Ukraine. Agricultural and industrial outputs plummeted after Ukraine gained independence from the Soviet Union in 1991, but had been picking up again before the fighting began. A "breadbasket," Ukraine is also a "basket case" in the wake of Russia's invasion. It would be difficult to make the case that Putin's Russia is profiting handsomely by taking Crimea and parts of eastern Ukraine. Instead, the intervention represents another economic burden for a Russia already saddled with economic sanctions and the impacts of low oil prices.

Ukraine paid a high economic cost for independence. Without large fossil fuel reserves to tap, Ukraine's economy has remained far more reliant on heavy industrial production than has Russia's. Previously the Soviet economy subsidized the provision of Russian oil, gasoline, natural gas, and uranium to Ukraine. As discussed earlier (**page 181**), Ukraine is now subject to the whims of Russia where oil prices and supplies are concerned. Ukraine's chemical plans and steel mills are guzzlers of natural gas, and Ukraine cannot afford to run out. Like the western European countries described earlier (**page 190**), Ukraine is looking to extract natural gas in its own territory through fracking of its shale deposits.

Many of the social problems that plagued Russia after the breakup of the Soviet Union also troubled Ukraine: increasing mortality, falling birth rates, and broken families are among them. In Ukraine, Moldova, Belarus, and Russia, increasing numbers of women fell victim to sex trafficking as they struggled to improve their lives (see Geography of Human Trafficking on **page 195**).

Chernobyl, the Type-Site of Nuclear Disaster

Sixty miles north of Kiev is **Chernobyl**, where an explosion at a nuclear power station in 1986 rendered parts of northern Ukraine and adjoining Belarus and Russia incapable of safe agricultural production for decades, possibly centuries, thereafter. The stricken reactor has been entombed by a gigantic shell designed to prevent any future leakage of radioactive material (**•Figure 5.34**). Encircling the station is an 18-mile (29-km) "exclusion zone" where farming is prohibited. Recent studies suggest that 100,000 to 200,000 people are still severely affected by the Chernobyl disaster and that 4000 deaths will ultimately be attributed to it. The toll on human health and life is well short of what had originally been predicted, but the psychological trauma to the region's inhabitants has been profound, and the three affected countries have spent large sums to provide benefits to Chernobyl's victims. Until recently, the Chernobyl disaster was a singular event. Now it is joined by Japan's Fukushima catastrophe in the world's experiences of nuclear power gone wrong.

• Figure 5.34 The sarcophagus under construction to contain Chernobyl's dangerous reactor 4.

Farming the Fertile Triangle

Most of the Fertile Triangle is within Russia, which still faces huge problems in transforming state-run into free-market farming. Throughout the second half of the 20th century, thanks to good soils, the Fertile Triangle was a global-scale producer of farm commodities such as wheat, barley, oats, rye, potatoes (Russia's leading staple food), sugar beets, flax, sunflower seeds, cotton, milk, butter, and mutton (see Figure 5.9). The overall high output, however, did not reflect high agricultural productivity per unit of land and labor. There were many inefficiencies in the state-operated and collective farms that dominated the agricultural sector during Soviet times. On the state-operated farms, workers received cash wages in the same manner as industrial workers. Bonuses for extra performance were paid. Workers on collective farms received shares of the income after obligations of the collective had been met. As on the state farms, there were bonuses for superior output.

171, 190

These methods, however, failed to provide enough incentives for highly productive agriculture. The patterns of agricultural failure parallel those in the industrial sector in the post-Soviet era. Farm machinery stood idle because of improper maintenance; because no individual owned the machine, the incentive to repair it was diminished. There were also shortages of spare parts. Poor storage, transportation, and distribution facilities, plus wholesale pilfering, caused alarming losses after the harvest. Young people deserted the farms for urban work, often living with the family in the countryside but commuting to their city jobs. With the shortage of younger male workers, the elderly, women, and children did an increasingly large share of the farm work.

In an effort to reverse declines in the agricultural sector, all of the countries of the Fertile Triangle are promoting land reform by privatizing collective and state farms and developing more independent farming. Privately owned farms generate over half of Russia's agricultural production, up from one-quarter in the early 1990s (**•Figure 5.35**). Russia has made revival of the agricultural sector one of its "national projects," but like almost all the countries in this region, Russia remains a net food importer.

Geography of Human Trafficking The Natashas

One of the outcomes of the collapse of communism and the dissolution of the Soviet Union has been the unleashing of long-restrained expressions of sexuality. Entrepreneurs in Russia and other former Soviet countries quickly adopted an axiom of Western advertising: sex sells. Russian television has surpassed even the spiciest fare on standard American television.

Many sexual awakenings in the region are applauded or accepted as good clean fun, but others have a decidedly dark side, and some send ripples far abroad. Far too many women of the former Soviet Union have become commodities (•**Figure 5.N**). Russian, Belarusian, Ukrainian, and Kazakh partners or brides can in effect be purchased by anyone in the world with access to the Internet. Dot-com agencies act as brokers for these New Age "mail-order brides." Some of the subsequent unions end happily, of course, but others collapse with physical abuse and heartache. Frequently, the woman ends up 359 abandoned and destitute in a strange land.

The sex trade is predatory and dangerous. For the purposes of sex, women (mainly) become victims of what is known as **human trafficking**, defined by the US State Department as "the act of recruiting, harboring, transporting, providing, or obtaining a person for compelled labor or commercial sex acts through the use of force, fraud, or coercion."[31] The State Department's annual Trafficking in Persons report is respected globally as one of the best resources on human trafficking (see the "TIP" reports at http://www.state.gov/j/tip/).

Often unwittingly and almost always out of poverty, women fall prey in a classic pattern, played out over and over, not only in this region but also around the world. Criminal gangs use false promises of conventional employment to lure young women into the trade. The typical victim is an 18- to 30-year-old Russian, Ukrainian, Belarusian, or Moldovan woman—often with children to support—who is approached by a trafficker promising her a job as a waitress, barmaid, babysitter, or maid in Europe. She pays a fee for travel and visa costs. Arriving in Bosnia and Herzegovina, the Czech Republic, or Germany, she is met by another trafficker who confiscates her passport and identity papers and then sells her for several thousand dollars to a brothel owner. She is escorted to her guarded apartment. There she is compelled to reside with several other women who have taken a similar journey, and she is informed of her real job: prostitute.

Berlin papers advertise the merchandise: "The Best from Moscow!" "New! Ukrainian Pearls!" The working women meet their clients in clubs or brothels or are driven to their homes. Of the rates charged for their services, typically less than 10 percent goes to the women. Their pimps and other gang members take the lion's share, and the women use their meager earnings to buy food and pay rent. Most end up in debt. Unable to pay off their debts, lacking identification papers, fearful of going to the police, and terrified the gang might harm family members back home if they try to escape, the women are trapped. Reports of depression, isolation, sexually transmitted disease, beatings, and rape abound.

This is a big business with striking demographics. The sex industry in Europe generates revenues of several billion dollars per year. Ukrainian sources estimate that since the country's independence in 1991, more than 100,000 Ukrainian emigrant women have ended up in the European sex trade. There is an overflow into other regions as well. These so-called "Natashas" work in Turkey, Israel, the United Arab Emirates, and other countries in the Middle East in similar deplorable circumstances. Many of the world's countries, including the United States, have human trafficking problems. The focus in this essay is on sex, but people of both genders, including children, are trafficked for a variety of reasons, including labor.

Joe Hobbs

• **Figure 5.N** Window shopping in Amsterdam's Red Light District. An estimated 85 percent of the working women here are trafficked, mainly from the former Soviet countries.

Joe Hobbs

• **Figure 5.35** Fluorescent lighting and artificial climate controls allow year-round harvests in Russian greenhouses.

Russia's Eastern and Northern Lands

Lake Baikal, the Pearl of Siberia

Lake Baikal, the largest freshwater lake in Asia, lies in a mountain-rimmed rift valley. It is the world's deepest body of freshwater, more than 1 mile (1.6 km) deep in places, and contains about one-fifth of the world's unfrozen freshwater (•**Figure 5.36**). Said to be the oldest lake in the world at 25 million years of age, the lake provides habitats to an estimated 1800 endemic plant and animal species. High endemism (species found nowhere else) like this is usually associated with tropical rain forests. "Nature as the sole creator of everything still has its favorites, those for which it expends special effort in construction, to which it adds finishing touches with a special zeal, and which it endows with special power," wrote Valentin Rasputin, a native of Siberia. "Baikal, without a doubt, is one of these. Not for nothing is it called *the Pearl of Siberia*."[32]

Jeff Schmaltz, LANCE/EOSDIS MODIS Rapid Response/NASA

• **Figure 5.36** Lake Baikal is the world's oldest and deepest lake. This NASA remote-sensing image shows the annual spring thaw, which progresses from south to north until the lake is usually ice-free in late June.

There was an environmental controversy over wastewater containing dioxin and sulfuric compounds discharged into Lake Baikal from the Baikalsk Pulp and Paper Mill beginning in 1966. The mill was shuttered in 2008 after environmentalists won a case arguing that the waste jeopardizes Baikal's unique natural history. In 2010, the then Prime Minister Putin signed a decree allowing the resumption of the discharge, which upset Russian conservationists, but the plant was closed once again in 2013. Public outcry also scrapped Moscow's plans to build a nuclear power plant several miles from the lake.

Proposed pipeline routes for exporting oil eastward from Irkutsk also threatened this natural wonder. Each route would have crossed a number of rivers flowing into Lake Baikal. Responding to fears that an earthquake in this seismically active region could result in calamity, President Putin in 2006 decreed that the proposed northern route should be moved farther north, away from Baikal's watershed. As major oil production comes online in the eastern Siberian oilfields around Irkutsk, there will be new environmental concerns.

Russia's Far East

The Far East is Russia's mountainous Pacific edge (see Figure 5.1). Most of it is a thinly populated wilderness in which the only settlements are fishing ports, lumber and mining camps, and the villages and camps of aboriginal peoples. Port functions, fisheries, and forest industries provide the main support for most Far Eastern communities, and until recently the output of coal, oil, and a few other minerals was small.

Most people here (primarily ethnic Russians) live in cities along two transportation arteries: the Trans-Siberian Railroad and the lower Amur River (including the diversified industrial and transportation center of Khabarovsk, a 5000-mile trip from Moscow via the Trans-Siberian Railroad). At the south, on the Sea of Japan, are two important port cities: Vladivostok, the largest Russian city east of Lake Baikal, and Nakhodka, the main commercial seaport of the Far East (•**Figure 5.37**). These ports are nearly ice-free and well positioned for trade in goods manufactured in Korea, China, and Japan. Seaborne shipments of cargo containers from northeast Asia to Scandinavia take 40 days, but the cargo can make the 6000-mile (9600-km) journey from Nakhodka to Finland by rail in only 12 days. Nakhodka is also home to the Kozmino oil terminal, where crude oil from the East Siberian Pacific Ocean pipeline is shipped to ports across North and East Asia.

• **Figure 5.37** Energy-hungry Asian countries are increasingly relying upon oil and natural gas produced in Russia's Far East to power their economies.

Before World War II, the Soviet Union and Japan held the northern and southern halves, respectively, of the large island of Sakhalin. The island has much pristine wilderness, with some forestry, fishing, and coal mining, but is particularly important for its offshore petroleum and natural gas. Russian energy officials describe it as a **"Second Kuwait,"** with about 1 percent of global oil reserves and enough natural gas to supply the US import market for about 25 years. Most of Sakhalin's natural gas will be exported to China, Japan, and South Korea.

World War II casts a shadow on modern relations between countries in the Far East. As the war drew to a close, the Soviet Union annexed southern Sakhalin and the Kuril Islands and expelled the half-million Japanese residents back to Japan. Russian control of these former Japanese territories, which the Japanese today call their "Northern Territories," continues to be a problem in relations between the two countries; despite attempts to break the impasse, disputes over control of the Kuril Islands has kept them from reaching a formal post–World War II treaty. Technically, Russia and Japan are 320 still at war!

Both countries are trying to put these differences aside in favor of economic cooperation, particularly in the energy realm. Japan negotiated contracts to buy natural gas from Sakhalin and to help build an oil pipeline from the eastern Siberian fields near Irkutsk to the port of Nakhodka. Russia is happy to have the foreign investment, and Japan, which has almost no oil, is happy to have the energy from a source outside the volatile Middle East. With its booming economy, China is also desperate for more energy, especially from closer and therefore cheaper sources. China is negotiating with Russia to obtain eastern Siberian oil and natural gas through a new pipeline (somewhat ostentatiously called the "Power of Siberia") that is currently under construction. Russia is strategically diversifying the geography of its oil exports to meet changing geopolitical circumstances, including Western sanctions. Until recently all of Russia's energy exports were geared toward western markets. Now Russia is opening export routes to the east, and these will be Russia's lifeline if Western sanctions become especially intense.

• **Figure 5.38** Karymsky is Kamchatka's most active volcano. A steep and symmetrical *stratovolcano* like Italy's Mt. Vesuvius, it periodically ejects lava flows from its 3-mile-wide (6-km-wide) caldera. Karymsky has been "hyperactive" since it stirred from quiescence some 500 years ago.

At the north, the Kuril Islands approach the sparsely populated mountainous peninsula of Kamchatka. Like the Kurils, Kamchatka is located on the Pacific Ring of Fire—the geologically active perimeter of the Pacific Ocean—and has 23 active volcanoes (**•Figure 5.38**). Soviet authorities pro- 25 tected Kamchatka for decades as a military area, and no significant development activities occurred there outside the small city of Petropavlovsk. A California-size land of rushing rivers, sprawling coniferous forests, and gurgling hot springs, Kamchatka is one of the last great wilderness areas on Earth, resembling the US Pacific Northwest landscape of a century ago. A struggle has been underway between those who want to develop Kamchatka's gold and oil resources and those who wish to set the land aside in national parks and reserves where ecotourism would be the only significant source of revenue.

Much to the surprise of environmentalists abroad, Russia's conservationists seem to have prevailed. Fossil fuel development will take place, but so will an ambitious effort to set aside seven wilderness tracts totaling more than 9400 square miles (24,400 sq km), or about three times the size of Yellowstone National Park in the United States. The main goal will be to have multiple uses from Kamchatka's rich salmon stocks—for food, sport, study, and profit—and also protect this resource from the kind of overfishing that has laid waste to the cod fisheries of the Atlantic and North Seas and to Russia's sector of the nearby Bering Sea. 112, 537

The Wild North

North and east of the Fertile Triangle lie enormous stretches of coniferous forest (taiga) and tundra extending from the Finnish and Norwegian borders to the Pacific (see Figure 5.5B).

These outlying wilderness areas in Russia may for convenience be called "The Northern Lands," although parts of the Siberian taiga extend to Russia's southern border.

The Arctic port of Murmansk (population 300,000, the world's largest city within the Arctic Circle) is located on a fjord along the north shore of the Kola Peninsula northwest of the similarly sized city of Arkhangelsk (or "Archangel"). It is the headquarters for important fishing trawler fleets that operate in the Barents Sea and North Atlantic. Murmansk is connected to Moscow by rail and has a major naval base and cargo port that is homeport to the icebreakers that escort cargo vessels and carry Western tourists to the North Pole and other high Arctic destinations (**Figure 5.39**). The harbor is open to shipping all year, thanks to the warming influence of winds associated with the North Atlantic Drift, an extension of the 98 Gulf Stream. Murmansk and Arkhangelsk played a vital role in World War II, when they continued to receive supplies by sea from the Soviet Union's Western allies after Nazi forces captured and closed the other ports of the western Soviet Union.

Murmansk and Arkhangelsk are western terminals of the **Northern Sea Route**, a waterway the Soviets developed to provide a connection with the Pacific via the Arctic Ocean (see **Figure 11.52**). Until recently, navigation along the whole length of the route was possible for only up to four months per year, despite the use of powerful icebreakers (including several nuclear-powered vessels) to lead convoys of ships. Areas along the route provide cargoes such as the timber of Igarka and the metals and ores of Norilsk. Some supplies are shipped north on Siberia's rivers from cities along the Trans-Siberian Railroad and then loaded onto ships plying the Northern Sea Route for delivery to settlements along the Arctic coast.

The Northern Sea Route is now open to international transit shipping, and if not for the capricious sea ice, it would be an attractive route. In recent winters, typical sea ice coverage in the Arctic Ocean has retreated substantially—one of the most accelerated trends in overall global warming, as described on **page 41**—and the Russian shipping industry is hoping this trend will continue. The distance between Hong Kong and all European ports north of London is shorter via the Northern Sea Route than through the Suez Canal, and Russia could earn valuable transit fees from this route. Russia is also anxious to develop the Northern Sea Route for its own exports; currently, Russian exports by land to Western Europe must pass over the territories of Ukraine, Belarus, and the Baltic countries, which collect transit fees from Russia.

Countries that have territories in the Arctic are making plans to exploit resources made more accessible by the forces of climate change. The race for Arctic riches is described on **page 538**. As a preview of that discussion, it is noteworthy that in 2007 Russians working in a mini-submarine planted the Russian flag on the seabed at the North Pole! The not so subtle message? Russia lays claim to the oil and gas wealth that may lie at the top of the world.

For over a century, there have been numerous proposals to facilitate trade and communication between Europe, Asia, and North America by constructing a transportation link across the Bering Strait between Siberia and Alaska. In 2015, a Russian plan called for the construction of a new superhighway that would originate in Europe, traverse the entirety of Russia, and cross the strait into Alaska. The highway would be paralleled by a railway branch from the Trans-Siberian Railroad and several oil and natural gas pipelines. China has also expressed interest in a rail link between its territory and the United States. Harsh climate conditions and the enormous costs involved—the 2015 Russian proposal could cost trillions of dollars—have always been severe impediments to these proposals. Neither northeastern Siberia nor western Alaska currently have any of the necessary infrastructure to support a direct transportation link, and it is unclear whether such a link would be an economically viable alternative to airplanes and cargo ships.

• **Figure 5.39** The Russian nuclear-powered icebreaker *Yamal* at the North Pole, August 2006. The former Soviet icebreaker fleet still escorts commercial ships on the Northern Sea Route but during the summer carries Western tourists such as these to the High Arctic. To be honest, we were 12 miles from the Pole (south of the Pole, of course) when I took this photograph. The North Pole was ice-free at this time, so we could not get down to touch the North Pole (see **page 41**). The captain steered the gigantic vessel to have the GPS device on the bridge indicate 90° north latitude, sparking loud cheers among his passengers. Twelve miles from the Pole, the captain "beached" the *Yamal*, and the passengers came down to spend some time on the ice.

The Caucasus

The Caucasian isthmus has been an important north–south passageway for thousands of years, and it was an important bridge for our ancestors coming out of Africa; this shows nicely in **Figure 3.2**. The population includes dozens of ethnic groups that have migrated into this region (•**Figure 5.40**). Most of these are small ethnic populations confined to mountain areas that became their refuges in past times. The republic of

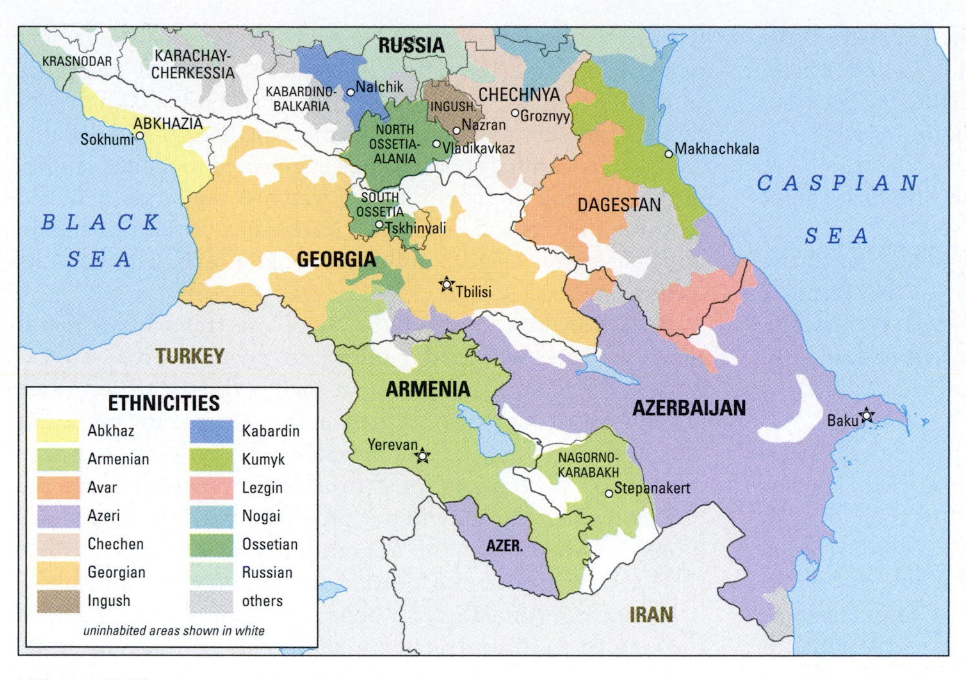

• **Figure 5.40** The Caucasus, rich in cultures.

Dagestan, within Russia on its southeastern border with Georgia and Azerbaijan, is by itself home to over 30 different ethnic groups; as we saw earlier, they do not always live comfortably 185–188 together. Ethnic Russians live primarily north of the Greater Caucasus Range; south of the mountains the largest ethnic groups are the Georgians, Azerbaijanis, and Armenians, each represented by an independent country. Throughout history, all have defiantly maintained their cultures in the face of pressure by stronger intruders.

The Armenian Genocide

The Armenian people historically occupied a large swathe of Anatolia (the Asian portion of Turkey), stretching far to the west of modern Armenia's borders. Between the World War I years of 1915-1918, a period of great political instability in the crumbling Ottoman Empire, Muslim Ottoman Turks annihilated an 273 estimated 1 million or more Christian Armenians. Most of the Armenians perished when they were uprooted from their homes and forced into "death marches" to concentration camps in the Syrian desert (•**Figure 5.41**). This **Armenian Genocide** grew out of Turkish fears that Russia would militarily aid the Ottoman Armenians to regain their ancient Anatolian homeland; such a loss of territory, and Russian presence within striking distance of the Ottoman capital of Constantinople, would be a severe threat to the empire's survival. After World War I, a proposed Armenian country in Anatolia never came to fruition, and the former Ottoman Armenian lands, including their sacred Mount Ararat, became part of Turkey. Turkey has never accepted responsibility for its role in the killings and officially denies the genocide took place. Animosity between Armenia and Turkey persists to this day: the two countries have no diplomatic relations, and the border between them is closed. Armenians in their capital city of Yerevan see Mount Ararat towering over Turkish land, a constant reminder of all they have lost.

Central Asia

The Shrinking Aral Sea

The five central Asian "stans" are mainly flat, with plains and low uplands, except for Tajikistan and Kyrgyzstan, which are spectacularly mountainous: summits in the Pamir and Tian Shan ranges reach over 20,000 feet (6000 meters) high. Central Asia is almost entirely a region of interior drainage. Only the waters of the Irtysh, a tributary of the Ob, reach the ocean; all the other streams either drain into enclosed lakes and seas or gradually lose water and disappear in the central Asian deserts.

Historically, many of this region's peoples were pastoral nomads who grazed their herds on the natural forage of the steppes and in mountain pastures in the Altai and Tian Shan ranges. Over the centuries, they slowly drifted away from nomadism. The Soviet government accelerated this process by forcibly collectivizing the remaining nomads and settling them in permanent villages. During the 20th century, Russians poured 220 into central Asia, mainly into the cities and the fertile plains of northern Kazakhstan. Meant to Russify this non-Russian area, they included farmers, administrative and managerial personnel, engineers, technicians, factory workers, and political dissidents banished by the Communists. Most of the Slavic newcomers lived apart from the local Muslims, as most of those who remain today still do.

• **Figure 5.41** The Armenian Genocide. Ottoman Turks uprooted over a million ethnic Armenians from their ancestral homeland between 1915 and 1918.

Today, most of central Asia's people live in irrigated valleys at the base of the great mountains in the region's south. There, the wind-blown loess and other soils are fertile; the growing season is long; and rivers flowing from mountains provide irrigation water. The principal rivers in the heart of the region are the Amu Darya (Oxus) and Syr Darya, both of which empty into the enclosed Aral Sea.

Only northern Kazakhstan, which extends into the black-earth belt, receives enough moisture for dry farming (unirrigated agriculture). Irrigation is needed everywhere else, providing water to mulberry trees (grown to feed silkworms), rice, sugar beets, vegetables, vineyards, and fruit orchards.

Cotton began to be cultivated here during the American Civil War, when US exports abroad virtually ceased. The Soviets later fostered a "plantation economy" of their own in central Asia, exploiting the region's potential to produce large quantities of this valuable crop (known for a time as **"white gold"**) for the central Russian textile mills. Cotton remains the region's major crop. Production flows to textile mills in Tashkent, Moscow (although at levels far below those of Soviet times), and locations outside the region.

There was a significant environmental price to pay for the development of central Asian agriculture. Driven by the need for foreign exchange from export sales, Soviet planners reshaped the basins of the Syr Darya and Amu Darya with 20,000 miles (32,000 km) of canals, 45 dams, and more than 80 reservoirs. As the area under irrigated cotton grew, the rivers and the Aral Sea became highly polluted with agricultural chemicals draining from the fields. Diversion of water from the rivers caused the Aral Sea to shrink rapidly in volume and area, virtually destroying a vast natural ecosystem and an important regional fishery. Since the 1950s, the Aral Sea has lost nearly 90 percent of its surface area and 80 percent of its volume, and has divided into three small lakes (see remote sensing images of the lake's transformation in •**Figure 5.42**). With assistance from the World Bank, Kazakhstan recently built a dam to contain the portion now known as the North Aral Sea, located completely within its borders. It is engineered to reduce salinity in the lake and divert any excess water to the larger, parched Aral Sea. The North Aral Sea's waters have risen faster than anticipated, and there is optimism about a revitalization of its fishing industry.

Both Kazakhstan and Uzbekistan use tremendous quantities of water to flood paddies to produce rice—an extraordinary land use decision in such a thirsty region. Rice cultivation is one reason Uzbekistan is desperate for water and accuses Kyrgyzstan of hoarding it. Kyrgyzstan insists it needs to store water for hydroelectric power because the downstream countries will not provide it with the fossil fuels it needs. Meanwhile, in a grandiose, Soviet-style project that is widely criticized outside the country, Turkmenistan is harnessing Amu Darya waters to create a new "Lake of the Golden Age" to irrigate an additional 1875 square miles (4860 sq km) of land. Progress has been slow, as much of the diverted water simply evaporates into the hot desert air before reaching the lake site, and the water reaching its destination is increasingly saline.

Another environmental problem—this one with international security dimensions—existed on the former island of Vozrozhdeniye, on Uzbekistan's side of the Aral Sea. In 1988, as Mikhail Gorbachev's Soviet Union hastened to build new bridges with the West, Soviet biological warfare specialists buried hundreds of tons of living anthrax bacteria, encased in steel canisters, on the island. As the Aral Sea continued to shrink, the island became a peninsula, raising fears that these more accessible biohazards could be exposed and blown into populated areas or be carried off by would-be terrorists.

NASA Images

• **Figure 5.42** NASA remote-sensing views of the Aral Sea in 2000 and 2014. The 2014 image shows that Kazakhstan's engineering works are beginning to revitalize the waters of its northern portion (at top).

Following a number of incidents of anthrax-tainted letters reaching US Senate offices and other facilities in the wake of the 9/11 attacks, the US government provided the funds needed to clean up "Anthrax Island"; all anthrax burial sites were decontaminated in 2002.

The central Asian countries are like those of the Middle East in many respects, particularly in their overwhelmingly Muslim populations, their traditions of pastoral nomadism and irrigated agriculture, and their petroleum endowments. The text now turns to that critical world region. The discussion of Islam in Chapter 6 provides especially useful background for further appreciation of central Asia and the Caucasus.

Study Guide

Summary

- From the Russian perspective, the other 14 countries of the former Soviet Union make up the region of the "Near Abroad," a special realm of policy and interaction. The three Baltic countries have struggled to divorce themselves from this orbit. Russia and the remaining countries except Georgia, Ukraine, and Turkmenistan, are members of the Commonwealth of Independent States (CIS). All have significant ties with Russia. Russia aspires to increase its influence on the former Soviet countries—sometimes by force, as seen in its 2008 invasion of Georgia and 2014 invasion of Ukraine.
- The area west of the Ural Mountains and north of the Caucasus Mountains is known as European Russia; the Caucasus and the area east of the Urals is Asiatic Russia. Siberia is the area between the Urals and the Pacific. Central Asia is the name for the arid area occupied by the five countries immediately east and north of the Caspian Sea.
- Russia and the Near Abroad span 11 time zones. Stretching nearly halfway around the globe, the region has formidable problems associated with climate, terrain, and distance.
- In the years of economic uncertainty after 1991, birth rates fell and death rates rose dramatically in Russia, leading to the demographic "Russian cross" and steep declines in population. Growing revenues from oil began to improve standards of living, however.
- Vast stretches of the region are at very high latitudes, making agriculture difficult. Much of the tundra and subarctic climate zones are underlain by permafrost. If this "permanently" frozen ground should thaw, large amounts of greenhouse gases would be released into the atmosphere, accelerating a trend of global warming. The subarctic taiga or coniferous forest acts as a carbon "sink" and helps offset carbon dioxide emissions, so there are fears about deforestation in Siberia. The most productive agricultural area is the steppe region south of the forest in Russia and in Ukraine, Moldova, and Kazakhstan.
- In early Russia, rivers formed natural passageways for trade, conquest, and colonization. In Siberia, the Russians followed tributaries to advance from the Urals to the Pacific. More recently, alteration of rivers for power, navigation, and irrigation became an important aspect of economic development under the Soviet regime.
- Meteors that have struck Siberia remind us of the potential for a devastating, world-changing impact that could come at any time.
- Large plains dominate the terrain from the Yenisey River to the western border of the country. The area between the Yenisey and Lena Rivers is occupied by the hilly Central Siberian Uplands. In the southern and eastern parts of the region, mountains, including the Caucasus, Pamir, Tien Shan, and Altai ranges, dominate the landscape.
- Slavic and Scandinavian peoples were prominent in the early cultural development that was the foundation for the Russian, Ukrainian, and Belarusian cultures. Although Slavs came to dominate economic and political life, the region has large and diverse populations of non-Slavic peoples.
- Early contacts with the Eastern Roman or Byzantine Empire led to the establishment of Orthodox Christianity in Russian life and culture. Although it was repressed after the Bolshevik Revolution, this and the region's other religions have been experiencing a rebirth since the breakup of the Soviet Union.
- From the 15th century until the 20th, the tsars built an immense Russian empire around the small nuclear core of Muscovy (Moscow). This imperialism brought huge areas under tsarist control, creating a Russian "land empire." Russian expansion extended to the Pacific, across the Bering Strait, and down the west coast of North America. Russia has often triumphed over powerful foreign invaders. These achievements have been due in part to the environmental obstacles the invaders faced, the overwhelming distances involved, and the defenders' willingness to accept large numbers of casualties and implement a scorched-earth strategy to protect the motherland.
- Policies of the Soviet Union permitted most ethnic groups to retain their own languages and other elements of traditional cultures. However, the Soviet regime implemented a deliberate policy of Russification in an effort to implant Russian culture in non-Russian regions. Millions of Russians settled in non-Russian areas, but local cultures were not converted to Russian ways.

- To meet the needs of its people, the Soviet Union became a collective welfare state. However, military and industrial superpower status was generally achieved at the expense of ordinary consumers.
- The sudden transition from a command economy to capitalism in Russia and other countries in the region led to a widening gap between the rich and the poor, with growing ranks of poor people. Organized crime and an underground economy emerged. Agricultural and industrial production fell dramatically. However, high oil prices in the early 2000s improved Russia's economic outlook. This risky overdependence on a single commodity set back Russia's economy when oil prices fell, threatening Vladimir Putin's plan to propel Russia to being the world's fifth largest economy by 2020. Russia's "Dutch Disease" broke out as overreliance on oil neglected manufacturing and the development of a diversified economy.
- There are three important realms of geopolitical concern in the region. Within Russia, the aspirations of non-Russian people like the Chechens and Tatars appear to pose a danger to the unity of the Russian Federation. Between Russia and the Near Abroad countries, there are challenging issues, including energy shortages and supplies, desires of Russians outside Russia to achieve their own rights and territories, stationing of Russian troops, and Islamist terrorism. Between this region and the rest of the world—the "Far Abroad"—there are concerns about the fate of Soviet-era nuclear materials, whether the oil-rich central Asian countries are more sympathetic to Russia or Turkey, and what should be done to stem the tide of narcotics and terrorism.
- The Russian government has said that it has the right and duty to protect Russian minorities in the other countries. Some observers contend that Russia is attempting to reestablish the Near Abroad as a buffer between Russia and the Far Abroad, or reoccupy portions of it in an effort to reconstitute parts of the Soviet empire. President Putin has ambitions to create a "Eurasia Union" that Russia would be the cornerstone of.
- Russia forcibly annexed Ukraine's Crimea Peninsula in 2014 and equipped pro-Russian rebels in eastern Ukraine to challenge Ukrainian sovereignty there. Russia has interests in maintaining "frozen conflicts" in Ukraine, Transnistria (within Moldova), Nagorno-Karabakh (within Azerbaijan), and Abkhazia and South Ossetia (both within Georgia). The West responded to Russia's incursion into Ukraine with economic sanctions that are difficult to strengthen because they would affect Europe's reliance on Russia's gas and oil imports. A "new cold war" between Moscow and the West may be underway.
- The Caspian Sea and adjacent areas make up one of the world's greatest petroleum and natural gas regions. Russia, Iran, and the West, particularly the United States, are vying for influence in this increasingly important area. Of particular concern to the rivals is how the energy supplies should be routed via pipeline to reach ocean terminals. There are similar questions about how Russian natural gas pipelines should be routed to reach western European markets.
- Because of nuclear weapons and other critical issues, relations between Western countries and the countries of the former Soviet Union, particularly Russia, are vital to global security.
- Nearly 75 percent of the people, and an even larger share of the cities, industries, and cultivated land of the region of Russia and the Near Abroad, are packed into the Fertile Triangle (Agricultural Triangle or Slavic Core), comprising one-fifth of the total area. The rest (mostly in Asia) consists of land where environmental obstacles limit settlement.
- Ukraine is one of the most densely populated and most productive areas of the region. Ukraine's industrial and agricultural assets were vital to the Soviet Union. Post-Soviet Ukraine relies more on heavy industry than most of the other successor countries, but the Russian invasion has impacted this sector by cutting off its raw materials in the east from the manufacturing plants in the west.
- Under Soviet rule, Belarus became an industrial power. Belarus's trade is now almost exclusively with Russia. Belarus and Russia technically formed a union state to more closely link the two predominantly Slavic countries, but political problems have prevented its implementation.
- Young women from Moldova, Ukraine, Russia, and other successor states of the Soviet Union are among the main participants in and victims of sex trafficking to Europe.
- In an effort to reverse the inefficiencies left by the state-run system, all of the countries are promoting land reform by privatizing collective and state farms and developing more independent or peasant farming. After a period of decline following independence in 1991, Russian agricultural production has improved.
- Russian industrial production plummeted after the Soviet Union dissolved. The industrialized area around Moscow has a central location within the populous western plains and is functionally Russia's most important industrial region.
- Most of the Russian Far East is a thinly populated wilderness with few settlements. Several small to medium-sized cities form a north–south line along the Trans-Siberian Railroad and the lower Amur River, the two main transportation arteries. The coastal ports of Vladivostok, Nakhodka, and Vostochnyy are coming to life as shipments of oil destined for Japan and other foreign markets have picked up. Oil from the eastern Siberian fields will flow through new pipelines to China and Vostochnyy. The island of Sakhalin is becoming an important oil and gas producer for the Japanese market.
- Russia's mineral-rich Northern Lands comprise one of the world's most sparsely populated areas, with coniferous forest (taiga) and tundra extending from the Finnish and Norwegian borders to the Pacific. Murmansk and Arkhangelsk are ports on the Northern Sea Route, a transpolar route for ocean shipping that can be kept open in winter with icebreakers. Russia is encouraged that melting of sea ice will make the Northern Sea Route a more permanent and profitable alternative to Eurasian overland shipping.
- The world's deepest lake, Lake Baikal, is a unique natural ecosystem recently threatened by effluent from paper and pulp mills. However, oil pipelines have been rerouted to reduce threats to the lake.

- The far southern Caucasus region borders Russia between the Black and Caspian Seas. It includes the Caucasus Mountains, a fringe of foothills and level steppes to the north, and the area south known as Transcaucasia. Russians and Ukrainians are the majority north of the Greater Caucasus Range. To the south, the important nationalities are the Georgians, Armenians, and Azerbaijanis, each represented by an independent country and maintaining their unique cultures. Numerous smaller ethnic groups also inhabit the region, which has been a cauldron of conflict since independence from the Soviet Union. One of the outstanding internal problems is the fate of Nagorno-Karabakh, a region claimed by Armenia but controlled by Azerbaijan.
- Soviet-era colonization of central Asia for a cotton plantation economy had pronounced and lingering effects on the environment. Most notable was the diversion of water for irrigated agriculture and the subsequent shrinkage of the Aral Sea. The post-Soviet countries of the region are locked in a critical competition over the region's water supplies from the Amu Darya and Syr Darya Rivers.

Review Questions

1. What events culminated in the creation of 15 independent countries from what had been a single superpower?
2. What are the main climatic belts and corresponding vegetation types of Russia and the Near Abroad? Which are the most and least productive for agriculture? What impacts do high latitude and other geographic variables have on economic activities and population distributions?
3. What are the region's major river systems?
4. Why did Russia's population decline to form the demographic "Russian cross"?
5. Why is Russia known as a land empire? How did it acquire its empire?
6. What significant conflicts have occurred in this region? Why has Russia so often won them? What was the geographic significance of Stalingrad in World War II? How did World War II impact Russian sentiment and nationalism today?
7. What were the perceived advantages of collectivized agriculture, and were they realized?
8. What major Soviet Communist projects changed the landscape of this region?
9. What were some of the successes and failures of efforts to industrialize the Soviet Union?
10. Why was post-Soviet Russia characterized as a "misdeveloped" country?
11. What were some of the results of Russia's "shock therapy" economic reforms and the subsequent conditions of the country and its people? What accounts for the recent turnaround in Russia's economy, and why was it vulnerable to sudden downdrafts? What is Russia's "Dutch Disease"?
12. Which of the non-Russian ethnic groups in Russia are particular security concerns to Russia today?
13. How does Caspian Sea oil reach markets abroad? What factors go into decision making about those exports?
14. Where, outside Russia, are Russian military troops stationed? What are the reasons for their presence? What events in 2008 and 2014 stirred new concerns about Russia's intentions? Why did Russia annex Ukraine, and how did the West respond to that incursion? What are the frozen conflicts that may bolster Russia's geopolitical interests?
15. What resources and boundaries are associated with the Fertile Triangle?
16. To what extent have Ukraine and Belarus been dependent on Russia or interested in developing further ties with Russia? Which are most and least like Russia? What problems in relations occurred between Russia and each of these counties prior to 2014?
17. What are the resources and economic development prospects of Sakhalin and Kamchatka? Why are the Kurils important in Russian-Japanese relations?
18. What major dispute between two groups is focused on Nagorno-Karabakh, and what is the source of that dispute?
19. What was the Armenian Genocide? Have the countries affected reconciled?
20. What environmental problems do the central Asian countries face, especially with water resources?

Key Terms + Concepts

Agricultural Triangle (p. 193)
Altaic language family (p. 164)
  Turkic subfamily (p. 164)
Armenian (p. 164)
Armenian Genocide (p. 199)
arms race (p. 171)
autonomies (p. 176)
Azerbaijani (p. 164)
barter (p. 174)
black-earth belt (p. 159)
Bolshevik Revolution (p. 168)
bourgeoisie (p. 169)
BRIC economies (p. 174)
Budapest Memorandum (p. 180)
buffer zone (p. 178)
Chechen (p. 164)
Chernobyl (p. 194)
chernozem (p. 159)
collective farm (*kolkhoz*) (p. 171)
Collective Security Treaty Organization (CSTO) (p. 178)
collectivized agriculture (p. 171)
command economy (p. 170)
Commonwealth of Independent States (CIS) (p. 150)
Commonwealth of Independent States Free Trade Agreement (CISFTA) (p. 179).

Community for Democracy and Rights of Nations (p. 185)
Cossacks (p. 167)
Deep and Comprehensive Free Trade Area (DCFTA) (p. 179)
Donetsk People's Republic (DPR) (p. 183)
"Dutch Disease" (p. 175)
economic shock therapy (p. 173)
Euromaidan (p. 182)
"Evil Empire" (p. 150)
Fertile Triangle (p. 193)
frozen conflict (p. 184)
Gazprom (p. 181)
Georgian (p. 164)
gigantomania (p. 171)
glasnost (p. 173)
"Golden Horde" (p. 166)
Gosplan (Committee for State Planning) (p. 170)
Great Volga Scheme (p. 163)
Group of Eight (G-8) (p. 188)
GUAM (p. 186)
hero projects (p. 172)
holodomor (p. 169)
human trafficking (p. 195)
humus (p. 160)
Islamic Movement of Uzbekistan (IMU) (p. 191)
Kazakh (p. 164)
kleptocracy (p. 174)
Kyrgyz (p. 164)
land empire (p. 167)
"loose nukes" (p. 190)
Luhansk People's Republic (LPR) (p. 183)
"misdeveloped country" (p. 174)
mollisols (p. 159)
Mongolic (p. 164)
Mongols (p. 166)
Near Abroad (p. 150)
"New Cold War" (p. 184)
Northern Sea Route (p. 198)
Novorossiya (p. 183)
oligarchs (p. 174)
Operation Barbarossa (p. 169)
Orange Revolution (p. 181)
pan-Turkism (p. 191)
Partnership for Peace (PfP) (p. 179)
perestroika (p. 173)
permafrost (p. 158)
Persian (p. 164)
podzols (p. 158)
proletariat (p. 169)
renationalization (p. 181)
"Russian cross" (p. 156)
Russian Revolution (p. 168)
Russification (p. 165)
sacred space (sacred place) (p. 170)
scorched earth (p. 168)
"Second Kuwait" (p. 197)
Second Russian Revolution (p. 173)
Shanghai Cooperation Organisation (SCO) (p. 192)
Slavic Core (p. 193)
soft power (p. 178)
spodosols (p. 158)
state farm (*sovkhoz*) (p. 171)
Tatar (p. 164)
Tatars (p. 166)
"Third Rome" (p. 165)
Turkmen (p. 164)
Tyvan (p. 164)
underground economy (p. 174)
Uzbek (p. 164)
Varangians (p. 165)
virgin and idle lands (new lands) (p. 159)
"white gold" (p. 200)
"Window on the West" (p. 167)
Yakut (p. 164)

Notes

1. Angela Stent, "Why America Doesn't Understand Putin," *Washington Post*, March 15, 2014.
2. Murray Feshbach, quoted in C. J. Chivers, "Putin Urges Plan to Reverse Slide in the Birth Rate," *New York Times*, May 11, 2006, pp. A1, A6.
3. http://science.nasa.gov/science-news/science-at-nasa/2008/30jun_tunguska/.
4. http://neo.jpl.nasa.gov.
5. On neo-Nazis in Ukraine, see "Ukraine Crisis: The Neo-Nazi Brigade Fighting Pro-Russian Separatists," *The Telegraph*, February 4, 2015.
6. For the transition to BICS, see Tod Lindberg, "An Elite Guide To Globalization," *The Wall Street Journal*, April 3, 2014, p. A15.
7. Quoted in Andrew K. Kramer, "For Investors, Russia's Putin is Good for Business," *New York Times*, September 28, 2011, p. B4.
8. Kathrin Hille, "Russia: Post-Soviet struggles bring on existential crisis," *Financial Times*, January 21, 2014, special issue, p. 2.
9. Mark Adomanis, "What Dutch Disease Is, and Why It's Bad," *The Economist*, November 5, 2014.
10. From Mikhail Blinnikov, *A Geography of Russia and Its Neighbors* (New York: Guilford, 2011).
11. Hille, op. cit., 2014, p. 2.
12. Quoted in C. J. Chivers, "For Putin and the Kremlin, a Not So Happy New Year," *New York Times*, January 3, 2006, p. A8.
13. The Phrase Finder, www.phrases.org.uk/meanings/31000.html.
14. Charles Clover, "Strife on the Edge," *Financial Times*, March 25, 2010, p. 8.
15. Stratfor, "Putin Brings Russia's Regions Back Under Its Control," May 16, 2014.
16. Clover, op. cit., 2010, p. 8.
17. Richard Weitz, "After Ukraine, Putin's Eurasian Union Could Be Dead on Arrival," *World Politics Review*, January 6, 2015, http://www.worldpoliticsreview.com/articles/14785/after-ukraine-putin-s-eurasian-union-could-be-dead-on-arrival.
18. http://www.nato.int/cps/en/natolive/topics_50349.htm.
19. Peter Finn, "Putin Threatens Ukraine on NATO—Russian Raises Issues of U.S. Missile Shield," Washingtonpost.com, February 13, 2008.
20. Blake Fleetwood, "Too Bad Ukraine Didn't Keep Its 2,000 Nuclear Weapons," *Huffington Post*, accessed June 30, 2014, from http://www.huffingtonpost.com/blake-fleetwood/too-bad-ukraine-didnt-kee_b_5235374.html.
21. 'The Ukrainian Question," *Economist*, November 29, 1999, p. 20.
22. David Klion, "Friend and Threat: Ukraine Crisis Confronts Belarus' Lukashenko with Russia Dilemma," *Briefing*, March 21, 2014.
23. Quoted in Clifford J. Levy, "Retreat of the Tongue of the Tsars," *New York Times*, September 13, 2009, p. wk3.
24. http://europa.eu/newsroom/highlights/special-coverage/eu_sanctions/index_en.htm.
25. "Ukraine Crisis: Russia and Sanctions," BBC News, December 19 2014, http://www.bbc.com/news/world-europe-26672800.
26. Nikola Vujinović, "Is Kosovo a Frozen Conflict?" University of Belgrade Faculty of Political Sciences, https://www.academia.edu/6713586/Is_Kosovo_a_Frozen_Conflict.
27. "The Economist Explains: What defines a frozen conflict?" October 23, 2014, Economist.com, http://www.economist.com/blogs/economist-explains/2014/10/economist-explains-19.
28. Caroline Wyatt, "Bush and Putin: Best of Friends," BBC news.com, June 16, 2001, http://news.bbc.co.uk/2/hi/europe/1392791.stm.
29. Hamid Golpira, "Iraq Smoke Screen," *Tehran Times*, November 20, 2008, http://www.tehrantimes.com/index_View.asp?code=182891.
30. Parag, Khanna, "The New Silk Road Is Made of Iron—And Stretches from Scotland to Singapore," *Quartz*, October 1, 2012, accessed March 5, 2015, from http://qz.com/6140/the-new-silk-road-is-made-of-iron-and-stretches-from-scotland-to-singapore/.
31. "Office to Monitor and Combat Trafficking in Persons," US Department of State, Diplomacy in Action, http://www.state.gov/j/tip/.
32. Valentin Rasputin, "Baikal," in G. Mikkelson, M. Winchell, and V. Rasputin (eds.), *Siberia on Fire: Stories and Essays* (De Kalb: Northern Illinois University Press, 1989), pp. 188–189.

Global Geoscience Watch

Go directly to the Russia Portal. Review Figure 5.24, a summary of post-Soviet multinational organizations in the region of Russia and the Near Abroad. Conduct a "Basic Search" on one of the following organizations mentioned in the figure: Eurasian Economic Union, Collective Treaty Security Organization (CSTO), Commonwealth of Independent States Free Trade Area (CISFTA), or the Baltic Assembly. Review news articles about one of these organizations, and write a brief report. What interest or agenda does the organization seem to advance? Are other organizations or countries mentioned, and what role do they play in the news story you read?

Online Resources

For access to MindTap and additional study materials visit www.cengagebrain.com. Read your textbook, take notes, complete activities, take practice quizzes and more.

Above: Jerusalem's Dome of the Rock is supercharged with *genius loci*, the spirit of place. Built as a Muslim shrine in 691 CE, it envelops the "Foundation Stone" that presumably occupied the Holy of Holies in the Jewish First and Second Temples. The Ark of the Covenant would have rested here within the First Temple. Joe Hobbs

Left: The Middle East and North Africa at night. Note especially the population concentrations along Egypt's Nile River, appearing as a white line curving across Africa's northeastern corner. The vast unlit wilderness of the Sahara stretches westward from there. C. Mayhew & R. Simmon (NASA/GSFC), NOAA/ NGDC, DMSP Digital Archive

6 The Middle East and North Africa

War is God's way of teaching Americans geography.

—AMBROSE BIERCE[1]

Welcome to the Middle East. What other part of the world outside your own is in the news every single day? What part of the world at times seems so threatening to your own? What part of the world seems to have so many problems for which no solutions seem to be in sight?

But there is also good news coming out of the Middle East, a region we often call the "cradle of civilization." This is a fountainhead of faiths, civilizations, and sciences. And there *are* solutions to the region's seemingly intractable problems. Peace and stability in the Middle East could even increase your security by diminishing motivations for terrorist attacks. This chapter will equip you with the geographical knowledge to talk about the misfortunes of war and components of peace in the Middle East.

Chapter Outline

Chapter Objectives

This chapter will enable you to

- Understand and explain the mostly beneficial relationships between villagers, pastoral nomads, and city dwellers in an environmentally challenging region.
- Know the basic beliefs and sacred places of Jews, Christians, and Muslims, and the political orientations of Sunni and Shi'a societies and nations.
- Recognize the importance of petroleum and natural gas to this region and the world economy, and the geographic challenges of transporting these fossil fuels.
- Understand the problematic issues of the Arab-Israeli conflict and the obstacles to their resolution.
- Learn about the promises of the Arab Spring, and the disappointments that followed.
- Consider the hydropolitical issues of the Nile and Mesopotamian river basins, and shortages of freshwater in this region's arid lands.
- Know what al-Qa'ida, ISIS, and other Islamist terrorist groups are and what they want to achieve.

The *Middle East* is a fitting designation for the places and the cultures of this vital world region: they really are in the middle. This is a physical crossroads where the continents of Africa, Asia, and Europe meet, and the waters of the Mediterranean Sea and the Indian Ocean mingle. Its peoples—Arab, Jew, Persian, Turk, Kurd, Berber, and others—represent the confluence of diverse and influential cultures. They have bestowed on humankind a rich legacy that includes the ancient civilizations of Egypt and Mesopotamia and the Abrahamic faiths of Judaism, Christianity, and Islam.

People tend to overlook contributions like these and associate the region only with war and terrorism. These negative connotations are often accompanied by muddled understanding. This chapter illuminates the region's geographical context and empowers you to think confidently about this complex and misunderstood region.

6.1 Area and Population

What is the Middle East, and where is it? It is a Eurocentric vernacular region invented by British colonial authorities who placed themselves in the figurative center of the world. They began to use the term before the outbreak of World War I, when the *Near East* referred to the territories of the Ottoman Empire in the eastern Mediterranean region, the *East* to India, and the *Far East* to China, Japan, and the western Pacific Rim. With *Middle East*, they designated as a separate region the countries and territories around the Persian Gulf (which has been called the "Arabian Gulf" by Arabs since the 1960s, a name not recognized by the UN or any non-Arab country). Gradually, the perceived boundaries of the region grew.

Sources today vary widely in their definitions of the Middle East. For some, the Middle East includes only the countries clustered around the Arabian Peninsula. This book adds North Africa to the Middle East to form the region of the Middle East and North Africa, often using the acronym MENA, which is popular in many regional studies. The North African peoples of Morocco, Algeria, and Tunisia share many traits with people to the east, but generally do not consider themselves Middle Easterners; they are, rather, from what they call the *Maghreb,* meaning "western land." The Middle East and North Africa is the region spanning a vast 6000 miles (9700 km) west to east, from Morocco in northwest Africa to Iran in central Asia, and a north–south distance of about 3000 miles (4800 km) from Turkey, on Europe's southeastern corner, to Sudan, which extends deep into Africa (•**Figures 6.1** and **6.2**).

In this book, the Middle East and North Africa region includes 21 countries, the Palestinian territories of the West Bank and Gaza Strip, and the disputed Western Sahara (•**Table 6.1**), occupying 5.6 million square miles (14.7 million sq km) and inhabited by about 530 million people. This area is about 1.8 times the size of the lower 48 United States and spans latitudes roughly equivalent to those between Boston, Massachusetts, and San Jose, Costa Rica (see •**Figure 6.3**).

These half-billion people are not distributed evenly across the region, but are concentrated in major clusters (•**Figure 6.4**). The pie chart illustrates that three countries have the lion's share of the region's population: Turkey, Iran, and Egypt, each with more than 75 million people. By looking at maps of precipitation and available water in the form of rivers, you can understand why people are clustered as they are. Where water is abundant in this generally arid region, so are people. Egypt has the Nile River, Turkey has the upper portions of the Tigris and Euphrates Rivers, and parts of Iran have bountiful rain and snow. Conversely, where rain seldom falls, as in the Sahara of North Africa and in the Arabian Peninsula, people are fewer.

The Middle East and North Africa have a moderately high rate of population growth compared to other world regions, but there is a considerable range between countries in these rates. Lebanon has the slowest rate of growth, at 0.9 percent, with

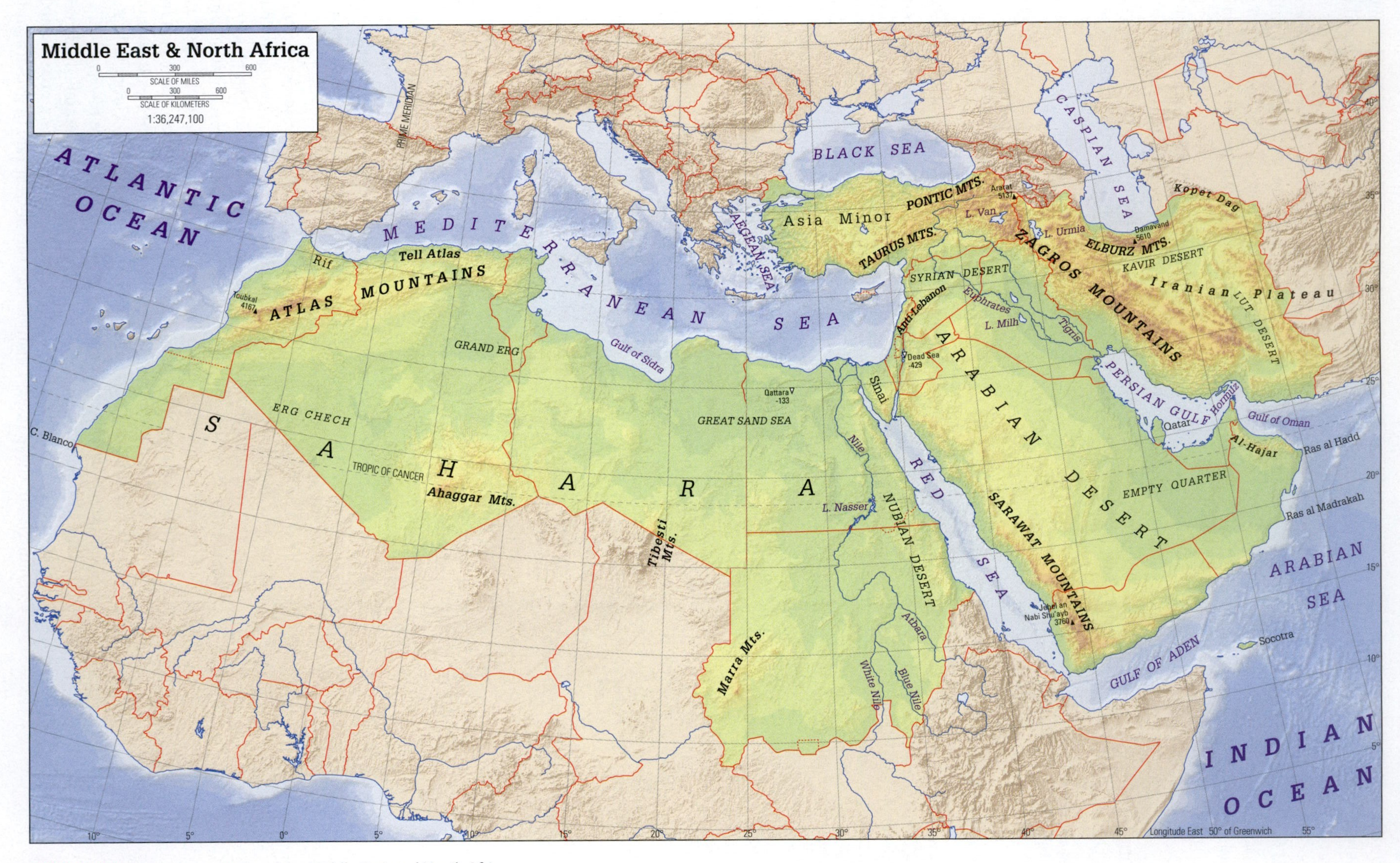

• **Figure 6.1** Physical geography of the Middle East and North Africa.

• **Figure 6.2** Political geography of the Middle East and North Africa.

Table 6.1 Middle East and North Africa: Basic Data

Political Unit	Area (sq mi, thousands)	Area (sq km, thousands)	Population (millions)	Rate of Natural Increase (%)	Urban Population (%)	Population Under Age 15 (%)	Agriculture Workers (%)	Per Capita GDP (PPP) ($US)	GDP ($US, billions)	Trade Balance ($US, billions)	Oil Production (million bbl/day)	Literacy Female (%)	Literacy Male (%)	HDI
Middle East and North Africa	**5416.1**	**14027.8**	**531.2**	**1.9**	**63**	**30**	**10**	**11,600**	**9,150**	**—**	**29.5**	**77**	**87**	**0.694**
Middle East														
Bahrain	0.3	0.8	1.3	1.3	100	21	—	29,800	61	8	*	92	96	0.815
Iran	630.6	1633.3	77.4	1.4	71	24	9	12,800	1,284	34	3.1	81	89	0.749
Iraq	169.2	438.2	35.1	2.6	71	40	3	7,100	505	32	3.1	73	89	0.642
Israel	8.1	21.0	8.2	1.6	91	28	2	36,200	268	-6	—	96	99	0.888
Jordan	34.4	89.1	7.6	2.4	83	34	3	6,100	79	-14	—	90	96	0.745
Kuwait	6.9	17.871	3.7	1.7	98	23	—	42,100	283	84	2.8	97	94	0.814
Lebanon	4.0	10.36	5	0.9	87	20	6	15,800	80	-16	—	86	93	0.765
Oman	82.0	212.4	4.1	1.8	75	22	1	29,800	163	24	0.9	73	87	0.783
Palestinian Territories	2.4	6.216	4.4	2.7	83	40	3	2,900	20	-6	—	93	98	0.686
Qatar	4.2	10.9	2.3	1	100	14	—	1,02,100	323	82	2.0	95	96	0.851
Saudi Arabia	830	2149.7	30.8	1.8	81	30	2	31,300	1,616	197	11.5	81	90	0.836
Syria	71.5	185.2	22	2.1	54	35	16	5,100	107	-6	*	74	86	0.658
Turkey	299.2	774.928	77.2	1.1	77	25	8	15,300	1,512	-64	*	98	99	0.759
United Arab Emirates	32.3	83.7	9.4	1.4	83	16	1	29,900	605	133	2.8	82	76	0.827
Yemen	203.8	527.8	26	2.8	29	42	9	2,500	106	-3	0.1	47	81	0.500
North Africa														
Algeria	919.6	2381.8	39.1	1.9	73	28	9	7,500	552	7	1.7	92	92	0.717
Egypt	386.7	1001.6	87.9	2.6	43	32	15	6,600	945	-28	0.7	66	81	0.682
Libya	679.4	1759.6	6.3	1.7	78	29	2	11,300	103	1	0.3	91	98	0.784
Morocco	172.4	446.5	33.3	1.5	59	28	14	5,500	254	-20	—	57	76	0.617
Sudan	718.7	1861.5	38.8	2.5	33	41	27	2,600	159	-2	*	63	81	0.473
Tunisia	63.2	163.7	11	1.3	66	24	9	9,900	125	-7	*	80	95	0.721
Western Sahara	97.2	251.7	0.6	1.5	82	27	—	—	—	—	—	—	—	—

* Less than 0.1. — Data not available or not applicable.
Sources: World Population Data Sheet, Population Reference Bureau, 2014; Human Development Report, United Nations, 2014; World Factbook, CIA, 2014.

Qatar and Turkey close behind at 1.0 and 1.1 percent, respectively. The highest is 2.8 percent in Yemen, followed by 2.7 percent in the Palestinian territories (consisting of the West Bank and Gaza Strip) and 2.6 percent in Iraq. These are some of the highest population growth rates in the world, on par with many countries in Africa South of the Sahara. Yemen's high population growth is related to its great poverty. In the Palestinian territories, rapid growth may be attributed in part to the Palestinians' poverty and probably also to the wishes of many Palestinians to have more children to counterbalance the demographic weight of their perceived Israeli foe. As we will see later in the chapter (on **pages 251–252**), there is an important demographic component to the Palestinian-Israeli conflict.

Between these extremes are countries with modest rates of population growth. The governments of many of these

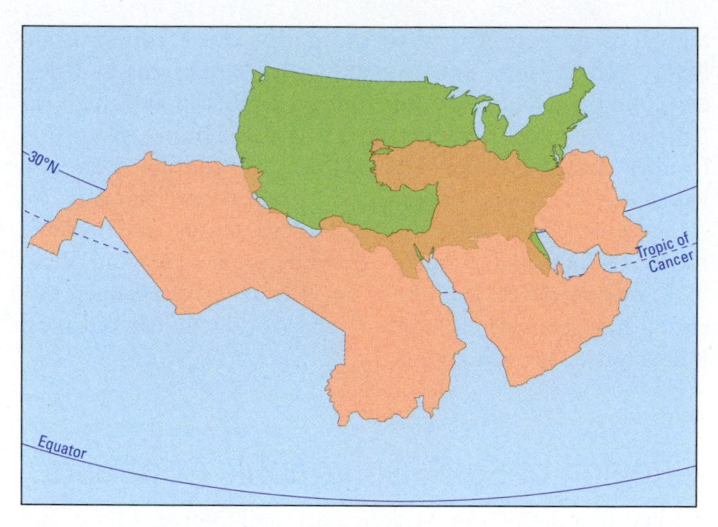

• **Figure 6.3** The Middle East and North Africa compared in latitude with the United States.

for example, is nonnative, mainly from South Asian countries. Many of these workers complain of dreams broken by poor living conditions and low wages, and in the worst cases, of physical and emotional abuse shared by victims of human trafficking. 195

The population growth rates of the Middle East and North Africa as a whole suggest that the region is moving out of the second and into the third phase of the demographic transition. 72 Death rates have been on the decline for decades, and recently birth rates have begun falling. The result is a declining rate of population growth. The average annual rate of population change for the 21 countries, the Palestinian territories, and Western Sahara was 1.9 percent in 2014. This is higher than the world average of 1.2 percent but is a significant decline from the high of 3.0 in 1980.

These numbers seem promising for progress: as we saw in Chapter 3, a lowering of population in the LDCs is a fundamental requirement for their sustainable development. On closer examination, however, we can see a problem: the youth bulge present in all but a handful of the Middle East and North African countries (illustrated well in the age structure diagram in •**Figure 6.5**). Sixty percent of MENA's people are less than 25 years old. The number of youth (people ages 15–24) in the region is estimated to grow until peaking at 100 million in 2035.

If there were jobs for such large numbers of people, there would be little reason for concern. Unfortunately, this region has the highest unemployment rate of all the world's regions (even higher than Sub-Saharan Africa)—over 10 percent, according to official statistics; the actual rate is likely considerably

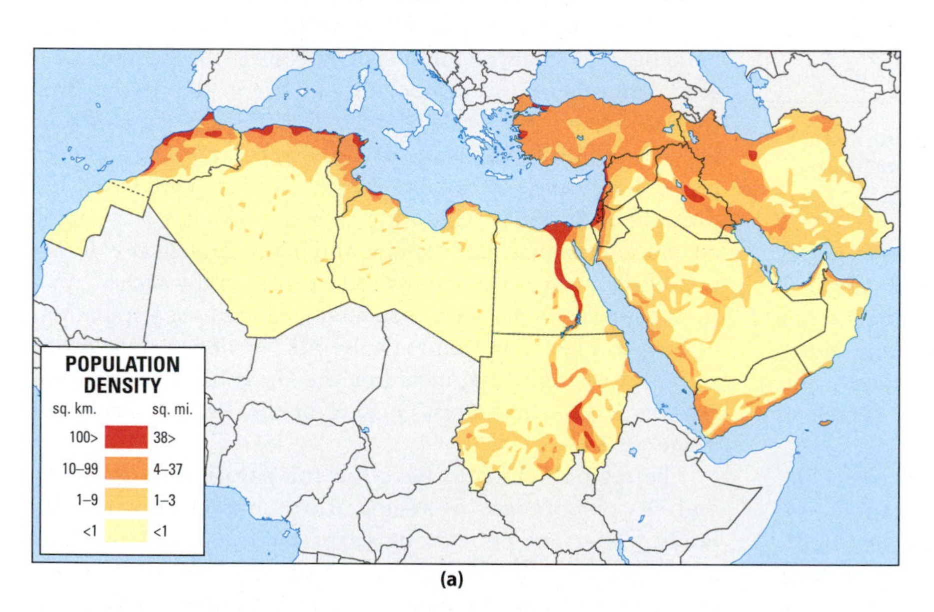

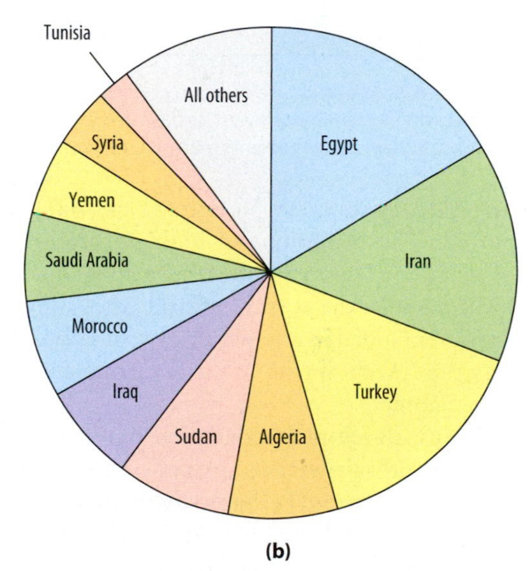

• **Figure 6.4** **(a)** Population distribution and **(b)** population pie chart in the Middle East and North Africa.

countries consider themselves too populous for their resources and economic bases, and have encouraged family planning. Birth rates have declined in these countries, which bodes well for their development.

At the other end of the spectrum, some oil-rich but population-poor countries of the Gulf nations have been encouraging their citizens to have more children. If there were more resident nationals, there would be less need to import foreign laborers and technicians, and these countries would be more self-sufficient in their development. Some oil-rich countries of the Gulf region have far more foreigners than citizens living in them; over 80 percent of the working-age population of the United Arab Emirates,

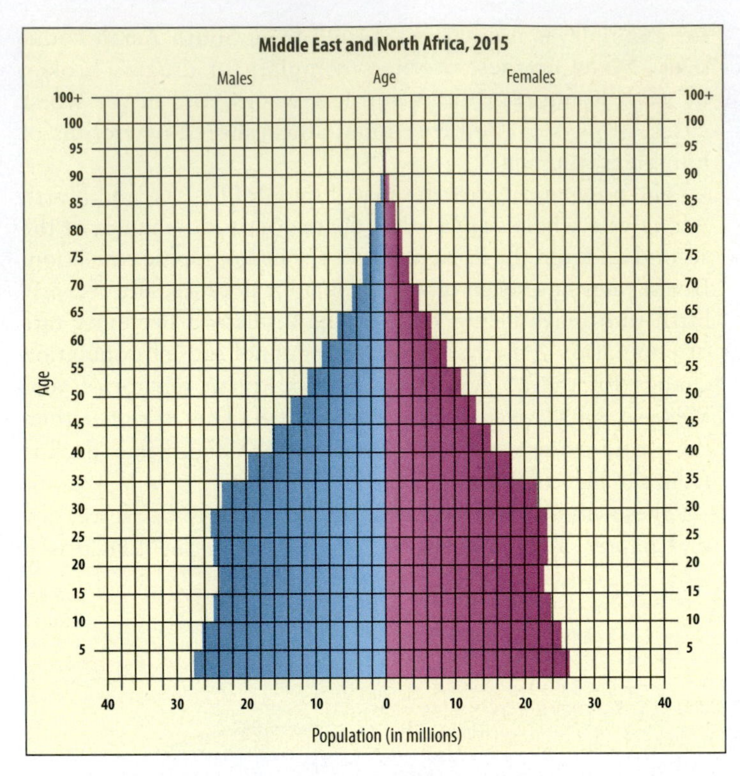

• **Figure 6.5** Age structure diagram of the Middle East and North Africa. Falling birth rates have narrowed its base somewhat, but the "youth bulge" is still prominent and represents many challenges.

higher. An estimated 30 percent of young people are unemployed. The conditions of high population growth and lack of economic opportunities have long made this region ripe for unrest. In the spring of 2011, the Middle East and North Africa exploded in a hopeful wave of popular revolutions. This was the "Arab Spring" that is introduced in **pages 255–259** of this chapter.

Many developing countries have economies largely dependent on subsistence agriculture, with low percentages of urban inhabitants. Perhaps surprisingly, however, the Middle East and North Africa have more urbanites than country folk. The average urban population among the 23 countries and territories is 63 percent. The most prosperous countries are also the most urban. Essentially city-states, Bahrain, Qatar, and Kuwait are 100 percent urban. The other oil-wealthy Gulf countries have urban populations over 70 percent. Consistent with its profile as a Western-style industrialized country without oil resources, Israel is 91 percent urban. At the other end of the spectrum, desperately poor Yemen and Sudan have urban populations of only 29 and 33 percent, respectively.

6.2 Physical Geography and Human Adaptations

The margins of the Middle East and North Africa are mainly oceans, seas, high mountains, and deserts (**Figure 6.1**). To the west lies the Atlantic Ocean; to the south, the Sahara and the highlands of East Africa; to the north, the Mediterranean, Black, and Caspian Seas, together with mountains and deserts lining the southern land frontiers of Russia and the Near Abroad; and to the east, the mountains on the Iran-Afghanistan frontier and the Baluchistan Desert, straddling Iran and Pakistan. These landscapes are arid plains and plateaus, together with large areas of rugged mountains and isolated "seas" of sand and "islands" or ribbons of water and fertility known as **oases**. Despite its environmental challenges, this region has given rise to some of the world's oldest and most influential ways of living.

A Region of Stark Geographic Contrasts

Aridity dominates the Middle East and North Africa (see the climate and biome maps of **•Figure 6.6**). At least three-fourths of the region has average yearly precipitation of less than 10 inches (25 cm), an amount too small for most types of dry farming (unirrigated agriculture). Sometimes, however, localized cloudbursts release moisture that allows plants, animals, and small populations of people—the Bedouin, Tuareg, and other pastoral nomads—to live in the desert. Even the vast Sahara, the world's largest desert, supports a surprising diversity and abundance of life. Plants, animals, and even people have developed adaptations of **drought avoidance** and **drought endurance**. Migrating to avoid drought is a coping adaptation that pastoral nomads and many animals use. Other organisms, such as trees, must endure drought by such adaptations as having very deep roots and small leaves. Populations of people, plants, and animals are all but nonexistent in the region's vast **sand seas**, including the Great Sand Sea of western Egypt and the Empty Quarter of the Arabian Peninsula (**•Figure 6.7**).

The region's climates have the comparatively large daily and seasonal ranges of temperature characteristic of dry lands. Desert nights can be surprisingly cool. Most days and nights are cloudless, so the heat absorbed on the desert surface during the day is lost by radiative cooling into the atmosphere at night. Summers in the lowlands are very hot almost everywhere, with temperatures regularly exceeding 100°F (38°C) daily for weeks or even months at a time. The average high temperature in July in Dubai is 105°F; in Riyadh, 108°F; and in Baghdad, 111°F. The Persian Gulf country Qatar is in these scorching ranks too. It raised a lot of eyebrows and led to arrests of the officials responsible for corrupt decision making when Qatar was awarded the 2022 World Cup; the soccer championship games normally held in summer.

Human settlements located near the sand seas often experience the unpleasant combination of high temperatures and hot, sand-laden winds, creating the sandstorms known locally by such names as *simuum* ("poison"), *sirocco*, and *haboob* (the latter, ushered in by an ominous, towering wall of sand, is also familiar to the people of southern Arizona; **•Figure 6.8**). Only in mountainous sections and in some places near the sea do higher elevations or sea breezes temper the

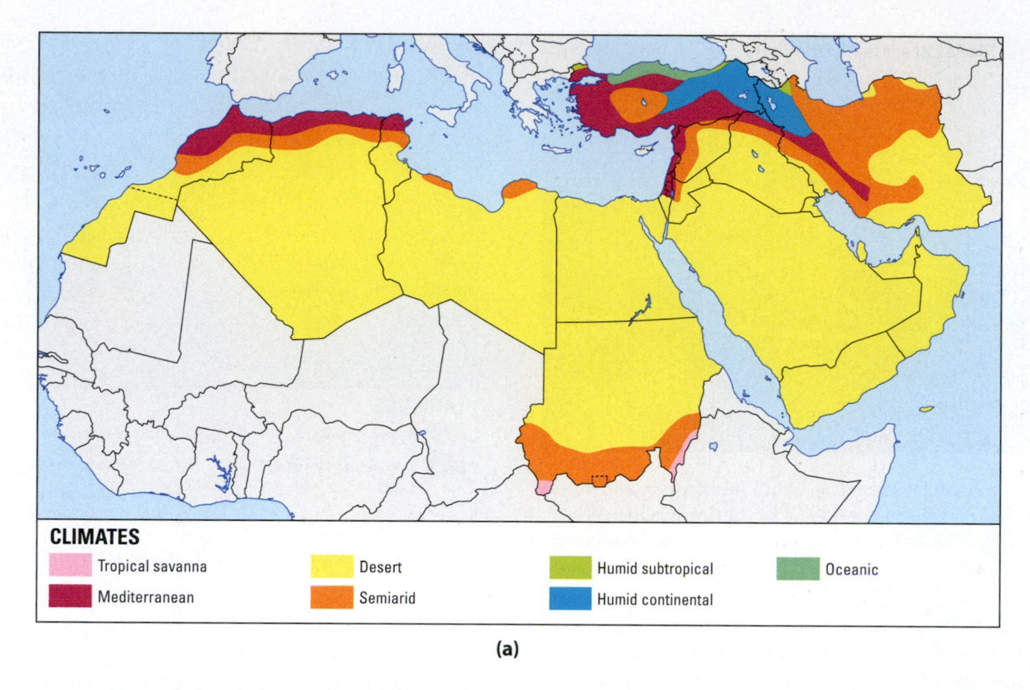

(a)

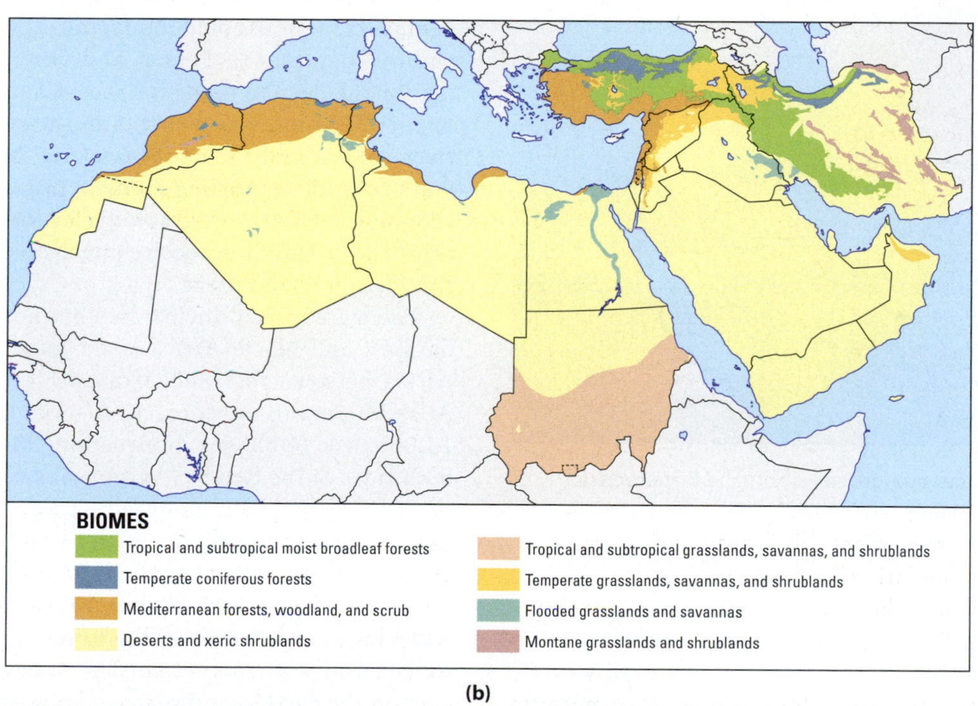

(b)

• **Figure 6.6** **(a)** Climates and **(b)** biomes of the Middle East and North Africa.

intense midsummer heat. The population of Alexandria, on the Mediterranean coast, explodes in summer as Egyptians flee from Cairo and other hot inland locations. In Saudi Arabia, the government relocates from Riyadh to the highland summer capital of Taif in summer, to escape the lowland furnace.

Lower winter temperatures bring relief from the summer heat, and the more favored places receive enough precipitation for dry farming of winter wheat, barley, and other cool-season crops. Land uses across the region are depicted in •**Figure 6.9**.

In general, winters are mild across the region. But cold winters and snowfalls occur in the mountainous areas of Iran and Turkey (there are even glaciers!). These locales are in the only zone of humid continental climate in the region (see **Figure 6.6a**). Only in the southernmost reaches of the region, notably in Sudan, do temperatures remain consistently high throughout the year. A semiarid climate and tropical grassland and savanna biome prevail there.

Except for extremely dry Libya and Egypt, most areas bordering the Mediterranean Sea receive 10 to 30 inches (25 to

Joe Hobbs

• **Figure 6.7** Seas of sand cover large areas in Saudi Arabia, Iran, and parts of the Sahara. This blacktop cuts through an edge of the Empty Quarter (al-Rub' al-Khali) in the United Arab Emirates. Sandstorms and even modest winds regularly bury this road.

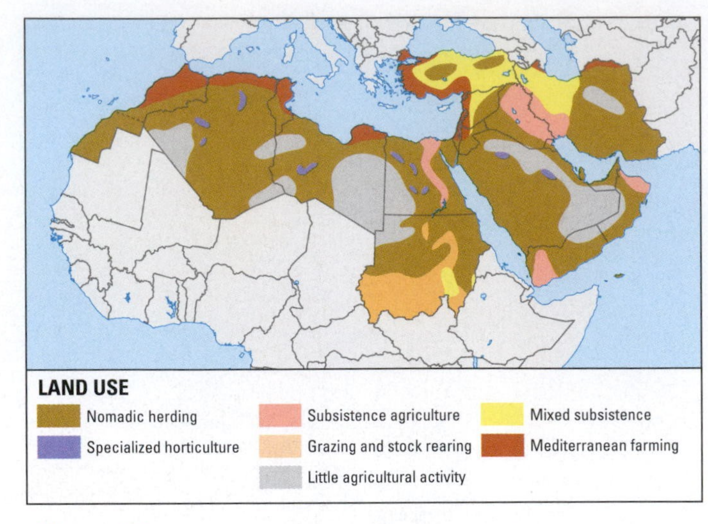

• **Figure 6.9** Land use in the Middle East and North Africa.

UniversalImagesGroup/Getty Images

• **Figure 6.8** A *haboob* is a wind-driven wall of sand that envelops everything in its path. Fine grains of sand penetrate even closed windows, leaving a gritty layer indoors. Cleaning up after a *haboob* like this one that swallowed the Al-Asad military base in Iraq in 2005 is a huge chore.

75 cm) of precipitation a year, falling almost exclusively in winter, whereas the summer is dry and warm—a typical Mediterranean climate pattern. Throughout history, some people without access to perennial streams have stored this moisture to make it available later for growing crops that require the higher temperatures of the summer months. The Nabateans, for example, who were contemporaries of the Romans in what is now Jordan, had a sophisticated network of limestone cisterns and irrigation channels to water their summer crops (•**Figure 6.10**).

In a few areas of the region, there is enough rainfall to support dry farming during the summer. There are some places that defy the usual perception of the region as bone-dry. The Black Sea side of Turkey's Pontic (Kuzey) Mountains, for example, is lush and moist in the summer, and tea grows well there (•**Figure 6.11**). In the southwestern Arabian Peninsula, a monsoonal climate brings summer rainfall and autumn harvests to Yemen and Oman, probably accounting for the Roman name for the area: *Arabia Felix* ("Happy Arabia"). Mountainous areas in the region, along with the river valleys and the margins of the Mediterranean, play a vital role in supporting human populations and national economies. Due to orographic (elevation-induced) precipitation, the mountains receive much more rainfall than surrounding lowland areas.

There are three principal mountainous regions of the Middle East and North Africa (see Figure 6.1). In northwestern Africa between the Mediterranean Sea and the Sahara, the Atlas Mountains of Morocco, Algeria, and Tunisia reach over 13,000 feet (4000 m) in elevation. Mountains also rise on both sides of the Red Sea, with peaks up to 12,336 feet (3760 m) in Yemen. These mountains rose with the tectonic processes that are pulling the African and Arabian plates apart at the northern end of the Great Rift Valley. The hinge of this crustal movement is the Bekaa Valley of Lebanon, where the widening fault line follows the Jordan River Valley southward to the Dead Sea (•**Figure 6.12**). This valley is the deepest depression on the earth's land surface, lying about 600 feet (180 m) below sea level at Lake Kinneret (also known as the Sea of Galilee or Lake Tiberias) and 1400 feet (400 m) below sea level at the shore of the Dead Sea, the lowest point of land on the planet. The rift then continues southward, through the very deep Gulf of Aqaba and Red Sea, before turning inland into Africa at Djibouti and Ethiopia, and continuing all the way down the Great Rift Valley of Africa (see **page 410**). 26, 410 Should you visit the Dead Sea one day, take a dip, and try to dive below the water's surface. It's nearly impossible: you will bob like a cork in its waters. With no outlet and a high rate of evaporation, the Dead Sea is one of the saltiest lakes in the world, with a salinity of 33 percent. Its North American counterpart, the Great Salt Lake in Utah, has a salinity of about 20 percent.

Joe Hobbs

• **Figure 6.10** The Treasury, a temple carved in red sandstone, probably in the 1st century CE, by the Nabateans at their capital of Petra in southern Jordan. The English poet Dean Burgen called Petra "a rose red city, half as old as time." The Nabateans built sophisticated networks for water storage and distribution. For scale, note the man at the lower right.

Joe Hobbs

• **Figure 6.11** Orographic precipitation is heavy on the Black Sea side of Turkey's Pontic (also known as Pontus or Kuzey) Mountains. Note that roofs are pitched to shed precipitation; in the region's drier areas, most roofs are flat. The crop growing here is corn.

In tectonic processes, one consequence of rifting in one place is the collision of the Earth's crustal plates elsewhere. There are several such collision zones in the Middle East and North Africa, particularly in Turkey and Iran, where mountain building has resulted. These are seismically active zones—meaning earthquakes occur there—and rarely does a year go by without a devastating quake rocking Turkey or Iran.

A large area of mountains, including the region's highest peaks, stretches across Turkey and Iran. The loftiest mountain ranges in Turkey are the Pontic, Taurus (• **Figure 6.13**), and Anti-Taurus, and in Iran, the Elburz and Zagros Mountains. These chains radiate outward from the rugged Armenian Knot in the tangled border country where Turkey, Iran, and the countries of the Caucasus meet. In Turkey, the glacier-fringed summit of the extinct volcano Mt. Ararat soars to 16,854 feet (5137 m) over the frontier between Turkey, Armenia, and Iran. Armenians resent that this mountain, which is sacred to them, fell into Turkish hands after World War I. Some people 199 believe that Mt. Ararat is where Noah's Ark lies, and many expeditions have sought it, for in Genesis it is written that the vessel came to rest "in the mountains of Ararat." (Incidentally, like so many things, the accounts of Noah are shared by all three Abrahamic traditions of Judaism, Christianity, and Islam.)

Extensive forests existed in early historical times in the Middle East and North Africa, particularly in these mountainous areas, but overcutting and overgrazing have almost eliminated them. Since the dawn of civilization in this area, around 3000 BCE, people have cut timber for construction and fuel faster than nature could replace it. Lebanon, described in ancient times as "an oasis of green with running creeks" and "a vast forest whose branches hide the sky," illustrates the result. If not for human impacts, forests would cover large parts of Lebanon today. On many mountainsides, the dominant tree would be the cedar, the tree that is the centerpiece of Lebanon's

Amiko Kauderer/NASA

• Figure 6.12 An astronaut aboard the Space Shuttle *Columbia* snapped this breathtaking photo of the northern end of the Great Rift Valley in 2002. At the bottom is the Red Sea, which traces the rift as it emerges from the African continent, and then splits into two arms. The shallow Gulf of Suez is the left arm. The deep Gulf of Aqaba, which continues the Red Sea rift, is framed by Egypt's Sinai Peninsula to the west and northwestern Saudi Arabia to the east. The Gulf of Aqaba points straight up toward the land portion of the rift. The rift's trough is clearly visible, descending northward to the lowest point on Earth (about 1300 feet below sea level) at the shoreline of the Dead Sea, bordered by Israel and the West Bank to the west and Jordan to the east. The proposed Two Seas Canal would allow the gravity flow of water from the Gulf of Aqaba to the Dead Sea to generate electricity and replenish the retreating waters of the Dead Sea, into which the depleted Jordan River flows.

Joe Hobbs

• Figure 6.13 The Taurus Mountains of southeastern Turkey.

flag. Historically this hardwood was highly valued throughout the Mediterranean Basin. The Phoenician civilization, with its base in Lebanon, grew wealthy by trading cedarwood. Ancient Egyptians were among those who prized the cedar. The solar boat of King Cheops, dating to the third millennium BCE, would carry this pharaoh's soul in the afterlife, and it had to be built of the best wood: cedar of Lebanon. Lebanon cedar was also used in building King Tutankhamen's funerary shrines and the First Temple in Jerusalem—the House of God on Earth and the abode of the Ark of the Covenant (**•Figure 6.14**). Today, 224 only a few isolated groves of cedar remain in Lebanon. Some of these are on the grounds of monasteries and survive because they grow in these sacred places. One grove is called "The Cedars of God."

As with the remnant cedars of Lebanon, the "primordial" or potential vegetation of Morocco's Atlas Mountains can be reconstructed by looking at the vegetation in the untouched sacred space surrounding saints' tombs.[2] This is another fine example of the importance of environmental perception in shaping cultural landscapes. In this case, 125 Islamic beliefs about the sacred space surrounding saints' tombs have protected the plants and animals living there, offering us a window into the environmental history of the Atlas Mountains.

Villager, Pastoral Nomad, Urbanite: The Ecological Trilogy

Who are the peoples who have created the cultural landscapes of the Middle East and North Africa? In the 1960s, the American geographer Paul English developed a useful model for understanding relationships between the three ancient ways of life that, although highly modified, still characterize the region today: villager, pastoral nomad, and urbanite. Each of these modes of living is rooted in a particular physical environment. Villagers are the subsistence

Joe Hobbs

• Figure 6.14 Ancient Egyptians believed that the solar boat of King Cheops of Egypt (c. 2500 BCE), builder of the Great Pyramid, would carry the pharaoh's spirit through the firmament. It had to be made of the best wood—cedar from Lebanon. The solar boat was interred next to the pyramid and now stands in a specially constructed museum.

farmers of rural areas where dry farming or irrigation is possible; pastoral nomads are the desert peoples who migrate through arid lands with their livestock, following patterns of rainfall and vegetation; and urbanites are the inhabitants of the large towns and cities, generally located near bountiful water sources but sometimes situated for particular trade, religious, or other reasons. Describing these ways of life as components of the **Middle Eastern ecological trilogy**, English explained how each of them has a characteristic, usually mutually beneficial pattern of interaction with the other two (• **Figure 6.15**).

The peasant farmers of Middle Eastern and North African villages (the **villagers**) represent the cornerstone of the trilogy. They grow the staple food crops such as wheat and barley that feed both the city dweller and the pastoral nomad of the desert. Neither urbanite nor nomad could live without them. The village also provides the city, often unwillingly, with tax revenue, soldiers, and workers. And before the

• **Figure 6.15** People, environments, and interactions of the ecological trilogy (here, villagers harvesting sugarcane in upper Egypt, Bedouin at camp in Egypt's Eastern Desert, and shoppers at the main gate to the historic city of San'a, Yemen). This relationship is generally symbiotic, although historically both urbanites and pastoral nomads preyed on the villagers, who are the trilogy's cornerstone.

THINK CRITICALLY: Who benefits most and who suffers most in the trilogy? Although villagers are the cornerstone, which of the three elements is the most dominant and powerful, and in what ways?

mid-20th century, villages provided **pastoral nomads** with plunder as the desert dwellers raided their settlements and caravan supply lines. Generally, however, the exchange is mutually beneficial.

The nomads provide villagers with livestock products, including live animals, meat, milk, cheese, hides, and wool, and with desert herbs and medicines. Educated and progressive **urbanites** provide technological innovations, manufactured goods, religious and secular education and training, and cultural amenities (today including films, music, and Internet resources).

There is little direct interaction between urbanites and pastoral nomads, although some manufactured goods such as clothing travel from city to desert, and some desert folk medicines pass from desert to city. Historically, the exchange was sometimes violent, as urban-based governments have sought to control the movements and military capabilities of the elusive and sometimes hostile nomads. Pastoral nomads once plundered rich caravans plying the major overland trade routes of the Middle East and North Africa. Later, some took to smuggling and other illegal pursuits. Governments did not tolerate such activities and often cracked down hard on the nomads they were able to catch. Even today you can witness Egyptian authorities trying to quell rebellious Bedouin in the Sinai Peninsula.

Later, in the 1970s, English wrote an article marking the "passing of the ecological trilogy." He noted that cities were encroaching on villages; villagers were migrating into cities and giving some neighborhoods a rural aspect; and pastoral nomads were settling down—thus the trilogy no longer existed. In reality, although the makeup and interactions of its parts have changed, the trilogy model is still valid and useful as an introduction to the major ways of life in the Middle East and North Africa. It is especially significant that a given person in the region strongly identifies as a villager, a pastoral nomad, or an urbanite. This perception of self has an important bearing on how these people of very different backgrounds interact, even when they live in close proximity. Urban officials may work in rural village areas, but they remain at heart and in their perspectives city people, and usually live apart from farmers. Extended families of pastoral nomads may settle down and become farmers, but they cluster in neighborhoods and continue to identify themselves by affiliation with the nomadic tribe. Many still venture into the desert to harvest wild resources on a seasonal basis, and they retain marriage and other ties with desert-dwelling relatives.

Let's take a closer look at each component of the ecological trilogy.

The Village Way of Life

Agricultural villagers historically represented by far the majority populations in the Middle East and North Africa; only in recent decades have urbanites begun to outnumber them. In the region's typically dry environments, the villages are located near reliable water sources with cultivable lands nearby. They are usually made up of closely related family groups, with many fields owned by an absentee landlord. Villagers typically live in closely spaced homes made of mud brick or concrete blocks. Production and consumption focus on a staple grain such as wheat, barley, or rice. As land for growing fodder is often in short supply, villagers keep only a small number of sheep and goats, and rely in part on nomads for pastoral produce. Residents of a given village usually share common ties of kinship, religion, ritual, and custom, and the changing demands of agricultural seasons regulate their patterns of activity.

Since the mid-18th century, these patterns of village life have been increasingly exposed to outside influences. European colonialism brought significant economic changes, including the introduction of cash crops and modern facilities to ship them. Improved and expanded irrigation, financed initially with capital from the West, brought more land under cultivation. Recent agents of change have been the countries' own government-supported doctors, teachers, and land reform officers. Modern technologies such as motor vehicles, gasoline-powered water pumps, radio, television, and—most recently—the Internet and cell phones have modified old patterns of living. The educated and ambitious, as well as the unskilled and desperately poor—motivated, respectively, by pull and push factors (see **Chapter 3**)—have migrated to urban areas. Improved roads and communications have in turn carried urban influences to villages, prompting villagers to become more integrated into the national society.

76

The Pastoral Nomadic Way of Life

Pastoral nomadism emerged as an offshoot of the village agricultural way of life not long after plants and animals were first domesticated in the Middle East (about 13,000 years ago). Rainfall and the wild fodder it germinates, although scattered, are sufficient resources to support small groups of people who migrate with their sheep, goats, and camels (and in some locales, cattle) to take advantage of this changing resource base (see Perspectives from the Field, **page 219**). In addition to selling or trading livestock to obtain food, tea, sugar, clothing, and other essentials from settled communities, pastoral nomads hunt; gather; work for wages; and, where possible, grow crops. Many now work in the tourist industry because Westerners hungry for insight into traditional cultures seek them out. Their multifaceted livelihood has been described as a strategy of **risk minimization** based on the exploitation of multiple resources so that some will support them if others fail.

Although renowned in Middle Eastern legends, pastoral nomads have been described as "more glamorous than numerous." It is impossible to get census figures on the number living in the deserts of the Middle East and North Africa, but they

Perspectives from the Field Way-Finding in the Desert

I have many interests within geography, but mainly I am a geographer of the Arabic-speaking Middle East, with an expertise on pastoral nomadism. I was among a handful of Middle East geographers trained at the University of Texas at Austin before regional studies fell out of favor. In Chapter 5, I lamented that Russian studies were likewise largely shuttered across the country (**page 149**). We are paying the price now: Where are the Middle East and Russia experts we desperately need?

The vehicle for my UT doctoral dissertation research was an 18-month journey with the Khushmaan Ma'aza, a clan of Bedouin nomads living in the northern half of Egypt's Eastern Desert, between the Nile River and the Red Sea. I studied their perceptions, knowledge, and uses of natural resources, and tried to understand how their worldviews and kinship patterns apparently helped them protect and sensibly use their scarce resources. I have worked with these people from the early 1980s to the present, most recently studying how they carefully protect acacia trees, which offer them many resources when drought prevents the growth of annual plants. I have always felt privileged to be invited into their world. They experience and perceive a world completely different from ours. They "read" the landscape and its clues with a set of skills I could never develop. In the following passage from my book *Bedouin Life in the Egyptian Wilderness*, I marvel at their way-finding abilities.

> The process by which the Khushmaan nomads have developed roots in their landscape, fashioning subjective "place" from anonymous "space," and the means by which they orient themselves and use places on a daily basis deserve special attention: these are essential parts of the Bedouins' identity and profoundly influence how they use resources and affect the desert ecosystem.
>
> The Bedouin have little need or knowledge of maps. On the other hand, when shown a map or aerial image of countryside they know, the nomads accurately orient and interpret it, naming the mountains and drainages depicted. Saalih Ali (**• Figure 6.A**) especially enjoyed my star chart. He would tell me what time the Pleiades would rise, for example, and ask me what time the chart indicated. The two were almost always in agreement.
>
> Many desert travelers have been astonished by the nomads' navigational and tracking ability, calling it a "sixth sense." The Khushmaan are exceptional way-finders and topographical interpreters able, for instance, to tell from tracks whether a camel was carrying baggage or a man; whether gazelle tracks were made by a male or female; which way a car was traveling and what make it was; which man left a set of footprints, even if he wore sandals; and how old the tracks are. Bedouins are proud of their geographical skills, which, they believe, distinguish them from settled people. Saalih told me, "Your people don't need to know the country but we do, to know exactly where things are in order to live." Pointing to his head, he said, "My map is here."
>
> The difference in the way-finding abilities of nomads and settled persons may be due to the greater survival value of these skills for nomads and to the more complex meanings they attribute to locations. Bedouin places are rich in ideological and practical significance. The nomads interpret and interact with these meanings on a regular basis and create new places in their lifetimes. Theirs is an experience of belonging and becoming with the landscape, whereas settled people are more likely to inherit and accommodate themselves to a given set of places.
>
> The nomads' homeland is so vast, and the margin of survivability in it so narrow, that topographic knowledge must be encyclopedic. Places either are the resources that allow human life in the desert or are the signposts that lead to resources. A Khushmaan man pinpointed the role of places in desert survival: "Places have names so that people do not get lost. They can learn where water and other things are by using place names." Tragedy may result if the nomad has insufficient or incorrect information about places: many deaths by thirst are attributed to faulty directions for finding water.*

Joe Hobbs

• Figure 6.A Saalih Ali, a Bedouin of the Khushmaan Ma'aza tribe, nearly 30 years after he took me in as his "wayfellow." He and his kin have brought me years of joy, and I credit Saalih for my professional life too: on a mountaintop he invited me to return for an extended stay with his people, and only then did I decide to pursue the PhD in geography. My dissertation and many of my publications are about his tribespeople and their perceptions and uses of desert resources.

*From Joseph Hobbs, *Bedouin Life in the Egyptian Wilderness* (Austin: University of Texas Press, 1989), pp. 81–82. Copyright © 1989 University of Texas Press. Reprinted with permission. For more information on Bedouin place-making, see Joseph Hobbs, "Bedouin Place Names in the Eastern Desert of Egypt," in Dawn Chatty (ed.), *Reshaping Tribal Identities in the Contemporary Arab World: Politics, Self-Representation, and the Construction of Bedouin History*, Nomadic Peoples Special Issue 18(2) (2014): 123–146.

are very small relative to the numbers of villagers and urbanites. Recent decades have witnessed the rapid and progressive settling down, or **sedentarization**, of the nomads—a process attributed to a variety of causes. In some cases, prolonged drought virtually eliminated the resource base on which the nomads depended. Traditionally, they were able to migrate far enough to find new pastures, but modern national boundaries now inhibit such movements. Some have returned with the rains to their desert homelands, but others have chosen to remain as farmers or wage laborers in villages and towns. In the Arabian Peninsula, the prosperity and technological changes prompted by oil revenues made rapid inroads into the material culture—and then the livelihood preferences—of the desert people; many preferred the comforts of settled life. Some governments, notably those of Israel and prerevolutionary Iran, were frustrated by their inability to count, tax, conscript, and control a sizable migrant population, and therefore forced the nomads to settle.

Pastoral nomads of the Middle East and North Africa identify themselves primarily by their **tribe**, not by their nationality. The major ethnic groups from which these tribes draw are the Arabic-speaking Bedouin of the Arabian Peninsula and adjacent lands, the Berber and Tuareg of North Africa, the Kababish and Bisharin of Sudan, the Yörük and Kurds of Turkey, and the Qashai and Bakhtiari of Iran. Members of a tribe claim common descent from a single male ancestor who lived countless generations ago; their kinship organization is thus a patrilineal descent system. It is also a segmentary kinship system, so called because there are smaller subsections of the tribe, known as clans and lineages, that are functionally important in daily life. Members of the most closely related families comprising the lineage, for example, share livestock, wells, trees, and other resources. Both the larger clans, made up of numerous lineages, and the tribes possess territories. Members of a clan or tribe typically allow members of another clan or tribe to use the resources within its territory on the basis of **usufruct**, or nondestructive mutual use. That arrangement has often proved to be lifesaving in an environment where water often is in dangerously low supply.

Although some detractors have depicted pastoral nomads as the "fathers" rather than "sons" of the desert, blaming them for wanton destruction of game animals and vegetation, there are numerous examples of pastoral nomadic groups that have developed indigenous and very effective systems of resource conservation. Most of these practices depend on the kinship groups of family, lineage, clan, and tribe to assume responsibility for protecting plants and animals.

The Urban Way of Life

The city was the final component to emerge in the ecological trilogy, beginning in about 4000 BCE in Mesopotamia (modern Iraq) and 3000 BCE in Egypt. Although they resembled villages in many ways, early cities were distinguished by their larger populations (more than 5000 people), the use of written languages, and the presence of monumental temples and other ceremonial centers. The early Mesopotamian city and, after the 7th century CE, the classic Islamic city, called the **medina**, had several structural elements in common (•**Figure 6.16**). The medina had a high surrounding wall built for defensive purposes. The congregational mosque and often an attached administrative and educational complex dominated the city center. Although Islam is often characterized as a faith of the desert, religious life has always been focused in, and diffused from, the cities. The importance of the city's congregational mosque in religious and everyday life is often emphasized by its large size and outstanding artistic execution.

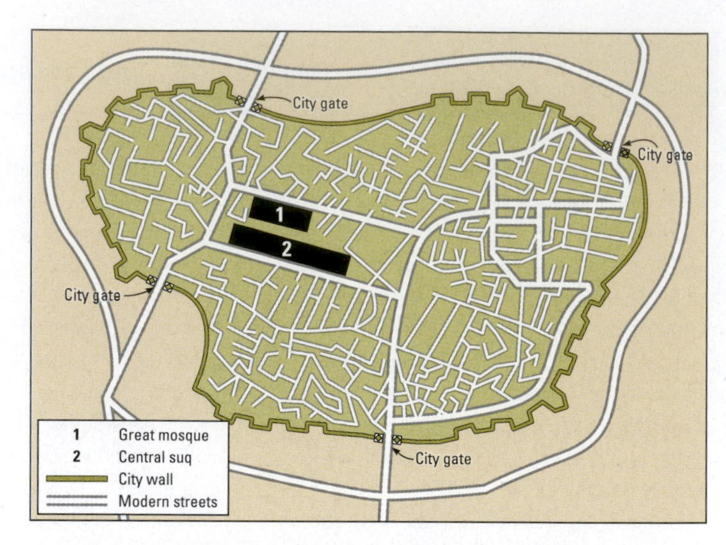

• **Figure 6.16** An idealized model of the classic medina, or Muslim Middle Eastern city. Figure 6.D shows the ethnic quarters that were also typical of the medina.

THINK CRITICALLY: Referring to the text, how does the medina differ ethnically, economically, and spatially from the modern Western city?

A large commercial zone, known as a ***bazaar*** in Persian and a ***suq*** in Arabic (and recognizable as the ancestor of the modern shopping mall), typically adjoined the ceremonial and administrative heart of the city (•**Figure 6.17**). Merchants and craftspeople selling various goods occupied separate spaces within this complex, and visitors to an old medina today can still find sections devoted exclusively to the sale of spices, carpets, gold, silver, traditional medicines, televisions and other electronics, and other categories of goods. Smaller clusters of shops and workshops were located at the city gates. The surviving historic suqs and bazaars are beguiling and endlessly fascinating places, partly because merchants always draw you in with a cup of tea or cold soda, sit you down, engage you in conversation, and, of course, work on the sale. Whether you buy or not, you are likely to enjoy the experience, especially if it is in one of those ancient markets as in Istanbul, Urfa, Isfahan, Cairo, or Marrakech. One of the oldest and most vibrant suqs was Aleppo's in Syria. A UNESCO World Heritage Site, it was a heartbreaking casualty of the Syrian civil war (more of these are related on **page 262**).

Joe Hobbs

• **Figure 6.17** The lanes of the Middle Eastern suq or bazaar (market) are busy, lively places. This one is in Cairo, Egypt.

Residential areas were differentiated as quarters, not by income group, but by ethnicity; the medina of Jerusalem, for example, still has distinct Jewish, Arab, Armenian, and non-Armenian Christian quarters (see •**Figure 6.D**, **page 225**). Homes tended to face inward toward a quiet central courtyard, buffering the occupants from the noise and bustle of the street. The narrow, winding streets of the medina were intended for foot traffic and small animal-drawn carts, not for large motor vehicles, a fact that accounts for the traffic jams in some old Middle Eastern and North African cities today and for the wholesale destruction of the medina, making way for modern throughways, in others.

The medinas that survive today are gently decaying vestiges of a forgotten urban pattern. Periods of European colonialism and subsequent nationalism changed the face and orientation of the city. During the colonial age, resident Europeans preferred to live in more spacious settings at the outer edges of the city, and later the national elite followed this pattern. In recent times, independent governments have adopted Western building styles, with broad traffic arteries cutting through the old quarters, and large central squares near government buildings. This opening up of the cityscape has spread commercial activity along the wide avenues, diluting the prime importance of the central bazaar as the focus of trade.

Rural-to-urban migration and the city's own internal growth contribute to a rapid rate of urbanization that puts enormous pressure on services in the region's poorer countries. Governments often build high-rise public housing to accommodate the growing population, contributing to a cycle in which the urban poor move into the new dwellings, only to leave their old quarters as a vacuum drawing in still more rural migrants. In Cairo, millions of former villagers now live in the "City of the Dead," an extraordinary urban landscape composed of multistory dwellings erected above graves—a last resort for the living poor with no other place to go. The overwhelmingly largest city, or primate city, so characteristic of Middle Eastern and North African capitals, thus grows at the expense of the smaller original city. Two of the primate cities of MENA, Cairo and Istanbul, are megacities, defined as having more than 10 million inhabitants (see **Figure 3.E** and **page 78**). Istanbul (along with Dubai) is also a "global city" (also known as a "world city" or "alpha city"), a city very important in the global economy. The region's major metropolitan areas and their populations are listed in •**Table 6.2**.

76 129 127

Much of the domestic rural-to-urban migration and subsequent urban gridlock and squalor could be prevented if governments invested more in the development of villages and smaller cities. The oil-rich countries with relatively small populations generally enjoy an urban standard of living equaling that of affluent Western countries. Modern industrial cities such as Saudi Arabia's Jubail and others founded on oil wealth were built virtually overnight, providing fascinating contrast with the region's colorful, complex ancient cities. With the recent resurgence of oil revenue into Saudi Arabia and the other Gulf countries, some of their cities boast the most elegant, sophisticated, and expensive amenities and architecture to be found anywhere in the world (see Problem Landscape, **page 222**). As we will see in the discussion of the Gulf region on **page 236**, international migrants, mainly from South Asia, have built and maintain these cities.

Table 6.2 Largest Cities in the Middle East and North Africa

City	Population
Cairo, Egypt	15.6
Tehran, Iran	13.5
Istanbul, Turkey	13.3
Baghdad, Iraq	6.6
Riyadh, Saudi Arabia	5.6
Khartoum, Sudan	5.1
Alexandria, Egypt	4.7
Ankara, Turkey	4.5
Kuwait City, Kuwait	4.3
Dubai, U.A.E.	3.9
Jiddah, Saudi Arabia	3.6
Aleppo, Syria	3.5

Population in millions.
Source: Demographia World Urban Areas, 2015.

Problem Landscape The Rise and Fall of Dubai's Artificial Islands

There are no constraints in the ultra-wealthy Persian Gulf emirate of Dubai, the preeminent "global city" in the Middle East. In a matter of decades, Dubai has transformed itself from a small trading port into a glitzy international transportation hub, business and financial center, and shopping capital. Dubai is literally over the top: at a height of 2717 feet (828 m), its Burj Khalifa, completed in 2010, stands as the world's tallest building (its primacy will fall to the 1000-meter-high Kingdom Tower, under construction in Jeddah, Saudi Arabia). The Dubai Mall is the world's largest, with over 1000 stores and an aquarium inside. In 2014, the mall received 80 million visitors—35 percent more than *all* of New York City. (Not to be outdone, the Mall of the Emirates, a few miles away, features a 300-foot-tall indoor ski slope with multiple trails serviced by snow machines; see **• Figure 6.B**.)

Dubai's land reclamation projects are this city-state's most unique and ostentatious symbols of its wealth. In 2001, Dubai began working with land and real estate developers to create a series of artificial islands in whimsical shapes including palm trees, elaborate crescents, and even a world map, all created from sand dredged up from the sea floor. The developers sold land to wealthy investors who would build mansions and resorts on the land they purchased. After construction began, conservationists argued that creating these islands was damaging the ecosystem by choking the waters with sediment, burying fragile coral reefs under sand, and threatening marine life. The World Wildlife Fund's annual reports showed Dubai as having the biggest ecological footprint in the world. The emirate also faced criticism about its excessive use of resources to create offshore vacation houses for the wealthy while the city's quarter-million foreign laborers lived in overcrowded tenements; worked long hours, often without pay, for weeks; and sometimes had their passports revoked so they would be unable to return home. Dubai's leadership pressed on, however, and the first island, Palm Jumeirah, was completed in 2007 at a cost of $12 billion. Within five years, two dozen hotels and a monorail had been built, and over 20,000 people were living on the island (**• Figure 6.C**).

William L. Stefanov/NASA

• Figure 6.C There seem to be no limits to the innovations that money can buy in the UAE's wealthy city-state of Dubai. These are Dubai's artificial islands Palm Jumeirah and The World, completed in 2008. Each required hundreds of millions of cubic meters of sand and tons of rock. The 300 "world" islands remain largely uninhabited. The better-built and protected Palm Jumeirah resort is on its way to housing 2000 villas, 40 luxury hotels, shopping centers, cinemas, and other facilities, and is meant to accommodate 500,000 people. Although the islands are touted as "being visible from the moon," an astronaut on the International Space Station captured this image from near-earth orbit.

Joe Hobbs

• Figure 6.B July in Dubai, one of the city-states of the United Arab Emirates. Wealth created initially on oil revenues has made it possible for some to buy or build almost anything, including a "ski resort" inside a shopping mall.

Dubai's most grandiose mega-project seemed to emerge from all the controversy as a success. But the fortunes of the island projects changed dramatically in 2008, when the global recession hit Dubai hard. The city's real estate prices plummeted 60 percent, and many investors pulled out of the projects. Suddenly, the islands' developers were faced with a glut of properties they could no longer sell profitably, and even the land that had been purchased remained undeveloped. As the economy contracted and construction workers were laid off in droves, work on the other islands was put on hold indefinitely. Two of the projects, Palm Deira and "The Universe," were cancelled outright.

To make matters worse, the islands were slowly receding back into the sea. Erosion was blowing away the sand above the waves, and the islands' foundations were sinking. This news drove away even more investors, and the developers were increasingly saddled with billions of dollars of debt. Except for Palm Jumeirah, the island projects sat desolate and uninhabited (save for the Greenland island within "The World" archipelago, where a villa owned by the emir of Dubai had been built).

By 2013, Dubai's economy had recovered sufficiently that buyers began expressing interest in island properties again. British billionaire Richard Branson purchased the Great Britain island of The World, and some other "European" islands are being developed into a collection of luxury villas. Still, The World was constructed at a reported cost of $14 billion (and significant ecological impact), and the majority of it remains undeveloped. Another island, Palm Jebel Ali, was built to be home for 250,000 people but remains completely vacant.

6.3 Cultural and Historical Geographies

Peoples of the Middle East and North Africa have made many fundamental contributions to humanity. During the Agricultural Revolution between 5,000 and 13,000 years ago, many of the plants and animals on which the world's agriculture is based today were first domesticated in the Middle East. 51 The list includes wheat, barley, sheep, goats, cattle, and pigs, whose wild ancestors were processed, manipulated, and bred until their physical makeup and behavior changed to suit human needs. The interaction between people and the wild plants and animals they eventually domesticated took place mainly in the well-watered **Fertile Crescent**, a vernacular region referring to an arc of land stretching from Israel to western Iran.

By about 6000 years ago, people sought higher yields by irrigating crops in the rich soils of the Tigris, Euphrates, and Nile River Valleys. Their efforts produced the enormous crop surpluses that allowed civilization—a cultural complex based on an urban way of life—to emerge in Mesopotamia (literally, in Greek, "the land between the rivers" Tigris and Euphrates) and in Egypt. Accomplishments in science, technology, art, architecture, language, mathematics, and other areas diffused outward from these centers of civilization. Egypt and Mesopotamia are thus among the world's great culture hearths. 53

Ethnicities and Languages

The Middle East and North Africa are sometimes mistakenly referred to as the "Arab world"; although the majority inhabitants are Arabs, this region has large populations of non-Arabs. An **Arab** is best defined as a person of Semitic Arab ethnicity whose ancestral language is *Arabic*, a **Semitic language** spoken by about 350 million of the region's people. Originally, the Arabs were inhabitants of the Arabian Peninsula, but conquests after their majority conversion to Islam took them, their language, and their Islamic culture as far west as Morocco and Spain.

The region is also the homeland of the **Jews**, whose definition today is complex. 108 Originally, Jews were both a distinct ethnic and linguistic group of the Middle East who practiced the religion of Judaism. It is possible for non-Jews to convert to Judaism (the convert is known as a proselyte, or "immigrant"), so a strictly ethnic definition does not apply today. Many Jews do not practice their religion but still consider themselves ethnically or culturally Jewish. Whatever debate exists about what defines a Jew, Jewish identity is strong and resilient.

It is important to recognize that although political circumstances made them enemies in the 20th century, Jews and Arabs lived in peace for centuries, and they share many cultural traits. Both recognize Abraham as their patriarch. Arabic is a Semitic language in the same **Afro-Asiatic language family** as *Hebrew*, which is spoken by most of the 6.1 million Jewish inhabitants of Israel (see the language map, **• Figure 6.18**).

There are other very large populations of non-Semitic ethnic groups and languages in the region. The greatest are the 65 million **Turks** of Turkey, who speak *Turkish*, a Turkic language, and the 55 million Persians of Iran, most of whom speak Persian (called *Farsi* in Persian) in the Indo-European language family. Both Persian and Arabic are written in Arabic script and so appear related, but they are not. Turkish was also written in Arabic script until early in the 20th century, but since then it has been written in the Latin alphabet. About 30 million **Kurds**—a people living in Turkey, Iraq, Iran, and Syria—speak *Kurdish*, which is also an Iranian language in the Indo-European family. Many people in North Africa, possibly 15 million, speak **Berber** and **Tuareg** (in the Afro-Asiatic language family).

The Abrahamic Faiths

Out of the Middle East came the closely related monotheistic faiths of Judaism, Christianity, and Islam. The region's human and political geographies are closely tied to these religions and

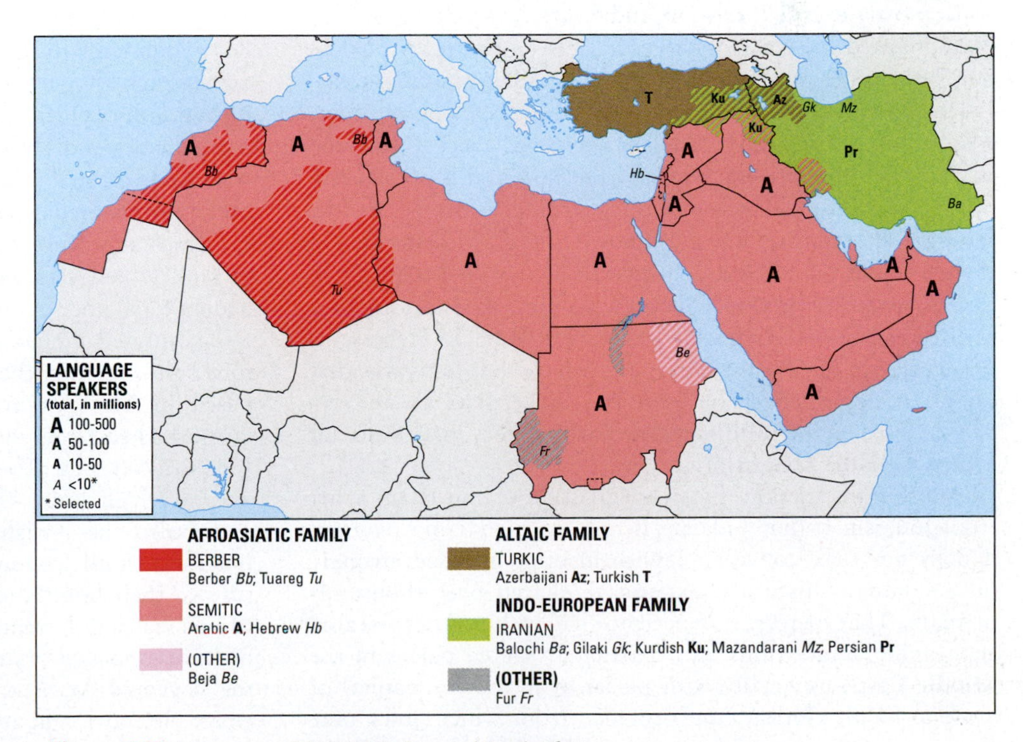

• Figure 6.18 Languages of the Middle East and North Africa.

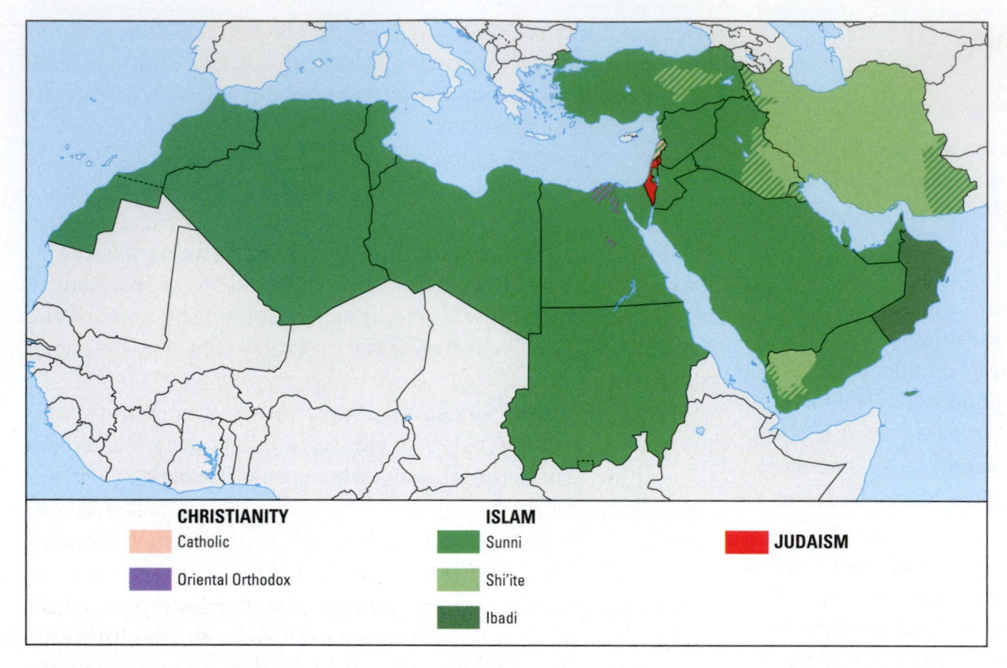

• **Figure 6.19** Religions of the Middle East and North Africa.

their holy places (• **Figure 6.19**). All three are Abrahamic traditions; they regard Abraham as their patriarch, and they share many fundamental beliefs. The first to emerge was the faith of Judaism.

The Promised Land of the Jews

Judaism, the first significant monotheistic faith as far as we know, is today practiced by about 15 million people worldwide, mostly in Israel, Europe, and North America. September 2015 marked the start of the year 5776 in the Jewish calendar, but unlike its kindred faiths—Christianity and Islam—Judaism does not have an acknowledged starting point in time. Also unlike Christianity, Judaism does not have a fixed creed or doctrine. Jews are encouraged to behave in this life in compliance with God's laws, which, according to the Torah (the Jewish holy scripture, which is known as the Old Testament in the Christian Bible) God delivered to Moses on Mount Sinai as a covenant with God's "chosen people." The coming of a savior known as the Messiah ("Anointed One" in Hebrew) is prophesied in the Torah; Christians believe that Jesus was that savior, as described in the New Testament; Jews do not recognize Jesus as the fulfillment of that prophecy and so do not accept the New Testament.

Another distinction that sets Christianity and Islam apart from Judaism is that Judaism is not a proselytizing religion; it does not seek converts. Jewish identity is based strongly on a common historical experience shared over thousands of years. That historical experience has included deep-seated geographic associations with particular sacred places in the Middle East—especially with places in Jerusalem, capital of ancient Judah (Judea), the province from which Jews take their name. Jewish history also has included unparalleled persecution. 108

Depending on one's perspective, the Jewish connection with the geographic region known as Palestine, the area now composed of Israel, the West Bank, and the Gaza Strip (see • **Figure 6.38**), is most significant on a time scale of either about 4000 years or about 100 years. According to the Bible, around 2000 BCE, God commanded Abraham and his kinspeople, known as **Hebrews** (later as Jews), to leave their home in what is now southern Iraq and settle in Canaan. God told Abraham that this land of Canaan—geographic Palestine—would belong to the Hebrews after a long period of persecution. The Hebrew Bible or Written Torah (the Christians' Old Testament) documents that the Hebrews did settle in Canaan until famine struck that land. At the command of Abraham's grandson Jacob, the Hebrews—known then as **Israelites** in reference to Jacob's descendants and to the geographical region of Israel and Judah—relocated to Egypt, where grain was plentiful. That began the long sojourn of the Israelites in Egypt, which, according to the Bible, ended in about 1200 BCE when Moses led them out on the **Exodus**, the journey toward Canaan.

According to Jewish history, the prophecy of Abraham was fulfilled when the Israelites settled once again in Canaan, their **Promised Land.** In about 1020 BCE, the Jewish King Saul unified the 12 tribes that descended from Jacob into the first united Kingdom of Israel. In about 950 BCE in Jerusalem—the capital of the kingdom enlarged by Saul's successor, David—King Solomon built Judaism's **First Temple.** He located it atop a great rock known to the Jews as *Even HaShetiyah,* the **Foundation Stone**, plucked from beneath the throne of God to become the center of the world and the core from which the entire world was created (see Geography of Sacred Space).[3] The Ark of the Covenant, containing the commandments that God gave to Moses atop Mount Sinai, was placed in the Temple's Holy of Holies. Believers and explorers have tried to find the real Mount Sinai and, of course, the Ark itself. 412

The united kingdom of Israel lasted about 200 years before splitting into the states of Israel and Judah. Empires based in Mesopotamia destroyed these states: the Assyrians attacked Israel in 721 BCE, and the Babylonians sacked Judah in 586 BCE. The Babylonians destroyed the First Temple (at which point the Ark of the Covenant disappeared) and exiled the Jewish people to Mesopotamia, where they remained until conquering Persians allowed them to return to their homeland. In about 520 BCE, the Jews who returned to Judah rebuilt the temple (the **Second Temple**) on its original site. A succession of foreign empires came to rule the Jews and Arabs of Palestine: Persian, Macedonian, Ptolemaic, Seleucid, and around the time of Jesus, Roman. Herod, the Jewish king who ruled under Roman authority and was a contemporary of Jesus, greatly enlarged the temple complex.

Geography of Sacred Space Jerusalem

The old city of Jerusalem is filled with sacred places and is one of the world's premier pilgrimage destinations. Over thousands of years, Jerusalem has been coveted and conquered by people of many different cultures and faiths. In the process, a place held sacred by one group has often come to be held sacred by a second and even a third.

The most important thing to notice on this map (**• Figure 6.D**) is that the location of the First and Second Jewish Temples is identical to that of the Muslim shrine known as the Dome of the Rock. We can assume that the great oblong rock within that shrine is the Foundation Stone around which the First and Second Temples were built (**• Figure 6.E**). Judaism, Islam, and Christianity are very close in their origins and many of their basic precepts, but their faiths are not conjoined or syncretized in practice, as, for example, the Maya and Catholic faiths have been in Mesoamerica (see **page 467**), or Buddhism and other traditions have been in East Asia (see **page 303**). In each faith, there has been a long tradition of quarreling over ownership of sacred space. Nowhere in the world does contested sacred space have more potential to ignite violence and even war than in Jerusalem (which means, in Hebrew, "City of Peace"). The later section on Regional Issues and Landscapes offers more insight into the role of Jerusalem's holy places in the region's turbulent history (see **page 253**). Figure 6.D depicts the major places sacred to Jews, Christians, and Muslims, and also the city's ethnic quarters; the walled old city is a classic medina.

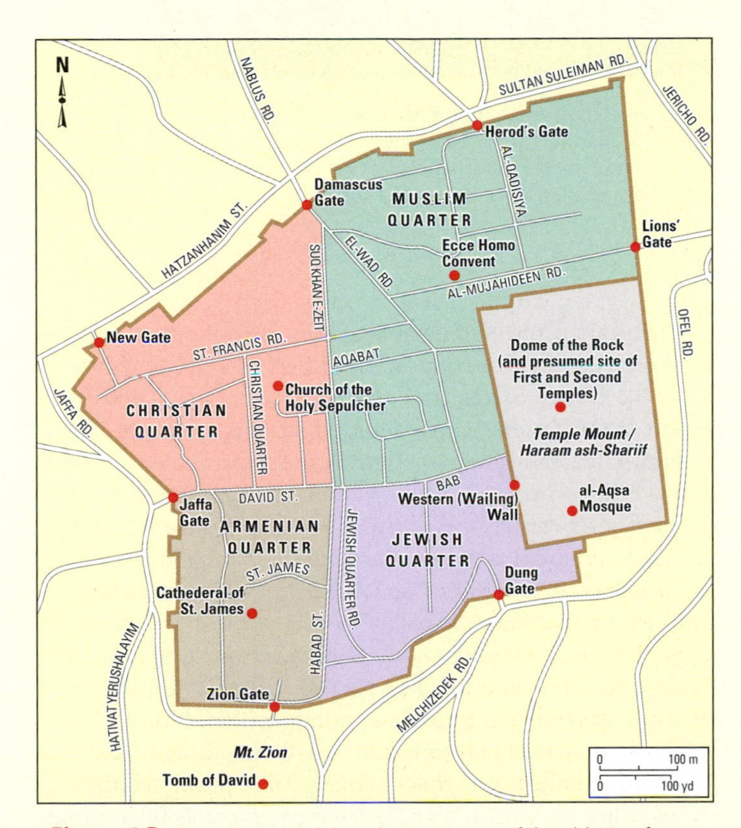

• Figure 6.D Sacred sites and the ethnic quarters of the old city of Jerusalem.

THINK CRITICALLY: How does this real city correlate with the model of the Middle Eastern medina discussed and depicted in Figure 6.16?

Marvin E. Newman/Getty Images

• Figure 6.E The rock occupying the floor of Jerusalem's Dome of the Rock is supercharged with religious meaning. It is presumably the "Foundation Stone" around which the Holy of Holies sections of the First and Second Jewish Temples were built. The Ark of the Covenant thus would have been stood atop the rock until Babylonians destroyed the First Temple in 586 BCE. Jewish tradition holds that God created the world at this spot and that Abraham tried to sacrifice Isaac here. Muslims believe that the Prophet Muhammad ascended into heaven from this spot. Completed in 691 CE, the Dome of the Rock is a gem of Islamic architecture.

THINK CRITICALLY: What would likely happen if an effort were made to build a third Jewish temple on this site as some have urged?

The Jews of Palestine revolted against Roman rule three times between 64 and 135 CE. The first revolt broke out as the profoundly monotheistic Jews refused to acknowledge the Roman emperor as a god. The Romans quashed these rebellions in a series of famous sieges, including those of Masada and Jerusalem. The Romans destroyed the Second Temple; a third has never been built. All that remains of the Second Temple complex is a portion of the surrounding wall built by Herod. Today, this **Western Wall**, known to non-Jews as the Wailing Wall, is the most sacred site in the world accessible to Jews (**Figures 6.D** and **6.20**). Some religious traditions prohibit Jews from ascending the **Temple Mount** above, the area where the temple actually stood, because it is too sacred. After the temple's destruction, that site was occupied by a Roman

Joe Hobbs

• **Figure 6.20** In this single image are Jerusalem's holiest Jewish and Muslim sites. At left, below the golden dome, is the Western Wall, which is all that remains of the structure that surrounded the Jews' Second Temple. The dome is the Muslims' Dome of the Rock shrine, presumably built exactly where the Jewish First and Second Temples had stood. At far right, with the black dome, is the al-Aqsa Mosque, another very holy place in Islam. The entire elevated platform is known to Jews as the Temple Mount and to Muslims as *al-Haraam ash-Shariif*, the Noble Sanctuary.

temple and then in 691 by the Muslim shrine called the **Dome of the Rock**, which still stands today (also in Figures 6.D and 6.20). The mostly Muslim Arabs know the Temple Mount as ***al-Haraam ash-Shariif***, the **Noble Sanctuary**. Supercharged with meaning, this place has, in modern times, often been the spark of conflagration between Jews and Palestinian Arabs. In the apocalyptic passages of the Books of Ezekiel and Daniel, the building of the Third Temple presages the coming of the Messiah to Jerusalem. Some Jews (and some Christians) want to build the Third Temple now to precipitate the end days. This would require the destruction of the Dome of the Rock, likely igniting a war between Jews and Muslims. Incidentally, the apocalyptic scenario ISIS has for Jerusalem is described on **page 245**.

The victorious Romans scattered the defeated Jews to the far corners of the Roman world. Thus began the Jewish exile, or **Diaspora**. In their exile, the Jews never forgot their attachment to the Promised Land. One Jewish prayer, recited during the weeklong Passover holiday, ends with the words "Next year in Jerusalem!" In Europe and Russia, Jews were subjected to systematic discrimination and persecution, and were forbidden to own land or engage in a number of professions. Known as **anti-Semitism**, hatred of Jews developed deep roots. This sentiment in part grew out of the long-simmering perception that Jews, rather than Romans, were responsible for the murder of Jesus Christ, and in part out of the fact that Christians were generally prohibited from practicing moneylending. Jews, however, were permitted to loan money to Christians, who resented this dependency on the moneylenders.

In the 1930s, anti-Semitism became state policy in Germany under the Nazi Party led by Adolf Hitler. Many German Jews fled to the United States, and others emigrated to Palestine in support of the **Zionist movement**, which, since the late 1890s, had aimed at establishing a Jewish homeland in Palestine, with **Zion** (a hill within Jerusalem, symbolizing the City) as its capital. Most Jews were not as fortunate as the Zionist emigrants. Within the boundaries of the Nazi empire that conquered most of continental Europe during World War II, Hitler's regime implemented its **"Final Solution"** to the "Jewish problem" by shipping to prison camps and eventually murdering all the Jews they could round up. The Nazi Germans and their allies killed an estimated 6 million Jews, along with millions of other minorities they deemed "inferior," including Roma (Gypsies) and homosexuals. It was this Holocaust that prompted the victorious allies of World War II, from their powerful position in the newly formed United Nations, to create a permanent homeland for the Jewish people in Palestine. A seemingly endless cycle of violence, discussed beginning on **page 249**, ensued.

Christianity: Death and Resurrection in Jerusalem

Nearly a thousand years after King Solomon built the Jews' First Temple, a new but closely related monotheistic faith emerged in Palestine. This was Christianity, named for Jesus Christ (*Christ* is Greek for "Anointed One," the equivalent of the Hebrew word for Messiah). Jesus, a Jew, was born near Jerusalem in Bethlehem, probably around 4 BCE. Tradition relates that when he was about 30 years old, Jesus began spreading the word that he was the Messiah, the deliverer of humankind long prophesied in Jewish doctrine. A small group of disciples accepted that he was the Messiah, the Son of God and a living manifestation of God Himself. They followed him for several years as he preached his message. He taught that the only path to God was through him and that faith in Christ as God's Son and as the redeemer of humanity's sins was the key to salvation.

Christianity is built upon a Jewish foundation, with the Hebrew Bible providing the common scriptures of Judaism and Christianity. Theologically, Christians would argue that Jews have the correct foundation but did not accept God's complete message, which continues beyond the Hebrew Bible to the new covenant, or the New Testament.

Jesus Christ's own teachings denied the validity of some Jewish doctrines, and around 29 CE, a growing chorus of Jewish protesters called for his death. Palestine was then under Roman rule, and Roman administrators in Jerusalem placated the mob by ordering that he be put on trial. He was found guilty of being a claimant to Jewish kingship, and Roman soldiers put him to death by the particularly degrading and painful method of crucifixion. The cornerstone of Christian faith is that Jesus Christ was resurrected from the dead on the third day after his crucifixion and later ascended into heaven. For Christians, this was fulfillment of the scriptures: Jesus had come to redeem humanity's sins through his own death on the cross and through his resurrection. Christians believe that he continues to intercede with his Father on their behalf and that he will come again on Judgment Day, at the end of time.

After a period of relative tolerance, the Romans began actively persecuting Christians. Nevertheless, Christian ranks and influence grew. The turning point in Christianity's career came after 324 CE, when the Roman Emperor Constantine embraced Christianity and established it as the official religion of the empire. The Christian Byzantine civilization that developed in the **"New Rome"** that Constantine established—Constantinople, now Istanbul, Turkey—created fine monuments at places associated with the life and death of Jesus Christ. These include the place long acknowledged as the center of the Christian world, Jerusalem's **Church of the Holy Sepulcher** (**•Figures 6.21** and **6.D**). This extraordinary, sprawling building, now administered by numerous separate Christian sects, contains the locations where tradition says Jesus Christ was crucified and buried. Incidentally, several of these Christian denominations have a long history of squabbling over turf within the church, and the key to this place is in the hands of a trusted neutral party: a Muslim family.

Christianity has seldom been the majority religion in the land where it was born; only until Islam arrived in Palestine in 638 CE was the region primarily Christian. From then until the 20th century, most of Palestine's inhabitants were Muslims, and since the mid-20th century, Muslims and Jews have been the major groups. Between the 11th and 14th centuries, European Christians dispatched military expeditions to recapture Jerusalem and the rest of the Holy Land from the Muslims. These bloody campaigns, known as the Crusades, resulted in a series of short-lived Christian administrations in the region.

Joe Hobbs

• Figure 6.21 Jerusalem's Church of the Holy Sepulcher, housing the locations where many Christians believe Jesus Christ was crucified and buried.

There are significant minority populations of Christians throughout the Middle East, including members of distinct sects such as the **Copts** of Egypt and the **Maronites** of Lebanon. Their share of the population has generally been declining, both because of emigration and lower birth rates than those of the majority Muslims. Christians have been persecuted in some countries (particularly where ISIS is present; see **page 263**), which is one of the push factors in their emigration. Nevertheless, the Middle East remains the cradle of their faith for Christians the world over, and Jerusalem and nearby Bethlehem are the world's most important Christian pilgrimage sites. Whatever your faith, you would surely find Jerusalem to be one of the most compelling places on Earth.

The sacred places and the archaeological sites are among the things you will enjoy if you visit the Middle East. Here is another that awaits you: the exceptional hospitality of the people. The obvious "don't go" places and peoples aside, you will feel the powerful pull of welcoming from Morocco to Iran, from Jews, Christians and Muslims, from Arabs, Turks, Iranians, and others. Driving with my wife, Cindy, across southern Turkey's frontier with Syria, I saw a wool tent and some people tending

life in EGYPT

Ahmed Sherif

Ahmed Sherif

As an Egyptian, I find that life in Egypt suits me very well. The style of living may look very different to Westerners, but it has its charm, and like any other country, it also has its negative side. I learned to adapt to it and I love it. Egypt is a third-world country with a very large population (around 90 million people in the year 2015), and there's poverty, ignorance, pollution, and corruption, among other negative points. But when you grow up and live here, you are also blessed with many positive sides that mostly overshadow the negative ones. From the social side, family ties are extremely important; neighbors and friends look after you; people care for each other and help each other in times of need—much more than in Western societies—because social welfare is quite poor, and services for the elderly are not adequate. Egyptians have a good sense of humor and are usually friendly people and very hospitable. It is customary to invite visitors and strangers into one's home and offer them meals, tea, and coffee as a sign of hospitality. No matter how poor or rich one is, it is very important for Egyptians to show hospitality to their guests, and it is considered rude by the visitors not to accept this hospitality.

Economically, Egypt is a poor country, even though it has all the necessary ingredients to become a very rich country. That's because it is badly managed and has been under a military rule by several dictators since 1952. It is a class-based society. There is a minority of very rich, and then there is the medium-sized middle class and then the very large poor class, of which many are uneducated. Because different occupiers colonized Egypt over many centuries, it has many different language schools. Arabic, of course, is the mother tongue, but English and French schools are also available, and many Egyptians who can afford an education can speak several languages. Tourism is also a factor accounting for why many Egyptians can speak other languages. Interacting with tourists helped many from the poor working class who work in tourism to pick up some words here and there, and it taught them a way to communicate with foreigners. The introduction of television in 1960 and the existence of many cinema theaters helped as well. Many programs and television series are broadcast in English. Later, when the Internet was introduced, it also helped increase the knowledge of English and French languages.

Politically, the country has passed through many different phases in its recent history. British colonialism left its marks on the society, and when the military took over in 1952 and ended the British rule and Egyptian monarchy, Egypt went into a socialist direction under President Gamal Abdel Nasser's rule. Under President Anwar Sadat's rule, Egypt changed course and went into an open door policy that leaned toward capitalism, and it continued to do so under President Hosny Mubaraks' rule. However, due to the bad management of the country by the various military regimes, and the corruption that expanded over the years, Egypt was in a very bad situation economically. This led to the revolution of January 25, 2011, against the rule of Mubarak. Egyptians had very high hopes for a better life and a chance to finally fix up the country and get rid of all the corruption. This unfortunately didn't happen, and the country fell into the arms of the religious fundamentalists (the Muslim Brotherhood) who hijacked the revolution and wanted to turn Egypt into a religious emirate that would become part of a bigger empire that would rule the whole Middle East under an Islamic Wahhabi umbrella. This did not suit the majority of Egyptians as they do not subscribe to the Wahhabi thought and are mostly moderate. On June 30, 2013, the moderate Egyptians went out into the streets all over Egypt in one of the biggest demonstrations in the history of mankind (estimated numbers between 15 to 30 million people), demanding the removal of President Morsi from power. This led to the overthrow of President Mohamed Morsi's regime, and the army took over power again, but with the blessing of a big majority of Egyptians, who preferred being ruled by a military dictatorship rather than a religious one, which would have led to a civil war. The goals of the January 25 revolution were not achieved, but a big change occurred in the Egyptian society: people now are no longer afraid to speak their mind and still have big hopes for a better future. Their leaders are also aware that if they do not deliver the basic needs of the people, they risk the same fate as their predecessors. At the end of the day, people power is the strongest.

Religion, however, is a very important part of the fabric of the Egyptian society. Since Pharaonic times, Egyptian have been known to be religious but also moderate and not fanatic. Jews, Copts, and Muslims have lived together in peace and harmony for centuries, until recently when politics got

sheep and goats. Although Arabic-speakers are rare in this area, I thought they might be Arabic-speaking nomads. They were. They invited us into their tent. As soon as we settled in, and even before they brought us tea, the women did something I had never seen before: they washed Cindy's feet. These people we had never met, and would never meet again, washed her feet! We know from the Bible that a host honors the guest with ritual washing of the feet. Here was that tradition of hospitality more than 2000 years on, in the same part of the world.

Our guest essayist has more to share with you about the traditions of hospitality, the importance of family, and other customs and interests shared by his fellow Egyptians and many people across the region. He is Ahmed Sherif, my friend and classmate from the American University in Cairo, writing to you about Life in Egypt above.

involved and caused rifts among the various different groups. However, there is still tolerance and coexistence among the majority of Egyptians.

Culturally, Egypt has a diverse mixture of citizens. There are those who live along the Nile valley, and are mostly into farming, trade, and business. And then there are the Bedouin tribes who live in the desert and live a nomadic life. Bedouin tribes differ in origins. Some are from African origins and live near the Sudanese borders with Egypt, like the Nubians, the Bisharin, and the Ababda; and then there are those who are descended from Arabs, like the tribes of North and South Sinai and those from the western desert oasis of Siwa near the Libyan borders. Some Bedouin tribes have their own language, like the Bisharin tribes who can speak "Tebdawy," a very old unwritten language, which is believed to predate Pharaonic times. But Arabic is the main language used among these tribes, although they have different traditions, wear different costumes, play different music, cook different food, and so on.

Egyptians, in general, are conservative people with Middle Eastern traditions. In many families, the father is the head of the house, and the wife is in charge of looking after the home. The man works and brings income to the family, and the wife looks after the children, cooks, does the washing, and attends to the family and the house. However, this is different in modern families. Many women now are educated and have jobs. Due to the economic situation, both women and men have to work to help bring an income to the family. The cost of living has gone up, and private education has become very costly. Many educated women have reached very high positions in various fields, and the mentality of traditional Egyptians is changing. Women are expected to play a bigger role in modern Egypt, but it will also take time before non-educated Egyptian men learn to accept this.

I myself am from the Nile Valley (Cairo). My ancestors were from Turkish origins (Ottoman Empire), and I was lucky to belong to a well-off, educated, modern family. I went to a French school in a suburb of Cairo called Maadi, then went to the American University in Cairo, where I graduated with a BSC in physics. I never used my degree in physics, but my ability to speak several languages helped me get a job with NBC News in Cairo. I also happened to have some knowledge of the German language thanks to my mother, who spoke it fluently even though she was Egyptian. Living in Cairo was fun at my younger age. I used a bicycle to go to school. The streets were empty, and we had a sporting club nearby, where we could go swimming, play football (soccer), tennis, and many other sports. Maadi also was home to many expats from various countries, so I was lucky to make friends with some of their children, and that helped increase my exposure to other languages. However, as I grew older, the population of Egypt kept on increasing, and more construction for housing kept appearing. This led to the disappearance of green areas around the city, replaced by cement blocks of houses and buildings. Traffic jams increased too, and there was more pollution, more garbage, and fewer jobs. Life became more difficult in Cairo and more stressful, so once I retired from my job, I moved to Sharm El Sheikh in Sinai. There the air is cleaner; nature is beautiful; and I'm lucky to have the beauty of the Red Sea corals and fishes right in front of my house. I can go diving and snorkeling, and the weather is great most of the year. Furthermore, I have the beauty of the Sinai desert, which is quite rich in fauna, flora, and antiquities. As I enjoy camping, I've been exploring the desert for many years and have been filming a lot of its beauty and its wildlife and historical sites, and I've made some documentaries about some of it. The city is multinational and is a famous tourist resort due to its location along the Gulf of Aqaba in the Red Sea. Because of tourism, many foreigners from various nationalities live and work here. This created a very nice atmosphere and increased the interaction among various cultures and nationalities. We all coexist happily in our little city. I also made friends with many Bedouin tribesmen and learned their way of living and their traditions, and in the process I learned to respect nature and appreciate its beauty and the simplicity of life. I spend my time snorkeling, diving, and camping in the desert.

Egypt may be a poor country, disorganized, going through tough times politically and economically, but what it has to offer in return is natural beauty, an incredible rich history with fantastic antiquities, a beautiful climate, access to two seas, the river Nile, and many desert oases. The culture is very rich and diverse, and it will leave an impression on anyone visiting it. We have a saying in Egypt: "He who drinks from the water of the Nile, always returns to it." I've travelled a lot around the world in my younger days, and now that I'm older and retired, I'm enjoying my life at home, which is Egypt. We call it "Umm al-Duniya," which means in Arabic "the Mother of the World."

Ahmed Sherif is a documentary filmmaker who has been on the front lines of Middle Eastern news for decades.

The Message of Islam

Islam is by far the dominant religion in the Middle East and North Africa; only Israel within its pre-1967 borders has a non-Muslim majority. Because of Islam's powerful influence not merely as a set of religious practices, but as a way of life, an understanding of the religious tenets, culture, and diffusion of Islam is vital for appreciating the region's cultural geography.

Islam is a monotheistic faith built on the foundations of the region's earliest monotheistic faith, Judaism, and its offspring, Christianity. Indeed, Muslims (people who practice Islam) call Jews and Christians the **People of the Book**, and their faith obliges them to be tolerant of these special peoples. Muslims believe that their prophet Muhammad was the very last and most important in a series of prophets, including Abraham, Noah, and Moses, who brought the Word of God to humankind (Muslims do not worship Muhammad, and orthodox Sunni Muslims believe that depicting him in any form is blasphemy). Thus they view the Bible as incomplete, but not entirely wrong—Jews and Christians merely missed receiving the entire message (just as Christians would insist that Jews missed the entire message). Muslims do not accept the Christian concept of the divine Trinity (God manifested in the form of the Father, His son Jesus, and the Holy Spirit) and regard Jesus as a prophet rather than as God.

Muhammad was born in 570 CE to a poor family in the Arabian city of Mecca. Located on an important north–south caravan route linking the frankincense-producing area of southern Arabia with markets in Palestine and Syria, Mecca was a prosperous city at the time. It was also a pilgrimage destination because more than 300 deities were venerated in a shrine there called the **Ka'aba** (the "Cube"; notably, at one corner is a meteorite said to have fallen from Heaven to this spot, showing Adam and Eve where to build an altar [**•Figure 6.22**]). Muhammad married into a wealthy family; the Umayyad clan of his wife Khadija led the caravan trade. Muslim tradition holds that when he was about 40 years old, Muhammad was meditating in a cave outside Mecca when the Angel Gabriel appeared to him and ordered him to repeat the words of God that the angel would recite to him. Over the next 22 years, the prophet related these words of God (whom Muslims call **Allah**) to scribes who wrote them down as the **Qur'an** (or **Koran**), the holy book of Islam.

During this time, Muhammad began preaching the new message, "There is no god but God (Allah)," which the polytheistic people of Mecca viewed as heresy. As much of their income depended on pilgrimage traffic to the Ka'aba, they also viewed Muhammad and his small band of followers as an economic threat. They forced the Muslims to flee from Mecca and take refuge in Yathrib (modern Medina), where a largely Jewish population had invited them to settle. There were subsequent skirmishes between the Meccans and Muslims, but in 630, the Muslims prevailed and peacefully occupied Mecca. The Muslims destroyed the idols enshrined in the Ka'aba, which became a pilgrimage center for their one God.

Zurijeta/Shutterstock.com

• Figure 6.22 In one of the most important *hajj* rituals, Muslim pilgrims circumambulate the *Ka'aba*, the black-shrouded cube at center, 7 times. The *Ka'aba* in Mecca is associated with the lives of the Prophets Abraham and Muhammad, and is the holiest site in Islam. Muslims the world over pray 5 times daily facing in the direction of Mecca and the *Ka'aba*. This photo was taken in 2012 before elevated walkways were built around the *Ka'aba* to accommodate more pilgrims.

The Ka'aba is Islam's holiest place, and Mecca and Medina are its holiest cities. Jerusalem is also sacred to Muslims. Muslim tradition relates that on his **Night Journey**, the Prophet Muhammad ascended briefly into heaven from the great rock now beneath the Dome of the Rock (the same rock Jews regard as the Foundation Stone). Nearby on the Temple Mount (*al-Haraam ash-Shariif*) is **al-Aqsa Mosque**, a sacred congregational site. The proximity and even the duplication of holy places between Islam and Judaism (see Figure 6.21) came to be the most difficult issue in peace negotiations between Palestinians and Israelis, as explained on **page 253**.

After Muhammad's death in 632, Arabian armies and traders rapidly expanded the geographic scope of their new faith (see Insights, **page 231**). Within 30 years, Muslims had conquered Egypt, the Sassanian Empire (based in modern Iran), and the southern provinces of the Byzantine Empire, including Damascus. Islam spread across northern Africa and across the Strait of Gibraltar into Europe; its advance was halted in 732 after a major defeat at Tours, France. Muslim armies continued expanding elsewhere, pushing into Sudan, Sicily, and Anatolia. In 1453, the Muslim Ottomans conquered the Byzantine capital of Constantinople, which would remain the center of their empire until World War I. In the 13th century, the Mongols, who ravaged the Middle East as they did Russia, converted to Islam across many of the lands they held. Through conquest and trade, Mongols brought their new faith into central Asia. One Islamic Mongol group, the Mughals, maintained a prosperous civilization across India from the 16th to 19th centuries. Arab traders who controlled the lucrative spice trade introduced Islam to ports across the Indian Ocean, from Zanzibar to India and to Malaysia, where the Sultanate of Malacca became the first Southeast Asian Muslim nation in the early 15th century. Today, the world's most populous Muslim country is not in the Middle East, and its 250 million people are not Arabs; it is Indonesia, 4000 miles (6400 km) east of Arabia.

302

Arab science and civilization flourished in Baghdad, immortalized in the legends of *The Thousand and One Nights*. Muslims made important advances, with discoveries in geography, mathematics, and astronomy (in fact, most of the names of the bright stars in the night sky, such as Betelgeuse, Rigel, Deneb, and Altair, are Arabic in origin). Many of these advances were recorded in Arabic, a language unfamiliar to contemporary Europeans, and had to be rediscovered centuries later by the Portuguese and Spaniards. Muslim scholars translated Greek and Roman texts, and if not for their efforts, many of these works would never have survived to become part of the modern European legacy.

Sunni and Shi'ite Muslims

A schism occurred very early in the development of Islam, and it persists today. The split developed because the Prophet Muhammad named no successor to take his place as the leader (Caliph) of all Muslims. Some of his followers argued that the person with the strongest leadership skills and greatest piety was best qualified to assume this role. These followers became known as **Sunni**, or orthodox, Muslims. Others argued that only direct descendants of Muhammad, specifically through descent from his cousin and son-in-law, Ali, could qualify as successors. They became known as **Shia**, or **Shi'ite**, Muslims.

The military forces of the two camps engaged in battle south of Baghdad at Karbala in 680 CE, and in the encounter Sunni troops caught and brutally murdered Hussein, a son of Ali and grandson of the Prophet Muhammad (Karbala is thus sacred to Shi'ites, as is nearby al-Najaf, where Ali is buried). The rift thereafter was deep and permanent. The martyrdom of Hussein became an important symbol for Shi'ites, many of whom still today regard themselves as oppressed peoples struggling against cruel tyrants, including some Sunni Muslims.

Today, only three of the region's countries, Iran, Iraq, and Bahrain, have Shi'ite majority populations (Azerbaijan is also majority Shi'ite). Iran's government is a Shi'ite theocracy. There are significant minority populations of Shi'ites in Yemen and Lebanon (see Figure 6.20) as well as in India and Pakistan. Conflicts between Shi'ites and Sunnis are central features in the region's modern political geography. When you talk with Muslims about these sects, you may find that they downplay the differences and stress the unity of Islam. Most importantly, the Qur'an represents the direct word of God for both Sunni and Shi'ite Muslims.

But then their paths diverge. For Sunni Muslims, there are other authoritative sources of religious guidance: the *Sunna*, or "Path" (authenticated traditions, words, and actions of the Prophet Muhammad), and the *Hadith* (each single tradition or narrative of the Sunna). Shi'ite Muslims do not accept these other sources in the same way as Sunni Muslims.[4] Many Sunni Muslims revile Shi'ites as heretical, especially because of their iconographic tendencies. Although they do not depict God, some Shi'ites depict a faceless Muhammad in addition to full portraits of Ali, Hussayn, and other prominent religious figures. This is anathema to Sunnis, who insist on unadornment and austerity in worship. Sunni Muslims frown upon Shi'ite pilgrimages to the tombs of Ali, Hussayn, and others; only pilgrimage to Mecca is permissible for Sunnis. Shi'ites seek the intercession of clerics, who are empowered to interpret God's will for them, whereas the Sunnis prize a direct and personal relationship with God. In the following pages, we will see a number of violent confrontations between Sunnis and Shi'ites in Syria, Iraq, Lebanon, and Yemen, but these grow more out of politics and historical legacies than because of differences in religious doctrine.

There is one group of Muslims who are neither Sunni nor Shi'ite: the Ibadis, whose doctrine emerged before the schism took place. Noteworthy for its tolerance of all faiths and for its disinterest in clergy and in the issue of succession to the Prophet Muhammad, Ibadi Islam is the majority sect in Oman. Ibadi Muslims are minorities in East Africa, Algeria, Tunisia, and Libya.

Whether in Arabia or Indonesia, whether they be Sunni Muslims or Shi'ite Muslims, all believers are united in support of the five fundamental precepts, or **Pillars of Islam**. The first of these is the **profession of faith**: "There is no god but God, and Muhammad is His Messenger." This expression is often on the lips of the devout Muslim, both in prayer and as a prelude to everyday activities.

The second pillar is **prayer**, required five times daily (three times for Shi'ites) at prescribed intervals. Two of these prayers mark dawn and sunset. Business comes to a halt as the faithful prostrate themselves before God. Muslims may pray anywhere, but wherever they are, they must turn toward Mecca. There also is a congregational prayer at noon on Friday, the Muslim Sabbath.

The third pillar is **almsgiving**. In earlier times, Muslims were required to give a fixed proportion of their income as charity, similar to the concept of the tithe in the Christian church. Today, the donations are voluntary. Even Muslims of very modest means give what they can to those in need.

The fourth pillar is **fasting** during Ramadan, the ninth month of the Muslim lunar calendar. Muslims are required to abstain from food, liquids, smoking, and sexual activity from dawn to sunset throughout Ramadan. The lunar month of Ramadan falls earlier each year in the solar calendar and thus periodically occurs in summer. In the torrid Middle East and North Africa, that timing imposes special hardships on the faithful, who, even if they are performing manual labor, must resist the urge to drink water during the long, hot days.

The final pillar is the ***hajj***, or **pilgrimage to Mecca**, Islam's holiest city. Every Muslim who is physically and financially able is required to make the journey once in his or her lifetime. A lesser pilgrimage may be performed at any time, but the prescribed season is in the 12th month of the Muslim calendar. Those days witness one of Earth's greatest annual human migrations as about 2 million Muslims from all over the world converge on Mecca (see **Figure 6.22**). Hosting these throngs is an obligation the government of Saudi Arabia fulfills proudly and at considerable expense. However, there has been some trepidation in recent years because of the security threat foreign visitors may pose to the host country and because accidents such as stampedes and tent city fires have cost numerous lives. Many pilgrims also visit the nearby city of Medina, where Muhammad is buried. Most Muslims regard the *hajj* as one of the most significant events of their lifetime. All are required to wear simple, seamless garments, and for a few days, the barriers separating groups by income, ethnicity, sect, and nationality are broken. Pilgrims return home with the new stature and title of *hajji*, but also with humility and renewed devotion.

All Muslims share the Five Pillars and other tenets, but they vary widely in other cultural practices related to their faith, depending on the country they live in; whether they are from the desert, village, or city; and their education and income. The governments and associated clerical authorities in Saudi Arabia and Iran insist on strict application of **Islamic law** (known as ***shari'a***) to civil life; in effect, there is no separation between church and state.

The Qur'an does not state that women are required to wear veils, but it does urge them to be modest, and it portrays their roles as different from those of men. Clerics in Saudi Arabia insist that women wear floor-length, long-sleeved black robes and black veils in public, that they travel accompanied by a male member of their families, and that they not drive cars. In Egypt, by contrast, Muslim women are free to appear in public unveiled if they choose. However, in most Muslim countries, conservative ideas about the role of women are still very strong: they should be modest, retiring, good mothers, and keepers of the home, although they can work. The Qur'an portrays women as equal to men in the sight of God, and in principle, Islamic teachings guarantee the right of women to hold and inherit property.

Most Muslim women argue that what others often see as "backward" cultural practices are, in fact, progressive. For example, they say, their modest dress compels men to evaluate them on the basis of their character and performance, not their attractiveness. They argue that segregation of the sexes in the classroom makes it easier for both women and men to develop their confidence and skills (**•Figure 6.23**). A married woman retains her maiden name. These apparent advantages can be weighed against the factors that make women generally subordinate to men in public and private affairs, and that fall short in utilizing the human capital of half the population.

Joe Hobbs

• Figure 6.23 Education of women in the Muslim countries has increased dramatically in recent decades, but strong traditions keep most women out of the workforce. This is a geographic information systems (GIS) class in the Department of Geography, United Arab Emirates University.

Powerful currents of Sunni "fundamentalism" flow throughout MENA and beyond. **Salafi Muslims**, or "Salafists," adhere to an interpretation of Islam they believe is closest to the faith's earliest tenets and social norms, and that most correctly follows *shari'a*. The roots of Salafi Islam are in 19th-century Egypt. The royal family of Saudi Arabia rules with the blessings of clerics representing another strict interpretation of Islam known as *Wahhabism*. Its roots are in 18th-century Arabia. Within Saudi Arabia, Salafism and Wahhabism have drawn closer together.[5] Wealthy Saudis send funds abroad to support "pan-Islamic" movements that are Salafi rather than Wahhabi in nature. Some of these and other Salafi groups advocate violent ***jihad***, literally meaning "struggle" but often translated in this context as "holy war." They use Islamic faith, culture, and history to legitimize their philosophy and actions. They frequently cite the Qur'an, the *Hadith* (sayings of the Prophet Muhammad), and the writings of earlier Islamic militants such as Ibn Taymiyya and Sayyid Qutb. I use the term *jihadi* or *jihadist* to refer to militants who use violence or terrorism to spread their interpretation of Islam. I adopt the Associated Press' definition of another important term, **Islamist**: "An advocate or supporter of a political movement that favors reordering government and society in accordance with laws prescribed by Islam." The AP advises reporters, "Do not use as a synonym for Islamic fighters, militants, extremists or radicals, who may or may not be Islamists."[6] Although not all Islamists are militant or terrorist, they all reject what they view as the materialism and moral corruption of Western countries and the political and military support these countries lend to Israel. Both Sunni and Shi'ite Muslims have advanced a wide range of Islamist movements, notably in Iran, Lebanon, Egypt, Sudan, and Algeria.

We will examine al-Qa'ida, ISIS, and other violent Islamist movements later in the chapter, but it is important to note here that very few Muslims around the world support such extreme interpretations of Islam: such groups have "hijacked" their religion, they say. In 9/11, the *Charlie Hebdo* attack, and other atrocities, a tiny minority of Muslims carried out terrorist actions that the vast majority of Muslims condemned. Islamic scholars and religious authorities point out that murder of civilians is prohibited in Islamic law and that the attacks have no legitimate religious grounds. Mainstream Islamist movements are not military or terrorist organizations and have distinguished themselves through public service to the needy. It is tempting to suggest, as many non-Muslims in the West have, that ISIS is not Islamic at all, but a "death cult" dressed in Islamic guise. But this is not the case. ISIS theology is "fundamentalist," drawing on the most basic precepts of the faith at the time of the Prophet, when beheadings, amputations, and enslavement and special taxes on non-Muslims were prescribed and practiced. Most Muslims have moved on with the times and see no place for these practices, but ISIS sees salvation and the fulfillment of prophecies in them; see **pages 244–245**.

6.4 Economic Geography

Overall, the Middle East and North Africa is a poor region; the average per capita GDP (PPP) for the 23 countries and territories is only US $9000 (see Table 6.1). This may seem surprising in view of the "rich Arab" stereotype, but only the small oil-endowed states around the Persian Gulf deserve the reputation for wealth; by most measures other than per capita wealth, only Israel is truly a more developed country (MDC). Israel's prosperity comes from its innovation in computer and other high-technology industries, the processing and sale of diamonds, large amounts of US and other foreign aid, and investment and assistance by Jews and Jewish organizations around the world.

Here we will examine the role of oil in the region's economies, and the lagging economic indicators and human conditions that for the region's authoritarian leaders should have served as the writing on the wall for the "Arab Spring" (see **page 255**).

"The Prize"

Vital to the industrialized countries as a source of fuels, lubricants, and chemical raw materials, petroleum is one of the world's most important natural resources. Daniel Yergin, recognized as the world's preeminent authority on oil, wrote the definitive book on this essential commodity. His title sums up what oil has come to represent in economic and geopolitical terms: *The Prize*.

A crucial feature of world geography is the concentration of about half of the world's proven conventional petroleum reserves (excluding those accessible by technologies such as fracking and the energy-intensive conversion of tar sands, for example) in the few countries that ring the Gulf. In the second half of the 20th century, this geographical quirk transformed a handful of inconsequential countries and sheikhdoms into some of the most critical locations in the global economic system (**• Figure 6.24**).

521, 522

By coincidence, the MENA countries rich in oil tend to have relatively small populations, whereas the most populous nations have few oil reserves (Iran is the exception). All together, the Gulf countries have about 800 billion barrels of proven reserves of crude oil (by comparison, Canada has 173 billion; the United States, 20 billion; and Mexico, 10 billion). Venezuela and Saudi Arabia are the world leaders in reserves; they have about 18 percent and 16 percent, respectively, of the proven crude oil reserves on the globe. Canada ranks third, with 11 percent (not in the form of easily extractable crude oil, but as tar sands; see **page 521**). The next four are all along the Gulf: Iran (9 percent), Iraq (8 percent), Kuwait (6 percent), and the United Arab Emirates (6 percent). Libya and Qatar also have significant oil reserves (**• Figure 6.25**). The world's largest oil consumer, the United States, ranks 13th in reserves, with just 1 percent. China, the second largest consumer, ranks 14th. It is important to distinguish between largest reserves and largest production: Saudi Arabia and the United States both produce about 11 million barrels of oil a day, but Saudi Arabia's reserves are far greater than that of the US (**• Figure 6.26**).

Natural gas is another significant fossil fuel resource in the region. Iran has MENA's largest reserves, with 16 percent of

• Figure 6.24 The Gulf countries.

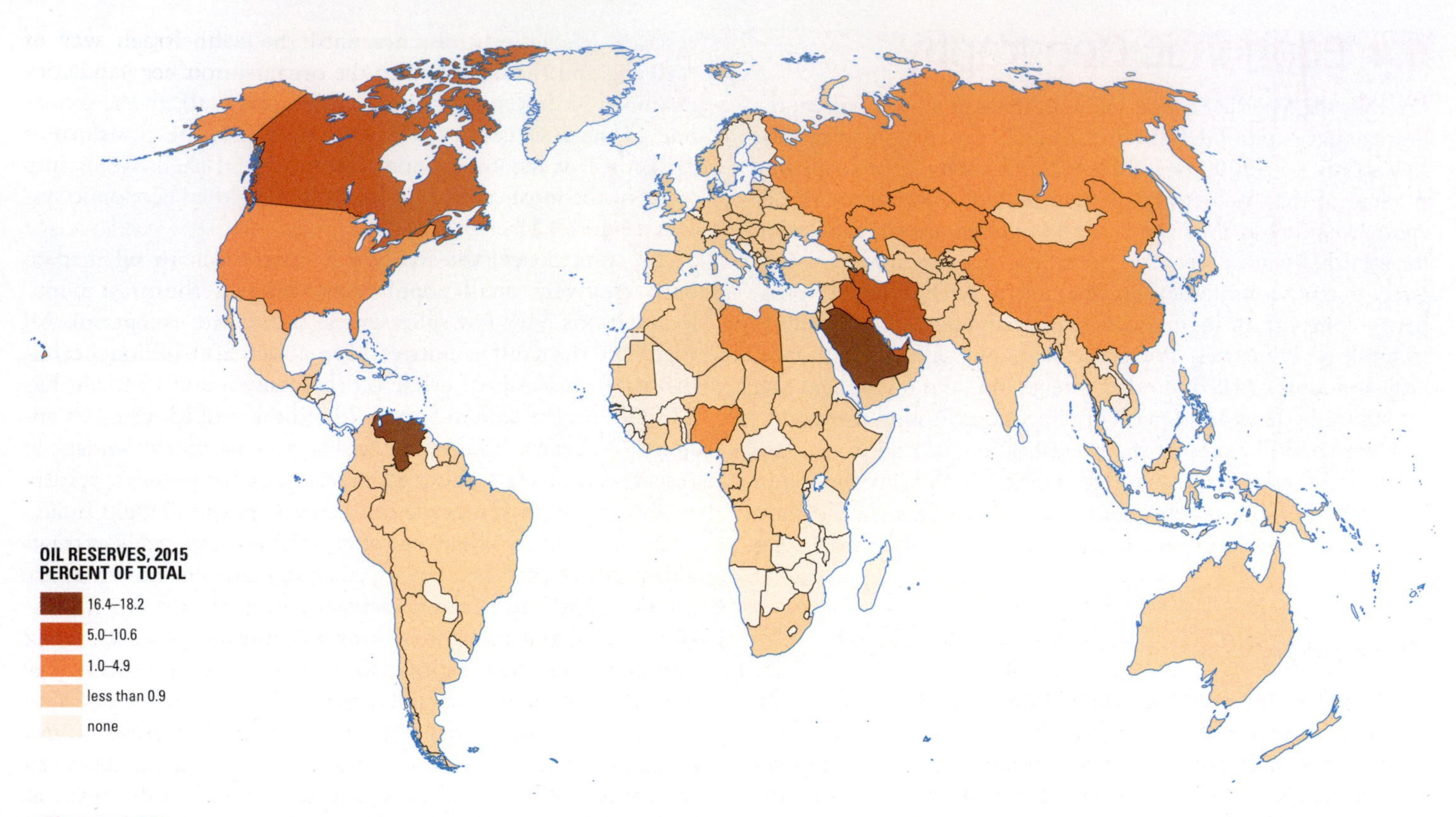

• **Figure 6.25** One-third of the world's proven oil reserves are in just two countries: Venezuela and Saudi Arabia.

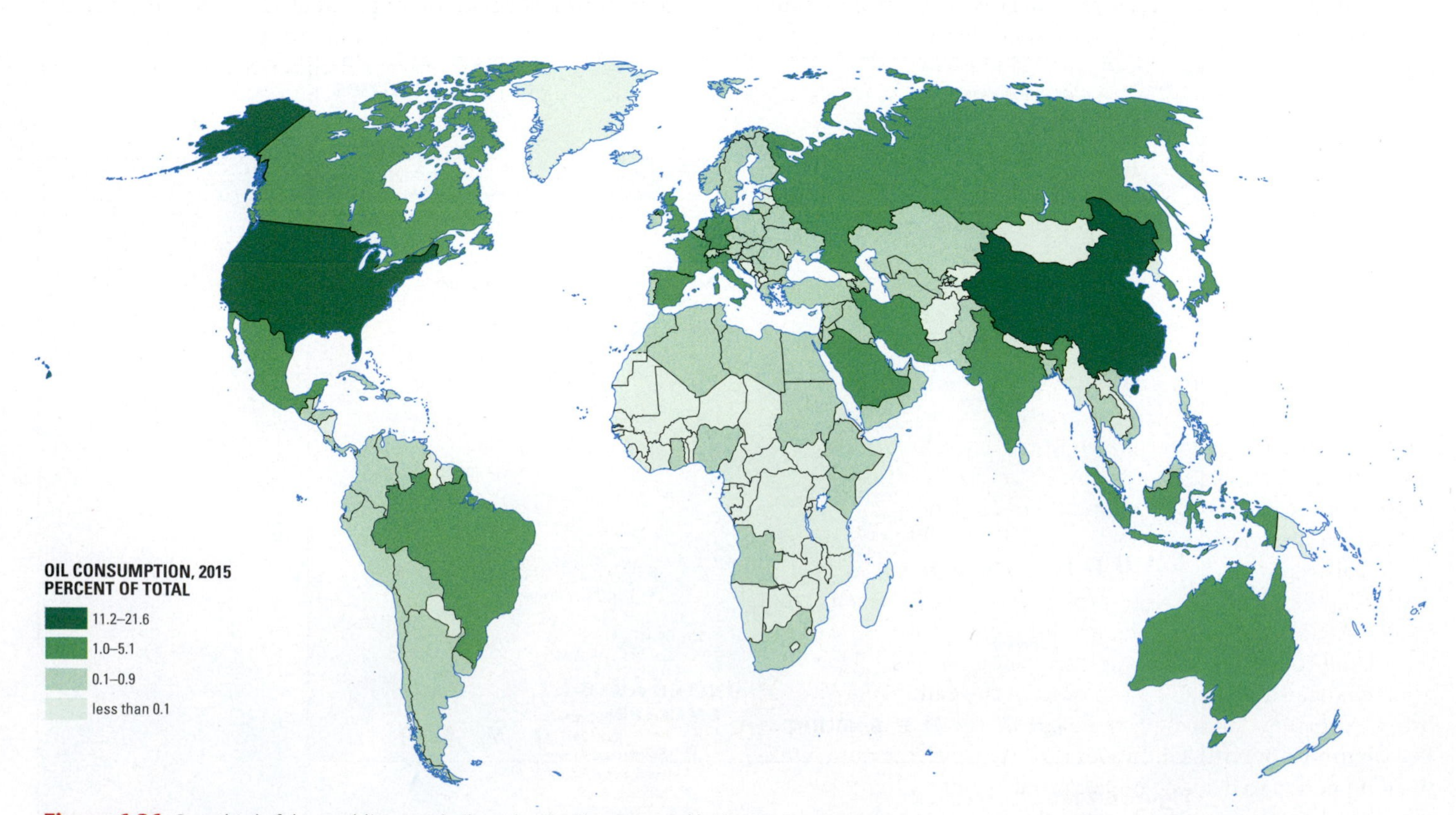

• **Figure 6.26** One-third of the world's annual oil production is consumed by just two countries: the United States and China.

Jamal Nasrallah/Corbis Wire/Corbis

• **Figure 6.27** Petroleum facilities in the interior wilderness of Arabia's Empty Quarter (*al-Rub' al-Khali*). Most of Saudi Arabia's production is from vast, shallow reservoirs of oil that are much closer to marine terminals on the Persian Gulf.

the world's total; only Russia has a greater supply. Qatar ranks just behind Iran, with 12 percent. Qatar (pronounced with an accent on the first syllable) essentially sits atop a vast reservoir of natural gas. This endowment has given this tiny country of 2 million an outsized influence on world affairs. Qatar is home to al-Jazeera, a highly influential media network. Qatar is often at odds with its Arab neighbors over foreign policy, for example, in backing Egypt's Muslim Brotherhood and in leaning toward Turkey's policies for the Middle East.

In the Gulf area, the great thickness of the region's oil-bearing strata and high reservoir pressures have made it possible to secure an immense amount of oil from a small number of wells (•**Figure 6.27**). The productivity of each well makes each barrel inexpensive to extract, and it is simple to increase or reduce production quickly in response to world market conditions. Gulf oil is thus cheap to produce unless expenditures to maintain huge military forces and infrastructure to defend it are factored in (in economic terms, unless the "external costs" of oil are internalized). Some analysts—including a former US Navy secretary—prefer to calculate the "real price" of oil with military expenditures calculated. If one accepts the premise that maintaining a costly military presence in the region is part of the United States' long-term strategy to maintain the flow of Gulf oil, that oil is no longer cheap by any measure.

Production, export, and profits of Middle Eastern and North African oil were once firmly in the hands of Western petroleum companies known as "the Seven Sisters." That situation changed after 1960, when most of the Gulf countries and other exporting nations formed a cartel, the **Organization of Petroleum Exporting Countries (OPEC)**, with the aim of taking joint action to demand higher profits from oil (•**Figure 6.28**). It changed again after 1972, when the oil-producing countries began to nationalize the foreign oil companies. OPEC was relatively obscure until the Arab-Israeli war of 1973, after which the organization began a series of dramatic price increases. In 1980, the organization's price reached $37 (US) per barrel, compared with $2 a barrel in early 1973; these were truly titanic price hikes.

These events had enormous repercussions for the world economy. Immense wealth was transferred from the more developed countries to the OPEC countries to pay for indispensable oil supplies. The skyrocketing cost of gasoline and other oil products helped cause serious inflation in the United States and many other countries and contributed to the 1973 **energy crisis** in the United States. The world's LDCs found that high oil prices not only hindered the development of their industries and transportation, but also reduced food production because of high prices for fertilizers made from oil and natural gas. In the Gulf countries, the oil bonanza produced a wave of spending on military hardware, showy buildings, luxuries for the elite, and ambitious development projects of many kinds. Per capita benefits to citizens were greatest in the Arabian Peninsula, where small populations and immense inflows of oil money made possible the abolition of taxes, the establishment of comprehensive social programs, and heavily subsidized amenities such as low-cost housing and utilities, including freshwater distilled from the salty Gulf in desalination plants.

The era of continually expanding OPEC oil production, sales, and profits seemed to come to an end in the 1980s. After 1973, the high price of oil stimulated oil development

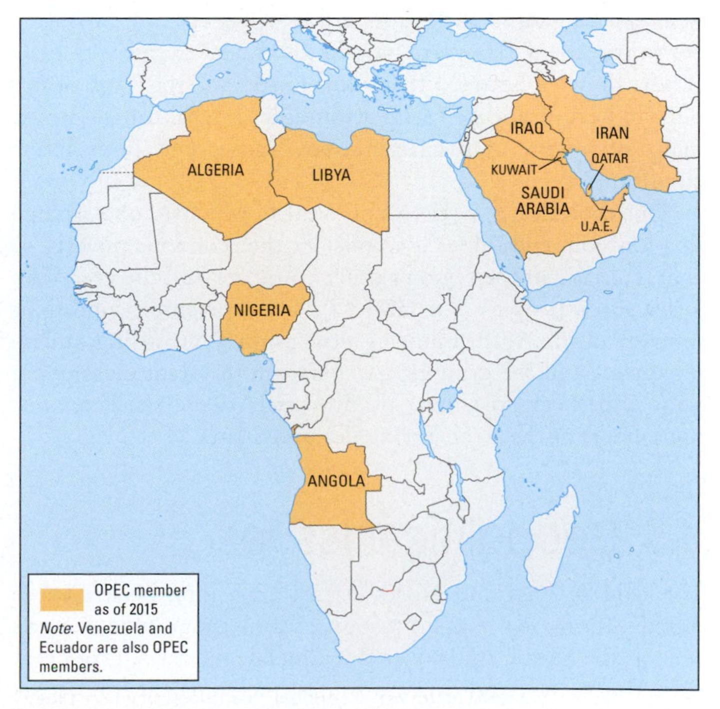

• **Figure 6.28** A map of current members of OPEC.

in countries outside OPEC. Oil conservation measures such as a shift to more fuel-efficient vehicles and factories were introduced. Substitution of cheaper fuels for oil increased. Coal replaced oil in many electricity-generating stations. Meanwhile, the world entered a period of economic recession due in part to high oil prices. Decreased business activity reduced the demand for oil. Profits of the world oil industry (and taxes paid to governments) were severely cut; large numbers of refineries had to close; and much of the world tanker fleet was idled.

Oil prices have risen and fallen ever since, with predictable consequences for producers and consumers. But the immense oil and gas reserves still in the ground guarantee that the Gulf region will continue to have major long-term influence in world affairs and will remain prosperous as long as these finite resources are in demand—especially in the established industrialized countries and in the booming economies of China and India. The great challenge for the oil-rich states is to lay the groundwork for future economies not based on fossil fuels. The Middle East is still in the middle, and as we will see, some small Gulf countries, like the United Arab Emirates, are capitalizing on their central geographic position in the midst of Europe, Africa, and Asia by developing seaports and airports and serving as financial centers.

There are, of course, other resources and industries in the region's economic geography—**remittances** (earned income sent home by guest workers) in the oil-rich countries; revenues from ship traffic through the Suez Canal; and exports of cotton, rice, and other commercial crops, for example—but oil dominates the region's economy and is central to the global economy. The oil-driven construction booms taking place on the Arabian Peninsula are responsible for the largest regional migration in the world, from South and East Asia to the Gulf, as seen in Figure 3.22. It is ironic that the Asian workers who build the Gulf's glitzy malls are not permitted entry into them. They are part of an almost exclusively male cohort shuttled between their workplaces and the large camps built to accommodate them. Although grateful for the work, many complain of discrimination, abuse, and poor living conditions.

This depiction of MENA's oil wealth, which is concentrated in a few countries, does not consider the economic poverty so characteristic of the region's other countries and the great majority of its peoples (see **Table 6.1**). The worsening conditions in many of the Arab countries—the poverty, crowding, unemployment, and the growing gap between the great mass of the poor and the increasingly wealthy rich—were the economic ingredients of the Arab Spring described later.

6.5 Geopolitical Issues

The Middle East and North Africa have long been vital in world affairs, with their resources and geographical assets such as the Suez Canal coveted by outside interests. From very early times, overland caravan routes, including the Silk Road, crossed the Middle East and North Africa with highly prized commodities traded between Europe and Asia. The security of these routes was vital. Geopolitical concerns are now focused on narrow waterways, access to oil, access to freshwater, and terrorism.

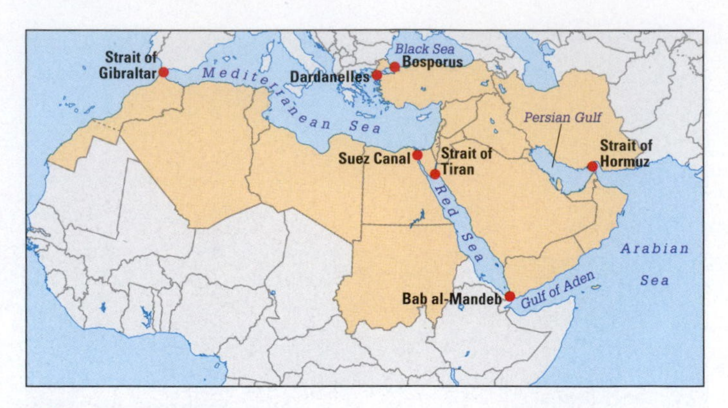

• Figure 6.29 Chokepoints in the Middle East and North Africa.

Chokepoints

One of the striking characteristics of the geography of the Middle East and North Africa is how many seas border and penetrate the region. In many cases, these bodies of water are connected to one another though narrow straits and other passageways. In geopolitical terms, such constrictions are known as **chokepoints**—strategic narrow passageways on land or sea that may be easily closed off by force or even the threat of force (**•Figure 6.29**). Chokepoints must be unimpeded if world commerce is to carry on normally. Keeping them open is therefore usually one of the top priorities of regional and external governments, especially the powerful ones. Likewise, closing them is a priority to a combatant nation or a terrorist entity seeking to gain a strategic advantage. Many notable events in military history and the formation of foreign policy in the Middle East focus on these strategic places.

One of the world's most important chokepoints is the Suez Canal, which opened in 1869. Slicing 107 miles (172 km) through the narrow Isthmus of Suez, the British- and French-owned Suez Canal linked the Mediterranean Sea with the Indian Ocean, saving cargo, military, and passenger ships a journey of many thousands of miles around the southern tip of Africa. Keeping the Suez Canal in friendly hands was one of Britain's major military concerns during World War II. That objective led to a hard-fought and eventually successful British and Allied campaign against Nazi Germany in North Africa, with a decisive end in the 1942 Battle of El Alamein, Egypt. Egyptian President Gamal Abdel Nasser's nationalization of the canal in 1956 led immediately to a British, French, and Israeli invasion of Egypt and a conflict known as the **Suez Crisis** and the **Arab-Israeli War of 1956**. Egypt effectively won that war when American President Dwight Eisenhower pressured the invading forces to withdraw, leaving the canal in Egypt's hands, where it remains (recently with enlarged capacity) as one of the country's main sources of revenue (**•Figure 6.30**).

• **Figure 6.30** The strategically vital Suez Canal zone saw bitter fighting in the Middle East wars of 1956, 1967, and 1973. The Sinai coastline is visible just behind the vessel, which is in a northbound convoy.

Joe Hobbs

Nearby, another chokepoint played a critical role in the most important Arab-Israeli war, the Six-Day War of 1967. One of the events that precipitated the war was President Nasser's closure of the Strait of Tiran (at the southern end of the Gulf of Aqaba) to Israeli shipping. Israel had won the right of navigation through the strait after its war of independence two decades earlier and would not accept its closure; that was one reason it attacked Egypt.

On either side of the Arabian Peninsula are two more critical chokepoints. One is exceptionally vital to the world's economy: the Strait of Hormuz, connecting the Persian Gulf with the Gulf of Oman and the Arabian Sea. Much of the world's oil supply passes through here in the holds of giant supertanker ships. Closure of the Strait of Hormuz would have devastating effects on the world's industrial and financial systems. Iran's plans to station Chinese-made missiles on the Strait of Hormuz and thus threaten international oil shipments led to a new level of US involvement late in the **Iran-Iraq War of 1980–1988**. A bloodbath initiated by the ambitions of Iraqi leader Saddam Hussein to capture yet more oil and achieve a historic Arab defeat over the Persian foe, the war ended in a geographic stalemate. That war also prompted a flurry of new oil pipeline construction designed to bypass the Strait of Hormuz chokepoint (see Geography of Energy, **page 238**). One of these pipelines, the Petroline route running east–west across the Arabian Peninsula, was built to bypass another important chokepoint, the Bab el-Mandeb, which connects the Red Sea with the Gulf of Aden and the Indian Ocean. Iran threatened to close the Strait of Hormuz when Western powers imposed sanctions against Iran for ramping up its nuclear program in 2012.

Turkey controls two more chokepoints that together are known as the Turkish Straits. The northernmost strait is the Bosporus, which cleaves the city of Istanbul into western (European) and eastern (Asian) sides, and the southern strait is called the Dardanelles (see Figure 5.26). Their security has long been critical to the successful passage of goods between Europe and Asia and even more so to the passage of vessels between Russia (and previously the Soviet Union) and the rest of the world. As discussed on **page 167**, throughout the 20th century one of the Soviet Union's constant strategic priorities was the right of navigation through the Turkish Straits.

Finally, the Strait of Gibraltar, connecting the Mediterranean Sea with the Atlantic Ocean, is also a chokepoint. Figure 6.26 shows how geographic good luck has given the lands of the Mediterranean Sea access to the rest of the world through Gibraltar; the Mediterranean is nearly a lake. Britain's insistence on maintaining control over its enclave of Gibraltar, decades after the decolonization of most of the world, is an excellent indicator of how critical this chokepoint is.

Access to Oil

The Suez Canal and other chokepoints, the cotton of the Egypt's Nile Delta, and the strategic location of the region were important during colonial times and have remained so ever since. But oil has been and will remain (as long as fossil fuels drive the world's economies) what keeps the rest of the world interested economically in the Middle East and North Africa. The region's oil flows to many countries, but most of it is marketed in Western Europe, China, and Japan. The United States also imports large amounts of Gulf oil but has a smaller relative dependence on this source than Japan and Europe do; about 20 percent of its imported oil came from the Gulf in 2014. However, the Gulf region is very important to the United States because of the great dependence of close American allies on Gulf oil and because of the importance of the oil as a future reserve. American companies are also heavily involved in oil operations and oil-financed development in the Gulf countries. Gulf "petrodollars" are spent, banked, and invested in the United States, contributing significantly to the US economy. Maintaining a secure supply of Gulf oil has therefore been one of the long-standing pillars of US policy in the Middle East.

The United States has long had a precarious relationship with the key players in the Middle Eastern arena. The US has pledged unwavering support for Israel while also courting Israel's oil-rich Arab foes, notably oil-rich Saudi Arabia. That Arab kingdom and its neighbors around the Persian Gulf possess 48 percent of the world's proven oil reserves. Thus they are vital to the long-term economic security of the Western industrial powers, China, India, and Japan. The United States and its Western allies made it clear they would not tolerate any disruption of access to this supply when Iraqi troops, under orders from Iraqi President Saddam Hussein, occupied Kuwait on August 2, 1990. US President George H.W. Bush "drew a line in the sand," proclaiming, "We cannot permit a resource so vital to be dominated by one so ruthless—and we won't." In what came to be known as the **Gulf War**, the United States and a coalition of Western and Arab allies mounted a massive array of military might that ousted the Iraqi invaders within months and secured the vital oil supplies for Western markets. When the United States invaded and occupied Iraq in 2003, ostensibly to extinguish Iraq's ability to develop and deploy **weapons of mass destruction (WMD)**,

Geography of Energy | Pipelines and Chokepoints in the Middle East

The oil-exporting countries of the Middle East, often with the financial support of the United States and other leading oil-consuming countries, have invested enormous resources to ensure the safe passage of oil to world markets. The Middle East's oil region is not landlocked like the oil-rich "stans" of central Asia, but some of the same kinds of geopolitical issues are involved in oil export 189 planning. These concerns often point to pipelines as the best means of routing oil and natural gas shipments (**• Figure 6.F**). Pipelines are attractive because they shorten the time and expense involved in seaborne transport, and they bypass chokepoints. But they are also vulnerable to disruption. Weapons as small as grenades can disrupt supplies, as in the Sinai Peninsula where militants have repeatedly ruptured a pipeline carrying Egyptian natural gas to Israel. A major challenge is how to route a pipeline so that it will not cross through a potential enemy's territory, and another is how to maintain friendships with the countries the oil crosses. Looking at the pipeline routes on Figure 6.F, you will get a feel for the compelling intersection of economics, military thinking, and the consequences of war in pipeline geography.

One of the first major pipelines in the region was the 1100-mile (1800-km) Trans-Arabian Pipeline (Tapline), leading from the oil-producing eastern province of Saudi Arabia to a terminal on the Mediterranean coast of Lebanon. It has the advantage of bypassing three chokepoints (Strait of Hormuz, Bab el-Mandeb, and Suez Canal). However, at the time it was completed (1950), it could not have been anticipated that it would cross two of the worst impending conflict zones in the Middle East: the Golan Heights of Syria (which fell to Israel in the 1967 war) and southern Lebanon (a major theater of the **Lebanese civil war** of the 1970s and the subsequent Israeli-Lebanese Shi'ite struggles). Tapline was knocked out early by its unfortunate geography.

Another Middle East conflict, the 1967 Six-Day War between Israel and its neighbors, led to the closure of the Suez Canal (for about a decade) and thus to the construction of two new pipelines to bypass Suez: one across southern Israel from the Gulf of Aqaba to the Mediterranean Sea and another across Egypt from the Gulf of Suez to the Mediterranean (thus its acronym, SUMED).

No Middle Eastern country has been more dependent on pipelines than Iraq, and none has so systematically experienced the liabilities of pipelines. The Iran-Iraq war of 1980–1988 had major impacts on pipeline geography. Saudi Arabia supported Iraq in the war and so tried to ensure it could get oil to market without being threatened by Iran; this meant bypassing the Strait of Hormuz. That led to the construction of Petroline (opened in 1981), the east–west pipeline across the Arabian Peninsula. Iraq built a pipeline to ship some of its oil south to Petroline, thereby bypassing the Strait of Hormuz as well. Used in conjunction with SUMED, Petroline allows Iraqi and Saudi oil to reach the Mediterranean coast without passing through the chokepoints of the Strait of Hormuz, the Bab el Mandeb, or the Suez Canal. Beginning in 1961, Iraq was also able to ship oil through pipelines almost due west through Syria to the Mediterranean Sea. But when it attacked Iran in 1980, Iraq lost that route because Syria was Iran's ally.

Iraq quickly responded to that setback by building a new pipeline leading almost due north to bypass Syrian territory and then making a sharp 90-degree turn westward through southern Turkey to the Mediterranean Sea. But in 1990, Iraq attacked Kuwait and threatened Saudi Arabia, which prompted a US-led counterattack. Saudi Arabia responded by closing the southern link with Petroline. Turkey, a NATO ally of the United States, reacted by closing the northern link. The United Nations responded by issuing strict controls on Iraqi oil exports. In sum, by his military adventurism, Saddam Hussein did almost everything imaginable to deprive his country of oil exports. It is difficult to explain his rationale; this is the man that also precipitated two wars that were ruinous for his country and his life.

• Figure 6.F Principal pipelines in the heart of the Middle East. Vulnerable chokepoints and volatile political relations have led to the construction, closure, and often indirect routing of many pipelines.

THINK CRITICALLY: How can Iraqi oil avoid the Strait of Hormuz? Why did Iraq stop using its pipelines crossing Syria, and how were new lines routed to avoid that country? What are the advantages of the SUMED pipeline?

including chemical, biological, and nuclear weapons, many critics insisted that this was just another example of America's determination to control Middle Eastern oil.

The Gulf War was not the first time the United States expressed its willingness to use force to maintain access to Middle Eastern oil. In 1979, in the wake of the Islamic revolution in 332 Iran, the Soviet Union invaded neighboring Afghanistan. US military analysts feared that the Soviets might use Afghanistan as a launching pad to invade oil-rich Iran. The United States deemed this prospect unacceptable, and President Jimmy Carter issued the foreign policy statement that came to be known as the **Carter Doctrine**: "An attempt by any outside force to gain control of the Persian Gulf region will be regarded as an assault on the vital interests of the United States of America, and such an assault will be repelled by any means necessary, including military force." "Vital interests" meant oil, and defending them with military force meant that the United States was willing to go to war with the Soviet Union, presumably nuclear war. The US began arming the *mujahidin* fighting Soviets in Afghanistan. In hindsight we can recognize that factions among these Islamist rebels would become al-Qa'ida and the Taliban.

Middle Eastern wars had already become proxy wars for the superpowers, with oil always looming as the prize. In the 1967 and 1973 Arab-Israeli wars, for example, Soviet-backed Syrian forces fought US-backed Israeli troops. American support of Israel in this war prompted Arab members of OPEC to impose an **oil embargo**, refusing to sell their oil to the United 534 States, thus precipitating the nation's first energy crisis. During the 1973 war, the United States put its forces on an advanced state of readiness to take on the Soviets in a nuclear exchange if necessary. All of these events illustrate that the Middle East and North Africa form, as Eastern Europe did, a shatterbelt—a large, strategically located region composed of conflicting states caught between the conflicting interests of great powers.

Access to Freshwater

The next war in the Middle East will be over water.
—Jordan's late King Hussein

Some of the most serious geopolitical issues in the Middle East and North Africa relate to **hydropolitics**, or political leverage and control over water. In this arid region, where most water is available either from rivers or from underground aquifers that cross national boundaries, control over water is an especially difficult and potentially explosive issue. An estimated 90 percent of the usable freshwater in the Middle East crosses one or more international borders.

Water is one of the most problematic issues in the Palestin- 249 ian-Israeli conflict (see **•Figure 6.31**). The average Israeli uses 62 gallons (235 l) per day, whereas the average West Bank Palestinian uses 20 gallons (76 l). (The World Health Organization calculates that 13 gallons, or 50 liters per person per day, is needed for minimal health and sanitation standards.) Under the terms of the Oslo II Accords, 80 percent of the water from the aquifers under the West Bank is sent to Israel, leaving the Palestinians with only 20 percent. The treaty also prohibits Palestinians from tapping into the Mountain Aquifer, the largest in the region (though they are permitted to drill a limited number of wells into other groundwater sources each year). As their population grows, the Palestinians become increasingly dependent upon Israel for their freshwater supply. When Palestinians exhaust their annual allocation of water, they must buy water from Israel, which is costly; up to 20 percent of a Palestinian household's income may be devoted to water costs. Many Israeli policymakers insist that water resources in the West Bank must remain under strict Israeli control, and on these grounds they oppose the creation of a Palestinian state in the West Bank.

In 2013, Israel, Jordan, and the Palestinian territory signed an agreement to build a pipeline that would send seawater from the Red Sea to the Dead Sea. Called the **Two Seas Canal**, this venture is intended to replenish the Dead Sea, which has been shrinking up to 3 feet (1 m) a year since the early 1980s from diversions of Jordan River water for agricultural purposes. Ecologists have questioned the usefulness of the pipeline, as the amount of water carried from the Red Sea would only be about one-eighth of what is needed to stabilize the Dead Sea, and have raised concerns about seawater damaging the Dead Sea's unique ecology.

Seawater would also be sent along the Two Seas Canal to a series of desalination plants, and the resulting freshwater piped to Jordan. Desalination of salt water has become an increasingly important way for Middle Eastern countries to add to their scarce freshwater supply, but it is very costly. Nearly 50 percent of Israel's drinking water now comes from desalinized seawater.

Water was a critical issue blocking a peace treaty between Israel and Syria before Syria descended into chaos. If Syria were to recover all of the Golan Heights (which it lost to Israel in the 1967 war), it would have shorefront on Lake Kinneret and therefore, presumably, rights to use its water. That prospect is unacceptable to Israel. This lake is Israel's principal supply of freshwater, feeding the National Water Carrier system that transports water south to Israel's population centers. The 1994 peace treaty between Israel and Jordan also gives Jordan access to about 12 percent of Lake Kinneret's water annually.

A useful way to think about the geography of hydropolitics is in terms of **upstream** and **downstream countries**. Simply because water flows downhill, an upstream country is usually able to maximize its water use at the expense of a downstream country (a situation described as a **zero-sum game**, where any gain by one party represents an equivalent loss to the other). However, this is not always true. Although a downstream Jordan River country, Israel is more powerful and could use the threat of force to wrest more water out of the system.

Historically, Egypt has also defied the norm of upstream countries exercising their power over downstream countries. Egypt is the ultimate downstream country, at the mouth of the great Nile River (**•Figure 6.32**). However, it has long been the strongest country in the Nile Basin and has threatened to use its greater force if it does not get the water it wants. In 1926, when the British ruled Egypt and many other colonies in Africa, 10 countries located on the Nile and its tributaries

• **Figure 6.31** Israel's National Water Carrier transports water from Lake Kinneret to Israel's thirsty cities. The Mountain Aquifer is also a major source of Israeli drinking water. The Two Seas Canal would generate hydroelectricity, bring desalinated water to Jordan and potentially stabilize the water level of the drying Dead Sea.

THINK CRITICALLY: Are the fresh water resources of the area distributed equitably to its peoples?

upstream of Egypt were compelled to sign the **Nile Water Agreement.** This guaranteed Egyptian access to 75 percent of the river's flow, even though barely a drop of the Nile's waters actually originates in Egypt. Sudan was given access to 25 percent of the water, and Ethiopia was given none. The treaty forbids any projects that might threaten the volume of water reaching Egypt (including from all the Nile's tributaries), prohibits use of Lake Victoria's water without Egypt's permission, and gives Egypt the right to inspect the entire length of the Nile to ensure compliance.

In recent years, however, one country after another has defied the treaty, calling it an outmoded legacy of colonialism. Tanzania flouted the treaty and built pipelines carrying Lake Victoria waters to thirsty towns and villages inland; Kenya is considering this as well. Uganda built its controversial Bujagali Dam on the White Nile, mainly for hydroelectricity production. With Chinese assistance, Ethiopia built the huge Tekezé Dam, for hydropower and irrigation, on a tributary of the Blue Nile. In order to use and sell hydropower, it is building an even larger dam, the Grand Ethiopian Renaissance Dam, on the Blue Nile itself. Sudan recently built the Merowe (Hamdab) Dam across the Nile, and with Chinese aid plans to add three more dams, two of which will drown more parts of Nubia, whose living people and archaeological legacy already lost much to the Aswan High Dam (see Regional Perspective, **page 241**). South Korea, China, and India have purchased or "grabbed" huge swaths of land in Sudan and Ethiopia to grow wheat and rice through irrigation. All of these projects will result in massive new drawdowns of Nile waters. Predictably, Egypt has condemned and challenged the upstream Nile countries. Egypt threatened to bomb Ethiopia's Renaissance Dam and called Kenya's stated intention to withdraw from the Nile Water Agreement an "act of war."[7]

66, 421

Meanwhile, Egypt's demands on Nile waters are increasing. Following on its ambitious Aswan High Dam, Egypt built the multibillion-dollar Toshka Canal, which transports water from Lake Nasser over a distance of 100 miles (160 km) to the Kharga Oasis of the Western Desert (see **Figure 6.32**). Proponents of the canal insist that it will result in the cultivation of nearly 2350 square miles (6100 sq km) of "new" land and provide a living for hundreds of thousands of people. Critics argue that it is a waste of money and that salinization and

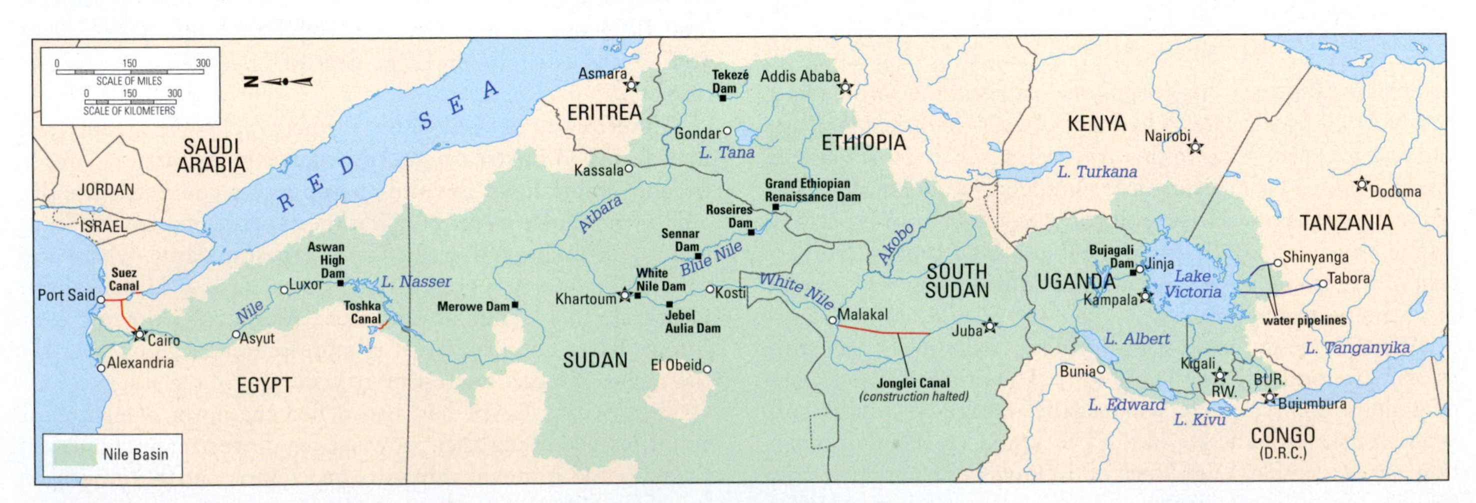

• **Figure 6.32** The Nile River and its tributaries are a vital source of water for about 200 million people in Africa. Many countries have erected dams to provide water for irrigation and to generate electricity. Note this map is oriented such that north is at the left rather than the top.

THINK CRITICALLY: Should downstream Egypt be permitted to exercise control over upstream Nile waters, or is it time to renegotiate the Nile Waters Agreement?

Regional Perspective The Aswan High Dam

Egypt's enormous Aswan High Dam (• **Figure 6.G**; and see location in Figure 6.32) is located on a stretch of the Nile known as the First Cataract (the Nile's many **cataracts** are areas where the valley narrows and rapids form in the river). Its reservoir, Lake Nasser, stretches for more than 300 miles (480 km) and reaches into northern Sudan. Like all giant hydrological projects, the Aswan High Dam has generated benefits and liabilities, and its construction was controversial. On the plus side:

- By storing water in years of abundant rain for use in years of low rainfall, the High Dam provides a constant water supply to the Egyptian Nile Valley's 11,000 square miles (28,500 sq km) of irrigated fields, thus extending the cultivation period.
- Year-round irrigation has significantly boosted crop production in Egypt.
- Because water levels can be adjusted, there are fewer threats to ships navigating the Nile.
- The dam provides hydroelectricity to Egypt's industries and citizens.
- Excess waters stored in Lake Nasser help Egypt to avoid the damaging impacts of drought.
- The dam can hold back overabundant water, helping Egypt to avoid the damaging impacts of flooding.

The Aswan High Dam has also had drawbacks:

- The dam has caused the water table to rise, making it harder for irrigated soil to drain properly. When farmers use too much water, standing water evaporates and leaves a deposit of mineral salts—a problem known as **salinization**—that causes soil to lose its productivity.
- Canal waters, which are now high throughout the year, are ideal breeding grounds for the snail that hosts the parasite that causes schistosomiasis (bilharzia), a debilitating disease affecting a large proportion of Egypt's rural population.
- Mediterranean sardine populations, now deprived of the rich river silt that nurtured their food supply of phytoplankton, have plummeted off the Nile Delta. The lucrative sardine fishing industry has declined by more than 90 percent.
- The dam prevents rich volcanic silt washed down the Nile from Ethiopia from reaching the floodplain below the dam. Without this "free fertilizer," Egyptian farmers must buy expensive artificial fertilizers.
- Because silt is no longer replenishing the soils of the Nile Delta, there is coastal erosion along the Mediterranean seafront. This is allowing seawater to destroy productive farmland.
- The Nubian people who lived in both the Egyptian and Sudanese areas now inundated by the reservoir behind the dam had to be relocated from their homeland to alien environments in both countries. Their economic and social adjustments in these areas were difficult.
- Many of the archaeological treasures of ancient Egypt were drowned by the reservoir (but others were relocated).
- If the dam burst because of an earthquake or a bombing, the lives of nearly 80 million people downstream would be at risk. One analyst wrote:

> If a bomb hit the top of the Dam, the water pouring out uncontrolled would make the Dam burst in its complete breadth within a few hours. The masses of water sweeping across the whole of Egypt (they would be at least the discharge of two years within hours or days) would cause utter devastation from Aswan to Damietta, destroy all cultivated land, and most probably about 99 per cent of the total population would lose their lives.[9]

Whenever a dam is built, there are questions about whether its advantages outweigh its drawbacks. When great dams are built in the LDCs—like the Aswan High Dam in Egypt or the Three Gorges Dam in China—such questions tend to be secondary to the symbolic importance of the dam. These massive engineering projects announce that a developing country is developed enough to master its environment and marshal the extraordinary engineering and other resources needed to build the dam. The dam announces that a country has "come of age."

349

Compliments of NASA/Earth Sciences and Image Analysis Laboratory at Johnson Space Center

• **Figure 6.G** Egypt's Aswan High Dam, on the First Cataract of the Nile River.

THINK CRITICALLY: Do the Aswan Dam's "pros" outweigh its "cons," or is the dam a colossal mistake?

evaporation will take a huge toll on the cultivated land and the country's water supply.

The familiar specter of population growth and the unknowable effects of climate change will play big roles in the hydropolitics of the Nile. About 200 million people live in the Nile Basin, a number predicted to grow to about 400 million by 2050. Every person's direct needs for water plus the food it produces mean growing demands on Nile waters and the potential for conflict. The solution? Lester Brown of Worldwatch proposes three steps: Reduce population growth; grow crops that are less demanding of water; and go back to the negotiating table for a new Nile Water Agreement.[8] See Geography of Water on **page 243** for option of securing "virtual water."

Turkey—the source of four-fifths of Syria's water and two-thirds of Iraq's—exercises its upstream advantage on the Tigris and Euphrates Rivers (**• Figure 6.33**). Turkey's position has long been that water in Turkey belongs to Turkey, just as oil in Saudi Arabia belongs to Saudi Arabia. Not surprisingly, downstream Syria and Iraq reject this position and are distraught by the diminished flow and quality of water resulting from Turkey's comprehensive **Southeastern Anatolia Project**. When completed, the project is expected to reduce Syria's share of the Euphrates waters by 40 percent and Iraq's by 60 percent. Further increasing the likelihood of serious future tension is a history of strained relations among Turkey, Syria, and Iraq, accompanied by the fact that no commonly accepted body-of-water law governs the allocation of water in such international situations. Turkish leaders have said they will never use water as a political weapon, but Turkey has already wielded water to its advantage. In 1987, for example, Turkey increased the Euphrates flow into Syria in exchange for a Syrian pledge to stop support of Kurdish rebels inside Turkey.

Terrorism

During the two terms of US President George W. Bush, almost all of the geopolitical issues related to this region—oil, economic development, trade, aid, the Arab-Israeli conflict, and more—were subsumed beneath the broader rubric of the president's declared **War on Terror**. Even his administration's 2003 invasion of Iraq was justified as a counterattack on a murky Iraqi hand in 9/11. The open-ended, unconventional War on Terror would destroy the enemies who attacked the United States on 9/11 and use preemptive strikes if needed to protect the US homeland against future attacks. President Obama far more frequently employed preemptive attacks on suspected terrorist targets, while avoiding the term *War on Terror*. One of his favorite tactics was the use of unmanned "predator drones" that fly quietly overhead and unleash deadly force on ground targets. Their use is controversial because they sometimes result in civilian casualties. In the MENA region, drone strikes have focused on al-Qa'ida in the Arabian Peninsula (AQAP) in Yemen, and further afield they have targeted presumed terrorists in Pakistan's frontier with Afghanistan, and in Somalia.[11]

319, 430

Since 9/11, Americans and their Western allies in particular have become accustomed to heightened security measures to provide protection against **terrorism**, which by the US State Department definition is "premeditated, politically motivated violence perpetrated against noncombatant targets by subnational groups or clandestine agents, usually intended to influence an audience."[12] The terrorists pursued by the United States are almost without exception Islamists, but here too President Obama tread carefully in his speech to avoid associating them with the religion of Islam.

Although nominally religious, the more radical of these movements have political and cultural aims, particularly the destabilization or removal of US and Israeli interests in the MENA region and abroad. In 1993, followers of the radical Egyptian cleric Sheikh Umar Abdel-Rahman bombed New York City's World Trade Center as a protest against American support of Israel and Egypt's pro-Western government. In an attempt to destabilize and replace Egypt's government, which it viewed as an illegitimate regime too supportive of the United States, another violent Egyptian Islamist group attacked and killed foreign tourists in Egypt in the 1990s (that effort to inspire Egyptians to rise up against their government backfired, as Egyptians recoiled from this slaughter of innocents). In the 1980s, members of the Shi'ite Muslim pro-Iranian **Hizbullah** ("Party of God") in Lebanon kidnapped foreign civilians to use as bargaining chips for the release of comrades jailed in other Middle Eastern countries. They killed American University in Beirut President Malcolm Kerr, shooting him in the back of his head outside his office. Hizbullah targeted him because he was an accessible, high-profile American (on a personal note, I mourned him as my mentor at the American University in Cairo, and as a friend of his family).

Within Israel and the autonomous Palestinian territories of the West Bank and Gaza Strip, Sunni Muslim Palestinian members of **Hamas** (an Arabic acronym for the Islamic Resistance Movement) and another organization called the al-Aqsa Martyrs Brigade carried out terrorist attacks on Israeli civilians and soldiers in a largely successful effort to derail implementation of the peace agreements reached in the 1990s

Joe Hobbs

• Figure 6.33 The Tigris and Euphrates rivers rise in Turkey, giving this non-Arab country control over a resource vital to the lives of millions of Arabs in downstream Syria and Iraq. This waterfall is on a tributary of the Tigris in far eastern Turkey.

Geography of Water | Virtual Water

Two British professors trying to figure out how Middle Eastern countries can conserve their scarce water supplies while also growing their economies came up with the concept of "virtual water."[10] The answer was to import certain foods, especially those demanding a lot of water to produce, rather than trying to produce them domestically. This would save the Middle Eastern countries a lot of virtual water, defined as the total volume of freshwater needed to produce and process a commodity or service, measured where the product was actually produced. It is "virtual" because it is an externality not factored into and not apparent in the production process. 83

The hidden amount of water used can be revealed by measuring the commodity's "water footprint," a component of the greater "ecological footprint" and "human footprint" introduced in Chapter 3. 55 • **Figure 6.H** shows, for example, that beef cattle require 15,000 liters of water per kilogram, or 4650 liters (1228 gallons) to produce a 300-gram (0.6 lb) steak. We have already seen how much grain goes into producing our meat, as we eat so high on the food chain (**page 82**); with the concept of virtual water, we can see how high we are drinking on the "water chain," and consider the pressure that demand places on natural ecosystems, agricultural systems, and human communities. We will see that the cutting of Amazon rainforest for cattle production is a major factor behind the destruction of Amazon rainforests. 486 There is a knock-on effect in each action: Amazon deforestation is a significant component of climate change. The virtual water concept also allows us to see how endangered or endowed individual countries are with respect to water and how they can make the best decisions for sustainable development. Countries like Libya and Saudi Arabia with nonrenewable aquifers cannot afford to drain their water resource to produce export crops—as Libya has been doing with its "Great Man-Made River," and Saudi Arabia formerly did with center-pivot irrigation of wheat (see **page 505** and the global aquifers map on **page 67**). Libya and Saudi Arabia might instead consider following the example of their neighbor Egypt, importing water-intensive commodities and even acquiring land abroad on which to produce them. Saudi Arabia has done so, joining Egypt and the UAE as the Middle Eastern "land grabbers."

Timm Kekeritz

• **Figure 6.H** By understanding the "water footprint" of commodities, it is possible to appreciate how water is traded virtually in the international marketplace. It takes 650 liters (172 gal) of water to produce 1 pound (500 g) of wheat, 2500 liters (660 gal) of water to produce enough meat for a burger, and a staggering 4650 liters (1228 gal) to produce a 300 g (11 oz) beef steak. 66

THINK CRITICALLY: What options do water-poor countries in the region have to secure the water-dependent foods they need?

between the Israeli government and the Palestine Liberation Organization (PLO).

Al-Qa'ida and ISIS

In 1998, the world began to hear about Osama bin Laden, a former Saudi businessman living in exile in Afghanistan, when his **al-Qa'ida** organization bombed US embassies in Kenya and Tanzania as part of its proclaimed *jihad* against American imperialism and immorality. Bin Laden's organization was also responsible for the bombing of the American naval destroyer *USS Cole* in Yemen's harbor of Aden in 2000. However shocking those assaults were, they pale in comparison to al-Qa'ida's attacks against civilian targets in the United States on September 11, 2001. In the most ferocious terrorist actions ever undertaken to that date, members of al-Qa'ida "cells" (small groups of agents) in the United States hijacked four civilian jetliners and succeeded in piloting two of them into the twin towers of New York City's World Trade Center and one into the Pentagon, home of the US Department of Defense, outside Washington. Passengers on a fourth plane aborted the terrorists' likely mission of destroying the White House or damaging the US Capitol, forcing the plane to crash in rural Pennsylvania. Altogether more than 3000 people, mainly civilians, perished in the 9/11 attacks (• **Figure 6.34a**). Bin Laden and his followers cheered the carnage as justifiable combat against an infidel nation whose military troops occupied the holy land of Arabia, where Mecca and Medina are located.

As the United States counterattacked, ousting al-Qa'ida from its bases in Afghanistan and cracking down hard on its leadership around the world, al-Qa'ida evolved into a far

Geography of Terrorism What Do al-Qa'ida and ISIS Want?

As a geographer of the Middle East, I was deeply interested after 9/11 in finding an answer to the al-Qa'ida question. Surprised that there was little attention to the organization's stated interests and the tenet of knowing the adversary well, I kept a close ear to the ground, listening to audio, watching video, and reading statements. I published those findings in *The Geographical Review* and share them briefly here. The bottom line for al-Qa'ida is that it wants to control geographic space, especially sacred space, and has not given up that effort.[13] With ISIS on the scene, we can ask the same question: What does ISIS want? And how, if at all, do the ambitions of these two terrorist groups differ?

To begin, it is useful to know what al-Qa'ida is. Al-Qa'ida ("the Base" in Arabic) is a transnational organization that seeks to unite Islamist militant groups worldwide in a common effort to achieve its goals. In a lawless region of Pakistan, bin Laden formed al-Qa'ida from a coalition of seven Islamist militant groups (three Egyptian, two Pakistani, one Bangladeshi, and one Afghan) called the "World Islamic Front for the Jihad against the Jews and the Crusaders." With his credentials that included being a *mujahid* ("holy warrior") against the Soviets in Afghanistan, bin Laden emerged as its leader.

Osama bin Laden called for the killing of American and other Western civilians. Ayman al-Zawahiri, al-Qa'ida's second in command, called for an escalation of such attacks with "the need to inflict the maximum casualties against the opponent, for this is the language understood by the West, no matter how much time and effort such operations take."[14] Rohan Gunaratna, a leading authority on the organization, regards al-Qa'ida as an extremely unusual terrorist group in the category he calls "apocalyptic," one that believes "it has been divinely ordained to commit violent acts, and likely to engage in mass casualty, catastrophic terrorism."[15] He warns that al-Qa'ida "will have no compunction about employing chemical, biological, radiological and nuclear weapons against population centers."[16]

But why? Why do these people want to kill Western, and particularly American, civilians? Al-Qa'ida leaders have been very explicit in their rationale. Osama bin Laden explained that the main reason for the terrorist attacks is to convince the United States that it should withdraw its military forces and other interests from the Islamic Holy Land. In the following statement, the "Land of the Two Holy Places" means Saudi Arabia. The two holy places are the Saudi Arabian cities of Mecca (site of the Ka'aba, and Islam's most sacred city) and Medina (Islam's second holiest place, where the Prophet Muhammad is buried). The Dome of the Rock is the shrine in Jerusalem, described earlier, containing the sacred rock from where the Prophet Muhammad was said to have ascended into heaven on the Night Journey.

> The Arabian Peninsula has never—since God made it flat, created its desert, and encircled it with seas—been stormed by any forces like the crusader armies spreading in it like locusts, eating its riches and wiping out its plantations . . . The latest and greatest of these aggressions, incurred by the Muslims since the death of the Prophet . . . is the occupation of the Land of the Two Holy Places—the foundation of the house of Islam, the place of the revelation, the source of the message, and the place of the noble Ka'aba, the *qibla* [holy direction] of all Muslims—by the armies of the American crusaders and their allies. We bemoan this and can only say "No power and power acquiring except through Allah" . . . To push the enemy—the greatest *kufr* [infidel]—out of the country is a prime duty. No other duty after Belief is more important than this duty. Utmost effort should be made to prepare and instigate the *umma* [Islamic community] against the enemy, the American– Israeli alliance—occupying the country of the two Holy Places . . . to the al-Aqsa mosque in Jerusalem . . . The crusaders and the Jews have joined together to invade the heart of Dar al-Islam—the Abode of Islam: our most sacred places in Saudi Arabia, Mecca and Medina, including the prophet's mosque and Dome of the Rock in Jerusalem, al-Quds.[17]

In what was arguably his most important policy statement, announcing the formation of the "Islamic World Front for the Jihad against the Jews and the Crusaders," bin Laden used the occupation of Islamic sacred space as the principal justification for war against the United States:

> In compliance with God's order, we issue the following *fatwa* [religious injunction] to all Muslims: The ruling to kill the Americans and their allies—civilians and military—is an individual duty for every Muslim who can do it in any country in which it is possible to do it, in order to liberate the al-Aqsa mosque and the holy mosque [in Mecca] from their grip, and in order for their armies to move out of all the lands of Islam, defeated and unable to threaten any Muslim.[18]

In an earlier *fatwa* (1996), bin Laden laid out these goals for al-Qa'ida: to drive US forces out of the Arabian Peninsula, overthrow the Saudi government, and liberate the holy places of Mecca and Medina. Overthrowing the Saudi government and the other autocratic dynasties of the Persian Gulf region, along with other secular and pro-Western regimes of the Middle East—notably Egypt's—is a theme that is very often articulated in al-Qa'ida ideology.

What is al-Qa'ida's ultimate goal? Is the organization satisfied now that US troops are no longer stationed in Saudi Arabia? If Americans return to the Kingdom to train Arab ground troops to take on ISIS, al-Qa'ida will feast again on the Crusader occupation narrative. What would be achieved if all Western interests were driven from the Islamic Holy Land and if all of the governments sympathetic to the West were overthrown? That would assist al-Qa'ida's ultimate goal, to reestablish the **Caliphate (*al-Khilaafa*)**—the empire of Islam's early golden age—and thereby empower a formidable array of truly Islamic states to wage war on the United States and its allies.[19]

With these goals in mind, what should the United States do? Al-Qa'ida's strategists clearly hope that terrorism will inflict unacceptable losses of American lives, forcing the United States to withdraw its presence in all forms from the Middle East. Al-Qa'ida thinks the United States has no stomach for sustained sacrifice, citing the withdrawal of US troops from Somalia after the loss of 18 American soldiers in Mogadishu in 1993 (the "Blackhawk Down" episode; see **page 430**), the withdrawal of US forces from Lebanon following the suicide bombing of the Marine barracks in Beirut in 1983, and the repositioning of American forces inside Saudi Arabia after the attack on the Khobar barracks in 1996.

Al-Qa'ida poses a dilemma for the United States. If the United States were to withdraw from the region, it might only reaffirm al-Qa'ida's belief that its adversary is weak and vulnerable, thus encouraging more attacks. And if al-Qa'ida's ultimate goal is to take on the United States once it has established an Islamic empire, there is no reason to believe that a unilateral withdrawal would bring a cessation of hostilities.

On the other hand, if the United States continues to conduct a war on al-Qa'ida and a broader "war on terror" in a host of Muslim countries, al-Qa'ida can use the American presence, and especially unintended civilian losses, to incite widespread hatred and violence directed against the United States. Al- Qa'ida used the US occupation of Iraq to argue its case that "the Crusaders" *are* waging a war against Muslims and the Islamic world. Many analysts argue that the 2003 US invasion of Iraq served as a catalyst to recruit uncounted numbers to the jihadist cause. Iraq became a theater where new militant groups could cut their teeth on American soldiers and civilian contractors, and their Iraqi allies. One of these was a well-organized insurgent group officially linked with al-Qa'ida, called **al-Qa'ida in Iraq**. Led by the Jordanian-born Abu Mus'ab al-Zarqawi until he was killed in a US drone strike in 2006, it later split with al-Qa'ida to become ISIS.

Al-Qa'ida's core leadership has been weakened by years of dogged hunting by US intelligence. Yet the organization continues to thrive, especially in lawless Yemen, where its affiliate al-Qa'ida in the Arabian Peninsula (AQAP) continues its quest for undetectable bombs that can bring down aircraft, and in lawless Syria, where its franchise is Jabhat al-Nusra and its leadership is called "the Khorosan Group" (**•Figure 6.I**).

ISIS (Islamic State in Iraq and Syria), also known as **ISIL (Islamic State of Iraq and the Levant)** and later as **IS (Islamic State)**, and to its enemies by its Arabic acronym DAISH, whose emergence is traced on **page 262**, is a "different beast." Although al-Qa'ida ranks have always been difficult to join, ISIS recruits new members openly, using social media and sophisticated Hollywood-inspired videos to reach potential recruits worldwide. That effort has paid off with legions of foreigners, including American, British, and other Western youth, travelling through Turkey to cross into Syria and join ISIS. Whereas Sunni-rooted al-Qa'ida goes to lengths to avoid alienating Shi'ites and other Muslim sects, Sunni-rooted ISIS considers every religious stripe but its own to be heretical and deserving of the most gruesome death sentences. ISIS beheaded not only Western journalists and aid workers but also Christian, Yazidi, Shi'ite, and even "apostate" Sunni men. Yazidi and other "infidel" women became objects of mass rape and imprisonment in brothels. Al-Qa'ida judged ISIS violence against prisoners of war and civilians to be unacceptable, and the two groups feuded openly over this issue.

For al-Qa'ida, establishing the Caliphate is a long-term goal; for ISIS, the Caliphate is a precondition for all events in its career. In June 2014, ISIS claimed that it had established the Caliphate as its new "Islamic State." The Caliphate is more a concept than an enduring historic reality, an empire of like-minded steadfast Sunni Muslims looking to their leader, the caliph, for political and religious guidance. Sunni Muslims acknowledge the early successors to the Prophet Muhammad as "rightly-guided" Caliphs who ruled the growing Islamic world from Medina, Damascus, and Baghdad. Although the Turkish-based Ottoman Empire labeled itself a Caliphate, most historic and modern Sunnis have not viewed it that way. ISIS followers recognize the first caliph to succeed the "rightly guided" caliphs as Abu Bakr al-Baghdadi, a former US prisoner of war in Iraq.

It is unlikely that al-Qa'ida will accept or pledge fealty to ISIS. Al-Qa'ida rejects ISIS's Abu Bakr al-Baghdadi as caliph. Al-Qa'ida sees a challenge in ISIS's declaration of its Caliphate, stretching across Syria and Iraq, with ambitions to take the same sacred heartland around Mecca that would be at the core of the al-Qa'ida Caliphate. The two organizations probably cannot have competing Caliphates, with the Holy Places of Saudi Arabia the prize of each. Al-Qa'ida has watched ISIS grow, acquire territory, and establish its Caliphate with breakneck speed (Figure 6.I). Al-Qa'ida, in contrast, takes its time, often planning an attack over a period of years. Al-Qa'ida has been losing ground to ISIS in the PR campaign to attract the hearts and minds of disaffected Muslims. Al-Qa'ida's likely gambit to reclaim its dominance of Islamist terrorism is to carry out a simultaneous mass-casualty attack on a number of Western targets.

ISIS has an even stronger apocalyptic vision than al-Qa'ida has. This group believes that by terrorizing civilian populations it can precipitate a military response by the "Crusaders" and hasten the end time with a decisive battle at Dabiq, near Aleppo in Syria.[20] This will be followed by another conflagration in Jerusalem and the appearance of Jesus. Borrowing from early Islamic traditions, ISIS thus has, like Christianity and Judaism, a millenarian narrative.

In March 2015, journalist Graeme Wood wrote an important piece about ISIS in *The Atlantic*, entitled "What ISIS Really Wants."[21] Arguing persuasively that ISIS's foes must know their enemy to engage them effectively, Wood rebuffs the much-espoused view that ISIS is not Islamic.

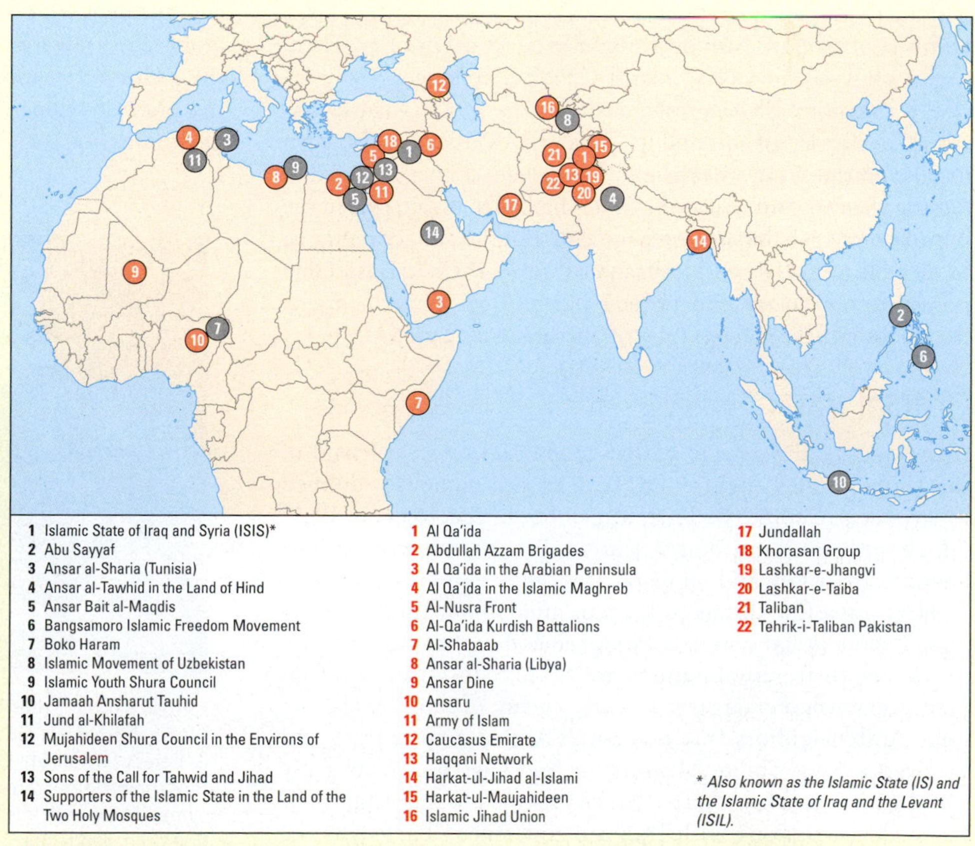

• Figure 6.I Although al-Qa'ida has many affiliate terror groups, the relatively newly formed ISIS has quickly acquired a number of pledges to its cause. Some of ISIS's affiliated terror groups originally supported al-Qa'ida but later switched allegiances.

AP Images/Hubert Boesl

(a)

AP Images/Charles Dharapak

(b)

• **Figure 6.34** Bookends to 9/11: **(a)** Lower Manhattan's skies filling with smoke, just before the second of the two World Trade Center towers collapsed. **(b)** Outside the White House on May 1, 2011, people celebrating President Obama's announcement of Bin Laden's death.

more geographically diffuse organization. It carried out or supported attacks in Indonesia, Morocco, Algeria, Tunisia, Turkey, Spain, Uzbekistan, Egypt, Saudi Arabia, Jordan, the United Kingdom, Yemen, Pakistan, and France (see **Figure 6.I**). A succession of al-Qa'ida videotapes and audiotapes promised an unrelenting and costly continuation of *jihad* against the United States and its allies. Even with Bin Laden dead, it will be many years before Americans and Europeans might emerge from the shadow of this threat (**Figure 6.34b**). Al-Qa'ida clearly will not hesitate to use the most devastating weapons, even against large numbers of civilians; this deadly goal is a tactical priority. There is also growing consensus that al-Qa'ida is no longer a solitary severe threat to US interests; instead, the *jihadist* cause has spread to smaller groups and to "lone wolves" that are also keen to take on the West. Yet there is little evidence that al-Qa'ida's message has any broad appeal, and it had virtually no role in the tumultuous Arab Spring (see **page 259**). But the danger remains; in fact, in terms of mass-casualty attacks, al-Qa'ida arguably poses a more "clear and present danger" to Western targets than does ISIS (see Geography of Terrorism on **page 244**).

Crucial Iran

Iran is a critical player in global geopolitics. Once a friend of the United States (before 1979), Iran was famously dubbed by President George W. Bush a member of the "Axis of Evil" (along with Iraq and North Korea). What went wrong? The answers are below and on **pages 270–272**.

Here are a few things to keep in mind about Iran in its regional geopolitical context. First, remember that Iranians are 223 not Arabs; their ethnolinguistic roots are Indo-European. Iran (and previously Persia) is a historic enemy of most Arabs. Its large Arab neighbors Iraq and Saudi Arabia despise Iran, and vice versa. Even Shi'ite Muslims in Iraq have not historically allied with Iran; their Arab ethnicity seemed to outweigh their faith where relations with Iran are concerned. That dynamic shifted as Sunni ISIS conquered territory in Iraq. Iraq Shi'ites welcomed Iranian fighters to take on ISIS.

Iran's identity as a Shi'ite state has had important effects on its relations with Arabs in several key situations. Iran claims sovereignty over the Gulf Arab country of Bahrain, where 70 percent of the population is Shi'ite. During the Arab Spring of 2011, while Saudi Arabia came to the aid of the minority Sunni monarchy in Bahrain, Iran appealed to the Shi'ites to overthrow Bahrain's government. There is a small minority of Shi'ites in Saudi Arabia, concentrated mainly in the kingdom's Eastern Province, and the Saudis constantly fear that with encouragement from Iran they might attempt to overthrow the monarchy (**•Figure 6.35**). ISIS has bombed Shi'ite targets in the Eastern Province to destabilize Saudi Arabia. Shi'ites are the poorest segment of Saudi society, and they complain of discrimination.

Joe Hobbs

• **Figure 6.35** Saudi Arabia's strict interpretation of Sunni Islam (Wahhabism) casts the Shi'ite minority of the country's Eastern Provinces as heretics. These Shi'ites have long been the country's poorest and most marginalized group. Already perceived as potentially destabilizing sympathizers with Iran, these Shi'ites could in fact be among the first to challenge the monarchy. These Shi'ites are at a market in the eastern province town of Qatif, where they were targeted in an ISIS bombing in 2015.

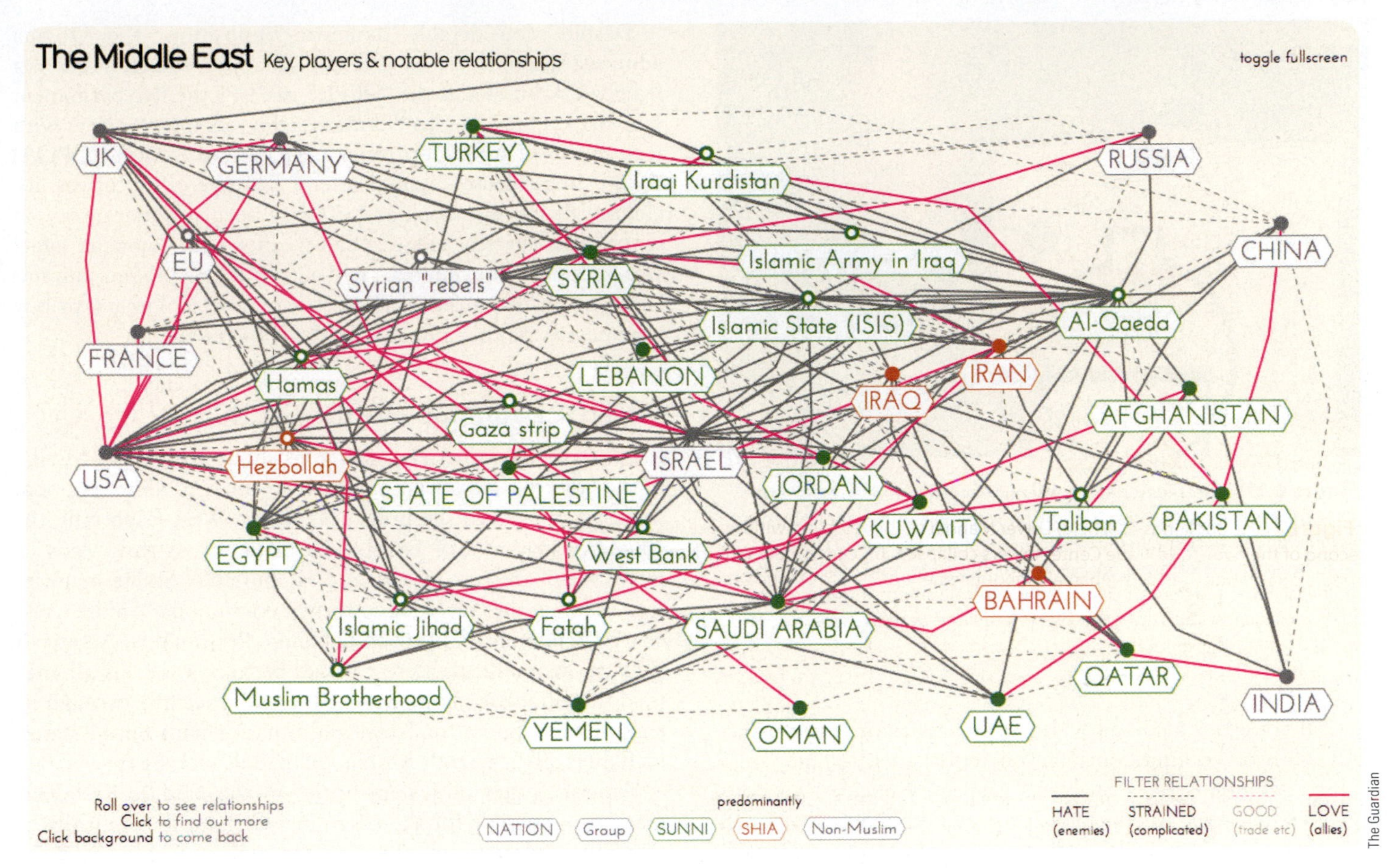

• Figure 6.36 *The Guardian* rendered this extraordinary interactive online graph of friends and enemies in the East. This is the site commentary: "As the US and UK are set to again commit to military involvement in the Middle East, this interactive visualizes the intricate, complex, and sometimes hidden relationships and alliances across the region. Its tangled, opened state is meant as a kind of visual joke, showing how its fabric defies simple solutions." You will benefit from exploring it at http://www.theguardian.com/news/datablog/ng-interactive/2014/sep/24/friends-and-enemies-in-the-middle-east-who-is-connected-to-who-interactive.

Be mindful of the confounding fact that non-Arab Iran has long been a supporter of Arab Syria's government. (In fact, the geopolitics of the Middle East can befuddle anyone; see **•Figure 6.36** and explore it dynamically on the website cited there if you can). In Syria, Sunnis are the majority and Shi'ites the minority. When (Arab) Iraq attacked Iran in 1980, the Arab nation of Syria allied itself with Iran. Why? In part because Syria's Sunni Muslim majority population is ruled by the al-Assad family and its associates who practice an offshoot of the Shi'ite faith known as Alawite Islam. Iran is also strongly allied with the militant Shi'ite Arabs of Hizbullah in Lebanon, and is able to support Hizbullah logistically because of strong ties with Lebanon's neighbor Syria. Iran also supports Hamas in the Gaza Strip, not because its members are Shi'ites (they are Palestinian Sunni Muslims), but because Hamas is set on Israel's destruction, which Iran would also like to see.

Iran's enmity toward Israel and its apparent interest in developing nuclear weapons places Iran at the center of geopolitical concern in Western capitals, particularly Washington. The US includes Iran on its list of "state sponsors of terrorism." Americans and Israelis share concerns about Iran's nuclear weapons potential because such weapons might find their way to terrorist groups or be delivered by Iran itself on its own missiles against Israel or another target (**•Figure 6.37**). Iran's development of nuclear weapons would also set off a regional arms race that would increase the size and the volatility of the Middle Eastern powder keg.

Iran insists that its nuclear program is aimed only at electricity generation and medical research, and denies that it is developing weapons. American and Israeli intelligence agencies, however, concluded that Iran was determined to have nuclear weapons. The United States, Israel, and the United Nations' **International Atomic Energy Agency (IAEA)** grew particularly watchful of Iran's acquisition of centrifuges, which are used to enrich uranium. The enrichment process is necessary for the production of fuel for civilian use in nuclear power production, which Iran is permitted to do, but enrichment can also be expanded for the production of nuclear weapons.

The United States weighed three difficult options in response to Iran's nuclear weapons threat. First, the United States could impose even harsher sanctions than it has already against Iran, especially to cut off investment and finance (the UN, EU, and

DON EMMERT/Getty Images

• **Figure 6.37** Israel regards an Iranian nuclear weapon as an existential threat. In September 2012, Israel's Prime Minister Benjamin Netanyahu used this illustration of a bomb to tell the UN General Assembly how close Iran was to having a nuclear weapon. He drew a red line beyond which Israel would not allow Iran to progress in weapon development, implying that Israel would bomb Iran's nuclear facilities if the line were crossed.

several countries also had levied sanctions against Iran). Second, with or without coordination with Israel, the US could launch military strikes on Iran's suspected nuclear weapons development sites. Israeli officials have said publicly that Israel (which has its own nuclear weapons, a fact it never acknowledges officially) regards the Iranian bomb as an "existential threat" and will prevent its development, presumably by a single surprise air strike like one Israel carried out against Iraq's nuclear facility in 1981.

Most analysts believe that a military strike against Iran would have far-ranging and dangerous consequences. Iran's nuclear program would be set back a few years, but the country would portray itself as a victim of US aggression, galvanizing support across the Muslim world (even among Iran's traditional enemies, the Sunni Arabs), perhaps inciting its Hizbullah allies in Lebanon and Hamas allies in the Palestinian territories to attack Israel, encouraging al-Qa'ida–style terrorism against the West and discouraging Iranians (particularly the younger ones) who favor peaceful relations with the United States. A military strike would probably also lead to a sharp reduction in oil exports from the Gulf, perhaps triggering a global economic crisis. Russia and China would meanwhile have an opportunity to enhance their geopolitical interests by offering diplomatic support and perhaps more to Iran throughout such a crisis. Both countries have considerable leverage in Iran.

Finally, the US could engage in diplomatic dialogue with Iran about its nuclear program, as it has done (largely unsuccessfully) with North Korea. The United States and Iran do share some common interests, notably stability in Iraq and Syria; Iran does not want to deal with strong new Sunni or Kurdish states that might arise from the ashes of sectarian wars in those countries, or with the tide of refugees these conflicts would send its way. Most importantly, as we will see now, the US and Iran are natural allies against ISIS.

319

Despite considerable domestic opposition, the Obama administration chose to give diplomacy a chance. Under a framework known as the **"P5+1,"** in 2015 the five permanent members of the UN plus Germany achieved an agreement with Iran called the **Joint Comprehensive Plan of Action (JCPOA)** that, if implemented, would offer a package of incentives, including lifting sanctions and offering security guarantees, in exchange for Iran's pledge to halt weapons development. Israel was skeptical and publicly fell out with the Obama administration, insisting that Iran was just using diplomacy to buy time before making a final "dash" for the bomb.

Axis of Evil/Axis of Resistance

Countering the American narrative about an "Axis of Evil," in the Shi'ite Middle East, a narrative emerged about regional powers opposed to the influences of the West (especially the United States) and of Israel. The members of this "Axis of Resistance" are Iran, Syria, and Hizbullah.[22] Shi'ite-majority Iraq is sometimes included to form a contiguous "Shi'ite Crescent" stretching from Iran to Lebanon. Presumably, America's primary geopolitical interest would be to oppose this alliance (and the alliance's main backer, Russia), especially through its traditional support for Israel and alliance with Sunni states, including oil-rich Saudi Arabia and people-rich Egypt.

America's geopolitical priorities in the Middle East have made it impossible for the US to maintain an ideologically or ethically consistent approach to the region. Throughout the region, the United States for decades had made "deals with the devil," helping to prop up authoritarian regimes like Egypt's that would maintain strategic and political interests in Middle Eastern oil, the region's chokepoints, and Israel's security, to name a few. Another interest was to prevent the ascendance of militant Islamists and terrorists. Many of the region's one-party states appealed to the United States for aid, saying that the alternative to their rule was that al-Qa'ida or other terrorists would find a footing. That was an effective ploy in Yemen, Egypt, and elsewhere.

Under George W. Bush, the US articulated an interest in bringing democracy to the Middle East. This endeavor, most significantly tied to the invasion of Iraq in 2003, had unintended consequences that undid many US strategic interests in the region. Regime change in Iraq introduced democracy that inevitably replaced Sunni dominance with Shi'ite dominance, setting the stage for an anti-American Sunni insurgency and the rise of ISIS. Taking down Qaddafi in Libya replaced a repressive but stable regime with a failed and fractured state. Democracy for the Palestinians empowered Hamas. Democracy for Lebanon empowered Hizbullah. When the Arab Spring of 2011 swept into Syria, President Obama said, "For the sake of the Syrian people, the time has come for President Assad to step aside."[23] The following year he said, "We have been very clear to the Assad regime . . . that a red line for us is we start seeing a whole bunch of chemical weapons moving around or being utilized."[24] When Syria actually did use chemical weapons, the US did not intervene and generally avoided any discussion of "regime change" to bring about democracy in Syria.

ISIS, with its ideological roots in Salafist Sunni Islam, has emerged as a profound and present danger to American interests in the Middle East and to America's regional allies, including Israel and Saudi Arabia. Who fled when ISIS carried out its blitzkrieg on Iraq in spring 2014? Sunnis of the Iraqi Army, trained and equipped at enormous expense by the United States. What countries and nonstate actors on the ground took up arms and confronted ISIS? Hizbullah, the Army of Bashir al-Assad in Syria, and units of Iran's army, along with Shi'ite members of the Iraqi Army and Sunni Kurdish fighters. The "Axis of Resistance" is an odd but inevitable bedfellow of the US in its effort to "degrade and destroy" ISIS. The US has been dealt a bad hand, in large part because of its own geopolitical priorities.

The following section offers more insight into the peoples and nations of this vital region. It begins with a continuation of the geopolitical theme, examining one of the world's most problematic, persistent, and influential conflicts, the one between Israel and its Arab neighbors.

6.6 Regional Issues and Landscapes

Israel and Palestine

The Arab-Israeli Conflict

The Arab-Israeli conflict persists as one of the world's longest lasting and most intractable disputes. It has not been resolved in part because the central issues are closely tied to such life-giving resources as land and water, and to deeply held and unyielding religious beliefs. You can be sure it will be in the news, even when it is overshadowed by other issues in the region like the fight against ISIS and the civil war in Syria. You will be rewarded when you understand these issues; among other things, you will have a strong foundation from which to follow the news about this conflict for years to come.

The Arab-Israeli conflict is above all a conflict over who owns the land—sometimes very small pieces of land—and is therefore of extreme interest in the study of geography. It is also a conflict that has repercussions far beyond the boundaries of the small countries and territories involved (• **Figure 6.38**). As long as it simmers or boils, there are other countries and entities that will use the unresolved Arab-Israeli conflict to advance their interests at the expense of others; both al-Qa'ida and Iran, for example, derive much beneficial propaganda value from it. A United Nations–sponsored group called the Alliance of Civilizations argues that the Palestinian-Israeli conflict is *the largest force behind global tensions.* Let's look at this in different terms: if it is resolved, the world should be a more peaceful place.

A geographic understanding of this conflict requires familiarity with the events leading up to the creation of the state of Israel and with the wars that have followed, particularly as they have rearranged the boundaries of nations and territories.

• **Figure 6.38** Israel, the Occupied Territories, and surrounding areas.

The Ottoman Empire, based in what is now Turkey, had ruled Palestine (the area now made up of Israel and the Palestinian territories) and surrounding lands in the eastern Mediterranean since the 16th century. After the British and French defeated the Ottoman Turks in World War I, they divided the empire's Arab provinces between themselves along the lines they had negotiated in the 1916 **Sykes-Picot Agreement** (• **Figure 6.39**). The British "mandate" (authority to establish a government) would be split into Palestine, Transjordan (modern Jordan), Iraq, and Kuwait; the French mandate would become Syria and Lebanon. Many of the boundaries you see on a modern map of the heart of the Middle East are based on these often arbitrary lines, created by European colonial powers competing amongst themselves for resources, with little

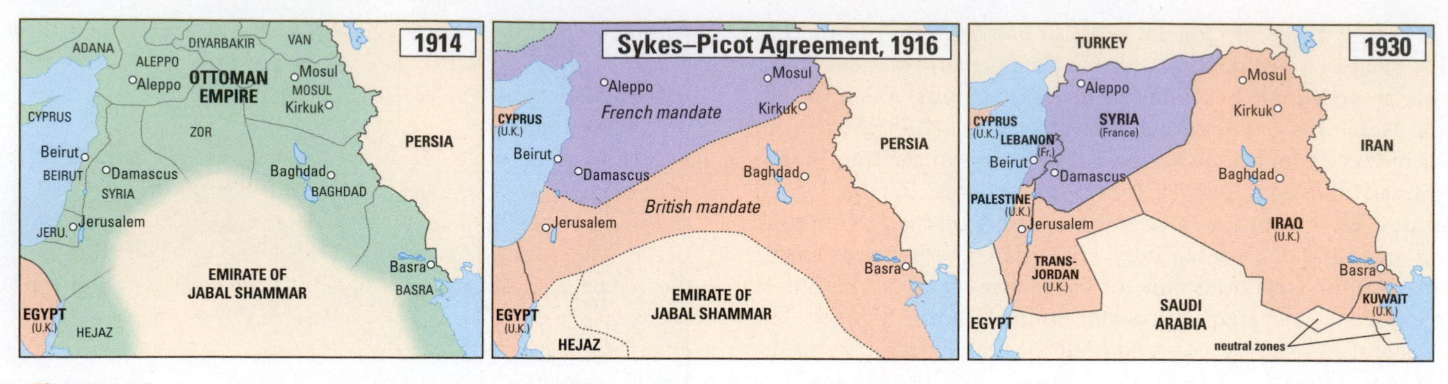

• **Figure 6.39** The British and French drew up the Sykes-Picot Line to divide the heart of the Middle East between them after the fall of the Ottoman Empire in World War I.

THINK CRITICALLY: Will these borders, particularly those of Iraq, survive the region's ongoing turmoil? What might the redrawn borders of Iraq look like?

regard for the Arabs and other peoples living within these new artificial borders.

During World War I, British administrators of Palestine had made conflicting promises to Jews and Arabs. They implied that they would create an independent Arab state in Palestine and yet at the same time vowed to promote Jewish immigration to Palestine, with an eye to the eventual establishment of a Jewish state there. The **Palestinians**—Arabs who historically formed the largest majority of the region's inhabitants—did not welcome the ensuing Jewish immigration and rioted against both the migrants and the British administration. Militant Jews attacked British interests in Palestine, hoping to precipitate a British withdrawal.

Placing themselves in a no-win position with these conflicting promises and under increasing pressure from both Jews and Arabs, in 1947 the British decided to withdraw from Palestine and leave the young United Nations with the task of determining the region's future. The United Nations responded in 1947 with the **two-state solution** to the problem of Palestine. It established an Arab state (which would have been called Palestine) and a Jewish state (Israel).

The two-state plan was deeply flawed. The states' territories were long, narrow, and fragmented, giving each side a sense of vulnerability and insecurity (• **Figure 6.40**). When Israel declared itself into existence in May 1948, the armies of the neighboring Arab countries of Transjordan, Egypt, Iraq, Syria, and Lebanon mobilized. In what Israelis call "the War of Independence" and Palestinians call "the Catastrophe" (*al-Nakba*), the smaller but better-organized and more highly motivated Israeli army defeated the Arab armies, and Israel acquired what have come to be known as its **pre-1967 borders** (the 1949 Armistice Agreement borders shown in • **Figure 6.40**). The boundary separating Israel from the West Bank would later come to be known as the "Green Line" (seen in Figure 6.44).

Prior to and during the fighting of this **1948–1949 Arab-Israeli War**, over 700,000 Palestinian Arabs chose or were forced to flee from the new officially Jewish state of Israel to neighboring Arab countries (• **Figure 6.41**). The United

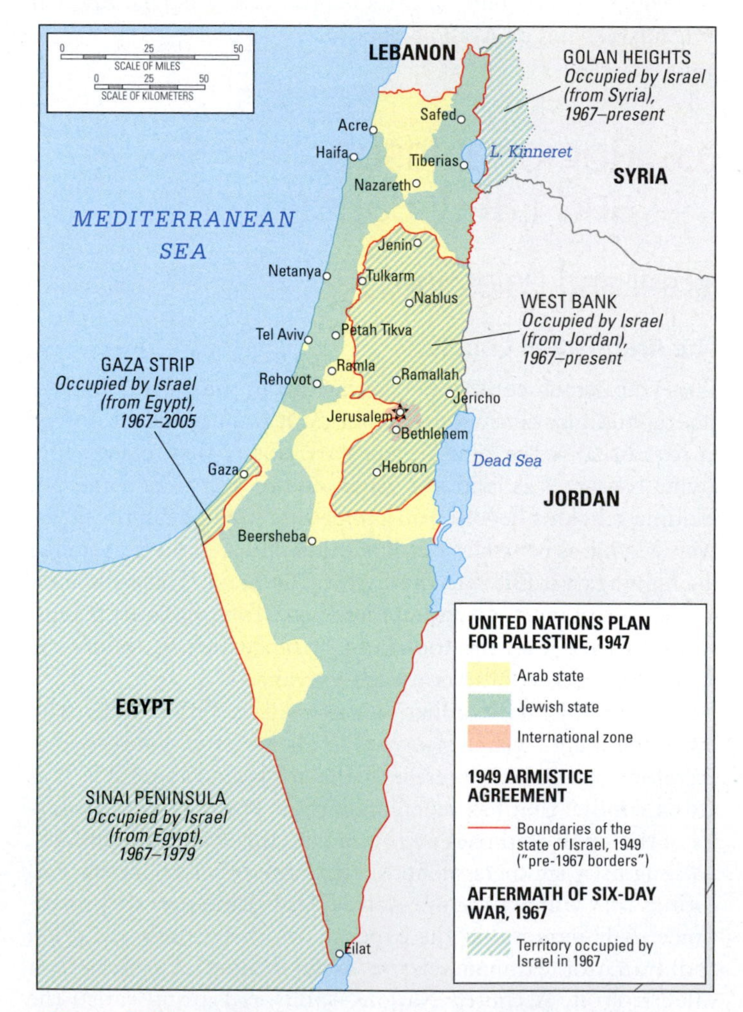

• **Figure 6.40** The 1947 UN partition plan for Palestine and Israel's original (pre-1967) borders. The 1948–1949 war, which began as soon as Britain withdrew from Palestine and Israel proclaimed its existence, aborted the UN plan and created a tense new political dynamic in the region.

Nations established refugee camps for these displaced peoples in Jordan, Egypt, Lebanon, and Syria. Little was done to resettle them in permanent homes, and both the local Arab governments and the refugees themselves continued to insist on

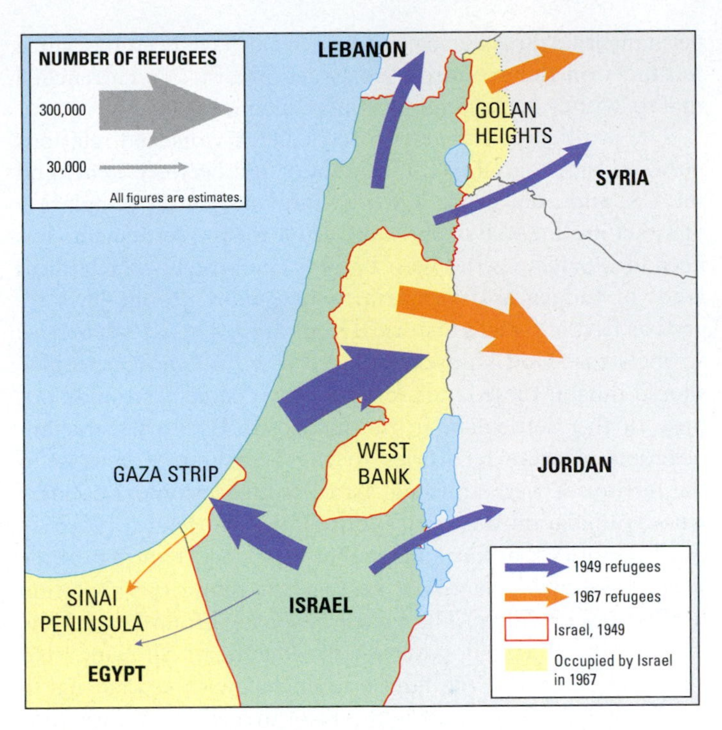

• **Figure 6.41** Palestinian refugee movements in 1948 and 1967. Many who fled in the first conflict had to relocate again in the second.

the right of the refugees to return to Israel and the restoration of their property there. This Palestinian **right of return** is one of the central problems of the modern peace process.

Other countries assumed control of those parts of the proposed Palestine that Israel did not absorb. Egypt occupied the Gaza Strip, a piece of land on the Mediterranean shore adjacent to Egypt's Sinai Peninsula that was inhabited mostly by Palestinian Arabs. Jordan occupied the predominantly Arab hilly region of central Palestine on the west side of the Jordan River, known as the West Bank, and the entire old city of Jerusalem, including the Western Wall and Temple Mount, Judaism's holiest sites. The Palestinian Arab state envisioned in the UN partition plan was thus stillborn.

The **Six-Day War** of 1967 fundamentally rearranged the region's political landscape in Israel's favor, setting the stage for subsequent struggles and the peace process. This conflict was precipitated in part when Egypt, by positioning arms at the Strait of Tiran chokepoint, closed the Gulf of Aqaba to Israeli shipping. Egypt's President Nasser and his Arab allies took several more belligerent but nonviolent steps toward a war they were ill prepared to fight. Israel chose to make a preemptive strike on its Arab neighbors, virtually destroying the Egyptian and Syrian air forces on the ground. Israel gave Jordan's King Hussein an opportunity to stay out of the conflict. However, feeling obliged to stand by the other Arab forces, Jordan went to war and quickly lost the entire West Bank and the historic and sacred Old City of Jerusalem. The entire nation of Israel was transfixed by the news that Jewish soldiers were praying at the Western Wall. The Israeli army (Israeli Defense Forces, or IDF) also seized the Gaza Strip, the Sinai Peninsula (shutting down the Suez Canal), and the strategic Golan Heights section of Syria overlooking Israel's Galilee region. These three pieces of land would henceforth be known to the world as the **Occupied Territories.** Israel had tripled its territory in six days of fighting (see **Figure 6.40**).

The Palestinian refugee situation became more complex as a result of the war. Israel took over the areas where most of the camps were located. Many people in the camps again took flight, and Palestinians fleeing villages and towns in the newly Occupied Territories joined them as refugees (see **Figure 6.41**). Some later returned to their homes, but an estimated 116,000 either did not attempt to return or were denied permission by Israel authorities to do so.

The **1973 Arab-Israeli War** was a multifront Arab attempt to reverse the humiliating losses of 1967. On October 6, 1973, the Jewish holy day of Yom Kippur, Egypt and Syria launched a surprise attack on Israel, with the hope of rearranging the stalemated political map of the Middle East. Egypt's army initially showed surprising strength, penetrating deep into the Sinai Peninsula and overturning Israel's image of invulnerability. Israeli troops soon reversed the tide and surrounded Egypt's army, but in the ensuing disengagement talks, Egypt won back the eastern side of the Suez Canal and by 1975 was able to reopen it to commercial traffic. In 1979, Israel agreed to return Sinai to Egypt under the US-sponsored **Camp David Accords,** and Egypt recognized Israel's rights as a sovereign state. The Gaza Strip, West Bank, and Golan Heights continued to be held by Israel. Subsequent Israeli administrations came to regard them as lands that rightfully belong to Jews or as cards to be played in the regional game of peacemaking.

Arabs and Jews: The Demographic Dimension

In addition to issues of land, water, politics, and ideology, the Palestinian-Israeli conflict is about sheer numbers of people. Each side has wanted to maximize its numbers to the disadvantage of the other. To realize the Zionist dream of establishing a Jewish state, Jews began immigrating to Palestine around the start of the 20th century. Jews made up 11 percent of Palestine's population in 1922, 16 percent in 1931, and 31 percent in 1946, on the eve of Israel's creation. In keeping with national legislation known as the **Law of Return**, the state of Israel has always granted citizenship to any Jew who wishes to live there. Following the 1948–1949 war, waves of new immigrants from Europe (**Ashkenazi Jews**) joined the Middle Eastern **Sephardic Jews** who had lived in Palestine and other parts of the Middle East (and until 1492, in Spain) since early times. 108

In the 1990s, more waves of Jewish immigration followed the dissolution of the Soviet Union and a change of government in Ethiopia, where an ancient Jewish group called **Falashas** had lived in isolation for thousands of years. In 2015, Jews made up 75 percent of the population of 8 million people living within Israel's pre-1967 borders. Palestinian and non-Palestinian Arabs are 21 percent of the country's population; they are Israeli citizens and residents, known as "Israeli Arabs." Palestinian Arabs without Israeli citizenship, along with a variety of other ethnic groups including recent Eritrean, Sudanese, and other Africans known as "infiltrators," who reach the country after arduous smuggling though northeastern Africa and the Sinai, comprise the balance of the population within Israel's pre-1967 borders.

Another 4 million people live in the West Bank, Gaza Strip, and Golan Heights—areas occupied by Israel in the wars of 1967 and 1973. About 300,000 of those are Jewish settlers, and most of the rest are Palestinian Arabs (mainly in the West Bank and Gaza Strip), **Druze** (members of a small offshoot of Shi'ite Islam, living mainly in the Golan Heights), and other Arabs. Another 200,000 Jewish settlers live in East Jerusalem, which is also part of the territories Israel conquered in 1967. There are approximately 6 million more Palestinians living outside Israel and the Occupied Territories. Jordan has the largest number (3 million), followed by Syria and Lebanon (with about 500,000 each). The remainder are in other Arab countries and in numerous nations around the world, comprising a sizable diaspora of their own that Palestinians call the ***ghurba***. In this respect, the Palestinians are in much the same situation as the Jews prior to Israel's creation.

After 1967, and especially after the election of its conservative **Likud Party** to power in 1977, Israel moved to strengthen its grip on the Occupied Territories through security measures and the government-sponsored establishment of new **Jewish settlements**. Most of these settlements were, and still are, spread through the West Bank (see **• Figure 6.42**). Much smaller numbers of Jews settled in the troubled Golan Heights, where about 20,000 lived in 2015. These are generally not frontier farming settlements, but communities inhabited by relatively prosperous middle-class Jews who commute to jobs in Israel. They were not built as temporary encampments, but as permanent fixtures meant to create what have been described as **"facts on the ground"**—an Israeli presence so entrenched that its withdrawal would be almost inconceivable.

The proliferation of Jewish settlements worsened relations between Israel and its Arab neighbors and between Israel and the US, and even drove a wedge into the Jewish population of Israel itself. Some of the motivation for the settlements has been ideological, as the West Bank is composed of the biblical lands of **Judaea and Samaria**, which many devout Jews regard as part of Israel's historical homeland. These Jews believe strongly that God Himself declared that his "chosen people" should inhabit these lands forever. Many Israelis strongly op- 224 pose further settlement in the Occupied Territories and annexation of these territories to the Jewish state because if the territories were annexed, Israel would become a country whose population was only about 60 percent Jewish. With an Arab minority that has a far higher birth rate than that of the Jews, Jews could eventually become the minority population in their own country. The Israeli advocacy group Peace Now leaked official Israeli government documents showing that about 40 percent of the land on which Jewish settlements in the West Bank were built is privately owned by Palestinians. There is also international pressure against the settlements because they violate Geneva Conventions on activities allowed in territories captured in war. The International Court of Justice has declared them illegal. **United Nations Resolutions 242 and 338** (passed after the wars of 1967 and 1973, respectively) called on Israel to withdraw completely from the Occupied Territories. The United States government routinely calls the Jewish settlements "an obstacle to peace" but is loath to push Israel hard about them.

The fate of the Jewish settlements in the Occupied Territories is one of the key issues in the resolution of the Palestinian-Israeli conflict. It is a difficult problem, but it is not insolvable.

The settlements and other aspects of Israeli occupation have fostered widespread and long-lasting Palestinian resistance against Israel. In 1987, Palestinians instigated a popular uprising in the Occupied Territories known as the ***Intifada*** (Arabic for "shaking"). Initially, they relied on rocks and bottles to engage Israeli troops, who answered with rubber and metal bullets. International media coverage of a Palestinian "David" fighting an Israeli "Goliath" did much to damage Israel's image abroad, and many Israelis came to question the justice and importance to security of Israel's continued occupation. Popular opinion and, by 1992, a new Israeli government led by the liberal **Labor Party** began to consider the previously unthinkable: giving the Palestinians at least some control over land and internal affairs in the Occupied Territories.

• Figure 6.42 This image captures the geographic realities of life for Israelis and Palestinians in the West Bank. The Israeli-built separation barrier cleaves a Palestinian village (with its mosque and minaret in the center) from a much more prosperous Jewish settlement. Inhabitants of these communities cannot mingle.

THINK CRITICALLY: What are the religious and political motivations of Jewish settlers in the Occupied Territories, and why does the US call the settlements "an obstacle to peace?"

Land for Peace

They loom large in the headlines and in world affairs, but a closer look reveals that Israel and the Occupied Territories are very small. Within its legally recognized, pre-1967 borders, Israel's area is only 8019 square miles (20,770 sq km), about the size of New Jersey or Slovenia (**• Figure 6.43**). Israelis often cite the small size of their country, nestled within the vastness of the Arab Middle East and North Africa, to highlight

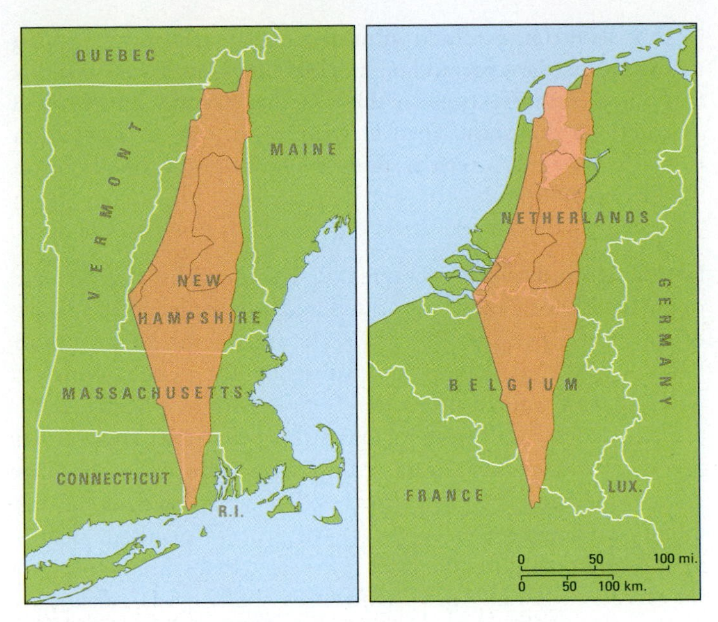

• **Figure 6.43** Israel and the Occupied Territories compared in size to New England and the Benelux countries.

their vulnerability on the world stage. If and when Palestinians acquire their country, Palestine will be even smaller; the West Bank is Rhode Island–sized, and Gaza is smaller than the District of Columbia. The question "Whose lands are these?" has always been at the heart of the conflict between Israel and its neighbors.

President Anwar Sadat of Egypt and Prime Minister Menachem Begin of Israel set the precedent in 1979: Arabs could make peace with Israel on the formula of **"land for peace,"** with Israel swapping Arab lands it occupied in 1967 in exchange for peace with its neighbors. The United States, Russia, and Norway initiated negotiations in the 1990s that made such a prospect appear possible. The process began in earnest in 1993, when Israel recognized the legitimacy of, and began to negotiate with, the **Palestine Liberation Organization (PLO)**, long recognized by the Palestinians as their legitimate governing body. In return, the PLO, under the leadership of Yasser Arafat, recognized Israel's right to exist and renounced its long-standing use of military and terrorist force against Israel. US President Bill Clinton orchestrated a historic handshake between Arafat and Israeli prime minister Yitzhak Rabin, and the leaders signed an agreement known as the **Gaza-Jericho Accord** (in its implementation, it came to be known as the **Oslo I Accord**, after the Norwegian capital where it was negotiated). It was designed to pave a pathway to peace by Israel's granting of limited **autonomy**, or self-rule, in the Gaza Strip and in the West Bank town of Jericho. Autonomy meant that the Palestinians were responsible for their own affairs in matters of education, culture, health, taxation, and tourism. Israel also pulled its troops out of these areas. Palestinians were allowed to form their own government, known as the **Palestinian Authority (PA)**, and they elected Yasser Arafat as its first president.

The accord established a five-year timetable for the resolution of much more difficult matters, the so-called **final status issues**. These included the political status of Jewish and Muslim holy places in Jerusalem and of the city itself, the possible return of Palestinian refugees (the "right of return"), the future of Jewish settlements in the Occupied Territories, and Palestinian statehood (independence). To this day, the final status issues are the ones that prevent the Palestinian-Israeli conflict from being settled.

In the wake of a deal known as the Wye Agreement, or the **Oslo II Accord**, three successive Israeli prime ministers (Yitzhak Rabin, assassinated by a Jewish settler in 1995, Benjamin Netanyahu, and Ehud Barak) promised to transfer more West Bank lands from Israeli to Palestinian control, and Arafat promised increased Palestinian efforts to crack down on Palestinian terrorists and so guarantee the security of Israelis in the Palestinian territories and in Israel. If finally implemented (it had not been as of 2015), this arrangement would have brought the total of West Bank lands under complete Palestinian control to 17 percent, leaving 57 percent completely in Israeli hands and 26 percent under joint control (• **Figure 6.44**).

On the Brink of Peace. During his final year in office in 2000, US President Clinton sought to solidify his legacy as peacemaker by brokering a historic final settlement between Israelis and Palestinians. PLO Chairman Arafat and Israeli Prime Minister Barak huddled with Clinton and his advisers in the presidential retreat at Camp David, a site chosen because of its historical significance in Middle East peacemaking. Over weeks of tough negotiations, mostly over the "final status" issues, the two sides came close to a deal. It would have included the following:

- The creation of an independent Palestinian country in the Gaza Strip and the West Bank.
- A "land swap": the Palestinian state would include 95 percent of the West Bank. The other 5 percent of the West Bank would be clusters of Jewish settlements that would remain under Israeli control. Palestine would receive 5 percent of Israel's land in exchange.

However, the talks broke down over two thorny issues:

- The status of Palestinian refugees abroad: Israel was unwilling to permit the "right of return" of the large numbers demanded by the Palestinians.
- The deal breaker, Jerusalem, which both Israelis and Palestinians proclaim as their capital: Israel agreed to have the Old City of Jerusalem divided, with Israel assuming sovereignty over the Jewish and Armenian quarters, and Palestinians assuming sovereignty over the Arab and Christian quarters (see Figure 6.D). These issues were not as problematic as those involving the places held holy by each side. President Clinton put an interesting compromise on the table: neither side would have sovereignty over the Jewish Temple Mount/Muslim Noble Sanctuary (*al-Haraam ash-Shariif*). Instead, these would belong to God. This proposal, which came to be known as "Divine Sovereignty," was rejected by both sides. Israel agreed to allow the Palestinians "soft sovereignty" over the Temple Mount/Noble Sanctuary, meaning

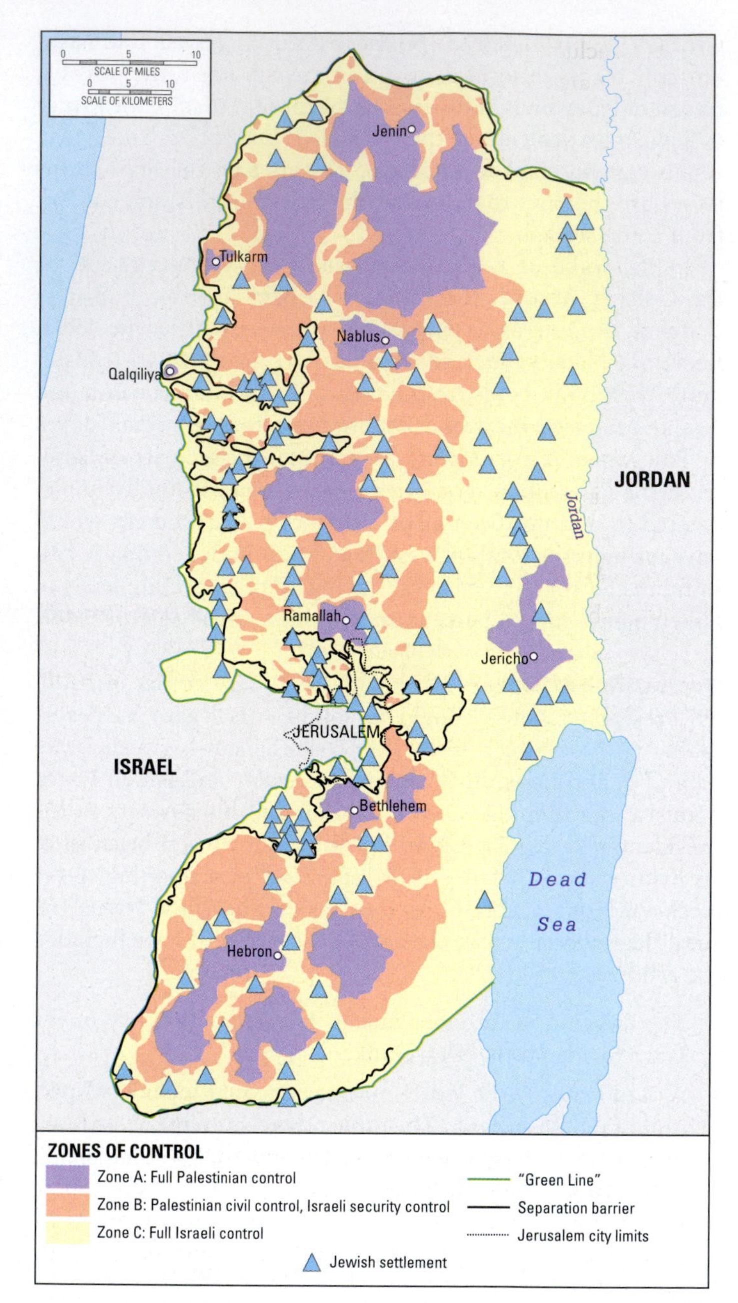

• Figure 6.44 The West Bank. Israeli and Palestinian areas of control were delimited in the peace process of the 1990s, but because of recurrent violence, almost all areas are effectively controlled by Israel. Also depicted are Jewish settlements, the completed and planned portions of the Israeli-built separation barrier, and the Green Line delimiting the internationally recognized border between sovereign Israel and the occupied West Bank.

THINK CRITICALLY: Is the Israeli-Palestinian conflict solvable? Use the Camp David negotiations under President Clinton (**pages 253–254**) as your reference. Where and in what circumstances could the Palestinians have independence? In geographic terms, how did the Camp David negotiations seek to address the settlements and international borders in an equitable way for Israel and the proposed Palestinian state? Geographically and politically, what did Camp David propose to resolve the difficult issues of Jerusalem's Old City?

that Palestinians could administer the area and fly a flag over it, but Israel would technically "own" it. Israel saw this point as essential because the First and Second Temples had stood on this spot. The Palestinians, however, insisted on full sovereignty over the Temple Mount and walked out on the negotiations.

Losing Ground Again. Tragically, within weeks, this historic opportunity for peace evaporated and was replaced by a state of war. On September 28, 2000, the right-wing Israeli Likud Party opposition leader (and subsequently prime minister) Ariel Sharon walked up to the Temple Mount under heavily armed escort. He meant to make the point that this sacred precinct belonged to Israel. Many observers claim he also meant to disrupt the peace process. Whether or not it was his intention, a disruption in the peace process resulted. Beginning that day and for the next few years, Palestinians took to the streets in the **al-Aqsa *Intifada*** to challenge Israeli forces with rocks, bottles, and guns. Hundreds of Palestinians and Israelis died.

Following the collapse of the Camp David talks and an escalation of suicide bombings of Israeli civilians, Israel chose to effectively suspend the peace process. Prime Minister Sharon branded Arafat an untrustworthy partner for peace negotiations, isolating him physically in the West Bank town of Ramallah and systematically dismantling the Palestinian Authority he headed. Israel also persuaded the United States to stop negotiating with Arafat, who died a forlorn figure in 2004, probably after being poisoned by radioactive polonium (an agent of assassination used elsewhere; see **page 181**). Israel and the United States insisted on the creation of the new post of Palestinian prime minister, hoping it would be filled by someone they could negotiate with. The United States led an effort joined by the other members of the so-called Quartet—the European Union, the United Nations, and Russia—to establish what they called a **"road map for peace"** between the Israelis and the Palestinian Authority. It was based largely on the 2000 Camp David formula and, if successful, would presumably have led to an independent Palestinian state consisting of the West Bank and the Gaza Strip.

Prime Minister Sharon, although resuming a dialogue with the Palestinian Authority after the *intifada* ended in 2005, decided unilaterally to withdraw Israeli settlers and troops from the Gaza Strip that year and to build an immense **separation barrier** walling the West Bank off from Israel (the barrier is mapped in Figure 6.44 and pictured in **• Figure 6.45**). Israel's stated goal in constructing the new barrier was to prevent Palestinian suicide bombers from reaching Israel. Noting that the security barrier did not follow the internationally recognized **Green Line** of Israel's pre-1967 border, but instead penetrated significant portions of the West Bank to pick up Jewish settlements there, Palestinians condemned it as a "land grab." The International Court of Justice concluded that the wall violated customary international law and several conventions on human rights to which Israel is a signatory.

Baz Ratner/Reuters

• **Figure 6.45** The Israeli-built separation barrier is most controversial in traditionally Palestinian-held East Jerusalem. The barrier here traces the line of Israeli-annexed East Jerusalem. On the Israeli side, the government has continued to build Jewish settlements, including Pisgat Zeev, photographed here in 2009.

On the Palestinian side, a political earthquake occurred in 2006 when a vote for parliamentary candidates gave the Islamist Hamas party control of the Palestinian Authority. Palestinians had grown disillusioned with the showy wealth of many elected Palestinian Authority officials and by the PA's failure to deliver an independent country or any other tangible benefits. Hamas, meanwhile, won hearts and minds through its network of social services, often providing what the Palestinian Authority failed to, including education, medicine, and food rations.

Hamas's stunning victory in these parliamentary elections added another layer of uncertainty and discord to the region's political landscape. In a subsequent power-sharing agreement, Palestinian president Mahmoud Abbas of the PLO retained his post, but a Hamas leader, Ismail Haniyeh, took office as the Palestinian prime minister presiding over the Hamas-dominated parliament. Because the Hamas charter does not recognize Israel and in fact calls for its destruction, Israel and the Quartet suspended financial aid and discontinued dialogue with the Palestinian Authority. These sanctions would be lifted on three conditions: the Hamas-led PA must recognize Israel's right to exist, renounce violence, and abide by previous agreements reached between Israel and the Palestinians.

As the Palestinian economy deteriorated, clashes erupted between Hamas and PLO factions led by Abbas's Fatah Party. Hamas emerged victorious in fighting to control the Gaza Strip in 2007. Expelled from the territory, Abbas withdrew his recognition of Haniyeh as prime minister, and the Palestinian Authority government was dissolved. Abbas's Fatah Party consolidated its hold in the West Bank and assumed total control of the PA. In the winter of 2008–2009, there was yet another conflict, the **Gaza War**, during which Israel sought to destroy as much of the weaponry and manpower of Hamas as possible. A second and more destructive Gaza War took place in summer 2014. With an effective US-supplied missile shield known as the "Iron Dome," Israel was able to fend off nearly all Hamas rockets launched toward Israel, and sustained few casualties. Palestinian casualties, however, including a high death toll among civilians as a result of Israeli air and ground operations, were quite high. Anti-Israeli sentiment was particularly strong in Europe, and it morphed into anti-Semitism on the political right. 96

Against this bleak backdrop President Obama charged his Secretary of State John Kerry with the responsibility of expediting a final status agreement between Israelis and Palestinians. Obama pressured Likud Prime Minister Benjamin Netanyahu to stop the construction of Jewish settlements in East Jerusalem and the West Bank. Netanyahu refused, and peace talks failed to get off the ground. They became technically impossible to resurrect in 2014 when the PA and Hamas announced a unity government. As Hamas did not recognize Israel and did not disavow terrorism, both the US and Israel had to stop negotiations. Abbas took the Palestinian case for statehood to the United Nations.

The Arab Spring

On December 17, 2010, in the Tunisian city of Sidi Bouzid, a 26-year-old vegetable vendor named Muhammad Bouazizi was shaken down for a bribe by a city inspector. Muhammad had paid the $7 to the inspectors many times before, but this time he protested. The inspector and her companions took apples from his rolling cart and confiscated his scale. The young man went to the local governor's office to lodge a complaint but was denied entry. Later in the day, he returned to the governor's office, shouting from the street, "How do you expect me to earn a living?" He poured paint thinner over his body and lit a match. Flames consumed his body.

This was the beginning of the Arab Spring.

Stirring from Stagnation

News of Bouazizi's self-immolation spread quickly throughout Tunisia and the Arab World by way of YouTube, Facebook, Twitter, and al-Jazeera television. This was an unfamiliar event in the region, quite unlike suicide bombings or anything known even in the worst of times.

The young man's death struck a nerve among much of Tunisia's largely young and disaffected population. Tunisian resentment against the rule of the autocratic President Zine al Abidine Ben Ali had been growing throughout his more than two decades in power. Ben Ali's family was estimated to have directly or indirectly controlled half of the country's economy.

Protestors poured into the streets of Tunisian cities, bearing signs reading "We are all Muhammad Bouazizi!" and calling for Ben Ali's resignation. Demonstrators soon numbered in the hundreds of thousands, and Ben Ali began to make political concessions. It was too late. Within a month of the start of what became known as the **Jasmine Revolution**, Ben Ali's own generals turned against him. Ben Ali and his family fled to Saudi Arabia. The first domino had fallen.

Protests soon erupted in a dozen other countries across the Middle East and North Africa. The experiences in each of these countries were diverse. We will focus on the most important country affected—Egypt—and identify some patterns underlying the dissent in all the countries. Later we will return to all the countries to see how they have fared since the Arab Spring.

The Pharaoh Falls

The next domino was the Middle East's demographic and political heavyweight, Egypt. There are several identifiable factors leading to this most important revolution in the region in half a century.

The first are two problems you know about: the general 81 trend of "people overpopulation," and MENA's own youth 211 bulge that is one of its manifestations. Over the past decade, the number of people of working age in Egypt, Jordan, Lebanon, Morocco, Syria, and Tunisia have grown by 2.7 percent, faster than any other world region except Sub-Saharan Africa.

Egypt has MENA's biggest youth bulge: 60 percent of Egypt's 88 million people are under 25. In Arabic the word *shabaab* means "young people" and represents that youth bulge. On the male side, it sometimes has the connotation of "bad boys." Bad or not, these boys have palpable frustrations. Without treading too much into psychosocial analysis, it is clear that the most basic desires and needs of young men that include courtship and marriage are extremely difficult to fulfill. Men are expected to have steady work and to provide a home and other amenities to their brides, and this is hard given the economic challenges they face.

Unemployment and underemployment were the second set of ingredients in Egypt's Arab Spring. Egypt's annual rate of population growth is 2.6 percent. At this rate, Egypt's population will double in 27 years. Will Egypt double the number of jobs for its people in 27 years? That prospect is remote. In recent decades, South Korea, China, and other successful developing countries nurtured the growth of private sector employment, whereas Egypt and other MENA countries generally stayed with a Soviet-style model of state-controlled enterprises that was unresponsive to people's needs and wants, 172 and failed to grow productive jobs. Outside the agricultural sector, 70 percent of working Egyptians are employed by the government. Private sector jobs are scarce.

Throughout MENA, the number of college graduates far exceeds positions to hire them. Egypt cranks out huge numbers of college graduates, but the quality of their education is questionable. All Egyptian college graduates are guaranteed jobs in the public sector. They are stuffed into already bloated bureaucracies where 10 people do the work that could be done by 1. Egypt and many other MENA countries suffer in this way from **underemployment** (the underutilization of labor). Their wages cannot possibly sustain a family, and most people "moonlight" with second or even third jobs. Under such circumstances, the visionaries of MENA are unlikely to ever be heard of, unless they flee with their brains to the West or the East.

Third, Egyptians suffered from government repression and a lack of individual freedom. I heard the same complaints in Egypt during the "Food Riots" of 1977 and on the eve of the Arab Spring. Egyptians had no control over their lives. They could not express dissatisfaction with government policies they were unhappy with, which were many. There was no real democratic political system. Every few years, Egyptians endured the sham of elections, in which an amazing 99 percent of Egyptians always chose President Hosni Mubarak's party to lead them. They had no freedom of the press. They lived under emergency law, which allowed authorities to arrest people without charges, detain prisoners indefinitely, and limit freedoms of expression and assembly. Their ranks were penetrated by the countless plainclothes secret police, the dreaded *mukhabarat* whose duty was to look for any sign of sedition. Egyptian security forces had torture down to a fine art (one of the reasons why the United States sent al-Qa'ida suspects here under the Bush and Obama programs of "extraordinary rendition").

The fourth factor in Egypt's uprising was the yawning gap between the haves and the have-nots, another problem of LDCs described in the book's introduction (see **page 54**). There was a veneer of prosperity. Egypt had a gleaming culture of conspicuous consumption for the relative few who could afford to consume. The gap between rich and poor grew widely in Mubarak's Egypt, and 20 percent of Egyptians were classified as poor at the time of the Arab Spring. For the wealthy and the well-connected, getting things done was easy: you just had to pay bribes and use your favors. To know someone with influence always helped. Corruption, known as *wasta*, was just the way it was. And if you didn't know anyone or could not afford to pay bribes, you could feel just like Muhammad Bouazizi did.

One of the opposition parties trying to organize in the last months of Mubarak's rule called itself Kifaya, meaning "Enough." Enough of all of this. The people of Egypt—not just the *shabaab*, but the older generations (including the *shabaab* I had seen rioting in 1977); girls and women; Muslims and Christians; the poor, the middle class, and even the well-to-do who wanted democracy and freedom of expression—were inspired by Tunisia's Jasmine Revolution to challenge the government—a government led by an 80-year-old man who dyed his hair jet-black and who was grooming his son to succeed him.

Egypt's revolution played out all over the country, but its epicenter was in Cairo's Tahrir ("Liberation") Square (**• Figure 6.46**).

Khaled Desouki/AFP/Getty Images

• Figure 6.46 Egyptians protesting *en masse* against the government of Hosni Mubarak in Cairo's Tahrir Square, February 2011.

Egypt's despised secret police used deadly force even when there was no need. Rather then disperse, over a period of 18 days the protestors held their ground in Tahrir and Egypt's other public squares and outside government buildings. Without arms, the demonstrators took casualties of dead and wounded. In the end, their people power prevailed. The army, an institution that Egyptians long revered, took up their cause. Mubarak offered political concessions to Egyptians, but it was too late. As the army assumed power, he fled from Cairo and was soon imprisoned and put on trial.

The Libyan Domino

A worse fate would befall Libya's dictator, Muammar Qaddafi. Libya was not like Egypt: with a population of just 6.3 million, Libya did not have an unmanageable youth bulge, nor was it propped up by American aid. With its considerable oil reserves, 3 percent of the world's total, Libya should have been a prosperous country. Instead, for more than 40 years, it was the toy of Muammar Qaddafi, an eccentric man (what other world leader would call for Switzerland to be abolished?) who ran the country as if it were his fiefdom, favoring and distributing money to those who were loyal to him. Tribal and ethnic politics were decisive. Qaddafi loyalists drew mainly from his own tribe on the central coast and from other tribes he favored in the center and west. King Idris I, whom Qaddafi overthrew in 1969, had favored tribes in the eastern region known as Cyrenaica, especially around the city of Benhgazi. Those tribespeople never forgave Qaddafi for ousting their leader, and Qaddafi made sure they never earned any graces from him. Qaddafi also treated the ethnic Berber tribes of Libya's western mountains (whose livelihood, but not ethnicity, is much like that of the Bedouin) as second-class citizens.

The people of Cyrenaica, along with the Berbers of the west, were inspired by the tide of the Arab Spring to rise up against Qaddafi, and a brief civil war emerged between them and Qaddafi loyalists. The opposition to Qaddafi was advantaged by the commitment of NATO air power, weapons, and intelligence, and it was only a matter of time until Qaddafi and 119 his power base would be eliminated. Qaddafi was captured and killed execution-style in October 2011, eight months after the uprising began. The scenario was now almost routine in the Arab World: a strongman ruler for life ruling no more, the people's freedom found at last, and almost universal joy. But, as elsewhere, the joy would be short-lived.

Syria's Minority Dynasty Challenged

The Syrian uprising began in the southern city of Deraa in January 2011 after another self-immolation. As in Bahrain and Libya, Syria's leaders opted for violent repression against demonstrators inspired by the Arab Spring; the Alawites controlled the Syrian military and brought all of its assets to bear on the rebels.

In Syria, the majority Sunni people (74 percent) long resented the rule of minority Shi'ite **Alawites**, making up 7 percent of population and led by President Bashir al-Assad, son of the late president Hafez al-Assad (see the map of Syrian ethnicities in **Figure 6.53**). In 1982, a brutal army assault against a Sunni Muslim Brotherhood uprising in the southern city of Hama resulted in tens of thousands of deaths. The 2011 revolt involved much broader participation in more cities. Arab Spring demonstrations in most of the Arab countries were loudest in the capital cities, but that was not the case in the Syrian capital, Damascus, where al-Assad's rule was most firm.

The rebels' challenge of taking down the Alawite regime in Syria was enormous. The Alawites controlled the key unit of the Syrian military, so it was difficult to secure a broad military uprising against the regime. NATO and the United States were far less inclined to become involved than they were in Libya; Assad's military assets were much greater. Western countries instead stepped up their economic sanctions against Syria. The Arab League moved to oust Syria from the organization and to impose its own sanctions. Turkey was more involved, providing shelter to forces of the **Free Syrian Army (FSA)**. Israel did not become involved. Nor for the time being did Iran, which would lose its only ally in the Arab World should the Syrian regime fall. The Alawites had free rein to brutally smash their opponents.

Bahrain: A Pearl Is Crushed

In Bahrain, a small but oil-rich Gulf island sheikhdom (Bedouin monarchy) linked to eastern Saudi Arabia by a causeway, there is a Shi'ite majority of 70 percent that has long been ruled by a Sunni monarchy. Like Egyptians, Bahrain's Shi'ites were inspired by the Jasmine Revolution to challenge King Hamad bin Isa al-Khalifa.

Bahrain was the Arab Spring's best example of how repression can backfire by fueling resentment. The mainly Shi'ite protestors began by simply expressing wishes for democracy, public participation, and justice. They asked that King Khalifa take his place in a new constitutional monarchy. The monarch could have responded as a benevolent despot, yielding some ground to his subjects.

Instead, he ordered government forces to crush the rebellion with deadly force. As the casualties mounted, demonstrators' cries changed from reasonable petitions to "Death for Khalifa!" Worried that Bahrain's Shi'ites might overthrow the monarchy and put Bahrain into the embrace of his adversary Shi'ite Iran, the king called on Saudi Arabian forces to drive across the causeway to Bahrain and helped put down the uprising. Government forces destroyed the demonstration's epicenter, Pearl Square, with the country's iconic pearl sculpture in the capital city of Manama, and turned it into a traffic intersection to ensure that crowds would not gather again. King Khalifa's opponents retreated, their hopes unrequited. Later we will consider Bahrain's future, especially in its geopolitical context. 261

Revolt in Yemen's Mountainous Redoubt

Although located on the Arabian Peninsula (see Figure 6.2), Yemen has only a small bounty of oil reserves. A stunningly beautiful country graced by mountains and deserts with a distinctive urban architecture looking like gingerbread houses, Yemen is cursed by poverty (•**Figure 6.47**). Loyalties are to clan

Joe Hobbs

• Figure 6.47 The palace of the Zaydi Imam Yahya in Wadi Dhahr, Yemen, is one of Yemen's architectural gems. The Zaydis, who are Shi'ite Houthis, ruled in Yemen for about 1000 years, until a revolution in 1962.

and tribe, and Yemen has no natural sense of cohesion as a nation-state. Until 1990, the country had been divided into North and South Yemen, and since 2007 an active southern secessionist movement has been increasingly active in efforts to regain the south's independence. Al-Qa'ida has a strong foothold here through its affiliate Al-Qa'ida in the Arabian Peninsula (AQAP; see **page 260**); Yemen is in fact the ancestral homeland of Osama bin Laden's clan.

With the Arab Spring, a widespread revolt against the longtime ruler Ali Abdullah Saleh erupted. After surviving an assassination attempt, President Saleh offered concessions to Yemeni protestors. The country's well-established opposition political parties pleaded with the young people to stop their protests. Yemen's *shabaab* ignored that advice. They are a force to be reckoned with; the demographic profile of Yemen's *shabaab* is very large. With a population of 26 million, Yemen has one of the world's highest population growth rates, 2.8 percent (with a doubling time of 25 years). Poverty, unemployment, and repression of dissent made Yemen ripe for the revolt that toppled Saleh in 2012.

Hallmarks of the Revolution

In the Arab Spring's aftermath, Middle East observer Fareed Zakaria recommended a book on his CNN show *GPS*. It was the United Nations' *Arab Development Report 2002*. Why recommend a decade-old book? To make the point that little had changed in those years. The chapter titles that so well summarized the region's conditions in 2002 did the same a decade later: "Development Not Engendered Is Endangered," "Bridled Minds, Shackled Potential," "As Development Management Stumbles, Economies Falter," and "The Curse of Poverty: Denying Choices and Opportunities, Degrading Lives."[25]

The Arab Spring rumbled through Tunisia, Egypt, Libya, Bahrain, Syria, and Yemen—described in some detail here—and in Sudan, Algeria, Mauritania, Morocco, Jordan, the Palestinian Territories, Lebanon, and Oman. These diverse countries shared several problematic traits: high population growth, unemployment and underemployment, lack of political representation, and oppression by authorities. Here are some of the other notable characteristics of the uprising and its aftermath in the countries of the Arab Spring:

- The revolutions were much facilitated by Facebook, Twitter, 24/7 satellite television news (notably from Qatar's al-Jazeera network), and other social media. Governments tried to crack down, bringing down the Internet and suspending cell phone service, but to no avail. By the time the authorities woke up, detailed instructions on how, where, and when to gather and protest in public spaces had been widely posted and tweeted.
- Although new media made their debut as tools of revolution in the Arab Spring, the more traditional geography of revolt in public spaces played a prominent role. In the traditional Middle Eastern *medina*, there was no large public space where large numbers of people could gather. However, spacious traffic hubs that doubled as public squares were established in the post-colonial modern Middle Eastern cities. In the Arab Spring, Cairo's Tahrir Square, Tripoli's Green Square in Libya (renamed Martyrs' Square in recognition of the lives lost there in protest), and Bahrain's Pearl Square served as central places where social media called upon demonstrators to gather. Many of the key events leading to success (or in Bahrain's case, the failure) of the revolutions took place in these locations.
- Women had an unprecedented strong role in the Arab Spring. In the conservative Muslim societies of the Middle East and North Africa, there is strict segregation of men and women (in some countries more so than in others) in public space. But in Tahrir Square and elsewhere, that convention was thrown aside as women and men stood shoulder to shoulder in protest.
- The traditional Islamic classification of sacred time played an important role in the Arab Spring. As often in Muslim history, Friday noon prayers and sermons incorporated politics with prayer. Prayer time gave way to protest time, often on such a huge scale that the authorities dreaded Fridays.

This traditional day of prayer was often tweeted, Facebooked, and otherwise announced as the "Day of Rage" for which demonstrators should prepare themselves.

- Remarkably, religion and militant Islamism did not otherwise feature prominently in the Arab Spring. Mainstream and radical Islamists alike were conspicuous by their absence. In Egypt, the Muslim Brotherhood, a well-organized political group that could have promoted itself and its agenda, kept a low profile. Throughout the region, no demonstrators evoked Osama bin Laden or al-Qa'ida. In fact, the Arab Spring succeeded nonviolently in doing what al-Qa'ida aimed to accomplish through terrorism: to overthrow the autocratic, Western-oriented regimes of the Middle East and North Africa. Although admitting to being awed, Bin Laden must have been dismayed by what he saw in the last months of his life.

The Arab Fall

What will take the place of the relative stability that the region's repressive, autocratic regimes guaranteed? What will happen to the dominoes that have yet to fall? What happened to the momentum of the Arab Spring? Surely this tumult must have given way to freedom and democracy, a new fluorescence of Arab culture, a genuine Renaissance that allowed the Arab World to race ahead and catch up on decades of potential lost to the ravages of autocracy, greed, and repression. Well, let's have a look.

On the whole there have been worrisome trends and incidents. The Arab Spring transitioned into an autumn, a period of danger and uncertainty that is likely to persist for years. Divisiveness and polarization based on major faith, minority sect, and tribal affiliation have been of special concern.

Let's start with post-Mubarak Egypt, where independent political parties were formed and reorganized after the Arab Spring. In 2012, Egyptians began voting in truly democratic parliamentary elections, which, if successful, would be a first for Egypt and would inspire many in the Arab World. Only one political party had the organization and manpower to galvanize the vote: the Muslim Brotherhood, a nonviolent, moderate Islamist party that had been outlawed for six decades. The Brotherhood easily beat the opposition candidate, Hosni Mubarak's former prime minister. After having no say for thousands of years, in June 2012 Egyptians were finally able to choose their leader. Muhammad Morsi of the Muslim Brotherhood was Egypt's first democratically elected president—ever.

Was Morsi able to make up for that lost time and be a benevolent, inclusive leader ready to unify Egypt's 88 million people and inspire millions more across the region? No. He was a divisive figure who immediately consolidated his power, dissolving parliament and bestowing unlimited powers upon himself. He drafted a constitution that enshrined Islamic law (*shari'a*) and precluded any role for the Christians who make up 10 percent of Egypt's population. Heedless of the Islamic tradition of giving the "People of the Book"—Jews and Christians—special dispensation, he looked the other way while Salafists and hoodlums burned churches. Violent sectarian clashes broke out between Muslims and Coptic Christians, leaving scores of people dead. Worse, sectarian clashes broke out between Sunni Muslim Egyptians and Sunni Muslim Egyptians. Until then, there had never been a sense of "us" versus "them" among Egypt's mostly homogeneous Muslims, except perhaps between the economic haves and have-nots, and violence and killings had never characterized Egyptian culture.

Anger mounted amongst the Egyptians who did not support Morsi and his Islamist agenda. Massive anti-Morsi and anti-Muslim Brotherhood demonstrations picked up steam in the summer of 2013, with the Egyptian Army actively fanning the flames of revolt. In a coup d'état, the Army deposed Morsi, reinstated martial law, banned the Muslim Brotherhood as a terrorist organization, and installed a military officer, Abdel Fattah al-Sisi, as Egypt's new leader. As was customary in the Mubarak years, he won a resounding 97 percent of the vote in 2013 presidential elections. Sisi said democracy in Egypt would be impossible for at least 25 years. What Egyptians know as the **deep state**, the all-pervasive repressive apparatus of the military security complex, was reinstated. All opposition was imprisoned or forced underground. Fareed Zakaria asked a secular, liberal Egyptian from Cairo who was involved in the uprising against Hosni Mubarak whether the current regime felt like a return of the old order. "Oh, no," he said. "This one is far more brutal, repressive and cynical than Mubarak's."[26]

Remember how the extremist jihadists were absent from the popular uprisings in Tahrir Square! The state's backlash against the Muslim Brotherhood and all perceived Islamist opponents lit a match under violent Islam. Ansar Bayt al-Maqdis, an al-Qa'ida affiliate operating from the northern Sinai Peninsula, stepped up a campaign of bombings of military targets across Sinai. Defecting from al-Qa'ida and pledging allegiance to ISIS in November 2014, this terrorist group set its sights on civilians and what remained of Egypt's vulnerable, valuable tourist sector.

The United States managed to become the enemy to both the Sisi and the Brotherhood factions. The US denounced the removal of Morsi as a military coup and at least temporarily suspended aid to Egypt. Egypt's peace treaty with Israel was tested on several fronts: Egyptian rioters sacked the Israeli embassy in Cairo; skirmishes broke out between Egyptian and Israeli forces on the Sinai frontier; and Libyan weapons began making their way overland through Egypt into the Palestinian Gaza Strip—until Sisi's regime demolished peoples' houses along this frontier and flooded the tunnels used to smuggle arms from Libya.

In Libya, the blowback of US-led NATO intervention was severe. Harvard's Kennedy School of Government called the mission "a model intervention of failure" and articulated its three lessons: "First, beware rebel propaganda that seeks intervention by falsely crying genocide. Second, avoid intervening on humanitarian grounds in ways that reward rebels and thus endanger civilians, unless the state is already targeting noncombatants. Third, resist the tendency of humanitarian intervention to morph into regime change, which amplifies the risk to civilians."[27] The NATO airstrikes that helped bring down the Qaddafi regime opened the door to jihadists, closed the door on American diplomacy in Libya, and had destabilizing effects on Egypt, Mali, and Algeria. We have seen how terrorists thrive in ungoverned or destabilized space in Yemen and elsewhere, and it happened in Libya too; one counterterrorism official described the country as "Scumbag Woodstock."[28] Tribal and regional competitions developed between Benghazi and the capital, Tripoli. A swirling mix of Islamists and anti-Islamists, tribal and regional competitions, regional political councils, and militias buffeted Libya even while a democratically elected House of Representatives tried to draft and ratify a constitution and resume oil production and export. Qaddafi had been a patron of the Tuaregs, many of whom fought for him as mercenaries. With Qaddafi gone, well-armed Islamist Tuaregs keen to establish an independent homeland in the Sahara carried their agenda back into Mali and Algeria. A collection of these Tuareg separatist groups and al-Qa'ida-linked jihadist groups —including al-Qa'ida in the Islamic Maghreb (AQIM)—gained control over northern Mali following the Arab Spring, thus prompting a French-led 431 intervention in January 2013. Libya provided refuge for Islamist fighters fleeing Mali and became a safe haven for terrorist training camps.

Tragic outcomes of the Libyan intervention included the murder of the American ambassador to Libya, Christopher Stevens, and later the closure of the US embassy in Libya. On the 11th anniversary of the 9/11 attacks, an excerpt from an American-made, inflammatory, low-budget movie depicting the Prophet Muhammad and Islam in a deeply offensive manner prompted anti-American demonstrations in several countries. Rioters in Cairo succeeded in breaching the outer perimeter of the American Embassy. What appeared to be an angry mob reacting spontaneously against the film attacked the US consulate in Benghazi, killing three Americans, including the ambassador. His gruesome death was attributed to a premeditated and well-organized assault by al-Qa'ida affiliate *Ansar al-Shari'a*. In July 2014, the State Department closed the American Embassy in Libya and evacuated all remaining US diplomatic personnel.

Six months later, the US "lost" Yemen as well. The American-backed government fought to contain the spread of AQAP, which has launched a number of attacks against the government in Sana'a, including several suicide bombings at President Hadi's inauguration. Tribal militants of various stripes have threatened Yemen's fragile infrastructure. Frequent militant attacks on energy facilities have hurt oil and natural gas production, and protests have grown over food, water, fuel, and electricity shortages.

One group of tribal militants, the Houthi, followed an unlikely path to control over Yemen. In 2004, the Houthis, who are Zaydi Shi'ites making up 40 percent of the country's population, began fighting Saudi and Yemeni forces in an effort to consolidate and spread their territorial holdings from the country's northwest. They have long resented being bottled up in the rugged, landlocked province of Saada, with few natural resources and no access to the sea. They wanted more political representation and a larger portion of the country's wealth.

In the political vacuum following President Saleh's ouster in 2012, the Houthis found opportunity and consolidated their gains. They branded themselves "Ansar Allah," the Partisans of God, fighting to reestablish their thousand-year reign of the Zaydi Imamate, which ended in 1962 (see Figure 6.47). They expanded out from Saada and captured the vital Red Sea port of al-Hudaydah in 2014. Hudaydah handles the bulk of the country's cargo imports and is the hub for Yemen's most agriculturally productive region. It is only 25 miles from the floating Ras Isa terminal, where Yemen's oil is loaded onto tankers. Houthis wanted control of the oil-producing region of Marib, but moving into that area would mean confrontation with their AQAP enemies. (AQAP, too, is taking advantage of the instability, expanding its territory by capturing new cities and districts in the wake of the weakness of the central government.) Houthis and their allies also fought for control of the southern port city of Aden, held by secessionists demanding independence from the north. In September 2014, Houthis seized the national capital of Sana'a, surprisingly not staging a coup, but settling for a power-sharing agreement (**• Figure 6.48**). But fighting continued to escalate. Seeking to avoid another "Benghazi"—which had become a political albatross for Democrats, especially presidential candidate Hillary Clinton who had been Secretary of State when that attack occurred—the State Department closed the Sana'a embassy.

The US, which had engaged in near-daily drone attacks against AQAP strongholds, thus lost its biggest listening post and strongest bulwark against AQAP in Yemen. Yemen's geopolitical situation became critical: as a failed state where AQAP could operate with impunity, it represented a threat to regional and US national security (remember that after the failed attempts to bring down planes with explosives concealed in underwear

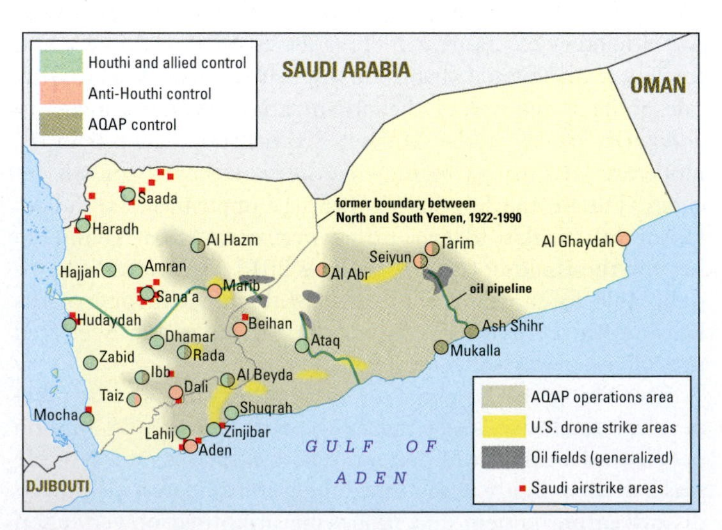

• Figure 6.48 Yemen fell into chaos after the collapse of the national government in 2012. Houthi rebels, secessionist factions, al-Qa'ida in the Arabian Peninsula terrorists, and surviving pro-government forces all battled for control of key cities. The Shi'ite Houthis reasserted control over large parts of the country, sparking a military confrontation with neighboring Sunni Saudi Arabia.

and printer cartridges, AQAP only strengthened its resolve to develop undetectable bombs). Regionally, the Houthis are seen as agents of fellow Shi'ite Iran, precluding any US-Houthi alliance, despite their mutual enemy in AQAP. Branding Houthis as terrorists, the Saudis and a coalition of other Sunni Arab countries launched airstrikes inside Yemen in 2015 to intervene on behalf of Yemen's Sunni tribesmen as they have in the past, despite concern about creating a broader Shi'ite backlash or further inciting jihadi terrorism. Yemen turned into a proxy war between Saudi Arabia and Iran for supremacy in the Middle East. If you are interested in seeing Yemen's web of friends and enemies, use the interactive graph cited in Figure 6.32.

In Bahrain, the government continued to marginalize and try to crush anti-regime forces. Elements of the Shi'ite opposition became more violent. Two groups, the Islamist Al Ashtar Brigade and the February 14 Movement, used improvised explosive devices (IEDs) against security forces and sought more sophisticated bombs. Iranian drones were seen operating in Bahrain's skies.

The one bright spot in the post–Arab Spring world was Tunisia. As in Egypt, an Islamist party, this one known as *Ennahda*, won the first election. Unlike Egypt's Muslim Brotherhood, Ennahda was willing to share power. It did not try to institute *shari'a* law, and it respected Tunisia's progressive stance on women's rights. Tunisia's different course from Egypt's after the Arab Spring may draw from some of its peoples' advantages, being more generally prosperous, urban, literate, and world-savvy, and with a more diverse civil society, than Egypt's. Ennahda ruled more effectively than the Brotherhood in Egypt and won a second election in 2014. Tunisia still has challenges, including a youth unemployment rate of 30 percent and a low-level Islamist insurgency that included attacks on Tunis' national museum and a beach resort in 2015. More Tunisians join ISIS than do citizens of any other Arab country. Overall, though, Tunisia shows that there is nothing in Islam or Arab society that makes it impossible for democracy to take root.[29] It simply has not; Freedom House reports that the MENA region is the world's least free, with 85 percent of the population living in not-free countries and only two countries, Tunisia and Israel, classified as free[30] (**•Figure 6.49**).

You have witnessed the Arab Spring, one of the region's most momentous periods in centuries, and the blowback of the "Arab Fall." These issues will be prominent in world affairs for a long time to come, and they help us refocus on some of the region's other problems. Problems between Israel and its Arab neighbors are a persistent feature of the Middle East's geopolitical landscape. In an odd twist, in the post–Arab Spring Middle East, with ISIS on the scene, the Arab-Israeli conflict appears as a set of problems that *can* be solved. With economic aid, the United States had effectively "bought" the peace between Egypt and Israel (excluding Iraq for military expenditures, these two countries consistently rank as the top two countries in the value of American foreign aid). With its economic rather than military clout, can the US bring more stability to this troubled, oil-rich part of the world? If not, how much instability will the United States tolerate without once again putting its boots on the ground?

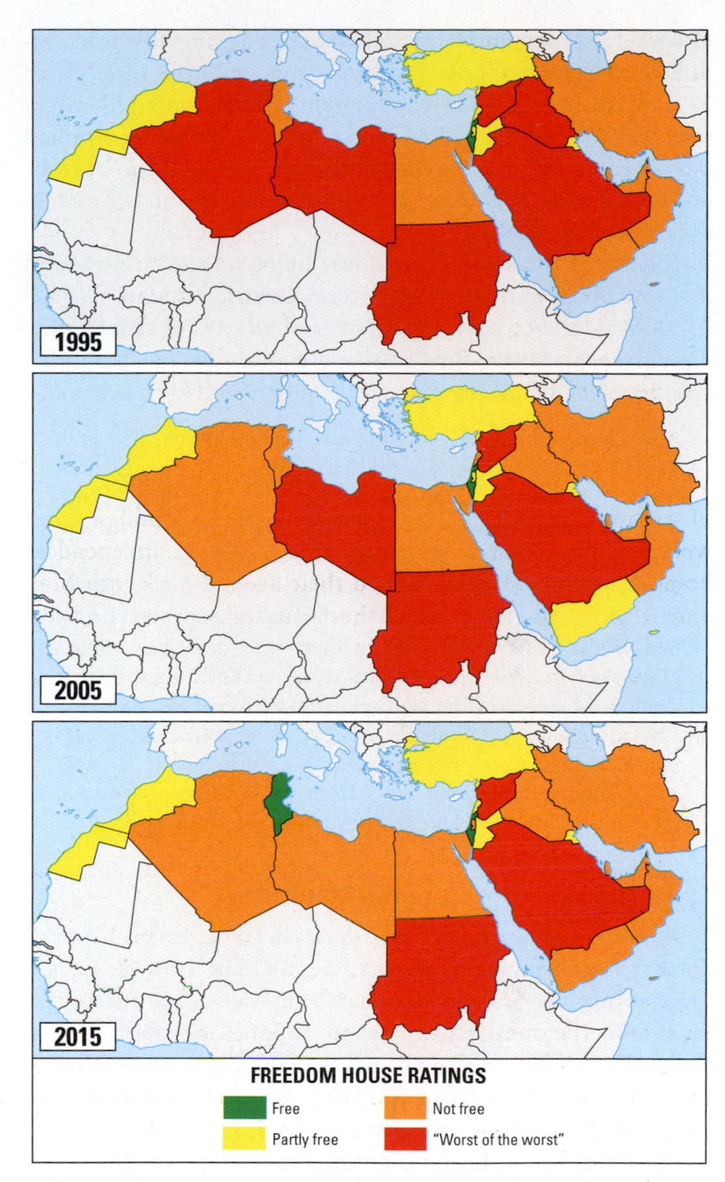

• Figure 6.49 Freedom House is a US-based organization that advocates for democracy and human rights around the world. It releases an annual report that rates countries as free, partly free, or not free based on a variety of criteria including the presence of an independent judiciary, free and fair elections, freedom of expression and the media, and gender equality. In the Middle East and North Africa, only Israel has consistently been rated a free country, though Tunisia's recent successes show that there are no inherent barriers to freedom within Arab societies.

THINK CRITICALLY: What does Hesham Melham mean by writing, "Every hope of modern Arab history has been betrayed?"

The United States would be loath to "lose" Bahrain to the rule of the majority Shi'ites, who could help Iran gain traction just miles from Saudi Arabia. Bahrain is also home to the US Fifth Fleet, whose ships patrol the Gulf to ensure safe passage of ships carrying oil through the Strait of Hormuz and to confront Iran if necessary. Following ISIS's rise, an anti-American Islamist regime could find solid footing in post-Qaddafi Libya, or more ominously in Yemen, on Saudi Arabia's doorstep. It is difficult to envision anything other than at least an ongoing low-level war waged by the United States against al-Qa'ida affiliates in Yemen. Those predator drones will be busy. With its oil wealth, Saudi Arabia represents the ultimate red line in the sand. Would the United States "allow" the Saudi Arabian monarchy to fall in a popular uprising? That scenario of revolution in Saudi Arabia is the worst nightmare of US foreign policy decision makers.

It is very difficult to find a silver lining on the landscape of the post–Arab Spring world. Fareed Zakaria laments that the region has failed to develop echoes in hallways of scholarship, diplomacy, and other fields. Lebanese journalist Hesham Melham has this disturbing view of the post–Arab Spring world:

> Arab civilization, such as we knew it, is all but gone. The Arab world today is more violent, unstable, fragmented and driven by extremism—the extremism of the rulers and those in opposition—than at any time since the collapse of the Ottoman Empire a century ago. Every hope of modern Arab history has been betrayed. The promise of political empowerment, the return of politics, the restoration of human dignity heralded by the season of Arab uprisings in their early heydays—all has given way to civil wars, ethnic, sectarian and regional divisions and the reassertion of absolutism, both in its military and atavistic forms. With the dubious exception of the antiquated monarchies and emirates of the Gulf—which for the moment are holding out against the tide of chaos—and possibly Tunisia, there is no recognizable legitimacy left in the Arab world.[31]

Syria: ISIS Emerges from the Maelstrom

With its crackdown on demonstrators in Deraa in 2011, Syria's descent into brutal civil war had begun. The carnage is on a scale reminiscent of the Second World War. More than three years into the conflict, more than 200,000 people had been killed. Atrocities including murder, torture, rape, and enforced disappearances have been carried out by multiple parties to the conflict. Early on, the government in Damascus was recognizable as the principal offender. The savagery with which the Syrian army took on armed and unarmed civilians was extraordinary. Soldiers seemed to have no compassion even for children and women. It was unrestrained carnage, and there appeared to be no way to bring it to a stop. As pressure on the al Assad regime mounted on the outskirts of the capital in August 2013, the Syrian army deployed the chemical weapon sarin, a nerve agent that typically kills within ten minutes by stopping the lung muscles from working. The images of civilian suffering from these attacks evoked outrage worldwide and also represented a crossing of that "red line" that the American 248 president had drawn one year earlier.

The United Nations and the Organization for the Prohibition of Chemical Weapons drew up a plan to rid Syria of its chemical agents and weapons, and that was essentially accomplished within a year. The regime, however, continued to use chlorine, ammonia, and other low-level chemical weapons. It also deployed a so-called "barrel bomb," a particularly insidious weapon (outlawed in international conventions) consisting of a barrel packed with high explosives and shrapnel. The Syrian Air Force dropped these into civilian gatherings, especially in Aleppo. Pilots often created civilian gatherings by dropping a bomb that attracted people desperate to free survivors from the debris; when there were enough of these first responders, the pilots dropped a second barrel bomb on them.

As many as a thousand rebel groups with 100,000 fighters emerged to take on the Assad regime—and one another. Who are these rebels? The **"Syrian Opposition"** is an umbrella term for groups opposed to the regime. Some are so-called "secular moderates." These are far outnumbered by moderate Islamists and by extremist jihadists. The rebel groups are deeply divided among themselves, with rival alliances fighting for supremacy. One of the most powerful is the moderate **National Coalition for Syrian Revolutionary and Opposition Forces**, backed by most Western and Gulf Arab states. Its members include the FSA, the first rebels to emerge as dissidents from the Syrian Army itself. That coalition's primacy is rivaled by the powerful but moderate Islamist alliance, the Islamic Front. Importantly, there is no nationally supported alternative to the current Syrian regime.[32]

Critical time passed as Syria's war raged unabated. President Obama received intelligence and counsel suggesting that the US should intervene more forcefully in the conflict, arming "moderate rebels" with the aim of bringing down the Assad regime. However, Obama came into office twice on the promise of drawing down US war efforts in the region, and he was loath to be pulled into the Syrian maelstrom. On what some observers called their "vacation from history," European powers watched from the sidelines. During the time between 2012 and 2014, ISIS gathered strength among Sunni insurgents, initially in Iraq. All but eradicated in its original form of al-Qa'ida in Iraq by US and allied Sunni forces by the time the US withdrew from Iraq in 2011, ISIS found an ideal breeding ground in war-torn Syria. It soon emerged as the most aggressive rebel force, and its success bred success.

Did the US miss an opportunity to prevent or blunt the growth of ISIS in Syria? Detractors of Obama's policy in the region argued that the president was "spooked by Iraq" and shunned intervention in Syria when the US truly needed to intervene there. According to this view, articulated by the Foundation for the Defense of Democracies, "the Obama Doctrine was to avoid using force," and the United States "didn't lose the war in Iraq until Obama pulled our troops out."[33] Furthermore, Obama decided *not* to use force in Syria after pledging to do so. This left allies, especially where the US has commitments as in Japan and Korea, questioning whether America would actually back them up in a crisis situation. 320

Partly through a highly effective social media and Internet campaign that included Hollywood-quality video production, ISIS drew in recruits from dozens of countries. ISIS appealed especially to disaffected youth, both men and women, from across the Middle East and from Europe and beyond. The "disaffected" are mainly from that "youth bulge" facing unemployment, underemployment, lack of opportunity, and especially lack of power. Joining ISIS brought immediate gratification in the individual's quest for power. Recruits received arms, training, and cash salaries supplied by donors in nearby Sunni countries, as well as large sums from bank loot

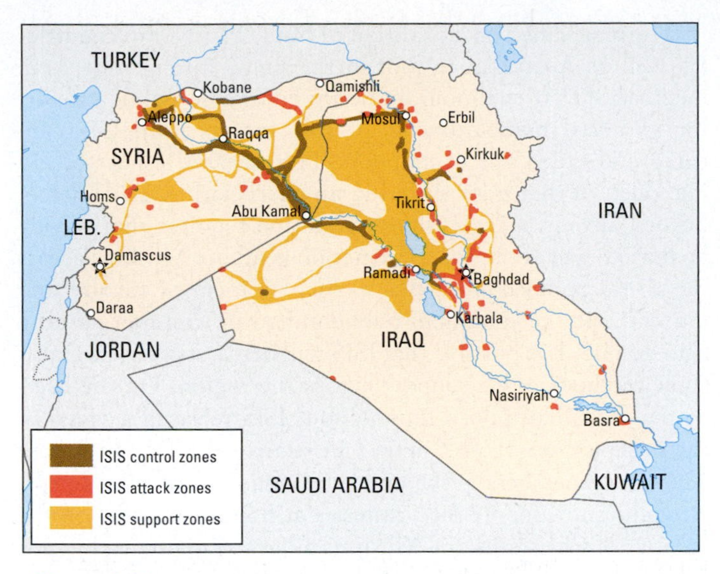

• **Figure 6.50** By 2015, ISIS was in control of a large area of both Syria and Iraq. Source: Institute for the Study of War.

THINK CRITICALLY: Why did ISIS take control of Iraq's mainly Sunni areas with relative ease? How does the area of ISIS control correlate with the depiction of Iraq's ethno-religious regions in Figure 6.53?

and sales of oil. ISIS has also attracted avowedly pious Muslims believing, at least initially, that it represented a "pure" 232 Islamic society governed by *shari'a* law. ISIS also offered its followers something al-Qa'ida (including its Syrian affiliate, the al-Nusra Front) could not: territory and lasting control over it (• **Figure 6.50**).

ISIS fighters broke down the defensive berms marking the Syria/Iraq border and poured into Iraq en masse in 2014. ISIS framed this in the anti-colonial and anti-infidel (*kufaar*) contexts they knew would appeal to many people in the region; one spokesman wrote on his Facebook page: "Today the borders of Sykes-Picot have ended, the Ummah [community of Muslims] is united beyond the imaginary lines created by the *kufaar*. The Islamic State is physically tearing down these markers of division between Iraq and Ash-Sham [Syria] and the believers are rejoicing, some of them brought to tears at this momentous and historic occasion."[34] ISIS's self-proclaimed caliph, Abu Bakr al-Baghdadi, demanded fealty (and strict religious adherence to his interpretation of Islam) of Muslims everywhere. He began to get it, not from the vast majority of Muslims who regarded his Caliphate as heresy, but from violent Islamist groups in a number of countries, including Abu Sayyaf in the Philippines; Tehrik-i-Taliban in Pakistan; Ansar Bayt al-Maqdis in Egypt; the Islamic Youth Shura Council in Libya (putting ISIS perilously close to Europe and the prized city of Rome); Jund al-Khilaafa, or Soldiers of the Caliphate, in Algeria; and Boko Haram in Nigeria. Some even broke ranks with al-Qa'ida to join ISIS (see **Figure 6.1**).

The Obama administration looked on this in horror; ISIS poured into Iraq right after the US had "wrapped up" the war in Iraq. As relatively small numbers of ISIS fighters assaulted and occupied city after city in northern Iraq, most notably Mosul (with a pre-ISIS population of 1.5 million), Iraq's army crumbled, its American-supplied weapons and armaments left to ISIS. Its Shi'ite fighters tore off their uniforms and tried to blend back in with the population. Many of the army's Sunni officers and recruits joined ISIS. ISIS had first taken hold in the absence of inclusive leadership that existed in Iraq after US troops withdrew. Iraq's nominally democratic but highly sectarian post-Saddam government greatly favored Shi'ites, ousting and marginalizing Sunnis decisively. Joining ISIS gave these disenfranchised Sunnis a degree of power they had not enjoyed since the reign of Saddam Hussein. In fact, once back in Iraq, ISIS mushroomed into such a powerful and effective military force in part because it was able to employ the very officers of the Iraqi army that the US had disbanded in 2003.

As ISIS gained control of more Syrian and Iraqi territory, it began to govern as a state, collecting taxes, providing services, and meting out its unique form of justice. It became apparent very quickly, however, that there are a number of very un-Islamic aspects of ISIS's Islamic Caliphate.

ISIS is zealous about levying *takfir*, the pronouncement of someone as an unbeliever, often translated simply as "excommunication." *Takfir* can be directed at an infidel or *kufr*, of which there are many subcategories, including polytheists (*mushriq*) such as Yazidis and Hindus; at the apostate or *murtaad* (those who abandon Islam in word, thought, or deed, including by converting to another religion); and at any Muslim who does not follow ISIS's austere vision of Islam (this includes the world's roughly 200 million Shi'ites, along with a great many Sunnis who do not embrace ISIS).

Unlike the historic Caliphates, this one denied all rights of the *dhimmi* (literally "protected people")—the "People of the Book," including Jews, Christians, Zoroastrians, and select other non-Muslims. We are witnessing the ethnic cleansing of Christians from eastern parts of the cradle of Christianity and of Yazidis (Kurdish speakers practicing a unique religion blending Zoroastrianism with early Mesopotamian beliefs) from their small hearth. ISIS announced that within the Islamic 138 State, Christians must pay a special tax (like one administered during the early Caliphates) or face death (not a choice the early Christians had to make). Iraqi Christians were barred from employment in public sector jobs. According to eyewitness accounts, nearly all of the Christian and Yazidi population in areas controlled or contested by ISIL were displaced as refugees. These were the lucky ones. Vast numbers of men—captured Alawite fighters, including Shabak and Turkmen Shi'ite civilians, Sunni Kurds, Yazidis, and Christians—were simply killed; executed, sometimes en masse; shot; or decapitated, and the heads of prominent men were mounted on posts. Women and girls as young as age 12 in these groups were subjected to systematic rape by ISIS fighters; many of them were put in brothels in cities taken by the group, and others were forced into marriage. Some women were sold in markets in the cities as sex slaves. There have been crucifixions of non-compliant Christians and Sunni Muslims. People have been tortured, burned alive, and buried alive. A group of Yazidi men had their eyes gouged out for refusing to convert to Islam, and then were then burned to death. ISIS has also killed Muslim

Polaris/Newscom

• **Figure 6.51** The cultural heritage of Mesopotamia and much of the Middle East is at great risk in ISIS-held areas. Here in 2015 ISIS militants are systematically destroying artifacts in the museum of ISIS-held Mosul in northern Iraq. In ISIS's extreme interpretation of Islam, all depictions of human figures are taboo. One of the ISIS iconoclasts at work in this museum explained, "These antiquities and idols behind me were from people in past centuries and were worshiped instead of God. When God Almighty orders us to destroy these statues, idols and antiquities, we must do it, even if they're worth billions of dollars."

clerics who disavow it, including more than a dozen Sunni Muslim leaders in Mosul.[35x]

In ISIS theology, all religious shrines are viewed as artifacts of idolatry (*shirk*) and are subject to destruction. Citing the Prophet Muhammad's destruction of idols in Mecca's Ka'aba, ISIS men smashed and drilled ancient Mesopotamian statuary in the Mosul Museum (**• Figure 6.51**). Not far from there, ISIS fighters bulldozed as much as they could of the ancient Assyrian city of Nimrud. Later, they leveled portions of ancient Palmyra in Syria. ISIS destroyed hundreds of Christian churches and Shi'ite mosques, especially of the Shabak and Turkmen ethnic groups, in equal measure. In Mosul they blew up the tomb of the Prophet Jonah (*Yunis*), revered there by both Christians and Muslims.

Such atrocities and the humanitarian issues associated with them stirred the US to action. Two events from around the same time in autumn 2014 are of particular note. First, ISIS pushed into Iraq's Sinjar district near the Syrian border, a predominantly Yazidi region. When Kurdish forces that had been resisting ISIS fell back, a large Yazidi population was left virtually undefended. ISIS killed as many as it could while thousands more sought refuge on Mount Sinjar. Worldwide news coverage of this humanitarian crisis brought Washington's war cabinet together. But perhaps the most decisive trigger for action was the series of terrifying ISIS videos depicting the beheading of American and British journalists and relief workers taken hostage by ISIS. American warplanes dropped palettes of relief supplies to the Yazidis on Mount Sinjar and opened a northern corridor for their escape into Turkey. Bombs and cruise missiles struck ISIS targets, initially only in Iraq. President Obama announced a strategy to "degrade and ultimately destroy" ISIS, and airstrikes widened to include Syria.

The plan called on a coalition of NATO and Arab countries (as well as Australia) to join in the fight. Jordan, one Arab member of that coalition, had only lukewarm public support for its aerial bombardments of ISIS targets until ISIS captured and killed one of its pilots. ISIS released a heartbreaking video of that pilot suffering an excruciating, cruel death, burned alive in a cage. Islamic tradition requires that a body be buried, and even proper cremation is taboo. Muslims across the world recoiled from this particular form of torture and murder, and the Jordanian public united in common purpose against ISIS. It is possible that ISIS will defeat itself as its loathsome methods erode support across the region. On the heels of the Jordanian pilot's immolation, ISIS released a video of the decapitation of 21 Coptic Christians on a Libyan beach. They were members of the poor working class of rural Egypt, struggling to support their families at home. Focusing on the identity of Christians and Muslims alike as Egyptians, Egypt's President Sisi responded quickly with bombings of ISIS targets around the Libyan city of Derna. The conflict in Syria and Iraq created one of the most serious humanitarian crises since World War II. Four million refugees have poured out of Syria into Jordan, Turkey, Lebanon, and beyond since 2011, bound especially for Western Europe (**• Figure 6.52** and **page 95**). International aid agencies erected gigantic camps in desert and other rural areas. As bleak as they are, these places offer safety, security, and food, with food provided mainly by the World Food Programme, to which the US is the largest donor.

Conditions are likely far worse for the refugees classified as internally displaced peoples (IDPs) inside Syria. Mid-2015 estimates put that population at about 8 million people. All told, approximately 12 million people—half of the country's population—were displaced. Probably 5 million people within Syria lived in areas under siege or hard to access. Both the government and some rebel groups have inflicted civilian suffering by blocking access to food, water, and health services.

76

What is going on, especially in ISIS-controlled parts of Syria? "Following takfiri doctrine, the Islamic State is committed to purifying the world by killing vast numbers of people," Graeme Wood writes. "The lack of objective reporting from its territory makes the true extent of the slaughter unknowable, but social-media posts from the region suggest that individual executions happen more or less continually, and mass executions every few weeks. Muslim 'apostates' are the most common victims."[36] We really know very little about what is going on inside Syria and ISIS-controlled Iraq. Early in the war, the Assad regime tried to keep foreign reporters out of the country, but a few sneaked in and were able to verify the authenticity of the video footage posted daily on YouTube. With the real hazard of capture and beheading, there are not many journalist eyes inside ISIS-held areas. A daring VICE News team managed to film some of the fighting, but described this as a dangerous get-in-and-out mission.[37] This media blind spot has consequences—we just don't know what they are.

With hindsight we can see that the Syrian Civil War began as one of the number of Arab Spring uprisings against an autocrat and grew into a brutal proxy war involving both regional and global powers. The Assad regime remained in power with the

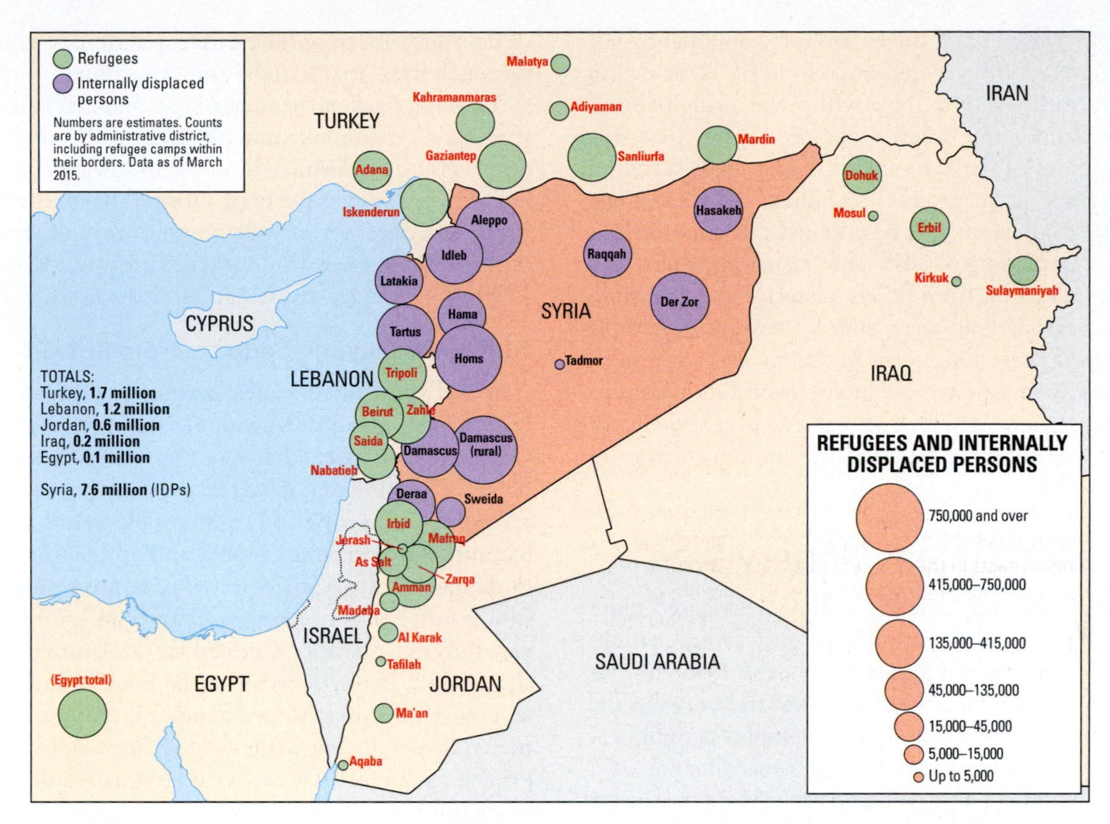

• **Figure 6.52** Years of war have left 8 million displaced persons within Syria and 4 million refugees outside the country.

Source: UNHCR.

support of Iran and Russia (including the deployment of conventional Russian military land, sea, and air assets), and on its western flank by help from Lebanon's Shi'ite Hizbullah. On the other side, elements of the Sunni-dominated opposition gained the support of Turkey, Saudi Arabia, Qatar, and other Arab states along with the US, the UK, and France. An odd feature of this conflict was the US finding common interests with its principal regional foe, Iran. With Iranian-supplied weapons, Iraqi Shi'ite militia, along with Iranian revolutionary guard troops, joined Iraqi troops to pry ISIS from cities it had occupied north of Baghdad.

The US is counting on three things to achieve its objectives to "degrade and destroy" ISIS. First is the political reconstruction of the Iraq government, with a system that is truly inclusive of the Sunni minority. The US forced the ouster of Iraq's sectarian president Nouri al-Maliki and called for a more truly representative government. The political landscape of Syria too will have to look completely different than it does now to be ISIS-free. The US is unlikely to garner maximum military support among regional powers, especially Turkey, unless it also commits to bringing down the Assad regime. Second is the military solution. The US sought to build a broad base of support across the region from the countries that have the most to lose to ISIS. But with tepid participation, the US bore the lion's share of the work and the cost. The US began its assault against ISIS with air strikes, in some places—notably Kobane, on the Turkish/Syrian border—backed by ground assaults by Kurdish *peshmerga* forces, and elsewhere in Iraq by Shi'ite units of the Iraqi Army and by Shi'ite militia, including the Badr Brigade. But military planners readily admitted that air strikes have limited utility, and that far more "boots on the ground" were required for success. Whose boots? The plan called for combat forces from the region itself, with US training, to do the fighting. The US slowly and steadily ramped up its "advisors" in Iraq. For Americans of a certain age, this hearkens back to the escalation of the war in Vietnam. They are mindful of "mission creep," in which limited military objectives are steadily widened to include growing commitments of men and matériel.

How long can the campaign to degrade and destroy ISIS last? Can it succeed, or will the Caliphate of the Islamic State emerge as a lasting entity? American military strategists have suggested it could be a "generational problem, a twenty year problem."[38] Council on Foreign Relations President Richard Haass likened the conflict to the Thirty Years War, a prolonged religious struggle that killed as many as a third of the people of Central Europe in the 17th century.[39] It is possible that the political boundaries of the Middle East, drafted a century ago along the Sykes-Picot lines of the European victors over the Ottoman Empire, will in fact be redrawn. The partitioning of Iraq into Sunni, Shi'ite, and Kurdish states is the most predictable outcome.

It is possible that an ongoing war against ISIS, particularly if led by the United States, will not defeat it, but will create countless numbers of radicalized ISIS-minded fighters. By engaging them in their *jihad*, their military opponents fall right into the ISIS playbook of creating a kind of "World Cup of Terrorism" in Mesopotamia. Nor will a war principally in the boots of the Saudis and other Arabs necessarily be a different scenario: many of these regimes are also long-standing targets of numerous jihadi groups, including al-Qa'ida. The US and its regional allies may at best be able to contain ISIS geographically. Opposition to ISIS may in time grow from within as people it controls tire of their austerity and brutality. Throughout the region, long-term alternatives and solutions will develop only if these long-term problems are addressed: education, especially of women; unemployment and underemployment; population growth; lack of opportunity; and lack of empowerment. Places to start include Syrian refugee camps in Jordan, Turkey, and Lebanon.

Iraq: The United States' "Pottery Barn"?

Iraq (population 35 million) occupies a broad, irrigated plain drained by the Tigris and Euphrates rivers, with fringing highlands in the north and east and desert in the west. Known since ancient times as Mesopotamia, meaning "the land between the rivers," this landscape of today's Iraq has a complex geopolitical history. It has been a cradle of civilization, a seat of empires, a target of conquerors, and in the 20th and early 21st centuries, a focus of oil development and political and military contention.

A critical feature of modern Iraq's geography is its ethnic and religious composition, with three major groups present: the Shi'ite Arabs, mainly in the south, who make up about 60 percent of the country's population; the Sunni Arabs, about 35 percent of the total and living mainly in the center (especially in the so-called Sunni Triangle); and the Kurds, mainly in the north, most of whom are Sunni Muslims and who represent 15 to 20 percent of the total (• **Figure 6.53**; see also the discussion of the Kurds on **page 269**). There are numerous smaller minorities, including Yazidi and Sunni Turkoman.

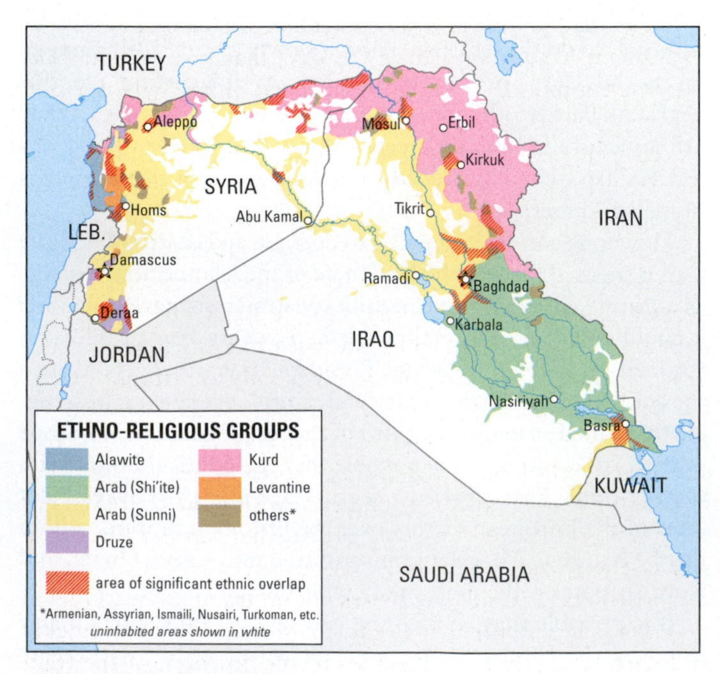

• **Figure 6.53** Ethno-religious groups across Syria and Iraq.

Looking back at the map of the Persian Gulf in Figure 6.24, you can see a fundamental geographic disadvantage of Iraq: it only has a tiny, 35-mile (55 km) window on the Gulf. Poor port facilities hamper exports of oil by tanker; this narrow access to the sea does not include any natural deepwater ports. One of the main reasons Iraq attacked Iran in 1980 and Kuwait in 1990 was to widen Iraq's access to the Gulf.

237

Gulf War I: Ousting Saddam from Kuwait

Iraq and the United States became entangled in 1990, when Iraqi forces invaded Kuwait—a country slightly larger than Connecticut in area that has the world's sixth largest oil reserves. Claiming that it was historically part of Iraq and calling it Iraq's "19th province," Saddam Hussein's regime attempted to annex Kuwait. Saudi Arabia and other Gulf oil states seemed in danger of invasion. Worldwide opposition to Iraq's aggression resulted in a sweeping embargo on Iraqi foreign trade imposed by the Security Council of the United Nations. Led by the United States, a coalition of 28 UN members (including several Arab states) staged a rapid buildup of armed forces in the Gulf region. The stage was set for what would come to be known as "the Gulf War," and later as "Gulf War I."

Beginning in January 1991, the coalition's crushing air war drove the Iraqi air force from the skies, severely damaged Iraq's infrastructure, and hammered the Iraqi forces in Kuwait and adjacent southern Iraq with intensive bombardment. In the ensuing ground war, the Iraqis were driven from Kuwait within 100 hours. Overall, an estimated 110,000 Iraqi soldiers and tens of thousands of Iraqi civilians were killed, whereas the technologically superior coalition forces suffered 340 combat deaths. As they withdrew from Kuwait, the Iraqis also made war on the environment, setting fire to hundreds of oil wells in Kuwait and allowing damaged wells to discharge huge amounts of crude oil into the Gulf.

A formal cease-fire to the Gulf War came with harsh conditions imposed on Iraq by the UN Security Council. UN-sponsored economic sanctions created shortages of food and medical supplies, bringing great hardship to the country's population. Iraq agreed to pay reparations for the damage its forces had caused, and it renounced its claims to Kuwait. Saddam Hussein's surviving forces then mercilessly put down rebellions by Shi'ite Arabs in southern Iraq and by Kurds in the north. To deter Iraq's military and provide a safe haven for the Kurds in the north and the Shi'ites in the south, the United Nations established "no-fly zones," where Iraqi aircraft were prohibited from operating. But Iraqi forces had no difficulties in mounting ground campaigns against the Shi'ites and Kurds. Coalition forces did not intervene on behalf of the Shi'ites and Kurds, who felt betrayed: the coalition had encouraged them to rise up against Saddam Hussein but then abandoned them and left them at the mercy of Saddam's vengeful forces. Iraq's Shi'ites, both in the countryside and in Basra and their other urban strongholds, fared especially badly.

One of history's great environmental and cultural tragedies was Iraq's systematic destruction after 1991 of the marshes

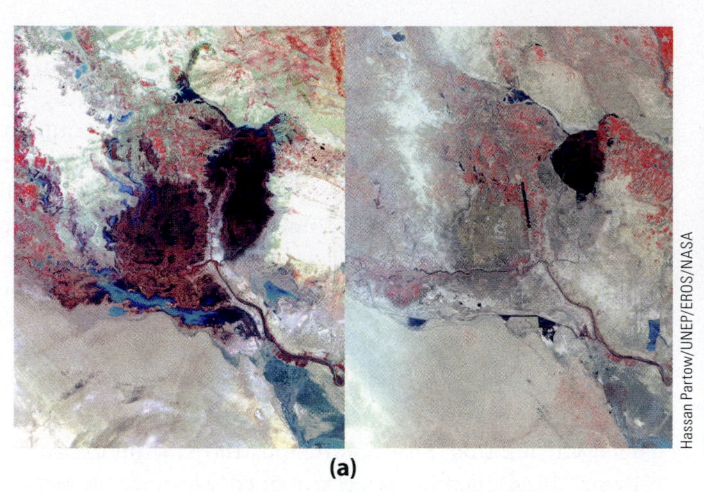

(a)

(b)

• **Figure 6.54** **(a)** Remote sensing of Iraq's southern marshes in 1971 and 2000. These are "false color" images; the red is the real-life green color of marsh vegetation. Iraqi engineers built embankments to cut off the flow of Tigris and Euphrates waters into these wetlands, laying waste to a unique ecosystem and the way of life of the "Marsh Arabs." **(b)** Marsh Arab life in the marshes before their destruction. Note the unique architecture of dwellings and community buildings, all made from marsh reeds. Some Biblical scholars have claimed that this marsh was the Garden of Eden, where Adam and Eve begat humankind.

adjoining the southern floodplains of the Tigris and Euphrates rivers (•**Figure 6.54a, b**). This area is the homeland of the ancient culture of the **Ma'adan**, or **Marsh Arabs**, who over several thousand years established a way of life seen nowhere else in the Middle East, using the resources of the marshland to sustain almost all their needs. To quell any opposition these Shi'ite Muslims might pose, the Iraqi regime destroyed their wetlands habitat and forced some 300,000 Ma'adan to flee. Engineers built massive embankments and canals to drain the marshes, and later set fire to the dry vegetation and to Ma'adan settlements.

After the Iraqi regime was taken down in 2003, the embankments were demolished, and restoration of the marshlands began. More than 50 percent of the marshes have come back to life, albeit with much less biodiversity than before. The water is much more saline now than in the past, and the fires changed the soil chemistry so that many of the soils are now claylike and unsuited for the vegetation that grew there previously. Less than one-third of Ma'adan who fled have returned to restore their traditional way of life. As extraordinary as their adaptation to this environment was, their life in the marshes was one of poverty, and many Ma'adan do not wish to return. The task of rebuilding communities in the marshes is an ongoing social challenge. Much damage was done to a culture whose social fabric and economy were woven into the reeds and waters of the marsh.

Throughout the 1990s, the regime of Saddam Hussein played a cat-and-mouse game with UN inspectors whose duty was to eliminate Iraq's potential to produce weapons of mass destruction. Iraq refused to allow full access by the UN inspectors. United States and allied forces responded with air strikes and more stringent economic sanctions. International support for the sanctions against Iraq weakened greatly, especially in the Arab world. President Saddam Hussein endured, much to the chagrin of his foes, and might have been able to hang on to power indefinitely if not for 9/11.

Gulf War II: Invasion and Occupation of Iraq

Immediately after the Gulf War during the presidencies of George H.W. Bush and Bill Clinton, the reasoning in Washington was that a unified Iraq under Saddam Hussein was better than the alternative: its possible fragmentation into the three entities of a Shi'ite south, a Sunni center, and a Kurdish north, some or all of which might be hostile to US interests in the region. Although unsavory, Saddam Hussein could be contained. After 9/11, however, under the George W. Bush administration, the downfall of Saddam Hussein became an urgent priority, ranking alongside the "War on Terror." The Bush White House asserted that Saddam Hussein had linkages with al-Qa'ida. There was no basis for this association, however, as al-Qa'ida had an unwavering contempt for Saddam.

Publicly, Washington built the case that Iraq was harboring and continuing to develop weapons of mass destruction that might be used against the United States and its allies. The United States was able to prompt further United Nations pressure on Iraq, which finally allowed UN weapons inspectors to resume their work. When these inspections failed to turn up weapons, the United States insisted that Iraq was hiding them and that only the use of force could remove this threat. It was necessary to effect "regime change" in Baghdad. To take on Saddam Hussein, President Bush assembled what he called a "coalition of the willing," consisting mainly of Britain, Australia, Denmark, and Poland. It fell far short of the coalition put together by his father for the first Gulf War. France, Germany, and other traditional allies of the United States opposed the impending war against Iraq, and even within the United States there was considerable dissent against what was perceived as a war of choice rather than one of necessity.

The US-led ground and air assault on Iraq (dubbed **Operation Iraqi Freedom**) that began in March 2003 led quickly and with relatively few casualties to the downfall of the Iraqi regime. With about 130,000 troops, concentrated mainly in the Sunni-dominated center of the country around Baghdad, the United States settled in for a long and troubled occupation of the country. The United States handpicked what it called the Governing Council of Iraqis, broadly representative of the country's ethnic composition, and ran many of Iraq's security and other affairs with its own Coalition Provisional Authority (CPA) until June 2004, when it turned over sovereignty to an interim Iraqi government.

The United Nations, with strong US support, organized a January 2005 election for a 275-member transitional national assembly. In this, Iraq's first-ever free election, Iraq's largest ethnic population unsurprisingly won the lion's share of the seats: the Shi'ites, who turned out in huge numbers to vote, won 48 percent for their main political party. Anticipating the Shi'ite success and condemning the election as an illegitimate US action, most Sunnis boycotted the vote. That made it possible for the main political party of the Kurds, a smaller ethnic group, to capture another large bloc of the assembly votes—26 percent. Iraq's Sunnis ended up with just 2 percent of the assembly seats. Having been the dominant force in Iraq for decades under Saddam Hussein (who was convicted of war crimes by an Iraqi court and hanged in 2006), the Sunnis now found themselves deprived of power and increasingly angry about the American occupation. The pro-US government meanwhile faced a perplexing array of challenges in cementing a constitution for the fragile country: how much autonomy the Kurds and Shi'ites might receive within the country, how oil revenues would be split among the geographic and ethnic regions, and what the role of Islam should be in national affairs, for example.

Sunni resistance against the United States grew. Using roadside IEDs, car bombs, small arms, and rocket-propelled grenades, Sunni militants were able to inflict American casualties on an almost daily basis. Predictably, al-Qa'ida used the US occupation of Iraq as a rallying cry for intensified resistance 245 against the United States. A well-organized insurgency officially linked with al-Qa'ida started up, under the leadership of Jordanian-born Abu Mus'ab al-Zarqawi. His al-Qa'ida in Iraq organization, which later morphed into ISIS, kept up a steady stream of bloody attacks on US targets even after al-Zarqawi was killed in 2006. Also in the crosshairs were the Iraqis perceived to be collaborating with the Americans, particularly those who joined the new Iraqi army. Sunni insurgents relentlessly targeted Shi'ites in particular, even twice bombing one of their holiest sites, the golden-domed Askariya Mosque in Samarra (**• Figure 6.55**).

These Sunni assaults on Shi'ite sacred space helped push Iraq into civil war. Increasingly, this internecine violence took the form of ethnic cleansing such as seen in the former Yugoslavia, where the cultural makeup of neighborhoods, cities, and regions was altered through violence and intimidation. The more militant anti-Sunni and anti-US Shi'ite factions, rallying chiefly behind a cleric named Moqtada al-Sadr and his Mahdi Army (Mahdi Militia), ascended. Moderate Shi'ite leaders such as Ayatollah Ali al-Sistani were no longer able to restrain a Shi'ite backlash. As Shi'ites retaliated against Sunnis, a ceaseless cycle of revenge killings ensued.

At home, the Bush administration found itself in an increasingly difficult position. Costly in lives and dollars—with thousands of soldiers and supporting civilians killed and a price tag of over $100 billion per year—the war grew increasingly unpopular among Americans, and the president's approval rating slid. Many opposition politicians, academics, and others spoke increasingly of Iraq as "Vietnam," referring to a previous war that proved untenable and unwinnable for the United States. 344 The Bush administration faced a dilemma: "Stay the course," as President Bush insisted, with about 140,000 US soldiers still on the ground in Iraq, or pull out (as the United States did in Vietnam) and risk Iraq's plunge into further chaos. Fearing anarchy, the Bush administration in 2007 opted for a "surge" in US troops to maintain order, even as its main ally, Britain, drew down its military presence (primarily in Shi'ite southern Iraq). The surge, which appears to have been an effective tactic in assisting Iraqi troops in securing the volatile Anbar province, took the number of US troops in Iraq up to nearly 170,000. By this point in the war, because of its violent tactics, al-Qa'ida in

• Figure 6.55 In February and June of 2006, Sunni militants—possibly belonging to al-Qa'ida in Iraq—bombed the holy site of the Imam al-Askari shrine in Samarra, north of Baghdad (seen here before and after the bombings). Shiite Islam's holiest sites outside Saudi Arabia are in Iraq. Sunni insurgents used the tactic of bombing these sites to sow sectarian strife in Iraq. Now-powerful Sunni ISIS has escalated these tactics to boost its strategic aim of expanding its Caliphate at the expense of Iraq's majority Shi'ites.

Iraq had become deeply unpopular among Iraqis, including the Sunnis who were intended to become al-Qa'ida's main supporters. "Moderate" Sunnis—the so-called "Sons of Iraq"—and US forces joined up to push the militants out of Anbar Province.

As a candidate for the presidential office, Bush's successor Barack Obama promised a speedy withdrawal of US troops from Iraq, effectively handing the country's problems back to the Iraqis. His Iraq policy helped Obama clinch the election in 2008. The troop drawdown began in 2009. In August 2010, President Obama announced that US combat operations in Iraq had ceased. He declared that Operation Iraqi Freedom was over, and that "Operation New Dawn" was beginning. All but some American military "advisors" were to be withdrawn from Iraq after 2011. To keep a foothold in the unstable Gulf region, the US military beefed up its forces in Kuwait. It would be especially vigilant toward Iran, which would surely seek to fill the vacuum left with the withdrawal of US forces.

The Iraq War that began in 2003 is now being viewed through historical lenses. The conflict was seen in a new light after the controversial release in 2010 of 400,000 Wikileaks documents about the war. These documents reinforced and fleshed out the major findings of the official US government investigation into the 9/11 attacks, establishing the following narrative. The Iraq war was a miscalculation by the United States. The claim that there were weapons of mass destruction in Saddam Hussein's Iraq was a costly intelligence failure. The allegation of Saddam Hussein's connection with al-Qa'ida was false. In the wake of 9/11, the Gulf War squandered American lives and military resources in Iraq while the United States lost its focus on Osama bin Laden and al-Qa'ida in Afghanistan and neighboring Pakistan. In invading Iraq, the United States lost most of the international sympathy and goodwill that grew out of 9/11. The United States inadvertently fulfilled many of Iran's ambitions in Iraq. And finally, the United States created enemies in Islamic countries who perceived that the United States was at war with Muslims. The occupation played into the hands of the conspiracy theories created by Osama bin Laden and other terrorists, and it sowed the seeds of the terrorist group ISIS by disempowering Iraq's Sunni Arabs and empowering its Shi'ites. 263 Could the US simply walk away and leave the problems to Iraq and its neighbors? Some analysts said "no," citing the so-called "Pottery Barn" analogy attributed to Secretary of State Colin Powell: you break it, you own it (or fix it).[40] The Obama administration, with Iraqi government urging, elected not to have a "Status of Forces Agreement" that might have left behind 10,000 US troops to help secure fragile Iraq and perhaps deter the spread of ISIS there. 344

About 4400 US soldiers and over 15,000 Iraqi soldiers were killed during the war. More than 100,000 Iraqi civilians died, and about 5 million Iraqis became refugees, with about half fleeing abroad (mainly to countries in the Arab World) and half displaced internally. 76

The Kurds

A mostly Sunni Muslim people of Indo-European origin, the Kurds are the largest of several non-Arab minorities in Iraq and the largest minority group in Turkey. They have long rebelled against Iraqi and Turkish authorities. Kurds today remain the world's largest ethnic group without a country. European powers promised an independent state of Kurdistan at the end of World War I, but the governments concerned never took steps to create it. The Kurdish question—"How are the political wishes of the Kurds to be accommodated?"—has geopolitical implications, with the region occupied by an estimated 28 million Kurds extending from Turkey and Iraq into Iran, Syria, and the southern Caucasus region (see **• Figure 6.56**); an estimated 2 million more Kurds live outside the Middle East.

• Figure 6.56 The Kurds are the largest ethnicity in the world without a country. In Iraq they have created a relatively prosperous de facto autonomous region. They have appealed for more backing from the US and other Western powers to preserve their gains there, especially by repelling ISIS.

In an effort to ward off Kurdish independence, the Kurds' neighbors have long tried to stifle Kurdish culture. The Kurds' political marginalization explains the title of a book about the Kurds entitled *No Friends but the Mountains*.[41] Turkish officials have long downplayed the identity of Kurds (who make up about 20 percent of Turkey's population), often referring to them as "Mountain Turks," and banned or restricted Kurdish-language media in Turkey until 2002. A Kurdish rebellion against Turkish rule, rising in part from the great poverty of the Kurds relative to the rest of Turkey's population, has smoldered and sometimes flared since 1984. The main Kurdish resistance to Turkish rule comes from the **Kurdistan Workers Party**, known by its Kurdish acronym **PKK**.

Under Saddam Hussein, Iraq tried to "Arabize" its Kurds by forcing them to renounce their ethnicity and sign forms saying they were Arabs. However, a no-fly zone and "safe haven" established after Gulf War I put them beyond Saddam Hussein's reach, Iraq's Kurds came to enjoy relative autonomy

Eddie Gerald/Alamy

• Figure 6.57 To take on their ISIS foe, Kurdish women warriors of the YPJ and YPG "People's Protection Units" are seen here recruiting fighters in the city of al Hasakah in the Kurds' de facto autonomous Rojava district in eastern Syria, July 2014. After declaring this area as part of "Western Kurdistan" in 2013, Kurdish leaders pledged to build a society guided by direct democracy, gender equity, and sustainability. Unveiled women fighters epitomize those priorities. According to an editorial in the journal *India Opines*, ISIS is terrified by the Kurds' military successes launched against it from Rojava. Rojava's flag colors are green, red, and yellow.

and prosperity. Even in the tumultuous years after the US-led invasion in 2003, the Kurdish region of the north escaped most of the bloodshed and economic chaos characteristic of the south. Kurdistan became a de facto independent state led by the Kurdish Regional Government (KRG) within Iraq, enjoying strong economic growth based on oil production centered around Kirkuk and on foreign direct investment (FDI). Syrian Kurds have also achieved a measure of de facto autonomy in Rojava, a region of northeastern Syria outside the control of the Syrian government. When ISIS swept across northeastern Syria and into northern Iraq, its strongest opposition came from Kurdish fighters belong to several groups, including the KRG's official military force known as the *peshmerga*, the PKK, and an offshoot of the PKK within Syria known as People's Protection Units, or YPJ, known especially for their fierce women fighters (**• Figure 6.57**).

The Pahlavis, the Ayatollahs, and the Youngsters of Iran

Iran, the easternmost country of the Middle East, was known as Persia until 1935 (see the heartland of the Middle East in **• Figure 6.58**). From the city of Persepolis, Persian kings ruled a great empire in the 5th century BCE. Even after Alexander the Great destroyed the city in 330 BCE, Persia's grandeur soon returned, and Persia was again the center of empires for another 800 years. Persia maintained its culture even after being conquered by Muslim Arabs in the 7th century, and became a center of literature and science in the Middle Ages. Some of the world's great architectural wonders were commissioned by Shah Abbas the Great in 16th-century Isfahan (**• Figure 6.59**).

• Figure 6.58 Political geography of the heartland of the Middle East.

©suronin/Shutterstock.com

• **Figure 6.59** Shimmering over the Naqsh-i Jahan Square in Isfahan, the Jame Abbasi mosque is a masterpiece of Persian architecture. Construction of the Meidan Emam complex, where this mosque stands, began in 1611 under the rule of Shah Abbas Safavi I. These architectural wonders are a UNESCO World Heritage Site.

Persian empires weakened in subsequent centuries as the Russian, British, and Ottoman empires rose around it, and by the beginning of the 20th century, it was a poor, undeveloped country. In the 1920s, a military officer of peasant origin, Reza Khan, seized control of the government and began a program to modernize Iran and free it from foreign domination. He had himself crowned as Reza Shah Pahlavi, the founder of the new Pahlavi dynasty. Influenced by the modernizing efforts of Mustafa Kemal Atatürk in neighboring Turkey, the new shah (king) introduced social and economic reforms, which continued when his young son Mohammed Reza Shah Pahlavi took the throne in 1941. In 1953, Iran attempted to nationalize its oil industry, a move opposed by Britain and the United States, the homes of the powerful companies that owned the lucrative Iranian oil fields. American and British intelligence agencies staged a coup to remove Iran's democratically elected prime minister and to solidify the absolute rule of the shah, who was seen as much friendlier to Western interests. In the 1970s, Iran played a leading role in raising world oil prices. Mounting oil revenues underwrote the shah's expanded program of modernization and Westernization and rapid industrial, urban, and social development that was not always welcomed by Iranian citizens.

In the countryside, the shah's government tried to upgrade agricultural productivity and rural standards of living through land reform and better technologies. Despite the fact that Iran was a cradle of plant and animal domestication, dry and rugged habitats make agriculture difficult over large parts of the country. Agricultural reforms generally failed, and poverty-stricken families from the countryside poured into Tehran and other cities. Along with students, populists, and others, this poor, devoutly Muslim group of new urbanites provided much of the support for the revolution that ousted the secular shah in 1978–1979. Dissent also grew because the regime spent vast sums on its military, mainly for US-made weapons rather than on civilian needs.

Until 1979, Iran and Saudi Arabia were the "twin pillars" of American foreign policy in the Middle East. The revolutionaries that overthrew the shah rejected American interests. The central figure among the revolutionaries was an **ayatollah** ("sign of God") named Ruhollah Khomeini, an elderly critic of the regime whom the shah had forced into exile in 1964. From Paris, Khomeini sent repeated messages to Iran's Shi'ites that helped spark the revolution. A furious tide of religious sentiment rose against corruption, police heavy-handedness, modernization, Westernization, Western imperialism, the growing divide between rich and poor, and the monarchy, which Shi'ite beliefs regarded as illegitimate. Growing numbers of people staged strikes and demonstrations in 1978, forcing the shah to flee the country and abdicate his throne. After hospitalization in the United States and life in exile in Panama and Egypt, the shah died in Cairo in 1981 and is buried there. Ayatollah Khomeini returned in triumph to Iran in 1979. His followers seized and held 52 US diplomatic personnel as hostages in Tehran for 444 days. US President Jimmy Carter proved unable to resolve this crisis, which contributed to his defeat in his 1980 bid for reelection. Authorities in Iran so reviled Carter that they did not release the hostages until the moment his successor, Ronald Reagan, actually assumed the presidency.

Under Khomeini (who died in 1989), Iran became an Islamic republic governed by Shi'ite clerics, who included a handful of revered and powerful ayatollahs as the head of the religious establishment, and an estimated 180,000 priests called **mullahs**. Khomeini served as the Guardian Theologian, an infallible supreme leader enshrined by the principle of **divine rule by clerics** (*velayet-e-faqih*). These religious leaders continue to supervise all aspects of Iranian life and perform many functions allotted to civil servants in most countries. They base their authority on Shi'ite interpretations of Islam (about 90 percent of Iran's 77 million people are Shi'ites). As noted earlier (page 231), Shi'ite Islam embraces the concept of an intermediating clergy, whereas Sunni Islam rejects it.

In contrast to their status during the shah's reign, when they were encouraged to take on larger public roles, women in post-revolutionary Iran lost many freedoms. Under the new regime, "polluting" influences of Western thought and media were purged. However, recent years have seen a general softening of restrictions on personal freedom of expression, and the Internet and other media have introduced broader worldviews to Iran's youth—a very important cohort, as about 55 percent of the population is under the age of 30 (• **Figure 6.60**). Young people led an unprecedented series of prodemocracy protests in 2002 and again in 2009. In subsequent incidents Iranians of different generations took to the streets to protest government hikes on gasoline prices; ironically, this oil-rich country has few oil refineries and needs to import 40 percent of its gasoline. The country is fortunate to have the world's fourth largest reserves of oil (nearly 10 percent of the world's total) and natural gas reserves that are second only to Russia's.

As gas-hungry Europe sought new sources after imposing sanctions for Russia's invasion of Crimea and eastern Ukraine, Iran proposed to fill the gap. But Iran would first have to throw

Majid/Getty Images

• **Figure 6.60** Iranian youth, particularly women, struggle to express their individuality and freedom under strict Islamic controls. In this encounter photographed in 2007, an Iranian policewoman in black garb on the left warns a young woman about her clothing and hair during a crackdown to enforce the Islamic dress code in Tehran. Warnings and arrests are routine during the annual pre-summer crackdown.

Joe Hobbs

• **Figure 6.61** The Hagia Sophia Istanbul is an apt symbol of Turkey as a bridge between East and West, and the civilizations and faiths of Christianity and Islam. It was built in 537 as an Orthodox Cathedral and with the "Fall of Constantinople" in 1453 it was converted to a great mosque. Like the state of Turkey itself it was "secularized" in 1937, and became a museum. Its construction materials came from many places in the Mediterranean Basin and beyond, and include imperial porphyry quarried on high mountainsides in the tribal territory of Egypt's Ma'aza Bedouin. Tracks created by the enormous weight of the wagons carrying porphyry across the Eastern Desert to Nile River ports remain etched on the desert floor.

off the yoke of Western sanctions, and that was possible only through successful negotiations. The most important backstory for the P5+1 talks is that diplomacy at last began to soften the hardened relations between the US and Iran that had existed since 1979—an issue that makes the region's Sunni-dominated countries, especially Saudi Arabia, furious with the US. Recent years have seen a tug-of-war between reform-minded President Hassan Rouhani and the less tolerant supreme religious leader, Ayatollah Ali Khamanei (Khomeini's successor), who has lifetime control over the military and the judicial branches of government. The country's younger and more Western-oriented sectors might win the day, and relations with the wider world—the United States in particular—could improve. The last large "emerging market" economy not integrated into the world economy, Iran may yet rejoin the global community.

Turkey: Where East Meets West

Like Iran, Turkey is a non-Arab demographic heavyweight (population 77 million people) and has long had an enormously important presence in the region (see **Figure 6.58**). The Turks who organized the Ottoman Empire beginning in the 14th century originated as pastoral nomads from central Asia, 164 where Turkic cultures still prevail. From the 16th to the 19th centuries, their empire, based in what is now Turkey, was a major power. Modern Turkey was created from the wreckage of the old empire after World War I. Its founder, Mustafa Kemal Atatürk, was determined to Westernize the country, raise its standard of living, and make it a strong and respected nation-state. He inaugurated social and political reforms designed to break the hold of traditional Islam and open the way for modernization and Turkish nationalism.

Islam had been the state religion under the Ottoman Empire, but Atatürk divorced church from state, and to this day Turkey is the only Muslim Middle Eastern country to officially separate them (• **Figure 6.61**). The wearing of the red cap called the *fez,* an important symbol of piety under the Ottoman caliphs, was prohibited, and state-supported secular schools replaced the all-pervasive religious schools. To facilitate public education and remove further traces of Islamic culture, Latin characters replaced the Arabic script. Slavery and polygamy were outlawed, and women were given full citizenship. Legal codes based on those of Western nations replaced Islamic law, and forms of democratic representative government were instituted, although Atatürk himself ruled in a dictatorial fashion.

Turkey's economic standing is near the top of the LDCs, close to MDC status. Not long ago, poverty and prominently rural characteristics marked Turkey. But the country embarked on a course of change that is modernizing agriculture, expanding industry, and raising standards of living. Its economic growth has been impressive. The value of output from manufacturing (mainly automobiles, televisions, textiles, metal industries, and weaponry) is greater than that from agriculture, which is considerable in this fertile, well-watered country.

Since the 1970s, Turkey has been engaged in a massive project to create a series of hydroelectric dams along the Tigris, Euphrates, and other rivers. Called the Southeastern Anatolia Project (known by its Turkish acronym as GAP), the dams that are its centerpiece have greatly increased Turkey's hydroelectric power, and the reservoirs that have filled upstream from them are feeding new irrigation lines that are expanding the amount of agricultural land in Turkey's most arid region. The project's

Joe Hobbs

• Figure 6.62 Hasankeyf, on the Tigris River. Important archaeological sites adjacent to the river here are to be inundated by a reservoir that will form behind a dam that is part of the Southeast Anatolia Project.

greatest technical feat is the Atatürk Dam on the Euphrates River: it generates about 5 percent of Turkey's electricity, and it irrigates the previously dry Harran Plain, which now produces over $1 billion of agricultural products annually. Long-term plans call for a total of 22 dams to be built in this region. Turkey hopes GAP will lead to increased industrialization in its poor southeastern area and create over 1 million jobs. There is an important political motivation here, as the beneficiaries should be the long-marginalized Kurds who are the ethnic majority in this region. The newly productive agricultural lands should provide big boosts to Turkey's wheat and cotton crops.

Like most great dam projects, including the Aswan Dam described earlier, GAP is controversial. It is being developed in the part of Turkey where a poor and restive Kurdish population is seeking development, human rights, and autonomy. Many Kurds do not have confidence in government promises that the project will boost their prospects and are actively campaigning against the project. Tens of thousands of Kurds have been relocated from areas inundated by reservoirs. In addition, Turkey's Tigris and Euphrates waters flow downstream into Syria and Iraq, raising serious disputes about downstream water allocation and quality (**• Figure 6.62**).

Is Turkey European?

Turkey is an "in-between" country. Economically, it is near the line between MDCs and LDCs. Culturally, it is between traditional Islamic and secular European ways of living. Part of Turkey—the section called Thrace, which includes the enormous city of Istanbul—is actually in Europe, and Turkey's institutions aspire to become more European. It has been a member of NATO since 1952. It is an associate member of the European Union and applied for full membership in 1987. The European Union, however, insists that it would first like to see Turkey become more European politically, for example, by diminishing the army's influence in government and by improving its human rights record. Turkey insists it has made huge strides toward satisfying the minimal democratic and human rights norms for EU member countries. It has abolished the death penalty and eased restrictions it imposed on the cultural expressions of its minority Kurdish population.

Turkey began membership negotiations with the EU in 2005, but the talks have repeatedly stalled. The EU argues that Turkey has still not gone far enough with democratic and human rights reforms. The Europeans probably fear the costs of absorbing Turkey's mainly low-income population into the union—a fear that may be justified, considering the EU's struggles with its own lower income countries like Greece, which have threatened the very existence of the EU. The Republic of Cyprus has blocked further EU negotiations with Turkey over the issue of Turkey's support for the breakaway Northern Cyprus region.

Many Turks feel that their membership in the European Union was denied because they are Muslims. It is indeed possible that behind the scenes, EU decision makers rejected EU membership for Turkey because they feared that Islamist terrorists would easily make their way from an EU Turkey into the heart of continental Europe, especially if Turkey were to join passport-free Schengenland. The European Union's rejection of Turkey contributes to the sentiment in the Islamic world that the West is waging a systematic campaign against Muslims. Turkish public opinion has become decidedly more anti-European, even as the prospect of joining the troubled European Union, with its sovereign debt crises and other problems, looks increasingly unattractive. A Turkey shunned by Europe may seek alliances eastward in Russia, India, and China.

...Or Is Turkey a Great Middle Eastern Power?

Turkey's standing in Middle Eastern affairs has changed dramatically in recent years. Not long ago, diplomatically and otherwise, the Arabs reviled Turkey, just as they had for centuries (remember that Turks are not Arabs, and the Ottoman Empire that subjugated much of the Middle East from the 16th century until World War I was Turkish). Turkey's closest ally in the Middle East was Israel, which is also reviled in much of the Arab world. That association made Turkey deeply unpopular among Arabs.

A single incident dramatically rearranged Turkey's regional priorities and its standing in the Arab Middle East. Israel has a blockade that prohibits any unauthorized access to the Palestinian-controlled Gaza Strip. On May 10, 2010, a "Gaza Freedom Flotilla" of ships carrying humanitarian aid and construction materials to the Gaza Strip was confronted at sea by Israeli commandoes. The Israeli forces boarded several of these ships belonging to the Free Gaza Movement and the Turkish Foundation for Human Rights and Freedoms and Humanitarian Relief. An altercation ensued, and a number of Turkish civilians as well as Israeli commandoes were killed. Turkey condemned Israel and demanded that Israeli apologize, pay compensation, and prosecute those responsible for the Turkish fatalities. Israel refused.

The incident accelerated an ongoing reshuffling of the Middle East's political landscape. Turkey's Prime Minister Recep Tayyip Erdoğan denounced the raid as Israeli "state terrorism." Turkey recalled its ambassador to Israel, suspended military trade with Israel, and otherwise indicated that Israel was no longer an ally. Turkey had long promoted a national policy of "zero problems with the neighbors," but Erdoğan now portrayed Israel as a regional problem and offered himself and Turkey as champions of Arab causes. Amidst the tumult of the Arab Spring in 2011, Turkey appealed to the United States to recognize it, rather than Israel, as the best ally of the US in the Middle East. Erdoğan visited the countries most transformed by the Arab Spring—Egypt, Tunisia, and Libya—and vowed to assist in their reformations. While the Muslim Brotherhood was in control of Egypt, he spoke of a new axis of power, based on Turkey's alliance with Egypt, the Arab World's strongest country. He indicated that Turkey would use military force if needed to advance the interests of its Arab allies and of itself—for example, in preventing the Republic of Cyprus from working with Israel to tap oil and gas reserves in the Mediterranean Sea. Turkey had close ties with gas-rich Qatar, and these two powers threw all their weight behind Egypt's Muslim Brotherhood. When the Brotherhood was ousted from power, Turkey retreated to the sidelines of Arab affairs.

Prime Minister (later President) Erdoğan shook up Turkey domestically. After coming to power in 2003, he began restraining the powers of the military. Turkey's constitution invests the military with the responsibility of protecting the country's secular government, and the army has used its powers to overthrow four elected governments since 1960. Erdoğan's growing power and popularity blunted any prospect that he would be ousted by the military. Turkey's economy improved markedly under his rule—most notably, it became a powerhouse of automobile manufacturing—and trade, transport, and communication with the Arab countries accelerated. As the leader of secular Turkey, Erdoğan has had to restrain his personal convictions. Prior to his election, he had described himself as a political Islamist and an advocate of Islamic law, or *shari'a*. Some analysts worry that he will violate Turkey's rigid separation of church and state, and advocate Islamist domestic and foreign policies.

As a NATO member with the region's largest army, Turkey was in the best position to advance the American-led coalition against ISIS. Initially, Turkey did not volunteer its forces in the fight and did little to slow the flow of Western recruits to ISIS through Turkey into Syria. Turkey did allow more than 1.7 million refugees from the crisis in Syria to cross its borders and settle in camps maintained by the Turkish government (see **Figure 6.52**). Not wanting to be overrun by ISIS, Turkey changed positions, allowing US warplanes to operate from its airbases, and using its own to strike ISIS targets inside Syria. In return, Turkey wanted to escalate its campaign against Kurdish separatists without condemnation from the West.

The Gulf Oil Region

GIS Helps Turn an Arabian Mirage into Reality

A futuristic city is taking shape on the desert sands of Abu Dhabi, one of the United Arab Emirates. An "arcology" project (a portmanteau combining architecture and ecology) called **Masdar**, meaning "the Source" in Arabic, this city has the extraordinary goal of being "carbon-neutral." That means that unlike any other settlement in the world, Masdar will minimize the carbon dioxide it produces and neutralize or offset the small emissions it does make (**•Figure 6.63**). Automobile use will be restricted, and the city's 45,000–50,000 residents will be required to use low-emissions public transport and clean-energy vehicles. Residents will work in nearby Abu Dhabi city or in the clean technology companies based in Masdar itself. Abu Dhabi's government is bearing the $20 billion price tag to build the city, which is supposed to become a magnet for domestic and foreign investment in high-tech

• Figure 6.63 Artist's rendition of Masdar, the carbon-neutral city under construction in a place built on fossil carbon fuels: Abu Dhabi in the United Arab Emirates.

THINK CRITICALLY: Does it make sense for a country dependent upon fossil fuel exports to research and build carbon-free infrastructure? Does Masdar portend impactful innovation, or is it a vanity project of a rich nation?

(a)

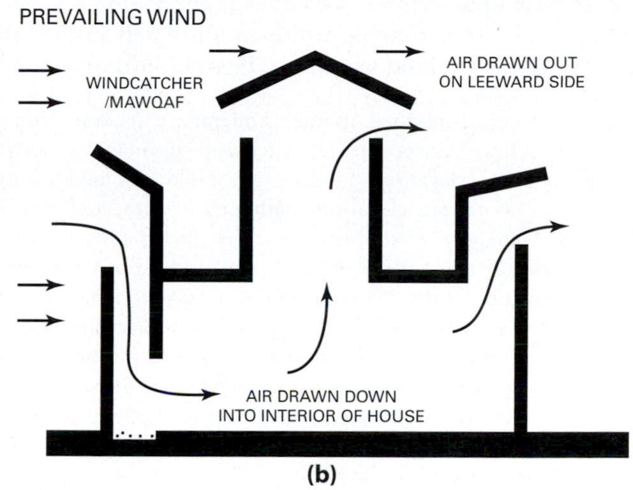

(b)

• **Figure 6.64** **(a)** Wind-catchers on the skyline of Yazd, Iran, and **(b)** a schematic of how they work. Known as *mawqaf* in the Arab countries and *badgir* in Iran, these superb features of traditional architecture served as natural air-conditioners around the Persian Gulf. Most have been replaced by modern systems. A few remain as valued heritage, and some new cities such as Masdar in the UAE pay tribute to them.

research and development firms. Masdar's first residents are students and faculty of the Masdar Institute, an educational center staffed by professors from the Massachusetts Institute of Technology (MIT). Its graduates are meant to help meet the world's growing demand for scientists and visionaries with knowledge of green technology, green cities, "smart" cities, and renewable energy. Abu Dhabi won a worldwide competition to headquarter the International Renewable Energy Agency (IRENA) at Masdar.

Energy for Masdar City is to be provided by renewable resources, especially photovoltaic energy, a solar tower, and other techniques of solar power that, as you might imagine, has abundant potential there in Arabia. Wind farms, geothermal energy, and the world's largest hydrogen power plant will contribute energy as well. Masdar will be zero-waste, with its waste products reused, recycled, or turned into yet more energy. Water will be provided by desalination of the very salty Gulf waters, an energy-intensive process that will also use solar power. Eighty percent of the water will be recycled, and gray water—water that has been used but does not contain human waste—will be used for irrigation.

Masdar is not entirely futuristic and transplanted; several of its arcology features are based on an understanding of historic adaptations to heat in Gulf region medinas. Those medinas countered the usual "urban heat island" effect of modern cities, where asphalt, glass and steel, and carbon emissions make the city warmer than adjacent rural areas. Narrow and short streets, no longer than 70 meters, helped maximize shade and move air up and away from street level. A unique device called a "wind-catcher," or *mawqaf*, directed air down into the bowels of homes and public buildings, serving as a natural air conditioner (• **Figure 6.64**). Masdar has incorporated these features, including a 45-meter-high *mawqaf*, making the city 27–36°F (15–20°C) cooler than the desert surrounding it.

In 2007, I was employed by a foundation in the UAE to come up with the first draft of an environmental master plan for the country. I visited each of the seven emirates and spoke with government officials, environmental nongovernmental organizations, CEOs, foreign workers, citizens, and others, to identify the country's environmental challenges, and then proposed specific methods for dealing with them. One of the most memorable visits was with Masdar's Chief Executive Officer Dr. Sultan al-Jaber. While we sipped tea, Dr. al-Jaber spoke at length about his vision of the futuristic city. As I prepared to leave, he said, "Come with me—I have something to show you." He led me to a conference room and opened the door. Inside, seated around a long table, were about 20 persons: engineers, urban planners, GIS technicians, and professors from a variety of fields at MIT. They were on a site visit to see whether they would commit to a five-year leave from MIT, relocating to Abu Dhabi with their families to help build Masdar. You know the expression "the smartest guy in the room"; this was a room full of the smartest guys in the room, and Dr. al-Jaber needed them to turn his zero-carbon vision into reality.

GIS has an important role in Masdar's urban planning and management. Using ESRI's ArcGIS software (see **page 17**), Masdar's GIS crew was able to carefully consider the city site's geography of sun angles, wind patterns, street widths, building density, and building heights. "Building a city like this has never been done before," GIS consultant Shannon McElvaney said, "and GIS is proving to be an absolutely critical tool."[42] ESRI's Derek Gliddon wrote of GIS and Masdar:

> Data layers contained in the geodatabase include information such as transportation, vegetation, drainage, structures, boundaries, elevation, biodiversity, buildings, and utilities, as well as terrain elevation, bathymetric data, and remotely sensed imagery. A sophisticated Web browser–based virtual city visualization and navigation tool is used to visualize the construction of the city over time. Construction managers can navigate anywhere in the city; "play" the project timeline; and identify spatiotemporal clashes, accessibility problems, and other logistical issues. On a fast-paced, high-density development, these issues are very important. ArcGIS introduced the spatial analysis and modeling necessary for the most efficient placement of facilities at the city. Water and sewage treatment plants, recycling centers, a solar farm, geothermal wells, and plantations of various tree species were placed using traditional planning principles modeled with ArcGIS. Questions—Is there enough physical space available? How much are the buildings shading each other? How much space is needed between a facility and the residents?—are modeled, and the best answer chosen through GIS.[43]

The global financial crisis derailed the city's ambitious construction schedule. Masdar opened for settlement a few years late and is now slated for completion by 2025. It won't be cheap to live there. Some critics have decried Masdar as a luxury development and the ultimate gated community. As a learning and living environment, it does have some global counterparts, including Japan's Tsukuba Science City and Saudi Arabia's King Abdullah University of Science and Technology. Whatever you call it, there is no place quite like Masdar.

This concludes our journey through the Middle East and North Africa. We now turn eastward into the realm of South and East Asia.

Study Guide

Summary

- The "Middle East" is a Eurocentric vernacular region developed by the British, who along with the French in the Sykes-Picot Agreement created many of the region's boundaries after defeating the Ottoman Turks in World War I.
- Misleading stereotypes about the environment and people of the Middle East and North Africa are common, as people outside the region often associate the region solely with military conflict and terrorism.
- The region has bestowed on humanity a rich legacy of ancient civilizations and the three great monotheistic faiths of Judaism, Christianity, and Islam.
- Middle Easterners include Jews, Arabs, Turks, Persians, Pashtuns, Berbers, people of Sub-Saharan African origin, and other ethnic groups who practice a wide variety of ancient and modern livelihoods.
- The Middle Eastern "ecological trilogy" consists of peasant villagers, pastoral nomads, and city dwellers. The relationships among them have been mainly symbiotic and peaceful, but city dwellers have often dominated the relationship, and both pastoral nomads and urbanites have sometimes preyed on the villagers, who are the trilogy's cornerstone.
- Arabs are the largest ethnic group in the Middle East and North Africa, and there are also large populations of ethnic Turks, Persians (Iranians), and Kurds. Islam is by far the largest religion. Jews live almost exclusively in Israel, and there are minority Christian populations in several countries.
- Population growth rates in the region are moderate to high. Sixty percent of the region's people are less than 25 years old. The number of youth (people ages 15–24) in the region is estimated to grow to 100 million in 2035. This "youth bulge" is a huge challenge for the region's development, and has played a major role in the Arab Spring and general discontent.
- Oil wealth is concentrated in a handful of countries, and as a whole, this is a developing region.
- The Middle East has served as a pivotal global crossroads, linking Asia, Europe, Africa, and the Mediterranean Sea with the Indian Ocean. These countries have historically been unwilling hosts to occupiers and empires originating far beyond their borders.
- The margins of this region are occupied by oceans, high mountains, and deserts. The land is composed mainly of arid and semiarid plains and plateaus, together with considerable areas of rugged mountains and isolated "seas" of sand.
- Aridity dominates the environment, with at least three-fourths of the region receiving less than 10 inches (25 cm) of yearly precipitation. Great river systems and freshwater aquifers have sustained large human populations.
- Many of the plants and animals on which the world's agriculture depends were first domesticated in the Middle East.
- Judaism, Christianity, and Islam are Abrahamic faiths that have coexisted rather peacefully, with political events of the

past century bringing them to blows, including over sacred places in Jerusalem.

- The split between Sunni and Shi'ite Muslims developed because Prophet Muhammad named no successor to take his place as the leader (Caliph) of all Muslims. Some of his followers argued that the person with the strongest leadership skills and greatest piety was best qualified to assume this role. These followers became known as Sunni, or orthodox, Muslims. Others argued that only direct descendants of Muhammad could qualify as successors. They became known as the Shi'ites. Three of the region's countries, Iran, Iraq, and Bahrain, have Shi'ite majority populations. There are significant minority populations of Shi'ites in Yemen and Lebanon.
- Many Sunni Muslims revile Shi'ites as heretical, especially because of their iconographic tendencies. Shi'ites seek the intercession of clerics, who are empowered to interpret God's will for them, whereas the Sunnis prize a direct and personal relationship with God.
- Powerful currents of Sunni "fundamentalism" flow throughout the region and beyond. "Salafists" adhere to an interpretation of Islam they believe is closest to the faith's earliest tenets and social norms and that most correctly follows *shari'a*, or Islamic, law. Islamists favor reordering government and society in accordance with *shari'a*, but not all of them are terrorists, meaning people who kill noncombatants.
- Since World War II, several international crises and wars have been precipitated by events in the Middle East. Strong outside powers depend heavily on this region for their current and future industrial needs. Known as the "Carter Doctrine," unimpeded access to Persian Gulf oil through military force if needed is one of the pillars of US foreign policy.
- The region of the Middle East and North Africa is characterized by a high number of chokepoints, strategic waterways that may be shut off by force, triggering conflict and economic disruption.
- Oil pipelines in the Middle East are routed both to shorten sea tanker voyages and to reduce the threat to sea tanker traffic through chokepoints, but are themselves vulnerable to disruption.
- Access to freshwater is a major problem in relations between Turkey and its downstream neighbors; Egypt and its upstream neighbors; and Israel and its Palestinian, Jordanian, and Syrian neighbors.
- Al-Qa'ida and affiliated Islamist terrorist groups aim to drive the United States and its allied governments from the region and to replace them with an Islamic Caliphate. Al-Qa'ida is an apocalyptic group that seeks to inflict mass casualties on its enemies, particularly on Americans in their home country.
- ISIS is another apocalyptic group that developed in Syria and Iraq in the wake of the 2003 US invasion of Iraq and subsequent withdrawal from the region. Unlike al-Qa'ida, ISIS targets include Shi'ite Muslims, and ISIS persecutes and executes a wide range of enemies. It has created what it calls the "Islamic Caliphate" in the heartland of the Middle East and with effective PR and military tactics has greatly expanded its area of control and range of influence.
- The United Nations Partition Plan of 1947 attempted a two-state solution to the dilemma Britain had created by promising land to both Arabs and Jews in Palestine. It envisioned geographically fragmented states, making each side feel vulnerable to the other. War prevented the plan's implementation. The Arab-Israeli conflict has continued from that time, with Israel gaining more territory and the indigenous Palestinians failing to acquire a country of their own. Principal obstacles to peace between Israelis and Palestinians are the status of Jerusalem, the potential return of Palestinian refugees, the future of Jewish settlements in the Occupied Territories, and Israeli construction of a security barrier that penetrates into portions of the Palestinian West Bank. Israel's eastern neighbor, Jordan, has a majority Palestinian population.
- The Arab Spring of 2011 was a wave of revolutions against autocratic regimes across the region, notably in Tunisia, Egypt, Libya, Syria, Bahrain, and Yemen. It has been followed by an "Arab Fall" that has seen even more authoritarianism and numerous civil wars and international entanglements.
- Masdar is a "carbon-neutral" city under construction in Abu Dhabi, one of the United Arab Emirates. The neighboring emirate of Dubai has capitalized on its geographical situation to become a hub for aviation, financial services, and other profitable ventures beyond fossil fuel exports. Dubai's wealth has transformed its landscape and seascape with towering buildings and with artificial islands threatened by poor design and rising sea levels.
- Two-thirds of the world's proven petroleum reserves are concentrated in a few countries that ring the Persian Gulf. Saudi Arabia controls more than one-fifth of the world's oil.
- Under Saddam Hussein, Iraq squandered its oil wealth on military misadventures, including invasions of Iran and Kuwait. The Kuwait invasion led to an enormous US-led counterattack that devastated the country's infrastructure and subjected it to years of economic sanctions. The US-led invasion of Iraq in 2003 resulted in Saddam Hussein's downfall and death, the eventual rise of ISIS, and an uncertain future for this ethnically diverse country (with large Shi'ite Arab, Sunni Arab, and Kurdish populations) engaged in civil war. The rise of ISIS in Iraq and Syria has presented the US with extremely difficult strategic challenges.
- Iran's oil revenues were used to modernize and Westernize the nation during the Pahlavi dynasty. Rapid social change and

uneven economic benefits precipitated a revolutionary Islamic movement, which forced the shah to abdicate and flee in 1979. Iran's conservative, theocratic Shi'ite government pervades all aspects of everyday life. Iran's young population is agitating for change, but the ruling clerics are resisting.

- Iran has had difficult relations with the United States, most recently over its alleged nuclear weapons program. US options for dealing with Iran's nuclear ambitions are limited: strengthen sanctions, strike militarily, or negotiate a settlement. The US and five other countries settled for the third option signing the Joint Comprehensive Plan of Action. Iran's Sunni Arab neighbors, including Saudi Arabia, are concerned about the US strengthening Iran's hand at their expense.
- Turkey was formerly the seat of power for the Islamic Ottoman Empire. It is a secular nation with membership in NATO and institutionally aspires to join the European Union. Turkey is at odds with the US over what to do about ISIS and for a time did not actively join the US-led coalition fighting ISIS. Turkey is blessed with freshwater and good soil resources, and with its GAP project has dammed Tigris and Euphrates watersheds, aiming to create the "breadbasket of the Middle East."

Review Questions

- What countries constitute the Middle East and North Africa? Which are the three most populous? What economic resources do the least and most populous countries have, and how do internal growth and international migration affect the countries' development?

1. What are the major climatic patterns of the Middle East? Where are the principal mountains, deserts, rivers, and areas of high rainfall?
2. Why can this region be described as a culture hearth? What major ideas, commodities, and cultures originated there?
3. What are the major ethnic groups, and in which countries are they found? What is an Arab? A Jew? A Turk? A Kurd? A Persian? A Muslim?
4. What are the components of the Middle Eastern ecological trilogy, and how do they interact? Which component is clearly the "cornerstone" of the trilogy, and why? What is the geographic pattern of the classic Middle Eastern medina?
5. What are the principal beliefs and historical geographic milestones of Jews, Christians, and Muslims? What traditions and beliefs do they share?
6. What is the difference between Shi'ite and Sunni Islam? How and where do these differences play out in political and other aspects of life in the region?
7. What are "Islamist" movements? Who are *jihadists*? What groups are clearly terrorists? What are Hizbullah, Hamas, al-Qa'ida, and ISIS, and what are they seeking and doing?
8. Where is oil concentrated in this region?
9. What is the Carter Doctrine?
10. What are upstream and downstream countries, and which are usually the more powerful? What are the exceptions to this pattern? What are the hydropolitical issues of the Nile basin? According to Lester Brown, what three steps can be taken to avert conflict over Nile waters?
11. What precipitated the various conflicts between Israel and its Arab neighbors? How did each of these conflicts rearrange the political map?
12. What lands did the peace process of the Oslo Accords yield to Palestinian control? What is the current situation in those areas? Who controls the Gaza Strip and the West Bank? What steps has Israel taken to ensure its security? What problems are associated with Jewish settlements in Palestinian territories?
13. What are the pros and cons of Egypt's Aswan High Dam?
14. What are virtual water and water footprints, and why are these important in this region?
15. What were essential elements of the Arab Spring, especially in Egypt, and what happened in the "Arab Fall?" What country evolved as the most successful after its Arab Spring uprising?
16. What is the ethnic composition of Iraq? How did that consideration cause the United States to avoid invading the country in 1991?
17. What happened when Saddam Hussein's troops invaded Kuwait? What happened when they invaded Iran? Why did the US invade Iraq in 2003, and what do historians now say about subsequent events there and in Syria? What is the "Pottery Barn" analogy of American responsibilities in Iraq? How is ISIS related to the war that began in 2003?
18. Where do Kurds live, and what are their prospects for autonomy or independence?
19. What were some of the grievances against the Pahlavi dynasty that led to revolution in Iran? What changes has the Islamic regime brought to the country?

20. What three options has the United States weighed in response to Iran's nuclear weapons program? Who are the "P5+1" and what is their role?
21. In what ways is Turkey an "in-between" country, and how is Turkey's role in the region and in the international arena changing? What was the Armenian Genocide and Turkey's role in it?
22. What are the objectives of the Masdar project?

Key Terms + Concepts

Notes

1. This quote, while most often attributed to Bierce, has also been attributed to Mark Twain, and its origins are murky; see http://quoteinvestigator.com/2014/05/19/geography/.
2. Ulrich Deil, Heike Culmsee, and Mohamed Berriane, "Sacred Groves in Morocco: A Society's Conservation of Nature for Spiritual Reasons," *Silva Carelica*, 49 (1980): 185–201.
3. An outstanding guide to the sacred places of Jerusalem, which includes this note on the Foundation Stone, is Martin Lev, *Traveler's Key to Jerusalem* (New York: Knopf, 1989).
4. "What Is Sunnah?" accessed March 7, 2015, from http://www.islaamnet.com/whatissunnah.html. See also http://www.bible-quran.com/sunna-hadith/.
5. Trevor Stanley, "Understanding the Origins of Wahhabism and Salafism," *Terrorism Monitor*, 3(14), July 15, 2005, accessed February 9, 2015, from http://www.jamestown.org/programs/tm/single/?tx_ttnews%5Btt_news%5D=528&#.VNpxpsbWtZU.
6. Abby Ohlheiser, "The Associated Press' New Definition of 'Islamist,'" *The Slatist*, April 5, 2013, accessed from http://www.slate.com/blogs/the_slatest/2013/04/05/_islamist_definition_changed_in_the_ap_stylebook_two_days_after_illegal.html.
7. Sudan and Ethiopia are weaker powers than Egypt, but they are upstream and have their hands on the tap. Egypt does not want military or diplomatic confrontations with Sudan that might lead an indignant Sudan to reduce the flow of Nile waters. You may want to keep your eyes on the troublesome issue between them involving the disputed "Hala'ib Triangle" east of Aswan (see **Figure 6.2**). In 1899, when Britain ruled both territories, it drew a political border along the 22nd parallel but three years later allowed Sudan to administer the triangular area north of this line. This helped accommodate the grazing needs of camel-herding Beja nomads who regularly range north of the political border, and disputes over the Halaib Triangle were minor. But now interests in exploiting the gold, petroleum, and tourism resources in the triangle has made competing claims of sovereignty a serious and potentially explosive issue that could spill over into the much more serious hydropolitical realm.
8. Lester Brown, "When the Nile Runs Dry," *New York Times*, June 2, 2011, p. A23.
9. Fouad Ibrahim and Barbara Ibrahim, *Egypt: An Economic Geography* (London: I.B. Tauris, 2003).
10. For more on virtual water, including how to get an interactive app to measure it for yourself, visit the Virtual Water Network at http://virtualwater.eu. For more on the water footprint, see http://www.waterfootprint.org/?page=files/Glossary. For the Global Footprint network, see http://www.footprintnetwork.org/en/index.php/GFN/page/earth_overshoot_day/.
11. An excellent source on the geographic distribution of drone strikes is http://drones.pitchinteractive.com.
12. From the US State Department website at http://www.terrorismanalysts.com/pt/index.php/pot/article/view/33/html.
13. Joseph J. Hobbs, "The Geographic Dimensions of Al-Qa'ida Rhetoric," *Geographical Review*, 95(3), 2003: 301–327; special issue on New Geographies of the Middle East.
14. Ayman Al-Zawahiri. "Why Attack America," in *Anti-American Terrorism and the Middle East*, Barry Rubin and Judith Colp Rubin, eds. (Oxford University Press, 2002), p. 133.
15. Rohan Gunaratna, *Inside al Qaeda: Global Network of Terror* (New York: Cambridge University Press, 2002), p. 93.
16. Op. cit., Gunaratna, op. 11.
17. Osama bin Laden, "Declaration of War (August 1996)," in *Anti-American Terrorism and the Middle East,* Barry Rubin and Judith Colp Rubin, eds. (Oxford University Press, 2002a), pp. 137, 139; Osama bin Laden, "Statement: Jihad against Jews and Crusaders" (February 23, 1998), in *Anti-American Terrorism and the Middle East,* Barry Rubin and Judith Colp Rubin, eds. (Oxford University Press, 2002b), p. 149; Osama bin Laden, "Al-Qa'ida Recruitment Video (2000)," in *Anti-American Terrorism and the Middle East*, Barry Rubin and Judith Colp Rubin, eds. (Oxford University Press, 2002c), p. 174.
18. Op. cit., Bin Laden, 2002b, footnote vii, p. 150.
19. Op. cit., Gunaratna, 2002, footnote ii, pp. 55, 89.
20. Graeme Wood, "What ISIS Really Wants," *The Atlantic,* 315(2), March 2015, pp. 78–94.
21. Ibid.
22. "Hezbollah Aims to Keep Lebanon as Part of 'Axis of Resistance,'" *Al-Monitor*, August 20, 2012, http://www.al-monitor.com/pulse/tr/politics/2012/08/lebanon-after-assad-as-seen-by-h.html#.
23. http://www.washingtonpost.com/politics/assad-must-go-obama-says/2011/08/18/gIQAelheOJ_story.html.
24. http://abcnews.go.com/blogs/politics/2013/08/president-obamas-red-line-what-he-actually-said-about-syria-and-chemical-weapons/.
25. United Nations Development Programme Report 2002. http://www.arab-hdr.org/publications/other/ahdr/ahdr2002e.pdf.
26. Fareed Zakaria, "Why Democracy Took Root in Tunisia and Not Egypt," *Washington Post*, October 30, 2014.
27. Alan Kuperman, "Lessons from Libya: How Not to Intervene," policy brief, Belfer Center for Science and International Affairs, Harvard Kennedy School, September 2013, http://belfercenter.ksg.harvard.edu/publication/23387/lessons_from_libya.html.
28. Daveed Gartenstein-Ross, "The Consequences of NATO's Good War in Libya," Foundation for the Defense of Democracies, May 8, 2014.
29. Op. cit., Zakaria, October 30, 2014.
30. https://freedomhouse.org/sites/default/files/01152015_FIW_2015_final.pdf.
31. Hisham Melham, "Barbarians at the Gate," *Politico*, September 18, 2014, accessed March 11, 2014, from http://www.politico.com/magazine/story/2014/09/the-barbarians-within-our-gates-111116.html#.VQA7NsbWtZV.
32. "Syria: The story of Conflict," BBC, November 11, 2014, http://www.bbc.com/news/world-middle-east-26116868.
33. Foundation for Defense of Democracy, 2014 Washington Forum, http://www.defenddemocracy.org/events/washington-forum-2014/.
34. Musa Cerantonio, Facebook page, accessed February 20, 2015.
35. "ISIL's Persecution of Religious Minorities in Iraq and Syria," testimony by Tom Malinowski, Assistant Secretary, Bureau of Democracy, Human Rights, and Labor, prepared as Opening Statement for House Foreign Affairs Subcommittees on Africa, Global Health, Global Human Rights and International Organizations and the Middle East and North Africa, Washington, DC, September 10, 2014.
36. Op. cit., Graeme Wood.

37. You can view this remarkable video at https://news.vice.com/video/the-islamic-state-full-length.
38. General Martin Dempsey in testimony on Capitol Hill, September 6, 2014.
39. Richard N. Haass, "The New Thirty Years War," Project Syndicate, July 21, 2014, https://www.project-syndicate.org/commentary/richard-n--haass-argues-that-the-middle-east-is-less-a-problem-to-be-solved-than-a-condition-to-be-managed.
40. The "Pottery Barn" debate is discussed in http://www.nytimes.com/2004/10/17/arts/17iht-saf18.html?_r=0.
41. John Bullock, *No Friends but the Mountains* (London: Oxford University Press, 1993).
42. Derek Gliddon, "Building an Oasis in the Desert," *Arc News*, 31(3), Fall 2009, 26–27.
43. Ibid.

Global Geoscience Watch

Go to the GREENR database and do a "Basic Search" on *Syrian War Refugees*. Take note of the dozens of news stories dealing with this humanitarian crisis. Next, review Figure 6.2 and write down the names of the five countries that have borders with Syria. Scan the news articles in more detail and develop a short report that includes as much information as you can gather to indicate how neighboring countries are responding to the crisis. What role is the United Nations playing? Feel free to expand your research beyond these news articles.

Online Resources

For access to MindTap and additional study materials visit www.cengagebrain.com. Read your textbook, take notes, complete activities, take practice quizzes and more.

Above: Boats on the New Year pilgrimage to the Perfume Pagoda, Vietnam. Joe Hobbs
Left: South and East Asia at night. Note especially (1) how the Himalaya separate the populous Indian subcontinent from the Tibetan wilderness; (2) the belts of urban-industrial activity in southeast China and in Japan; and (3) the contrast between the two Koreas. C. Mayhew & R. Simmon (NASA/GSFC), NOAA/NGDC, DMSP Digital Archive

7 South and East Asia

If a man takes no thought about what is distant, he will find sorrow near at hand.[1]

—CONFUCIUS (551–479 BC)

South and East Asia make up a great triangle that runs from Afghanistan in the west to Japan in the northeast and the island of New Guinea in the southeast (**•Figures 7.1, 7.2**). In this land area of about 8 million square miles (21 million sq km—a little more than twice the size of the United States [**•Figure 7.3**]), or less than one-quarter of the Earth's landmass, lives more than half of the world's population. The cultural makeup of these people is as diverse as the physical environments they live in. The world's highest mountain peaks, some of the longest rivers, and the Earth's most highly transformed and densely settled river plains are the settings for some of the oldest civilizations and the most modern economies. Just as this region has played a major role in human development over thousands of years, it is a major force in world affairs in the 21st century. Some experts say this will be known as the **Asian Century** for the ways in which the region will reshape the globe.

Chapter Outline

Chapter Objectives

This chapter will enable you to

- Appreciate China and India as the demographic and economic giants of early 21st-century Asia.
- Understand the geopolitical dimensions of the tensions between India and Pakistan, North Korea and the West, and Islamists and governments in Pakistan, and the factors that have given rise to regional insurgencies, including their relationships with faith and economic well-being.
- Appreciate how the partition of India and other legacies of colonialism created lasting problems in political relations, resource use and allocation, and industrial development.
- Learn about the cultures associated with rice, and the balance between precipitation, soils, crop varieties, and populations that have consistently allowed production to meet rising demand in the region.
- Appreciate the catastrophic reach of tsunamis and some of the region's other natural hazards.
- Consider the economic discrepancies between China's rural and urban populations, the forces behind large-scale rural-to-urban migration, and the urban planning that promises to address imbalances.
- Appreciate that in Korea different political and economic systems have produced dramatically different results for almost identical peoples.

The countries of eastern Eurasia were at one time referred to as "the Orient," from a Latin word meaning "facing the east" or "facing the sunrise." (The West was then known as the Occident, derived from a word meaning "facing the sunset.") But this terminology reflected the view from Europe and became associated with Western stereotypes and misconceptions about Asia. These terms have fallen out of favor, but in older literature you are sure to read about the Orient and its connotations of the exotic.

7.1 Area and Population

South and East Asia encompass the following subregions, which in some geography books are classified as separate regions (•**Table 7.1**; see also **Figures 7.1 and 7.2**):

- East Asia, which includes Japan, North and South Korea, China, Mongolia, Taiwan, and many near-shore islands
- South Asia, which includes Afghanistan, Pakistan, India, Sri Lanka, Bangladesh, the mountain nations of Bhutan and Nepal, and the island country of the Maldives
- Southeast Asia, which includes both the peninsula jutting out from the southeast corner of the Asian continent, on which are located Myanmar (Burma), Thailand, Laos, Cambodia, Vietnam, Malaysia, and Singapore, and the islands that ring this peninsula, which include the countries of Indonesia, the Philippines, Brunei, and Timor-Leste

Demographic Heavyweights of South and East Asia

Perhaps the most critical feature in the geography of South and East Asia is that this region is home to about 55 percent of the world's population. Two countries alone, India and China, together have 2.6 billion people (•**Figure 7.4a, b**), or 37 percent of the world total. Their extraordinary demographic weight reflects thousands of years of human occupation of productive agricultural landscapes, combined with 20th-century advances in health technology that reduced death rates. The highest population densities are found in the areas of most abundant precipitation or surface water and with the most productive soils; rivers and coastal plains are especially densely settled. Rugged mountains, almost waterless deserts, and (until recently) tropical rain forest vegetation have limited populations elsewhere.

Some of the world's highest urban densities are found in cities like Hong Kong and Singapore, whereas in Mongolia and Nepal population densities are extremely low (see **Figure 7.4a**). Singapore is one of the few nations to claim a 100 percent urban population; in contrast, only 17 percent of Nepal's and 15 percent of Sri Lanka's population are urban. A mere 1 percent of Mongolia is arable, whereas farmers in India and Bangladesh cultivate more than 50 percent of their lands.

Population Growth Patterns

It is impossible to generalize about population growth in South and East Asia. Growth rates range from about zero percent growth (ZPG) in Japan to a very high 2.7 percent per year in Afghanistan and Timor-Leste. The region reflects the general worldwide trends of poorer countries having higher growth rates and wealthier countries having lower growth rates. The region is demographically and otherwise recognizable as one of mainly less developed countries (LDCs), with some outstanding exceptions. Japan is a classic post-industrial country and, like

• **Figure 7.1** Physical geography of South and East Asia.

• Figure 7.2 Political geography of South and East Asia.

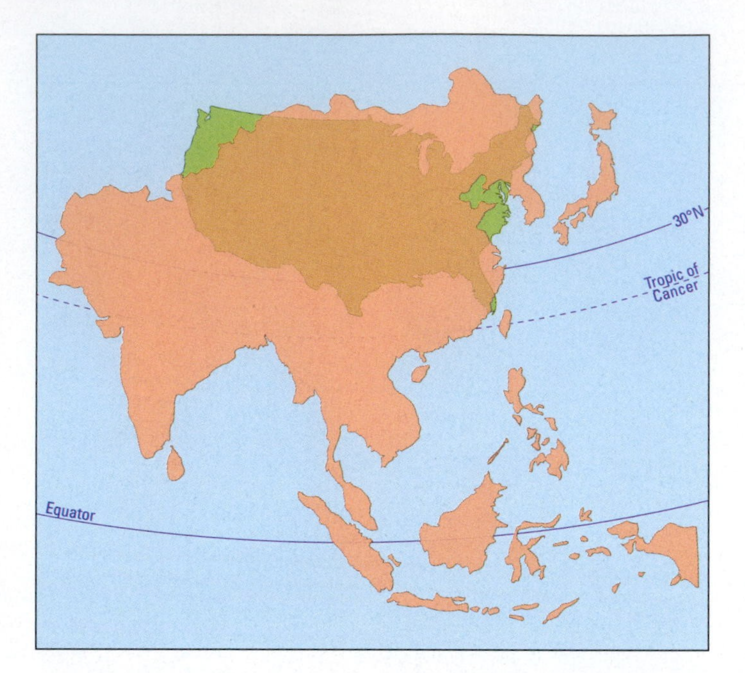

• **Figure 7.3** South and East Asia compared in area and latitude with the United States.

Italy, must worry about a declining and aging population in which young people will have to shoulder an increasing economic burden to support their elders. As discussed later in this chapter, there are other strong economies in the region with similar demographic issues.

94
359

One of the big anomalies in Asia's population equation is China. As a developing country with a per capita GDP (PPP) of just $9800, China might be expected to have a rather high population growth rate like India, which has a similar economic ranking. But in the 20th century, China's population was so alarmingly large and rapidly growing that its Communist leaders decided to restrain its growth drastically. China adopted a coercive "one-child policy" that has brought its growth rate to just 0.5 percent per year, representing a doubling time of 140 years (see Geography of Population, **page 288**). That is a remarkable turnabout from the situation in 1964, when China had a growth rate of 3.2 percent and a doubling time of 22 years.

The birth of a girl can be disappointing in Chinese and Indian societies. Some Chinese parents want to produce a son if their first child was a girl, and many girls are given to orphanages (China is the world's leading source country of adopted children, about 90 percent of them girls). Female abortion is increasingly associated with the desire to have a son. For firstborn children in China, about 108 boys are born for every 100 girls. In areas where parents can have more than one child, though, the surviving second child's gender is over 40 percent more likely to be a boy; the third child is nearly 70 percent more likely to be male. Although China has outlawed using ultrasound tests to determine the baby's gender (as has India), many couples, especially in rural areas, often find ways to have the procedure, and a second or third child that is found to be female is much more likely to be aborted. In

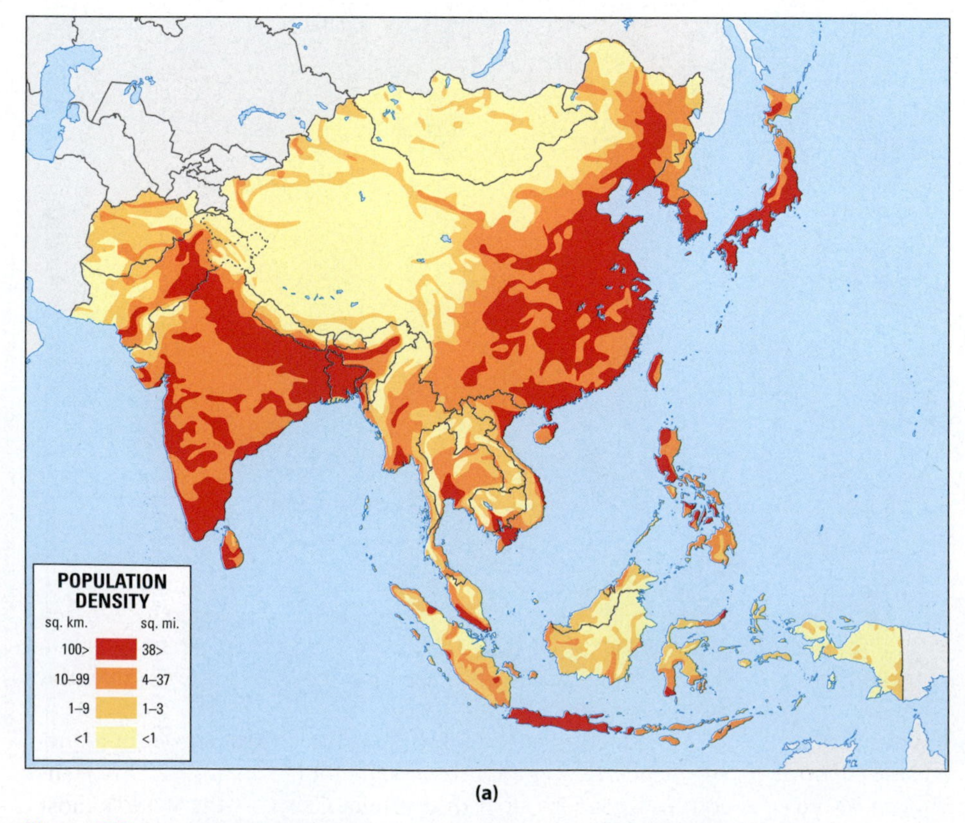

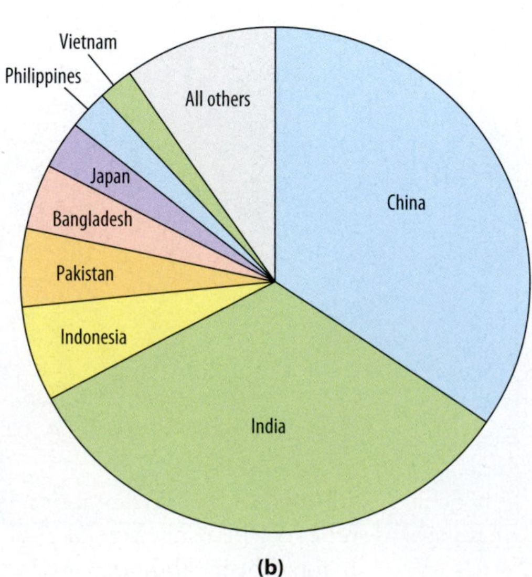

• **Figure 7.4** **(a)** Population distribution and **(b)** population pie chart of South and East Asia.

Table 7.1 South and East Asia: Basic Data

Political Unit	Area (sq mi, thousands)	Area (sq km, thousands)	Population (millions)	Rate of Natural Increase (%)	Urban Population (%)	Population Under Age 15 (%)	Agriculture Workers (%)	Per Capita GDP (PPP) ($US)	GDP ($US, billions)	Trade Balance ($US, billions)	Oil Production (million bbl/day)	Literacy Female (%)	Literacy Male (%)	HDI
South and East Asia	**8265.3**	**20407.1**	**3,950.7**	**1.1**	**45**	**25**	**14**	**7,900**	**41,074**	**—**	**7.4**	**79**	**88**	**0.646**
South Asia														
Afghanistan	251.8	652.2	31.3	2.7	24	46	27	1,100	61	−4	—	12	43	0.468
Bangladesh	55.6	144.0	158.5	1.5	26	29	15	2,100	535	−7	—	53	62	0.558
Bhutan	18.1	46.9	0.7	1.5	36	30	14	7,000	6	0	—	39	65	0.584
India	1269.3	3287.5	1,296.2	1.5	31	31	18	4,000	7,277	−165	0.8	65	82	0.586
Maldives	0.1	0.3	0.4	1.9	41	26	4	9,100	4	−1	—	99	99	0.698
Nepal	56.8	147.1	27.1	1.5	17	34	31	1,500	67	−6	—	57	75	0.540
Pakistan	307.4	796.2	194	2	35	38	25	3,100	884	−20	*	42	67	0.537
Sri Lanka	25.3	65.5	20.7	1.2	15	26	10	6,500	217	−7	—	98	97	0.750
Southeast Asia														
Brunei	2.2	5.7	0.4	1.3	76	25	1	54,800	32	7	0.1	94	97	0.852
Cambodia	69.9	181.0	14.8	1.8	20	31	33	2,600	50	−3	—	66	83	0.584
Indonesia	735.4	1904.7	251.5	1.4	50	29	14	5,200	2,554	13	0.9	90	97	0.684
Laos	91.4	236.7	6.8	2	34	35	23	3,100	34	−1	—	63	83	0.569
Malaysia	127.3	329.7	30.1	1.3	71	26	9	17,500	747	38	0.6	91	95	0.773
Myanmar	261.2	676.5	53.7	0.9	31	25	37	1,700	244	−2	*	90	95	0.524
Philippines	115.8	299.9	100.1	1.8	63	34	10	4,700	694	−20	*	95	95	0.660
Singapore	0.2	0.5	5.5	0.5	100	16	—	62,400	445	74	—	94	98	0.901
Thailand	198.1	513.1	66.4	0.4	47	18	12	9,900	990	13	0.4	92	96	0.722
Timor-Leste	5.7	14.8	1.2	2.7	30	42	5	21,400	8	0	0.1	—	—	0.620
Vietnam	128.1	331.8	90.7	1	32	24	18	4,000	509	8	0.3	92	96	0.638
East Asia														
China**	3696.1	9572.9	1,364.7	0.5	54	16	10	9,800	17,630	303	4.2	93	97	0.719
Japan	145.9	377.9	127.1	−0.2	91	13	1	37,100	4,807	−101	*	99	99	0.890
Hong Kong**	*	*	7.2	0.8	100	11	—	55,200	401	−32	—	99	99	—
Korea, North	46.5	120.4	24.9	0.5	60	22	25	1,800	40	−1	—	99	99	—
Korea, South	38.3	99.2	50.4	0.3	81	15	2	33,200	1,786	85	—	99	99	0.891
Mongolia	604.8	1566.4	2.9	2.3	68	28	16	5,900	30	1	*	98	97	0.698
Taiwan**	14.0	36.3	23.4	0.1	73	14	2	39,600	1,022	41	—	97	99	—

* Less than 0.1. — Data not available or not applicable. **China excludes Hong Kong and Taiwan.
Sources: World Population Data Sheet, Population Reference Bureau, 2014; Human Development Report, United Nations, 2014; World Factbook, CIA, 2014.

some provinces where the practice is widespread, three boys are born for every two girls. If this "gendercide" did not take place, this region would have an estimated 160 million more women than it currently has.[3]

In both China and India, this practice is having serious repercussions, reflected in a shortage of eligible brides. Rural Chinese women have been abducted and trafficked as virtual slaves to other parts of the country, where they become brides. On the positive side, the relative shortage of women in China has helped enhance their status, particularly in urban areas, where they are also becoming more influential in the workplace.

To reduce its poverty, India has made great strides in reducing its population growth, though not in China's coercive fashion. India is projected to overtake China as the world's most populous country by 2030, when it will have an estimated

Geography of Population | Changes in China's One-Child Policy

China's population of over 1.3 billion represents nearly one-fifth of the world's people and is increasing by 6.8 million each year. Population is a very serious matter for a less developed country whose area of about 3.7 million square miles (9.4 million sq km) is only slightly smaller than the United States but whose inhabitants outnumber the US population by about five to one. But a glimpse at China's age structure diagram, with its wide middle ages but tapering youth, reveals that China is one of the world's great success stories in terms of addressing "people overpopulation" (**•Figure 7.A**).

The drive for smaller families was motivated by the fact that China's population was surging while per capita food output was increasing at only a very modest rate. In 1976, when China's population was 930 million, the country's Communist regime instituted one of the most stringent and most controversial programs of birth control ever attempted. It culminated in the **One-Child Campaign**, which aimed to limit the number of children per married couple to one (exceptions have been made over time—for example, ethnic minorities and rural areas of certain provinces) (**•Figure 7.B**). To maintain compliance, the government tracks individual family birthing patterns—especially in the cities—and maintains surveillance through local authorities. It dispenses free birth control devices, free sterilization operations, free hospital care in delivery, free medical care and education for the child, an extra month's salary each year for compliant parents, and other favors and preferences to induce fidelity to the state's population policies.

Those who violate the norms are subjected to constant social and political pressure, denial of the privileges accorded compliant families, pay cuts, and fines. A woman who became pregnant after having had one child was pressured to have a free state-supplied abortion, with a paid vacation provided. Reasons for noncompliance with the one-child policy varied. In some cases,

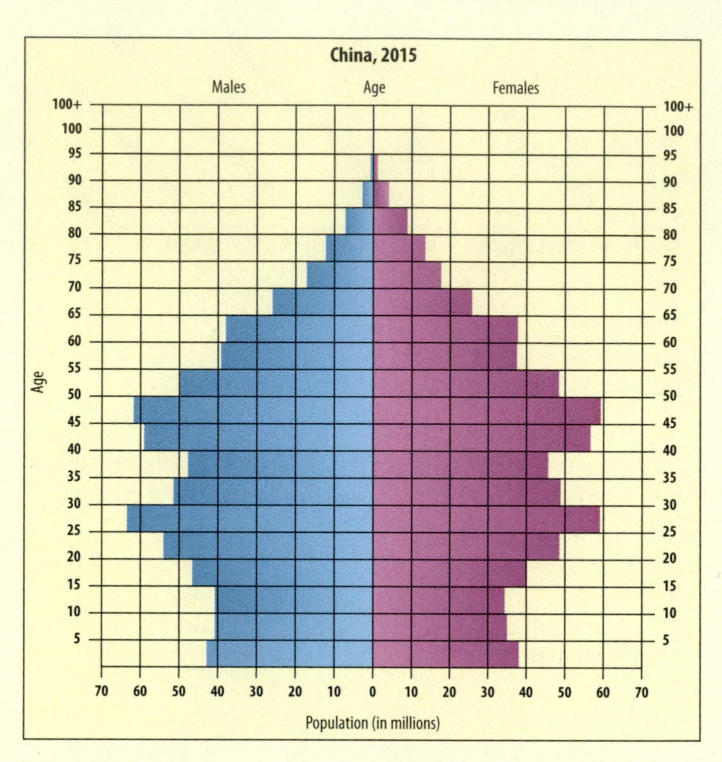

• Figure 7.A The age structure diagram of China (2015) reveals the country's success in slowing population growth.

THINK CRITICALLY: What else does it reveal? Compare the populations of females and males. Is there an imbalance, and if so, what might explain it? What would its implications be in coming years?

1.5 billion people. A wild card in India's population deck is HIV/AIDS. The number of Indians infected with HIV is estimated at 2.5 million, the highest worldwide outside of Africa. The potential for a **pandemic**—a disease that occurs in a large geographic area and affects a large portion of the population—in the world's second largest country is uncertain, and much will depend on how India's government chooses to fight the disease. About 80 percent of India's HIV infections are spread by heterosexual contact, and public health authorities have so far been shy to mount a strong campaign about sexual ethics.

408

7.2 Physical Geography and Human Adaptations

The physical geography of South and East Asia consists essentially of three concentric arcs, or crescents, of land: an inner western arc of high mountains, plateaus, and basins; a middle arc of lower mountains, hill lands, river plains, and basins; and an outer eastern arc of islands and seas (see **Figure 7.1**).

The inner arc includes the world's highest mountain ranges, interspersed with plateaus and basins. The great wall of the Himalaya, Karakoram, and Hindu Kush Mountains overlooks the northern part of the Indian subcontinent, while the Altai, Tian Shan, Pamir, and other towering ranges separate Afghanistan and China from the countries of central Asia. Between these mountain walls lie the sparsely inhabited Tibetan Plateau (averaging over 15,000 feet [4500 m] in elevation) and the deserts and uplands of western China's Xinjiang province.

The middle arc, between the western inner highland and the sea, is occupied by river floodplains and deltas bordered and separated by hills and low mountains. The major features of this area are the immense alluvial plain of northern India, built up through ages of meandering and deposition by the Indus, Ganges, and Brahmaputra Rivers; the hilly Deccan parents felt wealthy enough to pay the fines of about $6200 or more per extra child. Many rural dwellers thought their extra children would escape detection by authorities, who are most effective in the cities. Anticipating their years as elders, some couples wanted the "safety net" of additional children—especially boys. If a young urban couple has just one child and it is a girl, it is likely that she will marry and become, by Chinese tradition, such a part of the husband's family that she may not be available to care for her parents as they get older. China has traditionally used the son's family as the "old folks' home." On the other hand, rising property prices in cities may affect this preference for boys: parents have to provide a married son with an apartment, and that is becoming prohibitively expensive in cities.

With its coercive one-child policy, China's government aimed to decrease childbearing so that population would stabilize before reaching the projected 1.5 billion by 2025. But other processes in China's development began to make the one-child policy redundant. In recent decades, Chinese people have been becoming more urban, more affluent, and more educated. As we saw in Chapter 3 (**page 70**), all of these forces reduce birth rates, and that is what is happening in China. The fertility rate fell to 1.5, well below the natural replacement rate of 2.1.[2]

Chinese over 60 years old made up 16 percent of the population in 2015. This is projected to rise to 25 percent in 2020, with the working age population, ages 20–59, declining correspondingly. China strives to sustain robust economic growth, at least 7 percent annually, and that of course requires a robust work force. Worker productivity needs to increase, or the number of workers needs to increase, or both. In 2015, the Communist Party announced a change to the one-child policy, allowing couples to have two children. But getting to even that first child is going to be increasingly difficult for tens of millions of would-be grooms in China: their prospective brides were aborted or killed in infancy.

Joe Hobbs

351

• Figure 7.B Strong disincentives made one child the norm for Chinese couples, but with the country's population aging, couples can now have two children.

Plateau of peninsular India; the plains of the Irrawaddy, Chao Phraya, Mekong, and Red Rivers in the Indochinese Peninsula of Southeast Asia, together with bordering hills and mountains; the uplands and densely settled small alluvial plains of southern China; the broad alluvial plains along the middle and lower Chang Jiang (Yangtze River) in central China and the mountain-girded Red Basin on the upper Chang Jiang; the large delta plain of the Huang He (Yellow River) and its tributaries in northern China, backed by loess-covered hilly uplands; and the broad central plain of northeastern China, almost enclosed by mountains.

The outer arc is an offshore fringe of thousands of islands, mostly grouped in great **archipelagoes** (clusters of islands) bordering the mainland. On these islands, high interior mountains (including volcanoes) are flanked by coastal plains, where most of the people live. Most of the islands are in three major archipelagoes: the East Indies, the Philippines, and Japan. Sri Lanka, Taiwan, and Hainan are large, densely populated islands outside these archipelagoes. Between the archipelagoes and the mainland lie the South and East China seas and, to the north, the Sea of Japan (known in Korea as the East Sea). At the southwest, the Indian peninsula projects southward between two immense arms of the Indian Ocean: the Bay of Bengal and the Arabian Sea.

Climate and Vegetation

The region of South and East Asia is generally characterized by warm, well-watered climates, but winters can be quite cold in the north, and the western reaches of the region are very dry deserts (**• Figure 7.5a**). These are generally associated with predictable patterns of vegetation, which have been heavily modified by millennia of human use (**• Figure 7.5b**). Remember that the discussion of biomes and vegetation considers what would be growing in the absence of people. From this baseline, we will look at what people have done to the living landscape. The huge populations of this region over long periods of time,

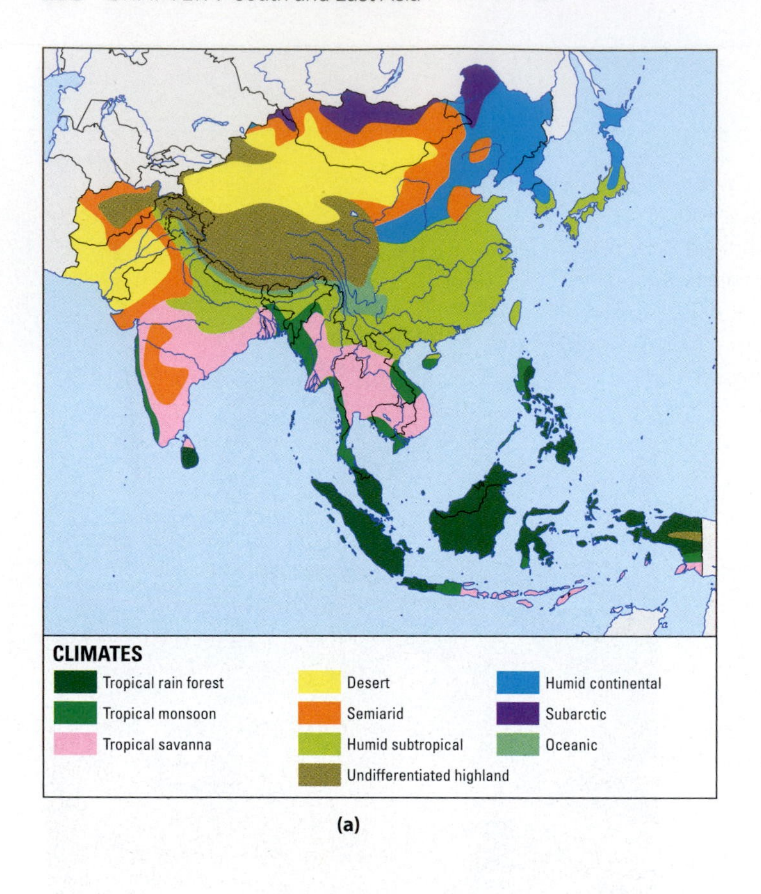

(a)

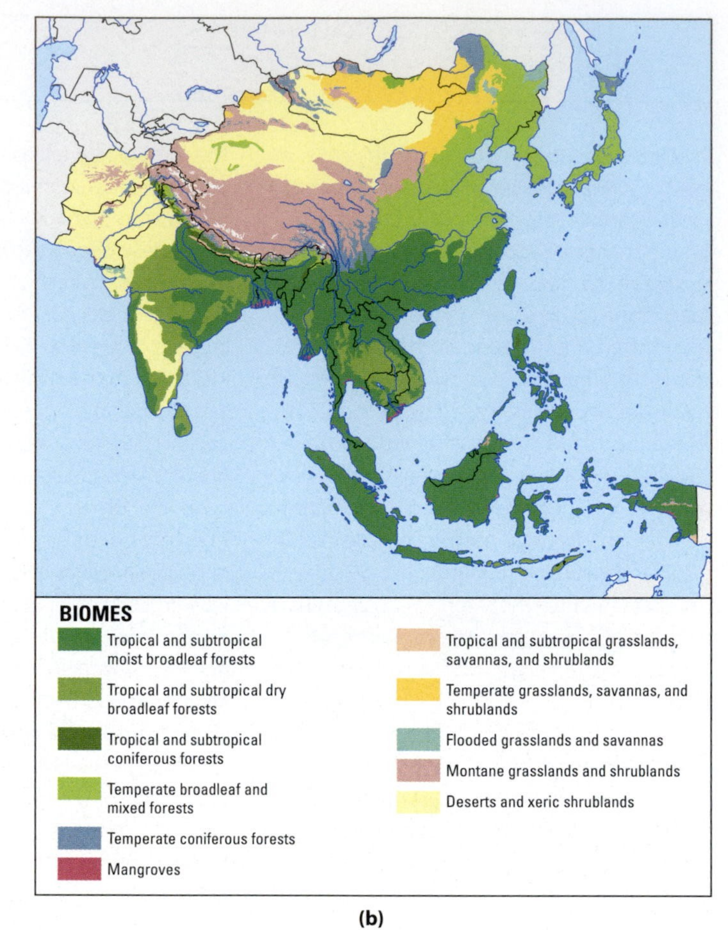

(b)

• **Figure 7.5** **(a)** Climates and **(b)** biomes of South and East Asia.

and the recent accelerated processes of urbanization, have transformed the natural environment profoundly, yet there are pockets of wild, pristine nature still to be found (see Perspectives from the Field, **page 291**).

Asia's tropical rainforests are vast, second only to the Amazon Basin in South America in size. Malaysia, Indonesia, and the Philippines have a tropical rainforest climate, but rainfall amounts in some surrounding areas in tropical monsoon and humid subtropical climates are high enough that the tropical rainforest biome has a greater range than the climate zone does. Tropical rainforests can be found beyond Southeast Asia in places like the west coast of India and parts of southern China and Taiwan. Asian rainforests are among the most endangered in the world; Malaysia and Indonesia are experiencing severe 334 deforestation, and nearly all of Bangladesh's original rainforests were cut down long ago to make room for its rapidly swelling population.

Central India and much of mainland Southeast Asia experience a tropical savanna climate, where the months-long pronounced winter dry season makes large-scale agriculture difficult. When the rains finally come, they can be so intense as to cause sudden widespread flooding. Unlike the tropical savanna climate zones in Africa, in South and Southeast Asia there are no widespread tropical grasslands; instead, the main vegetation found here are deciduous trees that drop their leaves during the dry season to conserve water.

Bangladesh and India share the world's largest mangrove forest, the Sundarbans. A 4000-square-mile (10,000 sq km) area of immense biodiversity at the mouth of the Ganges, it is home to several endangered species, most notably the Bengal tiger. Rapid population expansion has resulted in the loss of about half of the former extent of the mangroves since the 1950s; the establishment of several protected areas and a national park, starting in the 1970s, has slowed, but not halted, the loss of the ecologically valuable mangroves. In 2014, an oil tanker capsized in the Sundarbans, unleashing nearly 60,000 gallons of oil into the rivulets that snake their way through the forests toward the sea. Cleanup was slow and of limited effectiveness (local fishermen were encouraged to use their nets to remove the oil from the water), and conservationists feared that the oil would suffocate the roots of the mangrove trees, which are what holds the land in this waterlogged region together. Sea level rise also threatens the 34 Sundarbans. 330

Eastern China and southern Japan have a warm, muggy climate similar to the American South, but temperatures start

Perspectives from the Field The Lost City of Bangkok

The pace of landscape change in this region in recent decades has been truly breathtaking. We could say the same is true of all world regions. But the change is particularly striking in Asia as mushrooming cities and the activities of large numbers of people all across the land have replaced forests with the "human footprint." If you look at documentary footage of the Vietnam War in the 1960s and 1970s you will see extensive areas of tropical rain forest and hear descriptions of soldiers "fighting in the jungle." Travelling all through Vietnam since 2004, I have seen very little intact forest. In just my lifetime the landscape has changed profoundly.

My lifetime can also measure the changes brought about by burgeoning urbanization in Asia. As a kid I lived in the desert of Saudi Arabia. In 1968, my parents took me on a military transport plane to Bangkok, Thailand. This verdant paradise of a city was love at first sight for this animal-loving kid. Behind our hotel sprawled a vast lawn literally seething with reptiles and amphibians (many flew home to Saudi Arabia with me; others did not make it when my Dad saw the sides of my carry-on bag moving). This was a green, wet, amphibian of a city: canals ran every which way, and rain forest and fruit trees grew wherever they had an opportunity. It seemed like a forest with an enchanting city tucked into it (**•Figure 7.C**).

In 2006, I returned to Bangkok for the first time. How was it possible this was the same city? I looked for trees—any trees. And the canals—any canals. There were some remaining, including the famous "floating market," but most were gone, cut down and paved over in Thailand's rapid modernization, urbanization, and economic growth. Bangkok is often gridlocked, with some of the worst traffic anywhere on Earth (**•Figure 7.D**). One morning I had a 9:00 a.m. appointment across town. I took the locals' advice to leave the hotel at 6:00 a.m., arrive at my destination 30 minutes later, and just wait for the appointed time. Departing the hotel any later than 6:00 could turn it into a three-hour trip, or longer. During some long commutes, I had plenty of time to reflect: The verdant paradise of Bangkok was just a kid's memory.

Dean Conger/Documentary Value/Corbis

• Figure 7.C Bangkok in 1968.

Joe Hobbs

• Figure 7.D Bangkok in 2006.

falling quickly further north. Mongolia and northeastern China are adjacent to Siberia and can experience brutally cold winters. The vast Gobi Desert is far from the moderating influences of the ocean, and experiences a wide swing of temperatures, with an average winter temperature of 8°F (–13°C) rising to an average 75°F (24°C) in summer. West of the Gobi is the Taklamakan Desert, which lies in the rain shadow of several large mountain ranges and is one of the driest areas on Earth, with most of it receiving less than 1 inch (2 cm) of precipitation a year.

The Monsoons

The **monsoons** are the prevailing sea-to-land and land-to-sea winds that are the dominant climatic concern for people living in this world region (**•Figure 7.6**). They play a significant role in both wet and dry environments and are especially influential in the coastal plains and lowlands of South Asia, the peninsula and islands of Southeast Asia, and the eastern third of China. The ways in which the sea and land absorb **insolation** (heat from the sun) are different, causing the instability in air masses that creates monsoonal winds. If coastal waters and the adjacent coastline receive about equal amounts of warmth from the sun, the land becomes warm or hot much more quickly than the seawater. As a result, air over the land becomes unstable and starts to rise. This ascending air creates a low-pressure attraction for the air masses over the water, and marine winds begin to blow toward the land. In Asia, this is the **summer monsoon**, characterized by high humidity, moist air, and generally predictable rains. Because of the moisture carried by such air masses, these wind shifts are the sources of major

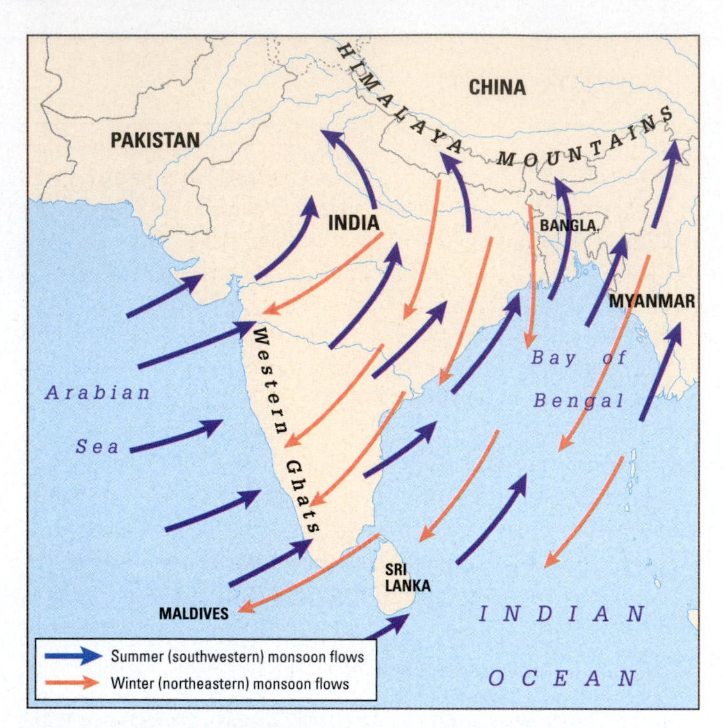

• **Figure 7.6** This map shows how the monsoons work, with prevailing winds blowing from the sea during the summer, bringing heavy rains, and winds blowing toward the sea in winter, bringing dry conditions.

Joe Hobbs

• **Figure 7.7** Torrential monsoon rains are regular and usually welcome features of land and life in much of the region. This is a flooded road in south central Sri Lanka.

rainfall in the late spring, summer, and early fall seasons. Agriculture and patterns of human activity are keyed to these incoming rains.

Hills or mountain flanks in the area force the moist air higher, where it cools and releases even more rain in the pattern known as **orographic precipitation**. The orographic effect of wringing out the moisture-laden atmosphere is most impressive along the giant mountain walls of the Himalaya and Karakoram, where tremendous rainfall and snowfall events can occur.

In the **winter monsoon**, the land loses its relative warmth while the sea and coastal waters stay warm longer. As a result, the wind shifts and air masses begin to flow from inland areas toward the sea. However, because there is hardly any moisture in the source areas over land for the winter monsoon, little moisture is picked up, and much less rain results. A monsoon climate, overall, is characterized by spring and summer precipitation and a long dry season in the low-sun (winter) cycle. Japan experiences a different pattern of winter precipitation related to the monsoon wind shifts. Here, the winter winds blow out of China and East Asia, sweep across the Sea of Japan, pick up moisture, and drop heavy, wet snow on the west coast of the Japanese islands.

The wet summer monsoon is of critical importance to the livelihoods of the people of India. By early May in India, the land has typically been parched by months of drought, and no thought can be given to farming until the monsoon rains come. The heat and humidity that build in the weeks leading up to the southwest monsoon can seem insufferable. Except in highland areas, May temperatures are in the low 100s (°F) during the day. (Nights can be uncomfortable too; I once showered throughout the night in a non-air-conditioned hotel room in Chennai (Madras) in a vain effort to cool off enough to sleep!) Anticipation and discomfort rise steadily until the monsoon finally "breaks," bringing great relief in the form of much needed rain and lower temperatures (•**Figure 7.7**). The monsoon rains don't always arrive at the same time each year; a native of India's Kerala state wrote about some of the problems coming with a late monsoon:

> The irregularity of the arrival of the monsoon can have significant influences on the morale of the society. When the monsoon is late the farmers have a tendency to get vexed. Despite the fact that it obviously will do no good, they often go on strike. When they do so, they incite bus drivers and students to go on strike with them. Schools are forced to shut down, and most transportation shuts down due to threats of stoning if anyone tries to go anywhere.[4]

Sometimes the onset of the rain comes with great ferocity. In some recent years, India, Bangladesh, and to a lesser extent Pakistan have experienced heavy early monsoonal rains, causing widespread flooding often resulting in significant loss of life. However, the overall trend is decreasing monsoonal rainfall, with increasing frequency of hot days. About 70 percent of India's groundwater replenishment comes from monsoon rains, and the agricultural impacts of less rain are especially pronounced in India's "breadbasket" states of Punjab and Haryana.[5] Inevitably, some scientific fingers point to global climate change as the culprit in poor monsoons.

41

Agricultural Adaptations

The wet and warm climates associated with the monsoons might be expected to offer excellent opportunities for agriculture. However, the high temperatures and heavy rains promote rapid leaching of mineral nutrients and decomposition of organic matter, so many soils are infertile. The lush vegetation

Insights Shifting Cultivation

Many Asian farmers, particularly in tropical rain forest regions, practice **shifting cultivation**, or land rotation between crop and fallow years (•**Figure 7.E**). The procedure typically begins with people clearing forest cover and then burning the dried vegetation during the dry season; this method is known as **slash-and-burn cultivation**). The area cleared for cultivation is known as a "swidden," and the term **swidden cultivation** is often used in place of shifting cultivation. Ash from the burn provides short-lived fertility to the soil in which farmers plant their subsistence crops. Most tropical soils lose their natural fertility quickly when they are cropped, and after two or three harvests must lie **fallow**, sometimes up to 15 years, before they will again produce a crop. During this period, the land reverts to wild vegetation.

The system of shifting agriculture is widespread in the tropics. Although it is not very productive, it does provide a bare living for people too poor to afford fertilizer or farm machinery (•**Figure 7.F**). Traditionally, this system has minimized soil erosion because most of the land is not in cultivation at any given time. Unfortunately, there has been a recent widespread trend for farmers to reduce the length of the fallow period in the cycle of shifting cultivation or to abandon it altogether. This is usually a direct result of "people overpopulation": with more mouths to feed, in many places there is simply not enough land to provide the sustainable method of letting it lie fallow.

Joe Hobbs

• **Figure 7.F** Part of the shifting cultivation cycle is harvesting crop residue (here, corn stalks) to use as fuel or fodder. This Vietnamese Hmong woman's backbreaking work is part of the greater fuelwood crisis scenario described on **page 68**.

The consequences of short-term overuse of the land are soil erosion and soil degradation. On the **lateritic soils** characteristic of many tropical regions, excessive exposure of the land to the sun's ultraviolet radiation and to repeated wetting and drying turns the ground to a bricklike, iron-rich substance called **laterite** (also called plinthite), which is virtually impossible to cultivate. In fact, this transformed soil is so firm and impermeable that people used it in the construction of some of Asia's greatest architecture, including the temples of Angkor Wat in Cambodia. 443, 486

When the best available lands are exhausted or denied to them, farmers are often marginalized to places unsuitable for farming, such as semiarid lands or slopes that are too steep to till without causing disastrous erosion. Such dilemmas of land use are typical of the world's tropics from Indonesia to West Africa and Amazonia.

① Clearing and burning vegetation
② Planting
③ Harvesting for 2 to 5 years
④ Allowing plot to revegetate 10 to 30 years

• **Figure 7.E** One of the first crop-growing techniques was a combination of slash-and-burn and shifting cultivation in tropical forests. This is a sustainable method if only a small portion of the forest is cleared. Soil fertility will not be restored unless each abandoned plot is left unplanted for 10 to 30 years.

cover in these low-latitude areas is deceiving; most of the system's nutrient matter is in the vegetation itself rather than in the underlying soil. 486 When farmers clear the tree cover, particularly in areas of tropical rain forest, they often find that the local soils will not support more than one or two poor harvests (see Insights, Shifting Cultivation, **page 293**). Agricultural and other land uses in South and East Asia are depicted in •**Figure 7.8**.

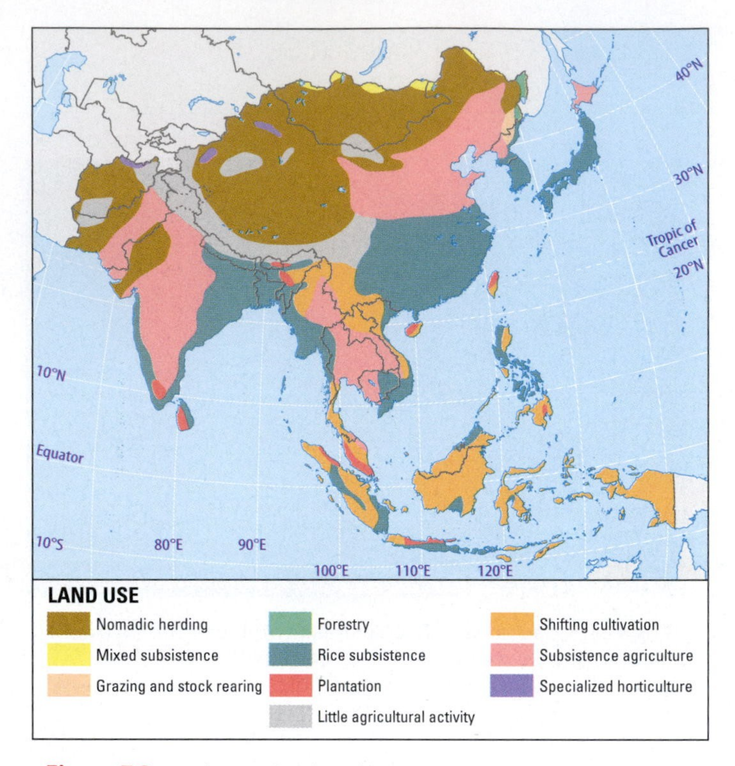

• **Figure 7.8** Land use in South and East Asia.

• **Figure 7.9** A Tay woman harvests wet rice on an irrigated terrace near Ha Giang, northern Vietnam.

The Importance of Rice

Agriculture in South and East Asia—with the principal exception of mechanized Japan—is characterized by the steady input of arduous manual labor. Shifting cultivation represents a precarious adaptation to local soil and climate conditions, capable of sustaining only small populations for brief periods in particular locales. At the other end of the spectrum of human adaptation to Asian lands is **wet rice cultivation**, capable of producing two or three crops per year and sustaining large populations over long periods of time. Rice is the principal component in the suite of cereal crops raised in the system of **intensive subsistence agriculture.** Rice is the premier staple of South and East Asia and the crop of choice in areas with adequate rainfall or where irrigation waters are available (see the very large areas of rice subsistence in **•Figure 7.8**, notably in Southern China). There are scores of different varieties of rice produced across this region.

Wet rice agriculture is practiced in both lowland floodplains and on carefully constructed upland **terraces** requiring enormous inputs of labor (the term **paddy**, also spelled **padi**, may refer to the rice itself or the field in which it is grown). Terraces have been built almost everywhere in Asia where there is a combination of sloping land, available irrigation water, and productive soil (**•Figure 7.9**). The dramatic stair-step terraces of the island of Luzon in the Philippines are the most famous, but dramatic cultural landscapes of this type can be found all across the region.

Terraces have the important function of preventing or dramatically slowing soil erosion, as the terrace walls allow heavy rainfall to run off without carrying topsoil away. In what the Chinese call "teaching water," people use canals and pipes to direct irrigation water from a source to the terrace, which acts like a kind of dam to hold the water in place. Wet rice is actually immersed in standing water for extended periods, creating an environment unfavorable to many crop pests. The productivity of wet rice agriculture can be exceptionally high, supporting as many as 500 people per square mile (200/sq km) on the rich volcanic soils of Java and Bali in Indonesia. This extraordinary output is a contributing factor to the high densities and overall populations of some of the Asian countries and regions, including Java. One anthropologist has observed that most of Asia's rice is eaten within walking distance of where it is grown.[6]

Some other forms of agriculture in South and East Asia are noteworthy. Where natural conditions are not suitable for irrigated rice, farmers grow grains such as wheat, barley, soybeans, millet, sorghum, or corn (maize). Large-scale commercial plantation agriculture, notably of rubber and oil palm, is common in Southeast Asia (note the area of plantation agriculture in **Figure 7.8**). Cash cropping has also been important for a long time in this area; rubber has been a major commodity in Malaysia since it was a British colony.

Are There Correlations between Agriculture and Culture?

Anthropologists have observed some unique cultural adaptations to opportunities for terraced wet rice. Group welfare takes precedence over individual interests, for example. Individual farmers cannot cultivate at will, but must work with the community to ensure that crops are staggered over time. This allows water to be shared fairly and efficiently, and gives some fields the chance to dry out while others are irrigated. This agricultural system also prevents the use of tractors and other mechanical technology. Over much of Asia today, just as in ancient times, wet rice cultivation is the product of intensive work by many human hands, mostly women's. Women spend much of their lives in very difficult labors, bent over to plant and harvest in hot and humid conditions.

Geographers and other social scientists have also observed a remarkable set of cultural adaptations to harsh environments on high elevation, steep slopes in the region, especially in Southeast Asia; see the Regional Landscape feature "Zomia," on **page 296**.

In some cases, population growth has outpaced the ability of even the most productive wet rice farming to feed the people, creating the classic scenario of "people overpopulation." 81 In Nepal, for example, high rates of population growth on limited areas of moderately sloped, terraceable land have 75 been a push factor behind migration. After the 1950s, malaria was largely eradicated in Nepal's lowland Terai region; this allowed millions of Nepalis to migrate there, clear the forests, and plant crops. For decades, the Terai served as a virtual "safety valve" for overpopulation in Nepal. Now that the Terai itself is mostly settled, there is nowhere for landless Nepalis to go but up. They clear and cultivate increasingly steep lands on which it is impossible to build terraces. Without terraces in place to hold the topsoil, heavy monsoon rains cause relentless erosion of their farmlands. Locally, the consequences are that the eroded plots cannot be cultivated again, and landslides occur downslope, often causing loss of 460 life. Earthquakes like the one in 2015 tear such slopes away, killing more.

Much farther away, in the lowlands of India and Bangladesh, the increased sediment load contributed by erosion in Nepal causes rivers to swell out of their banks and bring extensive flooding—again, with losses of crops and lives. This sequence of events, beginning with overpopulation in Nepal and ending with flooding in Bangladesh, has been described as the **theory of Himalayan environmental degradation.**[7] You can trace many of its components, for example how the deforestation leads to flood plain effects and siltation of reservoirs, in **• Figure 3.D** on **page 68**.

Where Asians Live

Many of the world's largest cities are in South and East Asia, including the greatest of all, Tokyo, which has a metropolitan population of 38 million. About 1.8 billion people, or 45 percent, of the region's total, live in urban settlements (see **• Table 7.2**).

Table 7.2 Largest Cities in South and East Asia

City	Population
Tokyo-Yokohama, Japan	37.8
Jakarta, Indonesia	30.5
Delhi, India	24.9
Manila, Philippines	24.1
Seoul, South Korea	23.4
Shanghai, China	23.4
Karachi, Pakistan	22.1
Beijing, China	21
Mumbai, India	17.7
Osaka-Kobe, Japan	17.4
Dhaka, Bangladesh	15.6
Bangkok, Thailand	15
Kolkata, India	14.6
Shenzhen, China	12
Tianjin, China	10.9
Chengdu, China	10.3
Nagoya, Japan	10.1
Lahore, Pakistan	10
Bengaluru, India	9.8
Chennai, India	9.7
Ho Chi Minh City, Vietnam	8.9
Hyderabad, India	8.7
Dongguan, China	8.4
Wuhan, China	7.5
Taipei, Taiwan	7.4
Hangzhou, China	7.2
Hong Kong, China	7.2
Chongqing, China	7.2
Ahmadabad, India	7.1
Kuala Lumpur, Malaysia	7
Quanzhou, China	6.7
Nanjing, China	6.1
Shenyang, China	6
Xi'an, China	5.9
Qingdao, China	5.8
Bandung, Indonesia	5.7
Pune, India	5.6
Singapore, Singapore	5.6
Surat, India	5.4
Suzhou, China	5.2

Population in millions.
Source: Demographia World Urban Areas, 2015.

Regional Landscape Zomia

Stretching from Vietnam in the east to India in the west, at elevations over 1000 feet (300 m), is a region known as **Zomia** (**• Figure 7.G**). Evocative of imaginary Asian realms like Shangri-La, the name *Zomia* comes from a word meaning "highlander" in Burma, Bangladesh, and India. Although first used in the scientific literature by a Dutch scholar, Zomia was popularized by Yale University professor James C. Scott in a groundbreaking book entitled *The Art of Not Being Governed.* Scott's thesis is that in China, Vietnam, Cambodia, Laos, Thailand, and Burma, upland peoples, including the Hmong, Akha, Karen, Lahu, Mien, and Wa, have found something perhaps all of us yearn for: an escape from tax collection and other unpleasantries of government rule. They have done so by choosing to live in the highest, most difficult, and least appealing lands of Southeast Asia—cold, steep, infertile places where no one else wants to live. They wish to be left alone, ungoverned and 464 undisturbed.

Scott's thesis is controversial, but after reading his book it is impossible to visit these places without thinking "I am in Zomia!", as I did on a journey in Ha Giang along the China-Vietnam frontier. In Ha Giang there is an extraordinary correlation between ethnic groups and elevations. At the lowest elevations, the ethnic Tay people enjoy a prosperous livelihood, growing wet rice (a Tay woman is picking rice in Figure 7.9). Higher in the mountains, ethnic Yao and Nung are clearly poorer than the Tay, growing dry rice and maize (corn) on slopes that they must maintain carefully with terraces to ward off erosion. Finally, on the steepest and most forbidding slopes near the mountaintops are the ethnic Hmong. The Hmong are the poorest of these ethnic groups, and it is easy to see why: it is all but impossible to productively farm these lands. The Hmong must resort to slash-and-burn farming, cutting the forest and planting maize without the benefit of terracing, which is impossible because the slopes are too steep. Without terraces, soils wash away and leave a desert-like environment in their wake. The Hmong move on and clear more forest. The result is breathtaking: an impoverished, lunar landscape where these people scour the hillsides for the last patches of forest to clear and plant (**• Figure 7.H**).

At first glance, it appears that the Hmong are the ultimate illustration of marginalization, where more powerful groups push weaker ones to the least hospitable lands (see **page 64**). If Scott is correct, however, the Hmong are marginalized by choice: they prefer to live in Zomia because no one else—including governments and other powerful entities—wants to be in such daunting places.

Many Hmong have resettled from Southeast Asia to the United States, especially in rural California and urban Minnesota and Wisconsin. Here too the Hmong have largely succeeded in being left alone, at least culturally. Their American neighbors are often baffled as to why Hmong have not assimilated as other immigrants have. Anne Fadiman's *The Spirit Catches You and You Fall Down* is an exceptional account of the culture clash between the traditional Hmong and modern America, especially in the realm of medicine. The book's focus is on a Hmong girl afflicted with epilepsy (which the Hmong attribute to evil spirits) and her parents' struggle to keep her out of the care of doctors with their powerful anti-seizure medications.

• Figure 7.G Zomia, a perceptual region comprised of the highland areas of Southeast Asia.

Joe Hobbs

• Figure 7.H Denuded landscapes of the Hmong at elevations up to about 5200 feet (1600 m) on the Dong Van Karst Plateau, northern Vietnam near the Chinese border.

In Villages

The main unit of Asian settlement has long been the village. Most of Asia's roughly 2 million villages are essentially groups of farmers' homes, although some villages are home to other occupational groups, such as miners or fisherfolk. Clusters of houses bunched tightly together are typical, and inexpensive and simple structures—often made of local clays and other building materials—are characteristic. Indoor plumbing continues to be the exception in village homes, but electricity lines have been extending steadily throughout Asia. So have cell phones and satellite dishes; it is not unusual today to see what appear to be very "traditional" and even quite poor Asians enjoying some of the latest innovations in global communications (**• Figure 7.10**). In South and East Asia, as elsewhere in the developing world, many people are making the jump directly from 19th- to 21st-centuries technology. 426 Wireless technology is especially appropriate where landlines and other infrastructure are difficult and expensive to install and where bureaucracies often slow down applications for access to utilities. In some countries, government officials decide what information the public cannot access online, creating one of those anachronisms that come with 351 rapid change in Asia.

The original site selection for Asian villages was usually closely related to natural conditions and to perceived spiritual circumstances as well (see Geography of Sacred Space, **page 298**). With consideration always given to the possibility of flooding monsoon rains, lowland villages tend to be situated on natural **levees** (raised riverbanks built up by deposition of sediments during floods), dikes, or raised mounds. Early villages in Indonesia were often built in defensible mountain sites. European colonists sometimes rejected traditional settlement patterns to serve their own interests. Dutch colonial administrators in Indonesia, for example, required that villages be built along main roads and trails in the lowlands, making it easier to exercise control, collect taxes, and draft soldiers or laborers for roadwork and other projects.

Joe Hobbs

• Figure 7.10 Across South and East Asia, satellite dishes sprout incongruously from traditional dwellings.

The most infamous example of foreign intervention in traditional rural settlements came during the Vietnam War in the 1960s. The American military wanted to make it more 343 difficult for insurgent Viet Cong forces to infiltrate South Vietnamese villages, where they might indoctrinate and blend in with civilians in their fight against the Americans. The military therefore forced millions of farming families to relocate into fortified communities known as **strategic hamlets**. The project was an utter failure. The insurgents managed to infiltrate these settlements. More importantly, the Americans violated one of the strongest precepts of Vietnamese culture: to live close to the graves of their ancestors. The US military lost the "hearts and minds" of 303 Vietnamese civilians in part because of the strategic hamlets debacle. Having been taught a painful lesson in Vietnam, today US intelligence agencies try to understand the culture of the enemy—for example, the Pashtun of Afghanistan's Taliban. In a program known previously as *Human Terrain* and now remarkably as *Human Geography*, social scientists with expertise in local cultures were imbedded with troops in Afghanistan and Iraq. Some of these scholars met gruesome ends when the enemy captured them and identified their intentions.

In Cities

South and East Asia is still a rural region, but its future, in terms of demographic weight, economic power, and cultural change, is focused on the cities (see the Regional Perspective on Asia, Motherland of Megacities on **page 299**, and **Table 7.2**). Some of the countries are already highly urbanized, notably Japan (91 percent urban), South Korea (81 percent), Taiwan (73 percent), and Malaysia (71 percent). Singapore and Brunei are effectively city-states. Slightly less than half of all Chinese citizens live in urban areas, yet China still has over 150 cities with populations greater than 1 million. The layouts and architectures of the cities span a continuum from the ancient, walled fortress city to the modern Western-style metropolis, and many, like Beijing in China and Lahore in Pakistan, accommodate both. Unfortunately, in the push for urban modernization, many traditional urban structures uniquely adapted to local environments have been sacrificed. Urban conservationists are struggling to protect some of the remaining old buildings, including those of Dhaka in Bangladesh and of Shanghai in China. Bestowing UN World Heritage status on such cities helps these efforts.

A striking pattern across the region is the steady stream of rural-to-urban migration, with people motivated by both push and pull factors to leave the relative poverty of the countryside in search of employment or even prosperity in the city. Many 76

Geography of Sacred Space The Korean Village

In North America, people are accustomed to communities laid out on a perfect grid of streets intersecting at right angles, with their homes and businesses symmetrically fronting those streets. But in large parts of Asia, settlement and residence considerations are far more varied, especially because they are tied to spirituality, aesthetics, and topography. Villages, and the homes in particular, tend to be arranged for maximum harmony with the natural and spiritual environments—a practice known as **geomancy**—and constitute a special realm that may be considered sacred space. The traditional Korean village illustrates these unique Asian qualities.[8]

From ancient times, Koreans developed ways of determining the most auspicious village sites. The principles of *feng shui* and of Confucianism played strong roles in how they adapted settlements and homes to geographic conditions. **Feng shui** (pronounced "fung-*shway*") is a traditional theory for selecting favorable sites for buildings, homes, and cities and for guiding many other urban and architectural planning decisions. It is essentially an ordering device for environmental planning and for daily living. *Feng shui* (the words mean "wind" and "water") attempts to balance **yin** and **yang** to achieve harmony. *Yin* (conceived of as negative, dark, and feminine) and *yang* (representing the positive, bright, and masculine) stand for all the dualistic, opposing forces that may be reconciled harmoniously, including female and male, night and day, moon and sun, cold and hot, and right and left. With cultural diffusion and growing interest in Asian lifestyles, *feng shui* principles are spreading in American suburban home architecture, as in houses built to conceal overhead beams and sharp protrusions (which are disharmonious) and provided with windows opening to the east (a harmonious situation).

The theory of *feng shui* and the Confucian view of nature have some common characteristics that affect site, settlement, and residence designs. Confucianism reflects the sociopolitical philosophy of the Chinese philosopher Kong Fuzi, known in the West as Confucius (see **page 305**). One of the main Confucian principles is that architecture and the lived environment should be in harmony with nature. This can be seen in the Confucian institutes for higher learning, called *seowon*, which in Korea peaked in the Joseon Dynasty, around 1400 CE. Each *seowon* had a lecture hall where Confucian scholars gathered to learn philosophy. The buildings were laid out in such a way that from the inside of the complex, the scholars could see distant landscapes. The scholars' minds and spirits were enhanced as they gazed at the mountains and contemplated nature's grandeur. They believed that the Earth's power was an invisible force that would help them become men of dignity and eminence.

The principles of *feng shui* and Confucianism are also apparent in the design of Korean villages. Each village is laid out to achieve harmony with the mountains—an important consideration in a land that is 70 percent mountainous. Rising behind many villages is a mountain, called *jinsan*, which is a spiritual focus for the villagers. *Jinsan* serves as the village symbol and guardian. Typically, a small stream flows in front of the village, and beyond it, there is a flat expanse of farmland. Beyond those fields is another mountain, called *ansan*, facing the village (**• Figure 7.I**).

These geographic conditions are economically advantageous to the community, providing it with productive fields and a good supply of water. But just as important, this situation is psychologically, spiritually, and symbolically beneficial to the villagers. The natural features are meant to embrace the village, as Koreans say a mother would hold a child. Korean historical records describe the traditional village as a place of comfort and serenity that safeguards against negative natural forces.

Symbolic guardians are associated with the topography. Ideally, there is the Green Dragon to the east, the White Tiger to the west, the Black Tortoise to the south, and the Red Bird to the north. These guardians, named for figures in Chinese astronomy, have varied manifestations. The Green Dragon, for example, could be a hill or a river to the east of the village, while the White Tiger could be a mountain or a road on the west side. One enters the village through a symbolic portal. For many villages, a pair of "spirit posts" stands at the foot of a nearby mountain. The posts mark the preliminary boundary to the outside, and they ward off evil spirits.

Not only is the village site selection process different from that in the West, but the settlement pattern is uniquely Asian too. Unlike in Western settlements, no more than two houses should be built side by side, they should not stand parallel to each other, and their gates should never face each other. The side street should curve naturally between the home lots. Overall, then, Koreans choose to emphasize the surrounding topography rather than geometric order in the construction of their communities and homes.

The traditional Korean home is adapted to the natural environment, embracing its surroundings while offering protection, comfort, and order. As in the *seowon*, the layout brings outdoor scenery into the house itself. Homes are also adapted to the environment in their construction materials. A thatched roof and earthen walls effectively block heat in the summer and insulate in winter, a perfect combination for Korea's sweltering summers and raw winters. The mulberry paper covering lattice windows diffuses sunlight, providing a pleasant natural light while absorbing sound and allowing some ventilation. The home's eaves are built so that sunlight penetrates deep into the house during the cold weather but never during summer—a rather recent innovation in the West, where it is known as "passive solar" design.

Joe Hobbs

• Figure 7.I The Korean village of Yangdong is nestled between mountains that are separated by a river and lush rice fields.

Regional Perspective Asia, Motherland of Megacities

A megacity is a city with more than 10 million inhabitants. We are accustomed to living in a world of such gigantic cities. Yet as recently as 1975, there were only three: New York City, Mexico City, and Tokyo. Today there are 34, with 20 of those situated in South and East Asia, including 9 of the top 10; New York is the only exception (see **Figure 3.E** for a map of all cities with more than 5 million inhabitants). Several of these Asian megacities—Tokyo, Jakarta, Manila, Seoul, Karachi, Dhaka, and Bangkok—are also 129 the primate cities of their countries. A number of region's megacities are also alpha or global cities with prominent roles in the global economic system, notably Hong Kong, Tokyo, Beijing, 127 Mumbai, and Kuala Lumpur.

It is impossible to make any generalizations about "the Asian megacity." They range from prosperous Tokyo to impoverished Dhaka, and from planned, vertical, tidy Shanghai to unplanned, sprawling, and in some areas squalid, Mumbai. In site and situation too, they span the continuum from the auspicious, chosen layout of the original Forbidden City in Beijing to the unfortunate but unavoidable placement of Tokyo atop a network of active tectonic faults.

Having described the Asian village as a sacred nurturing cradle, it is tempting to contrast it with secular fleshpots of the modern Asian city. But such a depiction of the Asian megacity would be wrong. As Edward Glaeser writes of the Asian city in his book *Triumph of the City*,

> Cities are at the apex of human endeavour. High-density cities are creative, thrilling and less environmentally destructive than sprawling car-based suburbs typical of America. Cities are passports from poverty. They attract poor people, rather than creating them. They are where humans are at their most artistically and technologically creative.[9]

Much of that creativity is going on in the slums of Asia and elsewhere. With necessity being the mother of invention, the needy in the world's "informal settlements," as slums are called, are coming up with ingenious solutions to local problems. They are also creating small enterprises that allow them to move up and out of the slums, or to become rather prosperous and remain in them. The effectiveness of the *zabaliin* in recycling Cairo's waste is the most well-known example of *in situ* success, but there are others in Rio's *favelas*, in the *Moti Jheel* area of Kolkata that was blessed by Mother Teresa's good works, and elsewhere. If you venture warily into these areas, expecting that the worst might happen, you are instead likely to be welcomed, invited into shops and homes, and encouraged to photograph people at work on their amazing variety of occupations.

Glaeser wants us to be optimistic about people of the slums, and have a better understanding of the choices they make:

> In poorer countries, people in cities also say that they are happier . . . Most people who've experienced both rural and urban poverty choose to stay in slums rather than move back to the countryside. As bad as conditions are in urban slums, they are better than the conditions they fled at home in the village . . . Over time, city dwellers' prospects can improve considerably. Unlike the hinterlands, urban slums often serve as springboards to middle-class prosperity . . . Urban poverty, despite its terrors, can offer a path toward prosperity both for the poor and for the nation as a whole. Brazil, China, and India are likely to become far wealthier over the next fifty years, and that wealth will be created in cities that are connected to the rest of the world, not in isolated rural areas. Cities and urbanization are not only associated with greater material prosperity.[10]

As for the sacred, one need not wander far from any spot in the Asian city to find people in prayer, whether in the neighborhood mosque or below the towering *goporums* of a South Indian Hindu temple complex. At the larger pilgrimage sites, many of the devotees are villagers who have travelled far to petition for better harvests or harmonious marriages of daughters and sons. Many of those daughters and sons will eventually move into the city.

Asian cities are hard-pressed to provide accommodation and services to their swelling populations, and in authoritarian countries (notably China), restrictions on internal migrations are imposed.

7.3 Cultural and Historical Geographies

South and East Asia have been home to some of the most important cultural developments of humankind in landscape transformation, settlement patterns, religion, art, and political systems. From Korea (not Gutenberg's Germany, as is widely thought) came the first movable printing type. From China came gunpowder, paper, silk, and china (porcelain). From India came the great faiths of Hinduism and Buddhism, and the concept of zero as a number, revolutionizing mathematics. Handmade textiles, artware, bone and leather and paper products, and precious metals and gems have flowed from Asia into global trade for millennia. Spices, foodstuffs, and exotic plants and products from the East Indies have been major commodities in Asian-European trade and commerce for more than five centuries. Some of the world's most important domesticated plants and animals originated in this region, including rice, cabbage, chickens, water buffalo, zebu cattle, and pigs.

These countries, and their cultures, histories and geographies, are a source of endless fascination. It is difficult for many people, the present author included, to dispute Ramachandra Guha's claim that "India is the most interesting country in the world."[11]

Ethnic and Linguistic Patterns

The South and East Asia region has tremendous ethnolinguistic diversity, given the region's huge population, vast area, and

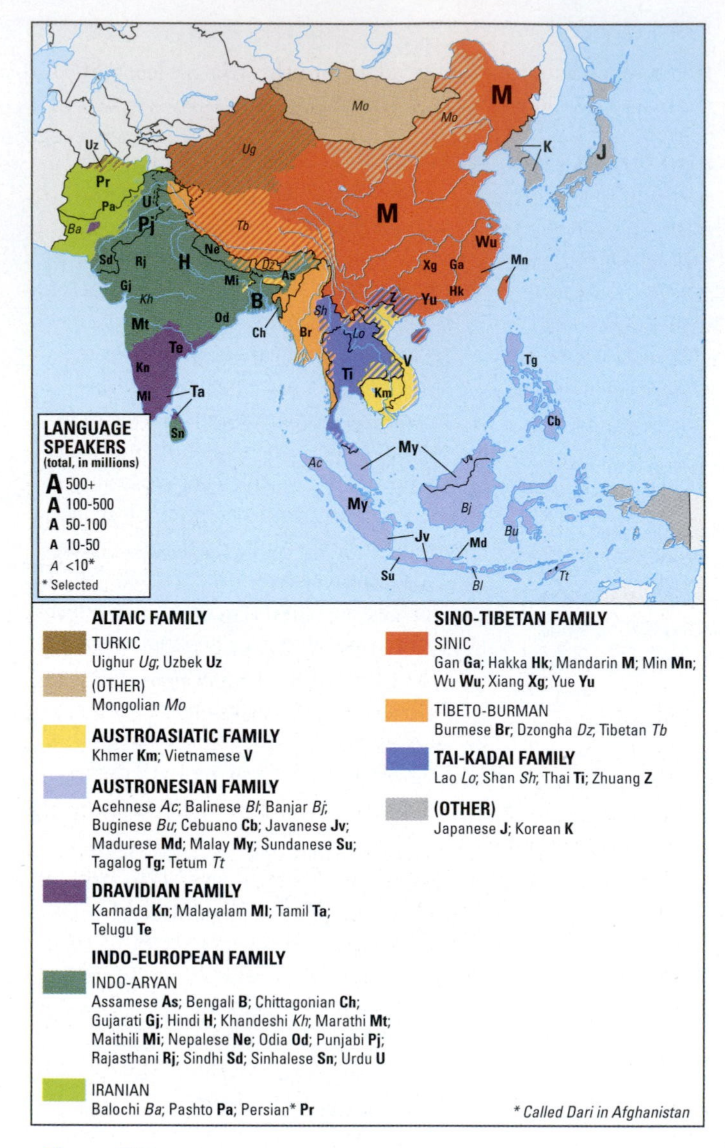

• **Figure 7.11** Languages of South and East Asia.

fragmentation from islands and mountain barriers (•**Figure 7.11**). British and American colonialism in India and the Philippines has resulted in English being one of the official languages there; for this reason many tech support call centers have been 313 established in these countries.

Languages

Only the Indo-European language family has more native speakers than the languages of the **Sino-Tibetan language family** (*Sino-* is a prefix that means "Chinese"). Mandarin Chinese, in the Sinic subfamily, has more primary speakers than any other language in the world by far, with about 1 billion people using it throughout northern and central China. Mandarin (specifically the Beijing dialect) serves as China's sole official language. Mandarin characters are transliterated into the Roman alphabet by a system called *pinyin*, which replaced several older systems in the late 1970s. This change is most notable when comparing maps of China made today and those made before 1980; major Chinese cities like Beijing, Tianjin, and Fuzhou were previously rendered as Peking, Tientsin, and Foochow, respectively. (Hong Kong is an exception, as its old spelling is still official; in pinyin it would be transliterated as Xianggang.)

Mandarin is spoken as a lingua franca in southeastern China, but people in this region natively speak a number of languages related to, but separate from, Mandarin (in the West, all these languages plus Mandarin are often referred to collectively as simply "Chinese"). **Wu** is the second-largest Chinese language, spoken by about 90 million people primarily around the city of Shanghai. **Min** is found in Fujian province, Hainan Island, and in Taiwan (though like mainland China, Taiwan's official language is Mandarin). **Gan** and **Xiang** are widely spoken in Jiangxi, Guizhou, and Hunan provinces. The Cantonese dialect of **Yue** is the primary language of Hong Kong, as well as the neighboring province of Guangdong. **Hakka** is also spoken in Guangdong and is widely used throughout the Chinese diaspora in Southeast Asia, including Taiwan, Singapore, and Malaysia. (A map showing Chinese provinces can be found in **Figure 7.65**.)

The Tibeto-Burman subfamily has hundreds of languages, most of which are spoken by only a few thousand people each, often in rugged, isolated areas. Only two tongues in this group are spoken by more than 1 million people: **Tibetan**, and the **Burmese** language of the predominant ethnic group of Myanmar.

Vietnamese and **Khmer** (the language of Cambodia) are the largest languages of the **Austroasiatic language family**, found mainly in Southeast Asia, which is also home to the **Tai-Kadai language family**, of which **Thai** (also called Siamese) is its primary tongue. Most Austronesian and Tai-Kadai languages are spoken by small minority ethnic groups, particularly in the uplands of Myanmar, Laos, Vietnam, and Yunnan and Guangxi provinces in southern China.

Over 300 million people speak languages in the **Austronesian language family**, which is found in peninsular Malaysia, across Indonesia and the Philippines, and all across the South Pacific. The Filipino dialect of **Tagalog** is the official language of the Philippines, along with English. **Malay**, known as *Bahasa* to its speakers, is the official language of Malaysia, Singapore, Brunei, and Indonesia, where they have codified different dialects. More people in Indonesia natively speak **Javanese** than Malay; with 100 million speakers, Javanese is the most widely used language in the world that is not an official language of any country.

Indo-European languages fill the bulk of the Indian subcontinent. On the Iranian branch, people in Afghanistan and Pakistan speak Dari (the local dialect of Persian) and Pashto. Nearly 1 billion people across Pakistan, India, Nepal, Sri Lanka, and Bangladesh speak languages in the **Indo-Aryan subfamily**. The largest of these is Hindi, spoken by almost 300 million people as a primary language throughout northern India. Bengali is close behind at over 200 million, used in Bangladesh and in eastern India, and third is Punjabi, spoken by 100 million people in the Punjab region that straddles India and Pakistan. Exact numbers are difficult to assign to

Indo-Aryan languages, as many are so similar that there is debate about whether they count as separate languages or dialects. For example, Hindi and **Urdu**, the official language of Pakistan, are often considered to be dialects of the same language (sometimes called Hindustani). Languages in this family are descendants of the ancient **Sanskrit** language, which, like Latin, is no longer a vernacular language but is still used extensively for liturgical purposes.

Around the fringes of the region, members of the **Dravidian language family** are found almost exclusively in southern India and northern Sri Lanka. Seventy million people speak **Tamil**; this includes about 2 million in Malaysia, where British colonists transported ethnic Tamils to work in 321 mines and on rubber plantations (**• Figure 7.12**). In the controversial Altaic language family, the Turkic Uighur language is spoken in the far western Chinese province of Xinjiang, and Mongolian is the primary tongue of the small Mongolic subfamily. The classification of the Japanese and **Korean** languages, spoken by about 125 million and 75 million people, respectively, is unclear. They appear to be language "isolates," unrelated to any other languages. Finally, the numerous **Papuan** languages of New Guinea will be discussed in 379 the next chapter.

Joe Hobbs

• Figure 7.12 This Tamil girl is from a village near Ipoh in West Malaysia. Making up about 10 percent of the Malaysian population, the Tamils are descendants of laborers brought from India to what was then called Malaya by the colonial British. Many of them worked on rubber plantations.

Ethnicity

Modern humans first arrived in South Asia at least 50,000 years ago, and within 10,000 years people had spread throughout the region. Agriculture reached Asia by 5000 years ago (and it likely developed independently even earlier in China), promoting the development of early civilizations such as the Harappan of the Indus River Valley in Pakistan and the Shang along the Yellow River in China. By the 4th century BCE, a number of small kingdoms appeared across Asia. Some grew into Asia's first empires.

Pressures from conquest and competition for land and resources triggered numerous migrations and ethnic mixing. Waves of migrating Indo-Aryan peoples pushed into India from Central Asia before the 5th century BCE, and later the invading empires of Alexander the Great and the Persians reached the vicinity of the Indus River. Among those displaced during these centuries were the Dravidians; currently living mostly in southern India, they once extended across the western parts of the subcontinent. The ancient Harappan people were probably ancestors of the Dravidians. In the 3rd century BCE, the Indo-Aryan Mauryan Empire, based along the Ganges River, expanded to control nearly the entire subcontinent. In the 1st and 2nd centuries CE, migrating Iranian peoples in turn pushed out the Indo-Aryans of the western mountain and desert areas, becoming the forerunners of current ethnicities including the Pashtuns and the Balochs.

In China, a succession of kingdoms such as Shang, Xia, and Zhou occupied territory around the Yellow River for nearly two thousand years. The Zhou Kingdom was overthrown by the Qin dynasty in the 3rd century BCE; the Qin were the first to unify China by bringing the southern tribes of China under northern rule (the name "China" is derived from Qin). The Han dynasty succeeded the Qin, ruling for four centuries and greatly expanding the territory under Chinese control into Korea, Vietnam, and far to the west. This is widely considered to be a Chinese "golden age," and the **Han Chinese** remain the dominant ethnic group in China to this day. There is also a large Han diaspora throughout Southeast Asia; they are economically prosperous minorities in several countries and are a majority in Singapore. Resentment against ethnic Chinese in Malaysia and other countries is palpable, and sometimes it erupts into violence.

The Qin and Han were expansionist and launched numerous military campaigns in southern China to enlarge their territories. Many of the peoples of southern China at this time were Austronesians, a group that originated on Taiwan about 8000 years ago. By about 2000 BCE, Austronesians had expanded out of Taiwan and colonized southern China and the islands of Southeast Asia (displacing the original Melanesian inhabitants) and later spread into the South Pacific. Austronesian 379 groups remain the majority ethnicities in Indonesia and the Philippines today, but Han migrations into southern China (and later Taiwan) absorbed or displaced most Austronesians in those areas; indigenous Taiwanese "aborigines" now only constitute about 1 percent of the island's population. This process of "Hanification" is still ongoing today in western China and Tibet. 348

Khmer (Cambodian) kingdoms dominated mainland Southeast Asia for a millennium until the Thai migrated from southern China in the 12th century. Successive Thai kingdoms, culminating in the establishment of Siam in the 18th century, gradually took over much of the area, except for the Burmese kingdoms in the Irrawaddy River valley to the west and for Indochina's easternmost areas. The northeastern coast, home of the Vietnamese people, was under almost continuous Chinese domination for over a thousand years until the 10th century, having lasting effects on the future Vietnamese nation. After freeing themselves from Chinese rule, the Vietnamese began a slow spread southward, conquering the empire of the Austronesian Cham people along the central coast and the Khmer-settled area of the southeast by 1800.

Farther north, the Korean people, possibly descendants of Siberian tribes, have lived in Korea for at least 3000 years. Korea as a political entity (previously known by other names like Goryeo and Joseon) has existed for over a thousand years, even during times of domination by Chinese dynasties, Mongols, and the Japanese. The origins of the Japanese are uncertain, but there is evidence they are descendants of indigenous hunter-gatherers and migrants from China that arrived in Japan about 2500 years ago. The Japanese spread throughout their archipelago, unifying the islands into one nation by the late 1800s, and displacing indigenous groups like the Ainu in Hokkaido and the Ryukyuan people in Okinawa.

Few peoples have had as much effect on world history as the Mongols of the northern steppes. Mongol peoples repeatedly invaded northern China, beginning in the 3rd century BCE, establishing their own kingdoms and gradually becoming absorbed into Chinese culture. Beginning around the year 1200, Mongol tribes united under the rule of Genghis Khan and his descendants and rapidly expanded out of Mongolia; within a century they conquered all of Korea, China, Central 166 Asia, Mesopotamia, Persia, and much of Russia and even attempted to subjugate Japan, Southeast Asia, India, Palestine, and Poland. This vast empire, the largest the world has ever known, began breaking down in the 1300s amid power struggles, rebellions, and the plague. The Mongol Empire was gone by 1400, to be succeeded by just a handful of small khanates across central Asia.

Religions and Belief Systems

The region of South and East Asia is, along with the Middle East, one of the world's two great hearths of religion. Hinduism, Buddhism, and related traditions of Confucianism and Daoism, collectively practiced or observed by an enormous 2 billion people (or more than a quarter of the world's population), originated in this region. The basics of these belief systems are introduced here. The precepts of Islam and Christianity, also practiced in the region, are described 223 in Chapter 6. The faiths with small numbers of adherents are described later in the contexts of particular countries. Religion is sometimes, unfortunately, at the root of or overlaps with domestic or international conflict, and these cases are also discussed in the appropriate places.

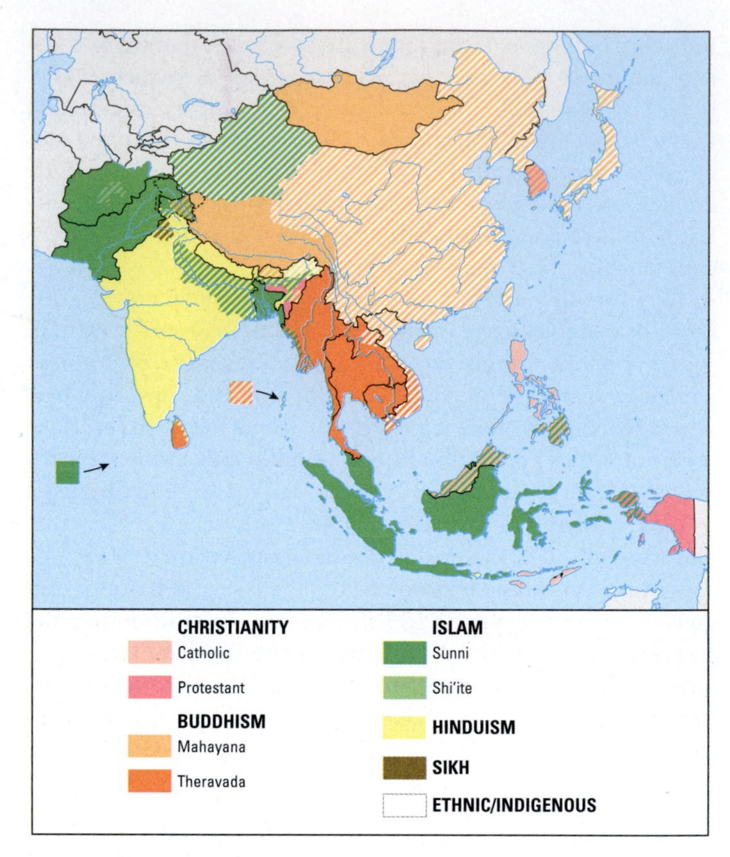

• **Figure 7.13** Religions of South and East Asia.

The geographic distribution of the region's faiths is depicted in • **Figure 7.13**. As this map reveals, Hinduism is dominant in India, although many other religions are practiced there as well. Indonesia's island of Bali is also predominantly Hindu. Islam prevails in Afghanistan, Pakistan, Bangladesh, Malaysia, and Brunei; additionally, over 100 million Muslims live in India. Indonesia is the world's largest Islamic country; 87 percent of the 251 million Indonesians are Muslim. Hinduism and Buddhism thrived in Indonesia until Islam arrived in the 15th century, and many aspects of these religions permeated Indonesia's unique Islamic culture. Various forms of Buddhism are dominant in Myanmar (Burma), Thailand, Cambodia, Laos, Tibet, and Mongolia. Most people in Sri Lanka, Bhutan, and Nepal are Buddhists or Hindus. The Philippines and Timor-Leste, with large Roman Catholic majorities, are the only predominantly Christian nations in Asia, although Christian groups are found in all other countries in this region. About one-third of South Koreans are Christian, and Korean missionaries work in many countries of the world.

The religious patterns of the Japanese, Chinese, Vietnamese, and Koreans are more difficult to describe. Among them, Buddhism, Confucianism, and Daoism have been influential, often in the same household.

Growing prosperity among the Chinese has been accompanied by a renewed interest in spiritual matters. China's government officially recognizes and even encourages five faiths (Buddhism, Islam, Daoism, Catholicism, and Protestantism) but discourages or forbids some sects and belief systems, including

Joe Hobbs

• **Figure 7.14** Falun Dafa (Falun Gong) is outlawed in China. In many countries, you will see followers sharing information about their movement with passersby. This posting is near the Sydney Opera House. After I took this photo, the Falun Dafa followers eagerly greeted Chinese tourists stepping off a tour, bus, telling them about the human rights abuses against their compatriots back home in China.

Tibetan Buddhism; the quasi-Buddhist movement known as **Falun Gong**, a practice of calisthenics and meditation that the government calls an "evil cult" (• **Figure 7.14**); and "house church" Protestantism. In Japan, religious affiliations often overlap, with an estimated 84 percent of the Japanese population sharing Buddhism with the strongly nationalistic religion of **Shintoism**, itself a blend of nature reverence, Japanese folklore, and Chinese ritual.

Other Asian religious customs include **ancestor veneration**, which is especially prominent among Chinese, Koreans, Japanese, and Vietnamese. Most homes in these countries contain small shrines with photographs and other memorabilia of deceased ancestors at which family members place fresh fruits and other offerings on a daily basis. It is crucial in these cultures that the ancestors never be forgotten (• **Figure 7.15**). Many members of hill tribes in Southeast Asia believe that natural processes and objects possess souls, a doctrine known as **animism.** Such indigenous animistic beliefs 392 have been incorporated into a number of mainstream religions, creating vibrant and highly varied belief systems by Western standards. This blending, known as **syncretism**, has parallels elsewhere in the world, for example, among the Maya of Central America. 467

Joe Hobbs

• **Figure 7.16** Worship at the Main Temple of the Holy See of the Cao Dai religion in Tay Ninh, southern Vietnam. Note the "All-Seeing Left Eye of God" at the upper right. Tay Ninh is like the Vatican City of the Cao Dai faith.

Joe Hobbs

• **Figure 7.15** Ancestor veneration is not a religion per se, but an important spiritual tradition that requires the living to remember and pay tribute to their forebears. This is a shrine to the deceased parents of an elderly woman in a southern Thai village.

The ultimate syncretic faith is the **Cao Dai** religion, which originated in Vietnam during the French colonial era and persists today despite constraints imposed by the communist Vietnamese government (• **Figure 7.16**). It recognizes and blends the precepts and iconography of Buddhism, Daoism, Confucianism, Judaism, Christianity, and Islam. Prophets and saints of the Cao Dai faith include Moses, Louis Pasteur, Victor Hugo, and Albert Einstein.

This kind of layering and mixing in the spiritual realm is typical of most belief systems in South and East Asia. Most of the faithful put stock in what many Westerners might dismiss as "superstitions."

Numerology, shamanism, amulets, horoscopes, and fortune telling have an important place in peoples' lives. Marriage, travel, site selection for homes and businesses, and countless other aspects of life are frequently influenced by auspicious and inauspicious omens. The number eight is auspicious in China and many other Asian countries (the Beijing Olympics began at 8:08:08 PM on 8/08/08); conversely, the number four is considered unlucky, similar to Western beliefs about the number 13. In Thailand, Laos, Cambodia, and Myanmar, the most visible sign of beliefs in a multifaceted spiritual realm is the "spirit house" (•**Figure 7.17**). The spirit house is placed in an auspicious spot of a home or business, where it serves as an abode for spirits that people petition for special favors or appease to avoid bad fortune.

Hinduism is a regionally varied faith and a belief system so complex that it is extremely difficult to summarize. However, it has the following characteristic elements. It lacks a definite creed or theology. It is very absorptive, encompassing an unlimited pantheon of deities (all of which, most Hindus say, are simply different aspects of the one God, Brahman) and an infinite range of types of permissible worship (•**Figure 7.18**). Most Hindus recognize the social hierarchy of the caste system, deferring authority to the highest (Brahmin) caste. They practice rituals to honor one or another of the principal deities (Brahman the creator, Vishnu the preserver, and Shiva the destroyer) and thousands of their manifestations (*avatars*) and lesser deities, attending temples to worship them in the form of sanctified icons in which the gods' presence resides. They believe in reincarnation and the transmigration of souls and revere many living things (see Insights, **page 305**). They aspire to be tolerant of other religions and ideas. They participate in folk festivals to commemorate legendary heroes and gods. To earn religious merit and to struggle toward liberation from the bondage of repeated death and rebirth, they make pilgrimages to sacred mountains and rivers.

Joe Hobbs

• **Figure 7.18** Hinduism has a complex and very colorful pantheon. This is Vishnu, preserver god, and his consort, Lakshmi, goddess of wealth and prosperity.

Joe Hobbs

• **Figure 7.17** Spirit House in Laos.

The Ganges is a particularly sacred river to Hindus, who believe it springs from the matted hair of the god Shiva, and who call it "Mother" and "Goddess." The pilgrimage to celebrate the festival of Kumbh Mela at Allahabad, at the confluence of the Ganges and Yamuna Rivers, drew over 80 million pilgrims in 2013. It was the largest pilgrimage, and probably the largest gathering of any kind, in human history. The most enduring Hindu pilgrimage destination is the Ganges city of Varanasi in the state of Uttar Pradesh. Many elderly people go to die in this city and to be cremated where their ashes may be strewn in the holy waters. Indian authorities have passed laws, so far ineffective, to clean up the Ganges, polluted in part by incompletely cremated corpses; many of the faithful poor cannot afford to buy the fuel needed for thorough immolation. Scavenging water turtles released into the river at Varanasi help dispose of the cadavers.

Sikhism is India's fourth largest religion, after Hinduism, Islam, and Christianity. India's 21 million Sikhs are concentrated mainly in the prosperous state of Punjab. A monotheistic faith, Sikhism was established in the 15th century by a man known as Guru Nanak. His religious reform movement was intent on narrowing the differences between Punjab's Hindus and Muslims. Sikhs regard men and women as equals and disavow the caste system recognized by Hindus. Sikh men are recognizable by their turbans, which contain their never-shorn hair; a bracelet called a bangle, worn on the right arm; and their ubiquitous surname Singh, meaning "lion." Some people in the US mistake them for Muslims and persecute them.

Insights The Sacred Cow

Traveling in India for the first time, you will do a double take and wonder "why" when you see a cow walking down or lying in the street. Many attitudes, beliefs, and practices associated with cattle make up the world-famous but often poorly understood "sacred cow" concept of India. There are well over 200 million cattle in India, more than 15 percent of the world total and the largest concentration of domesticated animals anywhere on Earth. India's dominant religion of Hinduism forbids the slaughter of cows but allows male cattle and both male and female water buffalo to be killed. Reverence for the cow is well founded. Indians favor cow's milk, ghee (clarified butter), and yogurt over dairy products from water buffalo. They value cows as producers of male offspring, which serve as India's principal draft animals. Both cows and bullocks provide dung, an almost universal fuel and fertilizer in rural India.

Reverence for cattle in general and the cow in particular pervades Hindu religion and mythology. The bull Nandi is associated with the Hindu god Shiva; the cow is linked to the goddess Lakshmi; and Krishna (an incarnation of the god Vishnu) was a cowherd. People allow cattle to roam the streets of Indian cities and simply accommodate them in vehicle and pedestrian traffic (• **Figure 7.J**). As a symbol of fertility, cows are associated with several deities. The mother of all cows, Surabhi, was one of the treasures churned from the cosmic ocean. Hindus honor cows at several special festivals. They use cow's milk in temple rituals. They believe that the "five products of the cow"—milk, curds, ghee, urine, and dung—have unique magical and medicinal properties, particularly when combined. In India, there are about 4000 *gaushalas*, or "old folks' homes," for aged and infirm cattle.

Remarkably, India is a major exporter of leather and leather goods. The Indian leather industry is mainly in the hands of Muslims, who do not share Hindus' restrictions on killing or eating cows. Muslims are forbidden to eat pork; therefore, pigs, a major food resource in many developing countries, are of little importance here. The main pork consumers are Christians, low-caste Hindus (some of whom are pig breeders), and tribal peoples.

Joe Hobbs

• **Figure 7.J** People and cows share street space throughout India.

Buddhism is based on the life and teachings of Siddhartha Gautama, known as the Buddha or the Enlightened One (• **Figure 7.19**). The Buddha was born a prince in about 563 BCE in northern India.[12] Although presumably born a Hindu, he came to reject most of the major precepts of Hinduism, including caste restrictions. When he was about 29 years old, he renounced his earthly possessions and became an ascetic in search of peace and enlightenment. The Buddha eventually settled for a "middle path" between self-denial and indulgence, and he meditated through a series of higher states of consciousness until he attained enlightenment (*bodhi*). From that point, he traveled, preached, and organized his disciples into monastic communities. In his sermons, he described the "Four Noble Truths" revealed in his enlightenment: life is suffering; all suffering is caused by ignorance of the nature of reality; suffering can be ended by overcoming ignorance and attachment; and the path to the suppression of suffering is the Noble Eightfold Path, made up of right views, right intention, right speech, right action, right livelihood, right effort, right-mindedness, and right contemplation.

An important Buddhist concept is the doctrine of **karma**, the sum of a person's acts and their consequences. A person's actions lead to rebirth, in which good deeds of the previous life are rewarded (for example, by which one could be reborn a human) and bad deeds punished (for example, by being reborn as a resident of hell). The goal of the practicing Buddhist is to attain **nirvana**, a transcendent state in which one is able to escape the cycle of birth and rebirth and all the suffering it brings.

Buddhism evolved into two separate branches: **Theravada** ("Way of the Elders") and **Mahayana** ("Great Vehicle"). (Mahayanists sometimes apply the term *Hinayana*, "Lesser Vehicle," in a derogatory fashion to the Theravada.) Theravada Buddhism is strongest in Sri Lanka, Myanmar, Laos, Cambodia, and Thailand. Theravada Buddhists claim to follow the true teachings and practices of the Buddha. Mahayana Buddhism originated in India and then diffused along the Silk Road to central Asia, Tibet, and China, and eventually into Vietnam, Korea, Japan, and Taiwan. Mahayana Buddhists accept a wider variety of practices than those espoused by the Theravada, have a more mythological view of the Buddha, and are interested in broader philosophical issues.

Confucianism is not as much a belief system as it is a sociopolitical philosophy that serves on its own or is blended with other religions, particularly among ethnic Han Chinese. It is based on the writings of Confucius (Kong Fuzi, or "Master Kong," who lived from 551 to 479 BCE), collected primarily in his *Analects*. The philosopher Mencius (c. 371–288 BCE) later became a major force in the widespread diffusion of Confucian thought throughout China.

The major tenets of Confucianism are embodied in a system of ethical precepts for the proper management of society, emphasizing honor of elders and other authorities, hierarchy, and

Joe Hobbs

• **Figure 7.19** The Buddha has many manifestations. This temple of the Theravada faithful is in Chiang Mai, Thailand.

education. Confucius viewed his philosophy as secular and never proclaimed his beliefs to constitute a religion; he was simply attempting to create a social contract between different classes central to Chinese government and society. A steady stream of Chinese emigrants diffused Confucian thinking to Korea, Japan, and Vietnam, where it became embedded in those other cultures.

Confucian educational priorities are strong in Chinese and many other Asian cultures. On the plus side, observers say, a firm emphasis on teaching helps create bright, competitive, and successful students; on the minus side, there is often too much emphasis on rote memorization and test taking, without enough critical thinking and creativity. In their educational reform programs, a number of Asian countries have identified critical thinking and problem solving as areas in which both professors and students need improvement. To meet this need, Vietnam, for example, has a government scholarship program sending large numbers of graduate students to US universities. The expectation is that students will be infused with these new skills and return home to help spread them as they become professors and managers. Our "foreign correspondent" for the *Life in Vietnam* feature, Quyen Nguyen, is earning her Masters Degree in Public Policy at the University of Missouri (see Life in Vietnam on **page 308**) and is on track to become a leader at home. Look around at your Asian classmates: many of them testify to the reputation for outstanding education that American universities have. Many are not on scholarships, but rather their families have put much of their savings into sending them to the United States.

Following the period of Chairman Mao's "Cultural Revolution," during which they were disavowed as backward and noncompliant with communist ideology, Confucian ideals are making a major comeback in China today. The government in Beijing sees them as having a positive influence at home and as making a uniquely Chinese contribution to world civilizations. China has invested large sums in creating "Confucius Institutes" at universities in the United States and elsewhere. Their mission statement is centered on the Confucian ideals of intellectual development, and these centers fund courses on Chinese language and life. Many universities, however, have pulled the plug on Confucius Institutes, citing them as tools of state propaganda and censorship.

Another important Chinese school of thought, second only to Confucianism and also widely blended with Buddhism among ethnic Han Chinese, is **Daoism** (sometimes spelled **Taoism**). Its philosophy comes from a body of work known as the *Daodejing (Tao-te Ching)*, or in English, *Classic of the Way and Its Power*, ascribed to Chinese Lao-tzu (Laozi), who lived in the 6th century BCE. Daoism encourages the individual to reject Confucian-style social conformity and seek to conform only to the underlying pattern of the universe, the "Way" (*Dao*, or *Tao*). To follow that way, the individual should "do nothing," meaning nothing unnatural or artificial. One can become rid of all doctrines and knowledge to achieve unity with the Dao and thereby gain a mystical power. With that power, the individual can transcend everything ordinary, even life and death. Daoists hold the simple earthly life of the farmer in high esteem.

There are about 25 million Christians in India, most of whom live in the southern and far eastern parts of the country; Buddhists, Jains, Parsis, and members of a variety of tribal religions make up the numerous smaller remaining religious minorities. Most of the estimated 7 million Buddhists are recent converts from among India's lowest castes. Numbering perhaps 75,000 and concentrated mainly in Mumbai, the famously entrepreneurial **Parsis** have attained wealth and economic power far out of proportion to their number. Their religion is the pre-Islamic Persian faith of **Zoroastrianism**, a monotheistic faith known mainly for its reverence of fire. Zoroastrians follow the teachings of the prophet Zarathustra, who was born around 600 BCE in Persia (modern Iran). He preached that people should have good thoughts, speak good words, do good deeds, and believe in a single god, Ahura Mazda ("Wise Lord"). Persia's Zarathustras, as the Zoroastrians are also known, fled Persia in two waves, first during the conquest by Alexander the Great around 300 BCE and then after Islam swept into Persia in the 7th century. They settled in China, Russia, and Europe, where they became assimilated with other cultures. In India, they maintained a separate existence until quite recently; now about one in three takes a spouse of another religion and abandons Zoroastrianism. Many scholars consider theirs an endangered religion, with fewer than 1 million followers worldwide.

To the extent they can with pressures on land use, the Parsis still use the Dakhma, or "Towers of Silence," to dispose of their dead. To prevent spiritual contamination of the corpse, it is left to the natural decomposing elements of the hot sun and vultures. In Mumbai, the Towers of Silence stand in a 54-acre wooded precinct in the city's most expensive real estate on Malabar Hill. The woods serve as habitat favorable to vultures, but recently their populations have plummeted because a drug given to cattle poisoned the vultures feeding on them. The Parsis tried unsuccessfully to breed vultures *in situ*. Most recently, they have installed powerful solar concentrators that accelerate the decomposition of bodies.

Like the Parsis, India's 6 million **Jains** also have influence beyond what their numbers suggest, as they control a significant share of India's business. They live mainly in the western state of Gujarat. **Jainism**, founded on the teachings of Vardhamana Mahavira (a contemporary of the Buddha, c. 540–468 BCE), is renowned for its respect for geographic features and animal life. Jains believe that souls are in people, plants, animals, and nonliving natural entities such as rocks and rivers. Jainism has taken the principle of nonviolence (*ahimsa*, also present in Hinduism) to mean they should not even harm microbes. Therefore, Jain worshipers often wear masks to prevent inhalation of microscopic organisms, and Jains cannot be farmers because they would have to destroy plant life and living organisms in the soil.

Effects of European Colonization

As in much of the developing world, Western colonialism reshaped many traditional geographic patterns in South and East Asia. By the end of the 15th century, Portugal and Spain had begun to extend economic and political control over some islands and mainland coastal areas of South and Southeast Asia. In the 18th and 19th centuries, the pace of colonization and economic control quickened, and nearly all of South and East Asia came under European rule, except for Afghanistan, Japan, Korea, Thailand, and most of China. By the end of the 19th century, the map of Asia was covered in names like British India, French Indochina, Portuguese Timor, and the Dutch East Indies. The United States acquired the Philippines from Spain after winning the 1898 Spanish-American War.

59

Western powers had their eyes set on controlling and profiting from trade with the vast market of China. They wrenched important trade and territorial concessions from China in two 19th-century conflicts known as the **Opium Wars**. Opium had long been part of Chinese traditional medicine, but the growing numbers of people addicted to opium led Chinese authorities to ban opium imports in 1800. Much of the opium came from poppy fields in the British colony of India, and British and other foreign merchants had gained huge profits from the captive market of Chinese addicts. In 1839, Chinese authorities in Canton burned chests of British-imported opium to demonstrate their resolve to stop the trade. Determined to keep control of trade in opium and other goods, Britain responded with overwhelming military force in the First Opium War. Defeated China was humiliated in the 1842 Treaty of Nanking, which guaranteed foreign access to Canton (now Guangzhou) and other ports, and ceded the critical port of Hong Kong to British control. China also lost the Second Opium War, which concluded with the 1858 Treaty of Tientsin (Tianjin). This treated opened more ports to foreigners and allowed foreigners to own property and proselytize Christianity throughout the country. Overall, this "time of unequal treaties," as the Chinese called it, was a degrading period in which China was forced to grant valuable economic and geographical concessions to European powers (**• Figure 7.20**). Japan too

168

• Figure 7.20 In this political cartoon published in 1900, colonial powers divide up China under the enraged but helpless Qing Dynasty. From left to right: Queen Victoria (Great Britain), Kaiser Wilhelm II (Germany), Czar Nicholas II (Russia), Marianne as the emblem of France, and a samurai soldier representing Japan.

life in VIETNAM

Photo provided by Quyen Nguyen

Quyen Nguyen

If our parents spent all of their childhood and adolescence in wartime, if our siblings, born in the 1990s, already knew how to use smart technology devices at a very young age, we, those who were born in the 1980s, witnessed a long transitional phase when Vietnam underwent postwar reconstruction and development.

I was born in a classic middle-rank family in the middle of Vietnam, where both my parents, who are from rural areas, struggled with the problems shared by the majority of Vietnamese, worked their way up to attend higher education, and built our family in the city. Living in Vietnam in a transitional time is interesting, and I will share some of the main aspects with you.

Education

We have 12 years for basic education including 5 years in primary school, 4 years in secondary school and 3 years in high school. Due to my parents' work and study requirements, I changed my primary school five times at four schools in three cities: in Hanoi (in the North), Quang Binh , and Hue (in central Vietnam). My mother, like any other Asian-tiger mother, sometimes regretted that I never got to be in a gifted class due to the short stays at any one school. However, I enjoyed my time, for I did not have to attend extra classes, got to mingle with new friends, and somehow won a number of competitions in sports and king chess at my new schools.

Things got serious when I entered high school, as we had to start preparing for college. Studying at a famous regional school for the gifted, which was already an advantage, I spent my time out of class in numerous extra classes ranging from Vietnamese, English, and math (some of us even took two extra classes) to physics and chemistry. In the senior year, my math class started at 5:00 a.m., and we normally had a very tight schedule for the whole year until we finished the college entrance exam in summer.

I finished my bachelor's degree in a national university in 2009, focusing on social sciences. The curriculum was designed so that, apart from technical knowledge of our major, we also took several classes in the Vietnamese political regime and its history, ideology, and practices.

Back in my parents' time, Russian was the popular language. However, English quickly became one of the main required skills, along with a rapidly increasing need for a quality graduate education and international experience. So a lot of us started to reach out, find a scholarship, and study abroad for a masters' or postdoctoral degree. Parents who can afford the expenses are also very willing to send their children abroad to the US, Australia, Europe, and other developed countries. This trend has become very popular in the last 10–15 years. My father sometimes told me that he still could not believe that he got to send his two children abroad for studying.

Health

As was the general practice at that time, I was born at my grandparents' home as a home-delivery case, with the help of a midwife. My mother recalled that if I had come out 30 seconds later, I would not have been alive, due to the mistake of the nurse and the lack of facilities. Fortunately, home birth no longer exists in a majority of places in Vietnam; the government and development agencies have been spending a lot of efforts to reach out to ethnic minority groups at remote and mountainous areas to end this practice.

My parents, who spent one-fourth of their lives living in war, hunger, and poverty, understood very well about the importance of nutrition in the development of a child. I was a tiny kid and got sick quite a lot. Because my mother did not have decent nutrition when she was pregnant

took advantage of China's weaknesses. Japan took Taiwan from China in 1895, Korea in 1905, Manchuria in 1931, and Hainan in 1939, and held these territories until its defeat in World War II.

Some Asian countries escaped domination by the Western powers during the colonial period. Thailand (historically known as Siam) formed a buffer between British and French colonial spheres in mainland Southeast Asia. Japan withdrew into almost complete seclusion in the mid-17th century but emerged in the late 19th century as the first modern, industrialized Asian nation and soon acquired a colonial empire of its own. Korea had come under Chinese and Japanese influence in its long history but was never a European colony. It is noteworthy that Japan, South Korea, and Thailand would emerge as very or relatively prosperous countries, whereas most of the former colonies lagged in economic development. Dependency theorists point to these discrepancies to affirm their argument that colonization hampered development.

59

Asia was an extraordinarily profitable region for the colonizers. Western nations extracted vast quantities of tropical agricultural commodities and in turn found large markets for their manufactured goods. Westerners also invested heavily in plantations, factories, mines, transportation, communication, and electric power facilities. Some of the region's most important cities, including Shanghai in China, Kolkata (Calcutta) in India, and Singapore, were developed mainly by Western capital as seaports serving Western colonial enterprises.

Western domination of Asia ended in the 20th century for a variety of reasons. The two world wars weakened the West's ability to conduct its colonial affairs in the region; Japan rose to great power status and challenged the West early in World War II; and effective anti-colonial movements arose in nearly with me, my parents took it very seriously later on. Generally, dairy products were seen to be an effective solution. My mom would send milk powder packs all the way from Russia; and my Dad, later on, would make me drink three glasses of milk per day, as well as have me frequently eat cheese and yogurt. There was no room for negotiation on this. Later we discovered that the majority of Vietnamese people, due to geographical history, do not have the enzyme that can transform one typical enzyme in dairy products. However, these products do help nutrition to some extent.

When it came to my brother, the conditions were much better. He was a healthy, happy baby. The public healthcare system has improved so that seeking a consultation from a good doctor is no longer a problem.

At home, we maintain a well-balanced diet with a lot of vegetables, fish, and meats. In 2013, the World Health Organization placed us last in the ranking of prevalence of an overweight population.

Technology

I still remember my first encounter with a computer in the second grade. We had several computers with the Windows 95 system. I remember it took me a long time to type my name, "Quyen," as I could not press the keyboard correctly. Before that, I saw my parents writing their long articles or graduate theses by hand. The Internet became popular in my high school time in the mid-2000s.

Within just a few years, we saw the influx of smartphones, tablets, and laptops, together with high-speed Internet connections. Vietnam, with its young population, is now among the developing countries with the highest potential in Internet applications, attracting numerous technology firms from all over the world.

Family Values, Culture, and Identity

Asian values teach children to respect and listen to adults, and parents have a certain amount of responsibility to take care of their children, even when they are over 18. For example, it is a common practice that the children take up the occupations of their parents. My dad is a journalist, and my mom is a teacher. Automatically, people assumed that I would either work at a news publishing house or at a university, and they were very surprised when it did not turn out to be so. Luckily, my parents have been very supportive on whatever decision I have made—unlike my experience with dairy milk.

They say Asians stick together. I find it completely true in various contexts. In a Vietnamese family, this communal culture brings us closely together. I have a very large extended family of aunts, uncles, and cousins, who live across the country. We frequently talk and visit each other several times a year. We help the others when in need and consider the act as a responsibility for our family.

If you ask about one thing that I am proud of about Vietnamese culture, I would say it is the food. An authentic Vietnamese dish like *pho* (noodle soup) looks simple and plain; but taking a closer look, you will feel the fulfillment and balance in the sweet soup and the fresh ingredients of meats and herbs. It requires a certain level of skill to make a good *pho*. This dish is a good representation of the true and classic portrait of the Vietnamese.

As a part of global integration, I sometimes question the Vietnamese identity, comparing it to the other neighboring countries that share a similar culture of rice-paddy agriculture. Thinking again, with the cultural background and a long history of war and reconciliation, I guess what we, as Vietnamese, must treasure is the legacy of discerning right from wrong, retaining fortitude, resilience, resourcefulness, and adaptability to rise and excel.

As we are transitioning, the country has been facing a lot of clashes in ideology and practices between the old and the young. I find myself to be absolutely fortunate to have experienced these drastic changes from the time I was born until now.

all areas subject to European control. In the decades after World War II, all colonial possessions in Asia gained independence. The last to revert were Hong Kong (returned by Britain to China in 1997) and Macau (returned by Portugal to China in 1999). Until this era of independence, however, 20th-century Asia was marked by revolution, war, turmoil, and an inability to gain an economic footing with the rest of the world.

7.4 Economic Geography

With some exceptions, South and East Asia is a dynamic region with some of the world's strongest and fastest-growing economies. Their economic performance is often described as "miraculous," given the region's tumultuous history. At the end of World War II, South and East Asia were poorer than Africa and Latin America. "The depleted states of East Asia, including Korea, Taiwan, and Singapore, with few sources of money and practically no existing industry, seemed especially hopeless," writes journalist Michael Schuman in his book *The Miracle*. "Yet it was in these countries, supposedly the bottom of the international economic barrel, that the Miracle was born."[13]

If not literally a product of divine intervention, what explains Asia's extraordinary economic success? Many of the answers—government policies that nurtured the growth of industries, and government investments in industrial growth rather than in public amenities and services, for example—are explained in the following pages in the context of individual countries, including Japan and South Korea. It is also useful to look at the level of the individual and family—like Quyen Nguyen and her family—to appreciate Asia's economic geography and economic success.

To generalize, Asians are savers, whereas Americans and Europeans are spenders. Leading up to the "Great Recession" of 2008–2009 Americans went deeply into debt to finance consumption, whereas Asians stepped up their savings and exported more goods to Western consumers.[14] To some extent, Asians are savers because they have to be: most Asian countries provide few of the "safety nets" such as retirement pensions and health care found in the West. And most important of all, savings help take care of the family, which is virtually sacred in many Asian cultures. It is always impressive to see how young people working even for low wages manage to save a significant share of those earnings as remittances to send home to parents and siblings. Needless to say, one of the enduring hopes in the West is that Asians will cut back on savings, loosening their purse strings especially to buy Western exports.

The economies of China and India have been surging, with their goods and services inundating the global economic system. Their clout, their accomplishments, and their ambitions are enormous. Some economists and equities marketers have taken to conflating the two countries and calling them "Chindia," a double-barreled engine of economic growth. It must be noted, however, that this growth is occurring against the backdrop of extraordinary poverty. Although the region as a whole has taken great strides at reducing poverty rates, with many hundreds of millions of people lifted out of poverty, hundreds of millions more remain desperately poor.

It is also troubling that against the backdrop of surging growth, the gap between rich and poor is growing in many countries, including China, India, and Indonesia.[15] In India, the overall economy has grown about 8 percent annually since 2000, yet in that same period, there has been almost no reduction in childhood malnutrition, which afflicts almost half of India's young. Two-thirds of India's people live on less than \$2 per day, which is the World Bank's benchmark for poverty; one-third live on less than \$1.25 a day, the definition of extreme poverty. About 70 percent of Indians live in rural areas, but agriculture produces only 18 percent of India's growth domestic product; no wonder then that rural people are poor, as they are a majority who live off a small piece of the economic pie. Poverty tends to be deepest in the rural areas, fueling the classic push factor of rural-to-urban migration.

In 2014, the IMF reported that China had overtaken the United States to become the world's largest economy, a position the United States had held since 1872 (•**Figure 7.21**).[16] It is important to recognize, however, that this is based on total output of goods and services and that China's accounts may have been exaggerated. China has a per capita GDP (PPP) of just \$9800 (compared with \$52,800 in the US), and 18 percent of China's people live on less than \$2 per day. "For the first time, you have this odd combination," an economist in Hong Kong wrote as China reached this milestone. "One of the world's largest economies is also one of the world's poorest economies."[17]

These concerns should not obscure the extraordinary, underlying story of the "economic miracle" of Asia. Against the odds, countries with few resources have become strong economic powers. Countries leveled by warfare have built

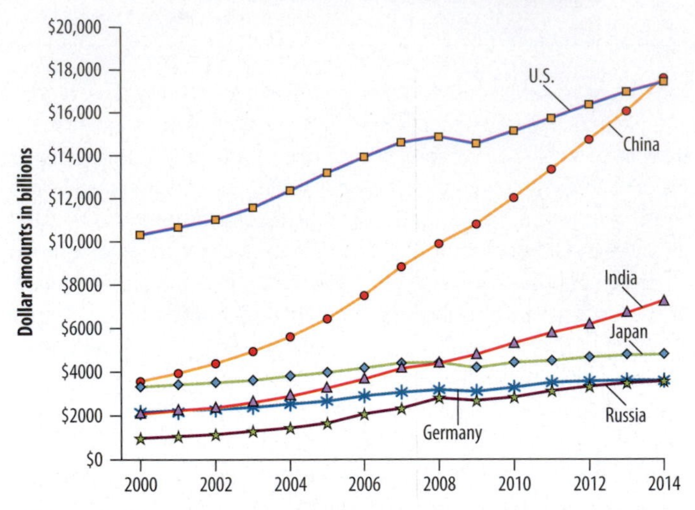

• **Figure 7.21** The sizes of the world's six largest economies from 2000 to 2014. The United States had been the world's largest economy for over a century until China overtook it in 2014.

economic strength from ruins. South and East Asia by IMF standards is home to the world's first, third, and fourth-largest economies (China, India, and Japan, respectively). The region is also home to the industrialized, export-oriented economies of South Korea, Taiwan, Hong Kong, and Singapore, which late in the 20th century came to be known as the **Asian Tigers** (or "Little Dragons"). Some economists consider the rapidly industrializing Southeast Asian countries of Thailand, Malaysia, Indonesia, the Philippines, and Vietnam the **New Asian Tigers**, or **Tiger Cubs**, following the path of rapid economic growth blazed by their neighbors. Indonesia's economy has been doing so well that some economists have wanted to include it as a second "I" in the exclusive BRIC club, which would become Brazil, Russia, India, Indonesia, and China, or BRIIC.

Japan was the first Asian country to develop modern cities and modern types of manufacturing on a large scale. It has been a major industrial power for more than a century, but like the US, Japan's economy is now primarily post-industrial, and is much more services- and knowledge-based. Less than one-quarter of Japan's GDP comes from industry, whereas half of China's does. China is much better supplied with mineral resources than is Japan, and combined with its large, cheap labor force, it leads Asia (along with Thailand, Malaysia, and Indonesia) in high **value-added manufacturing**—the process of refining and fabricating more valuable goods from raw or semi-processed materials. The biggest economic story in South and East Asia is the rise of China, and its potential fall. Some other highlights of that story are told on **pages 351–354**.

China's Surging Economy

The statistics portray China as a steamroller. With its system of state capitalism, China's economic growth soared

Bobby Yip/Reuters

• **Figure 7.22** Apple products are "designed by Apple in California" but assembled mainly in the "iPod City" of the Shenzhen SEZ. Foxconn factories there have introduced reforms to improve working conditions.

at an average annual rate of 9 percent between 1990 and 2014. China's exports (along with imports) increased about 2000 percent during those years, growing an average of 83 percent a year. China's share of the world's goods and services also increased 500 percent between 1990 and 2014 to 11 percent (compared with 11 percent for the EU and 8 percent for the US).

Before the 1990s, China was notable for making low-value-added products—cheap toys, for example—that the richer countries really had no interest in making. There were shirts and shoes as well, most of them manufactured in more than 20,000 factories in the special economic zone (SEZ) of 352 Shenzhen, adjoining Hong Kong. In the 1990s, China was making at least a little bit of almost everything, becoming the "workshop of the world." The country made especially rapid progress in manufactures and exports of information technology (IT) hardware. China is the world's largest manufacturer of personal computers and smartphones, sold by companies such as BlackBerry, Samsung, Lenovo, and Apple (**• Figure 7.22**). The roots of China's boom in IT may be traced to the dot-com bubble that burst in the United States in 2000. Many of the talented Chinese who worked in California's Silicon Valley at that time went back to China, taking their skills with them in 76 a kind of "reverse brain drain."

Advances in China's global status and in government economic policy have fueled the country's boom. China joined the World Trade Organization (WTO) in 2001, prompting a surge in foreign investment. China sweetens the deal for investors with incentives like subsidized loans, tax exemptions, and 50 percent discounts on land prices. The addition of Chinese brawn and brains created a nearly perfect investment climate. Chinese labor is much less expensive than is typical of most more-developed countries (MDCs), and China has a rapidly growing pool of university-trained engineers and other innovators. But China has suppliers and competitors in a region where market conditions and profit margins are constantly in flux. Everyone in this dynamic neighborhood is mindful of changing opportunities.

China's Economic Arena

China's economic boom is having enormous consequences for other countries, particularly its Asian neighbors. Japanese and Korean electronics companies are cutting costs by moving increasing amounts of production to China. Surging investment in China is linked to disinvestment elsewhere, especially in Southeast Asia. That region's Tigers and Tiger Cubs—notably Malaysia, Singapore, and Thailand—dominated the 1990s with manufactures and exports of medium- and high-tech industrial products. With its growing consumer appetite, China was buying such goods from these countries. But then the manufacturing shifted to China from those countries. South Korea's Samsung made China the main base for production of its flat screens and computers, for example. Japan's Toshiba makes televisions only in China. Dell recently moved some of its computer-making facilities from Malaysia to China.

With China's increasing gravity, Southeast Asian economies were at least temporarily relegated to supplying low-end products, especially food and raw materials, to China. In turn, they bought cheap Chinese manufactured goods, harming their own fledgling industries. They have been trying to counter this trend by developing higher-value niche products to meet demands in China. Thailand, for example, is boosting manufactures of health care products, and Singapore focuses on biomedical products and financial services. Indonesia has been able to help satisfy China's growing appetite for natural gas to fuel its southern industrial cities.

Cheap labor has not made China a permanent magnet for foreign investment, however. As affluence rises there, so do wages, and China's less prosperous neighbors will benefit from these trends. These figures are telling: in 2014, the average factory worker in Vietnam made $230 per month; in Indonesia, $300 per month; and in China, $650 per month. China's labor is no longer cheap. Although recently amended, the effects of China's one-child policy also means it will have fewer industrial workers in the future. Mindful of this demographic problem, farsighted multinational companies are boosting manufacturing investment in India, where the pool of young workers is likely to remain large and growing for decades to come.

In the long-dominant United States, there are rising fears of China's economic ascendance. Many Americans feel their country is becoming subservient to China because Chinese investments are funding US federal deficits and because the United States is losing ground to China's industrial might. China's rise has been steadily chipping away at US economic dominance in the region. In one country after another (including the most critical, Japan and South Korea), China has eclipsed the United States as Asia's most essential trading partner. The US-China relationship is growing in China's favor. Of China's exports, about 17 percent go to the United States (its largest single trading partner), but its imports from the United States lag far behind. This US trade deficit with China threatens to become a sore point in relations between the countries. The US government has passed punitive tariffs on Chinese imports such as tires and solar panels; attempts to place duties on other Chinese imports have failed for violating WTO regulations. Problems with tainted food, toys, and hardwood flooring have led to bans of some Chinese products in the United States.

The United States is also furious at China's reluctance to stop the counterfeiting or "piracy" of American products, particularly computer software and DVDs of Hollywood films. China is the epicenter of prolific Asian manufacture and trade in pirated products, but piracy is prolific throughout the region. In some countries like Malaysia, authorities try to crack down on sales of such goods on street corners, but the vendors are usually prepared to disappear at a moment's notice with their goods in tow, only to reappear moments later when police leave. In other countries like Vietnam, no one needs to run: there are countless permanent shops selling the latest music, movies, watches, handbags, and more (• Figure 7.23).

Another great concern in the United States is how many jobs and profits are lost to outsourcing, especially to India (see Insights, page 313).

Americans—and virtually everyone else in the world, now tied so closely together by the forces of globalization—need to be vigilant about a new aspect of China's economy: it is slowing down. From double-digit annual growth in 2010, China slipped to an annual growth rate of just 7.4 percent in 2014, the slowest since 1990, with future projections of 6 percent and even lower. Although 7.4 percent growth would be the envy of most countries, it may not be enough to ensure stability for China's government. As the population ages, the labor force is shrinking. Cheap capital, cheap labor, cheap energy, and cheap land were at the heart of the Chinese economic miracle, but each of these is becoming pricier. Inflation and unemployment are growing. The danger is, of course, social unrest. And the decline is already threatening the global economy. See What's Next for Industrialized China on page 351.

In the coming years, you will hear a great deal about the **Trans-Pacific Partnership (TPP)**. If negotiated successfully, it will be the largest trade agreement ever reached. Its members in discussions have so far included Australia, Brunei, Canada, Chile, Japan, Malaysia, Mexico, New Zealand, Peru, Singapore, the United States, and Vietnam. Conspicuous by its absence is China, which regards the TPP as an instrument of America and its allies' interests. American proponents of the TPP readily acknowledge that this free trade agreement (FTA) will strengthen the geopolitical hand of the US in Asia. The White House says it will "expand economic opportunity for American workers, farmers, ranchers, and businesses, boosting U.S. economic growth, supporting American jobs, and growing Made-in-America exports to some of the most dynamic and fastest-growing countries in the world."[20] Apart from a Wikileaks release of the environment chapter, few details about the negotiations have been made public, deepening the suspicions of groups normally opposed to FTAs. Calling it a "supersized and nuclearized NAFTA," and "NAFTA on steroids," these opponents say this FTA will strengthen multinational corporations, circumvent environmental standards, offshore domestic jobs to low-wage countries, increase inequalities, and squelch intellectual property (IP) and free speech. The battle lines are drawn, and the TPP is worth watching.

Joe Hobbs

• **Figure 7.23** Counterfeited goods made in China and elsewhere in Asia bear the brand names of US and other manufacturers but are sold to happy consumers at a fraction of the cost of the genuine product. DVD films typically sell for about a dollar, and CDs, for even less.

The Green Revolution

Most Asian countries do not want to improve their economies just by enhancing their outputs of services and manufactured goods. They also want to boost agricultural self-sufficiency and crop exports. Revolutionary changes in crops are at the heart of efforts to reshape Asia's agricultural economies. One of the most significant agricultural innovations in Asia during the past century is the development of new seed types and innovations in planting, cultivating, harvesting, and marketing of crops, efforts known collectively as the Green Revolution. 35 These developments are not exclusive to South and East Asia, but much of the growth and investment have taken place here.

Ever since 1960, when the International Rice Research Institute (IRRI) was founded in the Philippines, there have been efforts to use science to increase food yields (particularly rice) to stave off hunger and generate export income. Notable success has been achieved in breeding the new high-yield

Insights Outsourcing

Outsourcing, also known as **offshore outsourcing** or **offshoring**, is the flight of technology and other jobs from countries with high manufacturing and service costs to countries with low manufacturing and service costs. India is the world's largest recipient of these jobs, and the United States is the leading outsourcer. India's outsourcing industry was a natural development after 1990, when the country abandoned its Soviet-style, centrally planned economic model of self-sufficiency and opened its doors to world markets and economic liberalization.

In the United States, there is growing worry and even alarm over the trend of outsourcing jobs to India. Many of these jobs are in computer programming and a wide variety of technical support, from computer problem phone calls to X-ray diagnosis. Many are mundane, like bill processing and order taking. Increasingly, however, high-paying white-collar employment is being outsourced to India, in such fields as investment banking, aircraft engineering, and pharmaceuticals research. Western corporations like the United States' Cisco Systems, the world's leading maker of communications equipment, have relocated substantial numbers of their senior executives to India, where much of the firm's work is actually done. The top destination is the south-central city of Bangalore (officially Bengaluru), which has come to be known as **"India's Silicon Valley"** (**• Figure 7.K**).

India, which has earned a reputation for being the West's "back office," is an especially attractive venue for information technology outsourcing for several reasons. It has a large population of well-educated people, most of whom speak excellent English. Those who work the telephone call centers typically receive training in vernacular American English and strive to perfect their American accents while shedding their Indian and British pronunciations. The bottom line is another major advantage: Indian skilled labor is cheap. The average computer programmer in the United States earns $81,000 per year, whereas his or her counterpart in India earns $5500 per year (though this is partly because in recent years India has produced more programmers than it needs, lowering average salaries).

Finally, the Internet and superb telephone communications, ironically combined with India's physical distance from the United States, provide some remarkable opportunities. While people in the United States sleep, the day shift is on in India. An American doctor can X-ray a patient in the afternoon and by the early next morning have the analysis—performed overnight in India. On the downside, when an American consumer places a technical service call for a computer problem, he or she may be speaking with a very tired Indian employee working the graveyard shift. There are many accounts of fatigue and burnout among Indian and Filipino call center employees. In an extraordinary twist, outsourced Indian call center businesses are increasingly outsourcing their work to the Philippines!

What about the loss of American jobs due to outsourcing? There is an expression that one's job can be "Bangalored." Some projections are alarming. A former official of the US Federal Reserve described outsourcing as a "Third Industrial Revolution," predicting it would take jobs away from tens of millions of American workers in the coming years.[18] Outsourcing companies generally argue that reducing the costs of their services reduces inflationary costs for their consumers and improves efficiency and productivity. This rationale is, of course, cold comfort to the American worker whose job has moved offshore. Another trend that Americans fear is that with their well-educated work forces (including many graduates of US universities), India and other countries are moving further up the value chain. These countries are investing in education and research and development so that their information technology workforces will be innovating, and not just producing, goods that Western thinkers invented. The US is considering changes in its visa regulations that would encourage foreign students educated in STEM fields in the US to stay in the US and contribute to innovation there.

There are other surprising new twists in the economy of outsourcing. One is "reverse outsourcing," in which Indian and other companies hire American workers because their qualifications cannot be matched in India and other countries. Second, the pride in "Made in America" products has grown great enough, and the costs of intellectual capital, wages, raw materials, real estate, and energy in India and China have risen high enough that many American firms have begun to "onshore" work they had previously offshored. This countertrend, known as **insourcing** (also onshoring, inshoring, or homeshoring), is especially popular in the United States at a time when jobs are scarce. Insourcing can help reduce risks associated with supply chain delays, challenges of language and culture, and the "distance management" of overseas workers.[19]

Vivek Prakash/Reuters

• Figure 7.K It's not Egypt or the Louvre, and it's not your everyday image of India: This is the Media Center of Infosys, one of the world's leading IT firms, in Bangalore, India.

varieties of seed stock, and there has been a large upsurge of production in certain areas where the new strains have been widely introduced. Using a broad array of tools that range from conventional breeding techniques to the precise manipulation of the genetic components of a crop, scientists are not only able to produce a higher yield but also a crop more resistant to drought, flood, disease, or pests and which generates a higher amount of a desired nutritional component such as vitamin A. China, second only to the United States in the global biotechnology industry, has produced genetically engineered rice, corn, cotton, tobacco, sweet peppers, petunias, poplar trees, and more. It is useful to think of the Green Revolution today as having evolved into the "Gene Revolution."

Because the underlying premise of the Green Revolution is economic—more crops will be produced at lower cost for higher revenue—most countries in the region are racing to increase their biotechnological output, especially in an effort to catch up with China and ensure they do not lose positions in the global marketplace. South Korea, Japan, and India have large and well-funded biotechnology research programs, and Malaysia and Indonesia are hoping to build their own research centers as well. Japan long resisted production of GM food crops out of fears of possible health hazards, and to protect the country's traditional farming sector, but in 2013 approved 8 GM food and feed crops, excluding rice. The European-based No!GMO campaign is lobbying for "GMO-Free" zones in Japan, and maintains an atlas of these areas around the world.[19]

One of the most scrutinized GM crops is Golden Rice, currently under development by a multi-country project coordinated by IRRI (• **Figure 7.24**). Endowed with a gene from corn and another from a common soil bacterium, it is the only rice variety that produces beta carotene, which the body converts to vitamin A, as needed. If, as IRRI, the Bill & Melinda Gates Foundation, and other advocates hope, the crop can be made available to farmers around the world, it will contribute to reducing nutritional deficiencies and blindness dramatically. The Indian environmentalist Vandana Shiava called Golden Rice a "Trojan horse" designed to garner public support for all kinds of GMOs that would benefit multinational corporations at the expense of poor farmers and consumers. Responding to such allegations, Penn State biologist Nina Federoff said "It is long past time for scientists to stand up and shout, 'No more lies—no more fear-mongering,' we're talking about saving millions of lives here."[20] Typical of the GMO debate, the passions on both sides are intense. In 2013, Golden Rice opponents trampled one of the experimental field sites of Golden Rice in the Philippines.

For Asian farmers to capitalize fully on the Green Revolution, they must overcome many obstacles in addition to GMO concerns. Success requires levels of capital that are often beyond the means of peasant farmers, landlords, and governments. Such expenditures are needed for water supply facilities (for example, the tubewells that now number hundreds of thousands in the Indo-Gangetic Plain of the northern Indian subcontinent and that are severely depleting the groundwater; see **Figure 3.C** for a map of groundwater depletion), chemical fertilizers, and chemicals to control weeds, pests, and diseases. And as agriculture becomes more mechanized, considerable increases in the costs of machinery, fertilizers, and fuel have to be borne by farmers who have, in many cases, had little experience with the cash economy. Governments must improve transportation so that the large quantities of fertilizer required by the new seed varieties can be delivered in a timely fashion. Not only must there be these associated infrastructural changes to support this "revolution," but also the crop calendar of the farmer becomes much less forgiving because many of the newest seed grains demand more precise water, fertilizer, and cultivation requirements than traditional grains. Grain storage facilities, now subject to plundering by rats, must be improved.

There are other problems associated with the Green Revolution. Economic dislocations result when rice-importing countries become more self-sufficient, causing hardships for rice exporters. Large infusions of the agricultural chemicals

• **Figure 7.24** The IRRI puts a healthy and happy face on the development of "Golden Rice," which has been undergoing a strict series of development protocols.

THINK CRITICALLY: Where do you stand on the issue of this and other GMOs: Are they lifesavers or Trojan horses?

associated with high-yield varieties have negative repercussions on natural ecosystems. India's government subsidizes urea fertilizer, and farmers are naturally tempted to overuse this cheap product. Unfortunately, the overuse of urea degrades the soil, and fertilized soil demands much more water. There is also concern that development of a limited number of high-yield crop varieties grown in vast monocultures will dramatically reduce the genetic variability of crops, in essence interfering 35 with nature's ability to adjust to environmental changes.

7.5 Geopolitical Issues

In the Middle East and North Africa, and in Russia and the Near Abroad, principal geopolitical concerns focus on the production and distribution of energy resources. In South and East Asia, by contrast, some of the most serious geopolitical issues are prospects for what may be done with weapons created from a particular energy source: nuclear energy. There are also concerns about seafloor resources, Islamist terrorism, and the security of shipping lanes. Of these, seabed resources may prove to be the most critical. There are several discussions in this chapter of seemingly insignificant small islands that are extremely critical because of the fossil fuels and other resources that may lie in nearby waters.

The earlier discussion of the mushrooming economic clout of China and India raises a big-picture geopolitical reality: Asia is emerging as a center of gravity that will seriously challenge the century-long primacy of the United States in world 533 affairs. Japan long stood alone as a US-allied pillar of strength in Asia, but with China and even India surging ahead, regional might has been reshuffled. Eyes in the region and around the world will be focused especially on China, which some analysts fear may use its fast-increasing military power to enhance its fast-increasing economic clout. As China flexes its muscles, smaller countries—especially those who have historically experienced Chinese aggression—are calling on the US to extend its security umbrella over them. Geopolitics typically plays out in slow and subtle fashion, but the geopolitics of the Western Pacific is showing real-time in heated contests over territorial waters and the riches that may lie beneath them (see the Regional Perspective on **page 316**).

Nationalism and Nuclear Weapons

As we will see in **Section 7.6**, after its independence from Britain in 1947, India emerged as an avowedly secular democratic state. But after the 1998 victory of the Hindu nationalist Bharatiya Janata Party (BJP), hopes for new privileges emerged among India's Hindu majority. Among BJP supporters, there was hope for a "Hindu bomb," a counterweight to a long-feared "Islamic bomb" in neighboring Pakistan.

On May 11, 1998, much to the surprise of US and other Western intelligence agencies, India conducted three underground nuclear tests in the Thar Desert. With the blasts, India's government seemed to be staking India's claim as a great world power, exhibiting its military muscle to Pakistan and China, and increasing political support for the BJP. Initial reactions among India's vast populace were highly favorable; 91 percent of residents polled within three days of the event supported the tests. Outside India, there was alarm. India had defied an informal worldwide moratorium on nuclear testing that went into effect in 1996, when 149 nations (not including India and Pakistan) signed the **Comprehensive Test Ban Treaty** (also known as the **Nuclear Nonproliferation Treaty**, or **NPT**), which prohibits all nuclear tests.

The world's eyes turned quickly to India's neighbor to the west. Governments pleaded with Pakistan to refrain from answering India with nuclear tests of its own, arguing that Pakistan would have a public relations triumph if it exercised restraint: it would appear to be a mature and responsible power, whereas India would seem to be a dangerous rogue state. But Pakistan's leaders felt obliged by their population's demands for a tit-for-tat response to India's blasts and soon followed with six nuclear tests. India and Pakistan thus joined just five other nations—the United States, Russia, China, the United Kingdom, and France—in acknowledging that they possess nuclear weapons. (Israel does not acknowledge its arsenal. North Korea developed its weapons later.)

There is disagreement about what this regional and global shift in the balance of power means. Some analysts fear an escalating nuclear arms race that, perhaps ignited by a border skirmish in Kashmir or by a terrorist attack like the one in Mumbai in December 2008, could lead to the **mutually assured destruction** (**MAD**; that is, simultaneous launches result in both sides being "nuked") of Pakistan and India. Others, particularly in South Asia, argue that the weapons represent the best deterrent against such a MAD conflict, as they did for decades between the countries of NATO and the Warsaw Pact. However, few people anywhere argue against the view that the nuclear rivalry between India and Pakistan has taken an enormous economic toll in two nations that need to wage war on poverty.

In American geopolitical thinking, both India and Pakistan are **pivotal countries**, defined by several influential historians as countries whose collapse would cause international migration, war, pollution, disease epidemics, or other international security problems.[25] The disposition of their nuclear 491 arsenals is one of the main reasons for their being considered pivotal. If the government of Pakistan were to fall, Pakistan's nuclear weapons could come into the possession of a rogue element or a new government hostile to the West. One of those influential historians, Francis Fukuyama, declared that Pakistan is "the most dangerous place in the world."[26] India has much to fear from Pakistani scenarios. Either a right-wing or an Islamist government in Pakistan would be far more likely than the current regime to take India on in the disputed Kashmir region. Such a confrontation could set the stage for a nuclear exchange between Pakistan and India, not only decimating those countries but also sending economic shock waves around the world. India's recently surging economy has been a major force in India's recent diplomatic overtures to Pakistan; nothing discourages investment like war or the fear of it.

Problem Landscape | Geopolitical Hot Zones in the Western Pacific

Looking at the maps in • **Figures 7.L** and **7.N**, you can see the large marginal seas of the Pacific Ocean, named the East China Sea and the South China Sea. Those names suit China well, but not many of its neighbors. China, in fact, lays claim to about 90 percent of the waters of the South China Sea (enclosed within a maritime boundary known as the "Nine Dash Line," a reference to how this demarcation is usually portrayed on Chinese maps), and many of the countries also claiming those waters have been literally up in arms about the situation. What is China doing in those seas? And what are the risks posed by competing claims, not just between China and its neighbors but also between some of those neighbors themselves? Let's have a closer look at the most important locations and issues (see "By the Numbers," • **Figure 7.M**).

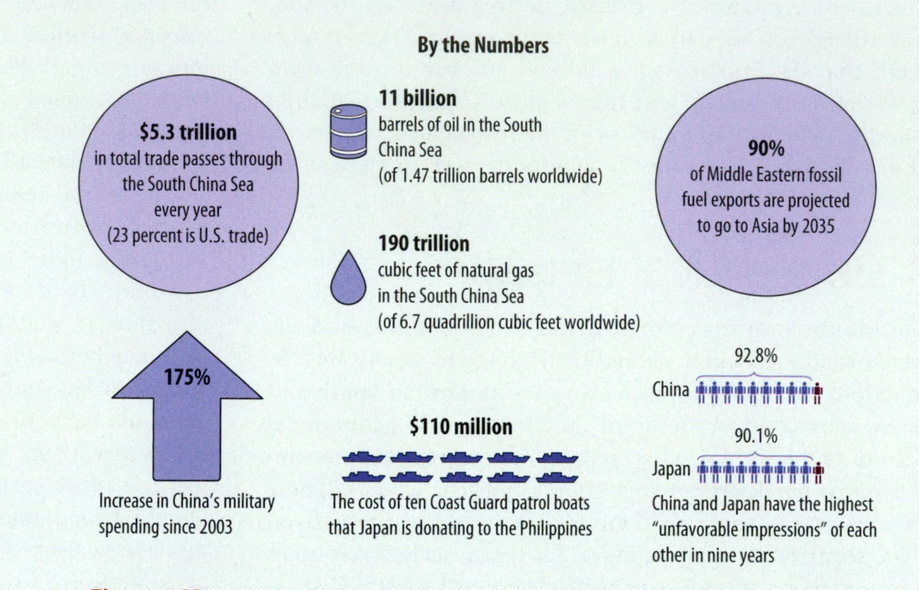

• **Figure 7.M** Energy resources in the Western Pacific are contested by a number of large and small powers. The stakes are high, and so is the potential for military conflict.

About 60 islands make up the Spratly Island chain, which lies in the South China Sea between Vietnam and the Philippines. They are an idyllic tourist destination where divers can hire luxury boats to explore the coral reefs and palm-lined beaches of remote atolls. However, the islands are much more significant for their strategic location between the Pacific and Indian Oceans. During World War II, Japan used the islands as a base for attacking the Philippines and Southeast Asia. Still more significant, as much as $1 trillion in oil and gas may lie beneath the seabed around the Spratlys.

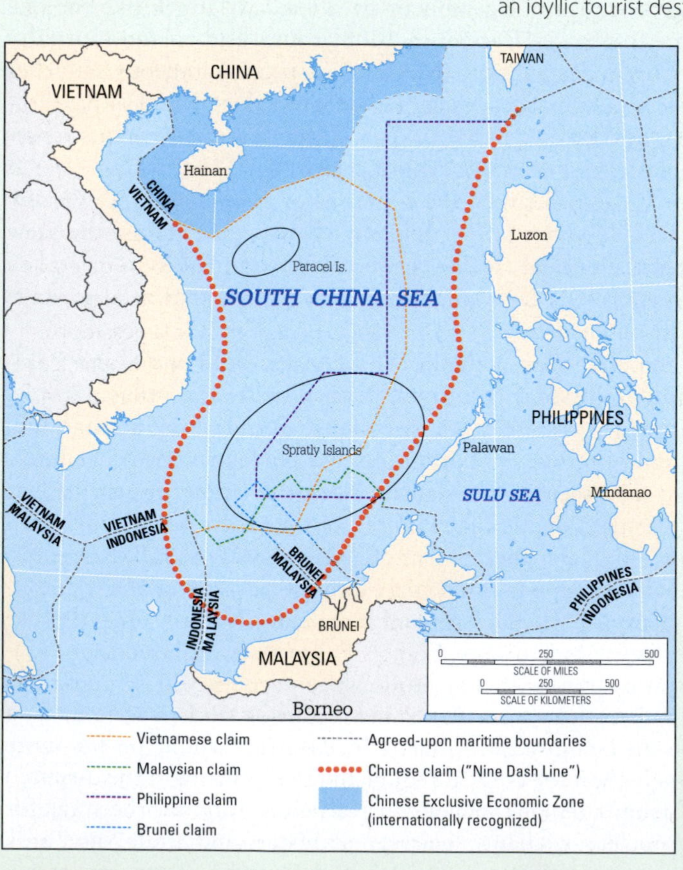

• **Figure 7.L** Maritime boundary disputes in the South China Sea. Boundaries not being contested (usually defined by a treaty between nations) are listed as "agreed upon."

Not surprisingly, many nations covet control of the Spratlys. Six countries claim some or all of the islands: China, Vietnam, and Taiwan claim sovereignty over all of them, and Malaysia, Brunei, and the Philippines claim some of them. During the Cold War, the competing nations felt it was too hazardous to push their claims on the islands. As the Cold War drew to a close, however, the situation became more volatile. All the contenders except Brunei placed soldiers, airstrips, and ships on the islands. In 1988, the Chinese navy invaded seven of the islands occupied by Vietnam, killing about 70 Vietnamese soldiers. In 1992, China again landed troops in the islands and began exploring for oil in a section of the seabed claimed by Vietnam. In 1995, China moved to expand its territorial designs on islands already claimed by the Philippines and built what it called "shelters," which look like fortifications to the Filipinos. Late in the 1990s, Malaysia occupied two disputed Spratly Islands and began building on one of them. In 2004, Vietnam began renovations of an airport on one of the islands, saying it was necessary to boost tourism there. China condemned the construction as a violation of its territorial sovereignty, an accusation it repeated as Vietnam negotiated natural gas and oil development with Indian and other foreign companies in the Spratlys.

The situation moved to the brink of all-out conflict in 2011, when Chinese ships sabotaged Vietnamese oil exploration vessels operating within what Vietnam claimed was its own economic zone in the South China Sea (which Vietnam calls the East Sea). Tensions boiled over again in 2014 when a huge Chinese portable oil rig began drilling amidst an oil field in waters claimed by Vietnam. There were large anti-Chinese demonstrations in Vietnam. Once again on the brink of conflict, China and Vietnam de-escalated the crisis and China withdrew its rig.

As China's need for energy grows, so does its assertiveness over the Spratlys, the nearby Paracel Islands, and the Senkaku Islands, among 327 others. In an effort to project its military power and secure more resources, including petroleum and fish, in 2015 China began constructing six artificial islands in the Spratlys. A runway to accommodate military planes was built on one of these islands, increasing the stakes of the ongoing disputes over sovereignty. Why can't these disputes be settled by normal treaty and other diplomatic instruments? As it turns out, international law is not well equipped to sort out these conflicting claims, even among presumed allies like Japan and South Korea, both protected by defense treaties with the United States.

Some 90 miles (145 km) between the shores of Japan and South Korea lie 34 small, inhospitable islands known as the Liancourt Rocks to outsiders, as Takeshima to the Japanese, and as Dokdo (or Tokdo) to the Koreans. South Korea has controlled the islands since 1956, but with only a handful of Korean civilians and coast guard personnel living on the islands, Japan has periodically asserted its right to them. Since the 1990s, these remote rocks have been the focus of a dispute between Japan and Korea, not because of any riches they contain, but because of a 1970s United Nations treaty known as the **Convention on the Law of the Sea**. This treaty would permit their sovereign power to have greater access to surrounding marine resources (• **Figure 7.N**).

The Law of the Sea began as an effort to apportion ocean resources as equitably as possible and to avoid precisely the kind of dispute that developed between Japan and South Korea. The treaty gives a coastal nation mineral rights to its own continental shelf, a territorial water limit of 14 miles (22 km) offshore, and the right to establish an **exclusive economic zone (EEZ)** of up to 230 miles (370 km) offshore (in which, for example, only fishing boats of that country may fish). The power that controls offshore islands such as Takeshima and Dokdo can extend its EEZ even further.

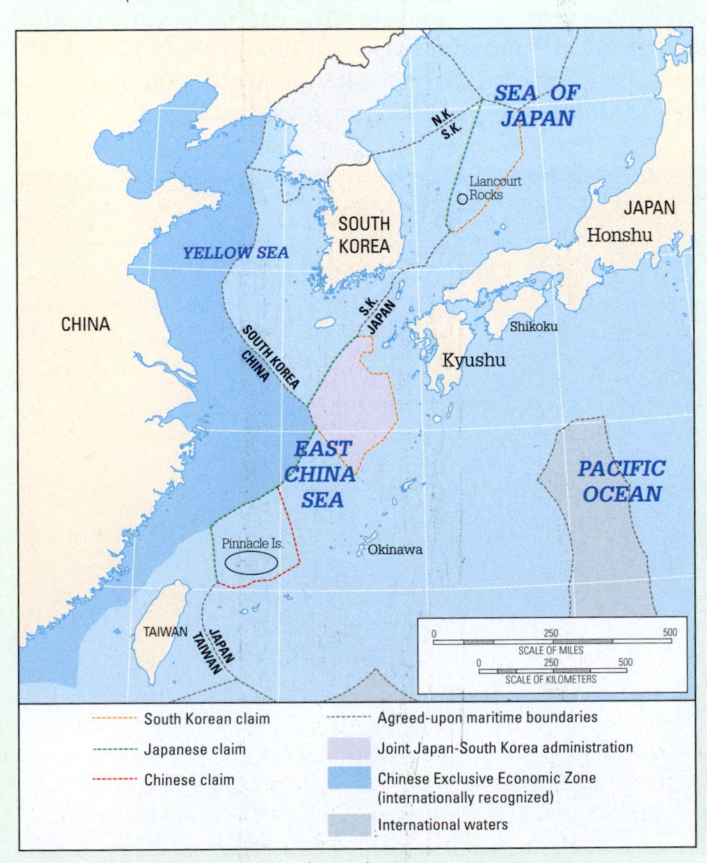

• **Figure 7.N** Maritime boundary disputes in the Sea of Japan (East Sea) and the East China Sea.

By early 1996, some 85 nations had ratified the treaty. South Korea ratified it late in 1995. Japan was preparing to ratify it in 1996, when news reached Tokyo that South Korea had plans to build a wharf on the islands. To avoid provoking Japan, North Korea, and China, South Korea avoided declaring an exclusive economic zone off its waters that would include the islands. Japan ratified the treaty. But fears that South Korea may build facilities and place more people on the islands as a step toward establishing an EEZ, and thereby exclude Japanese fishermen from the area, caused Japan to restate its claim to the islands in 1996. South Korean officials answered with military exercises near the islands, and South Korean civilians staged loud demonstrations outside Japan's embassy in Seoul. South Korea declared the islands a national monument in 2002. Long simmering, the issue boiled up again in 2006 when Japan announced it would send ships to survey the area, and South Korea answered that it would form a naval blockade against the Japanese vessels.

Even more important than Japan's competition with South Korea over territorial waters and islands is its contest with China. Both lay claims to the Pinnacle Islands in the East China Sea, known to the Japanese as the Senkakus and the Chinese as the Diaoyu Islands. Again, both are interested in staking out an EEZ in which oil reserves might be found. Despite being a signatory to the United Nations Convention on the Law of the Sea, China does not accept the UN definition of the EEZ beginning from a country's coastline; it insists on a wider zone that starts at the continental shelf.

The Senkaku, or Diaoyu, Islands are uninhabited, but their control is championed by nationalists on both sides and is therefore a potentially explosive issue. Even though China is Japan's largest trading partner and the largest recipient of Japanese foreign investment, the bitter legacy of World War II relations between these countries, the nationalistic pride on each side, and the potential prize of undersea resources threaten to unravel the much more important economic relationship the two have.

The big contest between China and Japan plays out in seemingly small incidents. During World War II, the Pinnacle Islands were occupied by the United States. In 1971, the occupation ended, and the US government turned over control of the islands to Japan, a move contested by China. In 2004, Chinese activists landed on one of the islands to stake China's claim there, but Japanese authorities promptly arrested them. Since then, there have been several encounters between Japanese and Chinese ships near the islands. The worst was the 2010 incident known as the "Senkaku Shock," which

Problem Landscape Geopolitical Hot Zones in the Western Pacific *(Continued)*

began when a Chinese fishing vessel rammed a Japanese patrol boat near the islands. Japan detained and refused to release the Chinese captain. China responded by halting exports of "rare earth" minerals needed in Japanese high-tech industrial products. By geological accident, and good fortune for China, many of the 17 rare earth elements are concentrated in 331 China. These are indispensible in cell phones, laptops, and many "green" technologies, like wind turbines and hybrid vehicles. They are also important in military technology, and the United States regards access to them as a matter of national security.

Japan's formal claim to the islands dates to 1895, when Japan declared that the islands had previously been *terra nullius*, or no one's property. But China insists that there is Chinese documentation about control of the islands dating to the Ming Dynasty in the 15th century. The islands are close to Taiwan, which also lays claim to the islands, so any move by mainland China to secure access to them could also precipitate conflict between the two Chinas.

"China has indisputable sovereignty over the South China Sea islands and adjacent waters," a Chinese Foreign Ministry spokesman asserted in 2011.[23] Such a bold claim makes smaller powers in the region fear that China could turn small islands into "permanent aircraft carriers" that China could use to dominate them militarily. China even used a tiny submarine to plant a Chinese flag on the bed of the South China

Pakistan since 9/11

It is ironic that Pakistan focuses so much on the Indian threat and nuclear weapons when the greater and immediate threat is within, in the form of Islamist insurgencies and rogue intelligence authorities. Pakistan's internal security problems proliferated after 9/11, when Pakistan became even more pivotal to US interests. Up to that point, the United States had slapped tough economic sanctions against Pakistan (and less severe ones against India) for the nuclear weapons tests. The United States pursued a diplomatic courtship with India, both as a counterweight against China (India's longtime foe in the region) and as a means of expressing displeasure with Pakistan (China's ally) for helping Afghanistan's Taliban harbor Osama bin Laden. But the United States and Pakistan did an abrupt diplomatic about-face and embraced one another after 9/11. Pakistani President Pervez Musharraf instantly dropped support for the Taliban and quietly allowed the United States to use its territory to prepare for the assault on the Taliban and al-Qa'ida in Afghanistan. In return, the United States forgave much of Pakistan's debt to the United States and lifted its sanctions against Pakistan. To avoid isolating India, the United States also lifted its post-nuclear test sanctions against that country.

The United States has dramatically strengthened military ties with both countries, recognizing Pakistan in 2004 as a major non-NATO ally—one of just 15 countries to have that designation as of 2015—and acknowledging India the same year as a "strategic partner." These designations opened the doors of both countries to major new shipments of American war matériel. But India would be favored: in 2007, Washington singled out India as its most critical strategic ally in the region by offering nuclear fuel and technology to New Delhi. That began a turnaround in US relations with India that blossomed with a series of bilateral agreements on trade and arms. The two countries are developing a strategic partnership based on converging geopolitical interests: what they share most is deep concern about China's might, especially its military muscle power. In South Asia, the proxies for great powers are lined up, with India in the US camp and Pakistan aligned with China.

Under US and allied attacks in Afghanistan after 9/11, Taliban and al-Qa'ida fighters retreated to and regrouped in western Pakistan, particularly in the semiautonomous Federally Administered Tribal Areas (FATA; see **Figure 7.39**). Here the people are mainly Pashtun, including many sympathetic to the causes of their Taliban ethnic kin and their al-Qa'ida spiritual kin (• **Figure 7.25**). This is the stronghold of the **Pakistani Taliban (Tehriq-i-Taliban Pakistan, or TTP)** forces, who are closely allied with the Afghan Taliban across their porous border. Pakistani government authority has

Joe Hobbs

• **Figure 7.25** The Pashtun of western Pakistan and Afghanistan are a tight-knit ethnic group with a long history of defiance against foreign rule. Most are strongly opposed to American interests in the region.

Sea. Not surprisingly, the United States took a position in this geopolitical chess match: in 2010, Washington declared that the South China Sea was of strategic importance to the United States. The United States is solidifying its position to offset China: it has an alliance with the Philippines and has drawn closer militarily to Vietnam. During his two terms in office, even while distracted by conflagration in the Middle East, US President Obama spoke repeatedly of America's "Pivot to Asia." In large part this new engagement was prompted by appeals from China's smaller neighbors, who fear China's growing military assertiveness. The US must be careful not to overplay its hand by fueling more military buildup in the region, however. "The danger is that 21st century Asia could go the way of 20th-century Europe," warned Council on Foreign Relations Chair Richard Haass.[24]

Clearly, an incident or miscalculation or a sustained naval offensive in these remote islands and their waters could trigger a much wider and more serious conflict in Asia, and draw in the United States with its defense pacts with Japan, South Korea, and the Philippines. The stakes are enormous, especially in view of the high value of trade in the Western Pacific (see "By the Numbers"). If the Law of the Sea cannot resolve competing claims, they could be considered in the International Court of Justice. There are great needs for instruments that could head off conflict: resource-sharing agreements, a crisis management system, and better military-to-military open lines of communication are among them.

long been kept at bay in the FATA, where it is extremely difficult to insert Pakistani forces on a sustained basis. Bowing to American pressure, but also because of growing domestic concerns, Pakistan's government periodically undertakes military operations in the FATA against Taliban and other insurgents. But these militants, including the Taliban-allied Haqqani Network, are so entrenched and well armed in the North Waziristan area of the FATA that it has been especially difficult to mount an effective military campaign against them there.

In 2014, an especially cruel TTP attack on a school in Peshawar (just north of the FATA), in which more than 140 children were massacred, may have been a watershed event in Pakistan's war on jihadists. Across Pakistan, people demanded a long-overdue, resolute crackdown on terrorism. Pakistan is very concerned about the drawdown of NATO troops in neighboring Afghanistan, fearing this will empower the militants in both countries and possibly allow the Pakistani Taliban to take and hold territory within Pakistan. 334

The United States is deeply unpopular in Pakistan, so the sustained insertion of US troops in Pakistan (in search of al-Qa'ida, for example) would be very destabilizing, perhaps even resulting in rebellion or a coup against Pakistan's government. This could lead to the rise of an anti-American administration in Pakistan and raise questions about the disposition of Pakistan's nuclear weapons. To avoid that scenario, the US military has relied heavily on predator drones to stalk and kill al-Qa'ida and Taliban suspects in Pakistan. A major source of anti-Americanism in Pakistan is the unintended killing of many civilians ("collateral damage") inflicted by these drone strikes.

Even those unmanned missions are politically risky for the United States, so imagine the risk involved in sending US troops on the 2011 mission that killed Osama bin Laden, 243 deep inside Pakistani territory. The fact that bin Laden had lived not far from Pakistan's capital, and just around the corner from a Pakistani military academy, raised many questions about the quality and the loyalty of Pakistani intelligence forces. Many analysts and politicians in the United States asserted that elements within Pakistan's intelligence service, the ISI (Inter-Service Intelligence Agency), had helped protect bin Laden. There are also well-founded suspicions that Pakistani intelligence and elements within the armed forces continue to support the Taliban both in Afghanistan and Pakistan. American military and intelligence communities often perceive Afghanistan and Pakistan as a single "theater" of operations and have a neologism for it disliked by both Afghans and Pakistanis: *AfPak*.

What Does North Korea Want?

There are also major concerns about nuclear weapons and terrorist alliances in Northeast Asia. Ever since suffering Hiroshima and Nagasaki, Japan has had an official policy never to develop or use atomic weapons. But Japan worries about the potential nuclear threat from three adversaries, all of which possess nuclear weapons: Russia, China, and North Korea. The Japanese feel that the West dismissed potential threats from Russia too readily when the Soviet Union dissolved. Japan still has territorial disputes with Russia, particularly involving four Kuril Islands (Kunashiri, Etorofu, Shikotan, and Habomai) lying just off the northeastern coast of Hokkaido, which the government of Josef Stalin seized at the end of World War II. To show that Russia was intent on maintaining its control, Russian President Medvedev visited one of the islands in 2010 and talked about raising living standards there. The Japanese were furious. 197

Japan does not favor the reunification of the Korean peoples, who, having been colonized and often treated with cruelty by the Japanese between 1910 and 1945, still resent the Japanese. The bigger issue in relations is that Japan is pondering the unthinkable and the officially disavowed—the development of nuclear weapons—to counter the threat to Japan posed by North Korea's nuclear weapons. With all of the instability in the region, at the very least Japan wants to beef up its Self-Defense Forces, which by postwar treaty were left with little power. By that treaty, the United States is the guarantor of Japan's military security. To protect Japan and project its power in the region, the US has 50,000 military personnel stationed in Japan, primarily on the southern island of Okinawa.

The world looks nervously at the troubled relations between North and South Korea and between North Korea and the West. A crisis flared in 1994 when North Korea refused to permit full inspection of its nuclear facilities by the International Atomic Energy Agency (IAEA). The country was suspected of separating plutonium that could be used in making nuclear bombs. Over time, the United States, South Korea, and Japan worked out an agreement in which North Korea would agree to freeze development of nuclear weapons in exchange for the others providing fuel oil and assistance in building nuclear power plants. Those nuclear reactors would be of the light-water variety, much less likely than North Korea's plutonium-based reactors to be "dual use" for both military and civilian purposes. The Clinton administration trumpeted the agreement as a success, but it had several flaws. One was that North Korea did not have to dispose of its existing nuclear fuel; it simply had to stow it safely away—meaning that it could quickly be reactivated.

President George W. Bush in 2002 proclaimed that North Korea was one of three countries making up an "axis of evil," 246 along with Iraq and Iran. Late that year, North Korea dropped a virtual bombshell on the United States by admitting—after being confronted with evidence collected by US intelligence agencies—that it did indeed have an active nuclear weapons program (developed, as it turns out, with Pakistani technical assistance). This admission followed earlier strident denials of such a program by North Korean officials.

North Korea's admission could have brought an end to the economic assistance in the form of food and fuel that the United States, South Korea, and Japan had been providing for a decade to the beleaguered country. What, then, was North Korea hoping to accomplish by revealing its nuclear weapons program? There are several possibilities, but it is often difficult to divine North Korea's intentions. The least likely is that North Korea was instigating a military confrontation with the United States and its allies South Korea and Japan. In any war scenario, those allies would obliterate North Korea. The United States, which has 30,000 military personnel stationed in South Korea, is obliged to defend South Korea in the event of a war with the North and would effectively wield its huge military advantage (•**Figure 7.26**). However, that victory would come at an enormous price for South Korea and possibly Japan. North Korea has a standing army of nearly 1 million soldiers, an impressive military arsenal including short- and medium-range missiles, a stockpile of chemical and biological weapons that it might not be shy to use, and, judging from underground tests in 2006, 2009, and 2013, presumably at least a few nuclear weapons that it could deploy. The casualties in South Korea would be enormous, whether from a conventional or an unconventional assault from the North.

It is much more likely that by disclosing its nuclear weapons program, North Korea sought guarantees that it could avoid war and also gain even more assistance from the West. For years, North Korea had only to mention the prospect of reactivating its nuclear weapons program to get more concessions from the United States and its allies—especially food aid during a succession of droughts and floods in the 1990s. In effect, North Korea was extorting money from the United States and its allies, which have anted up as a means of containing the rogue nation.

A series of on-again, off-again negotiations known as the Six Party Talks (held between North and South Korea, the United States, China, Japan, and Russia) culminated in 2007 with an agreement: the United States would help unfreeze North Korean funds in a Macau bank, remove North Korea from its list of countries supporting terrorism, lift trade sanctions in place since the Korean War ended in the 1950s, and, along with its negotiating partners, restart the flow of fuel oil to North Korea; in return, North Korea would close its nuclear plants and allow them to be inspected by international monitors. This agreement was never implemented, however. Talks broke down as North Korea continued testing missiles, and restarted its Yongbyon nuclear facility. North Korea has stated it is no longer bound by any of the Six Party Talks

Joe Hobbs

• **Figure 7.26** A little humor eases the tension for US troops serving along what may be the world's tensest border, the "no-man's-land" in the DMZ separating North and South Korea. Even a minor incident here could touch off a conflagration that would take a huge toll in human lives.

agreements, and will never return to the negotiating table. The United States hopes that the involvement of China and the other countries will make North Korea less likely to expand its nuclear program, but the regime of Kim Jung Un is unpredictable at best. Even China, its closest ally, is fatigued by the way in which North Korea repeatedly probes and tests the capabilities and patience of South Korea and the United States through missile launches, territorial incursions, and over-the-top rhetoric about annihilating its enemies.

Islands, Sea Lanes, and Islamists

315 Indonesia is another pivotal country from the US perspective. It is an ethnically complex nation whose integrity has at times been threatened by a variety of secessionist movements, all the way from Aceh (pronounced "a-chay") in the country's far west to Papua in the east (see **Figure 7.56**). The United States and other countries fear what would happen to vital international shipping lanes should several new and perhaps militant states emerge from any fragmentation of Indonesia. Washington is concerned that any anti-Western states that could emerge would threaten American interests in the country—notably its oil, natural gas, and copper resources and its location astride vital shipping lanes.

There have been scores of pirate attacks on vessels plying the strategic Strait of Malacca, a critical chokepoint between peninsular Malaysia and the island of Sumatra through which a quarter of the world's trade passes, including two-thirds of the world's shipment of liquefied natural gas and half of all sea shipments of oil (it ranks second only to the Strait of Hormuz 236 as an oil shipping lane). Security in the Strait of Malacca is a source of great concern to Japan, which imports 80 percent of its oil on ships that pass through the strait. Aceh's vast natural gas deposits and Papua's copper and gold are also seen as critical in the global economy; like Indonesia, many world powers are anxious to see that they stay in the "right" hands.

US intelligence agencies have kept a watchful eye on Southeast Asia, and Indonesia in particular, fearful that it might emerge as an Islamist terrorist hearth. Although Indonesia is the country with the largest number of Muslims, it is a secular rather than a religious state, and a moderate form of Islam prevails in politics and in everyday life. Still, some Indonesians resent the United States, perceiving it as anti-Muslim in the wake of 9/11, despite the generous amounts of foreign aid the United States provides to their country. What concerns Western intelligence authorities and Indonesia's moderate leadership are the more militant Islamist organizations that gathered strength after 9/11. One of these is **Laskar Jihad**, created in 2000 to mobilize Indonesian Muslims to fight Christians in Indonesia's Sulawesi and Muluku islands and Americans throughout the country. Another is **Jemaah Islamiah (JI)**, led by Abu Bakar Bashir, whose nominal role was preacher in an Islamic school. Western authorities are much concerned about such religious schools, known as *pesantrens*, in Indonesia. Like the *madrasas* of Pakistan and the Arab heartland of the Middle East, these are venues for both religious and secular learning, and some of the schools' curricula are imbued with virulent anti-Western content. With some exceptions, the Islamist threat emanating from Indonesia has subsided considerably in recent years, during which the country has enjoyed steady economic growth around 5 percent per annum.

This concludes an introduction that has set the stage for further exploration of land and life in the subregions and countries of South and East Asia. The journey continues in the Indian subcontinent.

7.6 Regional Issues and Landscapes

South Asia

South Asia has an enormous range of ethnic groups, social hierarchies, languages, and religions. This is the most culturally complex area of its size on Earth. Its cultures have roots that are deep in antiquity. Its modern engagements with the world are very important in the realms of security, economics, and population, among others.

Faith, Sectarianism, and Strife

The Indian subcontinent, depicted in **•Figure 7.27**, is one of the world's culture hearths. It is rich in a historical legacy and modern diversity of ethnicities, social practices, and faiths; India has no fewer than 23 official languages (with at least half a dozen more in official use at the state level), for example. Ironically, this cultural wealth often threatens the well-being and security of the subcontinent's peoples and those of nearby Sri Lanka.

There are major social and political divisions within the subcontinent that relate to religion. Religious differences between the two largest religious groups—Hindus (whose religion is described on **pages 304–305**) and Muslims (whose religion is described in **Chapter 6**)—are often troublesome. Hinduism was the dominant religion at the time Islam made its appearance in this region. Muslims are majorities in Pakistan (97 percent) and Bangladesh (90 percent), and they are a significant minority in India. Muslims make up a seemingly small percentage of India's population—13 percent—but they number nearly 170 million people, making India home to the world's third largest Muslim population, behind Indonesia and Pakistan.

Seldom in history have two large groups with such differing beliefs lived in such close association with each other as Hindus and Muslims in the subcontinent. Islam holds to an uncompromising monotheism and prescribes uniformity in religious beliefs and practices. Hinduism is monotheistic for some believers, but polytheistic for others, and asserts that a variety of religious observances is consistent with the differing natures and social roles of humans. The exuberant and loud celebrations of the Hindu faith are a striking contrast to the austere ceremonies of Islam. Islam has a mission to convert others to the true religion, whereas most Hindus regard proselytizing as essentially useless and wrong; one is born or reborn as a Hindu, so conversion is not an issue. Islam's concept of

• **Figure 7.27** Reference map of South Asia. In recent decades, some of India's cities and states changed their colonial-era names to more authentic Indian ones. Some city examples are Bombay (now Mumbai), Calicut (Kozhikode), Calcutta (Kolkata), Cochin (Kochi), Madras (Chennai), Pondicherry (Puducherry), Poona (Pune), and Trivandrum (Thiruvananthapuram). The state names of Uttaranchal (Uttarakhand) and Orissa (Odisha) have also been changed.

the essential equality of all believers is in sharp contrast with the inequalities of the caste system embedded in Hinduism (see **page 328**). Islam's use of the cow for food is anathema to Hinduism. And then there are religion-based politics, which often turn differences into discord and conflict.

In 18th-century India, the long-dominant Mughal Empire, led by Muslims, was warring with the Hindu-led Maratha Empire. The subcontinent was weak and unstable, which allowed the British East India Company and its private armies to occupy many key coastal cities; eventually the company's influence expanded until it ruled nearly all of India. In 1858, the East India Company was nationalized, and rule over India passed to the British crown. The "British Raj" primarily promoted Hindus to work in the civil service, and Hindus came to dominate India's economy. The Indian Congress was also primarily Hindu, and in the 1940s as discussions about Indian independence began, many Muslims feared being incorporated into a state with a Hindu majority. Muslims demanded a country of their own, and when independence came in 1947, the colony was partitioned into the secular but predominantly Hindu nation of India, and the Islamic nation of Pakistan (**•Figure 7.28**).

Originally, Pakistan constituted two separate parts (East and West) separated by over 1000 miles (1600 km) of Indian territory. The residents of East Pakistan, though Muslim, were culturally, ethnically, and linguistically different from the people of West Pakistan, and resented the west's political dominance and economic exploitation of their resources. Civil war between the two Pakistans broke out in 1971. India militarily aided East Pakistan; their combined armies defeated the West Pakistan forces, and East Pakistan became the independent country of Bangladesh.

Immediately preceding and following partition in 1947, violence broke out between Hindus and Muslims, and hundreds of thousands of lives were lost in wholesale massacres. More than 15 million people migrated between the two countries. Pakistan today has a large population known as **Mohajirs**, or "migrants," the Muslim immigrants and their descendants who poured into the two Pakistans from India when partition occurred. Particularly in Karachi and elsewhere in the southern province of Sindh, violent conflict frequently occurs between rival factions among the Mohajirs, and between the Mohajirs and other ethnic groups, including Pashtuns and Biharis. Large-scale migrations of ethnic groups like these are part of humanity's troubled experience.

The partition of colonial India precipitated all kinds of problems for the new countries. Some of them are still prominent today. Since 1947, India and Pakistan have been in conflict over the status of Kashmir, a disputed province straddling their northern border (**•Figure 7.29**). Before independence, Jammu and Kashmir was a princely state administered by a Hindu maharaja. About three-fourths of Kashmir's estimated population of 15 million is Muslim, which is the basis of Pakistan's claim to the territory. But under the partition arrangements, the ruler of each princely state was to have the right to join either India or Pakistan, as he preferred. Kashmir's Hindu ruler chose India, which is the legal basis of India's claim.

After partition, fighting between India and Pakistan led to the demarcation of a cease-fire line—the "line of control" that still separates the forces—leaving eastern Kashmir, with most

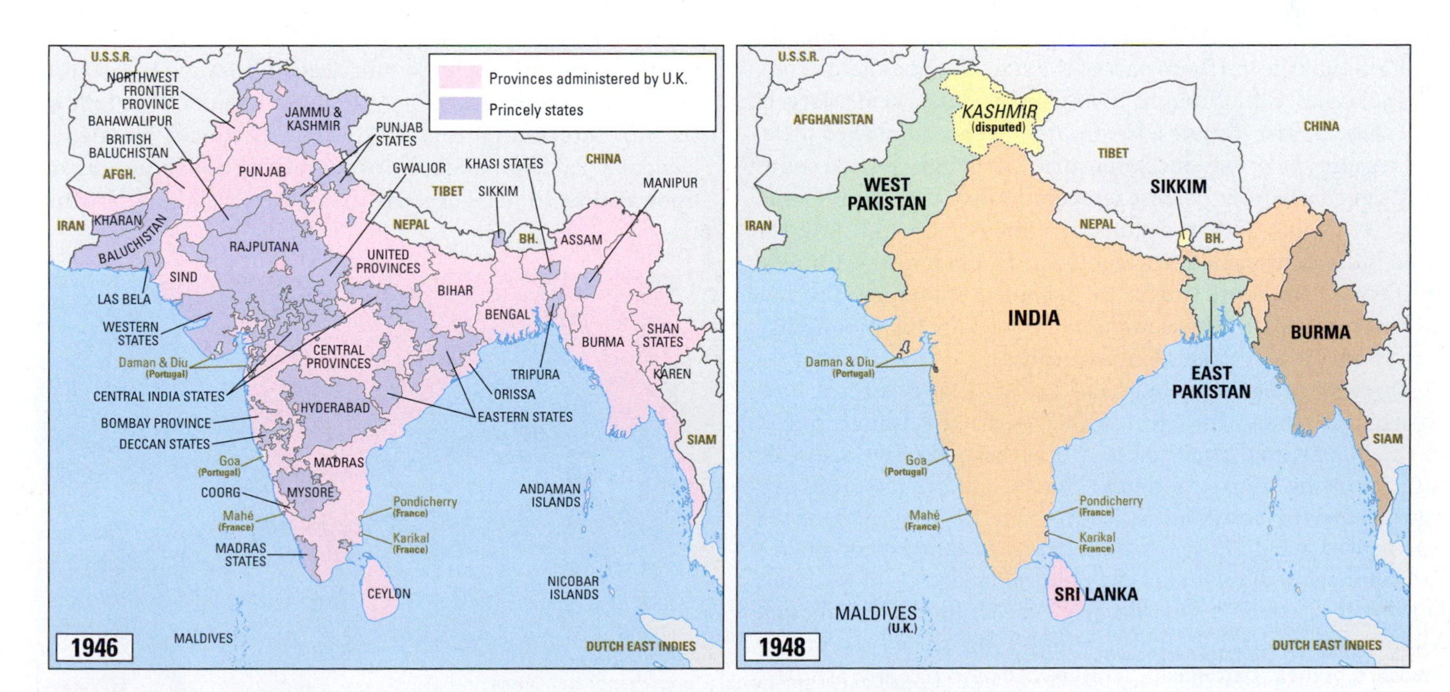

• Figure 7.28 Pre- and post-independence political units of South Asia. Before India and Pakistan were partitioned in 1947, they comprised the single unit of British India (sometimes called the Indian Empire). British India was made up of several provinces governed directly by the British and over 500 autonomous "princely states" controlled indirectly by Britain through a local Indian ruler. In 1948, Sri Lanka, Burma, and Sikkim also achieved independence from the UK (the Maldives remained a colony until 1962). Sikkim gave up its independence to join India as a state in 1975.

• **Figure 7.29** The political borders, areas of control, and religious composition of the disputed territory of Jammu and Kashmir.

of the state's population, in India, and the more rugged western Kashmir in Pakistan. Beginning in the mid-1950s, China pressed its own claims to remote northern mountain areas of Kashmir and occupied some of the border territories. Pakistan ceded part of the occupied territory to China and established friendly relations with its giant northern neighbor, but India rejected China's claims (relations between these countries remain chilly; see **page 327**). India's subsequent small-scale military actions failed to dislodge Chinese forces. India now holds about 55 percent of the old state of Kashmir; Pakistan, 30 percent; and China, 15 percent. (India has another border dispute with China over India's northeastern state of Arunachal Pradesh; China claims the northern part of the state but does not occupy it.) India and China fought a war in the early 1960s largely over these border disputes, but the conflict accomplished little, and tensions have remained ever since. Although the two countries signed a border defense cooperation pact in 2013, China sent troops into Indian-claimed portions of Ladakh in 2014 after India constructed a watchtower to observe the Chinese military on the other side of the disputed boundary. The improvement in US-India relations and especially their joint military cooperation irritates China greatly.

Three wars between India and Pakistan have effected little change in Kashmir. In 1965, conflict began in Kashmir, spread to the Punjab, and escalated to a brief but indecisive full-scale war involving tanks, airborne forces, and widespread air raids. Renewed hostilities in Kashmir in 1971, as part of the war in which India supported the revolt of Bangladesh against Pakistan, again did not alter the political landscape of Kashmir. During that conflict, the Siachen Glacier in the Karakoram Mountains, at about 20,000 feet (6000 m), earned the title of "world's highest battlefield" (**•Figure 7.30**). As recently as 2011, negotiations to demilitarize the glacier have failed, but the Indian and Pakistani governments are considering a proposal to make the Siachen Glacier an "international peace park." Without progress on this front, it is proving difficult for the two countries to normalize relations.

Kashmir's Muslim majority escalated the campaign for secession from India in 1989, and the strife has continued ever since. India and Pakistan came perilously close to the brink of war once again over Kashmir in 2002 after armed insurgents attacked India's parliament in New Delhi in a failed attempt to assassinate Indian politicians. India's government identified the assailants as members of two Muslim rebel movements (Lashkar-e-Taiba and Jaish-e-Muhammad) based in Kashmir. A far more spectacular attack—also attributed to these two groups and with possible links to al-Qa'ida—took place in December 2008. In India's commercial capital of Mumbai, scores of civilians were killed and injured in multiple terrorist strikes against a Jewish center, transport stations, and hotels and restaurants frequented by Westerners.

The upper portion of the Indus River and many of its tributaries are in Kashmir, and both India and Pakistan seek as much control as possible over these waters. The geopolitical contest for this water threatens to become very dangerous. The militant Lashkar-e-Taiba has threatened to blow up India's dams in its portion of Kashmir. The **Indus Waters Treaty** of 1960 prescribed exactly how much water each side could use (80 percent to Pakistan, 20 percent to India), and India's dams have so far not violated that treaty. Pakistan fears, however, that India could use its dams as a weapon to deprive downstream Pakistan of water. Pakistan has sent chilling warnings 239 to India, for example, in a newspaper editorial that read, "Pakistan should convey to India that war is possible on the issue of water and this time war will be a nuclear one."[27]

In the 2002 and 2008 incidents, Indian citizens and many government authorities laid blame on Pakistan's government. Foreign diplomats exerted tremendous efforts to avert war between the South Asian giants. Pakistan banned the militant groups after the 2002 assault (causing many Pakistanis to blame their government for "selling out" to the West following the 9/11 attacks), but the 2008 Mumbai attacks revealed how incapable Pakistan might be of preventing dangerous provocations against India. There are also questions about how much

• **Figure 7.30** Siachen Glacier in western Kashmir is known as the "world's highest battlefield." Here, Pakistani forces are lugging their weapons across the ice field in the height of summer.

control Pakistani leaders have over their own intelligence service, the ISI, which may contain rogue elements intent on continued support of the militants fighting India.

The Kashmir conflict has been very costly to both sides, with more than 65,000 people killed since 1965, about half of those since 1989. India's international reputation suffered because of the country's unyielding position on the region. Pakistan has steadily spent about a third of its budget on defense, mostly focused on a potential engagement with India over Kashmir. There would be economic benefits for both India and Pakistan if lasting peace were achieved. "Kashmir is a Paradise on Earth," wrote Samsar Kour.[28] The stunning snow-crowned peaks and flower-laden valleys of this western Himalayan region (more habitable than the Pakistani-held portion of Kashmir) once lured millions of Indian tourists and ranks of Western trekkers each year, and the tourism development potential is enormous.

Sectarian violence (known in the region as **communal violence**) between Hindus and Muslims is a frequent threat to India's social and political fabric. Contention over places sacred to both faiths sometimes ignites widespread violence. The 16th-century Babri Mosque in the city of Ayodhya in Uttar Pradesh (location in **Figure 7.29**) was a place of prayer for Muslims on a spot revered by the Hindus as the birthplace of the god-king Ram, an incarnation of the god Vishnu. Backed by Hindu nationalists in the provincial government, a huge mob of about 250,000 Hindu fundamentalists demolished the mosque in December 1992, with the intention of building a Hindu temple on the site. The ensuing communal violence between Hindus and Muslims throughout India, even among the cosmopolitan and usually tolerant urbanites of Mumbai, left thousands dead.

Ever since this incident, as a means of garnering votes from Hindu Indians, Hindu politicians have made veiled promises to proceed with construction of the temple. In 1998, the BJP platform included a pledge to build a Hindu temple atop the ruins of the mosque at Ayodhya. His apparent support for the temple helped sweep the BJP leader, Atal Bihari Vajpayee, into power as prime minister that year. A 2010 high court ruling divided the site, with two-thirds going to Hindus and one-third to Muslims. The conflict over this multireligious sacred space in India has parallels with the Temple Mount dispute between Israel and the Palestinians. 253 Politicians in India today are increasingly aware of the threat that Hindu-Muslim discord poses to the country's development. Indian politicians need to be attentive to their constituents; theirs is the world's largest democracy. Pakistan, in contrast, has alternated between democratically elected leaders and military-led dictatorships several times.

In India's Punjab state, where Sikhs make up 60 percent of the population and Hindus 36 percent, the 1980s and early 1990s were violent years during which about 20,000 people died in armed clashes. Sikh factions were intent on establishing their own homeland of **Khalistan**, prompted to do so in part by New Delhi's plans to divert water from Punjab. They challenged Indian authority and turned Amritsar's Golden Temple, the holiest site in the Sikh faith, into their military stronghold. In a controversial move to quell the revolt, Prime Minister Indira Gandhi ordered troops to storm the Golden Temple in June 1984. The resulting deaths and desecration led directly to Gandhi's assassination by her own Sikh bodyguards later that year. Indian authorities accused Pakistan of arming the Sikh militants.

Other trouble spots for India are the northeastern states of Assam, Manipur, Nagaland, and Arunachal Pradesh, an extraordinarily diverse region with more than 200 ethnic groups and followers of many different religions. The government has been investing more resources in these states, especially in infrastructure to facilitate trade with adjacent Myanmar. Several rebel groups have been active in this region for years, agitating for autonomy and independence from India. As many as 40,000 Maoist insurgents, known as **Naxalites**, claim to champion the causes of the rural poor and promise to carry violent rebellion into India's cities. Although governmental initiatives for reducing poverty and containing the Naxalites have had some successes in reducing their numbers since 2004, Naxalites continue to carry out deadly attacks. Naxalites are most active in parts of an area called the **Red Corridor**, a stretch of land in central India encompassing the states of Bihar and Jharkhand (two of the poorest states in India) and remote, rural areas of the states of Odisha, Chhattisgarh, and Telangana (**• Figure 7.31**).

Geographic patterns have made the Naxalite issue of great importance to the government in New Delhi. The Maoists are active in resource-rich areas that also happen to be areas where poor and marginalized ethnic peoples live; these tribal groups may be especially receptive to the Naxalite messages about disrupting India's economic and political systems. Much of the

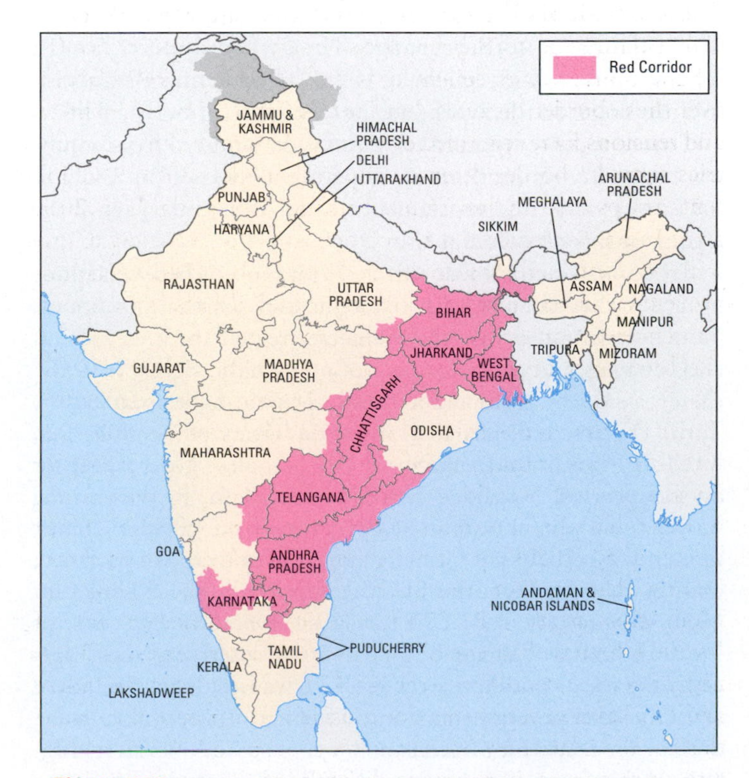

• Figure 7.31 India's "Red Corridor." Many of India's poorest areas lie within the Red Corridor, where the Naxalite insurgency is strongest.

Red Corridor overlaps with the distribution of many of India's most important natural assets, including coal, iron ore, bauxite, manganese, nickel, and copper, so controlling the Naxalite rebellion is a major national security issue for India. If India can provide more justice and economic security to the country's roughly 700 tribal groups, whose population is about 85 million, the Maoist alternative would be less attractive to these poor peoples. Following is an example of a legal resolution favoring indigenous interests.

Ancient tribal traditions were in deadlock with modern industry in a conflict over land use in the eastern state of Odisha. In 2008, a British mining company wanted to tap the state's rich deposits of bauxite (a major source of aluminum), iron ore, and coal, and develop mills near them. But the animist tribal people living in the forests atop those mineral deposits insisted that their identity is tied to the land, forest, and water. If they were displaced by mining, they said, they would lose their cultural and social identities. The mining and mills proceeded, and the tribe sued. The case made it all the way to the Indian Supreme Court in 2012. In what was considered a landmark ruling, the court said the native tribes had the final say on whether the mining operations should continue. The tribes voted unanimously against the project, and India's environment ministry closed the mining operations. 393

Known to early Arab seafarers as Seredib, the "island of Serendipity," Sri Lanka is a breathtakingly beautiful country of 20 million people. Sri Lanka has its own tense ethnic divide. Its two major ethnic groups, distinguished from each other by race, religion, and language, are the predominantly Buddhist **Sinhalese**, making up about 75 percent of the population, and the Hindu **Tamils**, about 12 percent. The lighter-skinned Sinhalese are an Indo-Aryan people who settled in Sri Lanka about 2000 years ago. The darker-skinned Tamils, whose main area of settlement is the Jaffna Peninsula and adjoining areas in the north, are descendants of early invaders and more recent imported tea plantation workers from southern India. The migrations of Tamil tea workers from India to Sri Lanka, and their subsequent travails there, are part of the legacy of British colonial rule. 307

Sri Lanka's civil conflict reflected antagonisms between ethnic groups and discontent with economic and political conditions; Tamils complained that the Sinhalese treated them as second-class citizens and deprived them of many basic rights. In 1983, these grievances prompted a Tamil uprising; fighters called the **Tamil Tigers** (officially the Liberation Tigers of Tamil Eelam, or LTTE) fought Sri Lanka's majority Sinhalese government for an independent homeland called Tamil Eelam. In the ensuing 25-year civil war, more than 70,000 people were killed and many more made homeless. Repeated peace talks and cease-fires failed, and the rebels held much of northern and eastern Sri Lanka for years (**•Figure 7.32**). The LTTE pioneered several methods of sowing terror that would later be used by Islamist terror groups, especially the use of suicide vests. The LTTE was gradually weakened after the international community declared it a terrorist group, and its procurement of funds and weapons became more difficult. Unrest in rebel-held areas made it particularly difficult to rebuild northern sections of the area hit hard by the December 2004 tsunami. After the failure of a cease-fire in 2006, government troops launched a final major offensive against the group, taking back all LTTE-held territory until the group had no choice but to surrender in 2009 (**•Figure 7.33**). In the years since the return of peace, Sri Lanka has successfully reopened doors to international tourism and investment. Sri Lanka now has the highest HDI and GDP (PPP) in South Asia. 338

Arguably, one of the victors in the Sri Lankan civil war was China, which sided with the government and financed the construction of the new port of Hambantota in the country's south. China's apparent generosity was in fact self-serving,

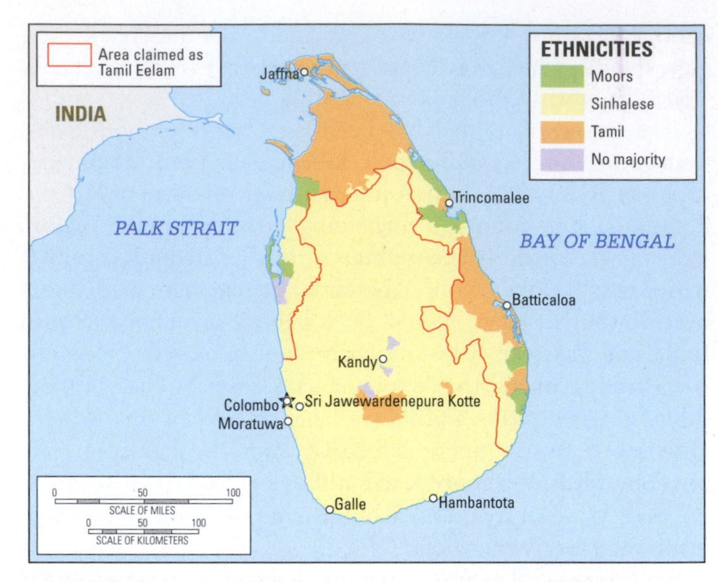

• Figure 7.32 Ethnicity in Sri Lanka. The heavily Tamil areas of the north and east fought a long civil war against the government to establish their own country, called Tamil Eelam, until 2009, when they were finally defeated. The city of Trincomalee would have been the capital of this new nation.

Joe Hobbs

• Figure 7.33 The war between Tamils and Sinhalese has ended in Sri Lanka, but Tamil protests against Sinhalese rule continue where they can be practiced. This is a demonstration in Sydney, Australia. The depiction is of the Tamil Tiger founder and leader, Velupillai Prabhakaran, who was killed just weeks before this demonstration in 2009. The red on the map on the right, like **Figure 7.38**, depicts parts of Sri Lanka held by the Tigers at their maximum advance. The Tamil flag, not fully in view, depicts a tiger jumping through a halo and crossed bayonets.

Insights China's String of Pearls

Some analysts, particularly in India, argue that China will use the Sri Lankan port of Hambantota to project its power into the Indian Ocean, where much of the world's oil and other valuable cargo is constantly in transit (see **Figure 2.11**). They see Sri Lanka as one pearl in China's **"string of pearls"** strategy to gain critical allies forming a maritime strand across the waters of South and East Asia (**• Figure 7.O**). The largest "pearl" in this part of the string is the Chinese-built port at Gwadar in Pakistan. India, which itself wants to project power across these sea lanes, views this as encirclement by China. Indians were unnerved when Chinese submarines docked in the friendly waters of Sri Lanka in 2014.

Keep in mind that China and India are enemies and have been ever since they fought a high-altitude border war in 1962. And Sri Lanka, nearly attached to India, has long benefited from Chinese aid (China backed the government in Colombo in its long war against the Tamil rebels, for example). Indians are suspicious of China portraying itself in benevolent terms; here is an Indian editorial that details their fears:

> China is attempting to disguise its strategy, claiming that it wants to create a twenty-first-century maritime Silk Road to improve trade and cultural exchange. But friendly rhetoric can scarcely allay the concern in Asia and beyond that China's strategic goal is to dominate the region.
>
> That concern is well founded. Simply put, the [maritime] Silk Road initiative is designed to make China the hub of a new order in Asia and the region of the Indian Ocean. Indeed, by working to establish its dominance along major trade arteries, while also instigating territorial and maritime disputes with several neighbors, China is attempting to redraw Asia's geopolitical map. . . .
>
> In reality, little distinguishes the maritime Silk Road from the "string of pearls". Though China is employing ostensibly peaceful tactics to advance the initiative, its primary goal is not mutually beneficial co-operation; it is strategic supremacy. Indeed, the [maritime] Silk Road is integral to President Xi Jinping's "China dream" ambitions, which entail restoring China's past glory and status.
>
> China, especially under Xi, has often used aid, investment, and other forms of economic leverage to compel its neighbors to deepen their economic dependence on—and expand their security cooperation with—the People's Republic. Xi's use of a $40 billion Silk Road Fund and the new China-sponsored Asian Infrastructure Investment Bank to develop the maritime Silk Road reflects this approach.
>
> Already, China is constructing ports, railroads, highways, and pipelines in the region's littoral states, not only to facilitate mineral-resource imports and exports of Chinese manufactured goods, but also to advance its strategic military goals. For example, China concluded a multibillion-dollar deal with Pakistan in order to develop the port at Gwadar, owing to its strategic location at the mouth of the Strait of Hormuz, which more than offsets the port's limited commercial potential. . . .
>
> Zhou Bo, an honorary fellow with the Peoples' Liberation Army Academy of Military Science, admits that China's mega-projects "will fundamentally change the political and economic landscape of the Indian Ocean," while presenting China as a "strong yet benign" power. This is important, because the new Asian order will be determined less by developments in East Asia, where Japan is determined to block the rise of China, than by events in the Indian Ocean, where China is chipping away at the longstanding dominance of India. . . .
>
> China's plans for the Silk Road combine economic, diplomatic, energy and security objectives in an effort to create an expansive network of linked facilities to boost trade, aid strategic penetration and permit an increasingly potent and active submarine force to play an expanded role. In the process, China aims to fashion an Asian order based not on a balance of power with the US, but on its own hegemony. Only a concert of democracies can block this strategy.[29]

This excerpt gives you excellent insight into India's thinking about its geopolitical struggles with India. You can be sure the Chinese perspective sounds very different.

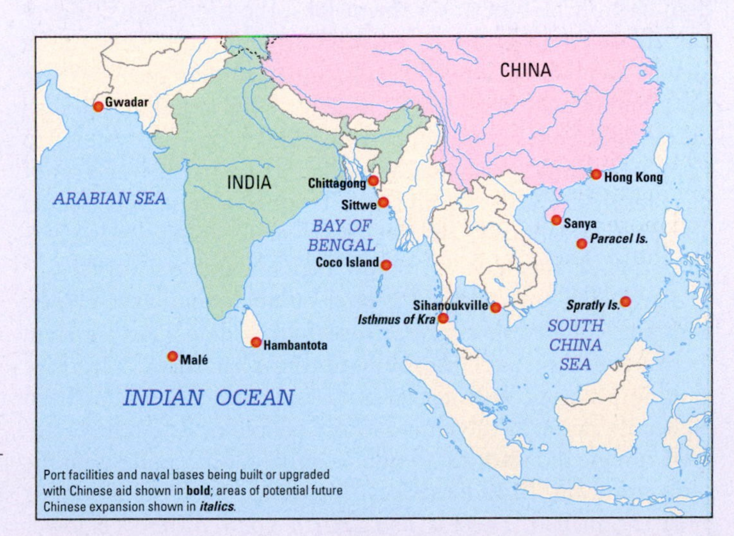

• Figure 7.O China's "string of pearls," stretching from Hong Kong to Pakistan.

THINK CRITICALLY: Is the "string of pearls" a stand of sensible commercial posts for China, or a strategic military placement in a Chinese geopolitical power play?

adding Sri Lanka as a gem on its militarily strategic "string of pearls" (see Insights, **page 327**).

The Caste System

One ancient Hindu belief is that every individual is born into a particular **caste**, a social subgroup that determines the individual's rank and role in society. Castes form a hierarchy, with the Brahmin caste at the top (comprising about 5 percent of all Hindus) and three others (Kshatriya, Vaisya, and Sudra) below. Caste membership is inherited and cannot be changed. Particular castes are associated with certain religious obligations, and their members are expected to follow traditional caste occupations. Marriage outside the caste is generally forbidden, and meals are usually taken only with fellow caste members.

At the bottom of the social ladder are the **Dalits**, or *scheduled castes*, once known as **untouchables**, accounting for about 20 percent of all Hindus. They are not part of the caste system, but are "outcastes," and according to Hindu belief, they are not reincarnated. The untouchables were so called because they traditionally performed the worst jobs, such as the handling of corpses and garbage; therefore, their touch would defile caste Hindus.

This ancient, rigid social system is changing. India's constitution outlawed the caste system in 1950, but a single decree could not undo thousands of years of tradition. Brahmin privileges are increasingly challenged and restricted by more laws. Indian law explicitly forbids recognition of Dalits as a separate social group. Confronting the fact that upper castes account for less than one-fifth of India's population but command more than half of the best government jobs, the Indian government has instituted an affirmative action program to allocate more jobs to members of lower castes. Indian leaders of high caste have championed the Dalits' cause for both moral and practical reasons. Politicians use caste divisions to their advantage to bring out voters, promising if elected to act in the interests of their respective large social blocs. In 1997, a Dalit was elected president of India for the first time in the country's history. Since then, Dalits have won more regional and local offices.

The Dalits are voting in greater numbers than ever before, with the conviction they can use India's democratic system to overcome ancient discrimination against them. They are making progress. Disintegration of the caste system has been especially rapid in the cities, where people of different castes must mingle in factories, in public eating places, and on public transportation. In India's vast rural areas, the caste system is more entrenched. There is also a north–south difference: lower caste southern Indians began agitating against the caste system early in the 20th century and kept up their struggle. In northern India, by contrast, there was less effort to tackle the caste system. Caste is so important in northern politics that it is said that voters don't cast their vote; they vote their caste. Remarkably, in large part because of the loosening of caste restraints, economic growth is much stronger in southern than in northern India.

Keeping Malthus at Bay

Poverty and human health are serious problems in South Asia, and the prospect of continued high population growth raises the question of whether growth in food supplies can avert the proverbial Malthusian crisis (see Chapter 3, **page 80**).

India's population has surged since independence, from 352 million in 1947 to 1.3 billion in 2014. In recent decades, the overall trend has been toward lower population growth rates. Fifty years ago, the average Indian woman had six children; now she has two or three, though that is still above the natural replacement rate that would keep the population steady. The population base is already so vast, and the base of its age structure diagram so wide (about 45 percent of the population is younger than 25), that even modest growth will add huge numbers. India is predicted to overtake China as the world's most populous country by 2028, when the UN projects that both countries will have 1.45 billion people. There is uncertainty about how high India's population will grow from there; estimates range from 1.5 to 1.9 billion. Most Indian decision makers think that even the low end of that range is too high for India's welfare. Unlike China, India is a democracy and cannot simply declare how many children a couple can have. Indian states, rather than the central government, establish laws and policies, and there is a wide range of decrees in the realm of population. Some states offer financial incentives, such as a $100 bonus if couples delay their first child's birth.

For many people, the name *India* invokes an image of grinding poverty, and with good reason: it is estimated that 32 percent of the population is "abjectly poor," defined as living on less than $1.25 per day, and almost half of India's children are malnourished. But perhaps in defiance of Malthus, an increased population has so far succeeded in producing more economic resources overall. A few people in India are very wealthy, and there is an emerging middle class of as many as 200 million people today; by 2030, India may have the world's largest middle class (**•Figure 7.34**). India's economy has grown impressively in recent years. An apparent problem is that the benefits of the growing economy are unevenly distributed. Most of the economic growth benefits those who are already better off, widening the gap between rich and poor.

Joe Hobbs

• Figure 7.34 This Indian family visiting the Red Fort in Agra is among the growing ranks of the subcontinent's middle class. Note the Taj Mahal gleaming in the distance.

The issue of population growth is also critical for Pakistan, which has never mounted an effective family planning program, mainly because of opposition from Muslim religious authorities, who regard birth control as an intervention against God's will (Pakistan's birth rate in 2014 was 28 per 1000, compared with 22 per 1000 in India). Muslim Bangladesh also has a high birth rate (20 per 1000), but it is much lower than it once was. In the 1970s, the average Bangladeshi woman had seven children, whereas now she has two, and contraception use nationwide has increased from 4 percent then to more than 50 percent now. South Asia's success story in bringing population growth under control is Sri Lanka, with a birth rate of just 18 per 1000 and an annual population growth of 1.2 percent.

Agricultural output in South Asia has increased since independence. Most notably, despite its huge and growing population, India had achieved self-sufficiency in grain production, with an 800 percent increase in output since 1960. Almost half a million "ration shops" sell subsidized food staples to the country's poorest people. The successes of agriculture in the subcontinent have been due mainly to the increased use of artificial fertilizers, the introduction of new high-yield varieties of wheat and rice associated with the Green Revolution, more labor provided by the growing rural population, the spread of education, the development of government extension institutions to aid farmers, and better irrigation. "Mother Nature," however, has the last word. Fifty-five percent of India's farmland is rain-fed rather than irrigated, and when summer monsoons fail to deliver rainfall, as they did in 2009, the yields of valuable rice, sugarcane, and other crops plummet. This causes immediate difficulties for people (in the form of higher food prices) and the nation (in terms of trade and economic 292 growth).

As the sale of wives and daughters suggests, one of the challenges confronting both rural and urban India is to improve the status of women. Despite a 1961 ban on dowries (money and gifts given by a bride's parents to the groom), the practice is still widespread. So is the killing of brides who do not provide enough dowry. The burden placed on the bride's family is one reason that many prospective parents choose to abort female fetuses, which can be detected with ultrasound tech- 286 nology. India banned this use of ultrasound in 1996, but the practice continues. About 113 boys are born in India for every 100 girls. As in China, female feticide and infanticide have been a part of India's patriarchal culture for centuries, primarily for financial reasons: a girl is seen as "other's wealth" who will benefit another family, but not her own. Education for girls has historically lagged in India, as many parents see no point in spending money on a child that will provide them with no benefits in the future. A solution to this problem has so far been elusive. A noticeable rise in female literacy rates in the southern state of Kerala seems to have been an answer, but increased literacy rates in other Indian states have had little effect. Because of the perceived financial burden, the rural poor families of India are thought to be the most likely to commit female infanticide, but gender ratios in the wealthiest and most urbanized of India's states, like Delhi and Haryana, are also among the most unbalanced. This is a complex cultural issue, and there may not be a "one size fits all" solution (**• Figure 7.35**).

Low-Lying Bangladesh and Maldives: Canaries in the Climate-Change Coal Mine?

Surrounded by India on three sides, Bangladesh is about the size of the state of Iowa or the nation of Greece—not an important fact until one considers that it is home to 158 million people. With its limited resources, overpopulated Bangladesh is South Asia's second-poorest country on a per capita GDP (PPP) basis, ahead of only mountainous Nepal.

One reason for Bangladesh's great population density is that its land is quite fertile, being located on the delta of several great rivers that drop their sediments as they discharge into the sea (see **Figure 7.37**). The Bangladesh portion of the Ganges-Brahmaputra Delta ranks, along with the Indonesian island of Java, as one of the two most crowded agricultural areas on Earth. This area is subject to catastrophic flooding. Massive tropical cyclones (hurricanes) ravage Bangladesh regularly in September and October. Deforestation upstream on the steep slopes of the Himalayas has increased water runoff and sediment load in the Ganges, also contributing to Bangladesh's flood problems.

Ironically, although there are concerns about climate change–related flooding in Bangladesh, there are even more

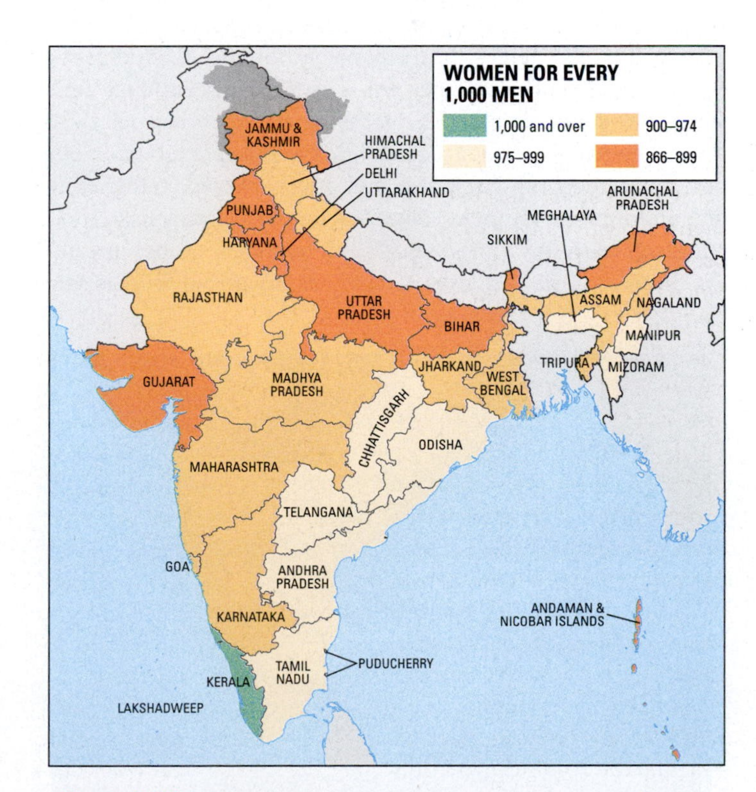

• Figure 7.35 India's gender ratios. Most of India's states report far more boys born than girls each year.

THINK CRITICALLY: What does this imbalance mean for Indian society and gender relations? India has been much in the headlines for violence against women.

worries about climate change–related water shortages and droughts. Climate change models point toward accelerating depletion of the so-called **"Water Towers of Asia"**—the glaciers and snow cover of the Himalaya, Karakoram, and other great mountain ranges that feed the region's 10 great rivers and scores of smaller ones. The perennial ice mass here is exceeded in volume only by the ice of the Arctic and Antarctic, earning the Tibetan Plateau and environs the name "The Third Pole." The region's roughly 18,000 glaciers, like most others in the world, are melting at a steady, relentless pace (**•Figure 7.36**). With temperatures on the Tibetan plateau rising much faster than the world average, a respected Chinese glaciologist predicts that two-thirds of them will be gone by 2050. These glaciers supply water to about 40 percent of the world's population! The prospect of their melting away is so alarming that long-standing enemies India and China are putting away some of their differences and collaborating in scientific research on climate change.

It is ironic that poor countries and poor people, often the most vulnerable to climate change, are generally the least responsible for greenhouse gas emissions. Bangladesh is a good example. The Intergovernmental Panel on Climate Change recognizes this as one of the world's most endangered areas and predicts a rise in sea level there of up to 23 inches (c. 58 cm) by 2100. Bangladesh is already suffering from rising sea levels attributable to climate change. About one-fifth of its land is less than 4 feet above sea level (**•Figure 7.37**). Projections suggest that by 2050 rising sea levels will cover 17 percent of Bangladesh's land, displacing 18 million people. But Bangladesh is responsible for just 0.3 percent—that is, three-tenths of 1 percent—of global greenhouse emissions. The country's position on the short end of the stick raises serious questions about accountability and responsibility. Some political leaders in Bangladesh and other affected countries are demanding that the rich polluting countries compensate poor countries with low emissions. "It's a matter of global justice," the director of Bangladesh's Center for Advanced Studies said.

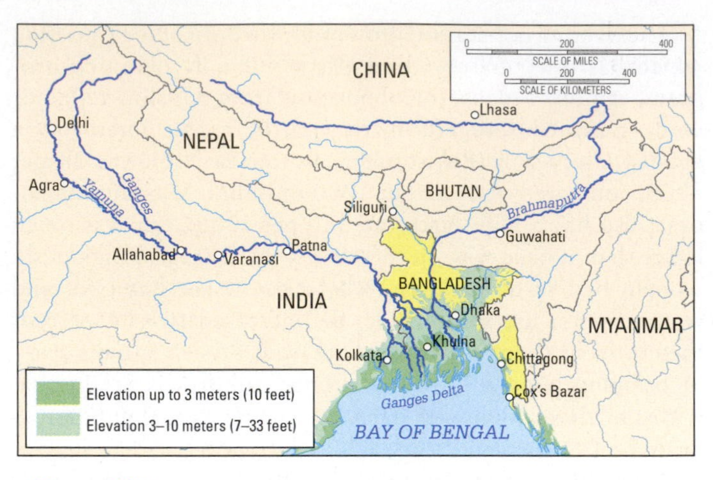

• Figure 7.37 The geographic situation of Bangladesh. Bangladesh is situated along the enormous delta made by the Ganges and Brahmaputra Rivers. The country's low elevation makes it very susceptible to storm surges and rising sea levels.

"These migrants should have the right to move to the countries from which all these greenhouse gases are coming. Millions should be able to go to the United States" [which produces about a quarter of world's greenhouse gases].[30]

The 26 atolls that make up the Maldives (population 400,000; capital, Malé, population 100,000), an independent country since 1965, appear to be the very essence of tropical paradise (**•Figure 7.38**). So prized are its palm-blessed and coral-fringed beaches that more than a million tourists, mainly Europeans, visit the country each year. More than 60 percent of the country's foreign-currency earnings come from tourism, with fishing and clothing providing most of the rest. Although not a rich country, the lucrative tourism industry combined with a small population has given the Maldives a poverty rate of just 1.5 percent, the lowest of all developing countries in Asia except for Thailand.

Joe Hobbs

• Figure 7.36 Melting of glacial ice in the Himalaya and other great mountain ranges of South Asia threatens future water supplies for people downstream. This is Gokyo Glacier, close to Nepal's Mt. Everest.

JBfotoblog/moment/Getty images

• Figure 7.38 The Maldives are endowed with a wealth of coral reefs and are threatened by sea level rise. Coral reefs are the most biologically diverse of all marine habitats, and provide important services to other ecosystems and to people.

Troubles could visit paradise. More than 80 percent of this country's very limited land area consists of limestone atolls less than 3 feet (90 cm) above the sea. If sea levels were to rise, as predicted by many common climate-change scenarios, the entire country could be submerged. 41

Afghanistan: Graveyard of Empires

Like Sudan and Turkey, Afghanistan is a country that is transitional between regions; some geographers place it in Central Asia or the Middle East. It is a land of limited resources, poor internal transportation, and little foreign trade. Afghanistan is one of only five landlocked countries in South and East Asia. As we have seen, most landlocked countries are disadvantaged 58 by their location and are poor. Afghanistan is indeed one of the world's poorest countries and also ranks high as a failed 59 state (seventh on that list).

Afghanistan is also a land of much conflict. It has repeatedly drawn the world's most powerful countries into conflict. Why? In large part because of its geographical situation: historically, it has occupied an important strategic location between India and the Middle East, and between India and Central Asia. Major caravan routes crossed Afghanistan, and a string of empire builders sought control of its passes. Afghanistan borders oil-rich Iran; this is the central geographical context of Carter 239 Doctrine discussed below and in Chapter 6.

"Afghanistan is poor in natural resources." That has been written about Afghanistan for a long time, but will be no more. The US Geological Survey (USGS) has been able to carry out enough field work in the war-torn country to determine that it may hold one of world's richest mineral troves, valued at nearly $1 trillion. "Afghanistan is a country that is very, very rich in mineral resources," said USGS Afghanistan project manager Jack Medlin; "its trove includes 60 million tons of copper, 2.2 billion tons of iron ore, 1.4 million tons of rare earth elements such as lanthanum, cerium, and neodymium, and lodes of aluminum, gold, silver, zinc, mercury, and lithium."[31] The rare earth elements are of particular interest because, until these discoveries, China held a near-monopoly 318 on them. These elements are rare because the processes creating them are. In Afghanistan, they are the products of the violent tectonic collision of the Indian Plate with the Eurasian Plate (see the map of plate tectonics in **Figure 2.1**). It will take time and elusive peace to have income from these resources, but they do represent hope for a country long dismissed as hopeless.

High and rugged mountains dominate Afghanistan. Most of the country's 31 million people live in irrigated valleys around the fringes of the mountains. Northern Afghanistan borders three of the five central Asian countries, and millions of people on the Afghan side are related to peoples of those countries. The most populous area is the northeast, particularly the fertile valley where the capital and largest city, Kabul (population 4.6 million; elevation 6200 ft [1890 m]), is located.

There are over a dozen ethnic groups that make up the Afghan state. About half of Afghanistan's inhabitants are Pashtuns (historically known as "Afghans," from which Afghanistan gets its name). Their language, Pashto, is closely related to Persian. Pashtuns also make up over 15 percent of neighboring Pakistan's population (despite that seemingly small number, there twice as many Pashtuns in Pakistan than Afghanistan) and are majorities in the Pakistani Federally Administered Tribal Areas (FATA) and the Pakistani province of Khyber Pakhtunkhwa.

The independent-minded tribal Pashtun people have always been loath to recognize the authority of central governments. In 1893, during the days of Britain's India colony, British authorities and the Afghan king drew a boundary called the Durand Line to separate their respective areas of influence. The line sliced right through traditional Pashtun lands, and Pashtun peoples have never accepted it (**•Figure 7.39**). This border area saw warfare among Pashtun tribes, tribal raids on British-controlled areas, and British punitive expeditions against the tribes. In those days, the city of Peshawar—on the Indian (now Pakistani) side of the Khyber Pass into Afghanistan—became the hotbed for British and Russian agents playing what came to be known as the **Great Game** of vying (and spying) for strategic influence in Central Asia. In a geopolitical scenario resembling the Carter Doctrine, the British were especially fearful that the Russians might use Afghanistan as a staging area to invade India, the "Jewel in the Crown" of Britain's empire. To this day, Peshawar feels like a wild, lawless frontier city (**•Figure 7.40**). It plays an important role in regional trade of weapons and drugs, especially heroin. For generations, the people of one of its outlying communities, Daraa, have specialized in replicating whatever foreign

• Figure 7.39 Ethnic Pashtun areas and international borders. "Pashtunistan" was cleaved in two by the Durand Line, which was established in 1893 and remains the boundary between Afghanistan and Pakistan today.

Joe Hobbs

• **Figure 7.40** Pashtun mujahadin in an Afghan refugee camp outside Peshawar, Pakistan. When I took this photo in 1985, these men were taking a break from fighting Soviet troops occupying Afghanistan. They and their sons could well have later become Taliban insurgents fighting American troops in Afghanistan. They bought their weapons in Daraa, the nearby factory and market for arms.

weapons a buyer desires, from old British muskets to modern anti-aircraft guns.

Afghanistan is overwhelmingly agricultural and pastoral. Its agriculture relies on traditional methods and produces low yields. The country is so mountainous and arid that only about 12 percent is cultivated (•**Figure 7.41**). The most successful crop has been the opium poppy, which routinely accounts for between one-third and two-thirds of the country's gross domestic product, and over 80 percent of the world's opium. In the context of the global drugs trade, portions of Afghanistan, Pakistan, and Iran have come to be known as the 156 **Golden Crescent**. Ranking second in global opium and heroin production is another locale in this world region, the Golden Triangle discussed on **page 342**. A succession of rebel armies and then Afghanistan's government when it was briefly under Taliban control used revenues from opium and heroin to obtain their arms. The Taliban and its insurgent allies, including the Haqqani Network, still do. Most of the production takes place near the southern cities of Kandahar and Lashkar Gah (where the rare earth minerals are also concentrated), where the Taliban have retrenched themselves most successfully.

Principally to reduce funding to the Taliban, and also to crack down on illicit drugs, the United States spent $7 billion on a poppy eradication campaign between 2004 and 2014. These efforts were fruitless, mainly because farmers cannot resist the higher prices opium fetches over conventional crops and because the investments in these conventional crops promised to farmers were never made. Destroying poppy fields 479 has the unintended effect of turning destitute farmers into vengeful fighters. One option not advocated, but very sensible in Afghanistan's situation, would be to allow the country to produce opium legally for pharmaceutical needs as a handful of countries, including Turkey, India, and Australia are allowed to do.

Throughout most of the 20th century, this highland country was remote from the main currents of world affairs. After the Islamic revolution in Iran in 1979, however, Afghanistan's location next to that oil-rich country made it once again the target of foreign interests. The Soviet military 239 intervention of 1979 and the ensuing devastation catapulted the country into world prominence. The USSR's motives for its invasion of Afghanistan may have included a desire to prevent by force the spread of Iranian-style Islamic fundamentalism into Afghanistan (which is 80 percent Sunni and 19 percent Shiite) or into the central Asian Soviet republics that were a vital buffer zone on the superpower's southern periphery. For its part, with the Carter Doctrine the United 178 States warned the Soviet Union that it would not tolerate further Soviet expansionism in the nearby oil-rich region of the Middle East.

The Soviet war in Afghanistan brought widespread killing and maiming of civilians, destruction of villages, burning of crops, killing of livestock, sabotage of irrigation systems, and sowing of land mines over vast areas. Soviet ground and air forces caused several million Afghan refugees to flee into neighboring Pakistan and Iran. Arms from various foreign sources

Thomas J. Abercrombie/National Geographic Creative

• **Figure 7.41** The Bamiyan Valley and the Koh I Baba Range of north-central Afghanistan. This photo conveys some of the country's rugged beauty and pastoral character.

filtered into the hands of the ***mujahidin***, the anti-Soviet rebel bands that kept up resistance in the face of heavy odds. The United States was one of the powers supporting the rebels and in that sense waged a proxy war against the Soviet Union in Afghanistan. In its waning years, the USSR recognized that it could not win its "Vietnam War," and its troops withdrew 344 from Afghanistan by 1989. It is estimated that of the 15.5 million people who lived in Afghanistan when it was invaded in 1979, at least 1 million died, 2 million were displaced from their homes to other places within the country, and 6 million fled as refugees into Pakistan and Iran. About two-thirds of these refugees have since returned to Afghanistan.

Support that the United States and moderate Arab states such as Egypt and Saudi Arabia lent to the *mujahidin* and sympathetic Arab fighters soon came back to haunt those countries in a phenomenon known as **blowback**. Emboldened by their victory against the Soviets, the most militant Islamists among the fighters turned their attention to the United States and its Middle Eastern allies. Chief among them was Osama bin Laden, who developed "the base" or "the database" (*al-Qa'ida* in Arabic) of thousands of Afghans, Arabs, and other anti-Soviet war veterans he could call on to wage a wider *jihad*. Bin Laden's organization trained an estimated 10,000 fighters in al-Qa'ida camps in Afghanistan, and from this unlikely, remote setting, Bin Laden devised spectacular acts of terrorism: an assassination attempt on Egyptian President Hosni Mubarak, the bombing of US military barracks in Saudi Arabia, the bombing of US embassies in Kenya and Tanzania, the 243 attack on the *USS Cole,* and the 9/11 attacks.

The *mujahidin* succeeded in overthrowing the Communist government of Afghanistan in 1992, but after that, rival factions among the formerly united rebels engaged in civil warfare. During this period, most other countries utterly neglected Afghanistan, inadvertently promoting the growth of militant movements (which tend to thrive in remote, underde- 260, 430 veloped, and ungoverned regions). By 1996, one of the rebel factions, the **Taliban** (backed by Saudi Arabia and Pakistan), gained control of most of the country, including the capital of Kabul. Proclaiming itself the sole legitimate government of Afghanistan, the Taliban (made up almost entirely of Pashtuns) imposed a strict code of Islamic law in the regions under its control and gained international notoriety for its austere administration. The Taliban removed almost all women from the country's workforce, forbade public education of girls, and outlawed "un-Islamic" practices such as dancing, trimming beards, watching television, keeping birds, and flying kites. In a parallel to the destruction of historical artifacts by ISIS in 264 Iraq, the Taliban received international condemnation in early 2001 when it destroyed two giant Buddha statues that had been carved into a mountainside in pre-Islamic times.

The Taliban continued to make advances against its opponents inside Afghanistan (particularly the Northern Alliance, a coalition comprised mostly of non-Pashtun ethnicities such as the Tajik, Hazara, and Uzbek, whose leader Ahmed Shah Massoud—to his people a hero known as "the Lion of Panjshir"—was assassinated just days before, and in apparent preparation for, the 9/11 attacks. By 2001, the Taliban controlled 95 percent of the country's territory. The neighboring central Asian countries, Russia, and even Iran grew increasingly fearful of the spread of the Taliban's extreme interpretation of Islam into their nations. Russia ended up supporting some of the rebels it had previously fought because they were now fighting the Taliban. Russia's once unlikely ally in supporting those rebels was the United States, which successfully lobbied the United Nations to levy economic sanctions against Afghanistan in 1999. The United States hoped that economic losses would pressure the Taliban into turning over bin Laden for prosecution, and it also announced a $5 million reward for information that would lead to bin Laden's capture or death.

The bin Laden bounty rose into the tens of millions of dollars after September 11, 2001. Named as the mastermind of the attacks against New York and Washington, bin Laden became the "world's most wanted" as US President George W. Bush evoked a Wild West vow to have him captured "dead or alive." In crafting its declared "war on terror" (which it promised to carry anywhere in the world necessary), the US administration identified the Taliban as al-Qa'ida's mentor and targeted both organizations for elimination. 242

Within a month of the attacks on the United States, American warplanes and special forces struck in **Operation Enduring Freedom** against Taliban and al-Qa'ida facilities and personnel around Afghanistan. The British military assisted in the air campaign, and most of the ground fighting was carried on by the Northern Alliance, the Taliban's rival and nemesis. The United States successfully persuaded Pakistan to drop its support of the Taliban and join the effort, and much of the military campaign was based on US-Pakistani cooperation. One after another of the Taliban's urban strongholds fell, including Kabul and the Taliban spiritual capital of Kandahar. Withering assaults from US warplanes crushed the tunnels and other hideouts used by al-Qa'ida guerrillas, and many al-Qa'ida prisoners were taken to the US naval base in Guantánamo Bay, Cuba. The elusive bin Laden, however, 476 along with Taliban leader Mullah Omar, slipped away. In a rugged border area called Tora Bora, the United States had bin Laden cornered for some days, but a decision to deploy local rather than American forces on the ground is now regarded as a tactical error that allowed bin Laden to escape into the FATA of nearby Pakistan. 318

The reconstruction of Afghanistan began almost as soon as the Taliban were driven from power. Billions of dollars in aid went into the building of highways and other infrastructure, hospitals, and schools. The numbers of Afghans enrolled in primary schools rose from 25 percent in 1999 to over 90 percent, and girls and women are again being educated. Progress has been slow because the country is so devastated, the security situation so tenuous, and the Afghan government so corrupt. Of all those billions of dollars in aid, some unknown amount has been siphoned off to line the pockets of those responsible for distributing it.

American forces and the Afghan government headed for many years by Hamid Karzai effectively took control of only a small area around Kabul, and NATO troops were stationed throughout the country, particularly in the troubled south

(•**Figure 7.42**). Much of the country remains in the hands of warlords whose allegiance is up for grabs. Taliban and even al-Qa'ida fighters operate in the rugged countryside, and sometimes target Kabul and other cities. One of the Taliban's deadliest tactics is the so-called "insider attack," using uniformed Afghan soldiers to kill American and other NATO personnel they work alongside. Taliban attacks have also been responsible for over three-quarters of all Afghan civilian deaths since 2010.

The mission of the war in Afghanistan—the longest in US history—changed substantially over the years. Initially, its purpose was clear: rid Afghanistan of al-Qa'ida and its Taliban supporters. With al-Qa'ida effectively displaced into Pakistan and elsewhere, the focus fell on the Taliban. American and NATO ground troops fought seemingly endless numbers of Taliban fighters, who fought with classic insurgent hit, run, and hide tactics. Western forces became engaged in a conflict that paralleled the long war of Soviets against the *mujahidin* in Afghanistan. The Taliban and other Afghans, including civilians, bore heavy losses, and the conflict cost more than 2200 American lives, the lives of over 1000 British and other NATO forces, and over $450 billion (through 2015). Like the war in Iraq, the Afghan war exacted massive numbers of injuries, some visible and others not, on American and 269 other NATO soldiers. Improvised explosive devices (IEDs), in particular, cost limbs and traumatic brain injuries, and many veterans suffer from post-traumatic stress disorder (PTSD). American soldiers often served multiple deployments in both Afghanistan and Iraq, compounding their suffering. The American public grew weary as well and anxious to quit the open-ended conflict.

Fulfilling pre-election promises, US President Barack Obama steadily drew down the number of American forces in Afghanistan. Looking at how the absence of a reserve force of American soldiers may have contributed to the rise of ISIS in Iraq, and at the recent spread of ISIS into Afghanistan, the US Department of Defense considered a long-term force deployment in Afghanistan. Meanwhile, other NATO forces also drew down or withdrew completely. NATO military analysts came to the conclusion that the fight against the Taliban was unwinnable. For the losses of life and retreats suffered in earlier conflicts by the British and Russians in Afghanistan, the country had come to be known as the "Graveyard of Empires." Historians began to add America to the list of empires bloodied by Afghanistan.

There are several geographic problems compounding the difficulty of conducting war in Afghanistan. Looking at Afghanistan in its regional context in **Figure 7.1**, you can see that its mountainous terrain makes supply logistics difficult, and that its landlocked situation would complicate those logistics. Appreciating these challenges, you are bringing the geographic "lens" to the region's geopolitical situation.

You or someone you know may be interested in educational initiatives for Americans to study "strategic" languages and cultures including those of Afghanistan; see the Geographic Spotlight on **page 355**.

CHRISTOPHER T. SNEED/UPI/Newscom

• **Figure 7.42** US Army soldiers meeting with leaders in Shabow-Kheyl, Afghanistan, April 8, 2009. The soldiers were assigned to Company D, 1st Battalion, 501st Parachute Infantry Regiment.

THINK CRITICALLY: Was the US-led war in Afghanistan winnable? Should it have been conducted after most al-Qa'ida fighters fled early in the conflict? Can the US sustain open-ended ground wars like these, and at what cost to those who fight? *(US Army photo by Sgt. Christopher T. Sneed)*

Southeast Asia

Deforestation of Southeast Asia

Southeast Asia today is composed politically of 11 countries: Myanmar (formerly Burma), Thailand, Laos, Cambodia, Vietnam, Malaysia, Singapore, Indonesia, Timor-Leste (formerly East Timor), Brunei, and the Philippines (see **Figure 7.2**). With the exception of Thailand, which was never colonized, all of these states became independent from colonial powers after 1946. These nations are fragmented both politically and topographically. Many of the countries are also culturally fragmented and have experienced strife between different ethnic groups. Outside intervention has sometimes complicated and worsened local discord, producing enormous suffering in the region and lasting trauma among many of the foreign soldiers who fought the determined inhabitants of Southeast Asia. The region also must contend with environmental problems related both to its economic progress and its lingering poverty, and to the

Geographic Spotlight Strategic Interests in US Education

In Chapter 1, we saw that geography is all about "getting it right" and "doing smart stuff" when it comes to understanding how the world works, and in making intelligent decisions about such critical issues as waging ground wars abroad. 3 Part of the problem in "doing smart stuff" is that the United States has been short on knowledge about the cultures, tribal allegiances, and other social complexities of Afghanistan and other countries, and in people fluent in some of the peoples' languages. The US government is playing catch-up, in part by funding undergraduate and graduate-level college students to study what it regards as "strategic languages" and become immersed for long periods in countries deemed critical to national security. See, for example, the Boren Scholarships, which "provide American undergraduate students with the resources and encouragement they need to acquire skills and experiences in areas of the world critical to the future security of our nation."[32] Boren Scholarships fund study of four of Afghanistan's languages: Pashto, Persian (Dari), Uzbek, and Turkmen. Scholarships like this can open many doors for students' life experiences and careers. They are also available for graduate study.

In the educational context, national security goes far beyond concerns with military thinking. Consider this definition of national security on the Boren Scholarship program overview: "It draws on a broad definition of national security, recognizing that the scope of national security has expanded to include not only the traditional concerns of protecting and promoting American well-being, but also the challenges of global society, including: sustainable development, environmental degradation, global disease and hunger, population growth and migration, and economic competitiveness." These concerns are virtually a definition of the field of geography. In a broad sense, your study of geography in this book is contributing to national security: geographic literacy is important, and for starters it does matter that you know where Ukraine is! 3

unequal usage of shared resources like the Mekong Basin waters (see Geography of the World's Great Rivers, **page 336**).

Near areas of dense settlement in Southeast Asia, there are still some large areas covered in primary forest, but they are shrinking as population expands and development progresses. Many scientists view the destruction of the region's tropical rain forest as an international environmental problem, especially because of its contributions to climate change. Primary (original) forests are much more ecologically rich than secondary-growth forests. By 1990, nearly all of Thailand's original forest had been lost; secondary forest covers about 30 percent of the country today. Poorer neighbor Myanmar has about a 45 percent forest cover. Vietnam was almost entirely forested up until World War II; by 2010, less than 1 percent of its original forest remained, though new plantings and secondary growth have filled over one-third of Vietnam's original forest area back in. Deforestation is advancing most rapidly in Malaysia. Officially, 59 percent of the land area is still forested. However, official statistics include palm oil plantations, so these figures should be evaluated carefully. (Only about 10 percent of Malaysia's remaining forests are considered "pristine.") Malaysia is now the world's largest exporter of tropical hardwoods. Virtually all of the primary forests in Peninsular Malaysia have already been cleared, and it is projected that commercial logging will decimate those remaining outside protected areas in Malaysian Borneo quite soon. Deforestation is also progressing quickly in Indonesia, where about half the land is forested (**• Figure 7.43**).

Although Southeast Asia's tropical forests are smaller in total area than those of central Africa and the Amazon Basin, they are being destroyed at a much faster rate. The main forces of deforestation are commercial logging for Japanese and Chinese markets, and clearing of forest to plant oil palm trees, from which palm oil is rendered (**• Figure 7.44**). Malaysia supplies over 80 percent of the world's palm oil. The demand for palm oil is soaring for use in biofuels and in household products, including Vaseline and Dove soap, and food, including Kit Kat candy bars and Nutella spread. Environmentalists fear the irretrievable loss of plant and animal species and the potential contribution to global warming, through both fires and the loss of the "carbon sink," that deforestation in this region may cause. 44

Remarkably, the destruction and burning of forests on the islands of Borneo and Sumatra have made Indonesia the world's fifth largest emitter of greenhouse gases, after China, the United States, India, and Russia. There is a new scientific consensus that a fifth of the world's greenhouse gas emissions is due to land use change, with Indonesia accounting for 60 percent of that (followed by Brazil at 14 percent and Malaysia at 7 percent); thus the cutting and burning of rain forest mainly to produce palm oil there is responsible for 4 percent of global greenhouse gas emissions. The conversion of peat bogs—some inhabited by Sumatran tigers and bears—to tree plantations for paper and pulp also releases carbon dioxide. As deep as 50 feet, and made up of decomposed trees and plants, the peat stores billions of tons of carbon dioxide.

Due to its large contributions to greenhouse gases through forest conversion, Indonesia is negotiating many projects for the Reducing Emissions from Deforestation and Forest Degradation (REDD) program described in Chapter 2 (see **page 44**). One major paper and pulp company working in Indonesia negotiated a REDD deal in which by protecting forests it receives carbon credits it sells on the global carbon exchange for large sums of money.

Environmentalists have put pressure on large companies like Cargill, Unilever, and Nestlé to stop buying palm oil produced unsustainably. Nestlé has responded favorably, promising a "zero deforestation" policy. Malaysia and Indonesia have

Geography of the World's Great Rivers The Mekong

The Mekong River, known as Water of Stone in Tibet, Great Water in Cambodia, Nine Dragons in Vietnam, and Mother of Waters (*Mae Nam Khong*, contracted to Mekong) in Laos and Thailand, is Southeast Asia's great river (**• Figure 7.P**). From its source on the Tibetan Plateau in the Chinese province of Qinghai, the Mekong snakes along a 2600-mile (4200 km) path through China's Sichuan and Yunnan provinces, forms much of the boundary between Laos and its neighbors Myanmar and Thailand, winds across Cambodia, and fans out across Vietnam's densely populated Mekong Delta region before empting into the South China Sea.

• Figure 7.P The Mekong River between Vientiane and Savannakhet, Laos. Its Laotian name means "Mother of Water."

Unlike the Chao Phraya in Thailand, the Red River in Vietnam, and the Irrawaddy in Myanmar, the Mekong is not a great magnet for cities and civilizations. The biggest city on its banks is Cambodia's capital city of Phnom Penh, with just over a million people. The main factor limiting more settlement along the Mekong has been the difficulty in navigating the river above Phnom Penh. Historically, very little trade has been carried along much of its length; the oceans and overland routes have provided superior opportunities.

The Mekong is not economically unimportant, however. It is one of the world's most prolific fisheries, yielding about 2 million tons of fish each year, or twice the annual North Sea catch. There are over a thousand species of fish in the Mekong, more than in any other river but the Amazon and the Congo. Unique creatures inhabit its waters, including about 70 Irrawaddy dolphins and the giant catfish that attain 10 feet (3 m) in length and 660 pounds (300 kg) in weight. Overfishing and the construction of dams have all but eliminated this superlative catfish throughout the Mekong Basin.

A big push is being made to capitalize on the Mekong's potential to produce hydroelectric power, with plans to build about 100 dams on the river and its tributaries. The first dam on the river itself (the Man Wan in China) was completed in 1993. China then went to work on five more, including the Nuozhadu Dam, one of the tallest dams in the world. Vietnam is building five dams on a Mekong tributary called the Dak Hoyt, which forms part of its border with Cambodia. Thailand has dammed its main tributaries to the Mekong, and Cambodia has two dams planned. Laos plans to build the most dams of all—twenty—and thereby become mainland Southeast Asia's biggest supplier of electricity. Laotian officials boast that their country will become "the battery of Asia." With funding from the World Bank—despite the criticism this organization has received for supporting large dams—Laos built the giant Nam Theun 2 dam. The World Bank insists it went to extraordinary lengths to minimize environmental impacts from this project.

These dams will nevertheless have enormous impacts, some of them unforeseeable. The fish catch will definitely fall sharply because the main reason for the Mekong's productive fishery is its regular flooding, which will be dramatically reduced. Fishermen are already complaining of smaller catches since dam building began. There are concerns that dams will ruin the ecosystem of Cambodia's Tonle Sap Lake, which is drained by a river that actually changes direction when downstream Mekong floodwaters inundate its basin. More than a million people make their living by fishing this lake. Inevitably, the downstream countries suffer many ill effects of dams. In this case, for example, Vietnam's Mekong Delta will be deprived of the fertile silt that helps make it an agricultural "breadbasket."

As always with shared river systems, there are serious questions about how the Mekong waters should be divided. Vietnam, Cambodia, Laos, and Thailand have joined to form the **Mekong River Commission**, whose biggest responsibilities are to fix the minimum amount of water each country must discharge downstream on the Mekong and its tributaries, and to agree on rules to ensure water quality. Both quality and quantity of water are threatened by upstream Myanmar and China, which have refused to join the Mekong River Commission. The downstream countries complain that China does not seem to care about what happens to them as a result of China's dam building and other uses of the Mekong.

240

passed laws to slow the destruction, with Indonesia taking the aggressive step of banning the use of fire to clear forests. However, enforcing such legislation has so far proved impossible. Hundreds of Indonesian and Malaysian companies, most of them large agricultural concerns (for example, palm oil plantations and pulp and paper companies) with close ties to the government, continue to use fire as a cheap and illegal method of clearing forests. The World Wildlife Fund estimates that

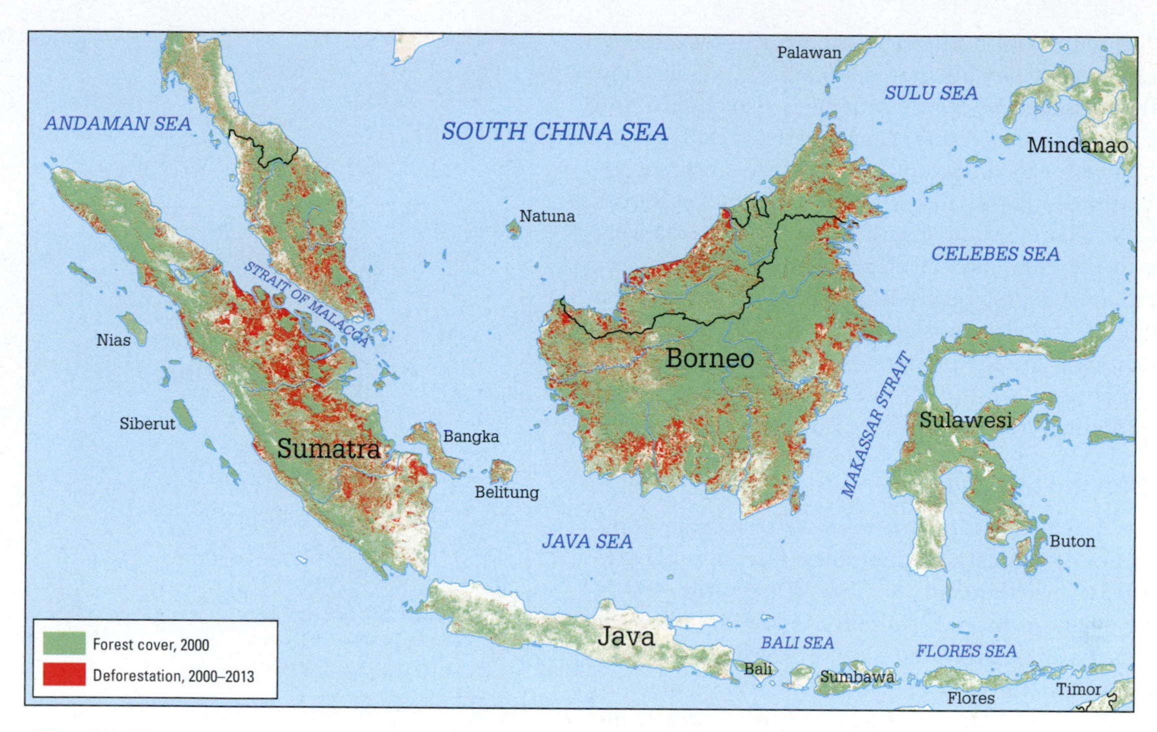

• Figure 7.43 In Southeast Asia, natural forest cover has been rapidly reduced by many human activities, especially commercial logging and farming.

Jason Isley - Scubazoo/Getty Images

• Figure 7.44 Over much of the island of Borneo and other areas in Southeast Asia, tropical rain forest is being torn out to make way for palm oil plantations. In the process, nature's factory of biodiversity is replaced by an impoverished human-made monoculture. Across the river, beyond the neat rows of oil palms, untouched rain forest is visible.

70 percent of Indonesia's exported trees are illegal, many of them removed by "timber barons" in the country's national parks while bribed park and police officials look the other way. The urgent need for construction materials after the devastating tsunami of 2004 (see **page 338**) further increased the pressure on the country's forests.

There is a growing movement in the consuming countries to boycott the tainted timber by selling only certified "good wood" and only palm oil that has been produced from sustainably harvested oil palms. The US company Home Depot, for example, sells only imported woods bearing the Forest Stewardship Council (FSC) logo. Even the source countries are coming on board the certification bandwagon. Malaysia began negotiations with the European Union in 2011 to become the first country to sign a voluntary agreement guaranteeing that all timber exports are harvested legally. Once the agreement is in effect, Malaysia would promise the European Union that its trees would not be cut from protected forests and that endangered tree species would not be harvested.

Deforestation in Southeast Asia has many actual and potential transboundary consequences. Periodically, widespread droughts attributed to the warming of Pacific Ocean waters by the phenomenon known as El Niño turn the annual July–October burning season into a conflagration of human origin. 458 Deliberately set fires race out of control over large areas of Sumatra and Borneo, resulting in a choking haze that shuts down airports, closes schools, deters tourists, and causes respiratory distress to millions of people throughout Southeast Asia. Even in years without the El Niño effect, less widespread but still severe smog crises strike Malaysia and Indonesia, particularly the island of Sumatra. Indonesian authorities fight back by revoking the licenses of plantation companies, but illegal activities continue.

Many of the plants and animals in these forests are **endemic species** (found nowhere else on Earth). Indonesia, which contains 10 percent of the world's tropical rain forests, is the world's second most important **megadiversity country** (after Brazil), with about 11 percent of all the world's plant species, 12 percent of all mammal species, and 17 percent of all bird species within its borders. Virtually all of Southeast Asia is part of a biodiversity hotspot identified by Conservation International (see **Figure 2.8**). Southeast Asia is also particularly significant in biogeographic terms because of the so-called **Wallace Line**—named for its discoverer, the English naturalist Alfred Russel Wallace—that divides it into two distinct ecozones (•**Figure 7.45**). Nowhere else on Earth is there such a striking local change in the composition of plant and animal species in such a small area as on either side of this divide. East of the line (separating the Indonesian islands of Bali and Lombok, for example), marsupials are the predominant mammals, and placental mammals prevail west of the line. Similarly, bird populations are remarkably different on either side of the line.

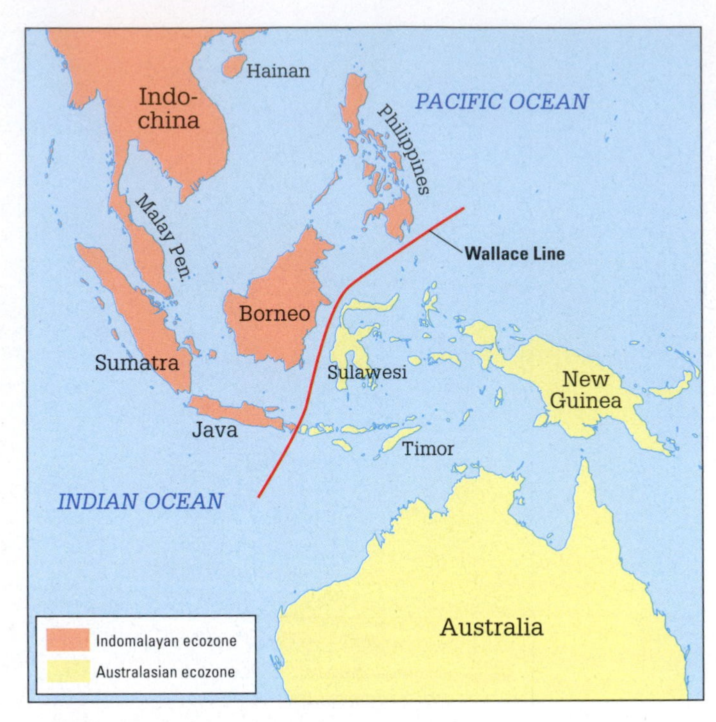

• **Figure 7.45** The Wallace Line. Flora and fauna on one side of the Wallace Line are remarkably different from those on the other side.

The Great Tsunami of 2004

On December 26, 2004, just off the northwestern coast of the Indonesian island of Sumatra in the Indian Ocean, a gargantuan "megathrust" earthquake measuring magnitude 9.1—with the force of more than 32 billion tons of TNT, or the equivalent of as many as a million nuclear explosions—struck at 6:58 AM local time. In the Java (Sunda) Trench, as much as 750 miles (1200 km) of fault line slipped 60 feet (over 18 m) along the subduction zone where the Indian Plate is sliding beneath the Burma Plate (a microplate along the edge of the greater Eurasian Plate; see **Figure 2.1**).

As the seabed of the Burma Plate instantly rose several meters vertically above the Indian Plate, the massive displacement of seawater created a series of tsunamis (*tsunami* is Japanese for "harbor wave"; the phenomenon is sometimes popularly but incorrectly referred to as a "tidal wave") that pulsed across the Andaman Sea and the Indian Ocean at up to 500 miles (800 km) per hour, striking coastlines around the Indian Ocean at heights up to 45 feet (14 m) (•**Figure 7.46**). On the open ocean, such waves are barely perceptible, but they break on coastlines with the force of giant storm surges.

This was the most cataclysmic natural event of modern times. The total number of dead exceeded 200,000. By far the greatest number of deaths (over 130,000) was in Indonesia, where Aceh's principal city of Banda Aceh was obliterated (•**Figure 7.47**). Sri Lanka lost 35,000; India, 10,000; and Thailand, over 5000. Almost all the deaths occurred within a mile of the shoreline and were caused by blunt force from debris and by drowning. An estimated one-third of the fatalities were children. There were reports of children running out to investigate fish stranded by rapidly retreating seas—a characteristic precursor to a tsunami—only to be struck seconds later by a wall of water.

As many as 2 million people were made homeless by the disaster. An unprecedented international relief effort, designed to

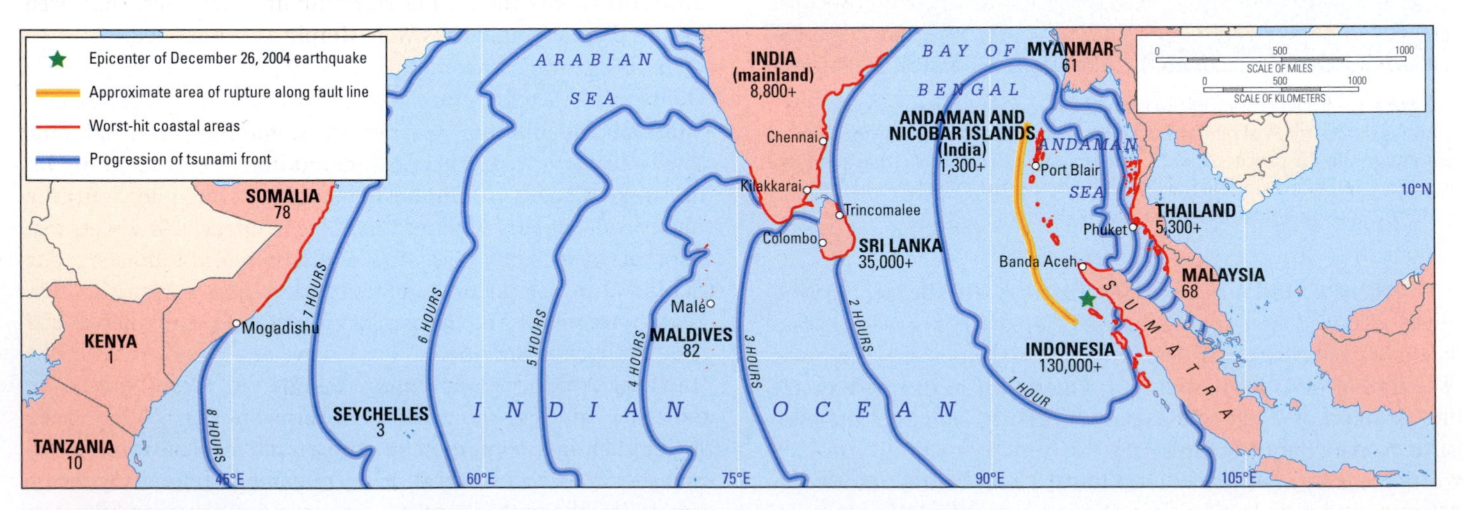

• **Figure 7.46** The earthquake and tsunami of 2004. The epicenter of the earthquake was located off the west coast of the Indonesian island of Sumatra. It created a tsunami of tremendous force and scope, inflicting massive suffering and loss of life.

• Figure 7.47 Banda Aceh before and after the tsunami, and as it appeared, rebuilt, in 2015.

feed and house these refugees and to contain the spread of epidemic diseases, began within hours. In many cases, it was clear that recovery efforts were profoundly successful, sometimes in unexpected ways. We have seen how the disaster hastened the war's end in Sri Lanka, and we will see how it accelerated reconciliation in Aceh. In India's Tamil Nadu state, there were new roads, housing and utilities, and jobs training. New homes were titled in women's names (because women were deemed more trustworthy than men), giving women an unprecedented rise in status. In the tsunami's wake, the social fabric was transformed. More than 10 years on, reconstruction of some of the hardest-hit areas is complete; Banda Aceh is rebuilt and thriving (see **Figure 7.47**).

326
347

Tsunami experts concur that countless thousands of lives would have been saved had the affected Indian Ocean region been equipped with the same kind of tsunami early warning system that was already in place in the Pacific Ocean.[33] That investment had not been made in the area affected, because historically about 90 percent of the world's tsunamis have occurred in the Pacific and because many of the nations afflicted lacked the financial resources for the wave sensors and other infrastructure required by the system. Installing an early warning system in the Indian Ocean basin became an urgent priority, and a system was up and running by 2006. Had this system been operational in 2004, Indonesia would only have had a few minutes' warning, but Thailand would have had two hours, and Somalia up to seven hours, to prepare for the disaster.

Japan's tragic 2011 tsunami may be seen in the remote sensing images on **page 21**. Its mechanism was similar: an upthrust earthquake along a section of underwater plates. Compared to the 2004 quake, its magnitude was slightly less (9.0), and the loss of life in wealthy, tsunami-prepared Japan far less (15,000). It was such a massive tsunami, however, that even Japan's engineering works were simply overwhelmed.

Misrule in Myanmar

Myanmar (population 54 million) is centered in the basin of the Irrawaddy River and includes surrounding uplands and mountains. The name of this country has generated some confusion for several decades. Britain conquered the area in a series of wars between 1824 and 1885. The British called it Burma (after the Burmese, the largest ethnic group in the country), and the country retained that name when it achieved independence in 1948. Its name is *Myanma* in the local language, which unlike "Burma" does not promote one ethnicity over any others. In 1989, the ruling military junta (see below) unilaterally changed the English-language name of the country to Myanmar. Although choosing a neutral name might be seen as a positive step for a multiethnic nation, many countries were slow to adopt the new name. The change was decreed without popular consent, and the United States, the United Kingdom, and others did not want to follow the dictates of a government they viewed as illegitimate. Some nations and media outlets, along with the United Nations, recognized the name change; others, like the United States, continued to use Burma. When relationships with the West began thawing after 2011, the name Myanmar started gaining increasing acceptance internationally.

Civil war has been almost constant in a country where ethnic minorities make up about 33 percent of the population (•Figure 7.48). Communism, targeted persecution, and ethnic separatism have motivated rebellions by ethnic Karens, Shans, and numerous hill peoples. The Muslim Rohingya ethnic group has been repeatedly targeted by the Buddhist central government; a major military action against the Rohingya in 1978 prompted over 200,000 to flee to Bangladesh and other Muslim countries. (The Rohingya crisis reescalated between 2012 and 2015, with more than 120,000 boarding ships to flee abroad and 90,000 internally displaced after conflict erupted with the neighboring Rakhine people.) After decades of unrest and rebellion, Myanmar's government reached cease-fire agreements with most of the ethnic rebel groups (except the Karens) in 1999 and gave these groups what it called **contingent sovereignty,**

• **Figure 7.48** Ethnicities in Myanmar.

offering them more civil rights and economic opportunities. The government also encouraged profits made by some of these groups in the drug business to be "invested" in national development.

The economy has been badly damaged by years of fighting and mismanagement. Once a country with the best health care, highest literacy rate, and most efficient civil service in Southeast Asia, a military coup in 1962 installed a government that promoted the **"Burmese Way to Socialism"**: a rigidly nationalistic, isolationist philosophy designed to promote national unity and governmental control over all sectors of the economy.

The results were disastrous. The military junta turned away most foreign aid and investment. Myanmar quickly became one of the poorest countries in the world (and remains so today, with a per capita GDP (PPP) of just $1700). It used to be the world's largest rice exporter; it now exports about 3 percent of the world total. Nearly 80 percent of the country lives without electricity, and mechanization in agriculture and construction is still a rarity. Until much-needed reforms were put in place in the early 2010s, about three-quarters of the economy was controlled by the black market.

In the 1990s, Myanmar began to slowly abandon socialism in favor of a free-market economy, and its economy finally began to grow, with China its main trading partner. Most of the fruits of this economic growth, however, lined the pockets of the ruling military. The **State Law and Order Restoration Council (SLORC)**, the military faction that seized power (from the previous military government) in 1988, allowed free elections in 1990. But this junta, which won just 10 of the nearly 500 parliamentary seats, refused to concede to the winning party led by Aung San Suu Kyi. The junta annulled the results and placed Suu Kyi under house arrest.

Myanmar remained one of the world's most repressive places to live. Access to the Internet was prohibited until 1999, and was strictly regulated afterward. Foreign journalists were banned. It was illegal to gather outside in groups of more than five. The few tourists that came were prevented from visiting most of the country, and Burmese citizens were strongly discouraged from having any "unnecessary contact" with tourists. Giant green billboards across the country proclaimed "Crush all internal and external destructive elements as the common enemy!"

In an apparent move to gain strategic depth from its own population and from any outside threats, the government in 2005 established a new capital at Naypyidaw, in the country's remote central region. The move did nothing to quell popular discontent with the regime. Widespread anti-government protests, with broad participation of Buddhist monks, broke out all over the country late in 2007. Authorities responded with deadly force, and the insurgency faded—at least temporarily.

In the wake of this political turmoil, Myanmar's isolated leaders turned a natural disaster into a humanitarian catastrophe. On May 2, 2008, a category 4 cyclone (hurricane) named Nargis slammed ashore across the Irrawaddy Delta. In this most agriculturally productive and densely populated region of Myanmar, more than 135,000 people lost their lives, and an estimated 2.5 million lost their homes and livelihoods. The country's isolationist junta (possibly unwilling to be seen as incapable of handling the disaster on its own) largely banned foreign aid agencies from providing medical, food, and other relief supplies. A US ship full of supplies off the coast was turned away, and cargo planes ready to fly to Yangon sat idle in their home airports when the military refused to issue visas to the relief workers. Foreign journalists were barred entry, and those already in Myanmar reporting on the situation were chased out of the country.

The Lady of Burma and Her Country's Turnaround

A much-publicized contest, followed around the world, raged for decades between the government and one of its citizens, the Nobel Peace Prize laureate Aung San Suu Kyi. She was placed under house arrest after the military overturned her victory in the 1990 election. In 1995, the government released "The Lady," as Suu Kyi is popularly known (**•Figure 7.49**). She then called for dialogue with the military junta and appealed for the parliament that was elected in 1990 to be convened. The government responded by again placing her alternatively under house arrest and in prison, triggering strong economic sanctions by the United States.

During her political exile, Aung San Suu Kyi called for would-be tourists to avoid Myanmar, which has abundant sites of cultural and environmental interest (**•Figure 7.50**). She argued that tourism revenue would help strengthen the repressive government. She was also successful in gaining support abroad for boycotts of US and other Western companies doing business with Myanmar. American corporations, including PepsiCo and ConocoPhillips, withdrew from Myanmar, and the US government banned new investments. The US government was displeased when the American firm Unocal built a pipeline from the offshore Yadana natural gas field to a port west of Bangkok, Thailand, where the gas is used to generate electricity. Human rights groups complained that slave labor was used to build the pipeline and that profits from gas sales helped prop up Myanmar's repressive government.

Joe Hobbs

• Figure 7.50 The huge golden Shwedagon Pagoda is the holiest Buddhist shrine in Yangon, Myanmar, attracting multitudes of the faithful and the tourists visiting the country. It is said that the relics or bones of three earlier buddhas and the eight hairs of Gautama Buddha are located within this spectacular golden architectural masterpiece.

Sean Gallup/Getty Images

• Figure 7.49 Aung San Suu Kyi is an international symbol of the unrelenting quest for democratic freedom. Like Nelson Mandela, she endured lengthy political detention, and like his followers, hers did not give up hope that she would emerge to lead a better country.

Other pipelines will serve the appetite of a country eager to consume Myanmar's energy and other resources: China. One carries Myanmar's offshore natural gas across the country to southern China. A parallel pipeline carries Middle Eastern oil offloaded on the coast of Burma and piped to southern China. This will increase China's energy security, in part by avoiding the congested chokepoint of the Strait of Malacca. Enjoying a 321 close relationship with the ruling junta, China long had almost exclusive access to Myanmar's natural gas, copper, and valuable gems, including rubies, sapphires, and jade.

Meanwhile, the United States began reevaluating its policies toward Myanmar, especially for an important geographical reason: its geographic *situation* is extremely important, as the country borders the two great powers of China and India. By withdrawing from Myanmar and imposing sanctions against it, the United States had been missing out on important commercial and strategic engagements with the country, pushing Myanmar into the arms of its rival China and enemy North Korea, and with sanctions effectively punishing the people rather than the government.

The US course change toward better relations with Myanmar has been helped by recent reforms made by the military government. In 2011, the government began cracking down on corruption, allowing foreign investment, easing restrictions on the Internet and media, releasing many political prisoners (including Aung San Suu Kyi), and promising free and fair elections. The reasons for the military government's pivot towards openness remain unclear, but the US approved of these changes and reestablished diplomatic ties with Myanmar in 2012. President Obama visited the country twice, indicating how important Myanmar is becoming for US economic and geopolitical interests. The biggest test of the military junta's promise to allow real democracy came in 2015, when Aung San Suu Kyi's National League for Democracy party won a landslide victory in national elections.

Another important milestone for Myanmar's reentry into the international community was its accession to the regional economic and political organization known as the **Association of Southeast Asian Nations (ASEAN)** in 1997. In an acknowledgement of Myanmar's recent progress, in 2014 it was selected to host the body's annual summit for the first time. The heads of state of all ASEAN members plus India, Japan, China, South Korea, Russia, and the United States met in Naypyidaw to discuss a variety of regional issues, especially the ongoing territorial dispute in the South China Sea.

Despite the country's recent strides, problems remain. Myanmar is still a very poor country fraught with ethnic tensions. Illegal weapons are smuggled into Myanmar in large numbers, most winding up in the hands of a variety of small ethnic groups demanding more autonomy (or to use against each other). Some of the cease-fire agreements from the 1990s started breaking down. Conflict erupted in 2015 between Myanmar's government and an ethnic Chinese minority called the Kokang in Shan State; for years the Kokang have been struggling to control an area rich in trafficked drugs, gems and other resources bound mainly for China. Sandwiched between contending powers and torn from within, for decades Myanmar has been a shatterbelt in miniature: "There is little danger that Burma will go Communist," wrote one observer in 1949, "but great danger it may go to pieces."[34]

Sex, Drugs, and Health in Southeast Asia

Along with Afghanistan, Southeast Asia has long been one of the world's main source areas for opium and its highly addic- 332 tive semisynthetic derivative, heroin. For decades, global concern focused on combating the drug trade originating in this region and on dealing with the problem of heroin addiction in the consuming nations. More recently, owing to the emergence of HIV/AIDS, the Southeast Asian industries of prostitution and tourism have woven drugs into a complex problem of global concern.

The region's drug production is centered in the so-called **Golden Triangle**, a vernacular region comprised of the borderlands where Laos, Thailand, and Myanmar historically exercised little control over their territories (see **Figure 7.48**). As in Afghanistan, the absence of strong government combined with rural poverty is conducive to drug production (**•Figure 7.51**). This Shan farmer's bottom line is typical in both regions: "We'd love to stop producing opium as soon as possible. But if you plant only coffee, you'll have nothing to eat."[35] Instability in Myanmar's rugged Kachin and Shan States bordering China provides especially fertile ground for poppy cultivation. Opium and heroin production in the region has risen and fallen over the years, but only Thailand has exerted successful eradication efforts. By 2015, Myanmar and Laos produced a combined 18 percent of the world's total opium output, up from 7 percent a decade earlier.

There is a global pattern to recognize here. When the nations of the Golden Triangle were most successful in eradicating

NYEIN CHAN NAING/EPA/Newscom

• Figure 7.51 Farmers in Myanmar's Shan State scrape raw opium oozing from slices in poppy pods. Heroin, which is derived from opium, has a hearth in Southeast Asia's Golden Triangle region.

opium production in the late 1980s, opium production spiked in the Golden Crescent of Afghanistan, surpassing the Golden Triangle's output in 1991. This phenomenon, in which cracking down on the problem in one place causes it to pop up in another, is called the **balloon effect**. As long as the demand for drugs remains high, production will meet it somewhere on the planet. Global drug trafficking routes are a cat's cradle, with traffickers taking advantage of globalization's conventional and unconventional ways of shipping goods. 478

In Asia, the US Drug Enforcement Administration and other international antidrug authorities focus their concerns on Myanmar, where refining of heroin from opium is done in remote laboratories. In northeastern Myanmar, armed ethnic minority groups have produced both heroin and a potent methamphetamine known as *ya baa*, meaning "crazy drug" in Thai. It is estimated that 1 billion *ya baa* tablets are produced here every year, making their way to users mainly in Thailand, China, and India.

Not all of the associated problems are exported. Mirroring the rise in raw opium production, addiction to cheap and potent heroin has been soaring in Myanmar. Some reports put the addiction rate of youths in Kachin state at 80 percent. Primary delivery is by needles shared in tea stalls and elsewhere; in Kachin's Internet cafes, patrons are advised not to shoot up while checking email.[36] Accompanying the heroin use is an AIDS epidemic, with the HIV infection rate among heroin users in Myanmar estimated at 40 percent.

The epidemic that began among heroin users spread quickly to Myanmar's sex industry. Recent estimates put the HIV infection rate among prostitutes at 30 percent, three times the rate in neighboring Thailand. Many young men with AIDS are shunned by their families and go to monasteries, where Buddhist monks offer them shelter, food, and prayer.

Myanmar's prosperous, cosmopolitan neighbor Thailand also has the scourge, with a 1.1 percent HIV infection rate among adults in 2014. This is a substantial reduction from the peak infection rate the country recorded in 1994. Unsafe sex is the main path of the virus's transmission, in a country where sex is a major industry. Thailand's liberal social climate has even given rise to the unique phenomenon of sex tourism,

Taimay Jones

• Figure 7.52 Bangkok's red-light district was a breeding ground for HIV/AIDS until the government and nongovernmental agencies aggressively promoted a safe-sex campaign.

with package tours catering to an international clientele (**•Figure 7.52**). Thailand's prostitution industry is estimated to bring in $4 billion or more each year, with prostitutes sending large sums as remittances each year to relatives in the villages from which they migrated. However, the spread of HIV infection among intravenous drug users and among prostitutes in the red-light districts of Bangkok and the popular beach resorts of Pattaya and Phuket threatens both the country's public health and its lucrative tourist trade.

There are striking differences in how this region's countries are dealing with HIV/AIDS. Thailand, regarded globally as a leader in efforts to combat the problem, has run an aggressive and increasingly successful anti-AIDS public awareness campaign, resulting in a dramatic decrease in the number of new HIV infections. Thailand distributes antiretroviral drugs to all AIDS patients at extremely low cost. Other Southeast Asian countries lag far behind. The government of the Philippines, where HIV/AIDS infections have doubled each year since 2008, has been unsuccessful in pursuing an anti-AIDS campaign because of opposition from the country's powerful and anti-contraception Catholic clergy. Church officials have denounced the government's condom-based anti-AIDS program as "intrinsically evil" and have been known to set boxes of condoms on fire. Similarly, in the conservative and largely Islamic nations of Indonesia and Malaysia, Muslim clerics denounce their governments' anti-AIDS campaigns as efforts to encourage promiscuity.

Vietnam Then and Now

The Vietnam War, which is such an important chapter in the American experience, had roots that preceded US interests in the region. This conflict profoundly affected Southeast Asia and has a lingering legacy there today. It is remarkable that Vietnam and the United States have reconciled to the extent that they are allies today.

The roots of the conflict began when France, believing it needed a strong colonial presence in Asia, invaded Vietnam in 1858. Southern Vietnam (an area the French called Cochinchina) was conquered in 1862, and Cambodia was annexed from Siam the following year. The Vietnamese called on China for help when France seized central Vietnam in 1884, but France defeated China in 1885 and annexed northern Vietnam soon afterward. France also occupied the various small Laotian kingdoms between 1893 and 1907. All of these areas were combined into the single colony of French Indochina, with its capital at Hanoi (**•Figure 7.53**). France's administration of Indochina, centered in Hanoi and Saigon, left a big cultural imprint. Its major economic success lay in opening the lower Mekong River area to commercial rice production, converting both Vietnam and Cambodia into important exporters of rice.

Japanese forces overran French Indochina in 1941, initiating five decades of warfare in the area. During World War II, a Communist resistance movement led by a Vietnamese nationalist named Ho Chi Minh carried on guerrilla warfare against the Japanese. When the French attempted to reoccupy the area after the end of the war, these forces fought to expel them. In 1954, Ho's forces destroyed an entrenched French garrison at Dien Bien Phu, in the northwestern mountains. With this defeat, France withdrew from Indochina, ending a century of colonial occupation.

Four countries came into existence with France's departure in 1954. Cambodia and Laos became independent non-Communist states (though Laos became Communist in 1975 after a North Vietnamese invasion). North Vietnam, where resistance to France had centered, emerged as an independent Communist state. South Vietnam gained independence as a non-Communist state that received many anti-Communist refugees from the north. Vietnamese Catholics were prominent among those who fled southward.

The partition of Vietnam into two countries was largely the work of US power and diplomacy. South Vietnam was from the outset a client state of the United States, whereas North Vietnam became allied first with Communist China and then

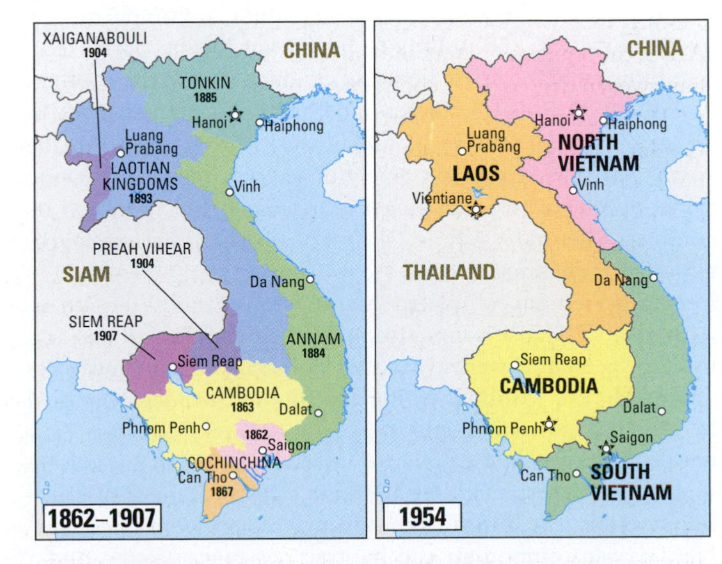

• Figure 7.53 Indochina's political evolution. The French colonized Indochina in stages in the latter half of the 19th century, consolidating their hold on the area by 1907. Four independent countries emerged after France withdrew from the area in 1954.

with the Soviet Union. Between 1954 and 1965, warfare in Vietnam gained momentum as the **Viet Cong**, an insurrectionist Communist force supported by North Vietnam, achieved increasing successes in the south in its bid to reunify the country. Increasing intervention by the United States in South Vietnam escalated in 1965 and eventually involved the commitment of half a million US military personnel. Why did the United States get involved on such a massive scale in a small country half a world away? Remember that those were Cold War days, and fighting communism was a foreign policy priority for the US. Vietnam was a "domino" that could fall.

172

North Vietnam responded to the US military escalation by fielding regular army forces against American and South Vietnamese troops. The United States never invaded North Vietnam, although American air strikes there were devastating, with more ordnance dropped on Vietnam than on all of Europe during all of World War II. American bombs, napalm (an incendiary explosive), and defoliants such as Agent Orange (dioxin) caused enormous damage to the natural and agricultural systems of Vietnam, destroying over 8400 square miles (nearly 22,000 sq km) of forest and farmland in an effort described in military parlance as **"denying the countryside to the enemy."** Agent Orange also left a legacy of birth deformities among Vietnam's postwar generation. One of Agent Orange's manufacturers, Monsanto, has recently begun agricultural development operations in Vietnam, where the government in 2014 approved the import of four GMO maize (corn) varieties developed by the biotech giant. Monsanto insists that no proof exists to link its product with birth defects, sparking angry reaction among the Vietnamese public.[37]

116

The United States avoided invading North Vietnam partly because of the risks posed by Chinese and Soviet support of the North. In limiting the theater of ground warfare, however, the United States found itself unable to expel from the South both the Viet Cong and the North Vietnamese army. Those combined forces were determined, skillfully commanded, increasingly better equipped by its allies, accomplished in guerrilla tactics, and willing to bear heavy losses. More than 3 million Vietnamese soldiers and civilians died in the conflict, and about 58,000 US soldiers and support staff perished. In 1973, almost all American forces were withdrawn from the costly war, which was extremely divisive at home as growing ranks of Americans questioned its rationale and protested its conduct. There are still 1627 Americans listed as missing in action in Indochina.

North Vietnam completed its conquest of South Vietnam on April 30, 1975, just hours after all remaining US civilians and military personnel were evacuated from Saigon. Saigon was renamed Ho Chi Minh City after the country's leader, who died in 1969 (most of its people still call it Saigon, however). Vietnam entered a period of relative internal peace, but there were ongoing conflicts, including Vietnam's 1979 conquest of Cambodia to oust the brutal Khmer Rouge regime, continued warfare against Cambodian and Laotian resistance groups, and a brief border war with China. Within its borders, the Communist government of the reunited Vietnam unleashed a repression against those who had assisted the Americans, including the rebellious hill tribes known collectively as Montagnards and members of the Cao Dai religion. This backlash caused a massive outpouring of refugees, including the "boat people," who risked much to flee. Most of the ethnic Vietnamese living in the United States today are those refugees and their descendants.

303

• **Figure 7.54** Vietnam today. Along with Laos and Cambodia, it forms the area called Indochina or the Indochinese Peninsula.

Vietnam (•**Figure 7.54**) has restored parts of its war-torn landscape through large-scale reforestation, agricultural reclamation, and nature conservation programs, aided by international organizations such as the World Wildlife Fund. Animal species new to science have been discovered in the rugged Annamite Range (Truong Son Mountains). Such natural treasures are attracting international ecotourists who provide much-needed revenue, and tourism in general is booming.

Since its embrace in 1986 of free-market reforms in the policy known as *doi moi*, Vietnam's economy has improved markedly. The poverty rate fell by half in the 1990s, and in recent years Vietnam has had Asia's second-fastest growing economy, after China's. The Communist government halted collectivized farming in 1988 and turned control of land over to small farmers under 20-year lease agreements. The results have been impressive; Vietnam since 1988 has become a major exporter of rice, third in the world after Thailand and India. This is a vulnerable commodity, however, especially because production is concentrated along the Mekong River, which is prone to flooding in the May–October rainy season and to the specter of rising sea levels due to climate change (see the Geographic Spotlight on page 346). Vietnam is also pinning hopes on exports of its proven reserves of 2.5 billion barrels of oil, especially to energy-hungry China, even as China has muscled in to extract oil in waters claimed by Vietnam (see page 316).

Vietnam and the United States restored full diplomatic relations in 1995 and have recently begun boosting military ties, as they share a joint interest in containing China. US businesses have been scrambling for consumers in this large, promising market of mainly young consumers. About 45 percent of the Vietnamese population of 91 million is under age 25. After the two countries signed a trade agreement in 2000, Vietnam began exporting shoes, finished clothing, and toys to the United States. Both its imports and its exports boomed when the country joined the WTO in 2007. The dual personalities of north and south are being perpetuated in the new economic climate. Most investment and infrastructural improvement focuses on the south, whereas the north is a more austere economic landscape; and northerners dominate the civil service in the capital of Hanoi (•**Figure 7.55**). Along with a growing gap between the rich and the poor, the differences between north and south could slow the country's overall development.

Indonesia: One Country, One People, One Language, 300 Ethnic Groups

The national credo of Indonesia, seen on banners and placards throughout the country, is "One country. One people. One language." The government's constitution, in an effort to promote national unity, officially recognizes four faiths: Islam (87 percent of Indonesia's population), Christianity (10 percent), Hinduism (2 percent), and Buddhism (1 percent). But the presence of some 300 different ethnic groups, combined with religious frictions, physical fragmentation, and economic problems, has made it difficult to attain peace, order, and unity in Indonesia. There is one official national language (Malay, known locally as Bahasa Indonesian), but more than 200 languages and dialects are in use.

Joe Hobbs

• **Figure 7.55** Street cacophony in Hanoi. Car ownership is on the rise in Vietnam, but for most people motorbikes are affordable and sensible. Who wants to drive a car on a street like this?

The largest ethnic group is the Javanese, who make up 40 percent of Indonesia's population. Java, an island the size of North Carolina, is volcanic (the famous Krakatoa volcano sits just offshore to the west), which provides the island with some of the most fertile soils on Earth. These soils, plus abundant tropical rainfall, allow for intensive rice cultivation; this abundance of food has led to Java becoming one of the most densely populated areas in the world. Nearly 150 million of Indonesia's 251 million people live there. Java is home to the country's capital of Jakarta, which with 30 million people is the world's second-largest urban area.

Various groups in the outer islands have resented the dominance of Jakarta and the Javanese, who have traditionally asserted their power in colonialist fashion over the larger, more resource-wealthy outer islands. Such animosities have escalated at times to armed insurrections. The Suharto regime (1968–1998) countered these militant expressions with a state ideology called ***Pancasila***. Aimed mainly at suppressing militant Muslim aspirations, *Pancasila* was a pan-Indonesian nationalist ideology designed to neutralize all ethnic identities.

For a time, Indonesia's government vowed to break with its repressive ways and appeased would-be separatists with promises of **liberal autonomy**, meaning that the provinces would have greater control over local administration and take more profits from the sales of local natural resources. With growing economic problems, pressure from nationalists, and accusations of corruption leveled against Suharto, however, the government shifted to using an iron fist against every Indonesian province that might aspire to follow the path the former enclave of East Timor took.

Now the independent country of Timor-Leste (population 1.2 million), East Timor was a Portuguese possession that Indonesia invaded and occupied after the collapse of the Portuguese colonial empire in the 1970s. The government declared East Timor to be an Indonesian province and violently suppressed an independence movement led by East Timor's Catholics. In the 24 years of occupation, about 20,000 Timorese were killed by Indonesian forces, and another 100,000 died from famine. By the late 1990s, Indonesia (which had been hit hard by the Asian financial crisis) could no longer afford its costly occupation and was facing strong international pressure to resolve the situation. In 1999, a new pragmatic Indonesian government allowed the people of East Timor to vote on whether they wanted autonomy or outright independence. The result was overwhelmingly in favor of independence, which was finalized in 2002.

Since its birth, this tiny young country has had many woes: a breakdown of law and order, interethnic violence, a failed 2006 coup attempt, and chronic unemployment. Over half of the country's exports come from a lucrative oil and natural gas field in the Timor Sea (which it shares with Australia), giving Timor-Leste a high GDP (PPP) of $21,400—yet its citizens remain among the poorest in the world: 40 percent of the population lives on $1.25 a day or less. All the oil and gas is sent to Australia by pipelines; there are no refineries or production

Geographic Spotlight Visualizing Climate Change with GIS

In 2010, I was invited by the government of Vietnam to advise the country's decision makers on how best to deal with the anticipated impacts of climate change. My focus had to be on adaptation rather than mitigation (see **page 43**); Vietnam produces only a tiny fraction of the world's greenhouse gases but has one of the world's longest and most densely populated coastlines. It is commonly ranked as one of the top 10 countries most threatened by climate change, particularly in the forms of rising sea levels and stronger typhoons (hurricanes; • **Figure 7.Q**). How could Vietnam adapt to these threats?

The best way to approach this question is by the basic geographical question: *Where* will the impacts of climate change take place? Using GIS, Vietnam's Ministry of Natural Resources and Environment (MONRE) has developed maps showing exactly where rising sea levels will inundate Ho Chi Minh City (the country's largest city) and the Mekong Delta (Vietnam's most densely populated region and one of the world's "breadbaskets," particularly for rice production). MONRE's Director Dr. Tran Thuc told me that IPCC forecasts for sea level rise were far too low because they did not take into account the melting of the world's great ice sheets in Antarctica and Greenland. Dr. Tran therefore commissioned maps depicting the impacts of 65-, 75-, and 100-centimeter (26-, 30-, and 39-inch) sea level rise by 2100—and told me he "fully expects" Vietnam to experience at least the 100-centimeter (1 meter) rise.

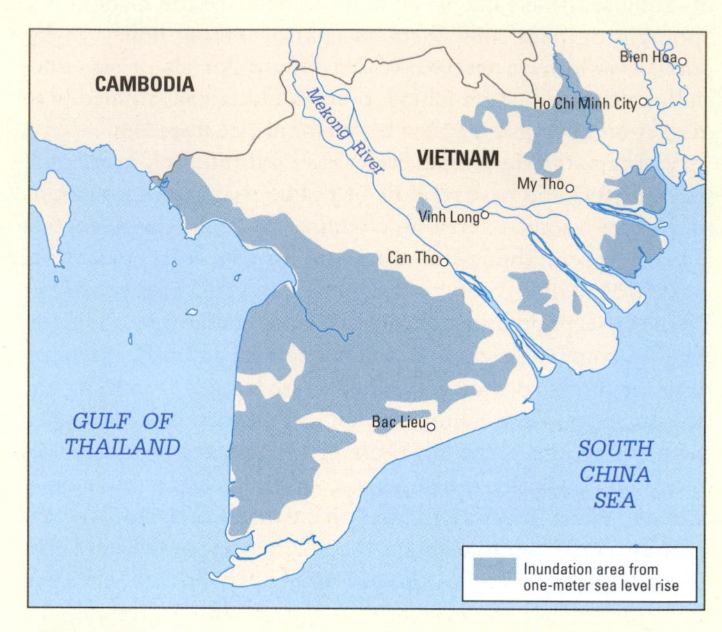

• **Figure 7.R** A 1-meter (39-inch) sea level rise would inundate much of the Mekong Delta, home of some of the world's most fertile farmlands.

• **Figure 7.Q** In recent years, Vietnam has experienced an unprecedented number of "super typhoons," equating to the category 4 and 5 hurricanes of the western hemisphere. I arrived in the coastal city of Hoi An hours before super typhoon Xangsane slammed ashore on October 1, 2006. I hunkered down with locals and took this photo three days later as the city's residents struggled to recover from extensive flooding and other damage. Many climate-change scenarios anticipate more frequent and severe typhoons in the country's future. Just as in Louisiana and in Bangladesh, higher sea levels would magnify the impacts of these storms. Engineering adaptation to climate change in the Mekong Delta must therefore consider sea level rise plus storm surge.

MONRE used a GIS method known as "statistical downscaling" to derive large-scale maps of the Mekong Delta and Ho Chi Minh City from small-scale global sea level models produced by the IPCC. The maps showing the 1-meter rise are breathtaking (the Mekong Delta map is shown in • **Figure 7.R**). Almost 40 percent of the Mekong Delta, and 25 percent of Ho Chi Minh City, would be underwater. About 9 million people—10 percent of Vietnam's population—would have to be relocated. These GIS visualizations are enormously useful, especially in helping Vietnamese leaders appeal for outside foreign aid to help build sea walls, dikes, and other infrastructure to adapt to climate change, and in illustrating the potential impact of climate change on global agriculture and urban infrastructure.

Much more sophisticated GIS work lies ahead. In order for sea walls to be built in just the right places and according to the correct specifications, very detailed large-scale modeling and mapping must be done. Vietnam can look to the Netherlands, where Dutch geographers are using a GIS-based process of "spatial planning" to map out exactly where engineering works must be built to stave off the rising waters of the North Sea.

facilities in Timor-Leste, so few jobs have materialized from the oil wealth. Over 60 percent of Timor-Leste's citizens work in agriculture; coffee is the country's only other significant export.

From the West... Of outstanding concern for Indonesia's unity is the province of Aceh (population 4.5 million), at the northernmost tip of Sumatra (**•Figure 7.56**). In 1976, the Aceh, who are a predominantly Muslim people of Malayan ethnicity, began seeking independence from Indonesia. The central government had made them a promise of autonomy upon Indonesia's independence from the Netherlands, but failed to keep it. The Acehnese insisted that the Jakarta government was too secular for their Muslim tastes and that too little of the revenue from sales of Aceh's abundant natural gas reserves remained in the province (which also has a wealth of oil, gold, rubber, and timber).

The secessionists' main voice was the **Free Aceh Movement** (known by the acronym **GAM**), which wanted to install a member of its own indigenous royal family as president of a free Aceh. Seeing Aceh as vital to the nation's unity and economic viability, and fearful that Aceh might inspire separatism elsewhere in the country, the government refused to discuss independence for the province. Violent clashes ensued, and the death toll grew to 12,000 between 1976 and 2004. But an unlikely arbiter appeared when the tsunami of December 2004 338 concentrated its unspeakable sorrow on Aceh (see **Figure 7.47**). The international community stepped in with a massive relief effort, and Indonesian authorities listened with unprecedented interest to Acehnese concerns. GAM dropped its demands for independence, and Indonesia agreed to allow GAM to participate in local elections. Nominally secular Indonesia allowed Aceh to become the only one of its 34 provinces to adopt Islamic *shari'a* law. Redefining itself as the Aceh Transi- 232 tional Committee, GAM laid down its weapons, and Indonesia scaled back its military presence in Aceh. The government gave amnesty to the rebels and, critically, pledged that 70 percent of the revenues from Aceh's natural resources would return to the provincial government.

Through a program called Aceh Green, founded by a former rebel, Aceh may emerge as a shining example of sustainable development. In one of Aceh Green's activities, hundreds of former rebels, who know the countryside very well, are being trained by the nongovernmental organization (NGO) Flora and Fauna International as rangers to look out for wildlife poachers and illegal loggers. If the rangers are successful in protecting the Ulu Masen rain forest, Aceh could net $26 million in carbon credits from the REDD funding mechanism (**•Figure 7.57**). 44

... to the East Progress in Aceh raised hopes for a similar future for Papua, at the extreme eastern end of the Indonesian archipelago (see **Figure 7.56**). The two provinces that comprise the Papua region are home to 3 million people of 200 different tribes speaking 100 different languages. Most are of Melanesian origin, and most are Christians who also have a strong substrate of indigenous beliefs. Collectively known as **Papuans**, they have little in common with the Javanese Muslims who control them from 2500 miles (4000 km) away in Jakarta. Their homeland on the western half of the island of New Guinea is of great importance to the economy of Indonesia because it contains the world's largest copper and gold mine, the Grasberg mine. The American company Freeport-McMoRan, which provides more taxes to Indonesia's treasury than any other foreign source, operates these mines.

• Figure 7.56 Indonesia, highlighting its Papua and Aceh regions. Timor-Leste achieved independence in 2002 after a quarter-century long attempt by Jakarta to incorporate it as an Indonesian state.

Courtesy of Goldman Environmental Prize

• **Figure 7.57** For his efforts combating illegal forest clearing and oil palm planting in Aceh, Rudi Putra is known as "Aceh's Green Warrior." He was nominated for the prestigious Goldman Environmental Prize in 2014.

• **Figure 7.58** The five autonomous areas of China.

The Free Papua Movement has been advocating independence for the region since 1965, and in 1971 it made a unilateral declaration of independence. This prompted intervention by the Indonesian military, and a low-level conflict has persisted since then. It is little wonder that Jakarta refuses to let its poorest but most resource-rich provinces go; the Dutch, coveting this wealth, initially refused to withdraw from the area when they granted independence to Indonesia in 1949. Pressure from Indonesia and the United States led the Dutch to relinquish control over Papua in 1963, and a UN-sponsored plebiscite on independence was held in 1969. The vote in favor of the region's becoming part of Indonesia was widely regarded as rigged. For many years Indonesia relocated people from other areas of the country to Papua to make the area friendlier to Jakarta. After the authoritarian Suharto regime fell in 1998, the subsequent, more permissive Indonesian government granted Papua a measure of autonomy in 2001. Indonesia has also promised that Papua will be given a greater share of the resource revenues it generates, but Papuans complain that these promises are unfulfilled.

China

Han Colonization of China's "Wild West"

China's growth as a land empire has involved both subjugation of people who are not ethnic Han and the colonization of those ethnic areas by ethnic Han. There are at least 56 non-Han ethnic groups in China, totaling about 100 million people living mainly in the west and southwest of the country. As the Soviets did in the USSR, China's Communist government granted at least token recognition of the distinctive identity and rights of five large minorities by creating **autonomous regions**. These are Guangxi, bordering Vietnam; Inner Mongolia (Nei Mongol) and adjacent, small Ningxia in the north; Tibet (Xizang); and Xinjiang in the far west (•**Figure 7.58**).

301 167

The Hui of Ningxia are Muslims of many different ethnicities who absorbed Chinese language and many Han customs. In contrast, the Uighurs of Xinjiang and the Tibetans have defiantly resisted becoming Chinese—and so Han colonization, or **Hanification**, focuses especially on their territories. The Chinese government aims either to assimilate these minority peoples into a broader, Han-based Chinese culture or to establish a strong enough Han demographic presence among them to dispel any hopes of autonomy or independence.

Chinese authorities have also initiated what they call the **Western Big Development Project** with the stated objective of improving locals' livelihoods enough to diminish their desire for ethnic and political separatism. Focused on the autonomous regions and four other provinces in which there are large minority populations, this project seeks to improve infrastructure like airports and roads and to boost output from factories and farms. The Uighurs, Tibetans, and other minorities generally fear that this "development" is just a cover for demographic, political, and economic domination by the Han.

Buddhism reached Tibet in the middle of the 8th century, and its arrival was followed by decades of conflict between Chinese and Indian Buddhists. The Chinese were defeated in this struggle and expelled from Tibet at the end of the 8th century. China regained loose control over Tibet in the 17th century, but Chinese authority vanished again with the overthrow of the Qing (Manchu) Dynasty in 1911. From then until the Chinese Communists' violent conquest of 1951, Tibet existed as an independent state, with its capital and main religious center at Lhasa.

After the Chinese completed roads to Tibet in late 1954, an increase in restrictive measures by the Communist government contributed to the rise of Tibetan guerrilla warfare. This culminated in a large-scale Tibetan revolt in 1959 and the flight of the **Dalai Lama** (the spiritual and political leader of Lamaism, or Tibetan Buddhism; •**Figure 7.59**) and many other refugees to India. Subsequently, the Chinese drove most of the monks from their monasteries, expropriated the large monastic landholdings, implemented socialist programs, and prohibited organized religion in the rest of China. Large-scale immigration by ethnic Han picked up; about half of Lhasa's population is now Chinese.

Harish Tyagi/EPA/Landov

• **Figure 7.59** The Dalai Lama is the enduring symbol of the nonviolent Tibetan quest for autonomy from China.

Beijing insists it is helping Tibet with major development investment, including the $4 billion Qinghai-Tibet railway constructed between Beijing and Lhasa. But Tibetan resistance to Chinese rule continues; some want stronger, even violent, action against China. From his place of exile in Dharamsala, India, and on frequent international trips, the Dalai Lama continues to appeal for Tibetan freedom while denying Chinese accusations that he is seeking Tibet's independence. He asks China to grant "genuine autonomy" to Tibet, meaning that Tibetans would have authority over most matters except foreign affairs and defense, which Beijing would still control. The Dalai Lama calls this the **Middle Way**. His peaceful cause has widespread support around the world. Beijing refuses to negotiate with the Dalai Lama, as it believes that requests for greater Tibetan autonomy are independence movements in disguise; the Chinese government even demands that the Dalai Lama publically apologize to China for what it sees as his attempts to divide the country. For many years, India had promoted the cause of Tibetan autonomy, but as its trade ties with China have grown more vital, its advocacy for Tibet has quieted. The future of the Dalai Lama tradition, and by implication Tibet's future, recently took an otherworldly detour when the Dalai Lama tested the waters of China's tolerance. In 2015, the Dalai Lama, who is believed to be the 14th reincarnation of a monk who lived in the 14th century, announced that he might not reincarnate, and therefore the identity of the next Dalai Lama is very much in question. Here is a situation you could only find in an authoritarian setting: China's government was enraged, saying only it could determine if the Dalai Lama is going to be reincarnated! The Dalai Lama is a "wolf in monk's robes," one government official fumed.[38]

China's government regards Xinjiang's Muslims, mainly from the 8-million-strong Uighur ethnic group who speak a Turkic language and are kin to the Kazakhs to the west, as the most problematic minority. Many Uighurs would like to pull at least a part of Xinjiang out of China's orbit to create an independent Chinese Muslim state. Two short-lived independent Uighur republics, both known as the Eastern Turkestan Islamic Republic, existed in 1933 and again from 1944 to 1949, and some Uighur separatists are fighting for one again. Beyond execution of those separatists deemed to be terrorists, China's response has been to press ahead with development projects and Han immigration.

At the outset of the Communist takeover of China in 1949, the Han population in Xinjiang amounted to 6 percent of the province's population. Today, the populations of Han and Uighur are nearly equal, and some 250,000 more Han Chinese immigrate to the cities (especially Urumqi and Yining) annually, enticed in part by relatively high salaries and other incentives. Many of these Han migrants are in Xinjiang only a short time; they stay long enough to make some money and then return to eastern China.

Human rights monitors report systematic Han abuses of the Uighurs. Scores of mosques have been torn down; Uighur literature has been burned; Muslim clerics have been forced into "reeducation" training; Islamic schools have been closed; public prayer is now forbidden; head scarves have been banned; and prodemocracy demonstrations are brutally quashed. Violent Uighur separatists strike back regularly with targeted killings and bombings of Chinese and their interests. China's struggle to maintain a grip on its ethnically distinct regions is reminiscent of Soviet efforts in non-Russian areas. 165

The Three Gorges Dam

The Chang Jiang ("Long River"), also known as the Yangtze River, has been central to China's identity and welfare for thousands of years. The river delivers precious water and fertile soils, creating environments that have been intensively settled and farmed, especially for rice and wheat. But this great river has also delivered tragedy in the form of devastating floods. Flooding in August 1998 affected 300 million people along the river and did an estimated $24 billion in damage, for example. Inevitably, questions arose about the advantages to be gained—especially in flood control, drought relief, and hydroelectricity production—if the river were to be dammed. The concept of a giant dam on the river dates to 1919, when it was proposed by Sun Yat-sen. His vision has been realized on a scale far grander than he could have imagined in the massive Three Gorges (Sanzxia) Dam, begun in 1994 and completed in 2009 (•**Figure 7.60**).

As with any large dam project, like the Aswan Dam seen earlier (**page 241**), the Three Gorges may best be evaluated by considering its pros and cons. On the plus side, this largest dam ever built—1.3 miles (2.1 km) wide and 610 feet (186 m) high—has created a reservoir 385 miles (620 km) long that will dramatically reduce the threat of downstream flooding. Stored waters will also help alleviate drought. The Three Gorges Dam will also improve navigation and therefore enhance trade, especially by connecting the burgeoning inland city of Chongqing to the world abroad. Historically, the Chang Jiang was able to accommodate freighters of 10-foot (3 m) draft all the way from the East China Sea to Wuhan. Above Wuhan, the shallower river was traditionally plied by smaller freighters to Yichang, situated just below the river's spectacular gorges. Yichang was a classic **break-of-bulk point**: the point at which cargo is unloaded and broken up into smaller units for delivery, thus minimizing transport costs. In this case, freight and passengers were

• **Figure 7.60** Floodwaters released from the Three Gorges Dam in September 2014.

THINK CRITICALLY: Do the "pros" of the Three Gorges Dam outweigh its "cons"?

offloaded and put on flat-bottomed junks and sampans drawn upstream through the gorges to Wanxian by human trackers who pulled and rowed the vessels. With the completion of the Three Gorges Dam project, vessels with much deeper draft are able to sail 1500 miles (2400 km) upstream, from Shanghai all the way to Chongqing in the Sichuan Basin (**•Figure 7.61**). Small vessels are able to pass over the dam in a ship elevator, a vast steel box that lifts the ship the height of the dam in 42 minutes. Most ship traffic uses a system of locks with five levels, taking three hours in transit. With these means, shipping to Chongqing has increased fivefold. Some observers say the increased commerce is in the process of transforming Chongqing into a "new Hong Kong." Chinese authorities view Chongqing as a new model for urban growth, as discussed on **page 354**.

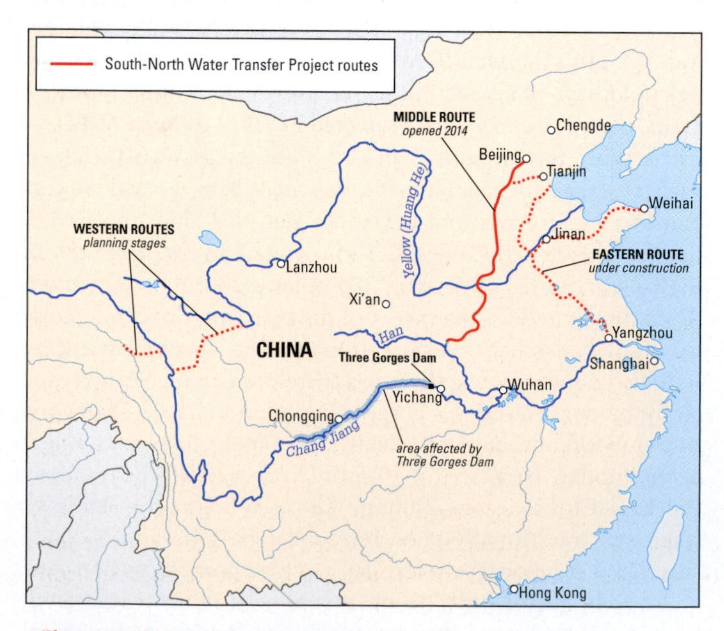

• **Figure 7.61** Locations of the Three Gorges Dam and the canals and pipelines of the South–North Water Transfer Project.

Most of the negatives are related to the vast reservoir forming behind the dam. The new 500 feet (150 m) deep lake has inundated 4000 villages, 140 towns, 13 cities, numerous archeological sites, and nearly 160 square miles (over 400 sq km) of farmland. Because their homes were to be drowned, as many as 1.4 million people were resettled. Many of the vacated settlements were quite ancient, and their inhabitants had deep cultural roots in them. The old villages were typically situated close to fertile farmlands that were also inundated. In the new, higher communities into which people were located, soil conditions are generally inferior.

The reservoir has had aesthetic consequences too, erasing a world-class wondrous wild river running through scenic canyons. The natural marvels of the three gorges drew 17 million Chinese tourists and hundreds of thousands of foreign visitors every year. There are also practical concerns about the silt that is building up behind the dam. It is possible that this increased soil load will have the ironic effect of causing flooding upstream from the dam, even as the dam prevents flooding below. There are fears of the almost unthinkable: that the dam could burst, causing unimaginable devastation and suffering downstream. Finally, there may also be seismic consequences of the dam. Central China suffered massive economic and human losses (about 70,000 dead and 5 million homeless) in the magnitude 8.0 Great Sichuan Earthquake of 2008. There is growing scientific evidence to suggest that this quake was triggered by the weight of a reservoir close to the fault on which the quake occurred. The Chinese government admits that the Three Gorges Dam has created seismic and landslide risks, and that its reservoir has been plagued by algae carpets, water contamination, and waste dumped by communities on its shores (though, in a positive twist, the dam has prevented about 10 million tons of garbage from flowing down the river and out into the Pacific).

Although recently admitting some environmental consequences, the government of China pursued the mighty Three Gorges Dam project without remorse and without tolerance for dissent within China. The enormous cost of the project, nearly $30 billion, was born solely by China. Like megadams elsewhere, the Three Gorges Dam is a statement that its builder is a great power to be reckoned with on the world stage.

The South–North Water Transfer Project

China has one of the world's lowest per capita water supplies and most uneven distributions of water. More than 40 percent of China's people live in the north, but only 23 percent of the country's water is there. Chinese engineers see water transfer on a massive scale as the way to redress this imbalance.

The massive **South–North Water Transfer Project** is a spectacular engineering project—more ambitious even than the Three Gorges Dam—that will move water from south to north through three canals (see **Figure 7.61**). One of its most difficult challenges is to blast through some of the giant mountains of the Himalayas to divert water from the upper reaches of the Yangtze River to the headwaters of the Yellow River. If completed, this feat will mark the first time in history that people have merged two river basins.

In contrast with their approach to the Three Gorges Dam, Chinese leaders view the South–North Water Transfer Project with some trepidation. There have been enormous cost overruns and years-long delays in construction. The project's concept originated with Chairman Mao and was approved long before China's present leaders came to power. Many of these leaders feel the project will have unfortunate, unintended consequences, both social and environmental. Like the Three Gorges Dam, this project will require people to relocate, and there are fears of social unrest. And the environmental consequences of changing the course of rivers and reallocating their waters are very difficult to foresee. Dai Qing, an environmental and political activist, had this to say of the South–North Water Transfer Project: "[It] will be considered a disaster within 20 years. Under Chairman Mao, we often heard the slogan 'Humans must conquer nature,' and that is still the mentality of so many people today—but gradually we are learning that human beings cannot conquer nature and that this philosophy inevitably leads to disaster."[39]

What's Next for Industrial China?

As described earlier (**pages 310–312**), China's economy has boomed. Yet even this economic dragon has weaknesses, and China's growth is slowing.

China's extraordinary recent successes are tempered by several concerns. One is freedom of expression. In 1989, the Chinese government cracked down brutally on prodemocracy demonstrations centered on Tiananmen Square in Beijing. Years later, there continue to be strictures on academic, media, artistic, and other freedoms that some observers believe are conducive to long-term economic well-being. China's government is one of few in the world (including those of Bahrain, Belarus, Cuba, Iran, Kuwait, North Korea, Pakistan, Qatar, Saudi Arabia, Syria, Turkmenistan, the United Arab Emirates, Uzbekistan, Vietnam, and Yemen) to censor the Internet, which is arguably one of the most important tools of economic development today. With what has become known as the "Great Firewall of China," the government requires all computers sold in China to be preinstalled with software blocking political and religious websites and pornography, and enabling the monitoring of personal communications and the websites visited by computer users. The government also monitors mobile phones, the use of which is growing 30 percent annually in China, to ensure that they too do not become instruments of 258 protest (as they did so profoundly in the Arab Spring of 2011). How sensitive is the government about unrest? Fearing the potency of a revolutionary symbol, China's government took to destroying jasmine plants in the wake of Tunisia's "Jasmine Revolution" (see **page 255**).

Another potential problem is that China's leaders have been breathlessly trying to feed what they call the "socialist market economy" with gigantic projects, thereby hoping to create jobs and stimulate economic growth. They reckon that the country needs to register at least 7 percent economic growth annually just to prevent massive unemployment and social unrest. Following on the massive infrastructure investment for the 2008 Olympic Games, one of the world's most impressive building booms ever is continuing: new subway systems, railroads, highways, bridges, sewage systems, dams, water diversion projects, a natural gas pipeline, a national electricity project, urban improvements, airports, industrial parks, and more. These efforts cost a lot of money, and the government is drawing down its national treasury to fund them. State banks are pouring billions of dollars into federal projects whose returns are uncertain.

A further hazard is that with its galloping growth China has been developing a classic economic bubble that will burst, probably with costly results. Historically, few economies have been able to maintain torrid growth for long, especially when that growth is fueled by speculative cash flows and questionable bank loans, as has been the case recently in China. The 2008 global credit crisis offers a bitter example of what can happen in these conditions, especially in the property sector. The problem is that China's breakneck development has funded mainly through domestic borrowing. All across the country, credit-fuelled property construction has created a landscape of unoccupied office and apartment blocks. Real estate accounts for one-fourth of China's economy, a higher proportion than the United States and Spain had before their property bubbles popped. Unfortunately, much 117, 529 of the collateral for the outstanding debt of the local governments that borrowed to fund the property boom is greatly overvalued government-owned property—mainly land. This may portend a precipitous chain reaction of collapsing valuations.

The unwinding of China's real estate sector could well send shockwaves through the global economy, starting with countries such as Australia, Brazil, Germany, and many in Asia that rely strongly on Chinese demand.[40] Prices of iron ore, copper, and other commodities used in the construction sector began to drop precipitously in 2014, as did oil, whose price is often a benchmark for anticipated global growth. China's colossal 176 manufacturing sector has far more supply than demand, a recipe for *deflation* within and beyond China.

The stage is set for a major reversal. As China's population ages, its labor force is shrinking. Cheap capital, cheap labor, cheap energy, and cheap land were at the heart of the Chinese economic miracle, but each of these is becoming pricier. Inflation and unemployment are growing, raising the specter of social unrest. Overall, China appears to be falling into the "development trap" described in Chapter 3 (**page 60**), in which an LDC can move into middle income status rather easily when commodity prices are rising, but find it much more difficult to advance to MDC status when commodity prices fall.

There is a solution for China's economic dilemma. The country is too dependent on exports and therefore on the whims of foreign consumers and market conditions abroad: recession in the US or Europe causes people affected to buy less of everything (including many things made in China), and that hits China hard. The solution is developing an economy based more on domestic household consumption and less on exports. So far China has been slow to develop a consumption-driven economy.[41] Household consumption represents only 34 percent of China's GDP, compared with 70 percent in the United States. For China's economy to hold up, urbanization combined with more domestic consumption on large scale is essential.[42] As many as 70 percent of Chinese may be middle class or above in the 2020s. That would represent a huge market of potential consumers.

There are environmental costs for China's spectacular growth. Due to generally lax environmental quality standards, China's skies and waters are often badly polluted. China has more entries on the World Health Organization's list of most polluted cities than any other nation. China's government acknowledges the problems. The State Environmental Protection Administration (SEPA) reported that one-quarter of the water in China's largest river systems was so polluted that it was dangerous for people and other living things to come into contact with or had "lost the capacity for basic ecological function." Polluted water was killing tens of thousands of people every year and reducing crop yields. SEPA warned that China's approach of "growth through industrialization" was pushing its environment "close to the breaking point."[43]

Having overtaken the United States as the world's largest producer of carbon dioxide, China (which gets 80 percent of its electricity by burning coal) presumably has the largest role in climate change. When the Kyoto Protocol on climate change was negotiated, China was exempted because it was an LDC. Technically this is still true, but China has surprised many by taking some responsibility for climate change. In its 2014 'great powers of climate change' agreement with the US, China vowed to reach a peak in CO_2 emissions around 2030. Thereafter, China would increase the non–fossil fuel share of all energy sources to around 20 percent by 2030. With that pledge, China also resolved to become the global leader in green technology. China sees this as an opportunity both to confront its own problems with climate change, apparently including both droughts and floods, and to create new jobs. To this end, China has quickly become the world's largest manufacturer of both wind turbines and solar panels, and aims to become the world's biggest builder of nuclear power plants. China has passed the United States to become the world's largest market for wind turbines, which are sprouting up on the Gobi Desert and elsewhere.

Addressing Inequities in China: Rich vs. Poor, Urban vs. Rural

There is concern that the benefits of China's economic growth have not been distributed evenly throughout the population. Prosperity and consumption have increased dramatically, but in a remarkably unequal manner. In 1980, China had one of the world's most even distributions of wealth, but that equation has changed dramatically, with the excesses of the flamboyant super-rich visible to the very poor. These inequalities are comparable to those that existed when the revolution brought Chairman Mao's Communists to power on the promise of redistributing wealth.

Remember that, despite its phenomenal growth, China is still relatively poor, with a per capita GDP (PPP) of $9800 (comparable to Turkmenistan) and agricultural (with a 46 percent rural population (see **Table 7.1**). The agricultural face of China is changing rapidly, though, as the cities boom (see the discussion of megacities on **page 299**).

China does have two very wealthy areas—the small Special Administrative Regions (SARs) of Macau and Hong Kong on its southern coast. Hong Kong's GDP (PPP) is nearly the same as that of the United States, and Macau's is $88,700—it would rank fourth highest in the world if it were a country. The British held Hong Kong until 1997, and Macau was a Portuguese colony until 1999. Both cities developed strong market-based economies and became major international commercial hubs during the 20th century. When they were retuned to China, it was done under the "one country, two systems" policy—Macau and Hong Kong would be part of China but would maintain their economies, foreign relations, politics, and judicial systems separately from Beijing. (Riots broke out in Hong Kong in 2014 when Beijing tried to implement changes to Hong Kong's electoral process to bring in leadership more to Beijing's liking.)

Most of China's economic activity and growth is in the east—its 11 coastal provinces account for 50 percent of the country's GDP. In 1978, in an experiment to open China up to the world economy, China created six **special economic zones (SEZs)**—five cities and one entire province—along the southeastern coast, designated to attract foreign investment and development without having to navigate the maze of bureaucratic red tape that hindered economic growth elsewhere in the country (**• Figure 7.62**).

Foreign corporations were eager to tap into the huge reservoir of cheap Chinese labor, and the SEZ experiment was a huge success—so much so that much of China is now open to foreign investment in a similar fashion. One of the results of the clustering of these cities, and the economic power they represented, has been a serious problem of regional imbalance. Coastal China has long been wealthier than the interior, and the success of the SEZs and the spread of that model to other eastern cities exacerbated that trend (**• Figure 7.63**). Even the rural areas of eastern China experienced much more growth relative to those in the west, especially because they were well linked to expanding markets and prospering cities.

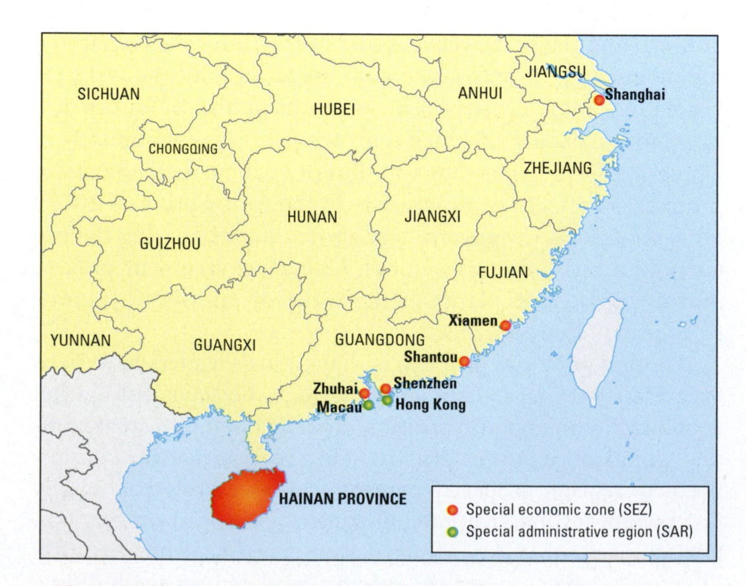

• Figure 7.62 The six original special economic zones, along with the special autonomous regions of Macau and Hong Kong that were incorporated into China in the late 1990s.

• Figure 7.63 The futuristic Pudong section of Shanghai is China's modern face to the world and is one of the country's SEZs.

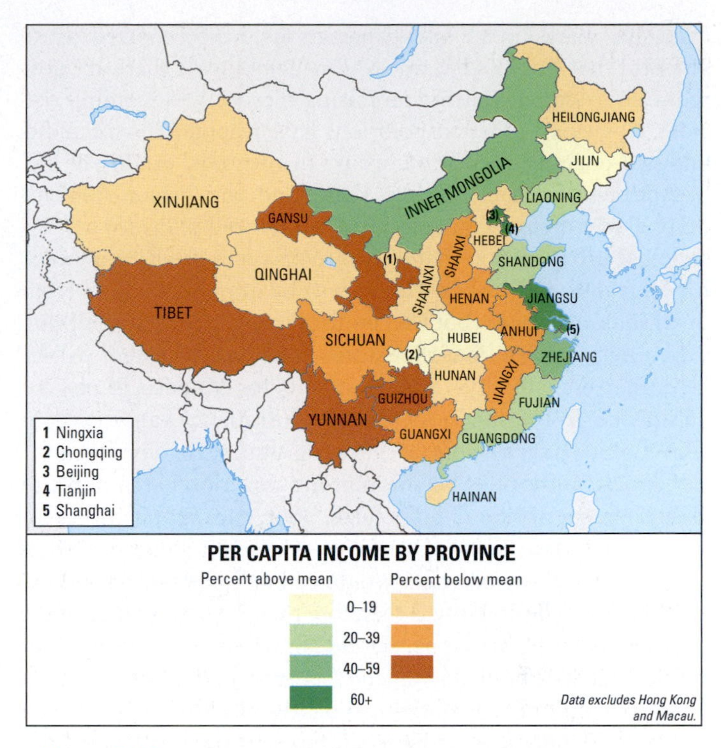

• Figure 7.64 The geographic distribution of Chinese incomes. The provinces along the coast tend to be much wealthier than those inland.

Meanwhile, people elsewhere in China were frustrated; they saw relatively little evidence of China's much touted economic prosperity. The new wealth bypassed most of western and southwestern China entirely. The northeastern provinces, an area known as Manchuria, was once China's industrial core, but most of its inefficient, Soviet-style heavy industry factories were shuttered during the market reforms of the 1990s, putting tens of millions out of work. Manchuria's proximity to Beijing and coastal cities has brightened its economic prospects somewhat, but as "China's Rust Belt" Manchuria has yet to fully partake in coastal China's prosperity (• Figure 7.64).

Predictably, large numbers of people have migrated from the less prosperous regions of the country to the promising urban centers of the east and south. As of 2015, 250 million migrant workers had left their villages—leaving them peopled mainly by children and the elderly—to work in construction or menial labors in the coastal cities, sending much of their meager monthly earnings of \$50 to \$100 home as remittances. The nonresident migrants provide cheap labor not only to factories but also to affluent urbanites who hire them as nannies and servants—you can imagine the clash in a society that was so equitable a generation earlier. People in the cities often discriminate against these migrants, known derisively as "outsiders," in education, housing, and jobs, treating them like illegal aliens. Many technically are illegal because they fail to sign work contracts that would both register them and require factories to pay fees associated with them to local cities. Police routinely "shake down" workers without papers, seeking bribes from the frightened migrants.

These problems ultimately are rooted in an antiquated bureaucratic mechanism that is badly in need of reform. Chinese cities are characterized by a two-tiered society of legal residents and nonresidents, created by a registration system known as ***hukou***. The system was designed in the 1950s mainly to discourage internal immigration. Residency status allows people access to better education and health care. But the number of residency permits is limited and nonresidents cannot afford to buy legal residence permits. In 2015, only 36 percent of China's urbanites held urban *hukou*.

What are conditions like for the rural migrants? Some of the factories—like the Foxconn complex in Shenzhen known as iPod City because these and other Apple devices are assembled there—have a good working environment, but others are **sweatshops** where working conditions border on the nightmarish. Management typically forbids strikes and independent trade unions, and does not enforce minimum-wage laws. Workers often have 18-hour workdays and take their brief rest in prisonlike dormitories. It is little wonder that many of the eastern boomtowns see as much as 10 percent annual turnover in the workforce, along with soaring crime rates. Even at Foxconn, excessive work hours and expectations were reported as so stressful that suicides became routine (employees there are now required to sign a contract promising they will not commit suicide). Apple and other corporations have responded with audits and still-ongoing efforts to improve the workers' quality of life.

Addressing problems like these and the urban resident/nonresident divide, China in 2014 announced plans for a "new style" of urbanization that would focus on making cities fairer for migrants. Full urban *hukou* will be granted to 45 percent of China's urbanites, but this will still leave about 200 million urban migrants without that status. The *hukou* system will be "eliminated gradually." The plan calls for 60 percent of China's people to live in cities by 2020—a statistic that would put China on a par with countries having similar income levels.[44]

Chinese authorities believe that urbanization is the ticket to greater prosperity for China's huge, poor, rural population. But they do not want the magnet cities of the east to absorb all of these people, many of whom would end up unemployed and living in slums. The strategy is therefore that the larger cities of the interior, such as Chongqing, should absorb much of the migrant traffic that has been flowing farther east and south. Already home to 8 million people, Chongqing (the wealthiest interior city in China) is poised to grow much more through its "one-hour economy circle plan." In this plan, 4 million rural residents will move into newly urbanized urban areas within an hour's drive of the city center. China hopes that millions of new jobs will be created in the long-neglected urban centers of the interior.

The normal processes of labor markets are helping the government achieve its goals. Wages have been rising sharply all over China. The government recently raised the minimum wage by 20 percent, and employers like Foxconn have granted even higher raises. Once pushed or pulled to the coastal cities, workers are now finding competitive jobs closer to home. The result is a sudden and surprising shortage of wage laborers in the coastal factories. Once-confident sweatshop managers now worry that the global marketplace will shift jobs to lower-wage 311 countries like nearby Vietnam.

There is much at stake for China's system of state-monitored capitalism. The government has a social contract with its people: it pledges that it will create and maintain a "harmonious environment" for them. Employment and economic growth need to be maintained at high levels if public unrest or even revolution are to be prevented. With a population of more than 1.3 billion, there is a lot of "people power" in China. Chinese leaders nervously watched the crowds in Egypt demanding and getting a change in government.

Taiwan and the Two Chinas Problem

Known by Westerners for centuries as Formosa (Portuguese for "beautiful"), Taiwan lies about 100 miles (160 km) off the southeastern China coast (•**Figure 7.65**). About the size of Switzerland, the island is home to 23 million people, nearly all of Chinese origin. Driven from the mainland in 1949, the Chinese Nationalist government fled to Taiwan with remnants of its armed forces and many civilian followers. Nearly 2 million Chinese were in this migration. Here, protected by American sea power, the government reestablished itself as the Republic of China (ROC, as opposed to the mainland People's Republic of China, or PRC) with its capital at Taipei.

The Nationalists, originally an authoritarian regime but with increasing elements of political democracy, were very successful in fostering industrial development. They united inexpensive Taiwanese labor with foreign capital to build one of Asia's first urban-industrial countries. By the 1990s, it had become one of the Asian Tigers. Its major exports are machinery, electronics 30 (notably computers, cell phones, modems, routers, and global positioning systems), metals, textiles, plastics, and chemicals, especially to mainland China and the United States. The population is 73 percent urban; its infant mortality rate is very low; and its population growth rate is the lowest in Asia except for Japan (see **Table 7.1**). The average ROC citizen is four times wealthier than the average PRC citizen. In almost all respects, Taiwan is clearly recognizable as an MDC. A major hurdle against even stronger growth has been a lack of energy resources. Some coal and natural gas exist, and some hydropower has been harnessed, but the island depends heavily on imported oil.

Taiwan's ROC continues its claim of being the legitimate government of China. Meanwhile, the PRC claims Taiwan as part of its own territory. The United States backed the Nationalist claim for a time, but during the 1970s the United States and the PRC developed closer relations. After years of opposition, in 1971 the United States supported the revocation of Taiwan's seat in the United Nations and its replacement by the PRC. In 1979, bowing to the wishes of the PRC, the United States withdrew its official recognition of Taiwan, instead recognizing China's claim of sovereignty over Taiwan. This **One China Policy** prevents the United States from having formal diplomatic relations with Taiwan. However, the US opposes the annexation of Taiwan to China by force, vowing to defend Taiwan from attack, just as it opposed Taiwan's call for a Nationalist effort to retake the Chinese mainland during periods of internal weakness there. The United States is a strong backer of Taiwan, supplying it with generous packages of weapons and economic aid.

Taiwan continues to resist mainland China's overtures for reunification, which include promises of broad autonomy, and is moving increasingly away from any type of reunion. That trend reflects Taiwan's ethnic makeup. Less than 15 percent of the island's people descend from the Nationalist refugees, who with their recent ties to the PRC are known as "mainlanders." The vast majority of the island's inhabitants are descendants of Han peoples who emigrated from China as many as four centuries ago, notably the Fujianese, who came from the southeastern coastal province of Fujian, and the Hakka, from Guangdong province. This majority is much less mainland oriented in political outlook. Despite discrimination by the Nationalists, these "indigenous" Chinese of Taiwan have enjoyed increasing political power.

The two countries must be cautious about their economic and political relations. Taiwan's economy has become very dependent on the PRC, where Taiwanese companies take advantage of low-cost mainland labor (about one-fifth the cost in Taiwan for the manufacture and assembly of products) and a huge potential market for their products. A recent survey found that if they were certain China would not attack their

• **Figure 7.65** East Asia: the economic powerhouses of Japan, South Korea, Taiwan, and coastal China.

island, 80 percent of Taiwanese would support declaring independence.[45] But any efforts to sever most ties with China and assert Taiwan's independence more forcefully could have large economic consequences and invoke military intervention from Beijing—and an American military response.

Japan and the Koreas

Postwar Japan

On the way up . . . Despite having lost its colonies and empire, Japan became an economic superpower after World War II. The nation's explosive economic growth after its defeat was one of the most remarkable developments of the late 20th century and, echoing the broader region's experience, is known widely as the Japanese "miracle." Japan watchers cite different reasons 309 for the country's economic success. Proponents of dependency theory argue that Japan, which was never colonized, escaped many of the debilitating relationships with Western powers that hampered many potentially wealthy countries. Some analysts 59 believe that the country's postwar economic miracle grew from an intense spirit of achievement and enterprise among the Japanese. Notably, many Japanese attribute this industrious spirit to Japan's geography as a resource-poor island nation. To overcome the constraints nature has placed on them, the Japanese people feel they must work harder and use their limited resources more efficiently. Still others attribute Japan's postwar achievements to its association with the United States following World War II. Some of the postwar US-imposed reforms worked well economically, and relations between the two countries since the war have also generally stimulated Japan's industrial productivity.

Economists also point to several unique features of Japanese management and employment to explain the country's meteoric postwar gains. One has been recruitment through an extremely challenging (some say brutal) educational system that emphasizes technical training. Although recent years have seen a move to a more lenient and more comprehensive education, a rigorous and stressful testing system still determines access to higher education. Many workers were hired right out of school and enjoyed guaranteed lifetime employment at a firm, with the company sometimes providing housing and recreational facilities to their employees.

One essential factor in Japan's postwar economic growth was a high level of investment in new and efficient industrial plants. Investment capital was made available when government policies cut expenditures on amenities and services such as roads, antipollution measures, parks, housing, and even higher education. Japanese industrial cities grew explosively and became highly polluted areas of dense and inadequate housing, with few public amenities and snarled transportation. The money "saved" by not being spent on amenities and services was made available for investment in industrial growth, but the costs to Japanese society were high.

Some analysts cite elements of Japan's political culture to explain the country's economic successes. One political party, the **Liberal Democratic Party (LDP)**, has governed Japan almost continually since the country regained its sovereignty in 1952. It is a conservative and strongly business-oriented and business-connected organization. The party has promoted Japanese exports with policies to keep the yen (the Japanese unit of currency)—and thus Japanese goods—inexpensive. The party has also cooperated closely with Japanese business in the development of new products and new industries. However, that very coziness between government and business—the "crony capitalism" seen in several other Southeast and East Asian countries—was partly to blame for the economic malaise that spread over Japan in the 1990s, and support for the LDP began to erode.

... and on the Way Down Despite all of these factors in Japan's economic favor, the Japanese miracle did not last. The peak of Japan's postwar success came in the 1980s. A powerful economic boom led to speculative rises in stock prices and land prices. At the end of the 1980s, the total value of all land in Japan was four times greater than that in the United States. The Tokyo Stock Exchange was the world's largest, based on the market value of Japanese shares. Then the **bubble economy** abruptly burst in the early 1990s, and real estate and stock prices fell by more than 50 percent. Japan had near-zero economic growth between 1991 and 2000 (in what has been called Japan's "Lost Decade"), and in the years since has only experienced slight improvement. Japanese industries whose growth had seemed unstoppable experienced an increasing loss of market share to US and European producers. Part of the reason the economy could not regain its footing was official Japanese commitment to prop up faltering industries and farms with huge subsidies. Because of the disproportionate political influence of small towns and rural areas, the Japanese tried to modernize remote villages and islands with expensive and underutilized public works projects. Due in large part to such projects, Japan is the world's most indebted country, with a debt-to-GDP ratio of 227 percent in 2014 (by comparison, the US debt ratio was 101 percent). Overall, the system succeeds in subsidizing its past at the expense of its future.

112,

Attitudes toward ethnic groups play a significant role in Japan's economic development, or more precisely lack of development. Japan has one of the world's most homogeneous populations; it is 98.5 percent ethnic Japanese. This lack of racial and ethnic diversity has had mixed results for Japan. Many observers believe that it has helped the country achieve a sense of unity of purpose, allowing the Japanese to persist through periods of adversity, especially the postwar years of reconstruction. However, the Japanese have also earned a reputation for intolerance of ethnic minorities. This attitude may be costly for Japan's future economic development. One way to increase productivity in the face of Japan's declining population would be to import skilled workers from abroad, but most people in the country remain averse to immigration. Inward-looking tendencies have also left Japan short of people capable of and interested in learning English, a language routinely sought by aspiring businesspeople in nearby South Korea and China.

93, 157, 497

There is definitely another upside to Japan's homogeneity. Capitalist Japan has had a remarkably egalitarian society, with about 80 percent of the population solidly middle-class. The richest third of the population has a total income just three times as great as that of the poorest third (compared with a fivefold differential in the United States). The Japanese have historically embraced the concept of ***wa***, or harmony, based in part on the principle of economic equality.

Japanese business culture also stands in the way of economic progress. Employment in Japan is very different from that of the United States. Many workers are not hired for a specific job at a company, but are transferred around to experience many different aspects of the organization. An employee's salary is not based on skill, but is tied to how long he or she has worked at the same company; if an employee leaves a firm to join another, he or she starts back at the bottom of the ladder in terms of pay and must work the way up all over again. Layoffs, though not illegal, are socially taboo, so in effect many employees have lifetime employment at their firms, even if their jobs have been eliminated. There are an estimated 25 million or more people, primarily middle-aged and older men, who collect paychecks without having an actual job to do. (When a job is cut, firms usually offer generous early retirement packages to get the employee to leave, but the employee is under no obligation to do so. In this case, the worker is sent to a so-called "boredom room" and given menial or meaningless tasks to do in the hope that he or she will eventually become demoralized and quit.) Japanese firms often have fairly low productivity levels, and the job market is very inflexible.

The government has long balked at many suggestions for economic reform because of the threat they pose to a socially harmonious Japan. There have been international appeals (and increasing domestic ones) for Japan to reform its economic system by allowing inefficient businesses to collapse rather than propping them up with expensive subsidies, and to resist the temptation to satisfy marginal rural populations with costly education and public works projects. Only in recent years have government officials seriously considered changing the seniority pay system to a merit-based one.

The legendary Japanese work ethic also has its advantages and drawbacks. It has helped the Japanese create a prosperous country, but it has also created a nation of workaholics beset with the same problems—stress, suicide, depression, and alcoholism—experienced by workers in many of the world's other MDCs. With their long hours at work, men are accustomed to spending little time with their spouses and children. It is said that Japanese men "commute to home from the office." There is a growing incidence of what Japanese call ***karoshi***, or death by overwork. The educational system encourages children to be highly successful but also to conform; critics say this stifles creativity and innovation. College graduates who are not hired during the annual recruiting season face difficult obstacles to entering the job market; many of them become "permanent temporary" workers who are not paid enough to find apartments in Tokyo and other very high-cost cities. The country's welfare system, pensions, and social security are inadequate, compelling workers to work harder for savings.

Women have not achieved parity with men in the workplace, and they complain increasingly of discrimination and sexism. In struggling to increase their footing in Japanese society, growing numbers of Japanese women are both working longer hours and marrying later, contributing to Japan's remarkably low birth rate (8 per 1000 annually, one of the lowest in the world). This trend and others account for Japan's declining population (with an annual population change rate of −0.2 percent), which stands out clearly in • **Figure 7.66**; note especially the tapering bottom of the country's age-structure diagram. Japan's falling birth rates will only increase the burdens of its workers. The population is both shrinking (it is expected to decline from 127 million in 2014 to 100 million in 2050 and 67 million in 2100) and aging (already one in four people were over age 65 in 2014, and 1 million people a year join those ranks). The Japanese of working age will face an increasing load of taxes and family obligations to meet the needs of older citizens.

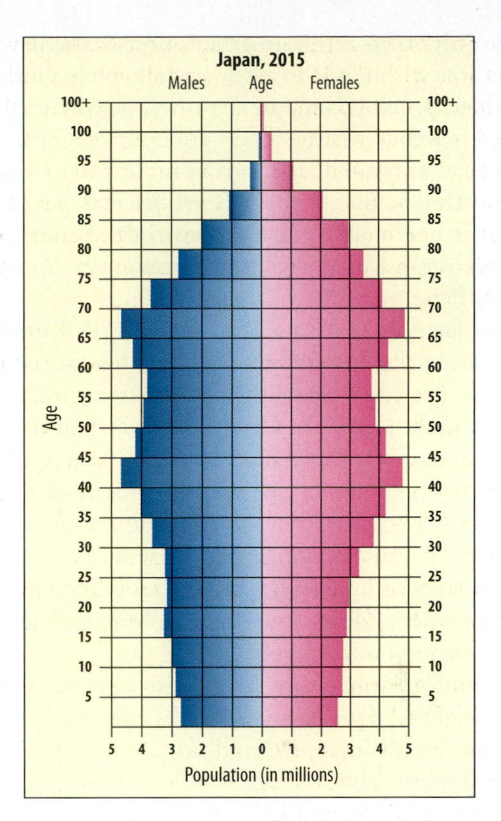

• **Figure 7.66** Many countries face problems of overpopulation, but Japan is losing population and is concerned about not having enough people to keep its economic engine running in the future.

North and South Korea: Night and Day

Just 110 miles (177 km) across the Korea Strait from Japan is the Korean Peninsula, roughly the size of Minnesota or Portugal. The two Koreas have occupied an unfortunate location in historic geopolitical terms: they are located in between larger and more powerful neighbors—China, Russia, and Japan—that have frequently been at odds with one another and with the Koreans. China and Russia have traditionally feared Japanese rule in Korea because it might serve as a springboard for invasion of their countries. Meanwhile, to Japan, Korea has been seen as a Chinese or Russian dagger aiming at its heart (see **Figure 7.65**). Throughout history, Korea was often a subject or vassal state of China or Japan as those two larger countries sought advantage over the other.

Despite external ambitions of conquest and division, from the late 7th century to the mid-20th century, Korea was a unified state, sometimes invaded and forced to pay tribute, but never destroyed as a political entity. The decline of Chinese power in the 19th century was accompanied by the rise of modern Japan. Korea came under increasing Japanese influence, and in 1910 Japan formally annexed Korea into its growing empire. A legacy of hostility and occupation continues to cast a shadow over relations between the Koreas and Japan, and disputes periodically emerge between them over issues such as control of islands and fishing rights (see **page 317**).

Japan lost Korea at the end of World War II, setting the stage for the division between North and South Korea. To understand Korea's split, it is important to consider the Korean War, sometimes in the United States called "America's Forgotten War." The conflict cost the lives of an estimated 2 million North Koreans, perhaps 300,000 South Koreans, and 33,629 Americans.

In the closing days of World War II, the Soviet Union entered the Pacific war as an ally of the United States against Japan. The two sides drew up plans to accept Japan's surrender on the Korean Peninsula, arbitrarily drawing a line at the 38th parallel. The Soviet Union would accept Japan's surrender north of that line and the United States south of that line. This was not meant to be a permanent boundary, but it became one. On either side of the line, the Soviet Union and the United States moved to set up governments that would be friendly to them. By 1948, the Soviet-style, Soviet-backed, Soviet-armed Democratic People's Republic of Korea was established in the north, while in the south, the Western-oriented Republic of South Korea was created under the auspices of the United Nations.

War came soon. On June 25, 1950, North Korean troops invaded the south and quickly took Seoul. Fearing Soviet expansionism, President Truman sent American forces in, and they soon fought under the United Nations flag with support from 16 other countries. General Douglas MacArthur served as supreme commander of the effort. By September, MacArthur's forces had pushed the North Korean advance all the way back to the Yalu River on the Chinese border. The general proposed using nuclear weapons to press forward from there, but the president refused, and MacArthur's protests over the issue cost him his job.

China, seeing enemy forces on its doorstep, reacted by sending huge numbers of forces across the border, driving the allies south across the 38th parallel and taking Seoul on January 4, 1951. Seoul was recaptured on March 15, and the battlefront stabilized generally along the 38th parallel. An armistice was signed at the border site of Panmunjom on July 27, 1953, by the Chinese, the North Koreans, and the United Nations command. It was only a cease-fire, and to this day, there has not been a full treaty or reconciliation between the Koreas.

The border between the two Koreas, often cited as the tensest boundary on Earth, still follows this armistice line. The **demilitarized zone (DMZ)** between them, 150 miles (240 km) long and 2.5 miles (4 km) wide, is a virtual no-man's land of mines, barbed wire, tank traps, and underground tunnels. The world's largest concentration of hostile troops faces off on either side of it. Remarkably, with the near absence of human activity, rare birds like the Manchurian crane and other endangered animal species have taken refuge in this narrow strip. The DMZ thus has the potential to become an international "peace park" like that taking shape along the old Iron Curtain in Europe.

119

The Korean armistice line divides one people, united by their ethnicity and language, into two very different countries (**• Figure 7.67**; also see **Table 7.1**). South Korea is a republic that endured years under a repressive military dictatorship before democratizing in 1987. It now has a modern, advanced capitalist economy heavily dependent on relationships with the United States and Japan. North Korea is a rigid and very tightly controlled Communist state that promotes what it calls ***juche*** (pronounced "*djoo*'-cheh"), or self-reliance. North Korea's leadership has used this political philosophy's three principles of political independence, economic self-sustenance, and national self-defense to limit economic and other exchanges with the outside world. North Korea's principal Cold War ties shifted between China and the Soviet Union. With the collapse of the USSR, China became North Korea's main ally and benefactor.

• Figure 7.67 The two Koreas are like night and day. A satellite image taken at dusk, with the Korean Peninsula near the photo's center, shows the profound economic distinctions. Prosperous South Korea is awash with urban and industrial lights; Seoul is the brightest. North Korea is nearly dark, with only the country's capital Pyongyang (near the west coast) visible. The islands of Japan sprawl across the right side of the image, with massive Tokyo visible at the far right.

THINK CRITICALLY: Much fun has been made of North Korea (think "The Interview" and Dennis Rodman's visits there). But what do you think can be done to accommodate North Korea in the world system?

From World War II until his death in 1994, the dictator Kim Il Sung, whom his citizens called "The Great Leader," governed North Korea. His successors, son Kim Jong Il (until 2011) and grandson Kim Jong Un (both dubbed "The Dear Leader"), have perpetuated the country's militaristic character. North Korea spends a staggering one-third of its gross domestic product on its military, compared with South Korea's expenditure of only 2.8 percent, China's 2 percent, Japan's 1 percent, and the US's 4 percent. North Korea's excessive expenditures on defense have been one cause of its severe economic decline. North Korea's economy is barely 2 percent the size of South Korea's. South Korea's per capita GDP (PPP)is $33,200 (similar to that of France), whereas North Korea's is just $1800 (comparable to Zambia).

Beginning in 1995, successive waves of flood and drought brought famine to North Korea. The natural calamities only contributed to the already plummeting economic health of the country that came in the wake of the collapse of the Soviet Union, formerly North Korea's main benefactor and trading partner. North Korea's almost impenetrable veil of secrecy made it difficult to calculate the losses, but estimates of the number of people who died in the famine range from 900,000 to 2.4 million, or up to 10 percent of the country's prefamine population. As one crop after another failed, North Korea gradually and reluctantly sought food aid from abroad. It coaxed emergency supplies of wheat from the United States in part by threatening to withdraw from its 1994 agreement with the United States to halt its nuclear weapons program (see the discussion on pages 319–321). Although food aid from the United States, China, and South Korea helped alleviate the immediate crisis, the poor conditions of North Korea's health system, drinking water supplies, and electrical generation have perpetuated malnutrition and disease.

Another physical contrast between the two Koreas is that most of the nonagricultural natural resources are in North Korea. All of the peninsula's major resources—coal, iron ore, some less important metallic ores, hydropower potential, and forests—are more abundant in the North than in the South. North Korea was thus originally the more industrialized state, featuring a typical Communist emphasis on mining and heavy industry. Viewed strictly in terms of resource potential for development, North Korea should have become the more prosperous power. Its decision to close its doors to the outside world and insist on strict state control of industry and agriculture provides a fascinating opportunity to appreciate how different political systems produce different economic results.

While the North focused on defense and strict socialism, the South took over industrial leadership in the 1970s and 1980s by developing a dynamic and diversified economy under a state capitalism approach rather than free-market and free-trade capitalism.[46] This was a time during which the government favored and strongly funded a few small companies that it nurtured into giant conglomerates called ***chaebols***. Hyundai is a good example: originally a small rice milk company, it diversified into trucks and buses. The government saw potential and supported Hyundai financially, supervising its growth into a world-class corporation renowned for the manufacture of automobiles and ships. Samsung, now one of the largest electronics companies in the world, followed a similar path. Altogether about 15 of these gigantic, interlocking, family-controlled conglomerates came to dominate South Korea's economy. Also aiding in the explosive development of *chaebol* industries were investments from Japan and the United States (as well as access to their markets) and the availability of inexpensive and increasingly skilled Korean labor. Finally, the factor South Koreans most frequently point to in their success has been an emphasis on education. They sell valued possessions if need be to afford their children's educations at home or abroad, and their culture prizes teachers and professors. South Korea has one of the world's highest numbers of Ph.D.s per capita.

With the world's 13th largest economy, South Korea is now among the top 10 countries worldwide in the production of automobiles, ships, steel, computers, and electronics. South Korea is the eighth largest trading partner of the United States, which has long subscribed to a "domino theory" that if South Korea's economy fell, so would Japan's and perhaps even its own. In 2012, the United States and South Korea entered into a free-trade pact that will even more strongly enmesh their economies.

In sum, South Korea is one of Asia's success stories, an Asian Tiger that made an extraordinary postwar recovery, especially by using brainpower to overcome its resource limitations, and that has enjoyed periods of rocketing economic growth. South Korea's rapid transition from LDC to MDC has brought some growing pains, seen especially in the social changes of more women in the workforce, higher divorce rates, and falling birth rates. South Korea, which is even more homogeneous than Japan, wants to maintain its cohesive society and does not invite foreigners to fill its impending shortage of productive young people. The government once encouraged family planning but is now asking its people to have more babies and to stop giving up their own for adoption abroad. A traditional preference for boys means there are fewer women than men in the Korea, and due to young women's migration to the cities for work, there is a pronounced shortage of eligible Korean women in the countryside. The result is a full-blown industry of bride shopping outside Korea, especially in China, Vietnam, and other parts of Asia. 289

The bride-seeking Korean farmer typically pays a broker for the full suite of services, from selecting his bride to the wedding ceremony. Prospective brides are lined up for his consideration—in Vietnam's Ho Chi Minh City, for example. He picks one and, with no courtship, marries her and takes her home to Korea. The Vietnamese and other brides are aliens in the homogeneous Korean society and are commonly ostracized by the grooms' extended families and by society at large. Most husbands and wives do not even speak one another's languages. There are many accounts of the wife's loneliness, depression, physical abuse, and even suicide. Mixed-race babies often suffer domestic abuse and bullying. Happily, there are also reports of successful unions and happy and cohesive families, against the odds.

Despite having some small, experimental free-trade zones, North Korea is a poor, economically isolated country almost frozen in time, intent on pursuing the command economy model that has failed in every other nation that employed it. Despite its motto of self-reliance, it depends on huge imports of food and energy, especially from Russia. Its principal exports are missiles and other military hardware and textiles, especially clothing. Its people, whose sentiments and hopes are unknown because they are not allowed contact with the outside world, periodically still hover close to famine. Only a select few are allowed access to the Internet, although the infrastructure is there: much of the country is wired with a fiber-optic intranet system. The North Korean leadership has a dilemma: If it gives up on *juche* and opens its doors, the economy might improve, but the people of this country, often called the **Hermit Kingdom**, will discover the bounties just across the border and might revolt.

Could the Koreans simply reunite as the two Germanys did? Right now, the prospect for Korean reunification appears to be very remote. 130 Outside observers generally believe that North Korea is not really interested in reunification because that would almost certainly doom the country's regime. For its part, South Korea is fearful of the enormous economic cost it would have to pay for absorbing its much poorer neighbor, just as West Germany did in absorbing East Germany. Reunification would probably begin with a huge stream of poor northerners moving south. The nearby great powers are also quietly pleased with Korea's lingering division. The two

Koreas provide dual buffer zones between the historical adversaries China and Japan. And if the two Koreas united, finally taking advantage of North Korea's strong natural resource base, China and Japan could suddenly have a rival great power to contend with.

Even with occasional diplomacy like the Six Party Talks (see **page 321**), there is much more concern with simply avoiding war than with reunification. Some incidents between the Koreas have been so serious that they threaten to spill over into war. For example, in 2011 North Korean forces sank a South Korean ship claimed to have crossed the Northern Limit Line (drawn hastily by the UN at the end of the Korean War) into North Korean waters. Forty-six South Koreans died. Soon after, North Korean artillery shelled the South Korean island of Yeonpeong, killing four. Because of treaty obligations, incidents in these places or in the DMZ could escalate tension between the United States and North Korea, and even be a tripwire for a great conflict (**•Figure 7.68**); if that were to happen, US troops could not avoid being involved, and Chinese troops might be brought in. Japan worries that it too would be drawn into any conflict between the Koreas. Some analysts fear that Korea may yet become a nuclear battlefield. The economic and political stability of these three small countries on the Western Pacific Rim is critical to the well-being of the global system.

Joe Hobbs

• Figure 7.68 Two South Korean soldiers keep a wary eye on their North Korean counterpart, just steps away in the Joint Security Area of the Panmunjom "truce village" in the DMZ. The low concrete slab running left to right on the left of the photo marks the border between the two countries. From time to time, a North Korean defector runs across the border, sometimes pursued by North Korean soldiers. Firefights involving both US and South Korean troops against North Korean soldiers have accompanied some of these incidents.

Study Guide

Summary

- South and East Asia include the countries of Japan, North and South Korea, China, Taiwan, Pakistan, India, Sri Lanka, Bangladesh, Bhutan, Nepal, Maldives, Myanmar (Burma), Laos, Thailand, Cambodia, Vietnam, Malaysia, Singapore, Indonesia, the Philippines, Brunei, Timor-Leste, and numerous islands scattered along the edges of this major continental bloc.
- This is the most populous world region, with 54 percent of the world's people. It includes the world's most populous countries, China and India. Population growth rates in the region vary widely, from zero growth in Japan to over 3 percent per year in Timor-Leste. China's one-child policy has been relaxed. Female feticide has created gender imbalances in both China and India.
- Three concentric arcs make up the broad physiography of the region: an inner arc of the high mountain ranges, the Himalaya, Karakoram, and Hindu Kush; a middle arc of major floodplains, deltas, and low mountains; and an outer arc of thousands of islands, including the archipelagoes of the East Indies, the Philippines, and Japan.
- Major rivers include the Indus, Ganges, and Brahmaputra in South Asia; the Irrawaddy, Chao Praya (Menam), Mekong, and Red in Southeast Asia; and the Chiang Jiang (Yangtze) and Huang He (Yellow) in East Asia.
- The region's climate types and biomes include tropical rain forest, savanna, humid subtropical, humid continental, steppe, desert, and undifferentiated highland. Shifting cultivation and wet rice cultivation are important forms of agriculture. Wet rice cultivation produces very high yields and is associated with dense human populations.
- Although South and East Asia have some of the world's largest cities (with 20 of its 34 megacities) about 55 percent of the region's population is rural. Many uniquely Asian traditions helped shape village settlement planning and home design.
- South and East Asia's ethnic and linguistic compositions are diverse. Major language families are Indo-European, Dravidian, Sino-Tibetan, Altaic, Austric, and Papuan.
- Major religions and sociopolitical philosophies of the Asia are Hinduism, Islam, Buddhism, Confucianism, Daoism, and Christianity. Sikhs, Zoroastrians, and Jains are significant minority faiths.
- Great Britain, the Netherlands, France, and Portugal were the most important colonial powers in this region. Most of these domains were relinquished by the middle of the 20th century, and the later British return of Hong Kong and the Portuguese return of Macao (both to China) closed the colonial period.
- China has the region's strongest economy, followed by Japan. Recent decades have seen strong economic growth among the Tigers and Tiger Cubs of the region, including South Korea, Taiwan, Singapore, Thailand, and Malaysia. China's labor,

long a magnet for investment, is becoming more expensive, creating ebbs and flows in regional manufacturing. India's well-educated and inexpensive labor force is contributing to its strong economic growth. There continue to be fears in the United States about the outsourcing of American jobs to India.

- The Green Revolution is a broad effort to increase agricultural productivity in dominant crops. Biotechnology has produced crops that are more drought and pest resistant and capable of creating much higher yields, but genetic engineering of crops has perceived risks. Profits and other benefits from the Green Revolution are not uniformly spread.
- There are several major geopolitical issues in South and East Asia. The traditional enemies Pakistan and India (both pivotal countries from the US point of view) now possess nuclear weapons. There are fears that destabilization in Pakistan might allow the weapons to fall into the wrong hands. Pakistan's support of the United States in its war on terrorism is very risky because a popular backlash could bring down the government. North Korea has nuclear weapons and has used them as bargaining chips to get more food aid and other assistance from the West.
- Afghanistan is one of the world's poorest nations. It is landlocked and has limited resources, poor internal transportation, and little foreign trade. Opium is its main export. Its location along the ancient caravan routes and adjacent to large oil reserves has made it the target of stronger powers. Backed by the United States, rebels succeeded in driving out a Soviet occupation force in the 1980s. That costly conflict was followed by a period of civil war during which outside countries neglected Afghanistan. The militant Islamist Taliban came to power and protected the presence and training of al-Qa'ida militants. Osama bin Laden and others planned anti-Western attacks, including 9/11, from Afghanistan. Al-Qa'ida was driven into hiding from there in US-led attacks following 9/11, and the Taliban were deposed from power. Reconstruction of the country is progressing, but a Taliban insurgency continues. There are questions about how far the US should draw down its troops there.
- The subcontinent is home to many different faiths, including Hinduism, Sikhism, Christianity, Islam, Buddhism, Zoroastrianism, and Jainism. Religious, ethnic, and other differences underlie serious and often violent conflicts. The main ones have been Hindus versus Muslims in India, Sikhs versus the government of India, and Tamils versus Sinhalese in Sri Lanka.
- Until 1947, what are now Pakistan, Bangladesh, and India formed the single country of India. For over a century, it was the most important unit in the British colonial empire—the "jewel in the crown."
- With independence, India was partitioned between avowedly Muslim Pakistan (which later divided into two countries, Pakistan and Bangladesh) and secular, primarily Hindu India. The partition had many consequences, including violence between Hindus and Muslims and their large-scale migrations, disputes over the sharing of Indus River waters, and resources and industries stranded on either side of the new partition lines.
- The most severe product of the partition has been lasting conflict in Kashmir. Several wars have been sparked or fought in the still volatile frontier province, which, despite its Muslim majority, was joined with India in 1947.
- Although a poor and rural country, India has many modern industries that, along with privatization of former state-run firms, have contributed to its substantial rate of economic growth in recent years. These include the software and computer industries of Bengaluru and the film industry of Mumbai. Pakistan and Bangladesh are much less industrialized. India and Bangladesh have been successful in lowering birth rates, but population growth is a serious issue for both countries. India has managed to feed its huge population.
- South Asia provides large numbers of workers to other countries, ranging from professional to semiskilled and unskilled. Remittances sent home by these workers are important to the South Asian economies. Many of the professionals do not return, contributing to the region's brain drain.
- Some of India's large minorities have had serious difficulties with the majority Hindu population. Hindu nationalism is a problem in India, fanning the flames of communal violence between Hindus and Muslims. In the 1980s, some of the Punjab's Sikhs agitated for an independent country.
- Pakistan's government faces challenges from its Pashtun population in the country's west and northwest and from strong Islamist sentiment throughout the country.
- Bangladesh and the Maldives are threatened by serious natural hazards from the sea. Strong typhoons (hurricanes) regularly devastate large parts of low-lying Bangladesh, and most of the Maldives would be inundated if sea levels were to rise by 3 feet (90 cm).
- Sri Lanka has a Buddhist Sinhalese majority and a Hindu Tamil minority. For more than 20 years, Tamil factions led a violent struggle to establish an autonomous homeland on the island's east, north, and west peripheries. That civil war has ended.
- A megathrust earthquake off the northeastern coast of Sumatra caused the great tsunami of 2004, which killed more than 200,000 people in 14 countries. A similar earthquake in 2011 triggered the tsunami that caused massive devastation on the coast of northeast Japan.
- Southeast Asia's tropical forests are being destroyed at a faster rate than others of the world, mostly for commercial logging for Japanese markets. In biodiversity terms, Indonesia is a megadiversity country whose forests are under intense pressure.
- Southeast Asia as a whole is among the world's poorer regions. The poorest countries—Myanmar, Cambodia, Laos, and Vietnam—have suffered from warfare and, in the case of Myanmar, from unwise political management.
- Drug production, drug abuse, and HIV/AIDS are serious problems in Myanmar. Neighboring Thailand has some of these problems but has dealt more effectively with them.
- Prodemocracy efforts in Myanmar, long thwarted by official repression, have made headway. Both China and the US are interested in the country's natural resources. The US has recently stepped up trade with Myanmar. The country continues to face many challenges from its ethnic minorities, and some, including the Rohingya, are experiencing repression.

- US forces succeeded the French withdrawal from Vietnam in the 1950s. American efforts to win a war against communism in Vietnam failed. Normalized relations now exist between the United States and Vietnam, and although Communist, Vietnam encourages private enterprise and has seen recent strong economic growth.
- Numerous dams are being constructed on the Mekong River. These will produce hydroelectricity but will damage the productive fishery of this basin.
- Much unrest in Southeast Asia is related to tensions between ethnic groups. Indonesia has dealt with issues of fragmentation. Catholic Timor-Leste has gained impendence from Indonesia, and the provinces of Aceh and Papua are seeking self-rule or independence.
- The majority ethnic Han are associated with the more densely populated humid east of China, and non-Han realms are primarily in the arid west. Assertive "Hanification" efforts aim to transplant Han culture and power to the west.
- The Huang He (Yellow River) and the Chang Jiang (Yangtze River) descend from the Tibetan highlands.
- The Three Gorges Dam on the Chang Jiang (Yangtze) is designed to control floods, provide hydropower, and allow transfer of water to the more arid north. The project has been both celebrated and lamented because of the scale of environmental change and social dislocation associated with it. Great engineering works are underway to divert water from south to north.
- China has modeled much of its most dynamic economic growth after capitalist rather than Communist economic and political models.
- Inexpensive labor helped fuel the booming industries of the coastal cities. There is still poverty in China, with the cities generally wealthier than rural areas. The most prosperous areas are in the southeast, along the east coast (including the Special Administrative Regions of Hong Kong and Macau, and several special economic zones). Regions in the west, southwest, and uplands have had less economic development. Jobs and wages are now growing in more areas, and labor shortages have developed in the coastal boomtowns.
- Taiwan (the Republic of China) continues to claim to represent the true government of China and has a tense relationship with the mainland People's Republic of China. Taiwan is heavily invested in mainland China's development. The United States does not have diplomatic relations with Taiwan but supports it militarily and economically.
- Japan experienced phenomenal economic growth after its crushing defeat in World War II, but it has recently suffered economic recession.
- Due to its relative poverty in natural resources, Japan must import most of its raw materials, most energy supplies, and a large share of its food.
- The quest for possession of offshore resources near islands is a source of tension between Japan and Korea, Japan and China, and China and Vietnam, along with other neighbors in the South China Sea.
- The hardworking Japanese are experiencing many symptoms of workaholism, and many complain about their low level of amenities relative to that of other prosperous countries.
- The Korean Peninsula has an unfortunate location in geopolitical terms, sandwiched between the greater powers of Russia, China, and Japan. Korea has had a turbulent recent political history, first as a colony of Japan and then as a land divided between the Communist North and the more dynamic and democratic South.
- There are marked physical and economic contrasts between North and South Korea, with North Korea being much more rugged and having more natural resources for industry. Capitalist South Korea is far more prosperous than Communist North Korea.
- North Korea has suffered famine recently but has been reluctant to open up to trade and assistance because North Koreans might revolt.
- There are prospects for reconciliation, but also for war, between North and South Korea. Reunification is not favored by the South because it would be too costly. China and Japan would apparently be happier to have a weaker divided Korea than a stronger unified Korea.

Review Questions

1. What countries make up South and East Asia?
2. Use the concept of the three arcs to trace the outlines of South and East Asia's geography. What are the largest plains and river valleys of the region? What are the most important mountain ranges? What are the major island groups?
3. What are the region's most populous countries? Where are the highest and lowest population densities? Which countries have the highest and lowest population growth rates? Why has China's one-child policy been relaxed?
4. What are the monsoons? What roles do they play in climates and human activities?
5. What types of environments are shifting cultivation and wet rice cultivation practiced in? What are some of the basic methods and land use considerations of both types of agriculture? What cultures are associated with different land uses? What are the promises and the drawbacks of the Green Revolution?
6. What are the components of Himalayan environmental degradation, and what are its downstream effects?
7. What is the geographical concept of Zomia, and what is James Scott's argument about the people who live in Zomia?
8. What roles have *feng shui* and Confucianism played in the site selection and design of homes, schools, and villages in parts of Asia?
9. What are some of the innovations and products that originated in South and East Asia and diffused from there?

10. What are the major linguistic and ethnic groups of South and East Asia? What are the principal religions and sociopolitical philosophies?
11. What were the major colonial powers in South and East Asia? What regions did they colonize? What were the "Opium Wars"?
12. What are the region's most significant economic trends?
13. Why are India, Pakistan, and Indonesia considered pivotal countries for US interests? Has the United States been forced to take sides in the traditional enmity between India and Pakistan?
14. Why are there far fewer women than men in some Indian provinces? What problems are associated with this phenomenon?
15. What is the caste system? How does it affect employment and social relations in India? What evidence of change in this ancient system is there?
16. Where is Ayodhya? What site there is critical in relations between Hindus and Muslims in India?
17. What are Pakistan's *madrasas*? Why is their curriculum of so much concern to the government of Pakistan and to the West?
18. What ethnic and political differences have given rise to strife in Sri Lanka?
19. How has Afghanistan's economic outlook changed recently? What "blowback" took place from early American policies in Afghanistan?
20. What is the "string of pearls" and what is the "maritime Silk Road"? What are China's geopolitical ambitions in the region?
21. What particular problems do Bangladesh and the Maldives have in terms of climate and climate change?
22. What caused the great tsunami of 2004, and what were some of its effects?
23. What countries are the largest producers and consumers of Southeast Asian tropical hardwoods?
24. What major changes are coming to the Mekong River basin?
25. Why is palm oil a major agent of landscape change in Southeast Asia?
26. What changes have been taking place in Myanmar?
27. What has been the course of US-Vietnam relations, and what are Vietnam's economic prospects?
28. What are the major political and resource-related issues faced by Indonesia? How has the country dealt with demands of people in Aceh and Papua?
29. What is the "social compact" that China has with its citizens, and what may threaten the continuation of China's economic boom?
30. What are the pros and cons of the Three Gorges Dam?
31. How is China dealing with ethnic and political situations in Xingiang and Tibet?
32. How can China's economy be characterized? What are the major threats to it? What is its "average" standard of living based on per capita GDP (PPP)? How effectively is the wealth distributed? Where is it concentrated geographically, and how is the government hoping to change that pattern?
33. Where are China's special economic zones? What does their distribution suggest about China's economic geography?
34. What are the major political attributes of Taiwan? What is the "One China Policy"?
35. What is Japan's ethnic makeup? What relations have the Japanese majority had with internal minorities and foreign groups at various times?
36. How has Japan been able to overcome its poverty in natural resource assets by commercial, military, or other means?
37. What are some of the features of Japanese society today, especially those related to the country's economy?
38. What precipitated the division of the Korean Peninsula into North and South Korea?
39. What are the major differences between North and South Korea in natural resources, political systems, and economic development?
40. What are the prospects for reunification of the Koreas? How do some outside powers and the Koreans themselves view reunification?

Key Terms + Concepts

ancestor veneration (p. 303)
animism (p. 303)
archipelago (p. 289)
Asian Century (p. 282)
Asian Tigers (p. 310)
Association of Southeast Asian Nations (ASEAN) (p. 342)
Austroasiatic language family (p. 300)
Austronesian language family (p. 300)
autonomous regions (p. 348)
balloon effect (p. 342)
blowback (p. 333)
break-of-bulk point (p. 349)
bubble economy (p. 356)
Buddhism (p. 305)
Burmese (p. 300)
"Burmese Way to Socialism" (p. 340)
Cao Dai (p. 303)
caste (p. 328)
chaebols (p. 359)
communal violence (p. 325)
Comprehensive Test Ban Treaty (p. 315)
Confucianism (p. 305)
contingent sovereignty (p. 340)
Convention on the Law of the Sea (p. 317)
Dalai Lama (p. 348)
Dalits (p. 328)
Daoism (Taoism) (p. 306)
demilitarized zone (DMZ) (p. 358)
"denying the countryside to the enemy" (p. 344)
Dravidian language family (p. 301)
endemic species (p. 337)
exclusive economic zone (EEZ) (p. 317)
fallow (p. 293)
Falun Gong (p. 303)
feng shui (p. 298)
Free Aceh Movement (GAM) (p. 347)
Gan (p. 300)
geomancy (p. 298)
Golden Crescent (p. 332)
Golden Triangle (p. 342)
Great Game (p. 331)
Haaka (p. 300)
Han Chinese (p. 301)
Hanification (p. 348)
Hermit Kingdom (p. 359)
Hinduism (p. 304)
hukou (p. 353)
"India's Silicon Valley" (p. 313)
Indo-Aryan language subfamily (p. 300)
Indus Waters Treaty (p. 324)
insolation (p. 291)

insourcing (p. 313)
intensive subsistence agriculture (p. 294)
Jainism (p. 307)
Jains (p. 307)
Javanese (p. 300)
Jemaah Islamiah (JI, or the Islamic Group) (p. 321)
juche (p. 358)
karma (p. 305)
karoshi (p. 357)
Khalistan (p. 325)
Khmer (p. 300)
Korean (p. 301)
Laskar Jihad (p. 321)
laterite (p. 293)
lateritic soils (p. 293)
levee (p. 297)
liberal autonomy (p. 345)
Liberal Democratic Party (LDP) (p. 356)
Mahayana Buddhism (p. 305)
Malay (aka *Bahasa*) (p. 300)
megadiversity country (p. 337)
Mekong River Commission (p. 336)
Middle Way (p. 349)
Min (p. 300)
Mohajirs (p. 323)
monsoon (p. 291)
summer monsoon (p. 291)
winter monsoon (p. 292)
mujahidin (p. 333)
mutually assured destruction (MAD) (p. 315)
Naxalites (p. 325)
New Asian Tigers (p. 310)
nirvana (p. 305)
Nuclear Nonproliferation Treaty (NPT) (p. 315)
offshore outsourcing (offshoring) (p. 313)
one-child campaign (p. 288)
One China Policy (p. 354)
Operation Enduring Freedom (p. 333)
Opium Wars (p. 307)
orographic precipitation (p. 292)
paddy (padi) (p. 294)
Pakistani Taliban (Tehriq-i-Taliban Pakistan, or TTP) (p. 318)
Pancasila (p. 345)
pandemic (p. 288)
Papuans (p. 347)
Parsis (p. 307)
pivotal countries (p. 315)
Red Corridor (p. 325)
Sanskrit (p. 301)
sectarian violence (p. 325)
shifting cultivation (p. 293)
Shintoism (p. 303)
Sikhism (p. 304)
Sinhalese (p. 326)
Sino-Tibetan language family (p. 300)
slash-and-burn cultivation (p. 293)
South–North Water Transfer Project (p. 350)
special economic zones (SEZs) (p. 352)
State Law and Order Restoration Council (SLORC) (p. 340)
strategic hamlet (p. 297)
"String of Pearls" (p. 327)
sweatshops (p. 353)
swidden cultivation (p. 293)
syncretism (p. 303)
Tagalog (p. 300)
Tai-Kadai language family (p. 300)
Taliban (p. 333)
Tamil (p. 301)
Tamils (p. 326)
Tamil Tigers (p. 326)
Taoism (p. 306)
terraces (p. 294)
Thai (Siamese) (p. 300)
theory of Himalayan environmental degradation (p. 295)
Theravada Buddhism (p. 305)
Tibetan (p. 300)
Tiger Cubs (p. 310)
Trans-Pacific Partnership (TPP) (p. 312)
untouchables (p. 328)
Urdu (p. 301)
value-added manufacturing (p. 310)
Viet Cong (p. 344)
Vietnamese (p. 300)
wa(p. 356)
Wallace Line (p. 338)
"Water Towers of Asia" (p. 330)
Western Big Development Project (p. 348)
wet rice cultivation (p. 294)
Wu (p. 300)
Xiang (p. 300)
yang (p. 298)
yin (p. 298)
Yue (p. 300)
Zomia (p. 296)
Zoroastrianism (p. 307)

Notes

1. From the Analects; see http://nothingistic.org/library/confucius/analects/analects27.html.
2. "Urbanization and Demographics Could Skew China's Economic Rebalancing," Stratfor, September 3, 2014, https://www.stratfor.com/analysis/urbanization-and-demographics-could-skew-chinaseconomic-rebalancing.
3. N. Ahmed, "Female feticide in India," *Issues Law Med*, 26(1), Summer 2010: 13–29.
4. "The Effect of the Monsoon on the Indian Society in Cochin, Kerala," 123HelpMe.com, May 30, 2015, http://www.123HelpMe.com/view.asp?id=150882.
5. D. R. Kothawale, J. V. Revadekar, and K. Rupa Kumar, "Recent Trends in Pre-Monsoon Daily Temperature Extremes over India, *Journal of Earth System Science*, 119(1), February 2010, pp. 51–65. Amit Bhattacharya, "16-Year Trend of Poor Monsoon in Punjab, Haryana," *Times of India*, http://timesofindia.indiatimes.com/india/16-year-trend-of-poor-monsoon-in-Punjab-Haryana/articleshow/43115393.cms.
6. John Reader, British anthropologist and traveler, cited in Fred Pearce, "Terraces: The Other Wonders of the World," *Eurozine*, March 2001, accessed July 11, 2007, from http://www.eurozine.com/articles/2001-03-01-pearce-en.html.
7. Jack D. Ives and Bruno Messerli, *The Himalayan Dilemma: Reconciling Development and Conservation* (New York: Routledge, 1989).
8. This discussion is based largely on two unpublished manuscripts, "Living Spaces of Korean Architecture" and "In Tune with Nature: Rural Villages and Houses in Korea," by the Korean geographer Sang-Hae Lee, of Sungkyunkwan University, and personal talks with him. I am grateful for his contributions.
9. Quoted in David Pilling, "Megacities," *Financial Times*, November 6, 2011, p. HH-1.
10. Edward Glaeser, *Triumph of the City: How Our Greatest Invention Makes Us Richer, Smarter Greener, Healthier, and Happier* (New York: Penguin Press, 2011), pp. 69–91.
11. Quoted in Pankaj Mishra, "Caste Adrift." *Financial Times*, January 23, 2011, p. 13.
12. Descriptions of Buddhism and Daoism are based on the *Microsoft Encarta Reference Library*, 2004.
13. Michael Schuman, *The Miracle* (New York: HarperBusiness, 2009).
14. D. Patrick Barta, July 27, 2009, "Asian Nations Revisit Safety Net in Effort to Bolster Spending," *The Wall Street Journal*, p. A2.
15. UNESCAP. October 3, 2014. "Poor-Rich Gap Growing in India, Asia-Pacific: UNESCAP," http://articles.economictimes.indiatimes.com/2014-10-03/news/54599647_1_income-inequality-gini-coefficient-poor-rich-gap.
16. Tim Worstall, "China's Now The World Number One Economy And It Doesn't Matter A Darn," Forbes, December 7, 2014.
17. Quote in Bob Davis, "Massive Population Lifts Nation's Growth." *The Wall Street Journal*, p. A12.
18. Alan S. Blinder, quoted in Anand Giridharadas, "India's Edge Goes beyond Outsourcing," *New York Times*, April 4, 2007, p. C4.

19. Patrick Dixon, "The Future of Outsourcing: Impact on Jobs," GlobalChange.Com, December 30, 2014, accessed February 23, 2015, at http://www.globalchange.com/outsourcing.htm.
20. The Bullet-Proof Patriot. June 7, 2015. "The Naked Statistics of Free Trade Agreements and the Trans-Pacific Partnership (TPP)" http://www.thebulletproofpatriot.com/2015/06/the-naked-statistics-of-free-trade-agreements-and-the-trans-pacific-partnership-tpp/
21. See http://www.gmo-free-regions.org/gmo-free-regions/maps.html.
22. Amy Harmon, "Golden Rice: Lifesaver?" *New York Times*, August 25, 2013, Sunday review, page 1
23. Quoted in Edward Wong, "China Navy Reaches Far, Unsettling the Region," *New York Times*, June 15, 2011, p. A11.
24. Richard Haas, "China's Maritime Disputes," A CFR InfoGuide, accessed March 2, 2015, at http://www.cfr.org/asia-and-pacific/chinas-maritime-disputes/31345#!/?cid=otr-marketing_use-china_sea_InfoGuideesen.
25. Robert Chase, Emily Hill, and Paul Kennedy, "Pivotal States and US Strategy." *Foreign Affairs,* January–February 1996, pp. 33–51.
26. Fukuyama, quoted in Martin Wolf, "The End of History Man," *Financial Times*, May 29, 2011, p. 3.
27. James Hookway, "Tensions Flare over Disputed Asian Sea," *The Wall Street Journal*, June 10, 2011.
28. Samsar Chand Kour, *Beautiful Valleys of Kashmir and Ladakh* (Mysore, India: Wesley Press, 1942), p. 2.
29. Brahma Chellaney, "India Is Feeling Increasingly Choked by China's 'String of Pearls,' *The Daily Star*, March 11, 2015, http://www.dailystar.com.lb/Opinion/Commentary/2015/Mar-11/290316-india-is-feeling-increasingly-choked-by-chinas-string-of-pearls.ashx.
30. Harris, Gardiner. March 28, 2014. "*Borrowed Time on Disappearing Land* Facing Rising Seas, Bangladesh Confronts the Consequences of Climate Change," *New York Times*, http://www.nytimes.com/2014/03/29/world/asia/facing-rising-seas-bangladesh-confronts-the-consequences-of-climate-change.html?_r=0.
31. Charles Q. Choi. September 4, 2014. "$1 Trillion Trove of Rare Minerals Revealed Under Afghanistan." Live Science. http://www.livescience.com/47682-rare-earth-minerals-found-under-afghanistan.html
32. See the Boren Scholarships site. https://www.borenawards.org/boren_scholarship
33. See UNESCO, *Assessment of Capacity Building Requirements for an Effective and Durable Tsunami Warning and Mitigation System in the Indian Ocean: Consolidated Report for the Countries Affected by the 26 December 2004 Tsunami* (Paris: UNESCO, 2005), accessed July 15, 2007, http://unesdoc.unesco.org/images/0014/001445/144508e.pdf.
34. Quoted in Lowell Dittmar, 2010. Burma or Myanmar? The Struggle for National Identity. World Scientific Publishing Company.
35. "Opium Poppy Cultivation Soaring in Myanmar: Boom Has Returned Nation to Being Major Player in Global Heroin Trade, *Straits Times Asia Report*, January 5, 2015, http://www.stasiareport.com/the-big-story/asia-report/myanmar/story/opium-poppy-cultivation-soaring-myanmar-20150105—sthash.ZFK1738A.I2iraMeC.dpuf.
36. Patrick Winn, "Syringes are Now Currency in Heroin-Addicted Myanmar," Global Post, October 28, 2014, http://www.usatoday.com/story/news/world/2014/10/28/globalpost-myanmar-heroin/18048031/.
37. See http://www.monsanto.com/newsviews/pages/agent-orange-background-monsanto-involvement.aspx)%5bcross.
38. Hannah Beech, "China Says It Will Decide Who the Dalai Lama Shall Be Reincarnated As," *Time*, March 13, 2015, accessed March 12, 2015, from http://time.com/3743742/dalai-lama-china-reincarnation-tibet-buddhism/.
39. Quoted in Jamil Anderlini, "A Blast from the Past," *Financial Times*, December 15, 2009, p. 11.
40. Jamil Anderlini, "China: Overborrowed and Overbuilt," *Financial Times*, January 29, 2015.
41. Minxin Pei, "China's Slowing Economy: The Worst Has Yet to Come," *Fortune*, January 21, 2015, accessed February 23, 2015, from http://fortune.com/2015/01/21/china-economy-growth-slowdown/.
42. "Urbanization and Demographics Could Skew China's Economic Rebalancing," Stratfor. September 3, 2014, https://www.stratfor.com/analysis/urbanization-and-demographics-could-skew-chinas-economic-rebalancing.
43. Jamil Anderlini and Mure Dickie, "Taking the Waters: China's Gung-Ho Economy Exacts a Heavy Price." *Financial Times,* July 24, 2007, p. 11.
44. "Urbanisation: Moving on Up. The Government Unveils a New 'People-Centred' Plan for Urbanization," *The Economist*, March 22, 2014, http://www.economist.com/news/china/21599397-government-unveils-new-people-centred-plan-urbanisation-moving-up.
45. G. John Ikenberry, "The Illusion of Geopolitics: The Enduring Power of the Liberal Order," *Foreign Affairs*, May/June 2014.
46. I thank the Korea Society for sponsoring my field study of Korea, on which much of this discussion is based. Professor Byong Man Ahn of the Hankuk University of Foreign Studies and Professor Taeho Bark of the Graduate School of International Studies at Seoul National University graciously provided much of the information on Korean economies and politics.

Global Geoscience Watch

First review Table 7.2, an overview of the largest cities in South and East Asia. Pick a city of interest listed in the Table to research in more depth. Then Click on the "World Map" menu item found on the GREENR database homepage and navigate to the country in which the city resides. Search within the country "Portal" for news about the city. Review several news articles mentioning your city. Look for interesting topics such as business deals, trade agreements, or government policies that involve your city. Also look for special problems the city might be facing such as poverty or environmental issues. Organize a short report on your city synopsizing what you researched.

Online Resources

For access to MindTap and additional study materials visit www.cengagebrain.com. Read your textbook, take notes, complete activities, take practice quizzes and more.

Above: A woman of Vanuatu preparing a dish of shredded coconut and taro root to be baked in taro and banana leaves. This trilogy of root crop, coconut, and banana is among the staples of the tropical Pacific diet. Joe Hobbs
Left: The vast Pacific realm at night. Sydney and the other cities of Australia's east coast stand out. C. Mayhew & R. Simmon (NASA/GSFC), NOAA/NGDC, DMSP Digital Archive

8

Oceania and Antarctica

Those stars are moving from place to place. Links from the stars down to the rock painting, and songline continues. All links together! That way it's connecting, whole lot of them, it's the right way. World is spinning! The stars moving around! Walk the moonlight.

—BILL HARNEY, WARDAMAN ABORIGINE, NORTHERN AUSTRALIA[1]

Covering fully one-third of the Earth's surface, **Oceania** (pronounced "oh-shee-*an*-ee-uh") is mostly water. The world's largest ocean, the Pacific, defines and dominates this region. It is bigger than all the Earth's continents and islands combined. Before World War II, the Western world spawned legends about this ocean and its islands as a kind of utopia. There has long been trouble in this paradise, however. On many islands, foreign traders, whaling crews, labor recruiters, and other opportunists exploited the indigenous peoples, reducing their numbers and disrupting their cultures. The military battles of World War II further shattered the idyllic qualities of many islands. Yet today, the Pacific mystique, which has been perpetuated in books and films, forms part of the allure for booming tourism development across the region. This chapter presents the cultural and natural geographies of Oceania, summarizes the impacts of colonial enterprises in the region, and discusses some of the obstacles to development confronting the countries of this watery realm.

Also covered in this chapter is Antarctica, an icy continent with no permanent human inhabitants.

Chapter Outline

Chapter Objectives

This chapter will enable you to

- Appreciate the economic prominence of larger, Europeanized Australia and New Zealand in a vast sea of small, mainly indigenous political units struggling to make a living.
- Recognize the associations between the physical geographies of islands, their typical social and political organizations, and their economic characteristics.
- Understand how minority control of most of the wealth generated by mineral and other resources has led to discontent and rebellion among majorities.
- Hear the concern expressed by low-lying island countries about the production of greenhouse gases in faraway industrialized nations.
- Understand the obstacles faced by indigenous people in winning legal recognition of ancestral claims.
- Appreciate the process by which Australia and New Zealand are loosening ties with their ancestral European homeland and strengthening their regional orientation.
- Learn about "unclaimed claims" to the continent of Antarctica.

8.1 Area and Population

Oceania includes Australia, New Zealand, and the islands of the southern and western Pacific (**•Figures 8.1** and **8.2**; **•Table 8.1**). Pacific countries and islands such as the Philippines, Japan, and the Galápagos Islands are not considered part of Oceania, nor are the vast stretches of nearly empty ocean in the eastern and northern reaches of the Pacific. Hawaii belongs to this region and is dealt with both here and in the Chapter 11 discussion of the United States. Australia and New Zealand 512 have strong political and economic interests in the smaller tropical islands and a similar insular character in many ways, and they share ethnic affiliations with the original inhabitants of the smaller islands. But they are also uniquely European in character and dwarf the other countries in economic and political clout. Therefore, these two countries, along with the continent of Antarctica, are examined in more detail later in the chapter.

Major Divisions of the Region

Geographers have long divided the Pacific Islands into three principal regions: Melanesia, Micronesia, and Polynesia. The islands of **Melanesia** (Greek for "black islands"), lying northeast of Australia, are relatively large. New Guinea, the largest, is about 1500 miles (2400 km) long and 400 miles (650 km) across at the broadest point. The western half of New Guinea is part of Indonesia and is discussed in Chapter 7. The eastern section is the independent country of Papua New Guinea. The other Melanesian countries are the Solomon Islands, Vanuatu, and Fiji. New Caledonia, also in Melanesia, is controlled by France.

Micronesia (Greek for "tiny islands") includes thousands of scattered small islands in the central and western Pacific, mostly north of the Equator. The Micronesian countries are Palau, the Federated States of Micronesia (FSM), the Marshall Islands, Nauru, and the western part of the nation of Kiribati (pronounced "keer-uh-*bahss*"). Micronesia also includes two United States territories, Guam and the Northern Mariana Islands.

Polynesia (Greek for "many islands") occupies a greater expanse of ocean than Melanesia or Micronesia. It is shaped like a rough triangle, with corners at New Zealand, the Hawaiian Islands, and remote Easter Island. Polynesia's independent countries are the eastern part of Kiribati, Tuvalu, Samoa, and Tonga. Possessions of other countries in Polynesia include French Polynesia and the French Wallis and Futuna Islands; Hawaii and American Samoa; Tokelau and the Cook Islands, administered by New Zealand; Britain's Pitcairn Island; and Easter Island (Rapa Nui), owned by Chile.

The typical Pacific island country (excluding Australia and New Zealand) has about 100,000 to 150,000 people in an area of 250 to 1000 square miles (c. 650 to 2600 sq km); consists of a number of islands; is poor economically; is an ex-colony of Britain, New Zealand, or Australia; and depends heavily on foreign economic aid. The total land area is 3.3 million square miles (8.5 million sq km), or about 90 percent of the size of the entire United States (**•Figure 8.3**). Without the three largest countries of Australia, New Zealand, and Papua New Guinea, the amount of land shrinks to just 38,000 square miles (100,000 sq km), slightly larger than the state of Indiana.

The People and Where They Live

The region's total population is 38 million (**•Figure 8.4a**). Aside from Australia's 23 million people, populations range from 7 million in Papua New Guinea to 10,000 in Tuvalu (**•Figure 8.4b**). The highest population densities are in the smallest island groups, notably the Tuamotu Islands in French Polynesia and the Ellice Islands of Tuvalu. Population growth rates vary widely, from the predictably low 0.7 in prosperous Australia and New Zealand, to a high of 2.9 percent in tiny Nauru.

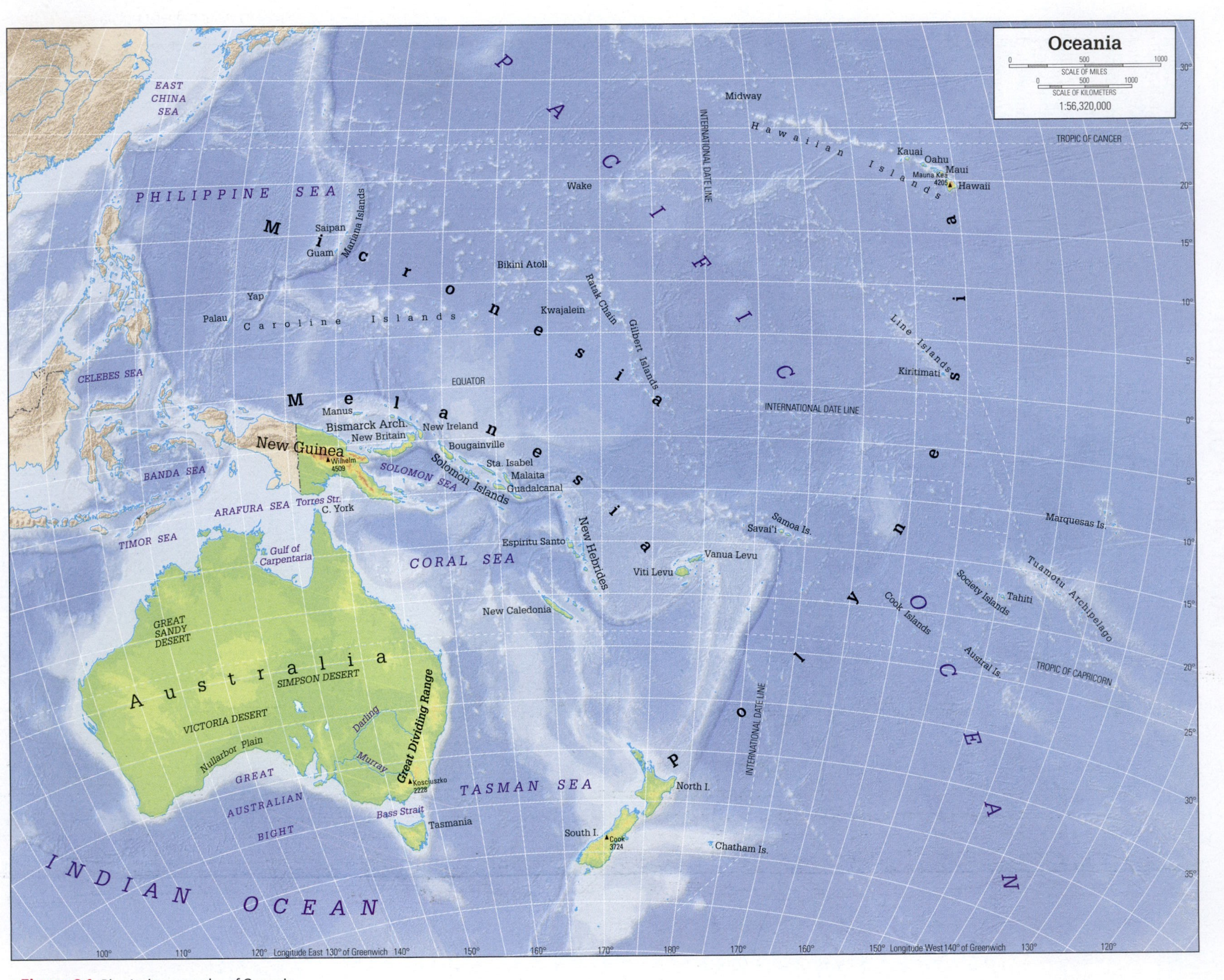

• **Figure 8.1** Physical geography of Oceania.

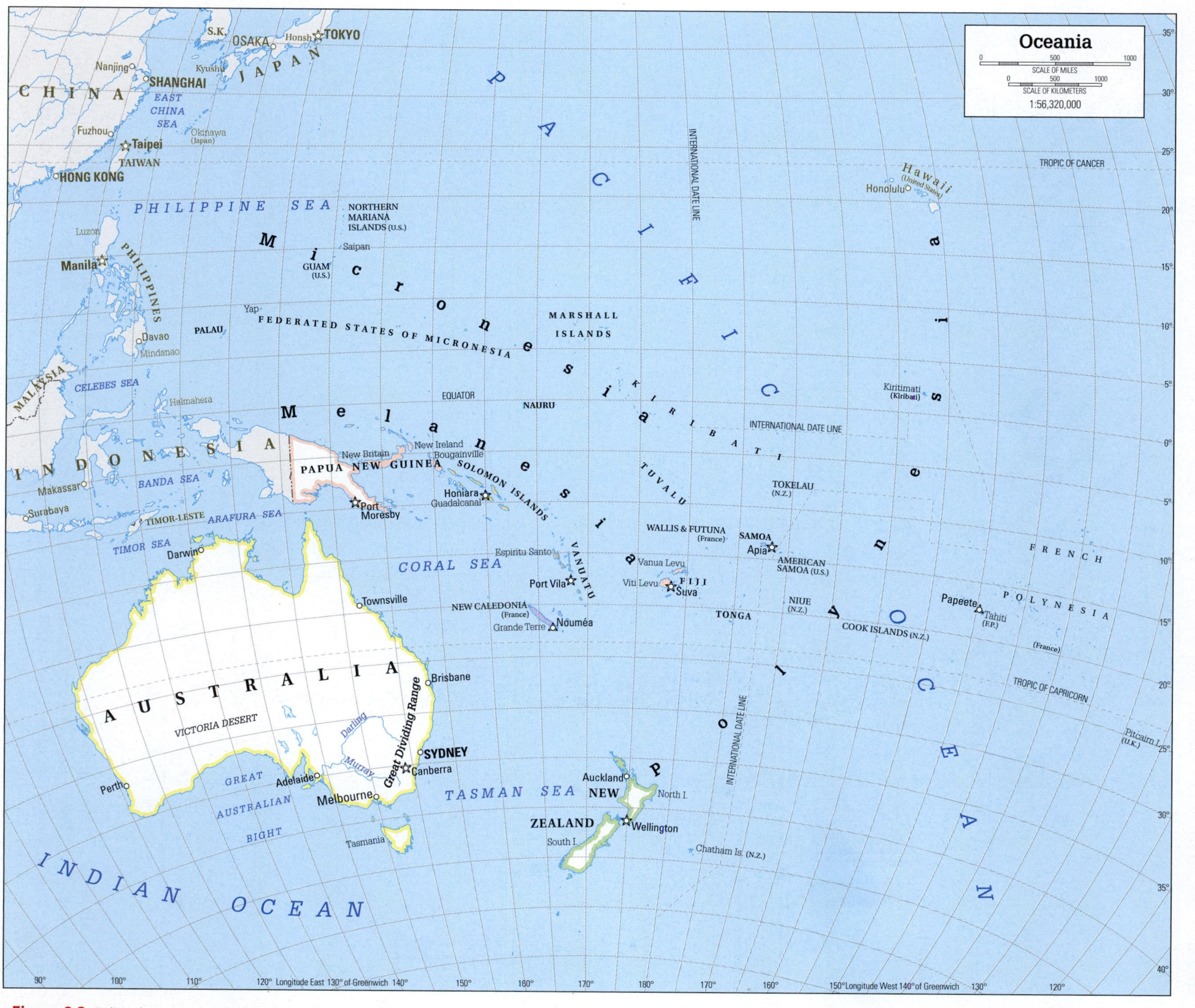

• **Figure 8.2** Political geography of Oceania.

Table 8.1 Oceania: Basic Data

Political Unit	Area (sq mi, thousands)	Area (sq km, thousands)	Population (millions)	Rate of Natural Increase (%)	Urban Population (%)	Population Under Age 15 (%)	Agricultural Workers (%)	Per Capita GDP (PPP) ($US)	GDP ($US, billions)	Trade Balance ($US, billions)	Oil Production (million bbl/day)	Literacy, Female (%)	Literacy, Male (%)	HDI
Oceania	**3306.8**	**8564.5**	**38.4**	**1.1**	**70**	**24**	**10**	**31,200**	**1,311**	**—**	**0.5**	**86**	**89**	**0.808**
Australasia														
Australia	2988.9	7741.3	23.5	0.7	89	19	4	43,000	1,100	5	0.4	96	96	0.933
New Zealand	104.5	270.7	4.3	0.7	86	20	4	30,400	158	0	*	99	99	0.910
Melanesia														
Fiji	7.1	18.3	0.9	1.4	51	29	13	4,900	7	−1	—	92	95	0.724
New Caledonia (France)	7.2	18.6	0.3	1.2	62	24	1	38,800	11	−2	—	96	97	—
Papua New Guinea	178.7	462.8	7.6	2.3	13	39	26	2,900	18	3	*	51	63	0.491
Solomon Islands	11.2	29	0.6	2.6	20	39	52	3,400	1	0	—	79	89	0.491
Vanuatu	4.7	12.1	0.3	2.4	24	39	64	4,800	0.6	0	—	—	—	0.616
Micronesia														
Guam (U.S.)	0.2	0.5	0.2	1.7	93	27	—	28,700	5	−1	—	99	99	—
Kiribati	0.3	0.7	0.1	2.1	54	35	24	6,400	0.1	0	—	—	—	0.607
Marshall Islands	0.1	0.2	0.1	2.5	74	40	14	8,700	0.1	0	—	93	93	—
Micronesia, Federated States of	0.3	0.7	0.1	1.9	22	34	14	7,300	0.3	0	—	88	91	0.630
Nauru	*	*	*	2.9	100	37	6	5,000	*	0	—	—	—	—
Northern Mariana Islands (U.S.)	0.2	0.5	*	1.1	89	26	2	13,600	0.7	0	—	96	97	—
Palau	0.2	0.5	*	0.5	84	20	3	10,500	0.2	0	—	90	93	0.775
Polynesia														
American Samoa (U.S.)	0.2	0.5	*	−0.4	87	25	—	8,000	0.5	0	—	97	98	—
Cook Islands (N.Z.)	*	*	*	−3	74	23	5	9,100	0.1	0	—	—	—	—
French Polynesia (Fr.)	1.5	3.8	0.3	1.1	51	25	2	26,100	7	−1	—	98	98	—
Niue (N.Z.)	*	*	*	0	41	—	24	5,800	*	0	—	—	—	—
Samoa	1.1	2.8	0.2	2.4	20	39	11	6,200	1	0	—	99	99	0.694
Tonga	0.3	0.7	0.1	2	23	37	18	8,200	0.5	0	—	99	99	0.705
Tuvalu	*	*	*	1.6	51	33	16	3,500	*	0	—	—	—	—
Wallis and Futuna (France)	*	*	*	0.3	—	23	—	3,800	*	0	—	50	50	—

* Less than 0.1. — Data not available or not applicable.
Sources: World Population Data Sheet, Population Reference Bureau, 2014; Human Development Report, United Nations, 2014; World Factbook, CIA, 2014.

There are some apparent problems of "people overpopulation," especially in Polynesia. Countries in Oceania have some of the highest emigration rates in the world, especially the FSM, Samoa, and Tonga. Emigrants head to Australia, New Zealand, and North America in search of work. These countries also suffer an intense "brain drain"; three-quarters of all college-educated people in Samoa leave, severely hampering development and public service in that country. Because birth rates are low, people in both Australia and New Zealand share the typical post-industrial fear of not having enough people to support the countries' economies and their aging populations. Both countries, however, are averse to liberal immigration policies that would boost their populations.

76

93, 157, 356, 496

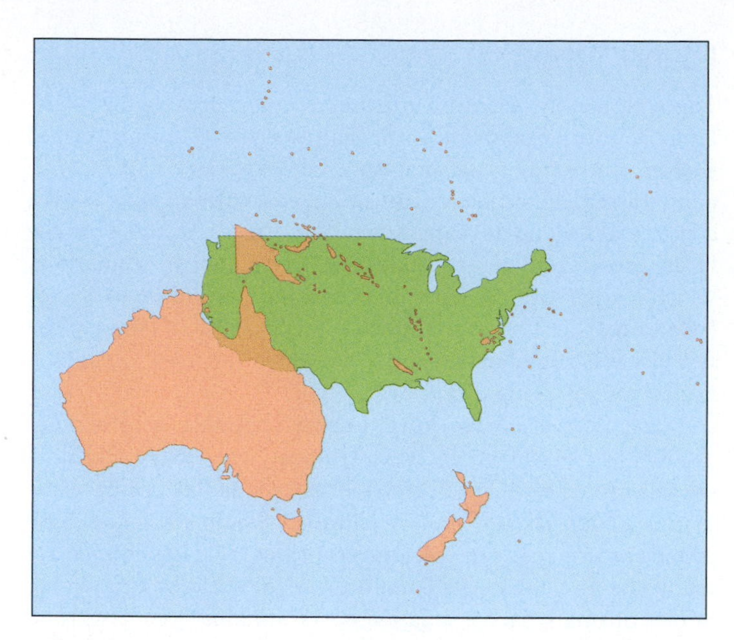

• **Figure 8.3** Oceania compared in area (but not latitude) with the United States.

8.2 Physical Geography and Human Adaptations

The climates, vegetation, landscapes, and cultural geographies of the islands scattered across the vast Pacific vary, but a few notable patterns emerge. The region's climates and biomes are shown in • **Figure 8.5a, b**; note that only Australia and the islands of the southwestern Pacific are large enough to have meaningful data depicted at this map scale.

Climates and Biomes

Outside of Australia and New Zealand, most of Oceania has tropical climates; the main exceptions are the higher elevations of New Guinea. In general, islands of Melanesia experience tropical rainforest climates, whereas Micronesia and Polynesia have more of a tropical savanna or tropical monsoon climate, with a significant dry season. On some high islands like New Caledonia and the main islands of Fiji, rainfall can vary significantly over a small area; the windward side of the island will receive abundant rainfall (and have a tropical rainforest climate), whereas the leeward side, in the rain shadow of the central peaks, usually receives much less (a tropical savanna climate).

Mangroves cluster around protected shorelines, as in southern New Guinea, and also may be found along the inner lagoons of low-lying atolls. Vegetation on low islands may seem lush, but the soils of these islands are not very fertile, and cannot support more than a handful of typical tropical vegetation species—palms and ferns, and drought-resistant scrub on islands having tropical monsoon or savanna climates. The tropical rainforest vegetation on the larger islands is more varied, and features the "canopy cover" found in rainforests elsewhere around the globe—many trees of roughly the same height growing so densely that their overlapping leaves block much of the sunlight from reaching the ground, which is dominated by scattered smaller trees, ferns, and shrubs.

Australia is marked by its dryness—its vast interior is dominated by a desert climate and sparse desert vegetation. Precipitation increases toward the margins of the continent. Southwestern Australia experiences a typical wet-winter, dry-summer Mediterranean climate (this is the center of Australian viticulture), and northern Australia has a tropical savanna climate, and the associated tropical grasslands that

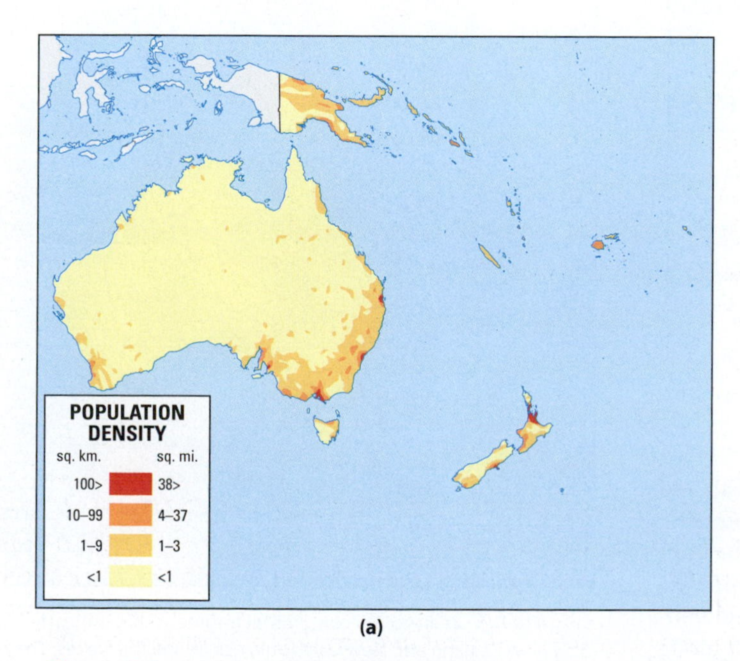

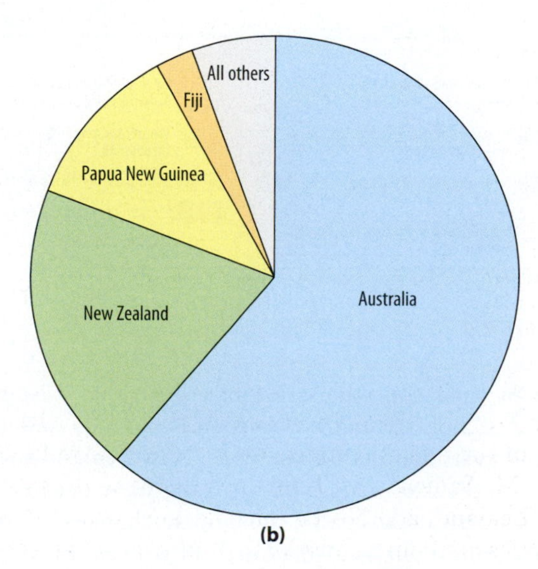

• **Figure 8.4** **(a)** Population distribution and **(b)** population pie chart of Oceania.

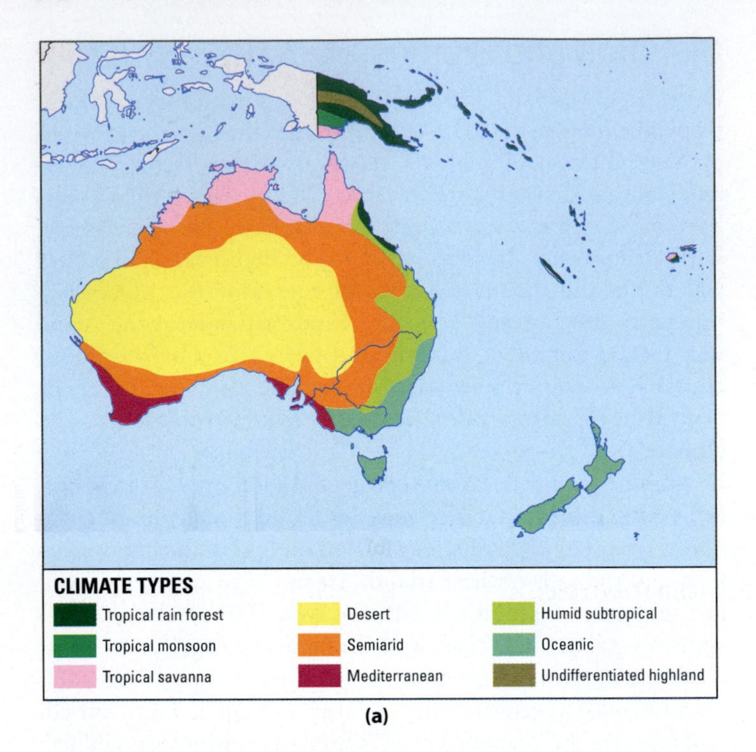

(a)

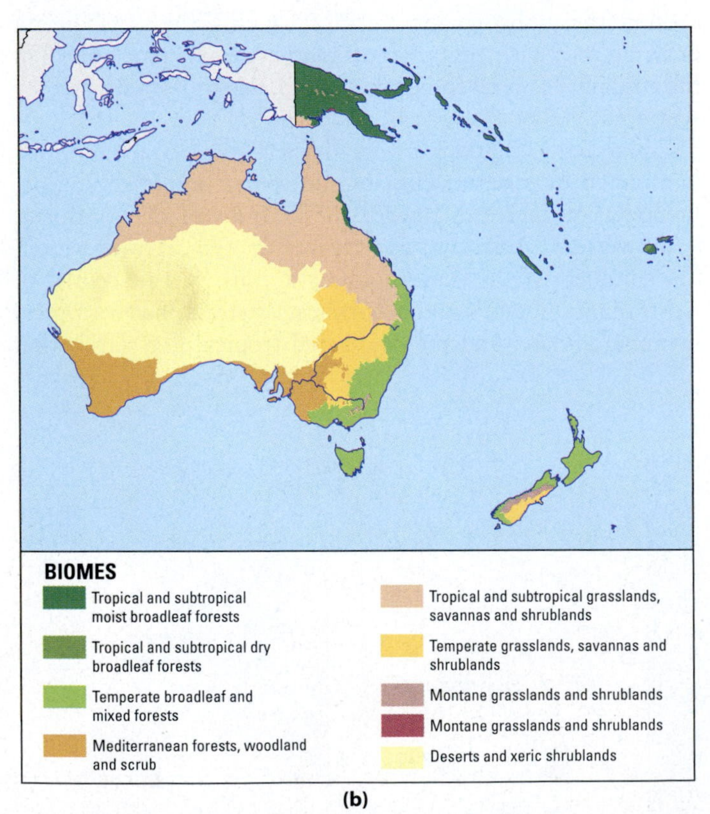

(b)

• **Figure 8.5** **(a)** Climates and **(b)** biomes of Oceania.

grow in such conditions. Eastern Australia, Tasmania, and New Zealand are more consistently humid, and temperate mixed forests dominate these areas. New Zealand and the populous southeastern coast of Australia have oceanic climates; farther north in Australia, the climate changes to humid continental, and there is a small area of tropical rainforest along the Queensland coast.

Island Types

Topographically, Oceania features three general types of islands. **Continental islands** are islands situated on continental shelves or are fragments that have been broken off of a continent. Tasmania and New Guinea are examples of islands on a continental shelf (in this case, Australia's). East of Australia lies the large, mostly submerged continental fragment of Zealandia, which broke off of Australia over 60 million years ago (•**Figure 8.6**). The main islands of New Zealand are among the few places Zealandia rises above the water line; New Caledonia is also part of Zealandia.

Some of these have lofty peaks, including several over 16,000 feet (4900 m) in New Guinea, whereas Australia is overall rather low. The region's other islands are categorized as either **high islands** or **low islands** (•**Figure 8.7**). Most high islands are the result of volcanic eruptions. The low islands are made of coral, a material composed of the skeletons and living bodies of small marine organisms that inhabit tropical seas.

These very different settings offer sharply different opportunities for human livelihoods and land uses (•**Figure 8.8**). Overall, however, outside urban, industrialized Australia and

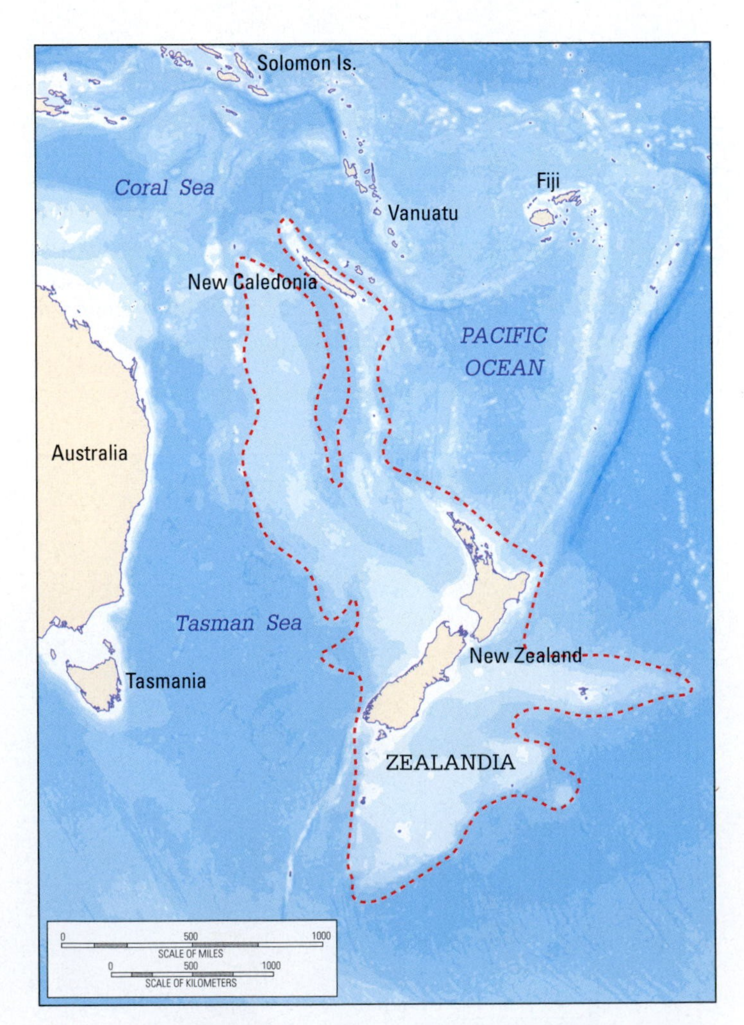

• **Figure 8.6** Zealandia, a continental fragment that is mostly submerged. Only New Zealand and New Caledonia rise above the waves. The ridges to the north are not part of Zealandia.

(a)

(b)

• **Figure 8.7** **(a)** Moorea Island, in Tahiti, is a classic and much romanticized high island. **(b)** Low islands, such as the one seen here, are most associated with the "coconut civilizations" of the Pacific. Low islands are extremely vulnerable to the dangers that would be posed by global warming.

New Zealand, the Pacific peoples are evenly divided between rural and urban, at 50 percent each. Some countries are highly urban (the Marshall Islands at 74 percent urban, Palau at 84 percent), but the Melanesian countries tend to be very rural: Vanuatu and the Solomon Islands are 20 percent urban each, and Papua New Guinea is the least urban country in the world, at just 13 percent. Rural people live in small farming or fishing villages, where they have long depended on root crops (such as taro and yams), tree crops (such as breadfruit and coconuts), ocean fish, and pigs (•**Figure 8.9**). Outside Australia, New Zealand, and Hawaii, there are only four cities with a population greater than 100,000: Port Moresby, Papua New Guinea; Suva, Fiji; Nouméa, New Caledonia; and Papeete, Tahiti (•**Table 8.2**).

Most of the region's high islands are volcanic in origin and have a familiar pattern. There is a steep central peak, with ridges and valleys radiating outward to the coastline. Permanent streams run through the valleys. A coral reef surrounds the island, and between the shore and the reef is a shallow lagoon. Many of the high volcanic islands are spectacularly scenic. Classic high islands in the region include the Polynesian groups of Hawaii, Samoa, and the Society Islands. If you have visited Hawaii, you will recognize the similarity with the beautiful Tahitian landscape in Figure 8.7a.

Some of the volcanic high islands of the Pacific comprise island chains. These are formed when the oceanic crust slides over a stationary geologic hot spot in the Earth's mantle,

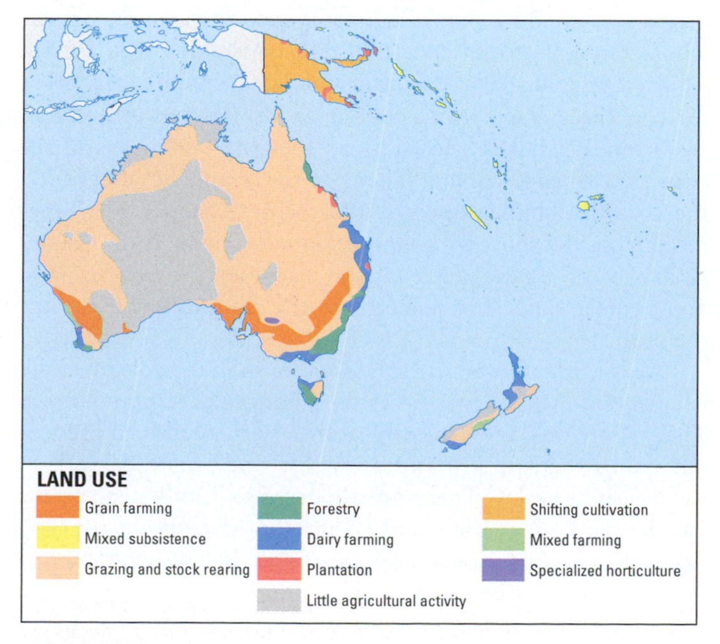

• **Figure 8.8** Land use in Oceania.

• **Figure 8.9** A market in Vanuatu. Root crops like those seen here are a staple in Melanesia.

Table 8.2 Largest Cities in Oceania

City	Population
Sydney, Australia	4
Melbourne, Australia	3.9
Brisbane, Australia	2
Perth, Australia	1.7
Auckland, New Zealand	1.3
Adelaide, Australia	1.1

Population in millions.
Source: Demographia World Urban Areas, 2015.

where molten magma is relatively close to the crust. As the crust slides over the geologic hot spot, magma rises through the crust to form new volcanic islands. The Hawaiian Island chain is an excellent example. The "Big Island" of Hawaii is now situated over the hot spot; it contains three active volcanoes (Kilauea, Hualalei, and Mauna Loa, the largest volcano on Earth) and is currently the chain's youngest member. In about 10,000 years the Lo'ihi seamount, currently about 3000 feet below the surface, will emerge as Hawaii's newest island (•**Figure 8.10**). The Pacific Plate of the Earth's crust is moving northwestward here; older volcanic islands that were born over the hot spot have moved to the northwest and have become inactive. Still older islands farther northwest in the chain have submerged to become **seamounts**, or underwater volcanic mountains.

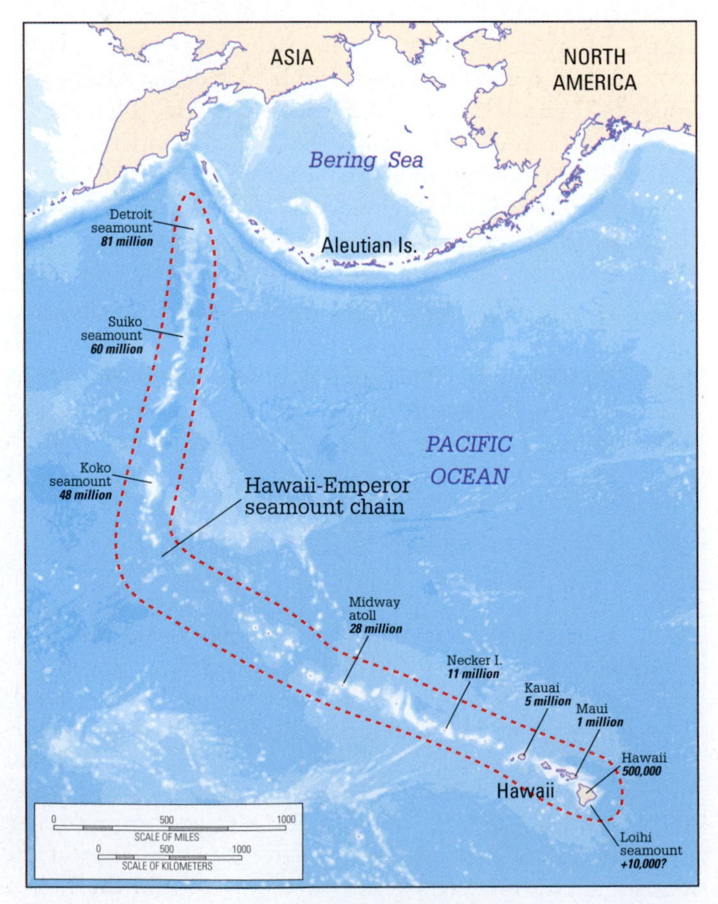

• **Figure 8.10** The hot spot in the Pacific has created a long chain of islands and seamounts as the Pacific Plate rode over it. The oldest islands in the chain are the northernmost. The island of Hawaii (the Big Island) is the youngest.

Although Hawaii occupies a mid-oceanic position, most of the region's volcanic and seismic activities are concentrated along the western side of the Pacific "Ring of Fire" (see Figure 2.1). A devastating series of earthquakes struck Christchurch, New Zealand's second largest city, in 2011. Read the dramatic firsthand account by Maori geographer Helen Hayward in her Life in New Zealand essay on **page 376**. The city has been rebuilding amidst a lively debate on whether the city center should be reconstructed as it was or fashioned into a new "smart city," employing the latest technology, to improve infrastructure, seismic safety, and quality of life.

The geographer Tom McKnight explained that the high islands' topography has historically influenced its cultural and economic patterns.[2] Each drainage basin formed a relatively distinct unit that was governed by a different chief. Within each unit, there were several subchiefs, each of them controlling a bit of each type of available habitat: coastal land, stream land, forested land, and gardening land. Because available resources were distributed relatively equally among chiefdoms, there were few power struggles or monopolies over resources. It is important to note that these are general rather than universal patterns of association between physical geography and human adaptations in the Pacific. Geographers reject the outdated notion of **environmental determinism**, which insisted that certain environmental conditions always correlate with specific culture traits.

Today, the main cities and seaports of Oceania, and the largest populations, are on the volcanic high islands and the continental islands. Their rich soils, typically of volcanic origin, support a diversity of tropical crops. These islands are generally more prosperous than the low islands.

241, 294, 408

The low islands are formed of coral. Most take the shape of an irregular ring surrounding a lagoon; such an island is called an **atoll**. Generally, the coral ring is broken into many pieces separated by channels leading into the lagoon, but the whole circular group is commonly considered one island. Charles Darwin devised a still widely accepted explanation for the three-stage formation of atolls (•**Figure 8.11**). First, coral builds what is known as a "fringing reef" around the edge of a volcanic island. Then, as the island slowly erodes, the coral reef builds upward and forms a barrier reef separated from the shore by a lagoon. Finally, the volcanic island sinks out of sight, and a lagoon occupies the former land area, whose outline is reflected in the roughly oval form of the atoll.

The low islands are generally smaller than the volcanic high islands and lack the resources to support dense populations. Their shorelines are typically dotted with the waving coconut palms that are a mainstay of life and the trademark of the South Sea isles. There are thousands of atolls across the Pacific; the Gilbert Islands of Kiribati in Micronesia are typical of the picturesque low islands idealized by Hollywood and travel brochures (see Figure 8.7b).

Despite their idyllic appearance, these islands pose many natural hazards to human habitation. The lime-rich soils are

(a) Volcanic island with fringing reef

(b) Slight subsidence, barrier reef

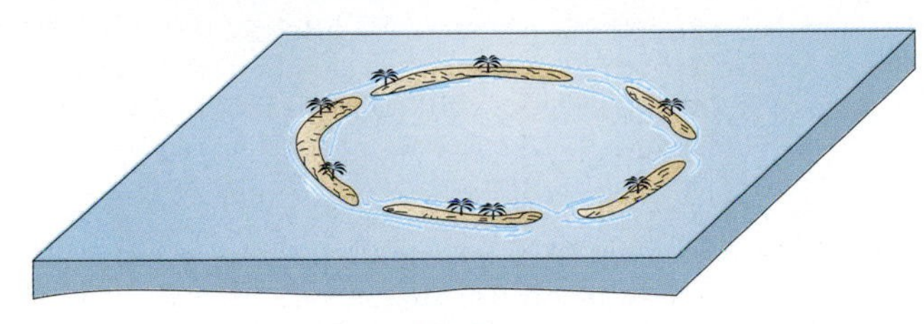

(c) Atoll

• Figure 8.11 Charles Darwin's explanation of the development of an atoll. **(a)** First, a fringing reef of coral is attached to the volcanic island's shore. **(b)** As the island subsides, a barrier reef forms. **(c)** With continued subsidence, the coral builds upward, and the volcanic center of the island finally becomes completely submerged, forming an atoll.

Joe Hobbs

• Figure 8.12 Pacific islanders know how to make many "appliances" and wares from coconut tree parts. This Fijian man is weaving a mat. Note the husked coconuts in the foreground.

often so dry and infertile that trees will not grow, and shortages of drinking water limit permanent settlement. Their low elevation above sea level provides little defense against storm 338 waves, tsunamis, and sea level rise due to climate change. Coastal regions of higher islands also suffer from these events; a 1998 tsunami killed 2100 people in northern Papua New Guinea, for example.

Before the intrusion of outsiders, the low-island economies were based heavily on subsistence agriculture (with strong reliance on coconuts), gathering, and fishing. Tom McKnight explained that sparse resources were distributed with relative uniformity among the low islands' populations.[3] The political and social units were smaller and less structured than those of the high islands. These resource-poor low islands have a long history of population-limiting events, both environmental and cultural.

Many low-island countries now rely heavily on modern commercialized versions of the coconut and fishing economy. The coconut is so important in their subsistence that many of these islands have developed what has been characterized as a **"coconut civilization."** Coconuts provide food and drink for the islanders, and the dried meat, known as copra—which is processed for coconut oil used in foods, cosmetics, and soaps—is the only significant export from many islands. The husks and shells of the nuts have many uses, as do the trunks and leaves of the coconut palms. People make baskets and thatching from the leaves, for example, and use timber from the trunk for construction and furniture (**•Figure 8.12**).

Why Are Oceania's Ecosystems So Vulnerable?

Island ecosystems in the Pacific region, like those across the globe, are typically inhabited by endemic species of plants and animals—those found nowhere else in the world. They result 443 from a process in which ancestral species colonize the islands from distant continents and, over a long period of isolation and successful adaptation to the new environments, evolve to become new species. Gigantic and flightless animals, like the moa bird of New Zealand and the giant tortoises of the Galápagos (in the Pacific) and the Seychelles (in the Indian Ocean), are typical of endemic island species. More than 70 percent of the native plant species of New Caledonia and Hawaii are endemic.

life in NEW ZEALAND

Helen Hayward

THINK CRITICALLY: In her account of life in New Zealand, how does Dr. Hayward draw from and share the deep well of indigenous Maori ties to the land?

Helen Hayward

Ko Aotearoa, ko te whenua
Ko Kurahaupo, ko Māmari me Tinana ngā waka o ōku tūpuna
Ko Te Kao tōku kainga, tōku here tangata
Ko Tawhitirahi tōku maunga, te iiringa kōrero no ōku tūpuna
Ko Pārengarenga tōku moana, te puna roimata mo rātou kua riro ki tua o te arai
Ko Te Awapoka tōku awa, te huarahi o ngā roimata ki te puna
Ko Pōtahi tōku marae, tōku tūrangawaewae
Ko Waimirirangi ko Haere Ki Te Rā, tōku wharehui, tōku whakaruruhau i ngā wā katoa
Ko Te Aupōuri tōku iwi, tōku mana, tōku tapu, tōku ihi
Ko Helen Hayward tōku ingoa
Rau rangatira mā, tēnā koutou katoa.

Aotearoa/New Zealand is the land. *Kurahaupo, Māmari*, and *Tinana* are the canoes of my ancestors. *Te Kao* is my home, my place of birth, *Tawhitirahi* is my mountain, upon whom is heaped the wisdom of my forebears. *Parengarenga* is my sea, the pool of tears for those who have passed on. *Te Awapoka* is my river, the path of tears to the pool. *Pōtahi* is my ceremonial hearth, my stronghold and place to stand. *Waimirirangi, Haere Ki Te Rā* is my meetinghouse, my shelter for all time. *Te Aupōuri* is my tribe, my dignity, my sacredness, and my strength. Helen Hayward is my name. Esteemed people, to one and all, greetings.

The *pēpeha* I've recited is a traditional practice in this part of the world. It precedes a new speaker's dialogue. The explicit and indigenous way of positioning oneself in time and space, and in relationship to others and landscapes, opens a space for genealogical links to be made and provides a place to "stand" and *kōrero*—talk.

The land where I first emerged into *Te Ao Mārama*, the World of Light, is situated along latitude 41° 00′ S and longitude 174° 00′ E. Aotearoa/New Zealand, a cluster of three main islands (*Te Ika a Maui*/North Island, *Te Wai Pounamu*/South Island, *Rakiura*/Stewart Island) and smaller offshore isles, lies in the south west of *Te Moana Nui a Kiwa*, the Pacific Ocean. Similar in size to Japan and the British Isles, *Aotearoa* has a land area of 271,000 square kilometers. Population-wise we are a small nation. On March 21, 2015, at 12.37:57 p.m., our Population Clock estimated we had 4,574,298 citizens and in the 2013 Census 14.9 percent of us, claimed Indigenous Māori descent or *Tangata Te Whenua*/People of the Land status.

As a settler-nation state, a British colonial past permeates our 21st-century education, social, economic, justice, and political systems. A realm of the British Commonwealth, we retain a constitutional monarchy, our Head of State being Queen Elizabeth II. Although holding no political influence the Queen has a representative on the ground, the Governor-General, Lt. Gen. Rt. Hon. Sir Jerry Mateparae, the second Māori to hold that office. Politically, we are governed by a parliamentary democracy led by a prime minister. Of the 120 seats in the House of Representatives, four are dedicated Māori seats.

Our nation's founding document, the Treaty of Waitangi (1840), is acknowledged on February 6 each year. As signatories to the treaty, Māori Rangatira chieftains of the day gave the British Crown rights to govern and to develop British settlement while guaranteeing Māori full protection of their interests and status, full citizenship, and religious freedom. Intended to create unity and certainty, major differences in the interpretation of the English and Māori texts and breaches of its principles triggered deep conflict and ongoing protest. In 1975, the Treaty of Waitangi Act established a permanent commission of inquiry—the Waitangi Tribunal—charged with making recommendations on claims brought by Māori regarding past actions or omissions of the Crown that potentially breached the tenor of the treaty. The prospect of cultural redress, reparations, and a Crown acknowledgment of and apology for alienation, confiscation, and theft of land has prompted *iwi*/tribes to initiate claims. Despite settlements, significant issues remain unresolved. The Tribunal has also been influential in gaining formal recognition for *te reo Māori*, the Māori language, as a *taonga*/treasure. The Māori Language Act (1987) enshrined Māori as an official language of Aotearoa. To the chagrin of some, the only other language with similar legal status is New Zealand Sign Language. English has a de facto authority mainly because of its widespread use.

Our physical isolation from the rest of the world has never stopped Kiwisa from having their say when things of import surface. A major turning point in our international foreign policy emerged from the 1960s as Kiwis rallied against nuclear atmospheric testing in the Pacific. The French government failed to observe the International Court of Justice ruling to cease testing at Moruroa[b] Atoll French Polynesia and lay the foundation for a more dedicated response from the then New Zealand government, who sent two Navy frigates into the test area to observe. The test site became the hub for ongoing anti-nuclear protest, with Greenpeace International taking a leading role. The 1970s and 1980s saw New Zealand public opinion against nuclear deterrents escalate to over 90 percent. A more "touchy" subject was the continued visits of US warships into our territorial waters and the US policy of neither "confirming nor denying" the presence of nuclear weapons on board. In early 1985, under a new Labour-led government and backed by overwhelming public support, *Aotearoa*/New Zealand declined a US request for the *USS Buchanan* (DGG14) to enter its waters.

In July 1985, French foreign intelligence agents under orders from the Direction générale de la sécurité extérieure (DGSE) bombed Greenpeace's flagship, the *Rainbow Warrior*, moored in our nation's largest city port of Auckland. Two explosions sank the vessel with one crewmember on board, Fernando Pereira. Relations between the two countries soured dramatically, with the UN mediating a settlement. The episode and the silence from other Western allies deepened our resolve for a nuclear-free Pacific. For prohibiting US warship entry into our waters, the US suspended military and security ties with *Aotearoa* in 1986, downgrading our status from "one of ally to one of friend." In 1987, The New Zealand Nuclear Free Zone, Disarmament and Arms Control Act was passed. Though abandoning atmospheric testing at Morurua in 1974, the French continued their testing regime underground. With the threat of the Comprehensive Nuclear-Test-Ban Treaty looming, France conducted its last detonation in January 1996 at Fangataufa Atoll close to Moruroa, and dismantling of the test site began. Nearly 30 years on, French authorities still guard the area, and both the US downgrade and our legislation remain.

The portent of our most recent and costliest natural disaster came in the early hours of September 4, 2010, in the form a 7.1 magnitude earthquake in our second largest city, Christchurch. For a country that sits astride tectonic plates, where in the North Island, the Pacific Plate subducts under the Australian Plate and the reverse occurs at the bottom of the South Island, volcanic and earthquake geological activity is inevitable. As a Christchurch resident, I experienced three of the major tremors of *Papatūānuku*/Earthmother's unborn child, Rūaumoko, in a suburban home, a low-rise building, and a new lecture block at *Te Whare Wānanga o Waitaha*/the University of Canterbury. Of the 11,000 plus quakes I felt, these particular events remain the most vivid. The 2010 arrival of a violent seismic P-wave sent a vicious jolt through the northwest side of the city. My brick house, set on piles, lurching in an impossibly frenetic way; my daughter's screaming; the intense noise—violent thunderclap meets freight train roar, followed by immense relief we were physically unscathed, with a house still intact—are some of the memory's bullet points.

On February 22, 2011, the second day of the first semester, I witnessed a strange spectacle outside my third-floor office window. The 100-year-old oak, as tall as the building I occupied, was visibly trembling. At 12.51 p.m., a 6.3-magnitude aftershock ripped through the region. With evacuation sirens blaring, staff worked hard to get students out, helping us make our way through corridors filled with a strange mist, the taste and smell of which registered as concrete dust. In a brief moment The Garden City, as it is called, had been leveled. In the city's outer reaches liquefaction, toppling chimneys and fences, and rock falls had also taken their toll. The event claimed 185 lives, the youngest being wee Baxtor Gowland, 5 months old. Miraculously, there were no fatalities on campus, though sadly one student died on the way to university. Like the rest of the community, we mourned this loss and those of faculty and students from the Kings Education School in the collapsed regional CTV building.

Mid-March, the university resumed teaching. Accompanied by safety crew and clad in boots, Hi-Viz jackets and hard hats, staff were given a strict 15 minutes to retrieve what teaching resources we could from battle-scarred buildings. Lectures went ahead in structures deemed safe, and in our large student car parks tents were pitched to become makeshift teaching spaces. Both staff and students adjusted to lessons under canvas and in community and church halls while portable on-site accommodation sprouted in open fields. For some students, the semester was completed in other New Zealand universities or in Australia. Supplying our on-campus food and caffeine needs was a huge marquee café, located in the Law car park, its name Intentcity 6.3.

In June, two 5.9- and 6.3-magnitude quakes put an end to afternoon lectures. Waiting in the Lecture Block C foyer after setting up to go live with education distance students in other parts of the country, I became aware that sound pressure was once again high enough to perceive the infrasound of the now familiar subterranean groan, and watched as large art installations suspended on wires swung in wide arcs and huge concrete columns "wavered." Cries from the lecture theatre where on-campus students had gathered sent me rushing in to see them emerge from under lecture stands—the "Drop, Cover, and Hold on" response in action. Overhead, screens showed my online students had heard the commotion. After seeing our students out, my teaching assistant and I returned to retrieve laptops; up above, the raft of concerned comments had grown exponentially, and my brief message was back: "Quake, lecture cancelled, all safe."

The UCSVA[c] formed after September 2010, and armed with spades, wheelbarrows, and gumboots, continued its sterling community work of clearing huge areas of liquefaction and providing cups of tea, hugs, and a kind word; when we could, staff joined them. The effort saw some 326,500 metric tons of liquefaction removed from back and front yards, and earned the community's gratitude and respect. Using Google Maps and data from Geonet[d] Paul Nicholls, a staff member of UC's Digital Media Group, developed the Christchurch Quake Map, an interactive animation site. The time-lapse map showing time, magnitude, and depth of all the earthquakes became a go-to online site to visit. As the CBD was gradually cleared, UC's HITLabNZ developed CityViewAR, a mobile augmented reality application allowing people to see how the city was before the earthquakes and building demolitions took place. It proved a useful tool as we became increasingly disorientated in a city we no longer recognized.

Four years on, many have yet to get their insurance claims resolved, and the rebuild continues. The road to Christchurch's recovery will be long and painful.

Life for now is Glenmāra, a 31-acre family vineyard in the *Wairau*/Marlborough region at the top of the South Island. The climate is ideal for grape growing. Long hours of sunshine, cool nights, and low annual rainfall combined with an extensive braided river system with free-draining stony, sandy loam soils covering very deep gravels, have contributed to the region's optimum conditions to account for 73 percent of NZ's grape production metric tons in 2013. Sitting a kilometer inland of *Te Koko-o-Kupe*/Cloudy Bay, Glenmāra grows Sauvignon Blanc, the region's leading wine grape, and Pinot Noir. The property also supports an olive grove; a small orchard with cherry, apple, peach, nectarine, feijoa, and berry fruits; and a vegetable garden. Throughout the year, labor is supplemented by WWOOFers,[e] often students on a gap year, mostly from Europe and the US; others have come from Japan, Mongolia, and Reunion Island, just off the coast of Madagascar. Right now, in March–April, Vintage 2015 is in full swing. In a massive logistical exercise, harvesters, gondolas, and trucks ply the millions of vine rows 24/7. Viewed from the air at night, the province looks like a sea of moving fairy lights. After a year tending vines, it all comes down to this moment, when winemakers determine final °Bx levels[f] and signal for harvesters to "tickle" the berries (grapes) off the clusters (bunches) for immediate transport to wineries. By the look of the fruit, it will be a great season.

Apart from wonderful wine, *Aotearoa*/New Zealand is home to many things, including hokey pokey ice cream, pineapple chunks/lumps (chocolate-coated, pineapple-flavored confectionery), *kūmara* (purple-skinned sweet potatos), *hāngi* (earth ovens), summer in December–February, jandals, rugby football, netball and cricket, the world's first commercialized bungee-jump enterprise, and kiwi-isms like "sweet as." Diverse landscapes of volcanic plateau, geothermal areas, mountains, fiords, glaciers, flat plains, and black and white sand beaches provide dramatic backdrops for movies like *Chronicles of Narnia*, the *Lord of the Rings* trilogy, *The Hobbit*, *The Last Samurai*, *The Piano*, and *Whale Rider*. And the land gives shelter to stands of magnificent native forest where trees have personal names, for example, *Te Matua Ngahere*/Parent of the Forest (3000–5000 years old) and *Tane Māhuta*/God of the Forest (2000 years old), and a habitat for unique birdlife such as the world's heaviest parrot, the *kākāpō* that, like a kiwi, can't fly.

Life in *Aotearoa*/NZ—well, it's rather like an unfinished musical score. Notes sound everywhere, in the excited whoop of a child who gets a *porotiti*[g] to hum and the exasperated sigh of a parent who doesn't. They pulse in the *haka* (war dance) *Te Kapa o Pango*, when our national rugby union team faces off before international matches and in the roar of the nation when we beat the Aussies at anything and in the collective groan of the country when we don't. They're heard in the 1982 Herbs anti-nuclear song "French Letter" that registered a small nation's self-determination and preparedness to risk it for humanity's sake and in the rich sonorous warble of the *kōkako* in the bush. They echo in the silence of a South Island city on February 22 each year as residents mourn the loss of 185 and their city's identity, and in the plaintive trumpet call of the Last Post in the dawn of an Australian and New Zealand Army Corps (ANZAC) remembrance day, when we recall all our war dead who died so far away from home. They reside in the flutter of the United Tribes of New Zealand and *Māori Tino Rangatiratanga*/ Sovereignty flags at the Waitangi Treaty Grounds on our national holiday. They murmur in the hush of a deep blue velvet Southern Hemisphere nightscape that shows *te ara whētu*, the star path my ancestors followed across the vastness of the Pacific Ocean to reach these shores. They even exist in the harsh biting staccato of statistics that report poverty and poor health exists among our nation's most precious resource—its people. And they stir in the moving *karanga* of a *kuia*, a female elder, who calls out across the courtyard of her *marae*, welcoming visitors forward, to bring with them their ancestors so that they too can be acknowledged. It is in the heart and the spirit of the land where the notes of this incomplete musical manuscript resound their strongest, for it cradles the blood and bone, the dreams, hopes, and aspirations of a young South Pacific nation trying to make amends for historical injustices and to meet the 21st-century demands of a population who wish to call this beautifully resonant land, home.

Dr. Helen Hayward is a Maori geographer, educationalist, and Viticulture Assistant living on New Zealand's South Island.

---

[a] Kiwis—informal term for New Zealanders; it is derived from our national flightless bird kiwi, order *Apterygiformes ratita*, which is endemic to New Zealand. Kiwifruit refers to the edible berry of the vine belonging to genus *Actinidia*.

[b] The atoll is part of the Tuamotu Archipelago. Indigenous Tahitians, the Maohi, have an ancestral name for it, *Hiti-Tautau-Mai*. Another historical name is *Aopuni*.

[c] University of Canterbury Student Volunteer Army.

[d] A collaborative partnership between the Earthquake Commission and GNS Science.

[e] Members of the Worldwide Opportunities on Organic Farms Association who work in exchange for food and accommodation.

[f] Short for Brix—a test for sugar and pH content, measured in degrees.

[g] Musical instrument and toy, a *porotiti* is a humming disc with two holes at its center, through which cord or string is looped. It is played by twirling the cord until the disc spins. Different materials (*pounamu* or New Zealand greenstone, wood, glass, shell, cardboard) create different sounds. Traditionally it was also used as a therapeutic tool, often over the chests of children with respiratory difficulties.

Because they have developed in the absence of natural predators and inhabit relatively small areas, island species are especially vulnerable to the activities of humankind. People have intentionally or accidentally introduced to the island alien plant and animal species, known as **exotic species** and including rats, snakes, sheep, pigs, cattle, and fast-growing plants, that often end up preying on or overtaking the endemic species. Habitat destruction and deliberate hunting also lead to extinction. For example, New Zealand's moas and the dodo birds of the Indian Ocean island of Mauritius are now—well, as "dead as dodos." Indigenous inhabitants of New Zealand set fires to hunt the giant ostrich-like moa and had already killed off most of the birds by the time the Māori people colonized New Zealand around 1350 CE. By 1800, the moa was extinct.

From one end of Oceania to the other, concern about such human-induced environmental changes has been mounting. Australia's Great Barrier Reef (the world's largest coral reef system) is facing threats from warming ocean waters, overfishing, and pollution runoff. The Hawaiian Islands have become known to ecologists as the "**extinction capital of the world**" because of the irreversible impacts that exotic species, population growth, and development have had on indigenous wildlife there. Environmentalists are also concerned about the impact of commercial logging on the island of New Guinea. On the eastern half of the island, there are accusations of "land grabbing" by Malaysian, Australian, American, South Korean, Chinese, and Japanese logging corporations. Wood is being "laundered" so that allegedly US and Chinese furniture are, in fact, goods made from New Guinea timber[4] (•**Figure 8.13**). About 22,000 plant species grow in New Guinea, fully 90 percent of them endemic. And ecologists consider nearby New Caledonia one of the world's biodiversity hot spots (not to be confused with the geologic hot spots like Hawaii) because of the ongoing human impact on its unique flora and fauna.

66, 421

36

Human-induced extinctions are not the only threats to island species. Volcanic eruptions, typhoons (hurricanes), and rising sea levels affect island plant and animal populations, sometimes in surprising ways. When Hurricane Iniki struck the Hawaiian Islands in 1992, for example, it destroyed numerous chicken coops on the island of Kauai. Today, as a result, feral chickens are thick on Kauai's grounds. Most animals find it difficult or impossible to evacuate in the event of a natural catastrophe, and an island's distance from other lands may hinder or prevent recolonization. So although island habitats have a higher percentage of endemic species than would be found on mainlands, environmental difficulties sometimes cause the number of different kinds of species (that is, species diversity) to be lower. Far more than natural hazards, however, it is the human agency that altered the Pacific Island ecosystems, in some cases completely transforming the natural landscape (see Insights, **page 380**).

8.3 Cultural and Historical Geographies

Their indigenous cultures having been greatly diminished in numbers and influence, Australia and New Zealand today are mostly European in culture, ethnicity, and language (•**Figure 8.14**). The region's other cultures, however, are overwhelmingly indigenous and quite diverse. Aside from Australia and New Zealand, with their majority European populations, and Fiji, New Caledonia, and Guam, each of which has populations about half indigenous and half foreign, approximately 80 percent of the Pacific's people are indigenous. The balance is 13 percent Asian and 7 percent European. Of the indigenous

Rachel Kramer/Getty Images

• **Figure 8.13** The luxuriant forests of New Guinea contain an exceptional wealth of plant and animal species. Here a logging road has been cut to access valuable hardwoods.

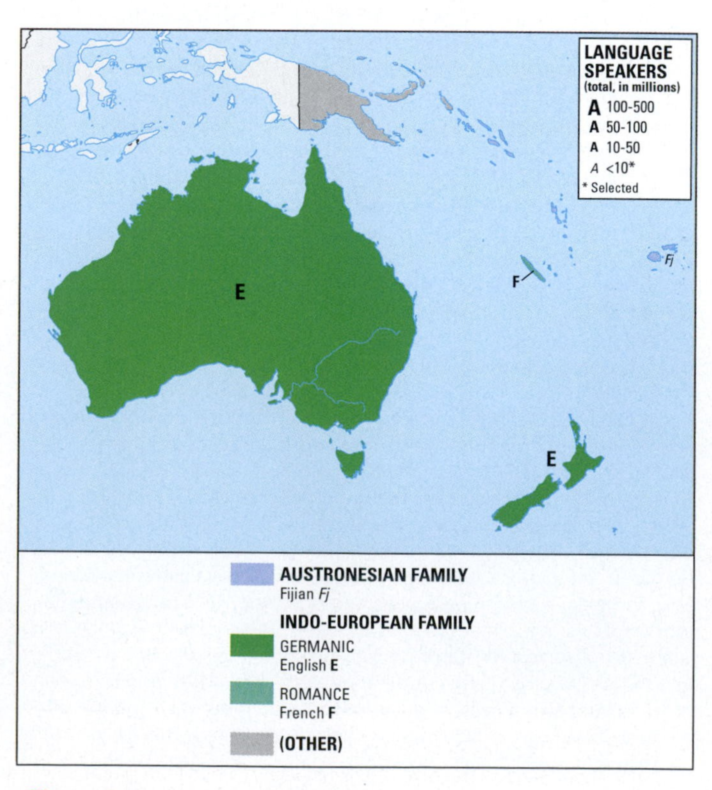

• **Figure 8.14** Languages of Oceania.

populations, Melanesians make up about 80 percent, with 14 percent Polynesian and 6 percent Micronesian.

The Indigenous Peoples of Oceania

Recent DNA analysis suggests that **Aborigines** arrived in Australia in a single major migration about 50,000 years ago (but archaeology may well push this date deeper into history). Some ethnographers refer to the Aborigines as a distinct Australoid race. However, they share racial characteristics with other indigenous groups in Asia, including the Mundas of central India, the Veddas of Sri Lanka, and even the Ainu of Japan. All of these groups probably descended from a common ancestral race in Asia, and the Aborigines developed their distinctive features over tens of thousands of years of living in Australia. The **Torres Strait Islanders**, a distinct indigenous group of Australians living mainly in Queensland, came to Australia from nearby Papua New Guinea.

During the Aboriginal migration to Australia, sea levels were much lower, and straits were narrower and easier to navigate. The settlers probably came to Australia across the land bridge that linked New Guinea and Australia until about 10,000 BCE (the historical continent made up of these two landmasses is known as Sahul). Once established in Australia, they developed a unique and enduring culture. Their **Aboriginal languages** are not clearly related to any languages outside the continent.

New waves of migrants known as **Austronesians** began a long process of diffusion across the western Pacific and eastern Asia after 3000 BCE. Their ancestral stock probably originated in 301 Taiwan. Their descendants migrated to the mainland and islands (initially the Philippines and Indonesia) of Southeast Asia. Their livelihoods were based on fishing and simple farming, especially of taro, yams, sugarcane, breadfruit, coconuts, and perhaps rice. Their domesticated animals included pigs and probably dogs and chickens, but they had no herd animals like cattle, sheep, or goats. From their Southeast Asian bases, over a period of several thousand years, the Austronesians embarked on voyages to Madagascar, New Zealand, Easter Island, Hawaii, and perhaps Chile (around 1400 CE), "island-hopping" all the way and accomplishing extraordinary feats of navigation in their outrigger canoes (•Figure 8.15). Their settlement of Polynesia came rather late in these adventures, within the last 1500 years. Today's Micronesians and Polynesians are mainly their descendants, but considerable mixing has occurred over the centuries.

These seafaring peoples spread their characteristic agriculture and a language called **Proto-Austronesian**—the ancestor of all modern Austronesian languages—across the Pacific. Most of the indigenous peoples of the region speak languages in the Austronesian language family. The Melanesian peoples, however, speak **Papuan languages**, which are not a distinctive linguistic family, but an amalgam of an extraordinary number of often unrelated and mutually unintelligible tongues. Today, there are about 700 Papuan languages. Some Torres Strait Islanders speak a Papuan language, and others speak an Aboriginal tongue.

Papua New Guinea is home to 850 languages (mainly Papuan but also Austronesian), or 20 percent of the world's total number of languages, making it by far the world's most linguistically diverse country, with an average of roughly one language per 5000 people. Vanuatu has fewer than 200,000 people and 105 identified languages, earning that country's population the distinction as the world's most linguistically diverse on a per capita basis, with roughly one language per

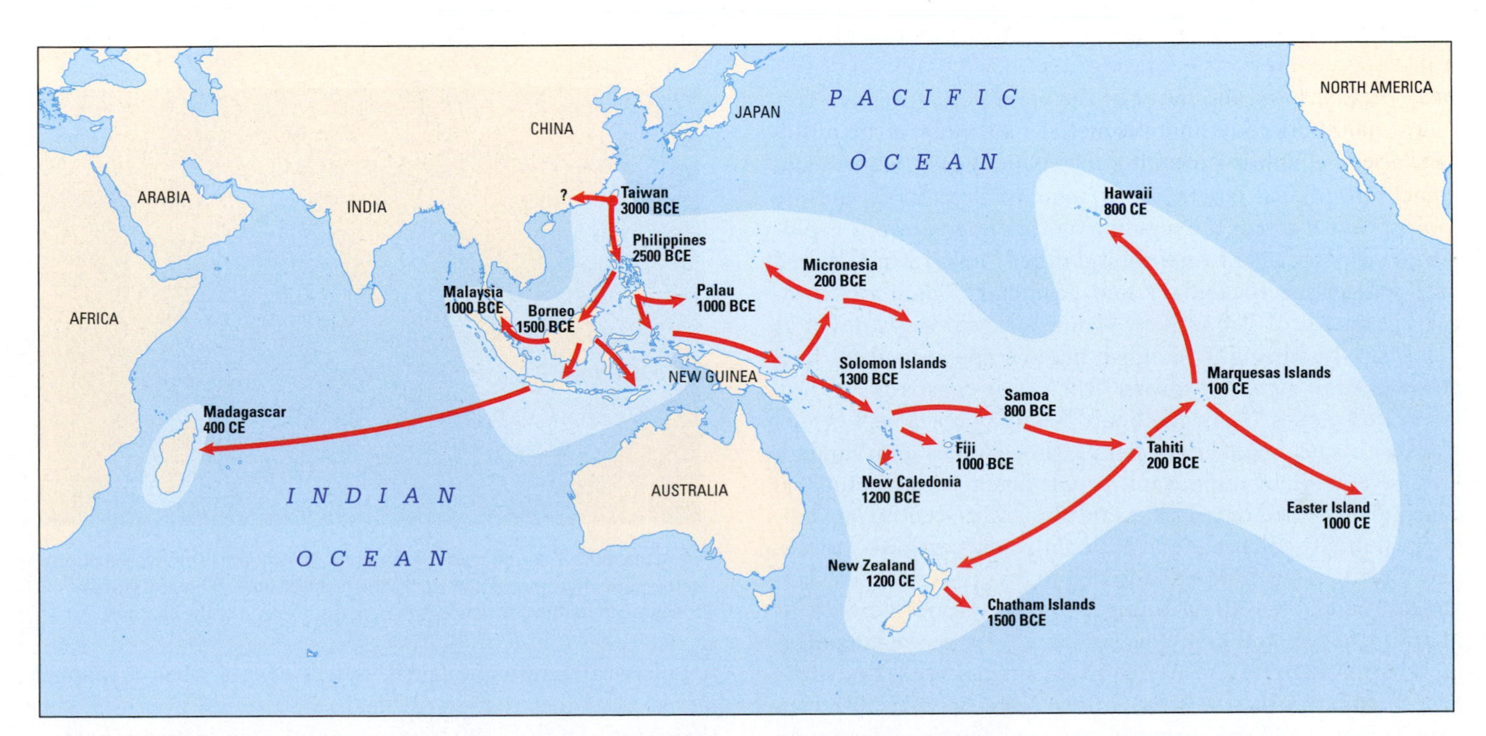

• **Figure 8.15** The origin and diffusion of the Austronesians. Austronesian peoples likely originated in Taiwan about 5,000 years ago. Over several millennia they spread throughout Southeast Asia and nearly all of the Pacific Islands. Some historians believe Austronesians may have reached the west coast of South America as well.

Insights Deforestation and the Decline of Easter Island: Too Simple a Parable?

"Easter Island is Earth writ small," wrote the naturalist and geographer Diamond. Archaeological and paleobotanical studies suggest that the civilization that built the island's famous monolithic stone statues (*moai*) destroyed itself through overpopulation and abuse of natural resources, Diamond argued (•**Figure 8.A**). Deforestation has serious repercussions wherever it occurs, but perhaps nowhere are these impacts so apparent as on Easter Island, one of the most remote islands in the world, lying 2200 miles off Chile's coast (see •**Figure 8.B**).

When the first colonists from eastern Polynesia reached remote Easter Island (known as Rapa Nui to locals) in about 1000 CE (although this date is uncertain), they found the island cloaked in subtropical forest. Plant foods, especially from the Easter Island palm, and animal foods, notably porpoises and seabirds, were abundant. The human population grew rapidly in this prolific habitat. The complex, stratified society that emerged on the island grew from an estimated 7000 to 20,000 people between 1200 and 1500 CE, when most of the famous statues were built. Apparently, in association with their religious beliefs, Easter Islanders erected more than 200 *moai* statues, some weighing up to 82 tons and reaching 33 feet (10 m) in height, on gigantic stone platforms. At least 700 more statues were abandoned in their quarry sites and along roads leading to their would-be destinations, "as if the carvers and moving crews had thrown down their tools and walked off the job," Diamond observed.

"Its wasted appearance could give no other impression than of a singular poverty and barrenness," the Dutch explorer Jacob Roggeveen wrote of the island on the day he discovered it—Easter Sunday 1722. Not a single tree stood on the island. The depauperate landscape bears testimony to the fate of the energetic culture that built the great statues. People were already exerting considerable pressure on the island's forests for fuel, construction, and ceremonial needs soon after their arrival. By 1400, people and the rats they introduced to the island caused the local extinction of the valuable Easter Island palm. Continued deforestation to make room for garden plots

Thomas J. Abercrombie/National Geographic Creative

• **Figure 8.A** Polynesian people who may have their origins in Taiwan erected at least 900 *moai* statues between 1200 and 1500 CE.

2000 people. There are fewer of the mainly Austronesian languages in Micronesia and Polynesia, and more of them are mutually intelligible. Reflecting the colonial past, English and French are official languages in some of the islands and are spoken widely across the region. Another lingua franca is **pidgin**, which consists of English and other foreign words mixed with indigenous vocabulary and grammar. Pidgin is the official language of Papua New Guinea, where it is known as Tok Pisin. Its kinship to English may be appreciated in these phrases: *kanu i kapsait* means "the canoe capsized," and *mi lukim dok* means "I saw the dog"[7] (•**Figure 8.16**).

Most of the languages of Papua New Guinea and Vanuatu are spoken only by a thousand people or fewer, with numbers continuing to decline. The establishment of central governments (initially colonial, then national), mass communications, and globalization have endangered many of the islands' languages; these forces draw younger people to learn more dominant languages, and many do not pass their native languages on to their offspring. At least a dozen languages are known to have become extinct in New Guinea since the mid-20th century, and over 100 are currently considered to be in danger. In 2011, the Papuan language of Laua had just one native speaker; by 2013, the language was gone. Laua had experienced what linguists call **language death**, which occurs when a tongue's last speaker dies. Language death occurs, on average, every two weeks. Of the 7000 languages of the world, about half are expected to die in this century, and many of these languages will be from the Melanesian islands.

Joe Hobbs

• **Figure 8.16** Can you read the language of Vanuatu? This is a cell phone carrier advertisement in pidgin. Try the first two lines. It helps if you say them out loud (the translation is in note 8 at the end of the chapter).

• **Figure 8.B** Easter Island is one of the most remote islands in the world.

and to supply wood to build canoes and to transport and erect the giant statues probably eliminated all of the island's forests by the 15th century. By then, people had hunted to extinction many terrestrial animal species and could no longer hunt porpoises because they lacked the wood needed to build seagoing canoes. Crop yields declined because deforestation led to widespread soil erosion. There is evidence that in the ensuing shortages, people turned on each other as a source of food. By about 1700, the population began a precipitous decline to only 10 to 25 percent of the number who once lived on this isolated Eden.[5] Today, about 6000 people live there, welcoming more than 50,000 tourists every year.

This is an exceptional story, an apparent Malthusian parable of people overusing their resources and suffering the consequences of doing themselves in, even eating one another. As it turns out however, there is probably more to the story. More recent archaeology post-dating that cited by Diamond suggests that Rapa Nui's palms were destroyed by rats eating their seeds, and not by human overuse. The island's human population continued to grow after the palms were exterminated, and probably crashed only after Dutch and other colonists inadvertently introduced killer diseases such as smallpox, and up to half the population was carried away by Peruvian slave traders in the 19th century. Finally, experiments seem to prove that the giant statues could be made to "walk" by their creators, coaxed along by ropes and muscle power.[6] Easter Island certainly tells the tale of how small island cultures and creatures can be wiped out—and that is a good parable for conservation.

Europeans in Oceania

Europeans began exploring the Pacific Islands during the 16th century, and their impact was especially profound after 1800. The initial Spanish and Portuguese voyagers were soon followed by Dutch, English, French, American, and German explorers. Many famous names are connected with Pacific exploration, including Ferdinand Magellan, Abel Tasman, Louis-Antoine de Bougainville, Jean-François de La Pérouse, and the most famous of all, British Captain James Cook, who undertook three great voyages in the 1760s and 1770s, only to be killed in 1779 by Hawaiian Islanders as a result of series of unfortunate interactions. Europeans were interested mainly in the high islands because of their relative resource wealth. They sought sandalwood, oyster shells, whales, and local people to indenture as whaling crews and as manual laborers. On island after island, European penetration decimated the islanders and disrupted their cultures. The intruders introduced sexually transmitted and other infectious diseases, alcohol, opium, forced labor, and firearms, which worsened the bloodshed in local wars.

The Europeans also came as Christian missionaries. Some ancient indigenous belief systems thrive in parts of the region, often nestled within Christianity and other recently founded introduced faiths (see Perspectives from the Field).

Overwhelmingly, the peoples of the Pacific adopted and continue to practice Christianity, with Methodists, Mormons, and Catholics the largest denominations (• **Figure 8.17**). Where the descendants of Asians that were brought to the Pacific by European colonial powers live, there are pockets of Hinduism, Buddhism, and Islam. Fiji, for example, is a distant outpost of India, with one-quarter of its population Hindu Indo-Fijians (• **Figure 8.18**).

Throughout the text, we have seen many and diverse impacts of colonialism, with more examples to come. In Oceania, Europeans created new settlement patterns, disrupted old political systems, and rearranged the demographic and natural landscapes. In Polynesia, for example, arriving Europeans typically sought shelter for their vessels on the lee side of an island. That safe place became the island's European port. According to Professor Knight, whatever chief happened to be in control of that spot typically, because of his association with the Europeans, became more powerful and wealthier than other island leaders. Eventually, he became high chief and often expanded his control to other islands; thus the first Polynesian kings emerged.

Europeans, North Americans, and East Asians changed the natural history of the islands by introducing exotic crops and animals, including arrowroot and cassava (manioc); bananas;

Perspectives from the Field In Search of John Frum

In 20th-century Melanesia, there was an extraordinary development in the history and geography of religion. A number of so-called **cargo cults** emerged in New Guinea and elsewhere in Melanesia, especially during World War II, as "Stone Age" people were exposed for the first time to chocolates, cigarettes, soft drinks, bottles, metals, manufactured cloth, and other material trappings brought by American soldiers and other Westerners. There were many variations among the cargo cults, but most were **millenarian movements**, maintaining that a supernatural event would trigger a new and more prosperous age for their followers. Intervention by white foreigners would end, and the locals would be mystically delivered the massive "cargo" of material possessions long denied them. Some of the believers built jetties and storehouses in anticipation of those events. Most withered with time, but one—the John Frum movement—endured with little change on Tanna, the southernmost island of Vanuatu (for location of Vanuatu, see Figure 8.2).[9]

As an undergraduate student of anthropology at the University of California–Santa Cruz, I was fortunate to be able to take a class on cargo cults taught by a visiting professor from New Zealand. I was captivated by his account of the John Frum "cult." It was not only a religion but a resistance movement against the colonial and missionary laws that would deprive locals of their culture (which they call *kastum*), including the vital ceremonial drinking of kava (**• Figure 8.C**). I wanted so much to meet these people who believed that one day a messianic American GI named John Frum would return to their village bearing endless quantities of cargo.

Joe Hobbs

• Figure 8.C The ritual and recreational drinking of kava (*Piper methysticum*) is an important feature of most Melanesian and some Micronesian and Polynesian cultures. This is a kava market in Fiji. Visible are the roots, which contain the plant's active compounds; powder of the root, from which the drink is prepared; and the kava beverage, ready to consume.

Joe Hobbs

• Figure 8.D Mt. Yasur, the "Lighthouse of the Pacific," is a perpetually active volcano at the southern end of the Vanuatu island chain. It is sacred to the local people. Its volcanic ash contributes to soil fertility beneficial for the cultivation of kava, coffee, and coconuts. But in the short term, continuous ash falls and acid rain are toxic to crops and livestock. Standing at the rim of its roaring caldera is exhilarating and scary. Yasur's lava bombs regularly kill visitors transfixed by this wonder of nature.

My wish came true in 2009. My wife, Cindy, and teenage daughters, Katie and Lily, and I made the long pilgrimage from Missouri to Tanna. On the island, we hired a local guide, who took us first by four-wheel drive to the foot of Mount Yasur, the active volcano whose nighttime glow led Captain Cook to discover the Vanuatu islands (named the New Hebrides by the British colonialists) in 1774 (**• Figure 8.D**). We hiked up to the cold, windswept rim of the volcano and peered

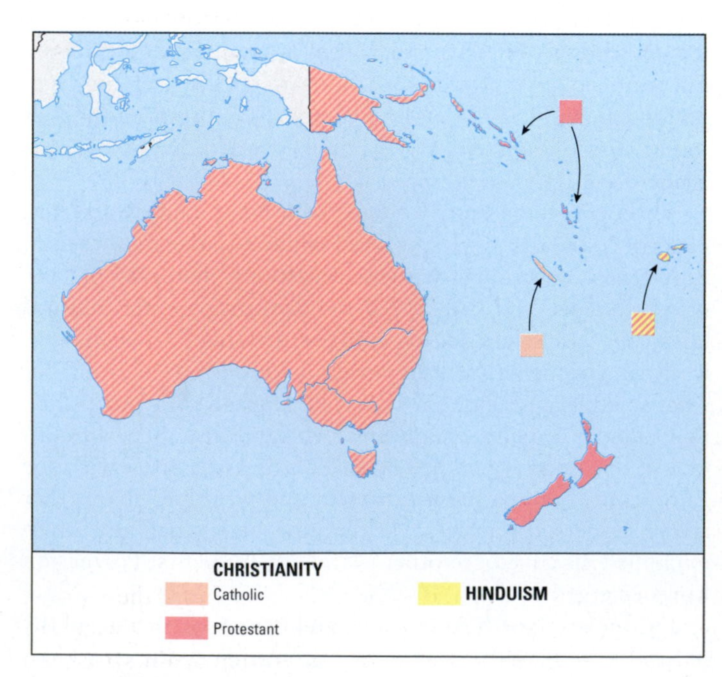

• Figure 8.17 Religions of Oceania.

Joe Hobbs

• Figure 8.18 The Hindu Temple of Sri Siva Subramaniya in Nadi, Fiji. The towering structure at left is known as a *goparum*. These structures are unique to South India's form of Hinduism. Their presence in Fiji, so far from India, is an element on the cultural landscape that tells you that some agent (in this case, British colonialism) brought South Indians to Fiji.

nervously down into its fiery mists. Every few minutes, Yasur would deliver a "lava bomb," a fiery red-hot projectile that shot far overhead and fell back into the crater. As it blasted off, each lava bomb issued a loud thud and shook the earth underfoot.

From the foot of Mount Yasur, we drove across a moonscape of recently made black earth, pocked with gurgling hot springs and stream vents. Just inside the edge of the forest, and still in the shadow of Mount Yasur, we arrived at Lamakara, one of Tanna's two villages where people honor John Frum. As in so many other religions, there had been a schism that split the approximate 6000 believers.

The headman of this village is Chief Isaak Wan Nikiau, and I was practically trembling with delight when he received us (•**Figure 8.E**). As we talked about John Frum, I learned many new things about the faith, most importantly that John Frum actually lived *inside* fiery Mount Yasur, from where he would emerge on that great day to meet his believers. Every February 15 since 1947, Chief Isaak said, all the villagers donned their homemade GI uniforms and with wooden models of rifles marched in formation to call upon John Frum to return. There were special martial ceremonies on each John Frum Day, but

Joe Hobbs

• **Figure 8.E** Chief Isaak Wan Nikiau of Lamakara, Tanna Island.

even on this ordinary day at dusk we saw men bring "the colors" down the flagpoles and carefully fold them (•**Figure 8.F**). These were flags of the United States, the US Marines, and Vanuatu.

I was very surprised to learn that the soft-spoken, humble Chief Isaak had visited the United States in 1995. He had been invited to speak about John Frum at a symposium on cargo cults. He travelled to New York, Washington, and Los Angeles. I wanted to know what this "man from the Stone Age" thought about flying high over the Pacific and seeing firsthand the great cities of the United States. He was "matter of fact" about those experiences, so

Joe Hobbs

• **Figure 8.F** Flying over the ritual parade ground of Lamakara are the flags representing the allegiances of John Frum's followers, including, from right to left, Old Glory, the US Marine Corps flag, and Vanuatu's national flag.

I returned to my many questions about him, his fellow villagers, and John Frum.

I hope that the people of Chief Isaak's village and those living all over Vanuatu will be able to recover soon from the devastation visited on them in March 2015, when Cyclone Pam took deadly aim on Vanuatu. With winds in excess of 150 miles per hour (240 km/h), Pam was the equivalent of a Category 5 hurricane and was the second strongest storm ever recorded in the Southern Hemisphere (exceeded only by Cyclone Zoe in 2002, which also struck Vanuatu).

tropical fruits such as mangoes, pineapples, papayas, and citrus; coffee and cacao; sugarcane; and cattle, goats, and poultry. They established extensive sugar plantations on some islands, notably in Hawaii, Fiji (where sugar is still the major export; see •**Figure 8.19**), and Saipan in the Marianas. The valuable Honduran mahogany trees planted by British colonists in Fiji early in the 20th century are only now beginning to mature, and competition for export profits from this resource has become intense.

Remarkably, until the middle of the 19th century, most of the Pacific was untouched by Western colonial powers; only Spain (which had governed the Marianas Islands since the 17th century) and Britain (which had held Australia since 1788) had a presence in the region. By the 1830s, many European merchants and missionaries were traveling to the Pacific, and the scramble for colonies began. Britain colonized New Zealand in 1840, and would later add Fiji, the Solomon Islands, the Gilbert Islands, and southeastern Papua New Guinea. Tahiti and New Caledonia became French colonies, and Britain and France administered the New Hebrides (Vanuatu) jointly. Germany colonized northeastern Papua New Guinea, the

Joe Hobbs

• **Figure 8.19** A train carries freshly cut sugarcane from the fields to a sugar processing plant on the northern side of Fiji's island of Viti Levu.

Bismarck Archipelago, and Samoa, but would lose these possessions after World War I. The United States made Hawaii a territory in 1898, took control of Guam from Spain after the Spanish-American War, and took eastern Samoa in 1900.

When Germany lost its Pacific territories at the end of World War I, Japan took control of German Micronesia, New Zealand took Western Samoa, and Britain and Australia took the German section of New Guinea and the Bismarcks. Phosphate-rich Nauru came to be administered jointly by Britain, New Zealand, and Australia.

With the outbreak of World War II, Japan quickly overran much of Melanesia. The ensuing ferocious battles between Japan and the Allies for control of New Guinea, the Solomon Islands, and much of Micronesia brought enormous changes to the region. (After World War II, the US took over Japan's former Micronesian holdings, including the FSM, Marshall Islands, and Palau.) Many of the cultures were in the Stone Age as the war began and by its end had seen the Atomic Age; Micronesia had become a nuclear proving ground. Everywhere, traditional peoples were drawn into cash economies and affected by other global forces originating far beyond their watery horizons.

Most of the colonists are gone, but some of the seeds of strife they planted are still germinating. Many conflicts in the region today reflect the demographic changes Europeans wrought to meet their economic interests. In many cases, the Europeans introduced laborers from outside the region because local peoples were too few in number or refused to work in the cane fields. Successive infusions of Chinese, Portuguese, Japanese, and Filipinos brought a polyglot character to Hawaii, with little apparent ethnic discord. But ethnic strife is ongoing in Fiji and the Solomon Islands, which became independent from Britain in 1970 and 1978, respectively.

The Solomon Islands' troubles date to 1942, when US Marines fought to push Japanese troops off the island of Guadalcanal. The American forces enlisted the support of thousands of people from the neighboring island of Malaita. After the successful offensive, many of these Malaitans elected to stay on Guadalcanal, where under British colonial rule they came to dominate economic and political life. The indigenous people of Guadalcanal resented their dominance and in particular what they saw as the loss of their lands to the Malaitans (who purchased the lands). Conflict between the groups erupted in 1999, with the majority Guadalcanalans murdering scores of Malaitans and driving tens of thousands off the island (most eventually returned to Malaita). An Australian-led peacekeeping mission arrived in 2003 to quell the unrest and put civil servants to work rebuilding the country's shattered economy. Thankfully, reconstruction has set the discord aside, but here again we see how the seeds of conflict were sown by colonial and other outside powers.

The violence in the Solomon Islands was inspired by ethnic unrest in Fiji. In the early 1900s, colonial British plantation owners imported mainly Hindu Indians as indentured laborers to work Fiji's sugarcane plantations. Today, indigenous Fijians now only slightly outnumber ethnic Indians (**Indo-Fijians**). Here too the indigenous inhabitants resent the dominance of the foreign ethnic group; ethnic Indians control most of Fiji's economic life. A coup led by ethnic Fijians in 2000 succeeded in ousting Fiji's ethnic Indian prime minister and throwing out a 1987 constitution that granted wide-ranging rights to ethnic Indians. Accused of lawlessness, the coup leaders were threatened with economic sanctions by the European Union, which pays higher than market prices for most of Fiji's sugar crop, and by the United States, Australia, and New Zealand.

The prospect of an economic crisis led the army to take control of Fiji's affairs, imprison the coup leader, and promise a general election and return to civilian government in 2002. The unrest caused a crash in tourism, Fiji's most valuable industry. Tourist numbers rebounded after a peaceful election that brought the indigenous Fijian political party back to power, but a military-led coup in 2006 temporarily diminished the vital industry again. Lack of progress toward democracy led to Fiji's ejection from the British Commonwealth of Nations, a rare action that Fijians felt quite ashamed of and were able to overcome with a democratic election in 2014 that restored Fiji to the Commonwealth. Ever resilient, tourism has bounded back again, with more than 600,000 visitors arriving yearly to enjoy a beautiful and interesting country troubled by its enduring ethnic fault line.

8.4 Economic Geography

A lack of industrial development characterizes the Pacific island economies, other than those of Australia and New Zealand, and the poverty typical of LDCs prevails in the region. Contributing to this underdevelopment are what could be described as tyrannies of size and distance. Many of the countries consist of little more than small islands of limestone sprinkled with palm trees and situated hundreds and often thousands of miles from potential markets and trading partners. Most of the countries suffer large trade deficits and must import far more than they can export.

Making a Living in Oceania

There are seven major economic enterprises in the Pacific region, only one of them (textiles) involving value-added processing of raw materials: exports of plantation crops (especially palm products), fish, and minerals; services for Western military interests; information technology; textile production; and tourism. Significant manufacturing outside Australia and New Zealand is limited mainly to Fiji. Until worldwide quotas on textile trading were lifted in 2005, giving an immediate advantage to China, Saipan in the Northern Marianas had a flourishing garment industry owned by South Korean companies and operated by Chinese laborers. Fiji also began exporting manufactured clothes after the mid-1990s, when Australia granted preferential tariff treatment to Fijian-made garments. Several island groups, like French Polynesia and the Cook Islands, make money by exporting exquisite sea pearls.

A few Pacific countries are beginning to participate, in rather unlikely and sometimes unscrupulous ways, in the information technology revolution that has generated so much global wealth since about 1990. One example is tiny Tuvalu in Polynesia. In 2000, an American entrepreneur agreed to pay the Tuvalu government $50 million to acquire Tuvalu's two-letter

Internet suffix, .tv. Although the new owner of the .tv domain has been busily profiting by selling Internet addresses, Tuvalu is delighted with its windfall. It used part of the money to finance the process of joining the United Nations, build new schools, add electricity and other infrastructure to outlying islands, and expand its airport runway. Vanuatu has enjoyed significant hard-currency revenue by permitting its country's phone code to be used for 1-900 sex calls. AT&T reports advise its customers to examine their bills for calls apparently (but not really) made to Vanuatu, billed from $9/minute and up. And Nauru and other island nations use telephone and digital technology to practice money laundering (more on this shortly).

• **Figure 8.20** The 10 square miles (26 sq km) of land that is Nauru have been cratered by the phosphate mining that once made Nauruans among the wealthiest people per capita on Earth.

TORSTEN BLACKWOOD/Getty Images

Mining takes place on only a few islands. New Guinea is mineral-rich, which is one reason why Indonesia refuses to grant 347 independence to Papua, on the western half of the island. Three-quarters of Papua New Guinea's exports are minerals (especially gold and copper) or petroleum. Copper is also found on Bougainville, an island in the Solomon Islands chain that belongs to Papua New Guinea. Until 1989, the world's largest copper mine, operated by the Australian mining company Rio Tinto, was on Bougainville. The mine provided half of Papua New Guinea's export revenue, but only a small fraction of the profits from the mine were invested in Bougainville. Many jobs at the mine went to people from Australia, New Guinea, and other islands, instead of to the people of Bougainville, who are ethnically distinct from New Guineans and had been pressing for independence or autonomy for many years. The mine also created many environmental hazards, and rebels of the Bougainville Revolutionary Army disrupted the mine's operations to get the government to act upon their concerns. A heavy-handed response of blockading the island led to an uprising against the government in 1988, which turned into a civil war lasting 10 years and resulting in about 20,000 deaths. This was the deadliest conflict in Oceania since World War II. After a 1997 cease-fire, Bougainville was granted more autonomy, and promised a referendum on independence to be held before 2020. The mine remains closed.

The 12,000 people of the tiny oceanic island of Nauru once enjoyed a high average income from the mining of phosphate, the product of thousands of years of accumulation of seabird excrement, or **guano**, which is used as fertilizer. Until recently, Nauru's economy depended entirely on annual exports of 2 million tons of phosphate and on revenues earned from overseas investments of profits from phosphate sales. This resource is nearly exhausted, however, and the phosphate mining has stripped the island of 80 percent of its soil and vegetation, turning it into a virtual lunar landscape. The island can no longer support its growing population. Nauru has one of the highest population growth rates in the world, at 2.9 percent, and many people in Nauru are now looking for a new island home (•**Figure 8.20**).

Many Pacific island countries, formerly including Nauru, have earned revenue through **money laundering**. These island entrepreneurs have undertaken offshore banking, in which they can operate loosely controlled banks anywhere but in their home islands. Many of these are **shell banks**, existing only on paper and specializing in turning money earned in illegal ways into money that looks legal. Although Nauru no longer launders money, like many Pacific Islands it has been sought out by businesses and individuals as a tax haven levying little or no taxes on enterprises.

After 2001, hard-up Nauru began earning revenue by serving yet another unlikely function: detention center for refugees from the war in Afghanistan. By 2004, more than 1000 Afghan refugees who were apprehended at sea in transit to Australia, where they hoped to gain asylum, had been diverted to Nauru. Australia paid the government of Nauru about $15 million yearly to hold the detainees while their cases were reviewed in the Australian courts, a component what it calls its **Pacific Solution** of sending would-be immigrants from a variety of countries to Nauru, in exchange for desperately needed hard currency (•**Figure 8.21**). 391

The most dynamic element of the Pacific economies is the rapid growth of tourism, with most visitors coming from North America, Australia, New Zealand, Japan, and increasingly China. Tourism is the largest industry in the Pacific region, providing substantial income and employment, and stimulating overall economic growth by encouraging foreign investment and facilitating trade.

martin berry/Alamy

• **Figure 8.21** A street protest by the Free the Children campaign in Sydney, Australia. The basketful of dolls represents the children detained in camps on Nauru and Christmas Island as part of Australia's "Pacific Solution" to its immigration problems.

THINK CRITICALLY: Considering US policy on immigration into the United States, or on its own merit, what do you think should be Australia's policies on immigration, on both ethical and economic grounds? See **pages 390–391** for more details on this issue.

The Pacific Islands continue to exert an irresistible appeal to people awash in the conveniences and congestion of the industrialized world. One online entrepreneur, hoping to capitalize on disenchantment with the modern world, built an e-community of people who would contribute funds to, and sometimes visit, his idyllic Fijian island (see http://tribewanted.com). His venture promising people an "off the grid experience" was so successful that it expanded to Sierra Leone, Mozambique, and Bali, with a "pilot project" in Myanmar.

Some of the countries have developed unique appeals for international tourists. In a bid to attract visitors by being the first country in the world to greet the new millennium, Kiribati decided unilaterally in 1997 to shift the International Date Line eastward by more than 2000 miles (3200 km). The impact of this decision stands out on any map depicting the date line (see Figure 8.2). Samoa in 2011 also moved the line eastward, in this case for the perceived economic advantages of being closer to the workdays of Australia and East Asia than to the United States.

French Polynesia (where Tahiti is located) is the prime destination for beach and resort vacations in the South Pacific, but there has been strong growth in ecotourism, archaeological tourism, and other tourism niches in destinations including Palau, Vanuatu, and Easter Island. Palau advertises itself as a paradise for scuba divers, including those wanting close encounters with sharks. Palau has used its shark tourism to make a convincing pitch about the tradeoffs between conservation and exploitation. Diver tourism contributes about 39 percent of Palau's $218 million annual gross domestic product. A 2010 research study by the Australian Institute of Marine Sciences and The University of Western Australia put some numbers together about the value of Palau's sharks. About 21 percent of the divers come specifically to see the sharks; therefore sharks contribute about 8 percent of GDP. There are roughly 100 sharks, so each is worth $179,000 annually to the country's tourism industry. Over its lifetime of about 16 years, each shark has a value of $1.9 million. Killed and sold only for the fins and meat, those sharks would be worth only $10,800.[10] A similar study has been done of the Maldives' sharks; each is worth $33,500 annually. These studies argue strongly for that principle of sustainable development that conservation and development *are* compatible (see **page 35**).

Finally, the military interests of outside powers, particularly the United States and France, generate much revenue for some of the Pacific economies, notably Guam, American Samoa, and French Polynesia. This military presence is a mixed blessing, however, because of its environmental impacts and because it perpetuates dependence on foreign powers. These are among the geopolitical concerns discussed next.

8.5 Geopolitical Issues

Once entirely colonial, Oceania today is a mix of units still affiliated politically with outside countries as well as others that have become independent. Since the end of World War II, the United States, Britain, Australia, and New Zealand have granted independence to most of their colonies in the region, but several still remain. Only France has insisted on holding on to all of its South Pacific possessions.

Why Are Foreign Powers Interested in the Pacific?

In some cases, colonial powers have postponed independence to would-be island nations. They reason that such territories are too small and isolated or are not economically viable enough for independence. Some islands remain dependent because they confer unique military or economic advantages on the governing power. For example, French Polynesia was, until recently, the venue for French atomic testing (see Regional Perspective, **page 388**, and the Life in New Zealand feature), and Guam and American Samoa are still useful to the United States for military purposes. Guam, which has been a US possession since 1898 and a US territory since 1950 (its residents are American citizens), is especially vital in US strategic thinking. Pentagon planners classify it as a **power projection hub** that can be used as a forward base that is five full days' sailing time closer to Asia than Hawaii is. Starting in 2003, it was used as a jumping-off point for American B-52 jets on bombing missions to Iraq, and it would be used in any US military action in Korea or elsewhere in East Asia. In the event of such hostilities, the United States would need Japan's permission to operate from bases in Japan, but no such approval would be necessary to act from Guam.

A generally unspoken but widely acknowledged US military objective in building up readiness in Guam is that the island would be critical to any confrontation with China, should that giant country ever become a belligerent power. China is well established in its "string of pearls" footholds across the Pacific and Indian Oceans (see **page 327**), and in a contest reminiscent of the Cold War, the US is positioning itself geostrategically to match China. Chinese officials were upset when, while visiting Australia in 2011, President Barack Obama resolved to station American troops and ships in Australia. "The US is a Pacific power and we're here to stay," he said.[11] One of the goals of the president's trip was to assert US access to the South China 316 Sea. Geopolitically, both Washington and Beijing perceive the Pacific region as the *Asia-Pacific* region. Russia wants a stake in this geopolitical arena too. At the 2014 Summit Meeting of the "G-20" countries in Australia, Vladimir Putin attended as Russia's head of state. Most of the other leaders snubbed him openly in his first major international appearance since the invasion of Crimea and the shoot-down of Malaysian Airlines 183 Flight 17 over eastern Ukraine. In a show of force and as a reminder of its regional interests, throughout the conference Russian warships steamed just outside of Australia's territorial waters.

Australia and New Zealand have Pacific-oriented defense and security agreements, somewhat precariously balanced with US interests in the region. In 1951, they joined the United States to form the **Australia, New Zealand, United States (ANZUS) security alliance**, and Australian troops supported American forces in the Vietnam War. However, in 1987 the liberal New Zealand government declared that it would henceforth 376 be a nuclear-free country (see Life in New Zealand essay on **page 376**). Legislation banned ships carrying nuclear weapons or powered by nuclear energy from New Zealand's ports. The United States refused to confirm or deny whether its ships violated either restriction, so New Zealand denied port access to them. This action strained relations between the US and New Zealand, and New Zealand was removed from ANZUS. New Zealand paid some heavy political and economic costs for its nonnuclear position; for example, it was pointedly left out of negotiations that led to the Australia-United States Free Trade Agreement (AUSFTA) that came into force in 2005.

Despite its small army (only 50,000 strong), Australia has begun to provide a selective security blanket over its interests in the region and elsewhere. In 1999, Australian troops were deployed to head up the United Nations peacekeeping force to oversee the referendum that gave independence to East Timor and its subsequent transition to full independence as Timor-Leste. Australia (and New Zealand) also lent troops and other support in the US buildup to and execution of the war against Iraq in 2003. Al-Qa'ida vowed revenge and apparently delivered it with a 2002 bombing in Bali, Indonesia, that killed many young Australian tourists. Australia's apparent concern about a long-term threat from North Korea also prompted it to join the United States in the development of the Strategic Defense Initiative missile shield program. 389 Australia is sometimes snubbed by its Southeast Asian neighbors in political and economic affairs because it is seen as something of a regional "police officer" or as a US "deputy."

Oceania's Environmental Future

Another set of geopolitical concerns in the Pacific region is environmental. The remote islands of Oceania, so far from the industrial cores of North America, Europe, China, and Japan, are actually the places most likely to suffer first and most from the most feared impacts of industrial carbon dioxide emissions. Along with the Maldives described earlier (**page 330**), the islands would be among the earliest and hardest hit victims of any rise in sea level due to climate change. There have 41 already been reports of unprecedented tidal surges on Pacific island shores. If the trend continues, the first Pacific Islands to be totally submerged would be Kiribati, the Marshall Islands, and Tuvalu (where the highest point is 15 feet above sea level), while Tonga, Palau, Nauru, Niue, and the Federated States of Micronesia would lose much of their territories to the sea. Long beset by troubles with atomic tests, the former residents of Bikini Atoll have reported flooding from the sea every year since 2011. There are already discussions about how **"deterritorialized states"**—those that have been drowned by the sea—might continue to have sovereignty and rights to underwater and other resources. Kiribati's 100,000 people could have most of their land inundated by a 3-foot (1 m) sea level rise by 2100. They have bought 6000 acres in Fiji to protect food supplies and possibly relocate. Fiji's president has said his islands will welcome climate refugees from its less fortunate neighbors. For its part, Fiji has begun moving people from 43 low-lying outer islands to its main islands of Viti Levu and Vanua Levu.

Most small Pacific island nations are among the 39 countries comprising the Alliance of Small Island States, which argued forcefully but unsuccessfully at the 1997 Kyoto Conference on Climate Change that by 2005 global greenhouse gas emissions should be reduced to 20 percent below their 1990 levels. What they did get was a pledge by most indus- 41 trialized nations, in the form of the Kyoto Protocol, to cut these emissions to at least 5 percent below their 1990 levels by 2012. Already claiming to be a victim of lost coastline, higher 45 storm surges, and more storms as a result of global warming, Tuvalu in the early 2000s lobbied other countries to join it in a lawsuit against the United States and Australia, both of which had signed the Kyoto Protocol, but not ratified it. Tuvalu made the case that these large countries' failure to ratify the Kyoto Protocol was a principal cause of global warming. Australia ratified in 2007, leaving the United States, Canada, Andorra, and South Sudan as the only countries currently not parties to the treaty. 524

Regional Perspective Foreign Militaries in the Pacific: A Mixed Blessing

The remote locales and small or nonexistent human populations of some Pacific Islands have made them irresistible sites for weapons tests by the great powers controlling them (•**Figure 8.G**). In the 1940s and 1950s, US authorities displaced hundreds of islanders from Kwajalein Atoll in the Marshall Islands to make way for intercontinental ballistic missile (ICBM) tests.

• **Figure 8.G** The greatly underestimated power of the Castle Bravo hydrogen bomb detonated on Bikini Atoll on March 1, 1954 irradiated numerous people downwind including relocated inhabitants, Japanese fishermen on a nearby trawler, and US military personnel. Of the 23 atomic weapons tested on Bikini, 20 were the enormous hydrogen bombs.

During those years, the United States also used nearby Bikini Atoll (namesake of the swimwear) as one of its chief nuclear proving grounds.

US government authorities relocated the indigenous inhabitants of Bikini to another island in 1946, promising they could return when the tests were complete. Over the next 12 years, there were 23 nuclear blasts on Bikini, out of a total of 67 conducted as **Operation Crossroads** in the Marshall Islands. The 15-megaton "Castle Bravo" hydrogen bomb detonation on Bikini in 1954 had the power of 1000 Hiroshima bombs, blowing a mile-wide crater in the island's reef (clearly visible in •**Figure 8.H**) and producing radioactive dust that fell downwind on the Rongelap Atoll, where children played in the dust "as though it were snow," one observer wrote.[12] The bomb's designers seriously miscalculated its power, estimating that it would produce only a 5- or 6-megaton explosion (one megaton is the equivalent of 1 million tons of TNT). For decades, the US Atomic Energy Commission claimed that a sudden shift in winds took the fallout to inhabited Rongelap, but recently revealed records proved that authorities knew of the wind shift three days prior to the test. The Marshallese today suffer a legacy of cancers and deformities linked to the weapons tests on Bikini and Eniwetok Atolls. So far, the United States has provided more than $60 million to compensate Marshall Islanders made ill by the blasts.

The Bikinians have never accepted their exile. But at the same time they worry about returning to their radioactive native soils. A repatriation of the Bikinians in the 1960s was found to be premature, as radiation persisted, and the nuclear nomads moved again. More recent tests have determined that Bikini has little ambient radioactivity and is essentially habitable, except for the cesium-137 that permeates the soil and becomes concentrated in fruits and coconuts. Bikinians do not want to go home to stay until they have assurances from the United States that the problem of poisoned produce can be resolved and that the island is completely safe. Along with the people of Eniwetok, they are seeking massive funding from the United States for complete environmental decontamination of their homeland and for further compensation for their travails.

Now that Bikini is at least safe to visit, some Bikinians are capitalizing on its formerly off-limits status. The island's lack of human inhabitants for more than 50 years has had the remarkable effect of rendering Bikini a true marine wilderness. In the late 1990s, it opened as a tourist destination, with naval ships sunk by the atomic blasts a highlight for many divers. The dive masters and tour guides are Bikinians, who welcome you to their island at www.bikiniatoll.com.

Like the United States, France has had a nuclear stake in the Pacific. In 1995, after a three-year hiatus, France resumed underground

Despite ratifying Kyoto, however, Australia still has a long way to go to meet its commitments. Australia's economy relies heavily on resource extraction; petroleum, coal, iron, gold, and other minerals comprise over half of the country's exports by value. Australia released over 30 percent more greenhouse gases into the atmosphere in 2013 than in 1990, whereas US emissions increased only 3 percent. In 2013, Australia's government eliminated several climate change programs and cut the position of science minister, and the following year Australia used its influence as host of the G-20 Summit to keep climate change off the agenda. Australia ranks higher than any other industrialized nation in greenhouse gas emissions per capita, and a UN report says Australia will not meet its targeted emissions reduction by 2020.

Reducing greenhouse gas emissions while minimizing the economic impact is a difficult task. Many countries, like France, Sweden, Japan, and New Zealand, did not meet their original Kyoto target emissions cuts. And between 2000 and 2014, China's emissions (unrestricted by Kyoto) increased a whopping 171 percent. As so often in the past, the peoples of the Pacific today feel vulnerable to the actions and policies of more powerful nations far from their tranquil shores.

8.6 Regional Issues and Landscapes

Australia and New Zealand

Becoming Less British, More Asian-Pacific

These two unusual countries (•**Figure 8.22**) are akin in population, cultural heritage, political problems and orientation, type of economy, and location. Australia and New Zealand are among the world's minority of prosperous countries. Australia's

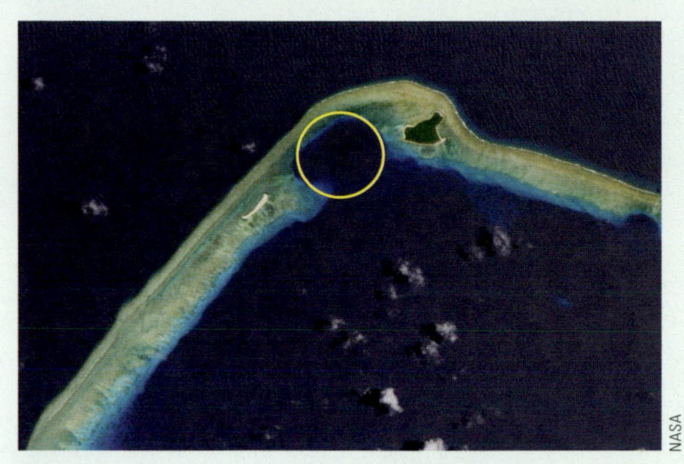
NASA

• Figure 8.H The Castle Bravo bomb, the largest nuclear device ever tested by the US, left this mile-wide crater in Bikini Atoll.

testing of nuclear weapons on the Mururoa ("Place of Deep Secrets") Atoll in French Polynesia. There was an immediate backlash from some national governments and from environmental and other organizations. France soon bowed to international pressure, ceasing its nuclear tests in the Pacific and signing and ratifying the Comprehensive Test Ban Treaty in 1996. The US has signed but never ratified the treaty, which has not taken effect because of reservations by the US and seven other countries.

Ironically perhaps, many islanders fear the retreat of French military interests in the region. Despite the nuclear tests, most inhabitants of French Polynesia have long favored continued French rule. Tourism and other local businesses contribute only about 25 percent of the territory's revenue; the remainder comes from French economic assistance, largely in the military sector. Many locals, whose parents and grandparents gave up subsistence fishing and farming for jobs in the military and its service establishments, are worried about their futures.

Similar fears stalk Palau, the Marshall Islands, and the Federated States of Micronesia (FSM), three independent nations carved from the former US Trust Territory of the Pacific. The United Nations awarded that territory, which includes the Bikini and Eniwetok Atolls, to the United States in 1946 as the world's first and only political-military zone. In 1986, the United States granted independence to the Marshall Islands and the FSM in an agreement known as the **Compact of Free Association**. In this agreement, the countries agreed to rent their military sovereignty to the United States in exchange for tens of millions of dollars in annual payments and access to many federal programs. The United States uses its leases to conduct military tests—for example, on Strategic Defense Initiative (SDI, also known as "Star Wars") technology. The United States also maintains a military presence in Palau and provides large cash grants and development aid to the nation, which became independent from the United States in 1994 but remains in free association with it.

In these cases, the American government handed over the monies with no strings attached. The funds have been embezzled and misspent, and there is little development to show for them. The United States has begun to reassess the payments, with an eye to their eventual curtailment. Many islanders fear that their economies will subsequently plummet. People in the Marshall Islands and the FSM are especially worried, because more than half of the gross domestic product of each country is comprised of economic aid from the United States. The Marshall Islands' leaders, however, are taking a more defiant stand, saying that the United States has no right to question what becomes of the aid money and pointing out that the United States gets much in return in the form of extensive military rights. The United States is hard-pressed here because the Kwajalein Missile Range in the Marshall Islands is the only reasonable place in the world where, Pentagon officials say, they can test many of the SDI program components.

per capita GDP (PPP) of $43,000 is higher than that of Britain, France, and Germany, although somewhat below that of the United States. Australia has been dubbed the "Lucky Country," and a recent World Bank study ranked Australia as the world's wealthiest country based on its natural resource assets divided by the total population. With China's strong appetite for minerals and other raw materials, resource-rich Australia has been particularly lucky until China's recent slowdown, with a gold rush and a coal boom to name just two developments. New Zealand is less affluent, with a GDP (PPP) of $30,400, but is still quite prosperous compared to most nations of the world. In both countries, there are relatively few people among whom to spread the wealth; Australia's 23 million people and New Zealand's 4 million together amount to only about 75 percent of the people living in California.

Both countries owe their prosperity to the wholesale transplantation of culture and technology from the industrializing United Kingdom to the remote Pacific beginning in the late 18th century. Australia (established originally as a penal colony for British convicts) and New Zealand are products of British colonization and strongly reflect the British heritage in the ethnic composition and culture of their majority populations. Despite their independence (for Australia in 1901 and New Zealand in 1907), both countries maintain close ties with Britain. Their peoples speak English, live under British-style parliamentary forms of government, and attend schools patterned after those of Britain. In both world wars, the two countries immediately came to the support of Britain and lost large numbers of troops on battlefields far from home (**• Figure 8.23**). In World War I, Australia lost 60,000 of the 330,000 soldiers it sent to fight in Europe, by far the highest proportion of dead among all of the Allied countries.

Both Australia and New Zealand belong to the British Commonwealth of Nations, and the head of state for both

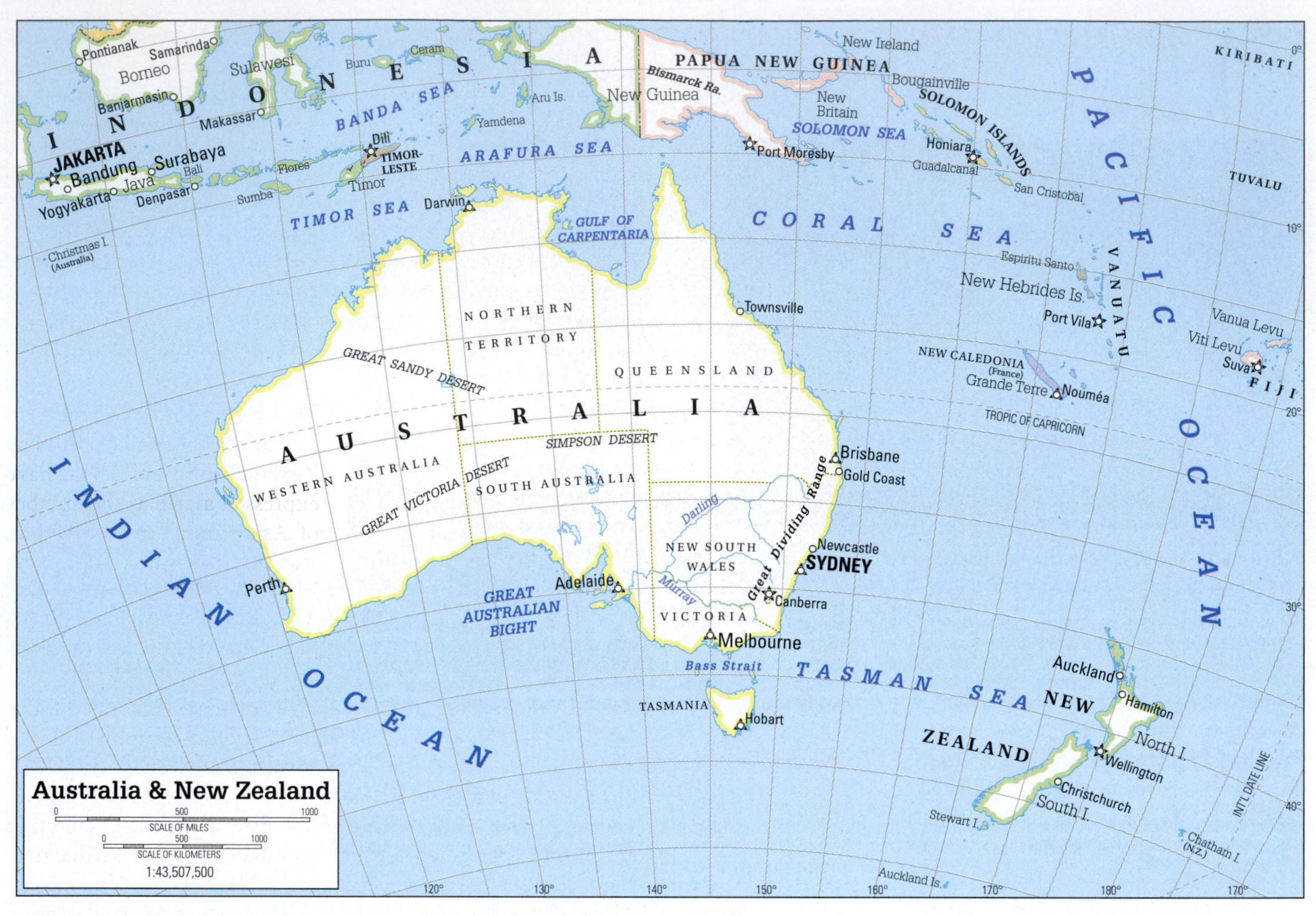

• Figure 8.22 Reference map of Australia, New Zealand, and nearby countries.

countries is the British sovereign. A barometer of Australia's more Pacific, less British orientation came in the form of a debate over whether that country should convert into a republic, which would allow Australians to have their own head of state (such as an elected president) and would sever its official ties with Britain. Although Australia would remain a Commonwealth member, the monarch would be removed from Australia's currency, and the Australian flag would be redesigned, removing the British Union Jack in the upper left corner. In a 1999 referendum on this issue, Australia voted to remain the constitutional monarchy it has been since 1901. This result was attributed to strong Australian affection for Queen Elizabeth II, and disputes among the pro-republican leaders over how a future government would be organized.

The monarchy-versus-republic debate led to something of an identity crisis for Australia and has focused more scrutiny on the sensitive issue of nonwhite immigration. Between 1945 and 1972, some 2 million people, mainly British and continental Europeans, immigrated to Australia. They were allowed in by a **"white Australia" policy** that excluded Asian and black immigrants because of fears of invasion from Asia and a desire to increase the country's population with white, skilled, English-speaking immigrants. That policy was dropped in 1973, and doors opened considerably to skilled nonwhites. About half of Australia's annual immigration quota (which changes each year; in 2015, the quota was 190,000) is Asian. Making up about 10 percent of Australia's population, Asians will form up to one-fourth by 2025, according to some projections. Australia has passed legislation requiring new immigrants to settle in cities other than Sydney (**•Figure 8.24**), where services threaten to be overwhelmed by growth. However, rich Chinese who can sink at least 5 million dollars into "qualifying investments" have a fast-track immigration status and the ability to settle where they will.

Recent years have seen a marked surge in racism and racial violence in Australia, targeted mainly at Asians and Middle Easterners (in 2010, the Indian government even put out an advisory to its citizens about travel to Melbourne). A growing tide of immigration from Indonesia, Afghanistan, Iran, Iraq, Syria, India, Pakistan, and Sri Lanka fueled a backlash prompted by fears of overpopulation, competition for jobs, and foreigners abusing Australia's generous welfare system. Opinion polls revealed that half of all Australians wanted to close the country off to immigration entirely, with less than a third believing immigrants were beneficial.

Joe Hobbs

• **Figure 8.23** The ANZAC (Australian and New Zealand Army Corps) War Memorial in Sydney, Australia, commemorates the sacrifice of thousands of young countrymen in World War I combat, including in the terrible Battle of Gallipoli at the Turkish Straits chokepoint (see **page 237**).

Prime ministerial candidates vowing to tighten restrictions against refugees and asylum seekers have fared well in elections, and in 2013 a "get tough" policy on illegal immigration took effect; after processing, intercepted migrants would either be granted temporary visas (instead of permanent residence), resettled in Papua New Guinea instead of Australia, or sent back home. The number of "boat people" (generally trafficked by Indonesian racketeers who charge a refugee as much as $7000 for the service) detained while struggling to reach Australia has been rising, from 6500 in 2010 to over 20,000 in 2013. As part of its "Pacific Solution," in addition to facilities in Nauru, Australia has opened more detention centers in Papua New Guinea and on its own Christmas Island, and has negotiated with Cambodia to permanently resettle detainees there—a policy critics decry as a "refugee-dumping deal."[13] These refugees are among a growing tide worldwide who have taken to sea on perilous journeys in search of a better life. In New 97 Zealand, some politicians have expressed alarm at the number of Asian immigrants arriving, and the anti-immigration New Zealand First political party won about 10 percent of the parliamentary seats in the 2014 election.

Ethnic tensions aside, Australia and New Zealand continue to forge important ties with their neighbors around the vast Pacific and with each other. Both countries adopted a free-trade pact in 1990 called the **Closer Economic Relations Agreement**, which eliminated almost all barriers between them to trade in farm and industrial goods and services. Both countries belong to the 21-member **Asia-Pacific Economic Cooperation (APEC)** group, which has agreed to establish "free and open" trade and investment between member states by the year 2020. Australia and the United States are the most assertive member states in arguing for the abolition of trade quotas and the reduction of tariffs imposed on imported goods. Along with New Zealand, both are members in the evolving and controversial Trans-Pacific Partnership (TPP), which would become the world's largest free-trade initiative. 312

Australia and New Zealand are seeking stronger roles in the economic growth projected for the Pacific Basin. Seven of Australia's 10 largest markets are in the Asia-Pacific region; China, Japan, the US, and South Korea are its leading trading partners (in that order). China, Australia, the US, and Japan are New Zealand's leading trade partners.

Australians refer to their Asian neighbors not as the Far East, but as the **Near North**. Australia's Northern Territory city of Darwin is well situated geographically to take advantage of new trading relations with the Near North. Darwin is closer to Jakarta, Indonesia's capital, than it is to Sydney, and Darwin promotes itself as Australia's "gateway to Asia." The city has established a trade development zone to provide manufacturing facilities and incentives for overseas companies and for Australian companies wanting to do business overseas. A deep-water port was built in the 1990s, and the transcontinental Darwin-Adelaide "Ghan" railway, which opened in 2004, is helping to fuel trade between Australia and its neighbors to

Joe Hobbs

• **Figure 8.24** Dedicated by Queen Elizabeth II in 1973 and registered as a World Heritage Site, Sydney's iconic Opera House is an architectural jewel. Its style is "expressionist modernism," involving innovative form and use of novel materials, in this case a series of concrete shells evoking sails that are covered by more than a million Swedish tiles.

the northeast. Australia's Asia-oriented perspective has already changed the nature of one of the country's export staples—beef—as ranchers have been shifting to breeds of cattle better suited for live shipment by sea to Indonesia, the Philippines, and Thailand.

Australia's Original Inhabitants Reclaim Rights to the Land

Both Australia and New Zealand have minorities of indigenous inhabitants. The native Australian Aborigines and Torres Strait Islanders have an extraordinarily detailed knowledge of the Australian landscape and its natural history, and complex belief systems about how the world came about. According to the Aborigines, the Earth and all things on it were created by the "dreams" of humanity's ancestors during the period of the **Dreamtime**. These ancestral beings sang out the names of things, literally "singing the world into existence." The Aborigines call the paths that the beings followed the **Footprints of the Ancestors** (or the **Way of the Law**; whites know them as **Songlines**). In a ritual journey called **Walkabout**, the Aboriginal boy on the verge of adulthood follows the pathway of his creator-ancestors. "The Aborigine clings to his native soil with every fiber of his being," wrote the ethnographer Carl Strehlow. "Mountains and creeks and springs and water holes are to the Aborigine not merely interesting or beautiful scenic features. They are the handiwork of ancestors from whom he himself has descended. The whole country is his living, age-old family tree."[14]

An estimated 300,000 to 1 million Aborigines and Torres Strait Islanders inhabited Australia when the first Europeans arrived in the 17th century. Colonizing whites slaughtered many of the natives and drove the majority into marginal areas of the continent. The Aborigines and Torres Strait Islanders today number about 600,000, living mainly in the tropical north of the country. Some carry on a traditional way of life in the wilderness, but most live in generally squalid conditions on the fringes of majority-white settlements.

As in the United States, newcomers to Australia forged a nation by dispossessing the ancient inhabitants of the land. And as in the United States, Australia has a long history of white racism, discrimination, and abuses against people of 507 color, who in Australia's case were the original inhabitants. Aborigines and Torres Strait Islanders were not mentioned in the 1901 constitution, not allowed to vote until 1962, and not counted in the national census until 1967. The most disadvantaged groups in Australian society, Aborigines and Torres Strait Islanders suffer from high infant mortality (four times that of the country's whites), low life expectancy (15 to 20 years fewer than that of white Australians), and high unemployment (officially 17 percent, although Aboriginal sources claim over 40 percent, versus 5 percent for whites). Like Native Americans, a disproportionately large share of Aborigine and Torres Strait Islanders fall prey to the economic and social costs of alcoholism. Statistics indicate they are 13 times more likely than non-Aborigines to end up in jail. Once in jail, they are more likely than non-Aborigines to commit suicide or be killed by guards. The Australian government is trying to improve the justice system for native Australians.

One of the largest issues of contention between Australia's indigenous people and the white majority (known to the Aborigines as "whitefellas") is land rights. European newcomers marginalized the native people, pushing them off potentially productive ranching, farming, and mining lands into special reservations on inferior lands. The Europeans used a legal 64 doctrine called ***terra nullius***, meaning "no one's land," to lay claim to the continent.

Under the 1976 Land Rights Law, aboriginal peoples were permitted to seek title to vacant state land ("Crown land") to which they could prove a historical relationship, but few succeeded. Following new legislation in 1992, the government began to return titles on a parcel-by-parcel basis—generally, in very marginal lands—to aboriginals who had argued their rights successfully. However, most aboriginals were unhappy with the terms of the returned titles, which allowed the native title to coexist with, but not supersede, the established Crown title. The government retained mineral rights to the land and could therefore lease native land to mining companies.

In 1993, after the longest Senate debate in the country's history, Australia's foreign minister pushed the **Native Title Bill** through Australia's parliament. The new legislation addressed the aboriginal peoples' major objections, giving them the right to claim land leased to mining concerns once leases expire, and allowing them to buy land to which they have proved native title. The Native Title Bill is of great consequence in the states of South Australia and Western Australia, where there is much vacant Crown land, and many Aborigines are able to file claims on it. Members of Australia's Mining Industry Council are unhappy with this legislation and worry that their mine leases will run out before the minerals do, which will require them to negotiate with the Aboriginal titleholders.

Potential Aboriginal claims to land expanded vastly in 1996 with another Australian High Court ruling on what is known as the **Wik Case**, named after the indigenous people of northern Queensland who initiated it. The court concluded that Aborigines could claim title to public lands held by farmers and ranchers under long-term "pastoral leases" granted by state governments. These lands comprise about 42 percent of Australia's territory.

What Aboriginal claims to these lands would mean was bitterly debated. White farmers and ranchers protest that they would be ruined economically if they had to share profits with Aborigines claiming land rights. Generally, however, Aborigines have insisted that their title rights would not mean running whites out of business, but would simply confer nominal recognition of their title and access to the land, especially for visitation to sacred sites. Subsequent legal clarification in Australian courts upheld the Aboriginal position, confirming that native peoples had rights to access their titled areas for traditional purposes, including to camp and perform ceremonies; visit and protect important places and sites; hunt, fish, and gather traditional resources; teach their laws and customs in situ; and even live "on" the area. White concerns about sharing profits have in most cases not been an issue.

Within the past two decades, more than 100 Aborigine and Torres Strait Islanders communities have won official land rights, typically in areas of little economic value (see Geographic Spotlight, page 394). However, there are important exceptions. On the northwest coast near Broome, the best site to build a liquefied natural gas (LNG) plant is on James Price Point, on traditional Aboriginal land. The opportunity to build the plant split Aboriginal opinions between those anxious to earn lucrative royalties from the development ($1.5 billion over 30 years) and those who wished to preserve the sacred landscape. An Aboriginal supporter of the development said, "These resources booms come along maybe every 50 to 100 years, big ones like this. So if we don't get positioned during this next stage, the chances are we're going to be locked out of the opportunities that are going to help build the economic basis for our families in the future. We can't sit back."[15] An Aboriginal opponent of the project, noting that James Price Point is the nexus of a sacred songline, where fossilized footprints on the beach are those of Emu-man, their ancestor in the dreamtime, said he was defending his "sense of country." Aboriginal supporters of the project, he said, were just "out for the money. They don't know how to connect to country anymore. They walk around with a dead feeling, these people, inside them."[16] The Australian Supreme Court took the side of the traditionalist Aboriginals in a 2013 ruling, blocking the LNG plant development.

In the Northern Territory, which has its own land rights legislation, vast tracts of land have been given back to Aborigines. Aborigines there make up 30 percent of the population and now control more than a third of the land. Aborigines also hold title to Australia's two greatest national parks, Uluru (Ayers Rock) and Kakadu, both in the Northern Territory. In a **co-management** arrangement that has become a model worldwide for management of national parks where indigenous people reside, the Aboriginal owners of the reserves have leased them back to the Australian government park system, which manages them jointly with the Aborigines.

Tourism programs in Uluru and Kakadu now highlight the cultural resources of the parks in addition to their natural wonders. At both parks, visitors can enjoy interpretive natural history walks that also emphasize Aboriginal culture and that are conducted by the land's oldest and most knowledgeable inhabitants. At Uluru, the Anangu Aborigines who own the rock hope that cultural awareness will cut down the number of tourists who climb the rock—estimated at 70 percent of the 400,000 annual total—because it is sacred to the Anangu (•**Figure 8.25**).

Joe Hobbs

• **Figure 8.25** A woman of the Anangu Aboriginal clan takes visitors on a walking tour around the base of Uluru (Ayers Rock), a site sacred to these people. In this national park program, visitors have the rare chance to learn about the meanings and uses of Aboriginal land. She is carrying a tool (not visible in this photo) that she uses to dig for grubs and other "bush tucker" (wild foods), as well as to plant crops.

Exotic Species on the Island Continent

Exotic species are nonnative plant and animal species introduced through natural or human-induced means into an ecosystem. Their impact tends to be pronounced and often catastrophic to native species, particularly on island ecosystems where resources and territories are limited. In Ecuador's Galápagos Island National Park, for example, cats, rats, pigs, goats, wasps, and other animals introduced by people have disrupted nests and food supplies and thereby reduced populations of endemic tortoises and other species.

Both inadvertently and purposely, people have introduced foreign plants and animals that have multiplied and severely impacted the country, sometimes in biblical plague proportions. English settlers imported the rabbit for sport hunting, confident in the belief that as in England, there would be natural checks on the animal's population. They were mistaken (•**Figure 8.26**). The rabbits bred like rabbits, eating everything green they could find, with ruinous effect on the vegetation. Observers described the unbelievable concentrations of the lagomorphs as "seething carpets of brown fur." In the mid-20th century, the government mounted a huge eradication effort, employing fences, snares, dogs, guns, fires, and numerous other means, and exhorting combatants with the slogan "The Rabbit War Must Be Won!" An introduced virus called "white blindness" eventually helped reduce the vast numbers, but the animal is still prolific throughout large parts of the country.

Paul Popper/Popperfoto/Getty Images

• **Figure 8.26** Rabbits were one of Australia's many human-caused scourges and are still a problem, but not nearly as much as in the 1920s, when people went on "rabbit drives" like these to chase, corral, and kill the declared vermin. Note the mounted rabbit drovers beyond the fence.

Geographic Spotlight Counter-mapping

The Aborigines' and Torres Strait Islanders' struggles for legal recognition of traditional land tenure is representative of a growing trend among indigenous peoples in many parts of the world. These cultures are increasingly enlisting the aid of geographers, anthropologists, and other social scientists to document, measure, and analyze traditional land claims. The objective is to produce stronger evidence of traditional land ownership that can be successfully presented in national and international court cases to win land rights. The methods often include **counter-mapping**, in which the accepted "truths" of colonial and other cartographies are challenged by mapping indigenous and local perceptions of place (**• Figure 8.I**). Issues such as these are discussed at international meetings like the International Forum on Indigenous Mapping. Among the contributors at these meetings are geographers fresh from the field with global positioning system (GPS) data and maps testifying to traditional land claims. Many belong to the active Indigenous Peoples Specialty Group of the Association of American Geographers.

Joe Hobbs

• Figure 8.I Australia's Aborigines say that spirits "dreamed" the world and all its features into existence. This is a rock painting of one of those creator beings, the *wandjina*, a deity of wind and storms.

Other remedies have been applied, with varying success, to correct numerous other problem species, including foxes, mice, water buffalo, cane toads, and the prickly pear cactus. Many of the problem populations, notably cats and water buffalo, are **feral animals**—domesticated animals that have abandoned their dependence on people to resume life in the wild. An estimated 12 million feral cats infest the country and are held responsible for causing the extinction or endangerment of 39 native animal species. One Australian politician promoted legislation that would kill all cats on the continent by 2020. He was not successful.

Exotic livestock such as sheep and cattle are also a problem in the Australian environment. These hard-hoofed imports cut into the soil, promoting erosion and desertification. Some range management scientists and conservationists believe that the future of Australian ranching lies with the country's 25 million kangaroos. These soft-footed marsupials are adapted to the Australian landscape and do not have such a damaging impact on the soil. Farmers and ranchers have traditionally eradicated about 2 million of them each year as pests. However, if more consumers at home and abroad could learn to appreciate kangaroo meat, there would be a strong incentive for ranchers to reduce their cattle and sheep herds and allow kangaroos to proliferate and be harvested. South Africa already imports about 20 tons of kangaroo meat yearly from Australia, and nearly 100 more tons go each year to countries including France, the Netherlands, Germany, and Russia, for total revenues of around $250 million for Australia. Kangaroo exporters are now eyeing China, which has a huge market for wild animal meats, and are waiting only for successful free trade agreements between the two countries. Kangaroo skin is also used to make athletic shoes. Australia has an annual cull (legal kill) of kangaroos deemed to be above the country's carrying capacity, and with a string of recent droughts taking an additional toll on the animals, export advocates say it makes more sense to develop a sustainable market for harvested healthy animals.

Australia is an unlikely exporter of another quadruped, the dromedary camel. Australia's nearly 1 million camels are feral descendants of those brought from South Asia and Iran early in the 20th century to provide transport across the arid continent. Australian camels are exported even to Saudi Arabia, where their wild temperaments suit them to racing. Oddly, camels are part of Australia's "carbon farming initiative," which allows the country to cull the feral camels and claim carbon credits in the global carbon emissions trading scheme to mitigate global warming. Every year a flatulent camel emits the equivalent of 99 pounds (45 kg) of methane, a greenhouse gas that is even more potent than carbon dioxide. The camel also eats about a ton of carbon dioxide–absorbing vegetation each year. On the carbon ledger, each camel's death earns an "emissions avoidance benefit" equal to about 15 tons of carbon dioxide.

From Australia, we move to the bottom of the world, to a continent twice its size: the white wilderness of Antarctica.

Antarctica: The White Continent

Antarctica is the world's fifth largest continent, with an area of 5.5 million square miles (14 million sq km), virtually filling the Antarctic Circle (**• Figure 8.27**). It is the setting of enormous human dramas in exploration, bravery, and foolhardiness as people have crossed the continent with dogsleds, on skis, on foot, and in airplanes and helicopters. Some, including all members of the British expedition led by Robert Falcon Scott, who had hoped to beat the Norwegian Roald Amundsen to

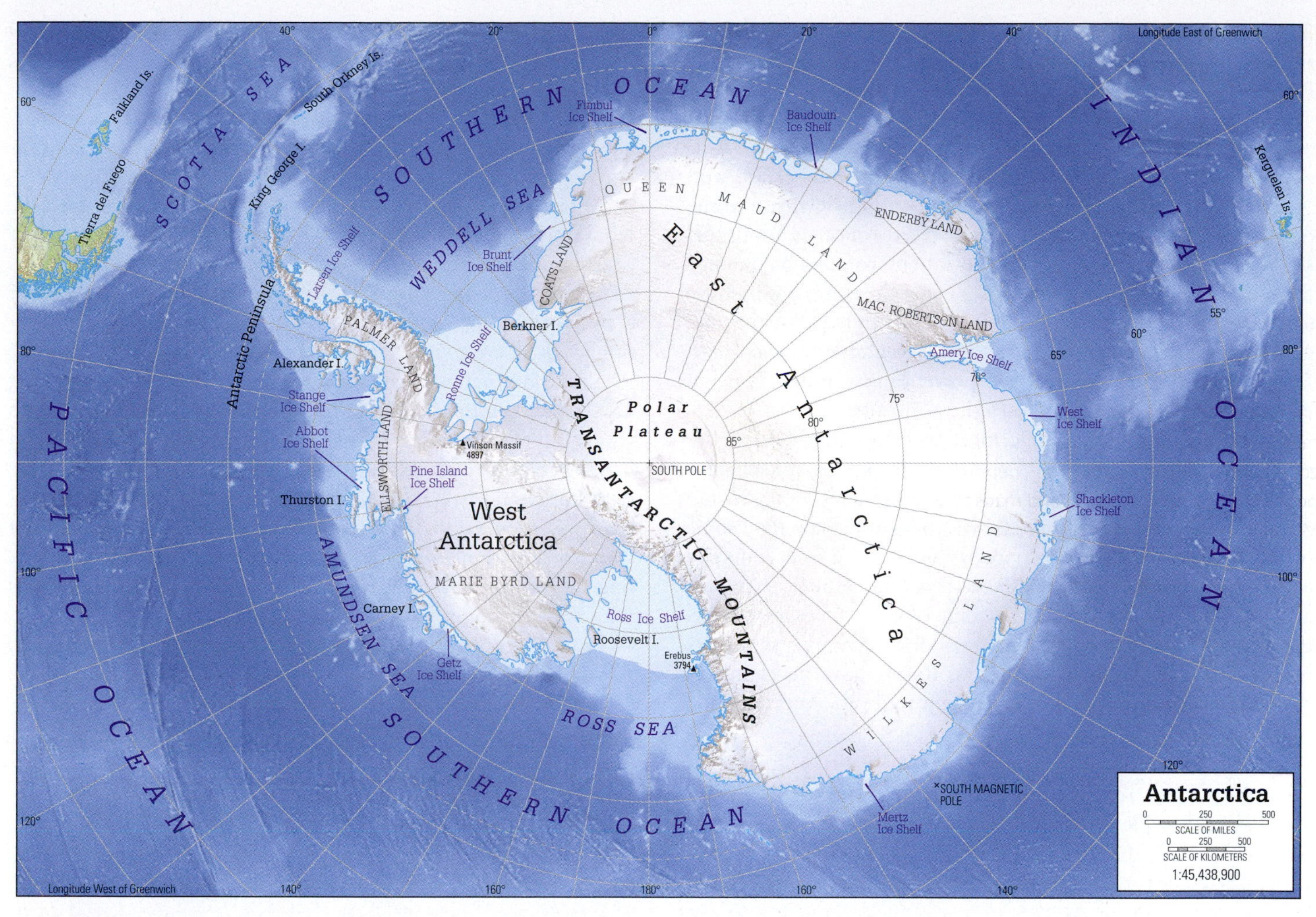

• **Figure 8.27** Reference map of Antarctica.

the South Pole in 1912 but arrived five weeks too late, did not return alive.

Antarctica's distinctiveness comes in part from its winter of darkness alternating with summer "whiteouts" caused by light refraction on snow and ice. Winter's months of darkness of the polar night are an astronomer's dream, and many scientific teams have set up observatories here. It is no surprise Antarctica is the coldest continent; in 1957, the lowest recorded temperature on Earth was recorded at the Russian Vostok research station: −128°F (−89°C). Antarctica is the world's windiest and driest continent (less than 10 inches [25 cm] of precipitation falls each year) and, during the summer, the sunniest as well—the South Pole on average receives more sunlight each year than the world's equatorial regions. It is also the continent with the highest average elevation, thanks to the glacial ice cap up to 10,000 feet (3000 m) thick that covers over 95 percent of Antarctica. About 70 percent of the world's freshwater is locked up in Antarctica's glaciers.

Antarctica has a prominent position in global concerns about climate change. The "ozone hole" in the Earth's stratosphere is concentrated seasonally over Antarctica, and scientists keep constant watch as it dwindles, thanks to the Montreal Protocol treaty to phase out CFCs. Records show that Byrd Station, an American research base in West Antarctica, has warmed by 4.3°F (2.4°C) since 1958. The Antarctic Peninsula has warmed by a similar amount, and large parts of the Larsen Ice Shelf fringing the eastern shore of the peninsula thinned rapidly and broke off in 2002. The loss of the ice shelves allows more water from melting land glaciers to reach the sea, contributing to the positive feedback of polar amplification and to rising sea levels worldwide.

In 2011, the US National Academy of Sciences published a report entitled *Future Science Opportunities in Antarctica and the Southern Ocean*. It identified key questions that should drive research in the region over the next two decades. Most of these questions focus on Antarctica's role in global climate change. Scientists see the southern polar region, like the northern polar region, as critical in understanding the global climate system and for creating models to predict future changes. Of particular concern is the Pine Island Ice Shelf, which is melting unusually rapidly. Scientists are using remote sensing to understand how relatively warm water is undercutting the ice and causing its melting. The fear is that accelerated melting will cause sea levels to rise more rapidly than anticipated.

There has never been any permanent human habitation on Antarctica. During the summer, about 5000 researchers live in the Antarctic at over 50 research stations maintained by

30 countries; the population shrinks to about 1000 during the long winter. Most research stations are on the Antarctic Peninsula or on one of the many small islands surrounding it. The US maintains the Amundsen-Scott South Pole Station, and also McMurdo Station on Ross Island, the largest of the Antarctic bases. In 2006, the US blazed a 1000-mile (1600-km) “ice highway,” the South Pole Traverse, from McMurdo to the South Pole. Some environmentalists voiced concerns about fuel spillage along the route and potential environmental damage from increased numbers of tourists using the road; others have said the road is largely environmentally sound, and using specialized land vehicles to bring supplies to and from the South Pole instead of cargo planes will prevent many tons of carbon dioxide from being released into the atmosphere.

Public, economic, and geopolitical interests in this distant part of the world are increasing. Expensive voyages by cruise ships and sightseeing flights bring about 40,000 visitors each year (up from 15,000 in 2000) to mingle with penguins and enjoy some of the wildest scenery on Earth (**•Figure 8.28**). Seven countries (Chile, Argentina, Australia, New Zealand, France, Norway, and the UK) have staked claims (some of which overlap) to portions of the continent as national territories. No other countries recognize these claims. The US and Russia have no claims in the Antarctic, but both reserve the right to do so. At present, all such claims are moot. The **Antarctic Treaty**, negotiated in 1959 and signed by 45 countries, including those laying claims, forbids any exploitation of the continent's natural resources until 2048. There has long been speculation about the potential for considerable resource wealth lying beneath the ice, but to date, no mineral deposits of significant economic value have been found.

Antarctica is the world's largest nature preserve and scientific laboratory. But even here there are pressures on resources. Some of Antarctica's waters are overfished, and illegal fishing remains a continual problem. China has announced plans to harvest over 2 million tons of Antarctic krill per year, alarming conservationists (krill are a vital part of the marine food chain). Japanese whalers have been active in the Antarctic for many years, taking 250 minke whales in 2014, although the UN has ruled Japan's whale hunts illegal. The US, EU, New Zealand, and Australia [137] have attempted to establish two marine reserves around Antarctica that would be safe from human exploitation, but the proposals were nixed by China and Russia in 2014. China did not want to constrain its rapidly growing fishing fleet, and Russia's opposition appeared to be motivated by political confrontations with Western countries after its invasion of Ukraine. [180]

From Antarctica, we move north (the only way to go) to Africa, another continent that boasts much wilderness and the continent that is the cradle of humankind.

iStockphoto.com/Mlenny

• Figure 8.28 Land, water, and ice on Antarctica's coast. The blue ice is part of a solitary iceberg that was “calved” from a nearby glacier. You are literally seeing the tip of the iceberg here. About 90 percent of an iceberg lies underwater, presenting a threat to navigation in more traveled maritime routes in the Northern Hemisphere. The underwater portion of an iceberg fatally damaged the *Titanic*.

Study Guide

Summary

- Oceania encompasses Australia, New Zealand, and the islands of the mid-Pacific lying mostly between the tropics. Tropical rain forest climates and biomes are most common, but Australia and New Zealand have several temperate climate and biome types.
- The Pacific Islands are commonly divided into three principal regions: Melanesia, Micronesia, and Polynesia.
- Oceania is ethnically complex, having been settled by peoples of various Asian origins through "island-hopping." Polynesia was the last to be populated. Papua New Guinea is the world's most linguistically diverse country. Language death threatens many tongues in Oceania. Christianity is the majority faith in this region.
- Although countless islands are scattered across the Pacific Ocean, there are three generally recognized types: continental islands, high islands, and low islands. Continental islands are either continents themselves (such as Australia) or were connected to continents when sea levels were lower (such as New Guinea). Most high islands are volcanic. Low islands are typically made of coral, a material composed of the skeletons and living bodies of small marine organisms that inhabit tropical seas.
- The island ecosystems of Oceania are typically inhabited by endemic plant and animal species—species found nowhere else in the world. Island species are especially vulnerable to the activities of humans, such as habitat destruction, deliberate hunting, and the introduction of exotic plant and animal species.
- Europeans began to visit and colonize the Pacific Islands early in the European Age of Exploration and brought mainly negative impacts to island societies. However, a steady process of decolonization has accompanied a recent surge of Western interest and investment in the region.
- There are ethnic conflicts, related mainly to maldistribution of income, between Malaitans and indigenous Guadalcanalans on the Melanesian island of Guadalcanal and between indigenous Fijians and Indo-Fijians in Fiji. Interest in securing more income from minerals has pitted the people of Bougainville against Australian corporate interests in Papua New Guinea.
- Aside from a few notable exceptions, the poverty typical of less developed countries prevails throughout most of Oceania.
- In general, the Pacific Islands' economic picture is one of nonindustrial economies. Typical economic activities include tourism, plantation agriculture, mining, and income derived from activities connected with the military needs of occupying powers. Several countries are profiting from offshore banking and telemarketing.
- Putting a value on the lives of sharks through shark tourism makes a persuasive case for ecotourism in Palau.
- During the 1940s and 1950s, the United States used the Bikini Atoll in the Marshall Islands as one of its chief testing grounds for nuclear weapons. Strong negative reaction arose throughout the region in the 1990s as the French resumed underground testing of nuclear weapons on the Mururoa Atoll in French Polynesia. That testing has since ceased. The United States relies on the region for testing of its Strategic Defense Initiative missile technology. Many inhabitants of French and American military zones are fearful of the economic repercussions of the withdrawal of foreign military presence.
- Some of the low-island countries are fearful that global warming might raise sea levels and inundate them, and Tuvalu considered legal recourse against the United States and Australia for failing to ratify the Kyoto Protocol (Australia later ratified).
- Australia and New Zealand are products of British colonization. The British heritage is strongly reflected in the ethnic composition and culture of the two nations' majority populations.
- Both Australia and New Zealand have indigenous inhabitants. Native Australians are known as Aborigines and Torres Strait Islanders, and the dominant indigenous group in New Zealand is the Māori.
- Australia has aggressive anti-immigration measures, including the "Pacific Solution" to divert would-be immigrants, especially from South Asia, to detention centers in Nauru and elsewhere.
- In Australia, one of the largest issues of contention between the Aborigines and the white majority is land rights. In 1993, the Native Title Bill addressed the Aborigines' major concerns and made several land rights concessions. New Zealand's Māori have won some concessions on resource use.
- Exotic species and human pressures have had catastrophic effects on native species and natural environments in Australia and New Zealand.
- Australia and New Zealand have been reorienting their focus toward the Pacific Rim and away from the United Kingdom. One expression of this change was the debate about whether Australia should become a republic, thus ending more than 200 years of formal ties with Britain.
- Antarctica is the world's coldest, driest, and least populated continent. There is no permanent population—only scientific researchers. The region's melting sea and land ice may reflect global warming, and the "hole" in the Earth's stratospheric ozone layer concentrates over Antarctica. Seven countries stake claims to the continent, but no other countries recognize these claims, and so far they have not had any economic significance.

Review Questions

1. What three principal subregions comprise Oceania? What are some of the countries and colonial possessions of each?
2. Describe the major differences between high and low islands. According to the geographer Tom McKnight, how did the physical characteristics of the islands relate to social and political developments? What are the typical demographic and economic qualities of each?
3. The small islands of Oceania appear to be idyllic but have some resource limitations on human habitation. What are these?
4. What is a geologic hot spot, and how does it relate to the Pacific Islands?
5. What unique ecological attributes do islands generally have? What are the major threats to them? What impacts do exotic species have upon endemic and other species in Oceania?
6. What is an atoll, and how does it form?
7. What are the major ethnic groups, languages, and religions of Oceania?
8. What are two notable ethnic conflicts on mineral-rich islands in Oceania?
9. How was Nauru able for a time to avoid the poverty typical of most Pacific Islands? How has Nauru's economy been diversifying now that its principal asset is dwindling?
10. What is Australia's "Pacific Solution" to its immigration problems?
11. What are the major economic activities in the Pacific Islands, excluding Australia and New Zealand? Why and where have military interests been prominent?
12. Why has Australia been dubbed "the Lucky Country"?
13. How is the US positioning itself geostrategically in the region to anticipate China's moves?
14. What explains the peculiar jogs in the International Date Line?
15. What fears do some of Oceania's countries have about global warming?
16. What features do Australia and New Zealand have in common? How are their identities changing?
17. What is the significance of the Dreamtime to Australia's Aborigines? What issues of land ownership concern them and the Torres Strait Islanders?
18. What advantages might kangaroo husbandry have over that of conventional livestock?
19. What are the major physical characteristics of Antarctica? What indications are there of climate change? What is the continent's political status?

Key Terms + Concepts

Aboriginal languages (p. 379)
Aborigines (p. 379)
Antarctic Treaty (p. 396)
Asia-Pacific Economic Cooperation (APEC) group (p. 391)
atoll (p. 374)
Australia New Zealand United States (ANZUS) security alliance (p. 387)
Austronesians (p. 379)
cargo cults (p. 382)
Closer Economic Relations Agreement (p. 391)
"coconut civilization" (p. 375)
co-management (p. 393)
Compact of Free Association (p. 389)
continental islands (p. 372)
counter-mapping (p. 394)
deterritorialized states (p. 387)
Dreamtime (p. 392)
environmental determinism (p. 374)
exotic species (p. 393)
"extinction capital of the world"(p. 378)
feral animals (p. 394)
Footprints of the Ancestors (p. 392)
guano (p. 385)
high islands (p. 372)
Indo-Fijians (p. 384)
language death (p. 380)
low islands (p. 372)
Melanesia (p. 367)
Micronesia (p. 367)
millenarian movements (p. 382)
money laundering (p. 385)
Native Title Bill (p. 392)
Near North (p. 391)
Oceania (p. 366)
Operation Crossroads (p. 388)
Pacific Solution (p. 385)
Papuan languages (p. 379)
pidgin (p. 380)
Polynesia (p. 367)
power projection hub (p. 386)
Proto-Austronesian (p. 379)
seamount (p. 374)
shell banks (p. 385)
Songlines (p. 392)
terra nullius (p. 392)
Torres Strait Islanders (p. 379)
Walkabout (p. 392)
Way of the Law (p. 392)
"white Australia" policy (p. 390)
Wik Case (p. 392)

Notes

1. Ray P. Norri and Bill Yidumduma Harney. 2014. "Songlines and Navigation in Wardaman and other Aboriginal Cultures," http://www.atnf.csiro.au/people/Ray.Norris/papers/n315.pdf.
2. Tom L. McKnight, *Oceania: The Geography of Australia, New Zealand and the Pacific Islands* (Englewood Cliffs, N.J.: Prentice Hall, 1995), p. 181.
3. Ibid.
4. "Interview—Violent corporate land grabbing in Papua New Guinea," December 1, 2013, http://farmlandgrab.org/post/view/22866-interview-violent-corporate-land-grabbing-in-papua-new-guinea#sthash.Vr6ry2Wx.dpuf.
5. All quotes and information to this point in the essay are from Jared Diamond, "Easter's End," *Discover, 16*(8), 1995, p. 64.
6. The newer information is related in Terry Hunt and Carl Lipo, *The Statues That Walked: Unraveling the Mystery of Easter Island* (Berkeley, CA: Counterpoint Press, 2012).
7. Jeff Siegel, "Tok Pisin," http://www.une.edu.au/langnet/definitions/tokpisin.html.
8. The ad's first two lines mean "Talk talk plenty—No more worry."
9. An excellent article on Tanna's John Frum Cargo Cult is Paul Raffaele's "In John They Trust," *Smithsonian*, February 2006.

10. David Jolly, "Priced Off the Menu? Palau's Sharks Are Worth $1.9 Million Dollars Each, Study Says," *New York Times*, May 2, 2011, p. A12.
11. Quoted in David Pilling, "How America Should Adjust to the Pacific Century," *Financial Times*, November 17, 2011, p. 9.
12. Nicholas Kristof, "An Atomic Age Eden (But Don't Eat the Coconuts)," *New York Times*, March 5, 1997, p. A4.
13. "A Cambodian Solution," *The Economist*, August 20, 2014, http://www.economist.com/blogs/banyan/2014/08/australia-s-asylum-seekers.
14. Quoted in Geoffrey Blainey, *Triumph of the Nomads* (Melbourne: Sun Books, 1987), p. 181.
15. Norimitsu Onishi, "Rich in Land, Aborigines Split on How to Use It," *New York Times*, February 13, 2011, p. 5.
16. Ibid.

Global Geoscience Watch

Once again, click on the "World Map" menu item found on the GREENR database homepage. This time use the "zoom" feature of the interactive Google Map to research more about at least two small islands in Micronesia. First use the hand tool to center the map over Micronesia, an area that includes thousands of small islands in the central and western Pacific, mostly north of the Equator. Use the "zoom" feature by clicking on the "+" button in the lower right until you see the names of islands that were previously not visible on the map. Write down the name of two islands to explore in more detail. Using the research skills you have been developing in other exercises, write a paragraph about each island. Be sure to indicate which of the five sovereign nations in Micronesia the island is part of, or is it one of the three US territories? What factors drive the economy of the island? Are there special problems that the inhabitants of the island are concerned about?

Online Resources

For access to MindTap and additional study materials visit www.cengagebrain.com. Read your textbook, take notes, complete activities, take practice quizzes and more.

Above: A majestic "tusker," or bull elephant, in Mana Pools National Park, Zimbabwe. This animal carries about 250 lbs (113 kg) of ivory, worth at least $1500 on the black market. Joe Hobbs
Left: Sub-Saharan Africa at night. This is the only world region where many communities are dark because they lack electricity; the true distribution of people may be seen in Figure 9.4. C. Mayhew & R. Simmon (NASA/GSFC), NOAA/NGDC, DMSP Digital Archive

9 Sub-Saharan Africa

I dream of the realization of the unity of Africa, whereby its leaders combine in their efforts to solve the problems of this continent. I dream of our vast deserts, of our forests, of all our great wildernesses.[1]

—NELSON MANDELA

For many people around the world, Sub-Saharan Africa is the most poorly understood or misunderstood of the world's regions. It seems to have so many countries, so much violence and disease, and so much poverty that it is simply too difficult or depressing to think about. Americans in general and our decision makers in particular have a lot to learn about Africa. Geographer Richard Grant writes, "Negative, outdated, and misplaced representations of Africa mean that many governments, for example, the United States, do not have clear and relevant Africa policies or a geographically informed citizenry that can debate and demand Africa policies."[2] Students from a variety of African countries studying at New York State's Ithaca College decided to help their American peers with a social media initiative called "The Real Africa: Fight the Stereotype."[3] They were stirred to act by their classmates' questions, including "Do you speak African?" and "What is Africa's flag?"

Chapter Outline

Chapter Objectives

This chapter will enable you to

- Understand what made Sub-Saharan Africa the world's poorest region, and how it is now finally breaking out of poverty.
- Know how the region came to have the world's greatest HIV/AIDS and Ebola outbreaks, and how these epidemics can be stopped.
- Consider the pressures on African wildlife and the unique approaches taken to protect the animals.
- Appreciate the diversity and richness of African cultures, and the role of ethnicity in the region's conflicts.
- See how Africans have leapfrogged communications challenges with mobile phones and are developing new IT hubs.
- Recognize why, after decades at the sidelines, Sub-Saharan Africa is considered important again in geopolitical affairs.
- Appreciate what corrupt leadership has done to impoverish people in countries with enormous oil and other natural resource wealth, and how good governance is taking its place.

Hollywood still perpetuates stereotyped images of the "Dark Continent," a self-contained, tribalized land of mystery. To the contrary, writes Dayo Olopade, Africa is the "Bright Continent," home to some of the world's fastest-growing countries and a growing middle class.[4]

This chapter will show that getting to know this bright continent is manageable, rewarding, and interesting. This region's diverse cultures and natural environments are sources of endless wonder. There are many reasons for hope and much to celebrate in the study of this fascinating region. Just ask the people: a Gallup survey of 50,000 citizens around the world found that Africans are the most optimistic.[5]

Geographers typically recognize the continent of Africa as having two major divisions: North Africa (the predominantly Arab and Berber realm of the continent, described in Chapter 6), and ethnically diverse Sub-Saharan Africa (**•Figures 9.1** and **9.2**; **•Table 1**). In his authoritative *Africa: Geographies of Change,* a book cited frequently in this chapter, Richard Grant argues that "Sub-Saharan Africa" is a Eurocentric term.[6] "Africa South of the Sahara" is less laden but is cumbersome to use, so this chapter often adopts Grant's solution of using "Africa" as shorthand for the most commonly used regional name, "Sub-Saharan Africa."

9.1 Area and Population

Sub-Saharan Africa has the second largest land area of all the major world regions described in this book. It covers 8.7 million square miles (22.4 million sq km) and so is more than twice the size of the United States (**•Figure 9.3**). Even with the terrible loss of lives due to AIDS, the rate of natural population increase in Sub-Saharan Africa is 2.6 percent per year, or about five times that of the United States. Africa's population of 920 million is the most rapidly growing in the world; half of all the persons born in the world between now and 2050 will be Sub-Saharan Africans. The lion's share of Africa's doubling of population by 2050 will be in a handful of countries: Nigeria will grow from about 177 million to 396 million then, when it will be the third most populous county in the world.[7]

The wild card in Africa's population deck is the **human immunodeficiency virus (HIV)**. The disease it causes—**AIDS (acquired immunodeficiency syndrome)**—has inevitably been identified by some as the Malthusian "check" to the region's population growth (see Geography of Disease, **page 408**).

The world's largest youth bulge, people aged 15–35, will be here, even more than in the Middle East, with all the political and people power that represents. The median age is now 20 (compared with 30 in Asia and 40 in Europe). There are different ways of viewing this young, growing population. Some see the youthful median age as a "demographic dividend," as large numbers of workers will be able to contribute to the region's economic growth. A similar demographic dividend boosted economic growth in Asia a generation ago.[9] As more people work, more money will circulate, and economies will grow. But population growth also means more people for whom to provide housing, educational and medical services, and food. Some observers fear that the Malthusian scenario will come about because of the gulf between population growth and agricultural output in these mainly agrarian societies. Since the 1960s, the population of Sub-Saharan Africa has grown at a rate of between 2 and 3 percent annually. In the 1980s, there were fears of widespread starvation across the continent as population growth accelerated faster than food production. As the severe African droughts ended by 1990, however, food production increased; by 2010, the region's population was 3.7 times larger than it had been in 1960, but food production was 4.3 times higher than in 1960.

A survey of more than 50,000 people in 34 countries revealed, however, that one in five Africans lacks food, health care, and clean water. Three-quarters of those surveyed believe their governments are not doing enough to reduce income

• **Figure 9.1** Physical geography of Sub-Saharan Africa.

inequalities.[10] Africa has the largest number and concentration of the world's fragile or failed states, as Figure 3.9 on **page 59** reveals. Africans are taking to the streets in increasing numbers to express their grievances, especially about the lack of jobs. However, one reason why more people are protesting is that governments are more democratic and open to dissenting views; on the whole, better governance is gaining ground across the region.

Despite the overall large population, much of the region is sparsely populated (**•Figure 9.4**). For now, the majority of people live in the countryside. Sub-Saharan Africa is the world's most rural region: only 37 percent of Africans live in urban areas. The lowest urban rates are found in Malawi, Burundi, Uganda, and South Sudan, all having urban populations between 16 and 18 percent. Typically of rural populations in the LDCs, there is widespread resistance to family planning. Many African parents want large families for several reasons: to have extra hands to perform work; to be looked after when they are old or sick; and in the case of girls, to receive the "bride wealth" a groom pays in a marriage settlement. Large families also convey status. However, there are signs of significant change in this pattern of population growth: although still high, birth rates have been dropping in every country in this region over the past two decades. We can anticipate that

• Figure 9.2 Political geography of Sub-Saharan Africa.

birth rates will drop, and population growth rates will stabilize as the region urbanizes. But life in villages is the rule for the majority. A typical rural home is a small hut made of sticks and mud, with a dirt floor, thatched roof, and no electricity or plumbing (**•Figure 9.5**). This is the only world region where large populations cannot be seen on the nighttime image from space that opens the chapter. However, the rate of urban growth is high: it is projected that over half of all Africans will live in cities before 2050 (see Geography of Cities, **page 434**). In 1990, there were 15 cities with populations over 1 million in Africa; by 2015, there were over 40 (see **•Table 9.2** for the largest cities in the region).

9.2 Physical Geography and Human Adaptations

Sub-Saharan Africa is both rich in natural resources and beset with environmental challenges to economic development. It is home to some of the world's greatest concentrations of wildlife and some of the most degraded habitats.

The Landscapes of Africa

Most of Africa is a vast series of plateaus with a typical elevation of more than 1000 feet (300 m). Near the Great Rift

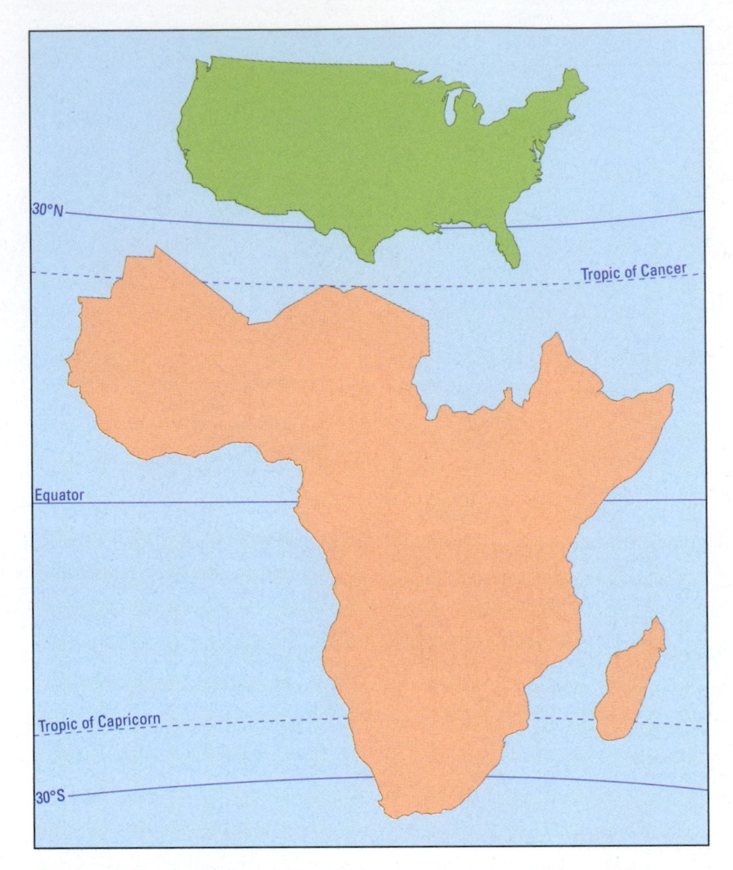

• **Figure 9.3** Sub-Saharan Africa compared in area and latitude with the conterminous United States.

• **Figure 9.5** Homes in African villages are typically made of local materials, including thatch, which, if replaced regularly, do a fine job of keeping rain out. Homes are usually elevated slightly to stave off the threat of flooding. Village roads are seldom paved. People live in close proximity with their livestock, in part to protect the animals from predators. This village is in northwestern Madagascar.

Valley in the Horn of Africa and in southern and eastern Africa, the general elevation rises 2000 to 3000 feet (600 to 900 m), with many areas at 5000 feet (1520 m) and higher (see Figure 9.1 and Regional Perspective). The highest peaks and largest lakes of the continent are located in this belt. The loftiest summits lie within 250 miles (400 km) of Lake Victoria. They include Kilimanjaro (19,340 ft/5895 m) and Kirinyaga (Mount Kenya, 17,058 ft/5200 m; •**Figure 9.6**), which are glaciated volcanic cones, and the Ruwenzori range (up to 16,763 ft/5109 m), a nonvolcanic massif produced by faulting and sometimes identified as the "Mountains of the Moon," described by Ptolemy as rising near the source of the Nile. Lake Victoria, the largest lake in Africa, is the third-largest lake in the world; it is relatively shallow, however, and the large numbers of people living on its shores are taxing its resources.

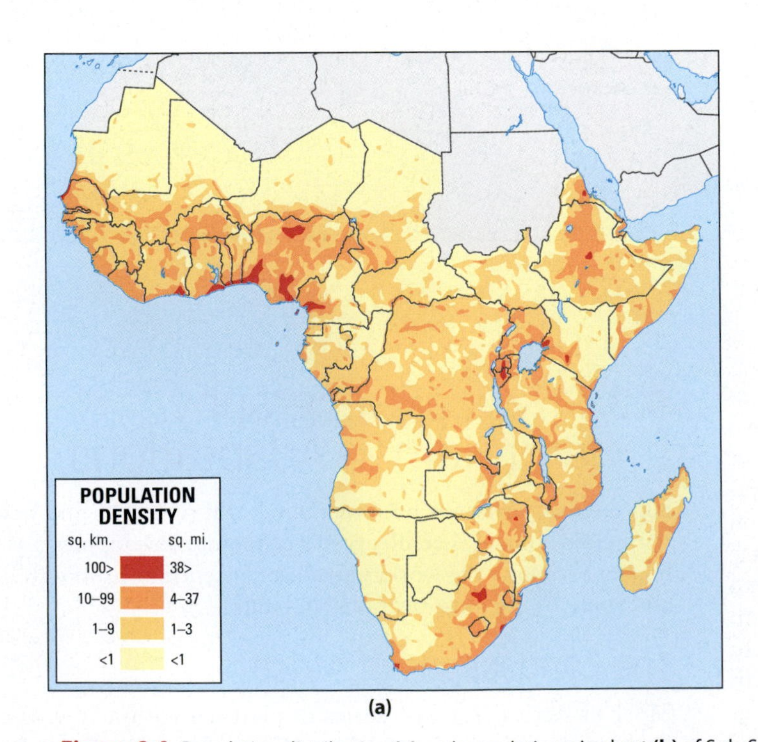

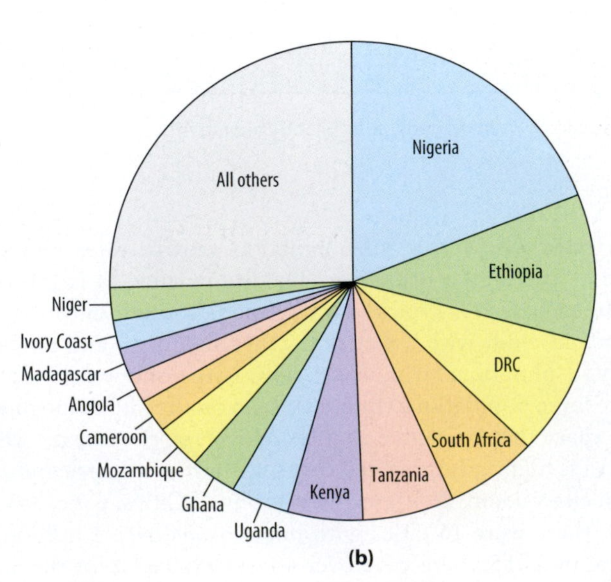

• **Figure 9.4** Population distribution **(a)** and population pie chart **(b)** of Sub-Saharan Africa.

Table 9.1 Sub-Saharan Africa: Basic Data

Political Unit	Area (sq mi, thousands)	Area (sq km, thousands)	Population (millions)	Rate of Natural Increase (%)	Urban Population (%)	Population Under Age 15 (%)	Agricultural Workers (%)	Per Capita GDP (PPP) ($US)	GDP ($US, billions)	Trade Balance ($US, billions)	Oil Production (million bbl/day)	Literacy, Female (%)	Literacy, Male (%)	HDI
Sub-Saharan Africa	**8655.2**	**22417.2**	**919.5**	**2.6**	**37**	**43**	**61**	**2,600**	**3,275**	**—**	**5.7**	**52**	**69**	**0.462**
The Sahel														
Burkina Faso	105.8	274.0	17.9	3.1	27	43	90	1,500	30	−1	—	21	36	0.388
Cape Verde	1.6	4.1	0.5	1.7	62	30	—	4,400	3	0	—	80	89	0.636
Chad	495.8	1284.1	13.3	3.3	22	49	80	2,500	29	1	0.1	25	45	0.372
Gambia	4.4	11.4	1.9	3.1	57	46	75	2,000	3	0	—	40	60	0.441
Mali	478.8	1240.1	15.9	2.9	35	48	80	1,100	27	0	—	20	36	0.407
Mauritania	396	1025.6	4	2.6	41	40	50	2,200	12	−1	—	51	65	0.487
Niger	489.2	1267.0	18.2	3.9	22	50	90	800	17	−1	*	15	43	0.337
Senegal	76	196.8	13.9	3.2	47	44	77	2,100	33	−3	—	29	51	0.485
South Sudan	248.7	644.3	11.7	2.4	17	42	—	1,400	23	—	0.2	16	40	—
West Africa														
Benin	43.5	112.7	10.3	2.7	45	43	—	1,600	19	−1	—	30	55	0.476
Ghana	92.1	238.5	27	2.5	51	38	56	3,500	109	−2	0.1	65	78	0.573
Guinea	94.9	245.8	11.6	2.7	36	42	76	1,100	15	0	—	30	52	0.392
Guinea-Bissau	13.9	36.0	1.7	2.5	44	41	82	1,200	2	0	—	42	69	0.396
Ivory Coast	124.5	322.5	20.8	2.3	53	41	68	1,800	71	5	*	46	65	0.452
Liberia	37.2	96.3	4.4	2.6	47	43	70	700	3	1	—	56	64	0.412
Nigeria	356.7	923.9	177.5	2.5	50	44	70	2,800	1,058	40	2.3	50	72	0.504
Sierra Leone	27.7	71.7	6.3	2.1	41	44	—	1,400	12	0	—	24	47	0.374
Togo	21.9	56.7	7	2.6	38	42	65	1,100	10	−1	—	47	75	0.473
West Central Africa														
Cameroon	183.6	475.5	22.8	2.7	52	43	70	2,400	67	0	*	64	78	0.504
Central African Republic	240.5	622.9	4.8	3.2	39	40	—	700	2	0	—	44	69	0.341
Congo Republic	132	341.9	4.6	2.8	64	42	—	4,800	28	4	0.3	78	89	0.564
Congo, Democratic Republic of	905.4	2345.0	71.2	3.0	34	46	—	400	55	0	*	57	77	0.338
Equatorial Guinea	10.8	28.0	0.8	2.2	39	39	—	25,700	25	7	0.3	91	97	0.556
Gabon	103.3	267.5	1.7	2.3	86	38	60	19,200	34	3	0.2	85	92	0.674
São Tomé and Principe	0.4	1.0	0.2	2.9	67	42	—	2,200	1	0		78	92	0.558
East Africa														
Burundi	10.7	27.7	10.5	3.2	16	45	93	600	8	0	—	61	73	0.389
Kenya	224.1	580.4	43.2	2.6	24	42	75	1,800	134	−10	—	84	90	0.535
Rwanda	10.2	26.4	11.1	2.3	17	41	90	1,500	18	−2	—	67	74	0.506
Tanzania	364.9	945.1	50.8	3.1	30	45	80	1,700	92	−6	—	62	77	0.488
Uganda	93.1	241.1	38.8	3.4	18	48	82	1,500	66	−2	—	57	77	0.484

Table 9.1 Sub-Saharan Africa: Basic Data (*continued*)

Political Unit	Area (sq mi, thousands)	Area (sq km, thousands)	Population (millions)	Rate of Natural Increase (%)	Urban Population (%)	Population Under Age 15 (%)	Agricultural Workers (%)	Per Capita GDP (PPP) ($US)	GDP ($US, billions)	Trade Balance ($US, billions)	Oil Production (million bbl/day)	Literacy, Female (%)	Literacy, Male (%)	HDI
Horn of Africa														
Djibouti	9	23.3	0.9	1.9	77	34	—	2,700	2	0	—	—	—	0.467
Eritrea	45.4	117.6	6.5	2.6	21	43	80	1,200	8	−1	—	—	—	0.381
Ethiopia	426.4	1104.4	95.9	2.1	17	43	85	1,300	139	−8	—	29	49	0.435
Somalia	246.2	637.7	10.8	3.2	17	48	71	600	5	−1	—	—	—	—
Southern Africa														
Angola	481.4	1246.8	22.4	3.2	59	48	85	6,300	175	41	1.9	58	82	0.526
Botswana	224.6	581.7	2	0.7	62	34	—	16,000	33	0	—	85	85	0.683
Lesotho	11.7	30.3	1.9	0.9	26	36	86	2,200	5	−1	—	95	83	0.486
Malawi	45.7	118.4	16.8	2.9	16	45	90	900	13	−1	—	68	81	0.414
Mozambique	309.5	801.6	25.1	2.9	31	45	81	1,200	29	−4	—	42	70	0.393
Namibia	318.3	824.4	2.3	2.2	38	36	16	10,800	23	−2	—	89	89	0.624
South Africa	471.4	1220.9	53.7	1.0	62	29	9	12,700	683	−4	0.1	92	94	0.658
Swaziland	6.7	17.4	1.3	1.6	21	38	70	5,700	8	0	—	80	82	0.530
Zambia	290.6	752.7	15.1	3.4	40	47	85	1,800	61	1	—	74	86	0.561
Zimbabwe	150.9	390.8	14.7	2.4	33	40	66	600	26	−2	—	87	94	0.492
Indian Ocean Islands														
Comoros	0.9	2.3	0.7	2.5	28	42	80	1,300	1	0	—	70	80	0.488
Madagascar	226.7	587.2	22.4	2.7	33	42	28	1,000	33	−2	—	61	67	0.498
Mauritius	0.8	2.1	1.3	0.3	42	21	9	16,100	23	−2	—	87	92	0.771
Mayotte (Fr.)	0.1	0.3	0.2	2.8	50	45	—	—	—	—	—	—	—	—
Réunion (Fr.)	1	2.6	0.9	1.2	94	24	13	—	—	—	—	—	—	—
Seychelles	0.2	0.5	0.1	1.1	54	22	3	25,900	2	0	—	92	91	0.756

* Less than 0.1. — Data not available or not applicable.
Sources: World Population Data Sheet, Population Reference Bureau, 2014; Human Development Report, United Nations, 2014; World Factbook, CIA, 2014.

The physical structure of Africa has influenced the character of its rivers. The main rivers of Africa rise in interior uplands and descend by stages to the sea. At some points, they descend abruptly, particularly at plateau escarpments, with rapids and waterfalls interrupting their courses. Although these often block navigation a short distance inland, they are also a great potential source of hydroelectric energy. There are major power stations on the Zambezi, Congo, Niger, Volta, and Blue Nile Rivers. But only about 10 percent of Africa's hydropower potential has been realized (compared to about 60 percent in North America). Many of the best sites are remote from large markets for power. In some cases, geopolitical considerations pose obstacles to dam construction. Downstream Egypt, for example, has expressed concern and even hostile rhetoric about dams and water diversions of the Nile and its tributaries by upstream Sudan, Ethiopia, Uganda, Kenya, and Tanzania (see **page 240** in Chapter 6).

Africa's Biomes and Climates

The Equator bisects Africa, so about two-thirds of the region lies in the low latitudes, having tropical climates and biomes; Africa is the most tropical of the world's continents (**•Figure 9.7**). Areas of tropical rain forest and tropical monsoon climates center on the great rain forest of the Congo Basin and extend along the shore of West Africa. The vast tropical grasslands and savanna that sprawl across much of Africa support most of the magnificent large mammals the continent is known for. South Africa has the most temperate climate on the continent, with Mediterranean-style conditions in the west and a more humid oceanic climate in the uplands of the east. Madagascar is highly varied climatically, with tropical rainforests in the northeast transitioning to deserts in the southwest.

Much of Africa is dry and drought-prone, and steppe and desert conditions are found in the Sahara, in Ethiopia and

Table 9.2 Largest Cities in Sub-Saharan Africa

City	Population
Lagos, Nigeria	13.1
Kinshasa, D.R.C.	11.5
Johannesburg, South Africa	8.4
Luanda, Angola	5.9
Abidjan, Ivory Coast	4.8
Nairobi, Kenya	4.7
Dar es Salaam, Tanzania	4.2
Accra, Ghana	4.1
Cape Town, South Africa	3.8
Kano, Nigeria	3.5
Dakar, Senegal	3.5
Durban, South Africa	3.4
Addis Ababa, Ethiopia	3.3
Ibadan, Nigeria	3.1
Younde, Cameroon	3
Douala, Cameroon	2.9
Pretoria, South Africa	2.9
Ouagadougou, Burkina Faso	2.7
Maputo, Mozambique	2.6
Bamako, Mali	2.5

Population in millions. Source: Demographia World Urban Areas, 2015.

Somalia, and across much of southwestern Africa. Total precipitation in the region is unevenly distributed; a few areas are typically saturated, whereas others are perennially bone dry. Even in many of the rainier parts of the continent, there is a long dry season, and wide fluctuations occur from year to year fluctuations in precipitation. One of the major needs in Africa is better control over water. In the typical village household, women carry water from a stream or lake or a shallow (and often polluted) well. Use of more small dams would help provide water storage throughout the year.

The Sahel, a broad belt of tropical steppe and savanna bordering the Sahara on the south, is especially vulnerable to **drought**, which is a naturally occurring climatic event in which rain fails to fall over an area for an extended period, often years. However, the Sahelian ecosystems have high **resilience**, meaning they are able to recover from the stress of drought and have mechanisms to cope with a natural cycle that includes periods of dryness and rain. When the rains do return, the arid and semiarid ecosystems of the Sahel come to life with a profusion of flowering plants, insects, and herbivores like gazelles, whose populations climb when foods are abundant. **Desertification** is the destruction of that resilience and the biological potential of arid and semiarid ecosystems. It is an unnatural, human-induced condition brought about by too many cattle and by faulty agricultural practices. Desertification has afflicted the Sahel periodically since the late 1960s, but people are taking steps to prevent it. In drought-prone Niger, for example, farmers have ceased the traditional practice of clearing trees before planting crops and are now protecting trees and growing crops under and between them. A key factor in the farmers' new concern for trees is that Niger's government released its long-held claim that all trees belonged to the state; now farmers recognize the trees as private property. New attitudes and practices like these, and a sustained period of good rains, have combined to produce a remarkable re-greening of much of the Sahel since 1985. That trend could change again, of course.

Mark C. Ross/National Geographic Stock

(a)

Robert Mackinlay/Photolibrary/Getty Images

(b)

• **Figure 9.6** Africa's highest mountains, both volcanic, are **(a)** Tanzania's Kilimanjaro, photographed from Kenya's Amboseli National Park, and **(b)** Kenya's Kirinyaga, also known as Mt. Kenya.

Geography of Disease HIV/AIDS and Ebola in Africa

AIDS is a global problem. The Earth has not seen a pandemic (very widespread epidemic) like this since the bubonic plague devastated 14th-century Europe and smallpox struck the Aztecs of 16th-century Mexico. AIDS had killed 39 million people worldwide by 2014, and there are 2 million new infections around the world each year. Sub-Saharan Africa is where the virus originated (apparently among chimpanzees, eaten as **bushmeat** and crossing the primate divide by some mutation); the first human case was reported in what is now the Democratic Republic of the Congo in 1959, and Africa today is the epicenter of this health crisis. The regional statistics are startling. In 2015, 70 percent of the world's estimated 37 million people infected with HIV lived in Sub-Saharan Africa.

The epidemic is most severe in southern Africa. In some countries, including Botswana and Lesotho, as much as one-fifth of the adult population is HIV positive (**•Figure 9.A**); in Swaziland, the rate is 27 percent. South Africa has more HIV cases (about 6 million) than any other country in the world. This scourge has been causing sharp reductions in life expectancy in Africa and, if not contained, would dramatically alter the region's population and development trends.

The good news is that HIV/AIDS is starting to be contained, and the HIV/AIDS scientific community is seeking total eradication of the virus and disease by 2030. The rate of new infections is down about 33 percent in the worst affected countries, mainly in southern Africa, since 2005. Africa saw fewer new HIV infections in 2011 than it did in 1991, when the disease was still ascendant, and the infection rate continued to fall from there. This is brightening the prospects for development in this least developed of the world's regions. The incidence of HIV infection is particularly high among the region's most educated, skilled, and ambitious young urban professionals, including teachers, white-collar workers, and government employees. These are the people, in addition to truckers and merchants, who travel the most, have higher incomes, and are therefore the most likely to have many sexual liaisons.

What is being done to fight the AIDS epidemic? African and other countries have enlisted a number of strategies to reduce their infection rates. One of the most effective has been ensuring that HIV-positive pregnant women receive **antiretroviral (ARV) drugs** to prevent passing the disease on to their infants. In Namibia, for example, ARV drugs have led to more than a 50 percent decline in the numbers of infected children since 2001. Namibia has also distributed ARV drugs to over 80 percent of its infected population,

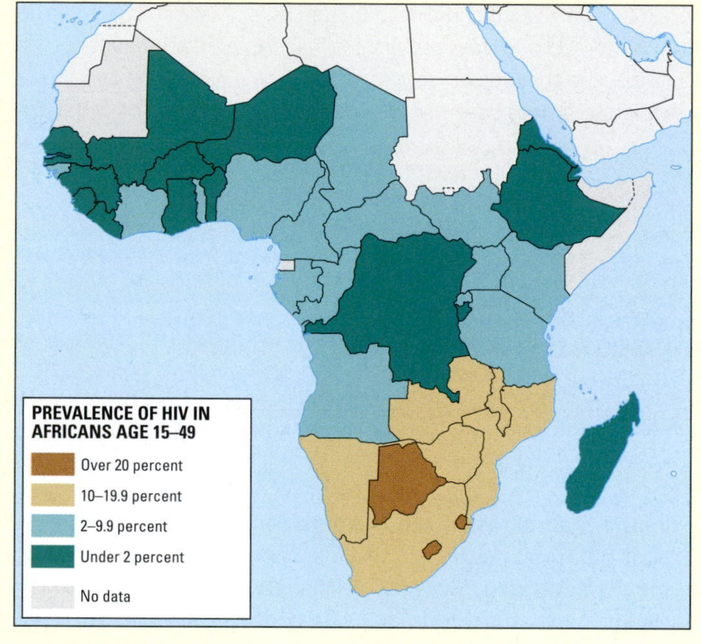

• Figure 9.A HIV/AIDS has had a devastating impact in Sub-Saharan Africa and is redrawing the demographic profiles of some countries in extraordinary ways.

The patterns of Africa's land use (**•Figure 9.8**) reveal that the most productive lands are on river plains, in volcanic regions (especially the East African and Ethiopian highlands), and in some grassland areas of tropical steppes (notably the Highveld in South Africa). Many soils of the deserts and regions of Mediterranean climate are poor. In the tropical rain forests and savannas, there are reddish, lateritic tropical soils that are infertile once the natural vegetation is removed and can support only shifting cultivation. Agriculture is discussed in more detail beginning on **page 419**.

486

Africa's Wildlife

Although cattle raising is widespread throughout the savannas, cattle are largely excluded from extensive sections both north and south of the Equator by the disease called **nagana**, which is carried by the tsetse flies that also transmit sleeping sickness to humans (see Insights, **page 411**). In tropical rain forests, tsetse flies are even more prevalent, and few cattle are raised, but goats and poultry are common (as they are in tsetse-frequented savanna areas).

Africa has the planet's most spectacular and numerous populations of large mammals. The tropical grasslands and open forests of Africa are the habitats of large herbivorous animals, including the elephant, buffalo, zebra, giraffe, and many species of antelope, as well as carnivorous and scavenging animals, such as the lion, leopard, and hyena. The tropical rain forests have fewer of these "game" animals (as Africans call them); the most abundant species here are insects, birds, and monkeys, with the hippopotamus, crocodiles, and a great variety of fish in the streams and rivers draining the forests and wetter savannas.

Although documentary films promote a perception outside Africa that the continent is a vast animal Eden, the reality is less positive. Human population growth, urbanization, and significantly lessening the death rate from HIV. About 15 million people, or 40 percent of the number of people worldwide living with HIV, were on these drugs in 2015. Scientific trials have demonstrated that early treatment with ARV drugs can prevent AIDS-related illness and death; testing for the HIV virus is therefore critical. Increased education (such as "HIV life skills" courses taught in schools), widespread testing and early detection, male circumcision, distribution of condoms and a vaginal microbicidal gel, and the use of mobile clinics to reach poor, isolated areas have also been effective at slowing the diffusion of HIV. Needle-exchange programs are most important in preventing transmission among drug users.

Wealthy countries and individuals outside of Africa have helped enormously through funding to combat and prevent the disease. The United Nations established the Global Fund to Fight AIDS, malaria, and tuberculosis in 2001. The United States gives more money to combat AIDS than all other donors combined. Apart from the UN, the US has its own effort to combat AIDS in 12 countries in Africa and two (Haiti and Guyana) in the Americas. In a program called **PEPFAR**, the President's Emergency Plan for AIDS Relief, the George W. Bush administration pledged over $45 billion to the effort. The Bush and Obama administrations spent most of those funds to put patients on life-lengthening ARV and other drugs. General "donor fatigue," the lingering impacts of the global Great Recession, and budget cuts have slowed international funding; the Obama administration for example could not afford the escalating costs of ARV treatment, and with its Global Health Initiative (GHI) shifted the focus to prevention and mother/child health.

In 2014, another deadly virus struck West Africa, and inspired global fears: **Ebola**. Like the AIDS virus, it was first identified in the Democratic Republic of the Congo in 1976, along the banks of its namesake Ebola River. It too made the jump to humans from the blood of bushmeat, especially fruit bats, chimpanzees, and forest antelopes in rainforest areas. Human-to-human transmission of the virus comes from contact with blood and other bodily fluids. Ebola symptoms manifest within three weeks and include sudden fever, weakness, muscle pain, and sore throat—much like those of flu. But in many Ebola victims, the symptoms become much worse, with vomiting, diarrhea, and internal and external bleeding; Ebola is known as a "hemorrhagic fever." Fatality rates are quite high—in the recent outbreak, between 50 and 60 percent.

The first case was reported in December 2013. Within 15 months, more than 10,000 people died, primarily in Guinea, Liberia, and Sierra Leone; Ebola also killed a few people in Nigeria, Mali, and the United States. This epidemic claimed the lives of more than five times the victims of all earlier outbreaks in Africa combined.[8] The economic costs were enormous too, with an estimated $33 billion in lost revenues up to November 2014 in the region affected.

Slowing down the spread of Ebola in West Africa required some significant cultural changes. People were advised to stop shaking hands, kissing, and embracing, and were sometimes quarantined in their homes while health care workers made their rounds to identify victims. The virus load in a person who has died of Ebola is quite high, and so traditional funerary practices were strongly discouraged. It had been normal to wash, touch, and kiss the deceased, and to braid the hair of a woman and shave the head of a man in preparation for burial. Hunting for bushmeat was banned. Changes in these practices, along with the enormous dedication and sacrifice of health care workers (many of whom died), did stem the tide of the epidemic. It peaked in spring 2015 and then declined steadily; while Liberia was declared Ebola-free in May 2015, 13 months after the first cases of the disease were detected in that country, a few new cases resurfaced several months later. There is no cure for Ebola, but an Ebola vaccine with 100 percent efficacy among a test group of thousands of people in Guinea in 2015 seemed to promise that the Ebola scourge would not race out of control again.

agricultural expansion are taking place in Africa, as elsewhere in the world, at the expense of wildlife. Hunting and competition with domesticated livestock also take their toll. Many species are safeguarded by law, especially in the protected areas that make up about 15 percent of the region. Such laws are difficult to enforce, however, and poaching has devastated many species. Still, Africa remains home to some of the world's most extraordinary and successfully managed national parks, including South Africa's Kruger National Park, Tanzania's Ngorongoro Crater National Park, and Kenya's Amboseli National Park (at the foot of Kilimanjaro; see Figure 9.6).

Elephants and rhinoceroses are Africa's largest herbivores and the most vulnerable to global market demand for wildlife products. In 1970, there were about 2.5 million African elephants living on the continent. Poaching and habitat destruction had reduced their numbers to less than 700,000 in 2014. Poachers slay over 20,000 African elephants every year for their tusks. They are supplying an insatiable demand for ivory in the regions of East and Southeast Asia; as the buying capacities of soaring numbers of Chinese and other Asians grow, so does their appetite for expensive ivory figurines used as gifts and religious objects, jewelry, hair ornaments, and chopsticks (**• Figure 9.9**). Ignorance is part of the problem; as many as two-thirds of Chinese ivory consumers have no idea about their impact on the animals, thinking that ivory grows back like fingernails.[11]

Poaching for ivory was already having a catastrophic impact on African elephants in 1989, when delegates of the signatory nations of the Convention on International Trade in Endangered Species (CITES) met to take action. The organization succeeded in passing a worldwide ban on the ivory trade, but the problem has persisted.

CITES won the ivory ban over the strong objections of several southern African states, led by Zimbabwe, that were

Regional Perspective The Great Rift Valley

One of the most spectacular features of Africa's physical geography is the **Great Rift Valley**, a broad, steep-walled trough extending from the Zambezi Valley (on the border between Zimbabwe and Zambia) northward to the Red Sea and the valley of the Jordan River in southwestern Asia (see Figure 9.1, **page 402**, and also Figure 6.12, **page 216**, and Figure 2.1, **page 26**). It marks the boundary of two crustal plates that are rifting, or tearing apart, causing a central block between two parallel fault lines to be displaced downward, creating a linear valley. Millions of years from now, this movement will tear much of southern and eastern Africa away from the rest of the continent and allow seawater to fill the valley.

The Great Rift Valley has several branches. Lakes, rivers, seas, and gulfs already occupy much of it. It contains most of the larger lakes of Africa, although Lake Victoria, situated in a depression between two of its principal arms, is an exception. Some, like Lake Tanganyika—at 4823 feet (1470 m), the world's second-deepest lake, after Russia's Lake Baikal—are extremely deep. Most have no surface outlet. Volcanic activity associated with the Great Rift Valley has created Kilimanjaro, Kirinyaga, and some of the other great African peaks, along with lava flows, hot springs, and other thermal features. Faulting along the Great Rift Valley in Ethiopia, Kenya, and Tanzania has also exposed remains of the earliest known ancestors of *Homo sapiens*.

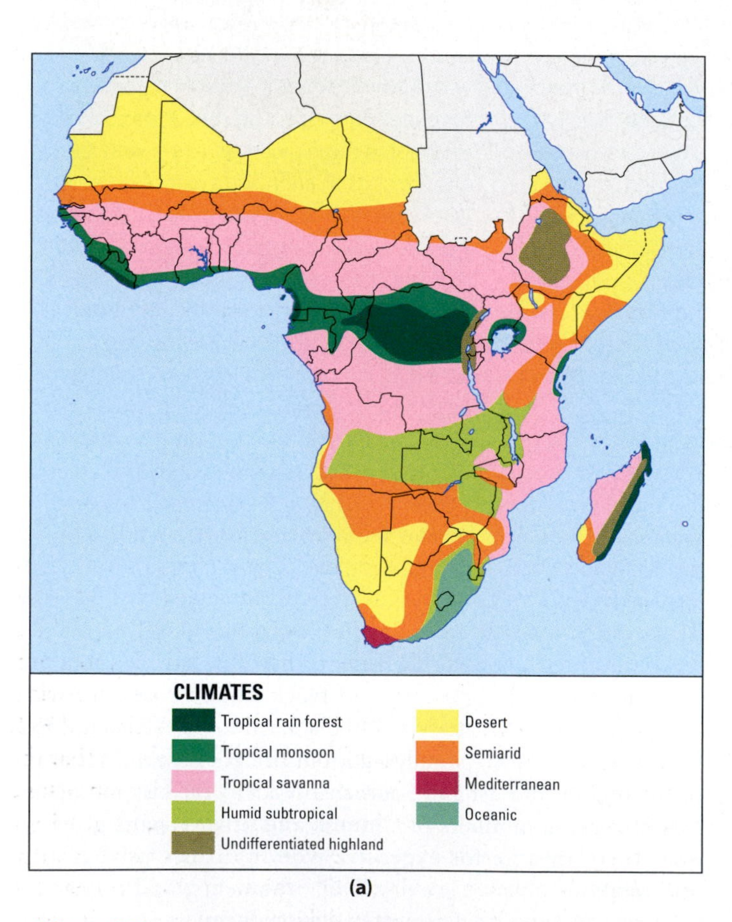

• **Figure 9.7** **(a)** Climates and **(b)** biomes of Sub-Saharan Africa.

experiencing what they saw as an elephant overpopulation problem. At the 1989 CITES meeting, these countries appealed for an exemption from the ivory ban so that they could earn foreign export revenue from a sustainable yield of their elephant populations. The majority of CITES members rejected this appeal, arguing that any loophole in a complete ban would subject elephants everywhere to illegal poaching. Since 1989, CITES has periodically allowed Zimbabwe, Namibia, Botswana, and South Africa one-time sales of ivory under rigorous monitoring as a reward for their positive wildlife policies. All continue to cull (kill) their "excess" elephants, with the meat going to needy villagers and crocodile farms.

If elephant ivory is like gold, rhinoceros horns are like diamonds. Men in the Arabian Peninsula nation of Yemen prize daggers with rhino horn handles (•**Figure 9.10**). Far more rhino horn is trafficked to Asia, where traditional medicines make wide use of it, including as an aphrodisiac and even a cure for cancer; it is a folk panacea. Western scientists deny the

Insights Africa's Greatest Conservationist

Not much larger than the common housefly, the tsetse fly of Sub-Saharan Africa packs a wallop. This insect carries two diseases, both known as **trypanosomiasis**, which are extremely debilitating to people and their domesticated animals. People contract **sleeping sickness** from the fly's bite, and cattle contract nagana. Since the 1950s, widespread efforts have been made to eradicate tsetse flies so that people can grow crops and herd animals in currently fly-infested wilderness areas (•**Figure 9.B**). Where the efforts have been successful, people have cleared, cultivated, and put livestock on the land. The results are mixed. Although people have been able to feed growing populations in the process of opening up these lands, they have also eliminated important wild resources and in many cases caused erosion, desertification, and salinization of the land. Where the tsetse fly has been eliminated, so has the wilderness. The diminutive tsetse fly thus may be characterized as a **keystone species**—one that affects many other organisms in an ecosystem. The loss of a keystone species—in this case, a fly that keeps out humans and cattle—can have a series of destructive impacts throughout the ecosystem. For its role in maintaining wilderness in Sub-Saharan Africa, some wildlife experts call the tsetse fly "Africa's greatest conservationist."

• **Figure 9.B** Strenuous and successful efforts to eradicate tsetse flies (inset) have opened vast new areas of Africa to human use. This is a tsetse fly trap with two components: a liquid chemical called "simulated cow's breath" that attracts the fly, and a pesticide-soaked tarp that kills it.

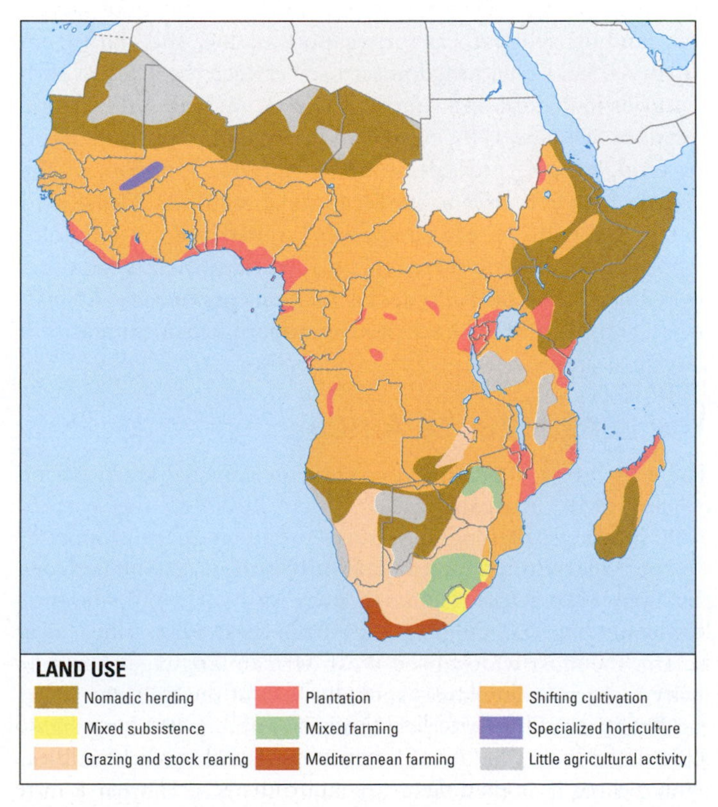

• **Figure 9.8** Land use in Sub-Saharan Africa.

medicinal efficacy of powdered rhino horn, noting that its makeup is mostly carbon, like that of your fingernails. These demands for rhino horn, and the current black-market value of tens of thousands of dollars per horn (roughly $45,000 per pound, more than gold, cocaine, or heroin), have led to a precipitous decline in population of black rhinoceroses in Africa. There were an estimated 65,000 black rhinos in Africa in 1982; in 2014, there were about 5000, but the good news is that the 2014 population is a doubling of the 1993 population. Along with captive breeding programs, aggressive anti-poaching efforts including shooting poachers on sight are in force to protect Africa's remaining rhinos. The tactic of tranquilizing the animal and using a chainsaw to dehorn it is widely used across the rhino's range, but the horn grows back at a rate of about 2 inches each year and poachers will take even a stub. Even rhinos mounted on display in European museums have had their horns poached!

Despite Africa's war on poaching, the Western Black Rhino became extinct in the wild in 2011, leaving three other subspecies with critically low numbers. No Northern White Rhinos exist in the wild, but five are in captivity. There is only one male of the species, living under heavily armed guard in a central Kenyan conservancy. Efforts are underway to promote natural or artificial insemination of the two females in his company.

• **Figure 9.9** Religious figurines made of African ivory are among the wares of this Hong Kong merchant.

• **Figure 9.10** Daggers are a nearly universal dress accessory for men in the Arabian Peninsula nation of Yemen. The most prized dagger handles are of rhino horn, a custom that has had a significant impact on African rhinos thousands of miles away. Even greater demand for rhino horn exists in South and East Asia.

9.3 Cultural and Historical Geographies

Sub-Saharan Africans' achievements and cultural contributions to humankind's legacy are enormous. The African continent was the original home of humankind. Recent DNA studies and archaeological finds that push dates back even further suggest that the first modern people (*Homo sapiens*) to inhabit Asia, Europe, and the Americas were descendants of a small group that left Africa across the Bab el-Mandeb sometime between 60,000 and 130,000 years ago (see **Figure 3.2**). After about 5000 BCE, indigenous people were responsible for agricultural innovations in four culture hearths: the Ethiopian Plateau, the West African savanna, the West African forest, and the forest-savanna boundary of West Central Africa. Africans in these areas domesticated important crops such as sorghum, yams, cowpeas, okra, and watermelons, and Africans were the first to domesticate cattle for milk and meat. From Africa, these diffused to populations in other world regions.

Civilizations and empires, some growing out of the original culture hearths, emerged in Ethiopia, West Africa, West Central Africa, and South Africa. In the 1st century CE, a Christian empire based in the Ethiopian city of Axum controlled the ivory trade from Africa to Arabia. Ethiopian tradition holds that a shrine in Axum still contains the biblical Ark of the Covenant and the tablets of the Ten Commandments, which disappeared from the Temple in Jerusalem in 586 BCE. Several Islamic empires, including the Ghana, Mali, and Hausa states, emerged in West Africa between the 9th and 19th centuries. All of these agriculturally based civilizations controlled major trade routes across the Sahara. They profited from the exchange of slaves, gold, and ostrich feathers for weapons, coins, and cloth from North Africa. Other kingdoms arose between the 4th and 18th centuries in central and southern Africa, including the Kongo kingdom, which was the hub of an interregional trade network for food, metals, and salt; and the Karanga kingdom of the 13th to 15th centuries in modern-day Zimbabwe. The skilled metalworkers of "Great Zimbabwe" mined and crafted gold, copper, and iron, and merchants traded these metals with faraway India and China. Prospects for elements of early African urban architecture to be adapted in modern urban contexts are described on **page 436**.

The Languages of Africa

There is great linguistic diversity in Sub-Saharan Africa (•**Figure 9.11**). Africans speak about 2000 of the world's 7000 languages, belonging to three broad language groups.

The **Niger-Congo language family** is the largest. It includes the many West African languages across a variety of subfamilies; many Nigerian languages are from these subfamilies, as is Fulani, the most widely used West African language. The majority of African languages spoken south of the Equator are in the **Bantu subfamily**, which comprises about 400 languages. Zulu is the most populous Bantu tongue, with over 10 million people using it as their first language. Swahili, another Bantu language, is spoken natively in Tanzania and has become a

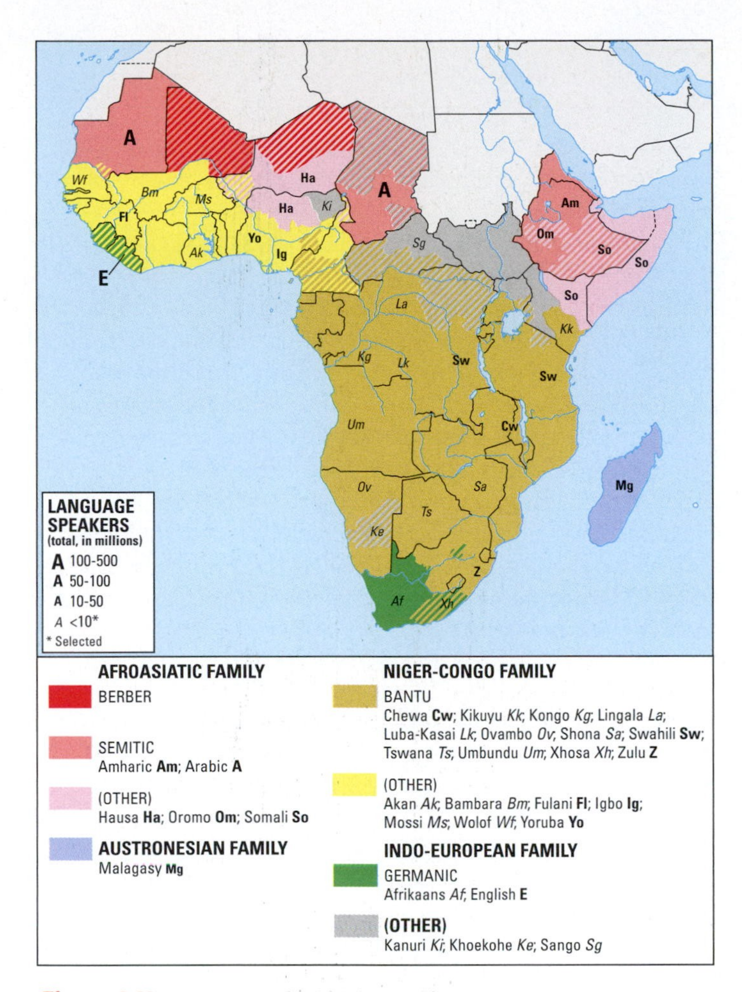

• **Figure 9.11** Languages of Sub-Saharan Africa.

lingua franca for much of eastern and southern Africa; some estimates put the number of people able to use Swahili as a second language as high as 150 million.

The Afro-Asiatic language family, spoken widely across the Sahara, the Sahel, and the Horn of Africa, includes Semitic languages (such as Arabic and the Amharic languages spoken in Ethiopia) and tongues in a handful of small subfamilies that 223 include languages such as Somali and Hausa. The Berber languages of the Sahara Desert are also Afro-Asiatic. Even some of the Niger-Congo languages, especially Swahili, have borrowed much from Arabic and other languages with roots elsewhere. The prominence of Arabic words in Swahili reflects a long history of Arab seafaring along the Indian Ocean coast of Africa; in fact, Swahili means "coastal" in Arabic.

The **Khoisan languages** of the **San** and related peoples in the western portion of southern Africa are nearly extinct. The largest, Khoekhoe, is spoken by fewer than 250,000 people in Namibia and Botswana. Recognizable by their prominent "clicking" sounds, these may be humankind's oldest languages.

Numerous small languages whose origins are unclear are found in the Central African Republic, South Sudan, and western Kenya. The people of Madagascar speak Malagasy, a language without African roots. Malagasy is an Austronesian 300 tongue that originated in Southeast Asia. The only place in Africa where Indo-European languages dominate is western South Africa, where the primary languages are English and Afrikaans, an offshoot of Dutch that was brought by the area's first European settlers in the 17th century.

Africa's list of lingua francas—the languages most likely to be recognized on the continent and those that Africans use to speak with the wider world—is provided by the six official languages of the African Union, the continent's chief supranational organization. They are English, French, Portuguese, Spanish, Swahili, and Arabic. In introducing Tanzania, our guest essayist Rhodes Makundi talks about language usage in various settings (see Life in Tanzania on **page 414**). Use of English is growing most rapidly, even in the formerly French-dominated "Francophone" countries. Countries that formerly were British colonies have a distinct advantage in the world of outsourced services: South Africa, Kenya, and Ghana, for example, have already opened call centers that they hope will rival those of India. Some Francophone countries also have call centers catering to French industries and clients. 313

The dominance of foreign languages in global mass media and digital technologies could endanger African languages, many of which are already in peril because so few people continue to speak them. As elsewhere in the world, many languages in Africa. including Khoisan, are threatened by language death. Many linguists, however, see computers as a tool for saving the 360 roughly 3000 endangered languages of the world. They argue that the computer will make it much easier for people to learn and preserve their traditional tongues. James McElvenny, a linguist at the University of Sydney, is developing software to revitalize vanishing languages. He gives young language learners portable electronic reference tools that provide the definitions and sounds of words that are otherwise no longer spoken. Dr. Tucker Childs, a linguist working in the West African nation of Sierra Leone, is using geographic information systems (GIS) to plot locational variations of Kim and other endangered languages. With these findings, linguists can decide where to focus their efforts.

Africa's Belief Systems

The religious landscape of Africa is complex and fluid. Spiritualism is extremely strong, but spiritual affiliations and practices are more interwoven and flexible than in most other world regions. It is not uncommon for family members to follow different faiths or for an individual to change his or her religious beliefs and practices in the course of a lifetime—or, as Rhodes Makundi put it, to practice more than 100 percent of one's beliefs.

Broadly, however, some dominant patterns of religious geography can be recognized (•**Figure 9.12**). Islam is the dominant religion in North Africa and the countries of the Sahel on the southern fringe of the Sahara. The **Ethiopian Orthodox Church**, closely related to the Coptic Christian faith of Egypt, makes Ethiopia an exception to the otherwise Islamic Horn of Africa region (•**Figure 9.13**). Islam is also the prevailing religion of the East African coast, where Arab traders introduced the faith. Muslims are a majority or strong

life in TANZANIA

Courtesy of Rhodes Makundi

Professor Rhodes Makundi

Tanzania (which used to be called Tanganyika but changed its name after uniting with Zanzibar in 1964) did not exist as a country before 1891. European imperialism and colonization resulted in the scramble for Africa in the latter half of the 19th century, leading to partitioning of the continent by the major European powers, mainly Britain, France, Germany, Belgium, Italy, Portugal, and Spain. Tanganyika was created out of a large landmass in East Africa and became what was known as German East Africa, which included present-day Rwanda and Burundi. To the north, the territory was bordered by present-day Kenya and Uganda, under the British; to the south by Mozambique, under the Portuguese; and to the southwest by Nyasaland (present-day Malawi), under the British. The Belgian Congo bordered Tanganyika on the west. For many years before the partition of East Africa, Arabs had established their rule in Zanzibar and were trading with tribes along the coast and the interior, but they also were engaged in the slave trade. The European colonizers wanted to spread Western civilization and Western culture in their colonized territories. This included the way of dressing, religion (Christianity as opposed to Islam, which was getting rooted along the trade routes and the coastal towns), education, and administration—after all, they referred to Africa as "The Dark Continent." After the First World War ended in 1918, Tanganyika changed hands, passing from the Germans to the British, and another era of colonialism began.

The 70–80 years of German and British colonialism had a major impact on the culture, economy, and religious life of the people of the territory called Tanganyika. Before colonization, the people were organized chieftainships, both small and large, weak and powerful. There were no national boundaries. Christianity was essentially unknown, and local beliefs abounded. The colonial order faced some resistance in trying to bring the different chiefdoms under one colonial administration, but because of their superior weaponry, the colonizers won. A new era began when these people started to demand independence peacefully, aiming to have a nation that was ruled by the natives. Tanganyika became independent in 1961 and later united with Zanzibar to become The United Republic of Tanzania.

At independence from the British, a young man (39 years old) called Julius Nyerere became the prime minister and later first president of the young nation. You cannot talk of Tanzania and life in Tanzania without this man. His 25 years of leadership had a lasting legacy on the political system, its people, and life in general. First, he worked hard to unite the people, a total of more than 120 tribes "to think" as one, rather than as different tribes. In this he was very successful! Second, he introduced "free" universal education to a country that at independence had only a few schools, no university, and only a few graduates from foreign colleges! Third, he introduced socialism—what he called African socialism, which blended socialism and a welfare state in a poor country. He aimed at satisfying the perceived needs of the people, particularly for education, clean water, and health through state-sponsored programs. However, much as the socialist economy he was trying to build appeared good on paper, in practice it could not withstand the forces of market economy sponsored by the powerful Western nations, the World Bank, the International Monetary Fund, and private creditors. So Tanzania in the 1980s changed and adopted a market economy, what became known as a liberalized rather than centrally controlled economy. This ushered in a new way of life for her people.

Each of the more than 120 tribes that make up Tanzania's population of 45 million has its own culture and dialect, but Kiswahili and English are the first and second languages, respectively, for official communication. However, although almost every Tanzanian speaks Kiswahili, only a small proportion of the population is actually fluent in both spoken and written English. Unlike in many other African countries, speaking your tribal language is kind of taboo and is unwelcome, especially when you are in the company of others who do not speak it. Tanzanians will prefer to speak Kiswahili, rather than English, unless they are talking to a foreigner who can't speak Kiswahili. If the foreigner speaks Kiswahili, they will rather converse in it than in English.

Tanzania is a very beautiful country, and its people are extremely friendly. We relate and mix with each other easily without bothering about the ethnicity or tribal affiliations. Try that in our neighboring countries, and it doesn't work. You can hardly find or hear of aversion or prejudice against foreigners in Tanzania. A visiting foreigner only needs to know a few Kiswahili words like *Jambo* ("hello") and *Habari* ("how are you"), and things are ripe for a conversation! Kiswahili developed originally in coastal areas of East Africa, stretching from Kilwa in Southern

minority in rural northern Nigeria and Tanzania, and there are minority Muslim populations in cities and towns across the continent. Christians are a majority most everywhere else in Africa, especially along the southern coast of West Africa and most of southern Africa, where European influence was strongest during colonial times. Both Christianity and Islam predominate in the cities, whereas traditional religions prevail or overlap with these monotheistic faiths in rural areas.

The dominant religious tradition of African Muslims is neither Sunni nor Shi'ite, but **Sufi Islam**. Sufism advocates a mystical approach to the faith that contrasts with the more legally inclined Sunni traditions. Sufis preach an emotional and intuitive close personal relationship with God. They also accept the roles of *skaykhs*, or "holy men," who act as intermediaries between the individual and God. Known as *marabouts* in West Africa and *wadaddos* in East Africa, these revered figures are believed to possess *Baraka* or supernatural blessings and mystical powers that give them special powers in this life and beyond; thus their tombs often become pilgrimage sites. The mystical aspects of Sufism have made it possible for pre-Islamic and local, traditional beliefs to be

Tanzania to north of Mombasa in Kenya. The language developed as a result of interactions between the natives of the coastal areas and traders from Arabia, India, and in later years Europeans. The vocabulary is rich in words originating from African languages, Arabic, Portuguese, some Spanish, English, and German, and it became a common lingua franca for traders travelling into the interior of East and Central Africa. It soon became a common language in trading centers in the central, south, north, and west of Tanzania, across Lake Tanganyika into Congo (the current Democratic Republic of the Congo), Malawi, Rwanda, and Burundi.

Most Tanzanians (about 80 percent of the population) live in rural villages and are involved in agriculture. Life in rural areas is more communal than in urban areas. The people are friendly and welcome any foreigner, whether they are from another village or another country, with open hands. The two verses below are from a poem[1] I wrote about life in my village, and they provide a good glimpse of her people's hospitality:

The village welcomed visitors from near and afar,
It wasn't how long they had walked to reach our village,
But how long they were staying with us,
You couldn't walk long and stay for a short period.

With some impoverished houses in some families,
Yet the doors were open to all kinds of visitors,
The occupants were full of smiles,
Guests treasured the magnitude of hospitality.

The countryside has its own beauty spots, unparalleled to any other on the African continent. We have more landmass that has been declared by law as national parks and game reserves than any other country in Africa. If you are not visiting the Manyara National Park to see the tree-climbing lions, you can go visit the Serengeti National Park to see all kinds of wildlife and hundreds of thousands of migrating wildebeests. Or perhaps you would wish to visit Olduvai Gorge, a place where human evolution led to the emergence of *Homo erectus* more than 500,000 years ago, and later to our species, *Homo sapiens*. This is where the first hominids walked and hunted in the savannas of Africa. Perhaps you might even like it better to be on top of the continent by climbing the snow-capped Mountain Kilimanjaro, the highest mountain peak in Africa, or visit the Selous Game Reserve, which is semi-explored, but still one of the wildest places on Earth, full of diverse wildlife and wilderness unparalleled to any other place in Tanzania and the world. Many other interesting places will suit your needs and desires if you are tired of the madding crowds of the city and you need to be in areas that have been saved from the curse of industrial activity and urban congestion—places that give the individual some peace of mind. Still, you can go to Zanzibar, where old traditions mix with the modern way of life, where a stone town borders modern-day high-rise buildings and Arab and African cultures have been blended to create the cradle of the Swahili culture.

Tanzania is politically a stable country following a democratic model of government. It is one of the few countries in Africa that have never been ruled by a military dictatorship and have not experienced civilian strife based on religious differences, political intolerance, or other causes that have been so common in other countries in Africa in the last 50 or so years. The economy of the country depends much on exportation of raw materials (cotton, coffee, etc.) and minerals (gold, diamonds, etc.), but recent discoveries of natural gas in large quantities could have a major impact on its economy. Tourism is a major economic activity that could surpass all others given good promotion and facilities to absorb and provide services to more tourists than we are currently handling. Religion-wise, the country is regarded as equally divided into Christians and Muslims. But if you ask an ordinary Tanzanian, "What is the percentage of the population that follow each of the religions?", you might as well get the answer "Fifty percent are Christians, 50 percent are Muslims, and the other 50 percent are none of the two," meaning among those who call themselves Christians or Muslims, there is also a large number who are not true converts and usually have strong traditional religious beliefs. Christians and Muslims live in harmony and marry without obvious restrictions.

I can now look back through the years and compare them with the present to predict the future, and what I see is a country whose development is quickly evolving to modernize. The majority of its people are still poor, but when I look at the smiles on their faces, even though they do not possess the material things that are taken for granted in the richer, developed countries of the world, I am convinced that with hard work and determination, their children's children will live a much better and richer life.

Rhodes Makundi is a professor at the Sokoine University of Agriculture in Morogoro, Tanzania.

[1] Makundi Rhodes, "Reflections: From the Village to the Peak of Knowledge" (*My Memoirs* in preparation, book draft, 2015).

syncretized with Islam across Africa.[12] Sufism is anathema to the rigid Salafist brand of Islam and the violent Islamist agendas that sometimes emerge from Salafism, as discussed on **page 232**.

Christians and orthodox (Sunni) Muslims have mounted efforts to dispose of "heretical" notions in African traditional belief systems, but these concepts are very resilient. Even where the large monotheistic faiths nominally prevail and their followers are devout, a strong substrate of indigenous beliefs has persisted and usually comfortably merged with the "official" 303, 467 faith—a syncretism also present in other cultures of the world. One example is the mingling of indigenous *vodun* ("voodoo") beliefs with Christianity and Islam in West Africa.

Indigenous African religions have many animistic elements, emphasizing that spiritual forces are manifested everywhere in the environment. Many spirits are tied to particular places in the landscape. Such beliefs contrast with the introduced Christian and Muslim dogmas that tend to see nature as separate from God and people as apart from and superior to nature.[13] Across the spectrum of African cultures, the use of mediums to contact spirits of deceased ancestors, creator beings, or earth genies that can intercede on one's behalf is

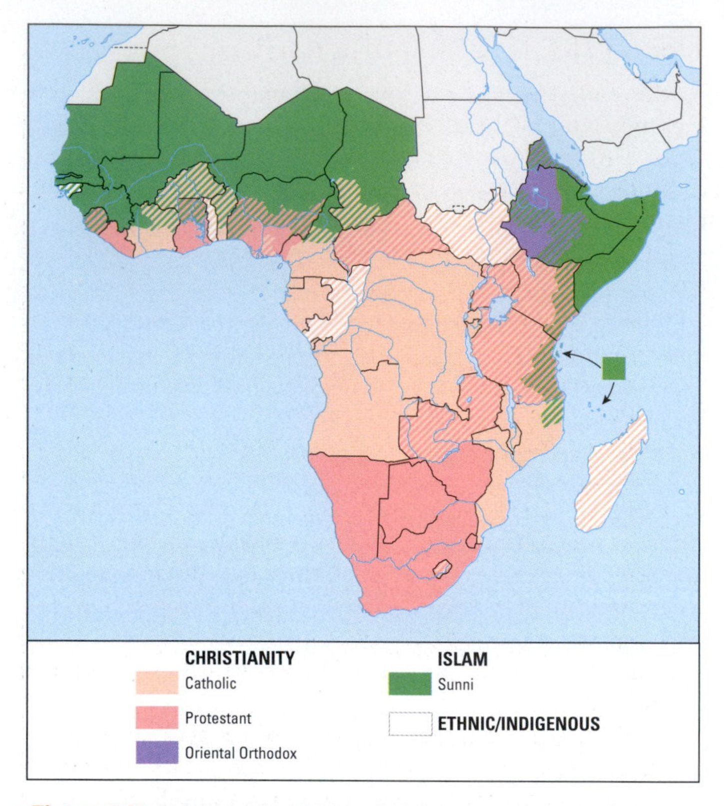

• **Figure 9.12** Religions of Sub-Saharan Africa.

• **Figure 9.13** The Ethiopian Church has been the dominant religion in Ethiopia since the 4th century. This priest's flock worships at the 12th-century Church of St. George in Lalibela. When my Mom and I visited Ethiopia in 1983, we were apparently the only tourists in the country! Ethiopia had a Marxist government that discouraged tourism and outlawed religion. The country's devout peoples defied the ban then, and freedom of worship is the norm for all denominations now.

Joe Hobbs

widespread. Reverence for ancestors is strong, reflected also in great respect for the elderly.

Outsiders have tried, seldom successfully, to give accurate names to indigenous African belief systems, practices, and personnel: ancestor worship, animism, *force vitale* ("life force"), living dead, and witch doctor among them. Although it would be gratifying to have precise names and descriptions for African beliefs, they are unique and can best be understood by careful reading, observation, and discussion on a case-by-case basis.

The Origins and Impacts of Slavery

Until about 1000 years ago, the cultures of Africa south of the Saharan desert barrier remained largely unknown to the peoples north of the desert. Egyptians, Romans, and Arabs had contact with the northern fringes of this region, and some trade filtered across the Sahara, but to most outsiders, Africa was *terra incognita*. Even at the opening of the 20th century, vast areas of interior tropical Africa were still little known to Westerners.

The tragic impetus for growing contact between Africa and the wider world was slavery. Over a period of 12 centuries, as many as 25 million people from Sub-Saharan Africa were forced into slavery, exported as merchandise from their homelands. The trade began in the 7th century, with Arab merchants using trans-Saharan camel caravan routes to exchange guns, books, textiles, and beads from North Africa for slaves, gold, and ivory from Sub-Saharan Africa. As many as two-thirds of the estimated 9.5 million slaves exported between the years 650 and 1900 along this route were young women who became concubines and household servants in North Africa and Turkey. Male slaves usually became soldiers or court attendants (some of whom eventually assumed important political offices). From the 8th to 19th centuries, about 5 million more slaves were exported from East Africa to Arabia, Oman, Persia (modern Iran), India, and China. Again, most were women who became concubines and servants.

The cruel but lucrative traffic in slaves provided the initial motivation for European commerce along the African coasts, inaugurating the long era of European exploitation of Africa for profit and political advantage. The European-controlled slave trade was the largest by far, particularly with the development of plantations and mines in the New World. Between the 16th and 19th centuries, the capture, transport, and sale of slaves was the exclusive preoccupation of trade between the European world and West Africa. Portuguese and Spaniards began the trade in the 15th century, and within 100 years English, Danish, Dutch, Swedish, and French slavers were also participating. The peak of the transatlantic slave trade was between 1700 and 1870, when about 80 percent of an estimated total of 10 million slaves made the crossing. In escape attempts, in transit, and in the famines and epidemics that followed slave raids in Africa, probably more than 10 million others died.

Slaves were a prized commodity in the **triangular trade** linking West Africa with Europe and the Americas. European 468, 514 ships carried guns, ammunition, rum, and manufactured goods to West Africa and exchanged them there for slaves. They then transported the slaves to the Americas, exchanging them for gold, silver, tobacco, sugar, cotton, rum, and tropical hardwoods to be carried back to Europe. As "raw material" and as the labor working in the mines and plantations of Latin America, the West Indies, and Anglo America, slaves generated much of the wealth that made Europe prosperous and helped spark the Industrial Revolution.

Although Europeans carried out the triangular trade, their physical presence was limited to coastal shipping points. Africans were the intermediaries who actually raided inland communities to capture the slaves and assemble them at the coast for transit shipment. West African kingdoms initially acquired their own slaves in the course of waging local wars. As the demand for slaves grew, these kingdoms increasingly went to war for the sole purpose of capturing people for the trade. As the exports grew, so did the practice of Africans keeping African slaves. Even after Britain and the other European countries abolished slavery, it continued to flourish in Africa. By the end of the 19th century, slaves made up half the populations of many African states.

Slavery has not yet died out in the region (and indeed, a modern-day form of slavery exists in the form of human trafficking elsewhere in the world; see **pages 195** and **359**). In Mauritania (where slavery was outlawed only in 1981), the practice remains widespread throughout the country; between 100,000 and 600,000 black Mauritanians are currently enslaved, primarily by lighter-skinned Arab and Berber Mauritanians (a similar situation exists in neighboring Mali, with about 200,000 southern black Malians enslaved by northern Tuaregs). Enslavement of children persists in West Africa. The typical pattern is that impoverished parents in one of Northern or Central Africa's poorer countries are approached by an intermediary who promises to take a child from their care and see that the boy or girl is properly educated and employed. This involves a fee (as little as $14) that, unbeknown to the parents, is a sale into slavery. The intermediary sells the child (for $250 to $400, on average) to a trafficker who sees that the child is transported to one of the region's richer countries. There, the child ends up working without wages and under the threat of violence as a domestic servant, plantation worker (especially on cacao and cotton plantations), or prostitute. Benin is the leading slave supplier and trafficking center; Gabon, Ivory Coast, Cameroon, and Nigeria are the main buyers of slaves. Slavery also exists in Sudan but has declined since South Sudan's independence from the north in 2011. During the years of fighting preceding independence, the common pattern was for northern slavers and allied rebel groups to capture mainly Christian Dinka tribespeople as slaves in raids or abductions in the south. In the 1990s, well-intentioned church groups such as Christian Solidarity International and other organizations in the West began paying hard cash to buy freedom for slaves in Sudan. Although many real slaves were freed, Sudanese profiteers also moved in with "counterfeit slaves," ordinary people hired to act as slaves until their "rescue" had been paid for.

The Impacts of Colonialism

In the 16th century, European colonialism began to overshadow and inhibit the growth of indigenous African civilizations. This is a diverse region, but one thing its peoples have in common is a sense that in the past they were humiliated and oppressed by outsiders.[14]

Portugal was the earliest colonial power to build an African empire. The epic voyage of Vasco da Gama to India in 1497–1499 via the Cape of Good Hope was the culmination of several decades of Portuguese exploration along Africa's western coast. During the 16th century, Portugal controlled an extensive series of strong points and trading stations along both the Atlantic and Indian Ocean coasts of the continent. European penetration of the African interior began in 1850 with a series of journeys of exploration. Missionaries like David Livingstone, as well as traders, government officials, and adventurers and scientific explorers undertook these expeditions. By 1881, when Africans still ruled about 90 percent of the region, foreign exploits had revealed the main outlines of inner African geography, and the European powers began to grab colonial territory in the interior.

Much of the "scramble for Africa" took place at the Conference of Berlin in 1884 and 1885, when various European powers established their respective spheres of influence in the region. By 1900, and continuing for more than half a century, Sub-Saharan Africa was a patchwork of European colonies (**• Figure 9.14**), and Europeans in these possessions were a privileged social and economic class.

At the outbreak of World War II in 1939, only three countries—South Africa, Egypt, and Liberia—were independent. The United Kingdom, France, Belgium, Italy, Portugal, and Spain controlled the rest. But after the war, mainly in the 1960s and 1970s, a sustained drive for independence was mounted. Ceasing to be a colonial region, Africa emerged with more than a quarter of the world's independent countries. This was a peaceful process in most instances, but bloodshed accompanied or followed independence in several countries.

Dependency theorists often point to Africa as a prime example of how colonialism created lasting disadvantages for the colonized. European colonization produced or perpetuated many of the enduring characteristics of underdevelopment. These included the marginalization of subsistence farmers, notably those who colonial authorities, intent on cash crop production, displaced from quality soils to inferior land. In addition, European use of indigenous labor to build railways and roads often took a high toll in human lives and disrupted countless families. Furthermore, the colonizers often corrupted traditional systems of political organization to suit their needs, sowing seeds of dissent and interethnic conflict. The Tutsi/Hutu divide in Rwanda and Burundi, which resulted in so much loss of life in the 1990s, is just one of the examples (see **page 437**).

The European colonial enterprise did have some beneficial impacts. The colonies, and the independent nations that succeeded them, were the beneficiaries of new cities and the transport links, built with cheap (or forced) African labor; new medical and educational facilities (often developed through

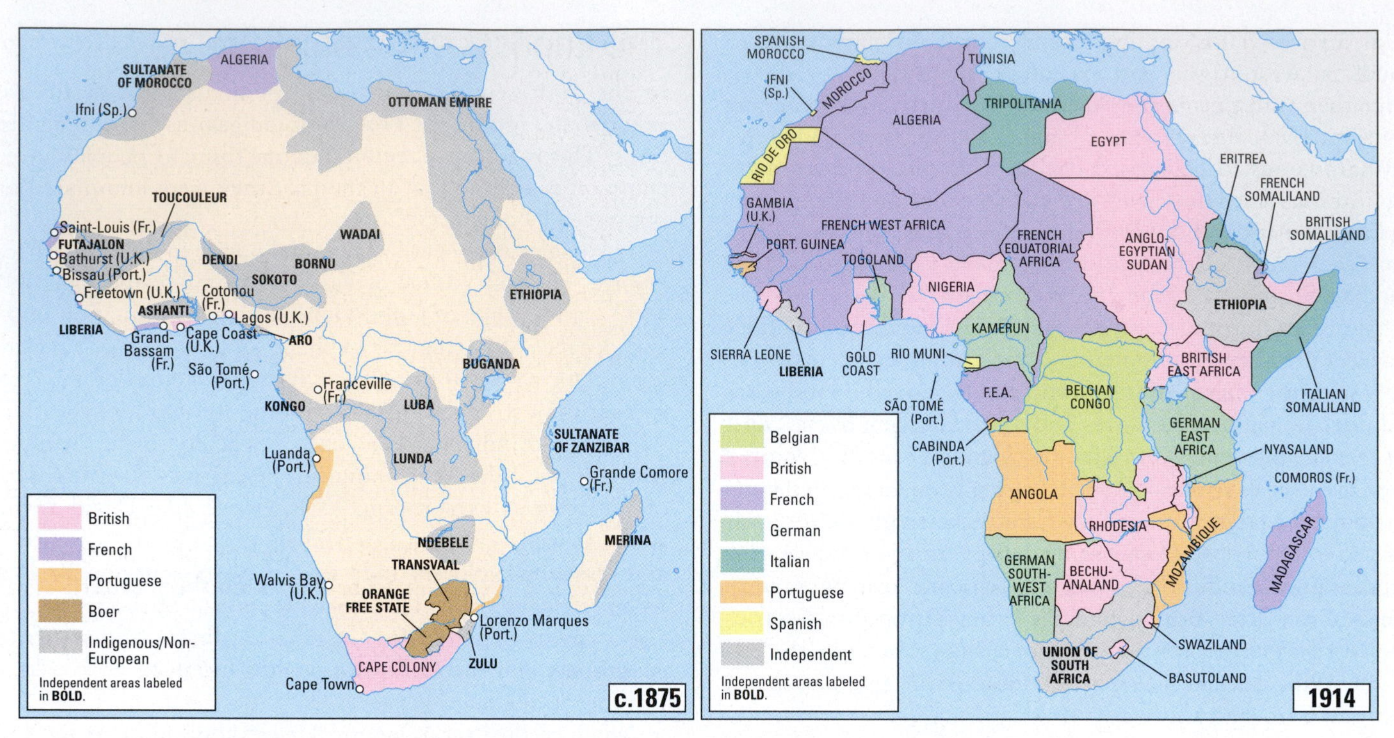

• Figure 9.14 In the 1870s, the European presence in Africa was mostly limited to several small coastal settlements. Less than half a century later, European powers controlled nearly the entire continent.

Christian missions); new crops and better agricultural techniques; employment and income provided by new mines and modern industries; new governmental institutions; and government-made maps useful for administration and planning. Such innovations were very helpful, but they were distributed unequally from one colony to another and were inadequate for the needs of modern societies when independence came.

One of the persistent problems in Africa's political geography is that many modern national boundaries do not correspond to indigenous political or ethnic boundaries. In most cases, this is another legacy of colonialism: occupying powers created arbitrary administrative units that were transformed into countries as the colonial powers withdrew. Nigeria is a good example of an ethnically complex, unnaturally assembled nation. On the political map, it appears as an integral unit, but its boundaries have no logical basis in physical or cultural geography. Some Nigerians still refer to the **"mistake of 1914,"** when British colonial cartographers created the country, heedless of its ethnic rifts (**•Figure 9.15**). The greatest divide is between poorer Muslim north and more prosperous Christian south, but there are at least 250 ethnic groups within the country. The British colonizers invested more in education and economic development in the south and built army ranks among northerners, whom they thought made better fighters; so the subsequent pattern is that the northern military leaders have tried to blunt the economic clout of southerners. Colonial administration and boundary drawing thus sowed seeds of modern conflict in Nigeria, a problem described in more detail in Section 9.6.

Although formal political colonialism has vanished, most countries still have important links with the colonial powers

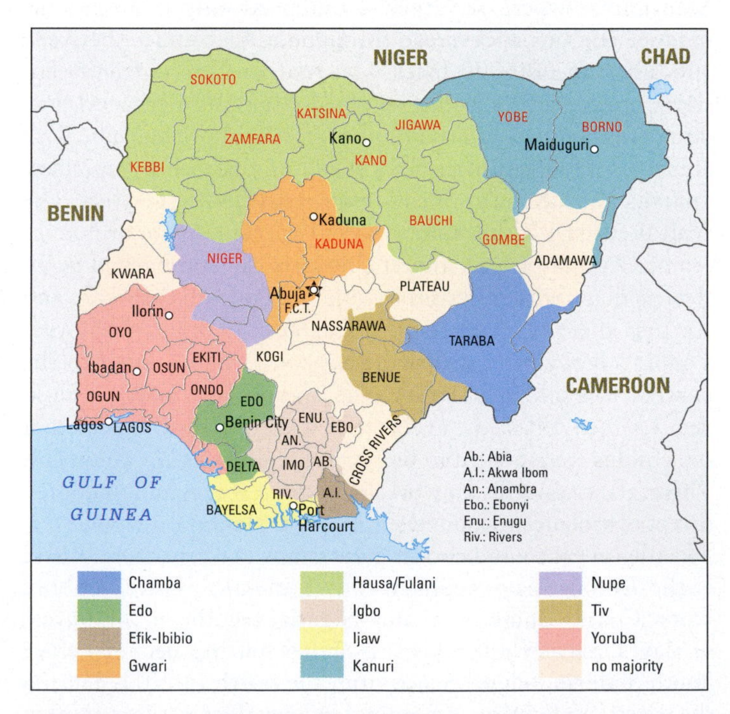

• Figure 9.15 Ethnic groups of Nigeria. States labeled in red are governed by *shari'a* law.

that formerly controlled them, and many foreign corporations that operated in colonial days still maintain a significant presence. France has a long history of post-independence intervention in the political and military affairs of its former African colonies. France is the only ex-colonial power to keep troops in Africa. France took steps to ensure that most of its former colonies trade almost exclusively with France and in turn supported national currencies with the French treasury. But France found this paternalistic approach to be extremely expensive and has been reducing its military presence and other costly assistance to its African clients. France nevertheless continues to deploy strike forces to aid local governments against insurgencies when it feels its interests on the continent are threatened, as it did in Mali in 2013.

9.4 Economic Geography

Sub-Saharan Africa ranks at the very bottom of every statistical indicator of global quality of life. Great poverty is characteristic of the region; 30 of the world's 35 poorest countries are there (see **Table 9.1**). Nearly half of this region's people subsist on less than $1.25 per day, according to the World Bank. But Africa is not as far behind as it has been, and the region is changing for the better. Sub-Saharan Africa not long ago was the world's worst region for conducting business because of corruption and bureaucratic obstacles, but with many conflicts ending and with better economic practices, the region's business climate is much improved. The number of significant conflicts was cut in half between the 1990s and 2000s. Governance is improving as corruption is being tackled. The so-called "cheetah generation" of new African leaders is more energetic and creative in business and in policy making than the retiring generation of post-colonial leaders. The "peace dividend" and improved governance is an attractive mix of factors that has led to a surge in foreign direct investment (FDI). Some financial authorities believe that Sub-Saharan Africa as a group should be just as attractive for investment as the BRIC (Brazil, Russia, India, and China) nations were, and have modified BRIC to BRICS, with the "S" representing South Africa. 174, 470

These factors have turned Africa's economic tide, as reflected in remarkable statistics. Despite the global economic downturn, Sub-Saharan Africa's economic growth of 5 percent annually in the past two decades was twice its pace in the previous two decades, placing it among the world's fastest-growing regions. As of 2015, 11 of the world's 20 fastest-growing economies were here, and between 2015 and 2020 the average African economy was projected to outpace its Asian counterparts in percent annual economic growth.[15] Private companies are creating jobs. Roads, bridges, dams, and other infrastructure are improving, in many cases through Chinese investment.

Debt is no longer the albatross it was around African necks, especially since the Millennium Development Goals (MDGs) began focusing energies on forgiving debts and making new IMF and World Bank loans less burdensome on the borrowers (see Millennium Development Goals on **page 433**).[16] And gone are the days when African governments were passive recipients of Western foreign aid. They have more choices, and more directions to turn to in receiving aid, especially eastward to China. China is Africa's largest trading partner (having replaced the US in 2009), and its investments are targeted, especially to the countries rich in natural resources. Investors in Africa are now bringing in more money than does foreign aid, and they are investing not only in natural resources but also in retail, financial services, tourism, transport, telecom, manufacturing, and construction. Information technology is attracting increasing attention, from the mobile phone systems (described on **page 426**) to call centers, to video gaming and entertainment. Hope City and Konza, discussed on **page 434**, aspire to be the capitals of Africa's IT boom.

There is an emerging middle class in the region; in other words, the number of people able to spend money on things other than life's bare necessities is growing. Much of this new economic activity is targeted at these rising middle classes, defined as those families earning more than $5000 annually.[17] Africa's economic growth is not just a story of the region's governments and external nations wielding tools of power. It also reflects the cultural attributes and the inherent capacities of individuals and communities endowed with historic traditions to persevere through struggles. Here is a list of the "durable strengths and resources" that African peoples and nations can tap into and strengthen to help position them better in global economic and other systems:

- Complex and resilient agro-pastoral systems
- Rich indigenous knowledge
- Water, land, forest, fisheries, animal, and mineral wealth
- Cultural, linguistic, and heritage assets
- Creative and talented populations and a connected diaspora
- Long history of resistance in the face of adversity (for example, slave trading, colonialism, and the struggle for independence)
- Strong collective structures at family and village levels
- Vigorous entrepreneurship in the informal economy
- Effective and vocal civil society
- Vocal and active women and women's groups
- Solitary among African governments and South–South partnerships
- New initiatives in African regional economic cooperation
- New impetus for African peacemaking and peacekeeping roles
- Growing democratization[18]

That's the big picture. Let's take a closer look at the region's economic challenges and the efforts and opportunities that may overcome them.

Subsistence Agriculture

Despite all this progress in the region, famine remains an all-too frequent visitor to Sub-Saharan Africa. Although the region's total food production is growing faster than the population, that increasing population is keeping the rise in food consumption per capita relatively low; it has grown by only 20 percent since 1960. Malnutrition is still a prevalent problem; nearly one-quarter of Africans are malnourished. To support growing populations, farmers across Sub-Saharan Africa have shortened fallow periods and pressed their lands to yield more crops. The result has been an unprecedented degradation 80 of the resource. Fully 75 percent of the region's farmland is severely low in the nutrients needed to grow crops, up from 40 percent a decade earlier. Studies suggest that at current rates, crop yields will fall as much as 30 percent by 2022. This projection does not even calculate the prospective disruptions caused by climate change. Most African farmers cannot afford fertilizers and are not familiar with the soil conservation techniques that would help reverse this ominous trend.

Over half of Africa's peoples practice subsistence agriculture (farming their own food but producing little surplus for sale) and pastoralism. Women do a large share of the farmwork—they produce 80 to 90 percent of Africa's food—in addition to household chores and the bearing and nurturing of children (•**Figure 9.16**). Mechanization is rare and fertilizers are expensive, so crop yields are low. In the steppe of the northern Sahel, both rainfall and cultivation are scarce (see **page 407**). The more dependable rainfall of the southern Sahel creates a major area of rain-fed cropping, with unirrigated millet, sorghum, corn (maize), and peanuts the major subsistence crops. In the tropical savannas south of the Equator, corn is a major subsistence crop in most areas, with manioc (cassava) and millet also widely grown. Corn, manioc, bananas, and yams are the major food crops of the rain forest areas.

Many peoples, particularly in the vast tropical savanna grasslands both north and south of the Equator, are pastoral. An increasing problem is that farmers often drive pastoralists from traditional grazing lands (this was a problem, for example, in the notorious Darfur conflict in Sudan and neighboring Chad, beginning in 2003). Confined to smaller areas in which to browse and graze, the nomads' cattle, sheep, and goats often overgraze vegetation and compact the soil. Access to drinking water for livestock and people is also a major problem.

Joe Hobbs

• **Figure 9.16** Mother and child in Zimbabwe. Women plant and harvest most of Africa's food and care for most of its children.

Most Africans who live by tilling the soil also keep some animals, even if only goats and poultry. Among some African peoples, livestock not only contribute to the daily diet but are also an indispensable part of customary social, cultural, and economic arrangements. Cattle are particularly important; traditional Maasai pastoralists of Kenya are probably the most famous African people with close dependence on cattle. They milk and carefully bleed the animals for each day's food and tend them with great care. The Maasai give a name to each animal, and herds play a central role in the main Maasai social and economic events through the year.

Madagascar is a good example of an African country in which cattle represent status, wealth, and cultural identity. In order to enhance their social standing, Malagasy livestock owners tend to want larger numbers of animals even more than they want better quality animals (•**Figure 9.17**). In the days-long funerary ritual known as the "turning of the bones," a Malagasy family commemorating a deceased ancestor sacrifices a large number of zebu cows for fellow villagers, and similar feasts accompany other important ritual dates. Due to such demands, the population of zebu cattle on the island is about 11 million. Their forage needs have grave consequences for Madagascar's rain forests and other wild habitats. People clear

Joe Hobbs

• **Figure 9.17** Zebu cattle are extremely important in the cultures and household economies of rural Madagascar.

the forests and repeatedly set fire to the cleared lands to provide a flush of green pasture for their livestock (the so-called "green bite"), causing a rapid retreat of the island's natural vegetation.

Commercial Agriculture and Marginalization

There is enormous potential to improve African agriculture. The region is a net food importer, yet 60 percent of the world's uncultivated arable land is in Sub-Saharan Africa. Much of the land is inherently difficult to farm, but cultivation methods, roads, food storage, and other infrastructure can be improved. Who will benefit from these improvements—the poor farmer, the wealthy landowners, or both?

Food shortages often result when governments promote the cultivation of cash crop primary commodities for export, instead of subsistence food crops for local people. Coffee, cacao, cotton, peanuts, and oil palm products, in particular, provide a means of gaining foreign exchange with which to buy foreign technology, industrial equipment, military armaments, and 64 consumption items for the elite. Secondary cash crops include sisal (grown for its fibers), pyrethrum (used in insecticides), tea, tobacco, rubber, pineapples, bananas, cloves, vanilla, cane sugar, and cashew nuts.

As commercial agriculture expands, small farmers whose land is owned by the government are experiencing disruption and dispossession. In a classic case of marginalization, small farmers are being driven off their land as it is sold to investors from China, South Korea, Libya, Saudi Arabia, and other countries. 64 In some cases, there are long-established villages located on what was sold to investors as "vacant" land. In this process of "land grabbing," the well-financed foreign land investors are able to secure irrigation water and other vital resources out 65 of reach to agricultural smallholders. None of the income from rice and sugarcane and other commercial products grown on these lands goes to local people. The consequences for those subsistence farmers who are not rendered landless include water shortages and reduced water quality. There are social stresses on them and on the marginalized, dispossessed former farmers who become dependent on internationally supplied and subsidized food aid. The process is creating a new source of landless, unemployed poor in African cities.

Sometimes the people push back successfully. When Madagascar's government was about to sell about half of the country's arable land to a South Korean firm, opposition to Madagascar's president swelled, and he was voted out of office.

International trading policies and patterns also disadvantage African farmers and render them less able to compete in world markets. One of the most disabling policies is the European Union's Common Agricultural Policy (CAP). The CAP's subsidies make it virtually impossible for farmers in African and other LDCs to compete with the low prices of sugar, dairy, cereals, and other produce exported worldwide 112 by large European agribusinesses. Reforms to the CAP system may eventually help level the playing field for African farmers. The termination of federal subsidies to American cotton farm- 472, 526 ers in 2013 should also help.

Overall, although cash crops are important sources of revenue in these poor countries, excessive dependence on them, as with other commodities, can harm a country's economy. The income they bring often goes almost exclusively to already prosperous farmers and corporations. The prices they fetch are vulnerable to sudden losses amid changing world market conditions. The plants themselves are susceptible to drought and disease. Finally, they are often grown instead of food crops, and cash crops do not feed hungry people.

Mineral Resources

Mining and mineral exports have long had strong impacts on the physical and social geographies of Sub-Saharan Africa. This is a well-endowed region with a remarkable one-third of the world's mineral resources, including 10 percent of the world's oil reserves, 40 percent of its gold, 65 percent of its diamonds, and 80 to 90 percent of its chromium and platinum (**•Figure 9.18**). Notable mineral exports include precious metals and precious stones (particularly diamonds), iron alloys, copper, phosphate, uranium, petroleum, and high-grade iron ore, all destined principally for Europe, the United States, and China.

Some of Africa's recent and current conflicts, in many cases fought by child soldiers, have not been about ideology or ethnicity, but simply about control over resources such as diamonds, oil, and other minerals. It is therefore of growing interest in the consuming countries that these commodities not be tainted by violence (see Problem Landscape, **page 423**).

Large multinational corporations, financed by investors in Europe, the United States, and China, do most of the mining in Africa. Mining has attracted far more investment capital to Africa than any other economic activity. Money is invested directly in the mines, and many of the transportation lines, port facilities, power stations, housing and commercial areas, manufacturing plants, and other elements in the continent's infrastructure have been developed primarily to serve the needs of the mining industry.

An economic activity as great as Africa's mining sector has major repercussions throughout societies and populations across the continent. Among major global migration flows, the huge circulation of migrants among Sub-Saharan African countries is outstanding; see this flow in Figure 3.22 on **page 76**. Many of these are people moving to where mining jobs and related services are located. Great numbers of workers in the mines are temporary migrants from rural areas, often hundreds of miles away. The recruitment of migrant workers has important cultural and public health effects. Millions of Africans who work for mining companies have come into contact with Western ideas as well as those of other ethnic groups and have carried these back to their villages. Unfortunately, migrant labor has also helped spread tuberculosis and HIV, the virus that causes AIDS. Most miners are single or married men 408 who spend extended periods away from their wives or girlfriends, and it is common for them to contract the virus from prostitutes and then return home to their villages, where they spread it still further. The HIV infection rate among migrant

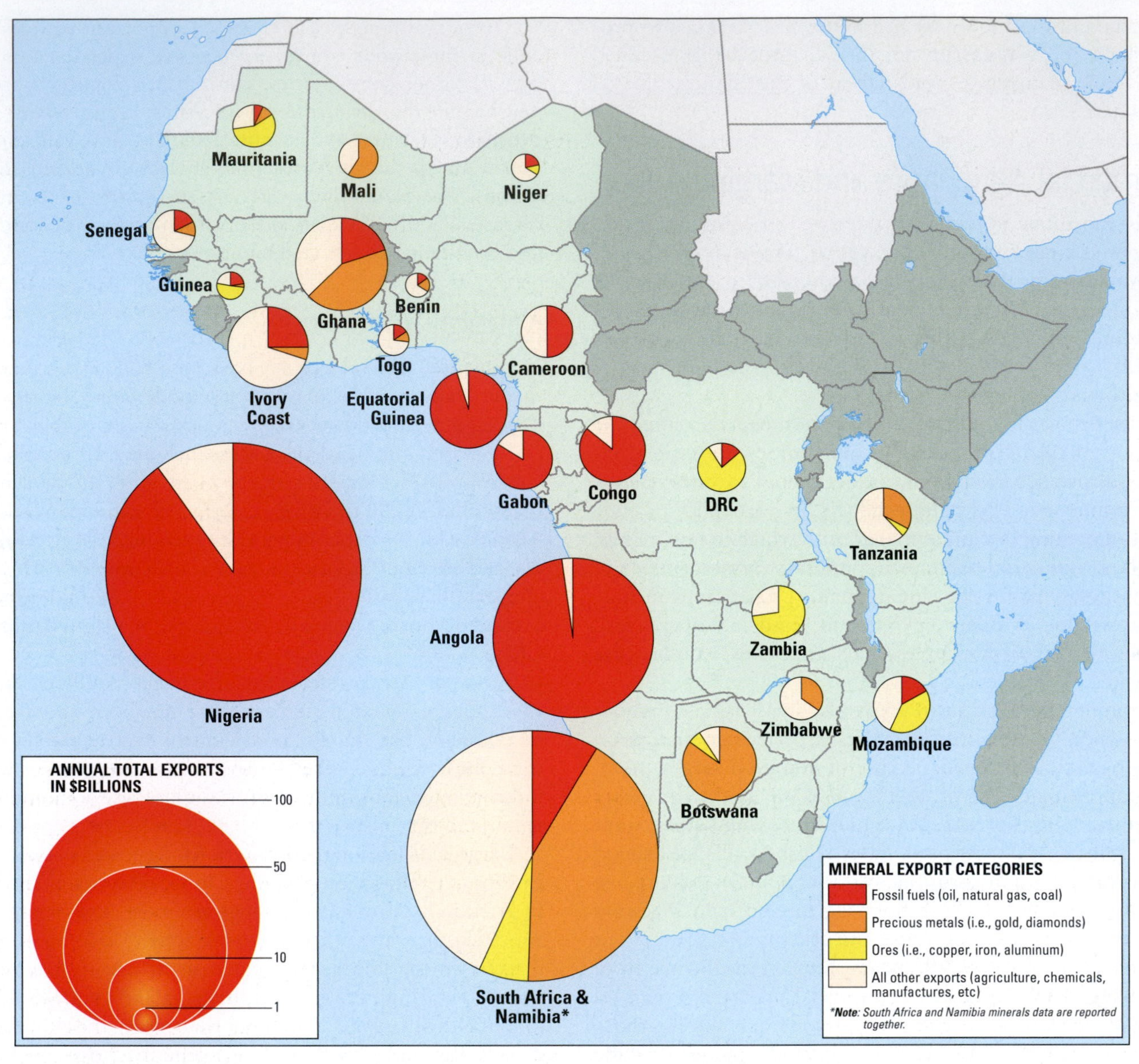

• **Figure 9.18** Many African economies are heavily dependent upon mineral exports, especially the oil-rich countries of West Africa.

miners working in South Africa was more than 25 percent in 2011—a high rate, but down from 40 percent just three years earlier.

Commodities: Boom! Crash!

Ever since colonial times, Africa's place in the commercial world has been mainly that of a producer of primary products, especially the cash crops and raw materials (particularly minerals) discussed above, for sale outside the region. In many nations, one or two products supply more than 40 percent of all exports. As recently as 2005, for example, Angola's oil and diamonds made up 99 percent of the country's exports, and in Kenya coffee and tea remain overwhelmingly important (•**Figure 9.19**). Such countries are vulnerable to international oversupply of an export on which it is vitally dependent; oversupplies cause prices to crash. But when demand for these exports is high, the money pours in.

Prior to a sharp downward trend beginning in 2014, the recent upward prices for commodities, driven especially by demand in China, India, and the older industrialized economies, have generally been a blessing for this resource-rich region. The ironic challenge is to prevent this blessing from becoming a mixed blessing or, worse yet, a curse (see Insights, **page 424**).

In Chapter 3 we considered the four poverty traps that impede development: conflict, natural resources, landlocked situations, and bad governance. These poverty traps converge in Sub-Saharan Africa to give this region the world's largest share of the "bottom billion," the poorest people on Earth. 60

Problem Landscape Cleaning Up the Dirty Diamonds

"Diamonds are a girl's best friend," sang Marilyn Monroe. *Diamonds Are Forever* proclaimed the title of a James Bond Film. Another film did not have such an endearing name: *Blood Diamond*. This film tells part of a sordid story about diamonds: as recently as 2004, as much as 15 percent of the world's annual production of rough diamonds was made up of **dirty diamonds** (also called "conflict diamonds" and "blood diamonds"), defined by the United Nations as "rough diamonds used by rebel movements or their allies to finance conflict aimed at undermining legitimate governments." Gems of such questionable origin financed at least three African wars. But diplomatic, nongovernmental, and business efforts have largely stemmed the tide of Africa's dirty diamonds.

In 2002, following four years of negotiations, 45 countries endorsed a UN-backed certification plan, called the **Kimberley Process**, designed to ensure that only legally mined rough diamonds, untainted by violence, reach the market. Rough diamonds must be sent in tamper-proof containers with a certificate guaranteeing their origin and contents. The importing countries (the biggest of which are China and India) must certify that the shipments have arrived unopened, and reject any shipments that do not meet the requirements. Only countries that subscribe to the Kimberley Process are allowed to trade in rough diamonds.

Corporate interests and consumer ethics have cleaned up the diamond trade. Diamond certification has been spearheaded by the world's leading diamond company (with about two-thirds of the market), the South African-based multinational De Beers. De Beers was embarrassed by a report that it had bought $14 million worth of diamonds from Angolan rebels in a single year. Perceiving that a public relations debacle could lead to a business disaster, as happened with an organized boycott against fur products in the 1980s, De Beers seized the initiative. In the name of Africa's welfare, but certainly also as a means of increasing demand and profit, De Beers introduced **branded diamonds**, certified as coming from nonconflict areas. Other diamond companies followed suit, especially in an effort to please Americans, who buy 40 percent of the world's diamonds.

Diamonds are not the only potentially "dirty" African mining products; "conflict metals" or "conflict minerals" are discussed on **page 439**. And outside the mining sector, there have been efforts to certify a legal trade in stockpiled and confiscated ivory, but as we have seen, these have so far proved unsuccessful.

Surging commodity prices of natural resources do not establish a sustainable foundation for development. In fact, dependence on primary commodity exports is associated with three problems that sustain underdevelopment: economic shocks related to volatile commodity prices, poor governance, and a high incidence of civil war.[19] These deter investment and the kind of economic diversification that ideally should include manufacturing and services.

The tide of globalization that swept the world between about 1980 and 2000 changed the structure of developing country exports profoundly. In 1980, 75 percent of LDC exports were primary commodities, but by 2000, 80 percent were manufactured goods. LDCs generally broke loose of their dependence on primary commodities—but not the African countries. Sub-Saharan Africa has been the last world region to be part of this transformation, and we need to see whether African countries can succeed in joining the global market for manufactures or secure a footing with profitable service sectors.

Yvette Cardozo/Alamy

• **Figure 9.19** The Kiambethu tea farm is just 20 miles from Nairobi in Kenya's beautiful highland country. Kenya's economy is highly dependent on exports of tea and coffee. Over-reliance on cash crops and other raw materials makes many Sub-Saharan African countries vulnerable to price swings in the global economy.

Economist Paul Collier did a statistical analysis of the relationship between primary commodity dependence and the risk of civil war. He found that in any given five-year period, countries not dependent upon primary commodity exports had only a 1 percent risk of civil war, whereas

Insights The Resource Curse

Many countries in Africa have been described as suffering from the **resource curse**, also known as the "paradox of plenty." This is a paradox in that a country with a great abundance of a valuable natural resource often experiences lower economic growth than countries without such an endowment. Nigeria provides a good example. More than $300 billion in oil revenues have poured into the country in the past three decades, yet 62 percent of Nigerians live below the poverty line (see **page 436** for more discussion of Nigeria's oil problems), a number barely changed from 1992. Primary commodity dependence is not just an African problem, but lingers in LDCs elsewhere. Venezuela's oil minister was quoted famously as saying that oil was not black gold; it was "the devil's excrement."[21]

What causes the resource curse? There are economic reasons: it is difficult to plan development when commodity prices (like Nigeria's oil) swing wildly up and down; oil prices, for example, can reach lofty heights and then plummet. These boom and bust cycles lead to risk taking, debt, and overinvestment. With busts, there are budget crises that hurt the poor. Oil-fueled growth does not always create abundant jobs; although oil can account for as much as 80 percent of a country's revenues, it often employs fewer than 10 percent of the workforce. Against this backdrop of economic inequality, the revenue earned from oil crowds out other sectors, such as agriculture, manufacturing, and services.

As we saw in Chapter 5 (**page 175**) the resource curse in Russia and other MDCs has been called the "Dutch Disease" because of circumstances following the discovery of natural gas in the Netherlands' waters in the 1950s. As large foreign exchange inflows poured in, local costs went up, so imports undercut domestic manufacturing and agriculture. All domestic industries except the fossil fuel industry went into a tailspin.

There are also political reasons for and consequences of the resource curse. A glut of money from the resource provides incentives for corruption and repression of dissent. Officials and company executives get rich, and the poor get poorer. Oil-producing LDCs often spend as much as 10 times what poorer LDCs do on their militaries, and are more likely to get involved in conflicts, both external and civil.

Is there light at the end of Africa's resource curse tunnel? Recent developments in some countries suggest there is. When commodity prices fall, the typical pattern is that the overall economy falls with them; for example, during an oil price fall in Nigeria in 1998–1999, Nigeria's currency lost 80 percent of its value, and the Nigerian economy crashed. What happened with the 2014–2015 oil price fall? Currency values fell by more than 10 percent in 10 African countries, but there were no catastrophic drops like Nigeria's 15 years earlier. The reason is that African economies are starting to diversify beyond natural resources to include manufacturing and services, and beginning to escape their curse. Successful countries include those like Botswana that are adding value to resources before exporting them—in that case, by doing processing, marketing, and sales of diamonds (previously these Botswanan gems were cut in London). Botswana is also diversifying into manufacturing and services and by this route achieved middle-income status in 2007.[22]

countries with primary commodity exports had a 20 percent risk.[20] In 1970, Africa had fewer civil wars than other developing regions. But in the period 1970–2000, although civil conflict declined in the other regions, it increased in Africa, so that by 2000 it had more civil wars than did any other region. Controlling large rural/agricultural regions, numerous rebel organizations looted revenues from primary commodity exports. Marauding across Zaire (now the Democratic Republic of the Congo) in 1997, rebel leader Laurent Kabila told a reporter that rebellion in Zaire was easy: all one needed was $10,000 (enough to hire a small army) and a satellite phone (for making deals on mineral extraction). Low costs and high revenues in a region marred by poverty made Africa attractive for rebel entrepreneurs. Unfortunately, the leaders of some of these rebel groups also wrested control of governments. These rebel leaders represented the very worst examples of "bad governance," which, although still prominent in the region, has typically become less sinister.

Manufacturing and Services

Manufacturing and services are growing steadily and beginning to chip away at Africa's dependence on primary commodities. Telecommunications, transportation, and finance are picking up. Economic diversification beyond raw materials is being fed by foreign direct investment. More FDI is going into resource-poor countries and into non-resource sectors like finance. Resource-dependent countries like Nigeria and Angola are working hard to diversify. Although government revenues still largely depend on oil in both countries, oil now represents just 15 percent of Nigeria's GDP, down from 38 percent of GDP in 2008. Non-mineral economic sectors including agriculture, manufacturing, and construction now make up a third of GDP in Angola, compared to just 1 percent in 2005.

There is also great potential for African nations to boost their services sectors. There are more English speakers in Sub-Saharan Africa than in Asia, and African time zones are closer to those of Europe and the US than Asian time zones are. Global telecommunications should therefore favor the kinds of outsourced services from Africa that several Asian countries have been providing (see Africa's New Talking Drums, **page 426**). Recognizing such opportunity in Ghana, where the official language is English, the American health insurance company Aetna hired 3000 people to process its paperwork. Botswana, where English is also the official language, has developed a number of financial and producer services, commercial banking, marketing, and product design. Service industries like these are especially appropriate for the

15 landlocked African countries like Botswana that are not well located to export manufactured goods. "Business parks" geared toward these services could be ideally located near the African universities that produce so many fluent and skilled graduates.[23] African countries that have been heavily dependent upon mineral exports but are now running low on reserves are almost compelled to boost services to keep their economies afloat.

Tourism falls in the service sector, and the region is building on its long history of offering the visitor the experience of a lifetime going on safari and experiencing local cultures. An estimated 1 in 20 jobs in Sub-Saharan Africa is related to tourism. About 35 million tourists visit the region each year, and an enormous infrastructure of goods and services supports them: the wider impacts of direct and indirect contributions of travel and tourism generate nearly $170 billion in revenue, or nearly 9 percent of the region's GDP. Growth of about 4 percent per year in the travel sector is anticipated through 2024.[24] A growing number of visitors are from within the region, reflecting the growth of Africa's middle classes. In recent years, South Africa has been the region's leading destination, with about 10 million visitors annually, followed by Namibia, Zimbabwe, Kenya, and Tanzania.

Trade Not Aid!

"Trade Not Aid!" is a growing mantra of African economics. But Africa's exporters are finding it difficult to get a solid foothold. One of the major obstacles to African economic development is that many countries outside the region have effectively closed their doors to African imports. These restrictions typically take the form of subsidies, high tariffs, or low quotas imposed on agricultural products (like cotton) or manufactured goods (like textiles). Potential importing countries, such as the United States, impose these restrictions to protect their own industries. Earlier, we saw how the subsidies of Europe's Common Agricultural Policy hamstring African farmers. 421 Government subsidies of American farmers likewise have disadvantaged African producers. Although the US is the world's leading exporter of raw cotton, several African countries, such as Burkina Faso and Ivory Coast, are rapidly gaining market share themselves. However, for years they were not playing on a level field in the global marketplace. The US government spent $2 billion each year from 1995 until 2012 to subsidize its 25,000 cotton farmers, prompting higher US production and exports, and consequently lower cotton prices and incomes for 472, 526 African farmers.

African middle classes are not yet large enough to consume what their countries could make, so most of the region's manufactured goods are produced for export. African producers have energetically been creating **export processing zones (EPZs)**, which are free trade areas offering incentives for international companies to invest. Incentives designed to boost an economy's competitiveness and reduce business entry and operating costs include import and export duty exemptions, streamlined customs and administrative controls, and income tax breaks. Recently 20 countries in the region have employed more than a million workers in 91 EPZs, producing clothing, electronic and light electrical goods, optical goods, watches, leather footwear, furniture and toys, and services. Although impressive numbers, these represent less than 3 percent of all the EPZs in developing countries; Asia and especially China with its similar SEZs dominates the field, followed by Latin America. 352 In many African examples, the lack of government commitments to follow through on the projects, bureaucratic obstacles, high up-front costs of building infrastructure, and the economic costs of providing tax breaks to foreign investors have raised questions about whether EPZs make economic sense.[25]

African EPZs are generally not providing stable employment with opportunities for sustainable growth and upward mobility. The most successful zones are those incentivized with Free Trade Agreements (FTAs) and special programs that guarantee minimum wages and other standards like the European Union's Everything But Arms (EBA) programs and particularly the American **African Growth and Opportunity Act (AGOA)**, passed by the US Congress in 2000 and reauthorized in 2015. AGOA reduced or ended tariffs and quotas on more than 1800 manufactured, mineral, and food items that could be imported to the US from about 40 countries in Africa. African exports to the United States have surged, especially in the clothing industry. Apparel made in Lesotho, Kenya, Mauritius, South Africa, Madagascar, Mauritania, and elsewhere in Africa is now sold in Target and Walmart stores (**•Figure 9.20**).[26] Under AGOA, the United States has also allowed tariff (tax) free imports from the group of countries comprising the **Southern African Customs Union (SACU)**: South Africa, Lesotho, Swaziland, Botswana, and Namibia. SACU is one of several trade pacts and organizations within Africa. These entities make it easier for international trade organizations to do business with Africa. In many cases, jobs and incomes for Africans have grown because of new opportunities created by trade pacts. AGOA, for example, helped tiny Lesotho establish a market niche in "sweatshop-free" garment production.

Addressing the 8th Millennium Development Goal ("Develop a global partnership for development"; see **page 434**), the European Union is also attempting to focus on Africa with a trade liberalization policy that would give African countries privileged status allowed by the World Trade Organization's Regional Economic Partnership Agreement (REPA) provision.[27]

Down on Debt

For decades, almost all of the countries of Sub-Saharan Africa were heavily in debt to foreign lenders. Young nations undertook costly development projects with borrowed money, particularly after the mid-1970s. Western financiers, planners, and contractors gave optimistic assessments of the benefits to be expected from such projects, and African leaders were ready to accept loans as a way to reap quick benefits from newly won independence. Countries soon found they had great difficulty in meeting even the interest payments on their debts, and their efforts to do so often resulted in further economic woes and destructive environmental effects. 64

Insights Africa's New Talking Drums: The Mobile Phone Revolution

There are more telephone land lines in the city of Tokyo than in all of Africa. So what? Who needs landlines? Not Africans. Without an infrastructure legacy, Africans are leapfrogging their telephone systems into the 21st century. Of all the world regions, Africa has been the most underserved by both mobile phone and Internet access. This represents an enormous market; Africa has the fastest-growing mobile phone market worldwide. Mobile phone network coverage was available to just 10 percent of Sub-Saharan Africa's people in 1999, but that grew to over 85 percent in 2014. In 2015, there were nearly 400 million users, up from 150 million users in 2008. There has also been explosive growth in broadband Internet coverage, made possible by recent projects laying tens of thousands of miles of undersea fiber optic cables.

Mobile phones are penetrating Africa far more quickly than the Internet, and African peoples are using them in many ways: to transfer money, buy and sell goods, and even turn on water wells. Cell phones are proving to be a critical tool in rural development. In Uganda, for example, farmers in remote areas are communicating with plant scientists in the nation's capital to report the incidence and spread of a disease infecting bananas. The farmers take photos, mark global positioning system (GPS) locations, write remarks, and then transfer all these data by mobile phone to the scientists. The scientists then prepare an attack plan against the disease and communicate disease treatment procedures back to the farmers. Using text messaging, villagers are also able to learn more about farming practices and other critical issues like health care. In time, web-enabled cell phones will make it much easier for millions of Africans to jump to the Internet without having to use computers. In 2015, only 20 percent of the handsets were data enabled, but that is expected to grow rapidly.

As this information suggests, cell phones are at the heart of a technological revolution in Africa's rural regions. Every phone owner needs to recharge the phone, but with most homes lacking electricity, that requires a trip to town and a recharge fee. The increasingly popular alternative is the purchase of an $80 Chinese-made photovoltaic power system that can recharge a phone and power a lamp. The investment pays for itself quickly. The solar panel is "green" in many ways, producing no greenhouse gases and also averting the need to buy candles, charcoal, batteries, wood, and kerosene. Perhaps the most impressive result of the purchase is a marked improvement in children's grades at school: they now have light for studying at night.

Whether by mobile phone or computer, in beginning to overcome the digital divide, African peoples are ramping up their development prospects. The World Bank reports that for every 10 percentage points of increase in high-speed Internet access, economic growth rises 1.3 percentage points. Internet connections are critical for the establishment of new businesses and the expansion of existing ones. Voice-over-Internet technology is making telephone calling more affordable for hundreds of millions of Africans and is also making African call centers more competitive with the established Indian market.

A valuable Internet tool created in Africa and used there and elsewhere mainly on cell phones is **Ushahidi** (ushahidi.com). Ushahidi (meaning "witness" or "testimony" in Swahili) is at the intersection of critical current events and their geospatial context. It originated as a means of tracking the violence accompanying the disputed 2008 presidential election in Kenya. Eyewitnesses sent text messages or wrote emails about what they saw, and each of these communications was "georeferenced" (placed on a known coordinate mapping system). With **georeferencing**, it was therefore possible to see in real time where the hot spots of violence were. Authorities, supporters, opponents, and others were able to act or make plans with this information. This kind of information can be thought of as **crowdsourcing**, or obtaining eyewitness accounts from a large group of people. 460

There are no limits to the possible applications of Ushahidi. It is particularly valuable as a means of reporting the fallout from human and natural disasters, and as a means of calling people together for social action, as it was in the 2011 Arab Spring protests and in the 2015 Nigerian elections, for which Ushahidi users created an "early warning system" against electoral violence. Its "CrisisNet" function, which Ushahidi calls the "firehose of global crisis data," is keeping real-time track of the fighting in Syria. 63 Have a look at the site to see what recent events Ushahidi has covered and to appreciate how inherently geographic this tool is. Having begun as "activist mapping," bringing together individual testimony with activist causes in their spatial context, Ushahidi—like Twitter and Facebook—now has no bounds and has a unique attribute: whatever Ushahidi is covering, it is doing so geospatially.

There is a widespread perception that Africa is always behind. Certainly much of the discussion in this chapter suggests that Africa has a lot of catching up to do. But Ushahidi is just one example of the surging innovation and creativity in African hands. As Pliny the Elder (23–79 CE) wrote, "There is always something new out of Africa."

Western donors began attaching more strings to loans as means of promoting democracy. African leaders were told that if they did not institute democratic reforms, hold fair elections, or improve bad human rights records, their countries would no longer be eligible for development aid or lending programs. Often the leaders made just enough concessions to win the aid without instituting real reform—a phenomenon known as **donor democracy**. (Other African countries simply turned away from the US and Europe and allied with the Soviet Union instead, which had no democracy requirements in its aid packages.)

In the West, increased foreign aid in the form of gifts or grants to Africa became controversial. Critics argued that foreign aid only increases dependence and deflects national attention away from problems underlying famine. International aid monies often free up funds from national treasuries to be spent on unfortunate uses. For example, when foreign money became available for famine relief, it allowed Ethiopia and Eritrea to spend more on weapons used in their war. Aid monies are often simply stolen, and monitoring by donors of how they are spent has been inadequate.

Rijasolo/Riva Press/Redux

• **Figure 9.20** Special programs like the US African Growth and Opportunity Act incentivize factory production in numerous African countries by reducing or eliminating tariffs. This is an AGOA-supported garment factory in Madagascar.

China's trade with Sub-Saharan Africa increased nearly 50-fold between 1999 and 2014; by 2010, China became the region's leading trade partner.

The growing value of China's aid rivals that pledged by Africa's traditional financiers in the West. There is an important difference, however: unlike the West, China has a "no strings attached" policy. It is not insisting on political reforms, is not expressing concern about environmental or social responsibilities, and (much to the chagrin of the West) is even engaging with repressive governments that the West is trying to isolate. At the same time, Western governments and financial institutions are increasingly alarmed that China is making rapid inroads into African markets at the expense of Western economies and geopolitical interests.

Since the 1990s, Western donors have proceeded more cautiously, more often funneling money through private and religious groups working in health care, education, and small-business development, using microcredit practices (see Try It, **page 428**). Foreign aid has been very effective in advancing primary and secondary education in the region. However, many African governments view universities as potential hotbeds of dissent and discourage investment in higher education. Africa's universities are crumbling, and the classic brain drain of African talent to Western institutions means there is that much less intellectual capital left at home to invest in the region's development. The development of the business parks proposed above could help reverse that brain drain.

76, 128, 129, 157, 175

Africa's untenable debt situation, and the slow flow of aid to the region, began to change around 2005. Two major factors were at work. First, in the West, awareness of the humanitarian costs of not directing more resources to Africa was growing. Media, political, and corporate stars kept up a steady appeal for governments and citizens of the MDCs to be more generous to impoverished Africa. Many efforts were directed specifically to meet the Millennium Development Goals. Second, just as it was becoming clear that the West was falling behind on its promised course to double aid to Africa, China entered the picture.

China Steps In

The second and more powerful shift in Africa's debt, trade, and aid picture came after 2006 with the meteoric rise of China's engagement with the region. In 2007, China pledged $20 billion in infrastructure and trade financing in the region.

Africans themselves look with some trepidation on China's advance. Some complain about China's record of not hiring locals, instead bringing in Chinese nationals for virtually all labor and management positions. In other cases, Chinese firms have reportedly mistreated African workers and paid them what have been described as "slave wages." Some Chinese projects have been documented as having detrimental environmental impacts. Although China is helping several African countries to build EPZs, just as China did to spur its own economic growth, China has also devised a unique way of profiting from these zones. China has taken advantage of the fact that AGOA-supported EPZs in Africa lack "rules of origins" provisions. Chinese and Taiwanese firms set up shop in the EPZs where they could enjoy duty-free access to US markets; they also quota-hop, avoiding punitive quotas on China's exports. In this light, many EPZs, especially in Lesotho, Madagascar, and Kenya, function virtually as transporting docks for foreign-sourced, fully assembled goods ready to go on to their final destination, tax free. In one Kenyan processing zone, nearly all of the 34 garment plants had Asian owners. A survey of Chinese and Taiwanese entrepreneurs found that "taking advantage of international trade agreements" was among their top five motives for investing and operating in Africa.[28]

China is unashamed about its interests in Sub-Saharan Africa: it needs and wants the region's raw materials—especially oil, iron ore, copper, and cotton—to feed its industrial appetite. To many Africans, China is building the same kind of exploitive mercantile relationship that characterized the colonial West: it takes Africa's raw materials, transforms them into value-added products, and sells them mainly to its own advantage. South Africa's president warned that Africa

427, 468, 484

Try it Microcredit

A promising alternative to conventional development based on large, top-down projects funded with nationwide borrowing and debt is **microcredit** (or microlending, microfinance), the lending of small sums to poor people to set up or expand small businesses. Typically, poor people have no collateral—assets to pledge against loans they seek—and therefore cannot borrow from commercial banks. This drives them into the arms of local loan sharks, who may charge huge interest rates of up to 10 to 20 percent *per day.* But in the interest of fostering development, microlending organizations have begun to encourage poor borrowers to form a small group that will cross-guarantee all members' loans. One member of the group may take out a loan of $25, for example, to help start up a small business such as a roadside restaurant. Only when that individual has paid it back may the next person in the group receive a loan. Peer pressure minimizes the default rate, and the accumulated history of timely paybacks increases the entire group's creditworthiness.

Microlenders generally prefer to lend to women because women are more likely to use the money in ways that will feed, clothe, and otherwise benefit their children, whereas men may be more likely to spend the money on alcohol or other frivolous pursuits. One fear in Africa is that microloans will not be repaid if the borrower or someone in the borrower's family contracts HIV or some other debilitating illness. Therefore, the loans are often extended only after the borrower receives some amount of public health training, for example, in the use of birth control and in general food preparation hygiene.

One microlending project that has received widespread and generally positive attention is *Kiva.* Many people find it attractive because they can lend small amounts and track the progress of the individuals they are helping (see kiva.org). Want to try it? There is a chance you would lose some or all of your investment of $25 or more, but the website states that the repayment rate has been 98.72 percent since the program began. You may want to at least explore the site to get a sense of how individuals and small groups around the world are seeking to improve their well-being through microborrowing.

risked becoming an economic colony of China. But desperate for assistance, even the wary are inclined to accept China's terms. The most typical arrangement is the **infrastructure-for-mineral swap**, in which the African country grants mining concessions to the Chinese, and the Chinese build railways, roads, schools, and hospitals. There is especially great appreciation for the schools—there are simply not enough of them in many African countries. China's model of state-led capitalism has proven attractive to many African governments. African officials go to Beijing to learn how to run state-owned companies more profitably.

It is useful to compare Chinese with American and other Western aid and interests. Most Chinese aid goes to transportation and storage equipment, energy generation and supply, communications, and upgrades to mining projects. Chinese government loans and related financing exceed those that the US extends to Africa. The priorities for the US, which is Africa's largest foreign aid provider, are in different directions: population/education/reproductive programs, health, agriculture, and food aid/security (**•Figure 9.21**). Are there "good guys" and "bad guys" that emerge from comparisons of Western and Chinese aid to Africa? Although the Chinese may appear to act in wanton self-interest, the West is not unsullied: A 2014 study by a group of UK and Africa-based NGOs concluded that Western countries are using aid to Africa as a smokescreen to hide the "sustained looting" of the continent. While each year Sub-Saharan Africa receives $134 billion in loans, foreign investment, and development aid, $192 billion leaves the region, creating a huge shortfall. According to the report, foreign multinational companies siphon $46 billion out of Sub-Saharan Africa each year, while $35 billion is moved from Africa into tax havens around the world annually. Meanwhile, African governments spend $21 billion a year on debt repayments. The report underscores that aid sent in the form of loans serves only to contribute to the continent's debt crisis.[29]

These discouraging findings about the ongoing looting of African resources suggest that even while African economies are making progress in joining the global economic system, outmoded external approaches that hamper the region's development are still in play. "The fundamental structure of relations remains intact," Richard Grant writes. "Simply put, improved economic performance may be due more to a new scramble for Africa than anything else. Many in the global North continue to understand Africa in an outdated and partial way as an inexpensive source for raw materials and/or as a rescue mission for their pop humanitarian impulses. Being caught in a loop between these positions has prevented the

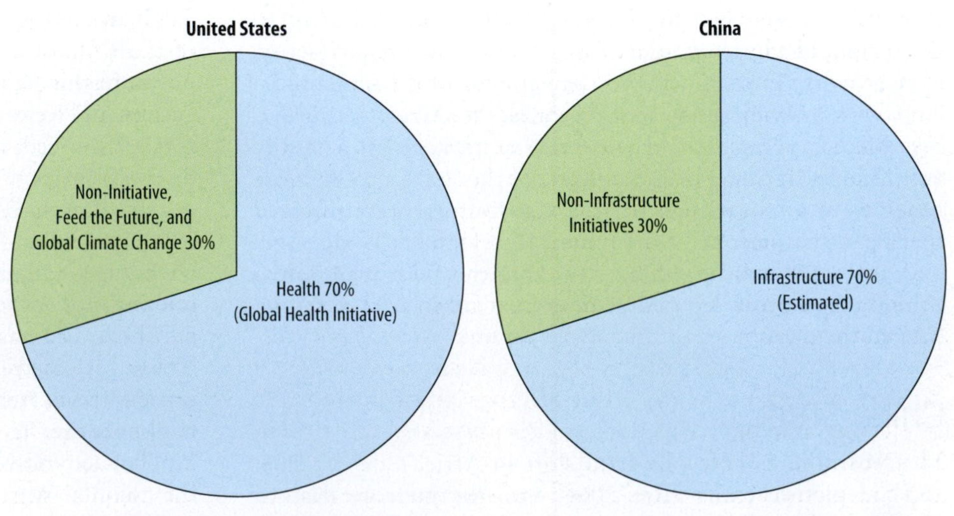

• Figure 9.21 There are significant differences in US and Chinese aid to African countries. The US **(a)** prioritizes health and social welfare, whereas China **(b)** targets infrastructure.

West from expanding its thinking about Africa, and prevents a balance of stories about Africa and Africans."[30]

Reshaping the African narrative from within, there are a growing number of African institutions designed to encourage economic growth and political stability. In 2002, the 53-member **African Union (AU)** was formed to replace the old and often inconsequential Organization of African Unity (OAU). One of the African Union's main goals is to implement self-inspection, or "peer review," to help promote democracy, good governance, human rights, gender equity, and development. It has formed an agency responsible for this monitoring, known as the **New Partnership for Africa's Development (NEPAD)**. NEPAD's premise is that improvements in such human affairs will improve the climate for international investment and assistance to Africa. The African Union has also created a 265-member (in 2015; just 235 in 2004) Pan-African Parliament to make regional laws.

Africa's Fragile Infrastructure: The Same Old Problems

Part of the reason that African countries have not made the leap to manufacturing—why many EPZs have not succeeded, for example—includes what the African Center for Economic Transformation calls "same old problems" hindering African development since the post-colonial era began: weak infrastructures, including deficiencies in telecommunications and power supply, unreliable and costly transportation services, and poor geographic situating of manufacturing centers.[31]

Africa has a huge infrastructure deficit, with 30 percent of its existing infrastructure, much of it built in colonial times to service mineral production, in dire need of rehabilitation (see Figure 9.19). Annual per capita expenditure on urban infrastructure is very low. Most African countries experience intermittent power cuts, and even South Africa, with the region's best infrastructure, imposes rolling blackouts to avoid widespread power disruptions. Overall, poor infrastructure reduces Africa's GDP growth by 2 percent per year, and the enormous funding to update infrastructure is out of reach from within.[32] In this context, it is easy to appreciate why China's focus on infrastructure-for-mining swaps is so compelling.

Geographically, manufacturing locations that are close to markets and suppliers, thereby economizing on transport costs, have a comparative advantage. Africa's many landlocked countries face prohibitive costs to export their manufactured goods. Coastal African countries in contrast are well-situated—much more so than Asian countries—for exports to European markets.[33] The Indian Ocean island of Mauritius initially was handicapped by its "tyranny of distance" from markets, especially because ocean cargo routes bypassed it, but shipping lanes were altered to accommodate Mauritius (see Figure 2.11 of the world's shipping lanes, page 38). (The **tyranny of distance** is the effect of geographical remoteness on shaping a country's identity, and is associated most often 384 with Australia and New Zealand.)

The region critically needs a good transportation network to enlarge market opportunities. Poor transportation is also often a contributing factor to famine. In Ethiopia, for example, the road network is so poor that it is extremely difficult to get food from the western part of the country, which often has crop surpluses, to the eastern part, which has chronic food shortages. On more than one occasion, it has proved cheaper to ship food from the United States to eastern Ethiopia than to truck it across this Horn of Africa country! Similarly, a World Bank study showed that it cost $50 to ship a metric ton of corn from Iowa 8500 miles (13,600 km) to the Kenyan port of Mombasa, but $100 to move it from Mombasa 550 miles (880 km) inland to Kampala, Uganda. Transportation problems contribute to the high costs of agricultural inputs like fertilizers, which in this region cost two to three times what they do in Asia. Poor communication, especially in digital technology applications, also hinders development. But as we have seen, that is changing quickly and for the better (see page 426).

To summarize this discussion of economic geography, it must be recognized that Africa's current development pathway is fraught with difficulties despite the region's overall high economic growth rates. Particularly vulnerable subregions are the Sahel, the Horn of Africa, and the Great Lakes area of East Africa, where there are still conflicts, unsustainable rates of natural resource exploitation, disease, rapid population growth, and the geographic handicaps of being landlocked. Here we need to turn to the region's geopolitical situation.[34]

9.5 Geopolitical Issues

Sub-Saharan Africa has waxed and waned as a theater of geopolitical interest since the end of World War II. The great powers were interested in the region during much of the 20th century. To boost their competing aims during the Cold War years, the Soviet Union and the United States played African 150 countries against one another, arming them with weapons with which to wage **proxy wars** (wars instigated or carried out by major powers that do not themselves participate). The superpowers also extended aid generously to many African nations, but not simply for altruistic reasons; as external powers always have, they coveted the continent's natural resource riches.

Great power concerns about Africa changed with the end of the Cold War around 1990, and the nature of conflict in Africa changed. The great powers withdrew support, but their weapons remained to prolong smoldering conflicts. The United States helped build the arsenals of eight of the nine countries involved in the Democratic Republic of the Congo conflict, for example. Cold War–era weapons like AK-47 assault rifles were "dumped"—sold at low prices—in Africa, where they were not considered obsolete and where they fanned the flames of conflict. Increasingly, fighting began to spread across international borders. Previously, the United States and the Soviet Union had maintained a kind of security balance that kept warfare from becoming internationalized, but that restraint no longer exists.

The scenario of regional or even Africa-wide wars was realized with Africa's first "world war," centered on the Democratic Republic of the Congo from 1996 to 2003 (see page 438). In the absence of regional or international powers to keep the peace, some countries (for example, Sierra Leone) simply imploded, fragmenting into fiefdoms run by factions that claimed to be revolutionaries but were essentially profiteers. Other national governments, such as Ethiopia's, spent vast sums on expensive military aircraft while their people suffered from malnutrition.

The end of the Cold War constricted (and in the case of the USSR, ruptured) aid pipelines. The United States cut its economic assistance to Sub-Saharan Africa by 30 percent between 1985 and 1992 and reduced it another 10 percent between 1992 and 2001. Why the waning concern? Early in 2001, former US President George W. Bush was quoted as saying, "While Africa may be important, it doesn't fit into the national strategic interests, as far as I can see them."[35] This was a widely held viewpoint in Washington at the time.

243

Then came September 11, 2001. Al-Qa'ida's attack on the United States and the US counterstrike against al-Qa'ida in Afghanistan raised fears that this Islamist terrorist organization might find new recruits and training grounds in Africa. Al-Qa'ida had already bombed US embassies in Kenya and Tanzania in 1998, and in 2002 struck again in the Kenyan city of Mombasa. The United States maintains a focus on **terrorism hot spots** in the region, including Kenya, Somalia, Djibouti, Niger, Chad, and Mali, all of which are regarded as real or potential training grounds for al-Qa'ida- and ISIS-affiliated groups. More broadly, the geographic patterns that emerge from terrorism's distribution across Africa are three discernible corridors: the East African corridor of al-Shabaab operations, particularly in rural areas; a second corridor of al-Qa'ida in the Islamic Maghreb (AQIM)–dominated activity based in northern Mali and extending into Mauritania, Niger, Burkina Faso, and Algeria; and a third corridor, with Boko Haram's reach from northern Nigeria into the neighboring countries of Chad and Cameroon (refer to the map in **Figure 6.I**).[36]

For many years following the violent deaths of US service personnel in Somalia in 1993 (the "Blackhawk Down" incident, in which al-Qa'ida was also involved), there was great reluctance to project US military power in African trouble spots. Even al-Qa'ida's simultaneous attacks on US embassies in Kenya and Tanzania in 1998 did not change American military commitments in the region, but the 9/11 attacks did. Since 2001, the United States has maintained a naval base at Camp Lemonnier in Djibouti, a tiny but strategically located country from where US military assets are projected against targets in nearby Somalia, Yemen, and other countries where Islamist militants are active. Unmanned drones equipped with laser-guided weapons have been the weapons of choice against suspected al-Qa'ida and other terrorist targets. Still, the US presence in Africa remains small; Lemonnier is its only African military establishment, and the entire continent represents just 5 percent of the Pentagon's annual budget.

260, 319

Terrorism is a darkness overshadowing many parts of "the bright continent." Of major concern are the nihilistic al-Shabaab, based in Somalia, and Boko Haram of Nigeria. Along with **al-Qa'ida in the Islamic Maghreb (AQIM)** (see **page 260** in Chapter 6), these groups have arisen and proliferated in geographic and socioeconomic conditions shared with other terrorist movements in the North Africa, the Middle East, and South Asia: porous borders and ungoverned spaces of failing or failed states; widespread poverty and alienation in undemocratic, repressive societies; lingering ethnic and sectarian conflict; outgunned national militaries; and abundant sources of funding from the region's natural resources. Smuggling of conflict diamonds has funded al-Qa'ida operations in the region, for example. Al-Qa'ida also uses West Africa as a way station in cocaine smuggling from South America to European markets. From 2007 to 2011, al-Qa'ida and its al-Shabaab affiliates controlling much of central and southern Somalia siphoned revenues from illegal acacia charcoal smuggled to Arabia and from smuggled ivory. Once again, Africa's natural resources are being plundered.

Al-Shabaab, "the Young Ones," a mainly ethnic Somali terrorist Islamist group, operates in the Horn of Africa and East Africa. Somalia has lacked a functioning government since 1991, and al-Shabaab has flourished in this lawless land. Initially, its goals were to oust Ethiopian troops and all other foreign presence from Somalia and to create an Islamic state in the country. After Ethiopia withdrew, its ambition grew to waging global *jihad* and establishing a global caliphate. In 2011, it pledged affiliation with the like-minded al-Qa'ida "Central" in Pakistan. Within the region, al-Shabaab also has ties with al-Qa'ida in the Islamic Maghreb (AQIM) and with Boko Haram.

243

After a three-year campaign dubbed "Operation Linda Nchi," a UN-backed African Union force ousted al-Shabaab from the Somali capital Mogadishu in 2011. But al-Shabaab retained control over a large part of southern Somalia, denying humanitarian access to this region in the grip of drought and famine. From southern Somalia, al-Shabaab has been able to infiltrate and terrorize neighboring Kenya. In proclaimed retribution for Linda Nchi, in September 2013 al-Shabaab gunmen seized control of Nairobi's Westgate Shopping Mall in a four-day siege that left 67 people dead. Many of those killed were Christians who had been singled out for execution. In northern Kenya, al-Shabaab employed similar tactics in a 2014 attack on quarry workers that left 37 non-Muslims dead, and in a 2015 assault on university students in which 147 Christians and converts to Islam died. Earlier terrorist attacks in and near the coastal city of Mombasa had already reduced tourism to Kenya's beautiful Indian Ocean coastline. The ongoing campaign elsewhere in the country led the US State Department to issue a travel warning to Kenya for its citizens, and many other countries issued similar advisories. Kenya had been one of Africa's leading safari destinations, but tourism there suffered; 1.7 million people visited Kenya in 2011, dropping to 1.4 million by 2013. Al-Shabaab promised that more Westgate Mall–style attacks would come.[37]

Elsewhere on the continent a number of groups carry the banners of Islamist terrorism. Often seizing the headlines is **Boko Haram**, meaning "Western Civilization Is Forbidden."[38] Founded within Nigeria's borders in 2002 and operating since 2009 mainly from the country's northeastern state of Borno, Boko Haram ranges freely across neighboring borders into Chad, Cameroon, and Niger. A Salafist Islamic group, its stated goals include the enshrinement of Islamic *shari'a* law in Nigeria and the establishment of an Islamic state, or Caliphate, in the region. Although 12 states in Nigeria's north adopted *shari'a* to varying degrees between 2000 and 2012, Boko Haram regards their interpretation of Islamic law as too 232 lenient. Boko Haram's strict interpretations of Islam echo those of ISIS in the heart of the Middle East, and in fact, after years of allegiance to and funding from al-Qa'ida via AQIM, in 2015 Boko Haram pledged allegiance 263 to ISIS. What places Boko Haram on the geopolitical radar screen is the prospect of it successfully capturing and holding ground and taking over or destroying Nigeria's economic lifeblood, the oil that it exports in abundance to the US and other consumers in the West.

AP Images

• Figure 9.22 In 2014, Boko Haram kidnapped more than 200 schoolgirls from the Nigerian town of Chibok, forcing them to convert to Islam, and raping and marrying many of them. International outrage came with the cry "Bring Back Our Girls!"

Nigerian army forces found it difficult to break Boko Haram's momentum. The group's strength grew enormously after 2011, when it was heavily armed with weapons spilling out of 257 fractured Libya, and its fighters received training by AQIM. As in Iraq under pressure from ISIS fighters, many Nigerian soldiers took flight from better-armed Boko Haram fighters using signature gruesome killing tactics: suicide and IED bombings, even of churches and mosques; assassinations of religious and political and traditional leaders (many of them Muslim); and kidnappings of foreigners, especially of women and children, avowedly in revenge for tactics used by the state against these militants. The group's atrocities included a mass kidnapping of more than 200 schoolgirls from the northeastern Nigerian city of Chibok in April 2014 (**•Figure 9.22**). These Christian girls were compelled to convert to Islam, and many were raped by or married off to Boko Haram fighters. The international campaign on Twitter and other social media crying out "Bring Back Our Girls!" evoked no compassion from Boko Haram.

Characteristically, France has sought to quell insurgencies in the "Francophone" countries that were its former colonies. In Mali, the Azawad National Liberation Movement, *Mouvement National Pour la Libération de l'Azawad*) (MNLA), a mainly secular Tuareg-dominated group whose ethnic kinspeople make up 10 percent of Mali's population, sought a political alternative to the Malian government. The MNLA seized control of northern Mali and in April 2012 declared the independence of the state of Azawad, with historic Timbuktu as its capital (**•Figure 9.23**). MNLA dominance was quickly challenged by the Islamist AQIM. Along with the Tuareg-dominated Islamist group Ansar Dine (AD), AQIM sought a *shari'a*-governed state by violent means, kidnapping foreigners in Mali and setting off car bombs in Timbuktu. While the MNLA fought for supremacy against AQIM and Ansar Dine, and the UN contemplated how to rid Mali of the "largest territory controlled by Islamic extremists in the world" (60 percent of the country's territory), France launched an intervention in January 2013.[39] French forces evicted the Islamist forces from Timbuktu, and MNLA leaders vowed they would fight alongside French and Malian troops in the future to prevent the imposition of an Islamist system on Mali.

Timbuktu's fate under the Islamists was heartbreaking. The city was established in the 12th century on the banks of the Niger River (the river's shifting channels has left the city miles away now). It occupied an important situation on Trans-Saharan trade routes for many centuries. Many civilizations left their mark on Timbuktu and endowed it with a wealth of cultural treasures; the entire old city is a UNESCO World Heritage Site. While AQIM rebels occupied the city in 2012–2013, they systematically destroyed much of this heritage, leveling 14 historic mausoleums and the El Farouk monument at the city's entrance. As did ISIS with Mesopotamian treasures in Iraq and Syria, these Islamists regarded the monuments as idolatry and blasphemy that had to be eradicated. A center of learning from the 13th through 19th 264 centuries, Timbuktu was a repository of Africa's written history, where scribes copied and authored innumerable works in the arts and sciences in a number of languages including Arabic, Fulani, Songhai, Bambara, Hebrew, and Turkish. AQIM also destroyed many of these irreplaceable works, the great majority of which had not been reproduced in any form.

There is consensus among analysts that the French intervention prevented Mali's government from falling to Islamists. Along with Boko Haram, those forces could have—and may still in the future—threaten the stability of Africa's most populous country, Nigeria. Stability will also be elusive in Chad and Niger with so many aspirational Islamist groups operating in the Sahel and North Africa. These include, in addition to those

Hector Conesa/Shutterstock.com

• **Figure 9.23** The Great Mosque in the Malian city of Djenné was built when Timbuktu thrived as a caravan crossroads and cultural center. Like much of historic Timbuktu, the Great Mosque of Djenné is a UNESCO World Heritage site. This distinctive adobe architecture of Mali is known as "Sudano-Sahelian."

discussed above, the AQIM offshoot **Movement for Oneness and Jihad in West Africa (MOJWA or MUJWA)** and Ansar al-Shari'a, dominated by Timbuktu's Arab tribe of Barabiche.

Although Boko Haram has so far not touched the oil of the country's southeast (see **page 436**), kidnapping threats and direct attacks on mining infrastructure have had serious economic impacts on regional development in the north. Western multinationals have scaled back operations in northern Nigeria and the broader Lake Chad region. Nigeria's war on terrorism, including its stated intention to "cripple" Boko Haram, has also affected infrastructure in Borno, with the closure of roads, border posts, and telecom systems, and the banning of commercial trucks—which are all vital to the business community. Nigeria's government also routinely detains women relatives of Boko Haram, a sensitive issue in their Muslim culture. With their tactics of **asymmetric warfare**, in which a weaker opponent uses unconventional means such as terrorism to exploit the enemy's vulnerabilities, Boko Haram and its affiliate Ansaru claim to be countering Nigeria's "Christian government."[40]

The Nigerian government's struggles to defeat Boko Haram played a large role in the defeat of incumbent President Goodluck Jonathan in Nigeria's March 2015 election. Much to the delight of everyday Nigerians and international monitors, the standing president conceded to the victorious Muhammad Buhari, who like most northern Nigerians is a Sunni Muslim. Buhari had promised that if elected he would defeat the Boko Haram insurgency, rein in Nigeria's rampant corruption, and reform Nigeria's economic injustices (see The Poor, Oil-Rich Delta of Nigeria on **page 436**).

So far in the big picture, Africans have proven resilient in the face of violent Islamism. The vast majority of people, including Muslims, recoil from the abhorrent tactics practiced by these groups. Culturally, few are predisposed to take up the terrorist cause. As we have seen, the dominant African sect of Islam is Sufism, which is more resistant than mainstream Sunni or Shi'a to radical interpretations.[41]

After its interests in counterterrorism (including blunting Somali piracy that occasionally threatens commercial traffic in the region's vital sea lanes) and advancing its neoliberal economic agenda in Africa, the United States has strong geopolitical interests in the region's oil-producing countries. Energy security in Africa became more critical for the US after 9/11 and the war in Iraq, which heightened the sense that Middle Eastern oil supplies were vulnerable. The United States is seeking more stable oil supplies and has pinned many hopes on Africa. Petroleum-rich Nigeria, Cameroon, Gabon, Congo, Angola, and Equatorial Guinea together in 2014 supplied 6 percent of the oil reaching world markets. However, except for Nigeria, reserves in these countries are not large, and their oil exports cannot fill in gaps for an extended period.

In the shift from unipolar to multipolar world power centers, increasingly the United States will be competing with China and other countries for these dwindling finite fossil reserves, and given current trends, the geopolitical landscape of Sub-Saharan Africa will see Asia on the rise and American clout in decline. Countries that are not pleased with Washington's wishes or demands will find partners elsewhere. 63 Some military analysts describe the emerging US-China rivalry in Africa as a modern Great Game, comparable to the British-Russian contest over central Asia in the 19th century. 332 Economic contests may have hidden military dimensions. Although China does not currently have a military presence in Africa (though some analysts caution that China is interested in turning Kenya's port city of Lamu into one of its "string of pearls" across the Indian Ocean), the International Relations and Security Network (ISN), 327 monitoring what it calls the "new era of resource imperialism," observes that that technicians who operate mineral-for-infrastructure concessions in Africa are also, "at least in the case of the Chinese, trained militiamen who could quickly assume a paramilitary role if required." The ISN's bottom line? "Geopolitical competition on the African continent has increased. And underpinning this competition is the desire to obtain and safeguard access to Africa's natural resources."[42]

Waxing and waning over the decades as a region of strategic geopolitical concern, Africa is now front and center on the geopolitical agendas of a number of countries. In Africa as elsewhere, the traditional framework of rich/poor, North/South relations is giving way to new forms of international cooperation shaped by the dynamics of globalization.[43] "There is an increasing competition among a large group of powers to engage Africa on a scale not witnessed since the original 'scramble for Africa,'" Richard Grant writes. "Important decisions are still made in London, Paris, Berlin, and Washington D.C.—capitals of 20th-century great powers—but equally consequential decisions are now made in BRICS capitals—Brasilia, Moscow, New Delhi, Beijing, and Pretoria. These days, there is not a middle-ranking power (rapidly developing powers with growing international influence) that is not deepening its ties with Africa."[44] Even Russia is re-engaging, picking up where it left off at the end of the Cold War, and between 2005 and 2012 wrote off billions of dollars of debt owed it by African countries. 60

The United States views South Africa as the region's economic powerhouse and as the key to regional stability. South Africa's military strength (which included nuclear weapons until they were decommissioned in 1990) contributes to its global significance. It is strategically important for its frontages on both the Atlantic and Indian Oceans, its possession of Africa's finest transport network (vitally important to many African countries that trade with and through South Africa), and its diversified mineral wealth.

Finally, HIV/AIDS and other infectious diseases in the region are also a geopolitical concern. Despite Africa's apparent remoteness from the United States, the two are linked by hundreds of air traffic routes traveled by thousands of people each day. Hundreds of HIV-positive Africans, perhaps most of them unaware of their virus, pass through US airport gateways daily. Some African strains (subtypes) of the virus are more easily transmitted by heterosexual contact than the strains prevalent in the United States, so the risk to the population at large is greater. There are other dimensions of the potential burden on the United States. As long as the epidemic rages in Africa, there may be large flights of "medical refugees" seeking asylum in the United States. Fears like these stalked American authorities and citizens in the recent Ebola outbreaks in West Africa. There could also be disease-related or even climate change–related political instability or civil wars that invite US military intervention in the region. How the world should respond to poverty and recurrent crises is one of the key questions facing the huge, resource-rich, problem-ridden but promising region of Sub-Saharan Africa. The following section examines some of these challenges in greater detail. 408

9.6 Regional Issues and Landscapes

Blueprints for Development in Africa: The Millennium Development Goals and Sustainable Development Goals

Despite substantial progress, Sub-Saharan Africa remains the world's last developed region. In 2000, the United Nations established the **Millennium Development Goals (MDGs)**. Targeting the world's poor, these had a special emphasis on Africa. Meant to be fulfilled by 2015, each of the eight MDGs listed here had a subset of specific activities with assigned target dates for their completion:

1. Eradicate extreme poverty and hunger.
2. Achieve universal primary education.
3. Promote gender equality and empower women.
4. Reduce child mortality rates.
5. Improve maternal health.
6. Combat HIV/AIDS, malaria, and other diseases.
7. Ensure environmental sustainability.
8. Develop a global partnership for development.

The leading industrialized "G-8" countries (at that time including Russia, which was later expelled) supplied most of the funding for implementing the MDG's blueprint for development. A large portion of these funds was used to pay down debts of the most heavily indebted poor countries (HIPCs)—a huge bonus for the debt-strapped region. Much more money than anticipated went into natural disaster relief, and in the end too little was left to actually invest in development. Nevertheless, progress was notable in maternal health and in primary education. 84

The MDGs expired in 2015 and were replaced by a UN-coordinated set of **Sustainable Development Goals (SDGs)** designed to guide international development policies through 2030, with emphasis on low-carbon technologies, green living, eco-friendly designs and communities, protection of biodiversity, and adaptation to climate change.[45]

The United Nations began constructing the SDGs' development agenda in 2013 by convening a high-level panel to brainstorm the broad outlines. To ensure that more than just a few eminent voices were heard, the UN "crowdsourced" input for the SDGs from millions of people in more than 100 countries, using digital media and mobile phone technology ("There are now more mobile phones than toilets worldwide," the UN call for input noted). The UN sponsored workshops in places where such technologies were not available. The clearinghouse for the crowdsourced data is the website http://www.worldwewant2015.org. On that site, you can explore the questions and responses in each world region; for example, on Africa the question was asked, "What would create lasting positive change in Africa?" The wealth of response is worth your time to explore.

The 17 Sustainable Development Goals that emerged from all these inputs are as follows:[46]

1. End poverty in all its forms everywhere.
2. End hunger, achieve food security and improved nutrition, and promote sustainable agriculture.
3. Ensure healthy lives and promote well-being for all at all ages.
4. Ensure inclusive and equitable quality education, and promote lifelong learning opportunities for all.
5. Achieve gender equality and empower all women and girls.

6. Ensure availability and sustainable management of water and sanitation for all.
7. Ensure access to affordable, reliable, sustainable, and modern energy for all.
8. Promote sustained, inclusive and sustainable economic growth, full and productive employment, and decent work for all.
9. Build resilient infrastructure, promote inclusive and sustainable industrialization, and foster innovation.
10. Reduce inequality within and among countries.
11. Make cities and human settlements inclusive, safe, resilient, and sustainable.
12. Ensure sustainable consumption and production patterns.
13. Take urgent action to combat climate change and its impacts (taking note of agreements made by the UNFCC forum).
14. Conserve and sustainably use the oceans, seas, and marine resources for sustainable development.
15. Protect, restore, and promote sustainable use of terrestrial ecosystems, sustainably manage forests, combat desertification, and halt and reverse land degradation, and halt biodiversity loss.
16. Promote peaceful and inclusive societies for sustainable development; provide access to justice for all; and build effective, accountable, and inclusive institutions at all levels.
17. Strengthen the means of implementation, and revitalize the global partnership for sustainable development.

Negative reviews of the MDGs were abundant, among them that too much emphasis was put on reducing problems by half; they should have been targeted for complete elimination, said critics. Efforts to meet the SDGs will be scrutinized and may be faulted as well. The Millennium Development and Sustainable Development Goals, however, need to be put in perspective and recognized as the vital beginnings of the process in which international efforts are confronting some of the world's most critical human and environmental problems. 83

Big, bureaucratic goals and programs like these risk bypassing the capabilities and know-how of African and other individuals and communities. To the list of "durable strengths and resources" that Africans have (see **page 419**) could be added ***kanju***, a term describing "the specific creativity born from African difficulty."[47] *Kanju* suggests that the best solutions are local—one of the tenets of sustainable development. In her book *The Bright Continent*, Dayo Olopade critiques Western-led efforts to assist Africa: "It turns out we have been throwing a party in an empty ballroom. One of the biggest problems with the world's longtime orientation toward Africa is a preference for interactions between governments, or between formal institutions, when the most vibrant, authentic and economically significant interactions are between individuals and decentralized groups."[52] *Kanju* often requires individuals and communities to bend or break the rules prescribed by those institutions in order to get things done.

Urbanization in Africa's Future

Africa is mainly rural—but not for long, and with guidance from Richard Grant this is a good time to take stock of the region's urban futures. Africa's rate of urbanization is unprecedented historically. We marvel at how industrialization brought about a sevenfold increase in London's population between 1800 and 1910. In just 65 years (1950–2015), the populations of Kinshasa, Congo, and Lagos, Nigeria, grew 45-fold![48] 110

Although urbanization generally corresponds with growth in agricultural productivity and with industrialization, neither of these factors is behind urbanization in Sub-Saharan Africa; the World Bank has re-characterized Africa's prognosis of "urbanization without industrialization" to "urbanization and growth of services bypassing the development of manufacturing."[49]

There are several factors that set African urbanization apart. First and most concerning is the explosive growth of slums, resulting mainly from rural–urban migration. An estimated 62 percent of city-dwellers live in slums. Many rural peoples bring their environment with them to the cities: Grant describes the "hermaphroditic landscapes" where "rural and urban features coexist in environmental, socioeconomic, and institutional terms."[50] Second, in many cases Africa's reliance on resource exploitation rather than diversified economies is driving urban expansion, as in oil-rich Gabon. Third, an unnumbered stream of migrants to cities are environmental refugees, including the leading edge of those long feared by those sounding the alarm about climate change (see **page 39**). A threefold increase in the number of storms, droughts, and floods between 1985 and 2015 has paralleled urban growth across Africa, and a connection likely exists. Finally, conflict and war have also driven people to the relative safety of cities. 76

Can Africa set the pace for urbanization in the LDCs?[51] Among LDC peers, African cities are uniquely hampered in lacking regional links and infrastructure within and between countries. Yet there are innovative urban projects underway in Africa, in some cases jump-started with financing from Chinese, Russian, and South African developers and speculators. These focus mainly on rebuilding central business districts and on building satellite cities to house workers outside metropolitan areas. Both employ "eco city" technologies such as wind and solar power generation and energy-efficient building design. Ghana's Hope City bills itself as "Africa's Silicon Valley" and is in friendly competition for the reputation of Africa's leading technological city with Kenya's Konza Techno City, trumpeted as "Silicon Savanna," a $10 billion government initiative to turn 5000 acres of savanna south of Nairobi into "the most modern city in Africa" (**•Figure 9.24**). Both urban projects are emulating the success of Bengaluru as India's information technology mecca. Spurred on by the Information Revolution, urban futures like these could help at least some African cities to overcome infrastructure handicaps, turning the "tyranny of distance" into the "death of distance" predicted by Frances Cairncross as an outcome of the Information Revolution, with supply chains leaping borders and oceans.[52] 313

Konza Technopolis Development Authority (KoTDA)

• **Figure 9.24** This is an artistic preview of Kenya's Konza Technopolis City, now under construction.

The most ambitious urban project in Africa is an "eco-city" known as "Eko Atlantic" on the Atlantic coast 8 miles (13 km) south of Lagos (Eko is the local name for Lagos). It is meant to offset the runaway urbanism of chaotic Lagos (population 13 million), an overcrowded and poor megacity that is the commercial capital of Africa's largest economy (•**Figure 9.25**). It is built on reclaimed land surrounded by a seawall ("the Great Wall of Lagos"). On its 6 square miles, 3000 buildings will be raised to accommodate 250,000 residents and 190,000 commuters to Lagos. The model city is supposed to boost employment, become a regional financial hub, become the "gateway to West Africa" and, its promoters say, be known as "an oceanfront city that will be one of the wonders of the 21st entury."[50] There are some distinctly unecological attributes of this eco-project. No environmental impact assessment was done before dredging began. Local blame the project for flooding and storm surges on nearby beaches. There are no provisions for affordable housing, and Eko Atlantic may become a blatant symbol of overconsumption by the few rather than a model for sustainable urbanism for the many.

There may be other pathways to genuine sustainable urbanism, however. Just as African peoples have leapfrogged telecommunications, bypassing landlines and adopting mobile phones, there is talk of their leaping to the green cities of the future. Green urbanism calls for adaptive, bottom-up inventiveness rather than top-down how-to instructions. As Glaeser informed us, the most flexible are people living in slums, who already have to devise ingenious ways to get hold of water, food, and transportation and to let hold of waste (see **page 471**). They tend to be highly efficient recyclers, reusers, and repurposers of water, waste, and plants; many have rooftop or structure-side farms and growing containers. A global NGO called Shack/Slum Dwellers International (SDI), which labels itself "A Global Network of the Urban Poor," trains people to set up waste management systems that collect, sort, and recycle materials.[54] In a number of countries "slum tourism" even offers relatively affluent foreigners a glimpse into how the other half (or more) lives, and encourages donations to local urban poor ("pro-poor") projects.

With the United Nations adopting Sustainable Development Goals to guide international development policy from 2015 to 2030 comes the opportunity to better align urban progress with the sustainable development agenda. What can Africans contribute to a new agenda for urban development? "Sustainable development offers a unique opportunity for cross-collaboration among nontraditional partners and for adopting a more holistic approach to the African development experience,"

Attilio Polo's Fieldwork/Getty Images

• **Figure 9.25** With more than 13 million residents, congested Lagos is the largest city in Sub-Saharan Africa. A new city called "Eco Atlantic" is under construction nearby. It is intended to relieve some of the crowding in Lagos and become West Africa's financial center.

writes Richard Grant, "emphasizing human well-being, social and economic inclusion and resilience in the face of climate change and greater environmental sustainability."[55] There were many environmental adaptations in early African cities that can be resurrected and used in modern urban contexts. Agriculture had a prominent place in indigenous African cities, and urban farming could be brought back to provide a green urban environment and supply additional foods to the population. Essentially, the sustainable city transplants components of the village into an urban environment. Courtyard architecture with rooms grouped around an open-sky courtyard maximized urban space, provided natural air conditioning, and was affordable. Benin City (in southern Nigeria) and other old African cities had roof technologies that allowed rainwater to be collected for drinking, cooking, and bathing. These are just a few examples of Africa's historic urban diversity that could feed its future.

The Poor, Oil-Rich Delta of Nigeria

Nigeria, often described as "Africa's keystone country," overtook South Africa in 2014 to become Sub-Saharan Africa's largest economy. How is it that 62 percent of Nigeria's 177 million people live in poverty? It looks like a classic 424 example of the resource curse.

Africa's largest oil producer, Nigeria has the 10th-largest proven reserves in the world. Most of the production is concentrated in the Niger River Delta, home to about 12 million mostly Christian people (•**Figure 9.26**). They belong to several different ethnic groups who have one thing in common: they say they have derived few benefits and have suffered greatly from oil development in their homeland. They complain that since the 1960s, thousands of oil spills have tainted their croplands and water, destroying their crops and fisheries, while the flaring off of natural gas has polluted their air and caused acid rain. They note that despite the enormous revenue generated from oil drilled on their land, little money has returned to the area; most goes to foreign oil companies and to the Muslim-dominated government in the north. Meanwhile, most of them live in palm-roofed mud huts; half of the Delta region lacks adequate roads, water supplies, and electricity. Schools lack books, and clinics have few medical supplies.

• **Figure 9.26** Political geography of Nigeria.

Where does Nigeria's oil money go? There is not enough transparency to answer this question. Incredibly, until 2011, it was illegal in Nigeria to publish official government data and statistics, including government accounts. Recently, $22 billion from the government's Excess Crude Account simply "went missing." An estimated $400 billion has been pilfered from Nigeria's treasury since the country became independent in 1960.

Grievances in the Delta region have deep and tragic roots. A 1967 independence movement for a country that would have been named Biafra, led mainly by the Ibo (Igbo) ethnic group, launched Nigeria into a three-year civil war. An estimated 1 million civilians died in this bloodbath. Some Ibo peoples of the Delta still agitate for an independent Biafra, and since 1990, several ethnically based organizations have been agitating for change. Ethnic Ogoni activists founded the **Movement for the Survival of the Ogoni People** (**MOSOP**), issuing a bill of rights in which they declared the right to a safe environment and more federal support of their people. One of the movement's leaders, a popular author, playwright, and television producer named Ken Saro-Wiwa, also called for self-determination for the Ogoni. Despite international condemnation, Nigeria's government executed him after a questionable murder trial.

Despite the return to civilian rule in Nigeria that came after Saro-Wiwa's death in 1995, dissent in the Delta continued to grow. Ogoni defiance and contempt of government and the oil companies spread to the region's other major ethnic groups, the Ibo, **Yoruba**, and Ijaw. The Ijaw (who make up about 5 percent of Nigeria's population) have been especially strident in demanding that some of the oil money be used to fund roads, electricity, running water, and medical clinics. Ijaw activists have spoken of secession and self-determination if their demands are not met. They have periodically registered dissatisfaction by seizing onshore stations and offshore rigs belonging to foreign oil companies, temporarily disrupting Nigeria's oil exports on both occasions. In 2006, a mainly Ijaw group calling itself the **Movement for the Emancipation of the Niger Delta** (**MEND**) escalated these tactics into frequent kidnappings of foreign oil workers. Opportunistic profiteers without well-established political views also joined in the action, and kidnapping became big business. An oil worker is typically freed unharmed upon payment of a ransom of several hundred thousand dollars.

This militancy in the Delta has periodically sent shockwaves through the world economy. Nigeria had touted itself as the safe alternative to Middle Eastern oil but has had to cut production by as much as 25 percent when tensions in the Delta and in the Boko Haram–plagued north have been particularly high. The resulting shortages on world markets helped send oil and gasoline prices to record levels. MEND promised greater disruptions when one of its leaders in 2014 vowed, "At the right time, we will reduce Nigerian oil production to zero and drive off our land all thieving oil companies."[56]

Questions of equity in a potentially rich country thus pose a threat to national, regional, and international stability. The peoples of the eight oil-producing states in the Delta have been promised 13 percent of all petroleum revenues (compared with 5 percent previously), and have been told that the region's foul gas flares will be shut off. But locals say the promised revenues have been embezzled while the flaring off of gas continues.

One of Nigeria's ironies is that in this oil-rich country there are continuous shortages of gasoline and electricity. Neglect, mismanagement, corruption, and theft are all responsible. One of the cruelest and most repetitive Nigerian news stories is the immolation of scores and even hundreds of poor Delta villagers who, in an activity known as "scooping" or "bunkering," illegally puncture gasoline or oil pipelines in an effort to sell the hydrocarbons, only to inadvertently ignite fires that consume them.

East Africa: No More Divisionism?

Episodes of bloodshed prompted by sectarianism and by rivalries dating to colonial times have punctuated East Africa's history. This section tells a complex and tragic tale of genocide. These details are related here because the potential for renewed conflict is high and because the world community vowed not to forget what happened here.

The small countries of Rwanda and Burundi have had tragic disputes between their majority and minority populations. About 85 percent of the population in Burundi and 90 percent in Rwanda are composed of the **Hutu (Bahutu)**, and most of the remainder of the two populations is **Tutsi (Watusi).** These were not originally separate tribal or ethnic groups; the peoples speak the same language, share a common culture, and often intermarry. The distinction between Hutu and Tutsi was instead based on socioeconomic classification. Historically, the Tutsi were a ruling class that dominated the Hutu majority. Tutsi power and influence were measured especially by the vast numbers of cattle they owned. Tutsi who lost cattle and became poor came to be identified as Hutu, and Hutu who acquired cattle and wealth often became Tutsi.

Colonial rule in East Africa attempted to formally polarize these groups, increasing antagonism between them. German and Belgian administrators differentiated between them exclusively on the basis of cattle ownership: Anyone with 10 or more animals was Tutsi, and everyone else was Hutu. Europeans mythologized the wealthier Tutsis as "black Caucasian" conquerors from Ethiopia who were a naturally superior, aristocratic race whose role was to rule the peasant Hutus. The colonists replaced all Hutu chiefs with Tutsi chiefs. These leaders carried out colonial policies that often imposed forced labor and heavy taxation on the Hutus. Education and other privileges were reserved almost exclusively for the Tutsis.

323

Ferocious violence between these two peoples marred Rwanda and Burundi intermittently after the states became independent in the early 1960s. After independence, the majority Hutus came to dominate the governments of both nations. In 1994, the death of Rwanda's Hutu president in a plane crash (in which Burundi's president also died) sparked civil war in Rwanda between the Hutu-dominated government and Tutsi rebels of the Rwandan Patriotic Front (RPF), based in Uganda. Hutu government paramilitary troops known as *Interahamwe*, still vengeful about decades of domination by Tutsis, systematically massacred Tutsi civilians with automatic weapons, grenades, machetes, and nail-studded clubs. Women, children, orphans, and hospital patients were not spared. An estimated 800,000 Tutsis and "moderate" Hutus (those who did not support the genocide) died. International media tracked these horrors, but outside nations did nothing to stop the violence.

Well-organized and motivated Tutsi rebels seized control of most of the country and took power in the capital, Kigali, in 1994. That precipitated a flood of 2 million Hutu refugees mainly into neighboring Zaire, now the Democratic Republic of the Congo (• **Figure 9.27**). Although the Tutsi-dominated government promised to work with the Hutus to build a new multiethnic democracy, Hutu insurgents, based mainly in eastern Zaire, continued to carry out attacks against Tutsi targets in Rwanda.

Rwanda now has a national unity policy aimed at reconciling Hutus and Tutsis; both occupy important government posts, and although the Tutsis still hold the political upper hand, they have encouraged the resettlement of Hutu refugees who fled the country during its troubles. In 2004, marking the 10-year anniversary of the genocide with a new policy meant to prevent its recurrence, Rwanda's government outlawed ethnicity. Now any Rwandan who speaks or writes of Hutus and Tutsis may be fined or imprisoned for practicing "divisionism." The country's past nevertheless continues to shape its future; for example,

Reuters/Landov

• **Figure 9.27** Hutu refugees swell a refugee camp near Goma, Democratic Republic of the Congo, in 1996. These Hutus fled from nearby Rwanda when Tutsis took control of that country, precipitating an enormous humanitarian crisis.

there is an ongoing defection of the faithful from Rwanda's majority Catholic Church to other religions, especially Islam but also smaller Christian denominations. Most of the country's worst massacres took place in Catholic churches with the apparent knowledge and even participation of church leaders.

Rwanda's catastrophe was mirrored in Burundi, where the Tutsi minority traditionally dominated politics, the army, and business. In 1993, there was a brief shift of power to the Hutu majority when Burundians elected the country's first Hutu president. Tutsis assassinated him in October 1993, and a large-scale Hutu retaliatory slaughter of Tutsis ensued. An estimated 300,000 died in a decade of violence. Burundi's Tutsis and Hutus now have a power-sharing agreement in which the presidency rotates between them.

The peace has been maintained by the first-ever deployment of the African Union's **African Standby Force**. To help ensure the stability needed for economic and political progress throughout the region, the African Union established this continental military force. It takes orders from the AU's Peace and Security Council. The force will, in theory, deploy peacekeepers or peacemakers to ensure that the map of Africa, so filled with deadly conflicts in the past, has far fewer trouble spots in the future.

Collective shame and embarrassment in the international community about the failure to stop the killing in Rwanda and Burundi led in subsequent years to numerous new anti-genocide measures. These include a United Nations "early warning" system to prevent genocide and a new forum for crimes against humanity in the **International Criminal Court**, established in 2002 in The Hague, Netherlands.

Africa's First World War

Just to the west of Rwanda and Burundi, the Democratic Republic of Congo (DRC) and its antecedents have long been a crucible of conflict. During the last quarter of the 19th century, the Congo Basin was virtually a personal possession of Belgium's King Léopold II, whose agents ransacked it ruthlessly for wild rubber, ivory, and other tropical products gathered by Africans. This lawless era inspired Joseph Conrad's famous novel *Heart of Darkness* (1902), in which the trader Kurtz, at the point of death, evokes the ravaged Congo region and the colonists' brutal treatment of Africans with the cry "The horror! The horror!"

In 1908, the Belgian government formally annexed the greater part of the Congo Basin, creating the colony of Belgian Congo. In 1960, Belgian Congo became the independent Republic of the Congo. During his tenure in office from 1965 to 1997, autocratic ruler Mobutu Sese Seko, who renamed the country Zaire in 1971, set a terrible standard for misrule. While amassing a personal fortune largely from embezzling foreign aid meant for development and health programs, Mobutu did little to develop the country's economy; in 1995, the country's GDP was less than half of what it had been in 1960. Ethnic tensions, political opposition, and public discontent with unemployment, inflation, and numerous human rights violations led to unrest in Zaire in the early 1990s.

After the fighting in Rwanda ended and a new Tutsi-led government was installed, as many as 2 million Hutus fled into eastern Zaire by 1996 to avoid possible government reprisals. Some Hutus used refugee camps in Zaire as a base to stage raids against Tutsis in Rwanda, and they soon clashed with native Zairian Tutsis as well. With aid from the governments of Rwanda and Uganda, the Zairian Tutsis organized into militias to defend themselves.

Seeing the Tutsi militias backed by other countries as a rebellion, Mobutu sent troops to subdue the Tutsis in 1997. This started an all-out war, with numerous anti-Mobutu factions joining the Tutsi militia groups. One of these factions was led by a non-Tutsi guerrilla named Laurent Kabila. With the assistance of troops from Angola (which also opposed Mobutu), Kabila's forces were able to overpower the weak Zairian army and quickly pushed westward through the huge country, capturing important cities with relative ease. In May 1997, Kabila occupied the capital of Kinshasa with little loss of life. President Mobutu fled Zaire on the eve of Kabila's triumph. Kabila then declared himself president of the country he now called the Democratic Republic of the Congo (DRC). 424

Unfortunately, little changed for the impoverished county under Kabila's rule. Like his predecessor, Kabila funneled much of the country's wealth and power into the hands of family and friends, and the DRC's economy continued to worsen. He obstructed international human rights investigations into the deaths and disappearances of thousands of Hutu refugees during the heady days of rebellion against Mobutu. Soon after his installation as president, Kabila expelled the foreign aides and fighters that had helped him achieve power, alienating Rwanda and other nations.

A year after Mobutu's government fell, Hutu forces based in the east launched successive devastating attacks on Tutsi interests in Rwanda, and Kabila's government did little to stop them. Although the Tutsis of Rwanda and the DRC had helped bring Kabila to power, they now turned against him, forming the backbone of a rebel alliance bent on bringing down Kabila's young regime. The unrest led to a broader conflict dubbed **"Africa's First World War"** that involved nine countries and 20 rebel movements, and resulted in the deaths of more than 5 million people (primarily civilians, mostly from disease and starvation), the deadliest conflict in the world since World War II. Tutsi militias backed by Rwanda, Burundi, and Uganda once again began battling government forces in eastern areas of the DRC, and within months they were poised to take Kinshasa. Kabila called for aid, and troops from Angola and Zimbabwe prevented a Tutsi takeover of the DRC. In 2000, UN troops were sent in to enforce a frequently violated cease-fire.

The turning point was the assassination of Laurent Kabila by one of his bodyguards in January 2001. His successor and son, Joseph Kabila, was seen by many as the reformer his father never was. The younger Kabila proved much more effective at governing and diplomacy, and soon came to an agreement with Uganda and Rwanda for those countries to begin removing their troops from the DRC. In 2003, the war was declared over, but pockets of terrible violence remained entrenched in some of the DRC's remote eastern areas for more than a decade.

The DRC epitomizes the persistent theme of Africa's plundered resources. Decades of chaos and misrule resulted in the

DRC being the world's poorest country despite its enormous mineral wealth; its GDP (PPP) is just $400. Conflict in the DRC has been enormously profitable for the warring factions, some foreign companies, and foreign governments; for example, Uganda and Rwanda looted much of the eastern DRC's gems, minerals, timber, agricultural produce, and wildlife, including elephant ivory from some of the country's national parks. Since 1998, most of the violence has been unabashedly about control over the Congo's vast mineral resources. Warlords and rebel proxies have been sustained by exploitation of minerals in the areas they controlled. In the east, where fighting to overthrow the government was centered, armed groups continue to control mines, sometimes in concert with the Congolese army.

You are probably part of this picture. There is a good chance that your cell phone or other electronic devices contain tantalum, which is extracted from coltan ore mined in this region. 423 American law requires that US companies perform "due diligence" to ensure that products made with "conflict minerals" (tin, tungsten, tantalum, and gold) that are sold in the United States be certified as free of conflict in the DRC and other conflict zones. This requirement affects sales of airplanes, computers, cell phones, television sets, and other goods. Seeking compliance, a handful of tech giants have voluntarily signed onto the Conflict-Free Sourcing Initiative (CFSI).[63] Validating that ores smelted for these companies do not come from conflict zones is extremely difficult, however.

Although peace has been negotiated in this area, the prospects for conflict and war continue. Now integrated into the Congolese army, ethnic Tutsi have been able to extend the area and the mineral wealth they control. Rival Rwandan Hutu militias who control their own mineral wealth have been carrying out reprisal attacks against Congolese civilians.

Further clouding prospects for peace are groups characterized as "pathological insurgents," including the **Lord's Resistance Army (LRA)**. Led by warlord Joseph Kony, the LRA was initially focused on rebellion against Uganda but evolved into a marauding force attacking and abducting people across northern Uganda, South Sudan, the DRC, and the Central African Republic. The cult leader Kony's atrocities against children earned him notoriety in the Invisible Children video "Kony 2012" that went viral, within days reaching half a million young adults in the US, generating much public concern and prompting the deployment of a small US military force to assist Uganda in confronting the LRA.[57] The LRA and other pathological insurgencies such as the Revolutionary United Front (RUF)'s in Sierra Leone are responsible for the region of Sub-Saharan Africa having the second highest number of casualties from subnational, internal (rather than international) terrorism, after Asia (including the eastern Middle East).[58]

The Galápagos Islands of Religion

Ethiopia is an ethnically and culturally diverse nation with a long, rich history.

About 45 percent of Ethiopians, including the politically dominant **Amhara** peoples, practice **Ethiopian Orthodox Christianity**, an ancient branch of Coptic Christianity that came to Ethiopia in the 4th century from Egypt. The entire area has had important cultural and historical links with Egypt, the Fertile Crescent, and Arabia. The Ethiopian monarchy based its origins and legitimacy on the union of the biblical King Solomon and the Queen of Sheba, who, tradition holds, gave birth to the first Ethiopian emperor, Menelik. Until a Marxist coup brought an end to the emperorship in the 1970s, Ethiopia's rulers were always Christian. Ethiopia has many outstanding Christian artistic and architectural treasures, including the 11 churches of Lalibela, carved from solid rock in the 12th and 13th centuries (•**Figure 9.28**). Most of the rest of Ethiopia's people are either Muslims (about 40 percent of the population) or members of Protestant, Evangelical, and Roman Catholic churches. There are still small numbers of Falashas, or Ethiopian Jews, in Ethiopia, a remnant of a very ancient and isolated Jewish population. The majority, about 100,000 people, have fled to Israel since the mid-1980s. Because this mountainous country has long served as an isolated refuge for such unique groups, it has been nicknamed the **Galápagos Islands of Religion**.

Perhaps the most unusual religion with roots in Ethiopia is that of the Rasta, or the Rastafari movement of the Jah People (called **Rastafarianism** by non-Rasta). This faith, which origi- 466 nated in the 1930s in the West Indies island of Jamaica, has the central doctrine that the Ethiopian Emperor Haile Selassie (1892–1975) was the earthly incarnation of Jah or Jehova (God). Ras Tafari was the Emperor's name before his coronation, and he was crowned on November 2, 1930, as the "King of Kings, Elect of God, and Conquering Lion of the Tribe of Judah."[59] Many poor Jamaicans gravitated to him for leadership and spiritual inspiration; he was the only black leader at the time to be recognized as legitimate in international circles. Their deep respect for him evolved quickly to reverence and worship, and the Rasta movement was born. The Rastafari faithful believe that blacks are the true children of Israel, and that on Judgment Day, Haile Selassie will call them to come home to Zion, which they identify with Africa rather than Jerusalem. During

Joe Hobbs

• **Figure 9.28** The churches of Lalibela, carved from volcanic rock in the 12th and 13th centuries, are among Ethiopia's many Christian cultural treasures. Note the men for scale.

his reign, Emperor Selaisse indulged the wishes of some Rastas to "repatriate" to Africa, allowing them to settle on his land in Ethiopia.

On April 21, 1966, Emperor Haile Selassie's plane touched down on the airport tarmac in Kingston, Jamaica. For the 200,000 Rastafaris who had gathered, God Himself had come for a visit. He told the assembled that they should not return to Ethiopia until they first liberated Jamaica. Haile Selassie's short stay bestowed new legitimacy to the Rasta faith, and Rasta culture flowered and diffused rapidly after that. One of its principal carriers was Bob Marley, whose emotive cries for freedom thrust reggae onto the international music stage.

Greg Hobbs

• **Figure 9.30** Scenes like this were rare during South Africa's apartheid period, but genuine reconciliation between the races is ongoing today.

Ethnicity, Colonialism, Strife, and Reconciliation in South Africa

Despite its location at the continent's southern extremity, South Africa has long been a global crossroads, and its diversity has been the source of both its wealth and its strife (**•Figure 9.29**). Visitors from Western Europe and North America will find in South Africa most of the institutions and facilities to which they are accustomed. However, the Europeanized landscape does not reflect the country's majority culture: whites represent only about 9 percent of the total population of 53 million. Seventy-nine percent of South Africa's population is black, composed of nine major ethnic groups (including the **Zulu** and **Xhosa** [pronounced "*khoh*-suh"]) and several smaller groups. Asians (primarily descendants of Indians brought to South Africa by the British in the 19th century to work on plantations) are 2 percent of the population. Most of the remainder are the **coloureds**, people who have mixed ancestry from a variety of ethnicities. Nelson Mandela, South Africa's first black president, envisioned all of these peoples living harmoniously in a multiracial, multicultural "Rainbow Nation" (**•Figure 9.30**).

• **Figure 9.29** Political geography of South Africa.

A great deal of work is underway to improve race relations and realize Mandela's vision of unity and equality for all. A long legacy of poisonous relationships is being overcome in a country where whites had exclusive control of political and economic systems for centuries. Following is a summary of how the country's racial structures—a critical issue in its modern geography—evolved.

Europeans first arrived in what is now South Africa in 1652, when the Dutch East India Company created an outpost near the Cape of Good Hope as a place for crews to rest and ships to resupply on their way from Amsterdam to the Dutch colonies in Asia. By the 1700s, the outpost had become a thriving colony. The Dutch drove out the indigenous San peoples, and took tens of thousands of blacks as slaves to work on the colony's outlying farms.

413

Great Britain took over the Cape Colony from the Dutch in 1806. Wanting more farmland and mineral wealth, the British greatly expanded the area of the Cape Colony by wresting lands inhabited by the Boers (Dutch settlers; Boer means "farmer") and by indigenous black tribespeople. Friction developed between many Boers and the British authorities, who imposed tighter administrative and legal controls than the Boers were accustomed to. The British abolished slavery throughout their empire in 1833, contributing to anti-British sentiments among the slaveholding Boers. Boer discontent resulted in the **Great Trek**, a series of northward migrations by which groups of Boers from the Cape Colony established new interior grazing lands and political units beyond British reach. These migrations resulted in the founding of the Orange Free State, Transvaal, and Natal as Boer republics. Britain annexed Natal in 1845 but recognized Boer sovereignty in the Transvaal and the Orange Free State in the 1850s.

Although Boer disaffection with Britain was growing, British settlers were coming to South Africa in increasing numbers, drawn by the area's rich farmland. The British colonies and Boer republics might have developed peaceably side by side if diamonds had not been discovered in the Orange Free State in 1867 and gold in the Transvaal in 1886. The discoveries set off a rush to these republics of prospectors and other fortune hunters and entrepreneurs from outside the region, particularly from Britain. The Boers felt threatened by the enormous numbers of British moving into their lands, and declared war in 1899. British troops defeated the Boer forces decisively, ending the war and annexing the Boer nations in 1902, but leaving behind a reservoir of animosity that still exists.

The Union of South Africa was organized in 1910 as a self-governing constitutional monarchy under the British Crown. (In 1961, South Africa became a republic outside the British Commonwealth.) Political life in the new country was dominated by **Afrikaners**, the majority European group in the country, made up mainly of descendants of the original Dutch Boer settlers and often at odds politically with the wealthier English-speaking minority.

The white minority created exclusive political dominance over South Africa's black majority. Fears of the large black majority having voting rights, and thus unacceptable political power over whites, led to the passage of laws barring blacks and coloureds from any representation in South African governance. Long-established social customs also discriminated against blacks; whites would not sell land to blacks, nor would whites teach them any skilled trades. One of the main reasons for all these restrictions was that South Africa's economy depended upon large numbers of low-paid black laborers to work in white-owned and -operated mines, factories, and farms. Whites reasoned that without education or representation, the black majority would maintain the status quo and not demand more pay or reforms, leaving the white minority prosperous. With virtually no economic opportunities of their own, black South Africans became almost totally dependent upon white payrolls for their often meager livelihoods. The few blacks with token leadership positions were often despised by the rest of the population, as those leaders benefited financially from the oppression most blacks were subjected to.

World War II created an economic boom in South Africa. As whites went off to the front lines to fight with their Western allies, many blacks migrated from the countryside to work as unskilled laborers in the rapidly expanding urban factories. This migration led to unprecedented (and, from the white perspective, unwelcome) racial mixing in South African cities. After the war, many low-skilled urban white laborers resented having to compete for jobs with blacks, who were paid much lower wages. This urban white economic insecurity gave rise to the National Party, dominated by Afrikaners, which vowed to codify racially based customs into law. After the National Party was swept into office (without blacks voting) in 1948, it passed a series of laws, known collectively as **apartheid**, mandating the geographic separation of South Africa's racial groups.

Apartheid transformed South Africa's cultural landscapes. Each South African was assigned to one of four races, and many Asians, coloureds, and blacks were uprooted from their homes and moved to special reserves established on the outskirts of cities or in marginal rural areas. Blacks received the brunt of these forced migrations. Most urban blacks were relocated into "townships" created for them on cities' fringes, like the notorious Soweto on the outskirts of Johannesburg. These townships became overcrowded and crime-ridden shantytowns. Blacks were not allowed to operate businesses without expensive permits, so Soweto and similar towns had few prospects for economic development. The government largely ignored pleas for better housing, services, and security in these areas. City hospitals, schools, transportation, labor unions, restaurants, and recreational areas were either completely segregated or did not admit blacks at all.

Blacks living in rural areas were exiled to the 10 so-called **homelands**, also known as **Bantustans**, established in economically and environmentally marginal rural areas of eastern South Africa (•**Figure 9.31**). These homelands were tribally based; for example, Bophuthatswana was created for the Tswana people, whereas Transkei and Ciskei were reserved for the Xhosa. The government declared many of these homelands to be independent, sovereign nations, and revoked the South African citizenship of the people forced to move there. These areas were small, geographically fragmented, and poverty stricken, and none were recognized politically outside of South Africa.

The apartheid system drew furious criticism from many other nations. Most of the world community ostracized South Africa and many countries, including the United States, imposed economic sanctions against it. Black unrest directed against apartheid and the general underdevelopment of the African majority became so widespread and violent by the 1980s that the government declared a state of emergency. Protests

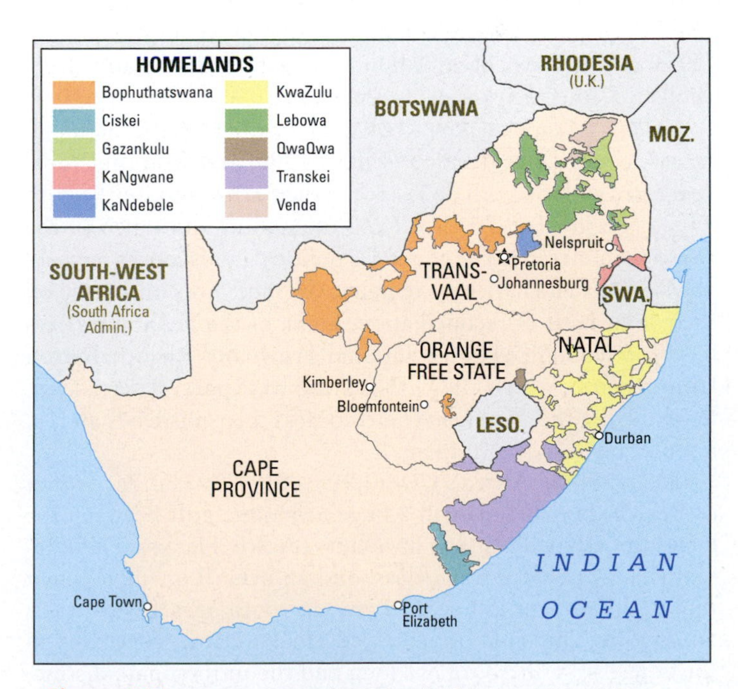

• **Figure 9.31** Between 1971 and 1981, South Africa established 10 "homelands" to segregate its native black population during apartheid. These homelands, often called Bantustans, had varying levels of autonomy, and some were declared independent nations, though no other countries ever recognized them. The Bantustans were reintegrated into South Africa upon the end of apartheid in 1994.

and strikes against apartheid were regularly met with excessive force by the police, and many political opponents of the white government were arrested and sometimes detained for years. Initially, factions like the Xhosa-dominated **African National Congress (ANC)**, led by Nelson Mandela, and the Zulu-dominated **Inkatha Freedom Party (IFP)**, led by Chief Mangosuthu Buthelezi, tried to create so much unrest in black areas that South Africa would become ungovernable. But political and philosophical differences between these rival organizations soon led to as much violence between the ANC and IFP as between them and the government.

By the late 1980s, South Africa was economically strangled by sanctions, ostracized from the international community, and wracked with political violence. There was growing antiapartheid sentiment among white South Africans, and the government began to see no alternative to reform, which finally began in 1989 when Frederik W. de Klerk became president. Urged on by white South African business leaders, antiapartheid activists, the powerful force of South African and international political opinion, the impact of international economic boycotts against South Africa, and his own convictions about the injustices and unworkability of apartheid, de Klerk took action. He launched a broad program repealing the apartheid laws, putting South Africa on the road to revolutionary political changes under a new constitution. Despite continuing violence, years of negotiations led to South Africa's first all-race national election in 1994. The African National Congress won the majority of seats in the new parliament, making Nelson Mandela the first black leader of South Africa. Mandela had been a political prisoner for 27 years and had been freed just four years earlier. This historic election ended white European political control in the last bastion of European colonialism on the African continent. The apartheid laws became null and void, the homelands were abolished, and their residents repatriated.

Nelson Mandela, now known as "Father of the Nation," served as president from 1994 until 1999, and won the Nobel Peace Prize along with de Klerk for their efforts to unite South Africa. He died in 2013, leaving the nation and much of the world in mourning. The country's political landscape is remarkably stable, but challenges remain. The races are continuing on their long path of reconciliation. Some of the healing process has been formalized in the national **Truth and Reconciliation Commission,** which allows those who were party to racial violence during the apartheid era to confess their misdeeds and, in a sense, be absolved of them.

Today, South Africa's GDP (PPP) is $12,700 (similar to that of Brazil), but there is still a huge economic gulf between the haves (mostly whites) and have-nots (mostly blacks). There remains a large black underclass, and an overall unemployment rate of 25 percent (50 percent among youth ages 15–24) fuels an ongoing epidemic of petty and violent crime, especially in the cities. Johannesburg has long had the unfortunate distinction of being the world's "murder capital." South Africa has the world's worst HIV/AIDS epidemic, a low level of education among workers, and relatively high labor costs that dampen foreign investment and economic growth. South Africa's commodity export-driven economy has made it susceptible to China's recent economic slowdown, and European consumption of South Africa's manufactured exports has also been lagging. While we have focused on BRICS and even included South Africa as the "S," some of these economies have recently been hit by emerging market downdrafts. The developing countries most susceptible to global financial shocks were labeled in 2013 as the "Fragile Five:" the BRIC countries Brazil and India, along with Turkey, Indonesia, and South Africa.[60]

312

60

There is a real need for land reform in South Africa, as the white minority still owns more than 70 percent of its productive land. A plan is in place to help reduce the imbalance by redistributing 30 percent of the country's farmland from white to black hands on a willing-seller, willing-buyer basis. Progress is slow, however; the original 2014 deadline for the transfers has been extended to 2025, and some landless blacks vow a takeover of white farms (as took place in neighboring Zimbabwe). A program is also underway to compensate the estimated 3.5 million blacks forcibly displaced by the government to the homelands between 1960 and 1982.

But there is good news coming out of South Africa as well. South Africa has an aggressive affirmative action program to help redress the economic imbalance between the races. The **Employment Equity Act** does not impose quotas, but requires employers to move toward "demographic proportionality" based on the national proportions of race and gender. Growing black wealth has boosted racial integration in neighborhoods, and over 10 million blacks are now in the middle and upper classes. South Africa's poverty level dropped from 26 percent in 2000 to 9 percent in 2015. No longer restrained by apartheid-era economic sanctions, South Africa has at the same time become a major investor in other African economies, buying banks, railways, cell phone networks, power plants, and breweries across the region. The economy is increasingly diversified across the primary, secondary, and tertiary sectors. South Africa was able to show its new face to hundreds of millions of people around the world in 2010 when it became the first African nation to host the World Cup. And there is progress toward Mandela's rainbow nation. The young South Africans born after 1994, who did not live through apartheid, are known as "Born Frees." They are often said to be colorblind and concerned far more with their economic welfare and the country's development rather than with racial issues.[68]

Madagascar and the Theory of Island Biogeography

Madagascar (in French, *La Grande Ile*) is the fourth largest island in the world, nearly 1000 miles (1600 km) long and about 350 miles (c. 560 km) wide. It lies off the southeast coast of Africa and has geological formations similar to those of the African mainland. Its distinctive flora and fauna include most of the world's lemur and chameleon species. These plants and animals are under tremendous pressure from people; Madagascar has 22 million inhabitants, many of them subsistence farmers who clear the island's forests to meet their needs.

The entomologist Edward O. Wilson and the biologist Paul Ehrlich have forecast that if tropical rain forests continue to be cut down at the present rate, a quarter of all of the plant and

(a)

(b)

• Figure 9.32 **(a)** The subsistence needs of a growing human population have had ruinous effects on Madagascar's landscapes and wildlife. Before people came some 2000 years ago, most of the island was forested, as in the Montagne d'Ambre National Park. **(b)** Today, less than 10 percent of the island is forested, and a characteristic landscape feature of its High Plateau is the erosional feature of lateritic soils known locally as lavaka.

animal species on Earth will become extinct by 2040.[61] They base their estimate on a model that correlates habitat area with the number of species living in the habitat. This **theory of island biogeography** emerged from observations of island ecosystems in the West Indies. It states that the number of species found on an individual island correlates with the island's area, with a tenfold increase in area normally resulting in a doubling of the number of species. If island A, for example, is 10 square miles in area and has 50 species, 100-square-mile island B may be expected to support 100 species.

What makes the theory useful in projecting species losses is the inverse of this equation: a tenfold reduction in area will result in a halving of the number of species; therefore, 1-square-mile island C can be expected to hold only 25 species. In applying the model, ecologists treat habitat areas as if they were islands. Thus, if people cut down 90 percent of the tropical rain forest of the Amazon Basin, for example, the theory of island biogeography suggests that they would eliminate half of the species of that ecosystem. Scientists caution that the theory is only a tool helping to make rough estimates; the actual number of species lost with habitat removal may be higher or lower.

As a rough guideline, the theory of island biogeography is useful in projecting and attempting to slow the rate of extinction 36 in the world's biodiversity hot spots, such as Madagascar. More than 90 percent of Madagascar's plant and animal species are endemic, occurring nowhere else on Earth. Extinction of species was well underway soon after people arrived on the island; the giant flightless elephant bird (*Aepyornis*) was among the early casualties. But human activities, particularly the clearing of forests to grow rice and provide pasture for zebu cattle, are eliminating habitat areas on the island at a faster rate than ever before. Meanwhile, scientists are anxious to learn whether some of Madagascar's remaining plants might be useful in fighting diseases such as AIDS and cancer. Already Madagascar's rosy periwinkle has yielded compounds effective against Hodgkin's disease and lymphocytic leukemia. Other species could become extinct before their useful properties ever become known. 36

How urgent is the task to study and attempt to protect plant and animal species in Madagascar? Scientists turn to the theory of island biogeography for an answer. Although people have lived on Madagascar for less than 2000 years, they have succeeded in removing 90 percent of the island's forest, setting the stage for some of the most ruinous erosion seen anywhere on Earth (**•Figure 9.32**). The theory of island biogeography suggests that in the process, they have caused the extinction of roughly half of the island's species. With Madagascar's human population on track to double in 26 years, and with pressure on the island's remaining wild habitats expected to increase accordingly, the task of conservation is extremely urgent.

Africa fits like a piece of Earth's jigsaw puzzle into neighboring South America. From Africa, we cross the Atlantic to explore that neighboring piece and the greater region of Latin America.

Study Guide

Summary

- Africa is the cradle of humankind, where hominids originated and from where they diffused. The region has seen many indigenous civilizations and empires and is ethnically and linguistically diverse.
- Africa's population of 920 million is the most rapidly growing in the world, but birth rates are declining.
- Sub-Saharan Africa has a relatively low population density overall, but a majority of this region's people live in a small

number of densely populated areas. Sixty-three percent of the region's people are rural.

- HIV/AIDS is taking a large toll, dramatically lowering life expectancy and projections for population growth, especially in southern Africa. But much progress is being made in the fight against the epidemic. ARV drugs are reaching 40 percent of the 37 million people infected worldwide. Ebola spread rapidly in West Africa in 2014, killing more than 10,000 people before it was brought under control.
- Most of Africa consists of a series of plateau surfaces dissected by prominent river systems.
- One of the most spectacular features of Africa's physical geography is the Great Rift Valley, a broad, steep-walled trough. The feature marks the boundary of two crustal plates that are rifting, or tearing apart.
- Although about two-thirds of the region lies within the low latitudes and has tropical climates and vegetation, Sub-Saharan Africa contains a great diversity of climate patterns and biomes, some resulting from elevation rather than latitudinal position.
- Between the late 1960s and 1985, the Sahel was subjected to severe droughts, which, in combination with increased human pressure on resources, prompted a process of desertification. Human and natural factors have helped restore the region recently.
- Africa's diverse wildlife is threatened by human population growth, urbanization, and agricultural expansion. Poaching of elephants and rhinos for tusks and horns, mainly for Asian markets, has decimated their populations.
- In the four culture hearths of Sub-Saharan Africa, early indigenous people were responsible for several agricultural innovations, including the domestication of millet, sorghum, yams, cowpeas, okra, watermelons, coffee, and cotton.
- The major language families of the region are Niger-Congo, Khoisan, and Afro-Asiatic. Malagasy, spoken only on Madagascar, is an Austronesian language.
- Islam and Christianity are the major faiths, but indigenous belief systems are often mixed with them or exist on their own in some locales.
- Slaves were "merchandise" in the triangular trade linking West Africa, Europe, and the Americas. About 10 million slaves crossed the Atlantic. Pockets of slavery still exist in the region.
- Most of Sub-Saharan Africa fell under European colonialism after the Conference of Berlin in 1884 and 1885.
- The majority of the people of Sub-Saharan Africa are poor, live in rural areas, and practice subsistence agriculture. Export crops include coffee, cacao, cotton, peanuts, and oil palm products.
- Frequent droughts, lack of education, poor transportation, and public health problems have hindered development in Sub-Saharan Africa. Many countries are under-industrialized and overly dependent on the export of a few primary agricultural and mineral products. But 11 of the world's 20 fastest-growing economies are in the region. Middle classes are growing.
- Land grabbing and other agents of marginalization negatively impact subsistence farmers in the region. European and American subsidies of cash crops have made it difficult for African farmers to compete in world markets.
- The export of minerals has had a particularly strong impact on the physical and social geography of Sub-Saharan Africa. The notorious trade in dirty diamonds has been largely cleansed, but certifying tantalum from the DRC is more difficult.
- The resource curse is also the "paradox of plenty"; a country like Nigeria with a great abundance of a valuable natural resource often experiences lower economic growth than countries without such an endowment.
- Mobile phones are helping with agriculture and other areas of development in Africa. English-speakers represent the potential growth of call centers.
- Export processing zones (EPZs)—free trade areas offering incentives for international companies to invest—are gaining momentum. Exports of clothing from Africa to the United States boomed after the US Congress lowered trade barriers on numerous products with the African Growth and Opportunity Act (AGOA).
- Many countries are heavily in debt to foreign lenders. Western economic and humanitarian assistance to the region slowed after the end of the Cold War but has picked up again since 9/11. China is the latest major economic power to engage in Africa and has "no strings attached" policies of aid and trade with the region.
- The plunder of African resources by external powers and even by African leaders is a historic and current problem.
- Although many countries have been under authoritarian governments since independence, there has been progress toward democracy. Serious political instability is characteristic of many African countries. Important links with the colonial powers that formerly controlled them remain strong in many countries of the region.
- The African Union is a supranational organization dedicated to solving African problems without the intervention of outside powers.
- Sub-Saharan Africa is receiving increased international attention because of its humanitarian problems, the global implications of its public health and environmental situations, problems in the management of its natural resource wealth, its oil reserves, and concerns about terrorism.
- Terrorism is growing, with al-Shabaab, Boko Haram, and al-Qa'ida in the Islamic Maghreb among the most dangerous groups.
- The Sustainable Development Goals are succeeding the Millennium Development Goals as a framework for international efforts to advance development in this region and others.
- There are innovative efforts to focus some of the region's rapid urbanization on sustainable city- and IT-driven models.
- The most significant mining development in West Africa has been Nigeria's emergence as a producer and major exporter of oil. There are serious conflicts among ethnic, religious, and

political groups in Nigeria, some resulting from the maldistribution of income from the country's oil wealth.

- Warfare between the Hutus and Tutsis of Rwanda and Burundi in the 1990s killed over 1 million people in total, but outside countries did nothing to stop the violence. The genocide was related to the later violence in the DRC.
- The DRC was the epicenter of "Africa's First World War" that killed 5 million people, the deadliest conflict on Earth since World War II.
- Ethiopia has been called the "Galápagos Islands of Religion." The Rasta faith of Jamaica has roots there.
- Racial segregation characterized South African life from 1652 onward, but it was systematized after 1948 under a body of laws known as apartheid.
- In 1994, Nelson Mandela was elected president of South Africa after an all-race election was held in the country. After years of conflict, this historic election ended apartheid and white European political control. Reconciliation between the races is continuing. The ranks of middle- and upper-class blacks are growing, but the gap between the haves and have-nots in South Africa is large.
- Madagascar has a wealth of endemic species under tremendous pressure by humans, who have cleared 90 percent of the original vegetation cover in the past 2000 years. The theory of island biogeography can be used to estimate the loss of species there.

Review Questions

1. Where are the five principal areas of population concentration in Sub-Saharan Africa?
2. What factors account for the high HIV infection rates in Sub-Saharan Africa? What is being done to fight the epidemic, and how successful have these efforts been? What is HIV/AIDS doing to life expectancy and the age structure profiles in countries where it is epidemic?
3. What is Ebola, and what cultural practices were changed to help bring the 2014 epidemic under control?
4. What are the principal climatic zones and biomes of Sub-Saharan Africa? What are the region's characteristic topographic features?
5. What is the difference between drought and desertification, and how are they linked in the Sahel?
6. What is a keystone species? Why is the tsetse fly called "Africa's greatest conservationist?"
7. What impact have consumers in Asia and Yemen had on rhinoceros and elephant populations in the region?
8. What are the main languages and religions of the region? How important are indigenous belief systems?
9. Where were the major slave trade routes in and from Sub-Saharan Africa? What were the components of the triangular trade?
10. Where were the major colonial possessions of various European powers in Sub-Saharan Africa? What did Europeans extract from the region? What was the "mistake of 1914" that impacted Nigeria?
11. What are the main features of African economies, and in what ways are they improving?
12. What are the "durable strengths and resources" that Africans can draw from to promote their development? What is *kanju*?
13. What is the significance of cattle in many cultures of Sub-Saharan Africa?
14. What are the region's major export crops? Why does dependence on them make a country vulnerable? How do foreign agricultural subsidies and land grabs impact African agriculture?
15. What steps are being taken to control the "dirty" trade in diamonds and tantalum?
16. What is the "resource curse," and what countries are afflicted by it?
17. How have cell phones revolutionized African communications, and what can their potential benefits include?
18. Why have clothing exports from Africa to the United States boomed in recent years? What countries are leading exporters?
19. What important differences exist between Chinese and Western approaches to foreign aid in Africa? What is an "infrastructure-for-mineral swap?" How does poor infrastructure impede development in the region?
20. What impacts did the end of the Cold War have on many African countries? Why has geopolitical interest been refocused on the region?
21. What are the region's main terrorism "hot spots" or corridors, and what groups pose the greatest threats?
22. What are the histories and the current manifestations of outsiders' "scramble for resources" in Africa?
23. What are the Millennium Development Goals and Sustainable Development Goals that would boost development in Africa?
24. What innovative approaches are being applied to the region's rapid urbanization?
25. What country has the greatest demographic, political, and economic clout in West Africa? What unique difficulties does this country face with respect to its oil wealth?
26. What was the original distinction between the Hutus and the Tutsis? What issues brought them into conflict in the 1990s?
27. What was "Africa's First World War," and what were its principal components and outcomes?
28. Why is Ethiopia known as the "Galápagos Islands of Religion"?
29. How did the modern nation of South Africa evolve? What are the main ethnic groups in South Africa? What were their traditional social and economic roles? How have these changed since the end of apartheid? How did apartheid affect the country's cultural geography?
30. How are deforestation and species loss measured in Madagascar?

Key Terms + Concepts

African Growth and Opportunity Act (AGOA) (p. 425)
African National Congress (ANC) (p. 442)
African Standby Force (p. 438)
African Union (AU) (p. 429)
"Africa's First World War" (p. 438)
Afrikaners (p. 441)
AIDS (acquired immunodeficiency syndrome) (p. 401)
al-Qa'ida in the Islamic Maghreb (AQIM) (p. 430)
Al-Shabaab (p. 431)
Amhara (p. 439)
antiretroviral (ARV) drugs (p. 408)
apartheid (p. 441)
asymmetric warfare (p. 432)
Bantu subfamily (p. 412)
Bantustans (p. 441)
Boko Haram (p. 430)
branded diamonds (p. 423)
bushmeat (p. 408)
coloureds (p. 440)
crowdsourcing (p. 426)
desertification (p. 407)
dirty (conflict, blood) diamonds (p. 423)
donor democracy (p. 426)
drought (p. 407)
Ebola (p. 409)
Employment Equity Act (p. 442)
Ethiopian Orthodox Christianity (p. 439)
Ethiopian Orthodox Church (p. 413)
export processing zones (EPZs) (p. 425)
"Galápagos Islands of Religion" (p. 439)
georeferencing (p. 426)
Great Rift Valley (p. 410)
Great Trek (p. 440)
homelands (p. 441)
human immunodeficiency virus (HIV) (p. 401)
Hutu (Bahutu) (p. 437)
infrastructure-for-mineral swaps (p. 428)
Inkatha Freedom Party (IFP) (p. 442)
International Criminal Court (p. 438)
kanju (p. 434)
keystone species (p. 411)
Khoisan language family (p. 413)
Kimberley Process (p. 423)
Lord's Resistance Army (LRA) (p. 439)
microcredit (p. 428)
Millennium Development Goals (MDGs) (p. 433)
"mistake of 1914" (p. 418)
Movement for Oneness and Jihad in West Africa (MOJWA or MUJWA) (p. 432)
Movement for the Emancipation of the Niger Delta (MEND) (p. 436)
Movement for the Survival of the Ogoni People (MOSOP) (p. 436)
nagana (p. 408)
New Partnership for Africa's Development (NEPAD) (p. 429)
Niger-Congo language family (p. 412)
Ogoni (p. 437)
PEPFAR (p. 409)
proxy war (p. 429)
Rastafarianism (p. 439)
resilience (p. 407)
resource curse (p. 424)
San (p. 413)
sleeping sickness (p. 411)
Southern African Customs Union (SACU) (p. 425)
Sufi Islam (p. 414)
Sustainable Development Goals (SDGs) (p. 433)
terrorism hot spot (p. 430)
theory of island biogeography (p. 443)
triangular trade (p. 416)
Truth and Reconciliation Commission (p. 442)
trypanosomiasis (p. 411)
Tutsi (Watusi) (p. 437)
tyranny of distance (p. 429)
Ushahidi (p. 426)
Xhosa (p. 440)
Yoruba (p. 436)
Zulu (p. 440)

Notes

1. Quoted in http://www.africansuccess.org/visuFiche.php?id=373.
2. Richard Grant, *Africa: Geographies of Change* (Oxford: Oxford University Press, 2015), p. 10.
3. Teo Kermeliotis, "'Africa Is Not a Country': Students' Photo Campaign Breaks Down Stereotypes," CNN, February 7, 2014, http://www.cnn.com/2014/02/07/WORLD/AFRICA/AFRICA-IS-NOT-A-COUNTRY-CAMPAIGN/INDEX.HTML.
4. Cited in Lydia Polgreen, "Home Improvement: 'The Bright Continent,'" by Dayo Olopade, *New York Times,* April 11, 2014.
5. Lydia Polgreen, "Misery Loves Optimism in Africa," *New York Times,* March 5, 2006, p. WK-1.
6. Op cit., Grant, p. 8.
7. Ibid., p. 17.
8. "Ebola Basics: What You Need to Know," *The Economist*, December 30, 2014, http://www.bbc.com/news/health-29556006; "Ebola: Mapping the Outbreak," *The Economist*, May 11, 2015, http://www.bbc.com/news/world-africa-28755033.
9. Op. cit., Grant, p. 18.
10. Ibid., p. 20.
11. http://theweek.com/articles/449437/tragic-price-ivory.
12. "The Destiny of Africa between Sufism and Salafism," *Invisible Dog*, 10, October 2012, http://www.invisible-dog.com/sufismo_salafismo_eng.html.
13. Robert Stock, *Africa South of the Sahara,* 2nd ed. (New York: Guilford Press, 2004), p. 41.
14. 14. Ibid.
15. Op. cit., Grant, p. 14.
16. Ibid., p. 16.
17. Ibid., p. 17.
18. A. Samatar, B. Wisner, R. Chitiga, T. Smucker, E. Wangui, and C. Toulman, "Agenda for Action," in B. Wisner, C. Toulman, and R. Chitiga (eds.), *Towards a New Map of Africa* (Sterling, VA: Earthscan, 2015), p. 331.
19. Paul Collier, "Primary Commodity Dependence and Africa's Future," World Bank, 2012, http://documents.worldbank.org/curated/en/2002/04/3030501/primary-commodity-dependence-africas-future-primary-commodity-dependence-africas-future.
20. Ibid.
21. Quoted in Moises Naim, "Oil Can Be a Curse on Poor Nations," *Financial Times*, August 19, 2009, p. 7.
22. Op. cit., Grant, p. 68.
23. Ibid., p. 24.
24. World Travel and Tourism Council, Travel and Tourism Economic Impact: Africa 2014. 2014. http://www.wttc.org/-/media/files/reports/economic percent20impact percent20research/regional percent20reports/africa2014.pdf.
25. Op. cit., Grant, p. 24.
26. African Center for Economic Transformation, "Africa's Export Processing Zones: It's Time to Lift the Game," August 17, 2012, http://acetforafrica.org/whats-new/post/africas-export-processing-zones-its-time-to-lift-the-game/.

27. Tiago Faia, *Exporting Paradise? EU Development Policy towards Africa since the End of the Cold War* (Newcastle: Cambridge Scholars Publishing, 2012), p. 186.

28. Adam Green, "China in Africa: Taking Advantage," *Financial Times*, December 13, 2102, http://blogs.ft.com/beyond-brics/2012/12/13/china-in-africa-taking-advantage/.

29. Mark Anderson, "Aid to Africa: Donations from West Mask '$60bn Looting' of Continent. UK and Wealthy States Revel in Their Generosity while Allowing Their Companies to Plunder Africa's Resources, say NGOs," *The Guardian*, July 15 2014, http://www.theguardian.com/global-development/2014/jul/15/aid-africa-west-looting-continent.

30. Op. cit., Grant, p. 25.

31. Op cit., African Center for Economic Transformation.

32. Op. cit., Grant, p. 24.

33. Op. cit., Collier.

34. Op. cit., Grant, p. 329.

35. Quoted in Ian Fischer, "Africans Ask If Washington's Sun Will Shine on Them," *New York Times,* February 8, 2001, p. A3.

36. Op. cit., Grant, p. 315.

37. "The State of al-Qaeda. The Unquenchable Fire: Adaptable and resilient, al-Qaeda and Its Allies Keep Bouncing Back," *The Economist*, September 28, 2013, http://www.economist.com/news/briefing/21586834-adaptable-and-resilient-al-qaeda-and-its-allies-keep-bouncing-back-unquenchable-fire.

38. Dan Murphy, "'Boko Haram' Doesn't Really Mean 'Western Education Is a Sin,'" *The Christian Science Monitor*, May 6, 2014, http://www.csmonitor.com/World/Security-Watch/Backchannels/2014/0506/Boko-Haram-doesn-t-really-mean-Western-education-is-a-sin.

39. Maha Hosain, "Mali: From Democratic Model to Terrorist Hot Spot," *Bloomberg Business*, January 13, 2013, http://www.bloomberg.com/bw/articles/2013-01-13/mali-from-democratic-model-to-terrorist-hot-spot.

40. Jacob Zenn, "Boko Haram's Evolving Tactics and Alliances in Nigeria," West Point Combating Terrorism Center, June 25, 2013, https://www.ctc.usma.edu/posts/boko-harams-evolving-tactics-and-alliances-in-nigeria.

41. Op. cit, Grant, p. 317.

42. "Colliding Geopolitics and African Resources," *International Relations and Security Network*, December 7, 2011, http://www.isn.ethz.ch/Digital-Library/Articles/Special-Feature/Detail/?lng=en&id=134747&contextid774=134747&contextid775=134743&tabid=1451531620.

43. J. Chissano, "The Current Geopolitical Dynamics in Africa and the Role of Partners like the EU," The European Center for Development Policy Management, September 2013, http://ecdpm.org/great-insights/new-impetus-africa-europe-relations/current-geopolitical-dynamics-africa-role-partners-like-eu/.

44. Op. cit., Grant, p. 309.

45. Op. cit., Grant, pp. 21, 329.

46. http://www.undp.org/content/undp/en/home/mdgoverview/mdg_goals/post-2015-development-agenda/.

47. Op cit., Polgreen.

48. Op cit., Grant, p. 21.

49. Op cit., Grant, p. 23.

50. Op cit., Grant, p. 24.

51. Op cit., Grant, p. 319.

52. "Land-Shackled Economies. The Paradox of Soil: Land, the Centre of the Pre-Industrial Economy, Has Returned as a Constraint on Growth," *The Economist*, April 4, 2015, http://www.economist.com/news/briefing/21647622-land-centre-pre-industrial-economy-has-returned-constraint-growth.

53. Op cit., Grant, p. 322.

54. For SDI, see http://www.sdinet.org.

55. Op cit., Grant, p. 333.

56. Elisha Balo-Gbobgo, "Nigeria's MEND Rebels Threaten Future Attacks on Oil Industry," *Bloomberg Business*, January 27, 2014. http://www.bloomberg.com/news/articles/2014-01-27/nigeria-s-mend-rebels-threaten-future-attack-on-oil-industry.

57. Kony 2012 may be viewed at https://www.youtube.com/watch?v=Y4MnpzG5Sqc.

58. Op. cit., Grant, p. 317.

59. This information comes from an excellent online description of the Rasta, http://en.wikipedia.org/wiki/Rastafarian#Haile_Selassie_and_the_Bible (accessed May 16, 2012). "The Galapagos Islands of Religion" was the title of a Marjorie Coeyman's *Christian Science Monitor* article on Ethiopia's religions, on March 30, 2000.

60. On the fragile five, see Landon Thomas, "Fragile Five' Is the Latest Club of Emerging Nations in Turmoil," *New York Times*, January 18, 2014.

61. Edward O. Wilson, "Threats to Biodiversity," *Scientific American,* 261(3), 1989, pp. 60–66.

Global Geoscience Watch

Go to the GREENR database and conduct basic research on the United Nations' Sustainable Development Goals (SDGs). Find at least two news stories related to the establishment of these goals. Considering all that you have learned in this chapter, write one to two paragraphs listing the SDGs you think are most attainable, which least likely to be realized. Do the articles you researched mention any potential adverse impacts of the newly established goals?

Online Resources

For access to MindTap and additional study materials visit www.cengagebrain.com. Read your textbook, take notes, complete activities, take practice quizzes and more.

Above: Disembarking a ferry at Lake Atitlan, Guatemala. Towering 11,598 feet (3535 m) over the lake is the active stratovolcano, Volcán Atitlán. Joe Hobbs

Left: Latin America at night, from space. The populous "rim" around the margins of South America is lit up. Note the dark vastness of the Amazon rain forest. Manaus, at the confluence of the Amazon and the Rio Negro, shines brightly. C. Mayhew & R. Simmon (NASA/GSFC), NOAA/ NGDC, DMSP Digital Archive

10 Latin America

Be brave. Take risks. Nothing can substitute experience . . . You have to take risks. We will only understand the miracle of life fully when we allow the unexpected to happen.

—Paulo Coelho, Brazilian lyricist and novelist[1]

The land portion of the Western Hemisphere south and southeast of the United States is commonly known as Latin America. This name reflects the importance of cultural traits inherited mainly from the European colonizing nations of Spain and Portugal, whose languages evolved from Latin. Native American cultures were flourishing when Columbus arrived in the Americas in 1492, but fell victim to the subsequent tide of European forces. The native cultures still persist to varying degrees, but "Latin" influences predominate today in the region as a whole.

This chapter will acquaint you with the varied indigenous and European cultures that have shaped the region. Latin America's exceptional topographic and environmental diversity are introduced as well. The economic prospects for this developing region are tied closely to relationships with the United States and China; the broader involvement of the US in affairs "south of the border" is also discussed. The must-read

Chapter Outline

Chapter Objectives

This chapter will enable you to

- Appreciate how topographic variety creates a predictable range of environmental conditions and livelihood opportunities, and what geographic traits correlate with core and peripheral regions.
- Learn about the accomplishments of indigenous cultures and how European conquest and colonization impacted indigenous peoples.
- Recognize the predominant ethnic patterns of the region and how ethnicity correlates with livelihood, wealth, and political power.
- Recognize the economic and social inequities in rural and urban settings that hamper development and contribute to dissent.
- Understand the obstacles Latin American economies face in transitioning from dependence on primary commodities to diversified manufacturing and services sectors, and the roles of free trade and fair trade.
- Understand how US interests have shaped the region's political and economic systems, and how the Latin American "ideology of fury" has pushed back against American influence.
- Examine human–environment interactions in the contexts of natural hazards, climate change, and economic development.

book for more in-depth knowledge of the geography of Latin America is *Latin America and the Caribbean: A Systematic and Regional Survey* (Brian Blouet and Olwyn Blouet, 7th ed., Wiley, 2015).

10.1 Area and Population

Latin America consists of a mainland region that extends from Mexico south to Argentina and Chile, together with the islands of the Caribbean Sea (•**Figures 10.1** and **10.2**; •**Table 10.1**). Its maximum latitudinal extent of more than 85 degrees, or nearly 5900 miles (9500 km), is greater than that of any other major world region. There are 38 countries and a handful of colonial possessions (primarily in the Caribbean) in the region.

Latin America has a land area of slightly more than 7.9 million square miles (20.5 million sq km) (•**Figure 10.3**). Latin America's two main subregions are neatly offset from one another geographically. The northern part, known as Middle America, includes Mexico, the countries of Central America and the islands of the Caribbean. South America protrudes much farther into the Atlantic Ocean than the Caribbean realm or Latin America's northern neighbor, North America. In fact, the meridian of 80 degrees west longitude (80°W), which touches the west coast of South America in Ecuador and Peru, passes through Pittsburgh, Pennsylvania. Brazil also lies less than 2000 miles (3200 km) west of Africa, to which it was joined in the supercontinent of Pangaea until around 110 million years ago. 25

Latin America's population of 618 million represents 9 percent of the world total. The population has uneven distributions and densities (•**Figure 10.4**). Most of Latin America's people are concentrated along two major geographic alignments. The larger of these two areas, known as "the rim" or "the Rimland," is a discontinuous ring around the margins of South America that is home to about two-thirds of Latin America's people. The second, generally a highland, extends along a volcanic belt from central Mexico southward into Central America.

Average population densities disguise the fact that most Latin American mainland countries reflect a basic spatial configuration: a well-defined population core (or multiple cores) with one or more outlying, sparsely populated **hinterlands** (peripheries). 13 The population density for the greater part of Latin America is quite low, averaging fewer than two people per square mile over approximately half of the entire region. The reason for this sparse density is that great stretches of rain forest, deserts, and mountains occupy large areas of the region.

Not every country, however, displays this pattern of core and hinterland. El Salvador and Costa Rica, with densely settled volcanic lands, and most Caribbean islands are different. The combined population of the Caribbean is about the same as Central America's. But these islands have heavy population densities, developed over a long period of growing populations in limited geographic areas. Continued high population growth rates on islands where steep slopes restrict farming and settlement compound the problems of "people overpopulation." 81

Very high rates of urban growth have characterized Latin America in recent times. Latin America became a majority-urban region in the 1960s; today the region is 78 percent urban (compared to the world average of 53 percent). The boom of urban population reflects the push factor of rural-to-urban migration, and especially the pull factor of "selective migration" related to the growth of urban jobs after World War II. Rural-to-urban migration beginning in the 1950s brought about "urbanization without industrialization," 75 similar to situations in Africa and the Middle East. 34, 434 Generally, manufacturing today employs about a third of the urban labor force, with the remainder in the service sector, both formal (e.g., government and retail jobs) and informal (e.g., street jobs). The region's largest cities are listed in •**Table 10.2**.

Booming urban populations also resulted from the population surge of stage two of the demographic transition. Many Latin American countries are in the second and 72

• **Figure 10.1** Physical geography of Latin America.

• **Figure 10.2** Political geography of Latin America.

Table 10.1 Latin America: Basic Data

Political Unit	Area (sq mi, thousands)	Area (sq km, thousands)	Population (millions)	Rate of Natural Increase (%)	Urban Population (%)	Population Under Age 15 (%)	Agricultural Workers (%)	Per Capita GDP (PPP) ($US)	GDP ($US, billions)	Trade Balance ($US, billions)	Oil Production (million bbl/day)	Literacy, Female (%)	Literacy, Male (%)	HDI
Latin America	**7946.2**	**20580.7**	**618**	**1.2**	**78**	**27**	**16**	**14,000**	**9,336**	**—**	**9.3**	**88**	**89**	**0.734**
Middle America														
Belize	8.9	23.1	0.4	1.9	45	36	10	8,800	3	0	—	84	75	0.732
Costa Rica	19.7	51.0	4.8	1.1	73	24	14	14,900	71	−6	—	96	96	0.763
El Salvador	8.1	21.0	6.4	1.4	65	30	21	7,500	50	−5	—	82	87	0.662
Guatemala	42.0	108.8	15.9	2.6	50	40	38	5,300	118	−6	—	—	—	0.626
Honduras	43.3	112.1	8.2	2.0	52	35	39	4,800	39	−3	—	80	80	0.617
Mexico	756.1	1958.3	119.7	1.4	78	28	13	15,600	2,143	−1	2.8	93	93	0.756
Nicaragua	50.2	130.0	6.2	1.9	56	33	31	4,500	29	−2	—	78	78	0.614
Panama	29.2	75.6	3.9	1.5	75	28	17	16,500	77	−7	—	91	92	0.761
Caribbean														
Anguilla (U.K.)	*	*	*	2.1	—	23	4	12,200	*	—	—	—	—	—
Antigua and Barbuda	0.2	0.5	0.1	0.8	30	24	7	18,400	2	0	—	98	99	0.774
Aruba (Neth.)	0.1	0.3	0.1	1.3	—	17	—	25,300	2	0	—	97	97	—
Bahamas	5.4	14.0	0.4	1.1	84	26	3	25,100	9	−2	—	96	95	0.789
Barbados	0.2	0.5	0.3	0.4	44	20	10	16,100	4	−1	—	99	99	0.776
Bonaire (Neth.)	*	*	*	—	17	—	—	—	—	—	—	—	—	—
Cayman Islands (U.K.)	0.1	0.3	*	2.1	—	18	2	43,800	2	—	—	99	99	—
Cuba	42.8	110.9	11.2	0.3	77	17	20	10,200	128	−9	—	99	99	—
Curacao (Neth.)	*	*	0.2	0.5	—	19	1	15,000	3	—	—	—	—	—
Dominica	0.3	0.8	0.1	0.5	67	22	40	14,300	1	0	—	94	94	0.717
Dominican Republic	18.8	48.7	10.4	1.5	67	31	14	9,700	135	−7	—	90	90	0.700
Grenada	0.1	0.3	0.1	0.8	39	27	11	13,800	1	0	—	—	—	0.744
Guadeloupe (France)	0.7	1.8	0.4	0.6	98	21	—	—	—	—	—	—	—	—
Haiti	10.7	27.7	10.8	1.9	53	35	38	1,300	18	−2	—	51	54	0.471
Jamaica	4.2	10.9	2.7	1.1	54	27	17	9,000	24	−3	—	91	84	0.715
Martinique (France)	0.4	1.0	0.4	0.4	89	19	—	—	—	—	—	—	—	—
Puerto Rico (U.S.)	3.5	9.1	3.6	0.3	99	19	2	16,300	64	—	—	94	94	—
St. Kitts and Nevis	0.1	0.3	0.1	0.6	32	22	—	16,300	1	0	—	—	—	0.750
St. Lucia	0.2	0.5	0.2	0.6	18	23	21	13,100	2	0	—	90	90	0.714
St. Vincent and the Grenadines	0.2	0.5	0.1	0.9	49	24	26	12,100	1	0	—	96	96	0.719
Trinidad and Tobago	2.0	5.2	1.3	0.3	14	20	4	20,300	42	3	0.1	98	99	0.766
Turks and Caicos (U.K.)	0.1	0.3	*	2.6	—	22	20	29,100	1	—	—	98	99	—
Virgin Islands (U.K.)	*	*	*	2.3	—	17	1	42,300	*	—	—	—	—	—
Virgin Islands (U.S.)	0.1	0.3	0.1	−0.6	95	18	1	—	1	—	—	—	—	—

(continued)

Table 10.1 Latin America: Basic Data (*continued*)

Political Unit	Area (sq mi, thousands)	Area (sq km, thousands)	Population (millions)	Rate of Natural Increase (%)	Urban Population (%)	Population Under Age 15 (%)	Agricultural Workers (%)	Per Capita GDP (PPP) ($US)	GDP ($US, billions)	Trade Balance ($US, billions)	Oil Production (million bbl/day)	Literacy, Female (%)	Literacy, Male (%)	HDI
South America														
Argentina	1073.5	2780.4	42.7	1.1	92	25	5	22,100	927	10	0.7	98	98	0.808
Bolivia	424.2	1098.7	10.3	1.9	67	35	32	5,500	70	2	*	86	95	0.667
Brazil	3300.2	8547.5	202.8	0.9	85	24	15	15,200	3,073	1	2.6	91	91	0.744
Chile	292.1	756.5	17.7	0.9	87	22	13	23,200	410	6	—	98	98	0.822
Colombia	439.7	1138.8	47.7	1.3	76	28	17	11,100	642	−1	1.0	93	93	0.711
Ecuador	109.5	283.6	16	1.8	63	31	28	10,600	182	1	0.5	90	93	0.711
Falkland Islands (U.K.)	4.7	12.2	*	—	—	—	—	55,400	*	—	—	—	—	—
French Guiana (France)	34.7	89.9	0.3	2.4	76	34	—	—	—	—	—	—	—	—
Guyana	83.0	215.0	0.7	1.3	28	36	—	8,500	5	−1	—	92	92	0.638
Paraguay	157.0	406.6	6.9	1.7	59	33	26	6,800	57	2	—	93	95	0.676
Peru	496.2	1285.2	30.8	1.5	75	29	25	11,100	376	−4	0.1	89	96	0.737
Suriname	63.0	163.2	0.6	1.3	70	28	8	12,900	9	0	—	90	95	0.705
Uruguay	68.5	177.4	3.4	0.5	94	22	13	20,500	69	−1	—	98	98	0.790
Venezuela	352.1	911.9	30.2	1.5	89	29	7	13,600	545	33	2.4	95	95	0.764

* Less than 0.1. — Data not available or not applicable.
Sources: World Population Data Sheet, Population Reference Bureau, 2014; Human Development Report, United Nations, 2014; World Factbook, CIA, 2014.

early third stages of the transition, clearly recognizable as less developed countries (LDCs) having relatively high (although declining) birth rates and low death rates due to advances in the spread of medical technologies. Many are progressing through the transition as population growth rates have fallen overall from 2.8 percent in 1980 to 1.2 percent today, the same as the world average. This regional rate of increase is down considerably, largely because the region's economies are generally improving.

Chile, with a population growth rate of 0.9 percent, is an example of a country in the third, or industrializing, stage of the demographic transition, with a low death rate and falling birth rate, hence a slowing population increase. Costa Rica, Argentina, and Uruguay are also in this category. The highest population growth rates tend to be in the least developed countries. Guatemala has the greatest growth rate in the region, at 2.6 percent; Honduras is second highest at 2.0 percent. Guatemala's underdevelopment, linked especially to the rural poverty of large indigenous populations, is reflected in its middling HDI rating of 0.628. At the other end of the HDI scale is Chile, with a score of 0.822. In addition to its low growth rate, Chile has a relatively small percentage of its population under age 15, a large urban population (87 percent), and a high per capita GDP (PPP) at $23,200—similar to that of Greece and Portugal.

The lowest per capita GDP (PPP) and lowest HDI ranking in all of Latin America belongs to Haiti. You would reasonably expect it to have the region's highest population growth rate, and it probably would if not for the scourge of HIV/AIDS. Haiti has the highest infection rate in the Western Hemisphere, affecting some 2.6 percent of adults. Haiti's annual population growth rate is just 1.9 percent. The region's lowest annual rate of population change is 0.3 percent in Cuba, caused more by strict Communist promotion of family planning than by general prosperity.

10.2 Physical Geography and Human Adaptations

Dramatic differences in elevation, topography, biomes, and climates characterize Latin America. Low-lying plains drained by the Orinoco, Amazon, and Paraná-Paraguay River systems dominate the north and central part of South America and separate geologically older, lower highlands in the east from the rugged Andes of the west. In Mexico, a high interior plateau

• **Figure 10.3** Latin America and the United States. Note the great longitudinal offset of North and South America.

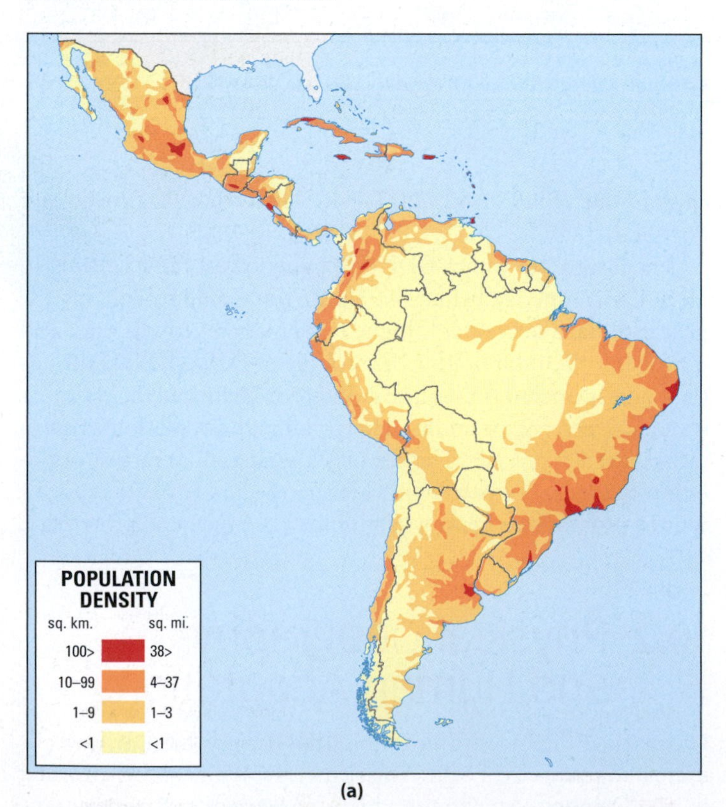

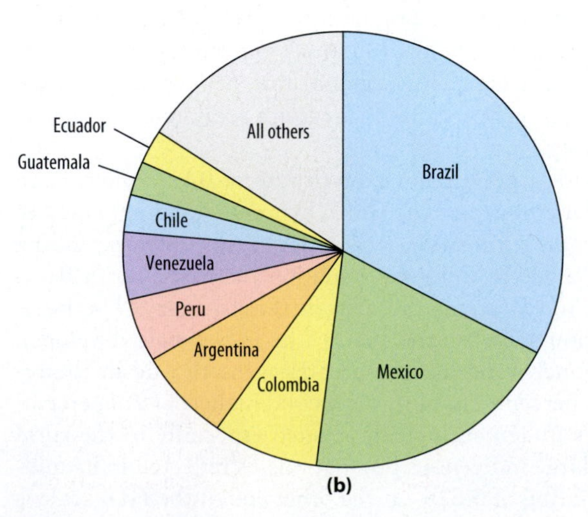

• **Figure 10.4** **(a)** Population distribution and **(b)** population pie chart of Latin America.

THINK CRITICALLY: Do you recognize the two major alignments of population along the Rimland and the highlands? According to the text, what factors favor habitation in these zones?

broken into many basins lies between north-to-south-trending arms of the Sierra Madre. High mountains, largely within the Sierra Madre and the Andes, form a nearly continuous landscape feature from northern Mexico to the southern tip of South America. Most of the smaller islands of the Caribbean Sea's West Indies are volcanic mountains, although some islands of limestone or coral are lower and flatter. The largest islands of the Caribbean have a more diverse topography, including low mountains.

Climates and Vegetation

The climatic and biotic diversity of Latin America is extraordinary (•**Figure 10.5**). Even within some single countries (Ecuador, for example), conditions range over a very short horizontal distance from sea level tropical rain forests to alpine tundra. The world's largest continuous expanse of tropical rain forest lies in the basin of the Amazon River system. Tropical rain forests 484 also grow in southeastern Brazil, eastern Panama, the western coastal plain of Colombia, on the Caribbean side of Central America and southern Mexico, and along the eastern (windward) shores of some Caribbean islands, particularly Hispaniola and Puerto Rico. Much of the original tropical rain forest vegetation of the islands has been lost to human activity, and even the vast Amazon forest is under considerable pressure (this problem, which is largely a reflection of Brazil's development, is described on **pages 482–484**).

On either side of the principal region of tropical rain forest climate, a tropical monsoon climate emerges from a more seasonally variable distribution of rainfall. Farther away from the Equator, the tropical savanna climate covers large portions of central Brazil, coastal Mexico, and Cuba. In Venezuela, this climate gives rise to the *llanos*, tropical grasslands interspersed with flooded grasslands. Southern Brazil and eastern Argentina experience a humid subtropical climate, with cooler winters than in the tropical zones. Here, this climate is associated mainly with the humid ***pampas***, the temperate grasslands that are analogous to the prairies of the American Midwest

Table 10.2 Largest Cities in Latin America

City	Population
São Paulo, Brazil	20.3
Mexico City, Mexico	20.1
Buenos Aires, Argentina	14.1
Rio de Janeiro, Brazil	11.7
Lima, Peru	10.7
Bogotá, Colombia	9
Santiago, Chile	6.2
Guadalajara, Mexico	4.6
Belo Horizonte, Brazil	4.5
Monterrey, Mexico	4.1
Medellín, Colombia	3.5

Population in millions.
Source: Demographia World Urban Areas, 2015.

(•**Figure 10.6**). On the Pacific side of South America, a small strip of Mediterranean or dry-summer subtropical climate in central Chile is similar to that of California. This is an ideal climate for wine production, and Chilean wines are now well established on world markets.

• **Figure 10.6** Wild horses on pampas grasslands, Patagonia, Argentina.

The humid climates of Latin America have a fairly predictable spatial arrangement. But the region's dry climates and biomes—desert and steppe—are the product of local circumstances such as rain shadows and offshore cold ocean currents.

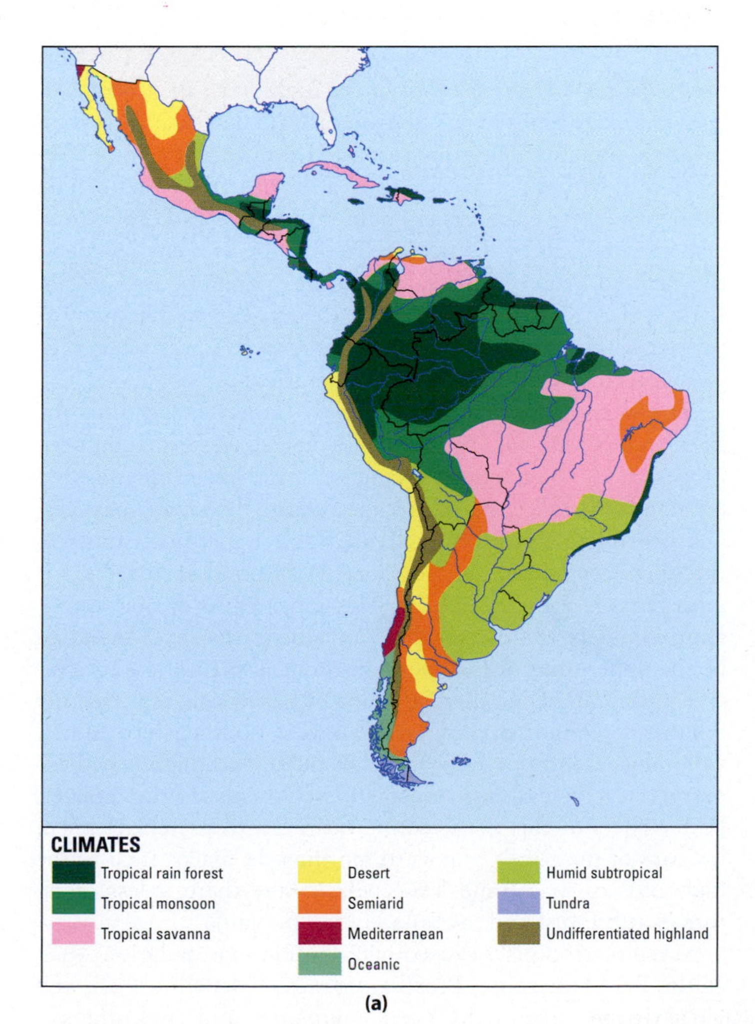

• **Figure 10.5** **(a)** Climates and **(b)** biomes of Latin America.

In northern Mexico, aridity is due partly to the rain shadow effect of high mountain ranges on either side of the Mexican plateau. High-elevation semiarid regions of Mexico are home to the world's largest concentration of the rare tropical coniferous forest biome. 32

In Argentina, the extremely high and continuous Andean mountain wall accounts for the aridity of large areas, especially the southern region of Patagonia. Here, the Andes block the path of the prevailing westerly winds, creating heavy orographic (mountain-induced) precipitation on the Chilean side of the border but leaving Patagonia in rain shadow. Farther north along the Pacific coast in Chile and Peru is the Atacama, the world's driest desert—even more arid than the Sahara. The Atacama lies in a "double rain shadow": to the east the Andes block humid air from the Amazon, and to the west the Chilean Coast Range blocks air from the Pacific Ocean. An offshore cold current and a semipermanent area of high pressure at these latitudes reinforce the dryness. Arid and semiarid conditions also prevail along the northernmost coastal regions of Colombia and Venezuela and in the poverty-stricken region called the Sertão in northeastern Brazil. 471

Elevation and Land Use

With respect to human adaptations, one of the most significant features of Latin America's physical geography is a series of highland climates arranged into zones by elevation. This zonation results from the fact that air temperature decreases with elevation at a normal "lapse rate" of approximately 3.6°F (1.7°C) per 1000 feet (300 m). At least four major zones are commonly recognized in Latin America (•**Figure 10.7**): the ***tierra caliente*** **(hot country)**, the ***tierra templada*** **(cool country)**, the ***tierra fría*** **(cold country)**, and the ***tierra helada*** **(frost country)**. Combined with the region's great latitudinal variations and local conditions, these zones define and limit land uses (•**Figure 10.8**). At the foot of the highlands, the *tierra caliente* is a zone embracing the tropical rain forest and tropical savanna climates. The zone reaches upward to approximately 3000 feet (900 m) above sea level at or near the Equator and to slightly lower elevations in parts of Mexico and other areas near the margins of the tropics. In this hot, wet environment, the favored crops are rice, sugarcane, bananas, and cacao. Latin America's black populations are concentrated in many of the *tierra caliente* zones, a legacy of the slave trade, when they were forced to work the region's plantations. 416

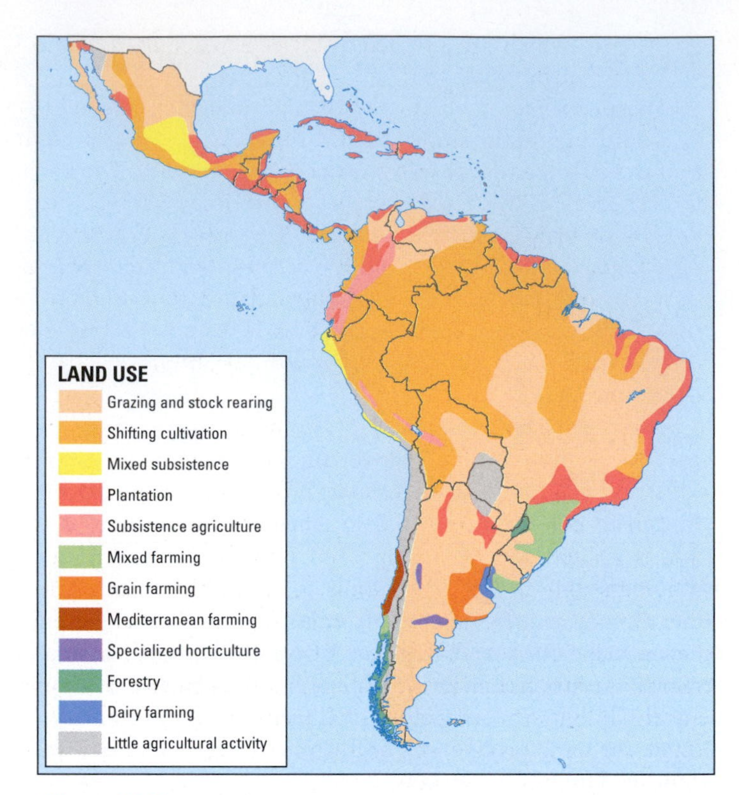

• **Figure 10.8** Land use in Latin America.

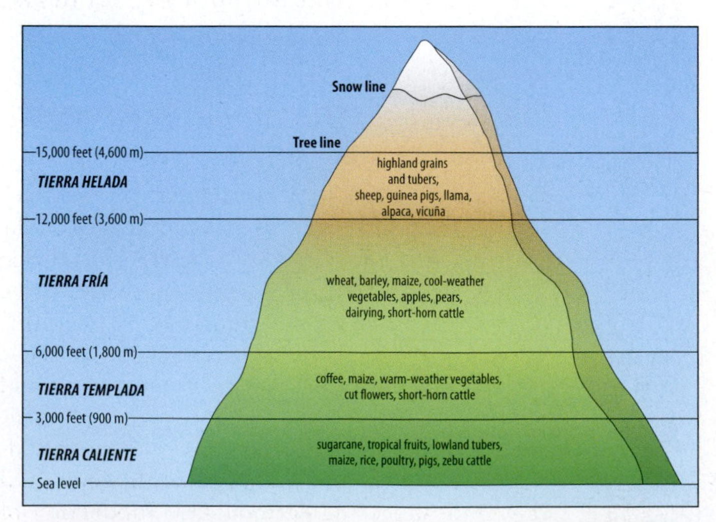

• **Figure 10.7** Altitudinal zonation in Latin America. Specific patterns of land use have evolved to take advantage of the region's diverse altitudinal zones. This graph shows elevations and general land use patterns, which may also be compared with land uses in Figure 10.8. Elevations shown here are for the Equatorial region and are somewhat lower toward the poles for the respective land uses.

The *tierra caliente* merges almost imperceptibly into the *tierra templada*, which flanks the rugged western mountain ranges and is the uppermost climate in the lower uplands and highlands to the east. Sugarcane, cacao, bananas, oranges, and other lowland products reach their uppermost limits in the *tierra templada*, but this zone of "eternal springtime" is most famous as coffee habitat. The upper limits of this zone—approximately 6000 feet (1800 m) above sea level—tend to be the upper limit of European-introduced plantation agriculture and modern commercial crops in Latin America. Densely inhabited sections occupy large areas in southeastern Brazil, Colombia, Central America, and Mexico. Although broadleaf evergreen (tropical rain forest) trees characterize the moister, hotter parts of this zone, coniferous evergreens replace them in parts of the zone's poleward margins. In places such as the highlands of Brazil and Venezuela, where there is less moisture, scrub forest and savanna grasses prevail.

Seven metropolises exceeding 2 million in population—São Paulo, Belo Horizonte, Brasília, Caracas, Medellín, Cali, and Guadalajara—are in the *tierra templada*, and two others—Mexico City and Bogotá—lie slightly above it. Others, like Rio de Janeiro, which are situated at lower elevations, have close ties with nearby residential and resort areas.

The *tierra fría* is comprised of high plateaus, basins, valleys, and mountain slopes within the great mountain chain that extends from northern Mexico to Cape Horn. By far the largest areas of this habitat are in the Andes, with significant areas also in Mexico (**•Figure 10.9**). Around the Equator, the *tierra fría* begins at about 6000 feet (1800 m) and extends upward to the **upper limit of agriculture** (represented by such hardy crops as potatoes and barley) and the **tree line** (the upper limit of natural tree growth) at about 12,000 feet (3600 m).

The *tierra fría* experiences frost and in many countries is the adverse habitat of a Native American economy based on subsistence agriculture. In a classic process of marginalization, European colonization of Latin America drove some Native American settlements upslope and into the *tierra fría* zone, although some major populations (notably the Inca of Peru) had already selected upland locations for their settlement long 421 before the arrival of Columbus. These upland settlements are most extensive in Ecuador, Peru, and Bolivia and are also present in Colombia, Guatemala, and southern Mexico. Most of them are rural villages based on subsistence farming of the cool weather crops listed in Figure 10.7. In some cases, especially in Bolivia and Peru, valuable minerals like tin and copper are mined on a large scale in the *tierra fría* as well as in the higher *tierra helada*.

The *tierra helada* zone lies above the other three and consists of the alpine meadows, known as ***páramos***, along with still higher barren rocks and permanent fields of snow and ice (**•Figure 10.10**). In the lower latitudes it generally lies between 12,000 feet (3600 m) and the lower edge of the snow line. It supports some grains and livestock (llama, alpaca, sheep) but is largely above the mountain flanks that are central to upland native settlement and agriculture.

Joe Hobbs

• Figure 10.10 The *páramo* at about 12,000 feet (c. 3600 m) in the Andes of Colombia. This habitat is often fog-shrouded, dripping, cool, and windy.

The rooftop of South America is Mount Aconcagua in Argentina, rising to 22,841 feet (6962 m) in the Andes. Its slopes, like many of the higher Andean reaches, are covered in glacial ice (**•Figure 10.11**). Andean glaciers are following the worldwide climate change trend of melting. As the glaciers melt, they take consistent, dependable water supplies with them—a problem that also threatens the "water towers of Asia" (see **page 330**). Some scientists predict that the Bolivian capital of La Paz, with its 2 million residents, will be the first large urban casualty of climate change. There are other natural threats as well.

Joe Hobbs

• Figure 10.9 Characteristic *tierra fría* landscape and agriculture in Andean South America. The lines of eucalyptus serve as shelter belts, protecting crops. Native to Australia, eucalypts flourish in the *tierra fría*, where local people use them for medicine, fuelwood, and construction. Their downside is their high water intake, and especially in parts of Peru, they are being eradicated for this reason.

Natural Hazards in Latin America

Adjoining a large section of the Pacific Ring of Fire and fronting two seasonal **hurricane** regions, Latin America is beset by natural hazards. The Pacific coast stretching from northern Mexico all the way to the southern tip of South America has a violent history of earthquakes and volcanic eruptions. On the map of global tectonic plates (Figure 2.1), you can see the Nazca Plate and South American Plate confronting one another, with tremendous geological consequences including the uplift of the Andes Mountain chain and those volcanoes and earthquakes. Subduction of the Nazca Plate under the South American plate generated the largest earthquake ever recorded, a titanic 9.5 magnitude event in May 1960, just off the coast of southern Chile. Known as the Great Chilean Earthquake or the Valdivia Earthquake,

• Figure 10.11 The spectacular glacial landscapes of the southern Andes in Torres del Paine National Park, Chilean Patagonia. At about 50 degrees south latitude, the ice line here is close to sea level.

this was a "megathrust" event like the 2011 earthquake that rocked eastern Japan. It created tsunami waves reaching 82 feet (25 m) high along Chile's coast and also affecting the Aleutian Islands, Hawaii, Japan, the Philippines, New Zealand, and Australia. Fifteen hours after the quake and 6200 miles (10,000 km) away, a tsunami 35 feet (11 m) in height obliterated parts of Hilo on the big island of Hawaii, killing 61 people. A total of 1600 people around the Pacific died.

Figure 2.1 also depicts an active tectonic area on the eastern side of the Americas: the Caribbean Plate is also in motion, causing earthquakes and a variety of associated tectonic features. You can see that one of this plate's boundaries passes through the island of Hispaniola, where the countries of Haiti and the Dominican Republic are located (see Figures 10.1 and 10.2, the Geography of Natural Hazards, **page 459**, and Try It, **page 460**).

Hurricanes, giant low-pressure systems that feed on warm waters of the Atlantic and Pacific Oceans, have sometimes devastating impacts. A single storm can wreck the economy of an island nation, as Hurricane Ivan did by wiping out Grenada's nutmeg crop in 2004. The monstrous Hurricane Mitch (category 5, highest on the Saffir-Simpson scale) made landfall in Honduras in 1998, and roamed for five days over that country and Nicaragua, El Salvador, Guatemala, and Mexico. Its ferocious winds and rainfalls of up to 2 feet (60 cm) brought floods, landslides, and storm damage—intensified by the region's widespread deforestation—that killed an estimated 15,000 people; inflicted enormous damage on property, crops, roads, power, and other infrastructure; and left disease epidemics in its wake. As if Mitch were not enough, the region had also suffered from drought and fires caused by El Niño in 1997 and 1998.

What is El Niño? It is best understood as a complete reversal of a normal climatic pattern. Normally, winds across the Pacific Ocean blow east to west, helping to promote the upwelling of cold, nutrient-rich water from the depths to the surface of the tropical eastern Pacific (**•Figure 10.12**). But every few years, usually beginning in December, the winds reverse direction, suppressing the upwelling water and thus raising the surface temperature of the water by as much as 20°F (11°C). Because of its common occurrence in December, that event has been dubbed **El Niño**, meaning "the Baby" in reference to the Christ child, whose birth is celebrated in December. Less prosaically, meteorologists know it as "El Niño Southern Oscillation (ENSO)."

El Niño conditions sometimes last a year or more, causing global climatic disruptions. These vary between events, but the typical El Niño pattern is illustrated in **•Figure 10.13**. Closest to the source of the condition, Peru experiences torrential rains. Unusually high rainfall also occurs in southern Brazil, the southeastern United States, western Polynesia, and East Africa.

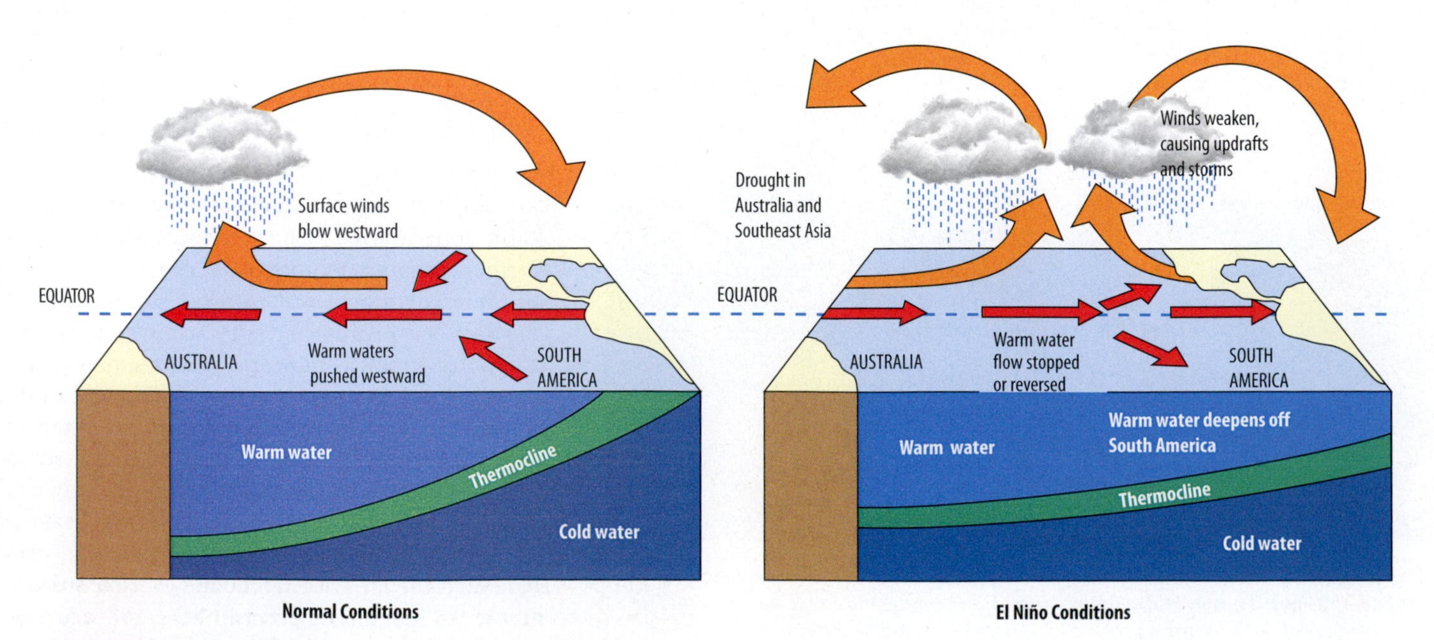

• Figure 10.12 Normal and El Niño conditions in the Pacific Ocean.
Source: From MILLER/SPOOLMAN, *Environmental Science*, 13E. © 2011 Cengage Learning.

Geography of Natural Hazards | Heartbreak in Haiti

On January 12, 2010, an earthquake with a magnitude of 7.0 struck Hispaniola on a strike-slip fault zone just 16 miles (25 km) from Haiti's capital, Port-au-Prince. In a more developed country better prepared with earthquake-resistant infrastructure, such a quake might not have produced extensive damage. But in Haiti, by far the poorest country in the Western Hemisphere, this temblor was catastrophic. It killed an estimated 315,000 people and made nearly a million homeless (**• Figure 10.A**).

A unique combination of geologic, historic, and political factors made Port-au-Prince especially vulnerable to seismic catastrophe. Not only is the city close to the fault zone on which the earthquake struck, but much of Port-au-Prince lies atop soft sedimentary rock that amplified the seismic waves. Then there were the urban population issues: far surpassing the metric of a primate city, the capital was home to one in three Haitians when the earthquake hit. Most of the city's housing stock was made up of **informal settlements**, a term often used interchangeably with *slums* but meaning the occupants lack legal claims to the land and that the housing is not in compliance with planning and building regulations. Much of the housing is fashioned from recycled materials. Meeting the requirements of earthquake resistance is far beyond the financial means of the people of Port-au-Prince, where two-thirds of the labor force lacks formal employment. Much of Port-au-Prince is a sprawling slum, described this way by journalist Amy Wiletnz: "a heart-tugging city that functioned on a complicated, hypercharged fuel of chaos, exposed wiring, pig slop, smog, gingerbread turrets, hot cooking oil, rum, cockfights and bougainvillea."[2]

Stocktrek Images/Getty Images

• Figure 10.A "Informal settlements" or slums, in many cases on steep hillsides as seen here, make up much of the housing stock of Port-au-Prince. They lack any kind of earthquake-resistant features, and the January 2010 quake reduced them to rubble.

The utter devastation of the 2010 earthquake gave Haitians the opportunity to refashion their urban geography. Assisted by geographers and other scientists from the University of Miami, a group of Haitian urban planners came up with a plan to decentralize Haiti, redistributing large numbers of people to smaller Haitian cities at safer distances from fault zones. The project will completely transform Haiti from a country dominated by the primacy of Port-au-Prince to one with a network of smaller urban "growth poles." These would be more suited to the economic assets Haiti could best develop: agriculture and tourism. Planners will locate schools, markets, and hospitals and health care centers in smaller towns and villages so that people will not need to leave these settlements and gravitate again to Port-au-Prince. Homes, stores, and offices will be constructed with earthquakes in mind and should grow over time in modular fashion from basic shelters into permanent fixtures.

Unfortunately, as a metropolis ripe for seismic ruin, Port-au-Prince is not alone. Other vulnerable cities, which seismologists right after the Haiti quake listed as "rubble in waiting," include Istanbul, Turkey; Karachi, Pakistan; Tehran, Iran; Lima, Peru; and Kathmandu, Nepal. All of these cities continue to swell from rural to urban migration, and all share a surplus of informal settlements built from such dangerous materials as brick and mud-brick. University of Colorado seismologist Roger Bilham discussed the seismic hazards of "an unrecognized weapon of mass destruction: houses."[3] Kathmandu moved off the waiting list when much of its brick and cement housing stock collapsed in a devastating magnitude 7.8 earthquake in April 2015, killing 9000 people and causing $10 billion in damage.

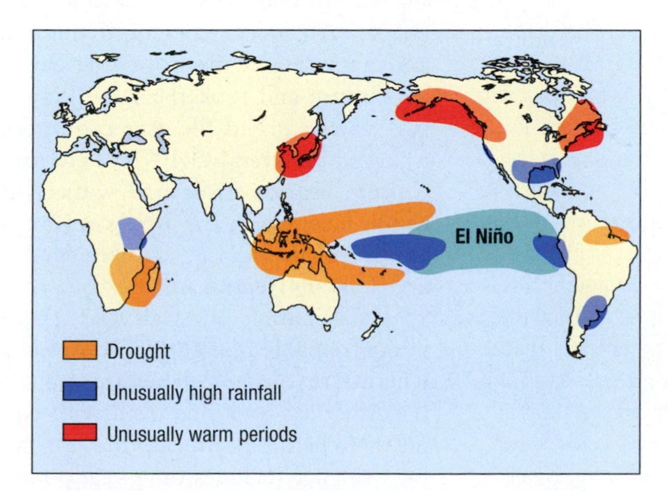

• Figure 10.13 The typical effects of El Nino on world climate.

Drought or unusually high temperatures settle in over northern Brazil, southeastern Canada and the northeastern United States, southern Alaska and western Canada, the Korean Peninsula and northeastern China, Indonesia and northern Australia, and southeastern Africa. When warmed, Pacific waters are far less rich in nutrients, so typically fish catches plummet, and food chains are disrupted, leading to massive wildlife losses. The most severe El Niño events on record occurred in 1982–1983 and in 1997–1998, but it is not unusual for three or four moderate events to occur every decade.

Economically, within Latin America El Niño losses are most immediate in Peruvian and Chilean fisheries. In normal years the nutrient-rich waters teem with anchovies, which sustain human diets (and are used in many pizzas) but mainly are processed into fertilizers and fishmeal for livestock. On the natural food chain, these swarms of fish support seabirds that produce their own fertilizers for human use; in the mid-19th century, Peru enjoyed an export "guano boom" facetiously dubbed the "Age of Manure," followed by an inevitable bust caused not by El Niño, but by Peruvian mismanagement of the resource. Aquaculture of shrimp in the coastal mangrove thickets of Mexico, Central America, and Ecuador produces frozen shrimp for markets in North America and Europe (see **Figure 2.6m**).[8] Productivity is good as long as the environment is clean. Many of the issues affecting salmon farming in Norway and the US are common to shrimp farming, including the rapid spread of disease in unclean, crowded waters and the application of antibiotics.

Try it Help Your Brother

You can personally help mitigate the threats posed by natural hazards around the world. This is possible because of the ways in which geospatial data are being "democratized" for the public good. There is a growing trend of crowdsourcing in which the public are asked to contribute their observations, ideas, and georeferenced information online to various projects; Wikimapia, for example, urges, "Let's describe the whole world!"

As discussed in the context of Ushahidi in Africa (see **page 426**), these projects often involve humanitarian and emergency situations, and one in particular asks you to help: **MapGive** (http://mapgive.state.gov, an initiative of the US State Department's Humanitarian Information Unit). After the 2010 earthquake that devastated impoverished Haiti, thousands of "digital do-gooders" answered the call to manage open-source geospatial data (non-copyrighted data that anyone can use or modify) at Open Street Map (OSM; http://www.openstreetmap.org). They traced roads, buildings, and refugee camps on fresh satellite imagery. In just two days they generated enough data to provide emergency workers on the ground with up-to-date maps. At the 2014 South-by-Southwest Festival in Austin, Texas, representatives of the US State Department invited the public to participate in a number of new MapGive projects, beginning with satellite imagery of war-torn South Sudan. On deck were humanitarian projects in the Democratic Republic of the Congo and Honduras.[4]

You can participate in projects like these, even if you've never mapped before. What can you do? The MapGive site presents you with this invitation:

> Join the community of online volunteer mappers. Make a powerful contribution with an Internet connection and basic computer skills, even if you do not live in the area you are mapping. Use satellite imagery to trace the outlines of roads, structures and land features for the creation of openly available data used to produce maps that help the humanitarian and development community. There are several ways to get involved and aid humanitarian crowd-sourced mapping initiatives.

The site provides you with a path signed "Learn to Map."

As for "Why Map?" the site responds:

> Quality geographic data helps empower organizations and communities to make important decisions across a range of environmental, economic and crisis management themes. For many places in the world, this information is incomplete or does not exist at all. Digital humanitarians map online to help give others the data they need to build a more sustainable future.

MapGive is proactive as well as responsive. In April 2013, fully two years before Kathmandu was stricken, the World Bank sponsored an OSM map-a-thon involving the public in 17 American cities. One participant wrote:

> Over the course of four hours, we traced building footprints from satellite imagery provided by the US State Department. These high-fidelity footprints are just one of many datasets needed for better disaster preparation, planning and mitigation in Nepal—a country that sits atop tectonic plates at high risk of earthquake.[5]

The work contributed by these volunteers that day was put to immediate good use when Kathmandu fell. You can see the results cartographically with a "before and after" slider at http://mapgive.state.gov/projects/nepal/. When you go online to help, you will be presented with the latest satellite imagery of the world's immediate and long-term place-based needs—for example flooding in Malawi, typhoon response in Vanuatu and Philippines, rural health care in Liberia—and you will be asked to contribute what time you can afford.

Tomnod (http://www.tomnod.com) is another organization seeking your help "to use satellite images to explore the Earth, solve real-world problems, and view amazing images of our changing planet [to] fulfill our purpose of seeing a better world."[6] Tomnod projects range all the way from mapping Swaziland in a malaria-eradication effort to locating the missing Malaysian Airliner MH370. What does "tomnod" mean, anyway? "Big eye," in Mongolian.

Are you interested in a related internship or career? A Silicon Valley start-up called PlanetLab is deploying "flocks" of "doves"—Earth-orbiting satellites—with the mission of "thinking about how space could be used to help humanity. We are concerned with the whole planet, as it changes. We are its stewards, accountable for its future and well-being. We aim to create commercial and humanitarian value with the market's most capable global imaging network." PlanetLab (http://planet-lab.org) is hiring and offering internships, and asking all of you to look upward where "There is a new human geography under construction above us, as well."[7]

10.3 Cultural and Historical Geographies

It is perhaps unfortunate that this region came to be known as "Latin" America because there were no Latins among its inhabitants before the end of the 15th century. In 1492, when the first Europeans arrived, the region was home to an estimated 50 to 100 million **Native Americans**, also known as **Amerindians** or, as Columbus (who thought he had landed in India) misidentified them, **Indians.** After their early migrations starting as long as 23,000 years ago from Asia, they spread throughout the Americas, developing many distinctive livelihoods and cultures. Some Native Americans practiced hunting and gathering while others developed agriculture and took the same kind of pathway to urban life and civilization followed by several Old World cultures. Culture hearths associated with civilization emerged in the Andes region of South America and in southern Mexico and adjacent Central America.

A fascinating question and a subject of ongoing scientific research is whether seafaring Polynesians reached the Americas and interacted with people there before Europeans did. Intriguing evidence includes the sweet potato, native to South and Central America, that was in some Polynesian diets several centuries before Europeans "discovered" the Americas and began the Columbian Exchange. Another tantalizing clue is the kayak design of the Chumash tribe of California. However, the chickens, rats, and certain technologies that accompanied the Polynesians everywhere else on their voyages are missing in the Americas. We know that Polynesians spread their cultures by amazing feats of seafaring in the Pacific, but if it occurred, this voyage across the vast, island-less, eastern stretch of that ocean would have been the most epic of all.

379

Civilizations Predating the Europeans' Arrival

Several indigenous civilizations flourished in the Americas before European contact (•**Figure 10.14**). In 1492, as they do today, the **Maya** inhabited southern Mexico, Belize, and Guatemala. They practiced simple agriculture based on maize (corn), squash, beans, and chili peppers and had by 200 CE developed a highly complex civilization in lowland tropical rain forests, tropical dry broadleaf forests, and highland volcanic regions. Some of the principal Maya settlements were in the Yucatán Peninsula, an area of tropical dry broadleaf forest with distinct wet and dry seasons. The Yucatán would have faced major water shortages for half the year, possibly precluding Maya settlement and civilization, if not for a bounteous supply of subsurface waters. The ground is a maze of dissolved cavities in the limestone partially filled with fresh water. These natural cisterns, known locally as *cenotes*, provided ancient and modern inhabitants with abundant water for all needs, including agriculture (•**Figure 10.15**). The Maya created monumental religious and residential structures, including pyramids, temples, and astronomical observatories. They built stone roadways through dense forest areas but notably had no wheeled vehicles, nor did they have sailing vessels, the plow, beasts of burden, and other tools and trappings associated with civilizations elsewhere. The Maya had highly developed systems of mathematics, astronomy, and engineering. (They also had a cyclical calendar; one cycle ended on December 12, 2012, a fact that sparked a rash of apocalyptic prose and films leading up to the date.)

• **Figure 10.14** Major Native American groups and civilizations in Latin America on the eve of the Spanish conquest.

Joe Hobbs

• **Figure 10.15** Some of the largest and longest-enduring settlements of the ancient Maya, including Chichén Itzá, were situated in the Yucatán Peninsula atop and near cenotes. These natural cisterns have had both practical and spiritual importance to the ancient and modern Maya. The water-giving god Chac dwelled in them, and the ancient Maya petitioned him to deliver rain when their crops withered. Chichén Itzá's Sacred Cenote was the most important of all. Maya offerings there included gold, jade, pottery and sacrificial men, women and children.

The Maya also had a hieroglyphic writing system that survives on stone monuments and in a handful of books. Regarding them as heretical, Spanish priests put almost all of the written volumes to the torch. On July 12, 1562, Bishop Diego de Landa destroyed at least 30 Mayan books in a bonfire outside a church in Mexico's Yucatán Peninsula. "They contained nothing in which there were not to be seen superstitions and falsehoods of the devil," de Landa later wrote. "We burned them all." Many Maya watched this event, which, de Landa recalled, "they regretted to an amazing degree and which caused them much affliction."[9] This was not like burning 30 ordinary books; they were rare and difficult to produce, and so only a few survive to this day.

The Maya civilization peaked around 900 CE and then began a slow decline. It is still unknown why this advanced civilization apparently abandoned its massive religious and urban centers and melted back into the forests of the Yucatán Peninsula and points south. Theories include natural disasters, agricultural failures, and revolts against ruling elites. When the Spaniards arrived, the great Maya cities were overgrown. Many can be seen in their restored glory today. Chichén Itzá in

Mexico's Yucatán Peninsula and Tikal in northern Guatemala (Figure 2.6a) are very significant sources of international tourism revenue for the respective countries.

North of the Maya realm, in the Valley of Mexico about 30 miles (40 km) from present-day Mexico City, arose Teotihuacán, the first true urban center in the Western Hemisphere. Its construction began about 2000 years ago, and by 500 CE it covered about 8 square miles (21 sq km) and had 200,000 residents, comparable to London a millennium later. The city was laid out on a grid aligned with stars and constellations that were central to the ceremonial timekeeping of the **Teotihuacános**. There were some 2000 apartment compounds, a 20-story Pyramid of the Sun, a Pyramid of the Moon, and a 39-acre (16,000-sq m) civic and religious complex called the Great Compound or Citadel (**• Figure 10.16**). The origins, language, and culture of Teotihuacános are still not fully understood, and it is hoped that future archaeology will shed more light on these extraordinarily creative people.

The **Aztecs,** who called themselves the Mexica (pronounced "meh-*shee*-kuh") and came later than the Teotihuacános, honored the earlier city as sacred space and a religious center when they took over the Valley of Mexico in the 1400s. They founded their capital, Tenochtitlán (now overlain by Mexico City), in 1325. With a population of about 250,000, Tenochtitlán had a sacred center surrounded by active marketplaces, all linked by stone-surfaced roads to other parts of the urban settlement. From this base, the Aztecs and their allies developed an empire stretching across central Mexico. The Aztecs left a legacy of much merit in metallurgy (which diffused to Mexico from the Andean region), woodworking, weaving, pottery, and especially urban design.

In South America, the **Inca** built an empire about 2000 miles (3200 km) long from northern Ecuador to central Chile. They controlled it for only about a century before 1532, when the Spanish conquistadores, under Francisco Pizarro, began their subjugation. The Inca's origins, dating to about 1200 CE, are not fully known. From their governing center in Cuzco, Peru, the Inca engineered a system of roads, suspension bridges, and settlements that connected and supported their empire. The Inca also achieved great skill in stonework, terrace construction, and irrigation networks (**• Figure 10.17**). Remarkably, they had neither paper nor a writing system, but kept mathematical records by knotted ropes called *quipus* (pronounced "*kee*-pooz").

• Figure 10.17 Machu Picchu, the legendary "lost city of the Inca" rediscovered by American archaeologist Hiram Bingham in 1911, has a dramatic site above Peru's Urubamba Valley.

• Figure 10.16 Teotihuacán's Pyramid of the Sun. People at lower left provide scale.

Still another group, the lesser-known **Nazca,** left a unique legacy on the landscape of South America about 2000 years ago. Their "Nazca lines" in the desert of southern Peru are a complex of nearly 200 square miles (520 sq km) of massive carvings in the sandy surface, depicting birds, insects, and fish, the smallest of which is an animal figure some 80 feet (23 m) in length. These designs can only be appreciated from the air (giving rise to the improbable theory that they were built as landmarks for "ancient astronauts"). No written records of the Nazca culture survive, so the purpose of their lines may never become known. Another of the less celebrated civilizations of South America is the **Chibcha** of what is now Colombia. The Chibcha lived in small agricultural villages rather than large cities, and their exquisite gold work is a major legacy of their culture. Speakers of Chibchan languages now live in the Andes of Colombia and Ecuador and as far north as Costa Rica. Beneath the dominant imported languages of Spanish and Portuguese, the linguistic substrate of Latin America is still rich.

Languages in Latin America

Linguists disagree on the exact history of and relationships between the indigenous languages of the Americas, but agree they are surprisingly complex. Probably over 50 indigenous language families exist in South America alone. Most of these languages today are small and isolated, spoken by just a few hundred to a few thousand people in small geographic areas. Still, there are several indigenous languages in Latin America that remain vigorous (**• Figure 10.18**).

In Guatemala, Belize, and southern Mexico, the 30 tongues of the **Mayan language family** are still widely spoken today by about 6 million people. Many are bilingual with Spanish,

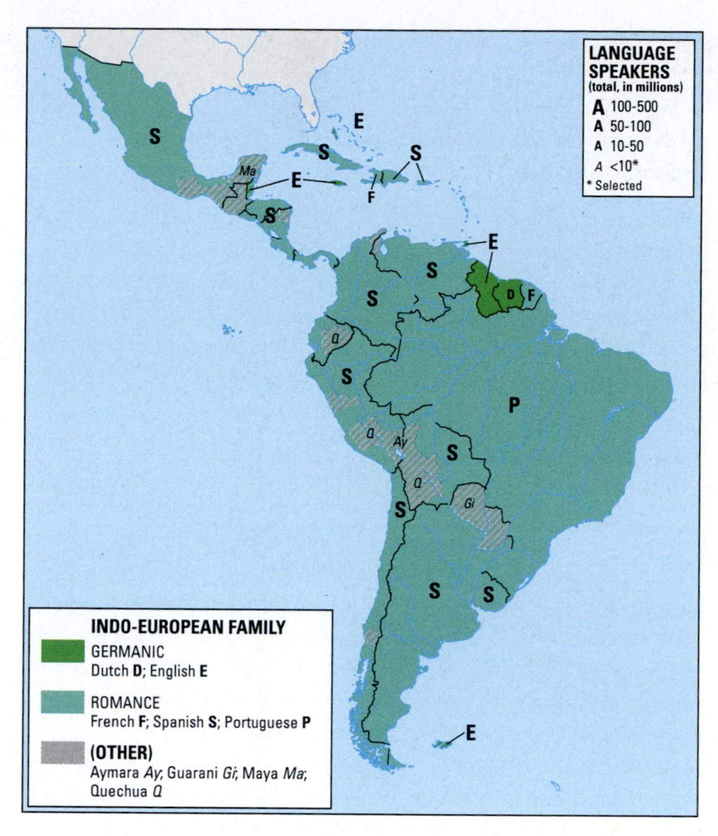

• Figure 10.18 Languages of Latin America.

but some Maya, particularly in highland Guatemala, never did mix with outsiders and are ethnically little changed from their forebears. The Maya of highland Guatemala tend to speak Mayan as their first language.

The most widespread indigenous language in Latin America is **Quechua**, spoken by about 10 million people. Quechua was the language of the Inca. It has status as an official language, along with Spanish, in Ecuador, Peru, and Bolivia. About 30 percent of Bolivians and 13 percent of Peruvians speak Quechua natively. An early contemporary of Quechua is **Aymara**, spoken by about 20 percent of Bolivians and 2 percent of Peruvians, primarily around Lake Titicaca. Aymara has official language status in Bolivia. Bolivia's president Evo Morales is of Aymara descent.

An interesting linguistic development has occurred in Paraguay. Just 5 percent of the citizens of that country are Native American (most of the rest are mestizo or European), but the original indigenous language of the area, **Guaraní**, is still widely used. Over 90 percent of Paraguay's population speaks it, including many non-indigenous people; more people speak Guaraní than Spanish. This is the only Native American language that is spoken by significant numbers of people outside a particular ethnic group. About 20 percent of Paraguayans, especially those living in the rural north of the country, speak only Guaraní; roughly 10 percent of Paraguay's population speaks only Spanish.

Some Latin American countries are strongly indigenous. More than half of all Bolivians (55 percent) are Native American. Peru (at 45 percent) and Guatemala (40 percent) also have large numbers of indigenous peoples. Most other Latin American countries have far smaller numbers (and the islands of the Caribbean have virtually none at all), but in the Amazon Basin and in Panama, scattered groups of lowland peoples maintain their traditional cultures. Remarkably, in 2011 remote sensing imagery of the upper Amazon region revealed 18 "uncontacted tribes," meaning that these isolated peoples have had little and perhaps no interaction with the wider world. The number of such groups detected by remote sensing has grown since then, as have the ethical and practical issues about whether they should be contacted; see Life in Amazonia on **page 464**.

European languages joined the linguistic mix in Latin America at the end of the 15th century. At first, only the colonizers spoke them, but they became more widespread as colonial administration progressed and expanded. The language distributions today reflect the pattern of colonial rule. Spanish, now one of the world's most spoken languages, or **megalanguages** (along with Chinese, English, Hindi, and Arabic), is the most widespread and is the prevalent European language throughout the region. Other European languages in widespread use are Portuguese, French, Dutch, and English. These languages are also spoken in the dependencies and former colonies of several other small Caribbean islands. The European languages generally became the sole official languages of both the colonial administrations and the independent countries that succeeded them. They serve as welcome lingua francas in countries like Mexico and Peru that have a great diversity of indigenous languages.

Another linguistic product of European colonialism in Latin America was the evolution of **Creole languages**. These are tongues that developed among the black slaves and indentured servants brought to work the plantations of the Caribbean islands and the Atlantic and Caribbean coasts of South and Central America. Their vocabularies come mainly from the European colonizers' languages. Sometimes a speaker of the parent language can recognize the Creole; for example, in Costa Rican Creole English, "*Mi did have a kozin im was a boxer, kom from Panama*" is easily enough understood as "I had a cousin from Panama who was a boxer."[10] But it is not always that simple. Like Pidgin in the South Pacific (see **page 380**), Creoles are distinct languages and have their own grammars.

The European Conquest

The arrival of Christopher Columbus and his three Spanish ships in 1492 brought more than a new language to the New World. It marked the beginning of profound changes in almost every aspect of life in what would become Latin America.

First came death, both deliberate and unintended. A series of expeditions to the New World convinced the Spaniards that there were riches worth pursuing in that far-off land. The Spanish conquistador Hernando (Hernán) Cortés landed on Mexico's coast near present-day Veracruz in 1519 and led his cohort of invaders inland, building alliances with ethnic groups opposed to the Aztecs and killing those who would not join

life in AMAZONIA FOR UNCONTACTED TRIBES

Rob Walker

Rob Walker with a poster illustrating his research.

THINK CRITICALLY: There are practical and ethical dilemmas about contacting uncontacted tribes. In your view, what is the best way forward with these groups: Should they be left alone, or are there methods of contact that would benefit them?

University of Missouri anthropology Professor Rob Walker has one of the most interesting and important research agendas a social scientist can have: studying and trying to improve the lives of uncontacted tribes in the vast Amazonian rain forest. I spoke with Dr. Walker about his work.

How did you get started with this work?

When I was an undergraduate I came across a newsreel about these people, and it really grabbed me. It's what got me into anthropology. I was just fascinated that there were these people out there almost completely disconnected from the rest of the world. I started working in the Amazon then, but I couldn't contact these people. It was really impossible. But with Google Earth and all the free satellite imagery out there, I am able to track down where exactly these people are. I'm getting images of their villages and then trying to track them.

What is the geographical pattern of their settlements?

They are on the margins of the Amazon Basin. They are in higher elevations at the top of the watersheds—generally upland areas not as productive as the lower places—that are way up in watersheds in places where others don't live. The closest water sources are little ephemeral streams that are not good for fishing. Soils aren't quite as good—these are not the ideal places to live. The people are getting pinched from both sides. Transport is on the larger rivers, and the people are getting forced away from them by the river traffic. They go to smaller rivers generally because people are not there, but now you have miners and loggers creeping in there. If you want to maintain your autonomy, you need to get away from these powerful outsiders. It looks like they've run away. Some of these places could be ancestral homelands, but for the most part it 296 looks like they've run away to get away from everyone else. They can't go to lower places because there are other people there, and they're going to be making contact in these areas.

By definition is it impossible to do field work with uncontacted tribes?

I would like to do field work with them, but it's difficult: you can't just go talk to these people. I've been working with people who have been recently contacted in the last 20–30 years in Brazil and Paraguay. There was a contact last year with six men who walked out of the forest. We interviewed them to find out what life was like before the pre-contact phase, and what they thought of us. I am asking people "what was it like when you made that transition? Did you want to make contact?"

What did you learn?

They have a view that there are the real people—themselves—and then there's everyone else, even including other speakers of their language (Pano). Traditional tribal warfare makes it difficult to create alliances with outsiders because they have had so many hostile interactions. It plays into their worldview. They have a warrior ethos. Making contact is a giant transformation. There is a certain amount of dissent among them about whether it is worth it to make contact. The older generations are more conservative. They've had a long history of persecution by the outside world, essentially being shot on sight, and the younger guys like these six are much more open and are interested in finding out what's going on. They want access to technologies like machetes and pots and pans, things to improve their lives. It's an interesting dynamic that's going on between old and young about what to do. And there is definitely a lot of fear. Their experiences with the outside world have been so negative. They are afraid of what's going to happen. They are walking on eggshells because they don't know what the future holds.

What about those who emerge and then re-isolate?

The classic examples are those who have had terrible experiences with the outside world—they were persecuted during the rubber boom, or made slaves of, and realized it was terrible and ran away. Some of these isolated groups we are working with now may in fact be ones that ran way from civilization hundreds of years ago. Those that have made contact usually stay that way. Life is fundamentally different for them. Very few actually go back if contact has been a relatively good experience. It's relative—you still have high mortality, but slavery has gone away. Relatively good, peaceful situations make for sustained contacts, and they don't go back to the forest. Once there is health care (critically important because of their lack of immunities), even meager care from the outside world, that will change their lifestyle and they won't go back.

him. Unfortunately for the Aztecs, Cortés's arrival seemed to fulfill the prophecy that Quetzalcoatl, a fair-skinned, bearded Aztec god, would return one day, approaching from the east. In Tenochtitlán, the Aztec leader, Moctezuma, received Cortés and his army as emissaries of Quetzalcoatl. Cortés and just 30 men took Moctezuma prisoner and were unopposed, apparently because of the widespread belief in Quetzalcoatl's return. Using a divide-and-conquer strategy that enlisted local forces, Cortés achieved a military victory over the Aztecs in 1521, destroying Tenochtitlán and laying the foundation for Mexico City in its ashes.

Spain soon dispatched new expeditions to conquer other peoples of Mexico and Central America. When conquistador Francisco Pizarro landed in Peru in 1531 with only 168 soldiers and 102 horses, he imprisoned the Inca leader Atahualpa, freed him for a huge ransom, imprisoned him again, and then had him hanged (he was to have been burned at the stake, but Pizarro modified the execution because Atahualpa agreed to be

How do governments view these uncontacted peoples?

In policy terms, most governments, even the United Nations, have documents that say in effect, "We should leave these people alone, they have a reserve, they will be fine if we don't do anything." That would be great if it were true. On the ground that's not what's happening. There are still external pressures going on, with people invading their lands. So governments that say "we won't allow contact" are allowing much more harmful things to happen—disease spreading and violence. I'm worried they will go extinct soon, and all we will have done is to sit back and watch it happen on the satellite images.

What are you able to do with remote sensing?

With the satellite images, we've found 30 groups so far where we can see their huts. We are tracking them through time. In one case, a group was there one year and gone the next in the 2015 image. This is the kind of thing that makes me worry: What if gold miners got in there and slaughtered them? On satellite images, you see gold mines 30 km away. There are as many as five bush planes a day flying into these gold mining areas. It's almost impossible to prevent—it's such a large area.

I want to get on the ground and figure out in certain areas what the major threats are and what we can do. But for now we can do a much better job with this satellite imagery to monitor not only them but also these external factors, like mining and deforestation. I want to turn this more into a conservation focus and ask whether these people are viable as uncontacted in the long term or not. I want to use these time-stamped images to tell people in the government, "Look, this is a real issue, you need to increase protection here, there are gold miners here next door, there is a road going through with lots of people."

The satellite imagery is a first step. A well-organized contact team would be a huge improvement over the disastrous contacts that have gone on for 500 years now.

What makes for a successful contact?

We have this group of Yanomamo. We have gold miners in this area. If we can't kick them out—and so far the government hasn't been able to do so—instead of running the risk that these gold miners will attack these people, it would be much better as soon as we know about them to take a health care team that is well trained and wants to stay on site a long time and do contact the right way. Once you have contact with that group and hear about gold miners, you contact federal police and they'll do something about it. When they are isolated, they are moving around, and we don't know what's going on. That's part of the risk here. We are not communicating with them, we don't know what's going on, and these external threats could swallow them up. Maybe that has happened in this Yanomamo case.

It must take a lot of coordination.

It's not an easy thing. Sometimes it works; sometimes it doesn't. That's where you need cultural translators who can say "we have your best interests at heart here." That takes some time to build that trust. Anthropologists and missionaries are good cultural translators. It takes a lot of buy-in from the government too. In a perfect world, FUNAI (Brazil's Indian agency) should do this, increasing indigenous participation and anthropological training, and also more health care training. Vaccines are scary to people, especially with their traditional beliefs about medicine and spirituality. FUNAI could make a perfect cultural translator team. And you want the language skills that the indigenous people bring. They've been through this situation before. You put this altogether in one team, and it would be perfect. They have some elements of that, but not as effective as they could be. A well-organized contact is in our minds, but we don't have a great example yet.

You wrote "isolated populations are not viable in the long term." What did you mean?

They are too small. A group of 50–100 people that is not growing is a group coming close to extinction. External threats are too big, and the groups, too small. You can see these numbers in the satellite imagery. In Acre (southwestern Brazil) there are 300 people in one settlement. That's probably a viable population. My expectation is that in the next couple of decades, there will be a lot of contacts going on. I'm just concerned that in that process a lot of groups will go completely extinct. Trade that off with a well-organized contact now to save these people from dying. Saving the culture is secondary. NGOs like Survival International say "leave these people alone to keep culture intact." I'm an anthropologist and saving culture is important to me. But it's more important to save actual people and individual lives. Our number-one priority should be to save people. If that means making contact and changing the culture that is the humane ethical thing, saving people, and if we save cultural traits that's great too. I'm worried about the smaller populations of Yanomamo. It would be better to have them with other Yanomamo with good health teams. But there are Yanomamo leaders that like it that smaller autonomous groups are out there. But they may not have thought through it. They want to maintain the Yanomamo lifestyle, but saving them should be the goal above saving their lifestyle.

baptized). Two years later, Pizarro's forces held the Incan capital Cuzco, and from that point the colonization of all of South America by the Spanish—and in Brazil, the Portuguese—progressed swiftly.

Disease brought death to the indigenous people on a vast scale. The Native Americans had never been exposed to, and hence had no natural immunities against, the host of diseases that the Spanish and the Portuguese introduced: bubonic plague, measles, influenza, diphtheria, whooping cough, typhus, chickenpox, tuberculosis, and smallpox, joined later by malaria and yellow fever that came with African slaves. The worst was smallpox, which killed between one-third and one-half of all Native Americans in affected areas. The diseases, famine, and enslavement that came with Spanish and Portuguese colonization brought what was probably one of humankind's largest population declines ever. The numbers are elusive but were certainly huge. The Spanish chronicler Bartolomé de Las Casas wrote that in Spanish-controlled Latin America, 40 million Native

Americans had died by 1560. The geographer David Clawson estimates that by 1650, 90–95 percent of the Native Americans who had come into contact with Europeans had died.

European settlement patterns emerged as the conquest proceeded. The development of ports as bases for subsequent penetration of the interior was the first major settlement task. Large coastal cities had not existed prior to European colonization. Many early ports eventually grew into large cities, and a few became major metropolitan centers. In some cases—notably Lima, Caracas, and Santiago—the main city developed a bit inland but retained a close connection with a smaller coastal city that was the actual port.

• Figure 10.19 Plaza and cathedral are at the center of most Latin American cities, and draw large numbers of people on a daily basis. This is the *zócalo* of Mexico City. From lower right to center are three male traditional healers, or shamans, practicing their trade.

Around the ports, agricultural districts developed, spread, and shipped an increasing volume of trade products overseas. Plants and animals domesticated by Native Americans would revolutionize diets and habits in Europe and elsewhere in the Old World: corn, tobacco, potatoes, cacao, and turkeys were among them. In time, on the other side of the Columbian Exchange, the European-introduced horses, cattle, sheep, donkeys, wheat, sugarcane, coffee, and bananas would revolutionize New World agriculture. In the second half of the 1500s, slave ships began to bring the Africans who provided the principal labor on the plantations created and owned by Europeans, primarily in the *tierra caliente* tropical lowlands with the generally fertile alluvial soils of the coastal plains.

The powerful lure of gold and silver stimulated deeper European penetration of the Andes and the Brazilian Highlands in South America. After capturing the stores of precious metals that had been accumulated by the indigenous peoples for ceremonial use, tribute, and trade, the foreigners opened mines and established market centers to service them. Some highland settlements became centers of new ranching and plantation enterprises. A few highland cities, of which the largest is Bogotá in Colombia, grew into large metropolises. Separated from the seaports by difficult terrain, these cities were eventually connected to the coasts by feats of great engineering skill. Some of the main seaports and highland cities became important centers of colonial government as well as economic nodes, and several are national capitals today, including La Paz, Bolivia; and Quito, Ecuador.

At the heart of each Spanish and Portuguese city established in Latin America, the newcomers placed their familiar combination of a plaza dominated by a massive Roman Catholic cathedral (**•Figure 10.19**). Latin American cities have a certain consistency and predictability about them.[11] You can expect to find that the core of the modern city (*ciudad*) is the old colonial city, with its gridiron pattern reflecting Spanish town planning (**•Figure 10.20**). The plan, formalized by Spanish Crown's 1573 Laws of the Indies, had a clear delimitation of core and periphery. Around the main plaza stand the church or cathedral, administrative buildings, and some businesses. Moving outward were homes of the wealthy, then middle- and lower-income peoples in their respective *barrios* or neighborhoods. At the town's edge lived the poor and Native Americans, near the cemetery, slaughterhouse, leprosy houses, gambling establishments, and brothels. Unlike most Middle Eastern historic medinas, the Latin American colonial cores remain vital in modern urban life, with the wide-open colonial plaza and church or cathedral adjacent to the modern central business district.

Roman Catholicism was introduced—actually, imposed—as the one and only acceptable faith in the New World, and today

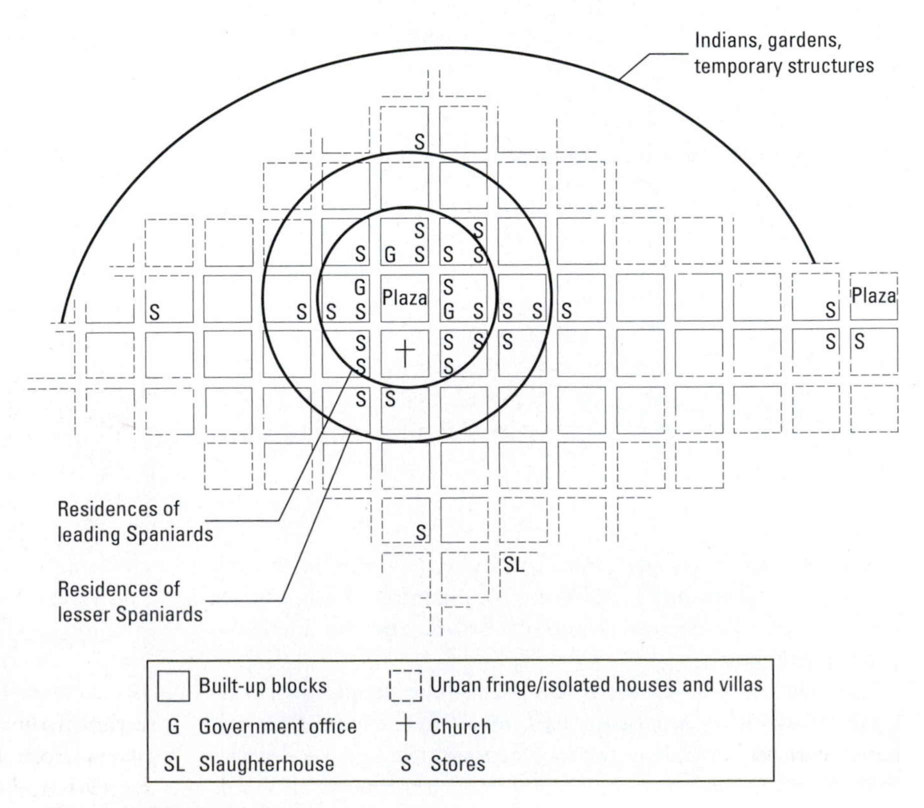

• Figure 10.20 Model of the Latin American colonial city.

80 percent of the region's people are Catholic (•**Figure 10.21**). Local cultures have contributed much to local expressions of Catholicism throughout the region. The "Lady of Guadalupe" for example, is a uniquely Mexican manifestation of Mary, the mother of Jesus. The British and Dutch brought their Protestant faiths to their possessions. Today, Protestant offshoots, notably Evangelical and Pentecostal religions—characterized by high-energy, participatory forms of worship and their appeal to disrupted families and the poor—are making inroads in traditionally Catholic communities throughout the region. Jamaica has embraced Rastafari, a faith championing black empowerment and holistic living dating to the early 20th century, 439 with unlikely links to Ethiopia.

Many indigenous beliefs have become syncretized with Catholicism, but some of them dating to **pre-Columbian times** (before 1492) remain alive among Native Americans in Latin America. The Catholic Maya of Guatemala, for example, still make offerings to the "Earth lords" who dwell in caves (•**Figure 10.22**). Innumerable icons testify to the blending of cultures and beliefs.

Ethnicity in Latin America

Despite the dominance in Latin America of culture traits derived from Europe, a large majority of the original European settlers or their descendants intermarried with Native Americans or blacks. Only Chile, Argentina, Uruguay, Brazil, Cuba, and Costa Rica have significant white European ethnic groups today (•**Figure 10.23a**). Some are descendants of European settlers from centuries ago, but most are families of immigrants from Europe who arrived relatively recently, during the 19th and early 20th centuries. Scattered districts in other countries are also predominantly European.

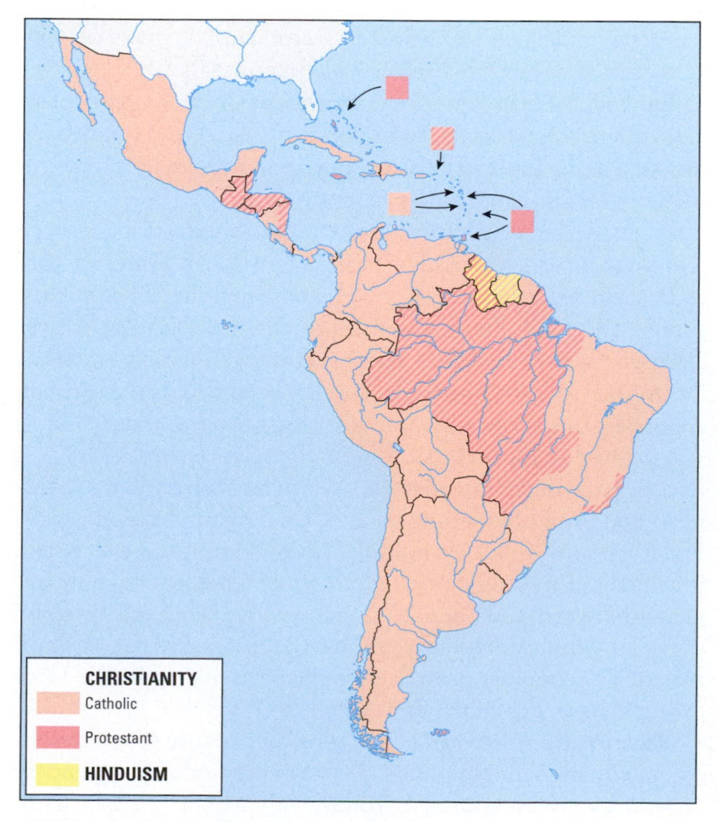

• **Figure 10.21** Religions of Latin America.

Joe Hobbs

• **Figure 10.22** Petitionary prayer to Earth lords, Christ, Mary, and Catholic saints for good crops. The participants in this cave ritual are Q'eqchi' Maya in Alta Verapaz, Guatemala.

All photos courtesy of Joe Hobbs

• **Figure 10.23** In addition to the Native Americans of Latin America, the leading racial types are **(a)** Europeans, **(b)** blacks, **(c)** mestizos, and **(d)** mulattoes.

Black Latin Americans of relatively unmixed African descent live in the greatest numbers on the Caribbean islands and along the Atlantic coastal lowlands in Middle and South America (**Figure 10.23b**). These are the areas to which African slaves were brought during the colonial period, mainly to work on sugar plantations. An estimated 3 to 4 million were sold in Brazil alone. Slavery was gradually abolished during the 19th century, although not until the 1880s in Brazil and Cuba. By that time, slavery had generated large fortunes for many owners of plantations and slave ships. The African peoples it introduced to the region have made cultural contributions that today provide a varied, vibrant, and important part of Latin American civilization.

416

Latin America has largely escaped many of the racial tensions and violence that have battered much of the world. This may be because a majority of Latin Americans are of mixed heritage, the most common being a mixture of Spanish and Native American that has resulted in a heterogeneous group known as **mestizos** (**Figure 10.23c**). A person of mixed African (black) and European (white) ancestry, sometimes with Native American blood as well, is known as **mulatto** (**Figure 10.23d**); in some locales, the term used is, like the linguistic term, **Creole**. It must be noted, however, that in most places, socioeconomic discrepancies are strongly associated with ethnicities in the region, with lighter skin associated with higher status. A thin stratum of people of mainly European origin often dominate government and business sectors, with mestizos and mulattoes making up the bulk of the middle and lower classes. Native Americans almost always have the lowest standing, and in some countries such as Guatemala, systematic exclusion and violence have been directed against them periodically. Economic inequities in Latin America are discussed on **page 470**.

Mestizos have not lost all traits of their Amerindian forebears. Although it can be hazardous to generalize about ethnicities, the mestizos' inheritance from their Native American roots includes the collective exploitation of land, the accumulation of lifelong personal bonds, the devotion to a whole series of supernatural beings (who have Christian names but non-Western origins), the ritualized remembrance of deceased relatives, and reverence for political leaders.[12] As will be seen later in the chapter, reverence for political leaders cannot be taken for granted in this region today.

Despite enormous impacts, some Native American cultures have proven remarkably resilient. Some, like the indigenous peoples of Nicaragua's Miskito coast, are actively pushing for independence (they call themselves the Communitarian Nation of Mosquitia). Many are technologically savvy, using the Internet and social media to elicit understanding and political and economic support in their struggles. Some are championed by Western celebrities and activists, but others, including some of the groups pushing back against oil development in the Ecuadorian Amazon, have managed to gain a footing without such assistance.

Some governments actively protect the lives and ways of life of Native Americans. Several countries, including Colombia, Peru, Bolivia, Ecuador, and Mexico, have amended their constitutions to expand indigenous rights. In these countries, Native Americans now have more rights over land, and the governments recognize communal rights over some resources. Brazilian law forbids mineral extraction by outsiders on aboriginal lands, but as reported by Dr. Walker, illegal gold mining has led to many bloody skirmishes between Native Americans and prospectors in several Indian reserves in the country. On the negative side of this protection, as the Native Americans see it, many resources of Brazilian and other national Indian reserves are placed under government control, and often indigenous peoples are forbidden from setting up their own businesses.

464

10.4 Economic Geography

Latin America is generally a region of middle-income LDCs. Only three countries in South America (Chile, Argentina, and Uruguay) and three Caribbean island nations (Trinidad and Tobago, Antigua and Barbuda, and the Bahamas) rank as MDCs based upon their per capita GDP (PPP) (see Table 10.1). Nine percent of the region's residents live in poverty. Both poverty and unemployment have diminished in recent years for the region as a whole, and the middle class is growing—but so is economic inequality.

62

In Chapter 3, we considered dependency theory, an economic argument that many of today's MDCs are successors to the colonizing powers who extracted raw materials from colonies and compelled the colonies to buy the colonizers' manufactured goods. The former colonies, now independent countries, remain dependent upon the MDCs and cannot participate in the global economy independent from them. Like other LDC economies, Latin American economies have long relied too much on the primary sector with exports of non-value-added goods like cash crops and minerals.

60

But today Latin America's dependency is not upon a former colonizer such as Spain or Portugal; the recent booms and slowdowns in Latin American raw materials exports are instead tied mainly to China's economy.

As in Africa and elsewhere, Chinese interests have been scouting out every possible prospect in Latin America that could help feed that nation's voracious appetite for minerals, energy, and food. Latin America's trade with China grew over 200 percent between 2002 and 2014, and China now stands as Latin America's leading trade partner. In some Latin American countries, this raises the same fears as in Africa: that China is seeking to develop a classic mercantile relationship with Latin America, taking in raw materials (both agricultural and industrial) and profitably selling back its own manufactured goods. But at the same time, the influx of Chinese capital is welcomed. It helped Latin America's economy grow 5 percent or more annually between 2003 and 2012. However, the economic slowdown in China in recent years has led to a slowing of Latin America's economic growth as well, particularly in the large export markets of Brazil and Argentina.

427

312

When Latin American economies lag because of conditions in China, the vulnerabilities of their commodities-dependent economies are accentuated. Today, only Mexico, Costa Rica, and Panama rely more on exports of manufactured goods than of commodities.[13] It is more important than ever to diversity the other Latin American economies by growing the manufacturing (secondary) and service (tertiary) sectors and obtaining a secure foothold in the global economic system. Here we will examine Latin America's historical dependence on commodities exports and what the region's economies are doing to move away from the shadow of dependency.

Agriculture and Mining: The Primary Sector

Farms in Latin America can be divided into two major classes by size and system of production: small holdings generally associated with subsistence agriculture and domestic consumption, and larger holdings associated with commercial agriculture (cash cropping) and production for export.

Minifundia (sing., *minifundio*) are the smaller landholdings focused on subsistence, that is, producing food for family use and for sale in local markets (**•Figure 10.24**). The crops most commonly raised, especially in Middle America, are beans and squash and, above all, maize (corn), which has long had religious significance in indigenous cultures. Many other crops are locally important, especially in the different climatic 456 zones of the highland regions. The *minifundio* farmer generally lacks funds to purchase larger and more fertile properties; he cultivates marginal plots of inferior soil fertility, often sharecropping (using his crops for rent) rather than owning. Individual landowners are often burdened by indebtedness, the threat of foreclosure of farm loans, and the fragmented nature of farms, which become smaller as they are subdivided through inheritance. Meanwhile, government promises to redistribute land generally go unfulfilled. Land ownership discrepancies contribute to the growing economic inequalities that trouble many Latin American societies (see Regional Perspective, **page 470**).

Large estates with a strong commercial orientation are known as ***latifundia*** (sing. *latifundio*). Also known as ***haciendas*** and plantations, these estates are owned by families or corporations (**•Figure 10.25**). Some have been in the hands

AP Images/ESTEBAN FELIX

• Figure 10.25 A plane fumigating a Chiquita banana plantation in Honduras. After Hurricane Mitch, Chiquita redeveloped its plantations in Honduras with denser crop placement requiring fewer workers; as a result many farmhands lost their jobs.

of the same family for centuries. The desire to own land as a form of wealth and a symbol of prestige and power has always been a strong characteristic of Latin American societies. Spanish and Portuguese sovereigns granted huge tracts to military noblemen who led the way in exploration and conquest. Such "land grants" in time became important properties for profitable ranching and mining, military garrisons, and more, and control over these lands was a major issue in wars and postwar settlements between Mexico and the young United States.

The value of commercial agriculture in national economies has declined in percentage terms as economies have become more diversified. But in many countries, more than half of all export revenue is still derived from farm products, mainly grown on the *latifundia*, and some countries rely heavily on one or two agricultural commodities.[13] Over-reliance on coffee and bananas in particular has whipsawed the economic fortunes of many countries, especially those in Central America such as Honduras and Guatemala that came to be known disparagingly as "banana republics." In colonial fashion, the American-owned **United Fruit Company** (**UFC**) owned vast tracts of land for the production of bananas and other produce, and had a powerful influence on local politics. Guatemala's president was ousted in 1954 in a CIA-led coup after the country developed land reform policies that were seen as a threat to the United Fruit Company (the CIA's then-Director Allen Dulles was a significant United Fruit Company shareholder). 473

As Latin American agriculture has become more commercial, the range of export crops has expanded beyond bananas and coffee. On a trip to the supermarket, you can find the ordinary and exotic fare of Latin America's commercial production: tomatoes, fruits, and vegetables from Mexico; bananas and coffee from Central and South America and the Caribbean; cantaloupes, vegetables, and snap peas from Central America; asparagus, onions, and quinoa (pronounced "*kin*-wa," protein-rich seeds from a flowering plant) from Peru; pears, apples, grapes, and wine from Chile's Mediterranean climate region; cut flowers from Colombia and the Caribbean

Joe Hobbs

• Figure 10.24 Weekly market in an Andean town, Ecuador.

Regional Perspective Inequities in Latin America

Land ownership mirrors the inequalities that plague most Latin American societies. As in many of the world's LDCs, there are great gulfs between the haves and the have-nots; but Latin America has the worst inequities of all the world regions. In some cases these chasms manifest themselves regionally between areas of prosperity, such as southern Brazil, and areas of poverty, as in the scrubby upland *sertão* region of northeast Brazil, whose very name evokes drought and rural despair (see Figure 10.34 for a map of Brazil). 62

In sporadic **land reform** programs—efforts to more equitably share quality farmlands among the population—governments reallocated some *latifundia* land to small farmers. A grassroots movement known as Sem Terra ("Without Land," or the Landless Workers Movement) effected change in Brazil's 1988 constitution, which states that that land must serve a social function and requires the government to "expropriate for the purpose of agrarian reform, rural property that is not performing its social function."[16] Few land reform schemes have been successful, however. In many countries, a very large share of the land is still in the hands of a small, wealthy, landowning class. Several Latin American countries—including Brazil, Colombia, Haiti, Honduras, Guatemala, and Suriname—have among the highest income inequalities in the world, surpassed only by a few African nations. In Brazil, for example, just 1 percent of the population owns 45 percent of the country's farmlands, and 3 percent own two-thirds. The rural poor have generally declined offers to assume government-owned mortgages for land ownership and have continued to migrate into Brazil's cities.[17]

The rich–poor divide is as glaring in Latin America's cities as it is in the countryside. Beneath the glitter of the great metropolises like Mexico City and Rio de Janeiro, with their forests of new skyscrapers, lie massive, squalid slums, known as ***favelas*** in Brazil and ***barrios*** in Spanish-speaking countries (**• Figure 10.B**). Such **shantytowns** or "informal settlements," which are practically universal in the world's LDCs, are filled with unemployed, underemployed, and sometimes undernourished people. *"Ni ni"* these people are called throughout much of the region: neither educated nor employed. 434

Joe Hobbs

• Figure 10.B Many of Latin America's poorest people live in ramshackle housing. This is the shantytown of Belen in Iquitos, Peru. The great Amazon River is just out of view to the left.

The sites and situations of some of these slums are remarkable. Some *favelas* border some of the most costly real estate in Rio de Janeiro and other cities, for example. More typically, they are situated in the least desirable, most hazardous areas of the urban landscape, such as steep hillsides or floodplains, making headlines when mudslides slam through them, taking scores of lives. Still, many people prefer their plight in the *favelas* to the conditions of the depressed rural landscapes from which they came. And as Edward Glaeser reminds us, urban poverty can offer a path toward prosperity both for the poor and for the nation as a whole. Urban life even outside the slums can be dispiriting. "Many Latin American cities are beastly places to live and work," Blouet writes.[18] But under the gritty veneer of pollution and crowding, life for most has improved, with longer life expectancies and the electronic and other amenities of the modern household.

The average GDP (PPP) per capita, as seen in Table 10.1, can give a misleading idea of average wealth because incomes are so unevenly distributed. In several Latin American countries, about 60 percent of all income is earned by the wealthiest 20 percent while the poorest 20 percent of populations receive 5 percent or less of the GDP—a much higher inequality rate than that of the United States. The top 10 percent of earners receive over 40 percent of all income while the bottom 10 percent live in poverty.[19]

There are some hopeful trends. In the big picture, globalization has helped redistribute wealth from the global "North" to the "South" as middle classes grow in the LDCs, as discussed in Chapter 3. As revenues rise overall, a key challenge is to ensure that they are spread among the population. Wealth disparities are falling in industrializing Mexico, and even Brazil is showing slow but steady improvement. There are risks in not spreading the wealth: social unrest and political instability can impede economic growth. On the surface, Latin America's economic disparities and related problems appear to have much in common with factors that culminated with the rage of the Arab Spring in the Middle East. One major difference is that in Latin America people have been able to democratically elect their leaders. Some, like Bolivia's Presient Evo Morales, have agendas for change that include the redistribution of wealth. 61 78 480 islands; and orange juice concentrate from Brazil.[14] Later in the chapter, we will explore Brazil's commercial agriculture in 484 more detail.

Latin America today is also a major producer and exporter of a number of valuable minerals; these are mapped in • **Figure 10.26**. Most of these minerals are exported in raw or semi-processed form, except for the Venezuelan aluminum and Brazilian and Mexican iron and steel. As happened with agricultural exports, with the Great Recession and China's recent economic slowdown, metals exports and revenues slowed down and posed special risks to the economies of Brazil, Peru, and Chile.[15]

Manufacturing and Neo-Liberalism

Like the agricultural commodity exports, these mineral exports have helped sustain economic growth in Latin America, especially because of growing demand in China and India. With the incoming revenues Latin American countries have invested in infrastructure—highways, power stations, water systems, schools, and hospitals—and more people have been employed.[20] But again, primary commodities exports subject Latin American economies to booms and busts, and only a diversified economy with healthy manufacturing and service sectors can provide long-term stability and economic growth, and escape the "re-424 source curse." New businesses in these secondary and tertiary sectors will grow the middle classes, whose appetite and capacity for consumption will further boost manufacturing and services.

Latin America's efforts to escape overreliance on the primary sector by boosting manufacturing have been fraught with difficulties and setbacks.

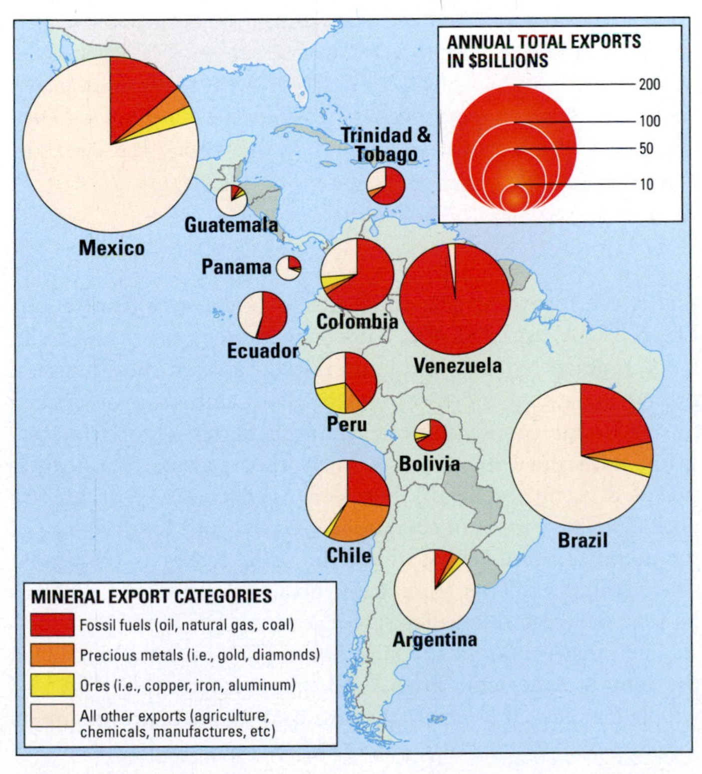

• **Figure 10.26** Exports of minerals and fossil fuels are very important to the economies of many Latin American countries.

In the early 1980s, a debt crisis swept through Latin America. Following the worldwide energy crisis of the 1970s, interest rates rose so high that Mexico, Brazil, and other debtor countries could not repay loans and were in danger of defaulting on them. The IMF and World Bank volunteered to help, but only if Latin American markets agreed to open their economies to competition and foreign investment. This was **neo-liberalism**, often referred to as "structural adjustment."[21] With neo-liberalism, earlier protectionism and the nationalist economic policies of "import substitution industrialization" (ISI) were rejected. The cornerstone of ISI was the introduction of high tariffs (taxes) on imported goods in order to promote industrialization while reducing imports of manufactured goods. The cornerstone of neo-liberalism was to reduce tariffs on imports to stimulate consumer demand, lower consumer prices, and promote economic growth. To bring in "hard currency," meaning monies like the dollar and euro that could be traded internationally, foreign investment was encouraged. Foreign investors could now buy or create companies to bring in capital in the form of hard currencies. Government-owned corporations were sold off (privatized) to generate revenue, but the most valuable sector of petroleum, wrested from control by foreign corporations like Shell and BP as early as the 1930s, remained in government hands: examples are Brazil's Petrobas and, in Mexico, Pemex—an infamously corrupt, state-run petroleum corporation.[22]

With the exception of the "Bolivarian socialist" countries of Cuba and Venezuela, most countries in Latin America took 480 the painful prescription of neo-liberalism. Economies were restructured, and a new era brought prosperity for some sectors and unemployment in others.[23] Overall, however, the region's economies moved away from commercial agriculture and raw materials and toward manufactured goods exported competitively into the world economy.

The region's strongest manufacturers are also Latin America's strongest economic powers. Three countries account for 80 percent of the region's manufacturing output: Brazil, Argentina, and Mexico. Below, we will look at Mexico's diverse and healthy industrial sector in the context of NAFTA.

Brazil deserves special attention. Brazil is the "B" in the dynamic (but slowing) BRIC economies and has the largest economy in South America and the seventh largest in the world. Brazil's manufacturing sector has experienced rapid expan-60 sion. Brazil makes early-stage industrial products like textiles and shoes, along with steel, machinery, chemicals, plastics, railway equipment, automobiles and trucks (Brazil is Latin America's leading automobile producer), ships, weapons, and aircraft. Chances are that if you fly much, you have travelled on Brazilian-built Embraer jets. Brazil vies with the Canadian firm Bombardier for world's third-place aircraft builder, after the EU's Airbus and America's Boeing. São Paulo, where Embraer is based, is the country's industrial core, producing nearly two-thirds of Brazil's manufactured goods.

Trade agreements exist to facilitate the profitable movement of such manufactured goods as well as agricultural produce. Generally—again, except for the "Bolivarian socialist" countries—Latin America has embraced trade agreements to facilitate its emergence on the world economic stage.

Regional Perspective The North American Free Trade Agreement (NAFTA)

In terms of both economic value and geographic area, the **North American Free Trade Agreement (NAFTA)**, composed of the United States, Canada, and Mexico, is the second largest trading bloc in the world after the EU. The essential goal in establishing this free-trade agreement, like most, was to reduce duties, tariffs, and other barriers to trade among the member countries, theoretically strengthening all the economies. Strictly speaking, however, NAFTA is not a pure free-trade agreement because it contains several restrictions and exclusions on certain agricultural products, such as dairy and sugar.[24]

NAFTA was already controversial before it went into effect in 1994, and it remains so today. Among US citizens, there were fears that factories would close down and jobs would flow south to Mexico. American environmentalists feared that the US environment would suffer because Mexican factories and other polluters, unable to afford the cleanup of their industries, would set new and lower standards that US companies could then match. Mexican officials feared that a surge in their consumption of American-made products would so increase their trade deficit with the United States that Mexico's monetary reserves would be exhausted.

And what has actually happened? Some US factories did close, in part because production was cheaper south of the border—notably in the ***maquiladora*** factories —but NAFTA was not the exclusive cause of the shutdowns. Manufacturers seek the lowest production costs, and they would have relocated to Asian countries if not to Mexico.[25] It is arguably much better for the American economy if the outsourced jobs remain in the neighborhood. Lower-cost manufacturing for American companies did shift to a host of countries outside NAFTA and the Western Hemisphere, but NAFTA rules buffered the impacts of those moves. Trying to maximize benefits for its three members in the global marketplace, NAFTA requires all its member countries to abide by **rules of origin** specifying that half or more of the components of any manufactured good must originate in Canada, the United States, or Mexico. This is designed to prevent a fourth country from using Mexican labor in the assembly of its own component parts. Critics of NAFTA say the agreement should have demanded that an even larger proportion of the component parts originate in the member countries to boost their economies.

American exports to Mexico were given a boost by NAFTA, as Mexican tariffs on American goods were reduced. The US exports more goods to Mexico than to any other country but Canada. NAFTA also increased investment in Mexico, especially by American firms. To attract foreign industries and boost exports of manufactured goods, the *maquiladora* offers tax breaks, cheap labor, and other incentives.

NAFTA did create more manufacturing jobs in Mexico, especially in the motor vehicles industry; Mexico is among the world's top 10 automobile producers. The geographical concept of *situation* was decisive in locating the *maquiladoras*: most were placed as close as possible to the great market of the United States—the principal destination for most of their output—and a few deep in areas like the country's eastern Yucatán region where labor costs are lowest. 94 There was a consequence of this strategy: the growth of *maquiladoras* along the US border contributed to the increasing disparity between the wealthier north and the poorer south, as few of these factories were located in the more agricultural south of Mexico.

This business location strategy was, however, enormously successful overall. Exports from Mexico to the United States have more than tripled since NAFTA went into effect in 1994. This bilateral trade is in Mexico's favor, with imports from Mexico exceeding exports to Mexico by 20 percent. Mexico is the third largest provider of imports to the United States, after China and Canada.

The *maquiladoras* have surged in growth and diversification since NAFTA went into effect; about 2800 of these factories, employing over 1 million Mexicans, are now operating in the US border region alone. Mexico's *maquiladoras* are evolving in the global marketplace. Originally established in the 1960s, the first *maquiladoras* produced jeans, T-shirts, and other textiles. After 2000, Mexico lost some of the competitive edge

Free-Trade Associations and Common Markets

In economic terms, globalization is the removal of barriers to trade by the lowering or elimination of tariffs and other restric- 60 tions on the movement of goods. Latin American economies became globalized beginning about 1960 when they formed or joined trading blocs including free-trade associations (FTAs), customs unions, and common markets, including some—notably, the North American Free Trade Agreement—with their giant neighbor to the north. The varying rules of trading blocs are described in note 26.

There are several free trade associations in Latin America, but the largest is the **Southern Cone Common Market**, also known as **Mercosur**, which was established in 1991 and is comprised of five nations in South America: Brazil, Argentina, Paraguay, Uruguay, and Venezuela. In value of products traded, this is the fourth largest trade group in the world, after the EU, NAFTA, and ASEAN. Mercosur reduced tariff and other long-standing barriers that had existed between the countries, promoting a huge increase in trade. Its members have worked to coordinate economic policies and a better negotiating position with the United States and other powers. (Coordination is difficult, especially between Argentina and Brazil; see note 27.) The people of this union can live and work in any of the member countries and have the same rights as citizens of those countries. This stipulation brought immediate relief to an estimated 3 million illegal workers.

The most economically significant FTA in the Americas is the **North American Free Trade Agreement (NAFTA)**, of which Canada, the United States, and Mexico are members (see Regional Perspective, page 472).

The United States is taking the lead in establishing a hemisphere-wide trade organization called the **Free Trade** represented by its inexpensive labor. Hundreds of *maquiladoras* closed, and hundreds of thousands of workers lost their jobs. Why? Mexico's low-tech industries could not match Chinese rock-bottom labor costs. Mexico itself became flooded with Chinese imports, even of pottery imitating Mexico's traditional wares. As personal electronics manufacture moved to China and Malaysia, and lower-cost textile production centers emerged in El Salvador, Mauritania and Vietnam, the *maquiladora* factories increasingly turned toward the manufacture of sophisticated medical equipment and electronics such as unmanned aerial vehicles (drones), mainly from components that are imported duty-free because they are assembled for export only (**•Figure 10.C**). There has also been growth in industries supporting the *maquiladoras*, such as small-scale retail and transportation services. Among the most recent developments are **energy *maquiladoras***. These are power generation plants—built by American firms just inside Mexico, where pollution standards are much lower than in the United States—that provide electricity to the US grid. Their power source, natural gas, comes from Texas, and Mexico supplies their coolant waters.

NAFTA gradually eliminated most Mexican tariffs on agricultural imports from the United States, with favorable results for American producers. The large, heavily subsidized American agricultural businesses can raise grain or livestock at prices below production cost. A pig, for example, can be raised and sold in Iowa for one-fifth of what it would cost to raise and sell it in Mexico. Cheap American agricultural products flooded the Mexican market, making it impossible for Mexican farmers to compete and driving many out of business; one-third of Mexico's 18,000 swine producers went out of business soon after NAFTA went into effect (on the positive side, Mexican companies have been able to turn the cheaper US pork into sandwich meats sold back at a profit to American consumers).

AP Images/Raymundo Ruiz

• Figure 10.C A technician repairing computer monitors in a maquiladora in Juarez, just across the border from El Paso, Texas. Assembly-for-export plants like these serve American companies and have boomed since NAFTA went into effect in 1994.

The United States had a strong if perhaps understated NAFTA goal of reducing Mexican immigration into the United States by boosting Mexico's prosperity. But with Mexican farmers losing incomes and livelihoods because of falling prices for corn, rice, beans, and pork, a new stream of poor southern Mexican immigrants was "pushed" into the United States. By about 2008, the hoped-for reduction of emigration to the United States did materialize, but for the "wrong" reason: the economic downturn in the United States made emigration from Mexico less attractive.

In sum, in the United States and Canada, NAFTA has been beneficial to farmers but not to factory workers, while the reverse is true in Mexico. Both the United States and Canada have trade deficits with Mexico, importing more from that country than exporting to it. Trade among NAFTA members continues to grow; US exports to Canada and Mexico increased by 34 percent (and its imports from those two countries by 60 percent) between the depth of the recession in 2009 and 2014.

Area of the Americas (FTAA). Like the political organization known as the **Organization of American States (OAS)**, it would include all 35 independent states of the Americas, including Cuba. Originally envisioned for establishment in 2005, this agreement is proving very difficult to achieve because some of its loudest critics in Latin America—Mercosur in general and Brazil in particular—insist that the United States is including too many restrictions in its vision of "free" trade. Brazil wants to see the United States drop the huge subsidies of many agricultural products and steel that keep American farmers and factory workers in business while shutting out 526 Brazilian products. Brazil argued successfully to the WTO that export subsidies to US cotton farmers did not conform to WTO rules: the US halted the cotton subsidies and paid Brazil 421 for lost revenues.

In FTAA negotiations, Mercosur members say they would reciprocate for American concessions by endorsing patent, copyright, and other **intellectual property rights (IPR)** protection of US industries. With the FTAA stalled, the United States has chosen to negotiate one-on-one trade agreements with members of the proposed FTAA; free-trade agreements have been signed with Chile, Peru, and Panama.

What have the results of neo-liberalism and free trade been for those countries following that path? Most of the countries' economies have been on an upward growth trajectory, with periodic setbacks most recently caused mainly by China's economic slowdown. Free trade benefits consumers with greater choices of goods and their prices. Educated and professional employees have prospered.

On the downside, countries can import too many goods relative to their exports, creating trade deficits that weaken already vulnerable currencies. Foreign factory owners can (and do) shift operations from Latin America to other regions where wages are lower, particularly in Asia. Creditors like the IMF and the World Bank—the "financial masters of the universe" as Blouet calls them—lay down rules that must be followed; if they are not, economic consequences are immediate.[28] International investors' "hot money" can be withdrawn on a whim, for example, when Wall Street money managers disavow the sometimes-darling "emerging market" stocks. Certain economic sectors can be wiped out, particularly if imports are "dumped," underpricing local producers—for example, as happened with US agricultural produce in Mexico and local 112 dairies and poultry plants in Jamaica.[29] These producers cannot compete with large-scale production by multinational corporations; exceptions are small growers of bananas in Ecuador and coffee in Colombia, who are getting a leg up thanks to the fair trade movement. And most worrisome of all, forces of globalization at work in Latin America are widening the gulf between rich and poor, even as middle classes grow, most noticeably in the cities.

Free trade is often not fair trade, especially in its tendency to undercut small-scale producers. Today's **fair trade movements** represent some degree of reconciliation between the hemispheric haves and have-nots. At many North American and European grocery stores, one may find bananas, pineapples, other fruits, chocolates, and coffees bearing "Fair Trade Certified" stickers. These are higher in price than the standard products, but consumers who buy them are generally motivated by a sense of social responsibility. Fair trade buyers deal directly with the farmer cooperatives they helped organize, avoiding brokers and intermediaries and guaranteeing higher prices for the farmers. The fair trade organizations also help the farmers establish schools and health clinics.

Remittances: Sending Money Home

More and more Latin American families, particularly from Mexico, Central America, and the Caribbean, have come to rely on having at least one member work abroad to help out the family economy. Among the largest sources of income for Latin American economies are the remittances, or earned savings, sent home by people working abroad, especially in the United States. In 2014, Latin Americans working in the United States sent home $62 billion. The workers typically send these payments home in periodic electronic transfers of $200 to $300 each. Mexico is the world's third largest remittance market (after China and India), with $23 billion infused yearly into the country from Mexican laborers in the United States.

About 54 percent of the 16 million immigrant workers originating in Latin America send monies home, in many cases to grandparents caring for the working parents' children; families often remain divided for many years while the father or both parents work abroad. Collectively, these remittances are so valuable—far surpassing the amount supplied to Latin America in foreign aid and foreign investment—that home countries are searching for ways to tax the income or otherwise find ways to funnel more of it into national development. Some critics say that Mexico and other Latin American countries have become so dependent on remittances that they neglect to invest their own resources in national development. Honduras is one such country: remittances amount to 17 percent of the country's GDP.

Tourism and the Service Sector

In economically advanced countries, the service sector (retailing, wholesaling, transportation, telecommunications, finance, medicine, education, administration, and so forth) provides approximately 80 percent of GDP. In the majority of Latin American countries, services account for over 50 percent of GDP, but official figures understate the service contribution because they do not account for the **informal sector**, including the "black market" and "under the table" transactions in cash that are not monitored or taxed by governments.[30]

Tourism, which belongs to the service sector, is a major regional economic asset to Latin America. The region's stunning archaeological and natural wonders and its diverse and rich cultures provide a wealth of choices for a variety of touristic pursuits. About 50 million foreign tourists visit the region each year, generating critical foreign exchange; only oil exports are more valuable.

During what many North Americans see as their too-long winters, they are bombarded by television and print advertisements of vacation in paradise: Mexico, Jamaica, Costa Rica, the Bahamas, and hosts of other destinations of the Caribbean. A wide variety of vacation experiences are available, from the hedonistic resort complexes that tourists seldom leave (one resort on Jamaica actually is called Hedonism), to the insular sea voyage that offers day trips ashore in exotic destinations, to whitewater and rain forest adventures of the ecotourist variety (• **Figure 10.27**).

The region benefits from the proximity of its year-round warmth to the wealthy and mobile and seasonally chilled

JOE HOBBS

• **Figure 10.27** International and domestic tourists flock to Mexico's "Maya Riviera," stretching along the Caribbean coast from Cancun southward to the border with Belize. These are passengers disembarking the Cozumel ferry in Playa del Carmen.

US and Canadian populations. Tourism revenues reflect that **distance-decay relationship:** the highest tourism receipts in Latin America flow to Mexico, the nearest neighbor to the wealthy countries, but tend to fall off for more far-flung destinations.

This overview of Latin America's economic geography has illustrated the growing engagement of these countries with the processes of globalization, particularly where trade is involved. As elsewhere in the world, we have seen how the major powers of the United States and increasingly China have boosted their commercial self-interests in the region. As we turn to geopolitics, we will see how sensitive the US is about its neighbors in the hemisphere, and how other powers have challenged American hegemony in the region.

10.5 Geopolitical Issues

The central geopolitical reality of Latin America is that it is in America's backyard. For better or worse, the United States has staked its geostrategic claim to the region. In 1823, US intentions for Latin America were formulated in a policy known as the **Monroe Doctrine**, after US President James Monroe announced that the United States would prevent European countries from undertaking any new colonizing activities in the hemisphere. Many subsequent interventions, wars, investments, and other US activities may be seen in light of this seminal policy. Even more consequential was the 1904 **Roosevelt Corollary** to the Monroe Doctrine, whereby the United States declared that it had the right to supervise the internal affairs of Latin American countries to ensure US national security.

Ever since the Monroe Doctrine was proclaimed, the United States has had a particularly firm hand in Central America and the Caribbean. Washington has dealt unapologetically and sometimes harshly with perceived threats to the United States from this region. Examples include actions to support governments against nationalist insurgencies, and to support insurgents against leftist governments: Nicaragua in the 1950s, Cuba and the Dominican Republic in the 1960s, and Grenada, El Salvador, and Nicaragua in the 1980s were the most significant instances.

One of the most aggressive interventions took place in 1954, when the US Central Intelligence Agency ousted the democratically elected Guatemalan leader Jacabo Árbenz in a coup. Guatemala's president had pushed for the redistribution of land in a country where 2 percent of the population owned 70 percent of the arable land. His subsequent branding as communist and his potential threat to United Fruit Company interests in Guatemala brought on the CIA intervention, which in turn led to a series of dictatorships and Guatemala's civil 469 war (1960–1996). That long war's main victims were the rural poor: Latino peasants and indigenous Maya peoples.

Guatemala was not the only Latin American country where the US sought "regime change." An even closer neighbor, Cuba, had long frustrated the American geopolitical vision for the region until a remarkable improvement in relations between the two countries in 2015.

Cuba and the US in Transition

Cuba and the United States have had a troubled relationship for many decades. A grave test of US hegemony in Latin America came with the **Cuban Missile Crisis** of 1962. Cuba is as close as 90 miles (144 km) from the United States. The frosty relations between the United States and Cuba's Marxist regime, led after 1959 by a firebrand revolutionary named Fidel Castro, were emblematic of the Cold War. Those tensions 119, 130 came dangerously close to war and even nuclear holocaust in 1962, when the Soviet Union began installing nuclear missiles in Cuba that were capable of striking Washington, New York, and other cities in the eastern United States (**•Figure 10.28**). As 150 it so often does today, remote sensing played a vital intelligence role. Photographs taken from an American U-2 reconnaissance aircraft provided undeniable evidence of missile launch infrastructure on the ground in Cuba. The pilot of that spy mission, Rudolf Anderson, was shot down by a Soviet-made surface-to-air missile. His death could have sparked a US-Soviet war, but President John F. Kennedy correctly concluded that the shoot-down order was given by a low-level Soviet officer in Cuba, and not by Soviet premier Khrushchev.

The missiles were ultimately withdrawn under intense pressure from the US government, and with a secret US-Soviet agreement: the Soviets would remove their nuclear missiles from Cuba, and the US would remove its own, aimed at the Soviet Union, from Turkey, as well as promise never to invade Cuba. Cuba, however, remained a storm center of political affairs in the Western Hemisphere. After several attempts to assassinate Castro—even with the unlikely ploy of an exploding cigar—and a disastrous effort by Kennedy's administration to overthrow him by military force in 1961 (in an incident known as the **Bay of Pigs invasion**), the US administration settled on a course of economic and political isolation of Cuba.

Despite the resumption of diplomatic and economic relations with most of the remaining Communist countries in the world—China most notable among them—for decades the United States remained steadfast in imposing sanctions on

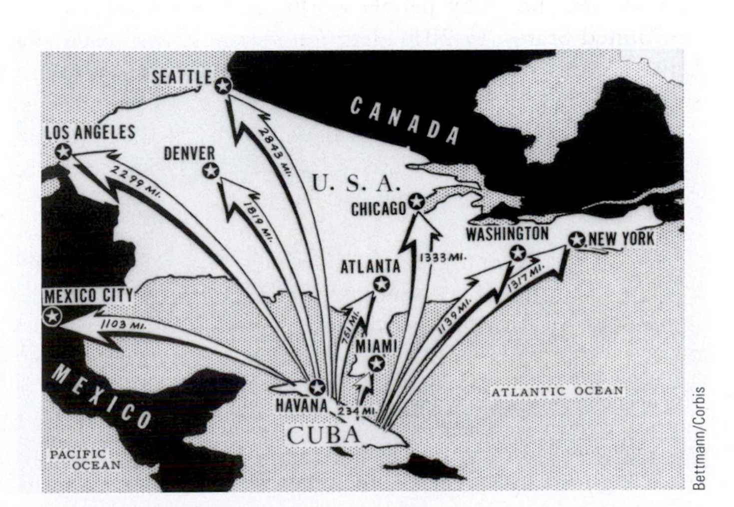

• Figure 10.28 The 1962 Cuban Missile Crisis brought the United States and the Union of Soviet Socialist Republics close to the brink of nuclear holocaust. This 1962 map shows the range that Soviet missiles would have if stationed in Cuba.

Cuba. This anomaly reflected domestic political realities in the United States, where a large population of **Cuban Americans** (including 800,000 in Florida alone) who fled Castro's Cuba tend to vote for the politicians who keep up the most pressure 517 against Castro. The Cuban Americans were against any relaxation of the virtual ban on American tourism to Cuba, which was in place because the "Trading with the Enemies Act" forbade Americans from spending more than $300 in Cuba. On the Cuban side, frozen relations with the United States long helped bolster a regime that was founded on the principle of opposition to US imperialism and that used American sanctions as an excuse for its poor economic situation. Cuba would, of course, like to see the country opened to American tourists, and American visitors would find much of interest in a country frozen in time (and filled with much-loved classic American cars; **•Figure 10.29**).

Gradually the door began to open. The failing health of Fidel Castro and the discovery of significant oil reserves off Cuba's shores led to unprecedented discussions among US lawmakers about resuming ties with Cuba, which had been broken after the 1959 revolution. One fear was that US businesses would lose important opportunities if the door remained closed. On the Cuban side, economic pressures were imperiling the regime: Russia did not carry on the Soviet legacy of subsidizing Cuba as its strategic partner in the Western Hemisphere, and plummeting oil revenues made it impossible for Cuba's neighbor Venezuela to keep propping up Cuba. Relations with the US could turn things around for isolated Cuba.[31]

President Barack Obama's administration opened a dialogue with Cuba. The two countries restored diplomatic ties and reopened embassies in their respective capitals in 2015. A number of issues would have to be resolved for fully normalized relations. The US economic embargo prohibiting trade with Cuba, in place since 1960, would have to be lifted. Cuba would like to see the United States withdraw from Guantánamo Bay, an enclave the United States secured in the 1903 treaty that addressed Cuba's fate following the Spanish-American War. The United States dutifully goes through the motions of paying Cuba the $4000 annual rent for the lease on Guantánamo, which it has used as a military base, and Cuba dutifully refuses to accept it. After 9/11, the United States established a prison camp for suspected al-Qa'ida members and other terrorists at the base. Its location outside the borders of the United States allowed the US government to suspend the usual legal and constitutional rights awarded to criminal suspects on US soil. Presidential candidate Obama vowed to shut down the prison at "Gitmo," but as president he failed to find an alternative to it.

Kamira/Shutterstock.com

• Figure 10.29 The lovingly maintained classic American cars of Havana's streets are artifacts of the era leading up to the freezing of Cuban-US relations in 1959.

The Panama and Nicaragua Canals

At the southern end of Central America, the narrow Isthmus of Panama separates the Pacific and the Atlantic oceans by a mere 50 miles (80 km). Figures 10.1 and 10.2 reveal that this is an excellent geographic situation for transcontinental shipment routes. In 1855, the Panama Canal Railway became the first transcontinental railway, and it remains in use today. This method of crossing, requiring a break of bulk of goods and passengers, would finally be superseded by the canal. Engineering a canal was, however, a great technical challenge: differences in sea levels of the Atlantic and Pacific Oceans required three sets of **locks**, the devices that allow a ship to move from one water level to another.

The United States began construction of the Panama Canal in 1904, and at a cost of $380 million and many lives lost to yellow fever, bubonic plague, and malaria, it was completed in 1914 (**•Figure 10.30**). The first crossing was made under a banner that read: "THE LAND DIVIDED; THE WORLD UNITED." The Panama Canal today is one of the world's most critical chokepoints: 14,000 ships a year, traveling 80 shipping routes and representing about 5 percent of the world's cargo volume, use

• Figure 10.30 The Panama and Nicaragua Canals.

the 50-mile (80-km) shortcut from the Atlantic Ocean to the Pacific (see Figure 2.11, **page 38**). A 14-day sea voyage between New York and San Francisco via the canal cuts 7800 miles (12,530 km) and 20 days from the trip. That translates into enormous cost and time savings. The United States is a major user; about one-seventh of all US foreign trade is carried through the Canal.

The Panama Canal was intended mainly to serve US commercial and strategic interests. For decades, it was sovereign US territory, carefully monitored by an American military presence, and the United States kept more than half of the transit fees charged to ships using it. In the late 1970s, however, the administration of US President Jimmy Carter accepted the Panamanians' complaint that US control of the canal was an outmoded vestige of colonialism. Carter negotiated the transfer of authority over the canal to Panama in a treaty that went into effect in 1999. Now the revenues from ships transiting the canal are exclusively Panama's.

From a strategic and economic defense standpoint, however, the United States is still deeply involved with the canal. The US Navy is concerned that its two access points (Balboa in the Pacific and Cristóbal in the Atlantic) are now controlled by a company with strong backing from China, the canal's second largest user. The United States has already effectively invoked the Roosevelt Corollary since returning control of the canal; in 1989, US troops invaded Panama and arrested its military dictator, Manuel Noriega, whom the Americans accused of money laundering and facilitating the trafficking of illegal drugs into the United States. The US **war on drugs** has been another central geopolitical issue of the region since the early 1970s (see Geography of Drug Trafficking, **page 478**).

Work on the canal has never ended. To meet anticipated demand, especially from surging exports of China's manufactured goods, the canal recently underwent the largest expansion ever, with the expectation that it would quadruple income from transit fees. The construction project aimed to overcome problems delaying and prohibiting the passage of many ships. To double capacity and accommodate larger cargo vessels up to the size known in the shipping trade as "post-Panamax," new channels with longer and deeper locks were added at each end of the canal. These and other measures doubled the canal's capacity. Thanks to growing trade, expanding ports, and residential and commercial development, Panama's economy has boomed in recent years.

Nicaragua wants to take advantage of its geographic situation to build its own profitable transcontinental canal, which, Nicaraguan authorities insist, will complement rather than challenge the Panama Canal. The idea of a shortcut between the Atlantic and Pacific Oceans had been around for a long time before the Panamanian site was chosen. In the early 1800s, Nicaragua was the first choice because of natural landscape features, especially Lake Nicaragua, that appeared better able to accommodate a crossing without massive construction. For decades, Nicaragua continued to consider building this canal, which would be longer, deeper, and wider than the Panama Canal, and (unlike Panama's even after the recent construction) would be able to accommodate the world's newest megaships. In 2014, Nicaragua decided to start the project with completion set for 2020 (see **•Figure 10.30**). This project is another in China's worldwide vision for securing strategic economic assets; China is putting $50 billion into what its contractor calls "the biggest [project] built in the history of humanity."[32] 327

Nicaraguan leaders tout the job-creating infrastructure to be associated with the canal, including two free-trade zones, two ports linked by a railway, and an international airport. The project faces great opposition, however, not from rival Panama, but from within Nicaragua itself. Protestors decry that Nicaragua is selling itself to China and that construction will force thousands of poor farmers from their land and have enormous environmental consequences, especially on Lake Nicaragua's vital freshwater resources.[33]

Access to Oil

The United States' geopolitical priorities around the world include secure access to petroleum resources. Latin America has a prominent place in this strategy. The US views Colombia, Venezuela, and Mexico as long-term counterweights to the volatile Middle East oil region. The United States obtains more of its oil from Latin America than from the Middle East, with Mexico, Venezuela, and Colombia the top providers.

Many American policies in the region are linked to oil. These include the war on drugs in the context of Plan Colombia, described below. Petroleum and related products represent about 70 percent of all legal Colombian exports by value. Washington's thinking has been that the best way both to secure this oil and to stem the drug tide is to make a multi-billion dollar investment in taking on Colombia's main rebel army, the Revolutionary Armed Forces of Colombia (FARC; see **page 478**) and another rebel group, the National Revolutionary Army (ELN). FARC is the quintessential **narcoterrorist organization**, deriving most of its funding from the profitable cocaine industry based mainly in the areas it controlled. In an operation related to Plan Colombia, dubbed **Plan Patriot**, US military personnel and civilian contractors working with Colombian forces have succeeded in wresting much control from FARC. The government-backed "paramilitary" forces that both fought these rebels and had their own stake in the drugs trade have also been weakened, and the security situation in Colombia has improved. Government peace talks aimed at the "demobilization" of FARC forces (and that also include discussions of land reform to address a root problem in Colombia's unrest) have taken place against the backdrop of continued FARC threats of violence.

Another major player in American geostrategic thinking is Venezuela, which has the world's largest petroleum reserves. Venezuelan-US relations are fraught with difficulty, however, as Venezuela is the region's principal opponent to Washington's vision for Latin America; see also **page 482**.

The Washington Consensus . . .

Between about 1980 and 2000, US administrations exercised a hands-on policy in Latin America. Through a combination of

Geography of Drug Trafficking The War on Drugs

From the Nixon administration of the early 1970s to the present, the United States has led a determined, expensive effort known as the **war on drugs**. The Nixon administration was the only one to focus significant efforts on combating demand for heroin, cocaine, and other drugs within the United States. All successive administrations have concentrated their efforts on the supply side, hoping to eradicate drugs at their sources, particularly in South America. Although it is far from over, the war on drugs has produced some definitive and sometimes predictable patterns.

One of the most consistent geographic patterns is the so-called "balloon effect," referring to the fact that when a balloon is pressed in one place, the air shifts elsewhere. In the case of drugs, concentrated eradication efforts in one place almost always shift production elsewhere, sometimes even from one continent to another. During the Nixon years, for example, US-funded efforts succeeded in all but eliminating production of opium poppies (the plant from which heroin is derived) in Turkey. With demand still high in the United States, Europe, and elsewhere, however, poppy production surged in a region where it had been only modest: the Golden Triangle of Myanmar, Laos, and Thailand. Increasing pressure on poppy producers in the Golden Triangle, the Bekaa Valley of Lebanon, and briefly in Afghanistan, then led to the eruption of the poppy industry in Mexico, Guatemala, and Colombia, oceans away from its Old World hearth.

332, 342, 343

The balloon effect is clearly visible in the geography of coca production within South America. Between 2001 and 2015, Colombia's efforts with the US-funded **Plan Colombia** succeeded in reducing coca production by two-thirds. (Coca leaves chewed and brewed as tea have a long history of beneficent traditional use, quite distinct from the dangerous recreational drug cocaine, which is produced through chemical processing of the leaves.) But overall coca production in Colombia, Peru, and Bolivia remained stable. Among those three, Colombian production dropped from first place to last, but production in Peru (now the largest producer) and Bolivia surged. In Colombia, those growers who remained adapted in innovative ways. They dispersed their farms into smaller and more widely scattered plots, often in new areas. They also practiced selective breeding to produce plants that contain more active alkaloids, are more resistant to defoliants, and can grow with more shade cover, thereby escaping detection from the air.

Plan Colombia is widely criticized by human rights groups for focusing more on counterterrorism (often employing paramilitary forces accused of abusive force) directed at leftist FARC and other guerillas than on drug control. These guerillas are deeply involved in the production and smuggling of cocaine, so disentangling these issues is challenging. Colombia's counterterrorism measures have had the unintended consequence of growing the rebel ranks. Coca cultivation generally coincides geographically with Colombia's poorest regions, the northeast and southwest. Plan Colombia's aerial spraying to eradicate coca often destroyed the food crops of peasants who, seeking vengeance, joined the rebellion. Aerial spraying was dropped from Plan Colombia in 2014 because of these problems.

There is an economic multiplier behind the balloon effect: when eradication is successful in one area, supply falls and even more value is added to the already high value of cocaine and heroin. Growers and traffickers are easily tempted by the money to be made under such market conditions. Distribution is much more profitable than production: in Colombia's producing areas, one kilogram (2.2 lb) sells for $2200; at Colombia ports, $5000–$7000; in Mexican/US border towns, $16,000; wholesale in the US, $24,000–$27,000 ($55,000 in Europe, and up to $200,000 in Australia).[34] "Cutting" or diluting the product with other ingredients brings purity down by roughly two-thirds from the source area to the American street. The traffickers' general business model is thus to sell their product for 10 times what it cost them to acquire it.

Seeking to increase revenues even more, the Mexican **cartels**—the illegal organizations controlling the production and distribution of drugs—now dominate trafficking and are penetrating the entire value chain. Plan Colombia's military tactics were largely successful in snuffing out the dominance of Colombian drug cartels in cocaine trafficking, mainly by sea. After the infamous Pablo Escobar, the Colombian "King of Cocaine" heading the Medellin Cartel, was killed in 1993, the geography of trafficking changed entirely, shifting to Mexican control. In predictable fashion, the balloon effect manifested itself in distribution as it had in production: where the Colombian cartels using maritime routes had reigned, now Mexican crime gangs took control of overland routes across the 2000-mile-long US-Mexico border.

The cartels have made other geographical adjustments. Previously reliant on middlemen in Central America to obtain the drugs, Mexican cartels are now producing and processing cocaine in South America. Due to the government's aggressive war on drugs in Mexico, they have moved their distribution centers southward to Central America (especially Guatemala and Honduras), where governments are weaker and systems even more corrupt (• **Figure 10.D**).

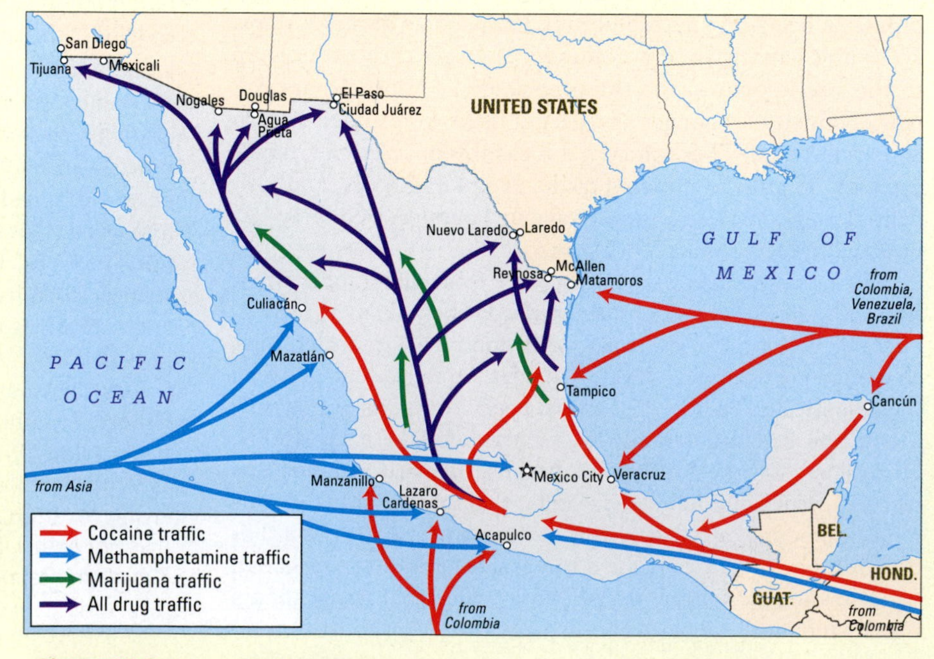

• **Figure 10.D** Drug trafficking routes.

Almost 90 percent of the cocaine bound for the United States is first brought by sea from Venezuela to Honduras, Guatemala, and El Salvador, where the Mexican cartels work with local youth gangs (gang violence has created a nightmarish existence for people in the Honduran capital Tegucigalpa and in other affected cities).[35] The Mexican cartels then move most of the cocaine overland to the United States. The cartels have also extended their trafficking to Europe, Africa, Asia (where growing wealth satisfies growing appetites), and Oceania, and are even involved in the final street dealings in US cities. By these methods, the Mexican cartels have become the biggest players in the global cocaine trade.

Who and where are these cartels, the illegal organizations controlling the production and distribution of drugs?[36] Now numbering about half a dozen, the cartels emerged in three geographic regions of Mexico (**• Figure 10.E**): western Mexico, from the Tijuana-San Diego border crossing to the Ciudad Juarez-El Paso border crossing south to Guadalajara; the eastern half of Mexico, including the Yucatán Peninsula; and the southern interior highlands between Mexico City and Guadalajara extending to the Pacific Ocean between Acapulco and Manzanillo.

The oldest, most powerful and still growing cartel is the Sinaloa Federation in the western region, controlling the flow of drugs across the US border from San Diego-Tijuana all the way to El Paso-Juarez. Under orders from notorious kingpin Joaquín "El Chapo" ("Shorty") Guzmán, twice-escaped from maximum security Mexican prisons, the Sinaloa Federation smuggles cocaine along western coastal routes, as well as marijuana grown in the rugged Sierra Madre Occidental. In these mountains, the group also cultivates opium poppies in one of the few New World areas to which it has been adapted. From the opium, it processes "black tar" heroin, marketed mainly in the southern and western US.

In the early 2000s, organized crime gathered momentum in the east, where Los Zetas emerged as the most prominent cartel. Los Zetas traffics drugs across the border from El Paso-Juarez all the way to the Gulf coast, with Laredo-Nuevo Laredo its most lucrative crossing.

In 2012, cartels in both the west and east were weakened by losses, leading to the emergence of crime groups in the southern region. Drug gangs there produce marijuana and opium/heroin in the interior highlands, as well as amphetamines made with precursor drugs imported from China.

Mexico's drug gangs have a tendency to splinter and "balkanize" as the relative power of these groups rises and falls. Violent turf wars periodically break out between rival cartels. This typically happens when a group controlling a specific smuggling corridor, known as a *plaza*, is weakened. Opportunistic rivals move in, and blood spills. Reprisals often involve gruesome murders, sometimes by beheading, that rival ISIS killings in their shock value. Between 2007, when Mexico's drug war began, and 2015, it is estimated that more than 160,000 people were murdered in the country, and that up to 55 percent of these homicides were drug war related. Although often involving civilians not related to the trade, cartel killings are, however, not by definition "terrorist" actions. Generally, the cartels try to avoid drawing more attention from law enforcement, and in many cases they have succeeded in corrupting the authorities to keep them at bay. Authorities have made some progress against some of the criminal activities perpetrated by these groups, which in addition to drugs often include theft of fuel, extortion, and kidnapping. However, there has been no progress in the main effort to reduce drug trafficking.

• Figure 10.E The distribution of the cartels across Mexico.
Source: Stratfor, 2015.

Overall, the drug war as it has been fought since the 1970s has not succeeded. What is to be done? Successful alternative proposals to the long-fought, conventional war on drugs typically have two elements. First, advocates of these other approaches argue that even more attention must be paid to the demand side, using education, treatment, and other means to wean addicts from drugs and discourage others from using them. The United States now spends about 45 percent of its drug war budget on treatment and prevention (only 6 percent is spent directly on targeting foreign sources). Conversely, in many European nations, treatment and prevention comprise only a small portion of their anti-drug budgets—13 percent in Britain (which has higher rates of cocaine and opiate usage than the US does), and just 1 percent in Sweden, for example.

The second element of success focuses on the supply side. Peasant farmers who grow drugs are generally not trying to get rich, but to secure a livelihood where jobs are scarce. Few crops offer the reliable market and certain returns that coca or poppies do, and **crop substitution** schemes to get farmers to switch to flowers or coffee almost always fail. Successful substitutions have occurred only in the context of so-called **alternative development** schemes, in which new crops are just one part of an overall effort to enhance the education, health, and livelihoods of rural communities. Alternative development in areas where coca is produced is not a component of Plan Colombia.

A successful war on drugs, won through efforts on the supply side, the demand side, or both, would have large environmental benefits, among others. Clearing of forest for new cultivation of coca has done enormous damage to South American forests, particularly on the steep eastern slopes of the Andes, where probably no other cultivation would ever be attempted. The processing of coca leaves to coca paste and finally powdered cocaine involves the use of highly toxic chemicals that make their way into water supplies, harming ecosystems and human health. A successful war on drugs could also help restructure US interactions with the countries of Latin America.

blunt US military interventions and the countries' own efforts, Latin America witnessed an overall transformation of political systems from authoritarian to multiparty democratic. In the thinking of successive US administrations, those greater political freedoms should have ushered in more prosperity. This was one component of the so-called **Washington Consensus**. In this political-economic philosophy, the United States would press the democratic governments of Latin America to expand neoliberalism by opening markets to international trade, reducing tariffs, expanding quotas, selling off state companies to private investors, allowing more foreign investment, cutting government spending, and reducing bloated bureaucracies. These reforms would bring the prosperity that would solve the region's problems. If serious problems did emerge, the United States could help; it did act, for example, to prevent the spread of a financial crisis that shook Mexico in 1994. But generally, the United States neglected Latin America and did so at its peril: into this vacuum stepped China, with its familiar pattern of economic aid and infrastructure, and even America's foes Russia (with military support of Nicaragua, Venezuela, and Cuba) and Iran (with military and other influence in Venezuela and Argentina).

. . . and Latin America's Pushback

After 2000 and especially after 9/11, aside from its boots on the ground in Colombia, the US government generally had a hands-off policy toward Latin America. The US was bogged down in the Middle East and Afghanistan, and insisted its most important geopolitical policy and action area would be Asia. Latin America fell down the list, and many Latin Americans themselves wanted a reduction of US influence in their region.

Washington watched from the sidelines while a succession of elections and polls showed a drift away from American wishes for the region. Widespread disillusionment with economic conditions led to successive elections of reformist, left-leaning and generally anti-Washington Consensus or anti-US leaders. First came Brazil's leftist labor union leader Luiz Inacio da Silva, known as "Lula," who vowed to fight the tide of globalization. Similar changes in administration took place in Argentina, Venezuela, Bolivia, Ecuador, and Nicaragua. In Venezuela, from 1999 until his death in 2013, President Hugo Chávez and then his successor, Nicolás Maduro, pursued assertive anti-American policies. Declaring that "the worst enemy of humanity is US capitalism," Aymara Native American and former coca grower Evo Morales became president of Bolivia in 2006 (**•Figure 10.31**). That same year, Ecuadorian President Rafael Correa was elected on promises to close a base used by US aircraft to bomb drug fields, reject free-trade deals, and spend more oil revenue on the poor. All of these leaders, Chávez most vocal among them, had in common a stated intention to reverse the US-backed push toward free trade and free enterprise, returning to state-owned companies and protected markets that would assist the poorest segments of their populations.

These leaders were generally supported by the indigenous Americans, who felt that the descendants of European colonizers of the New World, along with US and other multinational firms, were robbing them of both their natural resources and their political clout. The Native American majority of Bolivia actually succeeded in driving that country's president from office in 2003, setting the stage for the election of Morales, the first indigenous head of state in Bolivia since the European conquest. Morales championed the peasant Aymara cause by announcing that his government would nationalize all of Bolivia's valuable natural gas and other resources and use them for the benefit of the people. The so-called "ideology of fury" expressed by the Aymara in Bolivia rattled nerves among the elites in other countries of Latin America with large indigenous populations, including Ecuador, Peru, Paraguay, Guatemala, and even Mexico. It also rejected Washington's vision for the region.

Aizar Raldes/AFP/Getty Images

• Figure 10.31 Bolivian President Evo Morales in his Aymara Indian clothing.

In 2004, Venezuela and Cuba formalized their socialistic, anti neo-liberal, anti-US agenda with the creation of the **Bolivarian Alliance for the Peoples of our America** (its Spanish acronym is **ALBA**). ALBA calls for radical conflict to build a version of Simon Bolivar's "Gran Colombia," a bloc of Latin American states that would confront the "imperialist" United States; and an economic model of "Twenty-First Century Socialism" that would counteract every detail of the Free Trade Area of the Americas (FTAA). ALBA grew quickly to include nine other member countries: Antigua and Barbuda, Bolivia, Dominica, Ecuador, Grenada, Nicaragua, Saint Kitts and Nevis, Saint Lucia, and Saint Vincent and the Grenadines.

The following section explores more issues among the peoples and lands of the region, first in Middle America and then in South America.

10.6 Regional Issues and Landscapes

Middle America

Mexico's Stature

The federal republic of Mexico—officially, *Los Estados Unidos Mexicanos* (United Mexican States)—is the largest, most complex, and most influential country in Middle America (•**Figure 10.32**). The country's great population of 120 million makes Mexico by far the world's largest nation in which Spanish is the main language. Compared to most of the world's countries, Mexico is a spatial giant, but its large size tends to go unrecognized because it lies in the shadow of the United States. Mexico's area of 760,000 square miles (1.9 million sq km) is nearly eight times that of the United Kingdom, and its elongated territory would stretch from the state of Washington to Florida if superimposed on the United States. Its capital, Mexico City, is the world's 12th largest city in population, with 20 million inhabitants. Mexico is one of the most urbanized countries of Middle America, with about 78 percent of its people living in cities. In these and many other attributes, Mexico, in the Latin American context, lives up to its motto: "Higher and Further."

Mexico has been described as one of the globe's nine "pivotal countries," a country whose economic or political collapse could cause international refugee migration, war, pollution, disease epidemics, or other international security consequences. The country's stability is enormously important, particularly to its giant neighbor to the north. Mexico's natural and human capital is much greater than many outsiders realize. Mainly because of the wealth of its tropical forests in the south, Mexico ranks as the world's third most important megadiversity country, after Brazil and Indonesia. Mexico ranks highly among the world's producers of such commodities as metals, oil, gas, and sulfur, with oil the country's most important export. The Mexican oil industry is a government monopoly (carried on by a government corporation, Petróleos Mexicanos, or Pemex), and the United States is its largest foreign customer, taking in 90 percent of the country's oil exports. Mexican oil represented 9 percent of oil imported to the United States in 2014, making Mexico the second largest foreign supplier after Canada.

315, 321

• **Figure 10.32** Political geography of Mexico.

Mexico's oil will run out—some analysts say by 2030. The country is trying to diversify and grow its economy, the second largest in Latin America (after Brazil), away from reliance on commodities like oil. Mexico has made substantial gains in its development of manufacturing and component assembly industries, especially since the initiation of NAFTA. 474 The manufacturing sector is now Mexico's leading source of revenue, followed by oil and remittances from abroad. High-technology products represented 11 percent of Mexico's exports in 2014. A substantial portion of Mexico's industries is outside the *maquiladoras*: about one-third of all manufacturing employees in Mexico work in the thousands of factories located in metropolitan Mexico City.

There are both inseparable ties and divisive issues between Mexico and the United States. Huge disparities in wealth and power come between them. Mexicans have not forgotten that Mexico lost half its national territory to the *gringos* in the **Mexican-American War** (known to Mexicans as the **American Invasion**) in 1848. 511 Mexicans have enormous national and cultural pride and often feel misunderstood and underappreciated by their northern neighbors. In the next chapter, we explore immigration and other issues in relations between the United States and Mexico.

South America

Venezuela's Petroleum Politics

Oil resources in the coastal area around and under the large Caribbean inlet called Lake Maracaibo have been a huge blessing for Venezuela, the world's 11th largest oil producer. Oil makes up about 96 percent of the country's exports by value. The country was one of the founding members of the Organization of Petroleum Exporting Countries (OPEC) in 1960 and is still a member. 235 Oil revenues have fueled urbanization and industrialization and generated wealth. Yet Venezuela is rather like Nigeria, an oil-rich but resource-cursed country where the vast majority of people remain poor. 424, 436 Venezuela was ranked as the top spot globally with the highest "misery index" score in 2013, 2014, and 2015.[37]

The most prominent member in the lineup of left-leaning Latin American leaders elected in recent years was Venezuelan President Hugo Chávez. Invoking the revered name of Simón Bolívar (1783–1830), the Venezuelan-born independence leader who spearheaded a regional quest for liberation from Spain, he proclaimed to be undertaking a **"Bolivarian Revolution."** (Chávez had his hero exhumed and examined in 2010, ostensibly to determine how Bolívar died.) Chávez had his own ideas about liberation: he rewrote the constitution, put cronies on the supreme court, cracked down on the media, and jailed political foes. The Chávez agenda appealed to the poorest classes by promising a redistribution of wealth, drawn from the country's oil revenues, like that championed by Cuba's Fidel Castro (Chávez's strongest ally in the region). Chávez's political philosophy was not communism, but what he called "twenty-first-century socialism" (and others dubbed "Chavismo"), with the armed forces and the president controlling everything but a small and mainly foreign-owned private sector.

Venezuela's professionals—scientists, engineers, doctors, and businesspeople, for example—left Chávez's Venezuela in unknown numbers, probably hundreds of thousands, settling mainly in the United States and Canada. Meanwhile, there was a large influx of immigrants, including Chinese, Lebanese, Jordanians, Syrians, Haitians, and Colombians, drawn by high wages and new opportunities in Venezuela.

His power consolidated, Chávez used Venezuela's oil wealth to advance his political agenda across the Western Hemisphere. He built alliances with Latin America's leftist leaders and courted the center and right governments of the region to move his way. His main goal was to build a coalition of countries to counter US-led free-trade efforts in the region. He internationalized his pro-poor agenda, offering Venezuela's oil to Cuba and to select populations in other countries at little or no profit. He embarrassed US politicians by supplying cheap heating oil to hard-pressed American northeasterners in the cold winter of 2005.

Thanks largely to Chávez, old-fashioned, Cold War–style rhetoric between the United States and Latin America's leftist governments made a comeback. But while Venezuela condemned capitalism in general and the US in particular, it continued to sell its oil to the US uninterrupted. The country is enormously dependent on oil exports for revenue; despite the rhetoric, Venezuela could not afford to alienate its largest customer. 477

After Chávez died in 2013, his successor, Nicolás Maduro, vowed to continue his legacy. Maduro, however, lacks Chávez' charisma, and after Venezuela entered a deep recession in 2014, prompted by plummeting oil prices and hyperinflation, people took to the streets in massive anti-government demonstrations (**•Figure 10.33**). Among their complaints: doctors and other professionals earn the same wage as street cleaners, and some prefer to smuggle pasta and gas to Colombia.[38] Venezuela is falling apart at the seams.

Brazil, the Stirring Giant

Brazil is the giant of Latin America (see **•Figure 10.34**). It has an area of about 3.3 million square miles (8.5 million sq km)—about 5 percent larger than the conterminous United States

Juan Barreto/AFP/Getty Images

• Figure 10.33 Venezuelan protestors have repeatedly taken to the streets in opposition to the rule of Nicolás Maduro. The protestor at front here carries the scales of justice, tipped away from the people the government is blind to, and the image of a relative killed in an earlier protest. Anger in this 2014 demonstration focused on gun violence and other crimes, hyperinflation, and shortages of basic household goods.

• **Figure 10.34** Political geography of Brazil.

—and a population of 202 million as of 2014, second only to the United States in the Western Hemisphere.

Brazil has an increasingly important role in hemispheric and world affairs—so much so that it is lobbying for a permanent seat on the United Nations Security Council. Its growth is especially notable in its overall population, in the explosive urbanism that has made São Paulo and Rio de Janeiro two of the world's largest cities, in the rise of manufacturing, and in the diversification of export agriculture (**•Figure 10.35**). Brazil is the "B" in the BRIC economies, with the largest economy in South America and the seventh largest in the world. Its expanding domestic market and labor force have accelerated economic development in Brazil, but its population growth of nearly 1 percent per year also poses serious problems.

Brazil is also burdened with a large foreign debt owed to the International Monetary Fund and other international lenders for its ambitious and costly development projects launched after the mid-1970s. Over 40 percent of Brazil's annual government spending goes to debt reduction; just 4 percent is spent on health care. Paying down this crushing debt—over $450 billion—has prevented many improvements in social

• **Figure 10.35** Rio de Janeiro's harbor is known as one of the "natural wonders of the world." Charles Darwin wrote of its beauty: "Following a pathway, I entered a noble forest, and from a height of five or six hundred feet, one of those splendid views was presented, which are so common on every side of Rio. At this elevation the landscape attains its most brilliant tint; and every form, every shade, so completely surpasses in magnificence all that the European has ever beheld in his own country, that he knows not how to express his feelings."

services. An austerity program initiated in 2015 that mainly targeted already underfunded social services prompted widespread protests. However, Brazil has many things going for its future, including abundant and varied natural resources, many allies among the world's advanced economic powers, and relative internal harmony among its diverse ethnic and cultural populations.

Major investments in agriculture contributed to a rebound in the country's economy between 2010 and 2015. Brazil became one of the world's major breadbaskets and has the extraordinary ambition of overtaking the United States to become the world's largest food exporter by 2025. Brazil leads the world in exports of sugar, beef, poultry, coffee, orange juice, ethanol, and tobacco. Brazil's sugarcane production is used mainly to produce the biofuel **ethanol** for domestic consumption and exports. Brazil is recognized as having the world' first sustainable biofuels economy. Brazil and the US, which together produce 80 percent of the world's ethanol, are exploring ways to maintain and increase that share through cooperative research and development. Brazil may be able to help the United States get the kinds of secondary uses it does from its cane processed for ethanol: the cane waste, called *bagasse*, is burned in steam boilers to produce electricity used in ethanol production and for surplus sale back to the electricity grid. The US is also exploring better sources of biofuel; it is much more efficient and inexpensive to produce ethanol from sugar than from corn, as the United States currently does. 524

According to critics, major drawbacks to biofuel production are that it does nothing to enhance the welfare of Brazil's landless and rural poor. It puts more land in the hands of agribusiness and the elite—a classic illustration of the marginalization process. 64 It is also grown as a monoculture, cutting down on both natural biodiversity and crop diversity. Critics levy the same accusations at Brazil's soy industry. In the 1960s, Brazil became a major soybean producer and exporter, with production centered in the subtropical southeast and in a new district in the tropical Brazilian Highlands. Soybean production is now pushing into the Amazon Basin and is a major source of deforestation there. With the application of chalk and other additives, the infertile soils of the Amazon can be made more productive.

This expansion is in large part a response to growing demand for soy products in China, the world's largest soy consumer. Much of Brazil's soy is also bought by American-based companies and made into cattle feed consumed in China and Europe. Brazil is second only to the United States in soybean production, and in 2014 surpassed the US as the world's leader in soybean exports. To gain an even larger market share, Brazil has given up its long-standing opposition to genetically modified (GM) foods and is producing GM soybeans and other crops. 116 The country is trying to maintain a delicate balance between its North American markets, which have no problem with GM foods, and its European markets, which shun them. Brazil claims it can produce GM and non-GM foods to satisfy both. American farmers look nervously at Brazil's surging agricultural exports. Having won the case it made to the World Trade Organization that US farmers were too heavily subsidized, thus making Brazilian cotton and other exports uncompetitive, Brazil could undercut American agricultural exports on the world market. 421, 526

Brazil arguably has a "neocolonial" relationship with China, as many other Latin American and other developing countries do. Raw material and agricultural commodities— 427, 468 especially iron ore and soybeans—make up about 95 percent of Brazil's exports to China, and 98 percent of its imports from China are manufactured goods, including inexpensive cars.

Brazil's wealth of minerals feeds its rapidly expanding manufacturing sector and provide valuable export revenues. The Brazilian Highlands are very mineralized, with ores of iron, bauxite, manganese, tin, and tungsten. Brazil has about an eighth of the world's proven iron ore reserves and is the world's third largest producer and exporter of iron ore. Drawing from these reserves, Brazil is now in the top 20 of world steel exporters, thanks to China's voracious appetite for this important product.

For decades, Brazil's greatest resource deficiency was in energy. Coal and petroleum reserves were inadequate, threatening the country's continuing industrial development. Tropical hardwoods were widely used in place of fossil fuels in some manufacturing industries. Today, however, the country is in an enviable position with respect to energy: it is virtually self-sufficient, even meeting 90 percent of its needs for oil from domestic supplies. Large natural gas and petroleum reserves have been found offshore and in the remote central Amazon region. The proven offshore oil reserves are large, and some analysts suggest they may rival Saudi Arabia's. But drilling them will be a daunting task: they are very far offshore and are much deeper even than the Gulf of Mexico wells, and any accident would be difficult to contain. 522

The Amazon, Its Forest, and Its People

The Amazon is the world's greatest river by many measures. Although it is not the longest (its 3915 miles [6264 km] are surpassed in length by the Nile), by all other measures, the Amazon rules. It handles more volume than any other river. As much as one-fifth to one-fourth of all the world's available freshwater (excluding what is locked up in ice) is in the Amazon and its tributaries at any given moment. So much water flows from the mouth of the Amazon that it forces back the salty waters of the Atlantic as far as 200 miles (320 km) offshore. The river is so impressively large that early Portuguese explorers called it the **"River Sea."** Many creatures normally found only at sea have adapted themselves to its freshwater vastness, including dolphins, sharks, and rays.

The river drains the Amazon Basin, covering some 2.7 million square miles (4.4 million sq km), or about 10 times the size of Texas. About 60 percent of this basin is in Brazil, with the remainder sprawling into eight other countries. The basin is home to the world's largest remaining expanse of tropical rain forest and some of the world's most remote populations of indigenous peoples. It is also a region that promises, or seems to promise, prospects for economic development. The issues of pristine forest, Native American populations, and development are intertwined, often problematically. Here is a brief look at these problems.

The Amazon Basin rain forest is the world's largest storehouse of plant and animal species, the great majority of which have not yet been described by science. Ecologists argue that preservation of the rain forest's biodiversity is essential for ensuring the genetic variability that nature requires for change and evolution, and for helping ensure valuable future supplies 36 of food, medicines, and other resources for people. The forest also acts as a carbon sink to mitigate possible global warming due to excess greenhouse gas emissions. In economic terms, the amount of carbon that Amazon vegetation captures every year is worth over $13 billion in the global carbon trade. 44 Halting deforestation may be seen as the least expensive way to cut global carbon emissions; deforestation accounts for 20 percent of global carbon emissions. The challenge is to use mechanisms like the United Nations' REDD (Reducing Emissions from Deforestation and Forest Degradation) Program to persuasively make the case that conservation pays; the ecological services provided by the forests are enormously 44 valuable. The governments of Brazil and other Amazon Basin countries generally acknowledge the importance of the area's natural state and also argue that the area is so vast that portions of it can be reasonably developed to help advance their economies.

Since the 1970s, Brazil in particular has aggressively pursued development in five major areas of the Amazon: exploitation of iron ore and other minerals, resettlement of "excess" populations from Brazil's crowded southeast, farming and ranching, timber exploitation, and hydroelectric development. Each of these serves as an agent of deforestation. All are ultimately linked to Brazil's expanded road network in Amazonia. To begin its development of the Amazon, in the 1970s Brazil began construction of the **Trans-Amazon Highway** and its feeder system. The main line of the interoceanic highway runs roughly east to west from the Atlantic toward its ultimate, long-awaited destination: the Peruvian coast, finally reached in 2011. This road construction was part of an international project called the **Regional Initiative for the Infrastructure Integration of South America**. The final link crossing the Andes is intended to give Brazilian exporters an easier outlet to Asian markets. Principal feeder lines include the controversial BR-364 through the Brazilian states of Acre and Rondônia; BR-163, running north–south through the heart of the region; and BR-319, now under construction between the free-trade zone of Manaus and Rondônia. BR-319 is of particular concern to environmentalists because it runs through pristine forest to the "arc of deforestation" that was deforested as roads like BR-364 gave access to the forest (**•Figure 10.36**).

Here is how that arc was created: the first wave of settlers into Amazonia consisted mainly of these hardscrabble farmers and ranchers. Brazilian authorities advertised the Amazon as a "land without people for people without land." Brazil began transferring its landless poor from the overcrowded southeast of the country, up along the feeder roads into the wilderness of the Amazon where, with government cash subsidies in hand, they began a destructive cycle of land use. The process is a classic example of marginalization, with poverty fueling envi- 64 ronmental destruction.

• Figure 10.36 Deforestation in the Amazon Basin.

Road construction in Amazonia initiates a sequence of events, illustrated well by Highway BR-364 (seen in **•Figure 10. 37**). First, the main road is cut as a swath through the forest. Landless peasants, lured by cheap or free land, cut and burn the rain forest along the main road and the smaller lanes linked to it. Typical of farmers' experiences with tropical slash-and-burn agriculture elsewhere, they may be able to wrest only three to five years of corn, rice, or other crops from the soil before it is exhausted. They then move on to clear and 293 cultivate new lands—the typical pattern of shifting cultivation.

Tropical hardwoods are harvested and sold as well. About 40 percent of the wood cut from the Amazon is shipped overseas. China and the United States both import about 20 percent of Brazil's exported timber, and collectively the European Union nations take in 50 percent. In 2007, Brazil announced that it would open vast new tracts of the Amazon for large-scale logging but promised there would be strict monitoring to ensure that the resource was not overharvested, and that none of the 70 percent of the forest owned by the government would end up in private hands. Brazil also shifted some of the burden back to the consuming countries, insisting that they should buy timber only from these zones.

Brazil and other Amazon Basin countries have passed legislation to slow or halt illegal logging and other unlawful uses that contribute to the rain forest's destruction. Authorities in Brazil have slowed the illegal logging of valuable mahogany trees, but the problem has boomed in the Peruvian Amazon. At least 35 percent (and likely more) of the estimated 1.6 million cubic feet (45,000 cu m) of mahogany exported each year from Peru to the United States—where it becomes furniture, acoustic guitars, home decks, and coffins—is illegal. Despite US support for laws prohibiting the illegal logging of mahogany, enforcement has been lax, and legal and illegal mahogany is often imported to the United States in the same shipments.

If Amazonian lands used for farming and timber were abandoned and left to regrow, they would return to mature

Planet Observer/UIG/Getty Images

• **Figure 10.37** This sequence of satellite images shows the progression of Amazon forest reduction in Brazil's Rondônia state: the first shows intact forest in 1975; the second, Highway BR-364 recently cut in 1992; and the third, subsequent development along BR-364 and roads leading out from it.

tropical rain forest in a matter of decades. In Amazonia, however, the typical pattern is that large cattle operations move in to use the areas conveniently cleared for them by the farmers and loggers. As about 60 million cattle (one-third of Brazil's total cattle population) feed and tread over these Amazonian lands, the soils are exposed to excessive cycles of wetting and drying and to relentless bombardment by ultraviolet radiation. This turns the soil into a bricklike substance known as laterite, on which it is all but impossible for forest regrowth to occur. When ranchers have exhausted the land in this fashion, they move their cattle on to the next plots cleared by farmers. This **shifting ranching** follows the shifting cultivation; it is part of a cycle that is destructive to the rain forest.

293

The Amazon is gradually becoming urbanized in the process. Between 2000 and 2014, the population of Brazil's Amazon region grew by 33 percent to 17 million, the majority of these people living in cities and towns. There are now five cities in Amazonia with more than 300,000 residents—a threshold that is attracting Brazil's national retail chains. American-style shopping malls have begun sprouting in Amazonia.

These processes and the other uses of the forest, especially the boom in commercial soybean cultivation, have resulted since the 1960s in the removal of about 20 percent, or 1.6 million square miles, of the tropical rain forest cover of the Amazon Basin (see **Figure 10.36**). In recent years, the annual rate of deforestation in Brazil has slowed somewhat to about 2000 square miles (5000 sq km), or about the size of Delaware. National law in Brazil stipulates that 80 percent of every tract in the upper Amazon, and 50 percent in more developed regions, must remain forested, but there is little enforcement. The World Wildlife Fund projects that at current rates of deforestation, 55 percent of the Amazon forest will be gone by 2030. There is concern among ecologists that the Brazilian forest has reached a "tipping point," from where the vegetation will shift into a savanna biome with less biodiversity. Forest dieback has been accelerated recently by unprecedented droughts. These unusual dry spells are related to El Niño and to another weather phenomenon called the Atlantic dipole, which warms Atlantic waters. There were widespread forest fires in 2005 and again in 2010. A positive feedback loop is at work: with less foliage there is less moisture in the water cycle, making the local climate hotter and drier. Climate change also threatens indigenous cultures relying on hunting, gathering, and simple agriculture. Previously, their adaptation to environmental stress was to move, but that has become nearly impossible as other Amazonian interests stake claims to the forest.

At high environmental cost, and largely with Chinese financial assistance, more than 20 dams have been built or are scheduled for construction in the Amazon Basin to feed Brazil's appetite for electricity, almost 90 percent of which is produced by hydropower. The enormous Belo Monte Dam under construction on the Xingu River is opposed by a coalition of indigenous and environmental groups who argue that it will destroy pristine rain forest environment and that the 100,000 laborers and other outsiders it will attract will be detrimental to Indian cultures. The Indians themselves have taken action, as they are constitutionally allowed to do: in Brazil, Bolivia, and Peru, indigenous groups have the right of consultation on large infrastructure projects. In 2010, leaders of 13 tribes created what they called a "new tribe" of 2500 people to occupy the Belo Monte Dam site in an effort to stave off its development. The Indians' right of consultation held up the dam in Brazil's court system until 2012, when Brazil's Supreme Court ruled that indigenous tribes had been properly consulted before construction began. Work on the dam resumed.

Development in the Amazon has provoked conflicts with the region's indigenous populations on a number of other fronts. As Rob Walker described (see **page 464**), many indigenous groups, marginalized into areas seen originally as resource-poor, have suffered recently as gold and other riches have been discovered in their reservations. There is particular concern now about how fossil fuel production will affect the Kichwa, Achuar, Schuar, and other Native Americans. After 2004, 70 percent of Peru's lowland rain forest area, mostly inhabited by Native Americans, was zoned for oil and gas development, and indigenous advocates worry that exploration will introduce new diseases to the native populations. Several oil and gas pipelines have been built to transport energy from the Peruvian Amazon to coastal refineries and ports. There have been pipeline ruptures over the years that have caused serious contamination of local water supplies. In Ecuador, where oil already makes up half the country's exports by value, there are plans to double production, again mainly in areas where there are significant Native American populations.

In almost all cases, the Native Americans in the areas affected believe that little, if any, benefit will come to them from these developments. They are becoming activists, using a combination of threats against the oil interests and pleas for more benefits from the projects. Some of the native groups have powerful advocate allies abroad and have come to rely on the Internet to make their cause known; an example is the Kichwa people of Sarayaku (http://www.sarayaku.com), who live in one of the Ecuadorian oil blocks slated for development. Arguing that the neo-liberal agendas ruin national economies and have knock-on effects on local peoples, advocates of many Amazonian indigenous peoples have denounced existing and proposed free-trade agreements, including the Free Trade 472 Agreement of the Americas (FTAA).

Amazon rain forest destruction has become a *cause célèbre* for environmentalists around the world, who denounce the development efforts of Brazil and the other countries. Most of the region's governments and people see it differently, denouncing foreigners as hypocrites who have cut down their own forests and who are intent on stunting development of the rain forest countries. There are alternative development schemes, such as the commercial extraction of valuable exports like rubber and brazil nuts from intact tropical forests. These take place in areas like Brazil's **extraction reserves**, but they are very limited in economic impact relative to the conventional development enterprises in Amazonia, and their long-term prospects for growth are uncertain. Some Native American communities are developing their own ecotourist and cultural tourist enterprises, hoping both to demonstrate to governments that there are alternative economic uses for the rain forest areas and to gain valuable support and publicity from international visitors (• **Figure 10.38**). Sustainable development of the Amazon is elusive but, if achieved, would have enormous benefits for people and environments within and far beyond the forest biome.

The book continues, and ends, with the United States and Canada.

• **Figure 10.38** Are ecotourism and cultural tourism helping to protect indigenous people and natural environments in the Amazon? This community of Bora Native Americans in Peru's Amazon region promotes itself as a "genuine Indian village." Its residents wear native costume for visitors.

Study Guide

Summary

- Latin America lies south of the United States and includes Mexico, the Central American countries, the island nations of the Caribbean Sea, and all the countries of South America. It is called Latin America because of the post-15th-century dominance of Latin-based languages and the Roman Catholic Church. The term *Middle America* refers to Mexico, Central America, and the Caribbean.
- Latin American landscapes and livelihoods are correlated in altitudinal zones, which include *tierra caliente, tierra templada, tierra fría*, and *tierra helada*.
- Large areas of humid climates and biome types are in the lowlands of Latin America, with tropical rain forests occupying major river basins, especially in Brazil and Venezuela. Grasslands are found poleward of the tropics, and there is a zone of Mediterranean climate on the western coast of Chile.
- Dry climates and biomes, including desert and steppe, exist because of rain shadow and other factors. There are major desert regions on the west coast of South America and northern Mexico.
- Latin America's population is about 9 percent of the world's total, and its growth rate is above the world average. People are concentrated along two major geographic alignments. The largest is "the rim," or "the Rimland," a discontinuous ring around the margins of South America. The smaller, generally a highland, extends along a volcanic belt from central Mexico southward into Central America. Most countries have a recognizable core and hinterland.
- Earthquakes are a major threat to peoples living where tectonic plates meet, on both the Pacific and Atlantic sides of the region. A quake in 2010 leveled Haiti's capital, Port-au-Prince, and the country's urban geography is being remade to lessen impacts from future earthquakes.
- The public can contribute to prevention and reconstruction in hazard-prone areas of the world with crowdsourced and

georeferenced information. This chapter provides several opportunities to participate.

- Normally winds blow east to west and cause upwelling of cold and nutrient-rich water from the depths to the surface of the tropical eastern Pacific. But in El Niño conditions, winds reverse direction, blowing from west to east, suppressing the upwelling water and raising water surface temperatures, with pronounced environmental and economic consequences.
- Before the arrival of the Europeans in the late 1400s, Latin America had been home to advanced indigenous civilizations, notably the Aztecs, Maya, and Inca. There were many other Native American groups that did not develop urban livelihoods. All these cultures declined precipitously after Europeans arrived. Although some of the losses are associated with war and with mining and other land uses, the biggest killers were European diseases, especially smallpox.
- There are a number of "uncontacted tribes" in Amazonia presenting dilemmas to decision makers and social scientists about leaving them alone or intervening in ways that would benefit these Amerindians.
- The region's major ethnic groups today are Native Americans, mestizos, mulattoes, blacks, and Europeans. Political and economic systems still tend to favor people of European extraction, but non-Europeans have found increasing power through democratic political systems. Many reject the free-trade and capitalist systems that they believe favor ethnic Europeans.
- Latin America has been urbanizing rapidly, with more than two-thirds of its people now living in cities. Steady rural-to-urban migration continues, along with a steady stream of legal and illegal immigrants northward into the United States. The remittances of those workers are an important resource for Latin American economies.
- *Latifundia* and *minifundia* are the region's two major agricultural systems. The *minifundia* are small farms associated with subsistence agriculture. There is a strong plantation economy, especially in Central America, the Caribbean, and coastal Brazil. More than half of all export revenue for some countries comes from farm products, mainly grown on the *latifundia*, As in other LDCs, overreliance on a narrow range of exports makes these economies vulnerable.
- The fair trade movement raises prices for consumers of coffee, bananas, and other produce but helps ensure better wages for Latin American farmers.
- With land reform efforts faltering, ownership of farmland is concentrated in the hands of relatively few people, contributing to inequities that are a great problem in the region.
- Minerals have played a major role in the history, economic development, and trade networks of Latin America. The region's petroleum is of growing importance, especially in Mexico and Venezuela, and is a significant geopolitical concern for the US. The global commodities boom related especially to China's growth has helped Latin American economies in recent years.
- To reduce dependence on commodities exports, many countries after World War II embraced import substitute industrialization (ISI), introducing high tariffs on imported goods to promote industrialization while reducing imports of manufactured goods. With the debt crisis of the 1980s, they rejected ISI and moved to neo-liberalism, reducing tariffs on imports to stimulate consumer demand, lower consumer prices, and promote economic growth. The "Bolivarian socialist" economies of Venezuela and Cuba did not participate in these reforms.
- Three countries account for 80 percent of the region's manufacturing output: Brazil, Argentina, and Mexico.
- Free-trade associations (FTAs) have a strong role in the region's recent push away from commercial agriculture and raw materials and toward manufactured goods exported competitively into the world economy. The Latin American countries have formed or joined several FTAs. The largest in volume and value is the 1994 North American Free Trade Agreement (NAFTA) among Mexico, Canada, and the US. One of its goals is to promote economic growth in Latin America so that migration to the United States becomes less attractive to unemployed Latin Americans. Much of Mexico's economy is based on production in the *maquiladora* factories of the US border region. In the US and Canada, NAFTA has been beneficial to farmers, but not to factory workers, whereas the reverse is true in Mexico. NAFTA has been less beneficial to Mexico's mainly agricultural south, where farmers have a difficult time competing with subsidized US farmers. Mercosur is the leading FTA in South America.
- The Free Trade Area of the Americas (FTAA) FTA is facing opposition, especially from Brazil, which wants cuts in the subsidies of many agricultural products and steel that benefit American producers, but shuts out imports from Brazil.
- Tropical Latin America's proximity to temperate and affluent North America has helped make tourism one of the region's most important industries.
- US interests in Latin America dominate the region's geopolitical themes and issues. The Monroe Doctrine and its Roosevelt Corollary were designed to justify US activities and interventions in the region. Historical efforts have been directed at suppressing Communist or leftist interests in the region and at safeguarding passage through the Panama Canal. Modern interests focus on promoting trade, fighting drug trafficking, and guaranteeing secure access to oil in Venezuela and Colombia.
- Mexican cartels have acquired increasing control over drugs from source areas in South America and Mexico all the way to the retail trade on US streets.
- Mexico is the largest, most populated, and economically most developed of the Middle American nations. It is the largest Spanish-speaking country in the world.
- Panama's history over the past century has been shaped by US interests in the Panama Canal, which is now being enlarged. With Chinese funding, Nicaragua is building a transcontinental canal that will rival or complement the Panama Canal.
- In Venezuela, President Hugo Chávez and his successor Nicolás Maduro led a self-proclaimed "Bolivarian Revolution," ostensibly aimed at using the country's oil wealth to enhance the prospects for the poor. Anti-government protests accelerated

as the country's economy tanked. Along with Cuba, Venezuela has sought to blunt US influence in the region.

- The colonial Latin American city had a Spanish "gridiron" plan with a plaza and cathedral at its center. Typically the colonial city center remains at the center of the modern city.
- Brazil is the giant of South America. It is larger than the 48 lower United States but has more than 100 million fewer people. Settlement clusters along the coast, but Brazil is promoting settlement in the interior savanna and rain forest.
- Brazil's economy is the largest in South America. Booms and busts in commodities and a debt for development dating to the 1970s have been problematic. Beef and orange juice concentrate, along with soybeans for the Chinese market, are prominent in the agricultural export economy, sugarcane is grown mainly for Brazil's ethanol fuels, and recent oil finds will make up for the country's historic shortage of fossil fuels. Iron ore and steel are other important nonagricultural exports.
- Brazil has a large realm of tropical rain forest centered in the Amazon River system, which drains almost 3 million square miles (7.8 million sq km).
- The Amazon has a greater volume of water than any other river in the world. The Amazon Basin extends into nine countries, and about half of it lies in Brazil.
- Development of the Amazon region involves exploitation of iron ore and other minerals, resettlement of "excess" populations from Brazil's crowded southeast, farming and ranching, timber exploitation, and hydroelectric development. Each of these is an agent of deforestation. This development is controversial, in part because of the rich biodiversity of the rain forest and the forest's role in mitigating climate change. The Trans-Amazon Highway and its feeder roads are opening the Amazonian wilderness to development. Shifting cultivation is often followed by shifting ranching in a destructive land use cycle.
- Oil in both Ecuador and Peru is being extracted in the tropical lowlands draining into the Amazon River. Its development is controversial because many indigenous peoples live on lands slated for oil development, and they are generally opposed to the industry.

Review Questions

1. Where are the main areas of population concentration in Latin America?
2. What and where are the principal climate and biome types of Latin America?
3. What are the four elevation zones, known by their Spanish names? What human uses are generally associated with them?
4. Why was Port-au-Prince especially vulnerable to earthquake damage, and what plans are there to remake Haiti's urban geography?
5. How can you contribute spatial data to various humanitarian projects around the world, even if you have never made maps?
6. What causes El Niño? What effects are usually associated with it in particular regions?
7. What and where were the major Native American civilizations prior to the arrival of the Europeans?
8. What are the major pre-Columbian and modern languages of Latin America, and where are they spoken?
9. What are the principal faiths in Latin America?
10. What are the two main types of landholdings? What crops are associated with them? Why has land reform been an important issue in Latin America?
11. How are inequities in Latin America apparent in both rural and urban settings? What would help redress these imbalances, and what hopeful signs already exist?
12. What are the most important minerals exported from Latin America, and which are of particular interest to the US?
13. What is the Washington Consensus, and why have some Latin American countries pushed back against it?
14. What are the pros and cons of neo-liberalism and free trade for countries that have pursued these policies?
15. Which three countries account for 80 percent of manufactured goods in the region?
16. What are the region's most important existing and planned free-trade agreements? Why is the biggest, the proposed Free Trade Area of the Americas (FTAA) opposed by Brazil and other Mercosur countries?
17. What are "fair trade" products, and why do some consumers favor them despite their higher prices?
18. What drives the economies of towns on the Mexico-US border? What are energy *maquiladoras*? What resources do they use, where is their product exported, and why are they located where they are?
19. What considerations went into the construction of the Panama Canal? What steps are being taken to modernize the canal? What other country is building a transcontinental canal, and why?
20. What is the balloon effect, and what is its relationship to the "war on drugs"? How have Mexico's drug cartels capitalized on the drugs trade?
21. Which countries have been considered as "Bolivarian Socialist" and what are their views of the US?
22. What are the main components of the colonial Latin American city?
23. What are remittances? What role do they play in Latin American economies?
24. What attributes make Mexico a "pivotal country" in the world system and a significant power in its own right?
25. In what ways is Brazil the giant of South America? What is its economic situation in the region and globally? What roles do

soy and sugarcane production play in transforming ecosystems and in perpetuating economic inequities?

26. What exports from Brazil are particularly important today, and why?

27. How committed is Brazil to ethanol, and what interests do other countries have in this fuel source?

28. What is the Trans-Amazon Highway system? What role does it play in the region's development and deforestation?

29. What cycle of land use is particularly destructive to the Amazon rain forest? What is the "tipping point" scenario of environmental change in this forest biome?

30. Why are Native Americans of the Amazon Basin generally opposed to fossil fuel development there?

Key Terms + Concepts

alternative development (p. 479)
American Invasion (p. 482)
Amerindians (p. 460)
Aymara (p. 463)
Aztecs (p. 462)
barrio (p. 470)
Bay of Pigs invasion (p. 475)
Bolivarian Alliance for the Peoples of our America (ALBA) (p. 480)
"Bolivarian Revolution" (p. 482)
cartels (p. 478)
Chibcha (p. 462)
Creole (p. 468)
Creole languages (p. 463)
crop substitution (p. 479)
Cuban Americans (p. 476)
Cuban Missile Crisis (p. 475)
distance-decay relationship (p. 475)
El Niño (p. 458)
energy *maquiladoras* (p. 473)
ethanol (p. 484)
extraction reserve (p. 487)
fair trade movement (p. 474)
favela (p. 470)
Free Trade Area of the Americas (FTAA) (p. 472)
Guaraní (p. 463)
hacienda (p. 469)
hinterland (p. 449)
hurricane (p. 457)
Inca (p. 462)
Indians (p. 460)
informal sector (p. 474)
informal settlements (p. 459)
intellectual property rights (IPR) (p. 473)
land reform (p. 471)
latifundia (p. 469)
lock (p. 476)
MapGive (p. 460)
maquiladora (p. 472)
Maya (p. 461)
Mayan language family (p. 462)
megalanguage (p. 463)
Mercosur (p. 472)
mestizo (p. 468)
Mexican-American War (p. 482)
minifundia (p. 469)
Monroe Doctrine (p. 475)
mulatto (p. 468)
narcoterrorist organization (p. 477)
Native Americans (p. 460)
Nazca (p. 462)
neo-liberalism (p. 471)
North American Free Trade Agreement (NAFTA) (p. 472)
Organization of American States (OAS) (p. 473)
pampas (p. 454)
páramos (p. 457)
Plan Colombia (p. 478)
Plan Patriot (p. 477)
pre-Columbian times (p. 467)
Quechua (p. 463)
Regional Initiative for the Infrastructure Integration of South America (p. 485)
"River Sea" (p. 484)
Roosevelt Corollary (p. 475)
rules of origin (p. 472)
shantytown (p. 470)
shifting ranching (p. 486)
Southern Cone Common Market (Mercosur) (p. 472)
Teotihuacános (p. 462)
tierra caliente (hot country) (p. 456)
tierra fría (cold country) (p. 456)
tierra helada (frost country) (p. 456)
tierra templada (cool country) (p. 456)
Trans-Amazon Highway (p. 485)
tree line (p. 457)
United Fruit Company (UFC) (p. 469)
upper limit of agriculture (p. 457)
war on drugs (p. 477)
Washington Consensus (p. 480)

Notes

1. From http://www.goodreads.com/quotes/437387-be-brave-take-risks-nothing-can-substitute-experience.
2. Amy Wilentz, "The Dechoukaj This Time," *New York Times*, February 7, 2010, p. WK-12.
3. Quoted in Andrew C. Revkin, "In Quake-Threatened Cities, Quick Growth Invites Disaster," *New York Times*, February 25, 2010, p. A-1.
4. Association of American Geographers. *SmartBrief*, March 13 2014, http://www.aag.org/cs/pressroom/aag_smartbrief.
5. "Reducing Risk Before a Crisis, Nepal," *MapGive*, http://mapgive.state.gov/stories/reducing-risks.html.
6. Tomnod Loading Campaign, http://www.tomnod.com/FAQ.
7. https://www.planet.com/careers/#who.
8. Brian Blouet and Olwyn Blouet, *Latin America and the Caribbean: A Systematic and Regional Survey*, 7th ed. (Hoboken, NJ: Wiley, 2015; Kindle Edition), p. 122.
9. Quoted in Alfred M. Tozzer, "Landa's *Relación de las Cosas de Yucatán:* A Translation," in *Papers of the Peabody Museum of American Archaeology and Ethnology,* vol. 18 (Cambridge, MA: Harvard University Press, 1941), p. 78.
10. Bernard Comrie, Stephen Matthews, and Maria Polinsky, *The Atlas of Languages* (New York: Facts on File, 2003), p. 151.
11. This discussion of the colonial Latin American city draws from Blouet and Blouet op. cit., pp. 64–65, 163–164.
12. Simon Collier, Harold Blakemore, and Thomas E. Skidmore, eds., *The Cambridge Encyclopedia of Latin America and the Caribbean* (Cambridge: Cambridge University Press, 1985), p. 159.
13. Water consumption is a major issue in production of these commercial crops. Issues of blue water scarcity, water footprint, and virtual water are discussed in Mesfin M. Mekonnen et al., "Sustainability, Efficiency and Equitability of Water Consumption and Pollution in Latin America and the Caribbean," Sustainability, 2015, http://waterfootprint.org/media/downloads/Mekonnen-et-al-2015-WFA-LAC.pdf.
14. Ibid., p. 106.

15. "What China's Economic Slowdown Means for Latin America," *Stratfor*, April 23, 2014, https://www.stratfor.com/analysis/what-chinas-economic-slowdown-means-latin-america.
16. Friends of the MST, http://www.mstbrazil.org/content/need-basis-agrarian-reform.
17. Op. cit., Blouet and Blouet, p. 124.
18. Op. cit. p. 139.
19. Ibid., p. 112.
20. Op. cit., Blouet and Blouet p. 129.
21. This discussion draws on Blouet and Blouet, op. cit., p. 115.
22. Morris Beschloss, "Mexico Cancels PEMEX Oil Industry's Nationalization, Reverses Shrinkage." *The Desert Sun*, September 30, 2014, http://www.desertsun.com/story/money/industries/morrisbeschlosseconomics/2014/09/30/mexico-cancels-pemex-oil-industrys-nationalization-reverses-shrinkage/16480203/.
23. Blouet and Blouet, op. cit., p. 113.
24. Blouet and Blouet, op. cit., p. 79.
25. Ibid., p. 137.
26. Although member states of a *free-trade agreement* remove tariffs on goods moving between them, members have no policies to impose external tariffs, do not share common economic policies, and have no restrictions on trade with non-members. Unlike an FTA, *common market* members have common external tariffs to discriminate against imports from nonmembers, and members agree to allow free trade between themselves. Unlike a common market, a *customs union* generally does not allow free movement of capital and labor among member countries.
27. Mercosur faces internal opposition from Argentina, which is seeking trade protectionism to offset debt fueled by rising energy imports and outflows of foreign capital high levels of capital flight. The Argentine peso had been linked in value to the US dollar, making the peso a strong and stable currency. This made imports into Argentina cheap, but its exports, especially to neighboring Brazil (where the currency had lost value), became expensive. Brazil could not afford Argentinian vehicles and agricultural equipment. Argentine exports fell, and its currency collapsed. Unemployment and homelessness ensued. Brazil, meanwhile, wants to increase its exports and is urging Argentina to approve a free-trade deal between Mercosur and the European Union. Although a customs union, Mercosur allows the free movement of labor and capital across the member nations. See "Trade Tensions in the Southern Cone," *Stratfor*, March 28, 2012; and Blouet and Blouet, op. cit., p. 131.
28. Ibid., pp. 87, 161.
29. Ibid., p. 133.
30. Ibid., p. 133, 135
31. "The Geopolitics of U.S.-Cuba Relations," *Stratfor*, December 23, 2014, https://www.stratfor.com/weekly/geopolitics-us-cuba-relations.
32. "Nicaragua Canal Route: Atlantic-Pacific Link Unveiled," *BBC News*, July 8, 2014, http://www.bbc.com/news/world-latin-america-28206683.
33. "Nicaragua Canal Protest: Thousands Oppose Atlantic-Pacific Plan," *BBC News*, June 14, 2015, http://www.bbc.com/news/world-latin-america-33125526.
34. Amnesty International, "US Policy in Colombia," http://www.amnestyusa.org/our-work/countries/americas/colombia/us-policy-in-colombia.
35. "Changes in Cocaine Smuggling Tactics," *Stratfor*, March 11, 2014, https://www.stratfor.com/image/changes-cocaine-smuggling-tactics; "Peru Overtakes Colombia as Top Cocaine Producer," *NBC News*, July 31, 2012, http://worldnews.nbcnews.com/_news/2012/07/31/13045253-us-peru-overtakes-colombia-as-top-cocaine-producer?lite; "Bolivia's Burgeoning Cocaine Production," *Stratfor*, August 16, 2012, https://www.stratfor.com/analysis/bolivias-burgeoning-cocaine-production; and "Cocaine Complicates Peace Talks in Colombia," *Stratfor*, March 15, 2015, https://www.stratfor.com/analysis/cocaine-complicates-peace-talks-colombia.
36. Most of this discussion of the cartels is from these sources: Tristan Reed, "Mexico's Drug War: A New Way to Think about Mexican Organized Crime," *Stratfor*, January 15, 2015, https://www.stratfor.com/weekly/mexicos-drug-war-new-way-think-about-mexican-organized-crime; Scott Stewart, "Mexico's Cartels and the Economics of Cocaine," *Stratfor*, January 3, 2013, https://www.stratfor.com/weekly/mexicos-cartels-and-economics-cocaine; and Scott Stewart, "Agenda: Mexican Drug Cartels," *Stratfor*, January 3, 2013, https://www.stratfor.com/analysis/20110415-agenda-mexican-drug-cartels.
37. Steve H. Hanke, "The World Misery Index: 108 Countries," Cato Institute, January 22, 2015, http://www.cato.org/blog/world-misery-index-108-countries.
38. Girish Gupta, "Venezuela Is Slowly Coming Apart—and President Nicolas Maduro May Pay the Price," *Time*, February 27, 2015, http://time.com/3726757/venezeula-economic-collapse-nicolas-maduro/.

Global Geoscience Watch

Go to the GREENR database and search for information related to the proposed Nicaragua canal. The text mentions the internal political opposition to the project and environmental concerns voiced from many sides. Use the GREENR database to research the political and environmental concerns in more detail. Write at least two paragraphs that describe the internal and external obstacles construction of the canal faces.

Online Resources

For access to MindTap and additional study materials visit www.cengagebrain.com. Read your textbook, take notes, complete activities, take practice quizzes and more.

Above: Navajo horseman in the Monument Valley Navajo Tribal Park, Utah/Arizona borderland. Joe Hobbs
Left: North America from space, at night. Note the populous eastern half of the region and the extensive wilderness of the west and north. C. Mayhew & R. Simmon (NASA/GSFC), NOAA/ NGDC, DMSP Digital Archive

11 The United States and Canada

The genius of the United States is not best or most in its executives or legislatures, nor in its ambassadors or authors or colleges, or churches, or parlors, nor even in its newspapers or inventors, but always most in the common people.

—WALT WHITMAN[1]

The United States and Canada, together with Mexico and Central America, make up the continent of North America. Mexico and Central America are so linked with South America in culture, language, and tradition that they are best classified within the separate region of Latin America. The United States and Canada as a culture region is sometimes called "Anglo America," emphasizing their British colonial origins and their contrasts with Latin America. But that term is increasingly irrelevant today, as multiculturalism has become a hallmark of both societies. This chapter depicts a complex cultural mosaic overlaying an essentially British political, cultural, and economic foundation.

The island of Greenland is politically part of Denmark but is geologically part of North America, so it is also described briefly as part of this region, as is the small Atlantic island of Bermuda, a British possession (**• Figures 11.1** and **11.2**; **• Table 11.1**).

Chapter Outline

Chapter Objectives

This chapter will enable you to

- Appreciate the wide range of indigenous adaptations and ways of life in this region's diverse environments.
- Recognize Canada and the United States as countries shaped mainly by British influences but also by a wide range of foreign immigrant cultures, creating a complex ethnic geography.
- Understand how the United States and Canada acquired their vast land empires.
- View the prosperity of the United States and Canada in part as products of their great natural resource wealth.
- Trace the rise of the United States to a position of global economic and political supremacy and to recent rivalries from other powers.
- Follow the decline of traditional heavy industries in the process of deindustrialization and their replacement by high-technology and service industries.
- Evaluate the changing settlement patterns of the United States, including the depopulation of the Great Plains, the decline of small towns, and the movement of people back into cities.

11.1 Area and Population

Canada is the world's second largest country, after Russia. Its 3.85 million square miles (9.98 million sq km) makes it slightly larger in area than the United States' 3.79 million square miles (9.82 million sq km). (See **Figure 4.5** for a regional comparison with Europe.) Canada had a population of 35 million in 2014, compared with 320 million in the United States. Together, the two countries have 5 percent of the world's population, on 13 percent of its land surface.

Both the US and Canada were rural, agricultural lands until about 1920, when cities overtook the rural areas in total population. Today, this is the most urbanized world region; about 80 percent of both peoples reside in built-up areas. The largest cities in the region are listed in **•Table 11.2**.

The heaviest concentration of people in North America is found in a belt stretching from Virginia north into Maine that is known variously as the Northeastern Seaboard, the Northeast Corridor, and **megalopolis**. Five of the 20 largest cities in the United States—Washington, Baltimore, Philadelphia, New York, and Boston—are located within this belt, as are numerous mid-sized cities. **•Figure 11.3** depicts the population distributions for the US and Canada as well as Greenland.

The eastern half of the United States is heavily populated, with only a few exceptions; the average population density here is about 200 people per square mile (500 per sq km). Southeastern Canada is similarly densely populated, especially between the Great Lakes of Huron, Ontario, and Erie, and along the St. Lawrence River. This area is sometimes called the "Québec City-Windsor Corridor," named for its easternmost and westernmost reaches. This belt includes Toronto and Montreal, Canada's largest cities, as well as the national capital of Ottawa.

North and west of this area, the population thins out rapidly. The 25-inch (635 mm) isohyet (line of equal rainfall) that runs roughly from Winnipeg south to the Rio Grande marks the general divide between the heavily populated east and lightly populated west in both countries (the western US has a population density of just 60 people per square mile [150 per sq km]). On the plains and the dry intermontane areas, isolated pockets of urban development (like Denver, Phoenix, Calgary, and Edmonton) are separated by large expanses of sparsely populated land. West of the Cascades and the Sierra Nevada, population density rises again, although the Coast Mountains, extending right to the Pacific, hem in development on Canada's west coast north of Vancouver.

Canada's high, cold latitudes have a decisive impact on population distribution. About 90 percent of Canadians live within 100 miles of the US border. The majority of Canada's landmass has a population density of 0.4 people per square mile (1 per sq km) or less, as its vast northern areas are mostly uninhabited. Alaska too is quite empty, with about 1 person per square mile. With its vast icecap, Greenland is even more sparsely populated, with just 55,000 people living in an area of 836,000 square miles (2.17 million sq km), or 0.1 person per square mile (0.02 per sq km). Bermuda has 65,000 people in an area of just 20 square miles (52 sq km)—10,000 more people than Greenland, on an island over 4000 times smaller.

From the beginning of European settlement until the 1970s, this region's population grew rapidly. By 1800, about 5 million people lived in the United States and 500,000 in Canada. Birth rates had been high in both countries until the 1930s and 1940s (the Great Depression and World War II). After the war, birth rates surged again for about 20 years, in the so-called **"baby boom."** After 1964, birth rates plummeted; by 1970, they were lower than at any time in the past, and by 1995 the birth rate was barely half of what it had been in 1948. Birth rates have held relatively steady, but with a gradual downward trend, since then. Excluding immigration, population growth rates of the United States and Canada are now low (0.4 percent each).

72

• Figure 11.1 Physical geography of the United States, Canada, and Greenland.

Migration into North America

The burst of refugee migration into Europe aside, the US is the world's only MDC experiencing significant and sustained population growth. This is due mainly to immigration rather than natural increase; the US takes in 20 percent of the world's migrants each year. Both the US and Canada are best appreciated as nations of immigrants and of continuing immigration. Both countries have become very ethnically diverse, especially since the 1950s, when European migration to North America slowed. In Canada, 23 percent of the population is not solely of European ancestry; in the US that figure is 37 percent.

Immigration rates to the US have varied throughout its history. The first large wave of immigrants arrived between 1845 and 1860, mostly from Britain, Ireland, and Germany. The Civil War briefly stopped the flow of migrants to the US, but immigration surged during a sustained period from the 1880s until 1930, with a brief interruption during World War I; during this time over 25 million Europeans settled in the United States. Germans made up a quarter of that total, many of them settling

• **Figure 11.2** Political geography of the United States, Canada, and Greenland.

in the Midwest alongside Scandinavians and Eastern Europeans. The big cities of the Northeast took in large numbers of Italians, Poles, Jews, and Irish. We will explore America's ethnic origins, including the unwilling immigration of slaves as the ancestors of most African Americans, in more detail in Section 11.3.

During the Great Depression and World War II (1930–1945), immigration to the United States nearly came to a halt; fewer people moved to America during those 15 years than had come in the single year of 1924. A quota system was put in place in 1924 that gave preference to immigrants from northwestern European countries, restricted numbers from southern and eastern Europe, and essentially barred Asian immigration. When the nationality-based quotas were lifted in 1965, the nature of immigration to the US changed greatly. The numbers of European immigrants declined substantially, and those coming from Latin America and Asia surged. Legal immigration has remained relatively constant since the late 1980s, with about 800,000 to 1 million people per year.

Table 11.1 North America: Basic Data

Political Unit	Area (sq mi, thousands)	Area (sq km, thousands)	Population (millions)	Rate of Natural Increase (%)	Urban Population (%)	Population Under Age 15 (%)	Agriculture Workers (%)	Per Capita GDP (PPP) ($US)	GDP ($US, billions)	Trade Balance ($US, billions)	Oil Production (million bbl/day)	Literacy Female (%)	Literacy Male (%)	HDI
North America	**8403.8**	**21765.8**	**353.2**	**0.4**	**81**	**19**	**2**	**51,800**	**19,046**	**—**	**15.3**	**99**	**99**	**0.913**
Bermuda (U.K.)	*	*	*	0.5	100	17	1	86,000	5	—	—	99	99	—
Canada	3849.7	9970.7	35.5	0.4	80	16	2	43,100	1,579	-17	4.0	99	99	0.902
Greenland (Denmark)	836.3	2166.0	*	0.0	86	21	14	38,400	2	—	—	—	—	—
United States	3717.8	9629.1	317.7	0.4	81	19	2	52,800	17,460	-724	11.3	99	99	0.914

* Less than 0.1. — Data not available or not applicable.
Sources: World Population Data Sheet, Population Reference Bureau, 2014; Human Development Report, United Nations, 2014; World Factbook, CIA, 2014.

Table 11.2 Largest Cities in the United States and Canada

City	Population
New York, New York	20.1
Los Angeles, California	13.2
Chicago, Illinois	9.5
Dallas-Ft. Worth, Texas	6.9
Houston, Texas	6.5
Philadelphia, Pennsylvania	6
Washington, D.C.	6
Miami, Florida	5.9
Atlanta, Georgia	5.6
Toronto, Ontario	5.6
Boston, Massachusetts	4.7
San Francisco, California	4.6
Phoenix, Arizona	4.5
Riverside-San Bernardino, California	4.4
Detroit, Michigan	4.2
Montreal, Quebec	3.8
Seattle, Washington	3.6
Minneapolis-St. Paul, Minnesota	3.5
San Diego, California	3.2
Tampa-St. Petersburg, Florida	2.9
St. Louis, Missouri	2.8
Baltimore, Maryland	2.7
Denver, Colorado	2.7

Population in millions.
Source: US Census Bureau, 2015; Statistics Canada, 2015.

Immigration to Canada was similar to that of the US. European countries supplied the bulk of Canadian immigrants from the 1800s until the 1960s, with Britain, Ireland, and Germany leading the way. Like the US, Canada discriminated against non-European immigrants; for example, Chinese migrants were banned until 1947. That changed in 1967 when a "points" system was introduced that ignored national origin and instead gave preference to those with education and a skill set that would allow them to find jobs in the country. Unlike the US, Canada has a quota for immigrants, typically about 250,000 per year.

Countries of immigrant origin differ somewhat between the US and Canada. Half of all Canadian immigrants originate in South and East Asia, whereas only a quarter of US immigrants do. Unlike the US, Canada receives few migrants from Latin America. Nearly 30 percent of immigrants to the US are from Mexico, but less than 2 percent of Canadian immigrants are Mexican. In Figure 3.22 on page 76, you can see that the world's largest migration flow between individual countries is the steady stream from Mexico to the US.

What to Do About Illegal Immigration?

Issues of the status and future of illegal immigrants are contentious in the United States, but less so in Canada, where aversion to immigrants in general is rather muted. Numbers of illegal immigrants are much smaller in Canada than in the US, totaling only about 100,000. The number of illegal immigrants (also referred to as "undocumented immigrants" and "illegal aliens") living in the US was estimated by the Department of Homeland Security (DHS) at 11.3 million in 2013, down from the peak of 12.2 million in 2007. Just over half of the illegal immigrants are from Mexico, and one-quarter originate from elsewhere in Latin America and the Caribbean.

Many American citizens fear that an unstoppable tide of poor immigrants will take their jobs and bleed social and other

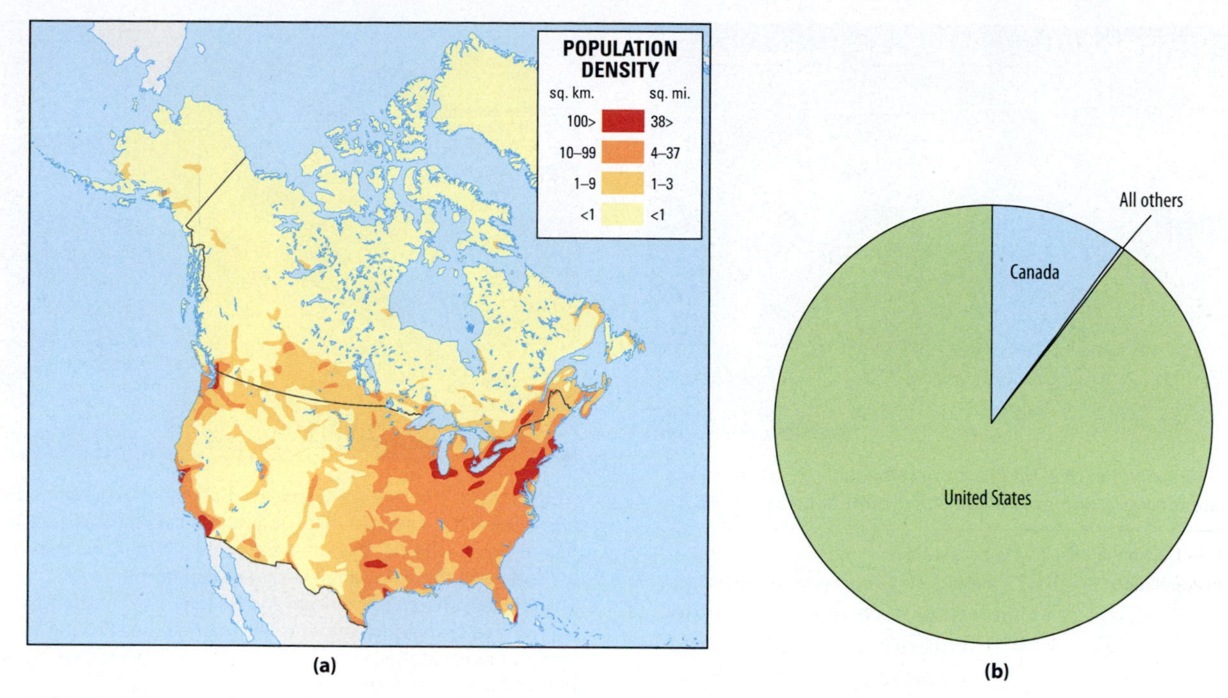

• **Figure 11.3** **(a)** Population distribution and **(b)** pie chart of the United States, Canada, and Greenland.

services. Others, particularly in the business community, argue that the low-wage immigrants who are undocumented workers are vital for the American economy, generally taking jobs shunned by most Americans while also contributing to the economy through their purchases (• **Figure 11.4**). Although they burden schools and other social services, most undocumented workers also pay federal taxes (a price they pay for using falsified Social Security and taxpayer ID cards) but are ineligible to receive Social Security benefits.

About 5 percent of the US workforce is classified as illegal, but contrary to stereotypes, most earn at least minimum wage, and many work outside agriculture, construction, and other labor-intensive industries. One-third own their own homes, and two-thirds are proficient in English.

• **Figure 11.4** In the US, Mexican farmhands with both legal and illegal status are essential where crops must be tended by hand. This field is in central California.

Joe Hobbs

The number of successful illegal entries to the US has been dropping steadily. In 2000, about 850,000 illegal migrants crossing the border were not detected; in 2014, that number was estimated at 300,000. How do they get into the US? Nearly half arrive legally on a temporary visa or from a guest worker program, and then fail to return home once their permit has expired. Others pay $10,000 or more to be smuggled by a "coyote" or "chicken rancher" across weak points in the US-Mexico border defenses. The majority of undocumented immigrants caught within the United States are returned to the Mexican side of the border, without prosecution. However, despite widely held (and deeply politicized) views, net migration of Mexicans into the US was near zero between 2010 and 2015; that is, the number of Mexicans leaving the US was about the same as those entering the US.

In 2014, an all-out immigration crisis occurred when about 60,000 migrants, mainly women and children from Honduras, El Salvador, and Guatemala, simply waded across the Rio Grande into the hands of immigration authorities in the erroneous belief that they would be permitted to stay in the US if they were able to endure their odysseys (• **Figure 11.5**). Most of the children were placed with sponsors or relatives across the US as their cases worked their way through immigration courts.

The United States does not have a firm or effective policy for managing illegal immigration. Passions run high over the immigration issue between those who support legalizing some illegal immigrants according to various criteria (a position known as **amnesty**) and those who favor strong methods of detecting and deporting illegal immigrants. The US Congress has often deadlocked over these issues and has not passed a comprehensive immigration bill since 1996. This has

AP Images/Eduardo Verdugo

• **Figure 11.5** Destination USA: Migrants from Honduras, El Salvador, and Guatemala pass through southern Mexico atop a northbound train, 2014.

Joe Hobbs

• **Figure 11.6** Did you know that any baby born in the United States, regardless of the circumstances, is automatically a US citizen? There is debate about whether this constitutional right (through the Fourteenth Amendment) should be amended to discourage immigration. Young Thai Nguyen Khang is American, and his parents Thai Thong and Doan Nguyen are Vietnamese students who earned their PhDs in architecture and economics at the University of Missouri. They are looking for permanent employment in the US.

led smaller jurisdictions to pass their own immigration laws. Arizona passed a law in 2010 requiring police to determine the immigration status of arrested persons having a "reasonable suspicion" of being in the US illegally. Conversely, over 30 large cities across the country have designated themselves **sanctuary cities**, where authorities will not cooperate with federal government efforts to identify illegal immigrants. Both the Arizona bill and sanctuary city policies have found popular support. These conflicting signals about immigration have led observers to describe US immigration policy as "schizophrenic"; as an exasperated US Border Patrol agent put it, "America loves illegal immigrants but hates illegal immigration."[2]

Measures are being taken to extend physical and other barriers that would deter new arrivals, including terrorists seeking easy entry into the United States. There are 18,000 Border Patrol agents, a huge increase from the 3500 of the mid-1990s. In the **Secure Fence Act of 2006**, the US Congress passed legislation calling for lengthening the existing 15-foot (4.5-m) border fences to 700 miles (1120 km). After 600 miles had been built, studies showed the number of border apprehensions had decreased to the lowest levels since the 1970s; however, critics charged that immigrants had simply found new ways over and around the physical barriers.

Some wealthy Chinese, Taiwanese, and South Korean parents have come up with an extraordinary strategy to give their children an advantage right out of the gate: have their babies in America. The Fourteenth Amendment to the US Constitution awards citizenship to any child born in the country (•**Figure 11.6**). In the practice of **birth tourism**, an expectant mother flies to the US explicitly for the purpose of having her baby (pejoratively called an **"anchor baby"**). The trip can cost thousands of dollars for travel and hospital bills, but parents expect the investment to be repaid many times over as the child is more likely to get into a good college and then to secure a well-paying job. Newborns of Latin Americans and others of poorer means also earn "birthright citizenship." In addition, at age 21 the citizen can petition for a grant of permanent residence in the US for her or his parents.

11.2 Physical Geography and Human Adaptations

The natural environments of the United States and Canada are remarkably diverse and include some of the most spectacular wild landscapes on the planet. They have presented people with a vast array of opportunities for land use and settlement.

Physical Geography of North America

North America's physiographic regions are depicted in •**Figure 11.7**. The ancient geological core of North America, with rocks up to 3 billion years old, is called the **Canadian Shield**, covering roughly the area from Nunavut south to Minnesota and northeast to Labrador—more than half of Canada's entire land area. The Canadian Shield was scoured by glaciers until about 10,000 years ago. The ice tore away the topsoil, leaving this huge, largely flat area unsuited for agriculture. Today, the area is dotted with many large lakes and thousands of smaller ones. The Canadian Shield surrounds Hudson Bay, which was the center of the ice sheets covering North America during the last glacial maximum. The immense weight of the ice pushed the land underneath it downward, and since the glaciers melted this land has been rising back up at a rate of about half an inch (1 cm) a year. This geological process, called isostatic rebound, is the same one affecting northern Europe.

Melting glaciers also created the **Great Lakes**, which together contain one-fifth of the world's freshwater. The Great Lakes are comprised of five lakes and four hydrological units. Lakes Michigan and Huron, despite their different names, are actually a single lake, and their water levels are the same (•**Figure 11.8**). Treated as a unit, Lake Michigan-Huron is the

• **Figure 11.7** Physiographic regions of North America.

Joe Hobbs

• **Figure 11.9** This is Horseshoe Falls, one of the two curtains of water making up Niagara Falls. This view is from the Canadian shore; the international boundary lies in the middle of the river.

world's largest freshwater lake by area; if considered separately, Lake Superior is the largest. The Great Lakes drain into the St. Lawrence River, spilling over the 167-foot (51-m) Niagara Falls between Lakes Erie and Ontario along the way (•**Figure 11.9**). The St. Lawrence Seaway is a system of locks and canals allowing ocean-going vessels to transit between the Great Lakes and the Atlantic Ocean. In the absence of navigable rivers this route has served as Canada's principal maritime shipping lane.

Southeast of the Canadian Shield lie the Appalachian Mountains and associated highlands (•**Figure 11.10**), with elevations up to 6684 feet (2037 m). Immediately east of the Appalachians lies the Piedmont ("foothills"), an area with a large population and good soils for farming. The Piedmont is fairly narrow, and its eastern edge is marked by the **fall line** (see Figure 11.7), a natural boundary where rapids and waterfalls commonly mark the "head of navigation" (the farthest point above the mouth of a river that can be navigated by ships) on many rivers; several important cities including Columbia, South Carolina; Richmond, Virginia; Washington, DC; and Philadelphia, Pennsylvania, are located along the fall line (see Figure 11.7).

East of the fall line is the Atlantic Coastal Plain, which is continuous with the Gulf Coastal Plain to the south. Forming a large crescent from eastern Texas up to the New Jersey shore, the coastal plains are characterized by a generally flat topography with sandy, largely infertile soils interspersed with more agriculturally productive river valleys. There are many important wetland areas along the coast and inland, such as the Great Dismal Swamp, the Okefenokee, and the Everglades (•**Figure 11.11**). North of the Gulf Coastal Plain lie the Interior Highlands, comprised of the Ozark Plateau and the Ouachita Mountains. This is the largest area of uplands between the

Joe Hobbs

• **Figure 11.8** A cargo vessel enters Lake Huron after transiting from Lake Erie through the St. Claire River. This is the American shore; the Canadian shore is just behind the ship.

Sean Pavone/Shutterstock.com

• **Figure 11.10** Autumn comes to the Appalachians in Smoky Mountains National Park, Tennessee.

• **Figure 11.11** Mangrove and other wetland species thrive in the estuary habitat of Florida's Everglades.

Art Wolfe Stock/AGE Fotostock

Appalachians to the east and the Rocky Mountains to the west. Despite the vernacular name "mountains," their highest elevation is only 2753 feet (839 m).

The Interior Plains of North America make up a vast, roughly triangular-shaped area stretching from Texas to Ohio to northern Alberta. Most of this area is in the watershed of the Mississippi River and its many tributaries. The prairies of the eastern sections, once dominated by long grasses with scattered wooded areas, have been largely transformed into built-up and agricultural land; only about 10 percent of the original ecosystem is intact. The western third of the region is called the Great Plains, a semiarid, mostly flat steppe dominated by short grasses. The Great Plains are higher in elevation than many of the peaks of the Appalachian Mountains; Denver, Colorado, on the far western edge of the steppe is known as the "Mile High City" as its elevation is 5280 feet (1609 m). A small, rugged area that includes the Black Hills (site of Mount Rushmore; •**Figure 11.12**) and the colorful region known as the Badlands disrupts the mostly level terrain of the western plains in the western Dakotas and eastern Montana.

The Great Plains end abruptly to the west at the Rocky Mountains, the highest area of a long chain of mountain ranges (collectively called the "Eastern Cordillera;" see **Figure 11.7**) stretching from Alaska to New Mexico. High and rugged with few easy passes, the Rockies (•**Figure 11.13**) are the source of many major rivers, including the Rio Grande, Colorado, Missouri, Fraser, Saskatchewan, and Columbia. The Continental Divide, an imaginary line separating the watersheds of the Atlantic and Arctic to the east, and the Pacific to the west, runs through the Rockies. The tallest mountain in the Rockies is Mount Elbert in Colorado at 14,440 feet (4401 m).

West of the Rockies is a geologically complex region of alternating wide valleys and low mountain ranges. This area is often called the "basin and range country." Rainfall is blocked by tall mountains to the east and west, creating rain shadow effects of arid conditions in the north and true desert in the south. This is the driest area of North America, but it is also home to the continent's two deepest lakes, Lake Tahoe and Crater Lake. Many of the region's plateaus are carved by canyons and dry valleys; the Grand Canyon (5000 ft/1500 m deep; •**Figure 11.14**) and the lowest point on the continent, Death Valley (282 ft/86 m below sea level), are located in this region. A prominent feature of the Great Basin is the Great Salt Lake in Utah, analogous to the Dead Sea with about 20 percent salinity. The Great Salt Lake is a remnant of the enormous, prehistoric Lake Bonneville, which all but dried up 14,000 years ago. Some of the ancient lake's shorelines can still be seen etched into the sides of mountains in the Wasatch Range.

A series of mountain ranges, known collectively as the Western Cordillera, parallel the Pacific Coast from Alaska southward to California. These ranges—including the Alaska Range, the Coast Mountains, the Cascades, the Sierra Nevada, and the Coast Ranges—are tall, rugged, and permanently glaciated in places. The tallest mountains in North

Joe Hobbs

• **Figure 11.12** The Black Hills are sacred to many Native American groups. While nearby Mt. Rushmore was being sculpted, Indian chiefs asked the sculptor Korczak Ziolkowski to carve an image of the Oglala Lakota warrior Crazy Horse in the Black Hills. Work began in 1948. When completed, this will be the largest sculpture in the world, exceeding the Mother Russia statue depicted on **page 171**.

Greg Hobbs

• **Figure 11.13** Autumn comes to the Rockies' Kenosha Pass in Colorado.

Joe Hobbs

• **Figure 11.14** The Grand Canyon. The Visitor Center complex on the South Rim is visible in the lower portion of the image. The muddied orange waters of the Colorado River can be seen in several places.

America (Denali, 20,320 ft/6194 m); in Canada (Mount Logan, 19,551 ft/5959 m); and in the contiguous United States (Mount Whitney, 14,505 ft/4421 m) are located in these ranges. Some of the ranges are separated by wide, fertile valleys that are good for agriculture, such as Oregon's Willamette Valley, California's Central Valley, and Alaska's Matanuska-Susitna Valley. In some cases these soils have been enriched by volcanic ash. It is a geographic twist of fate that some of America's most productive and most desirable places to live are in the crosshairs of deadly natural hazards (see Geography of Natural Hazards, **page 502**).

Stretching across northern Alaska and into the Canadian territories is the Arctic Coastal Plain, a flat, marshy area mostly underlain with a thick layer of permafrost. The Canadian Arctic Archipelago, consisting of over 30,000 islands, lies to the north of the mainland. East of the archipelago across the Davis Strait is Greenland, the world's largest island. A tiny island named Oodaaq off the Greenland coast is the world's northernmost point of land (at 83.66 degrees N), 450 miles (725 km) from the North Pole. About 80 percent of Greenland is covered in ice up to 10,500 ft (3200 m) thick. As we have seen, there is great concern about this ice as well as Antarctic ice melting and contributing to rising sea levels.

158

41, 395, 531

Climates and Land Uses

The United States, including Alaska and Hawaii, is more climatically varied than any other country in the world; all 11 climate types (plus undifferentiated highland) can be found within its borders (**•Figure 11.15a, b**). Canada's northern location

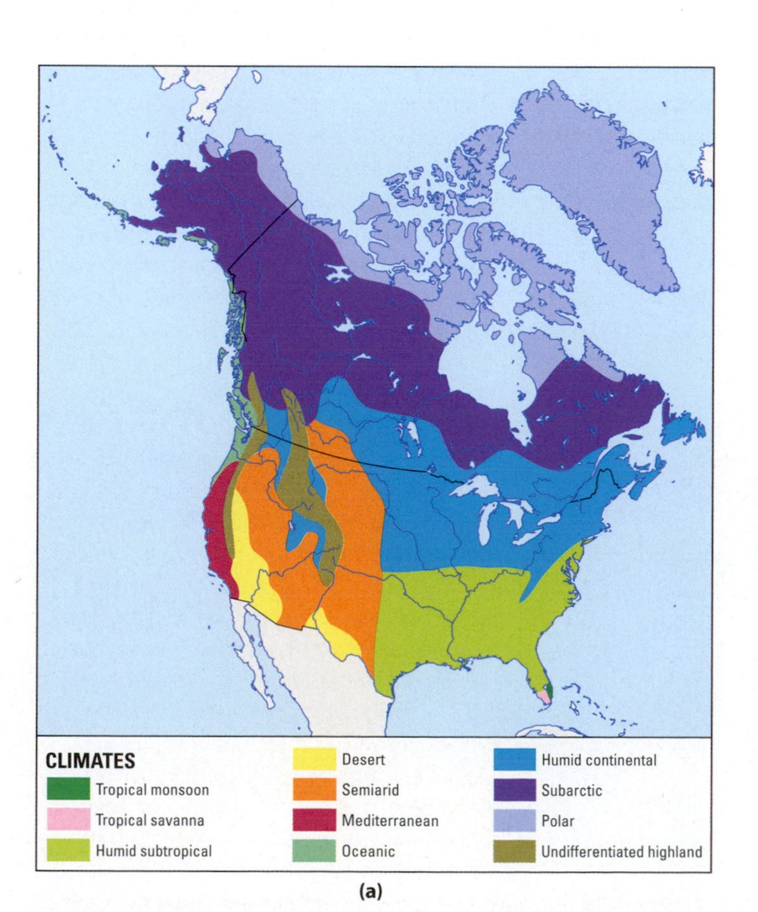

(a)

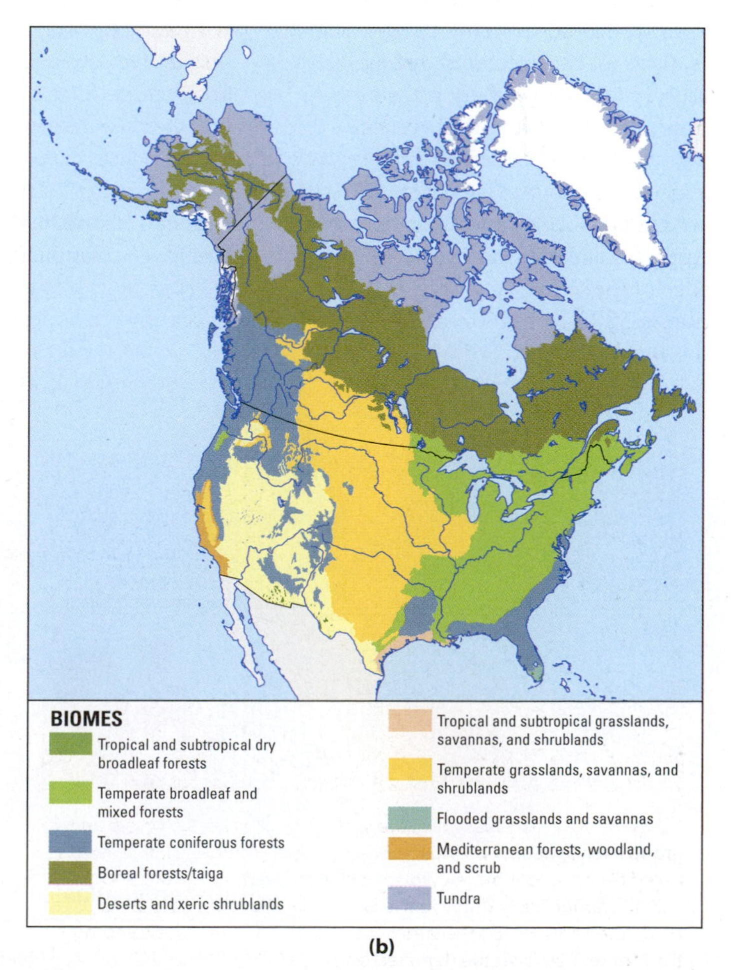

(b)

• **Figure 11.15** **(a)** Climates and **(b)** biomes of the United States, Canada, and Greenland.

Geography of Natural Hazards | Nature's Wrath in the United States

With its great size, its location astride a large range of latitudes, and its position atop a number of geologic fault lines, the United States bears the brunt of numerous and deadly natural hazards (**•Figure 11.A**).

The western edge of the continent lies on the eastern rim of the Pacific Ring of Fire, on a zone where two major plates of the Earth's crust collide (see Figure 2.1, **page 26**). Both earthquakes and volcanoes result from the tectonic processes involved. The most consequential quake in US history was the 1906 event of magnitude 8.0 on the San Andreas Fault near San Francisco, California, that killed about 3000 people. Large population centers have developed in areas all along the West Coast that are prone to great earthquakes. The pressure built up by tectonic movement is enormous and must eventually be released, so it is a question of "when" and not "if" these will occur. Some of the West Coast earthquakes will be catastrophic—particularly the megathrust variety that occurs along the 700 mile long Cascadia subduction zone from British Columbia to northern California, although the effects will be mitigated somewhat by strict building codes and emergency preparedness.[3] (It is not just the West Coast that experiences earthquakes; a 2011 quake centered in Virginia was felt throughout eastern North America, and a series of devastating quakes hit the New Madrid fault region of southern Missouri in 1811–1812.)

25, 374

The Pacific coasts must also be on guard for tsunamis, the waves that result when offshore earthquakes displace huge volumes of water. With a magnitude of 9.2, a 1964 Alaska quake was the second strongest ever recorded in the world. It damaged Anchorage extensively, and triggered a tsunami that devastated the small Alaskan town of Valdez and caused deaths and damage as far away as California and Hawaii.

The Aleutian Islands and the Cascades contain many active and dormant (potentially-active) volcanoes. An active volcano, Mount Rainier (close to Seattle) is among the planet's 17 "Decade" volcanoes, considered the most dangerous in the world. In 1980, Mount St. Helens in Washington blew up with an explosive force equivalent to 400 million tons of TNT. In 1912, the largest volcanic eruption of the 20th century occurred at Novarupta on the sparsely populated Alaskan Peninsula.

We have already considered why people choose to or must live in hazard-prone areas (see **page 28**).[4] Many Americans of the eastern United States think Californians are crazy to live in earthquake zones, but they regularly contend with their own menacing natural hazards. The US experiences more tornadoes than any other nation on Earth. Across a swath of the Midwest known as **"Tornado Alley,"** these intense vortices of very low atmospheric pressure cut paths of destruction, sometimes causing large losses of life and property (**•Figure 11.B**).

Successive days or even weeks of thunderstorm activity sometimes produce localized or widespread flooding. In the summer of 1993, weeks of heavy rains over portions of the Missouri and Mississippi River watersheds produced floods of historic proportions. Yet another precipitation-related natural hazard is the **blizzard**, a combination punch of heavy snowfall and high winds that strikes the United States—typically in the Midwest and Northeast—an average of 11 times each year.

Flooding and wind damage are typical results of the hurricanes that batter the East Coast and Gulf of Mexico regions between June and November. The most catastrophic in US history in terms of lives lost was a hurricane in 1900 that killed more than 8000 people in Galveston, Texas. The costliest storm in US history ($80 billion in cleanup and repairs) was Hurricane Katrina, which roared ashore as a category 3 storm in Louisiana in 2005. Katrina flooded 80 percent of New Orleans after the city's levees, which hold back the waters of the Mississippi River, failed (New Orleans lies 9 feet [3 m] below sea level), representing a problematic geographic site for a city that has an excellent geographic situation. In 2012, Hurricane Sandy, the largest Atlantic hurricane on record, struck New York City and New Jersey. This "superstorm" caused $60 billion in damage, largely from its **storm surge**—sustained coastal flooding from the storm's winds raising sea levels and pushing that excess water onto shore.

471

A more pernicious natural hazard, a kind of disaster in slow motion, is the drought that sometimes rakes the country, especially in the West. The worst in the collective American experience was the eight-year drought that produced the **Dust Bowl** of the 1930s, ruining crops and livelihoods and causing mass migrations of broken farming families to leave Oklahoma and adjacent areas, mainly to settle in the West. John Steinbeck told their stories in his epic novel *The Grapes of Wrath*. Recently, a drought lasting longer than the Dust Bowl years has afflicted the western and southwestern United States. Climatological records suggest that such lower precipitation may be the norm for this region. If so, local water supplies will not be able to sustain the region's large human populations. Problems in sharing Colorado River waters in the dry Southwest are discussed on **pages 544–546**.

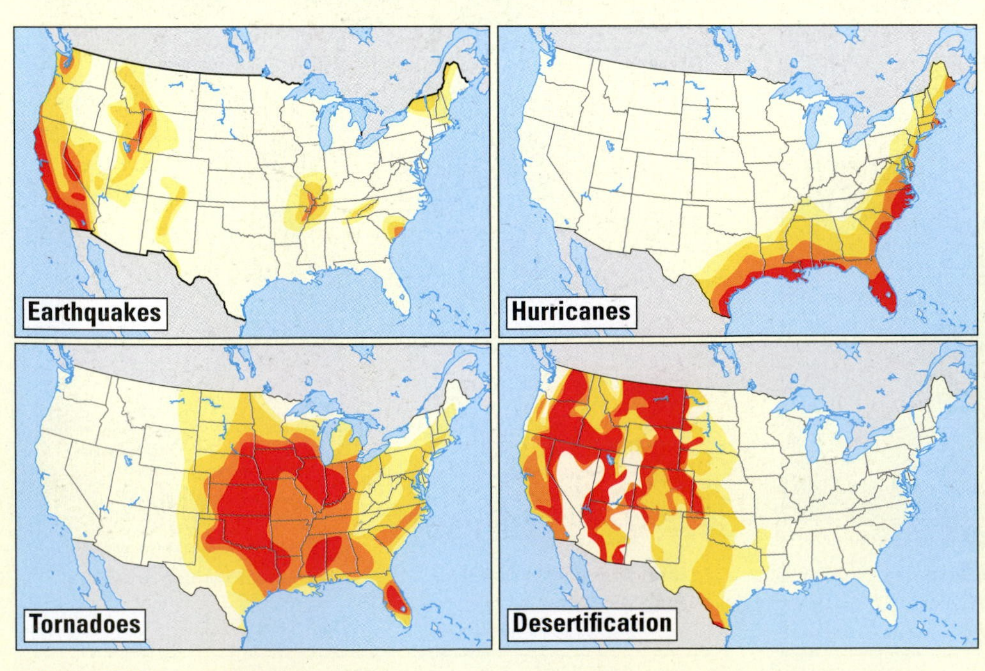

• Figure 11.A Selected natural hazards in the continental United States. Areas shaded red have the highest risk of experiencing that hazard.

National Oceanic and Atmospheric Administration (NOAA)

National Oceanic and Atmospheric Administration (NOAA)

• **Figure 11.B** The tornado measuring EF-5 (the strongest possible) that tore through Joplin, Missouri, on May 22, 2011, destroyed the city's only hospital (St. John's, seen here before and after the twister), complicating emergency relief efforts.

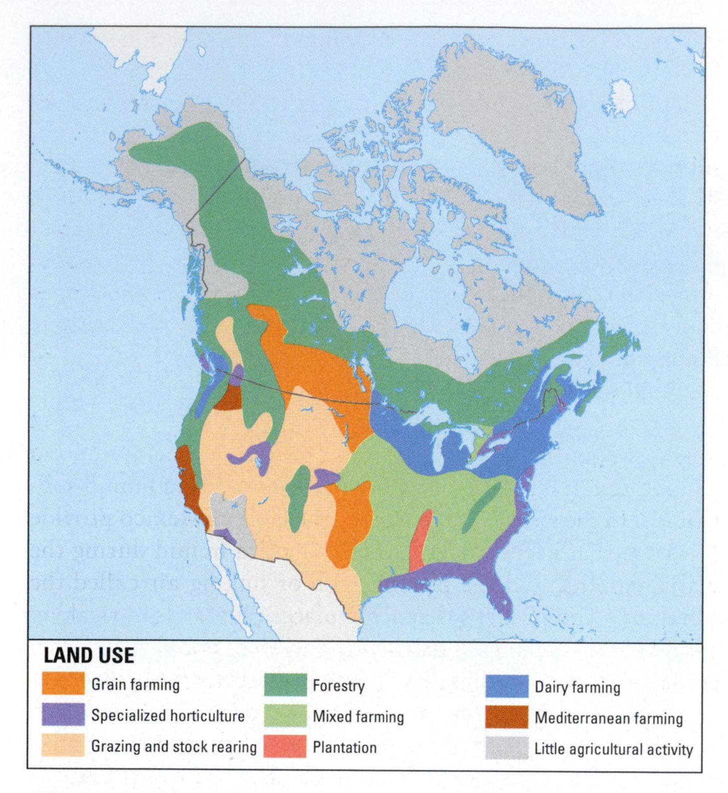

• Figure 11.16 Land use in Canada, the United States, and Greenland.

• Figure 11.17 Giant sequoias in King's Canyon National Park, central California.

makes it much less diverse. The wide range of land uses and corresponding economic opportunities afforded by this climatic variety helps make this world region very wealthy (**•Figure 11.16**; 520 for croplands see Figure 11.40, **page 523**).

The polar climate of Greenland, northern Canada, and Alaska offers little in the way of economic activity, and population is sparse. Traditional ways of life including fishing, whaling, and trapping predominate here, but thinning Arctic ice from climate change is threatening some of these traditions, and melting permafrost is forcing the relocation of some settle- 19 ments. In recent decades, oil production and tourism have be- 546 come increasingly important in the polar realm.

The subarctic climate to the south covers the bulk of Alaska and over half of Canada, largely coinciding with a vast boreal coniferous forest stretching virtually unbroken from Alaska to Labrador. The soils here are nutrient-poor and often waterlogged, so agriculture is impractical. Instead, hydropower, logging, and mining are the main economic activities. About half of Canada's annual timber production comes from the boreal forest, and the diamond mines in the Northwest Territories are among the world's largest.

Hugging the west coast from Alaska to northern California, the cool, rainy, foggy oceanic climate is favorable to the growth of large coniferous trees such as sitka spruce, western red cedar, Douglas fir, redwood (the tallest trees in the world, some towering over 350 feet [100 m]) and the massive sequoia (**•Figure 11.17**). Many of these woods are very valuable and in high demand; excessive logging has significantly reduced their 526 geographic ranges. The damp coasts of Washington and British Columbia have made them home to some of the few areas of temperate rainforest to be found anywhere in the world.

A pleasant Mediterranean climate occupies much of California. This is one of North America's most heavily populated areas, and it has a diversified economy. It is also a vitally important agricultural area; 17 percent of the total US agricultural output comes from California's $50 billion annual industry, located mainly in the Central Valley, which has some of the world's richest soils. California ranks at or near the top of production of an enormous number of crops, including 60 percent of US fruits and 70 percent of its vegetables. Extensive and prolonged drought became a major concern for the state's food growers, leading to mandatory water cutbacks in 2015. There were concerns of a **megadrought**, technically one lasting two decades or longer. California's climatic history is punctuated by megadroughts, some lasting longer than two centuries. 544

Semiarid conditions prevail over a large area stretching from the rugged Cascade and Sierra Nevada ranges across the Rockies to the Great Plains. Recreation, especially skiing and visiting national parks such as Yellowstone in Wyoming and Banff in Alberta, is a major draw (**•Figure 11.18**). Short grass, shrubs, and stunted trees supply forage for cattle ranching,

Joe Hobbs

• **Figure 11.18** Castle Geyser Basin in Yellowstone National Park is among the most active in the Old Faithful Geyser Basin.

the predominant form of agriculture. The Great Plains, once referred to as the "Great American Desert" before its agricultural potential was realized, has some of the world's finest soils and most productive farmlands, but is only lightly populated. Moister areas support wheat, and irrigation with center-pivot and other techniques enables production of a variety of crops including apples, cantaloupes, and potatoes (•**Figure 11.19**). Resource extraction is also important in this area, including mining of coal, oil, gold, and uranium (Canada produces one-fifth of the world's uranium).

The desert conditions of the Southwest lend themselves to recreational activities, tourism, and retirement. Irrigation supports agriculture in the Southwest; with its generous endowment from the engineering works described on **page 545**, Arizona is third largest US vegetable producing state. Small mining towns near active and abandoned silver and copper mines dot the landscape. Some, like Bisbee, Arizona, have evolved into artistic communities. The beautiful red sandstone landscapes of Sedona, Arizona draw artists and a variety of spiritual seekers.

Virtually all of the United States east of the Great Plains, and much of southeastern Canada, are located in the two humid climate zones: humid continental in the north, and humid subtropical in the south. The waters of the Gulf of Mexico provide moisture that keeps eastern North America humid during the warm months, and the phenomenon of sinking air called the "Bermuda High" keeps that air in place. This leads to the long stretches of hot, muggy days with only occasional relief from cooler Canadian air masses strong enough to push through. This clash between hot, humid air and cooler, drier air, especially in spring, leads to the development of intense thunderstorms and tornadoes across much of the Midwest and the South. The moist air also allows for ample snowfall during winter throughout the continental climate zone, and occasionally even in the subtropical zone.

About 80 percent of New England is forested, a remarkable turnaround from the 19th century, when over two-thirds of the original forest cover had been cleared for small agricultural plots and for use as firewood. After 1850, farmers gravitated toward the much richer soils of the Midwest, and hydropower and fossil fuels replaced timber as fuel. These shifts eased pressure on the forests, and secondary-growth trees have rebounded, though these forests are less diverse biologically than the primary forests.

Tallgrass prairie comprises much of the natural vegetation of the Midwest, giving way to mixed deciduous and coniferous forests farther east. The "Corn Belt" lies from the eastern Dakotas and Nebraska, through Iowa and into Illinois. The United States produces nearly half of the world's corn crop and over 30 percent of the world's soybeans. Dairy farming is prevalent in the upper Midwest. The Northeast's rocky soil makes it a generally less agriculturally productive region, but it is an important source of mushrooms and cranberries.

524

Joe Hobbs

• **Figure 11.19** Center pivot irrigation creates distinctive crop circles across southern Utah and other parts of the intermountain West and Great Plains regions of the US.

Tobacco is an important crop throughout the South (although China dwarfs the US, producing ten times as much tobacco), in addition to chicken, peanuts, and cotton. Florida is a major producer of citrus fruits including oranges, grapefruits, and tangerines. Half of all US sugarcane is grown in Florida, with most of the rest coming from Louisiana. Coal mines in the Appalachians, primarily in Pennsylvania, West Virginia, and Kentucky, produce about one-third of the total US supply.

Joe Hobbs

• **Figure 11.20** Na Pali coast, north end of Kauai Island, Hawaii.

520 The coal industry has been criticized for environmentally unfriendly procedures including using explosives in strip mining to access coal seams, and dumping the tailings into nearby valleys.

Southern Florida and Hawaii have tropical climates. South Florida's Atlantic side has a tropical monsoon climate, while its Gulf side (including the Everglades) has a tropical savanna climate. Hawaii is climatically complex for its small size; the northern coasts of islands tend to have a tropical rainforest climate, while the southern coasts, in the rain shadows of mountains, range from tropical savanna to semiarid (•**Figure 11.20**). The Big Island of Hawaii is an excellent example of undifferentiated highland: at sea level, tropical rainforest (on the east) and semiarid and desert (on the west) climates dominate. More temperate conditions prevail at higher elevations, and during winter on the peaks of Mauna Kea and Mauna Loa, the National Weather Service must occasionally issue blizzard 29 warnings!

The distribution of US and Canadian biomes as mapped in Figure 11.15b may be very different in the future. There is growing evidence that as global temperatures rise, agricultural and natural vegetation zones will shift poleward in this region, as they are doing elsewhere. Large-scale maps are already being re-drafted; for example, the 2012 United States Department of Agriculture map of hardiness zones for typical garden plants in the United States showed that many bands are a full zone warmer, and in some places two zones warmer, than when they were mapped in 1990.

11.3 Cultural and Historical Geographies

This discussion of the peoples of North America begins with the region's original, pre-European inhabitants, the Native Americans (in the United States) and **First Nations** (in Canada). The term "Indian" is still widely used in both countries, including on many official names and laws, and is not considered pejorative (in a 1995 survey, the US Census Bureau found that 50 percent of indigenous peoples preferred the term Indian, and only 37 percent used Native American; most of the rest had no preference and used the terms interchangeably). In neither country do the terms Indian, Native American, or First Nations include the **Inuit** peoples of northern Canada and Alaska, and the **Aleuts** of Alaska. Both groups were relative latecomers to the region. In Canada the Inuit are known as the **First Peoples**, while the Inuit and Aleuts living in Alaska are called **Alaska Natives**. The indigenous Greenlanders, known as the *Kalaallit,* are related linguistically and ethnically to the Inuit of Canada, Alaska, and Siberia.

Origins

The initial peopling of the Americas has been a subject of study for a century, and many details remain unknown or controversial. Recent archaeological and genetic findings indicate that humans crossed into North America about 23,000 years ago over a land bridge called Beringia that existed between Siberia and Alaska during the last glacial maximum. There is also growing evidence that small groups of people sailed from Beringia down the Pacific coast of North America around the same time or earlier; this is the "Coastal Migration Hypothesis."[5] These **Paleo-Indians** were the ancestors of all indigenous peoples in the Americas, except for the Inuit and some related groups, who arrived much later in a separate migration. (The possibility of some Polynesian groups reaching the Americas in more recent times is discussed in Chapter 10.) 460

These new arrivals and their successors heavily modified the landscape for their own benefit. Initially, according to the "Pleistocene Overkill Hypothesis," it is likely they overhunted the megafauna (woolly mammoths, mastodons, and others) that existed in North America at the time of their arrival, although a rapidly warming climate about 10,000 years ago may have played a role as well. In later millennia, the Paleo-Indians' 51 descendants made extensive use of fire to clear away brush, and in so doing changed many former woodlands permanently into grasslands. They domesticated numerous plants, and created modern corn by selectively breeding several types of maize.

Native American Civilizations

The Native American cultures that emerged ranged from nomadic hunter-gatherers to sedentary villagers who worked surrounding agricultural fields. Anthropologists have categorized North American indigenous peoples into ten groups based on geographic, linguistic, and cultural traits as they were at the time of European contact (•**Figure 11.21**). While each culture group contains dozens to hundreds of different peoples, some general trends are recognizable. While there were no extensive urban civilizations in North America like the Inca or Aztec, many tribes in the Northeast, Southeast, Plains, and Southwest culture areas settled in small villages and grew crops, fished, and hunted. The Iroquois Confederacy in present-day New York, for example, practiced agriculture, had a democratic political system, and lived in permanent small

• **Figure 11.21** Culture areas of Native Americans, First Peoples, First Nations, and Alaska Natives in Canada, the United States, and Greenland.

villages made up of wooden longhouses. The other culture areas were based primarily on hunting and gathering; in the food-rich Northwest and California culture areas these foraging tribes tended to be settled, while they were more nomadic in other regions where food was scarcer.

Indigenous peoples shared a number of cultural traits. Most Native Americans lived in bands (extended families of up to a few dozen people), either as nomads or in small villages. Nearly all of those who practiced agriculture grew the same basic crops, the "three sisters" of corn, beans, and squash. Like Australia's indigenous peoples they had creation myths, stories of how the world came to be and the place of their people 392 in it. Their religious practices varied considerably, but nearly all indigenous societies were animistic. Animals, forests, geological features, and the people themselves were tied together intimately in belief systems that emphasized the kinship between these diverse elements of life on Earth; there was a deep reverence for the natural world.

Complex, agriculture-based cultures with permanent settlements, divisions of labor, and stratified societies emerged independently in both in the modern southwestern and southeastern portions of the United States. In the Southwest, the **Ancestral Pueblo** (popularly known as the Anasazi) culture began around 550 CE, centered on the "Four Corners" region of southern Utah, northern Arizona, northwest New Mexico, and southwestern Colorado. The Ancestral Puebloans are best known today for their dwellings, called **pueblos**, which were

Joe Hobbs

• **Figure 11.22** Ancestral Puebloans' dwelling in Canyon de Chelly, Arizona.

interconnected residences and ceremonial centers made from stonework built into cliffsides or on the flat-topped mesas of the region (•**Figure 11.22**). After centuries of relative prosperity, in the 14th century these peoples abandoned their pueblos and the agriculture associated with them. The reasons are unknown, but it is likely that an extended drought that enveloped much of North America for several centuries beginning around 1300 was a factor.

A thousand miles to the east, a succession of cultures collectively referred to as the **Mound Builders** existed for millennia until the 16th or 17th century. The oldest existing, complex manmade constructs in the Americas are a series of earthen mounds created around 3500 BCE in modern-day Louisiana. The later Hopewell culture established an enormous trade network across the eastern half of what is now the United States, through which agricultural products, crafts, textiles, and new technologies (such as the bow and arrow) flowed. The Hopewell trade network and mound-building in these areas largely ceased around 500 CE upon the rise of the Mississippian culture, which created the only Native American city north of the Rio Grande: Cahokia (in modern Illinois), which at its peak around the year 1200 was home to 20,000 people. By the 14th century, Cahokia was abandoned; archaeological evidence suggests repeated flooding and other climatic events may have been the culprit.

European Impacts on Native Cultures

At the time of initial European contact in 1492, between 2 million and 18 million people lived in what is now the United States and Canada (these figures are vigorously debated, and there is no consensus on the number). As in Latin America, Native American cultures across North America were devastated by contact with Europeans. 463

Spanish conquistadores including Francisco Coronado and Cabeza de Vaca explored the interior of North America searching for rumored cities of gold, clashing repeatedly with the Native American peoples they encountered. The most destructive conquistador was probably Hernando de Soto, who from 1539 to 1542 led a 500-man expedition through much of what is now the southeastern United States "wrecking most everything it touched," writes Charles C. Mann. "[De Soto] managed to rape, torture, enslave, and kill countless Indians," but the most damaging aspect of his trek "was entirely without malice—he brought pigs."[6]

De Soto's expedition brought along hundreds of pigs and cattle for food. Once European animals entered the American ecosystem for the first time, their microbes quickly spread to other species, which then spread to the people. The pigs—some of which escaped into the woods and became the ancestors of today's feral "razorbacks"—hosted numerous deadly diseases that Native Americans had no resistance to. De Soto encountered Native Americans by the thousands on his journey. When French explorers visited the area 140 years later, they found it nearly uninhabited. In the wake of De Soto and other European contacts, indigenous nations suffered population losses between 30 and 90 percent, mainly from measles and smallpox epidemics. Entire cultures were wiped out. Disease, famine, and conflict reduced the pre-European population of millions to below 250,000 by 1890 in what is now the United States. Here is what happened.

Joe Hobbs

• Figure 11.23 The modern-day Lakota Sioux on the Pine Ridge Indian Reservation in South Dakota have a small museum whose theme is "The Indian Wars Are Not Over." This reflection on the broken treaty of Fort Laramie, which guaranteed exclusive Sioux use of the Black Hills, is on a museum wall.

Centuries of Conflict

The 17th through 19th centuries were a time of almost constant conflict between indigenous nations and British and American settlers. Colonial raids and land grabs and Indian reprisals, along with preemptive Indian raids and colonial retaliations, took thousands of lives. In the early decades of the 19th century, the United States signed peace treaties with several Native American nations, and negotiated to buy portions of their land. Prominent statesmen like George Washington and Benjamin Franklin had asserted that indigenous peoples, even if their societies were "backward," deserved the same treatment and justice as whites, and fair compensation for their lands; but these attitudes become the exception rather than the norm. These treaties were often flouted, and violent clashes erupted. In 1830 the Indian Removal Act authorized the expulsion of all Native Americans east of the Mississippi River who did not wish to assimilate into white American society. Thousands of Cherokees, Creeks, Seminoles, and others resisting assimilation were forced to migrate from their ancestral homes westward to a specially designated Indian Territory (now Oklahoma) in what came to be known as the **"Trail of Tears."**

Raids, skirmishes, and all-out wars marked American settlement of the interior. The US Army battled against various Native American tribes in nearly every state west of the Mississippi between roughly 1845 and 1890 (**•Figure 11. 23**). Most encounters were relatively small, with a handful to a few dozen people on each side. The deadliest and longest running clashes of the **Indian wars** were between the United States and the Apache in the Southwest and the various **Sioux** tribes of the northern plains. Federal troops fought the Apache until 1886, when the Apache war chief Geronimo was captured. In 1876, the Sioux handed the US Army its biggest defeat of the Indian wars at the Battle of Little Bighorn in South Dakota (commonly known as General George A. Custer's "last stand"), where over 250 US troops were killed (**•Figure 11.24**). The Sioux surrendered in 1890 after vengeful US cavalrymen massacred 200 Sioux at Wounded Knee, South Dakota. The Sioux surrender effectively ended the Indian wars.

Many tribes were resettled onto **reservations**. They resisted, and many of the Indian wars resulted from efforts to forcibly move them onto reservations. These small and fragmented territories were established to remove Native Americans from good farmlands they had previously occupied. Indians lived in isolation and poverty on the reservations largely without services or infrastructure.

Joe Hobbs

• Figure 11.24 General George Custer led his 7th Cavalry Regiment along with gold miners into the Black Hills lands designated by the 1868 treaty as belonging to the Sioux. Events escalated to the decisive 1876 confrontation on the Little Big Horn, where an Indian confederation led by Sioux Chief Crazy Horse killed Custer's men. Here lie their graves on "Last Stand Hill."

In the 20th century, when Native Americans were no longer a threat to white society, public attitudes toward indigenous peoples began changing. All Native Americans were made US citizens in 1924, and in 1934 the Indian Reorganization Act provided measures of self-rule on reservations.

A growing identity movement beginning in the 1960s brought Native American grievances into the political arena. The Indian Civil Rights Act was passed in 1968, and the federal government accredited tribal colleges. Populations are increasing; nearly 2 million Americans reported Native ancestry in 2010. Numerous programs across the country have been established to save indigenous languages from extinction. An ongoing issue has been the controversial use of offensive indigenous-related names in sports; some franchises, like the Washington Redskins football team, have come under intense pressure to change their names.

Despite progress, Indian reservations are still among the poorest communities of the United States. "The Res," as Native Americans call it, is often plagued by high rates of incarceration, alcoholism, drug abuse, depression, broken families, teen suicide, and unemployment. The Lakota's Pine Ridge Reservation in South Dakota (•**Figure 11.25**) has an annual per capita income of just $6300 (compared to the national US average of $42,600). Since 1988, about one-third of the over 500 federally recognized Native American tribes in the United States have attempted to bring in much-needed currency by establishing casinos on their reservations which, as semi-autonomous territories, are not subject to anti-gambling laws of the states they are located in. The **gaming industry** has been a mixed blessing for the Native Americans, with a wide variety of successes and failures in different places. In some, gambling revenues have supported community education and student scholarships, health care, and cultural centers, but in others, they have lined a few pockets and failed to benefit many members of the community.

Relations between First Nations and Europeans in Canada were not as bloody as those in the United States, but were still problematic. European diseases decimated indigenous communities in Canada. Armed white settlers were the main force behind the extirpation of bison, the primary food source for many Plains tribes, in Canada by the 1880s. Canadian authorities often reneged on treaties promising food aid to Indians who gave up their lands to whites, resulting in the starvation deaths of thousands. In 1885 in present-day Saskatchewan, several Plains tribes allied with the Métis (mixed-race people with European and First Nations ancestry) in rebellion against Canadian rule.

• **Figure 11.25** Lakota children at the Wounded Knee Massacre Site, Pine Ridge, South Dakota.

As in the US, indigenous peoples in Canada were placed onto reservations. Over 150,000 indigenous children were forced by the Canadian government into schools where conversion to Christianity and assimilation into white society were mandatory, and abuse, neglect, disease, and malnutrition were common. The federal and provincial governments sold quality land to white settlers that had been initially promised to First Nations as their reservations. In the 1950s, Canada's government forcibly relocated a group of Ungava Inuit from Hudson Bay 1200 miles (1920 km) north to inhospitable Ellesmere Island. The move, which caused these people enormous suffering, was meant to help confirm Canadian sovereignty over the island.

Official Canadian discrimination against First Nations peoples ended in the 20th century. They were given full citizenship in 1956 and finally allowed to vote in 1960. In 1991, the Canadian government officially apologized for its treatment of indigenous people, and has been continuing to grant land rights and financial support. Still, indigenous peoples remain the poorest segment of the Canadian population, with incomes about one-third lower than the population as a whole.

The situation for Canada's Inuit, who live in the far northern reaches of the country, is unique in the world. In 1999, Canada ceded one-fifth of its total land area to the Inuit, who called their new territory **Nunavut**, meaning "our land" (see **Figure 11.37**). Only about 30,000 people live in this vast area, and 90 percent of the funding to keep the territory viable, amounting to $500 million annually, comes from the federal government.

European Settlers and Settlements

Europeans and Europeanized cultures transformed the Americas over a period of five centuries (•**Figure 11.26**). Christopher Columbus never saw any of what would become the United States and Canada on his voyages. The first European to step foot on these shores was John Cabot, sailing for the English crown, in 1497. Soon numerous European explorers—including Giovanni Verrazano, Martin Frobisher, Jacques Cartier, Henry Hudson, and others—were sailing across the Atlantic. For European powers at the time, North America was less a destination in itself than an obstacle to be overcome. Europeans wanted trade with India and China, and they sought a direct trading route to Asia by sailing westward. Early exploration of North America was devoted to finding the theorized warm water **"Northwest Passage"** linking the Atlantic with Asia. The warm water passage Europeans hoped to find did not exist, however. 538

As European kingdoms continued to explore this new continent, many soon began establishing overseas colonies in North America, just as Spain and Portugal were doing farther south. The earliest colonists sought the new continent's resources; later, people escaping religious persecution joined the mix.

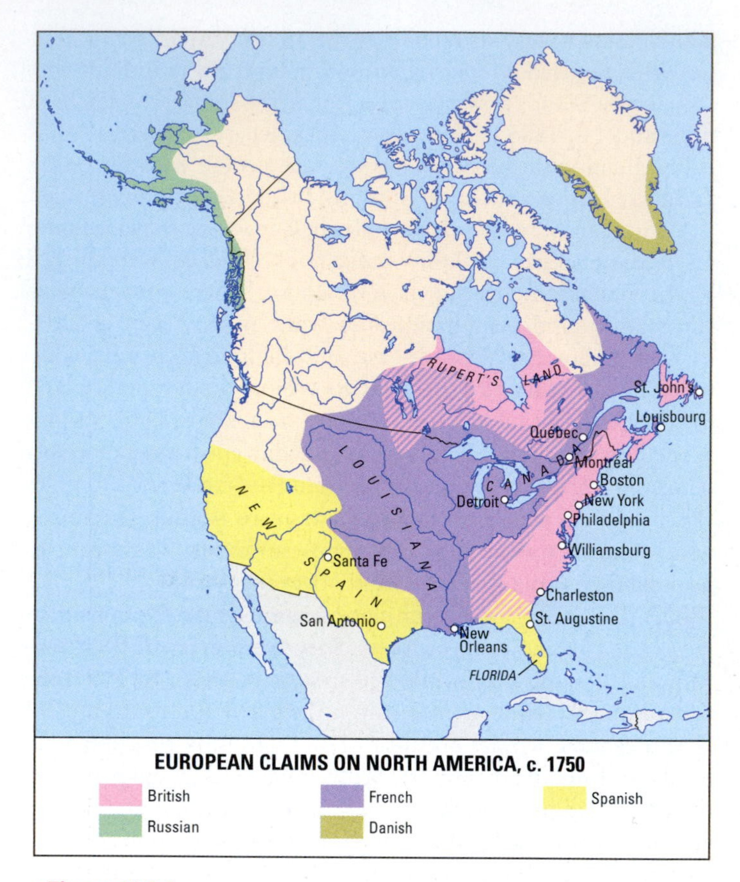

• **Figure 11.26** European colonization of North America.

The Spaniards, who were already settling in and exploiting the resources of Latin America and the Caribbean, were also the first to establish a permanent presence in North America, at St. Augustine in Florida in 1565. Spain also claimed most of what are now the south-central and southwestern parts of the United States. Being remote from colonial population centers, California initially did not interest the Spanish crown; finally in 1769, more than 200 years after Spain claimed the area, the Franciscan friar Junipero Serra (elevated recently to Catholic saint) founded San Diego as a mission to convert Native Americans to Catholicism. Spain established 20 more missions along the California coast during the following 30 years (•**Figure 11.27**).

France established its colony of Acadia in 1605, in what is most of today's Canadian Maritime provinces. The fortified settlement of Québec in 1608 was France's first successful establishment in the interior. Although the fur trade in Canada was lucrative for the French, there was no fur rush: only French Catholics were allowed to settle, and the Québec colony grew slowly. Although France claimed enormous amounts of territory in North America, it exerted control with strategically located forts rather than with large numbers of people. The most important French settlement was *La Nouvelle-Orléans*, founded in 1718 as the capital of the vast Louisiana colony. New Orleans quickly became a major port that asserted French control over the Mississippi River and much of the Gulf Coast. Acquiring New Orleans was a top priority for post-independence United States. It was also a key target of the British during the War of 1812.

In 1625, the Dutch established a fort on an island the local Lenape people called *Mannahatta*, buying the island for less than $1000. This was an ideal location for a trading port because of its good harbor with easy access to both the ocean and fur-trading posts farther north. New Amsterdam prospered, with a population of 9000 by 1664. In 1665, the entire New Netherland colony was transferred to English control.

England's initial colonial settlements were primarily located in the Chesapeake Bay area and New England. The first successful English colony in North America was founded in 1607 at Jamestown, Virginia, followed in 1620 by the founding of Plymouth, Massachusetts by a Puritan group who would later be known as the Pilgrims. The English hold on the East Coast was strengthened after it acquired the Dutch colonial area, giving London an uninterrupted stretch of shoreline from Maine to Virginia.

Britain, which was created in 1707 after England and Scotland united, defeated France in a series of wars over colonial territory, and acquired nearly all of the French colony of Acadia, renaming it Nova Scotia. This loss of territory was the beginning of French decline in North America. The British expelled over 10,000 French Acadians; many resettled in Louisiana (where over time "Acadian" morphed to "Cajun"). More warfare between France and Britain followed, notably with the Seven Years War (the North American theater of which became known as the French and Indian War). The war ended in 1763 with a treaty that turned over most of France's colonial holdings in North America (except Louisiana) to Britain. Britain then controlled nearly the entire continent east of the Mississippi River.

Territorial Development of the United States

Much like Russia and China, the United States grew as a "land empire" (•**Figure 11.28**). After the American Revolution, the 167 United States comprised 13 states along the Atlantic seaboard and a large unorganized area between the Appalachians and the Mississippi River. France sold its Louisiana colony to the

• **Figure 11.27** Spanish friar Junípero Serra established the first Franciscan mission in the Viceroyalty of Spain's California province here at San Diego in 1769.

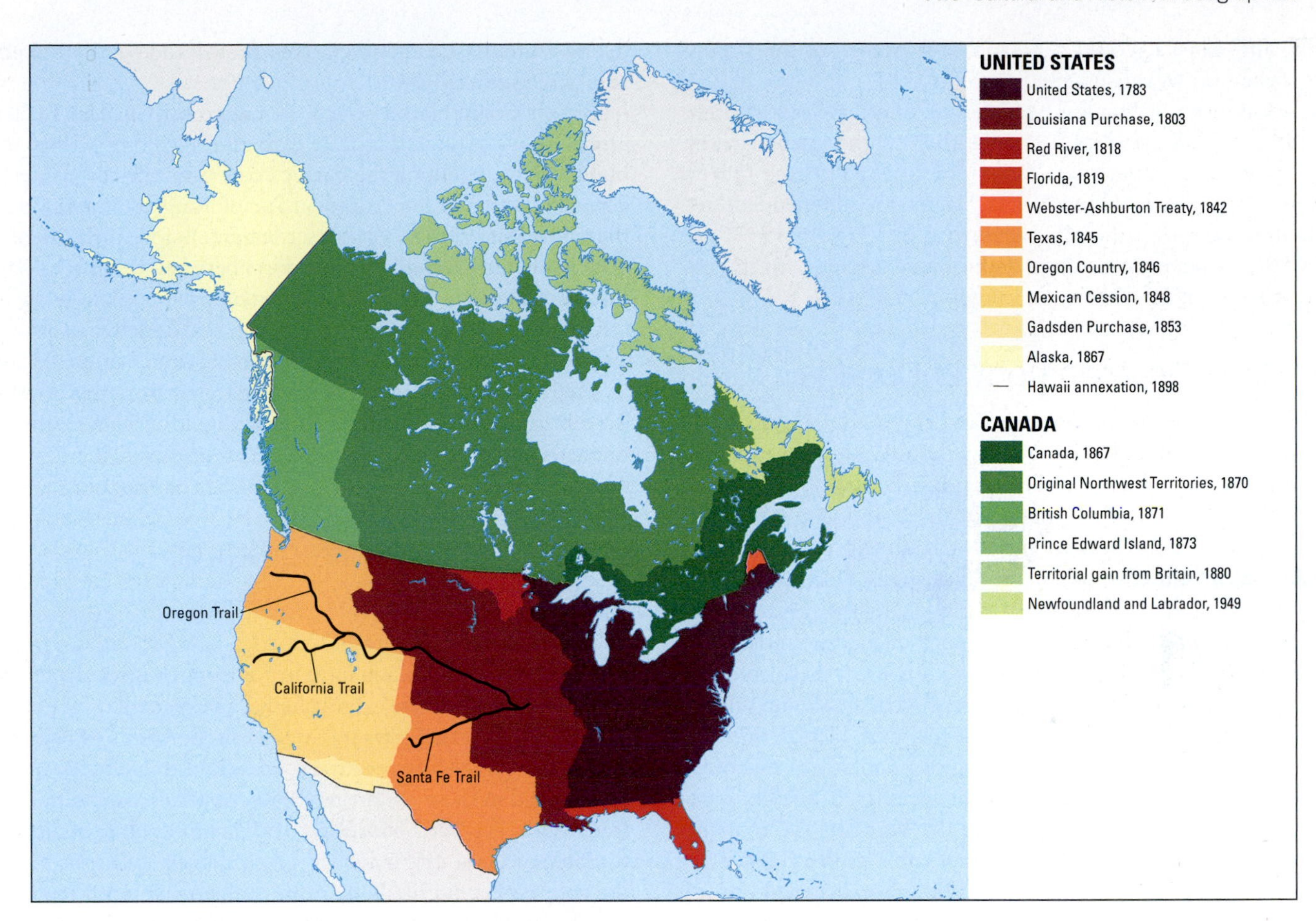

• **Figure 11.28** Territorial acquisitions of the United States and Canada.

US in 1803 for about 3 cents an acre. This **Louisiana Purchase** not only gave the US the vital port of New Orleans (and control of the Mississippi River), but also nearly doubled the size of the country. Much of this territory was unknown to European-Americans at the time, and in 1804 President Thomas Jefferson sent Meriwether Lewis and William Clark to explore the region. Lewis and Clark reached the Pacific Ocean in what is now Oregon; the "Oregon Country" (today's Pacific Northwest) was jointly administered by Britain and the US until 1846.

In 1819, a treaty with Spain transferred Florida to the US on the promise that the US would give up any claims to Spanish Texas, which many Americans wished to annex. By 1830 over 30,000 Americans had moved to Texas, which was then part of Mexico. When Mexico halted American immigration and outlawed slavery, the Americans of Texas began agitating for independence from Mexico. The American settlers were able to defeat the Mexican forces in 1836 and establish their own Republic of Texas. Texas was independent until 1845, when the insolvent republic gave up its sovereignty and became the 28th state of the Union so it could pay off its war debts.

The widespread American belief in **Manifest Destiny**—that the United States was fated to expand across the continent to the Pacific Ocean—led the American government to propose purchasing what were then Mexico's northern territories (today's American Southwest). The US particularly wanted California, fearful that Britain or France would colonize it and deprive the US of good Pacific harbors. Mexico refused the offer, and amid increasing border tensions around Texas, the two countries declared war on each other in 1846. The Mexican-American War lasted two years and ended with a decisive American victory. Mexico ceded California and the Southwest to the US, and later sold a strip of land (in the Gadsden Purchase) in southern Arizona to the US for it to build a railway through.

Sparsely inhabited by white settlers, the western half of the US was opened up with the development of several wagon routes (including the Oregon, California, and Santa Fe Trails; see **Figure 11.28**), transcontinental railways, and by the **Homestead Acts**, which in 1862 gave migrants 160 acres of land to settle on and farm provided they lived there for at least five years (there were later legal revisions). Settlement of the frontier west of the Appalachians in the latter half of the 19th century attracted over half a million homesteaders and their families from the eastern states, Canada, and Europe, particularly Britain, Germany, and Scandinavia.

In 1867, Secretary of State William Seward negotiated to buy the Russian America territory, renamed Alaska, for 2 cents an acre. Russia was unable to effectively govern this distant colony, and feared potential British conquest of the area; selling the land to the US halted British expansion westward. Some in the US at the time called the Alaska purchase "Seward's Folly," money wasted on a vast, frozen land with no economic benefit. These attitudes changed after the discovery of gold in the late

1890s, and Alaska would again prove its worth to the United 546 States when oil was discovered there in 1967.

The kingdom of Hawaii became an American protectorate in 1849. The US grew its influence there, culminating in the overthrow of the indigenous monarchy in 1893 during Queen Liliuokalani's reign. Hawaii was briefly an independent republic until the US officially annexed it in 1898. In 1993, the United States government issued an official apology for its role in overthrowing the Hawaiian kingdom.

Territorial Development of Canada

Canada had its own evolution as a land empire (see **Figure 11.28**). Britain's northern colonies of Newfoundland, Nova Scotia (which New Brunswick and Prince Edward Island would later be carved out of), Lower Canada (Québec), and Upper Canada (Ontario) remained loyal to the crown during the American Revolution. Britain also had nominal control of Rupert's Land, a vast northwestern area technically owned by the private Hudson's Bay Company that controlled much of the fur trade.

Outside of their loyalist sentiments, however, these colonies had little in common. Their populations were small and widely separated, and they had different levels and blends of English and French laws, language, religion, and social customs. "Canada" as we know it today did not yet exist. It was not until a botched American invasion during the War of 1812 that the 533 disparate northern colonies began to create a unified identity.

Development of the lands west of Ontario proceeded slowly. Most of the few European-Canadians here were fur traders, operating out of a handful of defensive forts. During the joint US-UK administration of the Oregon Country, many American settlers along the Pacific coast wanted America to own the entire region, which extended to 54°40′ north (leading to a rallying cry of "fifty-four forty or fight!"); Britain wanted the Columbia River and all lands north of it. In 1846, the boundary following the 49° parallel was extended to the Pacific, with a slight jog to the south through the Strait of Juan de Fuca to keep all of Vancouver Island in Britain's possession.

In recognition of the gradually developing Canadian identity, Britain appointed a governor-general for all of British North America in 1838, and in 1841 the colonies of Upper and Lower Canada were merged into the single colony of Canada. Growing numbers of Canadians wanted self-rule and the ability to control their taxes, yet most did not want to sever ties with Britain. After several years of talks, the Dominion of Canada was proclaimed in 1867 as an independent country, but one that still recognized the British monarch as its head of state. Keeping ties with Britain was in part a reaction to the recent American civil war; many Canadians feared the instability that a similar republican governmental system might bring them.

Four provinces—Ontario, Québec, Nova Scotia, and New Brunswick—made up the new country of Canada. Prince Edward Island initially rejected joining Canada, and stayed out of the confederation until 1873. Canada purchased Rupert's Land from the Hudson's Bay Company in 1870, reorganizing it as the Northwest Territories. Newfoundland voted to remain a British colony, largely over issues of taxation and protecting its maritime-based economy. Newfoundland finally joined the rest of Canada in 1949.

British Columbia became a Canadian province in 1871 only after Ottawa promised to take over all the colony's debts and to build a railway linking the Pacific coast with the eastern regions. The completion of the Canadian Pacific Railway in 1885 opened the Canadian Prairie to white settlement; before then, there were only a few farming villages and forts between the Rocky Mountains and Ontario. The quickly rising prairie populations gave rise to the provinces of Manitoba, Alberta, and Saskatchewan. Ownership of the Arctic islands was transferred from Britain to Canada in 1880. The Yukon Territory was separated from the Northwest Territories in 1898 after the Klondike gold rush brought thousands of prospectors to the area. A century later, the Northwest Territories shrank again upon the creation of Nunavut. 508

Peoples of the United States and Canada Today

So that is how mainly Anglo America came to be. But both the United States and Canada have become much more diverse since these foundations were established, and especially since changes to their immigration laws took effect in the 1960s (**•Figure 11.29**). About 200 non–Native American ethnolinguistic groups are represented in these countries! Both countries are multicultural. In Canada, the 1988 **Multiculturalism Act** acknowledged that Canada is a country made up of numerous cultures, and that the country would likely be more stable—avoiding the kind of racial tensions that wracked the United States—if those identities were recognized and encouraged (**•Figure 11.30**).

Europeans were the vast majority of immigrants to North America until the 1960s, founding and fostering what has been described as the **mainstream European American culture**.[7] For the United States, the ethnographer David Levinson describes that culture as one based on institutions such as the English language and British educational principles brought by the British colonists, combined with core American values that developed on US soil. These include English as the national language, religious tolerance, individualism, majority rule, equality, a free-market economy, and the idealization of progress.[8] Seventy-six percent of both Americans and Canadians identified as white in the most recent censuses (2010 in the US, 2011 for Canada).

For most of their histories, the United States and Canada (outside Québec) had a strong preference for majorities of northern European Protestants, to the general exclusion of everyone else. Many ethnic neighborhoods of large American cities were initially segregated to keep out "unwanted" European ethno-religious groups such as Italians and Jews. Some of these urban, working-class, white ethnic neighborhoods have survived to the present day, but the gradual assimilation of southern and eastern Europeans into the mainstream ethnic fabric, coupled with "white flight" out of large cities after World War II, has left most urban ethnic neighborhoods today dominated by non-European peoples. For example, the historically Czech 515 neighborhood of Little Bohemia in Omaha is now 50 percent Hispanic, and over 80 percent of the population in New York City's Little Italy is now Chinese.

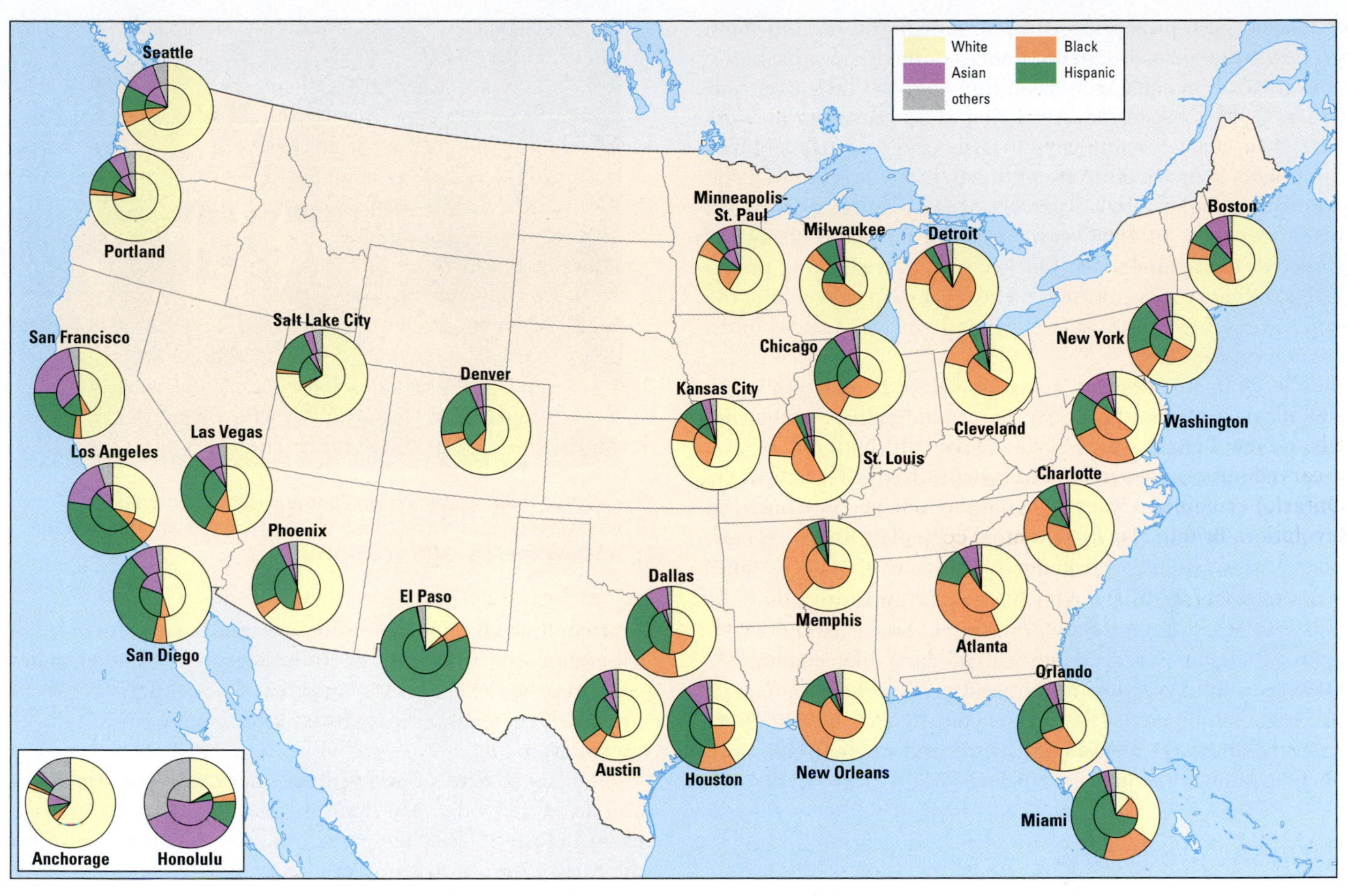

• **Figure 11.29** Ethnic composition of selected US metropolitan areas. The inner circle represents the city's center, and the outer circle represents the city's suburbs. The demographic shifts resulting from the Great Migration and "white flight" are especially noticeable in cities like Detroit and Cleveland.

Descendants of the original French settlers, the *Québécois* are about 20 percent of Canada's population and 75 percent of Québec's. Québec remained largely rural and agrarian until well into the 20th century, in part because of the legacy of the French "long lot" (or ribbon farming) system of allocating land. In this land use pattern, a farmer's plot has a narrow frontage along a river but stretches inland up to a mile, discouraging urban development (•**Figure 11.31**). Without an

• **Figure 11.30** This school outing to a Montreal museum provides a snapshot of Canada's multiculturalism. The exuberant boy at center is one of about a half million Sikhs living in Canada.

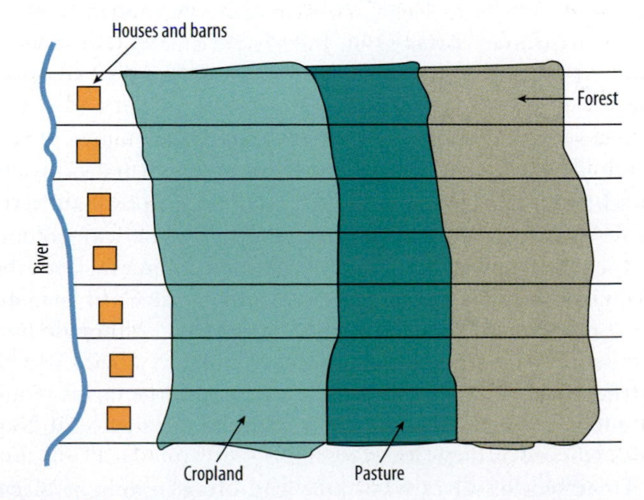

• **Figure 11.31** The distinctive French "long lot" system gave the farmer the advantage of access to a variety of agricultural habitats in a single plot, and access to a transportation corridor, either a river or road. The French transplanted this system from Europe to most of their New World holdings; long lots can still be seen on the landscapes of French Canada and Louisiana.

industrial base prompting rural–urban migration and immigration from abroad, Québec's society remained a bastion of conservative French Catholicism that changed little over time. When Québec began industrializing and urbanizing in World War II, it enacted many laws to discourage newcomers from diluting its French character. Although many people in Québec are now bilingual, only 8 percent speak English as their primary language. In 1995, unhappy with proposed changes to Canada's constitution that Québec saw as weakening provincial governance and culture, Québecers held a referendum that barely kept them in union with Canada (this is a focus of discussion on **page 536**).

The United States has no counterpart to the *Québécois*. The most distinctive European-descended culture in the US is that of the **Amish**, who number about 200,000 and are concentrated in Pennsylvania, Ohio, and Indiana. The Amish are Mennonites of Swiss-German origin who began moving to the US as early as the 1700s. Most speak English, although many of the "Old Order" Amish speak Pennsylvania Dutch (which is actually German). The Amish are not a uniform culture; beliefs and practices vary between groups, but these share some traits. Religion is very important in daily life, emphasizing modesty, humility, cooperation, and *gelassenheit* (submission to a higher authority). The Amish typically have large families, value community over individualism and manual labor over new technologies, and generally lead "plain" rural lives.

Joe Hobbs

• Figure 11.32 William Wesley's home in Florida's Eden State Park was inspired by plantations this prominent lumberman admired as he made his way home after the Civil War.

African Americans

African Americans (also referred to as "blacks," and up to the 1960s as "Negroes") have had a long and often difficult relationship with European Americans. Many of the US's decisive historical events, including the country's **Civil War** of 1860–1865, have reflected tensions between blacks and whites and invoked the legacy of slavery. The US Constitution had declared that "all men are created equal," and the Civil War confronted and resolved the question of whether slavery could persist in America. The Civil War also permanently established the United States as an indivisible nation with a sovereign government, rather than as a confederation of sovereign states.

Slavery was aptly described by former US Secretary of State Condoleezza Rice as "America's birth defect."[8] Its roots were broad: the English, Dutch, French, and Spanish brought slaves from Africa or their Caribbean and South American colonies to work the fields of their North American colonies. Slave labor was most useful on the large plantations of the Caribbean and the southeastern United States (**•Figure 11.32**). Although small numbers of slaves were taken farther north, Canada and the northern states did not have plantations or large farms requiring such abundant labor to run profitably; most slaves in these areas worked as domestic servants.

Treatment of slaves was harsh and often inhumane. Clearing land and maintaining large farms during excessive working hours was exhausting manual labor. Slaves were not allowed to learn to read or to gather in large groups. Punishments such as beatings and whippings were commonplace and were doled out for even minor infractions or for no reason at all other than to reinforce the slave master's authority. Occasional slave revolts prompted reprisals, even against slaves having no roles in the uprisings.

The United States outlawed the importation of slaves from abroad in 1808, but the domestic slave trade remained a vibrant industry. The widespread adoption of the cotton gin in the early 19th century, and the subsequent rise in demand for cotton products in the US and Europe, dramatically increased the demand for slave labor across the South. On the eve of the Civil War, there were about 4 million slaves across the South, and slaves made up majority populations of South Carolina and Mississippi. About one-third of white Southern families were slave owners, and nearly one-third of all slaves were owned by a small number of very wealthy plantation owners (the "1 percent" of their day).

Abolition of slavery progressed in the North early on; in 1780, Pennsylvania became the first state to abolish slavery, and over the next 20 years, all the other northern states followed suit. Most "free" blacks in these states, however, had many restrictions placed on them. They generally could not vote, serve on juries, or own property, and even in states where these activities were not prohibited, white social pressure usually marginalized black people.

The Civil War began in 1861 amid Southern fears that President Abraham Lincoln's 1860 election victory would prevent slavery from expanding into American territories and bring an end to it (and thus to profitable plantation economy) in the South. In 1863, well into the war, President Lincoln issued the **Emancipation Proclamation**, an executive order freeing all slaves in the 11 Confederate states. After the war ended in 1865, freedom for all slaves was codified in law as the **Thirteenth Amendment** to the Constitution. During the postwar period called **Reconstruction**, northerners tried to change the culture of the defeated South and promote the economic and political status of freed blacks. These attempts at reform largely failed, however; by 1877, former slave owners were once again ruling most Southern states. Lynching (extra-judicial killing) and other violence against blacks rose markedly during this time, at the hands of white supremacist groups such as the Ku Klux Klan (KKK) and unorganized mobs. Over 3000 blacks were lynched in the United States in the late 19th and early 20th centuries.

Postwar southern governments enacted numerous laws restricting the freedoms of their black populations. African Americans had a constitutional right to vote, but Southern states got around this by adopting "poll taxes" and literacy tests as prerequisites for voting. These effectively barred blacks, who were largely poor and illiterate, from participating in the political process. Laws enforcing racial **segregation**, collectively referred to as **Jim Crow** laws (named after a minstrel song stereotyping blacks), proliferated after Reconstruction, and many stayed on the books until the mid-1960s. States defended segregation under the doctrine of **"separate but equal"**—keeping whites and blacks apart was legal as long as public facilities for both races, such as schools, rail cars, and drinking fountains, were equal in quantity and quality. In reality, African American facilities were almost always inferior.

It was not until the 1950s that things began to change. A landmark 1954 Supreme Court ruling invalidated the "separate but equal" concept. Then a 1955 act of civil disobedience by an African American woman named Rosa Parks, who refused to give up her seat on a public bus in Montgomery, Alabama, to a white person, launched a wave of public protests and boycotts that eventually desegregated public transportation.

The movement toward equality reached its apex in the mid-1960s. In 1963, Dr. Martin Luther King Jr. led the **March on Washington**, in which he delivered his famous "I have a dream" speech, envisioning the end of racism, to about 250,000 people—mostly black, but with a strong white minority supporting them. This rally was a major influence on the **Civil Rights Act** of 1964 and the **Voting Rights Act** of 1965 during Lyndon Johnson's presidency. These laws officially ended all legal racially based discrimination, segregation, and disenfranchisement in the United States.

The legal battles to eliminate segregation have ended, but "hearts and minds" have been slower to change, and African Americans continue to struggle for equality. The 2008 election of Barack Obama as the 44th president of the United States marked a milestone in the nation's history, raising widespread hopes that remaining racial barriers in the United States would weaken and fall. However, more than 50 years after the landmark civil rights legislation, there is still a long way to go. The cry "Black Lives Matter!" emerged from a number of incidents in which white police officers shot and killed African Americans in illegal or legally questionable circumstances. A white supremacist's killings of nine people in a South Carolina church in 2015 seemed for that moment to crystallize the problem of race relations, and brought down the Confederate flag—seen by many as a symbol of the nation's segregationist past—in numerous public and commercial settings.

The geography of African American populations has changed substantially. Most African Americans lived in the rural South before World War I. Beginning around 1915 (and increasing substantially in the 1940s), in what is called the **Great Migration**, many blacks started moving out of the old Confederacy toward the cities of the Northeast and Midwest, where jobs were more plentiful. African Americans faced discrimination there too. Many white city dwellers did not want black neighbors, and African Americans were subjected to **redlining**, a practice (finally made illegal in 1977) in which minorities were prevented from getting loans to purchase homes or property in mainly white neighborhoods. African Americans ended up in poor, undesirable pockets of the city, particularly in the inner city, that came to be known as **ghettos**, plagued by problems of crime, joblessness, and substandard housing and services (•**Figure 11.33a**). Race riots plagued many cities across the nation from the 1920s through the 1960s.

The movement of about 6 million African Americans from south to north in the 20th century fundamentally changed American demographics and settlement patterns. Black culture changed from mainly rural to overwhelmingly urban, and many industrial cities changed from majority white to majority black in just a few decades. This turnover was caused by both the influx of southern African Americans and the massive **"white flight"** out of cities and into suburbs after World War II. Among the benefits of the GI Bill were low-interest mortgage loans for war veterans. Many vets opted to purchase a new, spacious single-family home in the newly developing (white) suburbs rather than continue to live in crowded cities with increasing minority populations. As African Americans moved into cities, whites moved out in far larger numbers. In recent years a "reverse migration" of African Americans back to Southern states has gained momentum.

Asian Americans

Asian Americans (generally known as "Orientals" until the 1960s) make up about 5 percent of the US population. The Census Bureau classifies "Asian" as anyone descended from people living in South, Southeast, and East Asia. Race relations between the Anglo majority and Asian groups in both the United States and Canada have generally been good in recent decades, but this has not always been the case.

With a population of over 3 million, ethnic Chinese are the largest Asian ethnicity in the United States (see •**Figure 11.33c**). About one-quarter of them live in New York City, where they make up the largest Chinese population outside Asia. Chinese immigrants were especially important in the construction and service sectors of the West Coast economies of both Canada and the United States between 1849 and 1882. Chinese laundries and the Chinese "coolies" who did the grueling and often fatal work of building western railways became the stuff of stereotype and legend. Many white workers resented the imported

Paul Fusco/Magnum Photos

(a)

Joe Hobbs

(b)

Joe Hobbs

(c)

• **Figure 11.33** Ethnic urban landscapes of the United States: **(a)** a black-run business in Los Angeles; **(b)** a Latino neighborhood in Chicago; **(c)** Chinatown in San Francisco.

Chinese laborers, as they would work for far less money than whites. This tension led to an American ban on Chinese immigration in 1882.

Even greater tensions affected ethnic Japanese living in America. Mining and construction companies turned to Japanese laborers after Chinese immigration ceased in 1882. As with the Chinese, many white workers felt threatened by cheaper Japanese labor and by the immigrants' success in running small businesses and farms. But the greatest hostility came after Pearl Harbor. Fearful that some Japanese Americans were secretly working as spies for imperial Japan, President Franklin Roosevelt authorized the forced internment of over 100,000 people of Japanese descent from California and other Western states into a number of concentration camps, where they remained until 1946, well after the war ended. No evidence was ever produced of Japanese Americans (most of them US citizens) collaborating with Tokyo; modern studies suggest their internment was officially sanctioned racism taken to an extreme. In the 1980s, the US government apologized for the Japanese internment and paid hundreds of millions of dollars in reparations to the survivors and their families.

More than half of the people in Hawaii claim some Asian ancestry, with ethnic Japanese the largest component. Asian minorities have intermarried with other ethnic groups in both the US and Canada, and their mixed offspring call themselves **Hapas**, a Hawaiian word meaning "half" that people of mixed Pacific Island and other ethnicities also use to describe themselves.[9]

The number of Indian Americans (not to be confused with American Indians) was small until the immigration reform of 1965; by the 2010 census, they numbered almost 3 million. Indian Americans have the highest educational attainment of all ethnic groups in the United States, compared with about one-fourth of the country as a whole. Following close behind with 2.7 million people are Filipino Americans; the Philippines is a predominantly English-speaking, Christian country, which has made assimilation into the broader American society easier for Filipinos than for many other immigrant groups. Today, there 344 are 1.6 million Vietnamese Americans. Few Vietnamese lived in the US until the end of the Vietnam War in 1975. Most of the initial wave of Vietnamese (often referred to as the "boat people") were refugees from conquered South Vietnam. Koreans also began moving to the US in large numbers in the 1970s; today there are 1.5 million Korean Americans. Many Koreans moved to inner cities, notably in Los Angeles, and started small businesses like convenience stores and laundries. Many of their locations had been previously African American owned, and racial tensions between the two groups occasionally flared.

Hispanic Americans

Unlike the white, black, Native American, and Asian categories (and the variants among them), **Hispanic** is not a racial term, but an ethnolinguistic one. Hispanics (also called **Latinos**) can be of any race; they are people whose ancestry is of Mexican, Central American, South American, Cuban, Puerto Rican, and other Spanish-derived cultures. By this definition, Brazilian Americans (with their historic ties to Portugal) are excluded, but informally they are sometimes grouped with those of Spanish origin.

Over 50 million people, or about 17 percent of the population in the United States, are Hispanic. Between the national census years 2000 and 2010, the number of Hispanics increased by 43 percent, compared with less than 5 percent for non-Hispanics. This large increase was the result of both high immigration levels (both legal and not) and high birth rates. In the mid-2000s, the natural growth rate of Hispanic Americans reached 3.7 percent, one of the highest in the world (comparable with growth rates of the fastest-growing African nations). That rate declined during the Great Recession, and by 2014 was slightly over 2 percent.

Hispanic settlement in the US has deep roots. The Southwestern states were part of Spain until 1821, and then Mexico until 1848, so some Hispanics in the area—especially in northern New Mexico—trace their ancestry back to colonial settlement. They call themselves *Nuevomexicanos*, and proudly distinguish themselves from later Mexican immigrants. The first wave of Mexican immigration came in the early 20th century when Mexican laborers, some fleeing from the Mexican Revolution, were recruited to work in the fields of California and Texas. During the Great Depression most of them were deported back to Mexico. Another wave came during World War II under the so-called **Bracero Program**, when Mexicans filled labor shortages resulting from the war effort (*bracero* means "manual laborer"). This program ended in 1964, but Mexican immigration continued to increase; Mexicans represented 5 percent of the total immigrant population in 1960, 15 percent in 1980, and 30 percent in 2000 (a proportion that has held generally steady since then). In 2010, 63 percent of Hispanics identified as being of Mexican origin, primarily mestizo (people of mixed European and Native American ancestry).

Although they are the largest single group of immigrants in the US, Mexican Americans are not evenly spread throughout the country. One-third live in the metropolitan areas of just five cities: Los Angeles, San Diego, Phoenix, Houston, and Chicago (see •**Figure 11.33b**). However, they are also the most rural of all immigrant groups: many of the counties of the Central and Imperial valleys of California, central Washington State, the Rio Grande Valley in Texas, and southwestern Kansas are over 50 percent Mexican; some, especially in Texas, are over 90 percent. The largest growth areas for Mexican immigrants since 2000 have been in the South: Georgia and North Carolina in particular have seen enormous growth in their Mexican populations, who find work in poultry plants, transportation, and the furniture industry.

Latinos of Cuban origin comprise an important demographic and political presence in the eastern US. Cubans began migrating to the US in the 19th century, and by the mid-20th century they numbered about 100,000 people. The pace of Cuban immigration quickened after Fidel Castro's takeover of the Cuban government in 1959. Initially, the US government admitted all Cubans as political refugees, and about half a million Cubans had fled Castro's regime by the late 1970s. Castro usually tried to prevent Cubans from leaving, but for a short time in 1980, let them go. About 120,000 Cubans left that year in an event called the **Mariel Boatlift**, named for the harbor from which the Cuban boats sailed. Castro put about 20,000 criminals in with the migrants to let them become America's problem. Americans were furious about this, and a backlash resulted: US policy changed in the 1980s so that if the Coast Guard intercepted Cuban migrants at sea, they were to be turned back to Cuba, but Cubans would be allowed to stay in the United States if they made it to American shores.

Two-thirds of Cuban Americans live in Florida. Most of the Cuban immigrants arriving after 1959 settled in Miami, fundamentally altering the city's demographics. In 1960, Miami was 5 percent Hispanic, and 80 percent non-Hispanic white. By the 2010 census, Hispanics were 70 percent of the city's people and non-Hispanic whites just 11 percent. Cubans make up half of Miami's Hispanic population; the other half includes a wide variety of national origins.

Visible Minorities in Canada

In Canada, people who are not of European or indigenous descent are known as "**visible minorities**." The Canadian government ostensibly uses the term to ensure fairness and equal opportunity in employment. The official definition, "persons, other than aboriginal peoples, who are non-Caucasian in race or non-white in color," has been criticized; the United Nations, for example, denounces "the assumption that whites are the standard against which anyone else is noticeably, visibly different."[10]

In Canada's 2011 census, 76 percent of Canadians reported European ancestry, and another 4 percent were "Aboriginals" (First Nations, Métis, and Inuit peoples). The largest non-European group in Canada is of Asian descent; those with ancestry from South, Southeast, and East Asia comprise 12 percent of the population. Much as in the United States, Chinese began migrating to western Canada in the mid-19th century; some established small businesses and farms, but most were laborers in fields and mines and on the Canadian Pacific Railway. Chinese and, later, Japanese and Indian immigrants faced significant hostility from white Canadians in British Columbia, where most Asian Canadians lived. Race riots wracked Vancouver in the early 1900s. British Columbia successfully lobbied Ottawa to ban Asian immigration in 1923 in a law that was not repealed until after World War II. Today, there are 1.5 million Chinese Canadians, 1.3 million Indian Canadians (belonging mainly to the Sikh religion; see **Figure 11.30**), and 700,000 Filipino Canadians.

Black Canadians are 3 percent of the population. Many blacks in Canada, especially those in the Maritimes, trace their origins to runaway American slaves who traveled north along the "Underground Railroad." Black immigration was banned in Canada for much of the latter 19th century and early 20th centuries. Once racial restrictions were eliminated, the country received a small but steady stream of migrants from the Caribbean, particularly Jamaica. Throughout their history, black Canadians have faced many of the same racial prejudices that American blacks have. Segregation was never mandated in Canada, but some schools and public facilities in

Ontario and other provinces were unofficially segregated until the 1960s.

Linguistic and Religious Patterns

English, French, and Spanish dominate this region's linguistic geography. English is the dominant language of both Canada and the United States, which share roots in colonial Britain (•**Figure 11.34**). Fifty-seven percent of Canadians and 80 percent of Americans use English as their primary language, while 76 percent of Canadians and 92 percent of Americans who speak a different language at home also know English. English and French are Canada's official languages. The United States has no official language, but 28 states have passed laws to make English official in those states. Efforts to pass a Constitutional amendment to designate English as the official language of the US have failed repeatedly.

French is the primary language of 21 percent of Canadians. The majority of these people are in Québec, where exclusively French speakers are 51 percent of the population. An additional 17 percent of Canadians and another 42 percent of Québecers are fluent in both French and English. Spanish-speakers are 12 percent of the population in the United States, where 10 percent of Spanish speakers have no working knowledge of English.

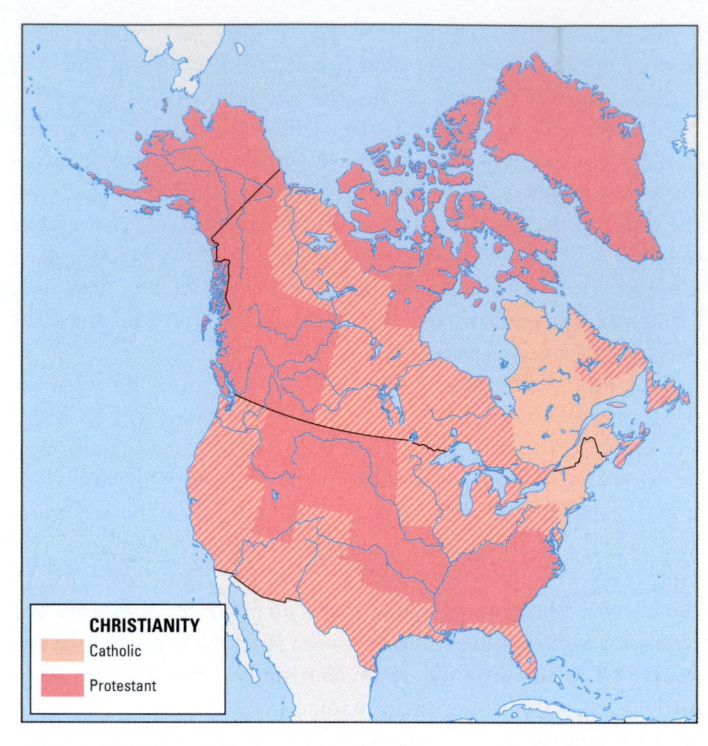

• **Figure 11.35** Religions of the United States, Canada, and Greenland.

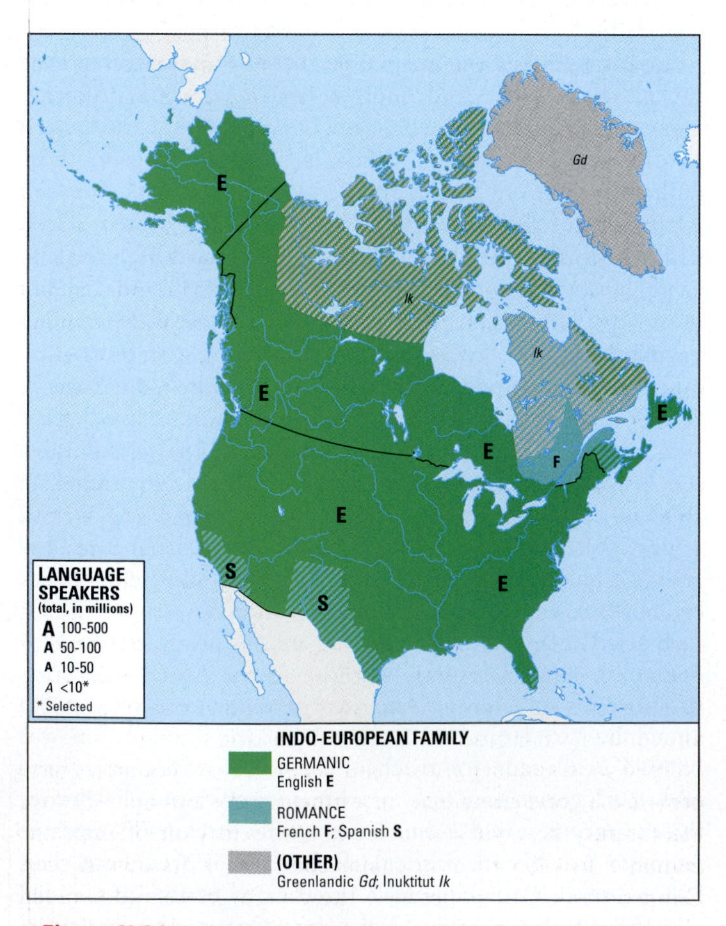

• **Figure 11.34** Non-indigenous languages of the United States, Canada, and Greenland.

Both countries have a diverse linguistic array. In the United States, over 30 languages are spoken by 100,000 people or more. After English and Spanish, Chinese, Tagalog (the language of the Philippines), French, Vietnamese, German, and Korean each have more than 1 million native speakers. Navajo is the most widely used indigenous language, with 170,000 speakers. In Canada, Chinese, Punjabi, Italian, German, Tagalog, Spanish, and Arabic are each spoken by more than 1 percent of the population.

Laws guarantee religious freedoms in both the United States and Canada, and both countries have a rich fabric of faith. Christianity predominates in both countries, as it has from the earliest days of European settlement (•**Figure 11.35**). Roman Catholicism is the largest single denomination in both countries today, for 39 percent of Canadians and 20 percent of Americans. All Protestant denominations combined represent 46 percent of the population in the US, and 27 percent in Canada.

About 6 percent of Americans practice non-Christian faiths. About 6 million Jews live in the United States, nearly the same number as in Israel. There are about 2.8 million Muslim Americans, and Islam is the fastest-growing religion in the US. Buddhism and Hinduism each have about 1.5 million adherents in the US. In Canada, where 11 percent of the religious population is non-Christian, nearly half that number practice Islam; there are about 1 million Canadian Muslims. Canada has about 400,000 Buddhists, Sikhs, and Hindus each, and about 300,000 Jews.

The United States is often considered an outlier among Western nations in its religiosity; but interestingly, 23 percent of both Americans and Canadians claim no religion, including those who are "spiritual" but do not belong to any organized

Joe Hobbs

• Figure 11.36 Protestant Evangelical churches have a tradition of hosting "revival" meetings, often in tents and over a period of several days, aimed at rekindling the spiritual dedication of churchgoers and attracting new members. This revival venue is in rural northwestern Florida.

church. Among those who are churchgoers, however, the two countries diverge. Canada has followed much of Europe in becoming increasingly secular over the last half-century. This is particularly notable in Québec: although most people identify as Roman Catholic, church attendance plummeted after the so-called "Quiet Revolution" of the 1960s. About 27 percent of Canadians attend religious services regularly. Attendance figures in the United States are also dropping, but more gradually; 46 percent of Americans attend services regularly. And although the number of Christian practitioners in the US is declining, the trends are not uniform. The population of Evangelical Protestants, who are much more inclined than other groups to attend services and who say that religion is an important part of their lives, declined only slightly between the censuses of 2000 and 2010 (**• Figure 11.36**). Catholics and Mainline Protestants have seen larger declines.

11.4 Economic Geography

The United States and Canada are very wealthy nations. Per capita GDP (PPP) figures for the countries are $52,800 and $43,100, respectively. Both countries belong to clubs representing the world's strongest economic powers: the Group of 8 (G-8) and the Group of 20 (G-20). Together, the United States and Canada generate an impressive 18 percent of the global gross national product.

As we have seen in Chapter 7, in 2014 the IMF reported that by one yardstick—total purchasing power parity (PPP) valuation—China had overtaken the United States to become the world's largest economy.[11] Subsequent investigations revealed however that China's statistics were exaggerated, leaving the US as the world's largest economy. Even if true, China in the top position would not necessarily be a bad thing for the US. The world economy is not a "zero sum game," where the advance of one economy comes at the expense of another. As China becomes wealthier and fulfills its ambition to shift from an economy based exclusively on exports to one sustained in good part by domestic consumption of imports, it will consume more American products, boosting the US economy.[12]

Generally, in the globalized economy, the United States may be seen as having built an economic structure of consumption, with consumer spending making up about 70 percent of the economy. By contrast, China has built its economy on production and export. The United States borrows money from other countries (it has the world's largest federal foreign debt—much of it to China—as well as the world's largest national debt), and it sells off assets to finance its consumption. The US has a trade deficit (for which China alone accounts for about one-third), exporting $1.6 trillion worth of goods while importing $2.3 trillion annually. Once the world's largest exporter of goods, the United States lost that position to China in 2013. The United States' share of global gross domestic production fell from 23 percent to 16 percent between 2000 and 2014. In the same period, the four leading emerging economies of Brazil, Russia, India, and China—the so-called BRIC—saw their share of global gross domestic production rise to a combined 29 percent.

Despite the rising economic competition, the US economy is powerful and has rebounded smartly since the Great Recession. These measures are indicative: the US dollar is the world's leading currency; the US is the world's second largest manufacturer, after China, with a fifth of global manufacturing output; the US is the largest producer of oil; the US has the world's leading financial markets; and due especially to immigration the US has the healthiest and most sustainable labor market among MDCs and the ability to attract and sustain more workers.[13]

Sources of the Region's Affluence

A number of geographic, political, and other circumstances have contributed to the development of the US and Canada as economically powerful countries. Both countries are among the five largest in area in the world, and a wide range of environmental settings has both allowed and stimulated full use of human ingenuity in economic development.

Although the US is the world's third most populous country, with so much land available, neither the US nor Canada is "people overpopulated." A large population represents both a big pool of prospective labor and talent to promote economic growth, and a vast market for the goods and services produced. Both countries developed mechanized economies early and under favorable conditions. Innovations continue to replace human labor with machine labor, generally increasing efficiency and productivity (with the downside of growing unemployment in traditional manufacturing sectors).

Peace and stability within and between these countries have provided a good climate for investment and economic development. Since the Civil War ended in 1865, the United States has managed to keep all of its major conflicts off its own shores. Cooperative relations between the United States and Canada also help. Both have an overall sense of internal unity despite episodes of poor race relations in the United States and the separatist movement in Québec. Both countries have a track record of continuity in political, economic, and cultural institutions. These stable conditions are conducive to economic growth. And most importantly, both have abundant natural resources.

An Abundance of Resources

This region, particularly the United States, has large endowments of some of the world's most important natural assets. The United States and Canada are the epitome of "neo-Europes," lands that European immigrants chose to settle largely because of their resemblance to the European environments and their potential for production of wheat, cattle, and 59 other vital agricultural products. The United States has more high-quality arable land than any other country in the world. A much smaller proportion of Canada is arable, but it still has more agricultural land than all but a handful of countries. Such abundance has helped these two countries become the largest food-exporting region of the world.

There is a strong geographic basis for American power and prosperity. Peter Zeihan calls the United States "The Accidental Superpower," and George Friedman uses the term "The Inevitable Empire," with both men arguing that geographic good fortune has endowed the country with an abundance of navigable waterways—17,600 miles of them—that lubricate the country's economy. Transporting goods by water costs 1/12 of shipping them overland. Of similar size to the US, China has only 2000 miles of navigable waterways. By more good fortune, many of those waterways overlay the Midwest, the world's largest patch of fertile ground. The US is also endowed with an abundance of natural harbors, including the world's three largest: Puget Sound, San Francisco Bay, and Chesapeake Bay.[14] "American geography dooms whoever controls the territory to being a global power," Friedman writes. "The Greater Mississippi Basin is the continent's core, and whoever controls that core not only is certain to dominate the East Coast and Great Lakes regions but will also have the agricultural, transport, trade and political unification capacity to be a world power—even without having to interact with the rest of the global system."[15]

The nation's physical geography does provide considerable advantages to the American economy: waterways and other endowments help make the US the world's most capital-rich country, with one of the lowest demands for capital because, unlike most of its MDC peers, it does not need to build an expensive internal trade infrastructure.[16] The economic geography of the US is one of diffused economic centers linked by transportation corridors, with a huge internal market and easy access to both of the world's great trading regions, Asia and Europe. The US has the geographic advantage of port infrastructure on both the Atlantic and Pacific and a strong (although aging) infrastructure linking them, so that in theory an economic downdraft occurring exclusively in either Asia or Europe would leave one coast an active trading zone.

These and other advantages allow the US federal government to maintain a hands-off, or "laissez-faire," economic system, contrasting with the government-planned economies even of capitalist, democratic Western European countries and with the "crony capitalism" of South Korea, for example, where the government invests heavily in favored firms. In the 359 free and competitive marketplace dominated by small and midsized businesses, American workers are more productive and efficient than most of their foreign counterparts. These smaller firms can go out of business (and be replaced) without dragging down the economy. In the liberal capital model, the market is more efficient than the government is in allocating resources. On the downside, laissez-faire economics are prone to boom-and-bust cycles.

The global economic system was built largely by the United States in the post–World War II system of monetary management known as the "Bretton Woods system," which established the IMF and the World Bank. The global economy depends overwhelmingly on the health of the US economy. The rapid growth of Asian economies (which Americans often perceive as threatening) has depended on American markets, and if the US economy is troubled, the Asian economies suffer the "contagion."

Oil and other Minerals

The United States and Canada are very rich in mineral resources (•**Figure 11.37**). Although a major mineral producer, the United States is also a major importer, especially of oil and industrial raw materials. Many of the ores easiest to mine and richest in content are now mostly depleted, and even large mineral outputs in the United States cannot meet the huge demands of the domestic economy. In many such cases, Canada has been able to assume the role of leading foreign supplier.

Abundant fossil fuel resources (coal, oil, natural gas) have been essential to the region's economies. Coal was the largest source of energy during the nineteenth-century industrialization of the United States. Today, the country leads the world in estimated coal reserves (about 25 percent of the total). Environmental concerns about the impacts of this most polluting of fossil fuels—and the leading source of greenhouse gas emissions linked to climate change—raise long-term questions about what the United States can do with its generous coal endowment. As it stands, the United States gets 39 percent of 47 its electricity from coal-fired plants, with natural gas supplying 27 percent and nuclear power 19 percent.

Recent years have seen impressive growth in proven reserves of fossil fuels in both Canada and the United States. Canada has about 11 percent of the world's proven oil reserves, and the US, about 2 percent. The huge US economy demands nearly 7 billion barrels a year—as much as the three next largest consumers (China, Japan, and Russia) combined. Although the US is the world's largest oil producer (as of 2015), this is far more than the country can produce, so the US imports nearly half the oil it consumes (44 percent). Its top five suppliers (in descending order) are Canada, Saudi Arabia, Mexico, Venezuela, and Iraq. The United States obtained 36 percent of its imported oil from Canada in 2015—more than from all the OPEC countries combined.

• **Figure 11.37** Distribution of fossil fuels across North America, with Keystone and Trans-Alaska Pipelines.

Think critically: Is it better for the Keystone XL project to be built so the tar sands will be refined in Texas rather than China? How does climate change figure into this debate?

Canada has an impressive fossil fuel endowment, with the world's largest reserves of **tar sands** (also known as "oil sands"). The potential output of this unconventional oil makes Canada third in the world (after Saudi Arabia and Venezuela) in proven oil reserves (Canada's conventional oil reserves are small). There is a catch, however. Tar sands are mineral deposits in which a thick liquid petroleum, called bitumen, is mixed with sand and clay (• **Figure 11.38**). Through a water-guzzling, energy-consuming

• **Figure 11.38** Extracting Athabascan tar sands.

process, the bitumen can be converted to petroleum. When conventional oil prices are high, the tar sands are economically attractive to exploit. The price surge in oil beginning in 2003 led to strong growth in production from Canada's Athabascan tar sands, with most of the output going to the United States. Collapsing oil prices in 2014 did not affect tar sand production—in fact, production increased—but as oil prices continued to drop, Canada's economy slid back into recession in 2015. Due to its reliance on tar sands, which are especially "dirty" in terms of carbon emissions, Canada also was forced to withdraw from the Kyoto Protocol in 2011; it could not possibly meet the Kyoto emissions reductions targets it ratified in 2004. 46

Canada's ability to refine tar sands into useable forms such as gasoline is limited. For years there has been a fierce debate on whether Canada's tar sand oil should be refined in the United States or China. The former option would involve shipping the coarse mixture of bitumen and other materials through a pipeline from their Canadian point of origin to refineries on the Gulf of Mexico. This would be done through the 1700-mile-long **Keystone XL Pipeline** (see Figure 11.37). Environmentalists have opposed the project on the grounds that the pipeline crosses portions of the critical Ogallala Aquifer in Nebraska (see Figure 3.C). An oil spill could taint these precious waters, which support about 27 percent of the irrigated farmland in the United States and is being depleted. There is an option to reroute the pipeline across less sensitive lands. However, opponents also warn that refining the "dirty" tar sands on the Texas coast would release enormous quantities of greenhouse gases into the atmosphere. Proponents appreciate the prospect of an estimated 20,000 US jobs that would be created by the Keystone XL construction effort. President Obama rejected the project in late 2015, but the highly politicized pipeline could be approved by a subsequent administration. Were this pipeline project to be cancelled, Canada would build a pipeline to a terminal on its Pacific coast, from where the tar sands fluids would be shipped to China for refining. Keystone XL proponents argue that whether the refining takes place in the US or China, the greenhouse gases will join Earth's atmosphere, and that US refining would be cleaner than China's.

There are other environmental risks in oil production. To access remote sources, producers are drilling technically challenging and environmentally risky reserves, including the Macondo Prospect in the US EEZ waters of the Gulf of Mexico. The Gulf is estimated to have 15 percent of 317 America's proven reserves. One of the US-owned and British Petroleum (BP)–operated offshore rigs there, the *Deepwater Horizon*, was drilling at a depth of 5000 feet (1524 m) below the Gulf of Mexico seafloor (at a combined depth of 10,000 feet [3048 m]) when an explosion took place in April 2010. Over the next three months, oil gushed virtually unabated into the Gulf of Mexico. This spill caused extensive damage to marine, saltwater marsh, and shoreline wildlife and to the fish, shrimp, oyster, and tourism industries of the Gulf, and restoration is ongoing.

Natural gas has become an increasingly important fossil fuel to these two economies. The United States is the world's largest consumer of gas and is a major importer, especially from Canada. The US ranks first in the world in output, producing 20 percent of the world's gas from its 4 percent of proven reserves worldwide. That equation means that the present high output cannot be sustained for long. Canada, with smaller but rapidly increasing proven reserves (from shale gas), produces about 4 percent of world natural gas output and is a major exporter.

In recent years many Americans have celebrated what they call **America's Energy Revolution**, thanks to an abundance of fossil fuels that were always there, but not always easy to access. Since the late 1990s, a spectacular development has been the exploitation of natural gas (and, to a lesser extent, oil) lying in shale deposits (see Figure 11.37). When energy prices were low, it was uneconomical to coax this energy from the ground, but as prices rose, so did the incentive to tap these hydrocarbons. The resources are obtained with innovative technologies including horizontal drilling and an energy- and water-intensive process known as **hydraulic fracturing**, or **fracking** (•**Figure 11.39**). Water, sand, and chemicals are pumped underground at high pressure through a well into shale deposits thousands of feet deep. This causes fissures and layers in the rock to crack. The sand particles keep the fissures open, allowing natural gas to flow up the well. Fracking is a special challenge and even a dilemma in drought-prone areas. In using enormous amounts of water, the process draws down on aquifer waters that are also needed for irrigation and other human uses. The gas can even leak into water reservoirs and even be lit as it comes out of users' kitchen faucets!

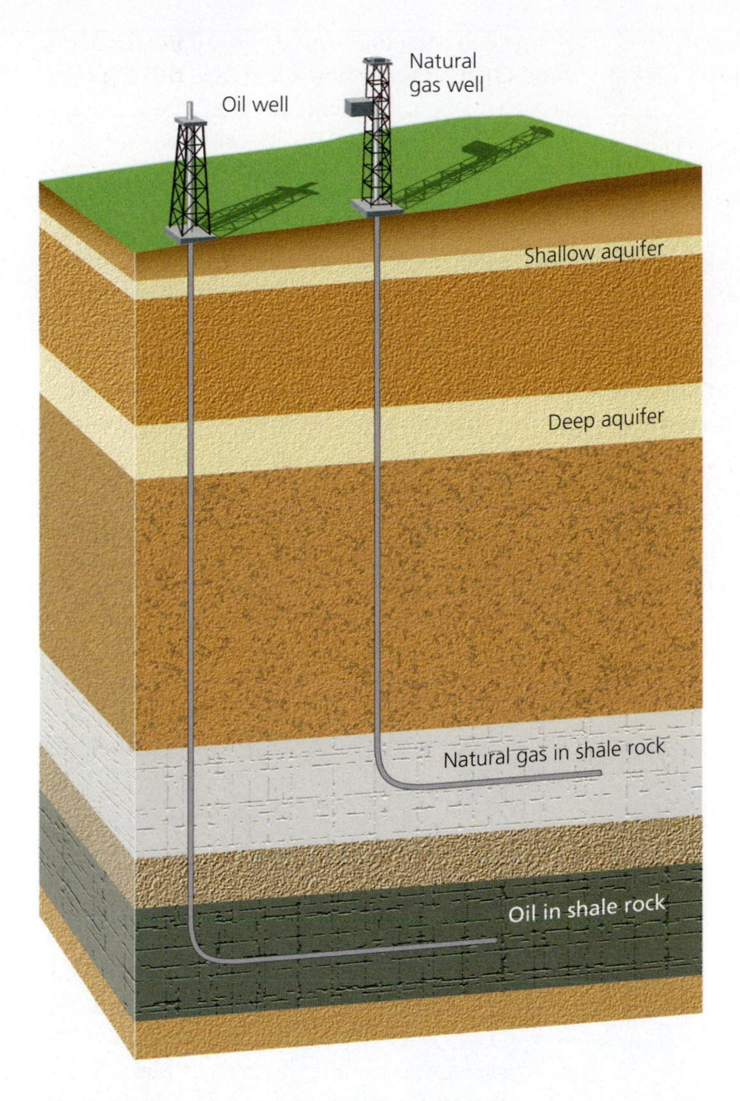

• **Figure 11.39** Horizontal drilling and hydraulic fracturing, or "fracking," are technologies that extract oil and natural gas held tightly in shale rock formations.

Some Canadian provinces and American states are particularly well endowed with frackable energy resources; Alberta, British Columbia, North Dakota, Texas, Arkansas, Oklahoma, Ohio, West Virginia, and Pennsylvania are among them (see Figure 11.37). Some of these states are in the "Rust Belt" of derelict manufacturing, and the shale gas industry helped revitalize their long-depressed economies. Tens of thousands of new jobs were created, not only for those who work directly to get the energy out of the ground but also in manufacturing the equipment and parts for the rigs and supplying services like food and lodging.

527

Exploitation of new reserves of oil in North Dakota and elsewhere decreased America's dependence on foreign oil imports, from 60 percent in 2005 to 44 percent in 2015. A windfall to the American consumer (especially in the form of lower gasoline prices) came in 2014–2015 with falling global oil prices due to surging US production, OPEC and especially Saudi restraint in oil production, and China's slowing economy. One downside of this oil glut was its immediate impact on the domestic oil industry and related services. Another was that it distracted attention and investment away from alternative energy development. It is easy to forget that fossil fuels are nonrenewable and that the future will have to depend on energy alternatives (see Geography of Energy, **page 524**).

Agriculture and Forestry in the US and Canada

Like most MDCs, the US and Canada transitioned from primarily rural to urban as the countries industrialized. Agriculture is still a great source of economic wealth in both countries, representing 5 percent of GDP in the US and 8 percent in Canada (• **Figure 11.40**).

Exporting about $115 billion in agricultural products annually, the US has an agricultural trade surplus; agricultural products represent 11 percent of American exports, and just 5 percent of its imports. One-third of American farm acreage is planted for export, and about one-third of farm income comes from exports. The principal markets for American and Canadian agricultural exports are China, Mexico, and the European Union. The prodigious exports of US produce are boosted in part by generous subsidies, which are controversial in the global marketplace (see Insights, **page 526**).

The demographics and technologies of agriculture have undergone profound change in both countries. In 1900, 40 percent of the US and Canadian workforce was in agriculture; by 2015, that share had fallen to less than 2 percent in both countries. There are about 2 million farms in the US, 87 percent of them operated by families. However, the top 10 percent of American farms produce over 70 percent of agricultural product sales. These are primarily farmlands owned by agribusinesses and large family corporations. In the 1930s,

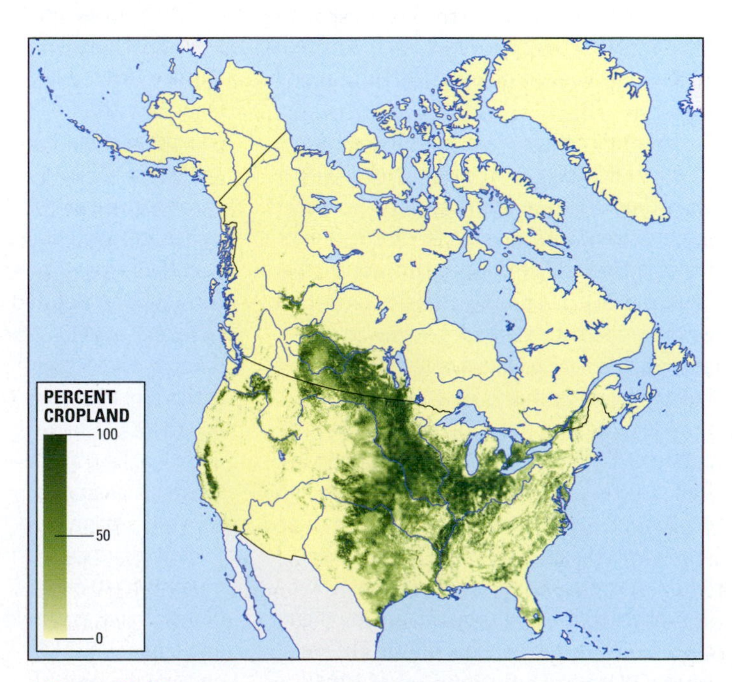

• **Figure 11.40** Croplands of North America.

Source: "Global Agricultural Lands: Croplands, 2000." NASA Socioeconomic Data and Applications Center (SEDAC).

Geography of Energy Energy Efficiency and Energy Alternatives in the United States

America's energy policy recently has been based on an "all of the above" strategy, exploiting conventional and alternative sources to feed the country's enormous energy appetite. In this mix, even the relatively dirty and dangerous energy sources are included, with the resolve to make them cleaner and safer.

Nuclear power is part of the mix. Like Canada, the US has large reserves of uranium, the fuel used for this source of electricity production. The United States is the world's largest producer of nuclear power and derives 19 percent of its electricity from that source. Nuclear energy has a difficult past and uncertain future in the US, however. In 1979, an accident occurred in the Three Mile Island nuclear power plant in Pennsylvania, releasing little radiation to the environment but spurring enormous public fear of nuclear power. The disastrous Chernobyl accident in 1986 in Ukraine and troubles at Japan's Fukushima reactors in the wake of the 2011 tsunami heightened the perceived peril of nuclear energy. Because of such fears as well as cost overruns and other problems, no nuclear power plants have gone online in the United States since 1978. The same concerns have troubled Canada, which gets 15 percent of its electricity from nuclear power. It too has not built a new nuclear power plant since the 1970s. Nevertheless, the external and other costs of rising fossil fuel use has forced a serious reevaluation of nuclear energy in both countries.

194, 339

83

The more benign sources of alternative energy—including windmills, photovoltaic panels, hydrogen fuel cells, and hydropower—meet just 13 percent of US energy needs. The United States did not ratify the Kyoto Protocol or other international legislation that would compel it to embrace alternative energies. Renewable energy sources have received only modest federal funding for research and development since the energy crisis of the 1970s. In 2009, however, as part of a large stimulus bill to help bring the United States out of the Great Recession, a hefty infusion of $10 billion in federal subsidies helped produced a boom in solar and wind power. Some states have acted independently of the federal government in adopting energy alternatives. California has led these efforts. The state's **Global Warming Solutions Act of 2006** requires that greenhouse gas emissions be cut to 1990 levels by 2020, by which time utilities will have to generate one-third of their energy from renewable resources.

45

Federal legislation has improved energy efficiency, especially in the vital transportation sector that accounts for about 28 percent of US energy needs. In 1975, in the wake of the energy crisis, Congress enacted the **Corporate Average Fuel Economy (CAFE) standard**, requiring that the fleets of passenger cars produced by all the auto companies achieve an average of 27.5 miles per gallon. The standard for light trucks (which include SUVs) was set lower, at 20.7 miles per gallon. President Obama's administration upped the CAFE ante to 54.5 mpg for cars and light trucks by 2025. Automakers have made great technical leaps to increase fuel efficiency, but they have had to reduce vehicle weight by using lighter components, and this may mean reduced safety for passengers.

Remarkable changes in the US auto and farming sectors have accompanied the rapid embrace of ethanol fuels, belonging in the biomass sector of alternative energies (**Figure 11.B**). Ethanol, made primarily from corn in the United States and typically mixed with gasoline to be sold as "super unleaded" fuel, helps reduce gasoline consumption. Boosted by federal ethanol subsidies of 45 cents per gallon (until the subsidy ended in 2011), more and more land in Midwestern states went into corn production (one-third of the total) and more corn (40 percent of the total) was turned into ethanol. The US became the largest ethanol producer in the world. Ten percent of the gasoline in the US is derived from ethanol. However, ethanol may not be a long-term solution. Much fossil fuel–based energy and fertilizer goes into its production; more energy may be used to create ethanol than the ethanol itself yields. Diversion of the corn crop into ethanol also raises food prices. Not only does corn as a human food source become more expensive, but so does the price of feed corn for hogs and cattle, leading to higher costs of pork, milk, and other produce.

484

Hydropower development has been so intense in the past that the two countries rank with China and Brazil as the world leaders in hydropower capacity and output. Canada gets 59 percent of its electricity from hydropower, and the United States, 7 percent. The countries'

Scott Olson/Getty Images News/Getty Images

• **Figure 11.B** Pictured here is stillage, a byproduct of corn ethanol production, which can be combined with other ingredients to make high protein livestock feed.

Joe Hobbs

• Figure 11.C Powerful wind turbines like these in an Illinois cornfield have transformed the rural landscapes of the Great Plains, the Midwest, and the West.

dams are controversial because of the landscape changes they produce and also because they often result in the inundation of indigenous peoples' lands, particularly in Canada. A coalition of Native American groups, environmentalists, and sport fishing enthusiasts has long been lobbying for dams in the Pacific Northwest to be **decommissioned** (heavily modified or removed) primarily for environmental reasons, especially to allow a dozen endangered salmon species to reach their traditional upstream spawning grounds. Wind-generated electricity would theoretically replace the hydropower lost when the dams are decommissioned.

Alternative energies with the least environmental impacts—wind and solar power—are becoming less expensive and more widely embraced. Wind resources in the US are enormous and have the potential to produce five times the country's total electricity production. Wind turbines are sprouting up across America's rural landscapes (**• Figure 11.C**) as farmers and ranchers have found wind energy to be profitable. The US Department of Energy's "Wind Powering America" initiative has provided $60 billion in capital investment to rural America, with the goal of producing 5 percent of the nation's electricity by 2020.[17] Farmers can lease land to wind developers, use the wind to generate power for their farms, or become wind power producers themselves. With "net metering," when a turbine produces more power than the farm needs at the moment, the extra power flows back into the utility system for others' use and turns the farmer's electric meter backward.

Solar power is the cleanest form of energy. The cost of rooftop panels dropped by half between 2011 and 2014, and the cost of producing electricity from solar power has become competitive with that of conventional sources like coal in several states.[18] The growth of the solar industry is impressive, increasing 17-fold between 2008 and 2014. Remaining capacity in the US is enormous, but challenges exist, especially in tying solar power into the electricity grid system. Giant "solar farms" in the southwestern US produce utility-scale output, supplying electricity to the grid feeding cities such as Phoenix and Las Vegas (**• Figure 11.D**).

Growing energy efficiency and energy alternatives in energy-hungry America represent considerable up-front costs, but long-term economic advantages. They also help confront global climate change by reducing greenhouse gas emissions. Some countries have been quicker to embrace the economics of green technologies; Germany and China, for example, are the leading producers and consumers of solar energy, but the US is narrowing the gap.

Joe Hobbs

• Figure 11.D In 2014, the Crescent Dunes Solar Energy Project in Nevada became the world's largest solar power plant. It generates about half a million megawatt hours of emission-free electricity, enough to meet the needs of about 75,000 households and to replace 290,000 metric tons of CO_2 emissions annually. A large circular field consisting of 17,170 heliostat mirrors focus sunlight on a central receiving tower where a conventional steam turbine produces the electricity.

Insights Trade Barriers: Subsidies and Tariffs

Many of the world's LDCs are convinced that the best way to develop is not to borrow money from the richer countries, but to trade with them. With their larger populations and lower labor costs, the LDCs should easily be able to churn out products at much more favorable prices than those sold by the MDCs. Buyers around the world should be snapping up those affordable products, pumping desperately needed money back into the developing countries. This exchange does not take place as freely as it could, however, because of two tools that the US, Canada, the EU countries, Japan, and other MDCs wield as much as possible to protect their own producers: subsidies 112 and tariffs.

Subsidies are expenses paid by a government to an economic enterprise to keep that enterprise profitable, even if the market would not. Tariffs are essentially taxes that a government imposes on foreign goods, basically for the same reason, to ensure that domestic products make it to market, even though they are not produced at competitive costs. Alone or in combination, subsidies and tariffs can make it all but impossible for a less developed country to join the world economy as a trading partner.

There are often struggles between the MDCs and LDCs over the issues of subsidies and tariffs, especially in meetings of the World Trade Organization (WTO). At these meetings, the LDCs are represented by a bloc, led by Brazil, that calls itself the **G-21 (Group of 21)** as a "play on numbers" against the affluent G-8.

Often bitter debate between these groups focuses on subsidies that the MDCs have used for decades to prop up certain sectors of their economies, especially agriculture. The subsidies were created to help farmers and the countries get through especially difficult times, including the **Great Depression** of the 1930s in the United States, when the livelihoods of so many farmers were threatened, and post–World War II Europe and Japan, when the specter of malnutrition loomed. Over the years, the critical need for subsidies passed, but the sectors that received them were loath to part with them. Federal supporters of subsidies found political capital in maintaining them. Today, farming subsidies are something of a sacred cow in many of the MDCs, and any effort to reduce or remove them would have domestic consequences—especially at the ballot box.

The G-21 countries, speaking for two-thirds of the world's farmers, demand that the MDCs reduce or eliminate such subsidies and tariffs or that the LDCs be compensated for the revenue lost to them. They threaten to begin erecting tariffs of their own (a proposal the MDCs say is bad for the developing economies) if the MDCs do not yield. The US subsidized its cotton farmers with over $4 billion a year, which had serious impacts on the economies of Brazil and several cotton-growing African nations: US-subsidized 421, 484 production drove world cotton prices down to 50 cents per pound, well below the African cost of 65 cents per pound, for example. The US ended its cotton subsidy (which was ruled illegal by the WTO) in 2013 after a decade of protests by Brazil, which argued that the US was depressing world cotton prices and which threatened sanctions against US interests. The US paid Brazil's cotton institute $300 million in return for Brazil dropping all cotton-related claims against the US.

there were about 7 million farms in the US, and the average plot was about 150 acres; today, the average plot is nearly 500 acres. The privately owned King Ranch in south Texas is the largest at over 800,000 acres.

Better land management practices and genetic modification 116 of crops have boosted crop yields. Today's farmers produce 262 percent more food with 2 percent fewer inputs (labor, seeds, feed, fertilizer, etc.) compared with 1950. Crop rotation (growing different crops in succession on the same land) and contour farming (planting crops across the slope of the land) help conserve water, protect soil, and boost yields. Just since 1982, erosion of cropland by wind and water has declined by half. Conservation tillage, which reduces erosion (soil loss) on cropland while using less energy, is used on two-thirds of America's farmland. More than half of America's farmers deliberately provide wildlife habitat. The Conservation Reserve Program, designed to reduce soil erosion and provide wildlife habitat, began in 1985. Farmers have enrolled more than 30 million acres in the program, which has restored more than 2 million acres of wetlands. Farm bill funding has prompted farmers and ranchers to create more than 2 million miles of buffer strips, creating wildlife habitat and improving soil, air, and water quality.[19]

Canada and the United States have abundant forest resources. Canada's forests are greater in area than those of the United States, but not as varied. Wood has been an essential material in the development of both countries, prompting the forest-rich LDCs, especially Brazil and Malaysia, to cry foul when American environmentalists denounce tropical deforestation. Even now, the United States cuts more wood for lumber 44 each year, produces more wood pulp, and produces more paper than any other country in the world. The United States both exports and imports large quantities of wood, with the exports going largely to Japan. US domestic industries can meet only two-thirds of their lumber needs, and their imports come mainly from Canada; half of the wood the US imports originates there. Canada is the world's largest exporter of wood (**•Figure 11.41**). The United States purchases 67 percent of Canada's timber exports; Japan is a distant second with 8 percent.

US-Canadian Economic Relations

Each of these countries is a vital trading partner of the other, although Canada is much more dependent on the United States than vice versa. In 2014, Canada supplied 15 percent of all US imports by value and took in 19 percent of all US exports, making Canada the leading country in total trade with the United States. That same year, the United States supplied 52 percent of Canada's imports and took in 76 percent of its exports. The general pattern of trade between the two countries is the exchange of Canadian raw and intermediate-state

Richard Kolar/ Animals Animals

• **Figure 11.41** Canadian timber in route to East Asia.

materials—such as ores and metals, timber and newsprint, and oil—for American manufactured goods.

Free-trade agreements designed to open markets and increase revenues for both countries have facilitated this exchange. Nevertheless, the countries have periodically violated the principles of free trade to protect their industries from one another. In the 1990s, there was a so-called "wheat war" between the two countries as they struggled for greater market share in an environment of high yields and low prices. US producers claimed that Canada was dumping (selling the product for less than its cost to produce it) its durum wheat on the US market. The United States also accused Canada of subsidizing timber (pricing this commodity artificially low) and dumping low-cost lumber in the US. In response, in 2003 the US imposed heavy new tariffs of 27 percent on softwood lumber imported from Canada. An agreement reached in 2006 lowered the tariffs and required Canada to price its lumber more competitively.

Despite such volleys occasionally fired on both sides of the border, the trend has been toward more cooperation and free trade. In 1965, the two countries negotiated a free-trade agreement for the production, import, and export of automobiles. This sparked rapid expansion of the auto industry into southern Ontario (Canada's most industrialized province), on the border with the US automobile-producing state of Michigan. When the auto industry suffered in Detroit, it also did so in neighboring Windsor, Ontario.

The two countries entered into the comprehensive **Canada-US Free Trade Agreement** in 1988, and in 1994 Canada, the United States, and Mexico enacted the North American Free Trade Agreement (NAFTA). Its goals and impacts are described in detail, especially as they affect Mexico, on **pages 474–475** in Chapter 10. NAFTA has brought a boom in trade between the three countries but also caused major industrial readjustments, generally characterized by the more labor-intensive firms (beginning with clothing) moving to Mexican locales from Canada's southeastern core region and from the industrial northeast of the United States.

Manufacturing and the Great Recession

Although raw materials contribute much to their wealth, the United States and Canada have become prosperous mainly because of machines and mechanical energy, complemented in recent decades by a boom in information technology (IT). From an early reliance on waterwheels powering simple machines, the United States increasingly exploited the power of coal-fired steam engines. Later came oil, internal-combustion engines, and electricity, all harnessed for greater and more efficient energy use.

In this energy-abundant and mechanized economy, the United States was able to take advantage of unique circumstances. Its abundant resources attracted foreign capital and stimulated domestic accumulation of capital through large-scale and often wasteful exploitation. There was also a labor shortage that attracted millions of immigrants as temporary low-wage workers but also promoted higher wages and labor-saving mechanization in the long run. A culture of striving for advancement emerged. This emphasis on productivity, dubbed the **Protestant work ethic** but shared by people of many faiths and ethnicities, characterized the struggle for economic success and the **"American Dream"** of achieving a comfortable life by working hard.

In the first part of the 20th century, wealth and jobs clustered where there were large industrial concentrations in Northeastern and Midwestern states. The brilliant industrialist Henry Ford located his innovative mass production of cars in Detroit, which brought down costs for consumers, and the American car culture grew. The Ford Motor Company would become one of the "Big Three" American automobile manufacturers, along with General Motors and Chrysler, both of which are also headquartered in Detroit, earning it the nickname **Motor City**.

The story of Detroit encapsulates the American experience with heavy manufacturing. A jobs mecca, the city boomed from the early 20th century to the 1960s. In addition to autos, other industries including heavy machinery, steel, and chemicals proliferated until about 1965, when the decline of American heavy industries began (**•Figure 11.42**). Many of these industries shifted overseas where labor costs were lower and regulations more lax. Increasing globalization and international trade brought more foreign-made products into the US, competing with American-made goods. Other factors were technological advances and increased automation, which required fewer workers. Factories for many industries started closing in Detroit and across the Midwest, and by the 1980s the region came to be known as the **Rust Belt**.

Growth in the service sectors has reduced the relative contributions of manufacturing in these two economies. Manufacturing now only accounts for about 10 percent of the US economy (down from as much as 30 percent in the early 1950s), and 11 percent of the Canadian economy, having fallen to the forces of globalization that moved traditional industries to countries with lower manufacturing costs. Mechanization and outsourcing have gutted the manufacturing and clerical jobs of unskilled workers in factories and offices and delivered them to

Joe Hobbs

• **Figure 11.42** The United States gained economic supremacy in the 20th century in large part because of its production of steel and other manufactured goods. It lost much of its production to other countries where labor and other costs were lower. South Korea's Posco steel mill, seen here, puts out a product that once epitomized the economy of Pittsburgh, Pennsylvania.

lower-wage workers in the LDCs. As these jobs moved to the LDCs, they lifted hundreds of millions of people out of poverty. Inequities grew in MDC economies but fell on the global scale. [62]

Despite representing only a small slice of the American economy overall, manufactured goods make up two-thirds of US exports by value, and the nation has the global lead in many high-tech products. Canada still has healthy automobile and aeronautical industries. Canadian manufacturing did not decline as precipitously as American manufacturing did, largely thanks to Canada's lower labor costs and the vast American market right next door.

Ironically, perhaps the best prospect for more growth in US manufacturing is investment from China, as that country looks to produce goods for the American market that are cheaper to produce in the United States than to export from Asia. Wages for Chinese and other foreign workers are rising, pushing manufacturing jobs back to the United States; this process is known as insourcing (or reshoring, that is, the reverse of outsourcing or offshoring; see **page 313**). Cars again provide a good example. Foreign auto companies like Mercedes, Honda, and Hyundai set up factories in the US, mainly in states such as Alabama, Kentucky, and Indiana, which offered lower labor and land costs than elsewhere in the country; these states also offered incentives like lower taxes to encourage the companies to establish manufacturing plants there. Other foreign-owned companies making chemicals, wind turbines, and jet engine parts have also established factories in the United States in recent years.

Services and Information Technology

The service (or tertiary) sector dominates the economies of both the United States and Canada. Services employ about three-quarters of the workforce in both countries. In the US, the largest single category of workers in the service sector is government employment. There are 2 million federal government employees and about 19 million people in state and local government; together they comprise 15 percent of all American workers. Professional and business services represent 12 percent; this broad category includes lawyers, scientists, accountants, human resources personnel, and engineers. Health care workers are 11 percent of the American workforce (including doctors, nurses, therapists, and paramedics). Ten percent are in sales—everything from cashiers to corporate purchasing agents to real estate brokers. The remaining 30 percent of American workers in service jobs run the gamut from educators and librarians to truck drivers, data processors, actors, bartenders, graphic designers, and financial analysts.

Some economists have suggested that some service sector jobs, particularly those in the STEM fields (science, technology, engineering, and mathematics), constitute a fourth economic sector, sometimes called the "knowledge economy." [63] Although technology has always been the engine of change for societies—the wheel was a major technological advancement—today we usually consider "technology" to mean electronic devices, especially those networked together. This required several innovations (particularly the invention of the microprocessor), but the most revolutionary was the development of the **Internet**. How it would transform the world could not have been foreseen during its early days.

The Internet, initially known as ARPANET, was originally created by California university researchers in the 1960s with funding from the US Department of Defense. The Internet was largely the domain of scientists and the US government until 1989, when the first ISPs (Internet service providers) were formed, allowing the general public access for the first time. The invention of the World Wide Web in 1991 and the opening up of the network to (previously forbidden) commercial use in 1995 led to a dramatic increase in the number of users. In 2000, the Internet was carrying half of the world's telecommunications; by 2007, it reached 97 percent. Companies like Microsoft and Apple were beneficiaries of a large upswing in computer sales, and many new companies such as Amazon, Google, and Facebook were founded to serve exclusively online audiences; they soon became as wealthy and influential as "old economy" corporations in the financial, oil and gas, and automotive industries (**• Figure 11.43**).

As telecommunications technologies have interconnected North America with the rest of the world, the United States has acquired global leadership in information technology (IT), whose industries encompass the Internet, computer hardware and software, electronics, semiconductors, telecom equipment, e-commerce, and computer services. Much of America's economic growth is concentrated in "smart cities" such as San Francisco, Seattle, Boston, and Washington, where the knowledge economy is driven by higher education levels and a larger supply of highly skilled workers. The United States hopes to [542]

Joe Hobbs

• **Figure 11.43** 1 Infinite Loop, Cupertino, California, is one of the most important tech addresses in the world. This is the Apple Campus, with a Chinese business delegation arriving to do business.

maintain its global leadership in design and innovation, even while other countries actually make products. As one American economist put it, "We want people who can design iPods, not make them."[20] High-tech industries increasingly built what came to be known as a **platform economy**, with companies like Dell operating headquarters in the United States but outsourcing virtually all work on its products to other countries.

In the 1990s, there was such a speculative frenzy about IT that a classic **economic bubble**, a surge in asset prices far above the fundamental value of the asset, developed. Tens of millions of Americans started investing in stocks for the first time, bidding up the values of companies—even companies that produced no tangible products—to lofty heights. There was talk of a "new economy" in which the old paradigms of corporate investment and growth no longer applied. However, this particular **tech bubble** burst in 2000. Clever or lucky investors withdrew their investments in time, but others lost fortunes.

The far larger **housing bubble** burst in 2007 and led to the **Great Recession** of 2007–2009, referring both to the largest economic downturn in the US since the Great Depression and to the ensuing global recession of 2009. The implosion began in the United States housing sector, where through so-called "subprime loans" millions of underqualified American borrowers bought homes they could not afford. The biggest asset in Americans' net worth is their homes, which declined sharply in value. Many went "underwater," meaning that they were worth less than the money owed on them. Unable to keep up on payments, many Americans who were lured by seductive credit terms into buying homes they could not afford lost their homes to foreclosure. As banks foreclosed on the unpaid properties, housing prices slumped, and a full-fledged crisis developed in the banking industry. This **credit crisis** in the United States spread quickly worldwide, sparking a sell-off in global stock exchanges, the widespread economic recession, and plummeting prices for oil and the other commodities that had supported booming economies.

The US dealt effectively with these crises, using assertive monetary and fiscal policies and reforming the banking sector.[21] The Bush administration used the Troubled Asset Relief Fund (TARP) to "bail out," or recapitalize, the broken banking system in 2008, especially institutions like Goldman Sachs and Morgan Stanley, judged "too big to fail" (so large and interconnected that their failure would spark a ruinous contagion in the global economic system). Instead of reining in spending as many European economies later did with their "fiscal austerity" approach during the 2010–2012 European Debt Crisis, in 2009 the Obama administration pumped billions of dollars of borrowed funds into stimulus. It was a risky recipe, but in bringing the US back from the brink of economic catastrophe it was arguably more effective than the European model.

Technically, the United States made its way out of the Great Recession by mid-2009, but recovery remained slow afterward, especially in the jobs market. Falling tax revenues meant huge cuts in city, state, and federal budgets. The US GDP declined by 3 percent in 2009, but resumed slow growth in 2010. The US unemployment rate peaked at 10 percent in 2010 (up from 4.6 percent in 2007); by 2015, it had fallen to under 6 percent. It must be noted, however, that these unemployment numbers do not count "missing workers," including those who are not actively seeking jobs. Canada's more peripheral role in international finance, and its larger reliance upon exports of raw materials and fossil fuels, allowed it to weather the recession better than the United States and Europe did. For decades, unemployment in Canada has almost always been slightly higher than in the US—in 2014, it had an unemployment rate of 7 percent—but peak unemployment in Canada reached only 8.3 percent during the recession.

Service sector productivity has remained roughly the same in recent decades, and only increased prosperity is growing the sector. Some economists see a "productivity paradox," arguing that the service sector actually hinders productivity. Economic development in the MDCs relies on consumer spending, but a lack of constant productivity improvements is restraining economic growth. Innovative information technology solutions have the most potential to solve the productivity paradox.[22] In a provocative essay entitled "How Technology Wrecks the Middle Class," two economists argue that we may have "mechanized and computerized ourselves into obsolescence."[23] The authors argue that advances in IT incentivize employers to substitute cheap computers for expensive labor, raising fears that workers will be displaced by machinery. Technological advances do not threaten employment, however; they polarize employment, with job growth concentrated in both the highest and lowest paid occupations, while jobs in the middle have declined. At the high end are the professional, managerial, technical, and creative occupations, which involve "abstract tasks" requiring critical thinking, problem solving, and creativity. At the low end are the manual jobs like food services, cleaning, and security that computers cannot replace. The middle of the labor market, where the routine task-intensive jobs are, is stagnant.

The authors foresee three categories of job types that will thrive in our technological future: professional and managerial jobs requiring college education or advanced training, jobs that are expanding as they use technology to improve their own productivity; manual task-intensive jobs that computers cannot do, such as cooking, cleaning, and truck driving, that are plentiful but pay poorly because so many people can do them; and middle-skill jobs that combine routine technical tasks with some adaptability, dexterity, and interaction with people. Jobs in this last category include skilled trades, customer service representatives, and medical paraprofessionals.[24] The middle-skill, middle-education, middle-wage jobs that are lagging will continue to be available, and may grow. These relatively well-paying paraprofessions demand some postsecondary vocational training, but not usually a four-year college degree.

Inequalities

62, 352, 424, 471

In the book, we have seen many examples of economic and social inequalities within and between countries. The distribution of national wealth in both Canada and the US is imbalanced. In America especially, there is a profound gap between the rich and poor that has widened significantly in recent years and has been at the root of many of the country's problems.

This polarization of job opportunities described above, with growth at the very top and the very bottom, has contributed to the country's historic rise in income inequality. The top 10 percent of Americans owned 75 percent of the country's wealth in 2014, representing the greatest inequity in the world's MDCs, according to the authoritative global wealth report by Credit Suisse.[25] Switzerland and Hong Kong are the only other members of this "very high inequality" group of MDCs. There are 12 LDCs in the same category, including Brazil, whose glaring inequalities were discussed in the previous chapter. America's top 1 percent held 20 percent of the country's wealth in 2014 (and also paid nearly half of the federal income taxes collected). At the other end of the spectrum, the bottom 90 percent owned less than one-quarter of the country's wealth. This level of wealth disparity was last seen in the 1920s, before the Great Depression dramatically changed wealth distribution. Although the US has always had a small cadre of the super-rich, inequity was much less pronounced from the 1940s through the 1970s than it is today. A variety of factors, including a rising stock market and changes in tax laws, allowed the wealth of the top earners to begin growing quickly in the 1980s. (Note that wealth and income are not the same; wealth is a person's net worth, which includes money in bank accounts, investments, houses, cars, etc., minus any debts owed.)

Large segments of the United States' population lack a "safety net" in the form of health insurance and other social services; even 5 years into the Affordable Care Act, or "Obamacare," about 10 percent of Americans had no health insurance. While real incomes of the poorest 50 percent in the US grew just 2.3 percent between 1988 and 2008, incomes of the top 1 percent of Americans grew at least 113 percent. In reaction to these trends, in 2011, a movement called Occupy Wall Street (later simply "Occupy") swelled to nationwide protests against the "1 percent." The top 1 percent of Americans includes top bankers and fund managers, heirs, sports figures, CEOs, movie stars, and tech entrepreneurs. There are other indexes of inequality; people of "Generation X" and "millennials" are not as wealthy as older generations, and men consistently make more money than women for the same job (a woman makes about 77 percent of what a man does, on average).

In relative terms globally, the United States is extremely well-to-do. But also in relative terms, the "average" American has been going nowhere economically for a number of years. Adjusted for inflation, the median income of the American family (with half making more, half making less) in 2014 was $52,000, roughly where it was in 1997, and is more than 7 percent below its 1999 peak. The gains of the boom years in the mid-2000s have been wiped out.

The poor are faring the worst. They were more than 45 million, or 14 percent of the population, in 2014. Twenty-two percent of American children are poor, as are one-third of households headed by single mothers. The poverty line for a family of four in the United States in 2015 was $24,250, and for a single person, $11,770 (**•Figure 11.44**). Geographically, there are a number of notable characteristics apparent in Figure 11.44.[26] There are 353 counties in the United States classified by the federal government as "persistently poor," meaning their poverty rate has been above 20 percent for three consecutive decades. Eighty-five percent of these are in rural areas of particular regions, including Hispanic communities along the Rio Grande Valley in Texas. Another poor region is the so-called "Black Belt" of the South, where "environmental racism" is a problem: poor and minority communities live in environments of toxic land, water, and air, in many places the byproducts of big industry. Within the Rust Belt the poverty associated with collapse of industries is most vividly apparent in Flint, Michigan, first a lumber town and later an automanufacturing giant. Flint has a poverty rate of 42 percent and ranks among the most dangerous cities in the US. Joblessness and crime sent people packing: Flint's population collapsed from 200,000 to 100,000 in the past 50 years. In the Northwest, as the lumber industry fell, poverty rose. Rural and isolated communities along with Indian reservations there and in the upper Midwest were hit hardest.

There are other ways to measure and characterize poverty in the United States. These include the "digital divide," the correlation between computer access and affluence. Only 63 about half of all households with incomes of less than $25,000 have Internet access. Being cut off from the Net is not trivial: it translates into lower-quality health services, education, and career opportunities.

Canada has a more equitable distribution of income than the United States, a smaller population (11 percent) living below the poverty line, and generally higher indexes of quality of life, although with generally lower incomes and higher unemployment rates than in the US. Canada has a more socialistic approach to its population than the United States does (for example, with taxpayer-funded universal health care) and tends to be more in line with Europe than with the United States on most social and political issues. Taken as a whole, however, 120 Canada and the United States make up a region where most people enjoy the "good life."

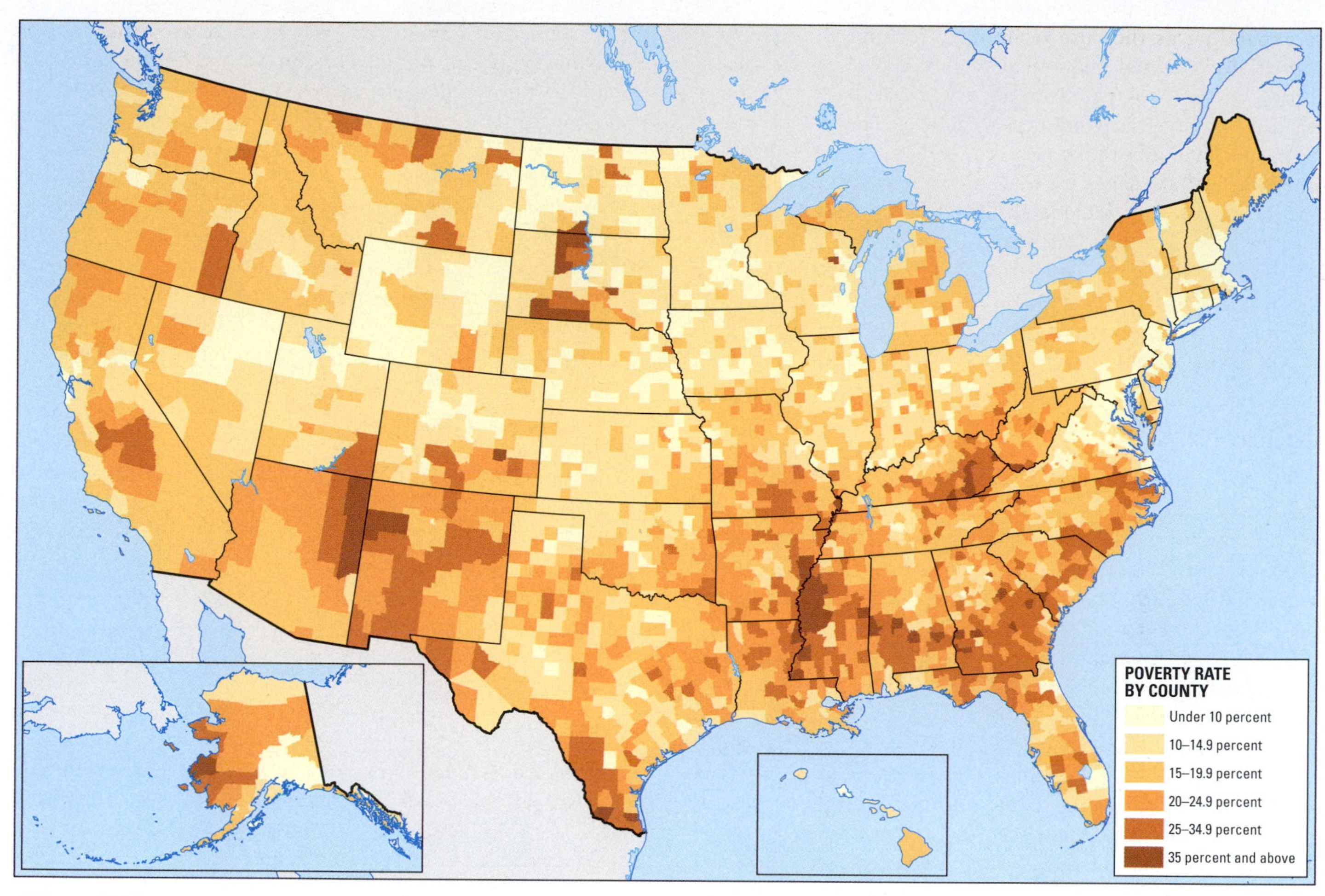

• **Figure 11.44** Poverty in America. High poverty rates are clustered in several areas of the country: the mostly white, rural counties of Appalachia; the mainly black Mississippi Delta region; the heavily Hispanic Rio Grande Valley in Texas; and the Indian reservations of the Dakotas, Arizona, and New Mexico. The highest poverty rate (55 percent) is found in Oglala Lakota County, SD, which is located within the Pine Ridge Indian Reservation. Wealthy suburban counties tend to have the lowest poverty rates; Douglas County, CO, Loudoun County, VA, and Hunterdon County, NJ, all have poverty rates of about 3 percent.

THINK CRITICALLY: What other geographic patterns of wealth and poverty stand out? How does your county fare?

Transportation Infrastructure

Both governments have sought to enhance national unity and economic strength with transportation networks spanning their vast landscapes (•**Figure 11.45**). The coasts of both countries were first tied together by heavily subsidized transcontinental railroads completed during the second half of the 19th century. Later, national highway and air networks improved transportation.

A landmark development in the United States was the growth of the **interstate highway system**, beginning in the late 1950s. Today, this system serves as the primary network for cargo trucking across the country. Although railways still carry the majority of goods transported across the country (the US arguably has the best railway network for carrying freight in the world), passenger rail traffic lags behind. The US has no high-speed rail lines on par with those in Europe, China, and Japan (Amtrak's high-speed Acela line from Boston to Washington has only a short stretch on which trains can reach 150 mph and is plagued with dangerous curves, freight traffic, and century-old track). More than 90 percent of the US rail network is owned by freight companies, and in part for safety reasons they do not want to share their tracks with passenger traffic. In many cases, old rails have been torn up and converted productively as "rails to trails," becoming bicycle and walking paths.

The stagnation or decline of US passenger railways is a direct reflection of the American love affair with the car. Personal prosperity (or at least the ability to take out a loan) and the freedoms associated with being on the road in one's own automobile diminished appetites to travel by rail and bus. The spread-out nature of American and Canadian settlement patterns (as opposed to the more concentrated European pattern) makes long-distance public transportation difficult to sustain in North America. Even public transportation usage in cities declined notably after World War II, and urban renewal projects in the 1960s brought highways to

• **Figure 11.45** There is an extraordinarily dense network of highways and railways in the United States and southern Canada. Much of this infrastructure is aging and in need of repair.

central business districts. Only in the 1980s and 1990s, once crime rates in cities dropped and gentrification took hold in many urban neighborhoods, did public transportation see a renewed upswing.

American infrastructure is imperiled. Many of the nation's highways, bridges, rails, and airports are poorly maintained. Americans travelling abroad are often stunned by the quality of airport infrastructure in Asia and the Middle East. These countries are building bigger and better airports to draw flights, investments, and visitors. In 2014 Vice President Joe Biden chose New York's La Guardia Airport to illustrate the dire condition of America's infrastructure, saying that if he blindfolded someone and took him to La Guardia, the person would think he was in "some third world country."[27] La Guardia is slated for major renovation at an astronomical cost.

The list of infrastructure woes is long. Harvard University researchers cite goods delayed at clogged ports; delayed and canceled flights that cost the economy $30–40 billion per year; average American stuck-in-traffic time of 38 hours per year, amounting to 5.5 billion hours in lost US productivity annually; wasted fuel; and a public health cost of pollution of about $15 billion per year. The average family of four spends about 20 percent of its household budget on transportation. Inequality issues are prominent here, as the poor cannot afford cars and are concentrated in areas lacking public transport.

How can these problems be solved? More spending and investment would help. Federal funding for American highways has been curtailed, but more taxes and government regulations are not the only way forward. Private-public partnerships can help, especially in boosting technologies, such as "smart" roads and rails with sensors to improve traffic flow, smart cars with accident-avoidance systems, and data analysis to facilitate air traffic. Oregon is running an experiment with a vehicle-miles-traveled fee to replace gasoline taxes for roads and bridges, which anticipates the spread of electric vehicles such as the Tesla that use roads, but not gas.[28]

11.5 Geopolitical Issues

Even with the meteoric economic rise of China, the collective strength of the European Union, and Russia's vision of its place at the heart of Eurasia, the United States remains the most important economic, military, and diplomatic power in the world. This section first examines the country's strategic relations with its northern neighbor and then with the rest of the world.

Relations between the United States and Canada

For many years after the American Revolution, which secured US independence from Britain, the political division of Canada and the United States—between a group of British colonial possessions to the north and the independent United States to the south—was accompanied by serious friction between the peoples and governments.

Several sources of antagonism existed. The northern colonies did not join the Revolution, and the British used those colonies as bases during the war. The **War of 1812** was fought largely as a US effort to conquer British Canada. Canada's emergence as a unified nation came partly as a result of US pressure. After the US Civil War (1861–65), during which more than 600,000 Americans died, the military power of the reunited nation seemed threatening to British eyes. In 1867, the British Parliament passed the **British North America Act**, bringing an independent Canada into existence. With that act, Great Britain sought to establish Canada as an independent nation capable of deterring US conquest of the whole North American continent. Relations between the two North American countries improved gradually thereafter.

Canada and the United States today are strong allies. But just as the United States overshadows Latin America in geo473political affairs, it overshadows Canada as well. Canadians, like Mexicans, often feel overlooked and underappreciated by the neighbors across the border. Canada maintains a rather low profile in international affairs, sometimes disagreeing with the United States on such critical issues as the Iraq invasion of 2003. Canada exerts a policy of relative independence and neutrality despite its membership in the North Atlantic Treaty Organization (NATO) and other international agreements.

The United States' Place in the World

Historically, the United States was not the same kind of empire-building power that many European countries were. The United States did have its overseas possessions—the Philippines, Puerto Rico, and Guam, for example—and still holds a few, but its power was historically projected more through trade, "soft power," and military engagement than by colonialism (**•Figure 11.46**).

Although America has nominally lost its superpower status, the American mind-set is hardwired to keeping the country number one. Geopolitical analyst George Friedman calls this arrogance: "Americans became convinced . . . that things can, will and should improve every day," he writes. "Americans came to believe that their wealth and security is a result of a Manifest Destiny that reflects something different about Americans compared to the rest of humanity. The sense is that Americans are somehow better—destined for greatness—rather than simply being very lucky to live where they do. It is an unbalanced and inaccurate belief, but it is at the root of American mania and arrogance."[29] The notion that the US is inherently different from and better than other nations is known as **American exceptionalism**. Political scientist Seymour Martin Lipset traces this mind-set back to the American Revolution and the Declaration of Independence: "The nation's ideology can be described in five words: liberty, egalitarianism, individualism, populism, and laissez-faire."[30] American Presidents John F. Kennedy and Ronald Reagan borrowed a phrase used by English colonists, and originating in the Biblical Sermon on the Mount, to describe their vision of America "as a city that is set on a hill" and exempt from historical forces that have affected other countries.[31]

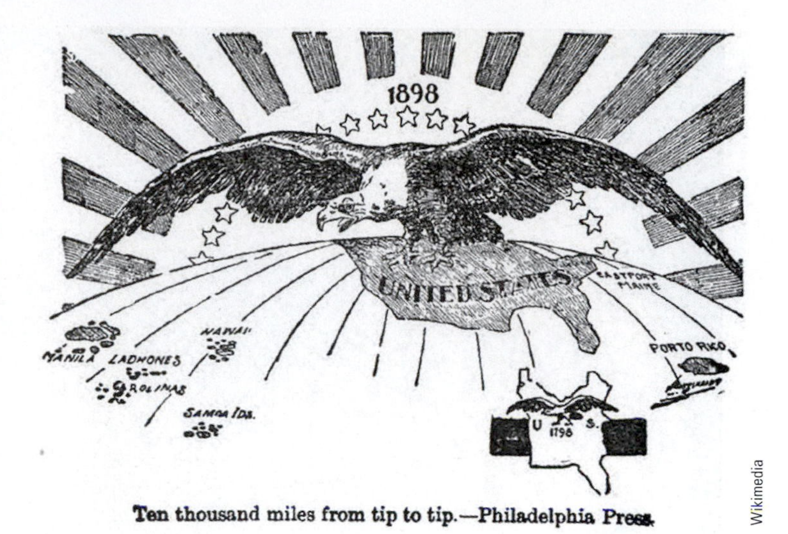

Wikimedia

• Figure 11.46 This political cartoon refers to the *Pax Americana*, the concept of "American Peace" projected through power over the Western Hemisphere and later, more broadly, over the rest of the world.

THINK CRITICALLY: George Friedman argues that "the greatest threat to the United States is its own tendency to retreat from international events." Do you see signs that the US is returning to its isolationist past? What evidence from the book supports your ideas that America is or is not moving this way?

Friedman asks what happens when something confronts this American vision: "What happens when something goes wrong? When one is convinced that things can, will and should continually improve, the shock of negative developments or foreign interaction is palpable. Mania becomes depression and arrogance turns into panic." Americans overreact when panicking, he argues: they did so in responding to the Japanese attack on Pearl Harbor, the Soviet launch of the Sputnik satellite, the Vietnam War, and the perceived Japanese economic takeover of the US economy in the 1980s. In none of these contexts was the US threatened to its core, he insists, but it responded with enormous energy, and not just militarily: from these engagements the US innovated and achieved unmatched leadership and technological supremacy.

The United States long enjoyed the geographic advantage of being situated far from the world's hot spots. In both world wars that ravaged Europe in the 20th century, the United States initially clung to **isolationism**—the view that those conflicts were someone else's and that the United States would do best to stay out of them. Only late in World War I, at the unpopular insistence of President Woodrow Wilson, did the United States enter the stalemated war on the side of Britain and France. The United States lost 116,516 soldiers in that war, but in helping

Joe Hobbs

• **Figure 11.47** A Japanese sailor pays respect to the 1177 American sailors who lost their lives on the battleship USS *Arizona* on December 7, 1941.

its allies secure victory, it also gained influence and importance on the world stage.

Again in World War II, the United States managed to stay out of the conflict for more than two years after Hitler's troops invaded Poland and ignited conflagrations across Europe. But Japan forced the United States into the war on the infamous day of December 7, 1941, with its surprise attack on Hawaii (**•Figure 11.47**). Again, US leadership and victory on both fronts of that war, which cost 405,399 American lives, helped the country achieve unprecedented strength in global affairs.

After World War II, the Cold War set in. US concerns about the spread of communism from the Soviet Union and China into newly independent countries—and the so-called "domino theory" that one after another of these countries might fall to communism—led to numerous and sometimes very costly American military interventions. The Vietnam War was the most significant of these. 172 343

The peaceful conclusion of the Cold War around 1990 was followed by a decade in which the United States sought a new role for itself on the world stage. There were military interventions to quell the conflict in disintegrating Yugoslavia and to try to deliver lawless Somalia from famine, but there was a new sense of security—and, in hindsight, complacency. There was no longer any defining framework, such as east–west relations, in international affairs; this was "the end of history" according to Francis Fukuyama. 63

Not unlike Pearl Harbor, the attacks of September 11, 2001, challenged the notion that geographic distance protected the United States. The events of that day represented a great 243 watershed in US geopolitical history. The United States under President George W. Bush developed a policy of **preemptive engagement** (also known as the "Bush Doctrine"): whenever and wherever the country perceived a threat to its security, it would take military action if necessary to defuse that threat. That action would be unilateral, if necessary; in its first term, 63 the Bush administration downplayed the need for diplomacy and partnership. The preemptive engagement policy was founded mainly on the premise that such actions would prevent potentially devastating terrorist attacks on the US homeland, perhaps with chemical, biological, or nuclear weapons of mass destruction (WMD). Such threats could emerge not only from transnational groups like al-Qa'ida but also from "rogue states," most of whom are on the United States' list of **official state sponsors of terrorism**, which includes Iran. As discussed earlier (**page 242**), the Bush administration justified the invasion of Iraq in 2003 on the grounds that Iraq was a state sponsor of terrorism and had WMD that it might use against the United States or its allies, especially Israel. Critics wrote of a "new American imperialism," proposing that US post-9/11 actions marked the beginning of a new era of aggressive global involvement for the country. Anti-Americanism rose abroad, even among the public of European allies of the US.

President Bush's successor, Barack Obama, was voted into office by a war-weary public who embraced his campaign pledge to withdraw American combat troops from Iraq and Afghanistan. He did so and drew distinctions from his predecessor in other ways; in ousting Qaddafi from Libya, for example, his strategy was to "lead from behind," allowing America's European allies do much of the heavy military lifting. The subsequent spiral of Libya into chaos and the emer- 259 gence of ISIS there and in the heartland of the Middle East suggested to many that "leading from behind" was not an effective policy and that a complete withdrawal of troops from Iraq was a big mistake. Under Obama's leadership, the US also brought itself close to the brink of military intervention in Syria, but then backed down. It appeared to many that Amer- 262 ica had become more reticent on the world stage, leaving allies such as Japan and Korea fretting about whether, if they were attacked, the US would come to their aid, as obliged by treaty. "The Obama doctrine is to avoid using force," wrote the Foundation for Defense of Democracies, with the mantra "We will not do ground wars."[32] Many Americans shared the view that 3 it should not be America's role to serve as "the world's policeman" and shoulder most of the burdens in situations where force might be called for.

According to the authoritative geopolitical analyst Saul Cohen, the United States has lost its position as the world's sole **superpower**, defined as an extremely powerful nation capable of influencing international events. Cohen argues that as of 2015, the world had no superpower, only a number of major powers with significant influence.[33] The US is, however, the strongest among these. It has not lost its capacity to play a decisive role in political, military, and economic events abroad. With its strong economy, military expenditures larger than those of the next seven countries combined, dominance of global popular culture, the world's best universities for graduate studies, and headquarters to many of the world's leading international organizations (including the United Nations, the International Monetary Fund, and the World Bank), the United States is unrivaled. Throughout the book, however, we have seen the shift to a multipolar world as rising powers like democratic Brazil and authoritarian China and Russia wield 63 growing influence. Between invading a neighboring country and positioning for regional military deployments, Russia and China, respectively, have seemed to emerge as threats to a "world order" led by the US. Military analysts concur that Russia poses the greater threat to the United States; China's economic survival (and thus its political system) depends upon trade with the US. George Friedman contends that China is on the way to doing itself in: it has "an unstable financial and economic system that will collapse under its own weight."[34] Despite its vast size, Russia has geographic disadvantages (dearth of warm water ports, indefensible borders, and high latitudes, for example) and for practical and political reasons is an ex- 155 pansionist power. Ominously, it has severed most of its ties to the world order and has passed up the so-called "off-ramps" of opportunities to de-escalate from the crisis it began by invad- 184 ing Crimea.

Some analysts, however, argue that the apparent challenges by Russia and China in fact make the world safer for the US and its allies. These undemocratic powers have nowhere near the collective military, economic, and diplomatic power represented by the Western alliance. Neither has a coherent ideological or strategic vision that offers an alternative to the "liberal order" of the West. "At most, China and Russia are spoilers," writes G. John Ikenberry in *Foreign Affairs*. "They do not have the interests—let alone the ideas, capacities, or allies—to lead them to upend existing global rules and institutions. . . . For them, international relations are mainly about the search for commerce and resources, the protection of their sovereignty, and, where possible, regional domination . . . In the struggle for world order, China and Russia (and certainly Iran) are simply not in the game . . . And the grand strategy (Washington) should pursue is the one it has followed for decades: deep global engagement. It is a strategy in which the United States ties itself to the regions of the world through trade, alliances, multilateral institutions, and diplomacy." Ikenberry adds, "Geography reinforces the United States' other advantages. As the only great power not surrounded by other great powers, the country has appeared less threatening to other states and was able to rise dramatically over the course of the last century without triggering a war. In fact, the United States' geographic position has led other countries to worry more about abandonment than domination. Allies in Europe, Asia, and the Middle East have sought to draw the United States into playing a greater role in their regions. The result is an 'empire by invitation.'"[35]

The main problem on America's geopolitical horizon appears to be that invited or not, and like it or not, the US is finding itself on a perpetual war footing. In 2015, the nation's top military officer, General Martin Dempsey, told the military to dig in for a long war. "Global disorder has trended upward while some of our comparative advantages have begun to erode," he said. "We are more likely to face prolonged campaigns than conflicts that are resolved quickly . . . control of escalation is becoming more difficult and more important . . . as a hedge against unpredictability with reduced resources, we may have to adjust our global posture."[36] One of those "comparative advantages" being eroded is the geographical advantage enjoyed by America's distance from other continents. Another is America's conventional military supremacy, which holds less weight in the threat environments of asymmetrical warfare and of "non-state actors" like ISIS, which can reach every American home through social media. 245

The Geopolitical Issues section of each regional chapter in this book has dealt with the United States' influence and roles in those areas. The details do not need to be revisited here. But our journey around the world has left us with six critical thinking questions about the US role in the world. Please consider these:

1. What can the US do to confront or contain ISIS, al-Qa'ida, and militant Islam?
2. How will the US deal with China as both a friendly trading partner and a potentially expansionist power in the western Pacific and Asia?
3. How can the US deal with the apparent expansionism of Russia and avoid a costly new Cold War or even "hot wars" involving NATO?
4. Can the US accommodate Iran's aspirations to be a nuclear weapons power without going to war?
5. Can the US avoid a confrontation with a weakened North Korea?
6. Will the US confront the consequences of climate change, which the Pentagon views as a critical security issue for the country?

Each of these questions involves significant challenges. Many raise the specter of costly and open-ended wars. Is America's military power its first or last resort of engagement on the world stage? The American people may have a decisive influence on future decision making. They reacted strongly to ISIS's brutality for example; a Quinnipiac University poll in 2015 showed that Americans, by a margin of two-to-one, supported deploying more US ground troops to fight and destroy ISIS.[37] But is it really necessary, in the way the Carter Doctrine proposed it was, for the US to fight in the Middle East in order to defend its vital interests in the region—which in Carter's time unabashedly meant its oil interests? Some say "no" because of 239 America's "energy revolution" described earlier. Among them 522 is Peter Zeihan, who views oil as the major reason in the past that the US needed to be involved with the world. With energy security, he sees America returning full circle to its traditional, pre-1945 strategy of prospering far from the ills of the world.[38]

11.6 Regional Issues and Landscapes

Canada

•**Figure 11.48** is the reference map of Canada, and is useful in the following discussion.

The Québec Separatist Movement

As we have seen, in Canada a culture clash exists between the mainly French heritage/French-speaking Québec and the predominantly British heritage/English-speaking remainder 513 of the country. Many of the distinctive and proud *Québécois*, or "Québecers," are pushing for devolution and independence from Canada, casting a long shadow over Canadian affairs.

In the 1960s, French Canadian dissatisfaction led to the formation of a separatist political party, the ***Parti Québécois (PQ)***, dedicated to the full independence of Québec from Canada. In 1976, the party gained control of Québec's provincial government, making French the province's sole language of government and commerce, and requiring that all immigrants to the province be educated in French. Separatists have been frustrated by the steady influx of non-French-speaking immigrants seeking jobs and settlement in Québec City and other cities along the Saint Lawrence Seaway. These immigrants tend to vote with the pro-English bloc or otherwise have no stake in Québec's bid for independence. Québec can ill afford to shut its doors to immigrants, who now make up more than 10 percent of the province's electorate. The province's historically high birth rates have declined, and future growth in the workforce will have to come from outside. We have seen how similar demographic and labor issues trouble a number of countries.

After two failed referenda in the 1980s, in 1995 the issue of Québécois independence was taken to provincial voters again. Those who favored staying a part of Canada very narrowly won, with just 50.5 percent of the vote. About 60 percent of French speakers voted in favor of independence, whereas English speakers and immigrants voted heavily in favor of Canadian unity.

Despite losing the bid for independence, Québec's separatists sparked a number of effects nationwide. In the 1995 referendum's wake, the federal government officially recognized Québec as a "distinct society" within Canada and made French an official language of Canada—along with English—across the nation (•**Figure 11.49**). Laws now stipulate that the

• **Figure 11.48** Principal features of Canada.

• **Figure 11.49** French and English multilingualism is characteristic of Canada, especially in the east.

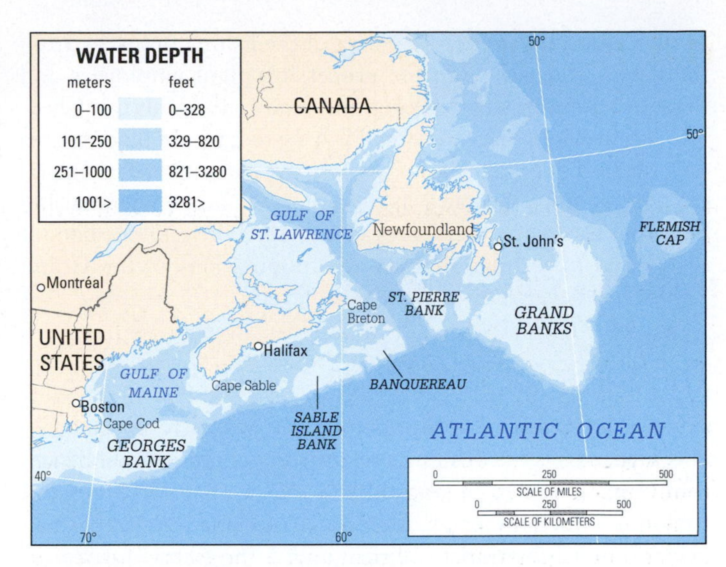

• **Figure 11.50** The Grand Banks.

children of French Canadians born outside Québec may have their children educated in French. Québec's French-only laws were challenged in court and were modified several times to be more lenient in the use of English.

The Parti Québécois continues to be stymied in its efforts to gain independence for Québec, in part because of its intolerance of Canada's increasingly diverse cultures. Mirroring issues in France and running against the grain of Canada's inherent tolerance, the PQ's candidate in 2014 declared multiculturalism a failure and argued that civil service employees should be banned from wearing "conspicuous religious symbols," including kippahs, turbans, and hijabs.[39] The PQ lost in a 2014 election that was essentially a referendum on a referendum; had the PQ won, it would have called for another referendum on independence. For the time being, the prospect of Québec's independence is remote, but separatist sentiments persist and will remain part of Canada's political landscape.

Overfished Waters

Canada's three eastern provinces of New Brunswick, Nova Scotia, and Prince Edward Island are known as the Maritime Provinces. The Maritimes plus Newfoundland and Labrador constitute Atlantic Canada. Historically the waters off Atlantic Canada teemed with fish, and the fishing industry was the region's economic backbone. Before the beginning of European settlement in the early 17th century, and possibly even before Columbus arrived in the Americas in 1492, European fishing fleets worked these waters.

Elevated portions of the sea bottom, known as **banks**, are located off the Atlantic coast from near Cape Cod to southeastern Newfoundland (•**Figure 11.50**). Shallow depths and the mixing of waters from the cold Labrador Current and the warm Gulf Stream favor the growth of plankton, the tiny organisms on which many fish feed. Early fleets fished mainly for bottom-dwelling cod, which were cured on land before being shipped to European markets (•**Figure 11.51**). Cod were said to have been so abundant "that you could almost walk across the ocean on their backs."[40] Vast schools of fish sustained Atlantic Canada's economy until the mid-20th century, when increasingly efficient fishing technologies began impacting this resource.

Worried about the effects of overfishing, in 1977 the Canadian government began enforcing a 200-mile (322-km) offshore jurisdiction prohibiting foreign fishing of the Grand Banks. But fish populations continued to decline, as increasingly larger vessels were able to haul in 200 tons of fish per *hour* for 24 hours a day, weeks on end. In 1968, over 800,000 tons of cod were brought in from the Grand Banks; by 1991, the number dropped to 100,000 tons. Alarmed, Canada altogether closed the Grand Banks to commercial fishing in 1992, setting off an economic shockwave throughout the Atlantic

• **Figure 11.51** Cod from the Grand Banks, harvested and dried before commercial bans went into effect.

provinces as jobs in every segment of the fishing industry, from the fishers themselves to the processing plant employees and those working in ship making and repair, suddenly vanished. The moratorium on commercial fishing remains in place to this day, but cod populations have not recovered in the nearly quarter century since the ban was first imposed. Some scientists believe that fishing trawlers' nets so damaged the seafloor habitat of fry and fingerlings that the cod populations of the Grand Banks will never return to previous levels.

The Race for the Arctic

Canada has a vast Arctic territory and wants to exploit it—and areas farther north of its territorial waters—to maximum advantage (see map in •**Figure 11.52**). "The true wealth of our nation lies north of the Arctic Circle," a Canadian spokesman said.[41] The US National Oceanic and Atmospheric Organization (NOAA) forecasts that the Arctic Ocean will be free of summer sea by 2040. As the climate changes and Arctic ice melts, more resources will become available to Canada and other countries of the Arctic Ocean rim.[42]

Seafaring explorers have long sought a Northwest Passage that would enable shipping across the top of North America. 509 Until recently, however, sea ice has made passage through Canada's arctic archipelago impossible except under extraordinary conditions. But maritime shipping would become routine if Arctic warming were to open up a reliable route. Canada's economy would benefit from the shipping fees charged for transit through the passage. This route would be the counterpart to Russia's Northern Sea Route (see **page 198**), allowing great time and cost savings: the distance traveled by ships between northeast Asia and Europe would be cut by 40 percent. Canada and Russia are also talking about opening the **Arctic Bridge**, an eight-day shipping route from Canada's port of Churchill, on Hudson Bay, to the northwestern Russian port of Murmansk (see **Figure 11.52**). Goods arriving in Churchill could be shipped quickly by rail south to ports on the Saint Lawrence Seaway that could otherwise be reached only after a 17-day sea voyage from Murmansk.

• **Figure 11.52** Recent dramatic reductions in sea ice cover and hopes of abundant resources have Arctic Ocean countries staking territorial claims and drawing future trade routes.

Other nations, including the United States, have rankled Canada by suggesting that the modern-day Northwest Passage should be classified as international rather than Canadian waters. Canada and the US are unlikely to have a serious showdown over this issue. But Canada is resolute in staking its claims: the country's prime minister said his country's attitude toward its Arctic territory was "use it or lose it" and announced plans for construction of more icebreakers, a new deepwater port on Baffin Island, and an army base at Resolute Bay.[43] Shipping routes are not the only prize in the polar realm: the Arctic Ocean's sea and seabed resources may be enormous. As Arctic Ocean waters warm, new fishing grounds will open. One-fourth of the world's undiscovered oil and gas reserves lie in the Arctic, according to US Geological Survey estimates. Five Arctic nations—Russia, Canada, Norway, Denmark, and the United States—are scrambling to stake their turf. The North Pole itself lies outside any country's EEZ, but Russia has actually used a submarine to plant its flag on the North Pole's seafloor 13,000 feet (3962 m) below the surface. Russia's claim is contested by Canada and Denmark, which have their own rationales for claiming the pole as theirs. The US does not 198 have a claim on the North Pole.

Russia's rhetoric about its claim to polar seabed resources is heating up. Without naming the country he was accusing, in 2010 Russia's President Medvedev said, "Regrettably we have seen attempts to limit Russia's access to the exploration and development of the Arctic's mineral resources. That's absolutely inadmissible from the legal viewpoint and unfair given our nation's geographical location and history."[44] The president signed a paper saying that the polar region must become "Russia's top strategic resource base" by 2020. The document included plans for beefing up military assets in the region in order "to ensure military security under various military political circumstances."[45] Canada and Denmark (through its Greenland colony) were not far behind in lodging their commitments to the Arctic. With so much at stake, negotiations and disputes among the Arctic Ocean countries will continue for years to come.

Greenland: A White Land

Greenland (officially *Kalaallit Nunaat*, Land of the Greenlanders; see Figure 11.2) was the site of the first European settlements in North America; Vikings landed there around the year 1000, but life was harsh, and the colonies died out within a

Joe Hobbs

• **Figure 11.53** Greenland's spectacular glacial landscape.

few centuries. Greenland has been under Danish control since 1814. In 1979, the Danes granted the island self-government while retaining control over its foreign affairs. Although part of Denmark, Greenland is not part of the European Union. Under the auspices of NATO, joint Danish-American air bases are maintained there, and the United States has a giant radar installation at Thule in the remote northwest.

Denmark subsidizes the island with about $600 million per year, or half of Greenland's annual revenue. There is some talk of independence among the Greenlanders, but severing the aid pipeline from Denmark and the European Union would make that a costly proposition. There is one way out of this: striking it rich with oil. Greenland may have done just that: there are estimates that the Greenland Sea may have large oil and natural gas reserves. However, significant challenges remain in exploring and tapping into these fossil fuels, and any production is still years away.

Greenland has nearly one-fourth the area of the United States, but about 80 percent of the island is covered by an ice-cap up to 10,000 feet (3000 m) thick (•**Figure 11.53**). Temperatures over Greenland are rising, and between 2003 and 2011 an average of about 50 cubic miles (c. 200 cu km) of ice melted each year. If the ice sheet were to melt completely—a prospect of the far distant future—climate models suggest the world's sea level would rise by 20 feet (6 m).

Greenland has a total population of 55,000, nearly all of whom live along the southwestern coast. The great majority of inhabitants are Inuit, with Danes making up about 15 percent 506 of the population. The leading means of livelihood are fishing, hunting, trapping, sheep grazing, and the mining of zinc and lead. Rare earth metals that are essential in the production of many electronic devices, as well as uranium, have also been discovered in Greenland.

318, 331

The United States

The book concludes with issues in the United States, the home country of most of its readers. •**Figure 11.54** is the reference map for these discussions.

The Changing Geography of American Settlement

Many of the issues discussed in the earlier section on economic geography are associated with changing patterns of settlement within the US. Four patterns are notable: As America has urbanized its rural, agricultural heartland has been depopulated, with some exceptions based on commodity booms; its people have moved South and West; cities have revitalized and attracted more people; and socioeconomic divides have widened within cities, forcing the relocation of poorer residents.

The United States became an urban-majority country around 1920. Today rural Americans make up only 20 percent of the population, living across three-quarters of the country's area. Their declining numbers are most pronounced across the Great Plains and in parts of Appalachia and the interior South (•**Figure 11.55**). Economic forces account for the dropping numbers. Most rural counties have economies based on agriculture, manufacturing, or resource extraction, and as mines and factories have closed and large corporations have bought up farmland, job seekers have migrated from these areas to metropolitan centers. About half of all rural counties have also experienced more deaths than births, further lowering their

• **Figure 11.54** Principal features of the United States. Alaska is shown at the same scale in Figure 11.63.

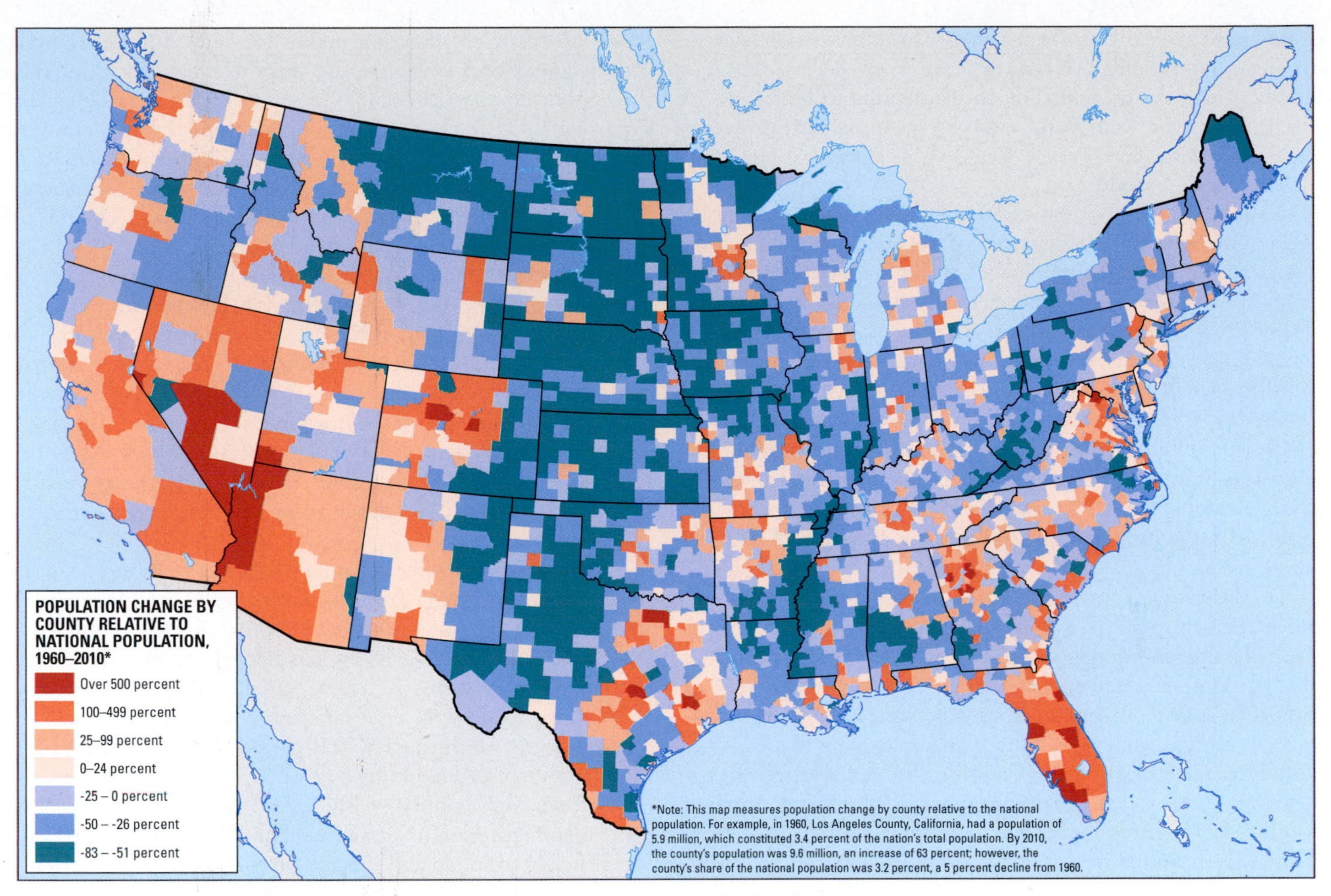

• **Figure 11.55** This map shows population movement trends in the US. Note especially the shift of population away from the Great Plains and rural areas of the eastern half of the country toward the West (especially around Las Vegas), Florida, and suburban areas surrounding cities such as Atlanta, Dallas, and Washington.

populations in recent years. Empty small villages—"ghost towns"—dot the landscape. Billings County, in western North Dakota, had 10,000 people in 1910, but a century later it was home to only 800 people—a remarkable 92 percent decline.

To slow down their population declines and attract new residents, rural communities are offering economic incentives and marketing the virtues of small-town living. Some have invited international immigrants, while others are enticing outsiders with tax breaks, low-interest loans, payoffs of student loans, and land giveaways to those who would build homes. Small towns are trying to attract doctors and lawyers. A program called "Project Rural Practice" offers subsidies to lawyers who relocate to rural South Dakota, for example. Some of these efforts are succeeding, but community leaders acknowledge the demographic and social challenges that lie ahead.

Not all rural counties have declined in population. Many situated near cities have become the domain of new suburbs and exurbs surrounding the central city, occasionally leading to clashes between the new, often wealthy migrants and the long-time residents who are loath to lose the character of their communities. Loudoun County in northern Virginia for example had a population of 20,000 in 1940, exploding to over 350,000 by 2014 as the urban area surrounding the nation's capital spread outward. Some in the county have called for it to be divided in two, to preserve the still largely rural western portion from the much more urbanized eastern half.

Since the 2010 census, a dramatic reversal of rural population loss has occurred in western North Dakota and eastern Montana. Fracking technology allowed the oil in the Bakken Formation to be tapped in the late 2000s. In the ensuing oil boom, thousands of people secured jobs in these rural counties. The town of Williston, North Dakota, grew from 14,000 people in 2010 to 25,000 in 2015. The population of McKenzie County, North Dakota, doubled during those years. Many counties in this region are seeing their populations rise for the first time in decades, and the sudden influx has created shortages in housing, medical care, and schools. If the global price of oil continues to fall and the oil boom ends, these communities may find the newcomers leaving as quickly as they came.

523

During the past century, the general trend in domestic migration (excluding international immigration) is that people have been moving from the old industrial and agricultural centers of the Northeast and Midwest to the South and West. Southern states including Florida, Georgia, North Carolina, and Texas have seen marked population gains as people from Michigan, Ohio, New Jersey, New York, and other northern states have moved into them. There is a climatic correlation to these migrations: Americans have been fleeing the cold northern climes and resettling in what came to be known in the 1970s as the **"Sun Belt,"** with its booming warm-weather cities including San Diego, Las Vegas, Phoenix, Houston, Atlanta, and Orlando.

California's demographics reflect this broad trend, along with some of the consequences of so many people seeking the good life. The "Golden State" was a magnet for migrants from all over the country; between 1940 and 1980, its population grew about 250 percent. Rising costs of living and a weakening economy reversed this trend by the early 1990s. California still takes in many immigrants from abroad, but in the last quarter century more people have moved out of California than into it from other states. The large population increases in other western states including Nevada, Colorado, Utah, Arizona, Idaho, Oregon, and Washington are partly the result of large numbers of Californians moving out of their home state.

On the national scale, the most notable evolution of American settlement patterns has been the repopulation of larger cities, reversing the trends of urban decay and population loss that had characterized many US cities, and slowing the trend of 527 suburban sprawl. American cities were wracked with racial violence and riots in the 1960s, high unemployment in the 1970s, and the crack cocaine epidemic of the 1980s. Cities were at the heart of a two-decade long crime wave. But suddenly, starting around 1990, crime rates began dropping significantly. New York City recorded over 2,200 homicides in 1990; by 2000, homicides fell to 673, and just 333 in 2014. Many other cities, including those with reputations for violence such as Washington, Los Angeles, and Miami, saw similar declines. This decline in crime (the reasons for which are still being debated), coupled with the strong economy of 1992–2007 (save for a brief recession in 2001), made cities attractive for employers and residents again.

As cities began rebounding, with more wealthy and educated people moving in and working in creative, high-paying service industries, tax revenues in city coffers also grew. These funds helped clean up neighborhoods and repair and rebuild infrastructure. Downtown revitalization has been transforming urban landscapes. A characteristic transformation is the **adaptive reuse** of older structures. In this process, abandoned commercial buildings are being converted to loft apartments and condos, and vacated mills and factories are being reborn as shopping malls, high-tech startup headquarters, and government offices, generating jobs and attracting people back. Adaptive reuse is often associated with **gentrification**, the process by which a low-cost, run-down neighborhood undergoes gradual physical renovation for residential or office space, resulting in increased property values and new homes and offices. Gentrification makes for safer and better-maintained neighborhoods, but also has the effect of making these neighborhoods unaffordable for many of the area's long-time residents. When they can no longer afford the city, many get displaced to distant housing that is more affordable but far less convenient for work and services (**• Figure 11.56**). There is 544 little they can do but protest as "the American Dream" retreats from their grasp.

Prominent among America's urban success stories are cities that already had white-collar job bases, educational institutions, government facilities, and abundant cultural and natural attractions: San Francisco, Portland, Seattle, Denver, Minneapolis, Austin, Raleigh, Washington, New York, and Boston are among them. They have succeeded in drawing in new residents, particularly young professionals of the Millennial generation (the more than 80 million born between the early 1980s and 2000, comprising the largest generation in American history).[46]

"Smart growth" is a key policy in many of America's urban success stories. Smart-growth policies began catching on in the 1990s. The Congress for New Urbanism (CNU), founded by a group of architects in 1993, has become an influential agent of smart urban growth. The CNU describes itself this way: "Working against the conventional, predominant, sprawl-oriented dogma of the post-WWII period, the group has worked for years to create buildings, neighborhoods, and regions that provide a high quality of life for all residents. We stand for the reconfiguration of sprawling suburbs into communities of real neighborhoods and diverse districts, the conservation of natural environments, and the preservation of our built legacy."[47] Smart growth has impacted urban sprawl, which has

Joe Hobbs

• Figure 11.56 Chicago's Pilsen area has been spiffed up in recent years. As the area becomes more expensive and attractive to urban professionals, its poorer, longtime, mainly Latino residents are driven out. With this mural, the less affluent residents condemn Pilsen's gentrification. Note on the left the sign reading "STOP GENTRIFICATION IN PILSEN!" Another sign equates the tax increment financing (TIF) for the neighborhood renovation with ethnic cleansing. Eastern European immigrants once lived in Pilsen but moved out during an earlier cycle of urban change.

been declining throughout the country since 1994, especially by discouraging cul-de-sac development and pushing for connected street grids.[48]

Meanwhile some cities, especially those hit hard by the decline of American manufacturing, have been struggling. 527 These include Detroit, Cleveland, St. Louis, Pittsburgh, and Buffalo, dubbed "Legacy Cities" for their historic role in growing America's economy.[49] But as housing costs skyrocket in the "smart cities," these former industrial cities are increasingly attractive options to cost-conscious residents and employers (the city of Columbus, Ohio even advertises its low cost of living on the platforms of Washington's subway system). Long-suffering Buffalo brought in the solar panel manufacturer Solar City. Cleveland's redeveloped North Coast Harbor is home to the Rock and Roll Hall of Fame, and medical research has a prominent role in the city's economy. Pittsburgh has redeveloped its waterfront and is working to shed its reputation as a grimy steel town.

The Big Apple

The United States' largest city, New York City, is centrally located on the Northeastern Seaboard's urban strip midway between Boston and Washington. An enormous harbor, an active business enterprise, and a central location within the economy of the colonies and the young United States made New York City the country's leading seaport. Superior access to the developing Midwest also gave New York unchallenged leadership among American cities in size, commerce, and economic impact.

New York City itself has a population of 8.5 million, and the metropolitan area is home to 20 million people. While the state of New York as a whole has been steadily losing population, the city is gaining people at an impressive pace: its population grew 6 percent between 2010 and 2014. The city is centered on the less than 24 square miles (62 sq km) of Manhattan Island. Manhattan is one of five "boroughs" (administrative divisions) of the city proper. The others are the Bronx, on the mainland to the north; Brooklyn and Queens, on the western end of Long Island; and Staten Island, to the south across Upper New York Bay. The whole city (excluding Staten Island) is held together in part by 722 miles (1155 km) of subway, parts of which are more than a century old.

New York is one of the world's most ethnically diverse cities. More than 35 percent of its residents are foreign-born, with the largest contingents coming from the Dominican Republic, China, and Jamaica. Their growing numbers more than offset the out-migration of longer-resident New Yorkers. Immigrants make up more than 40 percent of New York City's labor force, including more than half the employees working in restaurants and hotels, about 60 percent in construction, and nearly 70 percent in manufacturing. Most live in the outer boroughs, with two-thirds of them in Queens or Brooklyn. These boroughs today are ethnic stewpots where immigrants from numerous countries live cheek by jowl.

Size, financial importance, cultural complexity and variety, and national history have given New York City an uncontested position of urban prominence in the United States.

• **Figure 11.57** Manhattan's Times Square means a lot of different things to different people, but since a famous World War II photo of a kissing couple has been synonymous with romance.

Joe Hobbs

It is the country's media and financial capital, and has some of the world's greatest museums and other cultural assets. The heart of this alpha city is roughly the lower two-thirds of Manhattan (•**Figure 11.57**). The lower Manhattan Financial 127 District includes the imposing cluster of office skyscrapers in the Wall Street area, and it was the home of the twin towers of the World Trade Center until 9/11. This area's critical role in the global economy made it a prime target for al-Qa'ida's September 11 attacks (and an often forgotten smaller attack in 1993). In the footprints of the World Trade Center buildings is 243 a heart-rending memorial, and soaring symbolically 1776 feet above it is the Freedom Tower office building.

San Francisco, the City by the Bay

About 7 million Californians—roughly 20 percent of them Asian Americans—live in the San Francisco metropolis, the sixth largest urban area in the United States (see the map in Figure 1.6 on **page 10**). San Francisco—beloved as "The City" by its residents and neighbors—is located on a hilly peninsula between the Pacific Ocean and San Francisco Bay, but the metropolitan area includes counties on all sides of the bay

Joe Hobbs

• **Figure 11.58** San Francisco from the Golden Gate Bridge.

(•**Figure 11.58**). The climate is unusual. Marine conditions moderate San Francisco's temperatures from an average of 48°F (9°C) in December to only 64°F (18°C) in September. Local marine and topographic circumstances give rise to San Francisco's trademark fog, which can produce very chilly days even in the height of summer. Mark Twain is said to have quipped, "The coldest winter I ever spent was a summer in San Francisco."

San Francisco began as a Spanish fort (*presidio*) and a mission on the south side of the Golden Gate. It grew as a port for the gold-mining industry and then for the expanding agricultural economy of central California. The **Gold Rush of 1849** centered on the gold-bearing gravels of Sierra Nevada streams pouring into the Central Valley north and south of Sacramento. San Francisco, with its spacious natural harbor, was the nearest port to receive immigrants and supplies for the gold camps, and the city boomed during the race for gold.

Today's San Francisco Bay Area is a major domestic and international tourist destination, famed for its scenery, wines, arts, and other cultural attractions. In recent decades, high technology businesses have boomed from San Francisco to San Jose (50 miles south), in an area known as **Silicon Valley**. The rise of Silicon Valley as a center of high technology is due partly to its proximity to Stanford University, a center of computer research, development, and innovation (often in conjunction with the US Department of Defense) for decades; many of its graduates have gone on to found high-tech companies. Adobe, Apple, eBay, Facebook, Google, Intel, Netflix, Oracle, Symantec, and Yahoo are just some of the high-tech companies that are headquartered in this area.

The Bay Area has many of the characteristics described above in the changing demographics and economics of American cities. San Francisco became America's most expensive renter's market, with the median rent for a one-room apartment costing $3410 in 2015. Wealthy newcomers employed by high-tech firms can pay rents and buy homes in such areas, but many poor and middle-class people no longer can and have been priced out of the city. The new wealthy have used their influence to slow the construction of new housing units in order to preserve the character of the city and its surroundings. These conditions have created a genuine housing crisis in San Francisco and outlying areas. There are repercussions from such efforts to slow or freeze new urban development. A more accommodating housing market would allow the high-tech firms to hire more, which would create jobs for teachers, chefs, nannies, and others that service the high-tech workers. Housing regulations and shortages have similarly stifled economic growth in New York, Boston, Washington, and other cities.[50]

The Thirsty West

The American West has some very serious environmental challenges: dry climates, rough topography, and a lack of inland water transportation. Because settlement is clustered in places where water is adequate, the population pattern is oasis-like, with cities separated by great distances from each other by steppe, desert, and mountains. Prevailing dry conditions often menace the region with wildfire and droughts, ironically punctuated by floods and landslides. By 2015, the Southwest had been in drought for 15 years, a situation that climate scientists warned may represent "the new normal" for the region.

Water is therefore a critical issue throughout the western United States. Historically, settlements were small and self-sufficient in water. But with explosive suburbanization following World War II, huge numbers of people moved west, especially to California, in search of the American dream. Populations quickly surpassed local water supplies, and new sources had to be found or redirected from elsewhere. The region's older cities, such as Denver and Los Angeles, were able to grow because of federal support of giant dam, reservoir, and irrigation projects during the 20th century (for Los Angeles, large-scale waterworks included canals to bring water from distant Mono Lake and Owens Lake). But the realities of living in an arid environment have set in, and newer communities cannot meet their needs with large-scale manipulation of the environment. They must rely far more on water conservation and other ways of adapting to local conditions.

Management of Western waters is muddied by their classification: rather than being viewed as an integrated hydrologic system, in many cases a distinction is made between underground and riparian (river) waters.[51] Although Colorado state law recognizes the connection between tributary groundwater and surface water, California and Arizona laws do not. Prompted by the illusion that groundwater and surface water are separate "bank accounts," the result is a race to drill into what is left of those states' groundwater, intercepting water from rivers and taking water that would otherwise show up in surface streams in the future. About 60 percent of California's water and half of Arizona's came from underground sources in 2015. California does not require that groundwater extraction be measured at all; Arizona does, but its failure to recognize the integrated hydrologic system encourages overuse. Arizona used 90 percent of its available water in 2015, and is projected to use all of it by 2025 and be in a 10 percent water deficit a decade later.

The region's largest water source is the Colorado River, fed by its tributaries and snowmelt from mountain peaks (•**Figure 11.59**). It has been highly manipulated to suit human needs, in accordance with state and federal laws apportioning water rights to seven states. The 1922 agreement known as the **Colorado River Compact** allocated water use in an effort to

• Figure 11.59 The Colorado River Basin.

balance the present and future needs of states in both the Upper Basin (Colorado, Utah, Wyoming, and New Mexico) and Lower Basin (California, Arizona, and Nevada). This agreement and the later 1948 Upper Basin compact led to the construction of two major dams on the Colorado River: the Glen Canyon Dam in Arizona (completed in 1966, **• Figure 11.60**) and the Hoover Dam (completed in 1936) on the Arizona-Nevada border. Hoover Dam's Lake Mead (with the largest volume capacity of all American reservoirs) supplies a major portion of the water needs for about 25 million people. The Glen Canyon Dam and three smaller dams were built in response to a prolonged drought in the 1930s that demonstrated that during dry spells the states of the Upper Colorado Basin would not have the water allocated to them by the 1922 compact.

There are some popular misconceptions about these water-sharing compacts.[52] One is that the 1922 compact commission failed to consider the prospect of drought. In fact, the commission considered historical records that revealed the problem of very low water years, including a great drought in the 1890s that lasted into the early 20th century. Also contrary to widespread beliefs, Mexico (lacking its own storage facilities) is allowed to store waters in Lake Mead. Finally, Native Americans also have water rights; the Navajo Nation is apportioned Colorado River waters from the respective states in the Lower and Upper Basins where its lands are, and the Animas-La Plata Project delivers rightful waters to two Ute Indian tribes as well as non-native irrigators in Colorado.

Energy is an important component of the region's water system. Lake Powell (behind Glen Canyon Dam) and Lake Mead generate hydroelectricity for the region. Below the dams, a canal network known as the **Central Arizona Project (CAP)** pumps water from the Colorado River to sustain large, booming, and very thirsty cities of Phoenix and Tucson. About 90 percent of the energy required to move all that water is produced at the coal-burning Navajo Power Plant near Page, Arizona (**• Figure 11.61**). Considered a godsend for the Navajo Nation and to Arizona's growing economy when it was built, it is the object of much concern about its greenhouse gas emissions and their effect on air quality near some of America's greatest national parks, including the Grand Canyon.

Critics insist the Colorado River management system is inefficient and that the water crisis associated with it is a more a human-engineered one than a climatic one. Under the region's

• Figure 11.60 The Glen Canyon Dam, just upstream from the Grand Canyon, was constructed in the 1950s. Note the low water level in this photo. A full reservoir would cover the white strip of bank that extends behind the dam.

• Figure 11.61 The coal-burning Navajo Power Plant supplies about 90 percent of the electricity used for water distribution in the Central Arizona Project.

scorching sun, much water is lost to evaporation in the giant reservoirs of Lake Mead and Lake Powell. More makes its way into the porous sandstone, where it cannot be accessed for use. Investigator Abrahm Lustgarten backs a proposal to tear down the Glen Canyon Dam, reporting that if Lake Powell ceased to exist, 6 percent of the Colorado River's water would immediately be made available for use.[53] "A facility whose central purpose is to save water instead loses a mind-boggling amount of it," he wrote. But draining Lake Powell would deprive Upper Basin states of water and hydropower, and would impact programs to support fisheries in the Upper Basin.

There are some wasteful water uses in the Colorado River system. One is cotton farming in Arizona. Grown mainly around Phoenix, cotton is a thirsty crop whose production was made viable through the provision of federal subsidies until 526 2013 (**•Figure 11.62**). Provisions in early-20th-century water-use laws not only permit but compel cotton farmers and others to use more water than they need. Some farmers end up flooding their fields with water for no purpose at all except to maintain the levels of water allocated to them. They fear that if they use less, their water allotments will be cut in the future; this is the cynical "use it or lose it" proposition. Federal subsidies remain for corn, rice, wheat, and even notoriously water-hungry alfalfa. These subsidies rather than rational decisions about best crops for dry environments, including varieties that are being genetically engineered for drought tolerance, are decisive 117 in what farmers plant.

California is the last downstream power in the Colorado River system. Almost 20 million people in southern California drink this system's waters. California and other states have litigated water in federal and state courts ever since the Hoover Dam was completed in the mid-1930s. For decades, California has used more water from the river than was allocated to it in the 1922 Colorado River Compact, while the other six river basin states have drawn less. It took threats from the upstream states to actually cut off water to California to bring that state to the negotiating table. In 2003, California agreed to reduce its use of Colorado River water over a period of 14 years to allow the six upstream basin states to use their share. This agreement of course created unhappiness in California; water allocation is a classic zero-sum game in which any benefit to one party equals a loss to another.

Meanwhile, booming cities search desperately for new water sources. Las Vegas wants to divert water by pipeline from northern Nevada—a project opposed by people there and in neighboring Utah. St. George, a rapidly growing retirement and recreation community in southern Utah, wants to build a new pipeline from Lake Powell (which has been drying up)—much to the dismay of Nevada. New Mexico plans to build a channel diverting water out of the Gila River upstream from Arizona, but Arizona is counting on continuing to use that water. Colorado's state government is considering a plan to divert water all the way from the Missouri River, at the far eastern end of Kansas.

Joe Hobbs

• Figure 11.62 Cotton farms interspersed with subdivisions on the outskirts of Phoenix. Note at top center the development centered around a lake and canals. At far left is a dry "wash," or arroyo.

THINK CRITICALLY: Can Phoenix afford such "guilty pleasures" as this lakeside suburb in this bone-dry part of the country?

The Arctic National Wildlife Refuge

Alaska (from *Alyeska*, an Aleut word meaning "great land") (**•Figure 11.63**) makes up about 17 percent of the United States by area, but it is so rugged, cold, and remote that its total population was only 736,000 in 2014.

Alaska's Arctic Coastal Plain is an area of tundra along the Arctic Ocean north of the Brooks Range. Known as the North Slope, it holds the reserves of oil and natural gas that made Alaska the biggest fossil fuel–producing state in the United States for decades until some of the oil wells began to run dry. Oil supplies more than 80 percent of the state's revenues, most of which is invested in the state's **Permanent Fund**; the dividends from these investments are paid out to Alaska's citizens each year, usually in amounts over $1000.

Since 1977, North Slope oil has been transported southward through the Trans-Alaska Pipeline, an extraordinary feat of engineering snaking 800 miles (1287 km) across some of the most challenging environments on Earth, all the way to the port of Valdez on the south coast (**•Figure 11.64**).

Originally holding over 20 billion barrels of proven reserves, Prudhoe Bay's output peaked in 1988, when 20 percent of the country's oil was produced there, and has declined steadily since then. There are an estimated 2–4 billion barrels left. Attention has turned eastward to the **Arctic National Wildlife Refuge (ANWR)**, often spoken of as "Anwar," and the oil and natural gas beneath it. When the 12,500 square miles (32,000 sq km) of the refuge were set aside as wilderness in 1980, the fate of the oil beneath this tundra ecosystem was left for Congress to decide at some later date.

ANWR's oil has been the subject of a bitter and polarizing debate ever since. On the one side are the oil interests, Alaskan politicians, and almost

• **Figure 11.63** Reference map of Alaska.

all Republicans in Congress, who have argued that the United States must develop these reserves to give the country more independence from Middle Eastern and other imported oil. On the other side is a coalition of environmentalists, many Native Americans, and congressional Democrats who argue that oil production in ANWR will do irreparable damage to the coastal plain's unique environment and the indigenous people who depend on it.

Marc Shandro/Getty Images

• **Figure 11.64** The Trans-Alaska Pipeline delivers 720,000 barrels of crude oil per day from Prudhoe Bay in the north to Valdez, 800 miles (1300 km) to the south. Note how the pipeline is elevated so as not to melt the permafrost below.

The main Native American group involved is the Gwich'in, an Athabascan tribe of 7000 that derives much of its livelihood from exploiting caribou. The Gwich'in live mainly south of ANWR, but the 180,000-strong population of caribou known as the Porcupine Herd spends the winter in Gwich'in territory. The Gwich'in insist that oil drilling and related activity in ANWR will disrupt caribou calving, decimate the herd, and thus endanger the Gwich'in way of life. The indigenous perspective on ANWR is not united, however; an Inuit group called the Inupiat supports oil production there, especially because of the jobs it would bring them.

19

Environmentalists argue that habitats and native cultures would be despoiled to provide just six months of US oil needs, whereas the oil industry says that 20 years of critical US supply is in ANWR. The amount of oil in ANWR is uncertain; 10 billion barrels is one estimate, but initial exploration has revealed that many areas originally thought to have crude oil actually hold natural gas instead.

Our journey around the world has concluded in northern Alaska. May your own journeys bring you happiness and knowledge. And when you are thinking about courses, majors, minors, and certificates, don't forget geography!

One last thing: Do you want to draw your mental map of the world now? (See **page 11**.)

Study Guide

Summary

- The United States and Canada make up a region often known as Anglo America because of British influences. Both had numerous pre-European indigenous cultures and through immigration have been shaped by many others since 1492.
- Greenland is politically a part of Denmark but physically part of this region.
- The United States and Canada have large land areas of about the same size. Their combined population is about 355 million. Populations are clustered in both countries; one example is the American megalopolis of the Northeast, and about 90 percent of the Canadian population lives within 100 miles of the international border.
- Both are largely nations of immigrant origins. Due mainly to immigration, the US is the world's only MDC in the world with significant population growth. The US takes in 20 percent of the world's migrants each year, with Latin America the main source region. Net migration of Mexicans into the US was near zero from 2010 to 2015. The US lacks a consistent policy on how to deal with illegal immigrants and undocumented workers.
- The physiographic regions of the US and Canada are the Arctic Coastal Plains, Canadian Shield, Gulf and Atlantic Coastal Plains, Piedmont, Appalachian Highlands, Interior Highlands, Interior Plains, Eastern Cordillera, Intermountain Basin and Range and Plateaus, and Western Cordillera.
- The US experiences a number of natural hazards including volcanoes, earthquakes, tsunamis, floods, droughts, tornadoes, and hurricanes.
- The US is the most climatically diverse country in the world, with all 11 Köppen climate types. Climatic factors and climate change pose a number of risks and opportunities for both the US and Canada, including potential megadrought in California and the opening up of Arctic waters for navigation, fishing, and mineral exploration.
- Major resources include agricultural land, forests, coal, petroleum, natural gas, iron ore and other minerals, and waterpower.
- Human migrations into North America probably began about 23,000 years ago. Some indigenous groups developed livelihoods based on agriculture and permanent settlement, but most were at least partly nomadic and dependent on hunting and gathering. Great linguistic and cultural diversity existed among the 10 broad indigenous groups. All were well adapted to local environmental conditions, and all were disrupted profoundly by European contact. Native Americans are among the poorest populations in the United States but have fared better in Canada.
- Early European settlement displaced native populations in both countries, where waves of immigration continually changed demographic and ethnic characteristics. The French are the most significant minority in Canada. Hispanics and African Americans are the largest minority groups in the United States. Asians are prominent minorities in both countries.
- The United States built a land empire through warfare and purchases, including the Louisiana Purchase (from France) of 1803; the acquisition of Florida from Spain in 1819; the addition of Texas, the Southwest, and California as spoils of war with Mexico in 1845 and 1848; the Gadsden Purchase from Mexico of a strip of Arizona and New Mexico in 1853; and Alaska, bought from Russia in 1867. Canada's evolution as a land empire was less complex.
- While 76 percent of both Americans and Canadians identified as white in the most recent census years, both countries are highly multicultural.
- The cruel practices of slavery served mainly the plantation economies of the American South. Discrimination against African Americans has long outlived the abolition of slavery and even the advances of the Civil Rights era.
- In the Great Migration, African Americans moved from the South to the Northeast and Midwest, and with white flight many industrial cities in those regions changed from majority white to majority black.
- English, French, and Spanish dominate this region's linguistic geography.
- Christianity predominates in both countries. Roman Catholic is the largest single denomination in both, followed by Protestant denominations. About 6 percent of Americans practice non-Christian faiths. Secularism is stronger in Canada.
- Both the United States and Canada industrialized in the 19th and 20th centuries. The economies have changed with globalization as manufacturing became less important while employment in service industries and information technology grew. Economic inequalities grew in both countries, especially in the US.
- Canada and the United States are among the world's richest countries because of many factors. Both have large endowments of natural assets; both are large in area and population; and both industrialized fairly early. Innovations have increased their efficiency and productivity. Peace and stability at home have stimulated investment and economic development. Both have strong internal unity and political stability.
- Both countries have large fossil fuel endowments. Production of ethanol from corn, and wind and solar power, has grown in the United States, as has production of oil from tar sands in Canada. There is a debate about whether or not to route tar sand

products across the US via the Keystone KL pipeline. The US is experiencing an "energy revolution" largely through fracking, and is importing less oil than in recent decades. Nuclear energy production has stalled because of environmental and health considerations.

- The two countries have had generally good economic and political relations. Each is a major consumer of exports from the other.
- The United States is the world's strongest military and economic power (despite China's questionable economic leadership on a PPP basis, and despite arguments that the US is no longer a superpower). Despite trends of isolationism, its influence in world affairs grew in the wake of the two world wars of the 20th century. The terrorist attacks of September 2001 put US foreign policy on a new course that led to a more assertive interventionist foreign policy. Then, to other countries, the US seemed to become more reticent on the world stage.
- Atlantic Canada consists of the provinces of Newfoundland and Labrador, New Brunswick, Nova Scotia, and Prince Edward Island. Cod fishing was long the mainstay of its economy, but the Grand Banks have become so overfished that government regulation has restricted the harvest, causing much unemployment.
- The Ontario Peninsula is strongly British in origin and character, and Québec is strongly French, with 82 percent of the Québec population of French origin. Québec separatists have been a political presence for the past 40 years, and in 1995 the referendum for an independent Québec was defeated by less than 1 percent. But Québec has gained more provincial autonomy.
- As temperatures warm in the Arctic region, Canada is one of several nations capitalizing on new opportunities for trade and mineral exploitation. Several countries have claimed parts of the Arctic Ocean as sovereign territory, and some have conflicting claims to various areas, including the North Pole.
- Greenland is geologically a part of North America but is a self-governing property of Denmark. Greenland's foreign affairs are controlled by Denmark, which, along with the European Union, spends heavily on subsidies for the 50,000 inhabitants of the world's largest island. Scientists are watchful of the potential melting of the massive icecap that covers much of the island.
- About 40 percent of the world's corn comes from the Midwest. Corn production soared after 2006 as emphasis on ethanol production increased.
- Major demographic changes have taken place in recent decades in the United States. Rural areas in the Midwest have been depopulated, whereas southern and western cities have grown in population. Gentrification, adaptive reuse, and falling crime rates are revitalizing cities. People are moving from suburbs back into cities, pushing up urban housing costs. Suburbs are becoming older and more diverse.
- An alpha city, New York City is the media and financial capital of the US, and one of its most ethnically diverse cities. Its economic epicenter is Lower Manhattan.
- The San Francisco Bay Area is home to Silicon Valley, the dominant technology industry center in the United States. San Francisco is suffering a housing crisis.
- Settlement in the West is concentrated in well-watered areas that are, in effect, oases. The Colorado River and related waters are an integrated system that some states do not recognize in water use decisions. Several compacts apportion waters to upper and lower basis states. Thirsty crops like cotton and alfalfa have been favored by subsidies and other incentives in the arid Southwest.
- California's cities and farms have huge water requirements that cannot be met from local resources. California has long exceeded its allotment of Colorado River waters, and states upstream in the Colorado Basin have won court actions to increase their share and reduce California's.
- Alaska stretches north of the Arctic Circle and has only about 736,000 people. There is an ongoing dispute among Congress, the oil industry, environmentalists, and indigenous peoples about whether the oil reserves of northeastern Alaska's Arctic National Wildlife Refuge (ANWR) should be exploited.

Review Questions

1. What is the main reason for growth in the US' population?
2. What are the main traits and policies in the issues of legal and illegal immigration into the US and Canada? What role does the Fourteenth Amendment to the US Constitution play?
3. Where are populations of the United States and Canada concentrated?
4. What are their main topographic, climatic, and biotic zones and types of the region and the associated land uses?
5. What natural hazards threaten the United States? What new dangers and opportunities are posed by climatic factors in the Arctic and the American West?
6. What are some of the major indigenous cultures of the United States and Canada? How did these people live in their diverse environments? Which are classified as Native Americans, First Peoples, First Nations, and Alaska Natives? What were the causes of depopulation of a number of these cultures after 1492? What is life like on reservations today, and what makes Canada's Nunavut a unique territory?
7. What roles did colonization and immigration from Spain, Britain, France, and the Netherlands play in shaping the human geographies of the US and Canada? How did the French become such an important population in Canada, and how have their aspirations affected the country's politics?
8. How did the US and Canada evolve as land empires?
9. What roles did slavery, the Great Migration, and white flight have in shaping the cultural geographies of the US? What are the major characteristics of Asian Americans in the US and Canada and of Hispanics in the US?
10. What are the major non-indigenous faiths and languages of the United States and Canada?

11. What have been the major trends and developments in the economies of the United States and Canada? How has globalization affected them and the balance between their primary, secondary, and tertiary sectors?
12. What are the major characteristics of economic and political relations between the United States and Canada, historically and today?
13. Why has the US been described as the "Accidental Superpower" and the "Inevitable Empire?" What geographic characteristics favor its power?
14. What minerals and energy resources are especially important in the US and Canada? Where does the US import most of its oil from, and what has prompted its "energy revolution?" What energy alternatives has the US pursued and what factors encourage or discourage their adoption?
15. How important is agriculture in both countries' economies? What roles have tariffs and subsidies played between the countries and between them and other countries?
16. What is meant by the terms "Manifest Destiny," "Mainstream European American Culture," "Protestant Work Ethic," "American Dream," and "American Exceptionalism?"
17. How did the US economy get into and out of the Great Recession?
18. What are some of the characteristics of income and wealth inequality in the US?
19. What can be done about America's crumbling infrastructure?
20. What priorities have preoccupied US foreign policy in recent years? How has America's role in the world changed?
21. Why has fishing failed to provide a sustainable basis for a prosperous economy in Atlantic Canada?
22. What factors are behind Québec's push for independence? What are the prospects that this province might actually break away from Canada?
23. What new issues and potential conflicts have arisen as temperatures in the Arctic region have increased?
24. With what nation is Greenland politically affiliated? Why do most Greenlanders want to maintain these ties?
25. What is happening to Greenland's extensive ice cover? What explains that phenomenon? What would be the result of a complete melting of that ice sheet?
26. What are the four notable changing patterns of settlement within the US? What are small towns doing to lure back residents? How are the characters of the inner cities and the suburbs changing?
27. What factors give New York City its great prominence among American cities?
28. What are the economic and ethnic characteristics of California and its Bay Area?
29. Why is the term *oasis-like* used to describe urban settlement in the American West? What major problems and controversies have emerged involving the area's water supplies?
30. What are the main issues in the debate about ANWR?

Key Terms + Concepts

adaptive reuse (p. 542)
African Americans (p. 514)
Alaska Natives (p. 506)
Aleuts (p. 506)
American exceptionalism (p. 533)
American dream (p. 527)
America's Energy Revolution (p. 522)
Amish (p. 514)
amnesty (p. 497)
Ancestral Pueblo (p. 507)
"anchor baby" (p. 498)
Arctic Bridge (p. 538)
Arctic National Wildlife Refuge (ANWR) (p. 546)
Asian Americans (p. 515)
"baby boom" (p. 493)
bank (p. 537)
birth tourism (p. 498)
blizzard (p. 502)
Bracero Program (p. 517)
British North America Act (p. 533)
Canada-US Free Trade Agreement (p. 527)
Canadian Shield (p. 498)
Central Arizona Project (CAP) (p. 545)
Civil Rights Act (p. 515)
Civil War (p. 514)
Colorado River Compact (p. 544)
Corporate Average Fuel Economy (CAFE) standard (p. 524)
credit crisis (p. 529)
decommissioning (of dams) (p. 525)
Dust Bowl (p. 502)
economic bubble (p. 529)
Emancipation Proclamation (p. 514)
fall line (p. 499)
First Nations (p. 506)
First Peoples (p. 506)
fracking (p. 522)
G-21 (Group of 21) (p. 526)
gaming industry (p. 509)
gentrification (p. 542)
ghetto (p. 515)
Global Warming Solutions Act of 2006 (p. 524)
Gold Rush of 1849 (p. 544)
Great Depression (p. 526)
Great Lakes (p. 498)
Great Migration (p. 515)
Great Recession (p. 529)
Hapa (p. 516)
Hispanic (p. 516)
Homestead Act (p. 511)
housing bubble (p. 529)
hydraulic fracturing (*fracking*) (p. 522)
Indian wars (p. 508)
Internet (p. 528)
interstate highway system (p. 531)
Inuit (p. 506)
isolationism (p. 533)
Jim Crow (p. 515)
Keystone XL Pipeline (p. 522)
Latino (p. 516)
Louisiana Purchase (p. 511)
mainstream European American culture (p. 512)
Manifest Destiny (p. 511)
March on Washington (p. 515)
Mariel boatlift (p. 517)
megadrought (p. 504)
megalopolis (p. 493)
Motor City (p. 527)
Mound Builders (p. 507)
Multiculturalism Act (p. 512)
Northwest Passage (p. 509)
Nunavut (p. 509)
official state sponsors of terrorism (p. 534)
Paleo-Indians (p. 506)
Parti Québécois (PQ) (p. 536)
Permanent Fund (p. 546)
platform economy (p. 529)
preemptive engagement (p. 534)
Protestant work ethic (p. 527)
pueblo (p. 507)
Reconstruction (p. 514)
redlining (p. 515)
reservations (p. 508)
Rust Belt (p. 527)
sanctuary cities (p. 498)
Secure Fence Act of 2006 (p. 498)

segregation (p. 515)
"separate but equal" (p. 515)
Silicon Valley (p. 544)
Sioux (p. 508)
storm surge (p. 502)
Sun Belt (page 542)
superpower (p. 534)
tar sands (p. 521)
tech bubble (p. 529)
Thirteenth Amendment (p. 514)
"Tornado Alley" (p. 502)
Trail of Tears (p. 508)
"visible minorities" (p. 517)
Voter Rights Act (p. 515)
War of 1812 (p. 533)
"white flight" (p. 515)

Notes

1. http://www.brainyquote.com/quotes/keywords/united_states.html.
2. Quoted in Eilene Zimmerman, "Border Agents Feel Betrayed by Bush Guest-Worker Plan," *Christian Science Monitor*, February 24, 2004, p. 3; http://www.newyorker.com/magazine/2015/07/20/the-really-big-one.
3. http://www.nytimes.com/2015/04/14/science/californias-history-of-drought-repeats.html?_r=0.
4. See also www.nytimes.com/2014/03/29/us/governments-find-it-hard-to-restrict-building-in-risky-areas.html.
5. http://natural-history.uoregon.edu/research/paleocoastal-research-project/kelp-highway-hypothesis.
6. Charles C. Mann, *1491: New Revelations of the Americas Before Columbus* (New York: Vintage, 2006).
7. David Levinson, *Ethnic Groups Worldwide* (Phoenix, AZ: Oryx Press, 1998), p. 396.
8. Quoted in Richard W. Stevenson, "New Threats and Opportunities Redefine U.S. Interests in Africa," *New York Times,* July 7, 2003, pp. A1, A8. For more on how the Civil War addressed these two issues unresolved in the American Revolution, see James McPherson, "A Brief Overview of the American Civil War," at http://www.civilwar.org/education/history/civil-war-overview/overview.html.
9. http://www.npr.org/sections/codeswitch/2014/12/15/370416571/half-asian-half-white-no-hapa.
10. www.theglobeandmail.com/globe-debate/visible-minority-a-misleading-concept-that-ought-to-be-retired/article12445364/.
11. http://www.vanityfair.com/news/2015/01/china-worlds-largest-economy.
12. https://hbr.org/2011/06/what-the-west-doesnt-get-about-china.
13. https://www.stratfor.com/analysis/geopolitics-united-states-part-2-american-identity-and-threats-tomorrow.
14. Peter Zeihan, "The Accidental Superpower: The Next Generation of American Preeminence and the Coming Global Disorder," *Stratfor*, 2014, https://www.stratfor.com/analysis/geopolitics-united-states-part-1-inevitable-empire; see also Fareed Zakaria, "America's Prospects Are Promising Indeed," *The Washington Post*, November 20, 2014; and http://www.washingtonpost.com/opinions/fareed-zakaria-americas-prospects-are-promising-indeed/2014/11/20/eb91209a-70f3-11e4-ad12-3734c461eab6_story.html.
15. Friedman, op. cit., note 13.
16. Much of the discussion following draws from Friedman, op. cit, notes 13 and 14.
17. http://www.ucsusa.org/clean_energy/smart-energy-solutions/increase-renewables/farming-the-wind-wind-power.html#.Vav49XjWuPI.
18. http://energy.gov/eere/solarpoweringamerica/solar-energy-united-states.
19. http://www.fb.org/index.php?fuseaction=newsroom.fastfacts.
20. Quoted in Louis Uchitelle, "Goodbye, Production (and Maybe Innovation)," *New York Times,* December 24, 2006, p. BU4.
21. Nora Dunne, "How the American Dream Went Global: Interview with Fareed Zakaria," *Christian Science Monitor*, May 23, 2011.
22. Special Issue on "Innovative Information Technology in the Service Sector," http://www.inderscience.com/info/ingeneral/cfp.php?id=.
23. David Autor and David Dorn, "How Technology Wrecks the Middle Class," *New York Times*, August 25, 2013, *Sunday Review*, p. 6.
24. "Does Technology Really Wreck the Middle Class?" *Public Works Partners*, http://www.publicworkspartners.com/2013/08/28/techwrecksmiddleclass.
25. http://economics.uwo.ca/people/davies_docs/credit-suisse-global-wealth-report-2014.pdf.
26. http://www.msnbc.com/interactives/geography-of-poverty/index.html.
27. http://www.nytimes.com/2014/02/07/nyregion/biden-compares-la-guardia-airport-to-third-world.html; http://www.nytimes.com/2014/02/08/nyregion/some-see-third-world-as-too-kind-for-la-guardia.html.
28. This discussion of solutions draws from https://hbr.org/2015/05/what-it-will-take-to-fix-americas-crumbling-infrastructure.
29. Friedman, op. cit., note 14.
30. http://www.washingtonpost.com/wp-srv/style/longterm/books/chap1/americanexceptionalism.htm.
31. Mary Ali, *Through Three Miracles . . . Pulling the Sail in Together and Resetting "The Middle of the Political Spectrum"* (Pittsburgh, PA: Dorrance Publishing, 2013), p. 213.
32. https://twitter.com/followfdd/status/461860640219754497.
33. Saul Bernard Cohen, *Geopolitics: The Geography of International Relations*, 3rd ed. (Lanham, MD: Rowman & Littlefield, 2015).
34. Friedman, op. cit., note 14.
35. John G. Ikenberry, "The Illusion of Geopolitics: The Enduring Power of the Liberal Order," *Foreign Affairs*, May/June 2014.
36. http://www.defenseone.com/news/2015/07/d-brief-july-2/116824/.
37. Eric Pianin, "Most Americans Want Obama to Send Ground Troops to Battle ISIS," *The Fiscal Times*, March 4, 2015, http://www.thefiscaltimes.com/2015/03/04/Most-Americans-Want-Obama-Send-Ground-Troops-Battle-ISIS.
38. Zeihan and Zakaria, op. cit., note 14
39. Mathew Hays, "Québec's Separatist Movement Is Dead in the Water," *The Guardian*, April 9, 2014, http://www.theguardian.com/commentisfree/2014/apr/09/quebec-separatist-movement-independence-referendum.
40. http://www.fishwatch.gov/seafood_profiles/species/cod/species_pages/atlantic_cod.htm.
41. http://www.casr.ca/id-arctic-empires-4.htm.
42. http://www.theguardian.com/environment/2015/may/05/arctic-ice-retreat-scientists-climate-change.
43. http://www.casr.ca/id-arctic-empires-2.htm.
44. "Russian Leader: Defend Nation's Claim in Arctic," *Columbia Daily Tribune*, from Associated Press, March 18 2010, p. 11A.
45. Ibid.

46. http://www.thewire.com/national/2014/03/more-americans-moving-to-cities-reversing-the-suburban-exodus/359714/.
47. https://www.cnu.org/history.
48. http://www.citylab.com/commute/2015/06/us-sprawl-peaked-in-1994-and-has-been-declining-ever-since/395840/.
49. http://www.usnews.com/news/articles/2014/09/02/an-urban-revival-in-the-rust-belt.
50. http://www.vox.com/2015/6/15/8782235/san-francisco-housing-crisis.
51. http://mobile.nytimes.com/2015/07/19/opinion/sunday/how-the-west-overcounts-its-water-supplies.html; and Greg Hobbs, personal communication.
52. Greg Hobbs, personal communication and "Colorado River Compact Entitlements, Clearing Up Misconceptions"; "Upper Colorado River Basin Compact: Sharing the Shortage."
53. Much of the information for this section comes from the excellent series "Killing the Colorado," http://www.propublica.org/series/killing-the-colorado.

Global Geoscience Watch

Go to the GREENR database and do a "Basic Search" on "Keystone Pipeline." Click the "News" section coverage, and open up the entire list of news articles. Review Figure 11.37 again to see the states through which the proposed pipeline will pass. Find at least one article that mentions a state through which the pipeline will pass. Does the article discuss state-level impacts? Does the article seem to support or oppose the pipeline, and why? Write one or two paragraphs describing what you learned from your research.

Online Resources

For access to MindTap and additional study materials visit www.cengagebrain.com. Read your textbook, take notes, complete activities, take practice quizzes and more.

Glossary

1948–1949 Arab-Israeli War A war fought between Israel and the Arab countries comprising Egypt, Iraq, Syria, Lebanon, and Transjordan after Israel declared its independence as a Jewish country. Hundreds of thousands of Arab Palestinians chose or were forced to flee Israel during this war, many settling in the Gaza Strip and the West Bank.

1973 Arab-Israeli War A war launched against Israel by Egypt and Syria in an attempt to reverse the Arab losses of the 1967 war. Egypt succeeded in winning back the eastern side of the Suez Canal.

Aboriginal languages Indigenous languages of Australia that are not clearly related to any languages outside the continent.

Aborigines People indigenous to Australia that likely arrived in a single major migration 50,000 years ago. Europeans colonizing Australia in the 18th century slaughtered large numbers of Aborigines and drove them into marginal areas of the continent. Aborigines have long been Australia's most disadvantaged group and were not allowed to vote until 1962.

absolute (mathematical) location Determined by the intersection of lines, such as latitude and longitude, providing an exact point expressed in degrees, minutes, and seconds.

adaptation Measures designed to cope with and reduce the unavoidable impacts of climate change.

adaptive reuse New uses for older buildings and stores, often accompanied by a shift from decline to steady renewal in an urban neighborhood.

African Americans Americans of African ancestry.

African Growth and Opportunity Act (AGOA) A bill passed by the US Congress in 2000 that reduced or ended tariffs and quotas on many manufactured, mineral, and food items that could be imported from Africa.

African National Congress (ANC) The governing political party in South Africa since 1994. During the apartheid era, the ANC was dedicated to the nonviolent resolution of conflicts between ruling whites and the black majority. The ANC was banned until 1990. ANC leader Nelson Mandela became South Africa's first post-apartheid president in 1994.

African Standby Force The continental military force established by the African Union's Peace and Security Council for peacekeeping and peacemaking.

African Union (AU) An organization of most African countries formed to promote democracy, good governance, human rights, gender equity, and development.

"Africa's First World War" A widespread war involving nine African countries and 20 rebel movements between 1996 and 2001 that resulted in the deaths of more than 5 million people.

Afrikaners Dutch-speaking descendants of Dutch, French Huguenot, and German settlers in South Africa. They are the largest white population in South Africa; also known as Boers.

Afro-Americans See *African Americans.*

Afro-Asiatic language family A language family that includes Semitic languages such as Arabic and Hebrew.

Age of Discovery (Age of Exploration) The three to four centuries of European exploration, colonization, and global resource exploitation and trading led largely by European mercantile powers, beginning with Columbus at the end of the 15th century and continuing into the 19th century.

Age of Exploration See *Age of Discovery.*

age structure diagram (population pyramid) The graphic representation of a country's population by gender and five-year age increments.

Agricultural (Neolithic or New Stone Age or Food-Producing) Revolution The domestication of plants and animals that began about 13,000 years ago.

Agricultural Triangle See *Slavic Core.*

AIDS (acquired immunodeficiency syndrome) The disease caused by HIV. AIDS is a global pandemic that has killed over 25 million people since the first case was reported in Africa in 1959.

Al-Aqsa *Intifada* A Palestinian uprising, beginning in 2000, triggered by Israeli Prime Minister Ariel Sharon's visit to the Temple Mount in Jerusalem.

Alaska Natives The Inuit and Aleut peoples living in Alaska.

Al-Aqsa Mosque A congregational mosque that is a sacred Islamic site on the Temple Mount in Jerusalem.

Alawite A minority Shiite Muslim group, especially in Syria.

al-Haraam ash-Shariif See *Noble Sanctuary.*

Al-Qa'ida A terrorist group formerly led by Osama bin Laden that was behind several attacks against US interests, including the September 11, 2001, attacks.

Al-Qa'ida in Iraq An insurgent group arising in Iraq and affiliated with al-Qa'ida during the Iraq War, led by Abu Mus'ab al-Zarqawi until 2006. It is the forerunner of ISIS.

Al-Qa'ida in the Islamic Maghreb A Salafist and jihadist militant group operating in the Sahara and Sahel, designated by the US as a foreign terrorist organization.

Aleuts An indigenous people of the Aleutian Islands in Alaska.

Allah God in Islam.

Alliance of Small Island States A group of 39 island countries arguing for significant reductions in global greenhouse gas emissions.

almsgiving One of the Pillars of Islam, in which Muslims voluntarily donate a portion of their incomes to charity.

Al-Shabaab A mainly ethnic Somali terrorist Islamist group, operating in the Horn of Africa and East Africa.

Altaic language family A controversial language family that recent research suggests may not actually exist. Turkic and Mongolic languages are sometimes considered to be branches of Altaic, but there may not be any actual historic link between them.

alternative development An overall effort to enhance the education, health, and livelihoods of Latin American villages in an attempt to reduce the amount of cropland devoted to coca and poppies.

American Dream The concept of upward mobility in American society, associated in part with material possessions such as homes and cars.

American Energy Revolution The trend toward greater energy independence in the US, associated especially with fracking of shales for oil and natural gas.

American exceptionalism The notion that the United States is inherently different from and better than other nations.

American Invasion Mexican term for the Mexican War.

Amerindians See *Native Americans.*

Amhara The politically dominant ethnic group of Ethiopia.

Amish Peoples of Swiss-German origin in the United States who value community, religion, and manual labor; they typically live rural, "plain" lives.

amnesty In the US immigration debate, not persecuting illegal immigrants if they meet certain conditions and granting them permanent resident status.

Ancestral Pueblo A Native American culture that existed in what is now the southwestern United States from 550 CE to 1300 CE, known formerly as Anasazi.

ancestor veneration A widespread custom in eastern Asia, with shrines and offerings dedicated to preserving the memory of one's ancestors.

"anchor baby" A baby born in the United States to foreign parents, for the purpose of securing US citizenship.

animism The belief that natural processes and objects possess souls.

Antarctic Circle A line of longitude (65.56° S) commonly used to divide the southern middle latitudes to the north from the southern high latitudes to the south.

Antarctic Treaty An agreement, signed by 45 countries, forbidding any exploitation of Antarctica's natural resources until 2048.

Anthropocene Epoch An alternative name for the present geological epoch, usually termed the Holocene.

antiretroviral (ARV) drugs A medication for treatment of HIV infection.

anti-Semitism Anti-Jewish sentiments and activities.

apartheid The Republic of South Africa's former official policy of "separate development of the races," designed to ensure the racial integrity and political supremacy of the white minority.

aquaculture The cultivation of aquatic organisms such as fish for food.

Arab A person of Semitic Arab ethnicity whose ancestral language is Arabic.

Arab-Israeli War of 1956 See *Suez Crisis*.

archipelago A chain or group of islands.

Arctic Bridge A potential shipping route between Churchill, Canada, and Murmansk, Russia, across the Arctic Ocean.

Arctic Circle A line of longitude (65.56° N) commonly used to divide the northern middle latitudes to the south from the northern high latitudes to the north.

Arctic Cultural Geography Project National Science Foundation (NSF)–funded geospatial and cultural study of Inupiat traditional knowledge in Alaska.

Arctic National Wildlife Refuge (ANWR) A wildlife refuge in Alaska that contains 3 to 8 billion barrels of oil.

Armenian The language spoken in Armenia.

Armenian Genocide The deaths of an estimated 1.5 million Armenians between 1915 and 1918. Turkey and Azerbaijan have not accepted responsibility for their roles in the genocide.

arms race Usually associated with the competition between the United States and the Soviet Union, it refers to rival and potential enemy powers increasing their military arsenals—each in an open-ended effort to stay ahead of the other.

Ashkenazi Jews Jews originating in Eastern Europe.

Asia-Pacific Economic Cooperation (APEC) group A 21-member organization (including China, Australia, and the United States) that has agreed to establish free and open trade and investment among member countries by 2020.

Asian Americans Americans of Asian ancestry.

Asian Century A term for the 21st century, anticipating the large role many Asian countries may play in changing the globe.

Asian Tigers Several Asian regions including South Korea, Taiwan, Hong Kong, and Singapore that had rapidly growing economies in the late 20th century.

Association of American Geographers Organization dedicated to the study of the theory, methods, and practice of geography.

Association of Southeast Asian Nations (ASEAN) A regional economic association of 10 Southeast Asian countries.

asymmetric warfare A tactic used by a weaker opponent to use unconventional means such as terrorism to exploit the stronger enemy's vulnerabilities.

asylum People given permission to immigrate on the grounds that they would be harmed or persecuted in their country of origin.

atmosphere The layer of gases surrounding the Earth.

atoll Low islands made of coral and usually having an irregular ring shape around a lagoon.

Attila Line See *Green Line* (Cyprus).

Australia, New Zealand, United States (ANZUS) security alliance A defense and security agreement among the three countries, created in 1951. New Zealand has not been an active member since 1987 because of the country's anti-nuclear stance.

Austronesian language family The family of languages of most indigenous peoples of the Pacific Islands.

Austroasiatic language family A large language family of Southeast Asia that includes Vietnamese and Khmer.

Austronesians Original settlers of many locations in Southeast Asia and the Pacific beginning after 5000 BCE. Many of today's Micronesians and Polynesians are their descendants.

autonomous regions A region of a country granted a measure of self-rule without being fully independent of that country.

autonomy Self-rule, in the Peace Process with reference to Palestinians' rights to run their own civil (and some security) affairs in portions of the West Bank and Gaza Strip allocated to them in the 1993–2000 peace agreements.

ayatollah A revered and powerful cleric at the head of the religious establishment in Iran.

Aymara An indigenous language spoken by about 20 percent of Bolivia's population.

azimuthal A type of map projection in which the developable surface is based on a geometric plane.

Aztecs A Native American people that created an empire across central Mexico in the 15th century, which was destroyed by the Spanish in 1521.

Azerbaijani The language spoken in Azerbaijan.

"baby boom" A noticeable rise of the birth rate in the United States between the end of World War II and the mid-1960s.

"baby bounty" A financial incentive provided by the Italian government to families to have more than one child, to combat declining population.

balkanization The fragmentation of a political area into many smaller independent units, as in former Yugoslavia in the Balkan Peninsula.

balloon effect A term referring to the fact that if a balloon is pressed in one place, the air shifts elsewhere. This term is often used to describe areas where drugs are produced and shipped through; when drug supply or shipment is eradicated in one area, it will tend to surge in previously marginal locations.

Baltic languages A language group of Europe that includes Latvian and Lithuanian.

"banana war" Reference to the incidence in the late 1990s when the United States wanted to sell bananas grown in Latin America by US corporations to countries of the European Union (EU), but the EU members refused to buy them, as they wanted to protect their investments in banana farming in former European colonies in Africa and the Pacific.

bank In physical geography, an elevated area of the sea bottom.

Bantu subfamily A subfamily of the Niger-Congo language family, comprising about 400 languages in Sub-Saharan Africa.

Bantustans See *homelands*.

barrio A densely settled neighborhood in city space that is characteristically inhabited by migrants of Hispanic origin.

barter The exchange of goods or services in the absence of cash.

Basques A people, living primarily in Spain and France, who are ethnically and culturally unrelated to those of their host country majorities.

Bay of Pigs Invasion A failed attempt by the John F. Kennedy Administration to overthrow Fidel Castro by military force.

bazaar The central market of the traditional Middle Eastern city, characterized by twisting, close-set lanes, and merchant stalls.

Benelux The name used to collectively refer to Belgium, the Netherlands, and Luxembourg.

Berber An ethno-linguistic group of people living in northern Africa.

"big bang" The addition of ten primarily Eastern European countries to the European Union in 2004.

biodiversity See *biological diversity*.

biodiversity hotspots A ranked list of places scientists believe deserve immediate attention for flora and fauna study and conservation.

biogeography The distribution of living things on Earth.

biological diversity (biodiversity) The number of plant and animal species and the variety of genetic materials these organisms contain.

biomass The collective dried weight of organisms in an ecosystem.

biome A terrestrial ecosystem type categorized by a dominant type of natural vegetation.

biosphere The global ecological system, including all the relationships played out among the lithosphere, hydrosphere, and atmosphere. Also called the ecosphere.

"birth dearth" The declining population of many European countries from low birth rates.

birth rate The annual number of live births per 1000 people in a population.

birth rate solution Intentionally lowering birth rates to prevent an overpopulation crisis.

birth tourism A practice in which an expectant mother comes to the United States for the purpose of having her baby on US soil, making the baby an automatic US citizen.

black-earth belt An important area of crop and livestock production spanning parts of Russia, Ukraine, Moldova, and Kazakhstan. The main soils of this belt are mollisols.

blacks See *African Americans.*

blizzard A severe winter storm with heavy snowfall and high winds.

blood diamonds See *dirty diamonds.*

blowback Term associated with attacks on US forces by former *mujahidin* armed and trained by the United States to fight the Soviet occupation in Afghanistan.

Boko Haram An ISIS-affiliated Islamist terrorist group that operates primarily in northeastern Nigeria and surrounding areas.

Bolivarian Alliance for the Peoples of our America A group of several Latin American countries including Bolivia, Cuba, and Venezuela that opposes the economic policies of the United States in the region.

"Bolivarian Revolution" Hugo Chávez's term for his policy agendas after taking control of Venezuela.

Bolshevik Revolution The seizure of the Russian government by Vladimir Lenin's faction of the Communist Party in 1917.

boreal forest/taiga A type of forest primarily composed of coniferous trees that thrive in subarctic climates.

bourgeoisie In Marxist doctrine, the capitalist class.

Bracero Program An arrangement devised by the US government to import Mexican workers to fill labor shortages during World War II. *Bracero* in Spanish means "manual laborer."

brain drain The exodus of educated or skilled persons from a poor to a rich country or from a poor to a rich region within a country.

branded diamonds Diamonds produced by the De Beers company, certified as coming from nonconflict areas.

break-of-bulk point A classic geographic term describing a point in transit when bulk goods must be removed from one mode of transport and installed on another. A trainload of grain carried to a port for transshipment on cargo boats or barges is a common example.

BRIC economies An acronym for Brazil, Russia, India, and China, four countries that may have the world's largest economies by 2050; sometimes BRICS with "S" representing South Africa.

British North America Act An act of the British Parliament in 1867 uniting several British North American colonies to create the independent country of Canada.

bubble economy See *economic bubble.*

Budapest Memorandum An agreement made between the United States and Russia recognizing the independence, sovereignty, and borders of Ukraine.

Buddhism A religion based on the life and teachings of Siddhartha Gautama, known as the Buddha (or Enlightened One), who lived in the 6th century BCE in India.

buffer zone An area between conflicting powers, keeping them apart.

Burmese The principal language spoken in Myanmar.

"Burmese Way to Socialism" Rigid state control of the economy in Myanmar.

bushmeat In human diet, the meat of animals native to Africa, including chimpanzees, bats, and antelopes.

California Gold Rush The discovery of gold in California, leading to a population boom in that state.

Caliphate (al-Khilaafa) In Islamic history, a Sunni Muslim empire led by a legitimized heir to the Prophet Muhammad's mantle. In modern times, the political-religious goal of Salafist groups including al-Qa'ida and ISIS.

Calvinism A Protestant sect found in Britain.

Camp David Accords A 1979 agreement between Israel and Egypt that returned the Israeli-occupied Sinai Peninsula to Egypt in exchange for Egypt's recognition of Israel as a sovereign state.

Canada-US Free Trade Agreement A free trade agreement between Canada and the United States in 1988 that was the forerunner to NAFTA.

Canadian Shield The ancient geological core of North America.

Cao Dai A faith originating in Vietnam that blends elements of Buddhism, Daoism, Confucianism, Judaism, Christianity, and Islam.

cap-and-trade system An international system designed to reduce carbon emissions by having richer countries that produce lots of greenhouse gases pay poorer, energy-inefficient countries to cut their emissions. This allows the richer countries to offset their carbon emissions with the cuts in emissions made by the poorer countries. The poorer country is obligated to use the income from cap-and-trade to invest in energy-efficient and nonpolluting technologies.

carbon dioxide (CO_2) A greenhouse gas, the human production of which is very likely the cause of global warming.

carbon sequestration The capture and long-term storage of carbon dioxide. Usually refers to natural processes, but there are also artificial means.

carbon sink An area such as forest, farmland, or ocean that sequesters a large amount of carbon dioxide.

cargo cults Belief systems that appeared in New Guinea and on various Pacific islands during World War II when "Stone Age" peoples were exposed for the first time to Western technology and consumer goods.

carrying capacity The size of a population of any organism that an ecosystem can support.

cartels Illegal organizations controlling the production and distribution of drugs.

Carter Doctrine President Jimmy Carter's declaration, following the Soviet Union's invasion of Afghanistan in 1979, that the United States would use any means necessary to defend its vital interests in the Persian Gulf Region. The "vital interests" were interpreted to mean oil, and "any means necessary" interpreted to mean that the United States would go to war with the Soviet Union if oil supplies were threatened.

cartography The science, technology, and art of designing and making maps, the basic language of geography.

cash (commercial) crops Crops produced generally for export.

caste The hierarchy in the Hindu religion that determines a person's social rank. It is established by birth and cannot be changed.

cataracts Areas along the Nile River where its valley narrows and rapids form.

Catholic Republicans One side of "The Troubles" in Northern Ireland, those who wished the British Army to leave the area and for Ireland to be unified.

Celtic A language family that once spread throughout much of Europe, but is now restricted to extreme western portions of Ireland, Britain, and France.

Celtic Tiger A nickname for Ireland in the 1990s for the country's rapidly growing economy.

Central Arizona Project (CAP) A canal network bringing water from the Colorado River to the populous cities of Arizona.

Central European Free Trade Agreement (CEFTA) A free-trade agreement formed by countries in Eastern Europe, membership in which is a precursor to those countries gaining admittance into the European Union.

chaebols A selection of small companies in South Korea that were heavily funded by the government, eventually growing into giant conglomerates.

Chechen The language spoken in the Russian region of Chechnya.

Chernobyl The site in the Ukraine where, in April 1986, the worst nuclear power plant accident in history occurred. It is thought that approximately 5000 people died, and a zone with a 20-mile (32 km) radius is still virtually uninhabitable; 116,000 people were moved from the area, and cleanup continues to this day.

chernozem A Russian term meaning "black earth." It is a grassland soil that is exceptionally thick, productive, and durable.

Chibcha A small, pre-Columbian civilization in South America renowned for their exquisite gold works.

chlorofluorocarbons (CFS) A gas used in coolants and refrigerants that is a contributor to the greenhouse effect and also destroys ozone.

chokepoint A strategic narrow passageway on land or sea that may be closed off by force or threat of force.

choropleth maps A kind of thematic map that uses shades of color to represent the values of some phenomenon across statistical areas.

Christianity One of the world's three Abrahamic faiths, predominant in Europe and many former European colonial areas. Christianity has several major denominations, including Roman Catholicism, Protestantism, and Eastern Orthodox.

Church of England A Protestant sect that is the main church of England.

Church of the Holy Sepulcher Located in Jerusalem, this is the center of the Christian world, containing the places where tradition says Jesus Christ was crucified and buried.

"City of Light" A nickname for Paris, France.

civilization The complex culture of urban life.

Civil Rights Act A law passed in 1964 that ended all legal racially based discrimination in the United States.

Civil War A war between the northern and southern states of the United States from 1861 to 1865.

clean development mechanism (CDM) The process by which an Annex I (wealthier) country earns carbon emissions credits by investing in emissions-reducing projects in a poorer country.

climate The average weather conditions, including temperature, precipitation, and winds, of an area over an extended period of time.

climate change refugees People migrating from their home countries as a result of sea level rise or other harmful environmental changes brought on by climate change.

Closer Economic Relations Agreement A free-trade pact between Australia and New Zealand.

"coconut civilization" A term characterizing the importance of coconuts in the economies and diets of several low-island Pacific countries.

Cold War The tense political and military competition between the United States and the Soviet Union, and their respective allies, from the end of World War II until the collapse of the Soviet Union in 1991.

collective farm *(kolkhoz)* A large-scale farm in the former Soviet Union that usually incorporated several villages. Workers received shares of the income after the obligations of the collective had been met.

Collective Security Treaty Organization (CSTO) The Russian-initiated security alliance created in 1992 to replace the Warsaw Pact as a counterbalance to NATO.

collectivization The process of forming collective farms in Communist countries.

collectivized agriculture See *collective farm.*

colonization The European pattern of establishing dependencies abroad to enhance economic development in the home country.

Colorado River Compact A 1922 agreement between seven western US states that the Colorado River flows through, which determined how much river water each state may take.

coloureds A South African term for persons of mixed racial ancestry.

Columbian Exchange The transfer of plants and animals between Europe and the Americas after Christopher Columbus's voyages.

co-management A joint system of national park management in Australia between the Aborigines and the Australian government.

command economy A centrally planned economy typical of the Soviet Union and its Communist allies, in which the government, rather than free enterprise, determines the production, distribution, and sale of economic goods and services.

commercial crops See *cash crops.*

Common Agricultural Policy (CAP) Generous agricultural subsidies paid to farmers in the European Union, without which they would not be able to farm profitably.

Common Market An earlier name given to countries that make up the European Union. In 1957, an initial six countries combined to form the European Economic Community (EEC), and this supranational community has grown to have considerable economic and political importance in Europe. See *European Union.*

Commonwealth of Independent States (CIS) A loose political and economic organization of former Soviet republics (except for Estonia, Latvia, and Lithuania) formed in 1991.

Commonwealth of Independent States Free Trade Agreement (CISFTA) A free trade agreement between the members of the Commonwealth of Independent States.

Commonwealth of Nations A voluntary association of 54 countries around the world, most of them former British colonies.

communal violence Term used in India to refer to sectarian violence.

communism A type of national economy characterized by abolition of trade unions, abolition of private ownership in fields such as manufacturing and commerce, varying degrees of state ownership of agriculture, centrally planned and directed economies, and one-party dictatorial government.

Community for Democracy and Rights of Nations An organization comprised of Transnistria, Nagorno-Karabakh, Abkhazia, and South Ossetia, four breakaway areas whose independence is not recognized by most other countries.

Compact of Free Association A 1986 agreement between the United States and the Marshall Islands and the Federated States of Micronesia granting independence and subsidies to the two Pacific countries in exchange for renting their military sovereignty to the United States.

Comprehensive Test Ban Treaty A treaty signed by 149 nations in 1996 that prohibits all nuclear tests.

compromise projection A map projection that does not preserve any one metric, but is designed to distort all metrics roughly equally to achieve a map that "looks right."

conflict diamonds See *dirty diamonds.*

conformal projection A map projection that preserves shape.

Confucianism A Chinese sociopolitical philosophy based upon the writings of Confucius, either on its own or blended with other religions.

conic projection A type of map projection in which the developable surface is based on a cone.

coniferous trees Needleleaf evergreen trees; most bear seed cones.

consumption overpopulation The concept that a few persons, each using a large quantity of natural resources from ecosystems across the world, add up to too many people for the environment to support.

continental drift See *plate tectonics.*

continental islands Islands sitting on continental shelves, or fragments that have broken off a continent.

contingent sovereignty An agreement in Myanmar between the government and several minority ethnic groups, granting them more civil rights and economic opportunities.

Convention on the Law of the Sea A 1970s United Nations treaty permitting a sovereign power to have greater access to surrounding marine resources.

coordinate system A grid consisting of horizontal and vertical lines used to establish absolute location. On the Earth's surface, latitude and longitude lines form the coordinate system.

Copts A Christian sect in Egypt.

core location An area of the world with significant importance due to a central location relative to others.

cornucopians People who argue that humans have always been able to conquer food shortages in the past, thanks to technological innovations, and therefore will be able to in the future as well. Also called *technocentrists.*

Corporate Average Fuel Economy (CAFE) standard A 1975 US law stating that passenger cars produced by all auto companies must achieve an average of 27.5 miles per gallon. The mpg figures are periodically revised.

Cossacks Peasant soldiers of the Russian steppes who were among the first Russian settlers in Siberia in the 16th century.

counter-mapping Challenging the accepted "truths" of colonial and other cartographies by mapping indigenous perceptions of place.

credit crisis A financial crisis originating in the United States in 2008, after the housing bubble burst, that quickly spread worldwide, causing global selloffs in stock exchanges and a widespread economic recession.

Creole Tongues that developed among black slaves and indentured servants in the Americas, consisting mainly of vocabularies from the European colonizers' languages.

Creole languages See *Creole.*

crop substitution Efforts to get Latin American farmers to switch from growing illegal coca or poppies to legal flowers or coffee.

crowdsourcing Acquiring information from large groups of people.

Crusades A series of European Christian military campaigns between the 11th and 14th centuries aimed at recapturing Jerusalem and the rest of the Holy Land from the Muslims.

cryosphere The parts of the Earth's surface that are permanently frozen.

Cuban Americans Cubans who fled Castro's regime for the United States, and their descendants.

Cuban Missile Crisis The debacle between the United States and the Union of Soviet Socialist Republics in 1962 that put both countries on the brink of nuclear war. Soviet rocket launch facilities were being built in Cuba, presumably for possible nuclear strikes on the United States. President Kennedy demanded that the Soviets withdraw the missiles; they did.

cultural landscape The landscape modified by human transformation, thereby reflecting the cultural patterns of the resident culture.

culture The values, beliefs, aspirations, modes of behavior, social institutions, knowledge, and skills that are transmitted and learned within a group of people.

culture hearth An area where innovations develop, with subsequent diffusion to other areas.

cylindrical projection A type of map projection in which the developable surface is based on a cylinder.

Dalai Lama The spiritual and political leader of Tibetan Buddhism (Lamaism).

Dalits The bottom of the social hierarchy in India, not part of the caste system. Once known as "untouchables."

Daoism (Taoism) A Chinese school of thought, second only to Confucianism, originating in the *Tao-te Ching* in the 6th century BCE Daoism philosophy is commonly blended with elements of Buddhism.

Dayton Accord A peace agreement creating a joint multiethnic, democratic national government in Bosnia and Herzegovina, responsible for foreign and economic policies, while also recognizing a second tier of government charged with overseeing internal functions composed of the Federation of Bosnia and Herzegovina and the Republika Srpska.

death rate The annual number of deaths per 1000 people in a population.

death rate solution A catastrophic increase in death rates, seen as "nature's solution" to overpopulation.

debt-for-nature swap An arrangement in which a certain portion of international debt is forgiven in return for the borrower's pledge to invest that amount in nature conservation.

decommissioning (of dams) A major modification or removal of a dam primarily for environmental reasons.

Deep and Comprehensive Free Trade Area (DCFTA) A free trade agreement between Ukraine and the European Union. The failure of Ukraine's pro-Russian president to sign this agreement led to the 2014 Ukrainian Revolution.

deep state The pervasive, repressive military-security complex of Egypt.

deflation A sustained decrease in the price of goods that can result in widespread unemployment.

deforestation The removal of tree cover.

degrees Unit of measurement for longitude and latitude. There are 180 degrees of latitude (90 degrees north of the Equator, and 90 degrees south) and 360 degrees of longitude (180 degrees east of the Prime Meridian, 180 degrees west).

deindustrialization An economy shifting away from industries requiring large amounts of energy and labor toward an economy based on services and the production of high-tech goods.

delta The usually triangular-shaped alluvial area at the mouth of a river.

demilitarized zone (DMZ) The border between North and South Korea.

demographic transition The transition from high birth rates and high death rates to low birth rates and low death rates that has accompanied economic development in the MDCs and may be a model for population change in other countries.

demography The study of population.

"denying the countryside to the enemy" The use of explosives and defoliants such as Agent Orange by US troops to create enormous damage to the farms and forests of Vietnam.

dependency theory A theory arguing that the world's more developed countries continue to prosper by dominating their former colonies, the now independent less developed countries.

desert An area too dry to support a continuous cover of trees or grass. A desert generally receives less than 10 inches (25 cm) of precipitation per year.

desert and xeric shrubland Scant, bushy plant life occurring in deserts of the middle and low latitudes where there is not enough rain for trees or grasslands. The plants are generally xerophytic.

desertification The expansion of a desert brought about by changing environmental conditions or unwise human use.

deterritorialized states Low-lying countries that have been or will be completely covered by the sea as a result of climate change.

development A process of improvement in the material conditions of people, often linked to the diffusion of knowledge and technology.

development traps This theory argues that LDCs can move into middle-income status rather easily when commodity prices are rising, but find it much more difficult to advance to MDC status if commodity prices fall.

devolution The process by which a sovereign country releases or loses more political and economic control to its constituent elements, such as states and provinces.

Diaspora The dispersal and exile of the Jews outside Palestine beginning in the Roman Era.

diffusion The spread of cultural traits or other phenomena to other areas.

digital divide The divide between the handful of countries that are the technology innovators and users and the majority of nations that have little ability to create, purchase, or use new technologies. The divide also exists with countries and societies.

direct rule The policy of administering Northern Ireland directly by the central government of the United Kingdom.

dirty (conflict, blood) diamonds Rough diamonds used by rebel movements or their allies to finance conflict aimed at undermining legitimate governments.

distance–decay relationship The geographical concept that the interactions between places (and the people that occupy them) decline as the distance between them grows.

divine rule by clerics Iranian principle of the Guardian Theologian, an infallible supreme leader whose authority is based on Shi'ite interpretations of Islam.

Dome of the Rock A Muslim shrine on the Noble Sanctuary (Temple Mount) in Jerusalem, on the site where the two Jewish temples had previously stood.

domestication The controlled breeding and cultivation of plants and animals.

Donetsk People's Republic (DPR) A pro-Russian separatist region of eastern Ukraine which declared itself independent in 2014.

donor democracy A government that makes just enough concessions in holding elections or human rights to win foreign aid, without introducing any real reforms.

dot density map Type of map that uses dots to represent a stated amount of some phenomenon within a political unit.

doubling time The number of years required for the human population of a given area to double.

downstream countries A country that lies along a river that has already passed through at least one other country.

Dravidian language family A language family primarily found in southern India.

Dreamtime In Australian Aboriginal myth, the creation of the Earth and all things on it, created by the "dreams" of humanity's ancestors during the Dreamtime.

drought A naturally occurring climatic event when rain fails to fall over a given area for an extended period.

drought avoidance Adaptations of desert plants and animals to evade dry conditions by migrating (animals) or being active only when wet conditions occur (plants and animals).

drought endurance Adaptations of desert plants and animals to tolerate dry conditions through water storage and heat loss mechanisms.

Druze Members of a small offshoot of Shi'ite Islam, living mainly in the Golan Heights.

dry farming Planting and harvesting according to the seasonal rainfall cycle.

dumping Selling a product for less than the cost to produce it.

Dust Bowl An eight-year drought in the 1930s that ruined crops and livelihoods in the Great Plains, spurring many people to move west to California.

"Dutch Disease" A term used to describe how the Dutch currency became too valuable after production and export of natural gas in the Netherlands soared. This made the country's other products too expensive for other countries to import, and the economy sagged as industrialization fell and unemployment grew.

east longitude See *longitude.*

Eastern hemisphere All of Earth's surface eastward from the Prime Meridian to 180°.

Ebola A deadly disease with symptoms that are initially flu-like, but can progress to vomiting, diarrhea, and internal and external bleeding. An Ebola outbreak in West Africa in 2013–2015 killed over 10,000 people.

ecodevelopment See *sustainable development.*

ecological bankruptcy The exhaustion of environmental capital, leading to potential political and social crises.

ecological footprint The amount of biologically productive land needed to sustain a person's consumption and absorb wastes.

ecologically dominant species A species that competes more successfully than others for nutrition and other essentials of life.

economic bubble The sharply increasing valuations of companies or entire economic sectors to lofty heights that cannot be sustained, often leading to recessions when these bubbles "burst."

economic shock therapy Russia's economic transformation in the early 1990s from a command economy to a free-market economy. Overseen by Boris Yeltsin, this transformation was difficult for a country accustomed to government direction in all economic matters, thus the "shock."

El Niño (El Niño Southern Oscillation [ENSO]) A reversal of normal climatic patterns that occurs every few years, when the waters of the eastern Pacific Ocean become much warmer than usual, which can lead to global climatic disruptions.

Emancipation Proclamation An 1863 executive order by President Abraham Lincoln, freeing all slaves in the Confederate states.

emigrant A person moving away from a place.

Employment Equity Act A South African law that requires employers to move toward "demographic proportionality" in the workforce, based on national proportions of race and gender.

endemic species A species of plant or animal found exclusively in one area.

energy crisis The petroleum shortages and price surges sparked by the 1973 oil embargo.

energy maquiladoras Power generation plants built by American firms just inside Mexico that provide electricity to the US grid but are subject to less stringent pollution controls.

enosis The desire of the Greek majority in Cyprus for political unification with Greece.

environmental determinism The concept that the physical environment has played a sovereign role in the cultural development of a people or landscape. Also known as *environmentalism.*

environmental perception The concept that how people view the world and its landscapes and resources influences their uses of the Earth and therefore the condition of the Earth.

equal-area projection A type of map projection that preserves area.

Equator A line or latitude on the Earth's surface (0 degrees) halfway between the North and South Poles. The dividing line between the Northern and Southern Hemispheres.

equidistant projection A type of map projection that preserves distance from one specific point to all other points.

Eskimos See *Inuit.*

ETA (Basque Homeland and Liberty) A militant Basque group that carried out many terrorist attacks in Spain from the 1960s until 2011.

ethanol A biofuel made primarily from corn.

Ethiopian Orthodox Christianity See *Ethiopian Orthodox Church.*

Ethiopian Orthodox Church The Christian church of Ethiopia, related to the Coptic church of Egypt.

ethnic cleansing The relocation or killing of members of one ethnic group by another to achieve some demographic, political, or military objective.

Eurabia The nationalist perception in Europe that Muslims are changing the region's character.

Eurasia The world's largest landmass. Though Europe is often regarded as a continent unto itself, it is really much more of a large peninsula or subcontinent of western Eurasia.

euro Part of the authority of the European Union has been the institution of a new currency that has become "coin of the realm" since early 2002. Not all EU nations accepted the euro.

Euro-Fascism Europe's right-wing, nationalistic, anti-immigrant sentiments and movements.

Euromaidan The revolution that ousted Ukraine's pro-Russian president in 2014, resulting in the installation of a pro-Western government.

European core Primarily the countries of Western Europe, including Germany, France, and the United Kingdom.

European debt crisis A financial crisis beginning in 2010, when it appeared that Greece might default on its loans. Greece did not want wealthier European countries dictating its spending and taxation policies, whereas wealthier countries such as Germany did not want to spend their own taxes to bail out Greece and other heavy borrowers.

European Economic and Monetary Union An instrument of European Union economic integration bringing benefits of greater size and efficiency than individual EU countries can provide; the Euro is its common currency for many EU countries.

European Economic Community (EEC) An economic organization designed to secure the benefits of large-scale production by pooling resources and markets. The name has been changed to Economic Union. See also *common market.*

European Greenbelt A mosaic of national parks and other protected areas stretching across 18 countries.

European periphery Countries located in Northern, Southern, and Eastern Europe.

European Union (EU) A political and economic organization of 28 European countries (as of 2015). Formed in 1993, the EU has its own parliament and numerous supranational institutions.

Eurozone The countries that use the Euro as currency.

"Evil Empire" President Ronald Reagan's term for the Soviet Union in 1983.

exclusive economic zone (EEZ) An area up to 200 miles (320 km) offshore of a country, in which only that country can conduct economic activities such as fishing and mining.

Exodus The departure of the Israelites, led by Moses, from Egypt around 1200 BCE.

exotic species A nonnative species introduced into a new area.

export processing zones (EPZs) Free trade areas in African and other countries, offering incentives for international companies to invest there.

extensive land use A livelihood, such as hunting and gathering, that requires the use of large land areas.

external costs Consequences of goods and services that are not priced into the initial cost of those goods and services.

externalities See *external costs.*

"extinction capital of the world" A term used for the Hawaiian Islands because of the irreversible impacts that exotic species, population growth, and development have had on the native wildlife.

extraction reserve Areas of tropical rainforest in Brazil where valuable resources such as rubber are allowed to be commercially extracted.

"facts on the ground" The purpose of the Israeli settlements in the West Bank, to make the Israeli presence so entrenched that its withdrawal would be very unlikely.

fair trade movement The selling of commodities such as coffee through cooperatives that provide a living wage for the people in poorer countries growing that product, instead of through commercial producers.

Falashas An ancient Jewish group in Ethiopia, living in isolation for thousands of years.

fall line A zone of transition in the eastern United States where rivers flow from the harder rocks of the Piedmont to the softer rocks of the Atlantic and Gulf Coastal Plain. Falls and/or rapids are characteristic features. The fall line often corresponds with the head of navigation.

fallow A period of rest for soils. In tropical areas that have been farmed soils must be rested for a decade or more in order to revert to wild vegetation, thus restoring the soil's fertility.

Falun Gong (Falun Dafa) The quasi-Buddhist *qigong* movement combining calisthenics and meditation that China's government bans as an "evil cult."

fasting One of the Pillars of Islam, when Muslims are required to abstain from food, liquids, and indulgences, from first light to sunset, during the month of Ramadan.

fault A break in a rock mass along which movement has occurred. A break due to rock masses being pulled apart is a tensional, or "normal," fault, whereas a break due to rocks being pushed together until one mass rides over the other is a compressional fault. The processes of faulting create these breaks.

faulting The processes of rock crowding together or pulling apart along tectonic plate boundaries.

favelas Shantytowns (slums, informal settlements) in Brazil.

feng shui A traditional Chinese theory for selecting favorable sites for building homes, cities, and other urban or architectural structures.

feral animals Domesticated animals that have abandoned their dependence on people to resume life in the wild.

Fertile Crescent The arc-shaped area stretching from southern Iraq through northern Iraq, southern Turkey, Syria, Lebanon, Israel, and western Jordan, where plants and animals were domesticated beginning about 10,000 years ago.

Fertile Triangle See *Slavic Core.*

Final Solution German leader Adolf Hitler's plan for systematic extermination of Europe's Jews during World War II.

final status issues Issues deferred to the end of the Oslo Peace Process between Israel and the PLO. Finally dealt with at Camp David in 2000, they included the status of Jerusalem, the fate of Palestinian refugees, Palestinian statehood, and borders between Israel and a new Palestinian state. The negotiations broke down over these final status issues.

First Nations A term used by Native American groups in Canada, except for the Inuit and Aleuts, to refer to themselves.

First Peoples A Canadian term for the Inuit.

first (preindustrial) stage of demographic transition The stage characterized by high birth rates and high death rates, with little change in total population.

First Temple The Jewish temple built by King Solomon as the abode of the Ark of the Covenant. It was situated atop the Foundation Stone in Jerusalem. Built around 950 BCE, it was destroyed by Babylonians in 586 BCE.

fish farming See *aquaculture.*

Five Themes of Geography A list created by the National Council for Geographic Education and the Association of American Geographers, summarizing what geography is all about: location, place, human–environment interaction, movement, and region.

flooded grasslands and savannas A low-latitude grasslands in an area with marked wet and dry seasons.

flow map A type of map that uses arrows to indicate movement of people or goods from one place to another.

food chain The sequence through which energy, in the form of food, passes through an ecosystem.

Food-Producing Revolution See *Agricultural Revolution.*

Footprints of the Ancestors In Australian Aboriginal creation myth, a term for the paths humanity's ancestors walked.

foraging See *hunting and gathering.*

formal region See *region.*

Foundation Stone The great rock in Jerusalem upon which the Jews' First Temple was built.

fourth (post-industrial) stage of demographic transition The stage characterized by low birth rates and low death rates, with little change in total population.

fracking See *hydraulic fracturing.*

fragile states low-income countries characterized by poor governance and/or weak state capacity, leaving citizens vulnerable to a range of shocks.

"Frankenfoods" See *genetically modified foods.*

Free Aceh Movement (GAM) A secessionist movement in the Indonesian region of Aceh.

Free Syrian Army (FSA) Anti-Assad regime forces initially comprised of defectors from the Syrian Army.

Free Trade Area of the Americas (FTAA) A proposed free-trade area encompassing most countries of the Western Hemisphere.

frozen conflict A prolonged ethno-political conflict that falls short of all-out war.

fuelwood crisis Consequences of deforestation in the less developed countries brought about by subsistence needs such as cooking foods. It becomes increasingly difficult to acquire fuelwoods, which are usually collected by women.

functional region See *region.*

G-21 (Group of 21) An economic group of 21 LDCs strongly involved in discussions of subsidies and tariffs of agricultural goods at World Trade Organization meetings.

"Galapágos Islands of Religion" A term for the religious diversity of Ethiopia.

gaming industry The development of casinos and an economy dependent upon gambling for many Native American tribes.

Gan A language spoken in southern China.

Gaza-Jericho Accord See *Oslo I Accord.*

Gaza Wars Conflicts between Israel and Hamas during the winter of 2008–2009 and the summer of 2014.

Gazprom A state-owned Russian natural gas company.

genetically modified (GM) foods Products such as rice, corn, and tomatoes, that have been bio-engineered to be more productive and resilient. They are widely consumed in the United States but banned or stigmatized in European and some other countries.

genetically modified organism (GMO) foods See *genetically modified foods.*

gentrification The social and physical process of change in an urban neighborhood by the return of young, often professional, populations to the urban core. These people are often attracted by the substantial nature of the original building stock of the place and the proximity to the city center, which generally continues to have major professional opportunities. Although this process brings an urban landscape back into a primary role as a tax base, it does dispossess a considerable number of poor, often minority peoples.

geographic information systems (GIS) A computerized system for storing, analyzing, and displaying spatial data.

geography The study of the spatial order and associations of things. Also defined as the study of places, the study of relationships between people and environment, the study of spatial organization, and the study of Earth as the home of humankind.

geologic hot spot A small area of Earth's mantle where molten magma is relatively close to the crust. Hot spots are associated with island chains and thermal features.

geomancy The arrangement of homes and villages for maximum harmony with natural and spiritual environments.

geopolitics The struggle for space and power played out in a geographical setting.

georeferencing Placing geographic data onto a known coordinate mapping system.

Georgian The official language of the country of Georgia.

geospatial A term pertaining to the geographic location and characteristics of natural or constructed features and boundaries on, above, or below the Earth's surface, especially referring to data that are geographic and spatial in nature.

geo-engineering Using technology to intervene in the climate system in an effort to combat global warming.

Germanic languages A language family of Northern and Central Europe, including German, Dutch, Swedish, and English.

ghetto Impoverished areas of central cities where there are many problems, including high crime and lack of jobs.

ghurba The worldwide diaspora of Palestinians.

gigantomania A Soviet preoccupation with fulfilling huge quotas or implementing grandiose schemes.

glacial deposition In the process of continental and valley glaciation, the deposition of moraines that become lateral or terminal in the act of glacial retreat. This same process also leads to glacial scouring as moving ice picks up loose rock and reshapes the landscape as the glacier moves forward or retreats.

glacial scouring See *glacial deposition.*

glaciation A geologic and climatologic process in which great ice sheets form in the Arctic and Antarctic regions and advance toward the Equator.

glasnost Policies introduced by Russian leader Mikhail Gorbachev in the 1980s to allow a more democratic political system and more freedom of expression. See also *perestroika.*

global city (alpha city, world city) A city that is an economic and political power in the global economy.

globalization The spread of free trade, free markets, investments, and ideas across borders, and the political and cultural adjustments that accompany this diffusion.

Global Warming Solutions Act of 2006 A California law requiring the state's greenhouse gas emissions to be cut to 1990 levels by 2020.

Gold Rush of 1849 See *California Gold Rush.*

Golden Crescent A term for portions of Iran, Afghanistan, and Pakistan, referring to the large amounts of the world's opium supply being produced there.

"Golden Horde" Russian term for the Kazan Tatars in the 13th century.

Golden Triangle A region in the hinterlands of Myanmar, Laos, and Thailand that is the center of Southeast Asia's illegal opium and other drug production.

Good Friday Agreement An agreement between the Unionists and the IRA in 1998 to replace direct rule of Northern Ireland with the power-sharing Northern Ireland Assembly.

good governance Directing political energies at strengthening the economy rather than trying to cement power and keep down the opposition.

Gosplan (Committee for State Planning) The agency in Moscow that formulated the Soviet Union's national economic plans.

graduated symbol map A type of map that uses symbols such as circles scaled proportionally to the quantity of the attribute being mapped.

Great Depression The worst economic downturn in US history, beginning in 1929 and lasting through most of the 1930s.

Great Game A rivalry between Russia and the United Kingdom in the 19th and early 20th centuries over strategic influence in Afghanistan and other areas of central Asia.

Great Lakes Five interconnected lakes (Superior, Michigan, Huron, Erie, and Ontario) between the United States and Canada that together contain one-fifth of the world's freshwater.

Great Migration The migration of about 6 million blacks from the rural South to the cities of the Northeast and Midwest between 1915 and 1970.

Great Recession Between 2007 and 2009, the worst economic downturn in the US since the Great Depression. It was linked to the worldwide global recession of 2009.

Great Rift Valley As the result of tectonic processes, the splitting apart of Earth's crust extending from the Zambezi Valley in southern Africa northward to the Red Sea and the valley of the Jordan River in southwestern Asia.

Great Trek A series of migrations by the Boers of South Africa, who left the British-controlled Cape Colony to settle interior areas.

Great Volga Scheme The transformation of the Volga River in Russia in the 20th century to make the river more navigable and to supply hydroelectric power and water for irrigation.

Greek The principal ethnolinguistic group inhabiting Greece.

Green Line (Cyprus) A buffer zone dividing Cyprus between the Greek Cypriots and the Turkish Cypriots.

Green Line (Israel) A term referring to Israel's border with the West Bank as of 1967.

Green Revolution The introduction and transfer of high-yield crops (including those genetically modified) with mechanization, irrigation, and massive application of chemical fertilizers to areas where traditional agriculture has been practiced.

greenhouse effect The observation that increased concentrations of carbon dioxide and other gases in Earth's atmosphere cause a warmer atmosphere.

greenhouse gases Gases such as carbon dioxide and methane that trap heat from the sun in the atmosphere, increasing the Earth's temperature.

Greenwich Meridian See *prime meridian.*

gross domestic product (GDP) The value of goods and services produced in a country in a given year. It does not include net income earned outside the country. The value is normally given in current prices for the stated year.

gross domestic product purchasing power parity (GDP [PPP]) A measurement accounting for varying costs of living between countries, with gross domestic product converted to international dollars using purchasing power parity rates. An international dollar has the same purchasing power over GDP as the US dollar has in the United States.

gross national income (GNI) A measure of a country's wealth, counting gross domestic production plus income from abroad such as rents, profits, and labor.

gross national product (GNP) The value of goods and services produced internally in a given country during a stated year plus the value resulting from transactions abroad. The value is normally expressed in current prices of the stated year. Such data must be used with caution in regard to developing countries because of the broad variance in patterns of data collection and the fact that many people consume a large share of what they produce.

Group of Eight (G-8) The world's eight most economically powerful and politically influential countries: Canada, France, Germany, Italy, Japan, Russia, the United Kingdom, and the United States.

GUAM A regional grouping of Georgia, Ukraine, Azerbaijan, and Moldova, promoting economic integration, democratic reforms, and orientation toward Europe and away from Russia.

guano Seabird excrement that is also found in phosphate deposits, valued as a fertilizer.

Guaraní An indigenous language spoken by over 90 percent of Paraguay's population.

guest workers Migrants (mainly young and male) from less developed countries, who are employed, sometimes illegally, in more developed countries.

Gulf Stream The strong, warm ocean current originating in the tropical Atlantic Ocean that skirts the eastern shore of the United States, curves eastward, and reaches Europe with an appendage called the North Atlantic Drift.

Gulf War The 1990–1991 confrontation between a large military coalition, spearheaded by the United States, and the Iraqi forces that had invaded Kuwait. The allies successfully drove Iraqi forces from Kuwait. Earlier, the Gulf War had referred to the 1980–1988 war between Iraq and Iran.

Gypsies See *Roma.*

Hakka A language spoken in southern China.

hacienda A Spanish term for large rural estates owned by the aristocracy in Latin America.

hajj Pilgrimage to Mecca, one of the five Pillars of Islam.

Hamas Gaza Strip–based Palestinian Islamist party long dedicated to the destruction of Israel and opposed to the Palestinian-Israeli peace process. A major rival to the moderate al-Fatah party within the Palestinian Authority.

Han Chinese The original Chinese peoples who settled in North China on the margins of the Yellow River (Huang He) and who were central to the development of Chinese culture. Han Chinese make up more than 90 percent of China's population.

Hanification Efforts by the Chinese government to increase ethnic Han presence in China's outlying areas, particularly Tibet and Xinjiang.

Hapa The mixed-race offspring between Asian and other ethnic groups, living in the United States and Canada.

Hebrews A term for Jews in Biblical times.

hemisphere Half of the Earth's surface.

Hermit Kingdom A term for North Korea, indicative of its isolation.

hero projects The construction of dams, railways, factories, and other infrastructure, built by the Soviet government.

high islands Generally the result of volcanic eruptions, these are the higher and more agriculturally productive and densely populated islands of Oceania.

high latitude See *latitude.*

Hinduism A regionally varied, complex belief system that lacks a definite creed or theology but includes a belief in reincarnation, the reverence of many living things, an unlimited pantheon of deities, and an infinite range of types of permissible worship.

hinterland The region of a country that lies away from the capital and largest cities and is most often rural or even unsettled; it is often seen in the minds of economic planners as an area with development potential.

Hispanic A person of Mexican, Cuban, Central American, South American (except Brazil), and other Spanish-derived cultures.

Hizbullah Iranian-inspired and supported Shiite Lebanese political party holding many seats in Lebanon's Parliament. Broadly regarded as a terrorist organization and opposed to the existence of Israel.

Hmong-Mien subfamily A language family of Southeast Asia and southern China.

Hokan-Siouan language family A Native American language family in North America.

Holocaust Associated with the "Final Solution," Nazi Germany's attempted extermination of Jews, Roma (Gypsies), homosexuals, and other minorities during World War II.

holodomor The famine, engineered by Soviet leader Josef Stalin, that killed millions of Ukrainians in the 1930s.

homelands Former territorial units in South Africa reserved for native Africans (blacks). South Africa forced most of its black population onto these homelands, and declared some as "independent" republics, although none were recognized outside South Africa. Homelands were abolished in 1994. Also known as *Bantustans, native reserves*, and *national states*.

Homestead Act An 1862 US law that opened up much of the interior of the country for settlement. A pioneer family could claim up to 160 acres of farming land for a $10 fee.

homogeneous region See *region*.

housing bubble In the US, the speculative frenzy bidding up the costs of real estate purchased with unsustainable loans. Its "popping" was a major component of the Great Recession of 2007–2009.

hukou A registration system in China designed in the 1950s to discourage internal migration from the rural interior to the coastal cities.

Human Development Index (HDI) A United Nations–devised ranked index of countries' development that evaluates quality of life issues (such as gender equality, literacy, and human rights) in addition to economic performance.

human–environment interaction Study of the ways human beings use and change the natural environment.

human geography One of the two major branches of geography, primarily studying the world's peoples, their spatial relationships, and their usage of the natural environment.

human immunodeficiency virus (HIV) The virus that causes AIDS.

human trafficking Movement of people against their will or intention, largely for the sex trade and manual labor.

humid continental A climate with cold winters, warm to hot summers, and sufficient rainfall for agriculture.

humid subtropical A climate that generally occupies the southeastern margins of continents, with hot summers, mild to cool winters, and ample precipitation for agriculture.

humus Decomposed organic soil material. Grasslands characteristically provide more humus than forests do.

hunting and gathering A mode of livelihood, based on collecting wild plants and hunting wild animals, generally practiced by pre-agricultural peoples. Also known as *foraging*.

hurricane Large tropical storms that form over the warm waters of the Atlantic and Pacific Oceans.

Hutu (Bahutu) The majority population of Rwanda and Burundi. Ethnically, culturally, and linguistically the same as the minority Tutsi; the differences between the groups are socioeconomic, with the Tutsi a ruling class that dominated the Hutu majority.

hydraulic fracturing (fracking) A method of extracting natural gas or oil by pumping water, sand, and chemicals underground at high pressure, creating fissures in the rock that allow the natural gas to rise to the surface.

hydrologic cycle The process by which the sun evaporates seawater into water vapor that is later released as freshwater precipitation.

hydropolitics Political leverage and control over water.

hydrosphere All of the world's water features.

Ibo (Igbo) A major ethnic group in southern Nigeria.

Ice Age A period of widespread glaciation from about 2 million to 10,000 years ago.

ice cap A climate and biome type characterized by permanent ice cover on the ground, no vegetation (except where limited melting occurs), and a severely long, cold winter.

Ijaw A major ethnic group in southern Nigeria.

illegal alien A migrant moving to another country illegally.

immigrant A person moving into a place.

Inca An indigenous people of South America who created an empire running from modern-day Ecuador to Chile. The Spanish crushed the Inca Empire in 1532.

Indians See *Native Americans*.

Indian wars The clashes between the US government and Native American tribes throughout much of the 19th century.

"India's Silicon Valley" The city of Bengaluru (Bangalore) in India, which has become a major recipient of outsourcing by Western tech companies.

Indo-Aryan language subfamily A subfamily of the Indo-European language family, comprising many languages spoken in South Asia, such as Hindi, Bengali, and Urdu.

Indo-European languages One of the world's major language groups, comprising many languages spoken across much of Europe and South Asia.

Indo-Fijians Residents of Fiji who are descended from indentured laborers from India, brought to the islands by the British in the early 1900s.

Indus Waters Treaty A 1960 treaty between India and Pakistan, allocating how much water from the Indus River each country can use.

Industrial Revolution A period beginning in mid-18th-century Britain that saw rapid advances in technology and the use of inanimate power. It was widely associated with European colonialism and population growth.

industrialization The process where a society shifts from a primarily agricultural economy to one based on manufacturing goods and providing services.

informal sector A sector of a country's economy that is not monitored or taxed by governments. The "black market" is part of the informal sector.

informal settlements An area where residents lack legal claims to the land, and housing is not in compliance with regulations; also described as slums.

information technology (IT) The Internet, wireless telephones, fiber optics, and other technologies characteristic of more developed countries. Generally seen as beneficial for a country's economic prospects, IT is also spreading in the less developed countries.

infrastructure-for-mineral swaps Mining concessions by African countries to China, in exchange for schools, hospitals, and roads to be built in those countries by China.

Inkatha Freedom Party (IFP) South African political party with mainly Zulu membership. Founded by Zulu chief Mangosuthu Buthelezi, the IFP is a rival of African National Congress (ANC), long led by Nelson Mandela.

insolation Heat from the sun.

insourcing Keeping certain jobs within a country, or returning them to the country, instead of offshoring those jobs to another country.

intellectual property rights Legal claims to patents and copyrights.

intensive land use A livelihood such as farming, requiring use of small land areas.

intensive subsistence agriculture Agricultural practices built around the growing of cereals (especially rice) that require a steady output of arduous manual labor.

Intergovernmental Panel on Climate Change (IPCC) A United Nations panel, composed of 2500 atmospheric scientists from 130 countries, that has concluded that the evidence for global warming is "unequivocal."

internally displaced persons (IDPs) People dislodged and impoverished by strife in their home countries, but who have little prospect of emigrating from that country.

International Atomic Energy Agency (IAEA) A United Nations agency that carries out inspections of nuclear facilities.

International Criminal Court A United Nations forum for crimes against humanity, located in the Netherlands.

International Date Line A line roughly concurrent with the 180-degree line of longitude (with several deviations for political and practical reasons) where the beginning of one day and the end of another day meet. The date west of the International Date Line is one day ahead of the date east of the line.

international emissions trading See *cap-and-trade*.

Internet A vast, global network of interconnected computers, smartphones, and other electronic devices.

interstate highway system A massive infrastructure development in the United States that started in the 1950s and serves as the primary network for trucking cargo across the country.

Intifada The first Palestinian uprising against Israeli occupation, from 1987 to 1991.

Inuit A people living in Greenland and far northern Canada and Alaska, sometimes known as Eskimos to others.

Iranian A subgroup of the Indo-European language family. Persian and Tajik are Iranian languages.

Iran-Iraq War of 1980–1988 Launched by Iraqi leader Saddam Hussein in an effort to secure more Persian Gulf frontage and inflict a historic Arab defeat on the Persian foe, it ended in stalemate after great loss of life.

Irish Republican Army (IRA) A terrorist group based in Northern Ireland that launched many attacks in the United Kingdom that were designed to drive the British army out of Northern Ireland.

Iron Curtain A term used by Winston Churchill to describe the dividing line in Europe between capitalist democracies to the west and Communist authoritarian regimes to the east.

irredentism Movement by an ethnic group in one country to revive or reinforce kindred ethnicity in another country—often in an effort to promote succession there.

irrigation The artificial placement of water to produce crops, generally in arid locations.

isarithmic map A type of map that uses lines or bands of color to join the points of equal value of some phenomenon across a mapped area.

ISIS (Islamic State in Iraq and Syria), ISIL (Islamic State of Iraq and the Levant), IS (Islamic State) A Sunni Salafist "fundamentalist" Islamic militant group employing often cruel terrorist tactics to eradicate and intimidate members of almost all other religious groups. Originating in Iraq and Syria, where it has declared itself "the Caliphate," ISIS has expanded to a number of other countries.

Islam A monotheistic Abrahamic faith built upon the foundations of Judaism and Christianity. One of the world's major religions, it originated in the 7th century in what is now Saudi Arabia.

Islamic law (*shari'a*) Islamic interpretations of divine intention in legal matters, based on the Qur'an, the Hadith, and Islamic scholars' opinions in matters of jurisprudence.

Islamic Movement of Uzbekistan (IMU) A group dedicated to the overthrow of Uzbekistan's president and the creation of an Islamic state in the Fergana Valley of central Asia.

Islamists Mainstream and radical (and in some cases terrorist) Muslims dedicated to the principle that Islam provides guidance and solutions for all matters in life and society.

Islamophobia A fear of Islam and the Muslims that practice it.

isolationism The tendency of a country to stay out of the conflicts of other countries.

isostatic rebound The slow uplift of land that had been pressed down by the great weight of glaciers during the Ice Age.

Israelites A term for Jews, associated with the geographic area of Palestine, in Biblical times.

Jainism A small but influential religion of India renowned for its respect for geographic features and animal life.

Jains Practitioners of Jainism.

Jasmine Revolution The uprising in Tunisia against that country's leader in early 2011.

Javanese The language spoken by people on the Indonesian island of Java. Javanese is the largest language in the world that is not an official tongue of any country.

Jemaah Islamiah (JI, or the Islamic Group) A militant Islamist group in Southeast Asia.

Jewish settlements Communities of Israeli citizens settling in the West Bank to strengthen Israel's claims to the area.

Jews Practitioners of Judaism.

jihad Literally "struggle," the term that for Muslims can have a range of meanings, from expressing one's commitment to a personal conviction to all-out war against nonbelievers.

Jim Crow A collective term for laws in the United States enforcing racial segregation.

joint comprehensive plan of action (JCPOA) The agreement reached between Iran and the P5+1 in 2015, designed to prevent or delay Iran's development of nuclear weapons.

joint implementation (JI) A program that allows an Annex I (wealthier) country to earn carbon emission reduction credits by investing in emission reduction in another Annex I country.

juche North Korean principle of self-reliance.

Judea and Samaria Biblical lands that for Jews comprise the present-day West Bank.

Judaism The first significant monotheistic and Abrahamic faith, originating at least 4000 years ago. It does not have a fixed creed or doctrine, and it is not a proselytizing religion.

Ka'aba A Muslim shrine located in Mecca, toward which all Muslims pray, that is the centerpiece of the *hajj*.

kanju A term meaning "the specific creativity born from African difficulty." It suggests the best solutions to African problems can be found locally.

karma An important Buddhist concept of a person's acts and the consequences of those acts.

karoshi Japanese term for death by overwork.

Kazakh The principal language spoken in Kazakhstan.

keystone species A species that affects many other organisms in an ecosystem.

Keystone XL Pipeline A proposed pipeline to ship tar sand oil from Canada to refineries on the Gulf coast of the United States.

Khalistan A desired homeland for India's Sikhs.

Khmer The language spoken in Cambodia.

Khoisan language family The languages of the San people in southern Africa. These languages, nearly extinct, are among the world's oldest.

Kimberley Process A United Nations–backed certification plan for ensuring that only legally mined rough diamonds, untainted by violence, reach the market.

kleptocracy An economic and political system based on crime.

knowledge economy An economy based on innovation and services.

Köppen climate classification system (KCC) The most widely used system for categorizing climate types.

Koran See Qur'an.

Korean The language spoken in North and South Korea.

Kurdistan Workers Party (PKK) The main Kurdish resistance to Turkish rule.

Kurds The largest minority group in Turkey and the largest non-Arab minority in Iraq. Kurds are the world's largest ethnic group without a country.

Kyoto Protocol A treaty on climate change signed by 160 countries in Kyoto, Japan, in 1997, put into force in 2005 and extended to 2012. It required MDCs that ratified it to reduce their greenhouse gas emissions by more than 5 percent below their 1990 levels by the year 2012.

Kyrgyz The principal language spoken in Kyrgyzstan.

Labor Party The liberal political party of Israel.

land empire The establishment of colonies in a continental hinterland as opposed to overseas.

Land for Peace Part of the peace process between Israel and Arabs, with Israel swapping Arab lands it occupied in 1967 in exchange for peace with its neighbors.

land grabbing Wealthy countries buying land and water resources in poorer countries to protect their own food security and to turn profits.

land hemisphere The half of Earth's surface containing the largest possible area of land. The defined points of its center vary.

land reform In Latin America, the reallocation of land owned by the wealthy to small farmers.

landscape A portion of the Earth's land surface. Geographers are interested in the transformation of natural landscapes into cultural landscapes.

landscape perspective The understanding of the changes people have made on the natural environment over time.

language death The extinction of a language brought about by the death of its last speaker.

language isolate A language that has no known relatives.

large-scale map A map constructed to show considerable detail in a small area.

Laskar Jihad A militant Islamist group in Indonesia.

laterite A material found in tropical regions with highly leached soils; it is composed mostly of iron and aluminum oxides that harden when exposed and make cultivation difficult.

lateritic soils A soil characteristic of many tropical areas. Excessive exposure to the sun and repeated wetting and drying turns the soil into a bricklike substance nearly impossible to cultivate.

latifundia (sing. ***latifundio***) Large agricultural Latin American estates with strong commercial orientations.

Latino See *Hispanic*.

latitude A measurement that denotes position with respect to the equator and the poles. Latitude is measured in degrees, minutes, and seconds, which are described as parallels. Places near the Equator are said to be in low latitudes; places near the poles are in high latitudes. The poles are set at 90° in both hemispheres. The Tropic of Cancer and the Tropic of Capricorn, at 23.5°N and 23.5°S, respectively, and the Arctic Circle and Antarctic Circle, at 66.5°N and 66.5°S, respectively, form the most commonly recognized boundaries of the low and high latitudes. Places occupying an intermediate position with respect to the poles and the Equator are said to be in the middle latitudes. There are no universally accepted definitions for the boundaries of the high, middle, and low latitudes.

Law of Return An Israeli law permitting all Jews living in Israel to have Israeli citizenship.

Lebanese civil war The civil war in Lebanon in the 1970s, pitting the more prosperous and politically powerful Maronite Christians against the poorer majority Sunni and Shia Muslims. It resulted in the closure of the Tapline oil pipeline.

Lega Nord The "Northern League," a secessionist movement in Italy.

less developed countries (LDCs) The world's poorer countries.

levee A (usually artificial) raised riverbank designed to prevent flooding.

liberal autonomy See *autonomous regions*.

Liberal Democratic Party (LDP) The dominant political party of Japan since 1952.

life expectancy The number of years a person may expect to live in a given environment (typically defined as a country) and differentiated between women, who usually live longer, and men.

Liga Veneta An Italian organization wishing to create an autonomous region centered on the city of Venice.

Likud Party The conservative political party in Israel.

lithosphere The rocky portion of the Earth's surface.

location A concept central to all geographic analysis. Where something relates to all manner of influences, from climate to migration routes, and is a crucial component in trying to understand patterns of historic and economic development.

lock Devices that allow ships to move from one water level to another in a canal.

loess A fine-grained material that has been picked up, transported, and deposited in its present location by wind; it forms an unusually productive soil.

longitude A measurement that denotes a position east or west of the Prime Meridian. Longitude is measured in degrees, minutes, and seconds, with the Prime Meridian set at 0°, and 180° situated directly opposite the Prime Meridian. Meridians of longitude extend from pole to pole and intersect parallels of latitude.

"loose nukes" Stolen or unaccounted for radioactive materials, feared to be available on the black market for use by terrorists.

Lord's Resistance Army A rebel group based in Uganda and led by the warlord Joseph Kony.

Louisiana Purchase A large purchase of land by the United States from France in 1803 that significantly added to the size of the United States.

low islands Made of coral, these are the generally flat, drier, less agriculturally productive, and less densely populated islands of the Pacific World.

low latitude See *latitude*.

Luhansk People's Republic (LPR) A pro-Russian separatist region of eastern Ukraine that declared itself independent in 2014.

Lutheran Protestantism The dominant religion across Scandinavia and northern Germany.

Ma'adan (Marsh Arabs) Shi'ite Arabs living on marshlands in Iraq, creating a unique culture found nowhere else in the Middle East. Most of their habitat was destroyed by Saddam Hussein after the first Gulf War.

Maastricht Treaty of European Union The treaty that created the European Union in 1993.

Mahayana Buddhism One of the two branches of Buddhism, originating in India and spreading northward to Tibet and China.

mainstream European American culture A culture based on the English language and British educational principles brought by British colonists combined with core American values that developed on US soil.

major cluster of continuous settlement An area where all habitations lie no more than 3 miles (5 km) from other habitations in at least six different directions, and roads or railroads lie no more than 10–20 miles (16–32 km) away in at least three directions.

Malay The official language of Indonesia, Malaysia, Singapore, and Brunei. The Malay word for this language is *Bahasa*.

Malthusian scenario The model forecasting that human population growth will outpace growth in food and other resources, with a resulting population die-off.

mangroves A tropical saltwater marsh biome dominated by salt-tolerant mangrove tree species, often protecting vulnerable coastlines from erosion.

Manifest Destiny The opening and settlement of new frontiers across the United States from the Atlantic to the Pacific Oceans, believed to be preordained for Americans.

map A representation of various phenomena over all or a part of the Earth, either on a globe or a flat surface.

MapGive An initiative of the US Department of State's Humanitarian Information Unit allowing volunteers to contribute geospatial data to ongoing humanitarian efforts around the world.

map projection Mathematical ways to minimize the distortion that arises when the curved surface of the Earth is represented on a flat surface.

maquiladoras Operations dedicated to the assembly of manufactured goods, generally in Mexico and Central America, from components initially produced in the United States or other places. With the enactment of NAFTA in 1994, there was a massive expansion of *maquiladora* operations.

March on Washington A gathering of 250,000 people in Washington, DC, in 1963, where Dr. Martin Luther King Jr. gave his famous "I have a dream" speech, envisioning the end of racism in the United States.

marginalization A process by which poor subsistence farmers are pushed onto fragile, inferior, or marginal lands that cannot support crops for long and that are degraded by cultivation.

Mariel boatlift A mass exodus of Cubans to the United States in 1980. Cuban leader Fidel Castro placed 20,000 criminals among the 120,000 other migrants so Cuba would no longer have to deal with them.

Maronites A Christian sect in Lebanon.

Marshall Plan The plan designed largely by the United States, after the conclusion of World War II, by which US aid was focused on the rebuilding of the very Germany that had been its enemy in the war just concluded. Secretary of State George Marshall (who had been Chief of Staff of US Army from 1939 to 1945) was central to the plan's design and implementation.

Masdar A carbon-neutral city under construction in the United Arab Emirates.

mathematical location See *absolute location*.

Maya A Native American ethnic group inhabiting southern Mexico and northern Central America.

Mayan language family A language family of southern Mexico and Guatemala.

medina An urban pattern typical of the Middle Eastern city before the 20th century.

Mediterranean (dry-summer subtropical) A climate that occupies an intermediate location between a marine west coast climate on the poleward side and a steppe or desert climate on the equatorward side. During the high-sun period, it is rainless; in the low-sun period, it receives precipitation of cyclonic or orographic origin.

Mediterranean forests, woodland, and scrub The xeroyphytic vegetation typical of hot, dry summer, Mediterranean climate regions. Local names for this vegetation type include *maquis* and *chaparral*.

megacity A city with more than 10 million inhabitants.

megadiversity country A country containing a large percentage of the world's animal and plant species.

megadrought A naturally occurring drought that can persist for decades or centuries.

megalanguage One of the world's most-spoken languages, including Chinese, English, Spanish, Hindi, and Arabic.

megalopolis A chain of metropolitan areas with a population over 10 million.

Mekong River Commission A coalition of Vietnam, Laos, Cambodia, and Thailand, to determine rules to ensure water quality and quantity of the Mekong River.

Melanesia A group of relatively large continental islands in the Pacific Ocean, including New Guinea and New Caledonia.

mental map A term used to define the geographies encompassed in every individual's mind as a series of locations, access routes, physical and cultural characteristics of places, and often a general sense of the good or bad locales.

mercantile colonialism The historical pattern by which Europeans extracted primary products from colonies abroad, particularly in the tropics.

Mercator projection A cylindrical, conformal map projection intended to be used for navigation developed by Gerardus Mercator in 1569.

Mercosur Acronym for Southern Cone Common Market, an economic association among Brazil, Argentina, Paraguay, Venezuela, and Uruguay.

meridian See *longitude*.

mestizo In Latin America, a person of mixed European and Native American ancestry.

methane Greenhouse gas resulting from human activities (from rice paddies and the guts of ruminating animals like cattle).

Mexican-American War A war between the United States and Mexico in the 1840s, in which Mexico lost a significant portion of its territory to the United States.

Mezzogiorno A term for southern Italy.

microcredit The lending of small sums to poor people to set up or expand small businesses.

Micronesia The thousands of small and scattered islands in the central and western Pacific Ocean, mainly north of the Equator.

microstate A political entity that is tiny in area and population and is independent or semi-independent.

Middle Eastern ecological trilogy The model of mostly symbiotic relations among villagers, pastoral nomads, and urbanites in the Middle East.

middle latitude See *latitude*.

Middle Way The Dalai Lama's proposal for Tibetan autonomy. Under this plan, Tibet would have authority over most internal matters except defense and foreign relations, which would still be controlled by China.

migration A temporary, periodic, or permanent move to a new location.

millenarian movements In many religions, the belief of a major transformation of humanity, for example, with a "second coming" or apocalyptic event at the end time. See also *cargo cults*.

Millennium Development Goals (MDGs) Eight goals established by the United Nations in 2000, to be implemented by 2015, that would raise the quality of life for poor people around the world. In 2015, the MDGs were replaced with the Sustainable Development Goals.

Min A language spoken in southern China and Taiwan.

minifundia (sing. ***minifundio***) Small Latin American agricultural landholdings, usually with a strong subsistence component.

minute A unit of measurement equal to 1/60th of a degree.

"misdeveloped country" A term sometimes applied to post-Soviet Russia because of the rampant criminality and corruption in its society and its demographic trends of high death rates and low birth rates.

"mistake of 1914" Nigerian term for the creation of the country's modern borders by the British.

mitigation Measures taken to avoid the adverse impacts of climate change in the long term.

Mohajirs Muslim immigrants to Pakistan from India after the two countries were partitioned in 1947.

mollisol soils Thick, productive, and durable soils, such as the *chernozem*, whose fertility comes from abundant humus in the top layer.

moment magnitude scale A scale for measuring the intensity of earthquakes. The MMS replaced the Richter scale.

money laundering Methods of turning money earned in illegal ways into money that appears legal.

Mongolic A subfamily of the Altaic language family.

Mongols Conquerors of most of Russia and surrounding areas in the 13th century, originating from what is now Mongolia.

monoculture The single-species cultivation of food or tree crops, usually very economical and productive, but threatening to natural diversity and change.

Monroe Doctrine A policy established by US President James Monroe stating the United States would prevent European countries from establishing any new colonies in the Western Hemisphere.

monsoon A current of air blowing fairly steadily from a given direction for several weeks or months at a time. Characteristics of a monsoonal climate are a seasonal reversal of wind direction, a strong summer maximum of rainfall, and a long dry season lasting for most or all of the winter months.

montane grasslands and shrubland A biome of mountainous and highland areas (as in Tibet), with grassy and scrubby vegetation varying by elevation.

Montreal Protocol A 1989 international treaty to ban chlorofluorocarbons.

Moor A Muslim inhabitant of Spain.

more developed countries (MDCs) The world's wealthier countries.

Motor City A nickname for the city of Detroit, Michigan, from the concentration of automobile manufacturers in the area.

Mound Builders A collection of Native American culture groups, named for their creation of large earthen mounds, who created extensive trading networks and urban settlement patterns.

Movement for Oneness and Jihad in West Africa (MOJWA or MUJWA) An offshoot of Al Qa'ida in the Islamic Maghreb operating in countries of the Sahel.

Movement for the Emancipation of the Niger Delta (MEND) A mainly ethnically Ijaw group in Nigeria that has expressed dissatisfaction with the redistribution of Nigeria's oil wealth by kidnapping foreign oil workers.

Movement for the Survival of the Ogoni People (MOSOP) An ethnic Ogoni organization in Nigeria that petitions the government for a safe environment and more federal support for their people.

mujahidin Anti-Soviet rebel bands in Afghanistan during the Soviet invasion of the 1980s.

mulatto In Latin America, a person of mixed European and black ancestry.

mullah Religious leader in Iran's Islamic clergy.

Multiculturalism Act A 1988 Canadian law that officially established Canada as a multicultural society, recognizing and encouraging traditional identities.

multinational companies Companies that operate, at least in part, outside their home countries.

Muslims Adherents of Islam.

mutually assured destruction (MAD) The concept of nuclear deterrence: if one side in a conflict were to launch nuclear weapons, the other would. Both countries would be obliterated.

nagana See *trypanosomiasis.*

narcoterrorist organization Armed insurgent groups that are funded by the sale of illegal drugs.

National Coalition for Syrian Revolutionary and Opposition Forces A powerful Syrian "opposition group," backed by many Western and Gulf Arab states.

National Geography Standards A set of 18 standards drafted in 2012 by the National Council for Geographic Education to guide teaching of and literacy in geography.

nationalism The drive to expand the identity and strength of a political unit that serves as home to a population interested in greater cohesion and often expanded political power.

Native Americans The indigenous peoples of North and South America who originally migrated to the New World from Asia over a land bridge or along the Pacific coast during the Ice Age. Christopher Columbus misidentified them as "Indians" in 1492.

native reserves See *homelands.*

Native Title Bill Legislation passed by the Australian government in 1993, addressing many issues relating to Aborigines' rights to claim and buy land.

natural capital The natural resources and natural services that keep us and other species alive and support human economies.

natural hazard Elements of the natural physical environment that are hazardous to people.

natural landscape An environment without (or before) transformations caused by humans.

natural replacement rate The highest rate at which a renewable resource can be used without decreasing its potential for renewal. Also known as *sustainable yield.*

natural resource A product of the natural environment that can be used to benefit people. Resources are human appraisals.

natural services The processes provided by healthy ecosystems, including the natural processes of air and water purification and the renewal of topsoil.

Naxalites Maoist insurgents in India who claim to champion the causes of the rural poor.

Nazca An extinct ethnic group from Peru that left massive carvings in the desert two thousand years ago.

Near Abroad Russia's term for the other, now-independent former republics of the Union of Soviet Socialist Republics.

Near North Australian term for neighboring Asian countries.

neocolonialism The perpetuation of a colonial economic pattern in which developing countries export raw materials to, and buy finished goods from, developed countries. This relationship is more profitable for the developed countries.

neo-Europes Areas colonized by Europeans in lands similar climatically and geographically to Europe, creating "new Europes" with cultural similarities and ties to Europe.

neo-liberalism An economic policy of competition, foreign investment, and low tariffs to promote economic growth, often in the context of free trade agreements.

Neolithic Revolution See *Agricultural Revolution.*

Neo-Malthusians Supporters of forecasts that resources will not be able to keep pace with the needs of growing human populations.

New Asian Tigers A collection of Southeast and East Asian countries that have had rapid economic growth.

New Cold War A potentially emerging conflict between Russia and the countries of NATO.

new lands See *virgin and idle lands.*

newly industrializing countries (NICs) The more prosperous of the world's less developed countries.

New Partnership for Africa's Development (NEPAD) An agency of the African Union responsible for monitoring improvements in human rights across Africa.

"New Rome" A term referring to Constantinople as a center of Christendom in the late Roman period.

Niger-Congo language family A language family of central Africa.

Night Journey In Islam, the Prophet Muhammad's mystical journey from Mecca to Jerusalem, where he ascended briefly into heaven from the great rock beneath the Dome of the Rock.

Nile Water Agreement A treaty guaranteeing Egypt 75 percent of the Nile River's flow; it forbids any projects downstream that might threaten the volume of water reaching Egypt.

Nilo-Saharan language family A language family spoken in parts of Africa around the Sahara Desert.

NIMBY An acronym for "not in my back yard." Refers to the opposition of local residents to new developments in their area.

nirvana In Buddhism, a transcendent state in which one is able to escape the cycle of birth and rebirth.

nitrous oxide Greenhouse gas resulting from human activities, especially from the breakdown of nitrogen fertilizers.

Noble Sanctuary (*al-Haraam ash-Shariif*) What Muslims call the Temple Mount, where the Dome of the Rock and al-Aqsa Mosque are located.

nodal region See *region.*

nontimber forest products (NTFPs) Goods, other than trees, made from products found in forests, to prevent deforestation.

North American Free Trade Agreement (NAFTA) A free-trade agreement among Canada, the United States, and Mexico.

North Atlantic Drift A warm current, originating in tropical parts of the Atlantic Ocean, that drifts north and east, moderating temperatures of Western Europe. Also known as the *Gulf Stream.*

North Atlantic Treaty Organization (NATO) A military alliance formed in 1949 that included the United States, Canada, many European nations, and Turkey.

Northern Alliance The main domestic opponent of the Taliban in Afghanistan.

Northern Ireland Assembly The parliament of Northern Ireland, where Protestants and Catholics share power.

Northern Hemisphere The half of Earth's surface between the Equator and the North Pole.

Northern League An Italian organization calling for the creation of Padania as an autonomous area or independent country.

Northern Sea Route A waterway developed by the Soviets to facilitate Arctic Ocean commerce across the northern edge of Eurasia. Kept open with the aid of icebreakers, it may become ice-free with processes of climate change.

North European Plain The level to rolling lowlands extending from France westward through Germany to Poland. This area is rich in agricultural development, and is home to a number of major industrial centers.

north latitude See *latitude.*

North Pole A point at 90 degrees north latitude. The highest latitude in the northern hemisphere.

Northwest Passage A shipping channel through Canada's Arctic islands that may be used if global warming decreases sea ice enough to allow a reliable route.

Novorossiya A Russian term for eastern Ukraine, meaning "New Russia."

Nuclear Nonproliferation Treaty (NPT) See *Comprehensive Test Ban Treaty.*

Nunavut A Canadian territory created in 1999 as the homeland of the Inuit.

Oceania A term collectively referring to Australia, New Zealand, and the islands of the Pacific Ocean.

oceanic (maritime or marine west coast) climate Mainly moderate year-round climate characteristic of mid-latitude coastal regions (as in Europe) and of mountainous areas of tropical and subtropical regions.

Occupied Territories The territories captured by Israel in the 1967 War: the West Bank and Golan Heights captured, respectively, from Jordan and

Syria. Formerly occupied, the Gaza Strip has been relinquished to Palestinian control, as was Sinai to Egypt with the Camp David Agreement.

official state sponsors of terrorism A US list of countries that sponsor terrorism.

offshore outsourcing (offshoring) See *outsourcing*.

Ogoni An ethnic group in Nigeria.

oil embargo The 1973 embargo on oil exports imposed by Arab members of the Organization of Petroleum Exporting Countries (OPEC) against the United States and the Netherlands.

old-age support ratio The number of working-age adults for each older person.

oligarchs Literally "rule of the few," where a small number of wealthy businessmen control much of a state's production, especially in Russia.

one-child campaign See *one-child policy*.

one-child policy Official policy of the Chinese government, limiting families to one child.

One-China Policy The policy by which the US recognizes the People's Republic of China as the "only" China while allowing the US to maintain good relations with the Nationalist Chinese government in Taiwan.

Operation Barbarossa A German onslaught against the Soviet Union during World War II.

Operation Crossroads A series of 67 nuclear experiments by the United States in the Marshall Islands of the Pacific Ocean after World War II.

Operation Enduring Freedom The US military strike against Afghanistan that began in October 2001.

Operation Iraqi Freedom The US-led ground assault on Iraq that began in March 2003.

Opium Wars Two 19th-century conflicts in which Britain wrenched important territorial and trade concessions from China.

Orange Revolution The 2005 Ukrainian presidential election that brought pro-Western Victor Yushchenko to power.

Organization of American States (OAS) The 35 member political organization of every country in the Americas.

Organization of Petroleum Exporting Countries (OPEC) An international organization created to get higher profits from oil sales to its member countries.

orientation The relationship between direction on a map and the corresponding compass direction in reality.

original affluent society A description of hunter-gatherers, who are thought by some scholars to have enjoyed harmony with the natural world and suffered few social and psychological problems.

orographic precipitation Precipitation occurring when moist air is lifted upward over mountains or other high terrain.

Oslo I Accord An agreement signed by Israeli Prime Minister Yitzhak Rabin and Palestinian leader Yasser Arafat, designed to be a pathway to peace between Israel and the Palestinians.

Oslo II Accord Another agreement, not yet fully implemented, between Israel and the Palestinian Authority that would transfer more of the West Bank to Palestinian control in exchange for increased Palestinian efforts to crack down on terrorists and guarantee Israel's security.

outsourcing The flight of jobs and technology from countries with high manufacturing and service costs to countries with lower costs.

"P5+1" The five permanent members of the UN Security Council plus Germany. These countries developed a package of incentives to Iran in exchange for Iran halting its nuclear weapons development programs.

Pacific Solution Australian program for sending would-be migrants to Nauru and Papua New Guinea while Australian authorities review their cases.

Padania A term for northern Italy.

paddy (padi) The field in which wet rice is grown.

Pakistani Taliban (Tehrik-i-Taliban Pakistan, or TTP) Pakistan's main Taliban group, affiliated with the majority of Taliban groups across the border in Afghanistan.

Paleo-Indians The first people to inhabit the Americas, and the ancestors of all Native Americans today (except the Inuit and Aleuts).

Palestine Liberation Organization (PLO) The Palestinian military and civilian organization created in the 1960s to resist Israel and recognized in the 1990s as the sole legitimate organization representing official Palestinian interests internationally.

Palestinian Authority (PA) The Palestinian government created as a result of the Oslo I Accord.

Palestinians Arabs who formed the majority of inhabitants of Palestine prior to the creation of the modern country of Israel.

pampas Subtropical grasslands in South America.

Pan-African Parliament The lawmaking body of the African Union.

Pancasila A pan-Indonesian nationalist ideology designed to neutralize all ethnic identities.

pandemic A very widespread epidemic.

pan-Turkism The goal of uniting all Turkic peoples of Asia.

Papuan languages A collection of the 750 languages spoken on New Guinea and surrounding islands.

Papuans People of mostly Melanesian origin who live in New Guinea.

parallel A line of latitude, so named because all such lines run parallel to the Equator.

páramos Alpine meadows in Latin America.

Parsis An ethnic group in India that practices Zoroastrianism.

Parti Québécois (PQ) A separatist political party in Québec.

Partnership for Peace (PfP) A program of cooperation between 22 individual Euro-Atlantic partner countries (including Russia and Ukraine) and NATO.

pastoral nomads Desert peoples of the Middle East and northern Africa who migrate through arid lands with their livestock, following patterns of vegetation and rainfall.

Pentecostal churches About 170 Christian denominations, tracing back to the 20th century Holiness movement in Methodism, that focus on the power of the Holy Spirit and the gift of "speaking in tongues."

People of the Book Term used by Muslims to refer to Jews and Christians.

people overpopulation The concept that many persons, each using a small quantity of natural resources to sustain life, add up to too many people for the environment to support.

PEPFAR An acronym for President's Emergency Plan for AIDS Relief, in which the George W. Bush administration pledged over $45 billion to combat AIDS in Africa.

per capita GDP A measure of economic well-being found by dividing a country's gross domestic product by its population.

perceptual region See *region*.

perestroika Policies introduced by Russian leader Mikhail Gorbachev in the 1980s to allow a more market-oriented economy. See also *glasnost*.

peripheral location An area of the world with less importance because of a distant location relative to others.

permafrost Permanently frozen subsoil.

Permanent Fund A government surplus in Alaska created from oil revenues, the dividends of which are paid out to Alaska's residents each year.

Persian The language spoken by the majority of Iranians. The Persian word for this language is *Farsi*.

physical geography The subdiscipline of geography most concerned with the climate, landforms, soils, and physiography of the Earth's surface.

pidgin The official language of Papua New Guinea, a combination of English and foreign words mixed with indigenous vocabulary and grammar.

PIGS An acronym for Portugal, Ireland, Greece, and Spain, used in reference to the weaker economies of those countries as opposed to those of France and Germany.

pilgrimage (*hajj*) One of the five "Pillars" of Islam, requiring a Muslim to journey to Mecca once in his or her lifetime during the prescribed season of pilgrimage.

Pillars of Islam The five fundamental tenets of the faith of Islam, which are *profession of faith, prayer, almsgiving, fasting*, and the *hajj* (pilgrimage to Mecca).

pivotal countries Those countries whose collapse would cause international refugee migration, war, pollution, disease epidemics, or other international security problems.

place In geographic analysis of a given locale, the nature of place identity becomes a means of understanding people's response to that particular place. Determination of the environmental and cultural characteristics that are most frequently associated with a certain place helps establish that "sense of place" for a given location.

Plan Colombia A US-funded effort to reduce the amount of coca production in Colombia, intersecting with Plan Patriot to counter rebel movements associated with cocaine.

Plan Patriot An operation in Colombia involving US military personnel and civilian contractors targeting FARC to reduce drugs and secure oil reserves.

plates Giant slabs of rock in the lithosphere (the Earth's crust and upper mantle), the positions of which gradually change over millions of years through tectonic movements.

plate tectonics The dominant force in the creation of the continents, mountain systems, and ocean deeps. The slow but steady movement of these massive plates of the Earth's lithosphere has created the positions of the continents and major patterns of volcanic and seismic activity. Areas where plates are being pulled under other plates are called subduction zones.

platform economy Common in high-tech industries, with a corporation headquartered in the United States but most of the work on its products done overseas.

Pleistocene overkill hypothesis A hypothesis stating that hunters and gatherers of the Pleistocene Era hunted many species to extinction.

plinthite See *laterite*.

podzol Soil with a grayish, bleached appearance when plowed, lacking in well-decomposed organic matter, poorly structured, and very low in natural fertility. Podzols are the dominant soils of the taiga.

polar amplification The increasing warming of polar waters from melting ice. When sea ice melts, the darker ocean waters that emerge absorb more solar radiation, warming and contributing to more melting of sea ice.

polar climate Very high latitude climate in which no month has an average temperature higher than 10 degrees Centigrade (50 degrees Fahrenheit), as in Antarctica and Greenland.

Polynesia A large region in the eastern Pacific Ocean, primarily south of the Equator, filled with many small volcanic islands. The largest islands in Polynesia are those of New Zealand and Hawaii.

population change rate The birth rate minus the death rate in a population, typically measured in numbers per thousand.

population explosion The surge in Earth's human population that has occurred since the beginning of the Industrial Revolution.

population geography The study of the spatial distribution of human population.

population implosion A dramatic loss of population caused by low fertility rates or death rates that are much higher than birth rates.

population pyramid See *age structure diagram*.

population replacement level The number of new births required to keep a population steady. Generally calculated as 2.1 children per woman in MDCs.

post-industrial Description of an economy or society characterized by the transformation from manufacturing to information management, financial services, and the service sector.

potato famine The starvation of 10 percent of Ireland's population in the late 1840s, caused by a crop disease called blight.

potential evaporation (PE) A measure of how much water would be evaporated in a given area if the amount of water available were unlimited.

poverty trap Factors that impede the development of a poorer country. Bad governance, conflict, natural resources, and a landlocked situation are examples of poverty traps.

power projection hub In geopolitical terms, a strategic outpost from which a country could launch naval, air, and other forces against an enemy.

prayer One of the five "Pillars" of Islam, requiring the Muslim to pray at five prescribed times throughout the day.

pre-1967 borders The borders of Israel after 1949 and before the Six-Day War of 1967 in which Israel captured and occupied the Gaza Strip, West Bank, and Golan Heights.

precautionary principle The notion that it is appropriate to attempt to reduce or eliminate any risky practice.

pre-Columbian times North and South America before Christopher Columbus's first voyage in 1492.

precipitation Water falling from the sky to the surface in the form of rain, snow, sleet, or hail.

preemptive engagement A US policy developed after 9/11 that allowed the country to take unilateral military action if necessary to defuse threats to its security whenever and wherever those threats occur.

primary commodities Raw materials such as fruits and ores whose extraction or harvest needs little processing before use: examples are oil, rubber, bananas, and sugar.

primate city A city that dominates a country's urban scene and is usually defined as being larger than the country's second and third largest cities combined, or at least twice as large as its next largest city. Primate cities are generally found in developing countries, although some developed countries, such as France, have them.

Prime Meridian The line of zero degrees longitude, which passes through Greenwich, England. It separates the Western Hemisphere from the Eastern.

privatization The shift of ownership of economic activities from the state to nongovernmental.

profession of faith The utterance "There is no god but God, and Muhammad is His Messenger," which is one of the five "Pillars" of Islam and the fundamental tenet of the faith.

proletariat In Marxism, the industrial working class.

Promised Land Canaan (geographical Palestine), for the Israelites in Biblical times.

Protestant Reformation The schism in Western Christianity that began in 1517 when Martin Luther registered his objections to much of the dogma, rituals, and political organization of the Roman Catholic Church.

Protestant Unionists One side of "The Troubles" in Northern Ireland, those who wished Northern Ireland to remain unified with the United Kingdom and not unify with the rest of Ireland.

Protestant work ethic In the United States, the emphasis on work, productivity, and economic success in an individual's life. Despite its name, this ethic is shared by people of many faiths and ethnicities.

Protestantism One of the three major divisions of Christianity (along with Roman Catholicism and Orthodox Christianity), which began with the Protestant Reformation of the 16th century.

Proto-Austronesian The ancestor of all modern Austronesian languages.

proxy war During the Cold War, the situation created when the United States and the Union of Soviet Socialist Republics played client countries against one another, arming them with weapons, to boost their competing aims.

pueblo An interconnected series of residences and ceremonial centers composed mainly of rock, used by the Ancestral Puebloans and other indigenous cultures of the southwestern United States.

pull factors Emigration caused by so-called pull factors, as when an educated villager responds to a job opportunity in the city. People responding to pull factors are called selective migrants. Both push and pull forces are behind the rural-to-urban migration that is characteristic of most countries.

purchasing power parity (PPP) A method of comparing the real value of output between different countries' economies, considering factors such as differences in relative prices of goods and services.

push factors Emigration caused by so-called push factors, as when hunger or lack of land "pushes" peasants out of rural areas into cities. People responding to push factors are often referred to as nonselective migrants. Both push and pull forces are behind the rural-to-urban migration that is characteristic of most countries.

Quechua The most widespread indigenous language in Latin America, spoken by about 10 million people. Quechua was the language of the Inca.

Qur'an (Koran) The holy book of Islam, believed by Muslims to consist of the words of God recited to Prophet Muhammad through the Angel Gabriel.

Rastafarianism A religion, originating in Jamaica in 1930, that holds that Ethiopian emperor Haile Selaisse was the earthly incarnation of God.

Reconstruction A term used in the United States for the years 1865–1877, when the Confederate states that had been defeated in the Civil War were under control of the US Army.

Red Corridor The roughly contiguous region of eastern India where the Naxalite rebellion is prominent.

redlining A practice in the United States, now illegal, of denying minorities loans to purchase homes in majority-white areas of cities.

Reducing Emissions from Deforestation and Forest Degradation (REDD) A United Nations effort to create a financial value for the carbon stored in forests and to offer incentives for developing countries to reduce carbon emissions from forested areas and invest in low-carbon development.

reference map A type of map primarily depicting the locations of various features on the Earth's surface.

refugees Victims of severe push factors such as persecution, political repression, and war.

region A "human construct" that is often of considerable size, that has substantial internal unity or homogeneity, and that differs in significant respects from adjoining areas. Regions can be classed as formal (homogeneous), functional, or vernacular. The formal region, also known as a uniform region, has a unitary quality that derives from a homogeneous characteristic. The United States is an example of a formal region. The functional region, also called the nodal region, is a coherent structure of areal units organized into a functioning system by lines of movement or influence that converge on a central node or trunk. A major example is the trading territory served by a large city and bound together by the flow of people, goods, and information over an organized network of transportation and communication lines. Vernacular, or perceptual, regions are areas that possess regional identity, such as the Sun Belt, but share less objective criteria in the use of this regional name. General regions, such as the major world regions in this text, are recognized on the basis of overall distinctiveness.

Regional Initiative for the Infrastructure Integration of South America An international effort to improve the economies of South American countries by linking them together by road networks and other infrastructure projects.

relative location The location of a place defined by relationship to other places.

remittances Money earned by workers in a foreign country sent back to that worker's home country.

remote sensing Through the use of aerial and satellite imagery, geographers and other scientists have been able to get vast amounts of data describing places all over the face of the earth. Remote sensing is the science of acquiring and analyzing data without being in contact with the subject. It is used in the study of patterns of land use, seasonal change, agricultural activity, and even human movement along transport lines. This process relates closely to GIS. See also *geographic information systems (GIS).*

renationalization A process of reversing the trend toward private ownership of economic production in Russia during the presidency of Vladimir Putin.

renewable resource A resource, such as timber, that is grown or renewed so that a continual supply is available. A finite resource is one that, once consumed, cannot be easily used again. Petroleum products are a good example of such a resource, and because it takes too much time to go through the process of creation, they are not seen as renewable.

reservations Areas of land Native American tribes were forced onto by the US and Canadian governments.

resilience The ability of an ecosystem to recover from the effects of stress that is not too drastic, such as the stress of drought. A healthy desert or savanna ecosystem, for example, can recover from drought through adaptations to a natural cycle that includes extended periods of dryness and rain.

resource curse A paradox of plenty, when a country with a great abundance of a valuable natural resource experiences lower economic growth than countries without such abundance.

rifting The separation of tectonic plates on landmasses, producing very low land elevations.

right of return The Palestinian Arab principle that refugees (and their descendants) displaced from Israel and the Occupied Territories in the 1948 and 1967 wars be allowed to return to the region.

Ring of Fire A term referring to the numerous regions of volcanic and seismic activity on the edges of tectonic plates surrounding the Pacific Ocean.

risk minimization The exploitation of multiple resources so some resources will support a population if others fail.

"River Sea" A term referring to the enormity of the Amazon River.

"road map for peace" An effort led by the United States, the European Union, Russia, and the United Nations to reach a peace settlement between Israel and the Palestinians, intended to lead to an independent Palestinian state.

robotics The use of computer-controlled robots to perform what were human manual tasks, as on assembly lines.

Roma (Gypsies) Originating in India, they are one of Europe's largest ethnic minorities. Commonly referred to as Gypsies, they often move from place to place in caravans and have been discriminated against for centuries.

Roman Catholic Church The largest denomination of Christianity, led by the Pope in Vatican City.

Romance languages European languages that descended from Latin, such as Italian, Spanish, and French.

Roosevelt Corollary A 1904 policy in which the United States declared it had the right to supervise the internal affairs of Latin American countries to ensure US national security.

rules of origin A NAFTA requirement that half or more of all the components of any manufactured good must originate in Canada, the United States, or Mexico.

rural-to-urban migration See *urbanization.*

"Russian cross" The dramatic rise in death rates and fall of birth rates in Russia after the breakup of the Soviet Union.

Russian Revolution The overthrow of Czar Nicholas II of Russia in 1917 and the subsequent rise of the Bolsheviks.

Russification The effort, particularly under the Soviets, to implant Russian culture in non-Russian regions of the former Soviet Union and its Eastern European neighbors.

Rust Belt An area across the Midwest United States that was home to many heavy industries that, beginning in the 1960s, shifted overseas and to the southern US.

Saami (Lapps) The indigenous people of far northern Scandinavia.

sacred space (sacred place) Any locale that people hold in reverence, such as places of worship, cemeteries, and battlefields.

Salafi Muslims that adhere to an interpretation of Islam that they believe is closest to the faith's earliest tenets and social norms.

salinization The deposition of salts on, and subsequent fertility loss in, soils experiencing a combination of overwatering and high evaporation.

San The proper name for Bushmen, a major group of the Khoisan ethnolinguistic cultures living mainly in Botswana and Namibia. They were among the last people on Earth to practice hunting and gathering.

sanctuary cities Cities in the United States that do not cooperate with federal government efforts to identify illegal immigrants.

sand sea A virtual "ocean" of sand characteristic of parts of the Middle East, where people, plants, and animals are all but nonexistent.

Sanskrit An ancient language of India that is no longer used for everyday communication but is still widely used in religious rites.

scale The size ratio represented by a map; for example, a map with a scale of 1:12,500 is portrayed as 1/12,500 of the actual size.

Schengen Agreement A framework among most signatory countries of the EU allowing for free movement of people without need for visas, passports, or other border controls.

Schengenland A term for the countries participating in the Schengen Agreement.

scorched earth The wartime practice of destroying one's own assets to prevent them from falling into enemy hands.

seafloor spreading The creation of new crust from molten material rising from the Earth's mantle through ridges on the ocean floor.

seamount An underwater volcanic mountain.

second A unit of measurement equal to 1/60 of a minute, or 1/3600 of a degree.

"Second Kuwait" A term describing the Russian island of Sakhalin and the potential large reserves of oil and natural gas it has offshore.

second law of thermodynamics A natural law stating that high-quality, concentrated energy is increasingly degraded as it passes through the food chain.

Second Russian Revolution Events in Russia that precipitated the collapse of the Soviet Union in 1991.

Second Temple The second Jewish temple to stand on the Temple Mount, which was destroyed by the Romans.

second (transitional) stage of demographic transition The stage characterized by continued high birth rates but falling death rates, resulting in surging population growth.

sectarian violence Conflict between people of different religions, as in the Indian Subcontinent.

Secure Fence Act of 2006 Legislation passed by the US Congress to extend the fence on the Mexican border by 700 miles.

sedentarization Voluntary or coerced settling down of pastoral nomads in the Middle East.

segregation A policy of separation of races.

seismic activity Tectonic forces resulting in earthquakes.

semiarid A mainly dry climate, with sparse precipitation.

Semitic language A language family that includes Hebrew and Arabic.

"separate but equal" A legal doctrine in the United States that allowed for racial segregation, intended to provide separate facilities for whites and blacks that were equal in quality. The US Supreme Court struck down this doctrine in 1954.

separation barrier A wall that separates the West Bank from Israel, constructed to prevent Palestinian suicide bombers and militants from entering Israel.

Sephardic Jews Jews who historically inhabited Palestine and other parts of the Middle East and North Africa.

settler colonization The historical pattern by which Europeans sought to create new Europes, or "neo-Europes," abroad.

Shanghai Cooperation Organization (SCO) An organization composed of China, Russia, and several Central Asian countries designed to combat terrorism, separatism, and extremism in the region.

shantytown Areas of often squalid slums or "informal settlements" on the outskirts of large cities throughout the developing world.

shatterbelt A large, strategically located region composed of conflicting states caught between the conflicting interests of great powers.

shell banks A bank that exists only on paper and specializes in money laundering.

Shia, or Shi'ite The branch of Islam regarding male descendants of Ali, the cousin and son-in-law of Prophet Muhammad, as the only rightful successors to Prophet Muhammad.

shifting cultivation A cycle of land use between crop and fallow years that is needed to work around the infertility of tropical soils. After clearing tropical forest, a farmer may get only a few years of crops before there is no fertility in the soil and must move on to clear more land. In the meantime, forest reclaims the previously farmed plots. Where new lands are not available, fallow periods on old fields are reduced or eliminated, resulting in soil deterioration. Also known as *swidden* or *slash-and-burn cultivation*.

shifting ranching A destructive process in which ranchers raise cattle on areas of former rainforest originally cleared by farmers. When the land is exhausted, ranchers move their cattle on to the next plot of cleared land.

Shintoism The strongly nationalistic religion of Japan, a blend of nature reverence, Japanese folklore, and Chinese ritual.

Sikhism A monotheistic faith founded by the 16th century Guru Nanak and practiced mainly in India's Punjab, that advocates good deeds and the equality of all peoples.

Silicon Valley A region of the San Francisco Bay area known for its concentration of high-tech companies and workers.

Silk Road A mostly overland trade route that stretched from Venice to Xi'an in China.

Sinhalese The majority ethnic group of Sri Lanka, predominantly Buddhist.

Sinn Fein The political counterpart to the Irish Republican Army.

Sino-Tibetan language family A major language family of Asia, including Chinese and its variants.

Sioux A Native American tribal group with populations from the Midwest and Great Plains to the Rockies in the US and in southwestern Canada.

site The specific geographic location of a given place.

situation The accessibility of that site and the nature of the economic and population characteristics of that locale.

Six-Day War A 1967 war between Egypt, Syria, and Jordan against Israel, resulting in a dramatic Israeli victory and the Israeli occupation of the Sinai Peninsula, Gaza Strip, West Bank, and Golan Heights.

six essential elements of geography (1) The world in spatial terms; (2) places and regions; (3) physical systems; (4) human systems; (5) environment and society; (6) uses of geography.

slash-and-burn cultivation See *shifting cultivation*.

Slavic Core (Fertile Triangle or Agricultural Triangle) A large area of the western former Soviet region containing most of the region's cities, industries, and cultivated lands.

Slavic languages A language family, primarily in Eastern Europe, that includes Russian, Polish, and Czech.

sleeping sickness See *trypanosomiasis*.

small-scale map A map constructed to give a highly generalized view of a large area.

socialization State ownership of economic production.

soft power A persuasive approach to international relations that uses diplomatic, economic, or cultural tools (as opposed to warfare).

Solidarity An independent trade union in Poland, the creation of which was one of the conditions resulting in the withdrawal of Soviet control in Poland in 1989.

Songlines See *Footprints of the Ancestors*.

south latitude See *latitude*.

South–North Water Transfer Project An ambitious engineering project in China that would move water through a series of canals from the rainy south to the arid, more populated north of the country.

South Pole A point at 90 degrees south latitude, the highest latitude in the southern hemisphere.

Southeast Anatolia Project (GAP) An agricultural effort in Turkey aiming to double the amount of the country's irrigable farmland and generate hydropower.

Southern African Customs Union (SACU) A trade organization including South Africa, Lesotho, Swaziland, Botswana, and Namibia.

Southern Cone Common Market (Mercosur) See *Mercosur*.

Southern Hemisphere The half of Earth's surface between the Equator and the South Pole.

Soviet satellites Soviet-allied communist countries in Eastern Europe between World War II and 1989.

space In geography, the exact placement of locations on the face of the Earth.

spatial Geographers recognize spatial distributions and patterns in Earth's physical and human characteristics. The term *spatial* comes from the noun *space*, and it relates to the distribution of various phenomena on Earth's surface. Geographers portray spatial data cartographically—that is, with maps.

special economic zones (SEZs) In China, the urban areas designated after the 1970s to attract investment and boost production through tax breaks and other incentives.

spodosols Acidic soils that have a grayish, bleached appearance when plowed, lack well-decomposed organic matter, and are low in natural fertility. Also known as *podzols*.

state farm (*sovkhoz*) A type of collectivized state-owned agricultural unit in the former Soviet Union; workers receive cash wages in the same manner as industrial workers.

State Law and Order Restoration Council (SLORC) The military government that seized control of Burma in 1988 and changed the country's name to Myanmar.

storm surge A powerful wave in the ocean that is associated with landfalling hurricanes.

strategic hamlet Fortified communities in Vietnam that the US military forced millions of South Vietnamese farming families into, in a failed attempt to prevent Viet Cong forces from infiltrating South Vietnamese villages.

"String of Pearls" China's strategic placement of military and commercial bases across the Indian Ocean.

subarctic A high-latitude climate characterized by short, mild summers and long, severe winters.

subduction The process of one tectonic plate descending below another.

subduction zone See *plate tectonics*.

subsidies Payments made by governments to protect domestic producers, often impacting foreign producers.

Suez Crisis The British, French, and Israeli invasion of Egypt in 1956 after Egyptian president Nasser nationalized the Suez Canal. International pressure caused the withdrawal of the invading forces, and the canal stayed under Egyptian control.

Sufi Islam A branch of Islam, neither Sunni nor Shi'ite, that advocates a more mystical approach to the faith.

summer monsoon See *monsoon*.

Sun Belt A general term for the southern half of the contiguous United States, characterized by warm weather.

Sunni The branch of Islam regarding successorship to Prophet Muhammad as a matter of consensus among religious elders.

superpower A country with a far larger economy, stronger military, and more pervasive popular culture than most other countries. Most geopolitical analysts concur that no country has superpower status today, but that the US is the world's strongest power.

supranational organization An international organization in which member countries are united beyond the authority of any single national government and are planned and controlled by a group of nations.

suq A commercial zone adjoining the ceremonial and administrative heart of a Middle Eastern city. See *bazaar*.

sustainable development (ecodevelopment) Concepts and efforts to improve the quality of human life while living within the carrying capacity of supporting ecosystems.

Sustainable Development Goals (SDGs) A replacement for the UN's Millennium Development Goals, these 17 goals were established in 2015 to guide international adoption of sustainable development policies.

sustainable yield (natural replacement rate) The highest rate at which a renewable resource can be used without decreasing its potential for renewal.

sweatshops Factories in the LDCs infamous for low wages and long working hours. Part of the dark side of globalization: Chinese and other Asian companies in particular can undercut competition by underpaying and overworking employees.

swidden cultivation See *shifting cultivation*.

Sykes–Picot Agreement A secret agreement between Britain and France that in 1916 planned for the division of the Ottoman Empire's Arab provinces between themselves after World War I.

symbolization Representation of spatial phenomena on maps with the use of symbols such as lines, fill, shape, colors, and type.

syncretism The incorporation of some animistic beliefs into mainstream religions.

"Syrian Opposition" An umbrella term for the groups opposed to the rule of Syrian leader Bashar al-Assad.

Tagalog A major Austronesian language and ethnic group of the Philippines. Filipino language is the standard form of Tagalog and is one of the country's official languages.

Tai-Kadai language family . A tonal language of southern China and Southeast Asia that includes Thai and Lao, the chief languages of Thailand and Laos.

Taliban One of the Islamist mujahidin factions in Afghanistan fighting the Soviet invasion in the 1980s. Mainly of tribal Pashtun allegiance, the Taliban gained control over most of the country in the 1990s and instituted a very strict code of Islamic law. The Taliban were driven from power during the US-led Operation Enduring Freedom after 9/11 but regrouped as a significant insurgent force that took a significant toll of American and other NATO lives.

Tamil A language in the Dravidian language family of South Asia, and the ethnic group that speaks it primarily.

Tamils The minority ethnic group of Sri Lanka, primarily Hindu.

Tamil Tigers Tamil insurgents in Sri Lanka that wanted to establish an independent Tamil homeland. They were defeated by the Sri Lanka government in 2009.

Taoism See *Daoism*.

tar sands Mineral deposits in which liquid petroleum is mixed with sand and clay. Extracting the oil from tar sands is expensive and consumes large amounts of energy and water.

tariffs Tax penalties imposed on imports.

Tatar A Turkic language and ethnic group of about 5 million, mainly in the Russian republics of Tatarstan and Bashkortostan.

Tatars Nomads of central Asia who conquered much of Russia in the 13th century along with the Mongols.

tech bubble An economic bubble in the United States in the late 1990s, centering on high-tech products and services.

technocentrists (cornucopians) Supporters of forecasts that resources will keep pace with or exceed the needs of growing human populations.

tectonic forces Processes that derive their energy from within the Earth's crust and serve to create landforms by elevating, disrupting, and roughening the Earth's surface.

temperate grasslands, savannas, and shrublands Biomes composed mainly of short grasses. They occur in areas of steppe climate, a transitional zone between very arid deserts and humid areas.

temperate broadleaf and mixed forest Dominant forest type in areas with humid subtropical and humid continental climates.

Temple Mount The elevated platform in Jerusalem sacred to both Jews and Muslims. Two Jewish temples stood on the Temple Mount, and currently the Muslim Dome of the Rock and al-Aqsa Mosque are situated here.

Teotihuacános A Native American ethnic group that lived in central Mexico two thousand years ago, who were later replaced by the Aztecs. Their major site of Teotihuacán includes large pyramids.

terra nullius A legal doctrine declaring a location "no one's land," allowing for claims to that land.

terraces Carefully constructed rice paddies arranged in a stair-step fashion.

terrorism Premeditated, politically motivated violence perpetrated against noncombatant targets by subnational groups or clandestine agents, usually intended to influence an audience.

terrorism hot spot An area regarded by the United States as real or potential training grounds for terrorist groups.

Thai (Siamese) The principal tongue of the Tai-Kadai language family, spoken by a majority of people in Thailand.

thematic map A type of map showing the spatial distribution of one or more attributes across a given area.

theory of Himalayan environmental degradation The concept that overpopulation on steep slopes in the middle and high elevations of the Himalaya leads to excessive deforestation and erosion through shifting cultivation. This has many knock-on effects downstream, including flooding and siltation of hydroelectric reservoirs.

theory of island biogeography A theoretical calculation of the relationship between habitat loss and natural species loss, in which a 90 percent loss of natural forest cover results in a loss of half of the resident species.

Theravada Buddhism One of the two branches of Buddhism, found largely in Sri Lanka and Southeast Asia. Therevada Buddhists claim to follow the true teachings and practices of the Buddha.

third (industrial) stage of demographic transition The stage in which birth rates begin to decline and correspond with falling death rates, resulting in a leveling off and subsequent decline of population.

"Third Revolution" The concept that sustainable development will bring about a marked shift in human ways of interacting with the natural environment, so dramatic that it will be compared with the origins of agriculture and industry.

"Third Rome" A term used to describe Moscow's importance in Christian affairs (after Rome and Constantinople) in the 11th century.

Thirteenth Amendment An amendment to the US Constitution in 1865, ending the practice of slavery.

three-field system of crop rotation Growing cereals on an agricultural plot of land, then shifting to nitrogen-fixing legumes, and then letting the land lay fallow before returning to producing cereal crops.

Tibetan The language spoken in Tibet.

tierra caliente A Latin American climatic zone reaching from sea level upward to approximately 3000 feet (914 m). Crops such as rice, sugarcane, and cacao grow in this hot, wet environment.

tierra fría A Latin American climatic zone found between 6000 feet (1800 m) and 12,000 feet (3600 m) above sea level. Frost occurs and the upper limit of agriculture and tree growth is reached.

tierra helada A Latin American climatic zone above 10,000 feet (3048 m) that has little vegetation and frequent snow cover.

tierra templada A Latin American climatic zone extending from approximately 3000 to 6000 feet (914–1829 m) above sea level. It is a prominent zone of European-induced settlement and commercial agriculture such as coffee growing.

Tiger Cubs See *New Asian Tigers*.

tipping point In climate change, a time beyond which some changes are irreversible, likely bringing rapid acceleration in temperature and its related effects.

Tobler's First Law of Geography The Swiss-American geographer Waldo Tobler's axiom that "Everything is related to everything else, but near things are more related than different things."

Tornado Alley An area of the Midwestern United States where tornado activity is very common.

Torres Strait Islanders A distinct indigenous people of Australia with origins in Papua New Guinea.

trade barriers Restrictions on international trade, especially those created by MDCs to protect their industries from competition from cheaper goods produced in LDCs.

trade wars The use of barriers to restrict trade between countries to protect their interests and industries.

Trail of Tears The forced migration of numerous Native American tribes from the southeastern United States to what is now Oklahoma in the 1830s.

Trans-Amazon Highway A highway that, when completed, will link the Atlantic coast of Brazil with the Pacific coast of Peru.

transnational companies See *multinational companies*.

Trans-Pacific Partnership (TPP) A free-trade initiative between the United States, Australia, New Zealand, and several countries in Asia and South America.

tree line The upper limit (or northernmost limit) of natural tree growth.

trench Very deep linear features at the bottom of the ocean (e.g., the Mariana Trench off Japan).

triangular trade The 16th- to 19th-century trading links between West Africa, Europe, and the Americas, involving guns, alcohol, and manufactured goods from Europe to West Africa exchanged for slaves. Slaves brought to the Americans were exchanged for the gold, silver, tobacco, sugar, and rum carried back to Europe.

tribe In the Middle East, a group of kinspeople claiming common descent from a single male ancestor generations ago.

Tropic of Cancer A line of latitude located at 23.44°N. The general dividing line in the Northern Hemisphere between the tropics to the south and the mid-latitudes to the north.

Tropic of Capricorn A line of latitude located at 23.44°S. The general dividing line in the Southern Hemisphere between the tropics to the north and the mid-latitudes to the south.

tropical rain forest A low-latitude broadleaf evergreen forest found where heat and moisture are continuously, or almost continuously, available.

tropical savanna A relatively moist low-latitude climate that has a pronounced dry season.

tropical and subtropical coniferous forest A biome of needle-leaved evergreen forest in tropical and subtropical latitudes (as in mountainous Mexico and Guatemala).

tropical and subtropical grasslands, savannas, and shrubland A mainly grassy and scrubby biome in tropical and subtropical latitudes (as in the African plains populated with large herbivores).

tropical and subtropical moist broadleaf forest The world's most biodiverse biome, mainly a thick cover of broadleaf evergreen trees and including rainforests (as in Amazonia and central Africa).

The Troubles A period of strife in Northern Ireland marked by violence between Catholic Republicans and Protestant Unionists.

Truth and Reconciliation Commission A South African commission that allows those who were party to racial violence during the apartheid era to confess their misdeeds and, in a sense, be absolved of them.

trypanosomiasis A disease spread by the tsetse fly of Africa. Known as sleeping sickness in humans, and nagana in domestic cattle.

tsunami Large breaking waves created by undersea earthquakes.

Tuareg An ethnolinguistic group in northern Africa.

tundra A region with a long, cold winter, when moisture is unobtainable because it is frozen, and a very short, cool summer. Vegetation includes mosses, lichens, shrubs, dwarf trees, and some grass.

Turkic A subfamily of the Altaic language family, spoken mainly in western and central Asia.

Turkmen The principal language spoken in Turkmenistan, also referring to the Turkmen ethnic group.

Turks The predominant ethnic group of Turkey.

Tutsi (Watusi) See *Hutu*.

Two Seas Canal A proposed pipeline that would carry water from the Red Sea to replenish the Dead Sea, which has been shrinking since the 1980s, and supply hydropower and irrigation.

two-state solution Originally, the 1947 United Nations proposal to establish separate Jewish and Arab states in Palestine. Recently, this has also been the premise of the Oslo accords and the "Road Map to Peace."

tyranny of distance The effect of geographic remoteness in shaping a country's identity.

Tyvan The Turkic language of the Tuva people living along Russia's border with Mongolia.

underemployment The underutilization of labor.

underground economy The "black market" typical of the former Soviet region and many LDCs.

undifferentiated highland A climate that varies with latitude, altitude, and exposure to the sun and moisture-bearing winds.

undocumented workers People who are in the United States illegally but have obtained temporary or long-term employment.

uniform region See *region*.

United Fruit Company A US-based company that owned vast tracts of land in several Central American countries for the production of produce and that had a powerful influence on local politics.

United Nations Framework Convention on Climate Change (UNFCC) A treaty signed by 155 countries in 1992 acknowledging climate change as a problem. The United States signed the treaty after all binding actions such as carbon dioxide reductions were removed from it.

United Nations Resolutions 242 and 338 Resolutions passed by the United Nations after the Middle Eastern wars of 1967 and 1973, respectively, calling on Israel to withdraw from the Occupied Territories and for Arab states to recognize the sovereignty of Israel.

untouchables See *Dalits*.

upper limit of agriculture The highest elevation at which agricultural activities can be performed, at about 12,000 feet (3600 m) in the equatorial regions and increasingly low at heightening latitudes.

upstream countries Countries where a major river originates that have a large amount of control over that river's water, at the expense of countries farther down the river's course.

Uralic languages A language group that includes Finnish, Hungarian, and many Siberian languages.

urbanites In the Middle Eastern ecological trilogy, the educated people who provide manufactured goods and innovations, secular and religious educational instruction, and cultural amenities.

urbanization The population growth of cities, mostly from the movement of people from rural regions to built-up areas in the LDCs and from increasing industrialization in the MDCs.

Urdu The official language of Pakistan.

Ushahidi A nonprofit software firm and platform founded in Kenya. The software uses "crowdsourcing" for activist mapping of political, environmental, and other problems. It is a fine example of local people in an LDC developing a geospatial tool kit to address problems in development.

usufruct Nondestructive mutual use of resources by multiple clans or tribes located within the territory of another.

Uzbek The language of Uzbekistan.

value-added manufacturing The process of refining and fabricating more valuable goods from raw or semi-processed materials.

value-added products Finished products, worth much more than the raw materials they are created from.

Varangians Scandinavians who in the Middle Ages established trade routes, settlements, and political units along waterways between the Baltic and Black Seas.

vernacular region See *region*.

Viet Cong An insurrectionist Communist force in South Vietnam, supported by the North Vietnamese government, that aided the North's conquest of the South and reunification of the country during the Vietnam War.

Vietnamese The principal language spoken in Vietnam.

villagers In the Middle Eastern ecological trilogy, the peasant farmers that form the cornerstone of the trilogy.

virgin and idle lands (new lands) Steppe areas of Kazakhstan and Siberia brought into grain production in the 1950s.

virtual water The total volume of water needed to produce and process a commodity or service, measured where the product was actually produced. It is "virtual" because it is an externality not factored into and not apparent in the production process.

"visible minorities" Term used in Canada for people not of European, First Nations, or Inuit ethnicity.

volcanism Movement of molten material from the Earth's mantle, usually released through volcanoes.

Voting Rights Act A law, passed by the US Congress in 1965, ending official disenfranchisement of African Americans.

wa A Japanese concept of harmony based upon the principle of economic equality.

Walkabout A ritual journey for Australian Aborigines in which a boy on the verge of adulthood follows the pathway of his creator-ancestors.

Wallace Line A marked biogeographic division located in Indonesia, in mammals separating placental from marsupial.

War of 1812 A war between the United States and the United Kingdom, fought largely as a result of the United States attempting to annex Canada.

war on drugs A long-running US policy of reducing drug usage within the United States, mostly by attempting to eradicate drugs at their sources, especially in South America.

War on Terror An open-ended, unconventional war declared by President George W. Bush in 2001 against the Islamist groups that carried out terrorist attacks against the United States.

Warsaw Pact A military alliance, now dissolved, consisting of the Soviet Union and the European countries of Poland, East Germany, Czechoslovakia, Hungary, Romania, and Bulgaria.

Washington Consensus A political-economic philosophy by which the United States would pressure democratic governments of Latin America to open their markets to international trade, reduce tariffs, sell off state companies to private owners, and make other economic changes to bring prosperity to their region.

water footprint An indicator of freshwater use that looks at both direct and indirect water use of a consumer or producer; a component of the greater "ecological footprint" and "human footprint."

water hemisphere The half of Earth's surface containing the largest possible area of water. The defined points of its center vary.

"Water Towers of Asia" The glaciers and snowpack of the Himalaya, Karakorum, and other great mountain ranges.

Way of the Law See *Footprints of the Ancestors*.

weapons of mass destruction (WMD) Weapons designed to kill large numbers of people by chemical, biological, or nuclear means.

weather The atmospheric conditions prevailing at one time and place.

welfare state A national government that provides generous social services to citizens, paid for by high taxation rates.

west longitude See *longitude*.

westerly winds An airstream located in the middle latitudes that blows from west to east. Also known as westerlies.

Western Big Development Project A Chinese project to improve infrastructure and the economic livelihoods of residents of China's autonomous regions.

Western Hemisphere All of Earth's surface westward from the Prime Meridian to 180°.

Western Wall The most sacred site in the world accessible to Jews, it is a remnant of the Herodian-era wall that surrounded the Second Temple complex.

wet rice cultivation Rice harvests producing two or three crops per year that can sustain large populations over long periods of time.

"white Australia" policy An official Australian immigration policy until 1972 that excluded Asian and black immigrants.

"white flight" The mass migration of white Americans out of cities and into suburban areas after World War II.

"white gold" A term indicating how valuable cotton in central Asia was to the Soviet economy.

Wik Case A 1996 Australian high court ruling that Aborigines can claim title to public lands held by farmers and ranchers under long-term pastoral leases granted by state governments.

winter monsoon See *monsoon*.

"Window on the West" Russia's city of St. Petersburg, located on the Baltic Sea, giving Russia marine access to Europe.

"world city" A city that is a center of economic and political power in the global economy.

world regional approach Geography's tradition of understanding human–environment interaction on a regional basis.

Wu The second-largest language of China, spoken by 90 million people.

Xhosa One of the most populous ethnic groups of South Africa.

Xiang A language spoken in southern China.

Yakut A Turkic language spoken by the Sakha (Yakut) people of Siberia.

yang In *feng shui*, the masculine, bright, and positive.

yin In *feng shui*, the feminine, dark, and negative.

youth bulge A demographic situation in which a large percentage of a country's population is young, making competition for jobs, education, and land intense.

Yoruba One of Nigeria's three largest linguistic and ethnic groups, mainly in the southwest, including Lagos (the other two are the Hausa/Fulani of the north and the Igbo of the southeast).

Yue The Cantonese Chinese dialect that is the primary language of Hong Kong.

zero population growth (ZPG) The condition of equal birth rates and death rates in a population.

zero-sum game A term referring to a gain by one party resulting in an equivalent loss to another.

Zion A synonym for Jerusalem.

Zionist movement The movement, beginning in the 1890s, to establish a Jewish homeland in Palestine.

Zomia A term for the upland areas of Southeast Asia occupied by the Hmong ethnic group.

Zoroastrianism A monotheistic religion dating to 7th century Persia, practiced today mainly by the Parsis of western India.

Zulu One of the most populous tribal ethnic groups of South Africa.

Index

Note: Footnotes indicated by "n".

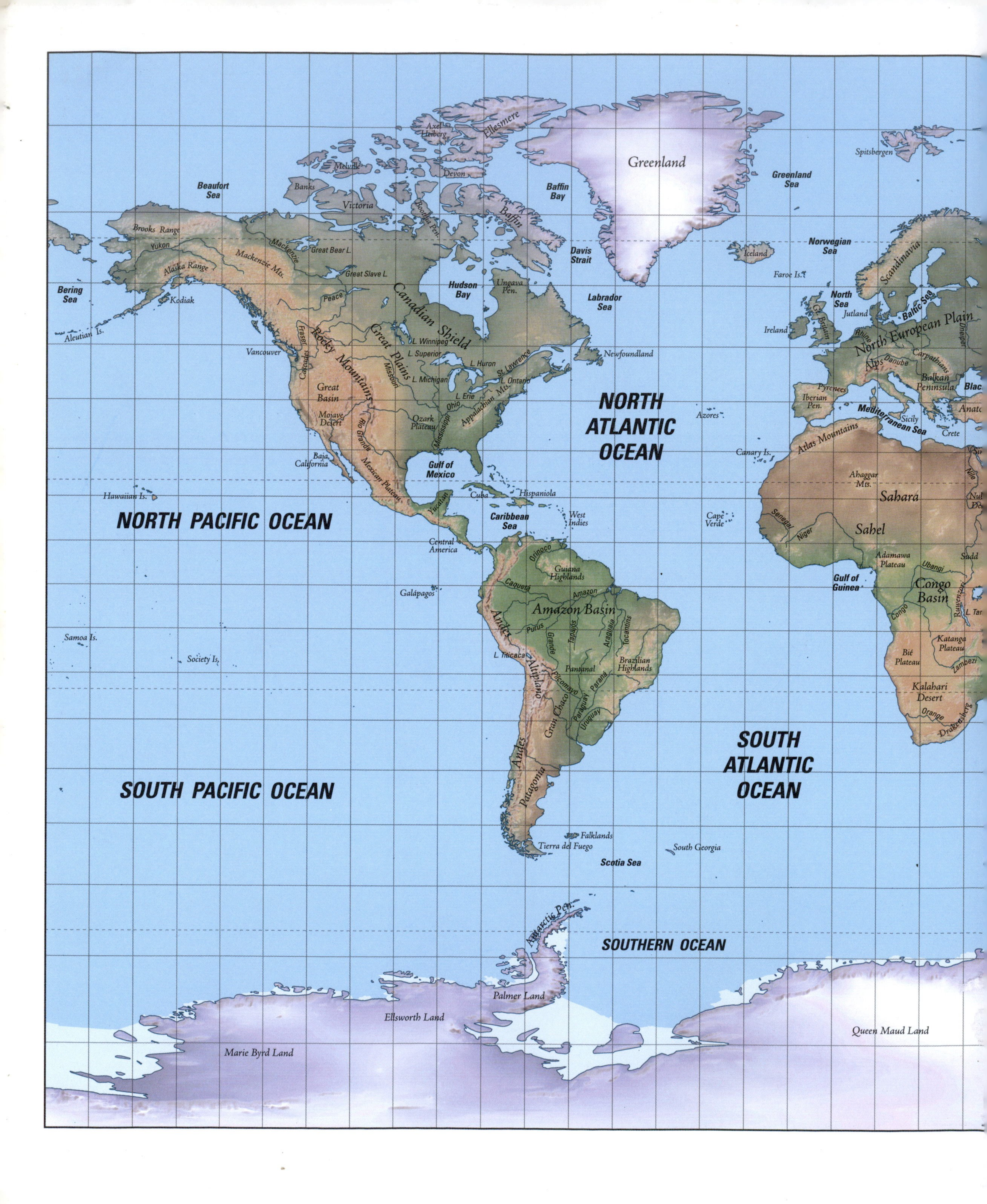

Greenland
Spitsbergen
Axel Heiberg
Ellesmere
Melville
Devon
Banks
Victoria
Baffin
Beaufort Sea
Baffin Bay
Greenland Sea
Norwegian Sea
Brooks Range
Yukon
Great Bear L.
Mackenzie
Mackenzie Mts.
Alaska Range
Great Slave L.
Davis Strait
Iceland
Faroe Is.
Bering Sea
Kodiak
Peace
Canadian Shield
Hudson Bay
Ungava Pen.
Labrador Sea
North Sea
Scandinavia
Jutland
Baltic Sea
Aleutian Is.
Fraser
Rocky Mountains
Great Plains
L. Winnipeg
L. Superior
Ireland
Gt. Britain
Rhine
North European Plain
Vancouver
Newfoundland
Cascades
Missouri
L. Huron
St. Lawrence
L. Michigan
L. Ontario
Danube
Carpathians
Alps
Great Basin
L. Erie
Appalachian Mts.
Ohio
Pyrenees
Balkan Peninsula
Iberian Pen.
NORTH ATLANTIC OCEAN
Mojave Desert
Ozark Plateau
Mississippi
Azores
Mediterranean Sea
Sicily
Rio Grande
Crete
Atlas Mountains
Baja California
Gulf of Mexico
Canary Is.
Mexican Plateau
Ahaggar Mts.
Nile
Hawaiian Is.
Cuba
Hispaniola
Sahara
Yucatán
NORTH PACIFIC OCEAN
Caribbean Sea
West Indies
Cape Verde
Senegal
Sahel
Niger
Central America
Orinoco
Adamawa Plateau
Ubangi
Guiana Highlands
Gulf of Guinea
Congo Basin
Galápagos
Caquetá
Amazon
Amazon Basin
Congo
Andes
Purus
Tapajós
Araguaia
Tocantins
Samoa Is.
Grande
Katanga Plateau
Bié Plateau
Society Is.
L. Titicaca
Brazilian Highlands
Zambezi
Pantanal
Altiplano
Pilcomayo
Paraná
Kalahari Desert
Paraguay
Gran Chaco
Uruguay
Orange
Drakensberg
SOUTH ATLANTIC OCEAN
Andes
SOUTH PACIFIC OCEAN
Patagonia
Falklands
Tierra del Fuego
South Georgia
Scotia Sea
Antarctic Pen.
SOUTHERN OCEAN
Palmer Land
Ellsworth Land
Queen Maud Land
Marie Byrd Land